STATISTICAL MECHANICS

The Theory of the Properties of Matter in Equilibrium

STATISTICAL MECHANICS

The Theory of the Properties of Matter in Equilibrium

by

R. H. FOWLER

SECOND EDITION

CAMBRIDGE UNIVERSITY PRESS

CAMBRIDGE

LONDON NEW YORK NEW ROCHELLE

MELBOURNE SYDNEY

CAMBRIDGE UNIVERSITY PRESS
Cambridge, New York, Melbourne, Madrid, Cape Town,
Singapore, São Paulo, Delhi, Tokyo, Mexico City

Cambridge University Press
The Edinburgh Building, Cambridge CB2 8RU, UK

Published in the United States of America by Cambridge University Press, New York

www.cambridge.org
Information on this title: www.cambridge.org/9780521093774

First published 1929
Second edition revised and enlarged 1936
Reprinted 1955, 1966
First paperback edition 1966
Reprinted 1980
Re-issued 2011

A catalogue record for this publication is available from the British Library

ISBN 978-0-521-05025-8 Hardback
ISBN 978-0-521-09377-4 Paperback

CONTENTS

CHAP.		PAGES
	Table of Common Physical Constants	x
I	Introduction	1–15
II	The General Theorems of Statistical Mechanics for Assemblies of Permanent Systems	16–76
III	Assemblies of Permanent Systems (cont.). The Specific Heats of Simple Gases	77–111
IV	Partition Functions for Temperature Radiation and Crystals. Simple Properties of Crystals	112–150
V	The General Assembly. Dissociation and Evaporation	151–186
VI	The Relationship of the Equilibrium Theory to Classical Thermodynamics	187–207
VII	Nernst's Heat Theorem and the Chemical Constants	208–235
VIII	The Theory of Imperfect Gases	236–274
IX	The Theory of Imperfect Gases (cont.)	275–291
X	Interatomic Forces	292–337
XI	The Electron Theory of Metals, Thermionics and Metallic Conduction. Semi-Conductors	338–436
XII	Electric and Magnetic Susceptibilities. Ferromagnetism	437–521
XIII	Applications to Liquids and Solutions	522–560
XIV	Assemblies of Atoms, Atomic Ions and Electrons .	561–583
XV	Atmospheric Problems	584–637
XVI	Applications to Stellar Interiors	638–657
XVII	Mechanisms of Interaction. Collision Processes .	658–699
XVIII	Chemical Kinetics in Gaseous Systems	700–719
XIX	Mechanisms of Interaction. Radiative Processes .	720–742
XX	Fluctuations	743–788
XXI	Recent Applications to Cooperative and other Phenomena	789–852
	Index of Authors quoted	853–857
	Index of Subjects	858–864

NOTE

For permission to reproduce the following figures thanks are due to the Council of the Royal Society of London, *Proceedings* (Figs. 2, 10·2, 10·3, 15, 18, 21, 66, 80, 84–94, 96); to Prof. Dr W. H. Keesom, *Physica* (Figs. 3, 37·1, 37·2); to the Editor of *Zeitschrift für Elektrochemie*, Deutschen Bunsen-Gesellschaft (Figs. 3·1–3·3, 8–9·2); to Herr S. Hirzel, Leipzig (Fig. 10); to Prof. M. Born and the Massachusetts Institute of Technology (Fig. 11); to Herr Julius Springer, *Handbuch der Physik* (Figs. 12, 36); to Messrs Benn [Ernest], Ltd. (Fig. 13); to Messrs Longmans, Green & Co. Ltd. (Fig. 14); to the Council of the Physical Society of London, *Proceedings* (Figs. 20, 38, 39, 67, 68); to the Editor of *Actualités Scientifiques et Industrielles* and MM. Hermann & Cie. (Figs. 32, 44); to Herrn Friedr. Vieweg & Sohn. A. G. *Lehrbuch der Physik* (Figs. 33–35); to the Editor of *Science*, U.S.A. (Fig. 53); to the Oxford University Press, *The Theory of Electric and Magnetic Susceptibilities*, J. H. Van Vleck (Figs. 54–57); to the Académie des Sciences, Paris, *Comptes Rendus* (Figs. 63 *a, b*); to the Editor of the *Physical Review* (Fig. 63 *c*); to Herr Johann Ambrosius Barth, *Annalen der Physik* (Figs. 64, 65, 69, 79); to the Council of the Faraday Society, *Transactions* (Fig. 72); to the Council of the Royal Astronomical Society, *Monthly Notices?* (Figs. 73, 77, 78); to Professor H. N. Russell and *The Astrophysical Journal* (Figs. 74 *a-c*); to the Council of the Cambridge Philosophical Society (Figs. 75, 76, 95); to the Council of the Royal Academy of Sciences, Amsterdam, *Proceedings* (Fig. 99); and to the Editor of the *Journal of Chemical Physics* (Figs. 100, 101).

PREFACE TO THE FIRST EDITION

My reasons for expanding the Adams Prize Essay for 1923–1924 into the present book are set forth in the introductory Chapter. Now that the book is finished it will be found I hope to be developed on a plan not too discreditable for 1926, but hardly one which would be adopted to-day. This is a fault hard to avoid, and I still hope that a systematic exposition of Statistical Mechanics, such as this book attempts to give, even if its tone is antiquated, may be of some value to students. I have therefore been at some pains to provide a reliable index of subjects. I hope that any matter which is treated in the book can be traced via the index with no more searching than is reasonable.

There remains only the pleasant task of thanking those who have helped me. The task is a heavy one, for without a number of collaborators the book could never have been finished. The greatest assistance has been given me by Prof. J. E. Lennard-Jones who has contributed Chapter x on a subject of which he is a master, and has also read many chapters in manuscript and proof. I could not have otherwise achieved a Chapter x of this completeness. Dr D. R. Hartree undertook for me the whole of the laborious calculations on which Chapter xvi is based, and provided similar material elsewhere in the book. Mr J. A. Gaunt wrote for me the greater part of the more elaborate discussion and development of Debye and Hückel's theory of strong electrolytes, and has read the whole book in proof. Mr W. H. McCrea has in the same way provided for me most of the material for the analysis of the specific heats of gases. To him also and to Dr L. H. Thomas I am grateful for reading proofs. I am deeply conscious that such merit as the book may have is largely due to the original work of these collaborators, started with the object of helping me. The contributions that I have gratefully taken from Mr H. D. Ursell and Dr P. M. Dennison stand in the same category. I have also benefited by Dr P. A. M. Dirac's criticisms of the last chapter, and Prof. J. E. Littlewood's mathematical assistance. Besides these primary helpers I have been generously given valuable information on various subjects by Dr S. Dushman, Prof. O. W. Richardson, Prof. A. Fowler, Prof. N. Bjerrum and Mr A. Egerton, to whom I offer my best thanks.

Of my obligations to Prof. C. G. Darwin I can make no adequate acknowledgment. The whole book is the outcome of my collaboration with him in which the revised method of approaching statistical theory was worked out.

Finally I must express my gratitude to the Cambridge University Press for their unfailing helpfulness and patience with a somewhat ruthless proof corrector.

R. H. F.

September 1928

PREFACE TO THE SECOND EDITION

I have been glad to take the opportunity provided by this edition to rearrange part of the subject-matter and to make other changes, described in the introduction, with the object of bringing the book more up to date. Numerous mistakes in the first edition have been pointed out to me by correspondents and reviewers to whom I am deeply grateful. I am particularly grateful for this help to Mr E. A. Guggenheim, Dr T. E. Sterne and Dr J. D. van der Waals, jr. I believe that all the important mistakes in the first edition have here been eliminated, but naturally one can scarcely hope that the second edition has escaped its own new crop. In preparing this edition I have received essential assistance from Dr G. B. B. M. Sutherland (Chapter II), Mr R. A. Buckingham (Chapter X), Mr E. A. Guggenheim (Chapter XIII) and Dr S. Chandrasekhar (Chapters XV, XVI). Without their help this edition would still be far from completion. For some very recent matter incorporated at the end I have to thank Prof. J. H. Van Vleck and Prof. H. Eyring; also Dr H. Grayson-Smith for invaluable help in reading the proofs.

When the first edition was written, Statistical Mechanics was still in process of being translated from classical to quantal language, and many of the features of this translation were still quite obscure. Since then this process has been completed, and no obscurities of principle remain. Developments in the near future seem likely to consist mainly of applications to more and more complicated models, which are designed to account for more and more subtle properties of matter in equilibrium. Some recent examples of such developments are considered in the new Chapter XXI of this edition.

R. H. F.

September 1936

TABLE OF THE VALUES IN C.G.S., CENTIGRADE AND ELECTROSTATIC UNITS OF THE COMMONER PHYSICAL CONSTANTS USED IN THIS MONOGRAPH

Charge on the electron, $-\epsilon$	$\epsilon = 4{\cdot}770 \times 10^{-10}$
Planck's Constant, h	$h = 6{\cdot}547 \times 10^{-27}$
Boltzmann's Constant, k	$k = 1{\cdot}371 \times 10^{-16}$
Avogadro's (Loschmidt's) Number	
Molecules per mole, N	$N = 6{\cdot}064 \times 10^{23}$
Molecules per c.c. in a perfect gas at 0° C. and 1 atmosphere	$2{\cdot}706 \times 10^{19}$
Mass of an atom of atomic weight 1 ...	$1{\cdot}649 \times 10^{-24}$
Mass of H-atom ... ... [atomic weight 1·0081]	$1{\cdot}662 \times 10^{-24}$
Mass of electron	$9{\cdot}035 \times 10^{-28}$
Radius of 1st Bohr orbit in H	$0{\cdot}528 \times 10^{-8}$
Bohr's magneton, μ_B (erg/gauss) ...	$\mu_B = 9{\cdot}174 \times 10^{-21}$
Mechanical Equivalent of Heat, J (erg/cal.)	$J = 4{\cdot}185 \times 10^{7}$
Gas Constant, R (cal./mole)	$R = 1{\cdot}986$
Velocity of light, c	$c = 2{\cdot}998 \times 10^{10}$
Electron-volt or volt	$1{\cdot}590 \times 10^{-12}$

These values are taken from Birge, *Reviews Mod. Phys.* vol. 1, p. 1 (1930); he has given supplementary discussions in *Phys. Rev.* vol. 40, pp. 230, 319 (1933); vol. 43, p. 211 (1933); vol. 48, p. 918 (1935), but no completely satisfactory set of numerical values can yet be given. [Atomic weight of H corrected according to Oliphant, Kempton and Rutherford, *Proc. Roy. Soc.* A, vol. 150, p. 241 (1935).]

CHAPTER I

INTRODUCTION

§1·1. *From the Introduction of* 1928. In attempting to study the physical state of matter at high temperatures on the lines suggested by the notice for the Adams Prize Essay for 1923–1924, it was at once apparent that the problem demanded all the available resources of present-day statistical mechanics. These have been somewhat increased in recent years, and the whole aspect of the kinetic theory of matter, at least in full statistical equilibrium, has been steadily altered by the development of the quantum theory. As a result there is no recent systematic exposition of the equilibrium theory of statistical mechanics,* envisaging throughout both classical and quantized systems, to which one may appeal in the further applications that it is proposed to make here. Prof. Darwin and I have been fortunate enough in recent years to have developed a method (new in this connection) which enables a systematic exposition to be undertaken with, we would submit, a sufficient degree of elegance. It has, at the same time, been possible to apply the results to a problem more immediately related to that proposed—that is to a theoretical study of the state of matter in stellar reversing layers and in the interior of gaseous stars.

These were of course the main problems with a view to which the essay was first written, but, for the reasons just given, it was thought best not to concentrate entirely on applications in the essay itself but to begin instead with the systematic survey of the equilibrium theory which was then needed and perhaps is still not superfluous. The essay, accordingly, from the first took the form of a monograph on the Equilibrium Theory of Statistical Mechanics. Originally the applications of the theory were mainly astrophysical, but it has been a simple matter to expand their scope. My object was to include all types of application of the equilibrium theory, so that, however inadequately, the monograph should cover the whole field. In the end, however, I have made no attempt to apply the theory to surfaces, or to liquids beyond the theory of dilute solutions; my knowledge of these branches of the theory is still too meagre to justify an exposition of them.

The standard results of the equilibrium theory have long been classical. They are here derived from the fundamental hypotheses in the systematic way mentioned above. The presentation here has been revised and to some

* More accurately, no such exposition existed in 1924. There were in 1928 at least two which should be mentioned: Herzfeld, "Kinetische Theorie der Wärme" (*Müller-Pouillets Lehrbuch der Physik*, vol. 3, part 2), and Smekal, "Allgemeine Grundlagen der Quantenstatistik und Quantentheorie" (*Encyclopädie mathematischen Wissenschaften*, vol. 5, part 3, No. 28).

extent remodelled, and now forms, I hope, a connected account of the greater part of the equilibrium theory of statistical mechanics, so far as this has yet been developed. In general, the theory and its simpler practical applications have been developed concurrently to avoid too continuous a sequence of unapplied theorems. The more complicated applications form the subject-matter of the later chapters.

At every stage the theory is developed for classical and quantized systems indiscriminately. It is therefore necessary from the start to be absolutely clear what is to be regarded, for the purposes of the theory, as the present logical position of the quantum theory. Though the quantum theory had its origin in Planck's statistical discussion of the laws of temperature radiation and in the breakdown of the theorem of equipartition, it should be regarded as a purely "atomic" theory—that is, a theory applying directly to individual atoms and other connected systems, but not *primarily* connected with the statistical behaviour of large collections of such systems. It is founded on the theory of spectra, and its laws must primarily be sought for by the study of the properties of individual atoms and molecules, and the interactions of pairs of such, or rather in those phenomena which can most certainly be referred back to such individual systems and interactions. Among these phenomena spectra stand first. The laws so derived for individual atoms, just as the laws for classical systems, are then at our disposal to use in discussing the statistical behaviour of large collections of such atoms and systems. If we can make use of them thus, the derived laws of temperature radiation and specific heats are then available for comparison with experiments on radiation or material systems in bulk. We thus ascertain whether the laws of atomic systems and the general hypotheses of statistical mechanics are adequate to account for such molar properties as we are able to compute. This complete divorce of the quantum theory from its historical setting seems to me to be essential to a grasp of its present logical position and to a properly proportioned view of the theory of statistical mechanics....

In these developments [therefore] we have deliberately used a non-historical deductive method. So far as possible the theory has been presented as a finished structure, with some attempt at logical completeness, not visibly constructed to fit the facts. Results have been deduced at each stage from the general theory, and checked by comparison with experiment.

Conforming to this method, the distribution laws for classical systems are derived by a limiting process from the similar laws for quantized systems. It is not difficult, I believe, to justify this somewhat unusual procedure, in which the laws for Planck's oscillator are fundamental, and the rest of the theory, quantized and classical, a generalization from this starting point. In the first place it is undesirable in a systematic exposition to regard both classical and quantized systems as fundamental. If we are so to regard one only it must be the latter, for we cannot derive the laws of quantized systems from those of classical systems. Secondly, it may at least be claimed that there is a gain in elegance and physical reality, for classical systems are the exception rather than the rule in atomic physics. This is not to say (of course) that we do not use classical mechanics, so far as we can, to derive the quantum mechanics of atomic systems by a process of generalization. But once the laws of quantum mechanics have thus been guessed, as they largely must be *before* we can discuss the theorems of statistical mechanics, quantized systems naturally come first. [In 1935 this attitude hardly needs apology, and from here on no further presentation of the point of view of 1928 will be given.]

§ **1·2.** *The generality of statistical theorems.* The equilibrium theory of statistical mechanics, as presented here or in any similar manner, is strictly a theory of the distribution of energy (and sometimes momenta) over systems, and of systems over phases, and derives these and other distribution laws by general arguments, making no reference whatever to the particular mechanisms of interaction which bring about the equilibrium between the individual systems and the different phases. If the fundamental hypotheses of the theory are accepted, there seems no escape from this conclusion. Thus, for example, Maxwell's law for the distribution of velocity among the molecules of a gas in statistical equilibrium with classical statistics, or the corresponding modifications with Fermi-Dirac or Einstein-Bose statistics, must be true *whatever be the laws of collisions between these and any other types of molecule in the gas.* The theorems of statistical mechanics thus appear to have something of the same generality as the laws of thermodynamics. They have necessarily less than the full generality of the latter, for they contemplate and refer to a particular molecular structure; granted this limitation, however, it seems that they must be granted also the universal character of thermodynamical theorems, with its advantages and disadvantages. The fact that a particular mechanism leads to a state of complete equilibrium in agreement with experimental facts is no evidence for the particular mechanism discussed. It is merely evidence that the laws of this mechanism have been correctly and consistently written down! Any other mechanism would give the same result.*

Particular mechanisms of interaction first become relevant in the study

* This is perhaps an overstatement. In the theory of imperfect gases, for example, we assume a mutual potential energy for each pair of particles and derive an equation of state depending on that potential energy. If the laws of classical mechanics are obeyed by the encounter, then the potential energy suffices to determine all its details as a mechanism for the exchange of energy and momentum. But these details are not relevant to the study of the equilibrium state itself and might conceivably be different without affecting it.

of *non-equilibrium* states, such as states of steady flow. Finally, of course, it is these mechanisms of interaction, for example between atoms and radiation or atoms and atoms in collision, that are of supreme interest; it must be regretfully admitted that the study of complete statistical equilibrium cannot *by itself* provide any information as to any particular process. It does, however, provide a rigid form to which all possible mechanisms whatever must conform; that is to say, any possible mechanism, left to act by itself, must set up and preserve the laws of statistical equilibrium. This idea, which is well known in the classical theory of radiation, has proved of great importance in general statistical mechanics, following a line of thought opened up by Klein and Rosseland.* It appears in general that a particular process can never be supposed to be able to act alone, unaccompanied by a corresponding reverse process; only the two together form a possible single mechanism. The next step forward from the purely equilibrium theory of statistical mechanics is obviously a systematic survey of possible mechanisms, working out the laws that they must observe in order to fit into the equilibrium theory and preserve, as they must, its distribution laws. We attempt to sketch such a survey in the concluding chapters of this monograph.

§ 1·3. *Scope of this monograph.* The scope of this monograph may now be more exactly indicated. At the close of this chapter we specify the fundamental assumptions on which the theorems of statistical mechanics are to be based. These are put on record in dogmatic form and all but the most superficial discussion of their foundations omitted. In Chapters II–IV we develop the equilibrium theory for all the types of matter commonly treated in this way—such as perfect gases, crystals, and any general body obeying classical laws. We include also a similar treatment of radiation, but exclude all cases in which dissociation or evaporation occur. Chapter III contains applications to the specific heats of gases, and the latter part of Chapter IV applications to the properties of simple crystals. The theory is generalized in Chapter V to include all types of dissociation and evaporation, and in Chapter VI the connection between the equilibrium theory of statistical mechanics and the laws of thermodynamics is considered in detail. We point out the close analogies which allow certain functions of the state of the bodies we discuss to be properly interpreted as the temperature and entropy of thermodynamics. This chapter concludes with criticisms of the commoner ways of introducing entropy into statistical mechanics, which, it is claimed, are either obscure or misleading, and certainly unnecessary.

Chapter VII returns to applications, now in the region of very low temperatures. Its subject is Nernst's heat theorem and the chemical constants

* Klein and Rosseland, *Zeit. f. Physik*, vol. 4, p. 46 (1921).

—entropy at the absolute zero. It is possible to obtain a clearer understanding of this theorem and of the chemical constants from the standpoint of statistical mechanics than in any other way. A comparison with experiment is also desirable at this stage of the theory, and can be most conveniently obtained in this field. The field of validity of Nernst's heat theorem and the precautions necessary in applying it to experimental data can now be accurately specified. In Chapter VIII we extend the general theory to include, so far as is possible, imperfect gases, allowing also for the possibility of electrostatic charges, and in Chapter IX apply the theory to a discussion of theoretical and semi-empirical equations of state. Chapter X, which was contributed to the first edition by Prof. Lennard-Jones, gives a general numerical survey of intermolecular forces so far as these can be derived by analysis of the equations of state of imperfect gases* and from the properties of allied crystals. It is interesting to find that one and the same law of force will account satisfactorily for so wide a range of properties.

Chapter XI attempts to cover the whole field of thermionic phenomena, so far as these can be related to states of equilibrium. The most important part is the theoretical formula for the vapour density of free electrons in equilibrium with a hot metal, including the effect of the space charge. This is of primary importance for further applications, because it involves the chemical constant of the electron and experiment confirms the theoretical value. This chapter also includes a formal account of the simpler parts of the theory of electronic conduction in metals and semiconductors. Its close relationship to thermionic theory excuses the fact that conduction cannot strictly be classed as an equilibrium property of matter. Chapter XII deals with the magnetic and dielectric phenomena of matter in bulk, the most important part being a semi-descriptive theory of ferromagnetism. Chapter XIII attempts to carry the theory on to describe the properties of liquids, but nothing is achieved beyond a development of the theory of dilute solutions including the theory of strong electrolytes. Chapters IX–XIII inclusive and the greater part of Chapter VIII are additions to the scope of the original essay.

Chapters XIV–XVI deal with applications of the theory to the high temperatures found inside and outside stars, the applications proposed by the examiners for the Adams Prize. In Chapter XIV the equilibrium theory of a gas of highly ionized atoms is developed as far as the methods available permit, including the effects of the sizes of the ions and their electrostatic fields. For many purposes approximate forms are necessary which may be expected to be qualitatively valid over wide ranges of conditions. Such forms are provided. These approximations are mainly required for Chapter XVI, which makes a start on the study of the properties of stellar material in the

* Evidence from viscosity is also used.

interior of a star. It would be out of place to carry these calculations to great detail or to trace their repercussions on Eddington's work in this monograph. In general they confirm the values of the physical constants of stellar matter which he uses, particularly for the larger stars. Chapter XV, meanwhile, has dealt with such problems of the atmosphere of a star as can be treated by means of the formulae of the equilibrium state—the more important are the elementary theory of the rise and decay of absorption lines with the rising temperature of the reversing layer and the theory of the rate of escape of molecules from an atmosphere. A summary is given of some of Milne's beautiful work on the calcium chromosphere, but here our connection with the equilibrium theory is getting very weak. The outward flux of radiation, which is an entirely trivial perturbation of complete equilibrium in the stellar interior, is now becoming the controlling feature.

The next group of three chapters, XVII–XIX, contains detailed studies of the laws to which actual mechanisms of interaction must conform in order to preserve the equilibrium laws. The laws of material collision processes between free atoms and molecules and free atoms and solid surfaces are discussed in Chapter XVII and applied in Chapter XVIII to the kinetics of homogeneous gas reactions. The laws of radiative processes are discussed in Chapter XIX. Chapter XX contains for completeness an account of the formal calculus of fluctuations, and applications of some of these theorems to the study of opalescence, Brownian movement, the shot effect and kindred phenomena. Chapter XXI gives an account of some miscellaneous very recent work which could not conveniently be incorporated in the other chapters; the most important sections give an account of cooperative phenomena, particularly the theory of order and disorder in alloys which has received such a successful start at the hands of Bragg and Williams* and Bethe.†

It will be seen that the content of this monograph is not *strictly* confined to equilibrium states of matter. We have ventured outside into regions dealing with steady rates of change (states of flow), but only where the application of the laws of the equilibrium state is immediate. When the changes are such, or the accuracy required is so great that the equilibrium laws can no longer be used without modification, as in the grand theory of transport phenomena in gases, the more advanced theory of electronic conduction in metals, or of the formation of stellar and nebular spectra, we must be silent. Nor can we do justice here to the phenomena of very high vacua. But where the direct application of the equilibrium laws themselves is relevant or sufficient, as in the study of unit mechanisms or in thermionics, we have endeavoured to press the theory forward.

* Bragg and Williams, *Proc. Roy. Soc.* A, vol. 145, p. 699 (1934).
† Bethe, *Proc. Roy. Soc.* A, vol. 150, p. 552 (1935).

§1·4. *The fundamental assumptions of statistical mechanics.* In my opinion any thorough discussion of the foundations of statistical mechanics is utterly unsuitable as an *introduction* to the study of this branch of theoretical physics. One might properly attempt to close this monograph with a chapter or chapters expounding the foundations as they now appear in quantum mechanics, thanks largely to the work of von Neumann.* But such an exposition to be of any value must be somewhat lengthy and would form a portion necessarily quite out of tone with the rest of this monograph. Statistical mechanics may really be regarded as consisting of two almost distinct subjects: the theory of the equilibrium properties of matter based on the usual assumptions as to the calculation of average states, assumptions which can be introduced in a way which makes them *a priori* eminently reasonable, and the deeper theory of these assumptions themselves. This deeper theory will be entirely omitted here, and would form the subject-matter for a substantial monograph by itself.

Though no thorough discussion of the foundations will be attempted here, it is none the less desirable to begin the exposition by considering shortly the usual bases, and specifying as clearly as possible the one selected, indicating the reasons for its choice.

There are two distinct starting points from which we may build up with equal success a theoretical model to represent the material systems of our more or less direct experience—the Gibbsian ensemble and the general conservative dynamical system. Of these the Gibbsian ensemble has perhaps the advantage in logical precision, in that the whole of the necessary assumptions can be explicitly introduced in the initial formulation of the "canonical" ensemble. For this reason it should perhaps be preferred, and is preferred by some theoretical physicists. But to others something more than success and logical rigour appears to be necessary for the acceptance of a model which is to account to our aesthetic satisfaction for the properties of matter. A certain "sanity", or physical reality, may be demanded in the initial postulates and in the details of the model, particularly in so far as they are to reproduce the well-known properties of matter. To these others the Gibbsian ensemble appears to be weak from this aspect, and they are led —in spite of logical and analytical incompleteness—to prefer the conservative dynamical system of many degrees of freedom as the more satisfactory model from which to derive (or attempt to derive) the properties of matter. This is the model, generalized from classical to quantum mechanics, which will be used in this monograph.

We have of course to deal in general with dynamical systems which are collections of large numbers of similar atoms, molecules, or electrons. It is

* von Neumann, *Mathematische Gründlagen der Quantenmechanik*, Berlin (1932).

convenient to introduce a consistent nomenclature for the whole collection and its constituent parts. We call the collection which composes the complete dynamical system *an assembly*. We call its constituent atoms etc., or any part of it which for the greater part of time has practically an independent existence, *a system*. The model we propose to use will then be called *an assembly of systems*. The motions and interactions of these systems are controlled by the laws of quantum dynamics, and the assembly as a whole is *conservative*.

We now ask the questions, will such an assembly attain in any sense a state of equilibrium and there exhibit permanent characteristics which can be identified by analogy with similar properties of matter determined experimentally? And if so, how are such permanent properties to be computed? If and only if these questions are answered can we claim to possess in any degree "a theory of the properties of matter in equilibrium". What we ordinarily call the observed properties of matter, e.g. the pressure of a gas, may be regarded as "short-time averages" of its instantaneous properties. It would be natural therefore to ask for the computed short-time average properties of our assembly and to require these to correspond to the observed properties of matter in equilibrium.

We have naturally no means of computing such time averages with full logical rigour short of a sufficiently detailed solution of the general dynamical equations of the assembly. Equally naturally we lack this information whether the assembly is classical or quantal.* It is necessary to assume the general form that the solution will take—the best known such assumption for classical assemblies was that of Maxwell, the assumption of quasi-continuity of path. It is extremely probable that this assumption is always untrue; it is, moreover, insufficient and at the same time unnecessarily restrictive for the purpose in hand. But some similar assumption must be made in its place. Its object was to entitle us to assert that the required time average properties may be correctly calculated as if they were averages over the whole phase space† of the assembly subject to the condition that the assembly has the proper energy, and perhaps momenta, and provided that the different elements of the phase space are "weighted" in the proper way. Even then it was necessary by an extra assumption or investigation

* It seems to be essential to possess an adjective which is the antithesis of "classical". Common usage gives us "quantum mechanical" for this purpose, a phrase which is really somewhat repugnant to English grammar. It is of course perfectly correct to use "quantum" itself for this adjective and it is habitually so used with the nouns theory, number, dynamics, mechanics. Preserving this usage, one should undoubtedly eliminate "quantum mechanical" either by substituting "quantum" everywhere or by use of some correct adjectival form. I have attempted in this monograph always to use "quantal" for this adjective, when I felt that "quantum" was inappropriate.

† See § 1·5.

(but no such was ever given) to identify the "long-time average" so calculated with the "short-time averages" which are of physical significance.

The modern quantum development is in many ways simpler than the classical which it has superseded. We shall assume that the reader is familiar with some standard exposition of quantum mechanics.* We may then enunciate the assumption replacing Maxwell's hypothesis of quasi-continuity of path as follows: *The observable equilibrium properties of any assembly are to be calculated by averaging over all the states of the assembly which are* **accessible** *under the given conditions.* All assemblies whose equilibrium states can be discussed are necessarily *enclosed* assemblies. An unenclosed assembly to which all space is accessible does not in general possess an equilibrium state about which anything significant can be said. Enclosed assemblies obeying the laws of quantum mechanics possess only discrete states of definite discrete energies. To any one of these energies belong a certain number of linearly independent wave functions for the assembly. *Each such distinct wave function represents a distinct state of the assembly.* The averaging required is averaging over all these states, without fear or favour, provided only that the assembly can occupy the state. With this meaning of "state" we are to average over all accessible states, assigning the same (unit) weight to each.

It is convenient to have a distinctive name by which to describe these distinct accessible states of unit weight, and they will be called *complexions*. This usage of the term complexion is a natural refinement of its traditional usage in statistical mechanics.

We can make no attempt in this monograph at any detailed justification of this procedure. Once this procedure has been adopted, the rest of the derivation of the equilibrium properties of an assembly is quite straightforward and can be carried through with logical rigour—simply for a simple assembly and with increasing difficulty and complication as the assembly gets more elaborate. But the critical step is so fundamental that we cannot pass it by entirely without comment, and therefore proceed to some such comment now by way of introduction.

The assignment of the weights for averaging is frequently spoken of as an assignment of *a priori* probabilities. If this is taken at its face value, the behaviour of the systems in our assembly must be according to the laws of chance, and cannot be controlled by dynamical (or any other determinate) laws in ordinary space and time. This is a possible hypothesis, but in the end hardly a satisfying one. It is equally repugnant to classical or to quantum

* For example, Frenkel, *Wave Mechanics*, I and II, Oxford (1932, 1934); Condon and Morse, *Quantum Mechanics*, New York (1929), or for a more fundamental treatment, Dirac, *Quantum Mechanics*, ed. 2, Oxford (1935).

mechanics. To avoid an appearance of definitely accepting this hypothesis we use the neutral word "weight" instead of the commoner "*a priori* probability", although in effect (though not in origin) they become synonymous terms. For the purpose of the ensuing calculations we require to know or assume the relative times (out of a long interval) during which the assembly occupies two given states. It is these which determine the relative weights. It is assumed here that these times are equal. Since each quantum state of a system of s freedoms corresponds to an element of classical phase space (of $2s$ dimensions) of extension h^s, this assumption reduces in the classical limit to assigning a weight to any element of phase space proportional to its extension. This procedure is consistent with, but by no means a deduction from, Liouville's theorem (§ 1·5).

It is usual to proceed from this basis by the calculation of values of maximum frequency of occurrence (most probable values) rather than average values. The results are, of course, identical—the mathematical machinery is not. Average values are as naturally as, if not more naturally than, most probable values identified with the observable properties of the assembly. They have besides an overwhelming advantage in ease and rigour of mathematical presentation; in particular the usual indiscriminate use of Stirling's theorem for large factorials can be entirely avoided.

No attempt has been made in these paragraphs to minimize the logical incompleteness of this development of statistical theory from the chosen starting point. It will be only too painfully apparent, and is of course the same for all variants of this development. I confess to a belief that it is unavoidable and may continue so for a long time. It can only be said that the assumptions made have a certain inherent plausibility and are justified by their success, and that, in spite of these lacunae, this starting point is, to some tastes, physically preferable to the Gibbsian ensemble.

One further step can be made, which, while it does not fill the logical gaps indicated above, yet goes a long way towards giving us confidence in our conclusions and warranting the belief that the average properties we calculate are really "normal" properties of the assembly, which it will always to our senses possess. This step is the calculation of *fluctuations*. Just as we calculate the average value of any quantity P, say, and find it is $\overline{P}$, so we can calculate the average value $\overline{(P-\overline{P})^2}$ of $(P-\overline{P})^2$. If we do this, we find that in all cases $\overline{(P-\overline{P})^2} \leqslant \overline{P}$. It follows that the average deviation of P from its average, and therefore normal, value $\overline{P}$ is of the order $\sqrt{\overline{P}}$, and if $\overline{P}$ itself is large the deviation is insignificant. We can interpret this by saying that out of any time interval only an insignificant fraction in general can be spent in states in which P differs effectively from its normal value $\overline{P}$. This very greatly consolidates the whole theory.

We have used here the word "normal" as a synonym for "short-time average" to recall the connection with the definition of "normal" adopted by Jeans.* In Jeans' language a property is a *normal* property of an assembly if that property is possessed by the assembly with negligible error in the whole of accessible phase space except for a negligible fraction. It is obvious that any "normal" property will be an equilibrium property of our assembly whose equilibrium properties are calculated by averaging over accessible phase space by the rule here adopted. Conversely any one of our averages is a normal property if its fluctuation is small. For all practical purposes the two methods of defining equilibrium properties are equivalent.

§1·41. *Concluding the Introduction of* 1935. In revising this monograph in 1933–5 it has been possible to retain the main framework. The old concluding chapter on quantum statistics has naturally been incorporated into the exposition from the start, but otherwise there has been no change of structure. A few new subjects such as ferromagnetics and semi-conductors have been added, but the main changes consist of revision and modernization of both theory and experimental data. Thus theory and experiment now agree about the specific heats of the simpler gases. But in 1928 there was no agreement at all even for diatomic gases due to the very slow rate of interchange of vibrational and translational energies, as a result of which the observed values did not refer to the equilibrium state. Again, there is now substantial agreement in the field of chemical constants and equilibrium constants for diatomic molecules due partly to improved data but here mainly to the improvements of theory which have enabled the states of such molecules in gaseous and solid phases to be correctly enumerated. Serious errors have been removed from the old account of the theory of metals and of dilute solutions which now, it is believed, survive in an acceptable form, and minor errors have been removed *passim*. It is too much to hope that these have all been detected and that no new ones have been introduced.

The concluding section of this chapter is obviously a survival from a preceding epoch, but has intentionally been left in place. As we have already insisted, no systematic examination of the foundations of statistical mechanics has been or should be undertaken in this introductory chapter. But to say no more than we have said hitherto is somewhat ruthless and the best compromise appears to be to give here a short account of the classical use of Liouville's theorem for classical assemblies in justifying (partially) our averaging rule for such assemblies.

§1·5. *Conservative classical dynamical systems.* We close this chapter by enumerating briefly the chief properties of conservative systems, which confirm the hypotheses of statistical mechanics. The state of the assembly,

* Jeans, *Dynamical Theory of Gases*, ed. 3, pp. 74 *sqq*.

which is the conservative system here in question, is fully defined by specifying the necessary N Hamiltonian coordinates q and N conjugated momenta p. It can be conveniently represented geometrically by a point in space of $2N$ dimensions, whose rectangular cartesian coordinates are the N p's and N q's. This space is called the *phase space* of the assembly (already referred to) and the point its *representative point*. The equations of motion of the assembly are

$$\dot{p}_s = -\frac{\partial H}{\partial q_s}, \quad \dot{q}_s = \frac{\partial H}{\partial p_s} \quad (s = 1, \ldots, N), \qquad \ldots\ldots(1)$$

where H is the Hamiltonian function. It is usually only necessary to consider assemblies in the formulation of which the time does not occur explicitly, so that H is the total energy E expressed as a function of the p's and q's. Through every point of the phase space passes a definite *trajectory* of the assembly satisfying (1), and confined of course to the surface $H = E$, constant, and perhaps to other surfaces defined by constant momenta as well.

We have mentioned above that no logical justification has ever been attempted for the assumption that the average values concerned in observations are equivalent to the long-time average properties of the assembly. Attempts, however, to justify identifying these long-time averages with averages over the accessible phase space, though far from successful, have led to interesting investigations, which tend to confirm the proposed choice of the weight for each element of phase space in this process of averaging. As the first of these we may cite Liouville's theorem.

Let τ be the density of a "fine dust" of representative points in any element of phase space. Then Liouville's theorem states that $D\tau/Dt = 0$, where D/Dt is the mobile operator of hydrodynamics (generalized), giving the rate of variation of τ for a given group of points as we follow them along their trajectories. Consider a fixed volume element in the phase space bounded by p_1, $p_1 + dp_1$; ...; q_N, $q_N + dq_N$, of extension $d\Omega$ $(= dp_1, \ldots, dq_N)$. The representative points crossing the face p_s, of area dS, have a component velocity $\dot{p}_s$ normal to that face, and so the rate of increase in $\tau d\Omega$ due to motion across this face is

$$(\tau \dot{p}_s dS)_{p_s}.$$

There is a similar rate of loss

$$(\tau \dot{p}_s dS)_{p_s + dp_s}$$

due to motion across the opposite face, and so a net rate of increase for this pair of faces

$$-\frac{\partial}{\partial p_s}(\tau \dot{p}_s)\, dp_s dS = -\frac{\partial}{\partial p_s}(\tau \dot{p}_s)\, d\Omega.$$

Summing for all the $2N$ pairs of faces, it follows that

$$\frac{\partial \tau}{\partial t} + \sum_1^N \left\{ \frac{\partial}{\partial p_s}(\tau \dot{p}_s) + \frac{\partial}{\partial q_s}(\tau \dot{q}_s) \right\} = 0. \qquad \ldots\ldots(2)$$

Hence
$$\frac{D\tau}{Dt}=\left\{\frac{\partial}{\partial t}+\sum_1^N\left(\dot{p}_s\frac{\partial}{\partial p_s}+\dot{q}_s\frac{\partial}{\partial q_s}\right)\right\}\tau, \qquad \text{......}(Def.)$$
$$=-\tau\sum_1^N\left(\frac{\partial\dot{p}_s}{\partial p_s}+\frac{\partial\dot{q}_s}{\partial q_s}\right)=0. \qquad \text{......(3)}$$

This is Liouville's theorem. If we consider it in terms of an element of phase space moving with the dust of representative points, it states that the extension of any such element is constant throughout the motion. This is of course easily proved directly. The rate of increase of volume is

$$\int_S\sum_1^N(l_s\dot{p}_s+\lambda_s\dot{q}_s)\,dS,$$

where dS is any element of the bounding surface and (l_s, λ_s) the direction cosines of its "normal". But this is equal to

$$\int_\Omega\sum_1^N\left(\frac{\partial\dot{p}_s}{\partial p_s}+\frac{\partial\dot{q}_s}{\partial q_s}\right)d\Omega$$

by Green's theorem, and so vanishes.

The content of Liouville's theorem relevant to the basis of statistical mechanics is that the density of a group of representative points remains constant along their trajectories. If at any time they are distributed with uniform density in the phase space, they will for ever have uniform density. There can therefore be no eventual crowding together of the points into favoured regions of the phase space. If "normal" properties are to be determined by averaging over the phase space, they must be properties true of almost all the phase space, and not properties of special regions, unless various regions are selectively weighted in this averaging. Again, the theorem suggests that no such selective weighting can be legitimate, for there is no natural crowding into one region rather than another, and therefore no excuse for selective weighting. In short, it suggests that the only reasonable choice of weight is the one actually made in statistical mechanics, namely a weight proportional to the extension of the region.

We can perhaps make this choice of weight clearer by a rather different presentation (of essentially the same argument). To each element of phase space $d\Omega$ we can certainly assign a time $t_{d\Omega}$ during which the representative point will lie in $d\Omega$ out of a total interval T, and can thereby define a function of position in the phase space

$$W(p_1, \ldots, q_N)\,d\Omega=\left\{\operatorname*{Lt}_{T\to\infty}\frac{t_{d\Omega}}{T}\right\}. \qquad \text{......(4)}$$

It seems reasonable to assert that this limit exists. It represents the "probability", defined as a limiting frequency ratio, that the representative point lies in $d\Omega$ at any specified epoch t, and is from its definition independent of t. This fits in with our physical preconception that such "probabilities" cannot depend on the epoch of observation.

The function W might be expected to depend on the particular trajectory chosen. No doubt it does so depend for any dynamical system, but it clearly cannot do so in any way which would make any difference to observable quantities, or consistency would vanish from physics. We therefore assume that there is some W, a definite one-valued function of position in the phase space, such that for any trajectory, or at least on the average for all trajectories,

$$KW\,d\Omega \quad (K \text{ constant})$$

represents the frequency ratio with which the representative point lies in $d\Omega$ for an arbitrary choice of epoch t. On this basis the frequency ratio with which the assembly has the property P is

$$\int_{\Omega_1} W\,d\Omega \bigg/ \int_{\Omega} W\,d\Omega, \qquad \text{......(5)}$$

where Ω_1 is that part of the whole phase space Ω in which P holds. Normal properties of the assembly are those for which (5) is effectively unity.

That such a W really exists for any actual assembly is largely a pious hope, but granted its existence its form can be fixed, and we can show that for any Hamiltonian assembly W may be taken to be independent of the coordinates and therefore may be put equal to unity. For since $KW\,d\Omega$ is the frequency ratio for the falling of the representative point in $d\Omega$, the total number in $d\Omega$ out of a "fine dust" of representative points will be effectively $K'W\,d\Omega$. This is the τ of Liouville's theorem, and by repeating his argument we find that

$$\frac{\partial W}{\partial t} + \sum_1^N \left\{ \frac{\partial}{\partial p_s}(W\dot{p}_s) + \frac{\partial}{\partial q_s}(W\dot{q}_s) \right\} = 0.$$

Therefore

$$\sum_1^N \left\{ \frac{\partial}{\partial p_s}\left(-W\frac{\partial H}{\partial q_s} \right) + \frac{\partial}{\partial q_s}\left(W\frac{\partial H}{\partial p_s} \right) \right\} = 0. \qquad \text{......(6)}$$

This is the general partial differential equation which any possible W must satisfy. Obviously a solution is $W = 1$.

The form of (6) shows that W is a *last multiplier** of the system of differential equations (1) which specify the trajectory, and it can be shown that the actual choice of last multiplier satisfying (6) can make no difference in statistical calculations. For if M and N are two last multipliers and a function f is defined by $f = N/M$, then $f = a$ is a uniform integral of the equations of motion (e.g. the energy integral itself), and N/M will be constant throughout the whole of the accessible phase space to which our calculations extend. The function f can therefore be absorbed into K and ignored in all calcula-

* See, for example, Forsyth, *Differential Equations*, ed. 3, § 174.

tions. Since $W = 1$ is one solution of (6) and the simplest, we may legitimately take

$$K\,d\Omega \quad (K \text{ constant})$$

to be the weight to be attached to the element $d\Omega$ in all statistical calculations. Since in any contact transformation the extension of any element of phase space remains unaltered,* the constant K is genuinely invariant and independent of the system of coordinates, provided only that they are Hamiltonian.

* See, for example, Boltzmann, *La théorie des Gaz*, vol. 2, p. 64. This invariance under a contact transformation is the most general assertion we have made about $d\Omega$. It includes its constancy during the motion, since the motion of any Hamiltonian system may be regarded as a succession of infinitesimal contact transformations. This alternative proof of Liouville's theorem is that used by Boltzmann, *loc. cit.*

CHAPTER II

THE GENERAL THEOREMS OF STATISTICAL MECHANICS FOR ASSEMBLIES OF PERMANENT SYSTEMS

§2·1. We shall establish in this chapter all the usual theorems of statistical mechanics for assemblies of permanent (non-combining and non-dissociating) systems, quantized or classical, which are in the highest possible degree independent of one another. These are, of course, the assemblies most amenable to exact treatment, about which most is known. They naturally include perfect gases and crystals, but it is convenient to postpone the actual calculations for crystals (and temperature radiation) to Chapter IV, though they are fully covered by the methods here developed.

The highest degree of independence is attained when it is sufficiently accurate to assume throughout the calculations that the energy of the assembly is the sum of the energies of the individual systems and contains no part depending on the coordinates of more than one such system. On this assumption, universally if sometimes tacitly made, some comment is needed. Such an assembly is, of course, an ideal limit to which an actual assembly may approximate but can never attain. For it is essential to the whole idea of an assembly that it should form a connected dynamical system with a single energy integral but not a number of separate ones. If, indeed, the energy were really entirely independent of such cross terms, which represent the interactions of the systems, the systems would never interact and the assembly would not be connected. We have therefore to assume that some such interactions do occur, but in this limiting case so rarely that their contribution to the total energy of the assembly may be neglected. They still suffice to preserve connection and ensure that only a single energy integral exists. This is an example, of course, of the general assertion underlying the whole theory that, while there must exist mechanisms of interaction, their mere existence is sufficient, their nature being irrelevant to the laws of equilibrium.

§2·11. *The analysis of weight.* We have stated in the introduction that in averaging to determine the equilibrium state we shall attach a weight unity to every distinct accessible state, that is complexion, of the assembly, each complexion being defined by a wave-function linearly independent of all others so used. This means that the average value $\bar{Q}$ of any quantity Q is to be calculated by the equation

$$\bar{Q}\Sigma_\sigma 1 = C\bar{Q} = \Sigma_\sigma Q_\sigma. \qquad \text{......(7)}$$

The summation Σ_σ is taken over all complexions, and Q_σ is the value of the quantity Q for that complexion of the assembly.

This formula can often be further rearranged. In the first place a number Ω_σ of different complexions may for many purposes possess indistinguishable properties, e.g. the same value of Q. We can then regroup the summations of (7) so that it reads

$$C\bar{Q} = \bar{Q}\Sigma_\sigma\Omega_\sigma = \Sigma_\sigma\Omega_\sigma Q_\sigma. \qquad \text{......(8)}$$

It will often be legitimate for conciseness of expression to speak of this group of complexions, between whose properties we cannot or do not care to distinguish, as a statistical state of the assembly and of Ω_σ as the weight of the statistical state σ.

We next recall that our assemblies are always to be regarded as collections of practically independent systems, between which to an approximation usually sufficient there are no interactions. When therefore we construct Schrödinger's equation which determines the wave-functions and energies of the states of the *assembly*, it can immediately be separated into equations for the distinct *systems*. The wave-function Ψ for the assembly can therefore be constructed out of products of the wave-functions ψ for the systems (see § 2·23). Now to any one value of the energy ϵ_r of an individual system there may belong a number ϖ_r of distinct wave-functions for the system, between whose properties we need not distinguish. We speak of ϖ_r as the *weight of the system* in the given state. It is convenient to use the term *degenerate* to describe states for which $\varpi_r > 1$ and *non-degenerate* for $\varpi_r = 1$. If the assembly consists of M such systems and in a particular state of the assembly there are $a_0, a_1, \ldots, a_r, \ldots$ systems in states of weight $\varpi_0, \varpi_1, \ldots, \varpi_r, \ldots$, and *if no special limitations of accessibility arise*, then it is clear that the weight Ω_σ of this group of complexions is given by the equation

$$\Omega_\sigma = \Pi_{(r)}\varpi_r^{a_r}. \qquad \text{......(9)}$$

We can thus (sometimes at least) attach the correct weight Ω_σ to the group of complexions by attaching suitable weights ϖ_r to the system states. But it must be remembered that such system weights have a statistical meaning *only when they are recombined by multiplication over all the systems in the assembly*. Strictly it is only the weights of the complexions that have a statistical application.

The same distinction was relevant in the older classical form of the theory. The classical average is taken over the whole of accessible phase space for the assembly attaching a weight

$$K(dp_1\ldots dq_s)_1 \ldots\ldots (dp_1\ldots dq_s)_M \qquad \text{......(10)}$$

to any element of phase space, K being constant. This can be achieved by attaching a weight

$$K' dp_1 \ldots dq_s \qquad \text{......(11)}$$

to any element of the phase space of each system, but the weights (11) only have a meaning when posterior remultiplication is assumed.

The conventional weight we attach to the cell or state of a *system* cannot be interpreted as proportional to the time which the representative point of the system spends in this cell, as can the weight attached to a cell or state of the assembly. Interpreted so it is definitely wrong. For if the weight for the cell of a classical assembly is given by (10), and if for simplicity we suppose that the energy function contains only square terms, then the average time spent by a selected system in a selected cell can be shown by integration to be

$$K''e^{-2jE_1}(dp_1...dq_s)_1,$$

where E_1 is the energy of the system in $(dp_1...dq_s)_1$, and K'' and j are constants. The conventional weight applied in this sense is wrong. We could if desired take this accurate value instead of the conventional weight, and on re-combining for the cell of the assembly we should obtain

$$K'e^{-2j\Sigma E_1}(dp_1...dq_s)_1 \ldots\ldots (dp_1...dq_s)_M,$$

or

$$K(dp_1...dq_s)_1 \ldots\ldots (dp_1...dq_s)_M$$

as before, since $\Sigma E_1 = E$, a constant. The factors e^{-2jE_1} are thus irrelevant. This digression should make clearer the extremely conventional use to be made of the weights of the system states. As we have said, it is only the weights of the complexions that have a direct statistical interpretation.

§2·2. *Weights of simple systems and the connection with classical phase space.* Let us start by considering the simple case of an ideal linear harmonic oscillator. If its mass is m and its classical frequency ν, its classical Hamiltonian function is

$$H = \frac{1}{2m}p^2 + (2\pi\nu)^2 \frac{m}{2} q^2 = E. \qquad \ldots\ldots(12)$$

The constant E is the total energy which, by solving the corresponding equation of Schrödinger

$$\frac{\partial^2\psi}{\partial q^2} + \frac{8\pi^2 m}{h^2}\left\{E - (2\pi\nu)^2 \frac{m}{2} q^2\right\}\psi = 0, \qquad \ldots\ldots(13)$$

is known to have one of the values

$$E = (n + \tfrac{1}{2})h\nu \quad (n = 0, 1, 2, \ldots). \qquad \ldots\ldots(14)$$

To each such value corresponds just one wave-function.

Now consider any possible orbit in the classical phase space whose equation is given by (12). The area enclosed by this ellipse is easily seen to be E/ν. Any such orbit is a classical possibility but quantum mechanics restricts the possible values of E/ν to $(n + \frac{1}{2})h$. Now an important feature of quantum mechanics is a correspondence in detail between classical and quantized systems by which the two become indistinguishable in the limit for large quantum numbers (or for $h \to 0$). Symbolically

$$\operatorname*{Lt}_{n\to\infty} \text{(Quantized system)} = \text{Classical system.}$$

It is convenient to refer to this as the *limiting principle.** By this principle it is easy to see that for large n, when the permitted orbits are close together but still separate off areas h, an element of phase space of extension h corresponds to each possible quantum state. The rule for averaging attaches the weight unity to each permitted value of E for this system. To conform to the limiting principle the rule should therefore attach the weight unity to an element h of phase space if and when the system can be treated as classical—that is the weight

$$\frac{dpdq}{h}$$

to the element of phase space $dpdq$.

This result is general. To conform to the limiting principle whenever we can treat a system of s freedoms as classical we must attach the weight

$$\frac{dp_1 \dots dq_s}{h^s} \qquad \dots\dots(15)$$

to the element of phase space $dp_1 \dots dq_s$. We shall not attempt to give a formal proof of this result.† It will be sufficient to illustrate it with some selected special cases.

§ **2·21.** *Weights of the states of isotropic oscillators in more than one dimension.* A two-dimensional isotropic harmonic oscillator is suitable for the next illustration of weight counting and of the limiting principle. Its classical Hamiltonian is

$$H = \frac{1}{2m}(p_x{}^2 + p_y{}^2) + (2\pi\nu)^2 \frac{m}{2}(x^2 + y^2) = E. \qquad \dots\dots(16)$$

* The content of this principle is distinct from that of the much wider Correspondence Principle (Bohr) which arose from it and which was finally superseded by the precise relationships of quantum mechanics. But these developments have retained the limiting principle unimpaired.

† In the older quantum theory these weights were justified (to some extent) by the theory of *adiabatic invariants*. [Bohr, *Proc. Camb. Phil. Soc.* Suppl. 1924 or *Zeit. f. Physik*, vol. 13, p. 117 (1923), Van Vleck, "Quantum Principles and Line Spectra", *Nat. Res. Council Bull.* (1926) or Born, *Vorlesungen über Atommechanik* (1925). The theory was begun by Ehrenfest.] The general form of the quantum conditions takes the form

$$J_r(a_1, a_2, \dots) = n_r h \quad (r = 1, \dots, s),$$

where the J's are action variables and the a's external parameters. If the a's are slowly varied during the motion, the J's remain in general invariant. [Burgers, *Com. Phys. Lab. Leiden*, Nos. 145–56 Suppl. 41*c*, *d*, *e* or *Proc. Sect. Sci. Amsterdam*, vol. 25, pp. 849, 918 (1916), p. 1055 (1917); Dirac, *Proc. Roy. Soc.* A, vol. 107, p. 725 (1925).] By varying the a's one should be able to transform "adiabatically" any one system of s freedoms into any other. Since an element of phase space is invariant for a contact transformation

$$\int dp_1 \dots dq_s = \int dJ_1 \dots dw_s,$$

where the w's are angle variables ranging from 0 to 1 along any classical orbit along which the J's are constant. The invariant quantum conditions therefore cut up the phase space into elements of invariable extension, and the assignment of any other weights than those in the text would be inconsistent with this intertransformability of the different systems, which extends even to different states of the same system. [Bohr, *Danske Vid. Selsk. Skrifter*, vol. 11, p. 24 (1918).]

The corresponding wave equation is

$$\frac{\partial^2\psi}{\partial x^2}+\frac{\partial^2\psi}{\partial y^2}+\frac{8\pi^2 m}{h^2}\left\{E-(2\pi\nu)^2\frac{m}{2}(x^2+y^2)\right\}\psi=0. \qquad ......(17)$$

This equation of course separates in x and y. The possible values of the energy are

$$E=(n+1)\,h\nu \quad (n=0,1,2,...), \qquad ......(18)$$

and the corresponding independent wave-functions are

$$\Psi=\psi_{n_1}(x)\,\psi_{n_2}(y) \quad (n_1+n_2=n), \qquad ......(19)$$

where $\psi_{n_1}(x)$, $\psi_{n_2}(y)$ are the wave-functions for an oscillator in one degree of freedom. To any value of E specified in (18) there correspond therefore $n+1$ distinct wave-functions, or $n+1$ distinct states. The weight of the nth state is therefore $n+1$. Now the classical phase space (x,y,p_x,p_y) enclosed by the condition $E=(n+1)\,h\nu$ is

$$\iiiint dx\,dy\,dp_x\,dp_y$$

extended over the region for which

$$\frac{1}{2m}(p_x{}^2+p_y{}^2)+(2\pi\nu)^2\frac{m}{2}(x^2+y^2)\leqslant E.$$

This is a familiar Dirichlet's integral, whose value is

$$\frac{\pi^2E^2}{\Gamma(3)}\frac{4m}{m(2\pi\nu)^2}=\frac{E^2}{2\nu^2}.$$

The extension enclosed between the surfaces $E=(n+1)\,h\nu$ and $E=nh\nu$ is therefore

$$\tfrac{1}{2}h^2\{(n+1)^2-n^2\}=h^2(n+\tfrac{1}{2}),$$

in agreement with the limiting principle, and the weight (15) attached to the classical phase space.

In the same way the number of distinct wave-functions which correspond to the energy value $E=(n+\frac{3}{2})\,h\nu$ for a three-dimensional isotropic oscillator is equal to the number of ways in which positive integral or zero values can be assigned to n_1, n_2, n_3 so that $n_1+n_2+n_3=n$. The weight of the nth state of such a system is therefore $\frac{1}{2}(n+1)(n+2)$. It may again be verified that

$$\int^{(6)}\!\!...\!\int dx\,dy\,dz\,dp_x\,dp_y\,dp_z$$

over the region for which

$$(n+\tfrac{1}{2})\,h\nu\leqslant\frac{1}{2m}(p_x{}^2+p_y{}^2+p_z{}^2)+(2\pi\nu)^2\frac{m}{2}(x^2+y^2+z^2)\leqslant(n+\tfrac{3}{2})\,h\nu$$

is $\frac{1}{6}h^3\{(n+\frac{3}{2})^3-(n+\frac{1}{2})^3\}$, in agreement with the limiting principle.

The number of distinct wave-functions belonging to the energy value

$E=(n+\frac{1}{2}s)h\nu$ for an s-dimensional isotropic oscillator is equal to the number of positive integral or zero solutions of the equation

$$n_1+n_2+\dots+n_s=n,$$

that is

$$\frac{(n+s-1)!}{n!(s-1)!}. \qquad \text{......(20)}$$

§ **2·22.** *Weights of the states of a rigid solid of revolution (diatomic molecule) with or without axial spin.* Another example is the rigid rotator with an axis of symmetry but no spin about that axis, which has important applications to the specific heats of gases. The motion of the centre of gravity separates and can therefore be ignored here.

Let A be the transverse moment of inertia of the molecule, and θ, ϕ the usual spherical polar coordinates of its axis. Then

$$p_\theta=A\dot{\theta}, \quad p_\phi=A\sin^2\theta\dot{\phi},$$

$$H=\frac{1}{2A}\left(p_\theta{}^2+\frac{1}{\sin^2\theta}p_\phi{}^2\right)=E. \qquad \text{......(21)}$$

The corresponding wave equation is

$$\frac{1}{\sin\theta}\frac{\partial}{\partial\theta}\left(\sin\theta\frac{\partial\psi}{\partial\theta}\right)+\frac{1}{\sin^2\theta}\frac{\partial^2\psi}{\partial\phi^2}+\frac{8\pi^2AE}{h^2}\psi=0. \qquad \text{......(22)}$$

The possible values of the energy are

$$E=\frac{h^2}{8\pi^2A}n(n+1) \quad (n=0,1,2,\dots), \qquad \text{......(23)}$$

and the corresponding wave-functions are the $2n+1$ spherical harmonics of order n, namely

$$P_n{}^m(\cos\theta)\,e^{\pm im\phi} \quad (m=0,1,\dots,n). \qquad \text{......(24)}$$

The weight of the nth state is therefore $2n+1$.

It is easy to show that the phase space integral

$$\iiint_0^\pi\int_0^{2\pi}dp_\theta dp_\phi d\theta d\phi,$$

subject to $p_\theta{}^2+p_\phi{}^2/\sin^2\theta\leqslant 2AE$, is equal to $8\pi^2AE$. We can thus again verify the limiting principle and the assignment (15).

When the molecule has axial spin let A, C be the transverse and axial moments of inertia, and θ, ϕ and χ the usual Eulerian coordinates. The angles θ and ϕ must be taken to fix the direction in space of one end of the axis, independently of any directions of rotation since the two ends of the axis may be different. Then

$$p_\theta=A\dot{\theta}, \quad p_\phi=A\sin^2\theta\dot{\phi}+C\cos\theta(\dot{\chi}+\dot{\phi}\cos\theta), \quad p_\chi=C(\dot{\chi}+\dot{\phi}\cos\theta),$$

$$H=\frac{p_\theta{}^2}{2A}+\frac{(p_\phi-p_\chi\cos\theta)^2}{2A\sin^2\theta}+\frac{p_\chi{}^2}{2C}=E. \qquad \text{......(25)}$$

The corresponding wave equation is

$$\frac{1}{\sin\theta}\frac{\partial}{\partial\theta}\left(\sin\theta\frac{\partial\psi}{\partial\theta}\right)+\frac{1}{\sin^2\theta}\frac{\partial^2\psi}{\partial\phi^2}+\left(\frac{A}{C}+\frac{\cos^2\theta}{\sin^2\theta}\right)\frac{\partial^2\psi}{\partial\chi^2}-2\frac{\cos\theta}{\sin^2\theta}\frac{\partial^2\psi}{\partial\chi\,\partial\phi}+\frac{8\pi^2A}{h^2}E\psi=0. \quad\text{......(26)}$$

The possible values of the energy are

$$E=\frac{h^2}{8\pi^2A}n(n+1)+\frac{\tau^2h^2}{8\pi^2}\left(\frac{1}{C}-\frac{1}{A}\right)\quad\begin{pmatrix}n=0,1,2,\dots\\ |\tau|\leqslant n\end{pmatrix}, \quad\text{......(27)}$$

and the corresponding wave-functions are of the form

$$\Theta(\cos\theta)\,e^{i\tau\chi+i\tau'\phi},$$

where τ, τ' are integers and $\Theta(\cos\theta)$ is a polynomial in $\cos\theta$ which is uniquely determined when n, τ and τ' are given, n being not less than the greater of

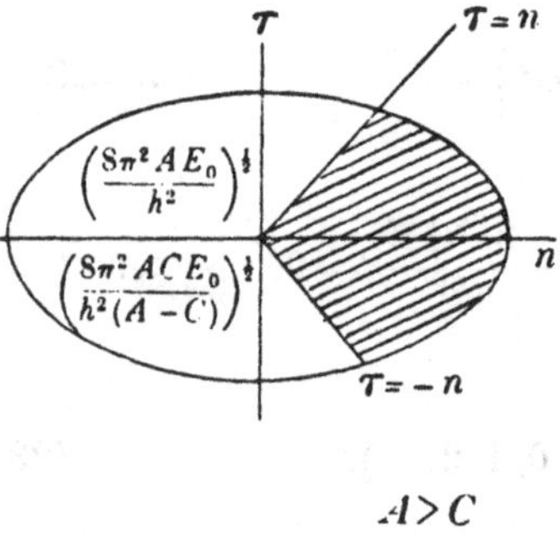

Fig. 1*a*.

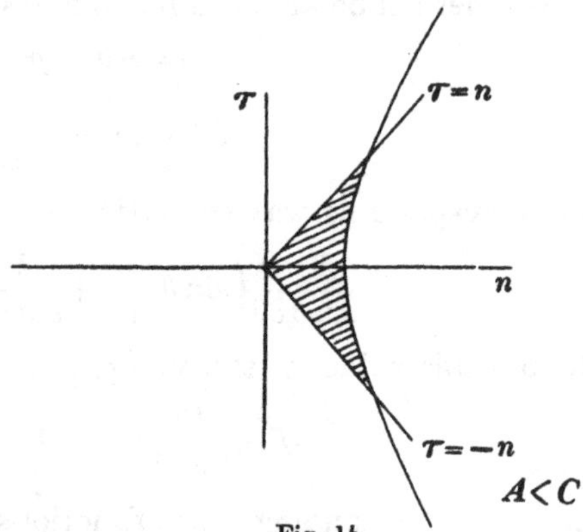

Fig. 1*b*.

$|\tau|$ and $|\tau'|$. Thus there are $2n+1$ distinct wave-functions for each value of E in (27), which is the weight of each of these states.

The verification of (15) and the limiting principle is in this case slightly more intricate and is a good illustration of how the general argument must go. The extension of phase space for which $H\leqslant E_0$ is

$$\iiint\!\!\int_{H\leqslant E_0}\int_0^{\pi}\int_0^{2\pi}\int_0^{2\pi}dp_\theta dp_\phi dp_\chi d\theta d\phi d\chi,$$

which can be shown without difficulty to be

$$\tfrac{32}{3}\pi^3(8A^2C)^{\frac{1}{2}}E_0^{\frac{3}{2}}.$$

We may compare this with the number of values of $E=E(n,\tau)$ given by (27) for which $E(n,\tau)\leqslant E_0$, each pair (n,τ) being counted $2n+1$ times. The relevant pairs (n,τ) are all points of positive integral n and integral τ for which $|\tau|\leqslant n$ and $E(n,\tau)\leqslant E_0$. They are therefore all points of integral coordinates which lie within the shaded region in Fig. 1*a* or 1*b* bounded by two straight lines and the conic (27). Since we are only concerned with asymptotic equalities for large n and τ, we may neglect the difference between $n(n+1)$ and n^2. By calculating the extension of the shaded area with the weighting factor $2n+1$ we find that the number of points of integral coordinates which it contains (with this repetition factor) is asymptotically

$$\tfrac{32}{3}\pi^3(8A^2C)^{\frac{1}{2}}E_0^{\frac{3}{2}}/h^3$$

as required by (15) and the limiting principle.

Other examples will appear at various stages throughout this monograph.

§ **2·23.** *The accessible states (complexions) of simple assemblies.* It is necessary next to consider in greater detail how to enumerate *complexions* in terms of the states of the component systems of an assembly. It is here and only here that an important divergence is possible between the classical and the quantal enumerations. We will consider first for simplicity an assembly of M permanent systems which interact only weakly (or occasionally) with each other. It is convenient to recall here in rather more detail than before the properties of Schrödinger's wave equation.

Let us consider first a single conservative isolated system whose classical Hamiltonian function would be $H(q,p)$, and energy integral

$$H(q,p)=E.$$

The variables q, p are to be taken as a shortened form of $q, q', q'', \ldots, p, p', p'', \ldots$, being any suitable set of Lagrangian coordinates and their conjugated momenta. Then in order to discuss such a quantum system we form Schrödinger's differential equation

$$\left[H\left(q, -i\hbar\frac{\partial}{\partial q}\right)-E\right]\psi=0, \qquad \ldots\ldots(28)$$

in which $\hbar=h/2\pi$. All the properties of the system are determined by the function ψ and the condition that ψ must be bounded* and one-valued in the q-space (configuration space) of the system. For an enclosed system the permissible values of E are necessarily discrete, though if the enclosure is a large one the values of E may lie so close together over some part at least of their range that they may be treated by a limiting process as a continuous distribution. These permissible values will be supposed to be enumerated by a suffix taking the values 1, 2, ..., so that no two values of E with different suffixes are equal. Corresponding to any possible value ϵ_σ of E, there exist one or more proper solutions ψ_σ of equation (28). The number of such independent solutions is denoted by ϖ_σ, the weight of this state. If the system is degenerate ($\varpi_\sigma \neq 1$), we can often reduce it to a non-degenerate one

* More correctly, that $\int|\psi|^2 dq$ taken over any finite part of the configuration space should exist, and that ψ should behave suitably at infinity, the precise behaviour there depending on the nature of the system.

($\varpi_{\sigma'}=1$) by including suitable perturbing fields, if it is convenient to do so for purposes of discussion.

Let us suppose next that our assembly is built up of two such similar systems with very weak interactions, so that to a first approximation the Hamiltonian of the pair is the sum of the separate Hamiltonians. Then the complete wave equation is

$$\left[H\left(q_1, -i\hbar\frac{\partial}{\partial q_1}\right)+H\left(q_2, -i\hbar\frac{\partial}{\partial q_2}\right)-E\right]\Psi=0. \qquad \text{......(29)}$$

It is obvious that the equation separates into two parts and that the permissible values of E are $\epsilon_\sigma+\epsilon_\tau$, and the corresponding solutions

$$\Psi=\psi_\sigma(q_1)\,\psi_\tau(q_2), \qquad \text{......(30)}$$

where ϵ_σ and ϵ_τ are possible values of E in (28) and ψ_σ, ψ_τ corresponding proper solutions. Such values of E are called characteristics or eigenvalues and the corresponding solutions characteristic functions, or eigenfunctions, or, in view of Schrödinger's interpretation, wave-functions. It is of the utmost importance to observe that in the limit of vanishing interaction the pair of systems is essentially degenerate except when $\sigma=\tau$ even if the single systems are not degenerate. For if $\sigma\neq\tau$ at least two wave-functions

$$\Psi_1=\psi_\sigma(q_1)\,\psi_\tau(q_2), \quad \Psi_2=\psi_\sigma(q_2)\,\psi_\tau(q_1),$$

obtained by permuting the individual systems, correspond to the eigenvalue $E=\epsilon_\sigma+\epsilon_\tau$. If the single systems are degenerate ($\varpi_\sigma, \varpi_\tau$), then obviously the total number of distinct wave-functions corresponding to $\epsilon_\sigma+\epsilon_\tau$ is $2\varpi_\sigma\varpi_\tau$ ($\sigma\neq\tau$) or ϖ_σ^2 ($\sigma=\tau$).

The argument is quite general. If the assembly consists of M weakly interacting systems, the complete wave equation is

$$\left[H\left(q_1, -i\hbar\frac{\partial}{\partial q_1}\right)+H\left(q_2, -i\hbar\frac{\partial}{\partial q_2}\right)+\ldots+H\left(q_M, -i\hbar\frac{\partial}{\partial q_M}\right)-E\right]\Psi=0. \qquad \text{......(31)}$$

To the eigenvalue

$$E=\epsilon_\sigma+\epsilon_\tau+\ldots+\epsilon_\omega, \qquad \text{......(32)}$$

no pair of the suffixes σ, τ, ..., ω being equal, there corresponds a set of $M!$ wave-functions obtained by permuting the systems 1, 2, ..., M among the suffixes of the eigenvalue σ, τ, ..., ω. A simple wave-function is

$$\Psi_1=\psi_\sigma(q_1)\,\psi_\tau(q_2)\ldots\psi_\omega(q_M). \qquad \text{......(33)}$$

For degenerate systems the total number of wave-functions corresponding to the eigenvalue (32) is

$$M!\,\varpi_\sigma\varpi_\tau\ldots\varpi_\omega.$$

If finally we consider the completely general eigenvalue

$$E=a_1\epsilon_\sigma+a_2\epsilon_\tau+\ldots+a_l\epsilon_\omega \quad (a_1+a_2+\ldots+a_l=M), \qquad \text{......(34)}$$

where the individual eigenvalues are equal in groups of $a_1, a_2, ..., a_l$, then there are sets of

$$\frac{M!}{a_1!\,a_2!\dots a_l!} \qquad\qquad \text{......(35)}$$

distinct wave-functions obtainable by permutation, of which

$$\Psi_1 = \psi_\sigma(q_1)\dots\psi_\sigma(q_{a_1})\,\psi_\tau(q_{a_1+1})\dots\psi_\tau(q_{a_1+a_2})\dots\psi_\omega(q_M) \quad \text{......(36)}$$

is typical. The total number of distinct wave-functions in the degenerate case is

$$\frac{M!\,\varpi_\sigma^{a_1}\varpi_\tau^{a_2}\dots\varpi_\omega^{a_l}}{a_1!\,a_2!\dots a_l!}. \qquad\qquad \text{......(37)}$$

This is a first and immediate generalization of the elementary enumeration used as an illustration in § 2·11. There we grouped together the complexions for which each separate system has let us say a given energy but may be in a state described by any one of ϖ corresponding wave-functions. Here we are grouping together not merely all such complexions, but also those in which the same numbers of systems, but not the same individual systems, have a given energy. When we form the group of complexions specified here and enumerated in (37), we are not concerned to discriminate between states which have the same distribution of energy over the component systems as a whole.

If to the assembly thus constructed we add another set of similar systems distinct from the set hitherto considered, we obtain obviously a new set of wave-functions similar to (36) each of which can be combined by multiplication with each one of (36) to give an independent wave-function of the complete assembly. We may not of course permute a pair of distinct systems, for we do not so obtain a solution of the wave equation of the assembly. Thus for an assembly of two distinct sets of similar systems the number of complexions corresponding to the eigenvalue

$$E = a_1\epsilon_\sigma + \dots + b_1\eta_{\sigma'} + \dots \qquad\qquad \text{......(38)}$$

is

$$\frac{M!\,N!\,(\varpi_\sigma)^{a_1}\dots(\rho_{\sigma'})^{b_1}\dots}{a_1!\dots b_1!\dots}. \qquad\qquad \text{......(39)}$$

All these formulae are identical with those that have long been classical in statistical mechanics. If the introduction of weak interactions between systems of the assembly or between the assembly and the outside world allows the assembly to pass from a state described by one of these wave-functions to any other, the formulae (37), (39) and similar generalizations must be used as in classical theory and we shall see no change will appear in the results. But this may not be so. It has been shown, principally by Heisenberg,* to whom we owe the first appreciation of the importance of this *exchange degeneracy*, that the wave-functions (36) of a set of similar

* Heisenberg, *Zeit. f. Physik*, vol. 38, p. 411 (1926). See also Dirac, *Proc. Roy. Soc.* A, vol. 112, p. 661 (1926).

systems, after reorganization into suitable linear combinations, necessarily divide into a number of groups, A, B, ..., S. These groups contain between them all the wave-functions belonging to all the eigenvalues, and they possess the extremely important property that *no interaction between the systems of whatever type or strength or between the systems and the outside world, so long as it is symmetrical in the coordinates of the similar systems can ever change the assembly from a wave-function of one group A to a wave-function of any other group B.* Thus if the assembly is originally represented by a wave-function of group A for one set of systems, it will for ever be confined to wave-functions of group A. Only these states are accessible.

§2·24. *Existence theorem for non-combining groups.** The general form of Schrödinger's equation for assembly of M similar systems to the zero order of approximation in which interactions of the systems and external perturbations are neglected, and from which the operator $\partial/\partial t$ has not yet been eliminated, is

$$\left[H\left(q_1, -i\hbar\frac{\partial}{\partial q_1}\right)+H\left(q_2, -i\hbar\frac{\partial}{\partial q_2}\right)+\ldots+H\left(q_M, -i\hbar\frac{\partial}{\partial q_M}\right)-i\hbar\frac{\partial}{\partial t}\right]\Psi^t=0. \quad \ldots\ldots(40)$$

The wave-function Ψ^t must here retain its time factor $e^{-iEt/\hbar}$; its presence is recorded by the affix t. The solutions of this equation are expressions such as (33) (with the time factor), or, since the equation is symmetrical in all the systems and linear in Ψ^t, any expression which can be derived from (33) by permuting the systems and taking linear combinations of any of these permutations. Any such wave-function we will denote for shortness by $\Psi^t(1,2,\ldots,M)$, the order of the suffixes of the systems being in general significant.

If equation (40) were exact, the assembly, once represented by a given wave-function, would remain so represented for all time. Owing however to interactions and perturbations this permanence does not exist and Ψ^t will pass from (approximately) one zero order form $\Psi_1{}^t$ to another $\Psi_2{}^t$ or rather to one of a number of other forms in a manner which can be best specified by certain probability coefficients. It is not necessary to enter into this here beyond observing that the exact Ψ^t has to satisfy exactly an equation of the form

$$\left[\mathbf{H}-i\hbar\frac{\partial}{\partial t}\right]\Psi^t=0, \quad \ldots\ldots(41)$$

* A proof not involving explicit appeal to group theory was first given by Hund, *Zeit. f. Physik*, vol. 43, p. 788 (1927). See also Heisenberg, *Zeit. f. Physik*, vol. 41, p. 239 (1927), and especially Wigner, *Zeit. f. Physik*, vol. 40, pp. 492, 883, vol. 43, p. 624 (1927). These detailed investigations of the structure of the non-combining groups are required for the theory of the structure of atomic and molecular spectra, but are not necessary to us here, where all that we require can be obtained very simply as suggested by Ursell, *Proc. Camb. Phil. Soc.* vol. 24, p. 445 (1928), amplified by Sterne, *Proc. Camb. Phil. Soc.* vol. 26, p. 99 (1930).

where **H** *includes the interaction and perturbation terms and is completely symmetrical in all the systems.*

Let P be any operation of permuting the systems in a given wave-function $\Psi^t(1,2,\ldots,M)$, and taking given linear combinations of these permutations, and let $P\Psi^t$ be the result of the operation. Then since **H** is completely symmetrical $P\Psi^t$ is also a wave-function. Suppose that at any given time $t=t_0$ $P\Psi^t\equiv 0$, the symbol $\equiv$ denoting identity for all values of the space variables $q_1, \ldots, q_M$. Then, since **H** is an operator in the space variables only in which the time enters if at all only as a parameter,

$$[\mathbf{H}]\,P\Psi^t=[\mathbf{H}]\,0=0,$$

so that $\partial(P\Psi^t)/\partial t\equiv 0$. If now we may assume that **H** and $P\Psi^t$ are analytic functions of t for all real values of t, it follows that we can prove, by repeated applications of this argument, that

$$\frac{\partial^n}{\partial t^n}(P\Psi^t)\equiv 0 \qquad \ldots\ldots(42)$$

for all n and therefore that $P\Psi^t\equiv 0$ for all time. This is all that is required to establish the existence of non-combining groups of wave-functions. We say that Ψ is *symmetrical* in the systems $\alpha_1, \alpha_2, \ldots, \alpha_\lambda$ if Ψ does not alter when any one of these numbers is interchanged. For example, if

$$\Psi(1,2,3,\ldots,M)=\Psi(2,1,3,\ldots,M),$$

then Ψ is symmetrical in the systems 1, 2. If

$$\Psi(1,2,3,4,\ldots,M)=\Psi(2,1,3,4,\ldots,M)=\Psi(1,3,2,4,\ldots,M),$$

then Ψ is symmetrical in the systems 1, 2, 3 and so on. But then $P_0\Psi=0$, where P_0 is the operation of interchanging 1 and 2 and taking the difference of these wave-functions, so that the wave-functions representing the assembly if they are ever symmetrical in any pair of systems must always be symmetrical in that pair. In the same way we say that Ψ is *antisymmetrical* in the systems $\alpha_1, \alpha_2, \ldots, \alpha_\lambda$ if Ψ changes to $-\Psi$ when any pair of these numbers is interchanged. It then follows as before that if Ψ is ever antisymmetrical in any pair of systems it is always antisymmetrical in that pair. The different non-combining groups can be specified by the various groups of systems in which they are symmetrical or in which they are antisymmetrical, but we need not examine the structure of these groups here.

§2·25. *The symmetrical and the antisymmetrical group.* From among the various non-combining groups two stand out, conspicuous for the simplicity of their properties and their mathematical form. One is the group of wave-functions which are symmetrical in *all* the systems. This group we shall call simply the symmetrical group S. The other is the group of wave-functions which are antisymmetrical in *all* the systems. This we shall call the antisymmetrical group A. These groups are unique in that for non-degenerate systems they alone contain at most one wave-function for any given eigenvalue $\Sigma \epsilon_\sigma$. There is always exactly one member of the S group, and there is one member of the A group if all the σ's are different or if all the a_l's are 0 or 1, and otherwise no member. This is easily verified if we observe that the antisymmetrical wave-function must take the form of the M-row determinant

$$\begin{vmatrix} \psi_\sigma(q_1)\,\psi_\sigma(q_2)\,\ldots\,\psi_\sigma(q_M) \\ \psi_\tau(q_1)\,\psi_\tau(q_2)\,\ldots\,\psi_\tau(q_M) \\ \ldots\ldots\ldots\ldots\ldots\ldots\ldots\ldots \\ \ldots\ldots\ldots\ldots\ldots\ldots\ldots\ldots \\ \psi_\omega(q_1)\,\psi_\omega(q_2)\,\ldots\,\psi_\omega(q_M) \end{vmatrix}, \qquad \ldots\ldots(43)$$

while the symmetrical wave-function is the same expression with all the signs taken positive.

These very simple enumerations for assemblies of non-degenerate systems can be generalized directly to degenerate ones. For example, if $\varpi_\sigma = 2$, the other ϖ's being 1 as before, there are just two alternative ψ_σ's which may be used in constructing Ψ. There are therefore in the A group clearly 1, 2, 1 or 0 wave-functions according as $a_1 = 0, 1, 2$ or $a_1 > 2$. In the S group there are $(a_1+1)!/a_1!$ or a_1+1 wave-functions instead of 1 as before. The general formulae can be given, but it is simpler to pass from non-degenerate systems to degenerate ones at a later stage. General formulae are therefore omitted here, and we shall confine attention for the present to non-degenerate systems.

We know as yet no *a priori* reason why wave-functions of only one group or of one group rather than another should be found in nature. To determine the proper group we must appeal to observation, and the proper group may vary from system to system. When the systems are electrons or protons (hydrogen nuclei) it is certain that the proper group is the *antisymmetrical*. For electrons this follows from the fact that the laws of interaction of electrons must embody Pauli's exclusion principle which is fundamental to the interpretation of spectra. According to this principle, as we know, two electrons in an atom may never possess the same set of quantum numbers, or as we should now say may never have the same wave-function. The group A is the only group of wave-functions of the assembly which possesses just this property, that it has no member whenever two systems have the same wave-function. Since the wave-functions for the electrons in any atom belong to the group A, one must suppose that this is due to the nature of the electron and that the wave-functions for the electrons in any other assembly will also belong to group A. For protons the evidence is not so extensive, as it depends only on the interpretation of the hydrogen band spectrum and the theory of the specific heat of hydrogen at low temperatures (§ 3·4). It is however sufficient to be convincing. More complicated systems such as atoms or molecules may require the use of either group A or group S, but discussion of such cases is postponed to Chapter v.

§ **2·26.** *The accessible states* (*complexions*) *of a simple assembly* (*concluded*). The foregoing considerations must now be applied and the enunciation of complexions completed. There are two types of simple assembly to consider for one of which the symmetry requirements modify the count, while for the other they make no difference.

Type I. Consider first an assembly containing a set of M similar linear harmonic oscillators, which have fixed positions in the assembly. They may for example be thought of as electrons suitably bound to the atoms of a solid. The wave-functions for electron p bound to atom α and in its n-quantum state of oscillation may be written $\psi_n{}^\alpha(x_p)$. When $a_0, a_1, \ldots, a_t$ are the numbers of oscillators with $0, 1, \ldots, t$ quanta of vibration respectively, we can make up

$$\frac{M!}{a_0!\,a_1!\ldots a_t!} \qquad \ldots\ldots(35)$$

distinct wave-functions of the type (36) belonging to the statistical state of the group. In all these wave-functions each electron belongs to the same atom and they are not antisymmetrical in the electrons. But by permuting the electrons p among the atoms α we can make up exactly one wave-function from each of those enumerated in (35) which is antisymmetrical in the electrons so that the enumeration in (35) is correct though it ignored symmetry requirements and treated the electrons as permanently attached to their own atoms. We shall discuss such short cut methods of enumeration more systematically at a later stage in § 5·11.

Type II. Consider next an assembly containing a set of M electrons (or other systems all similar) free to move about in the *same* enclosure of given volume. The wave-functions now lose their distinctive affix α. When now $a_0, a_1, \ldots, a_t$ are again the numbers of electrons or other systems in states of given quantum number* n, we can still make up unsymmetrized wave-functions of type (36) to the number (35). But now when we attempt to construct a wave-function for the assembly antisymmetrical in all the electrons, we can construct only one if $a_t \leqslant 1$ for all t and otherwise none; and if we attempt to construct a symmetrical wave-function for another type of system, we can always construct just one, as shown in § 2·25.

The difference arises from the loss of the distinctive affix α; physically the electrons or other systems now all belong to the same region of space. It is

* Here n can be taken to be an abbreviation for the whole set of quantum numbers $n_1, n_2, \ldots$ defining the state.

no longer physically significant to assert that an n-quantum electron is *here* and an n'-quantum electron *there*, whereas for a set of localized oscillators such statements are significant. The correctness of the two types of enumeration is best seen by considering simple examples, for instance one in which $M=3$, $a_0=2$, $a_1=1$, and wave-functions are required symmetrical in the three systems. In the first case the $3!/2!\,1!=3$ symmetrical wave-functions are

(i) $\psi_0{}^1(x_1)\,\psi_0{}^2(x_2)\,\psi_1{}^3(x_3)+\psi_0{}^1(x_2)\,\psi_0{}^2(x_3)\,\psi_1{}^3(x_1)+\psi_0{}^1(x_3)\,\psi_0{}^2(x_1)\,\psi_1{}^3(x_2)$
$+\psi_0{}^1(x_2)\,\psi_0{}^2(x_1)\,\psi_1{}^3(x_3)+\psi_0{}^1(x_1)\,\psi_0{}^2(x_3)\,\psi_1{}^3(x_2)+\psi_0{}^1(x_3)\,\psi_0{}^2(x_2)\,\psi_1{}^3(x_1)$,

(ii) $\psi_0{}^1(x_1)\,\psi_1{}^2(x_2)\,\psi_0{}^3(x_3)+\psi_0{}^1(x_2)\,\psi_1{}^2(x_3)\,\psi_0{}^3(x_1)+\psi_0{}^1(x_3)\,\psi_1{}^2(x_1)\,\psi_0{}^3(x_2)$
$+\psi_0{}^1(x_2)\,\psi_1{}^2(x_1)\,\psi_0{}^3(x_3)+\psi_0{}^1(x_1)\,\psi_1{}^2(x_3)\,\psi_0{}^3(x_2)+\psi_0{}^1(x_3)\,\psi_1{}^2(x_2)\,\psi_0{}^3(x_1)$,

(iii) $\psi_1{}^1(x_1)\,\psi_0{}^2(x_2)\,\psi_0{}^3(x_3)+\psi_1{}^1(x_2)\,\psi_0{}^2(x_3)\,\psi_0{}^3(x_1)+\psi_1{}^1(x_3)\,\psi_0{}^2(x_1)\,\psi_0{}^3(x_2)$
$+\psi_1{}^1(x_2)\,\psi_0{}^2(x_1)\,\psi_0{}^3(x_3)+\psi_1{}^1(x_1)\,\psi_0{}^2(x_3)\,\psi_0{}^3(x_2)+\psi_1{}^1(x_3)\,\psi_0{}^2(x_2)\,\psi_0{}^3(x_1)$;

in the second case the only symmetrical wave-function is

$$\psi_0(x_1)\,\psi_0(x_2)\,\psi_1(x_3)+\psi_0(x_2)\,\psi_0(x_3)\,\psi_1(x_1)+\psi_0(x_3)\,\psi_0(x_1)\,\psi_1(x_2).$$

The result of these restrictions is that in enumerating the accessible states of a group of similar non-degenerate systems which belong to one and the same locality in the assembly we have to replace the classical number of complexions

$$\frac{M!}{a_0!\,a_1!\ldots a_t!},\qquad\qquad \ldots\ldots(35)$$

which is still valid when the systems are separately localized, by 1 when the wave-functions are all confined to the group S, and by

$$1\quad [a_t\leqslant 1\ (\textit{all } t)],\qquad 0\quad [a_t>1\ (\textit{some } t)]$$

when the wave-functions are confined to the group A.

For degenerate separately localized systems the more general (39) still gives the correct number of complexions.

For simplicity of exposition we shall consider first simple assemblies for which the formulae (35) or (39) correctly enumerate the complexions. We proceed in the following sections to develop a general method for calculating average values by equation (8) for such assemblies. We return in § 2·4 to the more complicated calculations required to deal with assemblies in which (35) must be modified in the way detailed above. These modifications apply to any assembly containing a gaseous phase.

The distinction between these two cases is strictly apparent rather than real; it depends rather on a difference in the convenient first approximation than on anything more fundamental. We shall be able later on to trace the connection between these two limiting cases.

§ 2·3. *A simple case of Type I. An assembly of two sets of simple harmonic linear oscillators.* For simplicity of exposition we consider first this special case which will serve to bring out all the distinctive features of the problem and the method. Let us suppose that the assembly consists of two large sets of linear harmonic oscillators A and B, of total numbers M and N. As we have seen each oscillator has a series of stationary states of weight unity in which its energy takes the values $(n+\frac{1}{2})h\nu_1$ or $(n+\frac{1}{2})\epsilon$, and $(n+\frac{1}{2})h\nu_2$ or $(n+\frac{1}{2})\eta$. We may emphasize once again that in assigning individual stationary states and energies to the systems separately we tacitly assume that they are practically independent systems, each pursuing its own motion undisturbed for the greater part of time. This is essential to the energy specification and therefore essential to the treatment of assemblies composed of large numbers of practically independent systems. At the same time we must assume that exchanges of energy between the oscillators are possible and do occasionally take place, otherwise the systems will not form a connected assembly and obviously cannot possess unique equilibrium distribution laws. In the present very special case we may think of the exchanges of energy as effected by a few free atoms in an enclosure containing the oscillators—so few in number compared with the oscillators that we may ignore their energy altogether. (Later on we shall be able to include the energy of any number of such atoms or molecules in our discussion, as well as the energy of temperature radiation. The latter can then also be regarded as an agent of energy exchange.)

For the purposes of the proposed proof we must suppose that ϵ and η are commensurable and shall therefore suppose that they are integers with no common factor. This amounts to making a special choice of the unit of energy. The removal of this restriction by a limiting process will be considered at a later stage.

It is our object to determine the distribution laws of this assembly, that is, the equilibrium or average distribution of the oscillators among the various states of which they are capable. This is its only normal property of importance. A specification of this distribution—equilibrium or not—may, as we have said, be referred to as a specification of the *statistical state of the assembly*. This conveys correctly the idea that it is only the macroscopic state of the assembly that really interests us, not the microscopic state. If for example 47 systems A have the energy 5ϵ, we are not interested in which of the M systems these 47 may be. At the same time since the systems are localized a statement that such and such of the M systems are these 47 is significant.

An accessible statistical state of the assembly can be specified by any set of positive integers* (a_r, b_s) subject to the conditions

$$\Sigma_r a_r = M, \quad \Sigma_s b_s = N, \quad \Sigma_r r\epsilon a_r + \Sigma_s s\eta b_s = E. \qquad \ldots\ldots(44)$$

In the third of equations (44) we have omitted the $\frac{1}{2}$-quanta of energy on

* I.e. positive or zero. We use this convention throughout.

each oscillator (*the zero-point energy*), so that E is the energy of the assembly in excess of its minimum value, the zero-point energy. Only energy differences are physically significant (at least in non-relativistic theory), and for any assembly we can always assign the energy zero at will. Here we have assigned it so that the least energy the assembly can have is zero, but this is not always the most convenient choice. It is further to be understood in (44) that a_r denotes the number of systems of type A in the rth state of (extra) energy $r\epsilon$ and b_s the number of type B in the sth state of (extra) energy $s\eta$. The conditions (44) express the facts that there are M A's and N B's in all, and that the total energy is E. Since the assembly is of type I and the system weights are all unity, the number of complexions corresponding to this statistical state is

$$\frac{M!}{a_0!\,a_1!\ldots}\,\frac{N!}{b_0!\,b_1!\ldots}. \qquad \ldots\ldots(45)$$

The total number of complexions C is given by the equation

$$C=\Sigma_{a,b}\frac{M!}{a_0!\,a_1!\ldots}\,\frac{N!}{b_0!\,b_1!\ldots}, \qquad \ldots\ldots(46)$$

summed for all $a, b \geqslant 0$ subject to (44).

The equilibrium distribution laws for the assembly are to be obtained by averaging over all complexions. We can therefore find at once an expression for the average value $\overline{a_r}$ of a_r, or of any similar quantity, for we have, as in (8),

$$C\overline{a_r}=\Sigma_{a,b}\frac{a_r M!}{a_0!\,a_1!\ldots}\,\frac{N!}{b_0!\,b_1!\ldots}. \qquad \ldots\ldots(47)$$

The most important such quantity is $\overline{E_A}$, the average energy of the systems A. This is given by

$$C\overline{E_A}=\Sigma_{a,b}\frac{(\Sigma_r\, r\epsilon a_r)\,M!}{a_0!\,a_1!\ldots}\,\frac{N!}{b_0!\,b_1!\ldots}. \qquad \ldots\ldots(48)$$

Both these summations are of course over the same range of values as (46).

A rapid and powerful method of evaluating these sums is essential to the elegance of this development, and is provided by expressing them as contour integrals and evaluating the integrals by the method of steepest descents.

Now the general term of

$$(1+z^{\epsilon}+z^{2\epsilon}+\ldots)^M$$

expanded in powers of z is

$$\frac{M!}{a_0!\,a_1!\ldots}\,z^{\Sigma_r r\epsilon a_r}\quad(\Sigma_r a_r=M), \qquad \ldots\ldots(49)$$

and similarly of

$$(1+z^{\eta}+z^{2\eta}+\ldots)^N,$$

$$\frac{N!}{b_0!\,b_1!\ldots}\,z^{\Sigma_s s\eta b_s}\quad(\Sigma_s b_s=N).$$

By multiplying these series together it follows that the coefficient of z^E in

$$(1+z^\epsilon+z^{2\epsilon}+\dots)^M\,(1+z^\eta+z^{2\eta}+\dots)^N$$

is

$$\Sigma_{a,b}\frac{M!}{a_0!\,a_1!\dots}\frac{N!}{b_0!\,b_1!\dots},$$

summed for all positive values of the a's and b's subject to $\Sigma_r a_r = M$, $\Sigma_s b_s = N$ and also $\Sigma_r r\epsilon a_r + \Sigma_s s\eta b_s = E$, in short the conditions (44). *Thus C is the coefficient of z^E in*

$$(1-z^\epsilon)^{-M}\,(1-z^\eta)^{-N}. \qquad \dots\dots(50)$$

Similarly, observing the extra factor in (48), we find at once that $C\overline{E_A}$ is the coefficient of z^E in

$$(1-z^\eta)^{-N}\,z\frac{d}{dz}(1+z^\epsilon+z^{2\epsilon}+\dots)^M, \qquad \dots\dots(51)$$

for operating with $z\,d/dz$ introduces in the terms (49) just the required extra factor $\Sigma_r r\epsilon a_r$. The expression (51) can be put in the alternative forms

$$\left\{z\frac{d}{dz}(1-z^\epsilon)^{-M}\right\}(1-z^\eta)^{-N}, \qquad \dots\dots(52)$$

$$(1-z^\epsilon)^{-M}\,(1-z^\eta)^{-N}\left\{-Mz\frac{d}{dz}\log(1-z^\epsilon)\right\}. \qquad \dots\dots(53)$$

It thus appears that the required sums are the coefficients in certain simple power series, and, as we shall see, this conclusion is capable of immediate extension to general localized systems. Now the most convenient expressions for such coefficients are complex integrals taken round a contour enclosing the origin $z=0$. Thus we find

$$C=\frac{1}{2\pi i}\int_\gamma\frac{dz}{z^{E+1}}\frac{1}{(1-z^\epsilon)^M\,(1-z^\eta)^N}, \qquad \dots\dots(54)$$

$$C\overline{E_A}=\frac{1}{2\pi i}\int_\gamma\frac{dz}{z^{E+1}}\frac{-Mz\frac{d}{dz}\log(1-z^\epsilon)}{(1-z^\epsilon)^M\,(1-z^\eta)^N}. \qquad \dots\dots(55)$$

The contour γ may of course be any contour lying within the circle of convergence of these power series (radius unity) and circulating once counterclockwise round $z=0$.

We can determine a similar formula for $C\overline{a_r}$. We have

$$C\overline{a_r}=\Sigma_{a,b}\frac{a_r\,M!}{a_0!\,a_1!\dots}\frac{N!}{b_0!\,b_1!\dots},$$

$$=M\Sigma_{\alpha,b}\frac{(M-1)!}{\alpha_0!\,\alpha_1!\dots}\frac{N!}{b_0!\,b_1!\dots},$$

summed over all positive α, b subject to

$$\Sigma_r\alpha_r=M-1,\quad \Sigma_s b_s=N,\quad \Sigma_r r\epsilon\alpha_r+\Sigma_s s\eta b_s=E-r\epsilon.$$

This reduces at once to

$$\overline{Ca_r} = \frac{M}{2\pi i}\int_\gamma \frac{dz}{z^{E+1}} \frac{z^{r\epsilon}}{(1-z^\epsilon)^{M-1}(1-z^\eta)^N}. \qquad \text{......(56)}$$

These integrals are exact and hold for all values of M, N and E. They are, however, only physically significant when M, N and E are very large, since all assemblies that we can observe contain a very great number of systems. We therefore require primarily the asymptotic values of these integrals when M, N and E tend to infinity in fixed ratios. This means, physically, that we require the limiting properties of the assembly when its size tends to infinity without alteration of its *intensive* properties (constitution, etc.). The properties of the finite assembly will be shown to deviate only trivially from these limiting properties. These asymptotic values can be rigorously established here and in the general case by the method of steepest descents. It will tend to clarity first to sketch this method and the results, and to compare the results with those of other developments.

Consider the integrand on the positive real axis. It tends to infinity as $z \to 0$ and $z \to 1$, and somewhere between, at $z = \vartheta$, there is a unique *minimum*. For γ take the circle of radius ϑ and centre the origin. Then for values of z on γ, $z = \vartheta e^{i\alpha}$ say, $z = \vartheta$, $\alpha = 0$ is, when M, N and E are large, a strong *maximum* of the modulus of the integrand. Owing also to the fact that the differential coefficient of the integrand vanishes at $\alpha = 0$, the complex terms there are trivial and the whole effective contribution to the integral comes from very near this point. This remains true so long as there are no equal maxima elsewhere, which cannot occur when ϵ and η have no common factor. It remains true, moreover, when there are extra factors such as

$$-Mz\frac{d}{dz}\log(1-z^\epsilon),$$

and in effect such extra factors may be taken outside the sign of integration if in them we replace z by ϑ. The result is that ϑ is the unique positive fractional root of

$$\frac{d}{dz}\{z^{-E}(1-z^\epsilon)^{-M}(1-z^\eta)^{-N}\} = 0, \qquad \text{......(57)}$$

or

$$E = \frac{M\epsilon}{\vartheta^{-\epsilon}-1} + \frac{N\eta}{\vartheta^{-\eta}-1}. \qquad \text{......(58)}$$

By comparing (54) and (55) we find

$$\overline{E_A} = -M\vartheta\frac{d}{d\vartheta}\log(1-\vartheta^\epsilon),$$

$$= \frac{M\epsilon}{\vartheta^{-\epsilon}-1}. \qquad \text{......(59)}$$

Similarly,
$$\overline{E_B}=\frac{N\eta}{\vartheta^{-\eta}-1}, \quad \text{......(60)}$$

which satisfies the essential equation $E=\overline{E_A}+\overline{E_B}$. These equations determine the partition of energy among the two sets of oscillators in a large assembly.

It is already suggested by these formulae that ϑ is a function of the state of the assembly with the properties of *temperature*, and it turns out later that ϑ may be taken to represent the temperature on a special scale. It bears the relation
$$\vartheta=e^{-1/kT} \quad \text{......(61)}$$

to the absolute temperature T, where k is Boltzmann's constant. If we use this result in advance, then
$$\overline{E_A}=\frac{M\epsilon}{e^{\epsilon/kT}-1}, \quad \text{......(62)}$$

which is the familiar result due to Planck. In the same way, by comparing (54) and (56), we find
$$\begin{aligned}\overline{a_r}&=M\vartheta^{r\epsilon}(1-\vartheta^{\epsilon}),\\ &=Me^{-r\epsilon/kT}(1-e^{-\epsilon/kT}). \quad \text{......(63)}\end{aligned}$$

This result, in the more usual form
$$\frac{\overline{a_r}}{\overline{a_{r'}}}=\frac{e^{-r\epsilon/kT}}{e^{-r'\epsilon/kT}}, \quad \text{......(64)}$$

is also classical and due to Planck.

Logically the relation (61) must be deduced from the second law of thermodynamics, by which alone the *absolute* temperature scale can be defined. But we may anticipate this, if it is preferred, by asserting that in the limit in which $\nu\to 0$ and so $\epsilon\to 0$ the mean energy of a simple linear oscillator must be kT. This assertion then defines T on the basis of the theorem of equipartition of energy for classical systems which we have not yet proved (see § 2·62). It then follows from (59) that
$$kT=\operatorname*{Lt}_{\epsilon\to 0}\frac{\epsilon}{\vartheta^{-\epsilon}-1}=\frac{1}{\log 1/\vartheta},$$
which is (61).

We give the mathematical theorems in the next section, and shall there see that the full proof of (62)–(64) avoids all the apparatus of factorials approximated to by Stirling's theorem which disfigure the usual proofs. The use of this theorem is both cumbersome and, at least superficially, lacking in rigour, for it is often applied to 0! and 1!.

§2·31. *Application of the method of steepest descents.* We base the proof of the foregoing results on the following:

Theorem 2·31. *If*

(i) $\phi(z)$ *is a regular analytic function of z expressible in the form*

$$\phi(z) = z^{-\alpha_0}\{f_1(z)\}^{\alpha_1}\{f_2(z)\}^{\alpha_2}\ldots,$$

where the α's are positive constants, integral after multiplication by E, and the f(z)'s are power series in z which start with non-zero constant terms and have real positive integral coefficients and radii of convergence unity;

(ii) *Not all the indices in all the f(z)'s contain a common factor other than unity;*

(iii) $F(z)$ *is a regular analytic function with no singularity in the unit circle except perhaps a pole at* $z=0$;

(iv) γ *is a contour circulating once counter-clockwise round* $z=0$;

then

$$\frac{1}{2\pi i}\int_\gamma F(z)\,[\phi(z)]^E\frac{dz}{z} = \frac{[\phi(\vartheta)]^E}{[2\pi E\vartheta^2\phi''(\vartheta)/\phi(\vartheta)]^{\frac{1}{2}}}\{F(\vartheta)+O(1/E)\}, \quad\ldots\ldots(65)$$

where ϑ *is the unique positive fractional root of*

$$\frac{d\phi(z)}{dz} = 0. \quad\ldots\ldots(66)$$

We have not aimed at maximum generality, but only at a theorem sufficient for the purpose in hand. For example, the coefficients of the $f(z)$ need not be integers. It is sufficient to suppose that $f(z)\to\infty$ as $z\to 1$. Nor need the radius of convergence be unity for the purpose of the proof. But both these conditions are always satisfied by quantized systems.

Consider the function $\phi'(z)$ for real positive z. The equation

$$\frac{\phi'(z)}{\phi(z)} = \frac{-\alpha_0}{z} + \frac{\alpha_1 f_1'}{f_1} + \frac{\alpha_2 f_2'}{f_2} + \ldots \quad\ldots\ldots(67)$$

determines $\phi'(z)$. Consider the behaviour of $y = zf_1'/f_1$. This function y by (i) takes the value 0 for $z=0$ and steadily increases to $+\infty$ as z increases to 1. For if $f_1(z) = \Sigma\,\varpi_n z^n$, we have

$$y = \frac{\Sigma\, n\varpi_n z^n}{\Sigma\,\varpi_n z^n}, \quad zy' = \frac{(\Sigma\,\varpi_n z^n)(\Sigma\, n^2\varpi_n z^n) - (\Sigma\, n\varpi_n z^n)^2}{(\Sigma\,\varpi_n z^n)^2}.$$

The numerator of zy' is $\frac{1}{2}\Sigma\Sigma\,(n-n')^2\varpi_n\varpi_{n'}z^{n+n'}$, which is always positive. Therefore either $y\to\infty$ or $y\to A$, a finite limit, as $z\to 1$. But the latter is impossible as it implies that $f_1(z)$ is bounded as $z\to 1$, which is contrary to hypothesis. Thus the expression

$$\frac{\alpha_1 zf_1'}{f_1} + \frac{\alpha_2 zf_2'}{f_2} + \ldots$$

is zero for $z=0$ and steadily increases to $+\infty$ as $z\to 1$. It therefore takes the value α_0 once and once only, for $z=\vartheta$ say, which is then the unique root of (66). It follows also that $\phi''(\vartheta)>0$.

The method of steepest descents proceeds by making the contour γ pass through the col, $\phi'(z)=0$, in such a direction that the value of the integrand falls off *along* γ from a maximum value at the col at the greatest possible rate. This is here achieved (since $\phi(z)$ is real for positive real z) by taking for γ the circle $|z|=\vartheta$. On this circle (or any other) the maximum modulus of the integrand must occur for positive real z on account of (ii). It is easy to show rigorously that when E is large all parts of the contour except that in the immediate neighbourhood of $z=\vartheta$ make contributions exponentially small compared to this critical region. If we put $z=\vartheta e^{i\alpha}$, then when α is small

$$[\phi(z)]^E=[\phi(\vartheta)]^E\exp\{-\tfrac{1}{2}E\alpha^2\vartheta^2\phi''(\vartheta)/\phi(\vartheta)+KE\alpha^3+O(E\alpha^4)\},$$

where K is some function of ϑ. When E is large we may suppose that $E^{\frac{1}{2}}\alpha$ ranges effectively from $-\infty$ to $+\infty$, while α and all terms such as $E\alpha^3$ remain small. Thus

$$\frac{1}{2\pi i}\int_\gamma F(z)\,[\phi(z)]^E\frac{dz}{z}=\frac{1}{2\pi}[\phi(\vartheta)]^E$$
$$\times\int_{-\infty}^{+\infty}\{F(\vartheta)+i\alpha F'(\vartheta)+O(\alpha^2)+KE\alpha^3+O(E\alpha^4)\}\,e^{-\frac{1}{2}E\alpha^2\vartheta^2\phi''(\vartheta)/\phi(\vartheta)}\,d\alpha,$$

the error in taking the range of integration with respect to α infinite instead of some small number such as $kE^{-\frac{1}{3}}$ being exponentially small. Odd terms in α vanish on integration and

$$\int_{-\infty}^{+\infty}e^{-\frac{1}{2}E\alpha^2\vartheta^2\phi''(\vartheta)/\phi(\vartheta)}\,d\alpha=\left\{\frac{2\pi}{E\vartheta^2\phi''(\vartheta)/\phi(\vartheta)}\right\}^{\frac{1}{2}}, \qquad \text{......(68)}$$

$$\int_{-\infty}^{+\infty}\alpha^2e^{-\frac{1}{2}E\alpha^2\vartheta^2\phi''(\vartheta)/\phi(\vartheta)}\,d\alpha=O(E^{-\frac{3}{2}}),$$

$$\int_{-\infty}^{+\infty}E\alpha^4e^{-\frac{1}{2}E\alpha^2\vartheta^2\phi''(\vartheta)/\phi(\vartheta)}\,d\alpha=O(E^{-\frac{3}{2}}).$$

Hence the theorem.

The theorem applies at once to the assembly of two sets of linear oscillators with
$$\phi=z^{-1}(1-z^\epsilon)^{-M/E}\,(1-z^\eta)^{-N/E}\quad (M/E,\ N/E \text{ constant}).$$

All our conclusions hold with, in particular,

$$C=\frac{\vartheta^{-E}(1-\vartheta^\epsilon)^{-M}\,(1-\vartheta^\eta)^{-N}}{\{2\pi E\vartheta^2\phi''(\vartheta)/\phi(\vartheta)\}^{\frac{1}{2}}}\left[1+O\left(\frac{1}{E}\right)\right]. \qquad \text{......(69)}$$

Since we always suppose that the assembly is very large (E large) we shall in all formulae omit the factor $[1+O(1/E)]$ which is always present, and preserve only the limiting asymptotic form, which gives us all our results. We must be careful, however, not to overlook its presence in any formula in which the leading terms cancel.

The expression $E\vartheta^2\phi''(\vartheta)/\phi(\vartheta)$ can be simplified. Since $\phi'(\vartheta)=0$ it is equal to

$$E\left(\vartheta\frac{d}{d\vartheta}\right)^2\log\phi(\vartheta)=E\vartheta\frac{d}{d\vartheta}\left[-1+\frac{(M\epsilon/E)\,\vartheta^\epsilon}{1-\vartheta^\epsilon}+\frac{(N\eta/E)\,\vartheta^\eta}{1-\vartheta^\eta}\right],$$

$$=\vartheta\frac{\partial E}{\partial\vartheta}, \qquad \text{......(70)}$$

if E is regarded as a function of ϑ determined by the relation (58), that is,

$$E=\frac{M\epsilon}{\vartheta^{-\epsilon}-1}+\frac{N\eta}{\vartheta^{-\eta}-1}.$$

This relation (70) is valid generally.

Finally, condition (ii) is inessential. If it fails, *all* integrals such as C and $C\overline{E_A}$ are apparently β times as great as before, where β is the common factor in the indices of the $f(z)$. This is true because the function $F(z)$ is always composed of a selection of terms from the series in $\phi(z)$. From each of the β equal maxima of $|\phi(z)|$ on the circle $|z|=\vartheta$ we get a real, positive contribution, equal to the contribution from the main maximum on the positive real axis. But it is easy to see that this extra factor is only formal, and that C, for example, is unaffected by changing the unit of energy so as to insert or remove a common factor β. Thus no physical result is affected.

§ **2·32.** *Generalization to an assembly of any localized quantized systems and to any number of types of system.* Nothing in the preceding work depends essentially on the fact that we are discussing simple harmonic linear oscillators. Suppose instead we have two sets of *localized* systems A and B. Systems A are M in number; their sequence of stationary states has energies $\epsilon_0, \epsilon_1, \ldots, \epsilon_r, \ldots$, weights $\varpi_0, \varpi_1, \ldots, \varpi_r, \ldots$, and a statistical distribution specified by $a_0, a_1, \ldots, a_r, \ldots$. Systems B are N in number; their sequence of stationary states has energies $\eta_0, \eta_1, \ldots, \eta_s, \ldots$, weights $\rho_0, \rho_1, \ldots, \rho_s, \ldots$, and a statistical distribution specified by $b_0, b_1, \ldots, b_s, \ldots$. We assume for the present that all the ϵ's and η's are expressible as integral multiples of one basic unit of energy (and for simplicity not all expressible in the form $\alpha+r\beta$, $\beta>1$). Then in this case the total number of complexions representing this statistical state of the assembly is

$$C=\Sigma_{a,b}\frac{M!\,N!\,\varpi_0^{a_0}\varpi_1^{a_1}\ldots\rho_0^{b_0}\rho_1^{b_1}\ldots}{a_0!\,a_1!\ldots b_0!\,b_1!\ldots}, \qquad \text{......(71)}$$

summed for all positive a, b subject to

$$\Sigma_r a_r=M,\quad \Sigma_s b_s=N,\quad \Sigma_r\epsilon_r a_r+\Sigma_s\eta_s b_s=E. \qquad \text{......(72)}$$

We have here to form the functions

$$f(z)=\varpi_0 z^{\epsilon_0}+\varpi_1 z^{\epsilon_1}+\varpi_2 z^{\epsilon_2}+\ldots, \qquad \text{......(73)}$$

$$f'(z)=\rho_0 z^{\eta_0}+\rho_1 z^{\eta_1}+\rho_2 z^{\eta_2}+\ldots, \qquad \text{......(74)}$$

which from their special properties in the development of the theory we call *partition functions*. They are equivalent to the functions introduced by Planck under the name *Zustandsumme*, and are the transcription into the quantum theory of Gibbs' phase integrals.

Just as in § 2·3 it follows from the multinominal theorem that C is the coefficient of z^E in $[f(z)]^M [f'(z)]^N$. It follows at once that

$$C = \frac{1}{2\pi i}\int_\gamma \frac{dz}{z^{E+1}} [f(z)]^M [f'(z)]^N, \qquad \text{......(75)}$$

and, similarly, that

$$C\overline{E_A} = \frac{1}{2\pi i}\int_\gamma \frac{dz}{z^{E+1}} \left\{z\frac{d}{dz}[f(z)]^M\right\} [f'(z)]^N, \qquad \text{......(76)}$$

$$C\overline{a_r} = \frac{M}{2\pi i}\int_\gamma \frac{dz}{z^{E+1}} \varpi_r z^{\epsilon_r} [f(z)]^{M-1} [f'(z)]^N. \qquad \text{......(77)}$$

Theorem 2·31 applies to these integrals with $\phi = z^{-1}[f(z)]^{M/E}[f'(z)]^{N/E}$, provided that the partition functions converge for $|z| < 1$. If the sequence of energies can be expressed as here supposed and does not terminate, the series must converge for $|z| < 1$. For, if the system is of s degrees of freedom and non-degenerate, the ϖ's are all unity and the radius of convergence must be unity. If the system then degenerates until the energy depends on only u ($< s$) independent quantum numbers, the new partition function can be formed from the old by the grouping together of sets of terms whose energies are no longer distinct. This cannot alter the radius of convergence. We find, therefore, that ϑ is determined as the unique root of

$$E = M\vartheta\frac{d}{d\vartheta}\log f(\vartheta) + N\vartheta\frac{d}{d\vartheta}\log f'(\vartheta), \qquad \text{......(78)}$$

and

$$\overline{E_A} = M\vartheta\frac{d}{d\vartheta}\log f(\vartheta) = MkT^2\frac{d}{dT}\log f(T), \qquad \text{......(79)}$$

$$\overline{a_r} = M\varpi_r \vartheta^{\epsilon_r}/f(\vartheta) = M\varpi_r e^{-\epsilon_r/kT}/f(T). \qquad \text{......(80)}$$

There are similar formulae for systems B.

Obviously the restriction to two types of system is trivial. With any number of types of system the arguments are unaltered. We find for the total number of weighted complexions

$$C = \frac{1}{2\pi i}\int_\gamma \frac{dz}{z^{E+1}} \Pi_\tau [f_\tau(z)]^{M_\tau}. \qquad \text{......(81)}$$

There is a unique ϑ determined by

$$E = \Sigma_\tau M_\tau \vartheta\frac{d}{d\vartheta}\log f_\tau(\vartheta), \qquad \text{......(82)}$$

and

$$\overline{E_\tau} = M_\tau\vartheta\frac{d}{d\vartheta}\log f_\tau(\vartheta) = M_\tau kT^2\frac{d}{dT}\log f_\tau(T), \qquad \text{......(83)}$$

$$(\overline{a_r})_\tau = M_\tau(\varpi_r)_\tau \vartheta^{(\epsilon_r)_\tau}/f_\tau(\vartheta) = M_\tau(\varpi_r)_\tau e^{-(\epsilon_r)_\tau/kT}/f_\tau(T). \qquad \text{......(84)}$$

The partition function $f(z)$ is presumed to refer to the whole motion of the localized system. It should be observed that, in the important special case in which the motion splits up into two or more parts entirely independent of one another, the partition function $f(z)$ must factorize into functions of the same type, which refer separately to the independent motions. A particular case of this factorization occurs for the translatory motion and the internal motions and rotations of a free molecule, but in general free molecules are members of a gaseous phase and their complexions require an enumeration of type II.

We may properly comment at this stage on the properties of the parameter ϑ which, while mathematical in origin, is obviously fundamental in describing the state of the assembly, and should be identifiable by analogy with some physical property of the assembly. We have already stated in advance that ϑ measures the temperature. We can now see reason to justify this identification, though not of course the particular relation between ϑ and T. For ϑ *is a parameter helping to define the state of our assembly which must have the same value for all sets of systems in the assembly*. This is the precise property which distinguishes the temperature from other parameters and justifies the identification.*

It is natural at this stage to consider a few examples of special systems and construct their partition functions.

§ **2·33.** *Partition functions for two- and three-dimensional isotropic harmonic oscillators*. These are localized degenerate systems. The two-dimensional oscillator has (as we have seen in § 2·21)

$$\varpi_n = n+1, \quad \epsilon_n = (n+1)h\nu = (n+1)\epsilon,$$

and therefore,

$$f(z) = z^{\epsilon} + 2z^{2\epsilon} + 3z^{3\epsilon} + \ldots$$
$$= z^{\epsilon}(1-z^{\epsilon})^{-2}. \qquad \ldots\ldots(85)$$

By a different choice of energy zero, eliminating the zero-point energy, the extra factor z^{ϵ} is eliminated from every term and from $f(z)$. The three-dimensional oscillator has $\epsilon_n = (n+\frac{3}{2})\epsilon$ and $\varpi_n = \frac{1}{2}(n+1)(n+2)$ so that

$$f(z) = z^{\frac{3}{2}\epsilon}[1 + 3z^{\epsilon} + 6z^{2\epsilon} + \ldots + \tfrac{1}{2}(n+1)(n+2)z^{n\epsilon} + \ldots]$$
$$= z^{\frac{3}{2}\epsilon}(1-z^{\epsilon})^{-3}. \qquad \ldots\ldots(86)$$

The zero-point energy can again be eliminated. These are simple examples of weight counting. Thus in these two cases

$$\overline{E_A} = \frac{2M\epsilon}{\vartheta^{-\epsilon}-1} + M\epsilon, \quad \frac{3M\epsilon}{\vartheta^{-\epsilon}-1} + \tfrac{3}{2}M\epsilon, \qquad \ldots\ldots(87)$$

which fit in exactly with the expected requirements of two and three times the mean energy of a linear harmonic oscillator respectively.

* See, for example, Born, *Phys. Zeit.* vol. 22, pp. 218, 249, 282 (1921).

An isotropic harmonic oscillator of s degrees of freedom has a partition function

$$f(z) = z^{\frac{1}{2}s\epsilon}(1-z^{\epsilon})^{-s}, \qquad \text{......(88)}$$

and a mean energy s times that of a linear harmonic oscillator. In fact, for most purposes it is precisely equivalent to s independent simple harmonic linear oscillators.

§ **2·34.** *Partition functions for rigid rotators with or without axial spin.* The weights and energies for the states of a rigid rotator are given in § 2·22. Thus for a rigid rotator without axial spin and transverse moment of inertia A

$$f(z) = 1 + 3z^{2\epsilon} + 5z^{6\epsilon} + \ldots + (2n+1)\, z^{n(n+1)\epsilon} + \ldots \quad (\epsilon = h^2/8\pi^2 A). \qquad \text{......(89)}$$

This function can be used at once to determine the distribution laws and energy content of a set of such localized rotators in an assembly. The rotational energy of such a set will therefore be

$$\overline{E_A} = M\vartheta \frac{d}{d\vartheta} \log f(\vartheta)$$

and their contribution to the *specific heat*, C_{rot},

$$C_{\mathrm{rot}} = \frac{d\overline{E_A}}{dT} = Mk(\log\vartheta)^2 \left(\vartheta \frac{d}{d\vartheta}\right)^2 \log f(\vartheta). \qquad \text{......(90)}$$

If

$$\sigma = \frac{h^2}{8\pi^2 AkT} = \frac{h^2}{8\pi^2 A} \log 1/\vartheta,$$

and M refers to one gram-molecule so that $Mk = R$, the gas constant, then

$$C_{\mathrm{rot}} = R\sigma^2 \frac{d^2}{d\sigma^2} \log\left\{\sum_{n=0}^{\infty} (2n+1)\, e^{-n(n+1)\sigma}\right\}. \qquad \text{......(91)}$$

Practical applications of this formula to the specific heats of diatomic gases are made in §§ 3·3–3·4.

We find similarly for a rigid rotator with axial spin that

$$f(z) = \sum_{n=0}^{\infty} \sum_{\tau=-n}^{n} (2n+1)\, z^{n(n+1)\epsilon + \tau^2\epsilon'}, \qquad \text{......(92)}$$

where

$$\epsilon = \frac{h^2}{8\pi^2 A}, \quad \epsilon' = \frac{h^2}{8\pi^2}\left(\frac{1}{C} - \frac{1}{A}\right). \qquad \text{......(93)}$$

For an isotropic rotator $A = C$, and then $f(z)$ reduces to

$$f(z) = \sum_{n=0}^{\infty} (2n+1)^2 z^{n(n+1)\epsilon} \quad \left(\epsilon = \frac{h^2}{8\pi^2 A}\right) \qquad \text{......(94)}$$

In terms of the variables

$$\sigma = \frac{h^2}{8\pi^2 AkT}, \quad \sigma' = \frac{h^2}{8\pi^2 kT}\left(\frac{1}{C} - \frac{1}{A}\right) \qquad \text{......(95)}$$

these functions become

$$f(\sigma,\sigma')=\sum_{n=0}^{\infty}\sum_{\tau=-n}^{n}(2n+1)\,e^{-n(n+1)\sigma-\tau^2\sigma'}, \qquad \text{......(96)}$$

and when $A=C$

$$f(\sigma)=\sum_{n=0}^{\infty}(2n+1)^2\,e^{-n(n+1)\sigma}. \qquad \text{......(97)}$$

These formulae will be applied in §§ 3·3–3·4 to the specific heats of the more symmetrical types of polyatomic molecules, where asymptotic evaluations of the series will be given valid for small σ or large T.

A still more general model can be discussed—namely a rigid body with three unequal moments of inertia containing an internal spinning gyroscope representing electronic orbital or spin angular momentum.* Since however closed formulae for the energy as a function of the quantum numbers cannot be obtained, we shall not consider it further here.

§ **2·4.** *A simple assembly of type II, with two sets of systems confined to a common enclosure.* We shall now develop methods for calculating the equilibrium state of an assembly when the complexions must be enumerated in the manner detailed in § 2·26 for assemblies of type II. For simplicity we shall consider an assembly containing systems of two types A and B, M and N in number, all of whose states are at first non-degenerate, with eigenvalues of the energy

$$\epsilon_1,\ \epsilon_2,\ \ldots,\ \epsilon_l,\ \ldots,\quad \eta_1,\ \eta_2,\ \ldots,\ \eta_l,\ \ldots$$

respectively, all expressible as integers in terms of a suitable unit of energy. The number of systems in the states belonging to these eigenvalues will be specified as usual by

$$a_1,\ a_2,\ \ldots,\ a_l,\ \ldots,\quad b_1,\ b_2,\ \ldots,\ b_l,\ \ldots.$$

This set of numbers completely specifies a statistical state of the assembly.

In place of the expression (45) for the number of complexions corresponding to this statistical state, we can embody the rules laid down in § 2·26 by using the formal expression

$$\Pi_l\gamma(a_l)\,\Pi_l\gamma'(b_l). \qquad \text{......(98)}$$

This expression embodies the whole of the rules for non-degenerate systems if we define the γ's as follows:

(i) *The assembly wave-functions must be* **antisymmetrical** *in all the systems of a given set; then for these systems*

$$\gamma(0)=1,\ \gamma(1)=1,\ \gamma(a)=0\quad (a\geqslant 2). \qquad \text{......(99)}$$

Assemblies of such systems are said to obey **the Fermi–Dirac statistics.**

* Kramers, *Zeit. f. Physik*, vol. 13, p. 343 (1923); Kramers and Ittmann, *Zeit. f. Physik*, vol. 53, p. 553 (1929).

(ii) *The assembly wave-functions must be* **symmetrical** *in all the systems of a given set; then for these systems*

$$\gamma(0)=1,\ \gamma(a)=1\quad (a \geqslant 1). \qquad \text{......(100)}$$

Assemblies of such systems are said to obey **the Einstein-Bose statistics.**

It will not necessarily be true that the systems of various types in an assembly require the same γ's. We have therefore distinguished them in (98) by using γ and γ' for the two sets. For all systems yet known to occur in nature the γ's of either (99) or (100) are required. The formal development will be seen to be more general and to require only that the number of complexions for a given set of a's and b's (statistical state of the assembly) should be expressible in the factorized form (98). It may be noted that the classical expression (45) conforms to this form (apart from the constant factor $M!\,N!$) if we take

$$\gamma(a)=1/a!. \qquad \text{......(101)}$$

Assemblies of such systems will be said to obey **the classical statistics** when it is necessary to contrast them with the other types.

To find the total number of complexions C we have now to sum (98) subject to the conditions

$$\Sigma_t a_t = M,\quad \Sigma_t b_t = N,\quad \Sigma_t a_t\epsilon_t + \Sigma_t b_t\eta_t = E; \qquad \text{......(102)}$$

so that formally

$$C=\Sigma_{(a,b)}\,\Pi_t\,\gamma(a_t)\,\Pi_t\,\gamma'(b_t). \qquad \text{......(103)}$$

In § 2·3 we faced a similar problem, but there we were able to use the multinomial theorem so as to satisfy automatically two of the equations of condition. We then satisfied the third by the introduction of a selector variable z. Here no short cuts are possible, but we can still evaluate (103) by using *three* selector variables x, y, z, one for each of the conditions (102). The required expressions for C, and similarly for $C\overline{a_t}$, etc., will then be obtained as coefficients in triple power series which can still be expressed as multiple integrals and evaluated by the method of steepest descents.

Let us form the expression

$$\Sigma_{(a,b)}\,\Pi_t\,\gamma(a_t)\,x^{a_t}z^{a_t\epsilon_t}\,\Pi_t\,\gamma'(b_t)\,y^{b_t}z^{b_t\eta_t}, \qquad \text{......(104)}$$

where the summation is over unrestricted positive (and zero) values of all the a's and b's. This series can be partially summed at once. We write

$$g(xz^{\epsilon_t})=\sum_{n=0}^{\infty}\gamma(n)\,x^n z^{n\epsilon_t}, \qquad \text{......(105)}$$

$$g'(yz^{\eta_t})=\sum_{n=0}^{\infty}\gamma'(n)\,y^n z^{n\eta_t}. \qquad \text{......(106)}$$

The g-functions are defined entirely by the γ's and may be called *the generating functions* for the systems. Then the expression (104) reduces at once to

$$\Pi_t\,g(xz^{\epsilon_t})\,\Pi_t\,g'(yz^{\eta_t}). \qquad \text{......(107)}$$

Let us now return to (104) and from it select the coefficient of $x^M y^N z^E$. It is easy to see that this selects just those values of the a's and b's which satisfy all three conditions (102), and each coefficient so selected is one of the terms of (103). It therefore follows that C is the coefficient of $x^M y^N z^E$ in (104) and therefore also in (107). Using Cauchy's theorem three times over we therefore find

$$C = \frac{1}{(2\pi i)^3} \iiint \frac{dx\,dy\,dz}{x^{M+1} y^{N+1} z^{E+1}} \Pi_t g(xz^{\epsilon_t}) \Pi_t g'(yz^{\eta_t}). \quad \text{......(108)}$$

It is necessary to assume that all the radii of convergence of the power series concerned are non-zero. We shall verify later that this condition is satisfied.

We have next to construct a similar expression for $C\overline{a_r}$, where $\overline{a_r}$ is the average number of systems of type A in their rth state. By definition this average value is obtained by modifying (103) by the insertion of the extra factor a_r in every term. Thus

$$C\overline{a_r} = \Sigma_{(a,b)} a_r \gamma(a_r) \Pi_{t\neq r} \gamma(a_t) \Pi_t \gamma'(b_t). \quad \text{......(109)}$$

We then proceed to modify (109) as we modified (103). In place of (104) we form a similar unrestricted summation in which only the terms in a_r are different, by an extra factor a_r, thus:

$$\Sigma_{(a,b)} a_r \gamma(a_r) x^{a_r} z^{a_r \epsilon_r} \Pi_{t\neq r} \gamma(a_t) x^{a_t} z^{a_t \epsilon_t} \Pi_t \gamma'(b_t) y^{b_t} z^{b_t \eta_t}. \quad \text{......(110)}$$

This expression reduces at once to

$$\left\{ \sum_{n=0}^{\infty} n\gamma(n) x^n z^{n\epsilon_r} \right\} \Pi_{t\neq r} g(xz^{\epsilon_t}) \Pi_t g'(yz^{\eta_t})$$

which is

$$\left\{ x\frac{\partial}{\partial x} g(xz^{\epsilon_r}) \right\} \Pi_{t\neq r} g(xz^{\epsilon_t}) \Pi_t g'(yz^{\eta_t}). \quad \text{......(111)}$$

It follows at once that $C\overline{a_r}$ is the coefficient of $x^M y^N z^E$ in (111) and therefore that

$$C\overline{a_r} = \frac{1}{(2\pi i)^3} \iiint \frac{dx\,dy\,dz}{x^{M+1} y^{N+1} z^{E+1}} \left\{ x\frac{\partial}{\partial x} g(xz^{\epsilon_r}) \right\} \Pi_{t\neq r} g(xz^{\epsilon_t}) \Pi_t g'(yz^{\eta_t}). \quad \text{......(112)}$$

On comparing (108) and (112) we see that (112) contains in the integrand the extra factor

$$x\frac{\partial}{\partial x} \log g(xz^{\epsilon_r}).$$

If therefore, as we shall shortly establish, a theorem analogous to Theorem 2·31 applies to these multiple integrals, we shall find that

$$\overline{a_r} = \lambda \frac{\partial}{\partial \lambda} \log g(\lambda \vartheta^{\epsilon_r}), \quad \text{......(113)}$$

where λ, μ, ϑ determine the unique minimum of the integrand of C as a function of the real variables x, y, z.

Corresponding formulae for other average values can be derived in the same way. Thus

$$M=\Sigma_r \overline{a_r}=\lambda\frac{\partial}{\partial\lambda}\Sigma_r \log g(\lambda\vartheta^{\epsilon_r}), \qquad \text{......(114)}$$

an equation which in this simple assembly serves to determine λ in terms of M; similarly

$$\overline{E_A}=\Sigma_r \overline{a_r}\epsilon_r=\vartheta\frac{\partial}{\partial\vartheta}\Sigma_r \log g(\lambda\vartheta^{\epsilon_r}). \qquad \text{......(115)}$$

For systems of the other type,

$$\overline{b_r}=\mu\frac{\partial}{\partial\mu}\log g'(\mu\vartheta^{\eta_r}), \qquad \text{......(116)}$$

$$\overline{E_B}=\vartheta\frac{\partial}{\partial\vartheta}\Sigma_r \log g'(\mu\vartheta^{\eta_r}), \qquad \text{......(117)}$$

$$N=\mu\frac{\partial}{\partial\mu}\Sigma_r \log g'(\mu\vartheta^{\eta_r}). \qquad \text{......(118)}$$

It is obvious that these formulae can be extended at once to assemblies containing any number of distinct sets of systems.

§ **2·41.** *Degenerate systems. Special forms of generating function.* The equilibrium properties of the assembly have been shown to depend only on the function $\log g(\lambda\vartheta^{\epsilon_r})$ so far as concerns each set of systems. It is therefore now easy to remove the restriction to non-degenerate systems, by allowing the energies ϵ_r to become equal in groups of ϖ_r. If now the systems A have energies $\epsilon_1, \epsilon_2, \ldots, \epsilon_r, \ldots$ and weights $\varpi_1, \varpi_2, \ldots, \varpi_r, \ldots$, then

$$\overline{a_r}=\varpi_r\lambda\frac{\partial}{\partial\lambda}\log g(\lambda\vartheta^{\epsilon_r}), \qquad \text{......(119)}$$

$$M=\lambda\frac{\partial}{\partial\lambda}\Sigma_r \varpi_r \log g(\lambda\vartheta^{\epsilon_r}), \qquad \text{......(120)}$$

$$\overline{E_A}=\vartheta\frac{\partial}{\partial\vartheta}\Sigma_r \varpi_r \log g(\lambda\vartheta^{\epsilon_r}). \qquad \text{......(121)}$$

The special forms which these formulae take when the proper values of the γ's are inserted must now be recorded.

(i) *Assembly wave-functions antisymmetrical in the systems. Fermi-Dirac statistics.* For such systems

$$g(q)=1+q. \qquad \text{......(122)}$$

Thus

$$\overline{a_r}=\varpi_r\frac{\lambda\vartheta^{\epsilon_r}}{1+\lambda\vartheta^{\epsilon_r}}=\frac{\varpi_r}{1+1/(\lambda\vartheta^{\epsilon_r})}, \qquad \text{......(123)}$$

or in terms of T

$$\overline{a_r}=\frac{\varpi_r}{e^{\epsilon_r/kT}/\lambda+1}. \qquad \text{......(124)}$$

The equation which determines λ is

$$M=\Sigma_r \frac{\varpi_r}{e^{\epsilon_r/kT}/\lambda+1}. \qquad \text{......(125)}$$

It is important to examine the form taken by (125) when in general $\epsilon_r/kT \gg 1$, that is when $T \to 0$. It is clear that λ must have such a value that $e^{\epsilon_r/kT}/\lambda \to 0$ for the M states of lowest energy (degenerate states being counted multiply) while $e^{\epsilon_r/kT}/\lambda \to \infty$ for all higher states. If therefore we write

$$\lambda = e^{\eta/kT}, \qquad \text{......(126)}$$

then η is determined as a function of T by the equation

$$M = \Sigma_r \frac{\varpi_r}{e^{(\epsilon_r-\eta)/kT}+1}; \qquad \text{......(127)}$$

moreover η has the important property that as $T \to 0$

$$\eta \to \epsilon^*$$

given by the equation

$$M = \sum_{\epsilon_r \leqslant \epsilon^*} \varpi_r. \qquad \text{......(128)}$$

In many important cases we shall find that $\eta \simeq \epsilon^*$ over a long range of temperature, so that we shall frequently use (124) in the approximate form

$$\overline{a_r} = \frac{\varpi_r}{e^{(\epsilon_r-\epsilon^*)/kT}+1}, \qquad \text{......(129)}$$

a form which of course is exact if we allow ϵ^* to be a function of the temperature.

(ii) *Assembly wave-functions symmetrical in the systems. Einstein-Bose statistics.* For such systems

$$g(q) = 1/(1-q). \qquad \text{......(130)}$$

Thus

$$\overline{a_r} = \frac{\varpi_r}{-1+1/(\lambda\vartheta^{\epsilon_r})}, \qquad \text{......(131)}$$

or in terms of T

$$\overline{a_r} = \frac{\varpi_r}{e^{\epsilon_r/kT}/\lambda - 1}. \qquad \text{......(132)}$$

The equation which determines λ is

$$M = \Sigma_r \frac{\varpi_r}{e^{\epsilon_r/kT}/\lambda - 1}. \qquad \text{......(133)}$$

It is clear that we must always have

$$e^{\epsilon_r/kT}/\lambda > 1.$$

It follows therefore that as $T \to 0$

$$e^{\epsilon_1/kT}/\lambda \sim 1 + \varpi_1/M,$$

or when M is large, to sufficient accuracy

$$\lambda \sim e^{\epsilon_1/kT}, \qquad \text{......(134)}$$

$$\overline{a_1} \sim M, \quad \overline{a_r} \to 0 \quad (r>1). \qquad \text{......(135)}$$

(iii) *Enumeration of complexions for classical or localized systems.* It is interesting to verify that the more general formulae of this section reproduce the results of §§ 2·3–2·34 when we use the proper function $g(q)$. For such systems

$$g(q) = e^q. \qquad \text{......(136)}$$

Thus

$$\overline{a_r} = \lambda \varpi_r \vartheta^{\epsilon_r},$$

or in terms of T

$$\overline{a_r} = \lambda \varpi_r e^{-\epsilon_r/kT}. \qquad \text{......(137)}$$

The equation which determines λ is

$$M = \lambda \Sigma_r \varpi_r e^{-\epsilon_r/kT} = \lambda f(T), \qquad \text{......(138)}$$

where $f(T)$ is the usual partition function (in terms of T). Combining (137) and (138) we find

$$\overline{a_r} = M \varpi_r e^{-\epsilon_r/kT}/f(T), \qquad \text{......(139)}$$

which is the former result, (80). All other results which depend on the ratio of two integrals such as (108) are likewise unaffected. But from all the enumerations themselves we have omitted factors such as $M!N!$ compared with the similar enumerations of § 2·32 and all individual integrals such as (108) should be smaller by this factor than the corresponding (75).

It is interesting to verify that this ratio is preserved by the approximate evaluation of C by the method of steepest descents. By Theorem 2·31 and its analogue Theorem 2·42, which we shall shortly prove, the only term of importance in $\log C$ is log(integrand of C) evaluated at the col. Using (75) for C for localized systems, this is

$$M \log f(\vartheta) + N \log f'(\vartheta) - E \log \vartheta. \qquad \text{......(140)}$$

Using (108) with $g(q) = g'(q) = e^q$, it is

$$\lambda f(\vartheta) + \mu f'(\vartheta) - E \log \vartheta - M \log \lambda - N \log \mu. \qquad \text{......(141)}$$

But by (138) $M = \lambda f(\vartheta)$, $N = \mu f'(\vartheta)$, so that (141) reduces to

$$M + N - E \log \vartheta - M \log\{M/f(\vartheta)\} - N \log\{N/f'(\vartheta)\},$$

or to the required accuracy

$$M \log f(\vartheta) + N \log f'(\vartheta) - E \log \vartheta - \log(M!N!)$$

as we wished to verify.

§ **2·42.** *Proof of the result of* §§ 2·4, 2·41. We now give a proof of the results of §§ 2·4, 2·41 written out explicitly for three variables x, y, z. The proof however will be so arranged that it is easily seen to be general and to hold for similar integrals in any number of variables, so that the restriction to two sets of systems in §§ 2·4, 2·41 is immaterial and the results all hold for assemblies of any number of sets of systems as in § 2·32.

The form of the integrand of C and the analogous integrals is that of a triple (multiple) power series

$$\Phi = \Sigma_{abc} Q_{abc} x^a y^b z^c, \qquad \text{......(142)}$$

in which the Q_{abc} are all positive and the a, b, c (integers) start at negative values and run to $+\infty$. The domain of convergence of the series (142) in our actual problem must be examined, but for the moment need only be assumed to be non-zero in each variable. For our proof we require certain properties of Φ which are obtained in the following:

Lemma 2·42. *For real positive values of x, y, z the function Φ has an absolute minimum at λ, μ, ϑ which is the unique solution of the equations*

$$\frac{\partial\Phi}{\partial x}=\frac{\partial\Phi}{\partial y}=\frac{\partial\Phi}{\partial z}=0 \qquad \ldots\ldots(143)$$

in this domain.

(i) Since Φ is always positive, and since it may be assumed from the physical origin of the Q_{abc} that $\Phi\to+\infty$ as x, y, z tend to their boundary values (i.e. 0, ∞ or 0, 1) in any manner, Φ must have an absolute minimum value Φ_0 which it assumes at some points of the domain of real positive values x, y, z. At such a point λ, μ, ϑ equations (143) must of course be satisfied.

(ii) That λ, μ, ϑ is the *unique* solution of (143) in the real domain will follow at once if it can be shown that *any* stationary value of Φ must be an absolute minimum—that is that, if Φ_0 is any stationary value,

$$\Phi-\Phi_0\geqslant 0$$

for the whole domain, equality being only possible when $x=\lambda$, $y=\mu$, $z=\vartheta$. If we write $x=e^X$, $y=e^Y$, $z=e^Z$, then

$$\Phi=\Sigma_{abc}\,Q_{abc}\,e^{aX+bY+cZ},$$

and, by Taylor's theorem, for *any* stationary value Φ_0,

$$\Phi-\Phi_0=\frac{1}{2}\left[(X-X_0)^2\frac{\partial^2\Phi}{\partial X^2}+\ldots+2(X-X_0)(Y-Y_0)\frac{\partial^2\Phi}{\partial X\,\partial Y}\right], \qquad \ldots\ldots(144)$$

an expression in which all the partial differential coefficients are to be evaluated for some particular set of values of X, Y, Z. It is therefore only necessary to prove that the expression on the right of (144) is a positive quadratic form.

(iii) The proof of the lemma reduces therefore to the proof of the essential inequalities

$$\frac{\partial^2\Phi}{\partial X^2}>0,\qquad \begin{vmatrix}\dfrac{\partial^2\Phi}{\partial X^2} & \dfrac{\partial^2\Phi}{\partial X\,\partial Y}\\[2ex] \dfrac{\partial^2\Phi}{\partial X\,\partial Y} & \dfrac{\partial^2\Phi}{\partial Y^2}\end{vmatrix}>0,\qquad \begin{vmatrix}\dfrac{\partial^2\Phi}{\partial X^2} & \dfrac{\partial^2\Phi}{\partial X\,\partial Y} & \dfrac{\partial^2\Phi}{\partial X\,\partial Z}\\[2ex] \dfrac{\partial^2\Phi}{\partial X\,\partial Y} & \dfrac{\partial^2\Phi}{\partial Y^2} & \dfrac{\partial^2\Phi}{\partial Y\,\partial Z}\\[2ex] \dfrac{\partial^2\Phi}{\partial X\,\partial Z} & \dfrac{\partial^2\Phi}{\partial Y\,\partial Z} & \dfrac{\partial^2\Phi}{\partial Z^2}\end{vmatrix}>0.$$

[For more variables the series of inequalities is correspondingly extended.]

Firstly
$$\frac{\partial^2\Phi}{\partial X^2}=\Sigma_{abc}Q_{abc}a^2e^{aX+bY+cZ}>0,$$
for every term is positive. Secondly
$$\begin{vmatrix}\frac{\partial^2\Phi}{\partial X^2} & \frac{\partial^2\Phi}{\partial X\partial Y}\\ \frac{\partial^2\Phi}{\partial X\partial Y} & \frac{\partial^2\Phi}{\partial Y^2}\end{vmatrix}=\begin{vmatrix}\Sigma_{abc}Q_{abc}a^2e^{aX+bY+cZ} & \Sigma_{a'b'c'}Q_{a'b'c'}a'b'e^{a'X+b'Y+c'Z}\\ \Sigma_{abc}Q_{abc}abe^{aX+bY+cZ} & \Sigma_{a'b'c'}Q_{a'b'c'}b'^2e^{a'X+b'Y+c'Z}\end{vmatrix}$$
If we collect together all terms containing $Q_{abc}Q_{a'b'c'}$, we see that this determinant reduces to
$$\Sigma' Q_{abc}Q_{a'b'c'}e^{(a+a')X+(b+b')Y+(c+c')Z}\left\{\begin{vmatrix}a^2 & a'b'\\ ab & b'^2\end{vmatrix}+\begin{vmatrix}a'^2 & ab\\ a'b' & b^2\end{vmatrix}\right\}.$$
The terms in { } are formed of all possible permutations of the dashed and plain letters, and reduce to
$$ab'\begin{vmatrix}a & a'\\ b & b'\end{vmatrix}+a'b\begin{vmatrix}a' & a\\ b' & b\end{vmatrix}=\begin{vmatrix}a & a'\\ b & b'\end{vmatrix}^2.$$
The summation Σ' is over all possible values of a, b, c, a', b', c', the specified permutations being excluded. Since every term in Σ' is positive, the second condition is fulfilled. Finally, an exactly similar argument shows that
$$\begin{vmatrix}\frac{\partial^2\Phi}{\partial X^2} & \frac{\partial^2\Phi}{\partial X\partial Y} & \frac{\partial^2\Phi}{\partial X\partial Z}\\ \frac{\partial^2\Phi}{\partial X\partial Y} & \frac{\partial^2\Phi}{\partial Y^2} & \frac{\partial^2\Phi}{\partial Y\partial Z}\\ \frac{\partial^2\Phi}{\partial X\partial Z} & \frac{\partial^2\Phi}{\partial Y\partial Z} & \frac{\partial^2\Phi}{\partial Z^2}\end{vmatrix}=J\left(\frac{\frac{\partial\Phi}{\partial X}\ \ \frac{\partial\Phi}{\partial Y}\ \ \frac{\partial\Phi}{\partial Z}}{X\ \ \ Y\ \ \ Z}\right), \qquad \text{......(145)}$$
$$=\Sigma' Q_{abc}Q_{a'b'c'}Q_{a''b''c''}e^{(a+a'+a'')X+(b+b'+b'')Y+(c+c'+c'')Z}\begin{vmatrix}a & a' & a''\\ b & b' & b''\\ c & c' & c''\end{vmatrix}^2,$$
$$>0.$$
This completes the proof of the lemma which can obviously be extended to any number of variables.

The integrals which we desire to study asymptotically are all of the form
$$\frac{1}{(2\pi i)^3}\iiint\Phi\frac{dx\,dy\,dz}{xyz}. \qquad \text{......(146)}$$
We require in general only to evaluate the ratio of two such integrals in which the Φ's differ only in the *coefficients* of one (or more) of their component factors. On the contours of integration the maximum value of the modulus of the integrand occurs when all the variables are real and positive, and, as we shall see, if the contours are arranged to go through the real-value

minimum of Φ, it is only the contribution from this neighbourhood which need be considered.

Strictly speaking this neighbourhood might be only one of several making contributions of the same order. If certain relations are satisfied between the a, b, c, there might be other points on the contours at which the phases of all the terms are again equal, so that the same maximum value (of the modulus) of the integrand is repeated. The same difficulty occurred in § 2·31, when z was the only selector variable, where we showed that repetitions of the maximum are without effect on the value of the integral and can in fact be avoided by a proper choice of the unit of energy.

For the particular Φ's that occur here, built up of factors of the form $g(xz^{\epsilon_r})$, where

$$g(q) = 1+q, \quad 1/(1-q), \quad e^q,$$

it is not difficult to see that no repetitions of the maximum can occur except those which are identical with the repetitions of § 2·31 and therefore of no significance. To attain the maximum modulus of Φ each q, which is of given modulus on the contours of integration, must be real and positive. This will occur at points at which the amplitudes $\theta_x, \theta_y, \theta_z$ of x, y, z satisfy the relations

$$\theta_x + \epsilon_r \theta_z \equiv 0 \ (\mathrm{Mod}\ 2\pi) \quad (all\ r),$$
$$\theta_y + \eta_r \theta_z \equiv 0 \ (\mathrm{Mod}\ 2\pi) \quad (all\ r).$$

The first set of these equations asserts (i) that the ϵ_r are of the form $\epsilon_0 + n\zeta_r$, where n and ζ_r are integers, and (ii) that

$$\theta_z = \frac{2\pi s}{n}, \quad \theta_x = \frac{-2\pi s}{n}\epsilon_0 \quad (0 \leqslant s < n).$$

The second set asserts (i) that the η_r are of the form $\eta_0 + n'\zeta_r'$, where n' and ζ_r' are integers, and (ii) that

$$\theta_z = \frac{2\pi s'}{n'}, \quad \theta_y = \frac{-2\pi s'}{n'}\eta_0 \quad (0 \leqslant s' < n').$$

It is easy to see that if $n, n' > 1$ these relations may permit a number of subsidiary maxima. The energy zero can however always be defined so that $\epsilon_0 = \eta_0 = 0$. Subsidiary maxima therefore only occur for real x's, y's or q's. They are identical with the subsidiary maxima of § 2·31 and can be removed by a suitable choice of the unit of energy which will eliminate all common factors from n and n' and therewith all possibility of non-zero values of θ_z. Even if these subsidiary maxima are allowed to remain, every integral (146) can only exceed by a constant integral factor the value derived from the primary col on the real axes, and this extra factor is without significance.

We have not yet determined the general form of the functions such as $\Pi_r\, g(xz^{\epsilon_r})$ for systems of type II, but the ϵ_r for such systems always contain terms for the translatory motion of the system as a whole in an enclosure of volume V. We shall find that the integrals (146) with which we have to deal can always be cast into the form

$$\frac{1}{(2\pi i)^3}\iiint R\Psi^V \frac{dx\,dy\,dz}{x^{X+1}y^{Y+1}z^{E+1}},$$

where R and Ψ are independent of X, Y, E and also of V to a sufficient approximation, so that what we require is the asymptotic value of this integral as X, Y, E, $V\to\infty$ in fixed ratios. The function $\Psi^V/(x^{X+1}y^{Y+1}z^{E+1})$ with or without the factor R represents the former Φ and obeys the conditions of the lemma. Then the results which we have given in anticipation are consequences of the following

Theorem 2·42. If $\Phi(x,y,z)$ satisfies the conditions of Lemma 2·42 and the structural restrictions just discussed and if further when X, Y, E and V are large in fixed ratios Φ has the differentiable asymptotic form

$$\log\Phi \sim V\log\Psi+\log R-X\log x-Y\log y-E\log z, \qquad \text{......(147)}$$

where Ψ, R are functions of x, y, z independent of X, Y, E and V, and if γ is a contour circulating once counter-clockwise round the origin within the circle of convergence of each variable, then

$$\frac{1}{(2\pi i)^3}\int_\gamma\int_\gamma\int_\gamma \Phi\frac{dx\,dy\,dz}{xyz}=\frac{R(\lambda,\mu,\vartheta)\{\Psi(\lambda,\mu,\vartheta)\}^V}{\lambda^X\mu^Y\vartheta^E(2\pi V)^{\frac{3}{2}}}\left[\Delta^{-\frac{1}{2}}+O\left(\frac{1}{V}\right)\right], \qquad \text{......(148)}$$

where λ, μ, ϑ is the unique solution on the positive real axis of the equations

$$V\lambda\frac{\partial\log\Psi}{\partial\lambda}=X, \qquad \text{......(149)}$$

$$V\mu\frac{\partial\log\Psi}{\partial\mu}=Y, \qquad \text{......(150)}$$

$$V\vartheta\frac{\partial\log\Psi}{\partial\vartheta}=E, \qquad \text{......(151)}$$

Δ is a positive function of λ, μ, ϑ, dependent only on Ψ and the ratios of X, Y, E and V.

Corollary 1. *If I and I' are two such integrals for which* $\log\Phi$ *and* $\log\Phi'$ *differ only in the term* $\log R$, *then*

$$\frac{I}{I'}=\frac{R(\lambda,\mu,\vartheta)}{R'(\lambda,\mu,\vartheta)}. \qquad \text{......(152)}$$

Corollary 2. *If $\Phi(x_1,\ldots,x_n,z)$ satisfies analogous conditions, then*

$$\frac{1}{(2\pi i)^{n+1}}\int_\gamma\cdots\int_\gamma\Phi\frac{dx_1\ldots dx_n\,dz}{x_1\ldots x_n z}$$
$$=\frac{R(\lambda_1,\ldots,\lambda_n,\vartheta)\{\Psi(\lambda_1,\ldots,\lambda_n,\vartheta)\}^V}{\lambda_1^{X_1}\ldots\lambda_n^{X_n}\vartheta^E(2\pi V)^{\frac{1}{2}(n+1)}}\left[\Delta^{-\frac{1}{2}}+O\left(\frac{1}{V}\right)\right], \qquad \text{...(148a)}$$

where the notation is a natural extension of that of the main theorem.

The proof of this theorem and corollaries may now be easily completed. The extensions to Corollary 2 are obvious and will not be mentioned again. The integrand satisfies the conditions of Lemma 2·42 and therefore has a unique minimum on the real axes at λ, μ, ϑ which are given with sufficient accuracy by equations (149)–(151) whose roots depend only on the ratios of X, Y, E and V. [λ, μ, ϑ are *intensive* parameters.] When the three circles of integration are made to pass through λ, μ, ϑ respectively, the point λ, μ, ϑ itself is the only point in the domain of integration at which the integrand attains its maximum modulus in view of the structural restrictions on Φ. It remains only to show that the contribution of this neighbourhood itself is effectively of the order of $|\Phi(\lambda,\mu,\vartheta)|$ and actually to evaluate it. We write

$$x = \lambda e^{i\alpha}, \quad y = \mu e^{i\beta}, \quad z = \vartheta e^{i\gamma},$$

so that in the neighbourhood of the col (α, β, γ small)

$$R(x,y,z) = R(\lambda,\mu,\vartheta) + \left[i\alpha\lambda\frac{\partial}{\partial\lambda} + i\beta\mu\frac{\partial}{\partial\mu} + i\gamma\vartheta\frac{\partial}{\partial\vartheta}\right]R(\lambda,\mu,\vartheta) + O(\alpha,\beta,\gamma)^2,$$

$$\frac{\{\Psi(x,y,z)\}^V}{x^X y^Y z^E} = \frac{\{\Psi(\lambda,\mu,\vartheta)\}^V}{\lambda^X \mu^Y \vartheta^E}\exp\left\{-\tfrac{1}{2}\alpha^2 X - \tfrac{1}{2}\beta^2 Y - \tfrac{1}{2}\gamma^2 E\right.$$
$$\left. - \tfrac{1}{2}V\left[\alpha^2\lambda^2\frac{\partial^2}{\partial\lambda^2} + \dots + 2\beta\gamma\mu\vartheta\frac{\partial^2}{\partial\mu\,\partial\vartheta}\right]\log\Psi + V(\alpha,\beta,\gamma)^3 + O\{V(\alpha,\beta,\gamma)^4\}\right\}.$$

The quadratic form in the last exponential is equivalent to (144) with sign changed and is therefore essentially negative and reducible to the sum of three negative square terms. If X, Y, E and V are large, it follows by the arguments of § 2·31 that the variables α, β, γ in the quadratic terms may be supposed to range from $-\infty$ to $+\infty$ while all other terms remain small. When the quadratic form has been reduced to a sum of three squares by linear transformation, the exponential can be integrated with respect to all three variables, when it will be found that

$$\iiint_{-\infty}^{+\infty}\exp\left\{-\tfrac{1}{2}\alpha^2 X - \tfrac{1}{2}\beta^2 Y - \tfrac{1}{2}\gamma^2 E\right.$$
$$\left. - \tfrac{1}{2}V\left[\alpha^2\lambda^2\frac{\partial^2}{\partial\lambda^2} + \dots + 2\beta\gamma\mu\vartheta\frac{\partial^2}{\partial\mu\,\partial\vartheta}\right]\log\Psi\right\}d\alpha\,d\beta\,d\gamma = \left(\frac{2\pi}{V}\right)^{\frac{3}{2}}\Delta^{-\frac{1}{2}}.$$

Δ is the discriminant of this quadratic form with the factor $-\tfrac{1}{2}$ omitted. It is easily seen to be equivalent to $J/V\Phi$, where J is given by (145), and has already been shown to be positive.

Proceeding with the evaluation, the terms of odd order in α, β, γ vanish on integration and those of order $(\alpha,\beta,\gamma)^2$ or $V(\alpha,\beta,\gamma)^4$ leave an error term $O(1/V)$. The rest of the range makes a negligible contribution, hence the theorem.

The form in which we have proved Theorem 2·42 is not restricted to three variables. It extends at once to any number of variables with similar functions Φ obeying the analogous restrictions. We can therefore at once suppose all the results of §§ 2·4, 2·41 extended to assemblies of any number of types of systems. It is important to observe that in terms of the intensive parameters λ, μ, ..., ϑ the results for any set of systems are entirely independent of the constitution of the rest of the assembly.

§ **2·5.** *Structureless particles, moving in a volume V.* The important formulae of §§ 2·4, 2·41 become still more useful when the distribution of the ϵ_r for the free motion of a particle in an enclosure is explicitly introduced. Schrödinger's equation for a structureless particle of mass m in a field of potential energy W is

$$\nabla^2\psi + \frac{8\pi^2 m}{h^2}(E - W)\psi = 0. \qquad \text{......(153)}$$

An enclosure is sufficiently well represented by assuming that $W = 0$ inside the enclosure and $W \to \infty$ rapidly as we pass the walls.* If V has the form of a rectangular box of edges a, b, c, the determination of the ϵ_r is simple.

If ψ is to be a possible wave-function, it must be one-valued and bounded in V and vanish over the walls.† The possible forms of ψ are obviously

$$\psi_r = \sin\frac{s\pi x}{a}\sin\frac{t\pi y}{b}\sin\frac{u\pi z}{c}, \qquad \text{......(154)}$$

where s, t, u are positive integers (not zero), corresponding to the eigenvalue

$$E = \epsilon_r = \frac{h^2}{8m}\left(\frac{s^2}{a^2} + \frac{t^2}{b^2} + \frac{u^2}{c^2}\right). \qquad \text{......(155)}$$

There is only the one ψ_r for each eigenvalue. Any weight factors other than unity enter only in virtue of the internal structure of the systems.

On referring back to (120), (121) we see that the important series to be summed is

$$\Sigma_r \varpi_r \log g(\lambda\vartheta^{\epsilon_r}). \qquad \text{......(156)}$$

If $\lambda\vartheta^{\epsilon_r}$ or $\lambda e^{-\epsilon_r/kT}$ is less than unity for all ϵ_r the logarithm can be expanded in powers of λ, and the expression (156) can be rearranged in one of the forms

$$\textit{Antisymmetrical} \quad \sum_{j=1}^{\infty} \frac{(-)^{j-1}}{j}\lambda^j(\Sigma_r \varpi_r e^{-j\epsilon_r/kT}), \qquad \text{......(157)}$$

$$\textit{Symmetrical} \quad \sum_{j=1}^{\infty} \frac{1}{j}\lambda^j(\Sigma_r \varpi_r e^{-j\epsilon_r/kT}), \qquad \text{......(158)}$$

* For ions or electrons we thus ignore the effects of their charges. If the assembly as a whole is a neutral mixture, this is probably a valid but rough first approximation and is in common use.

† The boundary condition on the walls can be established thus: consider the wall $x = 0$, near which, as $W \to \infty$, $\psi'' \sim W\psi$. By well-known methods [e.g. Jeffreys, *Proc. Lond. Math. Soc.* vol. 23, p. 428 (1924)] it follows that $\log\psi \sim \pm\int^x W^{\frac{1}{2}}\,dx$. We must suppose that for a local boundary field the integral does not converge as $x \to 0$. Hence either $\psi \to 0$ or $\psi \to \infty$. The latter is impossible if ψ is to be bounded or even if only $\int|\psi|^2\,dx$ is to exist. Hence $\psi \to 0$ on the boundary.

according to the proper form of $g(q)$. Here therefore we have to sum $\Sigma_r e^{-j\epsilon_r/kT}$, which on using (155) breaks up into the product of three series of which

$$\sum_{s=1}^{\infty} \exp\left\{-\frac{jh^2}{8ma^2kT}s^2\right\} \qquad \text{......(159)}$$

is typical. This series is practically a ϑ-function and its value when $jh^2/(8ma^2kT)$ is small, as it is for all the important terms in (157) or (158) in all ordinary applications, can be obtained as accurately as may be required from the transformation theory. We have in fact*

$$\vartheta_3(v,\tau) = 1 + 2\sum_{s=1}^{\infty} e^{\pi i\tau s^2}\cos 2s\pi v,$$

$$= \sqrt{\left(\frac{i}{\tau}\right)} \sum_{n=-\infty}^{\infty} e^{-\frac{\pi i}{\tau}(v+n)^2}. \qquad \text{......(160)}$$

Putting $v=0$, and $\tau/i = jh^2/(8\pi ma^2kT)$, we find from (160), with great accuracy for small values of τ/i, that

$$\sum_{s=1}^{\infty} \exp\left\{-\frac{jh^2}{8ma^2kT}s^2\right\} = \frac{1}{2}\left\{\left(\frac{8m\pi a^2kT}{jh^2}\right)^{\frac{1}{2}} - 1\right\},$$

which is sufficiently nearly equal for all ordinary values of a and T and early values of j to

$$\frac{(2\pi mkT)^{\frac{1}{2}}}{h}\frac{a}{j^{\frac{1}{2}}}. \qquad \text{......(161)}$$

Thus†

$$\Sigma_r e^{-j\epsilon_r/kT} = \frac{(2\pi mkT)^{\frac{3}{2}}}{h^3}\frac{V}{j^{\frac{3}{2}}}. \qquad \text{......(162)}$$

For sufficiently small values of λ

$$\textit{Antisymmetrical} \quad \Sigma_r \log g(\lambda\vartheta^{\epsilon_r}) = \frac{(2\pi mkT)^{\frac{3}{2}}}{h^3} V \sum_{j=1}^{\infty} \frac{(-)^{j-1}}{j^{\frac{5}{2}}}\lambda^j, \qquad \text{......(163)}$$

$$\textit{Symmetrical} \quad \Sigma_r \log g(\lambda\vartheta^{\epsilon_r}) = \frac{(2\pi mkT)^{\frac{3}{2}}}{h^3} V \sum_{j=1}^{\infty} \frac{1}{j^{\frac{5}{2}}}\lambda^j. \qquad \text{......(164)}$$

The conditions under which the foregoing formulae hold are first that λ shall be sufficiently small for the expansion of the logarithms; on this condition we shall defer further comment. The other condition is that $h^2/(8\pi ma^2kT)$ shall be very small even when multiplied by any integer j which yields a significant term in the j-expansion. This condition asserts that the spacing of the energy value is very small compared with kT, and is fulfilled for all ordinary enclosures and ordinary temperatures; for if $a = 1$ cm., $T = 1°$ K., and m the mass of an electron, the value of this ratio is $1{\cdot}4 \times 10^{-11}$.

* Tannery and Molk, *Elliptic Functions*, vol. 2, pp. 252, 264.

† This result is really independent of the shape of the box. For a similar independence theorem see Weyl, *Math. Ann.* vol. 71, p. 441 (1911) or Courant, *Gött. Nachr.* p. 255 (1919); *Math. Zeit.* vol. 7, p. 14 (1920).

In order to combine these formulae into one valid for either statistics it is convenient to introduce the coefficient α_j which is $(-)^{j-1}$ in the antisymmetrical case and 1 in the symmetrical. We shall use α_j in what follows.

§ **2·51.** *The value of λ and the approximation to classical statistics.* The value of the foregoing expansions depends on the size of λ which must now be examined. By equations (120) and (163) or (164) for structureless particles the molecular density ν is given by

$$\nu = \frac{M}{V} = \frac{(2\pi mkT)^{\frac{3}{2}}}{h^3} \sum_{j=1}^{\infty} \frac{\alpha_j}{j^{\frac{5}{2}}} \lambda^j. \qquad \ldots\ldots(165)$$

It follows at once that λ is small if

$$\frac{\nu h^3}{(2\pi mkT)^{\frac{3}{2}}} \quad \text{or} \quad 5{\cdot}2 \times 10^{-21} \frac{\nu}{(TM^*)^{\frac{3}{2}}}$$

is small, where M^* is the molecular weight on the (chemical) oxygen scale. Thus even for molecular hydrogen at 1° K. and normal concentration, $\nu = 2{\cdot}7 \times 10^{19}$, λ is still less than 0·1 and the series in (165) reduces practically to its first term. *A fortiori* for heavier molecules and greater temperatures λ is still smaller, or as small up to higher concentrations. Thus in all applications to actual gases we may assume that λ is small and that (165) and similar series reduce to their first terms. The only exception will be electron gases at the concentrations at which one would expect to find free electrons in metals, about one per atom, $\nu = 10^{22}$. Such assemblies are still non-classical up to temperatures greater than 2000° K. The expansions are then valueless.

When λ is small the assembly is indistinguishable from a classical one, in which no account is taken of the symmetry requirements of the assembly wave-functions, so that the enumeration of wave-functions for localized or non-localized systems is the same. This limiting identity holds for any statistics in which $\alpha_1 = 1$. For if we use classical statistics we can employ the preceding analysis with $g(q) = e^q$ in which case our series for $\log g(q)$ reduce identically to their first term, $\alpha_1 = 1$, $\alpha_j = 0$ $(j \geqslant 2)$. Equation (165) reduces to

$$\nu = \frac{(2\pi mkT)^{\frac{3}{2}}}{h^3} \lambda \qquad \ldots\ldots(166)$$

and the distribution law (119) to

$$\overline{a_r} = \lambda e^{-\epsilon_r/kT}. \qquad \ldots\ldots(167)$$

These are the same results as we obtain for either symmetrical or antisymmetrical assembly wave-functions if we neglect all but the lowest power of λ in (165) and corresponding series.

Equations (166) and (167) are equivalent to Maxwell's and Boltzmann's distribution laws, as we shall show in §§ 2·6, 2·64. It is, however, hardly satisfactory to be content with only such a sophisticated derivation of these familiar classical laws. Having shown that quantum restrictions on accessibility are usually irrelevant for systems in the gaseous phase of an assembly, and that the spacing of the characteristic energies is very small compared with kT even down to 1° K., we are justified in deducing Maxwell's distribution law and similar classical theorems in a classical manner *ab initio*. Such deductions are still of great value in providing physical insight into the formulae, and we proceed to give them in the following sections. The reason why in general gaseous assemblies are effectively classical in spite of the quantum restrictions may be expressed thus, that the phase space of a system in the gas has so many cells of extension h^3, or the assembly so many accessible wave-functions, that it is extremely improbable that any pair of systems will attempt to occupy the same cell or to possess the same wave-function. It does not then matter how those configurations are enumerated in which two or more systems occupy the same cell.

We return to the exact quantum discussion in § 2·7.

§ 2·6. *Assemblies containing free molecules or other systems treated classically.* In order to discuss classical systems classically as simply as possible by the foregoing methods it is clear that some limiting process is essential, for we can only deal directly with a set of discrete commensurable energies. As in fact all motions are subject to the laws of quantum mechanics and all systems are really quantized systems, we are concerned in this limiting process only with questions of technical convenience. We shall now show that this can be done very simply. Questions as to the validity of this limiting process are postponed to the next chapter. For simplicity we shall suppose that the assembly consists of M localized systems of any type explicitly quantized, with partition function $f(z)$, and N atoms of mass m moving freely in a volume V whose energy is solely kinetic energy of translation. The whole discussion applies equally well to any number of types of classical and quantized systems; internal motions and rotations of the free atoms or molecules can be included among the latter.

The phase space for a free atom is specified by the six coordinates $p_1, \ldots, q_3$, and is divided up into small cells, 1, 2, ..., t, ..., of extension

$$(dp_1 \ldots dq_3)_t$$

and, by the rules of § 2·2, weight δ_t given by

$$\delta_t = \frac{(dp_1 \ldots dq_3)_t}{h^3}. \qquad \ldots\ldots(168)$$

To take account of the confinement of the systems to the volume V we start with the atoms in an external field of force of potential energy W which may finally be reduced to the local boundary field of the walls. Then there is an energy ζ_t associated with the tth cell given by

$$\zeta_t = \frac{1}{2m}(p_1^2 + p_2^2 + p_3^2)_t + W_t; \qquad \text{......(169)}$$

W is a function of q_1, q_2, q_3 only.

Consider an *artificial* assembly in which the cells are small and the energy anywhere in a cell constant and equal to ζ_t. Then all the ζ's and ϵ's can be supposed chosen so that they are commensurable and expressible as integers with the proper unit of energy. The artificial assembly is composed of systems with discrete energies only, and can be made to represent the actual one to any assigned standard of approximation. In the artificial assembly we have at once the partition function

$$h(z) = \Sigma_t \delta_t z^{\zeta_t}, \qquad \text{......(170)}$$

and the distribution laws are given at once by the old formulae of §§ 2·3–2·32. To obtain the distribution laws for the actual assembly we must proceed to the limit by making the extension of every cell tend to zero. We construct in fact any sequence of artificial assemblies for each of which we can determine the distribution laws, and which has the actual assembly as a limit ($\delta_t \to 0$ for all t). We must then *prove* that these laws have a unique limit, and that this limit represents the distribution laws of the actual assembly, evaluated, that is to say, after we have proceeded to the classical limit. This point is postponed to § 3·8. We really carry out some such process in any classical discussion of the classical distribution laws.*

Now by the definition of an integral, when $\delta_t \to 0$ (*all* t),

$$h(z) \to H(z) = \frac{1}{h^3}\int \dots \int e^{-\frac{\log 1/z}{2m}[p_1^2+p_2^2+p_3^2+2mW]}\,dp_1 \dots dq_3, \qquad \text{...(171)}$$

the integration being extended over all values of $p_1, \dots, q_3$. For an assembly in a volume V we again represent the walls by supposing that $W=0$ in V and that $W \to \infty$ rapidly near the wall. Then provided that the real part of $\log 1/z$ is positive ($|z| < 1$),

$$H(z) = \frac{V}{h^3}\iiint e^{-\frac{\log 1/z}{2m}(p_1^2+p_2^2+p_3^2)}\,dp_1 dp_2 dp_3,$$

$$= \frac{(2\pi m)^{\frac{3}{2}} V}{h^3(\log 1/z)^{\frac{3}{2}}}. \qquad \text{......(172)}$$

For the important point $z = \vartheta = e^{-1/kT}$ this yields

$$H(T) = \frac{(2\pi mkT)^{\frac{3}{2}} V}{h^3}. \qquad \text{......(172·1)}$$

* Cf. Jeans, *loc. cit.* chaps. III, V *passim*.

In the formulae for the distribution laws of the artificial assembly, $h'(z)$ and perhaps other differential coefficients occur. It is easily proved directly that $h'(z)$ has the limit $H'(z)$, etc. Thus the laws for the sequences of artificial assemblies have a unique limit which will be the laws given by the formulae of the preceding sections if we use (172) for the partition function of the free motion of the atoms. For example,

$$\overline{E_B} = N\vartheta\frac{\partial}{\partial\vartheta}\log H(\vartheta) = N\vartheta\frac{d}{d\vartheta}\log[\log 1/\vartheta]^{-\frac{3}{2}},$$

$$= \tfrac{3}{2}N/[\log 1/\vartheta] = \tfrac{3}{2}NkT. \qquad \text{......(173)}$$

$$b_t = \frac{N\delta_t\vartheta^{\zeta_t}}{H(\vartheta)} = \frac{N}{V}\left(\frac{\log 1/\vartheta}{2\pi m}\right)^{\frac{3}{2}} e^{-\frac{1}{2}m(\log 1/\vartheta)(u^2+v^2+w^2)}\, dp_1 \dots dq_3,$$

$$= \frac{N}{V}\left(\frac{m}{2\pi kT}\right)^{\frac{3}{2}} e^{-\frac{m}{2kT}(u^2+v^2+w^2)}\, du \dots dz. \qquad \text{......(174)}$$

This is Maxwell's Law.

Finally we observe that it is possible to replace $h(z)$ by $H(z)$ formally in C and the other integrals, if γ is fixed as the circle $|z| = \vartheta$, although the interpretation of the integral as a coefficient in a power series now fails, and the integrand is no longer single-valued. If $H(z)$ is taken to be real for real positive z, these formal integrals give all the correct results. It will be shown in § 3·8 that this is always true.

§ **2·61.** *Maxwell's distribution law with mass motion (classical treatment).* It is easy to extend the argument to the case of mass motion by introducing an extra variable for each of the additive integrals of our systems which is conserved in every interaction in the assembly. Consider for simplicity an assembly of any number of types of classical systems. Let ζ_t be the energy, and μ_t any component of the momentum, associated with the tth cell of the first set of systems. Then

$$C = \Sigma_{(a,b)} \frac{M!\,\delta_0^{a_0}\delta_1^{a_1}\dots}{a_0!\,a_1!\dots} \times \dots, \qquad \text{......(175)}$$

summed over all positive values of a, b, ... subject to

$$\Sigma_t a_t = M, \quad \Sigma_t b_t = N, \dots, \qquad \text{......(176)}$$

$$\Sigma_t a_t \zeta_t + \dots = E, \qquad \text{......(177)}$$

and

$$\Sigma_t a_t \mu_t + \dots = G, \qquad \text{......(178)}$$

where G is the total component of momentum of the assembly. There is a similar extra limitation for each component of momentum or angular momentum which is conserved. To sum (175) we introduce the partition function

$$h(z,x) = \Sigma_t \delta_t z^{\zeta_t} x^{\mu_t}, \qquad \text{......(179)}$$

and the value of C will be the coefficient of $z^E x^G$ in the double series

$$[h(z,x)]^M [j(z,x)]^N \ldots .$$

This is given by the double integral

$$C = \frac{1}{(2\pi i)^2} \iint \frac{dz\,dx}{z^{E+1} x^{G+1}} [h(z,x)]^M [j(z,x)]^N \ldots . \qquad \ldots\ldots(180)$$

Similarly, it is easily shown that we must have

$$C\overline{a_t} = \frac{M}{(2\pi i)^2} \iint \frac{dz\,dx}{z^{E+1} x^{G+1}} (\delta_t z^{\zeta_t} x^{\mu_t}) [h(z,x)]^{M-1} [j(z,x)]^N \ldots, \quad \ldots(181)$$

$$C\overline{E_A} = \frac{1}{(2\pi i)^2} \iint \frac{dz\,dx}{z^{E+1} x^{G+1}} \left\{ z \frac{\partial}{\partial z} [h(z,x)]^M \right\} [j(z,x)]^N \ldots, \qquad \ldots\ldots(182)$$

$$C\overline{G_A} = \frac{1}{(2\pi i)^2} \iint \frac{dz\,dx}{z^{E+1} x^{G+1}} \left\{ x \frac{\partial}{\partial x} [h(z,x)]^M \right\} [j(z,x)]^N \ldots, \qquad \ldots\ldots(183)$$

where $\overline{E_A}$ and $\overline{G_A}$ are the energy and component of momentum of the first set of systems. These double (and similar multiple) integrals can all be evaluated by Theorem 2·42. The resulting distribution laws can be reduced to those of the actual assembly by the limiting process of § 2·6.

The formal deduction of Maxwell's distribution law with mass motion by this method is very simple. We treat the case in which G and μ_t are the momenta, in the direction of q_1, of free molecules in a volume V, so that $\mu_t = (p_1)_t$. Then

$$h(z,x) \to H(z,x)$$

$$= \frac{V}{h^3} \iiint \exp\left[-\frac{\log 1/z}{2m} (p_1{}^2 + p_2{}^2 + p_3{}^2) + (\log x) p_1 \right] dp_1 dp_2 dp_3,$$

$$= \frac{(2\pi m)^{\frac{3}{2}} V}{h^3 (\log 1/z)^{\frac{3}{2}}} \exp\left\{ \frac{m(\log x)^2}{2 \log 1/z} \right\}. \qquad \ldots\ldots(184)$$

The distribution laws depend on *two* parameters ϑ and ξ which form the unique relevant root of the simultaneous equations

$$\frac{\partial}{\partial z}(z^{-E} h^M j^N \ldots) = \frac{\partial}{\partial x}(x^{-G} h^M j^N \ldots) = 0,$$

reducing in the limit to

$$E = M\vartheta \frac{\partial}{\partial \vartheta} \log H(\vartheta, \xi) + N\vartheta \frac{\partial}{\partial \vartheta} \log J(\vartheta, \xi) + \ldots, \qquad \ldots\ldots(185)$$

$$G = M\xi \frac{\partial}{\partial \xi} \log H(\vartheta, \xi) + N\xi \frac{\partial}{\partial \xi} \log J(\vartheta, \xi) + \ldots . \qquad \ldots\ldots(186)$$

We derive from (182) and (183)

$$\frac{\overline{E_A}}{M} = \vartheta\frac{\partial}{\partial\vartheta}\log H(\vartheta,\xi),$$

$$= \frac{3}{2}\frac{1}{\log 1/\vartheta} + \frac{m(\log\xi)^2}{2(\log 1/\vartheta)^2}, \qquad \text{......(187)}$$

$$\frac{\overline{G_A}}{M} = \xi\frac{\partial}{\partial\xi}\log H(\vartheta,\xi) = \frac{m\log\xi}{\log 1/\vartheta}. \qquad \text{......(188)}$$

Since $\overline{G_A}/M$ is the mean q_1-momentum per molecule of the first set, equal to mu_0 say, it follows from (188) and its analogues that the bulk-velocity u_0 must be the same in equilibrium for every set of systems in the assembly, and from (187) that the mean kinetic energy of translation per molecule is

$$\tfrac{3}{2}kT + \tfrac{1}{2}mu_0^2.$$

Finally, from (184)

$$H(\vartheta,\xi) = \frac{(2\pi m)^{\frac{3}{2}}V}{h^3(\log 1/\vartheta)^{\frac{3}{2}}}\exp\{\tfrac{1}{2}m(\log 1/\vartheta)\,u_0^2\}. \qquad \text{......(189)}$$

From (181) we find $\quad \overline{a_t} = M\delta_t\vartheta^{\zeta_t}\xi^{\mu_t}/H(\vartheta,\xi),$

$$= \frac{M}{V}\left(\frac{m}{2\pi kT}\right)^{\frac{3}{2}} dx\ldots dw\exp\left\{-\frac{m}{2kT}(u^2+v^2+w^2) + mu\log\xi - \frac{1}{2kT}mu_0^2\right\},$$

$$= \frac{M}{V}\left(\frac{m}{2\pi kT}\right)^{\frac{3}{2}} dx\ldots dw\exp\left\{-\frac{m}{2kT}[(u-u_0)^2+v^2+w^2]\right\}, \qquad \text{......(190)}$$

which is Maxwell's Law for this case. We may observe that the parameter ξ which arises from the second selector variable has a simple physical interpretation, for

$$\log\xi = u_0/kT,$$

where u_0 is the common bulk-velocity of all sets of systems in the assembly.

§2·62. *The theorem of equipartition.* The most important classical distribution law which we have not yet included is the theorem of equipartition. This is often stated as follows—*if we have any set of M classical systems in the assembly each of s degrees of freedom, whose energy (in Hamiltonian form) consists of the sum of t square terms ($s \leqslant t \leqslant 2s$), then in equilibrium the mean energy of the set is $Mt(\frac{1}{2}kT)$, or $\frac{1}{2}kT$ for each square term in the energy.* The present method enables us to give a very simple proof of this theorem, and to indicate its full range of validity, including, for example, the rotations of a rigid body, which some current proofs do not.

Suppose the equations of motion of the system do not contain the time explicitly. Then its Hamiltonian function is the energy and is the sum of (*a*) a homogeneous quadratic function of the *p*'s whose coefficients are functions of the *q*'s, and (*b*) a function of certain of the *q*'s (the potential energy). We will now suppose (1) *that the potential energy is a homogeneous* *quadratic function of $t-s$ of the q's whose coefficients may be functions of the other $(2s-t)$ q's,* (2) *that the coefficients of the quadratic p-terms are functions only of these $(2s-t)$ q's which do not contribute directly to the potential energy under* (1). We can then show that *the mean energy of the set is $Mt(\frac{1}{2}kT)$, which is the theorem of equipartition in its most general form.*

The classical partition function for these systems is

$$H(z)=\frac{1}{h^s}\int^{(2s)}\cdots\int e^{-(\log 1/z)\epsilon}\,dp_1\ldots dq_s, \qquad \ldots\ldots(191)$$

where ϵ is the energy in Hamiltonian form. The limits of integration of the $(2s-t)$ q's which provide no square terms in ϵ will be determined by the geometry of the system. Local boundary fields such as those defining the walls of a containing vessel can be regarded alternatively as geometrical constraints defining the limits of integration of certain q's. The other variables are to be integrated from $-\infty$ to $+\infty$. The homogeneous quadratic function of the p's can be expressed by a linear transformation as a sum of s squares with positive coefficients,

$$\alpha_1 r_1^2+\ldots+\alpha_s r_s^2.$$

We change the variables from $p_1, \ldots, p_s$ to $r_1, \ldots, r_s$ and integrate with respect to these from $-\infty$ to $+\infty$. Then

$$H(z)=\frac{1}{h^s}\int^{(s)}\cdots\int\left(\frac{\pi}{\log 1/z}\right)^{\frac{1}{2}s}\frac{\mu}{(\alpha_1\ldots\alpha_s)^{\frac{1}{2}}}e^{-(\log 1/z)W}\,dq_1\ldots dq_s$$

where μ is the Jacobian of the $(p—r)$ transformation, and W the potential energy. We can now find a linear transformation of $q_1, \ldots, q_{t-s}$ which casts W into the form

$$\beta_1 w_1^2+\ldots+\beta_{t-s}w_{t-s}^2.$$

We change the variables from $q_1, \ldots, q_{t-s}$ to $w_1, \ldots, w_{t-s}$. By hypothesis μ, the α's and the β's do not depend on the w's and are functions of the "geometrical" variables only. Integrating with respect to the w's we find therefore

$$H(z)=\left[\frac{\pi}{\log 1/z}\right]^{\frac{1}{2}t}\frac{1}{h^s}\int^{(2s-t)}\cdots\int\frac{\mu}{(\alpha_1\ldots\alpha_s)^{\frac{1}{2}}}\frac{\mu'}{(\beta_1\ldots\beta_{t-s})^{\frac{1}{2}}}dq_{t-s+1}\ldots dq_s, \qquad \ldots\ldots(192)$$

where μ' is the Jacobian of the $(q—w)$ transformation. The integral in $H(z)$ depends only on the geometrical limits, and is independent of z.

The mean energy for a set of M of these systems is

$$\begin{aligned}\bar{E}&=M\vartheta\frac{\partial}{\partial\vartheta}\log H(\vartheta),\\ &=-M\vartheta\frac{d}{d\vartheta}\log[\log 1/\vartheta]^{\frac{1}{2}t},\\ &=\tfrac{1}{2}Mt/[\log 1/\vartheta]=Mt(\tfrac{1}{2}kT),\end{aligned} \qquad \ldots\ldots(193)$$

which is the theorem stated. It is clear that the theorem cannot be true for non-relativistic Hamiltonian functions under conditions wider than those given here.

§ **2·63.** *Classical rotations.* A special case included in this proof is that of the rotations of a rigid body. General molecular rotations must therefore contribute $3(\frac{1}{2}kT)$ to the mean molecular energy, and pure transverse rotations of a body with an axis of symmetry and no axial spin $2(\frac{1}{2}kT)$. These results are in common use, and we shall refer in the next chapter to the classical values R, $\frac{3}{2}R$, $(R = Mk)$ here obtained for C_{rot}. We shall later want the complete expressions for these two partition functions, and it is convenient to insert the calculations here.

For the transverse rotations, moment of inertia A,

$$\epsilon = \frac{1}{2A}\left(p_\theta^2 + \frac{p_\phi^2}{\sin^2\theta}\right);$$

$$H(z) = \frac{1}{h^2}\iiiint \exp\left\{-(\log 1/z)\left(\frac{p_\theta^2}{2A} + \frac{p_\phi^2}{2A\sin^2\theta}\right)\right\} dp_\theta dp_\phi d\theta d\phi,$$

$$= \frac{1}{h^2}\frac{2\pi A}{\log 1/z}\int_0^\pi\int_0^{2\pi} \sin\theta\, d\theta\, d\phi,$$

$$= \frac{8\pi^2 A}{h^2 \log 1/z};$$

$$H(T) = \frac{8\pi^2 AkT}{h^2}. \qquad \text{......(194)}$$

For the general rotations of a rigid body, moments of inertia A, B, C,

$$\epsilon = \frac{1}{2A\sin^2\theta}\{(p_\phi - \cos\theta\, p_\psi)\cos\psi - \sin\theta\sin\psi\, p_\theta\}^2$$
$$+ \frac{1}{2B\sin^2\theta}\{(p_\phi - \cos\theta\, p_\psi)\sin\psi + \sin\theta\cos\psi\, p_\theta\}^2 + \frac{1}{2C}p_\psi^2;$$

$$H(z) = \frac{1}{h^3}\int^{(6)}\cdots\int \exp(-\epsilon\log 1/z)\, dp_\theta dp_\phi dp_\psi d\theta d\phi d\psi,$$

the limits of integration for θ, ϕ, ψ being $(0,\pi)$, $(0,2\pi)$, $(0,2\pi)$. The energy can be expressed in the integrable form

$$\frac{1}{2}\left(\frac{\sin^2\psi}{A} + \frac{\cos^2\psi}{B}\right)\left\{p_\theta + \left(\frac{1}{B} - \frac{1}{A}\right)\frac{\sin\psi\cos\psi}{\sin\theta\left(\dfrac{\sin^2\psi}{A} + \dfrac{\cos^2\psi}{B}\right)}(p_\phi - \cos\theta\, p_\psi)\right\}^2$$

$$+ \frac{1}{2AB\sin^2\theta}\frac{1}{\dfrac{\sin^2\psi}{A} + \dfrac{\cos^2\psi}{B}}(p_\phi - \cos\theta\, p_\psi)^2 + \frac{1}{2C}p_\psi^2. \qquad \text{......(194·1)}$$

Integrating with respect to p_θ, p_ϕ, p_ψ in that order, we find

$$H(z)=\frac{1}{h^3}\frac{(8\pi^3ABC)^{\frac{1}{2}}}{(\log 1/z)^{\frac{3}{2}}}\int_0^\pi\int_0^{2\pi}\int_0^{2\pi}\sin\theta\, d\theta\, d\phi\, d\psi,$$

$$=\frac{8\pi^2}{h^3}\frac{(8\pi^3ABC)^{\frac{1}{2}}}{(\log 1/z)^{\frac{3}{2}}};$$

$$H(T)=\frac{8\pi^2}{h^3}(8\pi^3ABCk^3T^3)^{\frac{1}{2}}. \quad \text{......(195)}$$

§2·64. *Boltzmann's distribution law (classical statistics).* This law is the complement of Maxwell's for classical systems, and was originally associated with the distribution in external fields of force. It can be stated generally as follows:

For any set of classical systems, the average numbers $\overline{a_1}, \overline{a_2}$ in any two equal elements of their phase space are in the ratio

$$\overline{a_1}:\overline{a_2}=e^{-\epsilon_1/kT}:e^{-\epsilon_2/kT}, \quad \text{......(196)}$$

where ϵ_1 and ϵ_2 are the energies of the systems in these elements of their phase space.

This law follows at once from previous theorems, for

$$\overline{a_1}=\delta_1\vartheta^{\epsilon_1}/H(\vartheta),\quad \overline{a_2}=\delta_2\vartheta^{\epsilon_2}/H(\vartheta),\quad \delta_1=\delta_2.$$

It applies of course to localized quantized systems in the form

$$\overline{a_1}:\overline{a_2}=\varpi_1 e^{-\epsilon_1/kT}:\varpi_2 e^{-\epsilon_2/kT}, \quad \text{......(197)}$$

also a consequence of preceding theorems. The law contains nothing not already given, but is inserted here formally for completeness.

Boltzmann's law has, of course, numerous important applications and important specialized forms. If we consider two elements of the *physical* space accessible to the systems, in which their dynamical state is the same, so that the Hamiltonian energy function differs only in the different values of W, we can integrate over all possible momenta and obtain

$$n_1:n_2=e^{-W_1/kT}:e^{-W_2/kT}. \quad \text{......(198)}$$

In (198) n_1 and n_2 are the average total numbers of systems without regard to their kinetic energy in equal volume elements of physical space. This leads at once to the density law for an isothermal atmosphere of perfect gases. Since n is proportional to ρ, the mass density of the gas, equation (198) can be written

$$\rho=\rho_0 e^{-(W-W_0)/kT}. \quad \text{......(199)}$$

If V^* denotes the gravitational potential in the atmosphere per unit mass (including any field of "centrifugal force") and m is the mass of a molecule, then

$$\rho=\rho_0 e^{-m(V^*-V_0^*)/kT}. \quad \text{......(200)}$$

This is the atmospheric density law, commonly known as Dalton's. Since V^* is the same for all molecules and m varies from molecule to molecule, equation (200) describes the well-known settling of the heavier molecules to the base of the atmosphere—the most prominent property of an atmosphere is statistical (i.e. isothermal) equilibrium.

The potentials here considered are primarily potentials due to bodies external to the assembly to which the systems of the assembly itself make no effective contribution. This restriction is removed in Chapter VIII.

We have no space to enter here into further atmospheric problems such as the nature of convective equilibrium and the rate of escape of molecules from the boundary of the atmosphere. Such problems belong more properly to the study of steady non-equilibrium states and require the explicit introduction of mechanisms of interaction, but escape is a border-line problem and of particular interest which we shall discuss in Chapter XV.

The validity of equations (196) and (197) is completely general, but that of (198) and (199) is not; they must be confined strictly to the field specified in their enunciation. For example, we cannot always apply (198) to elements of volume belonging to the system in different parts of the assembly which are different *phases* in the thermodynamical sense. More refined considerations are then necessary on which we embark in Chapter V. We must also be careful not to restrict in any way the range of the integrations with respect to the momenta. For example, if we apply Boltzmann's theorem to the number of *free* electrons in the neighbourhood of a fixed positive charge, we mean by free those which have sufficient kinetic energy to escape altogether. The relative numbers of these in two volume elements are not given correctly by (198). It is necessary to return to (196) and observe that the integration with respect to the momenta must be taken only over the region for which

$$(p_x^2+p_y^2+p_z^2)/2m \geqslant -W > 0.$$

Thus in this case the constant quantity is

$$\frac{e^{-W/kT}}{n}\iiint_{\Sigma p_x^2 \geqslant -2mW} e^{-(p_x^2+p_y^2+p_z^2)/2mkT}\,dp_x dp_y dp_z,$$

which reduces easily to give

$$n_1 : n_2 = e^{-W_1/kT}\int_{-W_1/kT}^{\infty} e^{-x}x^{\frac{1}{2}}dx : e^{-W_2/kT}\int_{-W_2/kT}^{\infty} e^{-x}x^{\frac{1}{2}}dx. \qquad \text{......(201)}$$

Equation (198) would of course continue to give the relative numbers of electrons *both bound and free*, were it not for the limitations imposed by the quantum theory on the bound electrons.

Though not strictly relevant to this chapter it is best to point out here that this classification of electrons into free and bound has another similar effect.* We have seen in equation (173) that the average kinetic energy of any separate classical system in a gaseous assembly is $\frac{3}{2}kT$, and we shall see later in § 8·2 that this is still true when there are forces acting between the particles. But here we arbitrarily classify an electron as free in a region where its potential energy is negative only when

$$\Sigma p_x^2 \geqslant -2mW > 0.$$

Its partition function in such a region is therefore no longer given by (172) but becomes instead

$$H(\vartheta) = \frac{V}{h^3}\iiint_{\Sigma p_x^2 \geqslant -2mW} e^{-(\log 1/\vartheta)(p_x^2+p_y^2+p_z^2)/2m}\, dp_x dp_y dp_z,$$
$$= \frac{2\pi V(2mkT)^{\frac{3}{2}}}{h^3}\int_{-W/kT}^{\infty} e^{-x} x^{\frac{1}{2}}\, dx.$$

Therefore in this region

$$\bar{E} = \vartheta\frac{\partial \log H}{\partial\vartheta} = kT\left\{\frac{3}{2} + \frac{(-W/kT)^{\frac{3}{2}} e^{W/kT}}{\int_{-W/kT}^{\infty} e^{-x} x^{\frac{1}{2}}\, dx}\right\}. \qquad \text{......(202)}$$

This effect illustrates the care necessary when such a classification has to be employed.

§ **2·65**. *Analogies with Gibbsian phase integrals.* The analogy with Gibbs' development can be clearly seen at this stage. The partition function for a molecule of a perfect gas in an external field of force, in which its potential energy is W, is, in terms of T,

$$H(T) = \frac{1}{h^3}\int^{(6)} \cdots \int e^{-\left\{\frac{1}{2m}(p_x^2+p_y^2+p_z^2)+W\right\}/kT} dp_x dp_y dp_z dx\, dy\, dz,$$

and the partition function for N such molecules (classical statistics) is this integral N times repeated, or $[H(T)]^N$.

Now this integral N times repeated is exactly Gibbs' integral† for this assembly of N molecules of a perfect gas over "an ensemble of such assemblies canonically distributed in phase". Gibbs defines a function ψ by the equation

$$e^{-\psi/\Theta} = \int_{\text{phases}}^{\text{all}} \cdots \int e^{-\epsilon/\Theta}\, dp_1 \ldots dq_n, \qquad \text{......(203)}$$

so that here $\psi = -N\Theta \log H(T)$. Gibbs' Θ is proportional to T, and his ψ is shown eventually to be equivalent to the thermodynamic potential $U - TS$, an equivalence established directly for our partition functions in Chapter VI. What we have done here may, if it is preferred, be regarded as a generalization of the Gibbsian phase integral so as to include quantized systems in the

* McCrea, *Proc. Camb. Phil. Soc.* vol. 26, p. 107 (1930).

† Gibbs, *Elementary Principles in Statistical Mechanics*, p. 33, eq. 92.

assembly. Our semi-logical dynamical foundation can be discarded, without altering the results, for the hypothesis of canonical distribution in phase.

We may observe here that whatever be the form of W the p_x, p_y, p_z integrations can be effected, giving

$$H(T)=\frac{(2\pi mkT)^{\frac{3}{2}}}{h^3}\iiint e^{-W/kT}\,dx\,dy\,dz. \qquad \text{......(204)}$$

Thus the partition function or phase integral splits into two factors for the kinetic and potential energies, which can always be discussed separately for classical systems. When $W=0$ except for boundary fields the factor for the potential energy reduces to the volume V. The form of both factors remains essentially Gibbsian.

§ 2·7. *Further theorems in quantum statistics for free particles in a volume V. Systems with internal structure.* After this classical digression we resume the development of the exact theory for free particles, and start by generalizing §§ 2·5, 2·51 to systems which, besides being free to move in a volume V, possess internal structure. It is a sufficient approximation to assume that the boundary fields do not affect the internal motions of the particle so that its translatory motion and internal motion and the corresponding Schrödinger's equation separate as if it were completely free. The eigenvalues of its energy and the corresponding weights are therefore

$$\epsilon_s=\epsilon_r+\epsilon_\tau,\quad \varpi_s=\varpi_\tau,$$

where the ϵ_r's are those of § 2·5 and the ϵ_τ, ϖ_τ's are the eigenvalues and weights for Schrödinger's equation for the internal motions. Any ϵ_r may be combined with any ϵ_τ. Therefore

$$\begin{aligned}Z&=\Sigma_s\,\varpi_s\log g(\lambda\vartheta^{\epsilon_s})=\Sigma_{r,\tau}\,\varpi_\tau\log g(\lambda\vartheta^{\epsilon_r+\epsilon_\tau}),\\&=\sum_{j=1}^{\infty}\frac{\alpha_j}{j}\lambda^j\,(\Sigma_\tau\,\varpi_\tau e^{-j\epsilon_\tau/kT})\,(\Sigma_r e^{-j\epsilon_r/kT}),\\&=\frac{(2\pi mkT)^{\frac{3}{2}}}{h^3}V\sum_{j=1}^{\infty}\frac{\alpha_j}{j^{\frac{5}{2}}}\lambda^j\,\Sigma_\tau\,\varpi_\tau e^{-j\epsilon_\tau/kT}. \qquad \text{...(205)}\end{aligned}$$

It is convenient to use a special symbol Z for this important sum.

The classical form of (205) or its limit for small λ is therefore given by

$$Z=\lambda\frac{(2\pi mkT)^{\frac{3}{2}}}{h^3}Vf(T), \qquad \text{......(206)}$$

where $f(T)$ is the partition function for the internal motion of the system (e.g. free molecule). It follows from (120) and (121) that in this case $M=Z$ and

$$\begin{aligned}\bar{E}&=kT^2\partial Z/\partial T,\\&=Z\{\tfrac{3}{2}kT+kT^2\partial\log f(T)/\partial T\},\\&=M\{\tfrac{3}{2}kT+kT^2\partial\log f(T)/\partial T\}. \qquad \text{......(207)}\end{aligned}$$

From (207) one can deduce the specific heat at constant volume $(d\bar{E}/dT)_V$. One sees that it is made up by addition of separate components from the translatory energy and the internal motions (including rotation).

It is perhaps of interest to give one example of the exact form (205) applied to systems with specified structure. Let us assume that the free systems are symmetrical rigid rotators without axial spin. Then using in anticipation the results of § 3·1 and assuming that for all important values of j and τ $j\epsilon_\tau \ll kT$, we find

$$\Sigma_\tau \varpi_\tau e^{-j\epsilon_\tau/kT} \sim \frac{1}{j}\frac{8\pi^2 AkT}{h^2}.$$

Then, with sufficient accuracy,

$$\Sigma_s \varpi_s \log g(\lambda\vartheta^{\epsilon_s}) = \frac{(2\pi m)^{\frac{3}{2}}(8\pi^2 A)(kT)^{\frac{5}{2}}V}{h^5}\sum_{j=1}^{\infty}\frac{\alpha_j}{j^{\frac{5}{2}}}\lambda^j.$$

The classical form or the limit of the exact result for small λ is therefore

$$\frac{(2\pi m)^{\frac{3}{2}}(8\pi^2 A)(kT)^{\frac{5}{2}}V}{h^5}\lambda,$$

as can be directly verified by combining (172) and (194).

§ **2·71**. *Space distributions of mass-points in external fields of force.* In the classical version of Boltzmann's distribution law we have obtained formulae for the average distribution of free particles in space under the action of an external field of force. The distribution laws of quantum mechanics however are primarily concerned only with distributions over the eigenvalues of the energy. At the same time these must imply some means of deriving the average number of systems "present" in a given volume element (not too small) of ordinary space.

The means required are provided by the properties of the wave-functions themselves. We interpret these wave-functions so that $|\psi_r|^2 dV$ is the probability that a given system with this normalized wave-function will be found in the volume element dV at any time. The average number n of molecules "present" in the volume element dV is therefore given by

$$n/dV = \Sigma_r \overline{a_r}\,|\psi_r|^2, \qquad \text{......(208)}$$

$$= \lambda\frac{\partial}{\partial\lambda}\Sigma_r \log g(\lambda\vartheta^{\epsilon_r})\,|\psi_r|^2. \qquad \text{......(209)}$$

We must now study the approximate forms of these wave-functions when the systems move in a field of force in which they possess the potential energy W.* We must restrict ourselves to wave equations for ψ which separate in the variables x, y, z, so that $W = w_1(x) + w_2(y) + w_3(z)$. Actually this restriction proves not to be serious since with a more general W we can always

* Mott, *Proc. Camb. Phil. Soc.* vol. 24, p. 76 (1928).

limit the assembly to physically small portions of the gas in which W is sensibly of this form. The equation for the x-factor in ψ, namely

$$\frac{d^2\psi(x)}{dx^2}+\kappa^2[\epsilon_s-w_1(x)]\,\psi(x)=0 \quad \left(\kappa^2=\frac{8\pi^2 m}{h^2}\right), \qquad \text{......(210)}$$

is then typical of all three factors. It must be solved with the boundary-conditions $\psi(0)=\psi(a)=0$. For convenience we shall assume that $w_1(0)=0$ and that $dw_1(x)/dx>0$, but these conditions are inessential. Since κ is very large we can apply the analysis developed by Jeffreys* for such problems. He has shown that if

$$\kappa \gg \int_0^x \frac{d^2}{dx^2}[\epsilon_s-w_1(x)]^{-\frac{1}{4}}\frac{dx}{[\epsilon_s-w_1(x)]^{\frac{1}{4}}}, \qquad \text{......(211)}$$

nen the two solutions of (210) approximate very closely to

$$[\epsilon_s-w_1(x)]^{-\frac{1}{4}}\exp\left\{\pm\kappa\int_0^x [w_1(x)-\epsilon_s]^{\frac{1}{2}}\,dx\right\}. \qquad \text{......(212)}$$

Since κ is large the condition (211) is satisfied for all but a very few of the possible ϵ_s except near a zero of $\epsilon_s-w_1(x)$, where there is a range very small compared with the total range $0, a$ in which the condition fails.

We now choose that solution vanishing at $x=0$ or $x=a$ which can also be made to vanish at $x=a$ or $x=0$. There are two cases. If $\epsilon_s-w_1(x)>0$ everywhere in the range the required solution is

$$[\epsilon_s-w_1(x)]^{-\frac{1}{4}}\sin\left\{\kappa\int_0^x [\epsilon_s-w_1(x)]^{\frac{1}{2}}\,dx\right\}. \qquad \text{......(213)}$$

The ϵ_s are then those values for which

$$\kappa\int_0^a [\epsilon_s-w_1(x)]^{\frac{1}{2}}\,dx=n\pi,$$

where n is an integer. If $\epsilon_s-w_1(x)=0$ at $x=x_0$ in the range we may take that solution which, when $\epsilon_s-w_1(x)<0$, approximates to

$$[w_1(x)-\epsilon_s]^{-\frac{1}{4}}\exp\left\{-\kappa\int_{x_0}^x [w_1(x)-\epsilon_s]^{\frac{1}{2}}\,dx\right\},$$

for this solution decreases very rapidly as x increases and may be taken to be zero for $x>x_0$, and so at $x=a$. When $x<x_0$ this solution has been shown to approximate to

$$[\epsilon_s-w_1(x)]^{-\frac{1}{4}}\cos\left\{-\tfrac{1}{4}\pi+\kappa\int_x^{x_0} [\epsilon_s-w_1(x)]^{\frac{1}{2}}\,dx\right\}.$$

These values of ϵ_s are determined by

$$-\tfrac{3}{4}\pi+\kappa\int_0^{x_0} [\epsilon_s-w_1(x)]^{\frac{1}{2}}\,dx=n'\pi,$$

* Jeffreys, *Proc. Lond. Math. Soc.* vol. 23, p. 428 (1924). An independent discussion has been more recently given by Kramers, *Zeit. f. Physik*, vol. 39, p. 828 (1926), and by other authors, in which numerical mistakes in Jeffreys' formulae have been corrected.

where n' is an integer. Since the period of the oscillating function is very short compared with the distances of appreciable variation in $w_1(x)$, the normalizing divisor for this wave-function is

$$N_s = \frac{1}{2}\int_0^{x_0} [\epsilon_s - w_1(x)]^{-\frac{1}{2}}\, dx,$$

and for the other type the same with x_0 replaced by a. The interval between two eigenvalues is given by

$$\kappa \int_0^{x_0'} [\epsilon_{s+1} - w_1(x)]^{\frac{1}{2}}\, dx - \kappa \int_0^{x_0} [\epsilon_s - w_1(x)]^{\frac{1}{2}}\, dx = \pi,$$

so that approximately, if $d\epsilon_s = \epsilon_{s+1} - \epsilon_s$,

$$\kappa N_s d\epsilon_s = \pi.$$

Thus the normalized wave-functions are approximately

$$\begin{aligned} \psi &= \left(\frac{\kappa\, d\epsilon_s}{\pi}\right)^{\frac{1}{2}} \frac{1}{[\epsilon_s - w_1(x)]^{\frac{1}{4}}} \sin G \quad \{\epsilon_s > w_1(x)\}, \\ &= 0 \qquad\qquad\qquad\qquad\qquad\qquad \{\epsilon_s < w_1(x)\}, \end{aligned}$$

where $\sin G$ oscillates very rapidly when either ϵ_s or x varies. There are similar factors in y and z.

We now insert these values in the series in (209), and average over a small volume element dV ($= dx\,dy\,dz$) so that factors such as $\sin^2 G$ may be replaced by $\frac{1}{2}$. We find for this series

$$\frac{(2m)^{\frac{3}{2}}}{h^3} \Sigma_r \log g(\lambda e^{-\epsilon_r/kT}) \frac{d\epsilon_s d\epsilon_t d\epsilon_u}{[\{\epsilon_s - w_1(x)\}\{\epsilon_t - w_2(y)\}\{\epsilon_u - w_3(z)\}]^{\frac{1}{2}}}$$

$$(\epsilon_r = \epsilon_s + \epsilon_t + \epsilon_u), \quad \text{......(214)}$$

summed over all r such that

$$\epsilon_s > w_1(x), \quad \epsilon_t > w_2(y), \quad \epsilon_u > w_3(z).$$

Using the substitutions $\epsilon_s = w_1(x) + \frac{1}{2}mu^2$, etc. and obvious approximations, this sum can be replaced by the integral

$$\left(\frac{2m}{h}\right)^3 \iiint_0^\infty \log g(\lambda e^{-\{W + \frac{1}{2}m(u^2+v^2+w^2)\}/kT})\, du\, dv\, dw. \quad \text{......(215)}$$

Therefore

$$\frac{n}{dV} = \left(\frac{2m}{h}\right)^3 \lambda \frac{\partial}{\partial\lambda} \iiint_0^\infty \log g(\lambda e^{-\{W + \frac{1}{2}m(u^2+v^2+w^2)\}/kT})\, du\, dv\, dw. \quad \text{...(216)}$$

The u, v, w are the exact analogues of the velocity components of the classical particle. Thus (216) gives us the space distribution law in its form integrated over the velocities. By returning to (214) and taking only those terms which correspond to specified velocity ranges we obtain the complete velocity-space distribution law in the form*

$$n(u,v,w,x,y,z)\,du\,dv\,dw\,dV$$
$$=\left(\frac{m}{h}\right)^3 \lambda\frac{\partial}{\partial\lambda}\log g(\lambda e^{-\{W+\frac{1}{2}m(u^2+v^2+w^2)\}/kT})\,du\,dv\,dw\,dV, \quad \text{......(217)}$$

which is the exact form of the Maxwell-Boltzmann distribution law. When λ is small the log reduces in all statistics to

$$\lambda e^{-\{W+\frac{1}{2}m(u^2+v^2+w^2)\}/kT}.$$

On carrying through the integrations in (216) we then obtain

$$\frac{n}{dV}=\nu=\frac{(2\pi mkT)^{\frac{3}{2}}}{h^3}\lambda e^{-W/kT},$$

which, λ being constant, is the classical result.

It is useful to express (217) in terms of the resultant velocity or total kinetic energy and the direction of motion. We can then put

$$u^2+v^2+w^2=c^2, \quad du\,dv\,dw=c^2\,dc\,d\omega,$$

where $d\omega$ is an element of solid angle, so that

$$\frac{n}{dV}=\left(\frac{m}{h}\right)^3\lambda\frac{\partial}{\partial\lambda}\log g(\lambda e^{-(W+\frac{1}{2}mc^2)/kT})\,c^2\,dc\,d\omega. \quad \text{......(218)}$$

On integrating over all directions, we find an expression for $n(c)\,dc$, the average number of systems per unit volume moving with velocities between c and $c+dc$,

$$n(c)\,dc=\frac{4\pi m^3}{h^3}\lambda\frac{\partial}{\partial\lambda}\log g(\lambda e^{-(W+\frac{1}{2}mc^2)/kT})\,c^2\,dc. \quad \text{......(219)}$$

Expressed in terms of energies ($\frac{1}{2}mc^2=\epsilon$) this reduces for the range ϵ, $\epsilon+d\epsilon$ to

$$n(\epsilon)\,d\epsilon=\frac{2\pi(2m)^{\frac{3}{2}}}{h^3}\lambda\frac{\partial}{\partial\lambda}\log g(\lambda e^{-(W+\epsilon)/kT})\,\epsilon^{\frac{1}{2}}\,d\epsilon. \quad \text{......(220)}$$

§ **2·72.** *Distribution of mass-points between different phases or enclosures.* The foregoing result can be obtained under more general conditions by a somewhat different method of treatment which does not contemplate in one survey the whole space V accessible to the systems, but starts instead by breaking it up into parts and treating each part as if it were a practically independent enclosure. The suitability of either procedure depends on what is the best type of idealization of the actual assembly to be investigated.

Consider for simplicity an assembly of two slightly connected enclosures in each of which the potential energy of the systems is constant. In one enclosure it may be taken to have the value zero, but in the other a different constant value W, which may of course be of either sign. In the former the

* A factor 8 drops out from (217) because only positive values of u, v, w were contemplated in (216), while the actual u, v, w may have either sign independently of each other.

eigenvalues are those already given in (155) with the corresponding spacing. We thus find, using the same groupings and transformations that lead to (220),

$$\Sigma_r \log g(\lambda \vartheta^{\epsilon_r}) = \frac{2\pi(2m)^{\frac{3}{2}}}{h^3} V \int_0^\infty \log g(\lambda e^{-\epsilon/kT})\, \epsilon^{\frac{1}{2}}\, d\epsilon. \quad \text{......(221)}$$

In the latter enclosure the wave equation is

$$\nabla^2 \psi + \kappa^2 (E - W)\psi = 0,$$

and the eigenvalues are given by

$$E - W = \frac{h^2}{8m}\left(\frac{s^2}{a^2} + \frac{t^2}{b^2} + \frac{u^2}{c^2}\right).$$

We therefore find instead of (221)

$$\Sigma_r \log g(\lambda \vartheta^{\epsilon_r'}) = \frac{2\pi(2m)^{\frac{3}{2}}}{h^3} V' \int_0^\infty \log g(\lambda e^{-(W+\epsilon)/kT})\, \epsilon^{\frac{1}{2}}\, d\epsilon. \quad \text{...(222)}$$

If we construct the usual expression for the number of complexions of this two-enclosure assembly, we have

$$C = \left(\frac{1}{2\pi i}\right)^2 \iint \frac{dx\,dz}{x^{X+1} z^{E+1}} \Pi_r\, g(xz^{\epsilon_r})\, \Pi_r\, g(xz^{\epsilon_r'}),$$

and the average numbers of systems in the two enclosures will be given by

$$\overline{M} = \lambda \frac{\partial}{\partial \lambda} \Sigma_r \log g(\lambda \vartheta^{\epsilon_r}), \quad \overline{M'} = \lambda \frac{\partial}{\partial \lambda} \Sigma_r \log g(\lambda \vartheta^{\epsilon_r'}).$$

On using (221) and (222) we see that these expressions are equivalent to (216) which we obtained by discussion of the whole assembly with a simple form for W.

We may note in conclusion that the dependence of (221) and (222) on T can be shown in a simple form by the substitution $\epsilon/kT = x$. We then find

$$\Sigma_r \log g(\lambda \vartheta^{\epsilon_r}) = \frac{2\pi(2mkT)^{\frac{3}{2}} V}{h^3} \int_0^\infty \log g(\lambda e^{-W/kT} e^{-x})\, x^{\frac{1}{2}}\, dx, \quad \text{...(223)}$$

which when $W = 0$ depends on T only through the outside factor and λ.

Whether or not the systems are practically classical in either of the enclosures will depend on the value of $\lambda e^{-W/kT}$. It may happen that λ is small (classical statistics) while $\lambda e^{-W/kT}$ is very large (tight-packed systems). This happens in applications to thermionics.

§ **2·73.** *Highly degenerate assemblies* (*of electrons*). For assemblies in which λ is not small, we require fresh means of evaluating the integrals of the last sections. Series expansions are now useless. As we have already shown in § 2·51 such assemblies in practice are only assemblies of electrons, so that we may confine attention to the generating function proper to the Fermi-Dirac statistics. It was moreover shown in § 2·41 that it is only for this statistics that λ can become large. We must also remember that the electron has a spin with two orientations, so that in the absence of external magnetic fields it is a degenerate system of weight 2 whatever its translational motion.

Before applying our theorems to assemblies of electrons we must recall that our assemblies have to be composed of practically independent systems, while electrons act on each other with long range fields. It is not possible that these long range fields should be entirely without effect on the eigenvalues of the assembly, but if the charges of the electrons are neutralized in each volume element by the charges of suitable associated positive systems, as in fact they are, it does seem reasonable to assume that the charges of the electrons can be neglected in constructing a valid approximation of zero order to the wave-function of the assembly. If we make this approximation we smooth out as it were the atomic structure of the charges, so that their only remaining effect from the point of view of the electrons is to create a region of uniform negative potential energy in which the electrons move almost freely. This is the model which (following Sommerfeld) we shall later apply to explain the leading features of metallic conductors* and of the interiors of ultra-white-dwarf stars.

For an assembly of electrons so treated we find on adapting (223) that

$$Z = \Sigma_r \varpi_r \log g(\lambda \vartheta^{\epsilon_r}),$$

$$= 2\frac{(2\pi mkT)^{\frac{3}{2}} V}{h^3} \frac{2}{\sqrt{\pi}} \int_0^\infty x^{\frac{1}{2}} \log(1 + \lambda e^{-W/kT - x})\, dx. \quad \text{......(224)}$$

If $W = 0$ for free space, then $W < 0$ for the interior of a metal. We shall write $W = -\chi_0$ and shorten the algebra by writing

$$\mu = \lambda e^{\chi_0/kT}. \quad \text{......(225)}$$

We can remove the logarithm if desired by integration by parts, so that

$$Z = 2\frac{(2\pi mkT)^{\frac{3}{2}} V}{h^3} \frac{4}{3\sqrt{\pi}} \int_0^\infty \frac{x^{\frac{3}{2}}\, dx}{1 + e^x/\mu}. \quad \text{......(226)}$$

For this group of electrons

$$\overline{M} = \mu \frac{\partial Z}{\partial \mu} = 2\frac{(2\pi mkT)^{\frac{3}{2}} V}{h^3} \frac{2}{\sqrt{\pi}} \int_0^\infty \frac{x^{\frac{1}{2}}\, dx}{1 + e^x/\mu}, \quad \text{......(227)}$$

$$\overline{E} = kT^2 \left(\frac{\partial Z}{\partial T}\right)_\lambda = kT^2 \left[\frac{\partial Z}{\partial T} + \frac{\partial Z}{\partial \mu}\frac{\partial \mu}{\partial T}\right] = \tfrac{3}{2} kTZ - \overline{M}\chi_0. \quad \text{...(228)}$$

We require a means of evaluating Z asymptotically for large μ. Consider the integral factor in Z in the form

$$I = \frac{2}{\sqrt{\pi}} \int_0^\infty x^{\frac{1}{2}} \log(1 + e^{\beta - x})\, dx \quad (\beta = \log \mu).$$

* Sommerfeld, *Zeit. f. Physik*, vol. 47, p. 1 (1928); *Naturwiss.* vol. 15, p. 825 (1927).

By breaking the range at $x=\beta$ this reduces to

$$\tfrac{1}{2}\sqrt{\pi}I=\int_0^\beta x^{\frac{1}{2}}(\beta-x)\,dx+\int_0^\beta(\beta-y)^{\frac{1}{2}}\log(1+e^{-y})\,dy$$
$$+\int_0^\infty(\beta+y)^{\frac{1}{2}}\log(1+e^{-y})\,dy.$$

In either of these integrals the logarithms can be expanded and the resulting series integrated term by term. Thus we get

$$\tfrac{1}{2}\sqrt{\pi}I=\tfrac{4}{15}\beta^{\frac{5}{2}}+\Sigma\frac{(-)^{s-1}}{s}\int_0^\beta e^{-sy}(\beta-y)^{\frac{1}{2}}\,dy+\Sigma\frac{(-)^{s-1}}{s}\int_0^\infty e^{-sy}(\beta+y)^{\frac{1}{2}}\,dy.$$

This is exact. When β is large the square roots in these integrals can both be replaced by $\sqrt{\beta}$ to give the dominant terms. The range of the first series of integrals can then be extended to infinity without sensible error. Thus

$$\tfrac{1}{2}\sqrt{\pi}I=\tfrac{4}{15}\beta^{\frac{5}{2}}+2\beta^{\frac{1}{2}}\Sigma\frac{(-)^{s-1}}{s}\int_0^\infty e^{-sy}\,dy,$$
$$=\tfrac{4}{15}\beta^{\frac{5}{2}}+\frac{\pi^2}{6}\beta^{\frac{1}{2}}.$$

The method can easily be extended to show that the error term is $O(\beta^{-\frac{3}{2}})$, or to give a general asymptotic expansion in powers of β^{-2}. We find therefore that

$$Z=2\frac{(2\pi mkT)^{\frac{3}{2}}V}{h^3}\frac{2}{\sqrt{\pi}}\left\{\tfrac{4}{15}(\log\mu)^{\frac{5}{2}}+\frac{\pi^2}{6}(\log\mu)^{\frac{1}{2}}+O(\log\mu)^{-\frac{3}{2}}\right\}.\quad\ldots(229)$$

Besides (229) we need only the distribution laws of §§ 2·71, 2·72 which here take the special forms

$$n(u,v,w,x,y,z)\,du\,dv\,dw\,dV=2\left(\frac{m}{h}\right)^3\frac{dV\,du\,dv\,dw}{1+e^{\{W+\frac{1}{2}m(u^2+v^2+w^2)\}kT}/\lambda},$$
$$=2\left(\frac{m}{h}\right)^3\frac{dV\,du\,dv\,dw}{1+e^{\frac{1}{2}m(u^2+v^2+w^2)/kT}/\mu},\quad\ldots\ldots(230)$$

$$n(\epsilon)\,d\epsilon\,dV=\frac{4\pi(2m)^{\frac{3}{2}}}{h^3}\frac{\epsilon^{\frac{1}{2}}\,d\epsilon\,dV}{1+e^{\epsilon/kT}/\mu}.\quad\ldots\ldots(231)$$

We postpone application of these formulae to Chapter XI where we shall examine more closely the conditions under which $\log\mu\gg 1$, required for the validity of (229); the modifications introduced by the relativistic variation of the mass will be considered in Chapter XVI.

§ **2·8.** *External reactions of the assembly.* In addition to the foregoing formulae for the distribution laws we require formulae for the average (equilibrium) values of the forces exerted by the assembly or its sets of systems on the bodies which control the external fields. The most important example is the formula for the pressure of a gas.

We may assume that the states of any system in the assembly (or of the assembly itself) are determined by solving Schrödinger's equation with a potential energy W which is itself a function of certain parameters $x_1, x_2, \ldots$, defining the positions of all the external bodies. The energies ϵ_r of the possible states of any system are then functions of $x_1, x_2, \ldots$. The weights ϖ_r are however constants, as they are necessarily merely the number of independent solutions for a given value of the energy and given x's. They could at most only change discontinuously for certain values of the x's, and if W is a continuous function of the x's this is impossible.

It will be observed that we have here classified the universe into two parts, the assembly in which we are interested, and the rest of the universe. This is an essential part of any discussion of external reactions and of course what part of the universe we call the assembly can always be chosen at our own discretion. Now the assembly and any one of the bodies which produce the external field also form together a quantum system, which if unperturbed is a *conservative* one. When therefore a relevant parameter x_s defining the position of this body is allowed to change by an amount δx_s, the system in the original assembly remaining in its original state r, energy must be conserved. The energy of the state will have increased by $\delta\epsilon_r$ and we express this conservation by saying that the system has done *work* on the external body to an amount $-\delta\epsilon_r$. This is more conveniently expressed by saying that the system exerts a generalized force component ξ_s^r on the external body such that in any infinitesimal displacement in which the system remains in its original state r (reversible displacement) the work done is $\xi_s^r\,\delta x_s$, where

$$\xi_s^r = -\frac{\partial \epsilon_r}{\partial x_s}.$$

There are similar generalized forces for other states and other parameters. In any state of the assembly it follows that the total generalized force X_s due to one set of systems and tending to increase the parameter x_s is given by

$$X_s = \Sigma_r a_r\left(-\frac{\partial \epsilon_r}{\partial x_s}\right).$$

The average value of this force which the assembly will exert in its equilibrium state is therefore

$$\overline{X}_s = \Sigma_r \overline{a}_r\left(-\frac{\partial \epsilon_r}{\partial x_s}\right). \qquad \ldots\ldots(232)$$

A detailed verification of these general considerations can be given, for which reference should be made to works on quantum mechanics.*

* For *adiabatic* (i.e. slow reversible) variations in quantum mechanics see Born, *Zeit. f. Physik*, vol. 40, p. 167 (1926). See also Born and Fock, *Zeit. f. Physik*, vol. 51, p. 165 (1928).

The equation (232) can be simplified. Using (119) it becomes

$$\overline{X_s} = \lambda \frac{\partial}{\partial \lambda} \Sigma_r \left(-\frac{\partial \epsilon_r}{\partial x_s} \right) \varpi_r \log g(\lambda \vartheta^{\epsilon_r}),$$

$$= \Sigma_r \left(-\frac{\partial \epsilon_r}{\partial x_s} \right) \varpi_r \lambda \vartheta^{\epsilon_r} \frac{g'(\lambda \vartheta^{\epsilon_r})}{g(\lambda \vartheta^{\epsilon_r})},$$

$$= \frac{1}{\log 1/\vartheta} \frac{\partial}{\partial x_s} \Sigma_r \varpi_r \log g(\lambda \vartheta^{\epsilon_r}). \qquad \text{......(233)}$$

In this form we can immediately extend the formula to an assembly of any number of groups of systems. The general result is obviously

$$\overline{X_s} = \frac{1}{\log 1/\vartheta} \frac{\partial Z}{\partial x_s} = kT \frac{\partial Z}{\partial x_s}, \qquad \text{......(234)}$$

where

$$Z = \Sigma_\tau \Sigma_r \varpi_r{}^\tau \log g^\tau(\lambda \vartheta^{\epsilon_r{}^\tau}), \qquad \text{......(235)}$$

the affix τ specifying the various groups of systems.

In the special case of free atoms or molecules in an enclosure in which the sole external field is the local boundary field of the walls equation (233) reduces in the limit to

$$p = kT \frac{\partial}{\partial V} \Sigma_r \varpi_r \log g(\lambda e^{-\epsilon_r/kT}) = kT \frac{\partial Z}{\partial V}. \qquad \text{......(236)}$$

This is the standard equation for the partial pressure exerted by any constituent of a perfect gas. It is most easily derived from (233) by regarding any small area ω of the wall of the enclosure as a piston free to move normally, whose position is fixed by the parameter x_1. Then $\overline{X_1} = p\omega$ by the definition of p, and $\omega\, dx = dV$.

We can apply equation (236) to a gas (or gaseous constituent) of free molecules using (205) for the sum Z. We then see at once that $p = kTZ/V$. By (121) of which (228) is a special case the average energy of the molecules is $kT^2 \partial Z/\partial T$. This includes both internal and translatory energy. If we ask only for $\overline{E_{\text{kin}}}$, the average kinetic energy of translation, we must operate with $kT^2 \partial/\partial T$ only on the ϵ_r terms in Z in (205) and not on the ϵ_τ. This means operating only on the $T^{\frac{3}{2}}$ factor. Therefore

$$\overline{E_{\text{kin}}} = \tfrac{3}{2} kTZ, \quad p = \tfrac{2}{3} \overline{E_{\text{kin}}}/V. \qquad \text{......(237)}$$

This result holds in all statistics, including the classical, for an ideal gas or gaseous constituent.

In the classical limit (or with classical statistics), equation (236) reduces to $p = kT\lambda \partial H(T)/\partial V$, where $H(T)$ is the partition function for the free molecules in an enclosure V. At the same time Z reduces to $\lambda H(T)$, and by (120) $Z = M = \lambda H(T)$. Thus

$$p = MkT \frac{\partial}{\partial V} \log H(T). \qquad \text{......(238)}$$

Since for ordinary free particles in a volume V $H(T)$ depends on V only through having V as a factor, it follows from (236) or (238) that

$$p\left(=\frac{kTZ}{V}\right)=\frac{MkT}{V}, \qquad \text{......(239)}$$

the classical equation of state of an ideal or perfect gas. This result can also be obtained at once by combining (237) and (173).

Equation (239) without the introduction of the variable T would have been obtained in the form

$$p=M/(V\log 1/\vartheta). \qquad \text{......(240)}$$

This equation can be used if desired for a preliminary definition of the relationship between ϑ and the absolute temperature, if the latter is defined so that for a perfect constant volume gas thermometer $p \propto T$. But as we have said before the absolute temperature can only be logically introduced with the aid of the second law of thermodynamics.

CHAPTER III

ASSEMBLIES OF PERMANENT SYSTEMS (*cont.*). THE SPECIFIC HEATS OF SIMPLE GASES*

§3·1. *The properties of perfect gases. Specific heats.* Further development of the general theory without some detailed application to experimental data would be somewhat arid. We pause here, therefore, to compare theory and experiment for perfect gases. Since actual gases are not perfect the properties of perfect gases cannot strictly be said to be observed. They must be obtained by extrapolation to zero concentration from the actual observations at ordinary concentrations. This presents no serious difficulty and introduces little uncertainty into the results. In this chapter we shall suppose that the necessary corrections have been made. The methods of doing this will be reviewed in Chapter IX.

We have seen in Chapter II that classical statistics can be used from which we have deduced or can deduce: (1) the equation of state $pV = MkT = RT$; (2) Maxwell's velocity distribution law; (3) formulae for C_V, for any given molecular model. It is hardly necessary to discuss the field of validity of the equation of state of a perfect gas. It is sufficiently a commonplace that $pV = MkT$ is accurately the limit of the actual equation of state for all permanent gases or gas-mixtures at all temperatures except very near to the absolute zero, when Fermi-Dirac or Einstein-Bose statistics must be used, with results given in § 3·73. For most of the simpler gases the equation of state is already very near to its limiting form at normal pressures of the order of one atmosphere even if the temperature is low. Maxwell's law enables us to calculate the numbers of events, such as collisions of a definite type, which occur per second per unit volume of the gas or per unit area of the surface of a wall. Results of this type are of great importance in surface phenomena and chemical kinetics and are obtained and used in Chapters XVII and XVIII. Here we shall be content to compare the theoretical and experimental values of the specific heats of perfect gases.†

From the definitions‡ of C_V and C_p, namely

$$C_V = \left(\frac{\partial E}{\partial T}\right)_V, \quad C_p = \left[\frac{\partial(E+pV)}{\partial T}\right]_p, \qquad \text{......(241)}$$

* I am deeply indebted to G. B. B. M. Sutherland for help in revising this chapter.

† The more important general authorities for the older experimental data used in these comparisons are: Partington and Shilling, *The specific heats of gases*, Benn (1924); Eucken, *Zeit. f. Physik*, vol. 29, pp. 1, 36 (1924); Lewis, *A system of physical chemistry*, vol. 3, chap. IV (ed. 1919); Jeans, *loc. cit.* chap. VII. These authors, especially the first two, contain a great quantity of well-digested information. It has, however, recently appeared that the rate of adjustment of the vibrational energy of some molecules is slow, so that many of the older measurements do not refer to the true equilibrium state. We shall refer to the more modern measurements when we discuss particular gases.

‡ See, for example, Planck, *Thermodynamik*, ed. 6, §§ 81, 82.

it follows at once that for perfect gases

$$C_p - C_V = R = 1{\cdot}98\,\text{cal./gm. mol.} \qquad \text{......(242)}$$

This relation is well known to be obeyed accurately so that it is only necessary to discuss C_V or γ $(=C_p/C_V)$ whichever is the more convenient. By (173), or more precisely (207), the contribution of the translational kinetic energy to C_V is $\frac{3}{2}R$ in all cases. Any excess of C_V over this value must come from internal motions of the atom or molecule, that is, from rotations of the molecule, from vibrations of the atoms in the molecule or from electronic rearrangements. Any defect of C_V below $\frac{3}{2}R$ must be due to quantum statistics and only occurs at extremely low temperatures.

§ **3·2.** *Monatomic gases.* A free atom possesses classical kinetic energy of translation and the internal energy of its electronic system. The energy step associated with the change from the normal state to the nearest excited state is very large, for all atoms ordinarily capable of existing as atoms in the free state, and the internal energy can contribute nothing to C_V except at

TABLE 1.

Observed values of γ for monatomic gases, corrected for deviation from the perfect gas laws.

Gas	Temp. ° C.	γ	Authority
He	18 − 180	1·660 1·673	} Scheel and Heuse*
Ne	19	1·64	Ramsay†
A	15 − 180	1·65 1·69	} Scheel and Heuse*
Kr Xe	19 19	1·68 1·66	} Ramsay†
Hg	275–356	1·666	Kundt and Warburg†

* See Partington and Shilling, *loc. cit.* chap. xv.

† Landolt and Börnstein, *Tabellen* (1923), No. 264.

very high temperatures. The energy required for this step varies for ordinary monatomic gases from 4 to 20 volts‡ (Hg, He), while kT in volts is $8{\cdot}60 \times 10^{-5}T$. Thus ϵ/kT is of the order $10^5/2T$, and $e^{-\epsilon/kT}$ is negligibly small for all terms in the partition function for the internal atomic energy, except the first (normal) term for which by definition $\epsilon = 0$, unless T is at least 10,000° K.§ The theory thus predicts $C_V = \frac{3}{2}R$, $C_p = \frac{5}{2}R$, $\gamma = C_p/C_V = \frac{5}{3}$ for

‡ A "volt" or "electron-volt" is here used as a measure of energy, denoting the energy acquired by an electron in falling freely through a potential difference of 1 volt. The "volt" is a convenient unit of energy in most problems concerned with atomic or molecular structure.

§ As the partition function is an infinite series, the argument is here incomplete, without the construction and examination of the partition function in detail. This forms the theme of Chapter XIV, and the exact investigation confirms the result used here.

all monatomic gases at ordinary temperatures. The experimental values of γ are in satisfactory agreement.

Many free atoms, for example thallium, not commonly experimented with as vapours possess however a normal state which is the lowest state of a multiplet. For such atoms the smallest excitation energy may be comparatively small, and the internal energy makes an important contribution to the specific heat. Though such cases are not yet of practical importance we shall assemble the formulae in § 3·7, as they are of general utility.

§ 3·3. *Diatomic gases at moderate temperatures.* In addition to the types of motion and energy content which they share with free atoms, diatomic molecules possess further types of motion. The atomic nuclei can rotate about their centre of gravity to a first approximation like a rigid body, and can vibrate along the line joining them to a first approximation like a simple harmonic oscillator. If the molecule is nearly rigid, so that the frequency of these vibrations is high, the rotations and vibrations are nearly independent of each other. Moreover, at fairly low temperatures the nuclear vibrations will not contribute to C_V for the same reason that the electronic structure does not contribute, and the whole extra motion reduces to the rotations of a rigid body. The non-vibrating molecule must indeed stretch under the centrifugal forces, but for stiff molecules of high vibrational frequency this effect will be small for moderate rotations—that is, at low temperatures. Just as for atoms the lowest electronic state of the molecule may be multiple with a small energy of excitation. Important examples actually occur among simple permanent gases (e.g. NO) and are discussed in §§ 3·71, 3·72.

Partition functions suitable for rotations when lack of rigidity and electronic excitation can be ignored were specified in § 2·34 and their classical form was given in § 2·63. It must be shown next that the quantum forms satisfy the limiting principle. The forms of (91), (96), (97) and similar functions for high temperatures ($T \to \infty$, $\sigma \to 0$) can be established by a variety of methods. Perhaps the simplest is to compare the sum, for example (91),

$$\sum_{n=0}^{\infty} (2n+1)\, e^{-\sigma n(n+1)}$$

with the corresponding integral

$$\int_0^{\infty} (2x+1)\, e^{-\sigma x(x+1)}\, dx.$$

It is easy to show, by breaking up the sum and the integral into two parts at the maximum of the integrand which is then monotonic in each part, that they differ at most by a term of the order of the largest term in the series.

This term occurs for the value of n nearest the root of the equation $(2n+1)^2 = 2/\sigma$ and is of order $\sigma^{-\frac{1}{2}}$. Hence

$$\sum_0^\infty (2n+1)\, e^{-n(n+1)\sigma} = \int_0^\infty (2x+1)\, e^{-x(x+1)\sigma}\, dx + O\left(\frac{1}{\sqrt{\sigma}}\right),$$

$$= -\left[\frac{e^{-x(x+1)\sigma}}{\sigma}\right]_0^\infty + O\left(\frac{1}{\sqrt{\sigma}}\right) \sim \frac{1}{\sigma}. \qquad \text{......(243)}$$

Similarly $$\sum_0^\infty (2n+1)^2 e^{-n(n+1)\sigma} \sim \int_0^\infty (2x+1)^2 e^{-x(x+1)\sigma}\, dx \sim \frac{\pi^{\frac{1}{2}}}{\sigma^{\frac{3}{2}}}; \qquad \text{......(244)}$$

$$\sum_0^\infty (2n+1)\, e^{-n(n+1)\sigma} \sum_{-n}^{n} e^{-r^2\sigma'} \sim \int_0^\infty (2x+1)\, e^{-x(x+1)\sigma}\, dx \int_{-x}^{x} e^{-y^2\sigma'}\, dy$$

$$= \frac{2}{\sigma}\int_0^\infty e^{-x(x+1)\sigma - x^2\sigma'}\, dx \sim \frac{\pi^{\frac{1}{2}}}{\sigma(\sigma+\sigma')^{\frac{1}{2}}}. \qquad \text{......(245)}$$

§ **3·31.** *More exact evaluations.* A more exact treatment of these relationships is not without importance. We have seen in § 2·5 that the series $\sum_1^\infty e^{-n^2\sigma}$ can be expressed in terms of Θ-functions, and an exact discussion can be given by using the transformation theory of these functions. The series

$$\sum_0^\infty e^{-(n+\frac{1}{2})^2\sigma}, \quad \sum_0^\infty (-)^n\, (2n+1)\, e^{-(n+\frac{1}{2})^2\sigma} \quad \text{and} \quad \sum_0^\infty (2n+1)^2 e^{-(n+\frac{1}{2})^2\sigma}$$

can also be expressed as Θ-functions, but the most important of these series $\sum_0^\infty (2n+1)\, e^{-(n+\frac{1}{2})^2\sigma}$, though closely allied, is not a Θ-function and the transformation theory does not apply. A special investigation* shows that

$$\sum_0^\infty (2n+1)\, e^{-(n+\frac{1}{2})^2\sigma} = \frac{1}{\sigma} + \frac{1}{12} + \frac{7}{480}\sigma + O(\sigma^2), \qquad \text{......(246)}$$

so that, for the partition function $f(\sigma)$,

$$\sum_0^\infty (2n+1)\, e^{-n(n+1)\sigma} = e^{\frac{1}{4}\sigma}\left\{\frac{1}{\sigma} + \frac{1}{12} + \frac{7}{480}\sigma + O(\sigma^2)\right\}. \qquad \text{......(247)}$$

The limiting form of this partition function when $T \to \infty$, $\sigma \to 0$ is therefore

$$\frac{1}{\sigma} = \frac{8\pi^2 AkT}{h^2},$$

in agreement with (94) as required by the limiting principle. By repetition of these arguments it is easily shown that this asymptotic relation can be differentiated any number of times. Thus

$$\sigma^2 \frac{d^2}{d\sigma^2} \log f(\sigma) = \sigma^2 \frac{f''f - f'^2}{f^2} \sim 1,$$

$$C_{\text{rot}} \to R \quad (T \to \infty), \qquad \text{......(248)}$$

* Mulholland, *Proc. Camb. Phil. Soc.* vol. 24, p. 280 (1928). This paper gives a general asymptotic expansion of the function in (246) for small σ.

again in agreement with the limiting principle. Using Mulholland's asymptotic expansion† we find more exactly for small σ

$$C_{\text{rot}}/R = 1 + \tfrac{1}{45}\sigma^2 + O(\sigma^3). \qquad \text{......(249)}$$

Thus the classical value is approached much more rapidly than one would have anticipated. It is obvious that

$$C_{\text{rot}} \to 0 \quad (T \to 0), \qquad \text{......(250)}$$

or more precisely that $\quad C_{\text{rot}} = O(\sigma^2 e^{-2\sigma}) \quad (\sigma \to \infty).$

Similar results are obtained from the more general expression (96) when $C \ll A$. There is then a range of values of T for which σ is practically zero but $\sigma'(\sim h^2/8\pi^2 CkT)$ still very large, so that only those terms of (96) are relevant for which $|\tau|$ has its least value. This least value is zero for the simple model of § 2·34, but to allow for electronic orbital or spin angular momentum the same formula may sometimes be usable as an approximation with values of $|\tau|$ proceeding by integers from a non-zero minimum value τ^*. The corresponding changes of n have no effect on the asymptotic form of the partition function, which is then

$$\frac{8\pi^2 AkT}{h^2} \quad (\tau^* = 0),$$

$$2\frac{8\pi^2 AkT}{h^2} e^{-\tau^{*2}\sigma'} \quad (\tau^* > 0).$$

Equation (248) holds unaltered for this range of values of T. In differentiating with respect to σ one need merely remember that σ'/σ is constant.

For all diatomic gases not containing a hydrogen atom A is at least as great as 10^{-39} gm. cm.^{2} and $1/\sigma$ at least $\tfrac{1}{4}T$. All ordinary values of T are "large" for such gases so far as concerns σ, and "small" for σ', since C is at most $A/10{,}000$. We shall therefore always find $C_{\text{rot}} = R$. The extra factor 2 agrees with the limiting principle, for it allows of the two possible directions of axial spin.

§ **3·32**. *Asymptotic formulae for* (96) *and* (97). The partition function $f(\sigma) = \sum_0^\infty (2n+1)^2 e^{-n(n+1)\sigma}$ being reducible to a Θ-function is easily evaluated. We find at once

$$f(\sigma) = -2e^{\frac{1}{4}\sigma}\frac{d}{d\sigma}\sum_{-\infty}^{\infty} e^{-(n+\frac{1}{2})^2\sigma}. \qquad \text{......(251)}$$

The series in (251) is $\Theta_2(0,\tau)$ if we put $\tau = i\sigma/\pi$. By the transformation theory‡

$$\Theta_2(0, i\sigma/\pi) = \frac{\pi^{\frac{1}{2}}}{\sigma^{\frac{1}{2}}}\left[1 + 2\sum_{n=1}^{\infty}(-)^n e^{-\pi^2 n^2/\sigma}\right].$$

† This calculation was incorrectly given in the introductory note to Mulholland's paper.
‡ Tannery and Molk, *Elliptic Functions*, vol. 2, pp. 252, 264.

It follows at once that

$$f(\sigma)=\pi^{\frac{1}{2}}e^{\frac{1}{4}\sigma}\left[\frac{1}{\sigma^{\frac{1}{2}}}+O\left(\frac{1}{\sigma^{\frac{3}{2}}}e^{-\pi^2/\sigma}\right)\right]. \qquad ......(252)$$

The corresponding formula for the rotational specific heat is

$$C_{\text{rot}}/R=\tfrac{3}{2}+O\left(\frac{1}{\sigma^5}e^{-\pi^2/\sigma}\right), \qquad ......(253)$$

the classical value $\frac{3}{2}$ being almost exact for small σ.

The partition function

$$f(\sigma,\sigma')=\sum_{n=0}^{\infty}\sum_{\tau=-n}^{n}(2n+1)\,e^{-n(n+1)\sigma-\tau^2\sigma'}$$

can be evaluated asymptotically, when σ, σ' are small in a fixed ratio, by a double application of the Euler-Maclaurin summation formula,* provided that $\sigma'>0$, that is $C<A$. It has been shown that

$$f(\sigma,\sigma')=\frac{\pi^{\frac{1}{2}}}{(\sigma+\sigma')^{\frac{1}{2}}}\left[\frac{1}{\sigma}+\frac{1}{\sigma+\sigma'}(\tfrac{1}{4}\sigma+\tfrac{1}{3}\sigma')+\frac{1}{(\sigma+\sigma')^2}\left(\frac{\sigma^3}{32}+\frac{\sigma^2\sigma'}{12}+\frac{\sigma\sigma'^2}{15}\right)\right]$$
$$+O(\sigma,\sigma')^{\frac{5}{2}}. \qquad ......(254)$$

Since σ/σ' is independent of T we may write $\sigma'=\beta\sigma$ so that

$$C_{\text{rot}}/R=\sigma^2\frac{\partial^2}{\partial\sigma^2}\log f(\sigma,\sigma\beta).$$

This formula gives after reduction

$$C_{\text{rot}}/R=\tfrac{3}{2}+\frac{1}{45}\frac{\sigma^2\sigma'^2}{(\sigma+\sigma')^2}+O(\sigma,\sigma')^3. \qquad ......(255)$$

Formulae (248) and (255) show how the corresponding quantum formulae approach their classical limits. It is interesting to observe that in both cases the approach to the classical limit is from above as $T\to\infty$. In all cases

$$C_{\text{rot}}\to 0 \quad (T\to 0).\dagger$$

§ 3·4. *Rotational specific heat of* H_2 *at low temperatures.* The predicted variation in the rotational specific heat—an increase from zero at low temperatures to classical values at high—has been observed for hydrogen (H_2 and recently D_2) alone among diatomic gases. The best observations for H_2 are shown plotted in Fig. 2. The specific heat C_V has approximately the normal value for a diatomic gas ($\frac{5}{2}R$) at ordinary temperatures of 300° K. and above, but falls steadily to $\frac{3}{2}R$ ($C_{\text{rot}}=0$) as T diminishes. For temperatures below 40° K. C_V and $\frac{3}{2}R$ are indistinguishable. This general behaviour

* Miss Viney, *Proc. Camb. Phil. Soc.* vol. 29, p. 142 (1933); mistake corrected, *ibid.* p. 407.

† Elaborate investigations of asymptotic forms analogous to these have been made by Kassel, *J. Chem. Physics*, vol. 1, p. 576 (1933).

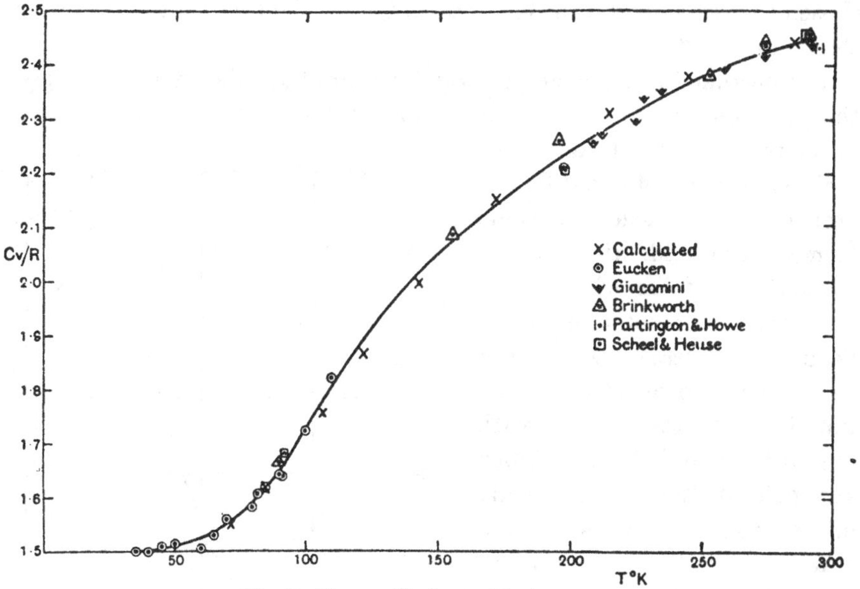

Fig. 2. The specific heat of hydrogen.

is completely accounted for if we may identify the variable part of C_V with C_{rot} and apply the foregoing theory. The values of the temperature for which the change occurs fit in with independent evidence as to the moment of inertia of the hydrogen molecule. This explanation, originated by Ehrenfest, has long been universally accepted.

It was only in 1927, however, that complete agreement *in detail* between theory and observation was obtained. Dennison* then showed that the data for the normal state of hydrogen, derived from the analysis of its band spectrum, yield precisely the observed values of C_V when properly applied.

As we have said, the straightforward application of quantum mechanics to the rotational states of a rigid rotator without axial spin yields the partition function (91), approximated to in (247),

$$f(\sigma) = \sum_{n=0}^{\infty} (2n+1)\, e^{-n(n+1)\sigma} \qquad \left(\sigma = \frac{h^2}{8\pi^2 AkT}\right). \qquad \text{......(256)}$$

This set of weights and energies is confirmed by the analysis of simple infra-red band spectra such as those of HCl, HBr, CO and CN,† and no theoretical modification is possible. No sufficient agreement, however, can be obtained with the observed curve of Fig. 2 for any value of A. If A is

* Dennison, *Proc. Roy. Soc.* A, vol. 115, p. 483 (1927). A full discussion of theories previous to the work of Dennison is given by Van Vleck, *Phys. Rev.* vol. 28, p. 980 (1926).

† R. H. Fowler, *Phil. Mag.* vol. 49, p. 1272 (1925); Kemble, *Zeit. f. Physik*, vol. 35, p. 286 (1925). The latter has numerous references to experimental data.

chosen to give a good fit at low temperatures, the best that can be done is shown in Fig. 3.*

The mistake in this attempted application of the theory lay in applying the partition function (256), which appears to be correct for a heteropolar, or more strictly heteronuclear, molecule formed of two different atoms, to a homonuclear molecule formed of two identical atoms, without examining whether the homopolar character is significant. It is necessary to consider in detail the forms of the wave-functions for the rotator with regard to their behaviour when any pair of identical parts are interchanged,† just as we examined wave-functions for the assembly in § 2·23 and the following sections. This has been done in detail with special reference to hydrogen by Hund.‡ In such a case, in which there are two like nuclei to consider, the wave-functions for the rotation break up into two distinct groups, those which are symmetrical and those which are antisymmetrical in the nuclei. Wave-functions for a rigid rotator without axial spin are symmetrical for n even in (256) and antisymmetrical for n odd. Further, if the two nuclei are absolutely indistinguishable there is no interconnection whatever possible between the symmetrical and antisymmetrical states and only one or other, but not both, can be expected to present itself in this universe. This is in beautiful agreement with observations on the band spectrum of helium (emitter He_2) which has for some time been recognized to possess only half the expected number of lines in each band, alternate lines being completely absent. The antisymmetrical rotational forms do not occur. The bands of H_2 on the other hand show alternating intensities and no missing lines. Both symmetrical and antisymmetrical rotational forms are present and the alternations are accounted for, as has been shown by Hori, § if the molecules with antisymmetrical wave-functions are three times as numerous as the symmetrical ones. This, however, is exactly what we should expect if

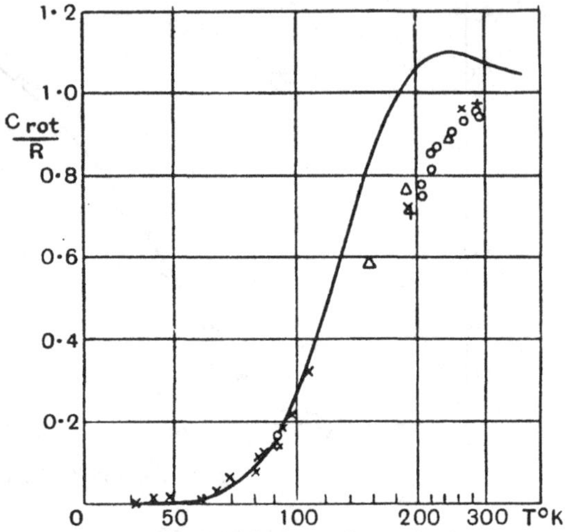

Fig. 3. The specific heat of a rigid rotator according to (256).

* Dieke, *Physica*, vol. 5, p. 412 (1925).
† Heisenberg, *Zeit. f. Physik*, vol. 38, p. 411 (1926); *ibid.* vol. 41, p. 239 (1927).
‡ Hund, *Zeit. f. Physik*, vol. 42, p. 93 (1927).
§ Hori, *Zeit. f. Physik*, vol. 44, p. 834 (1927).

the nuclei have spins like the electrons and if only wave-functions which are antisymmetrical for the nuclei can occur when account is taken both of their orientations and their rotations. Hori has also shown from the spectrum that the normal state of the molecule has no axial spin, so that the spin-free rotator is a legitimate model, and that its moment of inertia must be $4 \cdot 67 \times 10^{-41}$ gm. cm.^{2}.

These results require that we shall take for the partition function for the normal H_2-molecule not (256) but

$$f(\sigma) = \sum_{n=0,2,4,\ldots} (2n+1)\, e^{-n(n+1)\sigma} + 3 \sum_{n=1,3,5,\ldots} (2n+1)\, e^{-n(n+1)\sigma}. \qquad \ldots\ldots(257)$$

Partition functions such as this were therefore examined by Hund,* but were found to give curves bearing no resemblance whatever to the observations of Fig. 2. To apply the theory properly one further point must be made. While Hori's work leaves no doubt whatever that the possible states are correctly enumerated in (257), it is also assumed in (257) that interchanges between all the states of (257) take place freely, so that the distribution laws of the observed state are correctly given by (257) even at very low temperatures. But it is necessary to ask whether these interchanges can occur freely at ordinary and low temperatures; they must all be able to occur in a vacuum-tube discharge, but even then interchanges between the symmetrical and antisymmetrical rotational states are rare and no corresponding intercombination lines are observed in the spectrum of the discharge. They would *never* occur if the nuclei had no spin, and occur with a frequency proportional to the perturbation of the energy values by the nuclear spins—much less frequently than interchanges between the states of par- and ortho-helium. It is therefore reasonable to assume that in ordinary hydrogen gas interchanges between the symmetrical and antisymmetrical states only occur in times very long compared with the time of an experiment. The specific heat measurements are therefore not made on a gas in the true equilibrium state governed by (257) but in a metastable equilibrium in which the gas behaves like a mixture of different gases, one of them the symmetrical and the other the antisymmetrical molecules. By analogy with a terminology adopted for the states of the He atom the molecules with the antisymmetrical (rotational) states are called orthohydrogen and with the symmetrical parahydrogen.

We therefore proceed as follows. We introduce the functions

$$f_s(\sigma) = \sum_{n=0,2,4,\ldots} (2n+1)\, e^{-n(n+1)\sigma}, \qquad \ldots\ldots(258)$$

$$f_a(\sigma) = \sum_{n=1,3,5,\ldots} (2n+1)\, e^{-n(n+1)\sigma}. \qquad \ldots\ldots(259)$$

* Hund, *loc. cit.*

Then the rotational specific heat of the symmetrical gas (by itself) is given by

$$\frac{(C_{\text{rot}})_s}{R}=\sigma^2\frac{d^2}{d\sigma^2}\log f_s(\sigma),$$

and of the antisymmetrical gas (by itself)

$$\frac{(C_{\text{rot}})_a}{R}=\sigma^2\frac{d^2}{d\sigma^2}\log f_a(\sigma).$$

That of the actual 3 : 1 mixture will be

$$\frac{C_{\text{rot}}}{R}=\frac{\sigma^2}{4}\left\{3\frac{d^2}{d\sigma^2}\log f_a(\sigma)+\frac{d^2}{d\sigma^2}\log f_s(\sigma)\right\},$$

$$=\frac{\sigma^2}{4}\frac{d^2}{d\sigma^2}\log\{f_a{}^3(\sigma)\,f_s(\sigma)\}. \qquad \text{......(260)}$$

We have to replace the $(3f_a+f_s)$ of the true equilibrium by $(f_a{}^3f_s)^{\frac{1}{4}}$. The result of using (260) with $A=4{\cdot}64\times10^{-41}$ is the set of points shown by crosses in Fig. 2. The agreement with observation is all that can be desired.

In making these calculations an asymptotic formula similar to (247) is convenient. This is easily established, for the difference of f_a and f_s is a Θ-function which vanishes exponentially when $\sigma\to0$. Therefore with sufficient accuracy as $\sigma\to0$

$$f_s(\sigma)=f_a(\sigma)=\tfrac{1}{2}e^{\frac{1}{4}\sigma}\left\{\frac{1}{\sigma}+\frac{1}{12}+\frac{7}{480}\sigma+O(\sigma^2)\right\}. \qquad \text{......(261)}$$

It follows that the rotational specific heat approaches its classical value as $T\to\infty$ in exactly the same manner as for a simple rigid rotator, in particular from above.

Researches by Bonhöffer and Harteck* and by Eucken and Hillert† have shown that the two modifications of H_2, first postulated to explain the specific heat curve, do in fact exist. They have demonstrated that the slow change over from ortho- to para-hydrogen does take place even at low temperatures if sufficient time is allowed, with consequent changes in the thermal properties of the gas. In order to accelerate the change it is necessary to keep the gas at high pressure or better still to keep it over charcoal. At ordinary pressures and in glass-walled vessels the change from the metastable to the true equilibrium state requires times of the order of several days at least, just as was postulated for the explanation of the specific heat curve.‡

The metastable equilibrium of H_2 has further important consequences to which we must return in Chapter VII in discussing Nernst's Heat Theorem.

* Bonhöffer and Harteck, *Zeit. f. physikal. Chem.* vol. 4, p. 113 (1929).

† Eucken and Hiller, *Zeit. f. physikal. Chem.* vol. 4, p. 131 (1929).

‡ For a recent account of the many interesting features of the ortho-para hydrogen equilibrium see Farkas, *Light and Heavy Hydrogen*, Cambridge (1935).

Thanks to the discovery of heavy hydrogen two other diatomic molecules, HD and D_2, are now known whose rotational specific heats can be studied at low temperatures.

For HD the simple partition function (256) should yield the correct specific heat since this heteronuclear molecule has no rotational symmetry requirements. Fig. 3·1 shows this curve and the observations of Clusius and Bartholomé.* The agreement is excellent; in particular the existence of the maximum is clearly shown.

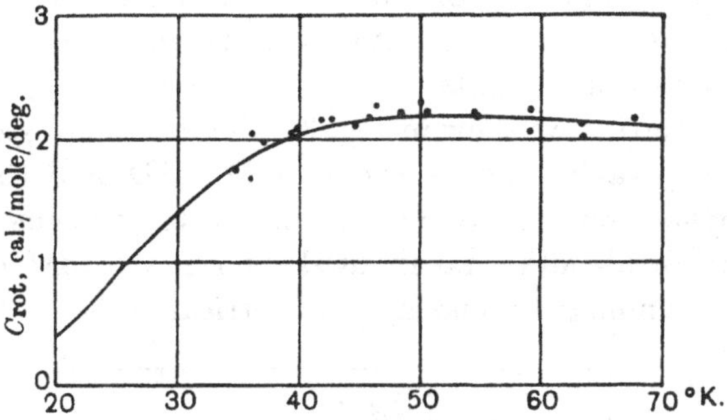

Fig. 3·1. The rotational specific heat of HD.

For D_2 symmetry requirements return. It is found that the nuclei have

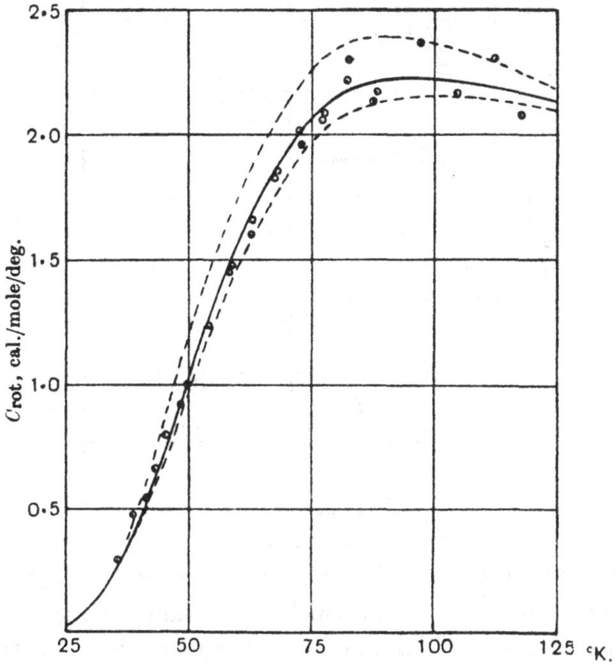

Fig. 3·2. The rotational specific heat of D_2.

* Clusius and Bartholomé, *Zeit. f. Elektrochem.* vol. 40, p. 524 (1934).

a spin unity and that the complete wave-functions must be symmetrical in the nuclei. The complete partition function is therefore now*

$$f(T)=6 \sum_{n=0,2,4,\dots} (2n+1)e^{-n(n+1)\sigma}+3 \sum_{n=1,3,5,\dots} (2n+1)e^{-n(n+1)\sigma}.$$

The moments of inertia of H_2, HD, and D_2 absorbed in σ are of course different, being in the ratios of $(m+m')/mm'$ or $\frac{1}{2}:\frac{2}{3}:1$. The observed value of the rotational specific heat will normally correspond to a constant mixture of the two sets of D_2 molecules in the ratio 2 : 1, instead of the 1 : 3 ratio for H_2. The theoretical value and the observations of Clusius and Bartholomé are shown in Fig. 3·2. The agreement is again excellent. The broken curves correspond to other spin weights.

As a final illustration we show in Fig. 3·3 the rotational specific heats of pure ortho- and para-H_2, ortho- and para-D_2, HD, and the metastable mixtures of ortho- and para-H_2 and D_2 drawn to the same temperature scale. The names ortho- and para- are used in both cases for the states which are respectively symmetrical and antisymmetrical in the nuclear *spins*.

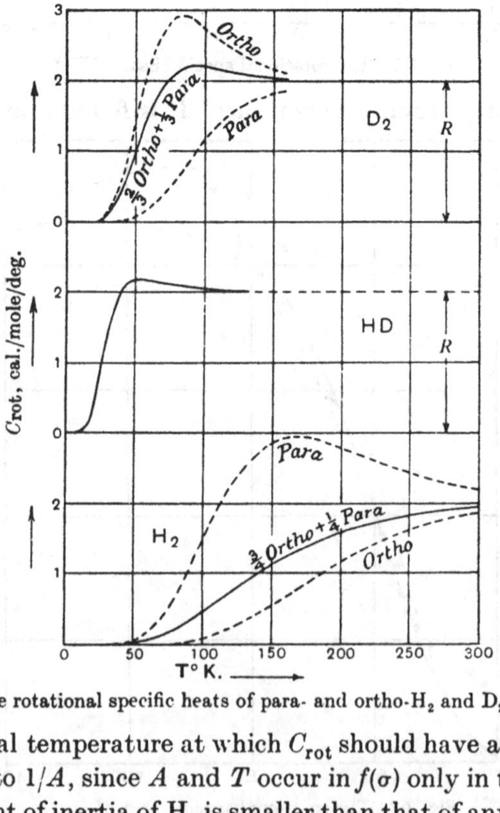

Fig. 3·3. The rotational specific heats of para- and ortho-H_2 and D_2 and of HD.

The theoretical temperature at which C_{rot} should have an assigned value is proportional to $1/A$, since A and T occur in $f(\sigma)$ only in the combination AT. The moment of inertia of H_2 is smaller than that of any other molecule

* For further explanation see § 5·31.

and C_{rot}/R is effectively unity for H_2 above 300° K. and effectively zero below 40° K. These temperatures can be divided by three for the heteronuclear HD though its moment of inertia is only increased by the factor $\frac{4}{3}$. We can therefore see at once that for example in O_2 C_{rot}/R must be unity as soon as the temperature reaches 300/16 or 20° K., the nuclei in O_2 being at least as far apart as in H_2. Observations of available C_{rot} are therefore impracticable for all diatomic molecules which contain no H-atom. Even the halogen hydrides have moments of inertia too large, that of HCl, for example, being five times* that of H_2, so that C_{rot} is normal above 60° K. The absence of other examples of this variation among diatomic gases is therefore in accordance with the theory.

This analysis has required a broadening of the classification of the states of a rotating homonuclear molecule with which we shall be further concerned later on. We have hitherto treated our atoms, still more our atomic nuclei, as structureless points, a treatment which has just proved inadequate. We may not even treat all nuclei as structureless mass-points. The proton—the H-nucleus—has two possible orientations and so must be assigned a weight twice as great as that which we have hitherto used for structureless points. The hydrogen molecule therefore should have a weight four times that of a structureless rigid rotator, due to the orientations of the nuclei. [In the normal state no further factor is introduced by the orbits or spins of the electrons.] This extra factor 4 however is reduced alternately to 1 or 3 by the requirements of symmetry in the nuclei, and therefore effectively to 2 at high temperatures. This reduction from 4 to 2 is the quantum analogue of the introduction of the symmetry number σ of classical statistics.†

§ **3·5.** *Diatomic gases. Vibrational energy.* The next approximation to a real molecule is to abandon the assumption of rigidity and allow for the vibration of the atomic nuclei along the line joining them. Besides the translations and rotations already dealt with, no other motion can contribute effectively to the partition function at temperatures below 10,000° K.

Let $b(T)$ be the partition function for the vibrational and rotational energy, referred to its own state of least energy as zero of energy. If the binding forces are very strong so that the molecule is nearly rigid and the frequency of vibration high, we may suppose to a first approximation that the vibrations and rotations do not affect one another; to this approximation $b(T)$ will factorize into

$$b(T) = r(T)\, v(T),$$

the partition functions for the rotations and vibrations respectively. To the same rough approximation we may suppose that the vibrations are like

* See Table 13, p. 223. Also Imes, *Astrophys. J.* vol. 50, p. 251 (1919); Colby, *Phys. Rev.* vol. 34, p. 53 (1929).

† Ehrenfest and Trkal, *Proc. Sect. Sci. Amsterdam*, vol. 23, p. 162 (1920). See also Chapter v.

those of a simple harmonic oscillator of frequency ν, but that as an energy χ will dissociate the molecule not more than p states are possible in which the molecule remains a molecule. Then

$$v(T)=\frac{1-e^{-ph\nu/kT}}{1-e^{-h\nu/kT}},$$

where $ph\nu$ is of the order of χ. [Actually this form is some way from the truth. The energies of the vibrational states tend to χ as a limit.] If χ and $ph\nu$ are fairly large compared with $h\nu$, there will be a considerable range of values of T for which $e^{-ph\nu/kT}$ is negligible compared with 1 even if $e^{-h\nu/kT}$ is not. For such temperatures

$$v(T)=(1-e^{-h\nu/kT})^{-1} \qquad \text{......(262)}$$

approximately, and in this region (262) will be an equally good approximation to more exact forms of the vibrational partition function. The contribution to the specific heat is then

$$C_{\text{vib}}/R=\left(\frac{h\nu}{kT}\right)^2\frac{e^{h\nu/kT}}{(e^{h\nu/kT}-1)^2}. \qquad \text{......(263)}$$

We shall write this $\quad C_{\text{vib}}/R=P(\Theta/T) \quad (\Theta=h\nu/k).$ ......(264)

When $h\nu/kT$ is large the contribution is zero. If ν is so large that $h\nu/kT$ is large at room temperatures, then for such diatomic gases we shall have

$$C_V=\tfrac{5}{2}, \quad \gamma=\tfrac{7}{5}.$$

Otherwise C_V exceeds $\frac{5}{2}$ by the amount given by (263). These predictions are in good general agreement with the facts shown in Table 2.

TABLE 2.

Observed values of γ for diatomic gases.

Gas	Temp. °C.	γ	Authority	Value of Θ $=h\nu/k$ from spectrum*
H_2	16	1·407	Scheel and Heuse†	6098
N_2	20	1·398		3358
	−181	1·419		·
O_2	20	1·398		2228
	−76	1·411		·
	−181	1·404		·
CO	18	1·396		3086
	−180	1·417		·
NO	15	1·38		2710
	−45	1·39		·
	−80	1·38		·
HCl	15	1·40	Partington‡	4208
HBr	—	—	—	3728
HI	—	—	—	3200

* Authorities for $h\nu/k$ are Imes, *loc. cit.* for the halides, Jevons, *Report on Spectra of diatomic molecules*, App. II (1932), down to HBr inclusive, and for HI, Salant and Sandow, *Phys. Rev.* vol. 37, p. 373 (1931). This value is provisional only, being derived from the Raman spectrum of the liquid.

† See Partington and Shilling, *loc. cit.* ‡ See *ibid.* Table B.

A proper study of the vibrational specific heat of any of these gases requires observations at rather high temperatures. A more accurate formula may then be necessary than that provided by the separate contributions of $r(T)$ and $v(T)$.

§ **3·51.** *Specific heat of* H_2 *at high temperatures.* A fairly satisfactory theoretical account of the specific heat of H_2 at high temperatures has been given by Kemble and Van Vleck, who discussed an elastic rotator with a particular law of force according to the classical quantum theory. They used values of A and ν_0, however, adjusted to fit the specific heat curve. From the point of view of *statistical* theory it is not necessary to go back to a mechanical model. It is sufficient to take the states of vibration and rotation of the normal H_2-molecule as enumerated directly from the band spectrum by Hori, to construct a semi-empirical partition function with their help and to use this function to evaluate C_{rot} and C_{vib}. This has been carried out successfully by McCrea.* At these higher temperatures the differences between symmetrical and antisymmetrical rotational states are unimportant, and we may use without serious loss of accuracy the one partition function

$$f(T)=\sum_{n=0}^{\infty}\sum_{j=0}^{\infty}(2j+1)\,e^{-E(n,j)/kT}, \qquad \text{......(265)}$$

where n is the vibrational and j the rotational quantum number and

$$\frac{E(n,j)}{hc}=A(n+\tfrac{1}{2})+(j+\tfrac{1}{2})^2B(n+\tfrac{1}{2})+(j+\tfrac{1}{2})^4\beta(n+\tfrac{1}{2}). \qquad \text{...(266)}$$

The functions of n, A, B, β are tabulated by Hori. It is easily verified that for our purpose β may always be neglected, and that at the temperatures concerned

$$\sum_{j=0}^{\infty}(2j+1)\,e^{-hcB(n+\frac{1}{2})(j+\frac{1}{2})^2/kT}=\frac{kT}{hcB(n+\frac{1}{2})} \qquad \text{......(267)}$$

with sufficient accuracy. It remains, therefore, only to compute the series

$$\sum_{n=0}^{\infty}\frac{1}{B(n+\frac{1}{2})}\,e^{-hcA(n+\frac{1}{2})/kT} \qquad \text{......(268)}$$

and its first two differential coefficients. For the values of T concerned not more than four or five terms are required, so that observed values are available and no extrapolation is needed. The results are given in the following table. Partington and Shilling state that the best representation of the observed total specific heat between 273° K. and 2273° K. is given by

$$C_V=4\cdot659+0\cdot00070T. \qquad \text{......(269)}$$

The agreement is satisfactory especially at the higher temperatures where it would fail but for these theoretical refinements.

* McCrea, *Proc. Camb. Phil. Soc.* vol. 24, p. 80 (1928), and since then by numerous other investigators in this field.

TABLE 3.

Specific heat C_V of H_2 at high temperatures.

Temp. °K.	C_V obs. (269)	C_V calc.
600	5·08	4·98
800	5·22	5·04
1000	5·36	5·16
1200	5·50	5·34
1600	5·78	5·72
1800	5·92	5·89
2000	6·06	6·05
2500	[6·41]†	6·37

† Extrapolated.

If it is desired to refer the energy values back to a model, it is necessary to assume a definite law of force between the nuclei. Such calculations have been made, for example, by Fues.* His result is quoted here for reference if required. For the law of force

$$E_{\text{pot}} = -E' + (2\pi\nu_0)^2 A\left\{\frac{1}{\rho} - \frac{1}{2\rho^2} + c_3(\rho-1)^3 + c_4(\rho-1)^4 + \dots\right\}, \qquad \dots\dots(270)$$

where $\rho = r/r_0$, r_0 is the equilibrium distance apart, and ν_0 the classical fundamental frequency of small vibrations, the energy values (constant terms omitted) are

$$E(n,j) = h\nu_0(n+\tfrac{1}{2})[1-\tfrac{3}{2}\kappa^2(1+2c_3)j(j+1)] + \frac{h^2}{8\pi^2 A}j(j+1)[1-\kappa^2 j(j+1)] - \frac{h^2}{8\pi^2 A}(n+\tfrac{1}{2})^2[3+15c_3+\tfrac{15}{2}c_3^2+3c_4], \qquad \dots\dots(271)$$

where

$$\kappa = h/4\pi^2\nu_0 A. \qquad \dots\dots(272)$$

§ **3·52.** *Other diatomic gases.* The same considerations can naturally be applied to explain the specific heats of other diatomic gases, but it is only recently that any agreement has been reached between theory and experiment for temperatures above room temperature. At these temperatures rotations are completely classical and it is then unnecessary as we have seen to take account of any requirements of nuclear symmetry. A sufficiently accurate formula for C_V, at least for a first survey, is provided by

$$C_V/R = \tfrac{5}{2} + P(\Theta/T), \qquad \dots\dots(273)$$

where P and Θ are defined in (263) and (264). More accurate comparisons can be made if desired by the methods of § 3·51.

Measurements of the specific heats of O_2, N_2 (air), CO and NO such as those recorded by Partington and Shilling for temperatures greater than 300° K. do not agree at all with (273) for the values of Θ given in Table 2.

* Fues, *Ann. d. Physik*, vol. 80, p. 367 (1926).

The reason for this is now understood. Except in H_2 the rate of conversion of translational or rotational energy into vibrational energy by molecular collisions is unexpectedly slow,* so that unless the time scale of the experiment is long compared with the time of adjustment of translational and vibrational energy, the true specific heat will not be measured, but rather a specific heat of a quasi-metastable state of the gas in which it behaves as if it had no vibrational degrees of freedom. Most of the earlier measurements were made by a velocity of sound method which measures C_p/C_V and whose time-scale is the period of the oscillation of the gas in the sound wave of the order of 10^{-3} sec. The measurements here successfully compared with the theory were obtained by a flow method of which the time scale was of the order of 1 second, a slow time scale which proves essential for the establishment of the true equilibrium state.†

Before passing to the discussion of the data, it is well to emphasize that we have here a second example of the care that is necessary in defining accessible states of an assembly. In the case of hydrogen at ordinary pressures and low temperatures, not in contact with any catalyzing surface, we can only specify the effectively accessible states of the assembly correctly by assuming that the hydrogen is a mixture of distinct gases orthohydrogen and parahydrogen. This specification is correct unless we experiment on a time scale of the order of weeks or months. In the case of ordinary diatomic gases analysed by sound waves of periods about 10^{-3} second or less at temperatures of the order of 300–500° K. there is effectively no interconnection between the different vibrational states; the accessible states of the assembly are those of a mixture of distinct gases each of given vibrational energy, and the specific heat C_V is given by $C_V/R=\frac{5}{2}$. If, however, we increase the experiment time to 1 second, the accessible states are effectively those of a single gas. If the accessible states are correctly specified, we can correctly treat either extreme case—the true equilibrium state or the perfect metastable equilibrium state—by the methods of statistical mechanics and the calculations so made may be compared with the results of experiments made on a suitable time scale. But the intermediate cases where there is a partial adjustment between different types of states can never be so treated. Their treatment requires always a knowledge of rates of interaction which is foreign to pure equilibrium theory, which from this point of view deals always with assemblies whose rates of interaction between states may all be classified as either infinitely fast or infinitely slow.

We pass on now to compare theory and experiment. Fig. 4 shows the

* This has been established theoretically by the work of Herzfeld and Rice, Zener and others; e.g. Herzfeld and Rice, *Phys. Rev.* vol. 31, p. 691 (1928); Zener, *Phys. Rev.* vol. 37, p. 556 (1931).

† The method is due to Blackett, Henry and Rideal, *Proc. Roy. Soc.* A, vol. 126, p. 319 (1930); first results described by P. H. S. Henry, *Proc. Roy. Soc.* A, vol. 133, p. 492 (1931).

observed values* for O_2, and air, which may be regarded as N_2 plus impurities, and the theoretical curves calculated by using (273) with the values of Θ given in Table 2. For O_2 a more exact theoretical curve is also shown (broken), calculated directly from the spectroscopic values of the

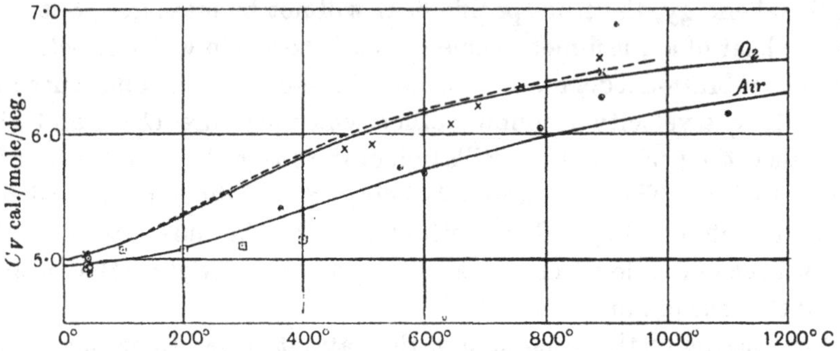

Fig. 4. The specific heat C_V of O_2 and air; observed values for O_2 × Henry, ⊡ Partington and Shilling; for air ⊙ Henry.

rotational and vibrational energies of the normal molecule. The agreement is for both molecules excellent. The older non-equilibrium values of C_V determined by a velocity of sound method are also shown for O_2 for comparison.

§ **3·6.** *Polyatomic molecules.* The only polyatomic molecules which could possibly show less than the classical rotational energy in accessible temperature ranges are those containing only one atom other than hydrogen and of these only CH_4, NH_3 and OH_2 need be considered. Low temperature observations are only practicable for CH_4 (methane) and values of C_V/R less than 3 have actually been recorded.† Owing to the identity of the four H-nuclei the rotational states of this isotropic rotator will break up into a set of non-combining groups, three in number and an elaborate analysis is necessary.‡ At high temperatures the non-combining sets of states are occupied by $\frac{5}{16}$, $\frac{9}{16}$ and $\frac{2}{16}$ of the molecules respectively. The complete partition function is of the form

$$f(\sigma) = 5f_1(\sigma) + 9f_2(\sigma) + 2f_3(\sigma),$$

where each $f(\sigma)$ contains a selection of the complete set of rotational states and as $\sigma \to 0$

$$f_1(\sigma) \sim f_2(\sigma) \sim f_3(\sigma) \sim \frac{1}{12}\sum_0^\infty (2n+1)^2 e^{-n(n+1)\sigma} \sim \frac{\pi^{\frac{1}{2}}}{12\sigma^{\frac{3}{2}}}. \quad \text{......(274)}$$

* P. H. S. Henry (*private communication*); modern high temperature work on CO has also been carried out by Sheratt and Griffiths, *Proc. Roy. Soc.* A, vol. 147, p. 292 (1934).

† Partington and Shilling, *loc. cit.*

‡ Hund, *Zeit. f. Physik*, vol. 42, p. 93; vol. 43, p. 778 (1927). The particular problem is correctly solved by Villars and Schultze, *Phys. Rev.* vol. 38, p. 998 (1931). The most profound account of the necessary enumeration is given by E. B. Wilson, *J. Chem. Physics*, vol. 3, p. 276 (1935).

At all temperatures in metastable equilibrium

$$C_{\text{rot}}/R=\sigma^2\frac{d^2}{d\sigma^2}[\tfrac{5}{16}\log f_1(\sigma)+\tfrac{9}{16}\log f_2(\sigma)+\tfrac{2}{16}\log f_3(\sigma)].$$

Calculations according to these formulae have been given by Villars and Schultze, but the observations are insufficient to provide an adequate test of the theory.

Turning now to higher temperatures, polyatomic molecules must be classified into two types, *linear* and *non-linear*, according as the equilibrium positions of all the nuclei do or do not lie on a straight line. For neither type is it necessary to take account of the refinements required by nuclear symmetry since the rotations are effectively classical, but for the linear type there are only two (classical) rotations while for the non-linear type there are three. It is of course useless at present to attempt the analysis for any but the simplest molecules and as among these there are good observations for CO_2, N_2O and C_2H_2 (linear type); OH_2, NH_3, CH_4 and C_2H_4 (non-linear) we confine attention to these. Owing to the considerable number of vibrational freedoms a convincing analysis is only possible when spectroscopic vibrational frequencies can be used with confidence. Neglecting for a first approximation the interaction of the rotations and vibrations we shall have

$$C_V/R=\tfrac{5}{2}+C_{\text{vib}}/R \quad \text{(Linear Type)}, \qquad \text{......(275)}$$

$$C_V/R=3+C_{\text{vib}}/R \quad \text{(Non-Linear Type)}. \qquad \text{......(276)}$$

§ **3·61.** *Specific heat of ammonia gas* (NH_3). Analysis of the infra-red and Raman spectra of gaseous ammonia* has shown that the molecule has the form of a regular pyramid with the nitrogen nucleus at its apex. There are four distinct fundamental frequencies of which two are double. It has been possible to identify with certainty three of these fundamentals with characteristic bands in the spectra, of wave numbers 950 and 3336 (single vibrations) and 1630 (double vibration). It seems highly probable that the other double vibration has a wave number 3300 approximately, but this is not yet completely established. If we make the rough approximation of using Planck terms (264) for each fundamental, then sufficiently nearly

$$\frac{C_{\text{vib}}}{R}=P\left(\frac{1361}{T}\right)+3P\left(\frac{4779}{T}\right)+2P\left(\frac{2335}{T}\right). \qquad \text{......(277)}$$

The observed values are shown plotted against the theoretical curve in Fig. 5. The agreement is rather poor. It suggests either that the fourth

* Sponer, *Molekülspektren* (1935).

frequency is higher than 3300 or that the observations have not recorded the full equilibrium value of the specific heat.

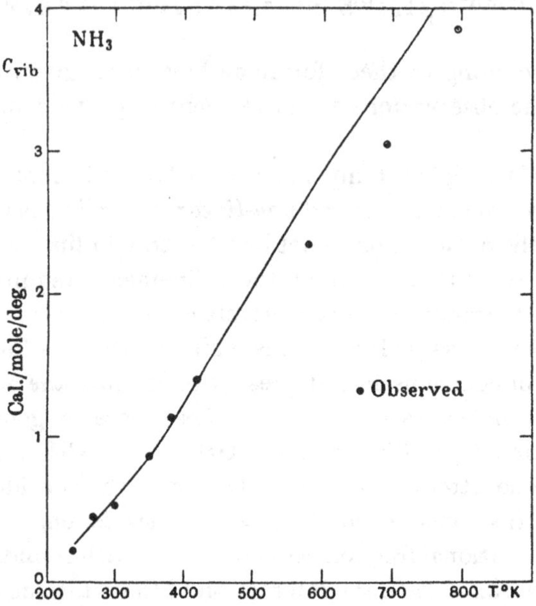

Fig. 5. The vibrational specific heat of NH_3.

Observations below 500° K. by Osborne, Stimson, Sligh and Cragoe, *Physikal. Ber.* vol. 5, p. 271 (1924); above 500° K. by Haber, *Zeit. f. Elektrochem.* vol. 20, p. 597 (1914).

§ **3·62.** *Specific heat of methane* (CH_4). We have already tacitly assumed that methane forms an isotropic rotator in which the H-nuclei lie in equilibrium at the corners of a regular tetrahedron with the carbon at the centre. The normal modes have been investigated by Dennison* and shown to consist of one single, one double and two triple vibrations, nine in all of four independent frequencies. The band structure, so far as it is known (which is not in great detail), agrees well with the assignment to these modes in order of the following wave numbers: 2915, 1520, 3014, 1304. The resulting specific heat is, using observed frequencies,

$$\frac{C_{vib}}{R} = P\left(\frac{4177}{T}\right) + 2P\left(\frac{2180}{T}\right) + 3P\left(\frac{4310}{T}\right) + 3P\left(\frac{1870}{T}\right). \quad \text{......(278)}$$

Observed and calculated values are given in Table 4. On the whole the two values are in good agreement and the excess of the observed values is certainly largely due to a stretching effect which has not yet been investigated. But the observations even after Eucken's discussion are hardly good enough for more serious comparison with theory.

* Dennison, *Astrophys. J.* vol. 52, p. 84 (1925).

TABLE 4.

Vibrational specific heat of CH_4, *cal./mole/deg.*

Temp. °K.	C_{vib}, observed	C_{vib}, calculated	Authority
193	0·07	0·04	1
218	0·09	0·10	1
243	0·16	0·20	2
297·5	0·62	0·58	3
397·7	1·98	1·74	3
481	3·34	2·91	3
573	4·9	4·23	2
673	6·3	5·60	2
773	7·6	6·85	2
873	8·8	8·00	2

(1) Heuse, *Ann. d. Physik*, vol. 59, p. 86 (1919).
(2) Eucken and Fried, *Zeit. f. Physik*, vol. 29, p. 41 (1924).
(3) Eucken and von Lude, *Zeit. f. physikal. Chem.* B, vol. 5, p. 413 (1929).

§ **3·63.** *Specific heat of water vapour* (H_2O). Attempts have been made to account in a similar way for the specific heat of H_2O, but without success. An analysis by McCrea* shows, however, that the very large specific heat at high temperatures must be accounted for by dissociation of the molecules of water into H_2 and O_2. There are symptoms also of an effect of polymerization at lower temperatures. The observational material is therefore not suitable for illustrating specific heat theory.

§ **3·64.** *Specific heat of carbon dioxide* (CO_2). Recent analysis of the infrared band spectrum and of the dielectric constant has finally established that the molecule CO_2 is a linear molecule with the oxygen atoms symmetrically placed so: O—C—O. There are three distinct fundamental frequencies ν_1, ν_2 (double) and ν_3, with wave numbers 1388, 667·5 and 2350 respectively.† The approximate specific heat formula is therefore

$$\frac{C_V}{R}=\tfrac{5}{2}+2P\left(\frac{960}{T}\right)+P\left(\frac{1989}{T}\right)+P\left(\frac{3367}{T}\right). \qquad \text{......(279)}$$

The values of the specific heats calculated for this model from (279) did not agree well with older observations at moderate and high temperatures. Recent observations by Henry‡ however have shown that the vibrational energy of CO_2 adjusts slowly to equilibrium with its translational energy just as for diatomic molecules not containing hydrogen, so that the older observed values were all too low. The agreement of Henry's observations with the theory is shown by Fig. 6. Besides the curve derived from (279)

* McCrea, *Proc. Camb. Phil. Soc.* vol. 23, p. 942 (1927).
† Sponer, *Molekülspektren* (1935).
‡ P. H. S. Henry (*private communication*).

the figure also shows a more accurate curve which includes the effect of the anharmonic character of the vibrations.*

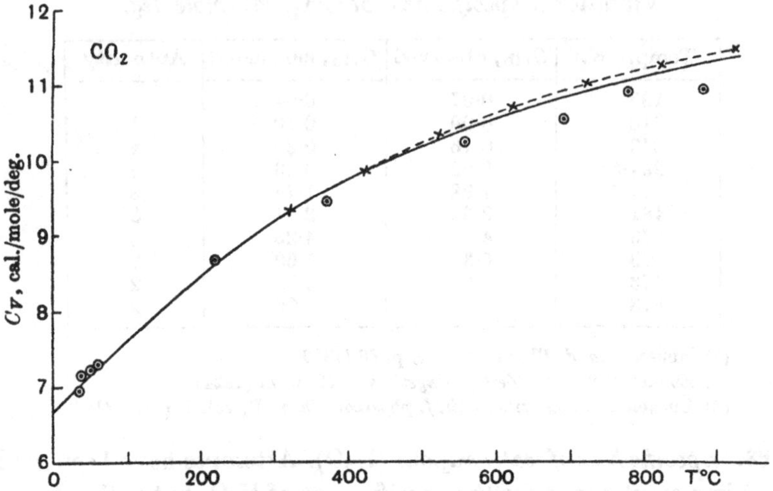

Fig. 6. The specific heat of CO_2; the continuous curve is calculated using Planck terms; the broken curve and crosses allowing for anharmonic terms; ⊙ Henry's observations.

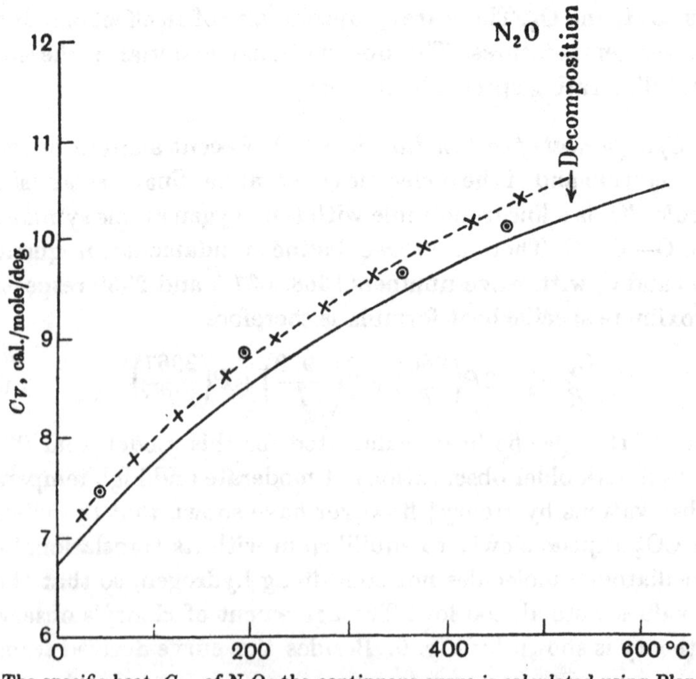

Fig. 7. The specific heat, C_V, of N_2O; the continuous curve is calculated using Planck terms, the broken curve and crosses allowing for anharmonic terms; ⊙ Henry's observations.

* Kassel, *J. Amer. Chem. Soc.* vol. 56, p. 1838 (1934).

§3·65. *Specific heat of nitrous oxide* (N_2O). The structure of this molecule has been shown to be linear but unsymmetrical (N—N—O) by the work of Snow* and Plyler and Barker† on its infra-red absorption spectrum; the latter workers give the wave numbers of its free fundamentals as $\nu_1 = 1285$, $\nu_2 = 593$ (double) and $\nu_3 = 2237$. Using these values we have

$$\frac{C_V}{R} = \tfrac{5}{2} + 2P\left(\frac{849}{T}\right) + P\left(\frac{1841}{T}\right) + P\left(\frac{3204}{T}\right). \quad \text{......(279·1)}$$

The specific heat has been measured recently by Henry‡ and his values are compared with the theory in Fig. 7. They are slightly greater than the theoretical values, but when the anharmonic effects are allowed for§ the agreement is very satisfactory.

§3·66. *Specific heat of acetylene* (C_2H_2). It has been known for some time that acetylene is a linear symmetrical molecule but the identification of the five normal frequencies of such a molecule has only recently|| been accomplished. Their values in wave numbers are $\nu_1 = 1974$, $\nu_2 = 3372$, $\nu_3 = 3288$, $\nu_4 = 730$ (double) and $\nu_5 = 605$ (double). The approximate formula for the specific heat can therefore be written

$$\frac{C_V}{R} = \tfrac{5}{2} + 2P\left(\frac{867}{T}\right) + 2P\left(\frac{1046}{T}\right) + P\left(\frac{2828}{T}\right) + P\left(\frac{4710}{T}\right) + P\left(\frac{4829}{T}\right). \quad \text{......(279·2)}$$

Unfortunately no accurate observations appear to have been yet made on the specific heat of acetylene over an extended range of temperatures, but Table 5 shows that the values calculated from (279·2) are in good agreement with the few observed values which exist.

TABLE 5.

Specific heat of acetylene, cal./mole/deg.

Temp. °C.	C_V calculated	C_V experimental¶
−70	6·62	6·86
18	8·43	8·34

¶ Heuse, *Ann. d. Physik*, vol. 59, p. 86 (1919).

§3·67. *Specific heat of ethylene* (C_2H_4). The structure of this molecule has been established as plane and symmetrical from spectroscopic evidence** and from a theoretical investigation†† of its structure using the quantum theory

* Snow, *Proc. Roy. Soc.* A, vol. 128, p. 294 (1930).
† Plyler and Barker, *Phys. Rev.* vol. 38, p. 1827 (1931).
‡ P. H. S. Henry (*private communication*).
§ Kassel, *J. Amer. Chem. Soc.* vol. 56, p. 1838 (1934).
|| Mecke, *Zeit. f. physikal. Chem.* B, vol. 17, p. 1 (1932); Sutherland, *Phys. Rev.* vol. 43, p. 883 (1933).
** Sponer, *Molekülspektren* (1935).
†† Penney, *Proc. Roy. Soc.* A, vol. 144, p. 166 (1934).

of directed valency. Consequently it possesses twelve normal frequencies all of which may be expected to contribute to the vibrational specific heat when sufficiently high temperatures are reached. The values for these frequencies (so far as it has been possible to determine them from infra-red and Raman spectra) are given in Table 5·1. The frequency ν_7, which corresponds to a twisting vibration of the two CH_2 groups relative to each other

TABLE 5·1.

*The normal frequencies of ethylene** (C_2H_4) *in cm.*$^{-1}$.

$\nu_1 = 2988$	$\nu_7 = ?$
$\nu_2 = 3019$	$\nu_8 = 1623$
$\nu_3 = 1444$	$\nu_9 = 950$
$\nu_4 = 1342$	$\nu_{10} = 1100$
$\nu_5 = 3107$	$\nu_{11} = 950$
$\nu_6 = 3256$	$\nu_{12} = 940$

* Authorities are given by Sponer, *Molekülspektren*. There is still some uncertainty regarding the values assigned to ν_9, ν_{10} and ν_{12} but the others are correct to within a few wave numbers.

about the C—C axis, is very difficult to determine since it is inactive as a fundamental both in infra-red absorption and in Raman scattering, and an insufficient number of combination bands are known as yet from which its value might be deduced. If now the specific heat of ethylene is known over a considerable range of temperature, then from the discrepancy between the experimental values and those calculated from the sum of the eleven Planck terms corresponding to the eleven known frequencies it should be possible to estimate the value of ν_7 provided it is not so high that its contribution in the observed temperature range is insignificant.

TABLE 5·2.

Specific heat of ethylene, cal./mole/deg.

Temp. °K.	Observed	Calculated *A*	Calculated *B*	Observer†
178·6	8·293	8·28	8·29	E. and P.
182·2	8·28	8·32	8·33	Heuse
192·8	8·45	8·44	8·44	E. and P.
205·2	8·54	8·61	8·60	Heuse
210·8	8·68	8·69	8·69	E. and P.
231·4	9·001	9·04	9·02	E. and P.
237·2	9·02	9·16	9·12	Heuse
250·9	9·332	9·43	9·38	E. and P.
272·1	9·809	9·88	9·81	E. and P.
291·2	10·14	10·33	10·24	Heuse
293·5	10·257	10·37	10·27	E. and P.
268·2	11·897	12·21	12·05	E. and P.
464·0	14·16	14·46	14·28	E. and P.

† E. and P. stands for Eucken and Parts, *Zeit. f. physikal. Chem.* B, vol. 20, p. 184 (1933); Heuse, *Ann. d. Physik*, vol. 59, p. 86 (1919).

Such an investigation has recently been carried out by Eucken and Parts* who have shown that the value of ν_7 is very probably somewhere between 750 and 800 cm.$^{-1}$. Their results are summarized in Table 5·2 in which the set of calculated values A was obtained by taking the values of the frequencies in Table 5·1 with $\nu_7 = 803$ cm.$^{-1}$, while those marked B were obtained with $\nu_7 = 747$ cm.$^{-1}$, $\nu_{10} = 1210$ cm.$^{-1}$, $\nu_{12} = 1040$ cm.$^{-1}$ and the values of the others as before. While the final assignment of these frequencies must be left to the spectroscopist,† it is clear that accurate specific heat data on polyatomic molecules may be of considerable service as a help to locating spectroscopically rather inactive frequencies of the above type.

§ 3·68. Summing up this survey we may say that the observed specific heats for the simpler gases are well reproduced by the theory, even when the approximation of using Planck terms is made, instead of the more precise observed vibrational energy levels. For gases with more complicated molecules there is no doubt that similar success can be achieved as soon as the requisite knowledge of the vibrational modes is available. An example of this is provided by the calculations for ethylene. For molecules which contain no hydrogen atoms the rate of adjustment of vibrational and translational energy may be very slow so that care may be necessary to ensure that observed specific heats refer to the true equilibrium state.

§ 3·7. *Specific heat contributions for atoms from excitation of the upper states of a multiplet.*‡ For such an atom as thallium in the vapour state, the normal electronic state is $^2P_{\frac{1}{2}}$, but the $^2P_{\frac{3}{2}}$ state, the upper state of the doublet, lies not far above. The weights of these states are 2 and 4 respectively and the energy difference $\Delta\epsilon = hc\Delta\nu$, where $\Delta\nu$ is the difference of term value in wave numbers. Since translational and internal motions are strictly separable, the partition function for such an atom merely contains the extra factor

$$2 + 4e^{-\Delta\epsilon/kT}$$

besides the partition function for a structureless mass-point. In general this extra factor for any doublet state is

$$\varpi_1 + \varpi_2 e^{-\Delta\epsilon/kT}, \qquad \text{......(280)}$$

or if the state is of multiplicity r

$$\varpi_1 + \varpi_2 e^{-\Delta\epsilon_1/kT} + \ldots + \varpi_r e^{-\Delta\epsilon_{r-1}/kT}. \qquad \text{......(281)}$$

For a gas of M atoms with normal state a doublet we therefore find from (280)

$$\overline{E}_{\text{exc}} = M \frac{\Delta\epsilon}{1 + (\varpi_1/\varpi_2)e^{\Delta\epsilon/kT}}. \qquad \text{......(282)}$$

* Eucken and Parts, *Zeit. f. physikal. Chem.* B, vol. 20, p. 184 (1933).

† Cf. a more recent discussion by Teller and Topley, *J. Chem. Soc.* p. 885 (1935).

‡ For spectroscopic notation see White, *Introduction to Atomic Spectra*, New York (1935); see also Chapter XIV.

This leads at once to a contribution C_{exc} to C_V given by

$$\frac{C_{exc}}{R}=\left(\frac{\Delta\epsilon}{kT}\right)^2\frac{(\varpi_1/\varpi_2)\,e^{\Delta\epsilon/kT}}{\{1+(\varpi_1/\varpi_2)\,e^{\Delta\epsilon/kT}\}^2}. \qquad \text{......(283)}$$

Examples of this function are shown in Fig. 7·1. It will be observed that

$$\frac{C_{exc}}{R}\sim\frac{\varpi_1}{\varpi_2}\left(\frac{\Delta\epsilon}{kT}\right)^2 \qquad \left(\frac{\Delta\epsilon}{kT}\ll 1\right), \qquad \text{......(284)}$$

$$\frac{C_{exc}}{R}\sim\frac{\varpi_2}{\varpi_1}\left(\frac{\Delta\epsilon}{kT}\right)^2 e^{-\Delta\epsilon/kT} \qquad \left(\frac{\Delta\epsilon}{kT}\gg 1\right). \qquad \text{......(285)}$$

In both limits C_{exc} vanishes and is only sensible when $\Delta\epsilon$ and kT are of the

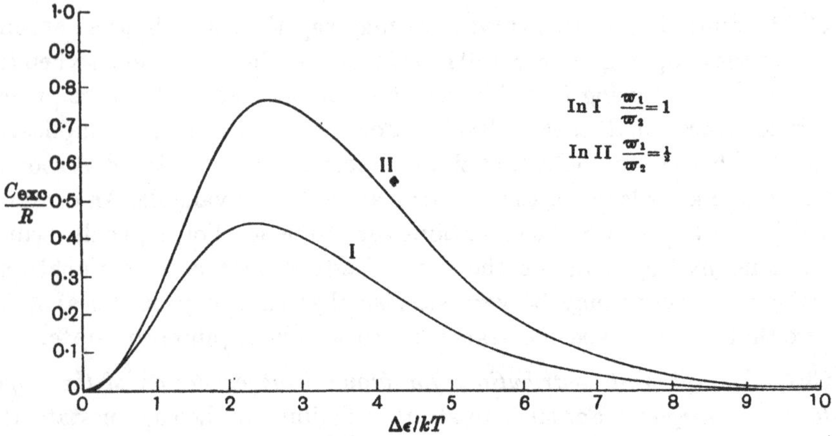

Fig. 7·1. The value of C_{exc}/R, equation (283), as a function of $\Delta\epsilon/kT$.

same order. When $\Delta\epsilon/kT\gg 1$ the upper state can be ignored. When $\Delta\epsilon/kT\ll 1$ the distinction between the two states can be ignored, and the combined states regarded as a single state whose weight is $\varpi_1+\varpi_2$.

There are no measurements on vapours with which these formulae may be compared. (See however § 21·6.)

§ 3·71. *Contributions from multiplet states of molecules.* NO. As a rough approximation these formulae might be applied to a molecule such as NO whose normal states belong to an electronic multiplet. The normal state of NO is $^2\Pi_{\frac{1}{2}}$, of the multiplet $^2\Pi_{\frac{1}{2},\frac{3}{2}}$. But the electronic spin and orbital angular momentum combine vectorially with the angular momentum of the rotating nuclei, so that the electronic states and the rotational states are not strictly independent. It is better therefore to use the more accurate partition function

$$f(\sigma)=\sum_{j=1}^{\infty}2je^{-\sigma j^2}+\sum_{j=2}^{\infty}2je^{-\sigma j^2-\Delta\epsilon/kT}, \qquad \text{......(285·1)}$$

where $\Delta\epsilon=124\cdot4$ cm.$^{-1}$. This may be re-written as

$$f(\sigma)=\sum_{j=1}^{\infty}2je^{-\sigma j^2}\{1+e^{-\Delta\epsilon/kT}\}-2e^{-\sigma-\Delta\epsilon/kT}, \qquad \text{......(285·2)}$$

and it can readily be verified that in subsequent calculations the last term may be neglected. Thus we may separate the rotational part $\sum_{j=1}^{\infty} 2je^{-\sigma j^2}$ from the excitational part $1+e^{-\Delta\epsilon/kT}$. The former can be proved* to lead to the formula

$$C_{\text{rot}} = R(1+\tfrac{11}{180}\sigma^2+\ldots), \qquad \ldots\ldots(285\cdot3)$$

while the latter has just been shown to give

$$C_{\text{exc}} = R\left\{\left(\frac{\Delta\epsilon}{kT}\right)^2 \frac{e^{\Delta\epsilon/kT}}{(1+e^{\Delta\epsilon/kT})^2}\right\}. \qquad \ldots\ldots(285\cdot4)$$

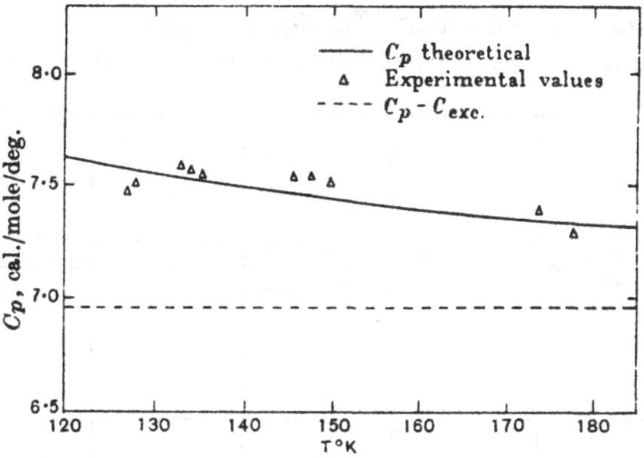

Fig. 8. The specific heat of NO at low temperatures.

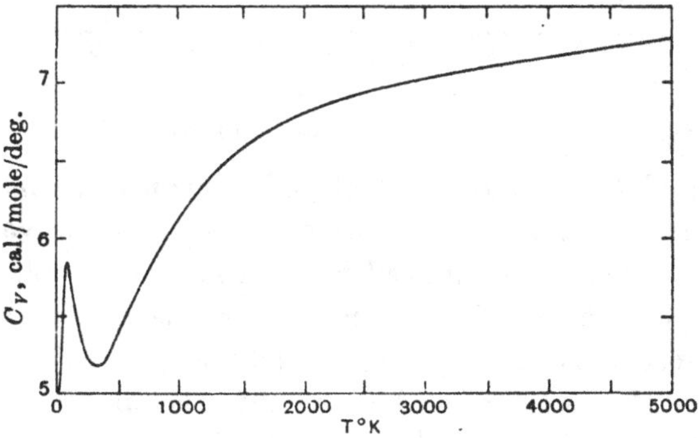

Fig. 8·1. The calculated specific heat, C_V, for NO for all temperatures.

The contribution from (285·4) begins to be appreciable about 30° K. and after rising to a maximum of approximately 0·8 cal./mole/degree near 75° K. it gradually drops off again until at room temperature it is barely observable. Eucken and d'Or† have recently made an investigation of the specific heat of NO between 130° K. and 180° K., and Fig. 8 shows that their results provide an excellent confirmation of this phenomenon of the

* Sutherland, *Proc. Camb. Phil. Soc.* vol. 26, p. 402 (1930).

† Eucken and d'Or, *Gött. Nachr.* p. 107 (1932).

"excitational specific heat" as calculated from (285·4). Johnston and Chapman,* whose results are shown in Fig. 8·1, have used exact spectroscopic levels for the calculation of the partition function.

§ 3·72. *Refined theory of the specific heat of* O_2. Oxygen is another molecule with an excitational specific heat. It exists in a triplet $^3\Sigma$ ground state, the three components of which have been termed by Mulliken the F_1, F_2 and F_3 coupling states respectively. The splitting between these three states depends on the rotational quantum number, but for a particular value of K the F_2 level is approximately 2 cm.$^{-1}$ higher than the F_1 and F_3 levels which form a very close doublet (Fig. 9). In this case it is not possible to separate the rotational from the excitational part of the partition function

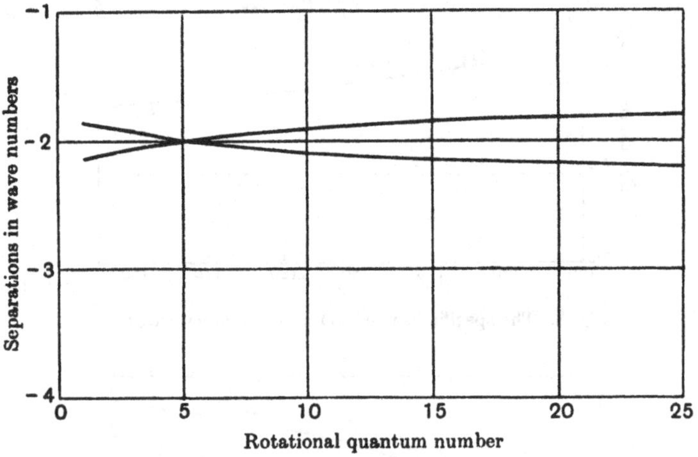

Fig. 9. The separations of the states of the oxygen triplet.

and so equation (285·4) cannot be employed. The only satisfactory method is that of computing the separate summations from the various energy levels. This is the method employed by Johnston and Walker† who used the energy levels deduced from spectroscopic data and then carefully computed the specific heat of O_2 between 0·02° K. and 10° K. Their results are illustrated in Fig. 9·1. It will be seen that the specific heat curve exhibits two maxima; the first very sharp one at 0·12° K. is due to the transition from all the molecules being in the F_3 state to an equilibrium between the numbers in the F_1 and the F_3 states, while the second smaller one is due to the next transition of the molecules up to the F_2 state 2 cm.$^{-1}$ higher. The succeeding increase in the specific heat is of course due to the rise of the rotational specific heat to its equipartition value. These anomalies are unfortunately unobservable.

* Johnston and Chapman, *J. Amer. Chem. Soc.* vol. 55, p. 153 (1933).

† Johnston and Walker, *J. Amer. Chem. Soc.* vol. 55, p. 172 (1933).

Lewis and Von Elbe* have detected another anomaly in the specific heat curve of O_2 at high temperatures due to a higher electronic state of the oxygen molecule. These experimenters determined the mean specific heat

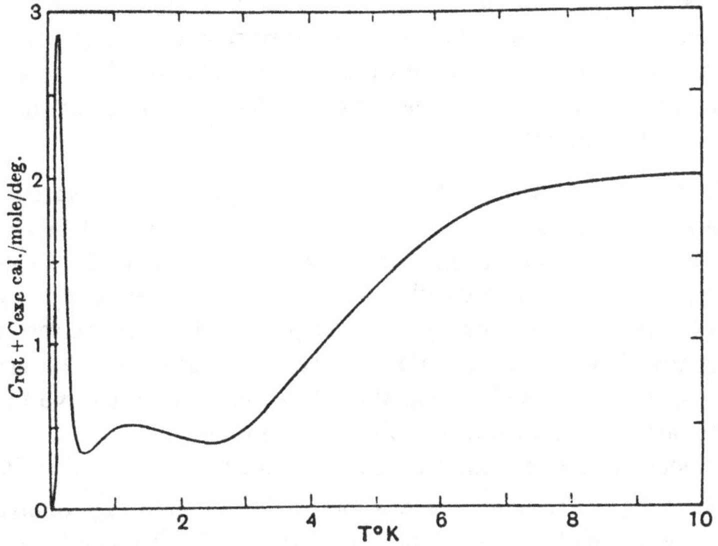

Fig. 9·1. The theoretical excitational plus rotational specific heat of O_2 at low temperatures.

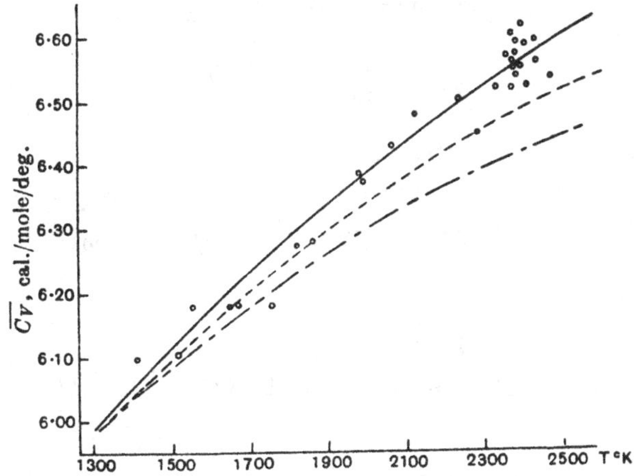

Fig. 9·2. The mean specific heat for O_2 $\overline{C_V}$ averaged from 300° K. to T° K.

○ Observations of Lewis and Von Elbe. Theoretical curves, — - — - without electronic excitation; - - - - with the observed excited state at 0·97 volt; —— with the derived excited state at 0·75 volt.

of oxygen by the explosion method up to 2500° C. They found that their results could not be accounted for by the sum of the translational, rotational and vibrational specific heats alone, even allowing for anharmonic correc-

* Lewis and Von Elbe, *J. Amer. Chem. Soc.* vol. 55, p. 511 (1933).

tions in the vibrational specific heat. They assumed therefore that the additional contribution must come from some higher electronic level, the position of which they calculated from the mean of a number of observations around 2400° K. to be 0·75 volt above the $^3\Sigma$ ground level. This level has since been determined spectroscopically by Herzberg* and others to be at 0·97 volt. The agreement between the experimental results of Lewis and Von Elbe and the theoretical values as calculated by Johnston and Walker† is illustrated in Fig. 9·2.

§ **3·73.** *Small corrections to the perfect gas laws for quantum statistics.* The properties of highly degenerate assemblies of electrons will be discussed later, but as we pointed out in § 2·51 a deviation from the behaviour of a classical perfect gas (λ not small) might be barely observable for H_2 and perhaps He at very low temperatures. We put the formulae on record here. The statistics obeyed by all actual molecular or atomic gases existing as such at low temperatures is the Einstein-Bose statistics (assembly wave-functions symmetrical) but the formulae apply to either case.

It was shown in § 2·8 that for all statistics $pV = \frac{2}{3}\overline{E_{\text{kin}}}$ so that it is unnecessary to discuss the equation of state. A discussion of $\overline{E_{\text{kin}}}$ for structureless mass-points will suffice. Equations (120), (121), (163) and (164) show that

$$M = \lambda \frac{\partial Z}{\partial \lambda}, \quad \overline{E_{\text{kin}}} = kT^2 \frac{\partial Z}{\partial T},$$

where

$$Z = \frac{(2\pi mkT)^{\frac{3}{2}}}{h^3} V \sum_{j=1}^{\infty} \frac{\alpha_j}{j^{\frac{5}{2}}} \lambda^j.$$

On working out $\overline{E_{\text{kin}}}$ explicitly we find

$$\overline{E_{\text{kin}}} = \tfrac{3}{2} MkT \left[1 - \frac{\alpha_2}{4\sqrt{2}} \frac{h^3 M/V}{(2\pi mkT)^{\frac{3}{2}}} + \left(\frac{3}{16} - \frac{2}{9\sqrt{3}} \right) \left(\frac{h^3 M/V}{(2\pi mkT)^{\frac{3}{2}}} \right)^2 + \dots \right]. \qquad \text{......(286)}$$

It follows that

$$\frac{C_V}{R} = \frac{3}{2} \left[1 + \frac{\alpha_2}{8\sqrt{2}} \frac{h^3 M/V}{(2\pi mkT)^{\frac{3}{2}}} - \left(\frac{3}{8} - \frac{4}{9\sqrt{3}} \right) \left(\frac{h^3 M/V}{(2\pi mkT)^{\frac{3}{2}}} \right)^2 + \dots \right]. \qquad \text{......(287)}$$

These formulae are valid so long as λ or, what is equivalent, $\dfrac{h^3 M/V}{(2\pi mkT)^{\frac{3}{2}}}$ is small. As T decreases the first effect of the quantum statistics is to increase or diminish C_V according as $\alpha_2 \gtrless 0$, that is according as the statistics is Einstein-Bose or Fermi-Dirac.

* Herzberg, *Nature*, vol. 133, p. 759 (1934); Ellis and Kneser, *Phys. Rev.* vol. 45, p. 133 (1934); Salow and Steiner, *Nature*, vol. 134, p. 463 (1934).

† Johnston and Walker, *J. Amer. Chem. Soc.* vol. 57, p. 682 (1935).

It is uncertain whether this calculated increase in C_V can be disentangled observationally from the corrections to the perfect gas laws due to intermolecular forces.

When T is exceedingly small these formulae fail for the Einstein-Bose statistics, and with them all formulae based on the approximation which replaces the discrete states in Z by a continuous distribution,* because this replacement is inadequate for the states of lowest energy which for this statistics and low temperatures control the distribution. It follows from (121) and (134) that ultimately as $T \to 0$ $\overline{E_{\text{kin}}} \to M\epsilon_1$, and $C_V \to 0$.

§ **3·8.** *The formation of partition functions by limiting processes.* We postponed from the preceding chapter a discussion of the difficult points in the formation of partition functions for classical systems by a limiting process. It will be sufficient to consider the partition function for the translatory motion of free atoms in an enclosure of volume V using classical statistics, which is a typical and important case.

In the earlier discussion, where no limiting process was involved, it was convenient to choose a unit of energy to fit the assembly, so that all the energies were measured by integers without a common factor. Here this implies a continual change of the unit of energy as we proceed to the limit, which is apt to obscure essential features. We therefore fix the unit of energy once for all and assume that for the artificial assemblies of the sequence the ζ_t are chosen so that $\zeta_t = \xi_t/\tau$, where the ξ_t and τ are integers, and τ changes from one sequence to another. As we proceed to the limit $\tau \to \infty$.

The partition functions for the atoms in the artificial assembly may still be taken to be

$$h(z) = \Sigma_t \, \delta_t z^{\zeta_t},$$

and we require the coefficients of z^E in, say,

$$[h(z)]^M \, [j(z)]^N,$$

and similar expressions. The powers of z in $h(z)$ are now fractional and $[j(z)]^N$ represents the partition functions for the rest of the assembly. If we write $x = z^{1/\tau}$, $h(x) = \Sigma_t \, \delta_t x^{\xi_t}$, etc., these coefficients are the coefficients of $x^{\tau E}$ in

$$[h(x)]^M \, [j(x)]^N,$$

and therefore

$$C = \frac{1}{2\pi i} \int_\gamma \frac{dx}{x^{\tau E+1}} [h(x)]^M \, [j(x)]^N. \qquad \text{......(288)}$$

On changing back to the variable z we find

$$C = \frac{1}{2\pi i \tau} \int_{\tau\gamma} \frac{dz}{z^{E+1}} [h(z)]^M \, [j(z)]^N, \qquad \text{......(289)}$$

* Uhlenbeck, *Over statistische Methoden in de Theorie der Quanta*, Inaug. Diss., Leiden (1927).

in which the contour $\tau\gamma$ means that the integral must now be taken τ times round the circle γ. We have to study the asymptotic form of (289) when $\tau \to \infty$.

The following arguments are incomplete in detail, but show how the use of $H(z)$ in place of $h(z)$ in all the integrals and the derived distribution laws may be justified. The details can be filled in without much trouble.

Let
$$h(z) = \Sigma_\xi \delta_\xi z^{\xi/\tau}, \qquad \text{......(290)}$$
where $h^3\delta_\xi$ is the extension of that set of points in 6 dimensions in which
$$\frac{\xi}{\tau} \leqslant \frac{1}{2m}(p_1{}^2 + p_2{}^2 + p_3{}^2) < \frac{\xi+1}{\tau} \quad \{(q_1, q_2, q_3) \text{ in } V\}.$$

Any other possible $h(z)$ will differ from this only by terms of order $1/\tau$. Then
$$h^3\delta_\xi = \frac{4\pi}{3}\{(\xi+1)^{\frac{3}{2}} - \xi^{\frac{3}{2}}\}\left(\frac{2m}{\tau}\right)^{\frac{3}{2}} V, \qquad \text{......(291)}$$
$$= 2\pi\left\{\frac{\Gamma(\xi+\frac{3}{2})}{\xi!} + \frac{\alpha}{(\xi+1)^{\frac{1}{2}}} + O[(\xi+1)^{-\frac{3}{2}}]\right\}\left(\frac{2m}{\tau}\right)^{\frac{3}{2}} V. \qquad \text{......(292)}$$

It is found that the dominant contributions come from the first term in { }, and
$$\Sigma_\xi \frac{\Gamma(\xi+\frac{3}{2})}{\xi!} z^{\xi/\tau} = \tfrac{1}{2}\Gamma(\tfrac{1}{2})(1 - z^{1/\tau})^{-\frac{3}{2}}.$$

Thus, approximately,
$$h(z) = \frac{(2\pi m)^{\frac{3}{2}} V}{h^3}\{\tau(1 - z^{1/\tau})\}^{-\frac{3}{2}}, \qquad \text{......(293)}$$

and any other $h(z)$ will likewise approximate to (293). In (289) z^E and $j(z)$ are uniform functions of z, and so is $[h(z)]^M$ when it is summed for all the τ circles. Thus we can write
$$C = \frac{1}{2\pi i}\int_\gamma \frac{dz}{z^{E+1}}[j(z)]^N Q(M, \tau, z), \qquad \text{......(294)}$$

where
$$Q(M, \tau, z) = \frac{1}{\tau}\sum_{r=0}^{\tau-1}[h(ze^{2\pi r i})]^M,$$
$$= \left\{\frac{(2\pi m)^{\frac{3}{2}} V}{h^3}\right\}^M \frac{1}{\tau}\sum_{r=0}^{\tau-1}\{\tau(1 - z^{1/\tau}e^{2\pi r i/\tau})\}^{-\frac{3}{2}M}. \qquad \text{......(295)}$$

In (295) it is supposed for definiteness that $am(z) = 0$ for real z and that $am(z)$ lies between $\pm\pi$ on the circle γ. The series in (295) can be expanded in the form
$$\frac{\tau}{\tau^{\frac{3}{2}M}\Gamma(\frac{3}{2}M)}\left[1 + \frac{\Gamma(\frac{3}{2}M+\tau)}{\tau!}z + \frac{\Gamma(\frac{3}{2}M+2\tau)}{2\tau!}z^2 + \ldots\right],$$
the other terms vanishing on summation. When $\tau \to \infty$
$$\frac{\Gamma(\frac{3}{2}M + r\tau)}{(r\tau)!} \sim (r\tau)^{\frac{3}{2}M-1}.$$

Therefore as $\tau \to \infty$

$$Q(M,\tau,z) \sim \frac{1}{\tau\Gamma(\frac{3}{2}M)}\left\{\frac{(2\pi m)^{\frac{3}{2}}V}{h^3}\right\}^M [z + 2^{\frac{3}{2}M-1}z^2 + \ldots + r^{\frac{3}{2}M-1}z^r + \ldots] = \frac{1}{\tau}q(M,z). \quad \ldots\ldots(296)$$

The factor $1/\tau$ will be irrelevant, as it will occur similarly in all the integrals. It can then be shown that we can work out all the distribution laws using $q(M,z)$ in the ordinary integrals in place of $[h(z)]^M$. The unique distribution laws so obtained in terms of $q(M,z)$ will be those obtained by proceeding to the classical limit before calculating average values.

In order to see that this form is the same as that obtained by proceeding to the classical limit afterwards as in § 2·6, we examine the form of $q(M,z)$ for large M. For real z it can be shown at once by comparing the series $z + 2^{\frac{3}{2}M-1}z^2 + \ldots$ with

$$\int_0^\infty x^{\frac{3}{2}M-1}z^x dx \quad \left(= \frac{\Gamma(\frac{3}{2}M)}{(\log 1/z)^{\frac{3}{2}M}}\right),$$

that the difference is negligible for large M. It follows that as $M \to \infty$

$$q(M,z) \sim \left[\frac{(2\pi m)^{\frac{3}{2}}V}{h^3(\log 1/z)^{\frac{3}{2}}}\right]^M = [H(z)]^M. \quad \ldots\ldots(297)$$

A closer discussion of the series for complex z shows that, for the purpose of substitution in the integral (294), $q(M,z)$ can be replaced by (297) for all values of z provided that $\log(1/z)$ is real for real z and that $|am(z)| < \pi$. This argument when the details are filled in, is the full justification of the procedure of § 2·6. It shows that the order of the operations $\tau \to \infty$ and $E \to \infty$ is indifferent, and that, with the proper convention as to z, $h(z)$ may be replaced by $H(z)$ in the integral for C. Similar arguments apply to the other integrals, to differential coefficients of $h(z)$ and $H(z)$, and to other classical systems.

A somewhat similar limiting process is required for sets of quantized systems when the energy quanta are incommensurable, but the arguments are simpler. We form a sequence of artificial assemblies with commensurable energies, whose limits are the actual energies of the real assembly. A procedure which uses the limiting partition functions with a properly defined range for $am(z)$ is justified by the same arguments. We shall not find it necessary to refer again to such limiting processes. We shall assume that, where necessary, they have all been carried out.

§ **3·9.** *Fluctuations.* We have hitherto ignored all questions of the fluctuations of a quantity P about its mean value $\bar{P}$. As we pointed out in the introductory chapter a proof that in general $\overline{(P-\bar{P})^2} = O(\bar{P})$ is essential to the completeness of the theory, to guarantee that an average property is one which the assembly may be expected actually to have. We shall show in Chapter XX that asymptotic formulae for the general fluctuations $\overline{(P-\bar{P})^s}$ can be calculated for all integral values of s by a simple extension of the analysis of this chapter. We shall not stay to consider these here, but content ourselves with proving that

$$\overline{(P-\bar{P})^2}=O(\bar{P}) \quad \text{......(298)}$$

in all the important cases that arise. We observe that

$$\overline{(P-\bar{P})^2}=\overline{P^2}-(\bar{P})^2. \quad \text{......(299)}$$

Consider a typical assembly of localized systems for which

$$C=\frac{1}{2\pi i}\int_\gamma \frac{dz}{z^{E+1}}[f(z)]^M[g(z)]^N\ldots, \quad \text{......(300)}$$

and consider the case $P=E_A$. Then, by the arguments of §§ 2·3–2·32,

$$C\overline{E_A{}^2}=\frac{1}{2\pi i}\int_\gamma \frac{dz}{z^{E+1}}\left\{\left(z\frac{d}{dz}\right)^2[f(z)]^M\right\}[g(z)]^N\ldots. \quad \text{......(301)}$$

If we evaluate this we find

$$\overline{E_A{}^2}=[f(\vartheta)]^{-M}\left(\vartheta\frac{d}{d\vartheta}\right)^2[f(\vartheta)]^M\left\{1+O\left(\frac{1}{E}\right)\right\}. \quad \text{......(302)}$$

We have also $$\overline{E_A}=[f(\vartheta)]^{-M}\left(\vartheta\frac{d}{d\vartheta}\right)[f(\vartheta)]^M\left\{1+O\left(\frac{1}{E}\right)\right\}. \quad \text{......(303)}$$

It is here necessary to include the O-terms, for the leading terms in the fluctuation will cancel. Thus, combining (302) and (303),

$$\overline{E_A{}^2}=[f(\vartheta)]^{-M}\vartheta\frac{d}{d\vartheta}\{\overline{E_A}[f(\vartheta)]^M\}\left\{1+O\left(\frac{1}{E}\right)\right\},$$

$$=\left\{(\overline{E_A})^2+\vartheta\frac{d\overline{E_A}}{d\vartheta}\right\}\left\{1+O\left(\frac{1}{E}\right)\right\}.$$

It follows that $$\overline{(E_A-\overline{E_A})^2}=O(\overline{E_A}^2/E)+O(\overline{E_A}),$$

$$=O(\overline{E_A}), \quad \text{......(304)}$$

which is the relation required.

The fluctuation of, say, $\bar{a}_r$ can be calculated in a similar way. We have

$$\overline{(a_r-\bar{a}_r)^2}=\overline{a_r(a_r-1)}+\bar{a}_r-(\bar{a}_r)^2,$$

and $$C\overline{a_r(a_r-1)}=\frac{M(M-1)}{2\pi i}\int_\gamma \frac{dz}{z^{E+1}}(\varpi_r z^{\epsilon_r})^2[f(z)]^{M-2}[g(z)]^N. \quad \text{...(305)}$$

Evaluating this we find

$$\overline{a_r(a_r-1)}=M(M-1)(\varpi_r\vartheta^{\epsilon_r})^2/\{f(\vartheta)\}^2\left\{1+O\left(\frac{1}{E}\right)\right\},$$

$$=\left(1-\frac{1}{M}\right)(\bar{a}_r)^2\left\{1+O\left(\frac{1}{E}\right)\right\}.$$

Thus
$$\overline{(a_r-\overline{a_r})^2}=\overline{a_r}-(\overline{a_r})^2/M+O\{(\overline{a_r})^2/E\},$$
$$=O(\overline{a_r}), \qquad \text{......(306)}$$

which is the relation required.

The method is quite general. For example, for assemblies of sets of systems confined to a common enclosure as in § 2·4 when quantum statistics is necessary we have similarly

$$C\overline{E_A{}^2}=\frac{1}{(2\pi i)^3}\iiint\frac{dx\,dy\,dz}{x^{M+1}y^{N+1}z^{E+1}}\left\{\left(z\frac{\partial}{\partial z}\right)^2\Pi_t\, g(xz^{\epsilon_t})\right\}\Pi_t\, g'(yz^{\eta_t}), \qquad \text{...(307)}$$

$$C\overline{a_r{}^2}=\frac{1}{(2\pi i)^3}\iiint\frac{dx\,dy\,dz}{x^{M+1}y^{N+1}z^{E+1}}\left\{\left(x\frac{\partial}{\partial x}\right)^2 g(xz^{\epsilon_r})\right\}\Pi_{t\neq r}\, g(xz^{\epsilon_t})\,\Pi_t\, g'(yz^{\eta_t}). \qquad \text{......(308)}$$

These formulae lead immediately to a repetition of (304) and (306). We shall not usually refer to such questions again except in Chapter XX, but shall in all cases leave it to the reader to supply such proofs as are necessary to establish the genuine normality of the equilibrium properties of the assembly. The omitted proofs are always extremely simple.

There is, however, one point of some importance which should not be overlooked. All our arguments can be used to determine $\overline{P}$, even when $\overline{P}$ is not large for values of E which are large enough to make other mean values such as $\overline{Q}$ effectively normal properties of the assembly. This does not in any way invalidate the calculation of $\overline{Q}$ or $\overline{P}$. It merely means that $\overline{P}$ itself is not yet, owing to its smallness and relatively large fluctuations, an effectively normal property of assemblies of this size.

CHAPTER IV

PARTITION FUNCTIONS FOR TEMPERATURE RADIATION AND CRYSTALS. SIMPLE PROPERTIES OF CRYSTALS

§ **4·1.** *Temperature radiation.* In addition to the energy of the material systems in our assemblies, there will be energy of radiation in equilibrium with the matter. It is desirable therefore to construct a partition function for this energy, to enable us to include it in a general discussion of equilibrium laws. This will be especially true of very hot assemblies, in which the energy of radiation is comparable to the energy of the matter. It is not without interest to observe that, if we treat the aether in any enclosure as an approximately independent dynamical system, obeying the laws of quantum mechanics, then Planck's well-known laws of temperature radiation follow at once from the equilibrium theory of statistical mechanics. This is in itself trivial, for of course the laws of the classical quantum theory and all later improvements were constructed to give it. What is of some importance is that we thus deduce Planck's law of temperature radiation as a theorem of the pure equilibrium theory, without appeal to any other fundamental principles or to the mechanisms of the processes of absorption and emission. Such an exposition was first attempted by Debye.*

§ **4·2.** *The normal modes of a continuous medium.* In order to construct a partition function for the energy of the aether, regarded as analogous to a material system, it is only necessary to find suitable coordinates by which to describe its motion and to apply the laws of quantum mechanics. This is easily done. We must start by analysing the number of degrees of freedom of a *continuous medium*—for the sake of generality this may be the aether or an idealized gas or elastic solid.† The gas, the aether and the elastic solid are ideally merely continuous media capable of transmitting respectively compressional oscillations only, transverse oscillations only or oscillations of both types.

The possible motions must all satisfy the general wave equation

$$\frac{\partial^2 \phi}{\partial t^2} = a^2 \nabla^2 \phi, \qquad \text{......(309)}$$

in which a is the velocity of propagation. In (309) ϕ has various meanings in the various problems—a velocity potential, a component of electric or magnetic force, the dilatation $\Sigma \partial u/\partial x$, or a component of the "molecular rotation" $\varpi_x = \frac{1}{2}(\partial w/\partial y - \partial v/\partial z)$, (u,v,w) being the velocity components of

* Debye, *Ann. d. Physik*, vol. 33, p. 1427 (1910).

† Jeans, *Dynamical Theory of Gases*, chap. XIV. Some minor oversights are here corrected.

the medium. For compressional waves there is only one independent type of solution. For transverse aether waves if the components (X,Y,Z) of the electric vector all satisfy (309) then so do the components (A,B,C) of the magnetic vector and, moreover,

$$\Sigma \frac{\partial X}{\partial x} = 0.$$

There are therefore here just two independent types of solution. For the torsional transverse waves of an elastic solid

$$\Sigma \frac{\partial \varpi_x}{\partial x} = 0,$$

so that there are just two independent types here also, or three in all for an elastic solid including the compressional waves.

Consider for simplicity an enclosure of the shape of a rectangular box

$$x=0, \quad x=\alpha,$$
$$y=0, \quad y=\beta,$$
$$z=0, \quad z=\gamma,$$

and a solution ϕ of (309). Let ϕ_0 be the value of ϕ for $t=0$. Then assuming the possibility of an expansion of ϕ_0 in multiple Fourier series we have

$$\phi_0 = \sum_{l=0}^{\infty} \sum_{m=0}^{\infty} \sum_{n=0}^{\infty} A_{lmn} \frac{\cos}{\sin} l\pi x/\alpha \frac{\cos}{\sin} m\pi y/\beta \frac{\cos}{\sin} n\pi z/\gamma. \quad \text{......(310)}$$

Similarly, if ϕ_0' is the value of $\partial\phi/\partial t$ for $t=0$, we can write

$$\phi_0' = \sum_{l=0}^{\infty} \sum_{m=0}^{\infty} \sum_{n=0}^{\infty} A'_{lmn} \frac{\cos}{\sin} l\pi x/\alpha \frac{\cos}{\sin} m\pi y/\beta \frac{\cos}{\sin} n\pi z/\gamma. \quad \text{...(311)}$$

Then it follows that the solution of (309) is

$$\phi = \sum_{l=0}^{\infty} \sum_{m=0}^{\infty} \sum_{n=0}^{\infty} \left\{ A_{lmn} \cos pt + \frac{A'_{lmn}}{p} \sin pt \right\} \frac{\cos}{\sin} l\pi x/\alpha \frac{\cos}{\sin} m\pi y/\beta \frac{\cos}{\sin} n\pi z/\gamma, \quad \text{......(312)}$$

where

$$p^2 = \pi^2 a^2 \left\{ \frac{l^2}{\alpha^2} + \frac{m^2}{\beta^2} + \frac{n^2}{\gamma^2} \right\}. \quad \text{......(313)}$$

In each of (310)–(312) there are eight possible terms and eight independent coefficients A, A' for given l, m, n.

We must now consider more closely the boundary conditions. For sound waves in a gas (ϕ velocity potential) we must have $\partial\phi/\partial n = 0$ on every boundary; that is, $\partial\phi/\partial x = 0$ at $x=0$ and $x=\alpha$, etc. This can only be effected by retaining only the cosine terms in ϕ, so that

$$\phi = \sum_{l=0}^{\infty} \sum_{m=0}^{\infty} \sum_{n=0}^{\infty} \left\{ A_{lmn} \cos pt + \frac{A'_{lmn}}{p} \sin pt \right\} \cos \frac{l\pi x}{\alpha} \cos \frac{m\pi y}{\beta} \cos \frac{n\pi z}{\gamma}. \quad \text{......(314)}$$

Thus to each lmn there corresponds here just one possible normal mode of the system—one degree of freedom. For compressional waves in an elastic solid (ϕ the dilatation) we must have $\phi=0$ over the boundary, only the sine terms can occur and there is just one normal mode as before.

The transverse aether waves and the torsional elastic solid waves are similar. For the former we may take $\phi=X$. We must then assume that the walls of our enclosure are perfect conductors, or energy will not be conserved in the assembly. Thus $X=0$ at $y=0$, $y=\beta$, $z=0$ and $z=\gamma$, which leaves the two terms of type

$$\begin{matrix}\cos\\ \sin\end{matrix}\, l\pi x/\alpha \sin\frac{m\pi y}{\beta}\sin\frac{n\pi z}{\gamma}.$$

There are similar terms in Y and Z. In X, Y and Z, however, the pure sine terms are impossible, or else the condition

$$\Sigma\frac{\partial X}{\partial x}=0 \qquad \text{......(315)}$$

cannot be satisfied. Of the one remaining term in each of X, Y, Z, two only remain independent when (315) is satisfied. There are thus 1, 2 or 3 normal modes per value of l, m, n in the three cases gas, aether, elastic solid.

We now return to (313), in which a can take different values for the different types of wave. Since $p=2\pi a/\lambda$, where λ is the wave length, we have

$$\frac{4}{\lambda^2}=\frac{l^2}{\alpha^2}+\frac{m^2}{\beta^2}+\frac{n^2}{\gamma^2}. \qquad \text{......(316)}$$

The number of normal modes with wave lengths λ satisfying $\lambda\geqslant\lambda_0$ is equal to 1, 2 or 3 times the number of points with integral coordinates inside an octant of the ellipsoid (316) with $\lambda=\lambda_0$, which has the volume

$$\frac{4\pi}{3}\frac{\alpha\beta\gamma}{\lambda_0{}^3}.$$

The number of normal modes with wave lengths between λ and $\lambda+d\lambda$ is therefore

$$4\pi\alpha\beta\gamma\frac{d\lambda}{\lambda^4}(1,2,3) \qquad \text{......(317)}$$

in the three cases, or

$$4\pi V\frac{d\lambda}{\lambda^4}(1,2,3), \qquad \text{......(318)}$$

where V is the volume of the enclosure. This result is really independent of the shape of the enclosure.* For the aether the number of normal modes with frequencies between ν and $\nu+d\nu$ is

$$\frac{8\pi V}{c^3}\nu^2 d\nu, \qquad \text{......(319)}$$

where c is the velocity of light.

Such a classical analysis into normal modes is easily translated into

* Weyl, Courant, *loc. cit.* p. 54.

quantum mechanics. To the same approximation as the classical analysis Schrödinger's equation for the system (crystal, aether) can be so transformed that it separates into a set of equations for simple harmonic oscillators one for each normal mode, and of the same frequency ν. It is therefore easy to construct the partition function for the system to this approximation.

§4·3. *The partition function for temperature radiation.* We shall assume that the zero of energy for the vibrations of the aether is the state in which every normal mode has its lowest possible quantum number.* The energy in any other state is then

$$h(n_1\nu_1+\ldots+n_r\nu_r+\ldots),$$

and each such state is of weight unity. The partition function is thus

$$\begin{aligned} R(z) &= \Sigma_{(n)}\, z^{h(n_1\nu_1+\ldots+n_r\nu_r+\ldots)}, \\ &= \Pi_r(1-z^{h\nu_r})^{-1}. \end{aligned} \qquad \ldots\ldots(320)$$

This factorization is typical of systems whose motions separate into independent parts, like the normal modes of a continuous medium controlled by linear partial differential equations. Convergency conditions are satisfied so long as $|z|<1$. To obtain an intelligible form of (320) we apply (319). Then

$$\log R(z) = -\frac{8\pi V}{c^3}\,\Sigma\,\nu^2 d\nu \log(1-z^{h\nu}). \qquad \ldots\ldots(321)$$

On proceeding to the limit $d\nu \to 0$ we obtain formally

$$\log R(z) = -\frac{8\pi V}{c^3}\int_0^\infty \nu^2 \log(1-z^{h\nu})\,d\nu. \qquad \ldots\ldots(322)$$

To evaluate the integral we can use the logarithmic series and integrate term by term. Then

$$-\int_0^\infty \nu^2 \log(1-z^{h\nu})\,d\nu = \frac{2}{h^3(\log 1/z)^3}\sum_1^\infty \frac{1}{n^4}. \qquad \ldots\ldots(323)$$

Since $\Sigma\, n^{-4} = \pi^4/90$, we find

$$\log R(z) = \frac{8\pi^5 V}{45c^3h^3(\log 1/z)^3}. \qquad \ldots\ldots(324)$$

Consider an assembly containing radiation and N material localized systems of partition functions $f(z)$. Then by the arguments of the preceding chapter

$$C = \frac{1}{2\pi i}\int_\gamma \frac{dz}{z^{E+1}}\,R(z)\,[f(z)]^N, \qquad \ldots\ldots(325)$$

$$C\overline{E_R} = \frac{1}{2\pi i}\int_\gamma \frac{dz}{z^{E+1}}\left\{z\frac{\partial}{\partial z}R(z)\right\}[f(z)]^N, \qquad \ldots\ldots(326)$$

where $\overline{E_R}$ is the average value of the energy of the radiation in the assembly. It follows that

$$\overline{E_R} = \vartheta\frac{\partial}{\partial\vartheta}\log R(\vartheta) \quad (\vartheta = e^{-1/kT}), \qquad \ldots\ldots(327)$$

* Any other choice of this zero introduces an infinite constant into the energy of the radiation which is without physical significance.

or
$$\overline{E_R}=\frac{8\pi^5 V}{15c^3h^3(\log 1/\vartheta)^4}=\frac{8\pi^5k^4}{15c^3h^3}VT^4. \qquad \text{......(328)}$$

This is the Stefan-Boltzmann law of total radiation with the usual theoretical value of Stefan's constant. To find the energy associated with any particular range of frequencies we write

$$R(z)=R_1(z)\,R_2(z),$$

where
$$\log R_1(z)=-\frac{8\pi V}{c^3}\nu^2 d\nu \log(1-z^{h\nu}).$$

Then by the usual arguments

$$C\overline{E_{R_1}}=\frac{1}{2\pi i}\int_\gamma \frac{dz}{z^{E+1}}\left\{z\frac{\partial}{\partial z}R_1(z)\right\}R_2(z)\,[f(z)]^N,$$

$$\overline{E_{R_1}}=\vartheta\frac{\partial}{\partial\vartheta}\log R_1(\vartheta),$$

$$=\frac{8\pi hV}{c^3}\frac{\nu^3 d\nu}{\vartheta^{-h\nu}-1}. \qquad \text{......(329)}$$

In the usual notation $E_\nu d\nu$ is the energy *density* in this frequency range, and we find

$$E_\nu=\frac{8\pi h\nu^3}{c^3}\frac{1}{e^{h\nu/kT}-1}, \qquad \text{......(330)}$$

which is Planck's law.

We can now introduce the energy of radiation by means of its partition function $R(z)$ into any discussion of equilibrium conditions. The limiting processes involved are of a simple type, for the final form of $R(z)$ is merely an analytical approximation to the partition function (320), which is that of a quantized system. No special investigation such as that of § 3·8 is necessary except to deal with incommensurable frequencies.

Though we do not take up thermodynamic relationships until Chapter VI, it is again convenient to record at once the thermodynamic consequences of the existence and form of the partition function $R(T)$. The radiation contributes $k\log R(T)$ to the characteristic function, and therefore by equation (588) we find

$$S_R=k\log R(T)+kT\frac{\partial\log R(T)}{\partial T},$$

$$=\frac{32\pi^5k^4}{45c^3h^3}VT^3=\frac{4}{3}\frac{\overline{E_R}}{T}, \qquad \text{......(331)}$$

$$p_R=kT\frac{\partial}{\partial V}\log R(T),$$

$$=\frac{8\pi^5k^4}{45c^3h^3}T^4=\frac{1}{3}\frac{\overline{E_R}}{V}, \qquad \text{......(332)}$$

for the entropy, S_R, and pressure, p_R, of radiation.

§ 4·31. *The equilibrium theory of radiation regarded as photons or light quanta.* For the sake of completeness we shall now show that Planck's law of temperature radiation may equally well be established by statistical arguments when an extreme particle view of radiation is adopted, provided we apply the Einstein-Bose statistics to an assembly of particles of the proper type and impose no restriction on the total number of such particles.

Treating the light quanta as particles of unrestricted number in an enclosure in which no forces act on the particles, we build up the corresponding factor for them in C as in § 2·4. Since their number is unrestricted no selector variable x is required. Since they obey the Einstein-Bose statistics $g(q) = 1/(1-q)$. Hence for these particles

$$Z = -\Sigma_t \varpi_t \log(1 - e^{-\epsilon_t/kT}), \qquad \text{......(333)}$$

where ϖ_t, ϵ_t are the weights and energies of the possible states of the particles in the given enclosure. If we restrict the summation to particles whose momenta lie in the ranges $p_1, p_1 + dp_1, p_2, p_2 + dp_2, p_3, p_3 + dp_3$, then just as in § 2·71 it is easy to show that

$$Z = -2\frac{V}{h^3}\log(1 - e^{-\epsilon/kT})\,dp_1 dp_2 dp_3. \qquad \text{......(334)}$$

The factor 2 remains as for electrons to allow for the two states of polarization of the light-particle. The deduction of (334) with $m^3 du\,dv\,dw$ in place of $dp_1 dp_2 dp_3$ given in § 2·71 is non-relativistic and to apply to the particles here considered we must remove this restriction; it is shown in § 16·31 that (334) remains true relativistically.

If we now introduce the specific properties of photons we have, for the relationship between energy, frequency and momentum, $\epsilon = h\nu$, where ν is the frequency, while the momentum is a vector of magnitude $h\nu/c$. For a photon therefore, when we change to polar coordinates for the momentum, we find

$$dp_1 dp_2 dp_3 = \frac{h^3}{c^3}\nu^2 d\nu\, d\omega, \qquad \text{......(335)}$$

where $d\omega$ is an elementary solid angle containing the direction of propagation of the photon. On integrating over all directions of propagation we find therefore that for photons in the frequency interval ν, $\nu + d\nu$

$$Z_\nu = -\frac{8\pi V\nu^2 d\nu}{c^3}\log(1 - e^{-h\nu/kT}). \qquad \text{......(336)}$$

The energy content $E_\nu d\nu$ of these light quanta per unit volume follows at once in the usual way, namely

$$E_\nu d\nu = \frac{kT^2}{V}\frac{\partial Z_\nu}{\partial T} = \frac{8\pi h\nu^3 d\nu}{c^3}\frac{1}{e^{h\nu/kT} - 1}, \qquad \text{......(337)}$$

which is Planck's law. The average number of light quanta of this frequency cannot be obtained by using the operator $\lambda\partial/\partial\lambda$, but follows at once from (337) by dividing by $h\nu$, and is

$$\frac{8\pi\nu^2 d\nu}{c^3}\frac{1}{e^{h\nu/kT}-1}. \qquad \text{......(338)}$$

§ **4·4.** *Applications to crystals.* The methods of § 4·3 can also be applied to crystals. We regard the whole crystal as a single Hamiltonian system, whose classical motion may, to a first approximation, be regarded as small oscillations about a position of equilibrium, which we then analyse into its normal modes. Each of these to this first approximation is an independent simple harmonic oscillation. This analysis will adequately represent the motion so long as the general run of the oscillations is small enough, that is so long as the crystal is not too hot. At greater violence of oscillation terms in the potential energy of higher order than the squares of the displacements must be introduced. The first effects of these on the energy and partition function can be investigated by perturbation theory. The quantum theory is a direct transcription of the classical theory.

Apart from these high temperature deviations from simple harmonic oscillations the construction of the partition function demands only an enumeration of the frequencies of the normal modes of the crystal consisting of a given number of molecules, and is precisely as accurate as the enumeration. It is difficult however to make an accurate enumeration, and various approximate enumerations have been given, more or less based on guesswork. The earliest was Einstein's* who suggested that for a crystal containing N atoms it was sufficiently accurate to take all the $3N$ frequencies equal. This still remains a valuable rough approximation. It was improved by Debye† who suggested that the $3N$ frequencies could be taken to be the $3N$ lowest frequencies of a *continuum* with the same elastic properties as the actual atomic crystal. This suggestion has proved of great importance and we shall give an account of Debye's theory in the somewhat more general form into which it was cast by Born.‡ The theory is very successful in accounting for observed facts, so successful that it has been strained beyond its natural range and facts which do not fit it have been thought to be anomalous and to require special explanations without due cause. It has only recently been realized thanks to the work of Blackman§ that Debye's suggestion for the

* Einstein, *Ann. d. Physik*, vol. 22, pp. 180, 800 (1907), vol. 34, pp. 170, 590 (1911).

† Debye, *Ann. d. Physik*, vol. 39, p. 789 (1912).

‡ Born, "Atomtheorie des festen Zustandes" (1923), *Encycl. Math. Wiss.* vol. 5, part 3, No. 25; Born and Göppert-Mayer, "Dynamische Gittertheorie der Kristalle", *Handb. d. Physik*, ed. 2, vol. 24, part 2, p. 623 (1933).

§ Blackman, *Proc. Roy. Soc.* A, vol. 148, pp. 365, 384 (1934), vol. 149, pp. 117, 126 (1935).

frequencies of the normal modes, even as elaborated by Born, may be a much less good approximation than it was formerly held to be. Considerable caution is therefore required before departures from Debye's theory can be held to indicate anything more than slight errors in the assumed distribution of the frequencies of the normal modes.

It is not possible here to enter in great detail into the equilibrium properties of crystals, in particular into the deduction of their properties from assumed lattice structures and laws of force. Some account of this part of the subject is given in Chapter X where the interatomic laws of force in gases and in crystals are brought into a common analysis. But some account of the very great success of statistical mechanics in the theory of the equilibrium properties of crystals should be given here, not entirely limited to specific heats. The field of application is of course far richer than for the permanent perfect gas, whose equilibrium properties are summed up almost completely in $pV = MkT$ and the form of C_V. The account we shall give in this chapter represents Debye's approximation. We shall also describe shortly the results of Blackman's investigation. Further refinements will be described at suitable places in Chapters XI, XII, the latest being collected together in Chapter XXI.

§ **4·5.** *The partition function for a crystal. Debye's approximation and its refinements.* In Debye's approximation and Born's generalization a crystal is supposed to be built up of a lattice of N congruent cells each of which contains s atoms, atomic ions or electrons. It is not normally necessary to include all the electrons in every atom among the s members of the cell. Those which are tightly bound are effectively tightly bound to a particular atom or molecular radical and belong to that atom or radical almost as in a gas; a partition function can be assigned to each atom, atomic ion or radical for its internal degrees of freedom contributed by these electrons. In general, however, these make no effective contribution to the properties of the crystal at relevant temperatures any more than to those of a gas, and for the same reason (§ 3·2). Important exceptions to this rule are provided by crystals containing paramagnetic ionic salts, such as the salts of the rare earths. These are discussed in Chapter XXI. Not all the electrons can properly be neglected in this way and it is necessary for generality to allow for some (or all) of the atoms being ionized, and for some few of their electrons having an independent existence in the cell. This description of the free electrons is very old-fashioned. It is however adequate for the moment and will be discussed further and modernized later on.

It has been shown by Born* that there are, as one would expect, $3s$ distinct sets of normal modes for a crystal of s structural units per cell. The

* Born, Born and Göppert-Mayer, *loc. cit.*

distribution of the *space* frequencies or wave numbers per cm. of each set τ_j ($j=1, 2, \ldots, 3s$) can be determined generally for large crystals. The wave numbers of each set are distributed uniformly in a three-dimensional space in which spherical polar coordinates represent wave number and direction of the wave normal for the corresponding wave. This is exact, but no use can be made of the result for calculating the partition function until the connection between wave number and time frequency ν is established. It is here that exact analysis is difficult and that the procedure suggested by Debye and in use hitherto has led to weaknesses in the theory recently pointed out by Blackman.* The refined form of Debye's procedure introduced by Born is equivalent to the following. The $3s$ sets of *space* frequencies divide necessarily into two groups, those whose *time* frequencies tend to zero for long wave lengths or small wave number of the normal mode, and those whose time frequencies tend to a non-zero limit. In general if the unit cell of the crystal is properly chosen there are just three of the first type for which

$$\nu_j = c_j \tau + \ldots \quad (j=1, 2, 3), \qquad \ldots\ldots(339)$$

where the c's are *the velocities of sound* and are functions of direction of the wave normal. The other $3(s-1)$ sets of frequencies may be called by distinction optical, and for them we have

$$\nu_j = \nu_j^0 + c_j \tau + \ldots \quad (j=4, \ldots, 3s). \qquad \ldots\ldots(340)$$

The ν_j^0 and the c_j depend on the crystal structure and the c_j are again functions of the direction of the wave normal. The ν_j^0 correspond to the frequencies peculiar to the crystal determining its anomalous reflections (*Reststrahlen*). Equations (339) and (340) as explicitly written here are only reliable approximations for small values of τ. The standard procedure however is to assume that it is sufficiently accurate to ignore the variation of ν_j with τ altogether in (340), and to use (339) as written without higher terms for all necessary values of τ. The number of acoustical modes lying within a given element of wave number space can be shown without difficulty to be

$$V\tau^2 d\tau d\omega, \qquad \ldots\ldots(341)$$

where V is the volume of the crystal. All these general results and approximations will be illustrated for a simple case in § 4·71 when we discuss Blackman's improvements.

On combining (341) and (339) we see at once that for the acoustical modes the number of the N frequencies of the jth set which lie in the (time) frequency range ν, $\nu+d\nu$ with their wave normals in the solid angle $d\omega$ is asymptotically equal for large N to

$$\frac{V}{c_j^3}\nu^2 d\nu d\omega \quad (j=1, 2, 3). \qquad \ldots\ldots(341\cdot1)$$

* Blackman, *loc. cit.*

The region of wave number space covered by permissible τ's is not a sphere; the true upper limit for τ and therefore for ν_j, ν_j^* say, is therefore in general a function of direction, as also are the c_j's. The strict balance account for the number of normal modes is therefore

$$V\int\frac{d\omega}{c_j^3}\int_0^{\nu_j^*}\nu^2 d\nu = \frac{V}{3}\int\frac{(\nu_j^*)^3}{c_j^3}d\omega = N. \qquad \text{......(342)}$$

It follows from (341·1) that, not too near ν_j^*, the *total* number dN of acoustical frequencies in the range ν, $\nu+d\nu$ is

$$dN = 4\pi V\left(\frac{1}{c_1^3}+\frac{1}{c_2^3}+\frac{1}{c_3^3}\right)\nu^2 d\nu, \qquad \text{......(343)}$$

in which the c's represent suitable mean values for direction. For isotropic crystals or quasi-isotropic mixtures there is no dependence on direction, and this reduces accurately to

$$dN = 4\pi V\left(\frac{2}{c_t^3}+\frac{1}{c_l^3}\right)\nu^2 d\nu, \qquad \text{......(344)}$$

where c_t and c_l are the velocities of torsional and compressional waves respectively—this is the result of § 4·2.

It is now easy to construct the partition function to this approximation. There is one new point that here first needs attention—the precise specification of the zero of energy. One might be tempted to define this as for the aether to be the state of lowest permissible energy in each normal mode. This specification, however, of the energy zero hardly goes deep enough, if variations in the volume of the crystal are taken into account, for such variations which vary the length of edge of the unit cell must alter the potential energy of the state of lowest permissible energy itself. This ambiguity can be avoided by taking the energy zero as the state of infinite separation at relative rest of all the constituent particles of the crystal, each particle separately being in some specified normal state. The energy of the crystal in its state of lowest permissible energy is then, say, $F_0(V)$. $F_0(V)$ is of course negative; its argument can usually be safely omitted. It is, however, only in the simplest case of an isotropic crystal subjected only to isotropic pressures or tensions that the argument V sufficiently defines its state. We return to the more general case in § 4·9. An entirely different and important type of generalization is introduced in Chapter v.

The energy of the state of the crystal specified by the quantum numbers $n_1, \ldots, n_r, \ldots$ is

$$F_0 + h(n_1\nu_1 + \ldots + n_r\nu_r + \ldots),$$

the weights unity (or at least all equal), and the partition function $K(z)$ given by

$$\log K(z) = F_0 \log z - \sum_{r=1}^{3sN}\log(1-z^{h\nu_r}). \qquad \text{......(345)}$$

This must of course be approximated to by the foregoing analysis of the frequency distribution. We then find at once

$$\log K(z) = F_0 \log z - V \sum_{j=1}^{3} \int \frac{d\omega}{c_j^3} \int_0^{\nu_j^*} \log(1 - z^{h\nu}) \nu^2 d\nu - N \sum_{j=4}^{3s} \log(1 - z^{h\nu_j^0}). \qquad \ldots\ldots(346)$$

It is usually sufficiently accurate to ignore variations of ν_j^* and c_j with direction and to define mean values $\overline{\nu_j}$ and $\overline{c_j}$ by the equations

$$\int \frac{(\nu_j^*)^3}{c_j^3} d\omega = \nu_j^3 \int \frac{d\omega}{c_j^3} = \frac{4\pi \overline{\nu_j}^3}{\overline{c_j}^3}. \qquad \ldots\ldots(347)$$

In virtue of (342)

$$\overline{\nu_j} = \overline{c_j} \left(\frac{3N}{4\pi V} \right)^{\frac{1}{3}}. \qquad \ldots\ldots(348)$$

If we now define three new constants of the crystal by the equations $k\Theta_j = h\overline{\nu_j}$ $(j = 1, 2, 3)$, and replace ν_j^* by $\overline{\nu_j}$ in (346), we find

$$\log K(z) = F_0 \log z - \frac{3N}{(\log 1/z)^3} \sum_{j=1}^{3} \frac{1}{k^3 \Theta_j^3} \int_0^{k\Theta_j \log 1/z} \log(1 - e^{-x}) x^2 dx$$

$$- N \sum_{j=4}^{3s} \log(1 - z^{h\nu_j^0}). \qquad \ldots\ldots(349)$$

Having constructed the partition function $K(z)$ we have at once the usual expression for the mean energy $\overline{E_K}$ of the crystal,

$$\overline{E_K} = kT^2 \frac{\partial}{\partial T} \log K(T). \qquad \ldots\ldots(350)$$

In terms of the absolute temperature T,

$$\log K(T) = \frac{-F_0}{kT} - 3N \sum_{j=1}^{3} \frac{T^3}{\Theta_j^3} \int_0^{\Theta_j/T} \log(1 - e^{-x}) x^2 dx - N \sum_{j=4}^{3s} \log(1 - e^{-h\nu_j^0/kT}). \qquad \ldots\ldots(351)$$

$$\overline{E_K} = F_0 + 3NkT^4 \sum_{j=1}^{3} \frac{1}{\Theta_j^3} \int_0^{\Theta_j/T} \frac{x^3 dx}{e^x - 1} + N \sum_{j=4}^{3s} \frac{h\nu_j^0}{e^{h\nu_j^0/kT} - 1}. \qquad \ldots\ldots(352)$$

The last result has been simplified after differentiation by an integration by parts. These are the complete formulae, but the $3s - 3$ frequencies can be grouped again into two classes: $3(p-1)$ infra-red frequencies not necessarily all different, and $3(s-p)$ ultra-violet, where p is the number of massive particles in the unit cell and $s-p$ the number of separated electrons. We can in most applications ignore the ultra-violet frequencies of the electrons altogether, and so find in all in (351) just $3p$ terms, pN being the total number of atoms in the crystal.

This view of the part played by the "free" electrons in the lattice is very old-fashioned, but none the less adequate here. In the modern electron theory of metals and crystals to which we return in Chapter XI we should rather regard these electrons as an electron gas in more or less free movement in the periodic field of the positive ions of the lattice. The gas, however, is an almost completely degenerate one and usually does not contribute appreciably to the specific heat. The result is then the same as for a lattice of high frequency. If we desire to calculate the actual contribution made by the electrons, if and when it is significant, it is of course essential to use the modern theory. Some beautiful examples of this electron contribution are already known, and will be discussed in Chapter XI.

If in (352) we ignore the differences between Θ_1, Θ_2 and Θ_3, and neglect the terms arising from the infra-red frequencies, we find approximately

$$\log K(T) = \frac{-F_0}{kT} - \frac{9NT^3}{\Theta^3}\int_0^{\Theta/T} \log(1-e^{-x})\, x^2\, dx, \qquad \text{......(353)}$$

$$\overline{E_K} = F_0 + \frac{9NkT^4}{\Theta^3}\int_0^{\Theta/T} \frac{x^3\,dx}{e^x-1}. \qquad \text{......(353·1)}$$

This is Debye's result, which we could recover directly by constructing a partition function according to Debye's theory. The complete *theoretical* result (352) is due to Born. The form was first suggested by Nernst. We have included in F_0 the zero-point energy of the lattice vibrations of $\frac{1}{2}h\nu$ per mode of frequency ν. It is sometimes convenient to have an explicit formula for this energy according to Debye's approximation, using all the simplifications in (353). In this approximation the number of modes with frequencies in the range ν, $\nu + d\nu$ is

$$\frac{9N\nu^2\,d\nu}{\bar{\nu}^3}. \qquad \text{......(353·2)}$$

The corresponding zero-point energy E_0 is therefore easily found to be given by

$$E_0 = \tfrac{9}{8}Nh\bar{\nu} = \tfrac{9}{8}Nk\Theta. \qquad \text{......(353·3)}$$

The two types of term in (351) and (352) are often referred to as Debye's terms and Einstein's terms respectively. The latter name arises from the early investigation by Einstein of specific heats of crystals already mentioned. Einstein's terms in (351) are obviously only significant when the $\nu_j{}^0$'s are effectively different from zero, so that it is allowable to ignore the terms $c_j\tau$. When $\nu_j{}^0 \to 0$ we must finally get an additional Debye's term. For example, in a two-atom lattice in which both types of atom are similarly situated and of approximately equal masses (e.g. KCl), it is obviously better to neglect Einstein's terms and treat the whole body as if it were built up of atoms of a single type. We shall thus get a better approximation to the corpus of normal modes.

§ **4·6.** *Debye's formula for C_V in theory and experiment.* The possible field of validity of Debye's formula (353) for the partition function of a crystal can be defined fairly closely. The formula can only be expected to apply to crystals built up out of atoms of one type, all of which are similarly situated in the lattice—that is, for elements crystallizing in the regular system, with extensions perhaps to nearly regular crystals and to simple compounds of atoms of nearly equal mass like KCl. The unit cell of the lattice may then be thought of as containing a single atom, and N is the number of atoms in the crystal or conglomerate. An examination of the facts shows a remarkable agreement with the theory in the expected region, as will now be described. The quantity directly observed is C_p while the theory gives C_V. The derivation of C_V from C_p is effected by formula (390) of a later section.

Let us introduce for shortness the notation

$$D(x) = \frac{3}{x^3}\int_0^x \frac{\tau^3}{e^\tau - 1}\,d\tau; \qquad \text{......(354)}$$

$D(x)$ is frequently called Debye's transcendent. Then

$$\overline{E_K} = F_0 + 3NkTD(\Theta/T), \qquad \text{......(355)}$$

$$C_V = 3Nk\left\{D\left(\frac{\Theta}{T}\right) - \frac{\Theta}{T}D'\left(\frac{\Theta}{T}\right)\right\}. \qquad \text{......(356)}$$

For large values of x

$$D(x) = \frac{\pi^4}{5x^3} - 3e^{-x}\left\{1 + O\left(\frac{1}{x}\right)\right\}. \qquad \text{......(357)}$$

This is easily proved by replacing $\int_0^x$ by $\int_0^\infty - \int_x^\infty$, expanding $1/(e^\tau - 1)$ in the infinite integral and integrating term by term. For small values of x, by direct expansion of the integrand,

$$D(x) = 1 - \tfrac{3}{8}x + \tfrac{1}{20}x^2 + O(x^4).$$

These relations can be differentiated. Therefore we have approximately

$$\overline{E_K} = F_0 + 3NkT, \qquad C_V = 3Nk\left(1 - \frac{1}{20}\frac{\Theta^2}{T^2}\right) \quad (T \to \infty), \qquad \text{......(358)}$$

$$\overline{E_K} = F_0 + \frac{3\pi^4}{5}Nk\frac{T^4}{\Theta^3}, \quad C_V = \frac{12\pi^4}{5}Nk\frac{T^3}{\Theta^3}\left(1 - \frac{15}{4\pi^4}\frac{\Theta^4}{T^4}e^{-\Theta/T}\right) \quad (T \to 0). \qquad \text{......(359)}$$

We make the following observations: (1) The relation $C_V = 3Nk$ is an example of the theorem of equipartition of energy and expresses the well-known laws of Dulong and Petit and Neumann and Regnault. It should hold to within $\frac{1}{2}$ per cent. for $T/\Theta > 3$. The law of Dulong and Petit states that the specific heat per gram-atom has approximately the same value (6·4 for C_p, 6·0 for C_V after correction) for all elements in the solid state. The law of Neumann and Regnault states that the specific heat per gram-molecule of a simple compound in the solid state is approximately equal to the sum of the specific heats of the corresponding solid uncombined components. These laws, their region or validity and the nature of the exceptions to them are well known. (2) The relation (359) states that C_V and $\overline{E_K}$ vary as T^3 and T^4 at low temperatures, with deviations of at most 2 per cent. so long as $T/\Theta < 1/10$. (3) The general form of (356) shows that C_V obeys a law of "corresponding states", being a function of the single variable Θ/T. (4) Equation (348) and the following definition show that Θ can be calculated from the velocities of sound and so from the elastic constants of the crystal. These four deductions are borne out by experiment. The nature of the agreement will now be examined more closely for the last three.*

TABLE 6.

The T^3-law for C_V at low temperatures.

Temp. °K.	C_V	$10^2 C_V^{\frac{1}{3}}/T$
Copper (Cu)		
14·51	0·0390	2·35
15·60	0·0506	2·37
17·50	0·0726	2·39
18·89	0·0930	2·40
20·20	0·1155	2·42
21·50	0·1410	2·42
23·5	0·22	2·57
25·37	0·234	2·43
27·7	0·32	2·47
Iron pyrites (FeS_2)		
27·5	0·1095	1·75
29·8	0·1385	1·74
32·9	0·179	1·72
35·8	0·232	1·72
38·3	0·295	1·74
42·2	0·402	1·75
46·7	0·530	1·74
51·7	0·712	1·74
54·7	0·844	1·74
56·9	0·952	1·73

Temp. °K.	C_V	$10^2 C_V^{\frac{1}{3}}/T$
Iron (Fe)		
32·0	0·152	1·67
33·1	0·177	1·70
35·2	0·244	1·77
38·1	0·288	1·73
42·0	0·325	1·64
46·9	0·522	1·71
Aluminium (Al)		
19·1	0·066	2·12
23·6	0·110	2·03
27·2	0·162	2·01
32·4	0·25	1·95
33·5	0·301	2·00
35·1	0·33	1·97
Beryl (BeO)		
76·8	0·202	0·765
78·1	0·219	0·773
79·3	0·226	0·769
80·3	0·223	0·756
82·6	0·236	0·750
84·9	0·274	0·766

The T^3-law at low temperatures has been accurately verified for elements and simple compounds. Typical examples are shown in Table 6. The figures of the third column should be constant when the T^3-law holds.

The law of corresponding states has also been found true for elements and simple compounds. Fig. 10† shows the observed values of C_V for eighteen

* The following statements of fact are based on Schrödinger, *Physikal. Zeit.* vol. 20, pp. 420, 450, 474, 497, 523 (1919), except where otherwise stated. Full references to the original literature will be found in this paper.

† Schrödinger, *loc. cit.*

substances plotted as functions of T/Θ, the value of Θ for each substance being chosen to give the best fit for that substance with the continuous curve which represents equation (356). The agreement is eminently satisfactory. The substances and temperature ranges are specified in Table 7 below.

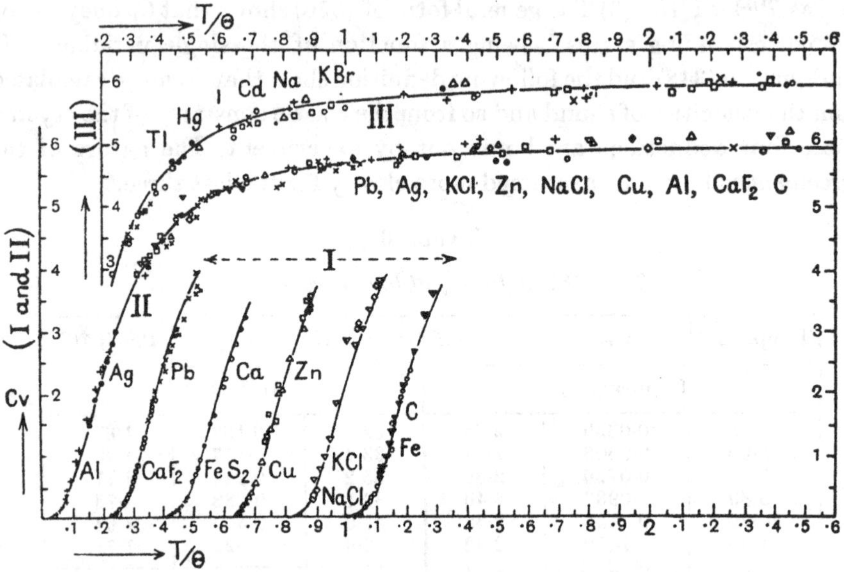

Fig. 10 The specific heats of various solids as functions of T/Θ.

TABLE 7.

Data for Fig. 10, *studying the law of corresponding states.*

Substance	Chemical symbol	Temperature range °K.	Θ	Points in Fig. 10 Curves I	II	III
Lead	Pb	14–573	88	×	×	
Thallium	Tl	23–301	96			○
Mercury	Hg	31–232	97			□
Iodine	I	22–298	106			×
Cadmium	Cd	50–380	168			+
Sodium	Na	50–240*	172			△
Potassium bromide	KBr	79–417	177			●
Silver	Ag	35–873	215		●	
Calcium	Ca	22–62	226	○		
Sylvine	KCl	23–550	230	▽	▽	
Zinc	Zn	33–673	235	□	□	
Rocksalt	NaCl	25–664	281	◊	◊	
Copper	Cu	14–773*	315	△	△	
Aluminium	Al	19–773	398		+	
Iron	Fe	32–95*	453	○		
Fluorspar	CaF_2	17–328	474	○	○	
Iron pyrites	FeS_2	22–57*	645	+		
Diamond	C	30–1169	1860	▾	▾	

* For Na, Cu, Fe, FeS_2, C_V rises *above* the curve after these temperatures.

The calculation of Θ from the elastic constants can be carried through very simply for isotropic solids and the value so calculated compared with the Θ of Table 7 which is chosen to give the best fit in Fig. 10. The elastic properties of an isotropic solid can be expressed in terms of two constants, the compressibility κ and Poisson's ratio σ. In terms of these the velocities of sound are*

$$c_l^2 = \frac{3(1-\sigma)}{(1+\sigma)\kappa\rho}, \quad c_t^2 = \frac{3(1-2\sigma)}{2(1+\sigma)\kappa\rho}, \qquad \text{......(360)}$$

where ρ is the density. Equation (344) shows that the required function is

$$\chi(\sigma)(\kappa\rho)^{\frac{3}{2}}, \quad \chi(\sigma) = \left\{\frac{1+\sigma}{3(1-\sigma)}\right\}^{\frac{3}{2}} + 2\left\{\frac{2(1+\sigma)}{3(1-2\sigma)}\right\}^{\frac{3}{2}}, \qquad \text{......(361)}$$

so that the theoretical mean value of Θ for use in Debye's formula is given by

$$\frac{1}{\Theta^3} = \tfrac{1}{3}\Sigma\frac{1}{\Theta_j^3} = \left(\frac{k}{h}\right)^3 \frac{4\pi V}{9N}\Sigma\frac{1}{c_j^3} = \left(\frac{k}{h}\right)^3 \frac{4\pi V}{9N}\chi(\sigma)(\kappa\rho)^{\frac{3}{2}}.$$

If V represents the volume of one gram-atom or gram-molecule, then $V\rho = M$, where M is the molecular weight, and

$$\Theta = \left(\frac{9N}{4\pi}\right)^{\frac{1}{3}} \frac{h}{k} \frac{V^{\frac{1}{6}}}{(M\kappa)^{\frac{1}{2}}\{\chi(\sigma)\}^{\frac{1}{3}}}, \qquad \text{......(362)}$$

$$= 3{\cdot}605 \times 10^{-3} \frac{V^{\frac{1}{6}}}{(M\kappa)^{\frac{1}{2}}\{\chi(\sigma)\}^{\frac{1}{3}}}. \qquad \text{......(363)}$$

The following examples are given by Born:†

TABLE 8.

Comparison of the values of Θ from specific heat data and from direct calculation from the elastic constants.

Substance	ρ	$\kappa \times 10^{12}$	σ	$\chi(\sigma)$	Θ Table 7	Θ Eq. (363)
Al	2·71	1·36	0·337	10·2	398	402
Cu	8·96	0·74	0·334	10·5	315	332
Ag	10·53	0·92	0·379	15·4	215	214
Pb	11·32	2·0	0·446	61·0	88	73

The agreement is excellent. There is, however, a difficulty that the calculations of Θ have been carried out with the values of the elastic constants determined at ordinary temperatures, whereas it has been argued‡ that it is the values of these constants at the absolute zero which are relevant, and the use of these would seriously damage the agreement. This objection, however, does not appear to me to be entirely valid. The elastic constants

* Love, *Theory of Elasticity*, ed. 3, p. 301.
† Born, *loc. cit.* p. 643.
‡ Eucken, *Verh. d. Deutsch. physikal. Ges.* vol. 15, p. 571 (1913).

and Θ are of course functions of V for an actual crystal, and the values required for these calculations are those corresponding to the actual volume of the crystal. The theory proceeds by imagining the crystal units at rest in equilibrium in their mean positions corresponding to a given crystal volume; this may involve large negative pressures over the boundary, which are theoretically unobjectionable. On this is superposed without change of volume (but with change of pressure) the heat motion corresponding to any given temperature, and at this temperature the observed C_V and the C_V calculated from the elastic constants for this volume should agree.* The observed values of C_V do not in fact therefore correspond to a sequence of C_V's at a single constant volume, the volume at absolute zero, but to a sequence of varying volumes which are the actual volumes of the crystal at each stage at atmospheric (or other small) pressure. These are compared with C_V, calculated for a genuinely constant volume, the volume at normal temperatures. The agreement in the Θ will largely depend on how the various parts of the theoretical curve are weighted in fitting it to the observations. There is a tendency to fit so that the deviations are of roughly constant absolute magnitude all along the curve. This will weight heavily the normal temperatures for which C_V is largest and would go some way to explain the satisfactory agreement in the table.

There is, however, a more precise and important point of divergence between theory and experiment, for the values of Θ derived from the T^3-law at low temperatures, equation (359), *should* agree with the values calculated from the elastic constants at low temperatures. There is some tendency for the T^3-law Θ's to be larger than those derived from the whole curve, but the increase is far smaller than the elastic constants require, and the matter remains not fully cleared up.

TABLE 9.

Comparisons of Θ derived by various methods.

Substance	C	Fe	Al	Cu	Ag
Θ from C-curve	1860	453	398	315	215
Θ from T^3-law	2230	455	385	321	—
Θ from elastic constants at 290° K.	—	—	402	332	214
Θ ditto, at low temperatures	—	—	488	344	235

§4·7. *Applications of formulae for C_V more accurate than Debye's.* Apart altogether from the basic approximations made in deriving (346), Debye's formula for C_V is admittedly derived from (346) by crude approximations, and should only hold at all strictly for isotropic bodies. With the same basic

* It is tacitly assumed in the theory that the elastic constants depend *directly* only on the volume.

approximations more accurate calculations are possible in the general case. Such calculations have been successfully carried out by Försterling.† When $|\xi| < 2\pi$ the function $\xi/(e^\xi - 1)$ can be expanded in the convergent power series

$$\frac{\xi}{e^\xi - 1} = 1 - \tfrac{1}{2}\xi - \sum_{n=1}^{\infty} (-1)^n \frac{B_n}{(2n)!}\xi^{2n}, \qquad \text{......(364)}$$

where the B_n are Bernoulli's numbers. If we now return to the original form of $\log K(z)$ in (346) we see that the part giving rise to Debye's terms gives an exact contribution to $\overline{E_K}$ of the form

$$Vh \sum_{j=1}^{3} \int \frac{d\omega}{c_j^3} \int_0^{\nu_j^*} \frac{\nu^3 d\nu}{e^{h\nu/kT} - 1}.$$

If $\Theta_j = h\nu_j^*/k$ as before, but now a function of direction, this becomes

$$kTV\left(\frac{kT}{h}\right)^3 \sum_{j=1}^{3} \int \frac{d\omega}{c_j^3} \int_0^{\Theta_j/T} \frac{\xi^3 d\xi}{e^\xi - 1}.$$

If we expand by (364) and differentiate to obtain C_V we find a more exact contribution

$$kV\left(\frac{k}{h}\right)^3 \sum_{j=1}^{3} \int \frac{d\omega}{c_j^3}\left[\frac{\Theta_j^3}{3} + \sum_1^\infty \frac{(-1)^n B_n (2n-1)}{(2n)!\,(2n+3)} \frac{\Theta_j^{2n+3}}{T^{2n}}\right].$$

Let us now assume that $\nu_j^* \propto c_j$. Then

$$\nu_j^* = c_j\left(\frac{3N}{4\pi V}\right)^{\frac{1}{3}}, \quad \Theta_j = \frac{h}{k} c_j\left(\frac{3N}{4\pi V}\right)^{\frac{1}{3}}.$$

We write

$$\sum_{j=1}^{3} \Theta_j^{2n} = \left\{\frac{h}{k}\left(\frac{3N}{4\pi V}\right)^{\frac{1}{3}}\right\}^{2n} \frac{1}{\rho^n} \sum_{j=1}^{3} \rho^n c_j^{2n}, \qquad \text{......(365)}$$

$$\gamma = \frac{1}{\rho}\frac{h^2}{k^2}\left(\frac{3N}{4\pi V}\right)^{\frac{2}{3}}, \quad \int\left(\sum_{j=1}^{3} \Theta_j^{2n}\right) d\omega = 4\pi\gamma^n \mathrm{K}_n. \qquad \text{......(366)}$$

Then the contribution to C_V is

$$3Nk\left\{1 + \sum_1^\infty (-1)^n \frac{B_n(2n-1)}{(2n)!\,(2n+3)} \frac{\gamma^n \mathrm{K}_n}{T^{2n}}\right\}. \qquad \text{......(367)}$$

Formula (367) replaces Debye's term in C_V. The integrands of the coefficients K_n can be shown to be rational functions of the measurable elastic constants of the crystal, so that the K_n can be calculated. The full formula for C_V is then

$$C_V = 3Nk\left\{1 + \sum_1^\infty (-1)^n \frac{B_n(2n-1)}{(2n)!\,(2n+3)} \frac{\gamma^n \mathrm{K}_n}{T^{2n}}\right\} + Nk \sum_{j=4}^{3p} \frac{(\Theta_j/T)^2 e^{\Theta_j/T}}{(e^{\Theta_j/T} - 1)^2}, \qquad \text{...(368)}$$

valid when $T > \Theta/2\pi$.

Försterling has used this formula in the most accurate comparison of

† Försterling, *Ann. d. Physik* (4), vol. 61, p. 577 (1920); *Zeit. f. Physik*, vol. 3, p. 9 (1920), vol. 8, p. 251 (1922).

theoretical and observed specific heats yet attempted on the basis of (346). Having calculated the first term entirely from elastic data, the correct number of extra terms of Einstein's type are introduced corresponding to the known lattice structure. The number of *different* Θ_j's allowable is also known from the structure and symmetry. These are then fixed to give the best fit possible between the observed and theoretical C_V, and the whole theory is checked by comparing these Θ_j's with the natural frequencies of the crystal determined by the method of *Reststrahlen*. Excellent representations of C_V are possible among other substances for NaCl, KCl, KBr, CaF_2 and SiO_2, and typical comparisons of the wave lengths of the natural frequencies derived from specific heats and optical measurements (*Reststrahlen*) are shown in the following table.

TABLE 10.

Comparisons of wave lengths of Reststrahlen *directly measured and deduced from specific heat curves.*

Substance	λ_0 optical 10^{-4} cm.	λ_0 from C_V
NaCl	66·7	64·5
KCl	78·0	77·0
CaF_2	53·1	51·0, 34·7*

* Optically inactive.

§4·71. *Blackman's discussion of Debye's theory for a simple crystal. Introductory.* The vital assumptions at the base of Debye's theory and all its refinements which we have indicated in the preceding sections concern the acoustical modes, and can be sufficiently illustrated by further examination of the simplest possible lattice, the simple cubic, with one atom per unit cell. Equation (353) should then be most nearly valid. In deriving it the vital approximation is that the normal modes of the system (with N unit cells in the lattice) have the frequency distribution of the continuum for $\nu < \bar{\nu}$, so that there are

$$\frac{9N}{\bar{\nu}^3}\nu^2 d\nu \qquad (\nu < \bar{\nu}) \qquad \text{......(368·1)}$$

modes with frequencies in ν, $\nu + d\nu$ for $\nu < \bar{\nu}$ and none for $\nu > \bar{\nu}$. This assumption was based on (339) and is certainly correct in form for small ν. It has been used, as we have seen in Fig. 10 and Tables 6–9, with great success right up to $\nu = \bar{\nu}$. If, however, we are more cautious and do not rely so implicitly on (339) we can merely assert that there will be some distribution law for the frequencies for a crystal of N cells, namely

$$g(\nu)\,d\nu$$

and that as $\nu \to 0$, $g(\nu) \sim \alpha\nu^2$. The form of $g(\nu)$ awaits more exact examination. We have thus replaced (353) by

$$\log K(T) = -\frac{F_0}{kT} - \int_0^{\bar{\nu}} g(\nu) \log(1 - e^{-h\nu/kT})\, d\nu. \quad \text{......(368·11)}$$

Though we cannot explicitly evaluate $g(\nu)$, evidence both theoretical and observational may be adduced to show that (368·1) is sometimes an inadequate approximation. An excellent example of a simple substance with a C_V which departs markedly from Debye's curve, most probably because (368·1) fails, is metallic lithium. The most accurate way of comparing obser-

TABLE 10·1.

The specific heat of lithium in cal./deg./gram atom, and the calculated values of Debye's Θ.

<table>
<tr><th>Temp. °K.</th><th>C_p</th><th>C_V</th><th>$\Theta\,(=h\bar{\nu}/k)$</th></tr>
<tr><td>15</td><td colspan="2">0·045</td><td>328</td></tr>
<tr><td>20</td><td colspan="2">0·095</td><td>340</td></tr>
<tr><td>25</td><td colspan="2">0·169</td><td>350</td></tr>
<tr><td>30</td><td colspan="2">0·273</td><td>356</td></tr>
<tr><td>35</td><td colspan="2">0·413</td><td>362</td></tr>
<tr><td>40</td><td colspan="2">0·573</td><td>367</td></tr>
<tr><td>45</td><td colspan="2">0·770</td><td>370</td></tr>
<tr><td>50</td><td>0·996</td><td>0·995</td><td>374</td></tr>
<tr><td>60</td><td>1·43</td><td>1·42</td><td>379</td></tr>
<tr><td>70</td><td>1·88</td><td>1·87</td><td>384</td></tr>
<tr><td>80</td><td>2·32</td><td>2·31</td><td>388</td></tr>
<tr><td>90</td><td>2·69</td><td>2·67</td><td>393</td></tr>
<tr><td>100</td><td>3·05</td><td>3·02</td><td>397</td></tr>
<tr><td>110</td><td>3·36</td><td>3·32</td><td>401</td></tr>
<tr><td>120</td><td>3·64</td><td>3·59</td><td>405</td></tr>
<tr><td>130</td><td>3·87</td><td>3·81</td><td>408</td></tr>
<tr><td>140</td><td>4·08</td><td>4·01</td><td>412</td></tr>
<tr><td>150</td><td>4·26</td><td>4·18</td><td>415</td></tr>
<tr><td>160</td><td>4·43</td><td>4·34</td><td>418</td></tr>
<tr><td>180</td><td>4·68</td><td>4·57</td><td>426</td></tr>
<tr><td>200</td><td>4·92</td><td>4·78</td><td>433</td></tr>
<tr><td>220</td><td>5·15</td><td>4·94</td><td>435</td></tr>
<tr><td>240</td><td>5·28</td><td>5·09</td><td>431</td></tr>
<tr><td>260</td><td>5·44</td><td>5·22</td><td>429</td></tr>
<tr><td>280</td><td>5·56</td><td>5·31</td><td>429</td></tr>
<tr><td>300</td><td>5·66</td><td>5·39</td><td>430</td></tr>
</table>

vations of C_V with Debye's theory is to calculate for each temperature what value of Θ will reproduce theoretically the observed C_V. Table 10·1 shows the measurements and analysis of Simon and Swain* for lithium, and it will be observed that Θ so calculated from the observations is very far from constant below 180° K.

Before one concludes that the figures of Table 10·1 indicate a failure of (368·1), one must examine alternative sources of failure for Debye's curve.

* Simon and Swain, *Zeit. f. physikal. Chem.* B, vol. 28, p. 189 (1935).

There are two known possibilities: (i) The electrons may contribute effectively to the specific heat especially at very low temperatures in such a way as to make the specific heat larger than is calculated from Debye's theory so that the derived Θ will diminish as T diminishes. This effect is examined in detail in § 11·55 and applied successfully to explain certain features of the C_V-curve for nickel. It appears that it cannot be responsible for the large effect here. (ii) The lithium ions in the lattice might have two electronic states of small energy difference, but otherwise so alike in their properties that the corpus of normal modes for the lattice is independent of the distribution of the ions over their pair of states, and this distribution is itself unaffected by the behaviour of neighbours so that it is identical with the similar distribution of gaseous atoms over a pair of states. The variation of this distribution will therefore make the contribution (283) to C_V. It has been shown by Simon and Swain that such a contribution with $\varpi_1/\varpi_2 = 1$, $\Delta\epsilon/k = 200$ when subtracted from the observed specific heat leaves behind a specific heat which is admirably represented by a Debye curve with $\Theta = 510$. The objection to this explanation is that there is no reason to believe that the simple lithium ion can possibly exist in the lattice with two such states with this energy difference and in spite of its empirical success the explanation must be discarded. We are driven back for an explanation on the probable failure of (368·1).

§ 4·72. *The normal modes of a linear lattice.* The further study of $g(\nu)$ may be introduced by starting with a linear lattice. We consider therefore a chain of $2N$ atoms, each of which may be displaced along the chain. Let the displacement of the nth atom be u_n and so small that only quadratic terms need be retained in the potential energy. We may further assume (merely for simplicity) that the potential energy due to the displacements depends only on the relative displacements of nearest neighbours. Then for such a system the total energy E is given by

$$E = \tfrac{1}{2}m \sum_{1}^{2N} \dot{u}_n^2 + \tfrac{1}{2}\alpha \sum_{0}^{2N} (u_{n+1} - u_n)^2. \qquad \text{......(368·2)}$$

This expression contains displacement terms for atoms 0 and $2N+1$ which do not belong to the system. It is simplest to include these and to impose as boundary condition a condition of periodicity that $u_n = u_{n+2N}$. The precise form of boundary condition imposed has no ultimate effect on the distribution of the normal modes.

Since we are concerned with simple harmonic oscillations classical equations of motion will suffice. These follow at once from the energy (368·2) and are

$$m\ddot{u}_n + \alpha(2u_n - u_{n+1} - u_{n-1}) = 0 \quad (n = 1, 2, \ldots, 2N). \qquad \text{......(368·21)}$$

Conforming to the periodicity condition let us now attempt a solution of equations (368·21) of the form

$$u_n = u' e^{i(2\pi\nu t + n\pi p/N)}, \qquad \text{......(368·22)}$$

where u' is a constant and p is an integer satisfying $-N < p \leqslant N$. There are exactly $2N$ distinct expressions of this type. The equations of motion then become

$$-4\pi^2\nu^2 m e^{i(2\pi\nu t + n\pi p/N)} + \alpha e^{i(2\pi\nu t + n\pi p/N)}(2 - e^{i\pi p/N} - e^{-i\pi p/N}) = 0$$

and are therefore all satisfied if

$$\nu^2 = \frac{\alpha}{\pi^2 m}\sin^2\frac{\pi p}{2N} = \nu_0{}^2 \sin^2\frac{\pi p}{2N}. \qquad \text{......(368·23)}$$

The frequencies are equal in pairs for equal positive and negative values of p. If we allow for this and write dN for the number of frequencies in the range ν, $\nu + d\nu$, we find

$$dN = \frac{4N}{\pi\nu_0}\frac{d\nu}{(1-\nu^2/\nu_0{}^2)^{\frac{1}{2}}}. \qquad \text{......(368·24)}$$

The velocity of sound along the chain is of course given by the equation $c = \lambda\nu$ for long wave lengths. It is easily verified from (368·22) and (368·23) that $c = \pi\nu_0 d$, where d is the lattice spacing. Thus alternatively

$$dN = \frac{4Nd}{c}\frac{d\nu}{(1-\nu^2/\nu_0{}^2)^{\frac{1}{2}}} = \frac{2L}{c}\frac{d\nu}{(1-\nu^2/\nu_0{}^2)^{\frac{1}{2}}}.$$

The normal modes of the one-dimensional continuum, or the modes of an actual continuum restricted to plane waves normal to a given axis, have a uniform distribution $dN = 2L\,d\nu/c$ which agrees with the distribution for the lattice until ν approaches ν_0. The $g(\nu)$ for the lattice does deviate markedly from that for a continuum as $\nu \to \nu_0$ and has there an infinite maximum, but no subsidiary maxima for smaller values of ν. No exact calculations have been made but it seems likely that specific heat curves based on (368·24) and on $dN = 2L\,d\nu/c$ restricted to $2N$ modes will not possess any striking differences.

§ 4·73. ***The normal modes of simple square and cubic lattices.*** In more than one dimension the $g(\nu)$ can be more exciting. Let us consider a two-dimensional square lattice of spacing d with $2N$ atoms in each row and column. Let us specify the atoms by the suffixes l, m $(1 \leqslant l, m \leqslant 2N)$ and denote their displacements along the rows and columns of the array by $u_{l,m}$, $v_{l,m}$ respectively. If now we assume that only the interactions between nearest neighbours are sensible, the oscillations along rows and columns separate and all the modes reduce to a double set of modes for a linear lattice. We only retain the essential nature of the two-dimensional array if we include interactions between next nearest neighbours as well, that is neighbours along the diagonals. The arrangement of the displacements is illustrated in Fig. 10·1. If the force constant for relative displacements of neighbours along the square edges is α and along the square diagonals is 2γ, we can at once write down the total energy for small oscillations in the form†

$$E = \tfrac{1}{2}m^* \Sigma_{l,m}\, \dot{u}^2_{l,m} + \dot{v}^2_{l,m} + \tfrac{1}{2}\alpha\, \Sigma_{l,m}\, (u_{l,m} - u_{l+1,m})^2$$
$$+ \tfrac{1}{2}\alpha\, \Sigma_{l,m}\, (v_{l,m} - v_{l,m+1})^2$$
$$+ \tfrac{1}{2}\gamma\, \Sigma_{l,m}\, (u_{l,m} + v_{l,m} - u_{l+1,m+1} - v_{l+1,m+1})^2$$
$$+ \tfrac{1}{2}\gamma\, \Sigma_{l,m}\, (u_{l,m} - v_{l,m} - u_{l+1,m-1} + v_{l+1,m-1})^2. \qquad \text{......(368·3)}$$

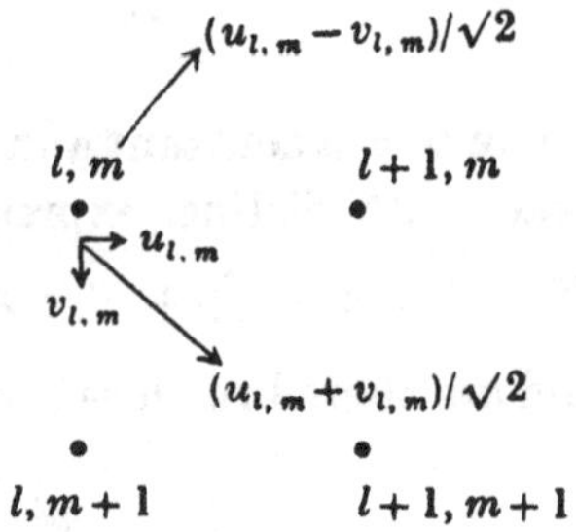

Fig. 10·1. Displacements of atoms in the oscillations of a square lattice.

This leads at once to the equations of motion

$$m^*\ddot{u}_{l,m} + \alpha(2u_{l,m} - u_{l+1,m} - u_{l-1,m}) + \gamma\{(u_{l,m} + v_{l,m} - u_{l+1,m+1} - v_{l+1,m+1})$$
$$+ (u_{l,m} + v_{l,m} - u_{l-1,m-1} - v_{l-1,m-1}) + (u_{l,m} - v_{l,m} - u_{l+1,m-1} + v_{l+1,m-1})$$
$$+ (u_{l,m} - v_{l,m} - u_{l-1,m+1} + v_{l-1,m+1})\} = 0, \qquad \text{......(368·31)}$$

$$m^*\ddot{v}_{l,m} + \alpha(2v_{l,m} - v_{l,m+1} - v_{l,m-1}) + \gamma\{(u_{l,m} + v_{l,m} - u_{l+1,m+1} - v_{l+1,m+1})$$
$$+ (u_{l,m} + v_{l,m} - u_{l-1,m-1} - v_{l-1,m-1}) + (v_{l,m} - u_{l,m} - v_{l+1,m-1} + u_{l+1,m-1})$$
$$+ (v_{l,m} - u_{l,m} - v_{l-1,m+1} + u_{l-1,m+1})\} = 0. \qquad \text{......(368·32)}$$

We can now attempt to find a solution of these equations with the periodicity conditions that both u and v are periodic in both l and m with period $2N$ by using the form

$$\left.\begin{aligned} u_{l,m} &= u' e^{i(2\pi\nu t + l\pi p/N + m\pi q/N)} \\ v_{l,m} &= v' e^{i(2\pi\nu t + l\pi p/N + m\pi q/N)} \end{aligned}\right\}, \qquad \text{......(368·33)}$$

where u' and v' are constants and p and q are independent integers satisfying $-N < p,\ q \leqslant N$. There are therefore $4N^2$ of these distinct forms. After an easy reduction it is found that all the equations of motion are satisfied if

$$\left.\begin{aligned} &\left[-\pi^2\nu^2 m^* + \alpha \sin^2\frac{\pi p}{2N} + \gamma\left(1 - \cos\frac{\pi p}{N}\cos\frac{\pi q}{N}\right)\right]u' + \left[\gamma \sin\frac{\pi p}{N}\sin\frac{\pi q}{N}\right]v' = 0 \\ &\left[\gamma \sin\frac{\pi p}{N}\sin\frac{\pi q}{N}\right]u' + \left[-\pi^2\nu^2 m^* + \alpha \sin^2\frac{\pi q}{2N} + \gamma\left(1 - \cos\frac{\pi p}{N}\cos\frac{\pi q}{N}\right)\right]v' = 0 \end{aligned}\right\}. \qquad \text{......(368·34)}$$

The period equation is therefore

$$\begin{vmatrix} -\pi^2\nu^2 m^* + \alpha \sin^2\frac{\pi p}{2N} + \gamma\left(1 - \cos\frac{\pi p}{N}\cos\frac{\pi q}{N}\right) & \gamma \sin\frac{\pi p}{N}\sin\frac{\pi q}{N} \\ \gamma \sin\frac{\pi p}{N}\sin\frac{\pi q}{N} & -\pi^2\nu^2 m^* + \alpha \sin^2\frac{\pi q}{2N} + \gamma\left(1 - \cos\frac{\pi p}{N}\cos\frac{\pi q}{N}\right) \end{vmatrix} = 0. \qquad \text{......(368·35)}$$

† The mass of the atoms has here been starred to distinguish it from the location parameter.

There are two roots for each given pair (p,q) so that there are $8N^2$ different normal modes in all, which is the correct number.

This determinantal equation for ν verifies at once for this simple case that the modes are uniformly distributed over (p, q)-space or, if

$$\phi = \pi p/N, \quad \psi = \pi q/N,$$

over (ϕ,ψ)-space, covering the square $-\pi < \phi, \psi \leqslant \pi$. This is a particular case of the theorem quoted in § 4·5. To determine $g(\nu)$ we have only to determine how much of (ϕ,ψ)-space corresponds to values of ν between ν and $\nu + d\nu$ for each of the two branches of roots of equation (368·35). There are N^2/π^2 roots per unit area of (ϕ,ψ)-space for each branch, but since the frequencies do not depend on the signs of ϕ and ψ we may confine attention to the positive quadrant, $0 \leqslant \phi, \psi \leqslant \pi$, and take $4N^2/\pi^2$ roots per unit area of that quadrant.

Even in this simple case the function $g(\nu)$ can scarcely be determined except by numerical computation, but certain of its properties can be derived at once analytically. Regarding ν as a function of ϕ and ψ we see at once that provided $\nu \neq 0$ $\partial\nu/\partial\phi$ vanishes if $\sin\phi = 0$ and $\partial\nu/\partial\psi$ vanishes if $\sin\psi = 0$. Therefore $\partial\nu/\partial\phi = \partial\nu/\partial\psi = 0$ whenever $\nu \neq 0$, $\sin\phi = \sin\psi = 0$, that is at the points

$$\phi, \psi = 0, \pi; \ \pi, 0; \ \pi, \pi.$$

[It is not excluded that there might be other stationary values of ν besides these.] At any such stationary point $\nu = \nu^*$ $g(\nu)$ must strictly speaking have an infinity since the area of (ϕ,ψ)-space enclosed between the curves of constant ν, ν' and ν^* tends to zero as $\nu' \to \nu^*$ at least as slowly as $|\nu' - \nu^*|^{\frac{1}{2}}$. The $g(\nu)$ curve has therefore in general more than one sharp maximum on it. These maxima are strictly infinities, but an infinity as such is not of great effect on C_V or on the other equilibrium properties which are all of the form $\int f g(\nu)\, d\nu$. It is rather the number of modes concentrated over a "physically small" range of ν than over the infinitesimal $d\nu$ which is significant.

When ϕ, ψ are both small (long waves) equation (368·35) factorizes and the two roots in ν^2 are approximately

$$\nu^2 = \frac{1}{4\pi^2 m^*}\,[(\alpha + 2\gamma)\,\phi^2 + 2\gamma\psi^2], \quad \frac{1}{4\pi^2 m^*}\,[2\gamma\phi^2 + (\alpha + 2\gamma)\,\psi^2]. \qquad \text{......(368·36)}$$

For each root the region of (ϕ,ψ)-space for which $\nu < \nu'$ is the positive quadrant of an ellipse—a quadrant of area

$$\frac{\pi^3 \nu'^2 m^*}{\sqrt{\{(\alpha + 2\gamma)\, 2\gamma\}}}.$$

The number of modes of both branches with frequencies between ν and $\nu + d\nu$ for small ν is therefore

$$\frac{16\pi N^2 m^*}{\sqrt{\{(\alpha + 2\gamma)\, 2\gamma\}}}\,\nu\, d\nu. \qquad \text{......(368·37)}$$

When $\psi = \pi - \psi'$ and ϕ and ψ' are small, the two roots are approximately

$$\nu^2 = \frac{1}{\pi^2 m^*}[(\alpha + 2\gamma) - \tfrac{1}{2}\gamma\phi^2 - \tfrac{1}{8}(\alpha + 4\gamma)\psi'^2], \quad \frac{1}{\pi^2 m^*}[2\gamma + \tfrac{1}{4}(\alpha - 2\gamma)\phi^2 - \tfrac{1}{2}\gamma\psi'^2]. \qquad \text{......(368·38)}$$

There are similar values for $\phi = \pi - \phi'$ and ϕ' and ψ both small. The frequency $\nu^2 = (\alpha + 2\gamma)/\pi^2 m^*$ is the greatest of the set. When $\phi = \pi - \phi'$, $\psi = \pi - \psi'$ and ϕ' and ψ' are both small, the two roots are approximately

$$\nu^2 = \frac{1}{\pi^2 m^*}[\alpha - \tfrac{1}{8}(\alpha - 4\gamma)\phi'^2 + \tfrac{1}{2}\gamma\psi'^2], \quad \frac{1}{\pi^2 m^*}[\alpha + \tfrac{1}{2}\gamma\phi'^2 - \tfrac{1}{8}(\alpha - 4\gamma)\psi'^2]. \qquad \text{......(368·39)}$$

We can now see clearly how the actual $g(\nu)$ can deviate widely from the $g(\nu)$ for a two-dimensional continuum which will of course be given by (368·37) for all ν if the elastic constants of the continuum are properly chosen. The distribution of frequencies for low frequencies and for high for the actual lattice are governed by two unrelated parameters while the distribution for a continuum is a one-parameter distribution. This is most easily seen if γ/α is fairly small. The low frequency distribution is then rather dense and controlled by the factor $1/\gamma^{\frac{1}{2}}$. The high frequency distribution on the other hand is given by the greater root of (368·38) and also by either root of (368·39) which here all coincide, neglecting γ. This distribution depends only on α and is effectively the same as for a linear lattice with the same force constant α when allowance is made for the extra degrees of freedom. A one-parameter curve such as Debye's cannot possibly therefore always represent effectively the actual two-parameter curve. The fact that Debye's curve is often so good in practice must be accounted as a happy accident.

By numerical calculations Blackman has constructed an approximate $g(\nu)$ curve for this square lattice when $\gamma/\alpha = 0{\cdot}05$. The contributions of the two roots of (368·35) are shown separately and also the total in Fig. 10·2. The infinities have been smoothed out by the numerical process adopted. Fig. 10·2 also shows the Debye distributions ($g(\nu) \propto \nu$) which contain the same number of modes $8N^2$, and (i) the same maximum frequency, (ii) the same low frequency density. The divergences are marked, particularly on account of the subsidiary maximum in $g(\nu)$ besides the expected maximum for the maximum frequency.

The three-dimensional lattice is naturally more complicated though it can be handled by the same methods. It presents a still greater variety of possible subsidiary maxima in $g(\nu)$. After this glimpse of the possibilities afforded by the two-dimensional discussion it is unnecessary to analyse it further. If the corresponding Θ which will give the same specific heat

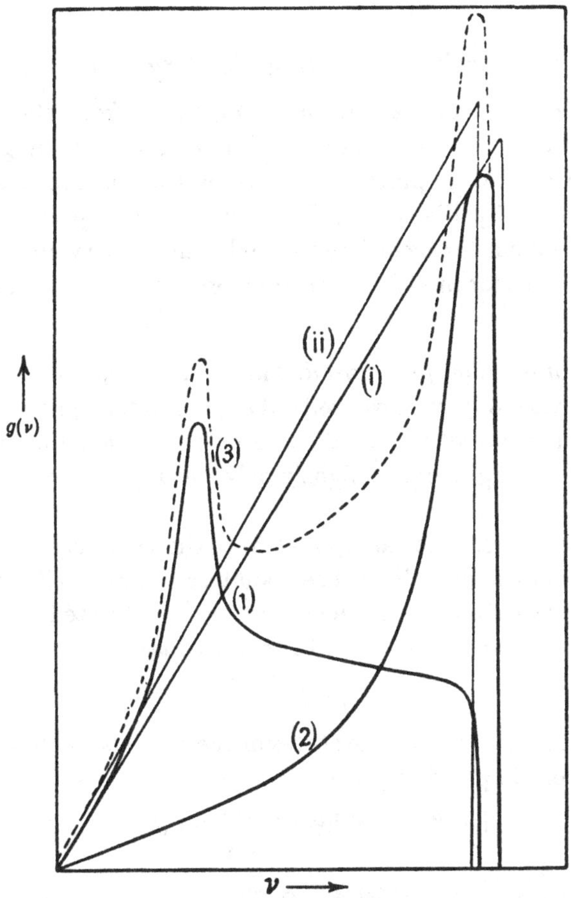

Fig. 10·2. Curves of $g(\nu)$ for a square lattice, after Blackman.

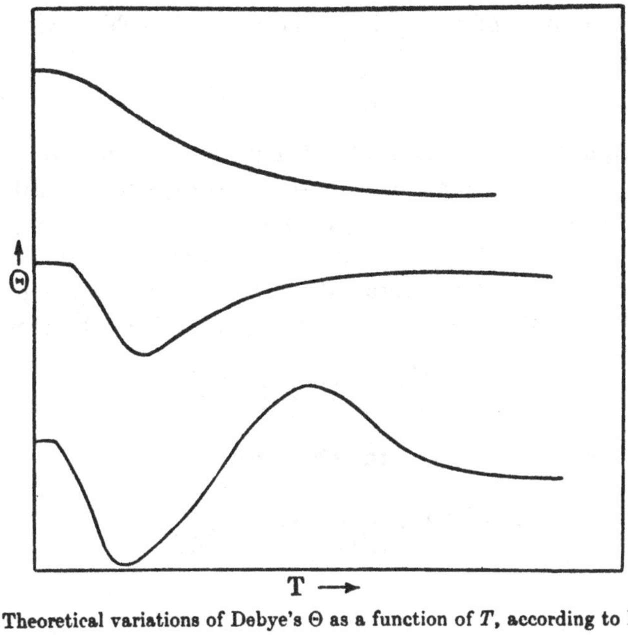

Fig. 10·3. Theoretical variations of Debye's Θ as a function of T, according to Blackman.

according to Debye's approximation is computed from the more exact theory as it was from observed values for Table 10·1, it will in general vary with the temperature. Typical variations of this nature are shown in Fig. 10·3 computed by Blackman. It is not, however, possible to discuss actual specific heat curves according to Blackman's theory until that theory has been worked out for models which are more genuinely representative of actual crystals.

§ **4·8.** *External reactions for a crystal.* The formulae of § 2·8 for the external reactions of the assembly were worked out explicitly for a gaseous assembly, but apply of course *mutatis mutandis* to a crystal. Since the most natural arrangement of the argument is slightly different here, we will give it again in full.

The crystal can be regarded as equivalent to the set of localized oscillators into which its movements have been analysed. The oscillators are not necessarily simple harmonic and need not be so restricted as in equation (345). The energy of any state (τ) of the crystal may be put in the form

$$F_0 + \Sigma_r \epsilon_r^{\tau}, \qquad \text{......(369)}$$

where the numbers (τ) specify the states of the various oscillators, and the partition function (in place of (345))

$$\log K(z) = F_0 \log z + \sum_{r=1}^{3sN} \log f_r(z), \qquad \text{......(370)}$$

where

$$f_r(z) = \Sigma_\tau \varpi_r^{\tau} z^{\epsilon_r^{\tau}}$$

is the partition function for the rth oscillator. In the state (369) the generalized reaction $\xi_t^{(\tau)}$ of the crystal on the external world is as in § 2·8 given by the equation

$$\xi_t^{(\tau)} = -\frac{\partial}{\partial x_t}[F_0 + \Sigma_r \epsilon_r^{\tau}]. \qquad \text{......(371)}$$

To find the equilibrium value $\overline{X_t}$ of this external reaction we could previously use $\overline{a_r^{\tau}}$, the average number of oscillators of type r in state τ. Here there is only one oscillator of each type, but $\overline{a_r^{\tau}}$, now a fraction, given by the old formula (80) with $M = 1$, still has the same physical meaning as before which is perhaps now more naturally described as the fraction of time that this oscillator spends in this state. It follows therefore just as before that

$$\overline{X_t} = -\frac{\partial F_0}{\partial x_t} + \Sigma_r \Sigma_\tau \overline{a_r^{\tau}} \left(-\frac{\partial \epsilon_r^{\tau}}{\partial x_t}\right), \qquad \text{......(372)}$$

which is equivalent to (232), and therefore that

$$\overline{X_t} = -\frac{\partial F_0}{\partial x_t} + \Sigma_r \frac{1}{f_r(\vartheta)} \Sigma_\tau \varpi_r^{\tau} \left(-\frac{\partial \epsilon_r^{\tau}}{\partial x_t}\right) \vartheta^{\epsilon_r^{\tau}}.$$

By comparison with (370) it is easy to see at once that this reduces to

$$\overline{X_t}=\frac{1}{\log 1/\vartheta}\frac{\partial}{\partial x_t}\log K(\vartheta). \qquad ......(373)$$

In terms of T (373) becomes

$$\overline{X_t}=kT\frac{\partial}{\partial x_t}\log K(T), \qquad ......(374)$$

and in particular

$$p=kT\frac{\partial}{\partial V}\log K(T). \qquad ......(375)$$

§4·81. *Equations of state for simple isotropic solids, deduced from Debye's partition function.* It is convenient to discuss at this stage other properties of solids which follow from Debye's form of the partition function. We consider in this section only the simplest case, in which with sufficient accuracy

$$\log K(T)=-\frac{F_0(V)}{kT}-9N\left\{\frac{T}{\Theta(V)}\right\}^3\int_0^{\Theta(V)/T}\log(1-e^{-x})\,x^2dx. \qquad ......(376)$$

In (376) we have shown Θ explicitly as a function of V; Θ must always vary with V according to the simple theory, unless $V^{\frac{1}{6}}\kappa^{-\frac{1}{2}}\{\chi(\sigma)\}^{-\frac{1}{3}}$ is independent of the volume. If the solid obeyed Hooke's law perfectly, so that for all displacements the stresses and strains were strictly proportional, this function would be independent of V, for Poisson's ratio σ would be an absolute constant, and the compressibility κ would be proportional to the linear dimensions. But this ideal case can never hold for actual solids, and a V-variation of Θ must be admitted.

By the general formula (375) it follows at once that

$$p=-F_0'(V)-3NkT\frac{\Theta'}{\Theta}D\left(\frac{\Theta}{T}\right). \qquad ......(377)$$

This is the *equation of state*. For many isotropic bodies it is sufficient to assume that

$$F_0(V)=-\frac{A}{V^{\frac{1}{3}m}}+\frac{B}{V^{\frac{1}{3}n}} \quad (n>m), \qquad ......(378)$$

where A and B are constants. This is the form of $F_0(V)$ when the atoms in a regular crystal act on one another with central forces consisting of a strong repulsive field of short range of potential β/r^n and a weaker attractive field of longer range of potential $-\alpha/r^m$, and the zero-point energy may be neglected. In the calculation of $F_0(V)$, by definition, the structural units are taken to be at rest in their mean positions. We cannot enter here into the calculation of A and B from the laws of force of individual atoms (see Chapter x). If V_0 is the natural volume of the solid at zero temperature and pressure, then

$$F_0'(V_0)=0. \qquad ......(379)$$

For the form (378) V_0 is therefore fixed in terms of the constants of the interatomic forces by the equation

$$\frac{mA}{V_0^{\frac{1}{3}m}} = \frac{nB}{V_0^{\frac{1}{3}n}}. \qquad \text{......(380)}$$

For small volume changes we may write

$$F_0'(V) = F_0''(V_0)(V - V_0),$$

$$= \frac{1}{9}\frac{n(n-m)B}{V_0^{\frac{1}{3}n+2}}(V - V_0), \qquad \text{......(381)}$$

and

$$F_0(V) = F_0(V_0) + \frac{1}{18}\frac{n(n-m)B}{V_0^{\frac{1}{3}n+2}}(V - V_0)^2, \qquad \text{......(382)}$$

$$F_0(V_0) = -\frac{n-m}{m}\frac{B}{V_0^{\frac{1}{3}n}}. \qquad \text{......(383)}$$

We may now introduce the usual coefficients of thermal expansion α and compressibility κ by the equations (definitions)

$$3\alpha = \frac{1}{V}\left(\frac{\partial V}{\partial T}\right)_p, \qquad \kappa = -\frac{1}{V}\left(\frac{\partial V}{\partial p}\right)_T. \qquad \text{......(384)}$$

Then

$$\frac{1}{\kappa V} = -\left(\frac{\partial p}{\partial V}\right)_T = F_0''(V) + 3NkT\frac{\partial}{\partial \Theta}\left\{\frac{\Theta'}{\Theta}D\left(\frac{\Theta}{T}\right)\right\}\Theta'. \qquad \text{......(385)}$$

At the absolute zero this reduces to

$$\kappa_0 = 1/V_0 F_0''(V),$$

and by (381) and (383) we have the relation

$$F_0(V_0) = -\frac{9V_0}{nm\kappa_0}. \qquad \text{......(386)}$$

This important relation between the compressibility and the lattice energy or heat of evaporation at zero temperature can be used to determine nm, or n when m is known as for an ionic lattice ($m = 1$). The actual comparisons are made with a calculated $F_0(V_0)$, after fixing B to give the right scale to the lattice. We shall take up the general question of the specification of interatomic forces in Chapter x.

On differentiating (377) with p constant we find that

$$\left[F_0''(V) + 3NkT\frac{\partial}{\partial \Theta}\left\{\frac{\Theta'}{\Theta}D\left(\frac{\Theta}{T}\right)\right\}\Theta'\right]\left(\frac{\partial V}{\partial T}\right)_p = -3Nk\frac{\Theta'}{\Theta}\left\{D\left(\frac{\Theta}{T}\right) - \frac{\Theta}{T}D'\left(\frac{\Theta}{T}\right)\right\},$$

$$= -\frac{\Theta'}{\Theta}C_V,$$

the latter reduction by (356). This can be reduced to the simple form

$$\frac{3\alpha}{\kappa} = \left\{-\frac{\Theta'}{\Theta}\right\}C_V. \qquad \text{......(387)}$$

We see at once that a body for which $\Theta' = 0$ should show no thermal expansion. Finally, from the thermodynamic relation

$$C_p - C_V = T\left(\frac{\partial p}{\partial T}\right)_V \left(\frac{\partial V}{\partial T}\right)_p, \qquad \ldots\ldots(388)$$

and from

$$\left(\frac{\partial p}{\partial T}\right)_V = \left\{-\frac{\Theta'}{\Theta}\right\} C_V = \frac{3\alpha}{\kappa}, \qquad \ldots\ldots(389)$$

obtained by differentiating (377) with V constant, we deduce

$$C_p - C_V = 9\alpha^2 TV/\kappa. \qquad \ldots\ldots(390)$$

This formula is of great practical importance in deducing C_V from observations of C_p. It is principally used in certain semi-empirical simplified forms.

These formulae appear to be in good agreement with observation for many solids. In particular (387) is satisfied if the value of $-V\Theta'/\Theta$ is about 2·3. A calculation of α and κ in terms of the interatomic forces would allow this value to be interpreted as fixing a relation between m and n.

§ **4·9.** *General equations of state of the general crystal. Specification of the unit cell.* We have hitherto, in discussing its equation of state, regarded a crystal as an isotropic solid, subjected only to changes of volume by isotropic (hydrostatic) pressures, which might be negative. The single geometrical variable necessary could then be the volume V as for a gas, and the single elastic constant the volume compressibility. This, however, is insufficient for even the simplest solid, since solids possess at least two independent elastic constants, Young's modulus and Poisson's ratio, and crystals of lower symmetry may possess many more, 21 in all. We are therefore led to re-formulate the calculations, to include the symmetry of the crystal structure and the general mechanical and perhaps electrical stresses to which a rigid body can be subjected.*

We start by specifying more closely than in § 4·5 the unit cell, by repetitions of which the crystal is constructed. The cell is of course not unique, but we suppose a definite choice has been made. The cell may then be taken to be a definite parallelepiped whose three concurrent edges are specified by the vectors $\boldsymbol{\delta}_1, \boldsymbol{\delta}_2, \boldsymbol{\delta}_3$. The volume of this cell is

$$\Delta = \delta^3 = \begin{vmatrix} \delta_{1x} & \delta_{1y} & \delta_{1z} \\ \delta_{2x} & \delta_{2y} & \delta_{2z} \\ \delta_{3x} & \delta_{3y} & \delta_{3z} \end{vmatrix}, \qquad \ldots\ldots(391)$$

and δ is called the lattice constant. Choosing any origin O in the basic cell, let $\mathbf{r}$ be the vector from O to any other point, usually to one of the other

* My knowledge of the atomic theory of crystals is derived almost entirely from Born, *Problems of Atomic Dynamics*, Cambridge, Mass. (1926), and "Atomtheorie des festen Zustandes", *loc. cit.*, *q.v.* The simple account, which I have tried to give here, not unnaturally follows Born's lectures closely.

atoms in the same cell. Then the vector distance from O to the congruent points in the other cells is written

$$\mathbf{r}+\mathbf{r}^l, \quad \mathbf{r}^l = l_1\boldsymbol{\delta}_1 + l_2\boldsymbol{\delta}_2 + l_3\boldsymbol{\delta}_3, \qquad \text{......(392)}$$

where l_1, l_2, l_3 are any integers positive, zero or negative. They are contracted to l and called the "cell-index".

The positions of the atoms or other units requiring separate consideration in the unit cell are denoted by the vectors from O,

$$\mathbf{r}_k \quad (k=1, 2, \ldots, s). \qquad \text{......(393)}$$

Any atom of the lattice is at a point specified by $\mathbf{r}_k{}^l$, where

$$\mathbf{r}_k{}^l = \mathbf{r}_k + \mathbf{r}^l. \qquad \text{......(394)}$$

The vector distance between any two atoms of the lattice is

$$\mathbf{r}_k{}^l - \mathbf{r}_{k'}{}^{l'} = \mathbf{r}_k - \mathbf{r}_{k'} + \mathbf{r}^{(l-l')} = \mathbf{r}_{k,k'}{}^{(l-l')}. \qquad \text{......(395)}$$

§4·91. *Crystal statics, or energies without heat motions.* We will next consider the form of the potential energy function when the atoms are at rest in their mean positions and act on one another with radial forces of potential energy $\phi_{k,k'}(r)$ for the pair of atoms of type k and k' at a distance r apart.

We require of course the potential energy per unit cell and have therefore to calculate the potential energy of each of the k units in this cell in the field of all the other units in the crystal, which is regarded as infinite in extent. We then obtain the total energy of the crystal, omitting surface effects, by summing these potential energies over the k units of the cell and multiplying by $\frac{1}{2}N$, where N is the number of cells in the actual crystal. Every term would be counted twice over by this summation, hence the factor $\frac{1}{2}$. We shall continue to call the total potential energy of the actual crystal F_0, and have therefore

$$F_0 = \tfrac{1}{2}N\,\Sigma_k\,\Sigma_{l,k'}\,\phi_{k,k'}(r^l{}_{k,k'}). \qquad \text{......(396)}$$

It is commonly convenient to use a different notation for the l and k summations, calling them S_l and Σ_k. We have therefore

$$F_0 = \tfrac{1}{2}N\mathsf{S}_l\Sigma_{k,k'}\,\phi_{k,k'}(r^l{}_{k,k'}). \qquad \text{......(397)}$$

In (397) the summation S_l runs from $-\infty$ to $+\infty$ in each of the three indices and the summation $\Sigma_{k,k'}$ twice over the s units of the basic cell. All terms for which $r=0$, that is, $k=k'$, $l=0$, are omitted.

Methods of evaluating these sums in terms of given atomic forces for the simpler cases are sketched in Chapter X, and we shall not consider them further here.

We should observe in passing that the form chosen for the unit of summation as a term depending on two atoms only is by no means general. More general forms are to be expected and are required to account for the more intimate properties of crystals. For example, if we have in the basic cell three non-collinear ions 1, 2, 3 with appreciably polarizable electronic structures, their mutual potential energy may be reduced to the form

$$\phi_{123}(\mathbf{r}_{12},\mathbf{r}_{23},\mathbf{r}_{31}),$$

at least approximately, but is *not* expressible in the form

$$\phi_{12}(r_{12})+\phi_{23}(r_{23})+\phi_{31}(r_{31}),$$

which is the form assumed in (397). Summations more complicated than (397) are then required, which we shall not formulate here.

Let us now consider the requirements of the equilibrium state. If we suppose that the vectors $\boldsymbol{\delta}_1$, $\boldsymbol{\delta}_2$, $\boldsymbol{\delta}_3$, $\mathbf{r}_1$, ..., $\mathbf{r}_s$ as specified refer to the equilibrium state with all units at rest in their mean positions, then the first order variations of F_0 must vanish for any small variations in position of any one (or more) of the constituents. Since all the cells are the same these conditions of course are not all independent, and it is easy to see that the independent conditions reduce to those in which every cell is submitted to the same variations. Such variations may be called *homogeneous displacements* and are of special importance.

Let us now pause a moment to consider the complete specification of the geometrical variables which describe the crystal, and which are entitled to enter into thermodynamical or statistical equations. It is no longer possible conveniently to use the actual volume; again, the displacements from the equilibrium state with the atoms at rest and undisturbed by external fields of force are practically always small in an actual crystal. It is therefore convenient to take as standard the undisturbed state of rest, and to refer all other states to that state. In any other actual state of the crystal (preserving the lattice repetition) the mean positions of the structural units can be just specified exactly by specifying the general homogeneous displacement which transforms the standard state to the required state, and this disturbance may usually be regarded as small. When external forces are acting, mechanical or electrical, they must therefore be supposed to be uniform over the extent of the crystal. Since the actual space variation of external fields is very small on the molecular scale this apparent restriction is a trivial one.

The general homogeneous displacements, which we are thus led to regard as a suitable geometrical specification of the crystal, contain terms for the displacement of one set of atoms relative to another which are at first sight not accessible to direct observation and control for the crystal in bulk and so not permissible statistical variables. Ideally, however, one can regard these displacements as directly measurable by X-ray methods, which can already place all the atoms in the basic cell at least in simpler cases. Moreover, in ionic lattices some at least of these displacements can be varied independently by external electrostatic forces. They cannot therefore all be ignored, and it seems a legitimate generalization to include them all in the statistical description of the crystal. As we shall see, they can be eliminated later from the partition function by the usual thermodynamic process as soon as it is desired to do so.

Disturbances of the crystal other than homogeneous only enter with its thermal motion.

The general homogeneous displacement consists of small vector changes in $\boldsymbol{\delta}_1, \boldsymbol{\delta}_2, \boldsymbol{\delta}_3, \mathbf{r}_1, \ldots, \mathbf{r}_s$, the lattice being rebuilt out of the cells so altered. In greater detail we write these changes as follows, where asterisks denote the new values. For the components of $\boldsymbol{\delta}_1, \boldsymbol{\delta}_2, \boldsymbol{\delta}_3$

$$\delta^*_{t,x} = \delta_{t,x} + \Sigma_y\, u_{x,y}\, \delta_{t,y} \quad (t=1,2,3); \qquad \ldots\ldots(398)$$

for the components of $\mathbf{r}_k$

$$x_k^* = x_k + u_{k,x} + \Sigma_y\, u_{x,y}\, y_k \quad (k=1,\ldots,s). \qquad \ldots\ldots(399)$$

The displacements then consist of a homogeneous strain of the whole of each cell defined by the tensor $u_{x,y}$, and then a displacement of the s elements of each cell by the vector $\mathbf{u}_k$. To exclude a rotation of the lattice as a whole as a rigid body the tensor $u_{x,y}$ must be symmetrical or

$$u_{x,y} = u_{y,x} \quad (\textit{all } x, y), \qquad \ldots\ldots(400)$$

and to exclude a translation of the lattice as a whole as a rigid body the sum of the s displacement vectors must vanish or

$$\Sigma_k\, u_{k,x} = 0 \quad (\textit{all } x). \qquad \ldots\ldots(401)$$

Let us now suppose that we can compute the value F_0^* of F_0 given by (397) under this homogeneous displacement, retaining terms in the displacements up to the second order. Then

$$\begin{aligned} F_0^* = F_0 &+ \Sigma_{k,x}\left(\frac{\partial F_0^*}{\partial u_{k,x}}\right)_0 u_{k,x} + \Sigma_{x,y}\left(\frac{\partial F_0^*}{\partial u_{x,y}}\right)_0 u_{x,y} \\ &+ \frac{1}{2}\left\{\Sigma_{k,x}\,\Sigma_{k',y}\left(\frac{\partial^2 F_0^*}{\partial u_{k,x}\,\partial u_{k',y}}\right)_0 u_{k,x}u_{k',y} + 2\Sigma_{k,x}\,\Sigma_{y,z}\left(\frac{\partial^2 F_0^*}{\partial u_{k,x}\,\partial u_{y,z}}\right)_0 u_{k,x}u_{y,z}\right. \\ &\left. + \Sigma_{y,z}\,\Sigma_{y',z'}\left(\frac{\partial^2 F_0^*}{\partial u_{y,z}\,\partial u_{y',z'}}\right)_0 u_{y,z}u_{y',z'}\right\}, \\ = F_0 &+ F_1 + F_2. \end{aligned} \qquad \ldots\ldots(402)$$

It is well to recall the meaning of the two zero suffixes. The suffix zero in F_0 and F_0^* denotes that there is no thermal agitation and that the terms refer to potential energy in an assigned configuration. The suffix zero in $(\;)_0$ denotes that the term is calculated for zero displacement from the standard configuration. By the conditions of equilibrium in the specified standard state

$$\left(\frac{\partial F_0^*}{\partial u_{k,x}}\right)_0 = 0, \quad \left(\frac{\partial F_0^*}{\partial u_{x,y}}\right)_0 = 0. \qquad \text{......(403)}$$

These equations are not all independent, for the change of potential energy must vanish identically when the crystal is moved as a whole or rotated as a whole in the absence of external forces. Thus

$$\Sigma_k \left(\frac{\partial F_0^*}{\partial u_{k,x}}\right)_0 \equiv 0, \qquad \text{......(404)}$$

and since in a pure rotation $u_{x,y} + u_{y,x} = 0$, etc.,

$$\left(\frac{\partial F_0^*}{\partial u_{y,z}}\right)_0 \equiv \left(\frac{\partial F_0^*}{\partial u_{z,y}}\right)_0. \qquad \text{......(405)}$$

The independent conditions (403) thus reduce to $3s+3$, which is the same as the number of independent displacements $(u_{x,y}, u_{k,x})$ subject to (400) and (401), and therefore just suffice to fix the standard equilibrium state. The numbers of independent variables and equations can of course be reduced by the crystal symmetry, but are reduced equally.

Suppose now that the crystal is again held in equilibrium but not in its standard state. This is possible if external mechanical or electrical forces exist, acting selectively on the different units of the crystal cell. We have then to consider a crystal element in a resultant state of homogeneous displacement. The selective forces on particular sets of atoms may naturally be thought of as being applied as body forces acting uniformly through the body of the crystal, but the ordinary mechanical stress tensor which primarily deforms the cell is in actual practice applied to the surface of the crystal. One may think at first sight that in a scheme which omits surface effects these surface applied stresses cannot be included, but this difficulty is only apparent. We find the same apparent difficulty in discussing imperfect gases in Chapters VIII and IX, which is considered in detail in § 5·7. We shall not go into details here, but by similar arguments the surface applied stresses can be seen to be transmitted uniformly through the solid by considering the actions across suitable sets of geometrical interfaces inside it, so that effectively the cells of the crystal are subjected to uniform body forces. In any such case of non-standard equilibrium under uniform homogeneous displacements we have therefore

$$U_{0,x,y} = -\frac{\partial F_0^*}{\partial u_{x,y}}, \qquad \text{......(406)}$$

$$U_{0,k,x} = -\frac{\partial F_0^*}{\partial u_{k,x}}, \qquad \text{......(407)}$$

where the functions U so defined, vanishing in the standard state, are the external stress tensor and the s external selective force vectors respectively. This identification follows at once from the requirements of the conservation of energy.

§ **4·92.** *Crystals with thermal agitation.* It is now possible to superpose on the distorted crystal the usual general thermal agitation, and so to construct its partition function. Into the details of this calculation we need not go. It requires calculations of F_0^* to the third order in the displacements in order to get the frequency spectrum of the crystal correct to the first order terms in the homogeneous displacements. The calculations are complicated, but can be carried through. With approximations of the same type as before we arrive at

$$\log K(z) = (F_0 + F_2)\log z - \frac{3N}{(\log 1/z)^3}\sum_{j=1}^{3}\frac{1}{k^3\Theta_j^{*3}}\int_0^{k\Theta_j^*\log 1/z}\log(1-e^{-x})\,x^2dx$$
$$-N\sum_{j=4}^{3s}\log(1-z^{h\nu_j^*}), \qquad \text{......(408)}$$

corresponding to (349), but now Θ_j^* and ν_j^* refer to the displaced lattice. If we anticipate Chapter VI as before and write (408) in terms of T, like (351), we have

$$\log K(T) = -\frac{F_0+F_2}{kT} - 3N\sum_{j=1}^{3}\frac{T^3}{\Theta_j^{*3}}\int_0^{\Theta_j^*/T}\log(1-e^{-x})\,x^2dx$$
$$-N\sum_{j=4}^{3s}\log(1-e^{-\Theta_j^*/T}), \qquad \text{......(409)}$$

where
$$\Theta_j^* = \Theta_j\{1 - \Sigma_{k,x} B^j{}_{k,x}u_{k,x} - \Sigma_{x,y} B^j{}_{x,y}u_{x,y}\}. \qquad \text{......(410)}$$

Strictly speaking, the averaging for direction of equation (347) must be carried out remembering that $B^j{}_{k,x}$ and $B^j{}_{x,y}$ $(j=1, 2, 3)$ are functions of direction. This will be indicated in the equations that follow.

The analysis is of course only carried through for small displacements, so that only the first order terms arising from the changes in the Θ's can be retained. When we expand (409) in powers of $u_{k,x}$ and $u_{x,y}$ and indicate the hitherto ignored directional averaging, the result can easily be reduced to the form

$$\log K(T) = -\frac{F_0+F_2}{kT} - 3N\sum_{j=1}^{3}\frac{T^3}{\Theta_j^3}\int_0^{\Theta_j/T}\log(1-e^{-x})\,x^2dx$$
$$-N\sum_{j=4}^{3s}\log(1-e^{-\Theta_j/T}) + \frac{1}{kT}\{\Sigma_{k,x}\mathbf{K}^\circ{}_{k,x}u_{k,x} + \Sigma_{x,y}\mathbf{K}^\circ{}_{x,y}u_{x,y}\}, \qquad \text{......(411)}$$

where, the bars denoting averaging for direction,

$$\mathbf{K}^\circ{}_{k,x} = NkT\left\{3\sum_{j=1}^{3}\overline{B^j{}_{k,x}\frac{T^3}{\Theta_j^3}\int_0^{\Theta_j/T}\frac{x^3dx}{e^x-1}} + \sum_{j=4}^{3s}B^j{}_{k,x}\frac{\Theta_j/T}{e^{\Theta_j/T}-1}\right\}, \qquad \text{......(412)}$$

$$\mathbf{K}^\circ{}_{x,y} = NkT\left\{3\sum_{j=1}^{3}\overline{B^j{}_{x,y}\frac{T^3}{\Theta_j^3}\int_0^{\Theta_j/T}\frac{x^3dx}{e^x-1}} + \sum_{j=4}^{3s}B^j{}_{x,y}\frac{\Theta_j/T}{e^{\Theta_j/T}-1}\right\}. \qquad \text{......(413)}$$

To the specified order of accuracy this is the complete partition function, replacing (351), and depending on all the geometrical parameters $u_{k,x}$ and $u_{x,y}$ instead of only on V. The temperature dependent terms have only been given to a first approximation in these parameters.

§**4·93.** *General applications.* By the general formula (374),

$$U_{x,y}=kT\frac{\partial}{\partial u_{x,y}}\log K(T), \qquad \text{......(414)}$$

$$U_{k,x}=kT\frac{\partial}{\partial u_{k,x}}\log K(T). \qquad \text{......(415)}$$

These are the average reactions of the specified types which the assembly exerts on the outside world, or the forces exerted by the surroundings reversed in sign, so that the external work done by the crystal in any specified displacement at the given temperature is

$$\Sigma_{x,y}U_{x,y}\,\delta u_{x,y}+\Sigma_{k,x}U_{k,x}\,\delta u_{k,x}. \qquad \text{......(416)}$$

The average energy $\overline{E_K}$ is given as before by

$$\overline{E_K}=kT^2\frac{\partial}{\partial T}\log K(T). \qquad \text{......(417)}$$

The forms of the average reactions are found on combining (406), (407), (414) and (415) to be

$$U_{x,y}=\mathbf{K}^\circ{}_{x,y}-\frac{\partial F_2}{\partial u_{x,y}}, \qquad \text{......(418)}$$

$$U_{k,x}=\mathbf{K}^\circ{}_{k,x}-\frac{\partial F_2}{\partial u_{k,x}}. \qquad \text{......(419)}$$

Anticipating the results of Chapter VI we may of course apply all the processes of thermodynamics to our crystal. In particular we can evaluate the displacements $u_{k,x}$ and construct a new function of the other variables and $U_{k,x}$ instead of the $u_{k,x}$ which has similar properties. This is a process of practical importance because all observable properties of a crystal are properties which it possesses for given values of the external forces $U_{k,x}$ and not for given values of the displacements $u_{k,x}$. As appears in Chapter VI $k\log K(T)$ for the crystal is the thermodynamic function known as Planck's characteristic function, and $-kT\log K(T)$ is the more usual *free energy*† *of Helmholtz*, F. The function we require here is

$$F^*=F+\Sigma_{k,x}U_{k,x}u_{k,x}, \qquad \text{......(420)}$$

expressed as a function of T, $u_{x,y}$ and $U_{k,x}$ after eliminating the $u_{k,x}$ from (420) by means of (419). It is as it were a partial transformation from the

† Also called *the work function* and denoted by A.

free energy of Helmholtz to the free energy of Gibbs. By the formation of the usual total variation we see that

$$u_{k,x} = \frac{\partial F^*}{\partial U_{k,x}}. \qquad \text{......(421)}$$

We have still to define the strength of the electric doublet **p** induced in the crystal by the displacements of type $u_{k,x}$, whether they are due to external electric fields, to thermal agitation or to the other displacements $u_{x,y}$. We have at once

$$p_x = N \Sigma_k \epsilon_k u_{k,x}, \qquad \text{......(422)}$$

where ϵ_k is the charge on the kth unit of the basic cell. If an external electric field of components $\mathbf{E}_x$, $\mathbf{E}_y$, $\mathbf{E}_z$ is acting, the forces $U_{k,x}$ are given by the equations

$$U_{k,x} = -N\epsilon_k \mathbf{E}_x. \qquad \text{......(423)}$$

The $u_{k,x}$ may be supposed eliminated from (422) also, with the help of (419).

We cannot pursue the theory of crystal structure in detail any further, but must content ourselves with general remarks. The general equilibrium properties of crystals and their relationships may be illustrated by the following diagram.†

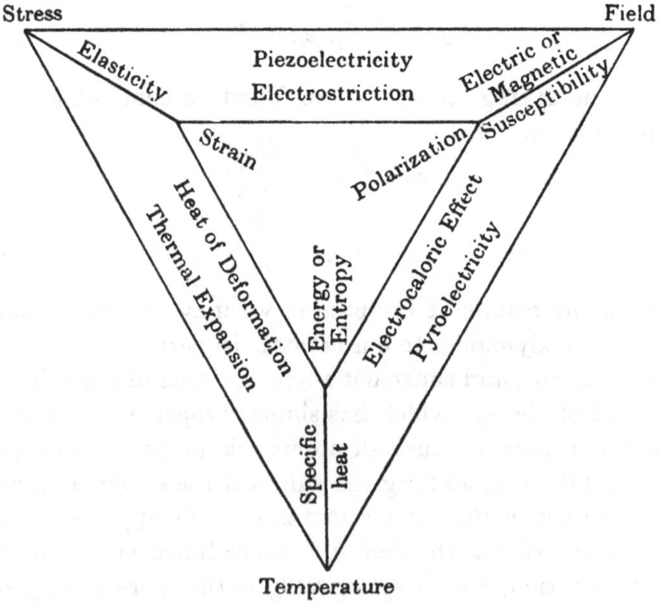

Fig. 11. The equilibrium properties of crystals.

The formal equations connecting these quantities can all be derived from the functions F or F^*, and can therefore be evaluated for given crystal structure and given atomic models by carrying through the calculations which we have indicated here. For example, the pyroelectric effect is a

† Modified here from Born, *loc. cit.* after Heckmann.

change of dipole strength in the crystal caused by heating it, and is of course coupled with the corresponding inverse electrocaloric effect which is a change of temperature following on the application of an external electric field. The pyroelectric effect appears of course as the production of equal and opposite surface charges on opposite faces of the crystal. The best known example is tourmaline in which the charges separate along the trigonal axis. This effect can be calculated by evaluating p_x, p_y, p_z for zero external stress tensor and zero electric field. The pyroelectric moment so calculated corresponds to that observed, but it is not what may be called the *true pyroelectric moment* which would naturally be that for zero external electric field and zero displacements $u_{x,y}$. The actual displacements $u_{x,y}$ for zero stress tensor themselves involve non-zero values of $u_{k,x}$ (independently of the direct temperature effect), and so produce a piezoelectric moment which is superposed on the true pyroelectric moment to give the observed value.

The point of chief interest to statistical theory is the dependence of all these parameters on the temperature. It is not difficult to see that the interconnections between the pyroelectric moment, the thermal expansions and the temperature are fixed by the coefficients $B^j{}_{k,x}$ and $B^j{}_{x,y}$ of equations (412) and (413). At low temperatures only the Debye terms, $j=1, 2, 3$, are important, and we find in addition to (359) for the energy, the relations

$$\text{Thermal expansions} \propto T^4, \qquad \qquad \text{......(424)}$$

$$p_x, p_y, p_z \propto T^4. \qquad \qquad \text{......(425)}$$

The proportionality between the excess energy content over the zero-point energy and the thermal expansions is in agreement with observation, and represents a law formulated on empirical grounds by Gruneisen, but the proportionality with the pyroelectric moment is not in agreement with observation. The observations suggest that p_x tends to zero rather like T or T^2. The explanation is still uncertain, but it is probably connected with certain other phenomena in crystals of low symmetry to which we shall now refer on account of their striking character.

§ **4·94.** *Some properties of strongly anisotropic crystals.* The investigations of Gruneisen and Goens* on single crystals of zinc and cadmium have shown that these hexagonal crystals are very strongly anisotropic, all their properties being markedly different along and across the hexagonal axis. In particular the frequencies of elastic vibration are very different in these directions. On following through the calculations for the coefficients of thermal expansion it is found that

$$\alpha_{\parallel} = \gamma_{11} p_{\parallel} + \gamma_{12} p_{\perp}, \qquad \qquad \text{......(426)}$$

$$\alpha_{\perp} = \gamma_{21} p_{\parallel} + \gamma_{22} p_{\perp}, \qquad \qquad \text{......(427)}$$

* Gruneisen and Goens, *Zeit. f. Physik*, vol. 26, pp. 235, 250 (1924), vol. 29, p. 141 (1924), vol. 37, p. 278 (1926).

where $$p_{\|}=f_1(T), \quad p_{\perp}=f_2(T), \qquad \text{......(428)}$$

$$\gamma_{12}=\gamma_{21}<0. \qquad \text{......(429)}$$

To a first approximation f_1 and f_2 are of Debye's type, but with very different Θ's. The γ's are elastic moduli, so that the p's are of the nature of "thermal pressures". The Θ's are as usual fixed by the elastic constants and serve to determine successfully the specific heats.

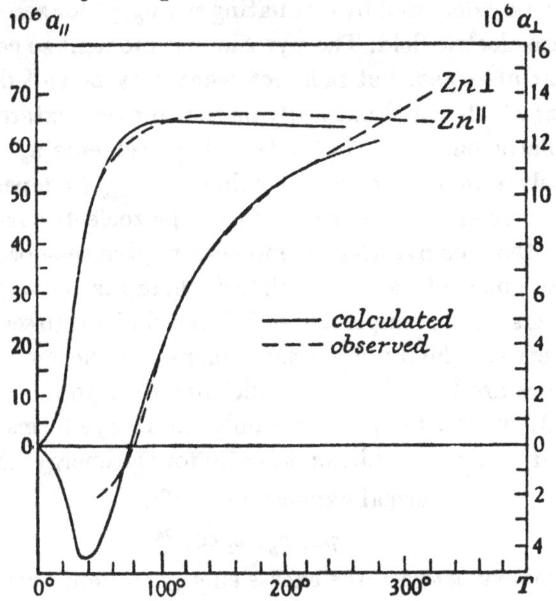

Fig. 12. The coefficients of thermal expansion $\alpha_{\|}$ and $\alpha_{\perp}$ of zinc crystals along and across the hexagonal axis.

The interesting special property of these crystals is that they show for certain temperatures a *negative* value of $\alpha_{\perp}$, that is a contraction across the hexagonal axis on heating. But this follows from the theory, for owing to the great difference in the Θ's f_1 at low temperatures has a sensible value while f_2 is still negligible. At higher temperatures both $\alpha_{\|}$ and $\alpha_{\perp}$ become positive, for the functions f_1 and f_2 become comparable and $\gamma_{22}>|\gamma_{12}|$. The observations and theoretical calculations for zinc are shown in Fig 12 and are in remarkable agreement.

In view of this result is seems possible that the observed variation of pyroelectric moment, which does not obey the T^4-law, arises from a combination of several rather different functions of the temperature which partly balance each other. The crystals concerned in pyroelectric effects are in general strongly anisotropic.

CHAPTER V

THE GENERAL ASSEMBLY. DISSOCIATION AND EVAPORATION

§5·1. *Introductory.* We now pass on to the general assembly in which gaseous atoms and molecules dissociate and combine, or evaporate from and recondense on crystals. Generally speaking we may say that we are now to investigate the equilibrium state of an assembly of practically independent systems, which no longer retain their individualities throughout the motion of the assembly, but are able to break up and recombine in such a way as to form any specified number of different types of independent systems. It will obviously be important to specify what we regard as the ultimate structural elements of the assembly—the indivisible systems which can combine, but never break up further. The formal exposition will be the same whatever the ultimate units are assumed to be, but the physical interpretation will be different. In the initial exposition we shall for definiteness suppose that the ultimate units are the atoms of chemistry. At a later stage the same formulae will be reinterpreted in terms of positive nuclei and electrons in the study of the properties of matter at very high temperatures. This interpretation is obviously the more fundamental, and will of course include the former as a special case, but in practice the two fields hardly overlap. In most assemblies hot enough to contain an appreciable equilibrium concentration of atomic ions and free electrons, the number of molecular species effectively present will be found to be always very small and generally zero.

Previous to the formulation of the laws of quantum mechanics any satisfactory discussion of the problems of dissociation from the standpoint of pure statistical mechanics adopted in this monograph was based on the work of Ehrenfest and Trkal.* These authors only discuss the dissociation problem on semi-classical lines; they assume, that is, that all degrees of freedom of a system are either "fully excited" so that the classical theory is adequate or "completely unexcited", in which case they may be ignored. This restriction is here removed so that the discussion is perfectly general, and subject only in its range of applicability to restrictions of a physical nature inherent in the problem. Thanks to the simplifications introduced by quantum mechanics the method of analysis of Ehrenfest and Trkal need no longer be followed. Actual matter however (for example, imperfect gases)

* Ehrenfest and Trkal, *Proc. Sect. Sci. Amsterdam*, vol. 23, p. 162 (1920). I do not know of any earlier discussion of dissociation or the kindred matters of chemical constants, "thermodynamic probability" and absolute entropy on the hypotheses of Chapter I which can be regarded as logically convincing.

will not always be comparable to an assembly analysed into practically independent systems. We shall make a first attempt in a later chapter towards the removal of this particular restriction.

For simplicity of exposition we begin by considering a gaseous assembly in which a single reaction $A^1 + A^2 \rightleftarrows A^1A^2$ can occur. This serves to bring out the essential point at which the analysis of Chapter II must be extended to take account of such reactions. We find that the expressions we require are still coefficients in multiple power series similar to those of § 2·4. After discussion of this simple case we give the analysis for the general gaseous assembly, and then the extension to assemblies containing crystals or other condensed systems for which we can construct partition functions. We include some simple examples of the interpretation of our formulae in terms of atoms, molecules and chemical reactions, but the main discussion of the results can only be profitably taken up after we have discussed the relationship between the equilibrium theory of statistical mechanics and the laws of thermodynamics. It will be shown that the selector variables each have their proper physical interpretation. Just as z corresponds to the temperature, so the other variables, one for each type of atom (or other structural unit), correspond to the partial potentials of thermodynamic theory.

§ 5·11. *Short cuts for enumerating complexions (accessible states) for a dissociating assembly.* Before enumerating the complexions of a dissociating assembly it is necessary to analyse its symmetry requirements in some detail. At the present time we can hardly specify with confidence what we should take to be the ultimate elements of which matter is built up. We are nowadays acquainted with electrons, protons, neutrons and positive electrons, all of which have some claim to be considered independent fundamental particles, though the last are seldom likely to be important in equilibrium problems. We even flirt with the idea of neutrinos. From what we already know about electrons and protons we may assert with some confidence that *the accessible states of all assemblies are those and only those which are antisymmetrical in all the electrons and also antisymmetrical in all the protons, account being taken of their spins.* If neutrons are admitted as a third fundamental particle, then the same rule may be expected to apply to them. If in fact any assembly may be regarded as built up of electrons, protons and neutrons and no other particles (besides light quanta which we are not considering at the moment), then these symmetry requirements should be and are sufficient for any enumeration of complexions. But for the sake of treating an assembly in a way which corresponds closely to its physical nature, it is inconvenient and of little value to regard it for example as an assembly of electrons and protons when it is in fact an assembly of permanent helium atoms or hydrogen molecules. We need to use these units themselves as quasifundamental systems with secondary derived symmetry rules.

Suppose such a complex system contains m electrons, n protons and o neutrons, or rather that some such analysis for definite values of m, n, o is a legitimate representation. Then since interchange of a pair of complex systems means interchange of $(m+n+o)$ pairs of electrons, protons and neutrons in each of which pairs the wave-function is antisymmetrical, *the wave-function of the assembly will be symmetrical in the complex systems if* $(m+n+o)$ *is even and antisymmetrical if* $(m+n+o)$ *is odd*. We have therefore at least these derived symmetry rules, and as we shall shortly see they are not only necessary but sufficient.

We shall therefore be able to avoid the difficulty or rather inelegancy mentioned above. When we have decided *a priori* from our knowledge of the properties of the secondary systems—atoms, molecules or nuclei—what systems can be properly regarded as the permanent population of the assembly in any particular problem, then we can appeal to the following

Lemma 5·11. *It is sufficient, in enumerating the complexions, that is the wave-functions characterizing the accessible states of the assembly, to construct formally, and so to enumerate, all those linearly independent wave-functions which have the correct symmetry properties in the "permanent" systems (regarded as wholes) of which the assembly is composed. The wave-function of any "permanent" system must have the correct symmetry properties in the electrons, protons and neutrons of which that system is composed, but the direct analysis of the complete assembly into the electrons, protons and neutrons of which it is ultimately compared may be omitted, for the number of complexions is not thereby affected.*

This simplifying lemma is easily established but as a formal proof is long and clumsy we shall be content with illustrative examples in which the argument is obviously general. It is a well-known property of Schrödinger's equation for an assembly that the neglect of the interactions between the constituent systems can never alter the number or the symmetry type of the various states, though of course it can alter the energies and therefore allow new degeneracies to enter. We may therefore proceed to enumerate states neglecting interactions when and where convenient.

For a first example consider an assembly of two electrons and two protons in a box, for which the conditions are such that the actual permanent systems may be assumed to be two hydrogen atoms. Let $\psi_\alpha(1,I)$ be the wave-function of the system electron 1 and proton I in the box, which describes a particular stationary state of $1+I$ neglecting the interaction with electron 2 and proton II. Under the conditions of the problem this is then the wave-function for a freely moving hydrogen atom in a given electronic state and a given state of translatory motion, and we need not specify it further. Let $\psi_\beta(2,II)$ have the obvious similar meaning. Then the assembly wave-functions describing the state, one atom in state α and one atom in state β, the atoms being "permanent", are

$$\psi_\alpha(1,I)\,\psi_\beta(2,II),\quad \psi_\alpha(2,II)\,\psi_\beta(1,I),$$

and the unique wave-function *symmetrical* in the two hydrogen atoms as wholes is

$$\psi_\alpha(1,I)\,\psi_\beta(2,II)+\psi_\alpha(2,II)\,\psi_\beta(1,I). \qquad \text{......(430)}$$

This wave-function is not antisymmetrical in the electrons and protons, but it corresponds to just one such wave-function, as the lemma requires. This wave-function is

$$\{\psi_\alpha(1,I)\,\psi_\beta(2,II)+\psi_\alpha(2,II)\,\psi_\beta(1,I)\}$$
$$-\{\psi_\alpha(2,I)\,\psi_\beta(1,II)+\psi_\alpha(1,II)\,\psi_\beta(2,I)\}. \qquad \text{......(431)}$$

Of course we may not argue that to (431) correspond two functions of type (430), namely the two terms in { } in (431), for to admit both these functions denies the permanency of the hydrogen atoms. We must then regard the assembly as one of electrons and protons and are driven back to the unique (431) itself.

For a second example let us suppose that an α-particle may be properly analysed into four protons and two electrons. It is far more likely that the correct analysis is into two protons and two neutrons but the proposed analysis will still serve equally well for illustration. Consider an assembly of eight electrons and eight protons for which we know that the "permanent" systems may be taken to be two α particles and four "free" electrons. The free electrons may of course be sometimes bound in the ordinary electronic orbits of He^+ or He. Let $\psi_\alpha(1,2,I,\ldots,IV)$ and $\psi_\beta(3,4,V,\ldots,VIII)$ represent wave-functions for the two α-complexes, each of proper internal antisymmetry in their own electrons and protons. Let $\psi_\gamma(5)$, $\psi_\delta(6)$, $\psi_\epsilon(7)$, $\psi_\zeta(8)$ be the wave-functions of the extra-nuclear electrons. The only wave-function Ψ for the complete assembly which is antisymmetrical in the extra-nuclear electrons, symmetrical in the α-particles, of the given grouping in the α-complexes, and representative of the specified state of the assembly (α-particles in states α and β, free electrons in γ, δ, ϵ, ζ), is

$$\Psi=\left\|\begin{matrix}\psi_\alpha(1,2,I,\ldots,IV)\,\psi_\alpha(3,4,V,\ldots,VIII)\\ \psi_\beta(1,2,I,\ldots,IV)\,\psi_\beta(3,4,V,\ldots,VIII)\end{matrix}\right\|\;\begin{vmatrix}\psi_\gamma(5) & \psi_\gamma(6) & \psi_\gamma(7) & \psi_\gamma(8)\\ \psi_\delta(5) & \psi_\delta(6) & \psi_\delta(7) & \psi_\delta(8)\\ \psi_\epsilon(5) & \psi_\epsilon(6) & \psi_\epsilon(7) & \psi_\epsilon(8)\\ \psi_\zeta(5) & \psi_\zeta(6) & \psi_\zeta(7) & \psi_\zeta(8)\end{vmatrix}$$

......(432)

In (432) $\|\ldots\|$ represents a *permanent* that is a determinant but with all its signs positive. It is easy to see that from any such function we can construct just one Ψ_0 antisymmetrical in all the protons by summation of the terms $P_p\Psi$ with the proper signs over all permutations P_p of the protons which make a significant change in Ψ (such as for example an interchange of *I* and *V* but not of *I* and *II*). We can then construct just one Ψ_{00} antisymmetrical in all the protons and all the electrons by a similar summation of the terms $P_e\Psi_0$ over all significant permutations P_e of the electrons. The order of these operations is irrelevant. Conversely from the given elementary ψ's of the electrons and protons we can obviously construct only one function with the groupings and symmetries of (432), which verifies the lemma.

For a third example let us take the same assembly of two α-particles and four electrons but analysed into four protons, four neutrons and four electrons—probably a more correct analysis—and let us specify the neutrons by the numbers *1*, *2*, *3*, *4*. Then the wave-functions for the two α-particles are $\psi_\alpha(I,II,1,2)$ and $\psi_\beta(III,IV,3,4)$ and for the four electrons $\psi_\gamma(1)$, $\psi_\delta(2)$, $\psi_\epsilon(3)$, $\psi_\zeta(4)$ as before. Then the only wave-function for the assembly with the right symmetries and groupings is

$$\Psi = \left\| \begin{matrix} \psi_\alpha(I,II,1,2)\,\psi_\alpha(III,IV,3,4) \\ \psi_\beta(I,II,1,2)\,\psi_\beta(III,IV,3,4) \end{matrix} \right\| \begin{vmatrix} \psi_\gamma(1) & \psi_\gamma(2) & \psi_\gamma(3) & \psi_\gamma(4) \\ \psi_\delta(1) & \psi_\delta(2) & \psi_\delta(3) & \psi_\delta(4) \\ \psi_\epsilon(1) & \psi_\epsilon(2) & \psi_\epsilon(3) & \psi_\epsilon(4) \\ \psi_\zeta(1) & \psi_\zeta(2) & \psi_\zeta(3) & \psi_\zeta(4) \end{vmatrix} \qquad \ldots\ldots(433)$$

This wave-function is already antisymmetrical in all the electrons of the assembly and we shall take no further notice of the electron factor. It remains to construct from (433) the unique Ψ_{00} which is antisymmetrical in all the protons and all the neutrons. It will be worth while to construct Ψ_{00} explicitly as it is not excessively complicated.

The wave-function Ψ_0 derived from Ψ by the necessary permutations of the protons (electrons ignored) is

$$\begin{aligned} \Psi_0 = {} & \psi_\alpha(I,II,1,2)\,\psi_\beta(III,IV,3,4) + \psi_\beta(I,II,1,2)\,\psi_\alpha(III,IV,3,4) \\ & + \psi_\alpha(II,III,1,2)\,\psi_\beta(I,IV,3,4) + \psi_\beta(II,III,1,2)\,\psi_\alpha(I,IV,3,4) \\ & + \psi_\alpha(III,I,1,2)\,\psi_\beta(II,IV,3,4) + \psi_\beta(III,I,1,2)\,\psi_\alpha(II,IV,3,4) \\ & + \psi_\alpha(III,IV,1,2)\,\psi_\beta(I,II,3,4) + \psi_\beta(III,IV,1,2)\,\psi_\alpha(I,II,3,4) \\ & + \psi_\alpha(I,IV,1,2)\,\psi_\beta(II,III,3,4) + \psi_\beta(I,IV,1,2)\,\psi_\alpha(II,III,3,4) \\ & + \psi_\alpha(II,IV,1,2)\,\psi_\beta(III,I,3,4) + \psi_\beta(II,IV,1,2)\,\psi_\alpha(III,I,3,4). \end{aligned} \qquad \ldots\ldots(434)$$

This Ψ_0 is antisymmetrical in all the protons. It is at the same time antisymmetrical in the neutrons *1*, *2* and *3*, *4*, and symmetrical for the interchange of the neutron pair *1, 2* with the pair *3, 4*. If we write Ψ_0 as $\Psi_0(1,2;3,4)$ to indicate this dependence, then

$$\Psi_{00} = \Psi_0(1,2;3,4) + \Psi_0(2,3;1,4) + \Psi_0(3,1;2,4) \qquad ......(435)$$

is the unique wave-function antisymmetrical in both protons and neutrons.

These last two examples serve to show that the precise nature of the fundamental particles into which the complex systems should be analysed makes no difference to the enumeration of complexions, provided of course that the wave-functions in terms of the complex systems themselves have the correct symmetry. In conformity with this principle it is possible in most statistical problems to treat atomic nuclei as permanent complexes. No matter what the correct analysis of nuclei into more fundamental particles may be, we may always proceed by satisfying the symmetry requirements of the assembly merely in all sets of equivalent nuclei and in the electrons (extra-nuclear). The symmetry rule for nuclei which holds universally so far as is known at present is that *the wave-functions must be antisymmetrical in all similar nuclei when the nuclei have an odd mass number and symmetrical when the mass number is even.* The mass number is of course equal to the number of protons and neutrons in the nucleus.

We have so far formulated the lemma for the analysis of "permanent" complex systems into electrons, protons and neutrons, but there is no need so to restrict it. It applies equally to the analysis of "permanent" complex systems X into any "permanent" secondary units Y and Z. For example in an assembly of chlorine atoms Cl^{35} the assembly wave-functions must be antisymmetrical in the Cl^{35} nuclei. But the assembly will often consist entirely of "permanent" molecules Cl_2^{35}. Applying the lemma, therefore, we see that the symmetry requirements of the assembly will be satisfied, and that the enumeration of wave-functions will be correct if we make the wave-function of each molecule *antisymmetrical* in its nuclei and then make the assembly wave-function *symmetrical* in the molecules.

As a simple example of this case we may take an assembly of four systems such as Cl^{35} atoms permanently combined in pairs, and consider the symmetry requirements of the nuclei only. If $\psi_\alpha(1,2)$, $\psi_\beta(3,4)$ are two wave-functions for pairs of nuclei, each antisymmetrical in their pair of arguments, then the single assembly wave-function corresponding is, by the rule,

$$\Psi = \Psi(1,2;3,4) = \psi_\alpha(1,2)\,\psi_\beta(3,4) + \psi_\alpha(3,4)\,\psi_\beta(1,2). \qquad ......(436)$$

The single corresponding wave-function for the assembly antisymmetrical in all the nuclei is again of the form

$$\Psi_0 = \Psi(1,2;3,4) + \Psi(2,3;1,4) + \Psi(3,1;2,4). \qquad ......(437)$$

The same analysis verifies that, for an assembly of four systems X in which the "permanent" systems are two X's and one X_2, the assembly wave-functions may be enumerated by being made antisymmetrical (or symmetrical according to requirements) in the two free X's; more generally in a dissociating assembly the assembly wave-functions may be correctly enumerated by fulfilling the symmetry requirements separately for each species of system present in the assembly, i.e. separately for each species of free atom or molecule which is one of the "permanent" systems. There is no need to fulfill the symmetry requirements for all atoms X say, whether they are free or combined, provided that we treat the combined pairs as "permanent" systems in the sense of this section.

§5·2. *The number of complexions and the equilibrium state of a gaseous dissociating assembly in which the single reaction $A^1+A^2 \rightleftarrows A^1A^2$ occurs.* In the proposed dissociating assembly the "permanent" systems in the sense of § 5.11 are the atoms or molecules A^1, A^2, and A^1A^2. For simplicity we can suppose that no other systems are present and that the total numbers of the systems A^1 and A^2, free or combined, are X_1 and X_2 respectively. Since dissociation and recombination is assumed to proceed freely, the number of molecules A^1A^2 present in an accessible state of the assembly may have any value whatever from zero to X_1 or X_2 whichever is the less, and all such states have to be included in the enumeration. It is convenient to refer to all the states in which there are M_1 free A^1's, M_2 free A^2's and N A^1A^2's with $M_1+N=X_1$, $M_2+N=X_2$ as an *example* of the assembly. We can sum the complexions for each example and then for all the examples to obtain the desired total.

To find the number of complexions for a particular example (M_1, M_2, N) we can proceed as in § 2·4 and form the g-function for each of the systems A^1, A^2 and A^1A^2, and with them construct the analogue of (107), namely

$$\Pi_t\, g_1(x_1 z^{\epsilon_t^1})\, \Pi_t\, g_2(x_2 z^{\epsilon_t^2})\, \Pi_t\, g_{12}(y z^{\eta_t}), \qquad \text{......(438)}$$

where ϵ_t^1, ϵ_t^2 and η_t are the sets of energies of the three systems, degenerate states being counted multiply. The number of complexions of this example of the assembly can then be obtained by picking out the coefficients of $x_1^{M_1} x_2^{M_2} y^N z^E$ in (438). In this way the proper symmetry requirements are satisfied for each set of systems separately, the systems being treated as "permanent", which gives the correct enumeration by Lemma 5·11. In order however to sum over all examples we must remove the independence of y and the x's, and use instead the expression

$$\Pi_t\, g_1(x_1 z^{\epsilon_t^1})\, \Pi_t\, g_2(x_2 z^{\epsilon_t^2})\, \Pi_t\, g_{12}(x_1 x_2 z^{\eta_t}). \qquad \text{......(439)}$$

The total number of complexions of the assembly is then given by picking out the coefficient of $x_1^{X_1} x_2^{X_2} z^E$ in (439). This coefficient is clearly the sum of all the coefficients of $x_1^{M_1} x_2^{M_2} (x_1 x_2)^N z^E$ in (438) for which the equations of condition $M_1 + N = X_1$, $M_2 + N = X_2$ are satisfied. Thus

$$C = \frac{1}{(2\pi i)^3} \iiint \frac{dx_1 dx_2 dz}{x_1^{X_1+1} x_2^{X_2+1} z^{E+1}} \Pi_t g_1(x_1 z^{\epsilon_t^1}) \Pi_t g_2(x_2 z^{\epsilon_t^2}) \Pi_t g_{12}(x_1 x_2 z^{\eta_t}). \quad \text{......(440)}$$

It will be seen at once that the difference between (440) and (108) consists solely in the introduction of the extra factor $\Pi_t g_{12}(x_1 x_2 z^{\eta_t})$, which allows for the extra states provided by the combined systems $A^1 A^2$. It follows at once, by the arguments of § 2·4, that

$$C\overline{a_r^1} = \frac{1}{(2\pi i)^3} \iiint \frac{dx_1 dx_2 dz}{x_1^{X_1+1} x_2^{X_2+1} z^{E+1}} \left\{ x_1 \frac{\partial}{\partial x_1} g_1(x_1 z^{\epsilon_r^1}) \right\} \Pi_{t \neq r} g_1(x_1 z^{\epsilon_t^1})$$
$$\times \Pi_t g_2(x_2 z^{\epsilon_t^2}) \Pi_t g_{12}(x_1 x_2 z^{\eta_t}). \quad \text{......(441)}$$

Further if $\overline{b_r}$ is the average number of molecules in their rth state, then similarly

$$C\overline{b_r} = \frac{1}{(2\pi i)^3} \iiint \frac{dx_1 dx_2 dz}{x_1^{X_1+1} x_2^{X_2+1} z^{E+1}} \left\{ x_1 x_2 \frac{\partial}{\partial (x_1 x_2)} g_{12}(x_1 x_2 z^{\eta_r}) \right\} \Pi_t g_1(x_1 z^{\epsilon_t^1})$$
$$\times \Pi_t g_2(x_2 z^{\epsilon_t^2}) \Pi_{t \neq r} g_{12}(x_1 x_2 z^{\eta_t}). \quad \text{......(442)}$$

Compared with (440) the integrals for $C\overline{a_r^1}$ and $C\overline{b_r}$ contain respectively the extra factors

$$x_1 \frac{\partial}{\partial x_1} \log g_1(x_1 z^{\epsilon_r^1}), \quad x_1 x_2 \frac{\partial}{\partial (x_1 x_2)} \log g_{12}(x_1 x_2 z^{\eta_r}).$$

In the same way it is obvious that integrals for $C\overline{M_1}$, $C\overline{N}$, $C\overline{E_{A^1}}$ and $C\overline{E_{A^1A^2}}$ can be constructed which contain respectively the extra factors

$$x_1 \frac{\partial}{\partial x_1} \Sigma_r \log g_1(x_1 z^{\epsilon_r^1}), \quad x_1 x_2 \frac{\partial}{\partial (x_1 x_2)} \Sigma_r \log g_{12}(x_1 x_2 z^{\eta_r}),$$

$$z \frac{\partial}{\partial z} \Sigma_r \log g_1(x_1 z^{\epsilon_r^1}), \quad z \frac{\partial}{\partial z} \Sigma_r \log g_{12}(x_1 x_2 z^{\eta_r}).$$

As is sufficiently obvious from the notation, $\overline{E_{A^1}}$ means the average energy of the free systems A^1 and $\overline{E_{A^1A^2}}$ the average energy of the combined systems A^1A^2.

These integrals can all be evaluated by Theorem 2·42. The unique saddle point λ_1, λ_2, ϑ of the integrand of the multiple integral is determined by equations which are most conveniently expressed in the notation

$$\frac{\partial \Sigma}{\partial \lambda_1} = \frac{\partial \Sigma}{\partial \lambda_2} = \frac{\partial \Sigma}{\partial \vartheta} = 0, \quad \text{......(443)}$$

$$\Sigma = Z - X_1 \log \lambda_1 - X_2 \log \lambda_2 - E \log \vartheta, \quad \text{......(444)}$$

$$Z = \Sigma_t \log g_1(\lambda_1 \vartheta^{\epsilon_t^1}) + \Sigma_t \log g_2(\lambda_2 \vartheta^{\epsilon_t^2}) + \Sigma_t \log g_{12}(\lambda_1 \lambda_2 \vartheta^{\eta_t}); \quad \text{...(445)}$$

Σ is thus log (integrand of C) evaluated at the saddle point. With the values $\lambda_1, \lambda_2, \vartheta$ determined by these equations

$$\overline{a_r^1} = \lambda_1 \frac{\partial}{\partial \lambda_1} \log g_1(\lambda_1 \vartheta^{\epsilon_r^1}), \qquad \overline{b_r} = \lambda_1 \lambda_2 \frac{\partial}{\partial(\lambda_1 \lambda_2)} \log g_{12}(\lambda_1 \lambda_2 \vartheta^{\eta_r}), \qquad \text{......(446)}$$

$$\overline{M_1} = \lambda_1 \frac{\partial}{\partial \lambda_1} \Sigma_r \log g_1(\lambda_1 \vartheta^{\epsilon_r^1}), \qquad \overline{N} = \lambda_1 \lambda_2 \frac{\partial}{\partial(\lambda_1 \lambda_2)} \Sigma_r \log g_{12}(\lambda_1 \lambda_2 \vartheta^{\eta_r}), \qquad \text{......(447)}$$

$$\overline{E_{A^1}} = \vartheta \frac{\partial}{\partial \vartheta} \Sigma_r \log g_1(\lambda_1 \vartheta^{\epsilon_r^1}), \qquad \overline{E_{A^1A^2}} = \vartheta \frac{\partial}{\partial \vartheta} \Sigma_r \log g_{12}(\lambda_1 \lambda_2 \vartheta^{\eta_r}). \qquad \text{......(448)}$$

For convenience of reference we shall rewrite these equations introducing explicitly weights for degenerate states. The equations (445)–(448) then become

$$Z = \Sigma_t \varpi_t^1 \log g_1(\lambda_1 \vartheta^{\epsilon_t^1}) + \Sigma_t \varpi_t^2 \log g_2(\lambda_2 \vartheta^{\epsilon_t^2}) + \Sigma_t \rho_t \log g_{12}(\lambda_1 \lambda_2 \vartheta^{\eta_t}); \qquad \text{......(449)}$$

$$\overline{a_r^1} = \varpi_r^1 \lambda_1 \frac{\partial}{\partial \lambda_1} \log g_1(\lambda_1 \vartheta^{\epsilon_r^1}), \qquad \overline{b_r} = \rho_r \lambda_1 \lambda_2 \frac{\partial}{\partial(\lambda_1 \lambda_2)} \log g_{12}(\lambda_1 \lambda_2 \vartheta^{\eta_r}), \qquad \text{......(450)}$$

$$\overline{M_1} = \lambda_1 \frac{\partial}{\partial \lambda_1} \Sigma_r \varpi_r^1 \log g_1(\lambda_1 \vartheta^{\epsilon_r^1}), \qquad \overline{N} = \lambda_1 \lambda_2 \frac{\partial}{\partial(\lambda_1 \lambda_2)} \Sigma_r \rho_r \log g_{12}(\lambda_1 \lambda_2 \vartheta^{\eta_r}), \qquad \text{......(451)}$$

$$\overline{E_{A^1}} = \vartheta \frac{\partial}{\partial \vartheta} \Sigma_r \varpi_r^1 \log g_1(\lambda_1 \vartheta^{\epsilon_r^1}), \qquad \overline{E_{A^1A^2}} = \vartheta \frac{\partial}{\partial \vartheta} \Sigma_r \rho_r \log g_{12}(\lambda_1 \lambda_2 \vartheta^{\eta_r}). \qquad \text{......(452)}$$

In the classical limit the equations defining λ_1, λ_2 and ϑ become

$$\lambda_1 f_1(\vartheta) + \lambda_1 \lambda_2 f_{12}(\vartheta) = X_1, \qquad \text{......(453)}$$

$$\lambda_2 f_2(\vartheta) + \lambda_1 \lambda_2 f_{12}(\vartheta) = X_2, \qquad \text{......(454)}$$

$$\lambda_1 \vartheta \frac{\partial}{\partial \vartheta} f_1(\vartheta) + \lambda_2 \vartheta \frac{\partial}{\partial \vartheta} f_2(\vartheta) + \lambda_1 \lambda_2 \vartheta \frac{\partial}{\partial \vartheta} f_{12}(\vartheta) = E. \qquad \text{......(455)}$$

The equations (450)–(452) become in the classical limit, in terms of T,

$$\overline{a_r^1} = \lambda_1 \varpi_r^1 e^{-\epsilon_r^1/kT}, \qquad \overline{b_r} = \lambda_1 \lambda_2 \rho_r e^{-\eta_r/kT}, \qquad \text{......(456)}$$

$$\overline{M_1} = \lambda_1 f_1(T), \qquad \overline{N} = \lambda_1 \lambda_2 f_{12}(T), \qquad \text{......(457)}$$

$$\overline{E_{A^1}} = \lambda_1 k T^2 \frac{\partial}{\partial T} f_1(T), \qquad \overline{E_{A^1A^2}} = \lambda_1 \lambda_2 k T^2 \frac{\partial}{\partial T} f_{12}(T). \qquad \text{......(458)}$$

In these classical equations $f_1(T)$, $f_2(T)$, $f_{12}(T)$ are of course the ordinary partition functions. The complete form of the *equation of mass-action* follows at once from (457) by eliminating λ_1 and λ_2. We find

$$\frac{\overline{M_1}\,\overline{M_2}}{\overline{N}}=\frac{f_1(T)f_2(T)}{f_{12}(T)}. \qquad \text{......(459)}$$

Thus the equilibrium constant of the law of mass-action is expressible at once, in the simplest way, in terms of partition functions. Equation (459) is also known as *the reaction isochore*.

It is extremely seldom that any equation more accurate than (459) is required. But when classical statistics is inadequate a similar equation of mass-action can still be obtained by eliminating λ_1 and λ_2 from the equations (451). Only the form of the eliminant need be recorded here. On employing (205) to evaluate (451) we see that (451) takes in all cases the form

$$\frac{\overline{M_1}}{V}=h_1(\lambda_1,T). \qquad \text{......(460)}$$

Equation (460) and the similar equations for λ_2 and $\lambda_1\lambda_2$ can be solved for λ_1, λ_2 and $\lambda_1\lambda_2$. The eliminant is therefore of the form

$$A_1(\overline{M_1}/V,T)\,A_2(\overline{M_2}/V,T)=A_{12}(\overline{N}/V,T); \qquad \text{(461)}$$

which is the generalized law of mass-action. The functions A_1, etc. reduce in the classical limit to $\overline{M_1}/f_1(T)$, etc., and could be made to play the part of the activities of classical thermodynamics.

Under special assumptions the functions h_1, etc. may depend in a specially simple way on T, with corresponding simplifications in the form of (461).

Integrals can obviously be constructed, which will enable us to evaluate fluctuations in these assemblies, and to prove all such relations as

$$\overline{(P-\overline{P})^2}=O(\overline{P}),$$

which guarantee the effective possession of normality. We need not stop over these points further here. Details will be found in Chapter XX.

It is clear that just the same analysis applies to an assembly in which the single reaction is $2A^1 \rightleftarrows A_2{}^1$. Equation (460) then becomes

$$\frac{(\overline{M_1})^2}{\overline{N}}=\frac{\{f_1(T)\}^2}{f_{11}(T)}, \qquad \text{......(462)}$$

where $f_{11}(T)$ is the partition function for the molecule $A_2{}^1$ taken over all its accessible states, but these states must now be enumerated with due regard to symmetry requirements in the nuclei, which are absent for A^1A^2.

§ **5·21.** *Special forms of the reaction isochore.* Let us consider the consequences of (459) and (462) in rather more detail in certain special cases.

Let us suppose that A^1 and A^2 are atoms of masses m_1 and m_2 whose states of least internal energy are of weight ϖ_1 and ϖ_2, all other states being irrelevant. Let us define the energy zero for the assembly to be that state in which all the atoms of both types are free in their states of lowest translatory and internal energy. It is no matter whether such a state is or is not an accessible state of the assembly. Then

$$f_1(T) = \frac{(2\pi m_1 kT)^{\frac{3}{2}} V}{h^3} \varpi_1, \quad f_2(T) = \frac{(2\pi m_2 kT)^{\frac{3}{2}} V}{h^3} \varpi_2,$$

while

$$f_{12}(T) = \frac{\{2\pi(m_1+m_2)kT\}^{\frac{3}{2}} V}{h^3} b'(T),$$

where $b'(T)$ is the partition function for the internal energy and rotational energy of the molecule A^1A^2. Relative to the assigned energy zero the normal state of lowest energy of the molecule will have an energy $-\chi$. It is therefore convenient to write

$$b'(T) = e^{\chi/kT} b(T)$$

so that $b(T)$ begins with a constant term, and is the partition function for the internal and rotational energy of the molecule referred to its own state of lowest energy as energy zero. If we express (459) in terms of average molecular densities $\nu_1 = \overline{M_1}/V$, etc., we find

$$\frac{\nu_1 \nu_2}{\nu_{12}} = \left(\frac{2\pi m_1 m_2 k}{m_1+m_2}\right)^{\frac{3}{2}} \frac{T^{\frac{3}{2}} e^{-\chi/kT}}{h^3} \frac{\varpi_1 \varpi_2}{b(T)}. \qquad \ldots\ldots(463)$$

It is convenient to collect the commoner approximate forms which (463) may be expected to assume, but in so doing we shall ignore all complications due to isotopic mixtures and non-combining sets of states. Such questions are examined in Chapter VII.

We may assume to this approximation that the rotations are independent of the vibrations and other internal motions so that $b(T)$ factorizes into $r(T)v(T)$. When $T \to 0$ $b(T) \to \varpi_{12}{}^0$, the weight of the molecular state of lowest possible energy. As T increases to a range in which $kT \gg \epsilon = h^2/8\pi^2 A$ the rotations become classical and

$$r(T) \sim 8\pi^2 AkT/\sigma h^2.$$

The symmetry number σ is 1 for the molecule A^1A^2 and 2 for $A_2{}^1$ or $A_2{}^2$, because half the rotational states drop out as in § 3·4 owing to symmetry requirements. In the early part of this range we shall often still have $kT \ll h\nu_0$, where ν_0 is the fundamental frequency of the nuclear oscillations. Then $v(T) \sim 1$ for a structureless harmonic oscillator. But if we include in $v(T)$ weight factors due to electronic or nuclear structure we shall have more generally $v(T) \sim \varpi_{12}$.

As T increases still further the higher vibrational states become significant, and so long as $kT \ll \chi$ and the effective states are nearly those of a simple harmonic oscillator we shall have sufficiently nearly

$$v(T) = \varpi_{12}/(1 - e^{-h\nu_0/kT}).$$

In the range (if any) in which $h\nu_0 \ll kT \ll \chi$ this gives approximately

$$v(T) \sim \varpi_{12} kT/h\nu_0.$$

Summing up this discussion we see that the reaction isochore takes successively the following standard simplified forms:

$$\frac{\nu_1 \nu_2}{\nu_{12}} = \left(\frac{2\pi m_1 m_2 k}{m_1 + m_2}\right)^{\frac{3}{2}} \frac{1}{h^3} \frac{\varpi_1 \varpi_2}{\varpi_{12}{}^0} T^{\frac{3}{2}} e^{-\chi/kT} \qquad (kT \ll \epsilon), \qquad \ldots\ldots(464)$$

$$= \left(\frac{2\pi m_1 m_2 k}{m_1 + m_2}\right)^{\frac{3}{2}} \frac{\sigma}{8\pi^2 hAk} \frac{\varpi_1 \varpi_2}{\varpi_{12}} T^{\frac{1}{2}} e^{-\chi/kT} \qquad (\epsilon \ll kT \ll h\nu_0, \chi), \qquad \ldots\ldots(465)$$

$$= \left(\frac{2\pi m_1 m_2 k}{m_1 + m_2}\right)^{\frac{3}{2}} \frac{\sigma \nu_0}{8\pi^2 Ak^2} \frac{\varpi_1 \varpi_2}{\varpi_{12}} T^{-\frac{1}{2}} e^{-\chi/kT} \qquad (\epsilon, h\nu_0 \ll kT \ll \chi). \qquad \ldots\ldots(466)$$

These formulae for the reaction isochore are of very well-known forms, of recognized validity. It is unnecessary to undertake a direct comparison of the whole of these and similar formulae with experiment, because, as we shall see in Chapter VII, these formulae and the general equation (459) really only differ from the formulae of thermodynamics by fixing the precise value of the constant factor—or in other words by fixing the "chemical constants". The ultimate test of the theory need therefore only be made by comparing observed and calculated values of these constants and to this we shall return in connection with Nernst's heat theorem in Chapter VII.

§5·3. *Gaseous assemblies with any numbers of components and reactions.* In view of the preliminary formulation of the simple case above, it is now only necessary to specify a notation suitable for the general gaseous assembly. Let the different types of atoms be denoted by the affix u, molecules by the affix v. Then the energy, weight and number of free atoms of type u in their rth state will be denoted by $\epsilon_r{}^u$, $\varpi_r{}^u$ and $a_r{}^u$. For molecules of type v the corresponding quantities are $\epsilon_r{}^v$, $\varpi_r{}^v$ and $a_r{}^v$. If A^u is the atomic symbol for the atom of type u, the chemical symbol for the molecule of type v will be

$$\Pi_u A^u{}_{q_v{}^u}.$$

It is assumed that the molecule of type v contains $q_v{}^u$ atoms of type u. All possible reactions may then be regarded as contained in the set

$$\Sigma_u q_v{}^u A^u \rightleftarrows \Pi_u A^u{}_{q_v{}^u} \qquad (v = 1, 2, \ldots, j),$$

or constructed out of members of this set. The actual sequence of reactions by which equilibrium is attained is without effect on that equilibrium, so long at least as reactions exist sufficient to preserve unimpaired the enumeration of accessible states.

Let the number of atoms of type u be X_u, there being t types of atom in all, and in any example the number of free atoms of type u, M_u, and of molecules of type v, N_v. Then

$$M_u + \Sigma_v q_v{}^u N_v = X_u \quad (u = 1, 2, \ldots, t). \qquad \ldots\ldots(467)$$

To preserve the correct atomic and molecular totals we require

$$\Sigma_r a_r{}^u = M_u \quad (u = 1, 2, \ldots, t), \qquad \Sigma_r a_r{}^v = N_v \quad (v = 1, 2, \ldots, j),$$

and therefore in general

$$\Sigma_r a_r{}^u + \Sigma_v \Sigma_r q_v{}^u a_r{}^v = X_u \quad (u = 1, 2, \ldots, t). \qquad \ldots\ldots(468)$$

To satisfy the energy equation we require also

$$\Sigma_u \Sigma_r a_r{}^u \epsilon_r{}^u + \Sigma_v \Sigma_r a_r{}^v \epsilon_r{}^v = E. \qquad \ldots\ldots(469)$$

To enumerate complexions subject to (467)–(469) we have now merely to form the analogue of (439) remembering that degeneracies have been admitted by the explicit introduction of the $\varpi_r{}^u$, $\varpi_r{}^v$. The required expression is obviously

$$\Pi_u \Pi_r [g_u(x_u z^{\epsilon_r{}^u})]^{\varpi_r{}^u} \Pi_v \Pi_r [g_v(x_1{}^{q_v{}^1} \ldots x_t{}^{q_v{}^t} z^{\epsilon_r{}^v})]^{\varpi_r{}^v}. \qquad \ldots\ldots(470)$$

The total number of complexions is then given by picking out the coefficient of $x_1{}^{X_1} \ldots x_t{}^{X_t} z^E$ in (470). We find therefore

$$C = \frac{1}{(2\pi i)^{t+1}} \int^{(t+1)} \cdots \int \frac{dz\, \Pi_u\, dx_u}{z^{E+1} \Pi_u x_u{}^{X_u+1}} \Pi_u \Pi_r [g_u(x_u z^{\epsilon_r{}^u})]^{\varpi_r{}^u}$$
$$\times \Pi_v \Pi_r [g_v(x_1{}^{q_v{}^1} \ldots x_t{}^{q_v{}^t} z^{\epsilon_r{}^v})]^{\varpi_r{}^v}, \qquad \ldots\ldots(471)$$

with similar integrals for $C\overline{a_r{}^u}$, *etc.*, containing extra factors similar to those in (441), (442) and the following expressions. The unique saddle point of these multiple integrals, $\lambda_1, \lambda_2, \ldots, \lambda_t, \vartheta$ is then defined by the system of equations

$$\frac{\partial \Sigma}{\partial \lambda_1} = \frac{\partial \Sigma}{\partial \lambda_2} = \ldots = \frac{\partial \Sigma}{\partial \lambda_t} = \frac{\partial \Sigma}{\partial \vartheta} = 0, \qquad \ldots\ldots(472)$$

where $\quad \Sigma = Z - X_1 \log \lambda_1 - X_2 \log \lambda_2 - \ldots - X_t \log \lambda_t - E \log \vartheta, \qquad \ldots\ldots(473)$

$$Z = \Sigma_u \Sigma_r \varpi_r{}^u \log g_u(\lambda_u \vartheta^{\epsilon_r{}^u}) + \Sigma_v \Sigma_r \varpi_r{}^v \log g_r(\lambda_1{}^{q_v{}^1} \ldots \lambda_t{}^{q_v{}^t} \vartheta^{\epsilon_r{}^v}). \qquad \ldots\ldots(474)$$

With the values of $\lambda_1, \lambda_2, \ldots, \lambda_t, \vartheta$ so determined we find

$$\overline{a_r{}^u} = \varpi_r{}^u \lambda_u \frac{\partial}{\partial \lambda_u} \log g_u(\lambda_u \vartheta^{\epsilon_r{}^u}), \quad \overline{M_u} = \Sigma_r \overline{a_r{}^u}, \qquad \ldots\ldots(475)$$

$$\overline{a_r{}^v} = \varpi_r{}^v \mu \frac{\partial}{\partial \mu} \log g_v(\mu \vartheta^{\epsilon_r{}^v}), \quad \overline{N_v} = \Sigma_r \overline{a_r{}^v} \quad (\mu = \lambda_1{}^{q_v{}^1} \ldots \lambda_t{}^{q_v{}^t}). \quad \ldots(476)$$

Analogues of (452) can easily be supplied. In the classical limit these equations take the form

$$\overline{a_r{}^u} = \lambda_u \varpi_r{}^u e^{-\epsilon_r{}^u/kT}, \quad \overline{M_u} = \lambda_u f_u(T), \quad \overline{a_r{}^u} = \overline{M_r} \varpi_r{}^u e^{-\epsilon_r{}^u/kT} / f_u(T), \qquad \ldots\ldots(477)$$

$$\overline{a_r{}^v} = \lambda_1{}^{q_v{}^1} \ldots \lambda_t{}^{q_v{}^t} \varpi_r{}^v e^{-\epsilon_r{}^v/kT}, \quad \overline{N_v} = \lambda_1{}^{q_v{}^1} \ldots \lambda_t{}^{q_v{}^t} f_v(T), \quad \overline{a_r{}^v} = \overline{N_v} \varpi_r{}^v e^{-\epsilon_r{}^v/kT} / f_v(T).$$
$$\ldots\ldots(478)$$

The laws of mass-action follow from (477) and (478) in the form

$$\frac{\overline{N_v}}{\Pi_u(\overline{M_u})^{q_v^u}}=\frac{f_v(T)}{\Pi_u\{f_u(T)\}^{q_v^u}}\qquad (v=1,2,\ldots,j).\qquad \ldots\ldots(479)$$

We can obtain also the energy distribution laws in the form

$$\overline{E_u}=\lambda_u kT^2\frac{\partial}{\partial T}f_u(T)=\overline{M_u}kT^2\frac{\partial}{\partial T}\log f_u(T),\qquad \ldots\ldots(480)$$

$$\overline{E_v}=\lambda_1^{q_v^1}\ldots\lambda_t^{q_v^t}kT^2\frac{\partial}{\partial T}f_v(T)=\overline{N_v}kT^2\frac{\partial}{\partial T}\log f_v(T).\quad \ldots(481)$$

Thus all the details of the distribution laws for such an assembly are the same as if it were not dissociating but had the numbers of the various species fixed at their equilibrium values.

For the reaction of the assembly on external bodies we find at once as in § 2·8 that

$$\overline{Y}=\Sigma_u\Sigma_r\overline{a_r^u}\left(-\frac{\partial\epsilon_r^u}{\partial y}\right)+\Sigma_v\Sigma_r\overline{a_r^v}\left(-\frac{\partial\epsilon_r^v}{\partial y}\right).$$

On using (475) and (476) we find easily that

$$\overline{Y}=\frac{1}{\log 1/\vartheta}\frac{\partial Z}{\partial y}=kT\frac{\partial Z}{\partial y},\qquad \ldots\ldots(482)$$

where Z is given by (474). In the classical limit this reduces to

$$\overline{Y}=\frac{1}{\log 1/\vartheta}\left\{\Sigma_u\overline{M_u}\frac{\partial}{\partial y}\log f_u+\Sigma_v\overline{N_v}\frac{\partial}{\partial y}\log f_v\right\}.\qquad \ldots\ldots(483)$$

In the case of local boundary fields this reduces of course to the ordinary equation of state for a "perfect" gaseous assembly (in which however the number of constituents is a function of the temperature),

$$p=\frac{kT}{V}\{\Sigma_u\overline{M_u}+\Sigma_v\overline{N_v}\}.\qquad \ldots\ldots(484)$$

§ 5·31. *A mixture of isotopes.* A simple example of the general dissociating assembly is a gaseous mixture of isotopes of a single element in which the relevant species are atoms A and molecules A_2 formed from the isotopes A^1 and A^2 whose relative abundances are D_1 and D_2, $D_1+D_2=1$. There are three distinct molecular species A^1A^1, A^1A^2 and A^2A^2, whose relative abundance must be determined. Equations (479), in an obvious notation, give here

$$\frac{\overline{N_{11}}}{(\overline{M_1})^2}=\frac{f_{11}}{(f_1)^2},\quad \frac{\overline{N_{12}}}{\overline{M_1}\,\overline{M_2}}=\frac{f_{12}}{f_1f_2},\quad \frac{\overline{N_{22}}}{(\overline{M_2})^2}=\frac{f_{22}}{(f_2)^2}$$

which combine to give

$$\frac{\overline{N_{11}}\,\overline{N_{22}}}{(\overline{N_{12}})^2}=\frac{f_{11}f_{22}}{(f_{12})^2}.\qquad \ldots\ldots(485)$$

The interesting case occurs when the atoms are practically all combined in molecules; we may then take with (485) the equations

$$2\overline{N_{11}}+\overline{N_{12}}=D_1X,\quad 2\overline{N_{22}}+\overline{N_{12}}=D_2X, \qquad \text{......(486)}$$

where X is the total number of atoms. Equations (485) and (486) are sufficient to fix the desired ratios.

It will shortly appear that in general $f_{11}f_{22}/(f_{12})^2$ is very nearly equal to $\frac{1}{4}$; writing it $\frac{1}{4}(1+\delta)$ we find on eliminating N_{11} and N_{22}

$$\delta(\overline{N_{12}})^2+X\overline{N_{12}}-D_1D_2X^2=0.$$

The relevant root is $X\{(1+4\delta D_1D_2)^{\frac{1}{2}}-1\}/2\delta$, since it must clearly reduce to D_1D_2X as $\delta\to 0$. Thus, to the first order in δ,

$$\overline{N_{11}}:\overline{N_{22}}:\overline{N_{12}}=D_1^2(1+\delta D_2^2):D_2^2(1+\delta D_1^2):2D_1D_2(1-\delta D_1D_2). \qquad \text{....(487)}$$

Since δ is small these ratios are very nearly $D_1^2:D_2^2:2D_1D_2$ which are the ratios one would find if the pairing of the molecules took place according to the laws of chance without preference for one or other structure. And in fact δ is generally so small that the departures of (487) from this law are not significant.

Let us now examine how the isotopic differences can enter into (485). A^1 and A^2 have different masses m_1 and m_2 and different nuclear spins with different orientational weights ρ_1 and ρ_2. These differences affect the molecular partition functions as follows:

(i) The translational factors $(2\pi mkT)^{\frac{3}{2}}V/h^3$ differ in the masses which are $2m_1$, $2m_2$ and m_1+m_2 respectively. This introduces a resultant factor

$$\left[\frac{(m_1m_2)^{\frac{1}{2}}}{\frac{1}{2}(m_1+m_2)}\right]^3$$

on the right-hand side of (485).

(ii) Electronic states are identical for all the molecular types and introduce no resultant factor.

(iii) Ignoring nuclear vibration, let us take the rotational states to be, with sufficient accuracy, those of a rigid rotator without axial spin. The rotational partition function must allow for the symmetry requirements of the nuclei, account being taken of their spins. In a region of high temperature, particularly if dissociation and recombination is occurring, there will be effective interchange between the states which are antisymmetrical in the rotations and those which are symmetrical, so that we have to deal with only a single set of rotational states.

We must next enumerate the number of orientational states, for a pair of nuclei A^1 say, which are respectively symmetrical and antisymmetrical in the nuclei. Let the nuclear spin wave-functions be $\phi_1, \ldots, \phi_{\rho_1}$ and $\psi_1, \ldots, \psi_{\rho_1}$

for the two nuclei. Suppose we take a pair of spin functions ϕ_s, ψ_s. From these we can form just one symmetrical combined function, $\phi_s\psi_s$ and there are ρ_1 such pairs. Suppose on the other hand we have a pair of spin functions ϕ_r, ψ_s $(r \neq s)$. Let us combine with these the pair ϕ_s, ψ_r. We can form from these two pairs, one symmetrical and one antisymmetrical, $\phi_r\psi_s \pm \phi_s\psi_r$ respectively, and there are $\frac{1}{2}\rho_1(\rho_1 - 1)$ such pairs. There are therefore $\frac{1}{2}\rho_1(\rho_1+1)$ symmetrical spin combinations and $\frac{1}{2}\rho_1(\rho_1-1)$ antisymmetrical.

We can now construct the complete rotational partition function $r(T)$, thus generalizing § 3·4. For a molecule whose wave-functions must be symmetrical in the nuclei we have

$$r(T) = \tfrac{1}{2}\rho_1(\rho_1+1) \sum_{n=0,2,4,\ldots} (2n+1)\, e^{-n(n+1)\sigma} + \tfrac{1}{2}\rho_1(\rho - 1) \sum_{n=1,3,5,\ldots} (2n+1)\, e^{-n(n+1)\sigma}$$

$$\left(\sigma = \frac{h^2}{8\pi^2 I_{11} kT}\right), \quad \text{......(488)}$$

where I_{11} is the moment of inertia of the molecule A^1A^1. If the wave-functions must be antisymmetrical in the nuclei, then

$$r(T) = \tfrac{1}{2}\rho_1(\rho_1-1) \sum_{n=0,2,4,\ldots} (2n+1)\, e^{-n(n+1)\sigma} + \tfrac{1}{2}\rho_1(\rho_1+1) \sum_{n=1,3,5,\ldots} (2n+1)\, e^{-n(n+1)\sigma}. \quad \text{......(489)}$$

Using the results of §§ 3·31, 3·4 we see that both (488) and (489) have the asymptotic form

$$r(T) \sim \tfrac{1}{2}\rho_1^2 \frac{8\pi^2 I_{11} kT}{h^2} \quad (T \to \infty), \quad \text{......(490)}$$

in agreement with the less detailed discussion of § 5·21. The asymptotic forms for the molecules A^2A^2 and A^1A^2 are similarly seen to be respectively

$$\tfrac{1}{2}\rho_2^2 \frac{8\pi^2 I_{22} kT}{h^2}, \quad \rho_1\rho_2 \frac{8\pi^2 I_{12} kT}{h^2}. \quad \text{......(491)}$$

The latter is immediate, for since there are no symmetry requirements for A^1A^2

$$r(T) = \rho_1\rho_2 \sum_{n=0,1,2,\ldots} (2n+1)\, e^{-n(n+1)\sigma}. \quad \text{......(492)}$$

If we now combine these results for a region in which the rotations are effectively classical we find that the rotational factors contribute the factor

$$\frac{1}{4}\frac{I_{11}I_{22}}{I_{12}^2}$$

to the right-hand side of (485).

(iv) We have still to examine the vibrational energy and the exact energy zero. The partition functions we have used have assumed that all the molecules may be taken to have zero energy when in their state of lowest translatory rotational and vibrational motion. This may not be exactly correct. If we take the usual energy zero as that state in which all the molecules are dissociated and all the atoms at rest in the lowest quantum state, then the lowest molecular states will have energies $-\chi_{11}$, $-\chi_{22}$, and $-\chi_{12}$ respectively, and these χ's will not be exactly equal. For example the zero-point energy $\frac{1}{2}h\nu$ in the vibrational motion will differ slightly from molecule to molecule. These differences in the energies of formation of the lowest states introduce the resultant factor

$$e^{(\chi_{11}+\chi_{22}-2\chi_{12})/kT}$$

into the right-hand side of (485).

If the vibrational quanta $h\nu$ are fairly large compared with kT, the vibrational partition functions introduce no further factors. But at higher temperatures for which $h\nu \ll kT \ll \chi$ the vibrational partition functions reduce as in § 5·21 to $kT/h\nu$. They then contribute the extra factor

$$\nu_{12}^2/\nu_{11}\nu_{22}$$

to the right-hand side of (485).

Collecting these results we see that (485) becomes

$$\frac{\overline{N_{11}}\,\overline{N_{22}}}{(\overline{N_{12}})^2} = \frac{1}{4}\left[\frac{(m_1 m_2)^{\frac{1}{2}}}{\frac{1}{2}(m_1+m_2)}\right]^3 \frac{I_{11}I_{22}}{I_{12}^2}\, e^{(\chi_{11}+\chi_{22}-2\chi_{12})/kT}, \qquad \text{......(493)}$$

or

$$\frac{\overline{N_{11}}\,\overline{N_{22}}}{(\overline{N_{12}})^2} = \frac{1}{4}\left[\frac{(m_1 m_2)^{\frac{1}{2}}}{\frac{1}{2}(m_1+m_2)}\right]^3 \frac{I_{11}I_{22}}{I_{12}^2}\,\frac{\nu_{12}^2}{\nu_{11}\nu_{22}}\, e^{(\chi_{11}+\chi_{22}-2\chi_{12})/kT} \qquad \text{......(494)}$$

according to the temperature range.

It may generally be assumed that the internuclear forces and distances in these isotopic molecules are unaffected by the changes of mass. When this condition is satisfied

$$I_{11} : I_{22} : I_{12} = \tfrac{1}{2}m_1 : \tfrac{1}{2}m_2 : \frac{m_1 m_2}{m_1+m_2}, \qquad \text{......(495)}$$

$$\nu_{11} : \nu_{22} : \nu_{12} = \sqrt{\frac{2}{m_1}} : \sqrt{\frac{2}{m_2}} : \sqrt{\frac{m_1+m_2}{m_1 m_2}}. \qquad \text{......(496)}$$

Further, the χ's will then only differ by the different zero-point energy of the nuclear vibrations so that

$$\chi_{11}+\chi_{22}-2\chi_{12} = \tfrac{1}{2}h(2\nu_{12}-\nu_{11}-\nu_{22}),$$

$$= h\nu_{12}\left[1-\frac{\sqrt{m_1}+\sqrt{m_2}}{\sqrt{2(m_1+m_2)}}\right]. \qquad \text{......(497)}$$

Equations (493) and (494) then reduce to

$$\frac{\overline{N_{11}}\,\overline{N_{22}}}{(\overline{N_{12}})^2} = \frac{1}{4}\frac{(m_1 m_2)^{\frac{1}{2}}}{\frac{1}{2}(m_1+m_2)} \exp\left\{\frac{h\nu_{12}}{kT}\left[1-\frac{\sqrt{m_1}+\sqrt{m_2}}{\sqrt{2(m_1+m_2)}}\right]\right\}, \qquad \text{......(498)}$$

or

$$\frac{\overline{N_{11}}\,\overline{N_{22}}}{(\overline{N_{12}})^2} = \tfrac{1}{4}\exp\left\{\frac{h\nu_{12}}{kT}\left[1-\frac{\sqrt{m_1}+\sqrt{m_2}}{\sqrt{2(m_1+m_2)}}\right]\right\}. \qquad \text{......(499)}$$

A compact formula for the δ of (487) is not available, but δ is easily computed in any actual case. For example for the isotopes of hydrogen at ordinary temperatures (498) applies and (H^1,H^2)

$$\delta = 0{\cdot}943e^{92/T} - 1 \sim 0{\cdot}27.$$

This is an extreme case. For the isotopes of chlorine, masses 35 and 37, at ordinary temperatures either (498) or (499) give

$$\delta \sim 10^{-4},$$

which is entirely trivial.* An exact comparison of the observed and calculated equilibrium for light and heavy hydrogen is recorded in Table 14, § 7·4.

§ 5·4. *A simple assembly with crystalline and gaseous phases. Evaporation.* We shall now show how to incorporate both crystals and gases into the same assembly. We shall start with a simple example, in which both the crystal and the gas may be supposed to be built up entirely of atoms of a single kind, the same in both phases; these atoms may further be treated as structureless massive points. In view of the discussions of the preceding chapters the necessary expressions for the number of complexions and for any average value can be rapidly constructed and no great detail is necessary.

Suppose that in a particular example of this assembly there are X atoms in all, N in the gaseous phase and P in the crystalline, and let us start by enumerating the number of complexions of the gas when its energy is F. We find as in § 2·4 that the number of complexions is the coefficient of $x^N z^F$ in $\Pi_r g(xz^{\epsilon_r})$. This enumeration of accessible states for the gas by itself has of course been made with due regard to symmetry requirements.

Let us now consider the crystalline phase by itself when it is composed of P atoms and has the energy U. We have constructed a partition function for such a crystal in § 4·5. Equation (349) gives the most general form there found. In this equation F_0 is the zero-point energy of the crystal, referred to a state of complete dispersion as the state of energy zero. F_0 is therefore the total heat of evaporation at zero temperature, and of the form $-P\phi_0$, where ϕ_0 is independent of P. It follows that the partition function $K(z)$ is of the form

$$K(z) = [\kappa(z)]^P, \qquad \ldots\ldots(500)$$

where $\kappa(z)$ is independent of P, for large P, to the approximation used. It may be assumed that in all cases the form (500) is a valid approximation for large P. The number of complexions for the crystal of energy U, containing P atoms, is the coefficient of z^U in $[\kappa(z)]^P$. The partition functions of Chapter IV were constructed without reference to symmetry requirements, since the atoms in a crystal are distinct localized systems. By

* A detailed study of the hydrogen isotopes has been published by Urey and Teal, *Rev. Mod. Phys.* vol. 7, p. 34 (1935); see also Farkas, *Light and Heavy Hydrogen*, Cambridge (1935).

the arguments of § 2·26 this yields a correct enumeration of the complexions.*

Now the crystal is merely a super molecule, and therefore by Lemma 5·11, in enumerating the complexions of an assembly containing a crystal (or crystals) and a gas, it is sufficient to satisfy the symmetry requirements for the atoms in the gas and for the crystal separately as if the free atoms and the crystal were permanent systems. The number of complexions for this example of the assembly in which there are N atoms in the gas with energy F and P atoms in the crystal with energy U is therefore

$$[\text{coeff.}\, x^N z^F \text{ in } \Pi_r g(xz^{\epsilon_r})] \times [\text{coeff.}\, z^U \text{ in } [\kappa(z)]^P].$$

The total number of complexions C_N for this example when the total energy of crystal and gas together is $E\,(=F+U)$ is therefore the coefficient of $x^N z^E$ in

$$[\kappa(z)]^{X-N}\,\Pi_r g(xz^{\epsilon_r}),$$

or, on applying Cauchy's theorem,

$$C_N = \frac{1}{(2\pi i)^2} \iint \frac{dx\,dz}{x^{N+1} z^{E+1}} [\kappa(z)]^{X-N}\,\Pi_r g(xz^{\epsilon_r}). \qquad \text{......(501)}$$

As before this number has been enumerated on the assumption that the free atoms and the bound atoms are distinct systems. If we allow for exchanges between the gas and the crystal (keeping N constant), then we must make the wave-functions of the assembly symmetrical or antisymmetrical in *all* the atoms bound and free together, and the number of distinct complexions is still given by (501).

In an assembly in which evaporation and condensation are proceeding the value of N is not fixed, as we have hitherto regarded it, but all values of N from 0 to X are possible. The total number of complexions summed for all examples is therefore given by

$$C = \sum_0^X C_N. \qquad \text{......(502)}$$

On combining (501) and (502) we find at once

$$C = \frac{1}{(2\pi i)^2} \iint \frac{dx\,dz}{x^{X+1} z^{E+1}}\,\Pi_r g(xz^{\epsilon_r})\,\frac{1-\{x\kappa(z)\}^{X+1}}{1-x\kappa(z)}. \qquad \text{......(503)}$$

Since the expansion of $\{x\kappa(z)\}^{X+1}/\{1-x\kappa(z)\}$ in powers of x begins with a term in x^{X+1}, it can contribute nothing to the integral and we may write

$$C = \frac{1}{(2\pi i)^2} \iint \frac{dx\,dz}{x^{X+1} z^{E+1}}\,\frac{\Pi_r g(xz^{\epsilon_r})}{1-x\kappa(z)}. \qquad \text{......(504)}$$

At first sight one might expect to be able to apply Theorem 2·42 to the evaluation of one or other form, but this cannot conveniently be done.

* A more detailed verification of this equivalence is given by Fowler and Sterne, *Rev. Mod. Phys.* vol. 4, p. 635 (1932), esp. Appendix I.

If we determine the saddle point of the integrand of (503) or (504) in the usual way and if, at the saddle point λ, ϑ, $\lambda\kappa(\vartheta) \leqslant 1-\delta < 1$, all the conditions of this theorem apply and from these and similar integrals we can determine all the average values for the assembly. But it is easy to see that this condition also implies that the crystal is a trivial part of the assembly, for the factor $1/\{1-x\kappa(z)\}$ then plays no part in determining λ, ϑ or the value of $\log C$ and the assembly is effectively the gas phase alone. For the crystal to matter it is essential that at the saddle point (determined as usual)

$$\lambda\kappa(\vartheta) \simeq 1. \qquad \text{......(505)}$$

But the conditions of Theorem 2·42 then break down† and the results (which remain true) are more easily determined by another method as follows.

§5·41. *Special methods for evaluating C.* We can legitimately use Theorem 2·42 to evaluate C_N, or any similar integral. We find that

$$\log C_N = \Sigma_r \log g(\lambda\vartheta^{\epsilon_r}) + (X-N)\log\kappa(\vartheta) - N\log\lambda - E\log\vartheta$$
$$+\log\left[\frac{\Delta^{-\frac{1}{2}}}{2\pi X} + O\left(\frac{1}{X^2}\right)\right], \qquad \text{......(506)}$$

where λ, ϑ is the unique root, in the real positive domain, of the pair of equations,

$$\lambda\frac{\partial}{\partial\lambda}\Sigma_r \log g(\lambda\vartheta^{\epsilon_r}) - N = 0, \qquad \text{......(507)}$$

$$\vartheta\frac{\partial}{\partial\vartheta}\Sigma_r \log g(\lambda\vartheta^{\epsilon_r}) + (X-N)\frac{\vartheta\kappa'(\vartheta)}{\kappa(\vartheta)} - E = 0. \qquad \text{......(508)}$$

In these equations Δ is a function of λ, ϑ, and the ratios only of X, N, E and V, the volume of the gaseous phase, and its variations when N varies are unimportant compared with those of the other terms in (506) and will be neglected.

The equilibrium state of the assembly is determined by (502) and similar equations such as

$$C\overline{N} = \sum_0^X N C_N, \qquad \text{......(509)}$$

$$C\overline{\alpha} = \sum_0^X \overline{\alpha_N} C_N. \qquad \text{......(510)}$$

In the last equation α is the quantity whose equilibrium value is desired and $\overline{\alpha_N}$ is the average value of α taken over all complexions of an example for given N. To evaluate these sums we start by determining the value N^* of N which makes $\log C_N$ a maximum. This value N^* must satisfy the equation

$$\frac{d\log C_N}{dN} = \frac{\partial \log C_N}{\partial\lambda}\frac{d\lambda}{dN} + \frac{\partial\log C_N}{\partial\vartheta}\frac{d\vartheta}{dN} + \frac{\partial\log C_N}{\partial N} = 0,$$

† The arguments of Theorem 2·42 were applied to this case unjustifiably in the first edition of this book. The corrected argument here given alters none of the results.

which reduces on using (507) and (508) to

$$\frac{d\log C_N}{dN}=\frac{\partial \log C_N}{\partial N}=-\log\{\lambda\kappa(\vartheta)\}=0, \quad \text{......(511)}$$

or

$$\lambda=1/\kappa(\vartheta). \quad \text{......(512)}$$

Combining this with (507) we find

$$N^*=\left[\lambda\frac{\partial}{\partial\lambda}\Sigma_r\log g(\lambda\vartheta^{\epsilon_r})\right]_{\lambda=1/\kappa(\vartheta)}. \quad \text{......(513)}$$

To determine the nature of this stationary value (or values) we observe that, since in C_N λ and ϑ are really only auxiliary functions of N satisfying identically

$$\frac{\partial\log C_N}{\partial\lambda}=\frac{\partial\log C_N}{\partial\vartheta}=0$$

it follows that

$$\frac{d^2\log C_N}{dN^2}=\frac{d}{dN}\left(\frac{\partial\log C_N}{\partial N}\right)=-\frac{1}{\lambda}\frac{d\lambda}{dN}-\frac{\vartheta\kappa'(\vartheta)}{\kappa(\vartheta)}\frac{1}{\vartheta}\frac{d\vartheta}{dN}. \quad \text{......(514)}$$

If we show that this is negative, then any stationary value is a maximum and must be unique. The functions $d\lambda/dN$ and $d\vartheta/dN$ are determined by equations obtained by differentiating (507) and (508) which yield

$$A\frac{1}{\lambda}\frac{d\lambda}{dN}+H\frac{1}{\vartheta}\frac{d\vartheta}{dN}=1,$$

$$H\frac{1}{\lambda}\frac{d\lambda}{dN}+B\frac{1}{\vartheta}\frac{d\vartheta}{dN}=\frac{\vartheta\kappa'(\vartheta)}{\kappa(\vartheta)}.$$

In these equations

$$A=\Sigma_r\frac{\lambda\vartheta^{\epsilon_r}d}{d(\lambda\vartheta^{\epsilon_r})}\frac{\lambda\vartheta^{\epsilon_r}g'(\lambda\vartheta^{\epsilon_r})}{g(\lambda\vartheta^{\epsilon_r})}>0,$$

$$B=\Sigma_r\epsilon_r^2\frac{\lambda\vartheta^{\epsilon_r}d}{d(\lambda\vartheta^{\epsilon_r})}\frac{\lambda\vartheta^{\epsilon_r}g'(\lambda\vartheta^{\epsilon_r})}{g(\lambda\vartheta^{\epsilon_r})}+(X-N)\frac{\vartheta d}{d\vartheta}\frac{\vartheta\kappa'(\vartheta)}{\kappa(\vartheta)}>0,$$

$$H=\Sigma_r\epsilon_r\frac{\lambda\vartheta^{\epsilon_r}d}{d(\lambda\vartheta^{\epsilon_r})}\frac{\lambda\vartheta^{\epsilon_r}g'(\lambda\vartheta^{\epsilon_r})}{g(\lambda\vartheta^{\epsilon_r})}.$$

The inequalities follow from the fact that all the functions $g(q)$ and $\kappa(\vartheta)$ are power series with positive coefficients. Solving these equations we find

$$\frac{1}{\lambda}\frac{d\lambda}{dN}+\frac{\vartheta\kappa'(\vartheta)}{\kappa(\vartheta)}\frac{1}{\vartheta}\frac{d\vartheta}{dN}=\frac{A\left(\frac{\vartheta\kappa'(\vartheta)}{\kappa(\vartheta)}\right)^2-2H\frac{\vartheta\kappa'(\vartheta)}{\kappa(\vartheta)}+B}{AB-H^2},$$

which is necessarily positive if $AB-H^2>0$. But this inequality follows at once from the form of the expressions A, B and H. It follows that (512) determines a unique maximum value of $\log C_N$ which of course will only be effective if the N^* so determined is less than X. The general nature of the argument shows that it will be possible to extend it to more complicated cases.

Having shown that C_N possesses a unique maximum at $N=N^*$ (provided $N^*<X$) it is easy to show that, in the neighbourhood of $N=N^*$, C_N can be cast into the form

$$C_N = C_{N^*}\exp\left\{-\frac{1}{2}\frac{(N-N^*)^2}{X}A\right\}\left[1+\frac{N-N^*}{X}\alpha+\frac{(N-N^*)^3}{X^2}\beta + O\left\{\frac{(N-N^*)^2}{X^2}\right\}+O\left\{\frac{(N-N^*)^4}{X^3}\right\}\right],$$

where A, α, β are independent of N and depend only on the ratios of the "large" variables X, V, N^* and E. We can then show, much as in § 2·42, that the immediate neighbourhood of the maximum contributes the whole of the dominant part of C, and thus establish the following

Theorem 5·41. *If C_N is of the form* (506) *and its maximum, determined by $\lambda\kappa(\vartheta)=1$, lies at $N=N^*<X$, then*

$$\log C = \log C_{N^*} + O(\log X). \qquad \text{......(515)}$$

If further $f(N)$ is any slowly varying function of N such that

$$f'(N)/f(N) = O(1/N)$$

and

$$R = \sum_0^X f(N)\, C_N,$$

then

$$R = f(N^*)\, C\left\{1+O\left(\frac{1}{X}\right)\right\}. \qquad \text{......(516)}$$

It follows at once from this theorem that the equilibrium value $\bar{\alpha}$ of any quantity α satisfies the equation

$$\bar{\alpha} = \alpha_{N^*}, \qquad \text{......(517)}$$

and is therefore to be determined by the following rule:

Rule C. Fix the value of N at its equilibrium value N and determine the average value of α for this example of the assembly. This value is equal to the true average value taken over all complexions of the evaporating assembly.*

§ **5·42.** *The vapour pressure of a simple crystal.* The first and most important application of this rule to (509) provides the relation

$$\bar{N} = N^*. \qquad \text{......(518)}$$

Combining this with (513) we have

$$\bar{N} = \left[\lambda\frac{\partial}{\partial\lambda}\Sigma_r \log g(\lambda\vartheta^{\epsilon_r})\right]_{\lambda=1/\kappa(\vartheta)}. \qquad \text{......(519)}$$

This is an equation for the vapour density in equilibrium with the crystal. Since in general classical statistics is sufficiently accurate, (519) reduces to

$$\bar{N} = f(\vartheta)/\kappa(\vartheta). \qquad \text{......(520)}$$

This equation may be cast into a more familiar form. Let us define the energy zero of the assembly to be that state in which all the atoms are condensed on the crystal and the crystal is in its own state of least energy. With this convention and for a simple atomic crystal $\kappa(0)=1$. In its state of least energy in the gas phase an atom possesses an energy χ, which is the heat of evaporation per atom at the absolute zero of temperature. [More strictly what we have thus assumed is that $P\chi$ is the work required to evaporate completely a crystal of P atoms, and that χ is independent of P. This assumption is correct for large P when we can ignore surface effects. The whole discussion would require modification when P gets small enough for surface effects to matter, for then $K(\vartheta)=K(P,\vartheta)\neq[\kappa(\vartheta)]^P$.] The partition function for the atom in the gas is then given by (see equation (172))

$$f(T)=\frac{(2\pi mkT)^{\frac{3}{2}}}{h^3}Ve^{-\chi/kT}. \qquad \ldots\ldots(521)$$

Again from (350) it follows that

$$kT^2\frac{\partial}{\partial T}\log\kappa(T)=\frac{\overline{E}_\kappa}{P}=\frac{1}{P}\int_0^T C_{sol}\,dT''.$$

On integrating again with $\kappa(0)=1$ and $Pk=R$ so that C_{sol} is the specific heat of one gram-molecule, it follows that

$$\log\kappa(T)=\int_0^T\frac{dT'}{RT'^2}\int_0^{T'}C_{sol}\,dT''. \qquad \ldots\ldots(522)$$

[A more elaborate discussion in Chapter VII shows that when $\kappa(T)$ is suitably defined this specific heat can be taken to be the specific heat at constant pressure $(C_p)_{sol}$.] In addition to these formulae we may apply to the vapour the equation of state for a perfect gas, $pV=\overline{N}kT$. Combining all these results we find

$$\log p=-\frac{\chi}{kT}+\tfrac{5}{2}\log T-\int_0^T\frac{dT'}{RT'^2}\int_0^{T'}(C_p)_{sol}\,dT''+\log\frac{(2\pi m)^{\frac{3}{2}}k^{\frac{5}{2}}}{h^3}. \ldots(523)$$

This is the most familiar form for the vapour pressure equation for a simple atomic vapour. It will be compared with experiment in Chapter VII when we shall have at our disposal the corresponding thermodynamic formula.

§5·5. *General assemblies of crystals and vapours.* It is now sufficiently clear how the formulae of §5·4 can be extended to the general case of an assembly with a gaseous phase and any number of crystalline phases each of which consists of a crystal built up out of P molecules of one of the species present in the gaseous phase.

We start by considering an example of the assembly in which there are s different types of crystal effectively present, containing $P_1, \ldots, P_s$ molecules respectively.* The nth of these crystals may be assumed to be composed of molecules whose formula is $\Pi_u A^u{}_{q_n{}^u}$. If the total number of atoms of the uth species in the assembly is X_u, the number free and combined in the gaseous phase will then be $X_u - \sum_{n=1}^{s} q_n{}^u P_n$. By modifying (471) to include the crystalline partition functions we find that the number of complexions $C_{P_1,\ldots,P_s}$ belonging to this example is given by

$$C_{P_1,\ldots,P_s} = \frac{1}{(2\pi i)^{l+1}} \int^{(l+1)} \cdots \int \frac{dz\, \Pi_u\, dx_u}{z^{E+1}\, \Pi_u\, x_u{}^{X_u - \Sigma_n q_n{}^u P_n}}\, \Pi_u\, \Pi_r\, [g_u(x_u z^{\epsilon_r{}^u})]^{\varpi_r{}^u}$$
$$\times\, \Pi_v\, \Pi_r\, [g_v(x_1{}^{q_v{}^1} \ldots x_l{}^{q_v{}^l} z^{\epsilon_r{}^v})]^{\varpi_r{}^v}\, \Pi_n\, [\kappa_n(z)]^{P_n}. \quad \ldots\ldots(524)$$

This can be evaluated in the usual way and the total number of complexions is given by

$$C = \Sigma_{P_1,\ldots,P_s} C_{P_1,\ldots,P_s}. \quad \ldots\ldots(525)$$

It will not be necessary to write out the rather laborious algebra of the next stages. It is clear that we shall find that the terms of (525) have a unique maximum at the values of $P_1, \ldots, P_s$ which satisfy the s equations

$$\Pi_u \lambda_u{}^{q_n{}^u} \kappa_n(\vartheta) = 1 \quad (n = 1, \ldots, s), \quad \ldots\ldots(526)$$

provided that the values of P so determined leave the expressions $X_u - \Sigma_n q_n{}^u P_n$ positive. Otherwise certain of the assumed crystals are not really present. It can then be shown that average values for the assembly can be determined by fixing the P's at the values satisfying (526) and determining average values for this example. We find at once for example from (476) and (526) that

$$\overline{a_r{}^n} = \left[\varpi_r{}^n \mu \frac{\partial}{\partial \mu} \log g_n(\mu \vartheta^{\epsilon_r{}^n})\right]_{\mu = 1/\kappa_n(\vartheta)}, \quad \ldots\ldots(527)$$

$$\overline{N_n} = \left[\Sigma_r\, \varpi_r{}^n \mu \frac{\partial}{\partial \mu} \log g_n(\mu \vartheta^{\epsilon_r{}^n})\right]_{\mu = 1/\kappa_n(\vartheta}. \quad \ldots\ldots(528)$$

In the classical limit, always sufficiently accurate,

$$\overline{N_n} = f_n(T)/\kappa_n(T) \quad (n = 1, \ldots, s). \quad \ldots\ldots(529)$$

An equation identical in form with (520) therefore holds for every component of the gaseous phase which is also effectively present in the assembly in a crystalline phase of its own.

§5·51. *Law of mass-action in the presence of crystalline phases.* The general laws of mass-action in a purely gaseous assembly (in which of course the gas is "perfect") are given by equations (479). These can be rewritten in a more useful form which does not make explicit reference to free atoms which may not be effectively present, or may not occur in the simplest

* It is now more convenient to specify the example by the numbers of molecules in the crystals than by those in the gaseous phase.

stoichiometric formula for the reaction. Suppose that a particular reaction is specified by the stoichiometric formula

$$\sum_{n=1}^{p} Q_n B_n = 0, \qquad \text{......(530)}$$

in which B_n stands for any one of the atoms or molecules present in the assembly and Q_n for the number of these molecules concerned in the reaction. We shall take Q_n positive for the products resulting from the reaction taken in a specified direction and negative for the reactants that are consumed. If B_n has the chemical formula $\Pi_u A^u{}_{q_n{}^u}$, then of course

$$\Sigma_n Q_n q_n{}^u = 0 \quad (\textit{all } u). \qquad \text{......(531)}$$

It follows at once from (531) that equations (479) can be combined to give

$$\Pi_n (\overline{N_n})^{Q_n} = \Pi_n \{f_n(T)\}^{Q_n}. \qquad \text{......(532)}$$

The crude form of the law of mass-action of course asserts merely that

$$\Pi_n (\overline{N_n}/V)^{Q_n} = F(T),$$

a function of T only.

If certain of the molecules have a crystalline phase present we apply the corresponding equations (529). The $\overline{N_n}$ for all such molecules can thereby be removed from (532) and the $f_n(T)$ on the right replaced by $\kappa_n(T)$, which depends only on T. If $\Pi_n{}'$ denotes a product over constituents which have no crystalline phase present and $\Pi_n{}''$ over those which have, then (532) becomes

$$\Pi_n{}' (\overline{N_n})^{Q_n} = \Pi_n{}' \{f_n(T)\}^{Q_n} \Pi_n{}'' \{\kappa_n(T)\}^{Q_n}, \qquad \text{......(533)}$$

and the crude form

$$\Pi_n{}' (\overline{N_n}/V)^{Q_n} = F(T). \qquad \text{......(534)}$$

These are familiar formulae of physical chemistry.

§ **5·52.** *General remarks on the nature of statistical equilibrium.* In concluding this discussion of the most general assemblies which we have yet handled one should include the energy of the radiation in the assembly which we have not yet done explicitly for dissociating assemblies. We observe that it merely needs the inclusion of the factor $R(z)$ in every integrand to take complete account of the equilibrium temperature radiation present. Equations such as (472)–(474) in so far as they determine T as a function of E are of course altered, but obviously no distribution law of the assembly expressed as a function of T. So far as the laws of dissociative equilibrium are concerned it makes no difference whether radiation is or is not explicitly included. The equilibrium will be the same whether exchanges of energy take place by radiation or by collisions alone, but of course this does not imply that the steady state remains unaltered when the assembly is subjected to radiation of a different temperature from outside. We must of course explicitly include $R(z)$ when the energy E (or the entropy Σ, see § 6·3) of the assembly is under discussion as a function of T.

A further application open to these methods is to include explicitly partition functions for adsorbed films or surface phases in general in the integrand of C, but this will only be taken up in this monograph shortly in the supplementary Chapter XXI. These omissions will not affect our study of the equilibrium properties of bulk phases.

§ 5·6. *Mixed crystals*. We have hitherto in this chapter considered only simple crystals composed of one of the atoms or molecules in the gaseous phase. The analysis given for this case is formally correct even if the atom or molecule in the crystal has a number of states of internal motion or nuclear orientation, but if explicit account is to be taken of such motions and orientations the notation of Chapter IV can sometimes be departed from and the notation of this chapter refined with advantage, especially with a view to extensions to mixed crystals. We must attempt to take account of such formations constructed out of two or more components of the gaseous phase—possibly mixable in all proportions and possibly not—for important examples commonly occur. One such is a metal in equilibrium with an atmosphere of evaporated electrons which will be discussed in Chapter XI. Another example, which we shall examine here, is a crystal formed of a mixture of isotopes. In this application owing to the great similarity of isotopic molecules the general formulae can be considerably simplified. An interesting extension to metallic alloys is considered in Chapter XXI.

The necessary *formal generalization* to mixed crystals is due to Schottky* and is very simple. It is sufficient for the present to consider a crystal of two components evaporating separately without interaction in the gas phase. Such interaction can obviously be included when required.

We may formally suppose that we can construct a partition function $K_{P,P'}(z)$ for the crystal containing P, P' systems of the two types—this partition function to take account of every possible state of the crystal including all relative rearrangements of the two types of system. Actually of course only large values of P and P' will prove to be important and then only in a ratio which deviates only slightly from some fixed value. For an assembly containing such a crystal and a vapour phase of $X-P$, $X'-P'$ systems respectively we have

$$C_{P,P'} = \frac{1}{(2\pi i)^3} \iiint \frac{dx\,dx'\,dz}{x^{X+1} x'^{X'+1} z^{E+1}} x^P x'^{P'} K_{P,P'}(z)\, \Pi_r g(xz^{\epsilon_r})\, \Pi_r g'(x'z^{\epsilon_r'}). \quad \ldots\ldots(535)$$

* Schottky, *Ann. d. Physik*, vol. 78, p. 434 (1925).

$C_{P,P'}$ can be evaluated in the usual way in terms of the usual parameters λ, λ' and ϑ. Its maximum can then be determined by the equations

$$\partial C_{P,P'}/\partial P = \partial C_{P,P'}/\partial P' = 0,$$

which give $$\log\lambda + \frac{1}{K}\frac{\partial K}{\partial P} = 0, \quad \log\lambda' + \frac{1}{K}\frac{\partial K}{\partial P'} = 0. \qquad \text{......(536)}$$

These equations effectively determine P^* and P'^* in terms of λ, λ' and ϑ, just as in § 5·41 equation (511) determines N^* and thereby P^*. By obvious though complicated generalization of that section we can proceed to show that $P^* = \overline{P}$, etc., and to determine all the details of the equilibrium state by extending and using Rule C.

These equations though formally sufficient are naturally of no value unless the actual structure of $K_{P,P'}$ as a function of P and P' can at least be approximated to. The evaporating electron problem can be carried through because then the second constituent, the metal ions, need not be regarded as evaporating at all and the problem simplifies to a study of the partition of electrons between their states in two different phases, metal and vacuum. For the crystal of mixed isotopes, the calculations can be carried through because the similarity of isotopes allows $K_{P,P'}$ to be constructed explicitly.

§ **5·61.** *Approximate evaluation of* $K_{P,P'}(z)$. Let us first return to a simple crystal composed of P similar systems and suppose that they have all a unique internal state so that from the point of view of crystal structure they may be regarded as massive points. If for such a crystal we construct a partition function as in Chapter IV, we shall have only the acoustical modes of motion to consider and shall obtain a result of the form $[\kappa_a(z)]^P$ approximately.† In enumerating the states of such a crystal we can proceed simply as in Chapter IV, where we regarded all the constituent atoms as permanently assigned to a fixed location in the lattice, or we can allow for complete rearrangements, but in the latter case we must make the resulting wave-function symmetrical (or antisymmetrical) in all the atoms and the enumeration of complexions is unaffected.

Suppose, however, we now attempt to generalize this result to a crystal of atoms with internal structure and more than one relevant state of internal motion or nuclear orientation, but assume that the external properties of the atoms are so insensitive to the internal states in question that to a sufficient approximation the acoustical normal modes are unaffected. Such a crystal then has two sets of (practically) independent motions: the acoustical modes with partition function $[\kappa_a(z)]^P$ approximately, and the internal

† It is convenient to distinguish a crystalline partition function $\kappa(z)$ which refers only to the energies and states of the *acoustical* modes of motion by the suffix a, $\kappa_a(z)$.

motions of a group of P localized systems with partition function $[f(z)]^P$ (see § 2·26), where $f(z)$ is the partition function for the internal motions of a single atom at its lattice point. Since the two sets are independent the partition function for the complete set of states of the crystal is as in § 2·7

$$[\kappa_a(z)f(z)]^P. \qquad \text{......(537)}$$

These arguments are of course only another aspect of those of Chapter IV, less accurate in form, but giving greater insight into the physical origin of the terms.

Now let us suppose that we have a crystal made up of P massive points of mass m and P' others of mass m' arranged in some definite manner among the $P+P'$ lattice points. There is then just one linearly independent wave-function for the crystal in a state in which the oscillations in each of the acoustical modes have given quantum numbers—just one whether we ignore symmetry properties or whether we make the wave-function have the proper symmetry by permuting the points of mass m over their own P lattice points and the points of mass m' separately over their own P' other lattice points. If, however, we interchange any one of the masses m with a mass m', we shall obtain a significantly different crystal with slightly different normal modes and energy levels. If mixing of the m, m' mass-points can occur indifferently in all fashions, there will be $(P+P')!/(P!\,P'!)$ different crystals possible each with slightly different normal modes and energies. In general therefore there will be a different partition function for each of these crystals, differing in internal arrangement of the masses, which we shall denote by $K_j(z)$ $[j=1, 2, \ldots, (P+P')!/(P!\,P'!)]$. It follows that

$$K_{P,P'}(z) = \Sigma_j K_j(z). \qquad \text{......(538)}$$

When the masses m, m' are nearly equal and the forces exerted on each other by systems of the two types are also the same or nearly the same, then no matter how the two sets of particles are mixed the normal modes and energies of the crystal in any arrangement will be very nearly equal and we shall have

$$K_j(z) = K(z) = [\kappa_a(z)]^{P+P'}$$

to a close approximation. It follows then that

$$K_{P,P'}(z) = \frac{(P+P')!}{P!\,P'!}[\kappa_a(z)]^{P+P'}. \qquad \text{......(539)}$$

Now suppose finally that the two sets of systems have different sets of possible states of internal motions and nuclear orientation, giving rise to the partition functions $f(z)$, $f'(z)$ respectively. If as before these internal motions may be assumed to be effectively independent of the acoustical modes, then to every one of the states enumerated in (539) there corresponds a set of internal states which are those of P localized systems with partition function $f(z)$ and P' with $f'(z)$. The total internal motions therefore have a partition function $[f(z)]^P[f'(z)]^{P'}$ and the complete partition function for the whole mixed crystal (perfect mixing) is given by

$$K_{P,P'}(z)=\frac{(P+P')!}{P!\,P'!}[f(z)]^P[f'(z)]^{P'}[\kappa_a(z)]^{P+P'}. \qquad \text{......(540)}$$

Still more general mixed crystals are considered in § 21·2.

§ **5·62.** *A simple illustration of* (540). It may be desirable before going further to illustrate the enumeration of complexions summarized in (540) by a simple example. Equation (540) for example asserts that a crystal of 3 molecules at lattice points a, b, c, two of one kind A and one of another kind B, in a given vibrational state, when the A's possess a unique internal state and the B's one of weight 2, possesses

$$\frac{3!}{2!\,1!}1^2 2^1=6$$

complexions, symmetrical (or antisymmetrical) in the systems A, and separately in the systems B. This is easily verified in this simple case by explicit construction of the wave-functions which are, in an obvious notation, just the six functions

$$\begin{gathered}
\{A_a(1)A_b(2)+A_a(2)A_b(1)\}B_c{}^1(3),\\
\{A_a(1)A_b(2)+A_a(2)A_b(1)\}B_c{}^2(3),\\
\{A_b(1)A_c(2)+A_b(2)A_c(1)\}B_a{}^1(3),\\
\{A_b(1)A_c(2)+A_b(2)A_c(1)\}B_a{}^2(3),\\
\{A_c(1)A_a(2)+A_c(2)A_a(1)\}B_b{}^1(3),\\
\{A_c(1)A_a(2)+A_c(2)A_a(1)\}B_b{}^2(3).
\end{gathered}$$

In this counting the crystal wave-function is symmetrized separately in the A's and B's here regarded as distinct systems. In certain other cases such as when the A's and B's represent non-combining sets of states of the same molecule—e.g. ortho- and para-hydrogen, the crystal wave-functions may need to be symmetrized over the whole of the A's and B's, exchanges being allowed between all the systems. It is easily verified that this makes no difference to the enumeration of complexions. We still obtain in this example just six wave-functions which are then the permanents

$$\left\|\begin{matrix} A_a(1) & A_a(2) & A_a(3)\\ A_b(1) & A_b(2) & A_b(3)\\ B_c{}^1(1) & B_c{}^1(2) & B_c{}^1(3)\end{matrix}\right\|,$$

$$\left\|\begin{matrix} A_a(1)\ \dots\\ A_b(1)\ \dots\\ B_c{}^2(1)\ \dots\end{matrix}\right\|,\quad
\left\|\begin{matrix} A_a(1)\ \dots\\ B_b{}^1(1)\ \dots\\ A_c(1)\ \dots\end{matrix}\right\|,\quad
\left\|\begin{matrix} A_a(1)\ \dots\\ B_b{}^2(1)\ \dots\\ A_c(1)\ \dots\end{matrix}\right\|,\quad
\left\|\begin{matrix} B_a{}^1(1)\ \dots\\ A_b(1)\ \dots\\ A_c(1)\ \dots\end{matrix}\right\|,\quad
\left\|\begin{matrix} B_a{}^2(1)\ \dots\\ A_b(1)\ \dots\\ A_c(1)\ \dots\end{matrix}\right\|.$$

§5·63. *Properties of an ideal mixed crystal of two components.* We have now shown that for an ideal mixed crystal we can use the expression (535) in which $K_{P,P'}(z)$ is evaluated by (540). On evaluating $C_{P,P'}$ we now find

$$C_{P,P'} = \frac{(P+P')!}{P!\,P'!} \frac{[\lambda f(\vartheta)]^P [\lambda' f'(\vartheta)]^{P'} [\kappa_a(\vartheta)]^{P+P'}}{\lambda^X \lambda'^{X'} \vartheta^E} \Pi_r\, g(\lambda\vartheta^{\epsilon_r})$$
$$\times \Pi_r\, g'(\lambda'\vartheta^{\epsilon_r'}) \left[\frac{\Delta^{-\frac{1}{2}}}{(2\pi X)^{\frac{3}{2}}} + O\left(\frac{1}{X^{\frac{5}{2}}}\right)\right], \quad \text{......(541)}$$

where λ, λ' and ϑ are determined by the usual equations and Δ is a slowly varying function as in (506) whose variations with P, P' will be neglected. Since P and P' are both large for a significant mixed crystal we may employ Stirling's theorem and find with sufficient accuracy

$$\log C_{P,P'} = (P+P')\log[(P+P')\,\kappa_a(\vartheta)] + P\log\frac{\lambda f(\vartheta)}{P} + P'\log\frac{\lambda' f'(\vartheta)}{P'}$$
$$+ \Sigma_r \log g(\lambda\vartheta^{\epsilon_r}) + \Sigma_r \log g'(\lambda'\vartheta^{\epsilon_r'}) - X\log\lambda - X'\log\lambda' - E\log\vartheta + O(\log X). \quad \text{......(542)}$$

Applying Rule C we find that equations (536) characterize the equilibrium state of the assembly and reduce to

$$\lambda f(\vartheta)\,\kappa_a(\vartheta) = \frac{\overline{P}}{\overline{P}+\overline{P'}}, \quad \lambda' f'(\vartheta)\,\kappa_a(\vartheta) = \frac{\overline{P'}}{\overline{P}+\overline{P'}}, \quad \text{......(543)}$$

so that
$$\{\lambda f(\vartheta) + \lambda' f'(\vartheta)\}\,\kappa_a(\vartheta) = 1. \quad \text{......(544)}$$

This last equation is the analogue of (512) and the remaining independent equation

$$\frac{\lambda f(\vartheta)}{\lambda' f'(\vartheta)} = \frac{\overline{P}}{\overline{P'}} \quad \text{......(545)}$$

gives the equilibrium ratio of the constituents in the mixed crystal.

The density of the mixed vapour phase in equilibrium is given by the equations

$$\overline{N} = \lambda\frac{\partial}{\partial\lambda}\Sigma_r \log g(\lambda\vartheta^{\epsilon_r}), \quad \overline{N'} = \lambda'\frac{\partial}{\partial\lambda'}\Sigma_r \log g'(\lambda'\vartheta^{\epsilon_r'}),$$

where λ, λ' and ϑ are determined by the three equations which result from (544), (545) and the three equations fixing the saddle point of (541) when P and P' have been eliminated. It is found that the vapour can always be regarded as classical. Then the last equations reduce as usual to

$$\overline{N} = \lambda f_v(\vartheta), \quad \overline{N'} = \lambda' f_v'(\vartheta), \quad \text{......(546)}$$

where $f_v(\vartheta)$, $f_v'(\vartheta)$ are the partition functions for the systems in the vapour phase.

If
$$\frac{f_v(\vartheta)}{f(\vartheta)} = \frac{f_v'(\vartheta)}{f'(\vartheta)}, \quad \text{......(547)}$$

then
$$\frac{\overline{N}}{\overline{N'}} = \frac{\overline{P}}{\overline{P'}} = \frac{X}{X'}, \qquad \ldots\ldots(548)$$

and
$$\overline{N} + \overline{N'} = \frac{f_r(\vartheta)}{f(\vartheta)\,\kappa_a(\vartheta)} = \frac{f_r'(\vartheta)}{f'(\vartheta)\,\kappa_a(\vartheta)}. \qquad \ldots\ldots(549)$$

The equilibrium ratio of the constituents is then the same in both phases and the total vapour density is the same for any mixed crystal as for a simple crystal of either component. When (547) does not hold, the vapour density depends essentially on the abundance ratio $D:D'$ $(D+D'=1)$. Since in general the overwhelming majority of the systems will be in the crystal, we shall have

$$\frac{\overline{P}}{\overline{P'}} \cong \frac{X}{X'} = \frac{D}{D'}.$$

We then find
$$\overline{N} + \overline{N'} = \frac{f_r(\vartheta)}{f(\vartheta)\,\kappa_a(\vartheta)}\,D + \frac{f_r'(\vartheta)}{f'(\vartheta)\,\kappa_a(\vartheta)}\,D'. \qquad \ldots\ldots(550)$$

This equation can easily be generalized to a mixed crystal of any number of components. It asserts that the equilibrium vapour density of the mixed crystal is equal to the mean value of the vapour densities in equilibrium with the simple crystals, weighted in proportion to the abundance ratios in the mixed crystal.

§ **5·64.** *Crystals with imperfect mixing.* We have already discussed the limiting case of perfect mixing. In the other limiting case we must assume that the components are so antagonistic that all the states of a mixed crystal have energies much greater than states in which each component forms a separate simple crystal. In such a case only the simple crystals will form in the assembly. Between these two extremes every intermediate degree of mixability is possible, and it is convenient, if only in a rough and ready way, to be able to construct partition functions for models which might imitate such crystals.

The simplest way in which a limitation of mixing can be specified is to require that the first type of system must be grouped together in blocks of n systems each and the other type in blocks of n', the blocks being then able to mix perfectly. This is a possibility that might well be realized in real crystals. The total number of possible arrangements is then

$$\frac{(P/n + P'/n')!}{(P/n)!\,(P'/n')!}$$

instead of $(P+P')!/(P!\,P'!)$ for a crystal of P, P' systems respectively. This number of arrangements is the same as the number for a crystal in which $1/n$ of the systems of the first type and $1/n'$ of the others mix in all fashions while the remaining systems of both types remain permanently attached to their own lattice points, and take no part in the mixing. It is therefore convenient to introduce *two mixing coefficients* α, α' so defined that the number of effectively possible arrangements of the systems over the lattice points is

$$(\alpha P+\alpha' P')!/(\alpha P)!\,(\alpha' P')!. \qquad \text{......(551)}$$

If $f(z)$, $f'(z)$ are the partition functions for the internal motions, etc. of the systems in the crystal and if the systems are still sufficiently alike for the approximate form of the partition function for the acoustical modes to be independent of the mixing and of the form $[\kappa_a(z)]^{P+P'}$, then the partition function for the whole mixed crystal will be of the form

$$\frac{(\alpha P+\alpha' P')!}{(\alpha P)!\,(\alpha' P')!}[f(z)]^P\,[f'(z)]^{P'}\,[\kappa_a(z)]^{P+P'}. \qquad \text{......(552)}$$

The succeeding arguments determining $C_{P,P'}$ and the equilibrium properties of the assembly follow the usual course. We find

$$\lambda=\left(\frac{\alpha\overline{P}}{\alpha\overline{P}+\alpha'\overline{P'}}\right)^{\alpha}\frac{1}{f(\vartheta)\,\kappa_a(\vartheta)}, \qquad \text{......(553)}$$

$$\lambda'=\left(\frac{\alpha'\overline{P'}}{\alpha\overline{P}+\alpha'\overline{P'}}\right)^{\alpha'}\frac{1}{f'(\vartheta)\,\kappa_a(\vartheta)}, \qquad \text{......(554)}$$

and therefore, with the classical approximation for the vapour and the assumption that $\overline{P}/\overline{P'}\cong X/X'$,

$$\overline{N}=\frac{f_v(\vartheta)}{f(\vartheta)\,\kappa_a(\vartheta)}\left(\frac{\alpha X}{\alpha X+\alpha' X'}\right)^{\alpha}, \qquad \text{......(555)}$$

$$\overline{N'}=\frac{f_v'(\vartheta)}{f(\vartheta)\,\kappa_a(\vartheta)}\left(\frac{\alpha' X'}{\alpha X+\alpha' X'}\right)^{\alpha'}. \qquad \text{......(556)}$$

When at least one α is zero there is no mixing, and equations (555) and (556) both reduce to an equation of the ordinary type $\overline{N}=f_v(\vartheta)/f(\vartheta)\,\kappa_a(\vartheta)$ for the vapour density of a simple crystal. The total vapour density is then the sum of the two vapour densities for the separate crystals and has approximately twice the value of the vapour density when $\alpha=\alpha'=1$, when there is perfect mixing.

§5·7. *Internal stresses.* We have so far defined the pressure only in such a way that $p\,dS\,dn$ is the average work done by the boundary field on the assembly when an element dS of the boundary is moved a distance dn normal to itself. Then the pressure (or other stress) so far refers entirely to the relationship of the assembly to the outside world. The general formula for such a stress for an assembly containing a vapour and a crystal phase has the form

$$\overline{Y}=\frac{1}{\log 1/\vartheta}\left\{\Sigma_u\overline{M_u}\frac{\partial}{\partial y}\log f_u(\vartheta)+\Sigma_v\overline{N_v}\frac{\partial}{\partial y}\log f_v(\vartheta)+\overline{P}\frac{\partial}{\partial y}\log\kappa(\vartheta)\right\}, \qquad \text{......(557)}$$

the classical approximation being used as in (483) for the vapour phase. Pressures at points or across areas inside the assembly, or in general internal stresses are as yet undefined. Definitions of these quantities will often be required.

The average *stress* per unit area across any imaginary surface inside the assembly is defined to be *the resultant force per unit area exerted by systems on side A on systems on side B, together with the momentum transferred per unit area per second from A to B by systems crossing the surface from A to B* and *B to A*. The stress exerted by side A on side B is necessarily equal **and** opposite to the stress exerted by B on A so long as the forces obey Newton's third law.

The stress so defined is of course in general a symmetrical tensor of the second rank of nine components p_{xx}, p_{yy}, p_{zz}, p_{yz} $(=p_{zy})$, p_{zx} $(=p_{xz})$, p_{xy} $(=p_{yx})$. For any gas (or fluid) in equilibrium it reduces necessarily to the simple form $p_{xx}=p_{yy}=p_{zz}$, $p_{yz}=p_{zx}=p_{xy}=0$. This simple isotropic stress per unit area, always normal to the surface across which it acts, is defined to be *the pressure*. The verification of this simplification and the explicit calculation of the pressure across any internal surface is immediate for perfect gases. The forces between the systems are negligible, so that, for example, p_{xx} is the rate of transfer of x-momentum across a unit surface normal to the x-axis. Molecules with the x-component of velocity between u and $u+du$ carry x-momentum mu across the surface, and the number of such molecules crossing unit area in time dt is the number of such molecules in a volume $u\,dt$. Both other velocity components are entirely irrelevant, merely fixing the shape of the volume $u\,dt$. Thus by Maxwell's law the number is

$$\frac{Nu\,dt}{V}\left(\frac{m}{2\pi kT}\right)^{\frac{1}{2}} e^{-\frac{1}{2}mu^2/kT}\,du,$$

and

$$p_{xx}=\frac{Nm}{V}\left(\frac{m}{2\pi kT}\right)^{\frac{1}{2}}\int_{-\infty}^{+\infty} u^2 e^{-\frac{1}{2}mu^2/kT}\,du,$$

$$=\frac{Nm}{V}\left(\frac{m}{2\pi kT}\right)^{\frac{1}{2}}\left(\frac{2kT}{m}\right)^{\frac{3}{2}}\frac{\pi^{\frac{1}{2}}}{2}=\frac{N}{V}kT.$$

Similar values are found for p_{yy} and p_{zz}. For p_{zx}, which is the rate of transfer of z-momentum across unit area normal to the x-axis, we find similarly

$$p_{zx}=\frac{Nm}{V}\,\frac{m}{2\pi kT}\int_{-\infty}^{+\infty}\int_{-\infty}^{+\infty} uw\,e^{-\frac{1}{2}m(u^2+w^2)/kT}\,du\,dw,$$

$$=0.$$

This calculation is of course merely one version of the classical pressure calculation of the Kinetic Theory.

We have so far considered the case of a single set of systems and calculated the "partial" stress which they cause. Obviously for perfect gases the stresses are additive, and reduce as above to an isotropic pressure given by the equation

$$p=\frac{kT}{V}\{\Sigma_u \overline{M_u}+\Sigma_v \overline{N_v}\}. \qquad \text{......(558)}$$

In (558) the values of $\overline{M_u}$ and $\overline{N_v}$ are *local* values in the neighbourhood of the internal point considered. If there are only boundary fields, then the pressure is constant throughout the gas and equal to the pressure on the boundary previously otherwise defined. This result of course continues to hold generally, e.g. for imperfect gases, in which connection we return to it in Chapter IX.

The surface across which we calculate the stress may be an interface between two phases. It follows at once that the pressure on a solid phase is equal to the pressure in the surrounding gas phase evaluated at its surface.

We are now in a position to examine the form of (557) for boundary fields. By definition of the external pressure p, $Y\,dy=p\,dV$, and from the geometry of the assembly $dV=dV_s+dV_g$, where V_s, V_g are the volumes of the solid and gaseous phases. Hence

$$Y\,dy=p\,dV_s+p\,dV_g.$$

But by (557)

$$Y\,dy=p_s dV_s+p_g dV_g,$$

where p_s and p_g are the expressions obtained by applying the general pressure formulae of §§ 2·8, 4·8 to the two phases in their average state as if they were separate assemblies. Such an application will therefore always give the correct equal pressures in the various phases.

We may notice also that the equilibrium state of statistical mechanics as calculated on our general hypotheses is a state of mechanical equilibrium of the matter in the assembly. No element of it has any mass motion relative to the enclosure containing the assembly, and in the absence of external fields the pressure is everywhere constant.

The final result of these paragraphs may be expressed by saying that the stresses on any volume element or any phase of the assembly may be calculated by applying the general laws to this element or phase with its average constitution and energy as if it were itself a separate assembly.

It is possibly more significant to start by *postulating* that the general laws of statistical mechanics apply not merely to the assembly as a whole but also to its constituent elements and phases as if they were separate assemblies with their average constitution and energy. We can then work backwards and deduce the constancy of the pressure and the existence of mechanical equilibrium.

§5·8. *Dissociative equilibrium in an external field of force.* The formulae of the preceding sections refer to gaseous (and other) assemblies subject to no external fields of force, except the local boundary fields. This restriction can easily be removed for the gaseous part of the assembly, for which alone it is of importance. We will suppose for simplicity that all the gaseous components are confined to the same volume V, and that classical statistics may be used. Then the V-factor in the partition functions must be replaced by $V_\alpha(T)$, where

$$V_\alpha(T) = \iiint_V e^{-W_\alpha/kT}\, dq_1 dq_2 dq_3; \qquad \ldots\ldots(559)$$

W_α is the potential energy of the atom or molecule of species α in the external field. The equilibrium laws for the *whole* gaseous part of the assembly are then unaltered in form. It is however now necessary to consider also the equilibrium laws for any physically small element δV_x of the assembly, since the distributions are no longer uniform in space.

The function $V_\alpha(T)$ is strictly the partition function for the potential energy, and has all the properties of a partition function (see Chapter VIII). If $\overline{n_x{}^u}$, $\overline{n_x{}^v}$ are the average numbers of the atoms u or molecules v in the volume element δV_x, then, by general formulae such as (477) and (478),

$$\overline{n_x{}^u} = \overline{M_u}\, e^{-W_u/kT}\, \delta V_x / V_u(T), \qquad \ldots\ldots(560)$$

$$\overline{n_x{}^v} = \overline{N_v}\, e^{-W_v/kT}\, \delta V_x / V_v(T). \qquad \ldots\ldots(561)$$

Besides being obtainable directly from the properties of $V_\alpha(T)$, these formulae can of course be obtained from (477) and (478) by summing or integrating for all variables except the positional coordinates q_1, q_2, q_3. For the whole gaseous part of the assembly the laws of mass-action (479) take the form

$$\frac{\overline{N_v}}{\Pi_u (\overline{M_u})^{q_v{}^u}} = \frac{V_v(T)}{\Pi_u \{V_u(T)\}^{q_v{}^u}} \times \frac{F_v(T)}{\Pi_u \{F_u(T)\}^{q_v{}^u}} \quad (v = 1, 2, \ldots, j), \qquad \ldots\ldots(562)$$

in which the F's are the partition functions without their volume factors. For the volume element δV_x they take the form, after reduction by (562),

$$\frac{\overline{n_x{}^v}/\delta V_x}{\Pi_u \{\overline{n_x{}^u}/\delta V_x\}^{q_v{}^u}} = \frac{F_v(T)}{\Pi_u \{F_u(T)\}^{q_v{}^u}}\, e^{-(W_v - \Sigma_u q_v{}^u W_u)/kT} \quad (v = 1, 2, \ldots, j). \qquad \ldots\ldots(563)$$

In general, therefore, the equilibrium constant might be expected to be a function of position in the gas, varying from place to place according to the equation (563). In actual fact there is no such variation in any known conditions. For there will be no variation provided

$$W_v - \Sigma_u q_v{}^u W_u = 0, \qquad \ldots\ldots(564)$$

that is, *provided the potential energy of any system or group of systems in the field is unaltered by dissociation and recombination.* But this proviso is in general satisfied in actual conditions. Actual fields are usually combinations of gravitational (including inertial) and electrostatic fields. Let Φ be the gravitational and Ψ the electrostatic potential. Then

$$W_v = m_v\Phi + \epsilon_v\Psi, \quad W_u = m_u\Phi + \epsilon_u\Psi, \qquad \text{......(565)}$$

where m_v, m_u and ϵ_v, ϵ_u are the masses and charges of the systems v, u. But total mass and total charge are conserved by dissociation and recombination. Therefore

$$m_v = \Sigma_u q_v{}^u m_u, \quad \epsilon_v = \Sigma_u q_v{}^u \epsilon_u,$$

and $$W_v - \Sigma_u q_v{}^u W_u = \{m_v - \Sigma_u q_v{}^u m_u\}\Phi + \{\epsilon_v - \Sigma_u q_v{}^u \epsilon_u\}\Psi = 0.$$

Thus in all such cases

$$\frac{\overline{n_x{}^v}/\delta V_x}{\Pi_u\{\overline{n_x{}^u}/\delta V_x\}^{q_v{}^u}} = \frac{F_v(T)}{\Pi_u\{F_u(T)\}^{q_v{}^u}}; \qquad \text{......(566)}$$

and the equilibrium constants for all volume elements of the gas are the same, and the same as that for the whole gas, *without* external fields.* The total amount of dissociation, however, in the whole gas will in general have been altered by the external field.

This constancy does not necessarily hold for magnetic systems in magnetic fields, which require further consideration (see § 12·8).

It is interesting to verify that the equilibrium laws of the assembly still contain the laws of mechanical equilibrium under the influence of external fields (gravitational, inertial, or electrostatic). Equation (558) still gives the pressure in the gas if $\overline{M_u}/V$ is replaced by $\overline{n_x{}^u}/\delta V_x$ and $\overline{N_v}/V$ by $\overline{n_x{}^v}/\delta V_x$. As functions of position $\overline{n_x{}^u}$ and $\overline{n_x{}^v}$ are given by (560) and (561) and W_u and W_v by (565). Hence

$$p = \frac{kT}{\delta V_x}(\Sigma_u \overline{n_x{}^u} + \Sigma_v \overline{n_x{}^v}),$$

$$dp = -\frac{1}{\delta V_x}(\Sigma_u \overline{n_x{}^u}\, dW_u + \Sigma\, \overline{n_x{}^v}\, dW_v),$$

$$= -\frac{1}{\delta V_x}\{(\Sigma_u m_u \overline{n_x{}^u} + \Sigma_v m_v \overline{n_x{}^v})\, d\Phi + (\Sigma_u \epsilon_u \overline{n_x{}^u} + \Sigma_v \epsilon_v \overline{n_x{}^v})\, d\Psi\},$$

$$= -\rho\, d\Phi - \sigma\, d\Psi. \qquad \text{......(567)}$$

Since ρ and σ are the density of mass and charge in the gas, equation (567) is the usual equation for fluid equilibrium.

* For the independence of gravitational forces see Gibbs, *Collected Papers*, "Thermodynamics", pp. 144, 171; for the extension to electrostatic forces, Milne, *Proc. Camb. Phil. Soc.* vol. 22, p. 493 (1924).

CHAPTER VI

THE RELATIONSHIP OF THE EQUILIBRIUM THEORY TO CLASSICAL THERMODYNAMICS

§**6·1.** In the preceding chapters we have obtained all the distribution laws of the equilibrium state of any assembly for which we can construct partition functions. Except for the extension to imperfect gases which is the subject of Chapter VIII, this includes all types of assembly commonly treated in statistical mechanics. We have obtained all these distribution laws without any reference to thermodynamical ideas except to specify the exact relation between ϑ and the absolute temperature. where it will be seen that such an appeal is logically essential. It is fair to claim this feature for a merit in the present method of exposition. The ideas of thermodynamics are entirely foreign to the foundations of statistical mechanics which are mainly dynamical. The proper course is to prove that the laws of thermodynamics are true for the assemblies of statistical mechanics if we use suitable analogies to interpret their properties.* Such proofs are given in the succeeding sections, and it will be seen that the direct introduction of the laws of thermodynamics in this way is satisfactorily simple. We definitely discard Boltzmann's hypothesis relating entropy to a probability too often ill-defined, and introduce the entropy in just the classical way in which it is introduced into ordinary thermodynamics.

When the true relationship between the equilibrium theory of statistical mechanics and thermodynamics has thus been made apparent by showing that our assemblies in equilibrium are thermodynamic systems, it is natural to enquire into other methods of exposition in which an early introduction of entropy plays a leading part. This is the more natural, since many writers have contributed such expositions, and it cannot be maintained that logical clarity has often been achieved. Detailed criticism is out of place, but for completeness a short survey is included.

§**6·2.** *Temperature.* We have already in anticipation identified ϑ with the temperature on some empirical scale, but we may conveniently recapitulate the argument here. The legitimacy of the identification depends solely on the possession by ϑ of properties strictly analogous to those assigned to the "empirical temperature" in a rational formulation of the foundations of thermodynamics.† The basic fact of thermodynamics is that the parameters specifying the state of any body can be so chosen that when-

* This is made abundantly clear by Gibbs, *Elementary Principles in Statistical Mechanics*, chaps. IV and XIV.

† See, for instance, Max Born, *Physikal. Zeit.* vol. 22, pp. 218, 249, 282 (1921).

ever any two bodies are in equilibrium in thermal contact one of the parameters for each body has a common value for both. This parameter is defined to be the empirical temperature, but of course on an arbitrary scale, and any convenient body may be chosen for thermometer. On the statistical side we have shown in Chapters II–V that when two assemblies or sets of systems in an assembly can exchange energy, so that there is one common energy total for the whole, then their equilibrium states are defined in terms of a common value of a parameter ϑ. The analogy is exact, and we are therefore logically justified in identifying ϑ with the empirical temperature as defined in thermodynamics.

It was shown further that it was possible to prove that $\vartheta = e^{-1/kT}$ by postulating the properties of perfect gases, using a perfect gas for thermometer, and asserting that the temperature shall be proportional to the pressure of the perfect gas at a constant volume; but this appeal to the properties of an ideal substance is illogical (though often convenient) and inessential. In thermodynamical theory the *absolute temperature* is defined in connection with the second law, and can only be defined in this way. We wish to show that the assemblies of statistical mechanics obey the laws of thermodynamics (or from our assumptions to prove the laws of thermodynamics), and so we must not postulate a knowledge of the absolute temperature but define it in connection with entropy, just as we do in classical thermodynamics.

§ **6·3.** *Entropy and absolute temperature.* In classical thermodynamics the "heat" dQ taken in in any small change is defined to be the increase in internal energy plus the external work done by the assembly.* Thus for our general assembly

$$dQ = dE + \Sigma_t \overline{Y}_t dy_t. \qquad \text{......(568)}$$

The second law of thermodynamics asserts that *there exist functions T and S of the state of the assembly such that $T = f(\vartheta)$, where ϑ is the empirical temperature, and*

$$dQ = T\,dS. \qquad \text{......(569)}$$

Except for a certain arbitrarily assignable constant multiplier and an additive constant these functions are unique.

For the assemblies contemplated in § 5·3 $\overline{Y}$ and E are given by (482) and (472). These equations, written explicitly, give

$$E = \vartheta \frac{\partial Z}{\partial \vartheta}, \quad X_j = \lambda_j \frac{\partial Z}{\partial \lambda_j}, \quad \log 1/\vartheta\, \overline{Y}_t = \frac{\partial Z}{\partial y_t}, \qquad \text{......(570)}$$

where Z is given as a function of ϑ, the λ_j's and implicitly of the y_t's by (474). Consider the function Σ given by (473) in the form

$$\Sigma = Z - \Sigma_j X_j \log \lambda_j - E \log \vartheta. \qquad \text{......(571)}$$

* Born, *loc. cit.*

Then using (570)

$$d\Sigma = \left[dZ - \Sigma_j \frac{\partial Z}{\partial \lambda_j} d\lambda_j - \frac{\partial Z}{\partial \vartheta} d\vartheta\right] + \log 1/\vartheta\, dE,$$

$$= \Sigma_t \frac{\partial Z}{\partial y_t} dy_t + \log 1/\vartheta\, dE,$$

$$= \log 1/\vartheta[\Sigma_t \overline{Y}_t dy_t + dE],$$

$$= \log 1/\vartheta\, dQ.$$

Thus $\log 1/\vartheta\, dQ$ is a perfect differential and our assembly obeys the second law of thermodynamics. The postulated functions T and S exist and are defined by the equations

$$1/kT = \log 1/\vartheta, \qquad \text{......(572)}$$

$$S - S_0 = k\Sigma. \qquad \text{......(573)}$$

The constant k, Boltzmann's constant, is of course fixed by fixing the scale interval between two standard temperatures, and S_0 is essentially undetermined. It is almost always convenient to put $S_0 = 0$.

We may note here that S is closely connected to C, the total number of complexions which represent accessible states of the assembly. It is easy to show further that this total number of complexions does not differ significantly from the number of complexions which represent *nearly average states* of the assembly. The value of C for the general dissociating assembly defined by (471) can be evaluated at once by Theorem 2·42, Corollary 2 and is then found to satisfy

$$\log C = \Sigma + O(\log X). \qquad \text{......(574)}$$

Terms of order $\log X$ are negligible compared with the main terms of order X in Σ. Combining (573) and (574) with $S_0 = 0$ we obtain the fundamental relationship

$$S = k \log C, \qquad \text{......(575)}$$

where k is Boltzmann's constant. We may describe this relation by saying that *the entropy is equal to k times the logarithm of the integrand of C evaluated at the saddle point* (λ, ϑ). We may also express it by saying that *the entropy is equal to k times the logarithm of the number of complexions belonging to states which differ only insignificantly from the equilibrium state.*

We have so far considered only a gaseous phase, but the same results hold true for a crystalline assembly or for one containing both a gaseous and crystalline phase. To begin with a simple crystal of P molecules, the equations (350) and (373) of Chapter IV show that equations (570) for E and $\overline{Y}_t$ continue to hold for a Z given by

$$Z = P \log \kappa(\vartheta). \qquad \text{......(576)}$$

It follows that (572) and (573) remain true if now

$$\Sigma = P \log \kappa(\vartheta) - E \log \vartheta. \qquad \text{......(577)}$$

Finally for the most general mixed assemblies considered at the close of Chapter v we can apply Rule C of § 5·41 and evaluate all their equilibrium properties (including the enumeration of the complexions belonging to states insignificantly different from the average state) after first fixing the number of molecules in each crystal at its equilibrium value $\overline{P}$ and the number of each type of atom (free or combined) in the gaseous phase at $\overline{M_u}^{\dagger}$. Though we have not yet given them explicitly, it is easy to verify that for such an assembly equations (570) hold good in the form

$$E = \vartheta \frac{\partial Z}{\partial \vartheta}, \quad \overline{M_u}^{\dagger} = \lambda_u \frac{\partial Z}{\partial \lambda_u}, \quad \log 1/\vartheta \overline{Y_t} = \frac{\partial Z}{\partial y_t}, \qquad \text{......(578)}$$

where $$Z = \Sigma_u \Sigma_r \varpi_r^u \log g_u(\lambda_u \vartheta^{\epsilon_r^u}) + \Sigma_v \Sigma_r \varpi_r^v \log g_v(\lambda_1^{q_v^1} \dots \lambda_t^{q_v^t} \vartheta^{\epsilon_r^v})$$
$$+ \Sigma_\tau \log K^\tau_{(\overline{P_\tau})}(\vartheta). \qquad \text{......(579)}$$

In (579) Σ_τ denotes summation over the different species of crystal present and $K^\tau_{(\overline{P_\tau})}(\vartheta)$ the partition function for the τth species containing its equilibrium number $\overline{P_\tau}$ of molecules, or if a mixed crystal its equilibrium number $\overline{P_\tau}$, $\overline{P_\tau'}$, ... of all the molecules it contains. It follows at once that (572) and (573) are true with

$$\Sigma = Z - \sum_{u=1}^{t} M_u^{\dagger} \log \lambda_u - E \log \vartheta. \qquad \text{......(580)}$$

It will be recalled that $\overline{P_\tau}$ and $\overline{M_u}^{\dagger}$ are determined by equations of the form $\partial\Sigma/\partial\overline{P_\tau} = 0$ so that they need not be varied in forming $d\Sigma$.

It is evident by inspection of (571), (577) and (580) that the contributions to Σ of the gaseous and the crystalline phases are additive. This is further true of the different components of the gaseous phase. For

$$\overline{M_u}^{\dagger} = \overline{M_u} + \Sigma_v q_v^u \overline{N_v},$$

and therefore the free atoms of type u contribute to Σ

$$\Sigma_r \varpi_r^u \log g_u(\lambda_u \vartheta^{\epsilon_r^u}) - \overline{M_u} \log \lambda_u - \overline{E_u} \log \vartheta \qquad \text{......(581)}$$

and free molecules of type v

$$\Sigma_r \varpi_r^v \log g_v(\mu \vartheta^{\epsilon_r^v}) - \overline{N_v} \log \mu - \overline{E_v} \log \vartheta \quad (\mu = \lambda_1^{q_v^1} \dots \lambda_t^{q_v^t}). \qquad \text{......(582)}$$

These forms hold whether or not there are crystalline phases present. They contain redundant variables and either $\overline{M_u}$ or λ_u may be eliminated by means of (475) and (476); $\overline{E_u}$ may also be eliminated by expression in terms of ϑ and $\overline{M_u}$ or λ_u.

On account of their importance we now summarize these results in the form of contributions to the *entropy* in the following theorem, assembling therein the various important forms.

Theorem 6·3. *The entropy defined by* (573) *with* $S_0 = 0$ *of an assembly consisting of (ideal) gaseous and crystalline phases is the sum of the following contributions made by each component in the gaseous phase and by each crystalline phase:*

For each gaseous component, with partition function $f(\vartheta)$, *present to the average number* $\overline{M}$ *with average energy content* $\overline{E}$ *and parameter* μ

$$k[\Sigma_r \varpi_r \log g(\mu\vartheta^{\epsilon_r}) - \overline{M} \log \mu - \overline{E} \log \vartheta], \qquad \text{......(583)}$$

which can be expressed, in the classical limit, in either form

$$k[f(\vartheta)\{\mu - \mu \log \mu\} - \overline{E} \log \vartheta], \qquad \text{......(584)}$$

$$k[\overline{M} \log\{f(\vartheta)/\overline{M}\} + \overline{M} - \overline{E} \log \vartheta]. \qquad \text{......(585)}$$

For each pure crystal present with an average number of molecules $\overline{P}$, *partition function per molecule* $\kappa(\vartheta)$, *and average energy* $\overline{E}$

$$k[\overline{P} \log \kappa(\vartheta) - \overline{E} \log \vartheta]. \qquad \text{......(586)}$$

For each ideal mixed crystal of two components present to the average numbers $\overline{P}$, $\overline{P'}$ *with partition functions per molecule* $f(\vartheta)\kappa_a(\vartheta)$, $f'(\vartheta)\kappa_a(\vartheta)$, *and average energy* $\overline{E}$

$$k[\overline{P} \log f(\vartheta)\kappa_a(\vartheta) + \overline{P'} \log f'(\vartheta)\kappa_a(\vartheta) - \overline{E} \log \vartheta$$
$$+ (P+P') \log(P+P') - P \log P - P' \log P']. \qquad \text{...(587)}$$

For radiation $\quad k[\log R(\vartheta) - \overline{E}_R \log \vartheta] = k\left[\dfrac{32\pi^5}{45c^3h^3} \dfrac{V}{(\log 1/\vartheta)^3}\right]. \qquad \text{......(588)}$

Formulae for S are so important that we shall record here for reference further forms obtained by elimination of $\overline{E}$ and substitution of T for ϑ. It will be observed that (583)–(588) and all the succeeding forms are homogeneous functions of degree unity in the extensive variables V, $\overline{M}$, $\overline{P}$ and $\overline{E}$.

Alternative forms for the contributions to the entropy S. *Gaseous component.*

$$k\left[Z - \mu \log \mu \frac{\partial Z}{\partial \mu} - \vartheta \log \vartheta \frac{\partial Z}{\partial \vartheta}\right] \quad (Z = \Sigma_r \varpi_r \log g(\mu\vartheta^{\epsilon_r})); \qquad \text{...(589)}$$

$$k\left[Z - \mu \log \mu \frac{\partial Z}{\partial \mu} + T \frac{\partial Z}{\partial T}\right] \quad (Z = \Sigma_r \varpi_r \log g(\mu e^{-\epsilon_r/kT})); \qquad \text{...(590)}$$

or in the classical limit,

$$k\left[f(T)\{\mu - \mu \log \mu\} + \mu T \frac{\partial f(T)}{\partial T}\right]; \qquad \text{......(591)}$$

$$k\overline{M}\left[\log \frac{f(T)}{\overline{M}} + 1 + T \frac{\partial \log f(T)}{\partial T}\right]. \qquad \text{......(592)}$$

Pure crystal.
$$k\overline{P}\left[\log\kappa(\vartheta)-\vartheta\log\vartheta\frac{\partial\log\kappa(\vartheta)}{\partial\vartheta}\right]; \qquad \text{......(593)}$$
$$k\overline{P}\left[\log\kappa(T)+T\frac{\partial\log\kappa(T)}{\partial T}\right]. \qquad \text{......(594)}$$

Formulae (593) and (594) hold also for the contribution by any set of N localized systems of partition function $\kappa(\vartheta)$ if $\overline{P}$ is replaced by N.

It is to be particularly noticed that Theorem 6·3 determines explicitly the dependence on M.* This is because, for the assemblies discussed, M can be made (by dissociation, etc.) to vary reversibly—that is, a sequence of natural equilibrium states can be found, in which M varies. A great part of the controversies about entropy in statistical mechanics has centred round the determination of the variation of S with M, ignoring just this point, that such variations can only be relevant, and therefore determinable, when M can change reversibly. This has naturally led to great confusion of thought,† only avoidable in some way equivalent to the foregoing.

The forms of (583), (586) and (587) suggest at once that the connection between partition functions and the functions of thermodynamics can be simplified by the use of *Planck's characteristic function* as the primary thermodynamic quantity. Ψ is defined in thermodynamics by the equation
$$\Psi=S-E/T; \qquad \text{......(595)}$$
it is thus merely a modification of the Helmholtz's free energy‡ F. It has the properties
$$E=T^2\frac{\partial\Psi}{\partial T},\quad S=\Psi+T\frac{\partial\Psi}{\partial T},\quad Y_t=T\frac{\partial\Psi}{\partial y_t}. \qquad \text{......(596)}$$

In terms of Ψ the foregoing theorem can be rewritten as follows:

Theorem 6·31. *Any particular species of free molecule contributes to the characteristic function*
$$kM\{\log[f(T)/M]+1\}; \qquad \text{......(597)}$$
any pure crystal
$$kP\log\kappa(T). \qquad \text{......(598)}$$
Any one of the equivalent forms (583)–(594) *has an analogue here.*

We may conveniently recall here the further property of Ψ as a thermodynamic potential, that for any assembly at given temperature and given volume (or generally given parameters y_t) the equilibrium state is determined by the equation
$$d\Psi=0 \qquad \text{......(599)}$$
for all relevant variations. It is easily verified by differentiation of Ψ that

* Strictly $\overline{M}$. But since it never matters whether we are dealing with a fixed M or an average $\overline{M}$ we shall usually omit the bar in future over symbols representing numbers of systems of a given type unless the context requires it for clarity.

† See Ehrenfest and Trkal, *loc. cit.*, for a critical exposition of this confusion.

‡ See Planck, *Wärmestrahlung*, ed. 5, p. 127 (1923). This function is also called the work function and sometimes denoted by A.

the general laws of mass-action, *etc.*, which we have determined directly in Chapter v are equivalent to this equation.

§**6·4.** *The increasing property of entropy.* We are now in a position to complete our account of the thermodynamic properties of our assemblies by showing that the function S which we have identified with the entropy possesses the characteristic *increasing property*. We have to show that, with such conventions for the arbitrary constants that $S'+S''=S$ when an assembly in equilibrium is separated into two by an ideal workless process such as the closing of an ideal door, then on junction by a similar process

$$S'+S''\leqslant S; \qquad \ldots\ldots(600)$$

S' and S'' are the entropies of the two assemblies before junction or after separation and S the entropy of the combined assembly. Since the joining together of two gaseous assemblies is essentially irreversible, only a convention can settle the values of the various entropy constants. A sufficient convention is to take $S_0=0$ in (573)—that is, to take S as given completely by Theorem 6·3.

To establish (600) it will be sufficient to consider a single type of assembly, for example the most general gaseous one. The proof for other types is similar. If then we distinguish all quantities referring to the two separate assemblies by single and double primes, we have

$$S'/k=Z(V',\vartheta',\lambda_1',\ldots,\lambda_t')-\Sigma_j X_j'\log\lambda_j'-E'\log\vartheta',$$

$$S''/k=Z(V'',\vartheta'',\lambda_1'',\ldots,\lambda_t'')-\Sigma_j X_j''\log\lambda_j''-E''\log\vartheta'',$$

$$S/k=Z(V'+V'',\vartheta,\lambda_1,\ldots,\lambda_t)-\Sigma_j(X_j'+X_j'')\log\lambda_j-(E'+E'')\log\vartheta.$$

In formulating the terms in Z in these expressions we have used the fact (see (163) or (221)) that each of the terms in Z contributed by the states of each type of free system depends on the particular enclosure only through the volume V which enters as a multiplying factor. Thus

$$Z(V'+V'',\vartheta,\lambda_1,\ldots,\lambda_t)=Z(V',\vartheta,\lambda_1,\ldots,\lambda_t)+Z(V'',\vartheta,\lambda_1,\ldots,\lambda_t).$$

But it was shown in Theorem 2·42 that ϑ', λ_1', λ_2', ..., λ_t' define the unique minimum of the function S', so that

$$S'(\vartheta',\lambda_1',\ldots,\lambda_t')\leqslant S'(\vartheta,\lambda_1,\ldots,\lambda_t),$$

equality being only possible when

$$\vartheta'=\vartheta,\ \lambda_1'=\lambda_1,\ \ldots,\ \lambda_t'=\lambda_t.$$

Similarly $$S''(\vartheta'',\lambda_1'',\ldots,\lambda_t'')\leqslant S''(\vartheta,\lambda_1,\ldots,\lambda_t),$$

equality being only possible when

$$\vartheta''=\vartheta,\ \lambda_1''=\lambda_1,\ \ldots,\ \lambda_t''=\lambda_t.$$

We see therefore at once that $S'+S''\leqslant S$, equality being only possible when

$$\vartheta'=\vartheta''=\vartheta,\ \lambda_1'=\lambda_1''=\lambda_1,\ \ldots,\ \lambda_t'=\lambda_t''=\lambda_t.$$

Since ϑ and the λ's as determined by (472) are intensive parameters, the necessary and sufficient conditions are

$$\vartheta' = \vartheta'', \quad \lambda_1' = \lambda_1'', \ldots, \lambda_\iota' = \lambda_\iota'', \qquad \ldots\ldots(601)$$

which are equivalent to asserting that, if there is no change of entropy on junction, the separate assemblies must have had equal temperatures and concentrations, or rather temperatures and partial potentials (§ 6·5).

§ 6·5. *The physical meaning of the λ's.* The parameters $\lambda_1, \ldots, \lambda_\iota$, which we have been led to introduce by the nature of the mathematics, play such a natural part in the preceding discussions that one is led to expect them to possess a natural physical interpretation, just as ϑ may be interpreted as the temperature. This is the case, and we can relate them in a simple way to the *partial potentials* of the various constituents in the assembly, as is in fact already obvious from the form (591). Just as ϑ was identified with the temperature because it is a parameter which has the same value for every component part of the assembly, so λ_u must be equivalent to the partial potential of the uth constituent because it is a parameter which has the same value for every component part of the assembly *in which the uth constituent occurs.*

Thermodynamic partial potentials μ_u may be defined by the equation†

$$dE + \Sigma_t Y_t dy_t = TdS + \Sigma_u \mu_u dM_u^*; \qquad \ldots\ldots(602)$$

in forming this variation we are to suppose that the total masses‡ M_u^* of the various constituents in our assembly or in any phase or part of it are varied, as well as the temperature and the geometrical parameters y_t. If we form the variations of (571) in this manner we obtain

$$TdS = dQ - \Sigma_u kT \log \lambda_u dX_u.$$

Now X_u is the total number of atoms of type u and M_u^* is the mass of the uth independent constituent in gram-molecules. Therefore

$$X_u = RM_u^*/k,$$

and

$$\mu_u = RT \log \lambda_u. \qquad \ldots\ldots(603)$$

This is the desired relation between λ_u and the corresponding partial potential. If we evaluate λ_u by means of (166) for a structureless particle in the classical limit we find

$$\mu_u/RT = \log \nu_u - \tfrac{3}{2}\log T + \log \frac{h^3}{(2\pi m_u k)^{\frac{3}{2}}}, \qquad \ldots\ldots(604)$$

which is consistent with the usual value of μ_u for a perfect gas.§

† Bryan, *Thermodynamics*, p. 152.

‡ The masses are here denoted by M_u^* to avoid confusion with the numbers of free atoms M_u or the total in the gaseous phase $M_u^\dagger$ of § 6·3.

§ Bryan, *loc. cit.* p. 120.

If in a similar manner we imagine an addition to the assembly not of free atoms, but of $dN_v{}^*$ gram-molecules of the molecule v, then

$$dX_u = Rq_v{}^u dN_v{}^*/k,$$

and the partial potential μ_v of the vth molecule is given by

$$\mu_v = RT\,\Sigma_u q_v{}^u \log \lambda_u,$$

$$\mu_v/RT = \log \nu_v - \log\{f_v(\vartheta)/V\}, \qquad \text{......(605)}$$

which is also consistent with the usual thermodynamic value. We can thus see that the partial potential of the molecule v in the equilibrium state is equal to the sum of the partial potentials of its constituent atoms. This is the usual relation between partial potentials necessitated by the existence of a chemical reaction.† It is another aspect of the usual theorem that the partial potential of any constituent must be the same in all phases in which it occurs.

Finally let us consider the more general case of an assembly containing a mixed crystal dealt with in § 5·6. The contribution to the entropy by the mixed crystal is

$$S_\kappa = k[\log K_{P,P'}(\vartheta) - \overline{E_\kappa} \log \vartheta].$$

It follows at once that

$$TdS_\kappa = dQ_\kappa + kT\left[\frac{1}{K}\frac{\partial K}{\partial P}dP + \frac{1}{K}\frac{\partial K}{\partial P'}dP'\right];$$

by equations (536) this reduces to

$$TdS_\kappa = dQ_\kappa - kT[\log \lambda dP + \log \lambda' dP'],$$

where λ and λ' are the partial potentials of the constituents in the gaseous phase in equilibrium with the crystal. The relationship between the λ's and the thermodynamic μ's is thus repeated.

§ **6·6.** *Thermodynamic necessity for the invariance of the weights.* The preceding discussion, establishing the existence of S, has proceeded on the assumption that

$$\overline{Y_l} = \Sigma_u \Sigma_r \overline{a_r{}^u}\left(-\frac{\partial \epsilon_r{}^u}{\partial y_l}\right) + \Sigma_v \Sigma_r \overline{a_r{}^v}\left(-\frac{\partial \epsilon_r{}^v}{\partial y_l}\right),$$

$$= \frac{1}{\log 1/\vartheta}\frac{\partial Z}{\partial y_l}. \qquad \text{......(606)}$$

This however is only true because the weights $\varpi_r{}^u$, $\varpi_r{}^v$ do not depend upon the y's and are therefore what are called *adiabatic invariants*. Since in the present theory the weights are by definition integers specifying the number of solutions of Schrödinger's equation for a given characteristic energy they must vary discontinuously if at all, and in fact can never vary. Equation (606) is therefore justified. But in earlier presentations of the subject this

† Gibbs, *Scientific Papers*, "Equilibrium of Heterogeneous Substances", pp. 67–70.

was not yet clear and it was of some importance to establish the invariance of the weights by showing that such invariance was also necessary for the existence of the entropy.* On account of its historical importance the argument is reproduced here for a classical gaseous assembly in dissociative equilibrium.

When the weights depend on the y's (606) fails and (classically) must be replaced by

$$\overline{Y}_t = \frac{1}{\log 1/\vartheta}\frac{\partial Z}{\partial y_\tau} - \frac{1}{\log 1/\vartheta}\left[\Sigma_u \Sigma_r \lambda_u \frac{\partial \varpi_r^{\,u}}{\partial y_\tau}\vartheta^{\epsilon_r^u} + \Sigma_v \Sigma_r \lambda_1^{q_v^1}\ldots\lambda_l^{q_v^l}\frac{\partial \varpi_r^{\,v}}{\partial y_\tau}\vartheta^{\epsilon_r^v}\right].$$

Therefore the general form of $\log 1/\vartheta\, dQ$ will be

$$dS(\vartheta,\lambda_1,\ldots,\lambda_l,y_1,\ldots) - \mu_1(\vartheta,\lambda_1,\ldots,\lambda_l,y_1,\ldots)\,dy_1 - \ldots, \quad \ldots\ldots(607)$$

where

$$\mu_1(\vartheta,\lambda_1,\ldots,\lambda_l,y_1,\ldots) = \Sigma_u \Sigma_r \lambda_u \frac{\partial \varpi_r^{\,u}}{\partial y_1}\vartheta^{\epsilon_r^u} + \Sigma_v \Sigma_r \lambda_1^{q_v^1}\ldots\lambda_l^{q_v^l}\frac{\partial \varpi_r^{\,v}}{\partial y_1}\vartheta^{\epsilon_r^v}.$$

We may suppose that the independent parameters that define the state of the assembly are S, y_1, y_2, ... Then the conditions that (607) should be a perfect differential contain the equations

$$\left(\frac{\partial \mu_1}{\partial S}\right)_{(y)} = \left(\frac{\partial \mu_2}{\partial S}\right)_{(y)} = \ldots = 0 \quad \ldots\ldots(608)$$

for all values of S (or ϑ) and the y's. They must also hold for all X's and however many types of atom or molecule are present in the assembly.

Consider first the simplest case in which only the uth type of atom is present in number X_u. Then

$$\mu_1 = X_u \Sigma_r \frac{\partial \varpi_r^{\,u}}{\partial y_1}\vartheta^{\epsilon_r^u}/f_u(\vartheta)$$

and (608) reduces to $\partial\mu_1/\partial\vartheta = 0$. Now μ_1 is of the form $\Sigma\, a_n \vartheta^n/\Sigma\, b_n \vartheta^n$, and therefore if it is a function of the y's only we must have

$$\Sigma\,(b_n\mu_1 - a_n)\,\vartheta^n = 0 \quad (all\ \vartheta)$$

or

$$\mu_1 = \mu_1(y_1,\ldots) = a_n/b_n \quad (all\ n, y_1, \ldots).$$

That is to say, we must have

$$\frac{\partial \varpi_r^{\,u}}{\partial y_1} = \kappa_u \varpi_r^{\,u}, \quad \kappa_u = \frac{\partial}{\partial y_1}\log \varpi_r^{\,u}, \quad \ldots\ldots(609)$$

where κ_u depends only on the y's. Thus the $\varpi_r^{\,u}$ may have a common factor $\omega_u(y_1,\ldots)$ and the resulting change in S would be $-X_u \log \omega_u$. This would not affect any physical result for such assemblies. Similar arguments hold for assemblies of a single type of molecule.

* Bohr, *Proc. Camb. Phil. Soc.* Suppl. p. 17, and previously Ehrenfest, *Physikal. Zeit.* vol. 15, p. 660 (1914), have attempted to show that the weights must be adiabatic invariants. It will appear in the course of this section that, though their results are effectively correct, their arguments hardly went sufficiently deep.

Returning now to the general gaseous assembly we should have

$$\mu_1 = \Sigma_u \lambda_u f_u(\vartheta) \frac{\partial}{\partial y_1} \log \omega_u + \Sigma_v \lambda_1^{q_v^1} \dots \lambda_t^{q_v^t} f_v(\vartheta) \frac{\partial}{\partial y_1} \log \omega_v,$$

and the extra terms in $\log 1/\vartheta dQ$ reduce to

$$-\Sigma_u \lambda_u f_u(\vartheta)\, d\log \omega_u - \Sigma_v \lambda_1^{q_v^1} \dots \lambda_t^{q_v^t} f_v(\vartheta)\, d\log \omega_v$$

or

$$-\Sigma_u X_u\, d\log \omega_u - \Sigma_v \lambda_1^{q_v^1} \dots \lambda_t^{q_v^t} f_v(\vartheta)\, d\log \frac{\omega_v}{\Pi_u \omega_u^{q_v^u}}.$$

The ω_u, ω_v being functions of $y_1, y_2, \dots$, these extra terms can in general only be a perfect differential if

$$d\log \frac{\omega_v}{\Pi_u \omega_u^{q_v^u}} = 0 \quad (all\ v). \qquad \dots\dots(610)$$

For the extra terms are of the form

$$\Sigma_v A_v(\vartheta, y_1, \dots)\, dB_v(y_1, \dots)$$

and can only be a perfect differential if for $y_1, y_2, \dots$

$$\Sigma_v \frac{\partial A_v}{\partial \vartheta} \frac{\partial B_v}{\partial y_1} \equiv 0.$$

Since $\partial A_v/\partial\vartheta \neq 0$ and the A_v's have in general no special relations between them, these relations can only be satisfied if $\partial B_v/\partial y_1 = 0$ ..., which are the equations (610). It follows, therefore, that the weights can contain factors dependent on the $y_1, y_2, \dots$ provided that

$$\omega_v = \alpha_v \Pi_u \omega_u^{q_v^u},$$

where the α's are absolute constants. Thus the non-invariant factors ω_u, if they exist, must persist with the atom through every combination into which the atom can enter. They are therefore without significance in any physical problem and can be omitted without loss of generality, and apart from these trivial factors the adiabatic invariance of the weights is established. The conclusion holds good for the most general assemblies so far discussed.

§ **6·7.** *The inverse relation between specific heats or average energies and the weights and energies of the states of an individual system.* In the foregoing chapters we have shown how to determine the average energy and specific heats of any system or set of systems, when the weights and energies of the permitted states of one system are known so that the partition function may be constructed. We have had in fact the relations (classical statistics)

$$\bar{E} = MkT^2 \frac{\partial}{\partial T} \log f(T),$$

$$C_v = Mk \frac{d}{dT} T^2 \frac{\partial}{\partial T} \log f(T), \qquad \dots\dots(611)$$

where

$$f(T) = \Sigma_r \varpi_r e^{-\epsilon_r/kT}.$$

It is desirable at some stage to pause and attempt to answer the question whether given $\bar{E}$ or C_V, and so the form of the function $f(T)$, we can deduce the ϖ_r and ϵ_r, and in particular whether such values, if they can be derived, are unique. It appears that the theoretical answer is "yes". The ϖ_r and ϵ_r can be derived and are unique, with the exception of trivial constant factors, which arise as the constants of integration in determining $f(T)$ from (611). The practical importance of this answer is (as will be seen) limited, but its theoretical importance is still considerable.

The most convenient variable to work with here is τ where $\tau = 1/kT$. Then

$$C_V = Mk\tau^2\left(\frac{d}{d\tau}\right)^2 \log f(\tau), \quad \bar{E} = -M\frac{d}{d\tau}\log f(\tau),$$

$$f(\tau) = \sum_0^\infty \varpi_r e^{-\epsilon_r \tau}.$$

A knowledge of C_V determines $f(\tau)$ except for trivial constants of integration; the problem is to derive from this $f(\tau)$ the ϖ_r and ϵ_r. For this purpose we express $f(\tau)$ as a Stieltjes' integral in the form

$$f(\tau) = \int_0^\infty e^{-\tau\epsilon}\, dw(\epsilon), \qquad \text{......(612)}$$

where $dw(\epsilon)$ represents the weight corresponding to the energy ϵ or the range $\epsilon, \epsilon + d\epsilon$, and it is indifferent whether we are concerned with quantized systems or with classical. In all cases $w(\epsilon)$, the integrated weight, is an increasing function in the wider sense, which has simple isolated discontinuities if the system is quantized or contains a quantized part.

Now it has been shown that (612) can be inverted so as to express $w(\epsilon)$ as an integral of $f(\tau)$.* It may be supposed that $w(\epsilon)$ is a monotonic increasing function of ϵ for all values of ϵ considered, which has only a finite number of simple discontinuities or steps ϖ_i in any finite interval. The function $w'(\epsilon)$, derived from $w(\epsilon)$ by the removal of the discontinuities, has a differential coefficient Ω which exists and is continuous, except perhaps at a finite number of points in any finite range, and is bounded in any finite range. These conditions will be referred to as conditions W. Then we have the following

Theorem 6·7. *If $w(\epsilon)$ satisfies conditions W, and if $\sum^\infty \varpi_i e^{-\tau\epsilon_i}$ and $\int^\infty e^{-\tau\epsilon}\Omega\, d\epsilon$ converge for $\tau = \gamma_0$, and if*

$$f(\tau) = \int_{-a}^\infty e^{-\tau\epsilon}\, dw(\epsilon), \qquad \text{......(613)}$$

* For a proof see R. H. Fowler, *Proc. Roy. Soc.* A, vol. 99, p. 464 (1921). A deeper version of the same theorem has been given by Burkill, *Proc. Camb. Phil. Soc.* vol. 23, p. 356 (1926); *Proc. Lond. Math. Soc.* vol. 25, p. 513 (1926).

then $f(\tau)$ is a holomorphic function of τ in the half-plane $R(\tau) > \gamma_0$, and

$$\tfrac{1}{2}\{w(\epsilon+0)+w(\epsilon-0)\}-w(-a)=\frac{1}{2\pi i}\int_{\gamma-i\infty}^{\gamma+i\infty} f(\tau)\,e^{\epsilon\tau}\frac{d\tau}{\tau}, \quad \text{......(614)}$$

where $\gamma > \gamma_0$, $\gamma > 0$, and the infinite integral in (614) *is evaluated as*

$$\underset{\Gamma\to\infty}{\mathrm{Lt}}\int_{\gamma-i\Gamma}^{\gamma+i\Gamma}.$$

It is clear that the theorem theoretically must apply to the physical problem in hand, and therefore, given $f(\tau)$, the ϖ_r and ϵ_r (and any classical part) are uniquely determinate. But it is not practically applicable unless our knowledge of $f(\tau)$ is so precise that we know it not merely numerically but formally, as a function of the complex variable τ. This difficulty can be turned by a method due to Schwarzschild,* but as it is still doubtful whether even so the method would give results of practical interest, we shall not pursue it further.

In the case, however, of radiation (or the linear harmonic oscillator) we believe that we do know the exact form of $f(\tau)$. It is then of interest to see that the weights must be of the form assigned. For if we demand that

$$\bar{E}=M\frac{h\nu}{e^{h\nu/kT}-1},$$

so that
$$f(\tau)=C(1-e^{-h\nu\tau})^{-1} \quad (C\ \textit{constant}), \quad \text{......(615)}$$

then
$$\tfrac{1}{2}\{w(\epsilon+0)+w(\epsilon-0)\}=\frac{C}{2\pi i}\int_{\gamma-i\infty}^{\gamma+i\infty}\frac{e^{\epsilon\tau}}{1-e^{-h\nu\tau}}\frac{d\tau}{\tau}.$$

Now $\quad (1-e^{-h\nu\tau})^{-1}=1+e^{-h\nu\tau}+\ldots+e^{-ph\nu\tau}+e^{-(p+1)h\nu\tau}/(1-e^{-h\nu\tau}).$

Choose p so that $p<\epsilon/h\nu<p+1$. Then

$$\begin{aligned}&\tfrac{1}{2}\{w(\epsilon+0)+w(\epsilon-0)\}\\&\quad=C\left\{\sum_0^p\frac{1}{2\pi i}\int_{\gamma-i\infty}^{\gamma+i\infty}e^{(\epsilon-rh\nu)\tau}\frac{d\tau}{\tau}+\frac{1}{2\pi i}\int_{\gamma-i\infty}^{\gamma+i\infty}\frac{e^{[\epsilon-(p+1)h\nu]\tau}}{1-e^{-h\nu\tau}}\frac{d\tau}{\tau}\right\},\\&\quad=C(p+1+J).\end{aligned} \quad \text{......(616)}$$

By an application of Cauchy's theorem it is easily shown that $J=0$. It follows that the weights must be C for $\epsilon=rh\nu$ and zero for all other ϵ. No other scheme of weights can be admitted.

If alternatively we demand the average energy

$$\bar{E}=M\left\{\frac{h\nu}{e^{h\nu/kT}-1}+\tfrac{1}{2}h\nu\right\},$$

then by simple integration

$$f(\tau)=\frac{Ce^{\frac{1}{2}h\nu\tau}}{1-e^{-h\nu\tau}} \quad (C\ \textit{constant}). \quad \text{......(617)}$$

* See Fowler, *loc. cit.* p. 470.

The weight function can be evaluated by the same arguments and we find that the only admissible weights are weights C for $\epsilon=(r+\frac{1}{2})h\nu$ and zero for all other ϵ.

If we accept the general basis from which we have developed the equilibrium theory of statistical mechanics in this monograph, then the experimental laws of temperature radiation may be taken as demanding that $f(\tau)$ for a simple linear harmonic oscillator shall be given by (615). It then follows that the weight function must be determined by (616), that is, *must* be that assumed by Planck. It is of course possible to call in question the general basis, but not, accepting this basis, Planck's assumption.*

A direct analysis of the Stefan-Boltzmann law by this method is also not without interest; the value of $f(\tau)$ is

$$f(\tau)=C\exp(\sigma V/\tau^3).$$

From this it follows by further applications of Cauchy's theorem that

$$w(-0)=0, \quad w(\epsilon)=C\left[1+\frac{\sigma V}{1!}\frac{\epsilon^3}{3!}+\frac{(\sigma V)^2}{2!}\frac{\epsilon^6}{6!}+\frac{(\sigma V)^3}{3!}\frac{\epsilon^9}{9!}+\dots\right].$$

The weight function therefore must have a discontinuity C at $\epsilon=0$. The state of zero energy must have a non-zero weight. For other energies the function is too complicated to give us much information, but the non-classical nature† of the weight function is already evident.

§6·8. *Entropy and thermodynamic probability.* It is not in general our purpose in this monograph to attempt critical discussions of alternative presentations of statistical mechanics, but rather to develop the theory on a single consistent plan in a manner as logical as possible. It is not, however, possible at this point to pass by entirely in silence other methods of introducing entropy into statistical mechanics, without giving some indication of why the very strict analogy to classical thermodynamics has been preferred here.

Entropy is usually introduced into statistical mechanics by means of Boltzmann's hypothesis relating it to probability. We cannot do more than abstract the various arguments here; for the best systematic account the reader should refer to Planck.‡ It will be assumed that he is familiar with Planck's account. Boltzmann's hypothesis is based in general on the fact

* This discussion contains the whole substance of Poincaré, *J. de Physique*, vol. 2, p. 5 (1912), but much simplified and rendered more rigorous by the use of the machinery of the present methods. The hypothesis and conclusions are essentially the same.

† See the classical paper by Ehrenfest, *Ann. d. Physik*, vol. 36, p. 103 (1911), where just this point is established by reasoning essentially the same. The present methods again allow of great simplification.

‡ Planck, *Wärmestrahlung*, 3rd Abschnitt, ed. 5 (or later).

that on the one hand the assembly tends to get into its most probable state (which is equivalent to the average state with which we work here), while on the other hand its entropy tends to increase, so that a functional relation between the entropy and the probability W of a state

$$S = f(W)$$

may be postulated by a legitimate analogy. The analogy is of the same type as that by which we have postulated functional relationships between ϑ and T and the λ's and the μ's. The argument then proceeds somewhat as follows. Suppose we can assign the numerical value of W for the probability of the state of any assembly. If therefore we have two such assemblies which are entirely independent, then by a fundamental principle of probability the joint probability is the product of the two separate probabilities or

$$W_{12} = W_1 W_2. \qquad \ldots\ldots(618)$$

On the other hand, by the second law of thermodynamics the joint entropy is the sum of the separate entropies (with suitable adjustments of the additive constants) and so

$$S_{12} = S_1 + S_2. \qquad \ldots\ldots(619)$$

The functional relationship must then be

$$S = k \log W, \qquad \ldots\ldots(620)$$

k being a universal constant. We have still to assign a definite way of specifying W for any statistical state of any assembly. For use as W in this connection the quantity "thermodynamic probability" is introduced and defined to be equal to the number of complexions corresponding to the specified state. This number W is then made a maximum subject to the condition of constant energy—the assembly is then in its most probable state—and the maximum value of $k \log W$ so obtained is equated to the entropy S. The entropy S so defined has been shown in § 6·3* to agree with the entropy of classical thermodynamics, and in general possesses the fundamental increasing property. It should be observed that there are two quite distinct steps in the argument after W has been equated to the thermodynamic probability. In the first the determination of the maximum fixes the most probable state of the assembly *by itself*. In the second the assembly is related *to the outside world* by determining its entropy by (620). Finally, the absolute temperature scale is introduced by the relation $\partial S/\partial E = 1/T$.†

Unfortunately, there is much to be criticized in this argument. In the first place there is some vagueness as to what precisely is happening in the

* A slight extension of § 6·3 is required to show that W_{max} and W_{tot} are effectively the same. The correctness of S is easily verified directly in simple cases.

† This abstract is intended to do proper justice to the argument described, which is in any case elegant and attractive. The reader should supplement it by reference to Planck (*loc. cit.*) at least.

building up of one assembly out of two to yield equations (618) and (619). The addition of entropies can usually only be realized by some form of thermal contact, and is then true in general only when the temperatures are equal. But both these conditions require that the assemblies shall not be independent. So it is only possible to give a meaning to (619) by making (618) invalid or at best not necessarily valid, which destroys the *a priori* nature of the argument. For example, it is argued that $S_{12} = S_1 + S_2$ follows directly from the second law. It is hard to see how this can be derived except from the equation $dQ_{12} = dQ_1 + dQ_2$ which is true neglecting surface interactions. This then leads to $dS_{12} = dS_1 + dS_2$, and so to (619), only if the temperatures are equal.* It is not maintained that this criticism could not be turned, but it disposes at least of Planck's example of a composite assembly consisting of any body on the Earth's surface and an enclosure of temperature radiation on Sirius.

In the second place, the probability W of any statistical state of an assembly requires of course precise and careful definition in such a way that (618) holds *a priori*. The definition of W actually used, thermodynamic probability or number of complexions, makes W a large integer and not a probability at all (which must be a proper fraction); thus (618) cannot be maintained by any appeal to the theory of probability, for, frankly, that theory is irrelevant. On the other hand, genuine probabilities such as the ratio of the number of complexions representing any statistical state to the total number of complexions do not lead (straightforwardly) to the right result.

It is well known that actually the "thermodynamic probability" does lead to the right value of the entropy, and it is perhaps worth while to pause and enquire how the criticism just formulated is to be satisfied. If we take the genuine probability, in so far as we have to calculate the most probable state for given energy the total number of complexions C is constant, we are concerned only with the equation

$$S' - S'' = k \log(W'/W''), \qquad \text{.....(621)}$$

i.e. with ratios of W, and so the argument is unaffected. But when in the second step we attempt to determine the value of the entropy itself from (620) with $W_{\max}$, we find in all cases the trivial result $S = 0$. It is a simple consequence of the arguments leading to (575) that $W_{\max}/C$ or

$$W(\text{average state})/C$$

is always effectively unity, expressing the fact that the possession of the average or most probable state is a normal property of the assembly.

* It is just at this step that an important logical obscurity enters. We have no right to derive $S_{12} = S_1 + S_2$ from $dS_{12} = dS_1 + dS_2$ without an *explicit* recognition that we have made a special choice of the additive constants in the entropies.

It is thus clear that the straightforward process is useless, and if we are to retain a relation between entropy and probability we must find a way of justifying the omission of the denominator C. As long as we consider the assembly as a whole this is impossible, for C depends on T and cannot be ignored when variations of temperature are contemplated. In fact the usual arguments are attempting an impossibility, for they attempt to determine the entropy of an assembly, which determines its relation to the outside world, by consideration of the assembly by itself. The difficulty can be overcome, therefore, only by considering the assembly in question as part of a much larger assembly. Consider, for example, a group of M localized systems immersed in a bath of a very much larger number N of other localized systems. The statistical state of the whole is specified as usual by $a_0, a_1, \ldots, b_0, b_1, \ldots$, the numbers of systems in the permitted states or cells. The probability of the statistical state specified by $a_0, a_1, \ldots$ is

$$W(a_0, a_1, \ldots) = \left(\frac{M!\, \varpi_0^{a_0} \varpi_1^{a_1} \ldots}{a_0!\, a_1! \ldots}\right)\left(\Sigma_b \frac{N!\, \rho_0^{b_0} \rho_1^{b_1} \ldots}{b_0!\, b_1! \ldots}\right) \Big/ C,$$

where Σ_b denotes summation over all b's such that $\Sigma_s \eta_s b_s = E - \Sigma_r \epsilon_r a_r$. Comparing C and Σ_b we see at once that

$$C = \frac{1}{2\pi i}\int_\gamma \frac{dz}{z^{E+1}} \{f(z)\}^M \{g(z)\}^N,$$

$$\Sigma_b = \frac{1}{2\pi i}\int_\gamma \frac{dz}{z^{E-E_a+1}} \{g(z)\}^N.$$

Since N is very large compared to M, Theorem 2·31 shows that approximately

$$\frac{\Sigma_b}{C} = \frac{\vartheta^{E_a}}{(f(\vartheta))^M},$$

where ϑ is fixed by

$$\frac{d}{d\vartheta}\frac{(g(\vartheta))^N}{\vartheta^E} = 0.$$

Thus

$$W(a_0, a_1, \ldots) = \frac{M!\, \varpi_0^{a_0} \varpi_1^{a_1} \ldots}{a_0!\, a_1! \ldots} \frac{\vartheta^{E_a}}{\{f(\vartheta)\}^M}. \qquad \ldots\ldots(622)$$

Equation (622) shows that the true probability W is proportional to the thermodynamic probability provided E_a, the total energy of the M systems, is fixed, but even now not otherwise unless the temperature of the large assembly is practically infinite, $\vartheta = 1$. If we wish to maintain the relation between the entropy and probability, it seems as if we can only justify the use of "thermodynamic probability" by considering a group of systems as part of a very large assembly at a very great temperature. If we do this, then it is easily verified that $W_{12} = W_1 W_2$ for any two parts of such an assembly (whether or no the temperature is very great) and the first difficulty is also turned. This treatment would be perhaps artificial but not illogical.

We can cast (622) into an alternative, more illuminating form. For the natural contribution of the M systems to the entropy of the assembly in equilibrium is

$$k\{M \log f(\vartheta) - E_a \log \vartheta\},$$

which we may call $S_{\max}$. Therefore

$$W(a_0, a_1, \ldots) = \frac{M!\, \varpi_0^{a_0} \varpi_1^{a_1} \ldots}{a_0!\, a_1! \ldots} e^{-S_{\max}/k}, \qquad \ldots\ldots(623)$$

which shows at once that $W_{\max} = 1$ as before, and that, defining entropy *via* probability, we arrive at

$$W/W_{\max} = W = e^{(S - S_{\max})/k}. \qquad \ldots\ldots(624)$$

It is only in some such sense as this that a meaning can be assigned to Boltzmann's hypothesis, but then it must be noted that it survives in Boltzmann's own form

$$S - S' = k \log W/W'. \qquad \ldots\ldots(625)$$

The whole development has thus become rather clumsy, for the entropy itself is a function only of the group and its temperature, but has to be derived for the group in relation to an infinite assembly at infinite temperature. It would appear to be much better, if a direct definition of entropy is required in terms of complexions and not *via* the classical form $dQ = TdS$, to abandon all reference to the theory of probability and define the entropy simply as k times the logarithm of the number of complexions, that is by (575). This definition must then of course be justified by direct comparison with classical thermodynamics, not by the *a priori* arguments which we have been criticizing here. It is a definition which has much to recommend it, especially as it enables the definition of entropy to be extended to cover non-equilibrium states.

As yet no reason has been given against the introduction of entropy by (625) or in the more general way just suggested, if such a way is still preferred. It must be observed, however, that there is no hope of a logical definition of *absolute entropy* by such an equation as (625). This is as it should be. Much has been written of absolute entropy in the belief that in this way a basis could be found for Nernst's heat theorem. We shall show in the next chapter that this theorem takes its natural place in the equilibrium theory of pure statistical mechanics, and can be formulated without reference to entropy at all, still less to absolute entropy.* Even if entropy is defined by (575)

* Even in the 5th edition (p. 119) of his *Wärmestrahlung* Planck says: "As opposed to [Boltzmann's hypothesis (625)] we assign to the entropy S a quite definite absolute value. This is a step of essential import, which can only be justified by the verification of its consequences. It leads, as we shall see later, of necessity to the quantum hypothesis, and thereby on the one hand, for radiant heat, to a definite law of distribution of energy for black radiation and on the other for the heat of solids to Nernst's heat theorem." It should be remembered that Planck is thinking

there is still nothing really *absolute* about it. We can never know whether our analysis of complexions is really complete. At one stage we may omit all reference to nuclear spin and its orientation. Then we generalize the count to include it. But why should this be the last step?

There is, however, yet a further difficulty in the introduction of entropy in this way, logically founded on its increasing property. The identification of S and $k \log W$ is based on an analogy, correct enough so far as it goes, but insufficiently deep. For it is tacitly assumed that the entropy is the only function of the state of the assembly which has this increasing property. This, however, is untrue, for $\Sigma = S + bE$, where b is any constant, also has the same increasing property, and we have no *a priori* reason for preferring one value of b to any other. If we take the general value Σ for the entropy we find for the relation between T and ϑ

$$\frac{1}{T} = \frac{\partial \Sigma}{\partial E} = k \log 1/\vartheta + b,$$

which does not determine a unique temperature scale. In fact we can only see that $b = 0$ by a direct appeal to the second law $dQ = TdS$, which will only hold with $b = 0$. The use of functions with the increasing property can apparently never lead to precise results without an appeal to dQ. If this appeal has to be made in any case, the method of approach by the increasing property loses any possible advantage over the classical method adopted in this monograph.

The difficulties pointed out above occur for the simplest non-dissociating assemblies, and render unsatisfactory, even for these, the introduction of entropy *via* Boltzmann's hypothesis. When we come to general dissociating assemblies the difficulties become still more pronounced, because the logical determination of the proper dependence of the entropy on the number of systems of any type present is almost impossible by this method. These difficulties have been pointed out by Ehrenfest and Trkal* and we need not stress them here. They again arise in what is virtually an attempt to determine the dependence of S on N or M without reference to any reversible method by which N or M may be supposed to be varied, and as such are doomed to failure.

§**6·9**. *Position of Boltzmann's hypothesis in the present theory.* The very general form of (624) or (625) makes these equations important instruments of investigation especially in complicated assemblies for which explicit

primarily of the fact that in the classical theory the entropy of a solid would not remain finite as $T \to 0$, and the requirement that it should remain finite demands the quantum theory. Even allowing for this, it is impossible to accept his statement fully. It will be maintained in this monograph that whatever the practical convenience of the idea of absolute entropy (often great) it is of no theoretical importance whatever.

* Ehrenfest and Trkal, *loc. cit.*; Fowler, *Phil. Mag.* vol. 45, p. 497 (1923).

forms for the partition function may be difficult or impossible to construct. It is important, therefore, to fix their position in the theory as here developed, a position of course not that of a fundamental hypothesis but of a general theorem, when our definition of the entropy S is suitably extended. We have shown in equation (575) that for the whole assembly $S = k \log C$. We can extend our definition of S, in conformity with the usages of classical thermodynamics, so as to apply to any physically separable part of the assembly by the convention that the entropy of the part may be calculated as in § 6·3 as if the part were a separate assembly with specified energy and configuration. This convention is obviously self-consistent. Special adjustments of S_0 are of course implied to make the entropy additive in equilibrium. We then recover (625) as a general theorem, valid at least for any small part of a large assembly, if ratio of probabilities means simply ratio of representative complexions.

In order to provide *a priori* for a connection between entropy and probability we were logically compelled to discuss a part of a very much larger assembly, and have formulated (625) in this connection. The large assembly, however, is irrelevant to the truth of (625) as a general theorem in statistical mechanics. It is easy to verify that (625) applies at once to any specified configurations of the whole assembly or of the whole of any parts into which we choose to divide it. We can calculate the entropy of each part as specified,* and it follows at once by (575) that for each part $S = k \log C'$. The total number of weighted complexions representing the specification is therefore $\Pi(C')$ taken over each part and therefore the probability is proportional to $\exp(\Sigma S/k)$. When the specified state is the equilibrium state for the whole assembly ΣS reduces to the usual S and $\Pi(C')$ effectively to C. For the ratio of the probabilities of two such specifications we have therefore $\exp\{(\Sigma S - \Sigma S')/k\}$ which is (625). We summarize this conclusion in the following

Theorem 6·9 (***Boltzmann's hypothesis***). ***If the entropy S of any assembly or its part is defined as in* § 6·3 *(with suitable additive constants), and if as usual the probability W of any specification of the assembly is proportional to the number of representative complexions, then***

$$S - S' = k \log W/W'. \qquad \text{......(626)}$$

The theorem as proved refers only to specifications of the assembly or its parts in which molar variables alone are concerned. If the specifications become so detailed as to be molecular, then our S ceases to have a meaning. We shall not be concerned to use S in such cases, but if an extended definition to cover such cases is required it can obviously be provided by the equation

* Subject to the conditions specified each part will be in its own equilibrium state.

$S = k \log C$, which we have already seen to be valid in all cases in which a thermodynamic S exists.

Problems arising in complicated systems such as liquids or far-from-perfect gases, particularly in connection with fluctuations, can sometimes be handled with the help of (626) and the theorems of classical thermodynamics, when a direct treatment would fail for lack of power to construct or handle the complicated partition function. A well-known example is the theory of density fluctuations and the opalescence of liquids near their critical point.* A formula is required for the relative frequency of volume fluctuations of specified range in a given small element of volume. This frequency is given of course by the W of (626), which can be used in the form

$$W = \alpha e^{\Sigma(\Delta S)/k} \quad (\alpha \text{ constant}). \qquad \ldots\ldots(627)$$

If, however, the volume fluctuations may be thought of as isothermal—legitimate certainly for a small element of volume out of the large assembly—this can be cast in a form which is easier to use. For then, since

$$S = (E - F)/T,$$

where F is Helmholtz's free energy (work function), and since for the volume element and the rest of the fluid $\Sigma(\Delta E) = 0$, we have

$$W = \alpha e^{-\Sigma(\Delta F)/kT}, \qquad \ldots\ldots(628)$$

where $\Sigma(\Delta F)$ denotes the maximum work that the assembly can be made to do in returning isothermally and reversibly to its equilibrium state.

We should not close this chapter without a reference to the ideas put forward by G. N. Lewis,† with the help of which he propounds a generalized form of thermodynamics incorporating the laws of fluctuations which are more usually regarded as a deduction only from statistical mechanics. It would however take us too far afield to describe his arguments here.

* Einstein, *Ann. d. Physik*, vol. 33, p. 1275 (1910). See also Chapter xx.

† G. N. Lewis, *J. Amer. Chem. Soc.* vol. 53, p. 2578 (1931).

CHAPTER VII

NERNST'S HEAT THEOREM AND THE CHEMICAL CONSTANTS

§7·1. *The vapour pressure equation of classical thermodynamics.* As an introduction to the main subject of this chapter we must compare the formulae of classical thermodynamics and statistical mechanics for the vapour pressure equation (§ 5·42). Later on we must do the same for the reaction isochore (§ 5·3). For convenience of reference we start by giving the thermodynamical formula for the vapour pressure and its deduction, which enables us to make clear the necessary assumptions and approximations.

Consider an assembly containing originally a gram-molecule of a condensed substance, at a temperature T_2 and pressure p_2 which is its natural vapour pressure. The vapour is assumed to be an ideal gas so that for one gram-molecule of vapour $pV = RT$. (i) Allow the condensed phase to evaporate completely at temperature T_2 and pressure p_2. This is a reversible process in which the energy absorbed is Λ_2 and the gain of entropy Λ_2/T_2. (ii) Now expand the vapour isothermally and reversibly, doing work, to pressure p_1. The heat absorbed is $RT_2 \log(p_2/p_1)$, equal to the work done; the energy does not change and the gain of entropy is $R\log(p_2/p_1)$. (iii) Lower the temperature at constant pressure p_1 until condensation starts, supercooling not being allowed to occur. The assembly loses energy $\int_{T_1}^{T_2} C_{p_1} dT'$ and entropy $\int_{T_1}^{T_2} C_{p_1} dT'/T'$. Here C_{p_1} is the specific heat of the vapour at constant pressure p_1, but since the vapour is ideal C_{p_1} is independent of p_1 and may be written simply C_p. (iv) Compress the vapour isothermally until it is entirely condensed at temperature T_1 and pressure p_1. The assembly loses energy Λ_1 and entropy Λ_1/T_1. (v) Heat up the condensed phase from T_1, p_1 to T_2, p_2 keeping the pressure at its equilibrium value for each temperature, so that evaporation is always just unable to occur. The assembly gains energy $\int_{T_1}^{T_2} (C_p)_{\text{sol}} dT'$ and entropy $\int_{T_1}^{T_2} (C_p)_{\text{sol}} dT'/T'$. In this integral $(C_p)_{\text{sol}}$ is the specific heat of the condensed phase under the conditions specified, that is at a pressure always equal to the vapour pressure for the temperature, and the integral must include all the heats of transition (melting is one such transition) if any transition points occur in the range. The cycle is reversible and the entropy change vanishes so that

$$\frac{\Lambda_2}{T_2} - \frac{\Lambda_1}{T_1} + R\log\frac{p_2}{p_1} - \int_{T_1}^{T_2} [C_p - (C_p)_{\text{sol}}]\, dT'/T' = 0.$$

This expression obviously reduces to an equality between two expressions one depending only on T_1, the other only on T_2; each must therefore be constant and we find, dropping suffixes, that

$$\log p = -\frac{\Lambda}{RT} + \int^T \frac{C_p - (C_p)_{\mathrm{sol}}}{RT'}\,dT' + i'', \qquad \text{......(629)}$$

where i'' is some constant. The energy changes also must vanish so that

$$\Lambda_2 - \Lambda_1 = \int_{T_1}^{T_2} [C_p - (C_p)_{\mathrm{sol}}]\,dT'. \qquad \text{......(630)}$$

Classical thermodynamics has nothing to say as to whether the integral in (630) should converge as $T_1 \to 0$. But in fact it does so converge since $(C_p)_{\mathrm{sol}} = O(T^3)$ and $C_p = O(1)$. It follows that, letting $T_1 \to 0$,

$$\Lambda(T) = \int_0^T [C_p - (C_p)_{\mathrm{sol}}]\,dT' + \Lambda_0. \qquad \text{......(631)}$$

On using this value of $\Lambda(T)$ in (629) we obtain

$$\log p = -\frac{\Lambda_0}{RT} - \frac{1}{RT}\int_0^T [C_p - (C_p)_{\mathrm{sol}}]\,dT' + \int^T \frac{C_p - (C_p)_{\mathrm{sol}}}{RT'}\,dT' + i''.$$

On rearranging the integrals by integration by parts this becomes

$$\log p = -\frac{\Lambda_0}{RT} + \int^T \frac{dT'}{RT'^2}\int_0^{T'} [C_p - (C_p)_{\mathrm{sol}}]\,dT'' + i'''. \qquad \text{......(632)}$$

It is convenient to break up the specific heat of the vapour into two parts $(C_p)_0$ and $(C_p)_1$ of which $(C_p)_0$ is a constant part (constant over the temperature range which may be in question) and $(C_p)_1$ a variable part. It is a quantum result that this can always be arranged so that the double integral converges when the outer lower limit is zero, and the inner integrand is $(C_p)_1 - (C_p)_{\mathrm{sol}}$. The term in $(C_p)_0$ integrates and gives

$$\frac{(C_p)_0}{R}\log T$$

besides contributions to i'''. We are therefore finally left with

$$\log p = -\frac{\Lambda_0}{RT} + \frac{(C_p)_0}{R}\log T + \int_0^T \frac{dT'}{RT'^2}\int_0^{T'} [(C_p)_1 - (C_p)_{\mathrm{sol}}]\,dT'' + i. \qquad \text{......(633)}$$

The constant i is commonly known as *the chemical constant*, but for reasons that will appear would be better known as *the vapour pressure constant* of the vapour in question. In deriving (633) we have assumed, besides the first and second laws of thermodynamics, merely that the vapour is an ideal gas. The trivial differences between $(C_p)_{\mathrm{sol}}$ at zero pressure or one atmosphere or for the sequence of natural vapour pressures can also be ignored so that throughout our analysis of vapour pressures we shall use for $(C_p)_{\mathrm{sol}}$ the specific heat of the condensed phase at some standard pressure, usually one atmosphere. With these assumptions together with the quantum assumptions of convergence of the integrals at the lower limit equation (633) is a classical thermodynamic relation. About the actual values of Λ_0 and i classical thermodynamics has nothing to say. The value of i will differ according to the value assigned to $(C_p)_0$, the constant part of C_p; this may be conveniently varied according to the temperature range, and in all statements about i $(C_p)_0$ must be specified.

§ **7·11.** *The vapour pressure equation of statistical mechanics.* In § 5·42 we obtained the corresponding statistical formula (equation (520)) which can be written in the form

$$p = \frac{kT}{V}\frac{f(T)}{\kappa(T)}. \qquad \text{......(634)}$$

From this we derived equation (523) which is repeated here for reference:

$$\log p = -\frac{\chi}{kT} + \tfrac{5}{2}\log T - \int_0^T \frac{dT'}{RT'^2}\int_0^{T'} (C_p)_{\text{sol}}\, dT'' + \log\frac{(2\pi m)^{\frac{3}{2}} k^{\frac{5}{2}}}{h^3}. \qquad \text{......(523)}$$

This equation is the expanded form of (634) when the systems in both gas and crystal are effectively structureless massive points so that for example $\kappa(0) = 1$. We see at once that (523) and (633) are equivalent, since for atoms representable by such systems

$$(C_p)_0 = \tfrac{5}{2}R, \quad (C_p)_1 = 0.$$

The essentially new contribution made by the statistical treatment is the evaluation of i:

$$i = \log\frac{(2\pi m)^{\frac{3}{2}} k^{\frac{5}{2}}}{h^3}, \quad (C_p)_0 = \tfrac{5}{2}R \qquad \text{......(635)}$$

(*structureless monatomic systems in both vapour and gas*).

In establishing this interpretation of (634) we have made use of the relationship

$$\log \kappa(T) = \int_0^T \frac{dT'}{RT'^2}\int_0^{T'} (C_p)_{\text{sol}}\, dT'' \quad [\kappa(0) = 1], \qquad \text{......(636)}$$

an equation which requires a more exact study, now to be undertaken. The volume V of the crystal containing a fixed number of atoms is not itself fixed, but is a function of the temperature. In order to construct $\kappa(T)$ we have to suppose that V has been given an assigned value, and the value chosen must be specified before (636) has a precise meaning. The necessary precision is provided by the following

Lemma 7·11. *Let the state of zero energy of a crystal (containing one gram-molecule or P atoms) be defined to be its state at zero temperature and pressure. Let the exact partition function for the crystal, $[\kappa(V,T)]^P$, be a function of V and T only. Let the specific heat of the crystal C_p be measured in such a way that at each temperature of measurement the pressure p acting on the crystal* *is constant during measurement and has a definite value $p(T)$ subject to the restriction that $p(0)=0$. Then*

$$\log\frac{\kappa(V(T),T)}{\kappa(V(0),0)}=\int_0^T\frac{dT'}{RT'^2}\int_0^{T'}C_{p(T'')}dT''+\frac{1}{RT}\int_0^T p(T')\frac{dV(T')}{dT'}dT'. \qquad\ldots\ldots(637)$$

In this equation $V(T)$ is the volume which the crystal must have at temperature T when the pressure acting on it is $p(T)$.

The proof is as follows. From the thermodynamic definition of $C_{p(T)}$

$$C_{p(T)}=\left(\frac{dH}{dT}\right)_p=\left(\frac{dE(V(T),T)}{dT}\right)_p+p\left(\frac{dV(T)}{dT}\right)_p.$$

It follows easily by integration and use of (350) that

$$RT^2\frac{\partial}{\partial T}\log\kappa(V,T)=E(V(T),T)=\int_0^T C_{p(T')}dT'-\int_0^T p(T')\frac{dV(T')}{dT'}dT'.$$

Moreover by (375)

$$\frac{d}{dT}\log\kappa(V(T),T)=\frac{\partial}{\partial T}\log\kappa(V,T)+\frac{p(T)}{RT}\frac{dV(T)}{dT}.$$

Combining these equations and simplifying the p, V terms, we find

$$\frac{d}{dT}\log\kappa(V(T),T)=\frac{1}{RT^2}\int_0^T C_{p(T')}dT'+\frac{d}{dT}\frac{1}{RT}\int_0^T p(T')\frac{dV(T')}{dT'}dT'.$$

On integrating this equation we obtain (637).

An immediate consequence of this theorem is that

$$\log\frac{\kappa(V(T),T)}{\kappa(V(0),0)}=\int_0^T\frac{dT'}{RT'^2}\int_0^{T'}(C_p)_{p=0}dT''. \qquad\ldots\ldots(638)$$

It is easily verified that if C_p, measured at some pressure or pressures other than zero is used in (638) in place of $(C_p)_{p=0}$ the error committed will be trivial so long as the work done in the expansion of the crystal against this pressure is small compared with RT. This error is entirely trivial in vapour pressure applications, and equation (523) is justified.

Lemma 7·11 is more useful in the following extended form:

Lemma 7·11 (*extension*). *Let the partition function for a mixed crystal of P molecules in all (P is the number of molecules in one gram-molecule) be*

$$N_P[\kappa_a(V,T)\{f_1(T)\}^{D_1}\{f_2(T)\}^{D_2}\ldots]^P \quad (D_1+D_2+\ldots=1).$$

In this form N_P is independent of V and T, $\kappa_a(V,T)$ is the partition function for the acoustical modes of the crystal, and $f_1(T)$, $f_2(T)$, ... are the partition functions for the orientations and internal energies of the various molecules in the lattice. Let the reference state of zero energy be chosen so that $\kappa_a(V(0),0)=1$. Let the molecular states of lowest energy in $f_1(T)$, $f_2(T)$, ... have weights and *energies $\varpi_0^1, \varpi_0^2, \ldots, \epsilon_0^1, \epsilon_0^2, \ldots$ respectively. Then if $(C_p)_{\rm sol}$ is the specific heat of the crystal measured always at the pressure of its own vapour and the trivial work term is neglected*

$$\int_0^T \frac{dT'}{RT'^2}\int_0^{T'} (C_p)_{\rm sol}\, dT'' = \log\{\kappa_a(V,T)\,[f_1(T)]^{p_1}\,[f_2(T)]^{p_2}\ldots\} + \frac{D_1\epsilon_0^1 + D_2\epsilon_0^2 + \ldots}{kT} - D_1 \log \varpi_0^1 - D_2 \log \varpi_0^2 - \ldots \quad \ldots\ldots(639)$$

The proof is simple and will be omitted. The extension shows that if the crystalline partition function in its most general form is $[H(T)]^P$ then

$$\int_0^T \frac{dT'}{RT'^2}\int_0^{T'} (C_p)_{\rm sol}\, dT'' = \log H(T) + (\textit{terms in } \Lambda_0,\, i). \quad \ldots(640)$$

It remains to evaluate explicitly the contributions to $\log p$ arising from $\log f(T)$. In all cases there must be the terms given in (521) for a structureless system, namely

$$\log\left\{\frac{(2\pi mkT)^{\frac{3}{2}} V}{h^3}\right\} - \frac{\chi}{kT}.$$

To this a term $\log f_v(T)$ must be added when there is any structure, where $f_v(T)$ is the partition function for any rotational, vibrational, orientational or electronic states of the system that may be relevant in the vapour phase. If we choose to regard only the translational motion as contributing to $(C_p)_0$ then, by arguments similar to those of Lemma 7·11,

$$\log \frac{f_v(T)}{f_v(0)} = \int_0^T \frac{dT'}{RT'^2}\int_0^{T'} (C_p)_1\, dT''.$$

The definition of χ is such that $f_v(T)$ always starts with a constant term. The partition function $f_v(T)$ can often be analysed further (at least approximately) into rotational and vibrational parts. When any one of these factors is effectively classical, the contribution to $\log f_v(T)$ takes the form $s \log T + \textit{const.}$, where $2s$ is the number of square terms in the classical part of the energy. We can absorb these terms into $[(C_p)_0/R] \log T$ by a new choice of $(C_p)_0$ and the extra constant into i. The rest of $f_v(T)$, $n_v(T)$ say, necessarily continues to satisfy

$$\log \frac{n_v(T)}{n_v(0)} = \int_0^T \frac{dT'}{RT'^2}\int_0^{T'} (C_p)_1\, dT''.$$

The principal results are summarized below. Of course Λ_0/R and χ/k are identical.

Vapour Pressure Formulae: Pure Crystals

$$\log p = -\frac{\Lambda_0}{RT} + \frac{(C_p)_0}{R}\log T + \int_0^T \frac{dT'}{RT'^2}\int_0^{T'} [(C_p)_1 - (C_p)_{\rm sol}]\, dT'' + i. \quad \ldots\ldots(633)$$

Crystalline partition function, $[\kappa_a(T) f(T)]^P \quad [\kappa_a(0) = 1, f(0) = \varpi_0].$

I. *Monatomic vapours and crystals. Atoms with internal structure arising from nuclear orientations and electronic multiplets*

Gaseous partition function,

$$\frac{(2\pi mkT)^{\frac{3}{2}}}{h^3} V e^{-\chi/kT} f_v(T) \quad [f_v(0)=\rho_0];$$

$$\frac{(C_p)_0}{R}=\tfrac{5}{2}; \quad i=\log\frac{(2\pi m)^{\frac{3}{2}} k^{\frac{5}{2}}}{h^3}\frac{\rho_0}{\varpi_0}. \qquad ......(641)$$

If $(C_p)_1$ is neglected in (633), then i is slightly variable,

$$i=\log\frac{(2\pi m)^{\frac{3}{2}} k^{\frac{5}{2}}}{h^3}\frac{f_v(T)}{\varpi_0}. \qquad ......(642)$$

II. *Diatomic molecules in vapours and crystals*

(*a*) *Non-classical rotations in vapour—as above, equation* (641).

(*b*) *Classical rotations in vapour.*

Gaseous partition function,

$$\frac{(2\pi mkT)^{\frac{3}{2}}}{h^3}\frac{8\pi^2 AkT}{\sigma h^2} V e^{-\chi/kT} n_v(T) \quad [n_v(0)=v_0],$$

where A is the transverse moment of inertia and $\sigma=1$ for molecules XY, $\sigma=2$ for X_2;

$$\frac{(C_p)_0}{R}=\tfrac{7}{2}; \quad i=\log\frac{(2\pi m)^{\frac{3}{2}} k^{\frac{5}{2}}}{h^3}\frac{8\pi^2 Ak}{\sigma h^2}\frac{v_0}{\varpi_0}. \qquad ......(643)$$

III. *Polyatomic molecules in vapours and crystals*

(*a*) *Linear molecules—as for* II (*b*).

(*b*) *Non-linear molecules.*

Gaseous partition function,

$$\frac{(2\pi mkT)^{\frac{3}{2}}}{h^3}\frac{8\pi^2(8\pi^3 ABC)^{\frac{1}{2}} k^{\frac{3}{2}}}{\sigma h^3} V e^{-\chi/kT} n_v(T) \quad [n_v(0)=v_0],$$

where A, B, C are the principal moments of inertia, and σ, the symmetry number, can be evaluated by a study of the symmetry requirements of the atomic nuclei;

$$\frac{(C_p)_0}{R}=4; \quad i=\log\frac{(2\pi m)^{\frac{3}{2}} k^{\frac{5}{2}}}{h^3}\frac{8\pi^2(8\pi^3 ABC)^{\frac{1}{2}} k^{\frac{3}{2}}}{\sigma h^3}\frac{v_0}{\varpi_0}. \qquad ......(644)$$

If f vibrational degrees of freedom in the gaseous molecules of frequencies $\nu_1, \ldots, \nu_f$ with $2f$ square terms are effectively classical, their contribution may be extracted from $(C_p)_1$. We must then add

$$\text{to } \frac{(C_p)_0}{R}, \quad f; \quad \text{to } i, \quad \log\frac{k^f}{h^f \nu_1 \ldots \nu_f}. \qquad ......(645)$$

§7·12. *The vapour pressure equations concluded. Mixed crystals.* It remains to discuss mixed crystals but we shall confine attention to ideal ones in which the components mix perfectly. It has been shown in equation (550) for two components forming such a crystal in the abundance ratio $D_1 : D_2$, $D_1 + D_2 = 1$, that

$$p = \frac{kT}{V}(\overline{N_1} + \overline{N_2}) = \frac{kT}{V}\left[D_1 \frac{f_v^{\,1}(T)}{\kappa_a(T) f^1(T)} + D_2 \frac{f_v^{\,2}(T)}{\kappa_a(T) f^2(T)}\right]. \quad \text{...(646)}$$

Equation (646) is easily generalized to any number of components. We find

$$p = \frac{kT}{V}\left[\sum_{s=1}^{r} D_s \frac{f_v^{\,s}(T)}{\kappa_a(T) f^s(T)}\right]. \quad \text{......(647)}$$

It follows at once from the definitions of f, f_v and κ_a that

$$p = \sum_{s=1}^{r} D_s p_s, \quad \text{......(648)}$$

where p_s is the vapour pressure of a pure crystal consisting of the sth component only.

Equation (648) is exact (under the rather stringent assumptions we have used for ideal mixed crystals). It is however inconvenient to use, and since in any application made where the components are different isotopes the differences of the p's among themselves are only of the order of one part in seven, we can replace (648) by the approximation

$$\log p = \sum_{s=1}^{r} D_s \log p_s \quad \text{......(649)}$$

with an error of not more than 1 per cent. in p.* We shall therefore use (649) as the statistical result for mixed crystals.

Equation (649) gives us at once

$$\log p = \log\left\{\left[\frac{kTf_v^{\,1}(T)}{V}\right]^{D_1}\left[\frac{kTf_v^{\,2}(T)}{V}\right]^{D_2}\cdots\right\} - \log\{\kappa_a(T)[f^1(T)]^{D_1}[f^2(T)]^{D_2}\cdots\}.$$

Lemma 7·11 (extension) evaluates the second term. The first term can be similarly evaluated, for we can cast $\log\{kTf_v^{\,s}(T)/V\}$ into the form

$$-\frac{\chi_s}{kT} + \frac{(C_p^{\,s})_0}{R}\log T + j_s + \log n_v^{\,s}(T),$$

where χ_s is the heat of evaporation of a molecule of the sth type at the absolute zero, $(C_p^{\,s})_0$ the constant part of the specific heat at constant pressure, j_s a constant and $n_v^{\,s}(T)$ the partition function factor for the rest

* If $p_s = \bar{p}(1+\alpha_s)$, $\Sigma\, D_s\alpha_s = 0$ and $p = \exp(\Sigma\, D_s \log p_s)$ instead of $p = \bar{p}$, then the error Δp satisfies $\Delta p/p \simeq \frac{1}{2}\Sigma_s D_s \alpha_s^{\,2}$.

of the (non-classical) energy—all for the sth type of molecule. This can then be still further reduced to

$$-\frac{\chi_s}{kT}+\frac{(C_p{}^s)_0}{R}\log T+\int_0^T\frac{dT'}{RT'^2}\int_0^{T''}(C_p{}^s)_1\,dT''+j_s+\log n_v{}^s(0).$$

Now consider the (ideal) mixed vapour. We have at once $\Sigma_s D_s(C_p{}^s)_0=(C_p)_0$, $\Sigma_s D_s(C_p{}^s)_1=(C_p)_1$, where $(C_p)_0$ and $(C_p)_1$ refer to the mixture at the given (fixed) composition. The χ_s's are unlikely to differ much among themselves, but if we write $\Sigma D_s\chi_s=\chi$, then χ is the mean heat of evaporation for the mixture of the assigned composition. Absorbing the $1/T$ terms from the crystalline partition functions in χ we thus find as before

$$\log p=-\frac{\chi}{kT}+\frac{(C_p)_0}{R}\log T+\int_0^T\frac{dT'}{RT'^2}\int_0^{T'}[(C_p)_1-(C_p)_{\text{sol}}]\,dT''+i,$$

where now
$$i=\Sigma_s D_s\left\{j_s+\log\frac{(v_0)_s}{(\varpi_0)_s}\right\}. \qquad \text{......(650)}$$

It is obvious from the form that

$$i=\Sigma_s D_s i_s, \qquad \text{......(651)}$$

where i_s would be the vapour pressure constant in the equation for a pure crystal of the sth type of molecule, in equilibrium with its own pure vapour.

We can summarize this discussion in the following statement.

Vapour pressure formulae. Ideal mixed crystals

Calculate the specific heats $(C_p{}^s)_0$, $(C_p{}^s)_1$, $(C_p{}^s)_{\text{sol}}$, *and the vapour pressure constants* i_s *for each type of system, forming separately a pure crystal and vapour. Then for the mixed crystal*

$$(C_p)_0=\Sigma_s D_s(C_p{}^s)_0,\quad (C_p)_1=\Sigma_s D_s(C_p{}^s)_1,\quad (C_p)_{\text{sol}}=\Sigma_s D_s(C_p{}^s)_{\text{sol}},\quad i=\Sigma_s D_s i_s, \qquad \text{......(652)}$$

where the abundances of the types are given by the D_s, $\Sigma_s D_s=1$.

§**7·13.** *Basis of the comparison between theory and experiment.* (i) *Use of diatomic or polyatomic formulae for i.* Formulae (633) and (641)–(645) allow us considerable latitude in the manner in which a comparison between observed and theoretical values of i may be made for other than monatomic vapours. For monatomic vapours the only choice lies between the exact (641) and the approximate (642) which can be used when $(C_p)_1$ is small and $f_v(T)$ does not vary much over the temperature range used. For other gases the monatomic formulae can always be used so long as we retain the whole rotational, vibrational and electronic energy in $(C_p)_1$ and use observed values for $(C_p)_1$. We can obtain in this way further direct tests of the monatomic formula for i unaffected by specific heat *theory*.

This procedure is however often impracticable or inconvenient and we use one of the other formulae with $(C_p)_0>\frac{5}{2}R$. It must be remembered however in comparing such forms with experiment that they are no longer independent of specific heat theory, and errors in that theory will reveal themselves as errors in i. Consider for example the rotational term. The formula (633) for $\log p$ strictly contains a term which we may write

$$\left[\int_0^T \frac{dT'}{RT'^2}\int_0^{T'} C_{\text{rot}}\,dT''\right]_{\text{obs}}, \qquad \ldots\ldots(653)$$

reducible exactly to $[\log\{r'(T)/r'(0)\}]_{\text{obs}}$, where $r'(T)$ is the "observed" partition function exactly reproducing the observed specific heats. But when for example we take the rotations as classical for $T > T_0$ and use (643) for i we replace this by a theoretical expression which of course varies with T in very nearly the correct manner for $T > T_0$ but which may not exactly reproduce the residual constant in (653). Assuming that the monatomic formula for i is correct we are by such a comparison testing specific heat theory rather than the theory of vapour pressure constants.

(ii) *Changes of phase in the condensed form.* Changes of phase in the condensed form may be regarded as automatically allowed for by equation (633) when we use proper observed values of C_{sol}. Where the term

$$\int_0^T (C_p)_{\text{sol}}\,dT$$

occurs, the inclusion of any latent heat terms must be understood. In terms of what are normally called specific heats, this integral must be interpreted to mean

$$\int_0^T (C_p)_{\text{sol}}\,dT' \quad (T < T_t),$$

$$\Lambda_t + \int_0^T (C_p)_{\text{sol}}\,dT' \quad (T > T_t),$$

where T_t is a transition temperature and Λ_t is the latent heat of transition, which will usually be positive as written above.

Trouble, however, will arise if we have extrapolated back to zero an apparently well-determined curve of specific heats through an unknown transition point. This can of course only happen at low temperatures at which, apart from changes of phase, $\int_0^T (C_p)_{\text{sol}}\,dT'$ will be given sufficiently closely by AT^4. If, however, there was really a transition point (Λ_t) at T_t, then the correct value of the integral for $T > T_t$ is not AT^4 but

$$AT^4 + \Lambda_t + (A' - A)\,T_t^4.$$

In this $A'T^4$ is the form of the integral for $T < T_t$. We may assume that $\Lambda_t > 0$ and perhaps that $A' < A$, since the transition is likely to be to a more stable and therefore more rigid form. It seems to be the case, however, that A' and A, or more generally the specific heats on either side of the transition point are very nearly equal, and therefore that the sign of the error, which we will call μ_t, is the same as that of Λ_t. Instead of the correct value of

$$\int_0^T \frac{dT'}{RT'^2}\int_0^{T'} (C_p)_{\text{sol}}\, dT''$$

we shall therefore be using

$$\int_0^T \frac{dT'}{RT'^2}\int_0^{T'} (C_p)_{\text{sol}}\, dT'' - \int_{T_t}^T \frac{\mu_t dT'}{RT'^2}.$$

The right-hand side of our formula for $\log p$ is therefore too *large* by

$$\frac{\mu_t}{R}\left(\frac{1}{T_t}-\frac{1}{T}\right).$$

The term in $1/T$ does not matter. It is absorbed in Λ_0, which is in practice an adjustable constant. The observed value of i will therefore be too *small* by μ_t/RT_t, which may be expected to be positive.*

(iii) *Monatomic vapour pressure constants* via *dissociation equilibria*. To obtain as many examples of monatomic vapour pressure constants as possible, it is important to recognize that a knowledge of vapour pressures and dissociation equilibria for diatomic gases such as the halogens provides us at once with a direct determination of the vapour pressure constant of the atom into which theoretical uncertainties as to the structure of the free molecule do not enter. Expressed in partition functions we have

$$\overline{N_{11}} = \frac{f_{11}(T)}{\kappa(T)}, \quad \frac{(\overline{M_1})^2}{\overline{N_{11}}} = \frac{f_1{}^2(T)}{f_{11}(T)},$$

and therefore $$\overline{M_1} = f_1(T)/\{\kappa(T)\}^{\frac{1}{2}},$$

an expression into which no reference to the molecular form enters. Since

$$\frac{\overline{M_1}}{V} = \left\{\frac{(\overline{M_1}/V)^2}{\overline{N_{11}}/V}\cdot \overline{N_{11}}/V\right\}^{\frac{1}{2}} = \left\{\frac{\mathrm{K}_p p}{kT}\right\}^{\frac{1}{2}},$$

where K_p is the equilibrium constant, and p the *diatomic* pressure, $\overline{M_1}/V$ can be "observed" and the *monatomic* vapour pressure constant directly determined. In practice vapour pressures and equilibrium constants are not observed for the same temperatures and an extrapolation of one or other is needed by the theoretical formula. This will involve a knowledge of the specific heats of the molecule over the range of extrapolation, but will involve the molecule in no other way. The same result can easily be obtained from the thermodynamical forms. In practice the analysis will be carried out separately to determine $2i(X)-i(X_2)$ from $\log \mathrm{K}_p$, see § 7·4, and $i(X_2)$ from $\log p$. These are then added together and halved, to give the entry in the table.

* For this discussion see Cox, *Proc. Camb. Phil. Soc.* vol. 21, p. 541 (1923).

§ 7·2. *Comparison of observed and theoretical monatomic vapour pressure constants.* For numerical comparisons it is usual to rewrite (633) so that p is measured in atmospheres and the log's are $\log_{10}$. In that case the monatomic (641) becomes

$$i = -1{\cdot}587 + \tfrac{3}{2}\log_{10} m^* + \log_{10} \rho_0/\varpi_0, \qquad \text{......(654)}$$

where m^* is the chemical atomic weight, defined by taking the value of m^* for the (average) oxygen atom to be 16·000. The following table contains all the reliable comparisons of which I am aware. The table is followed by comments explaining and justifying the values of ρ_0 and ϖ_0 adopted for each class of atom; accepting these values the agreement between theory and observation is all that could be desired.

TABLE 11.

Monatomic vapour pressure constants.

Vapour	Atomic weight	ρ_0/ϖ_0	i calculated	i observed and range	Authorities for i obs.
Hg	200·6	1	1·866	1·83 ± 0·03	(1)
Cd	112·4	1	1·488	{1·45 ± 0·1 1·57	(1) (6)
Zn	65·37	1	1·135	1·21 ± 0·15	(9)
Pb	207·2	1	1·887	{1·8 ± 0·2 2·27 ± 0·36	(1) (7)
Mg	24·32	1	0·49	0·47 ± 0·2	(9)
Ne	20·20	1	0·37	0·39 ± 0·04	(10)
A	39·88	1	0·813	0·79 ± 0·04	(1)
H_2 ($T<40°$)	2·015	1	− 1·132	− 1·09 ± 0·02	(2)
Na	23·0	2	0·756	{0·63 0·78 ± 0·1 0·97	(3) (8) (4)
K	39·1	2	1·10	{0·92 1·13	(3) (4)
Tl	204·4	2	2·18	2·37 ± 0·3	(9)
Cl	35·46	4	1·44	1·53 ± 0·2	(5)
Br	79·92	4	1·87	2·00 ± 0·2	(5)
I	126·9	4	2·17	2·19 ± 0·2	(5)

(1) Egerton, *Proc. Phys. Soc. London*, vol. 37, p. 75 (1925).
(2) Eucken, *Zeit. f. Physik*, vol. 29, p. 1 (1924).
(3) Edmonson and Egerton, *Proc. Roy. Soc.* A, vol. 113, p. 533 (1927).
(4) Zeidler, *Zeit. f. physikal. Chem.* vol. 123, p. 383 (1926).
(5) These values have been obtained as explained in the preceding section from the observed values of $2i(X) - i(X_2)$ in Table 14 and of $i(X_2)$ in Table 13.
(6) Lange and Simon, *Zeit. f. physikal. Chem.* vol. 134, p. 374 (1928).
(7) Harteck, *Zeit. f. physikal. Chem.* vol. 134, p. 1 (1928).
(8) Ladenburg and Thiele, *Zeit. f. physikal. Chem.* B, vol. 7, p. 161 (1930).
(9) Coleman and Egerton, *Phil. Trans.* A, vol. 234, p. 177 (1935).
(10) Clusius, *Zeit. f. physikal. Chem.* B, vol. 4, p. 1 (1929).

The errors given in the table must be used with caution in estimating the reliability of the observed values. For instance for Hg the experimental evidence indicates that i probably lies between 1·80 and 1·86 *apart from concealed errors in the extrapolated specific heats used.* Owing to difficulties of extrapolation the real uncertainty in i may well be greater than the range given.

We proceed now to justify the values of ρ_0/ϖ_0 used.

For Hg, Cd, Zn, Pb, Mg and A the lattices of the crystals are atomic, and the atoms themselves in their vapour phase in ^1S states. All other states lie so high that they do not contribute to the partition function at the relevant temperatures. No weight factors other than unity can then arise except from nuclear spins. Nuclear spin will introduce an equal weight factor into both ρ_0 and ϖ_0 and can therefore be ignored. Isotopic mixtures have strictly to be considered, but since the foregoing remarks hold for each pure species no weight factors enter when equations (652) are applied to the mixture.

H_2 is a special case. Below 40° K. all its molecules are in their lowest possible rotational states, so that in spite of its form it is effectively monatomic both in the vapour and the solid phase. Its possible rotational states owing to its high degree of external symmetry may be assumed to be effectively the same both in the vapour and the solid. It possesses however a mixture of almost non-combining ortho- and para-states, so that it must be treated here as usual as a mixture of two gases, parahydrogen in a $^1\Sigma$ electronic state, rotation quantum number zero and weight 1, and orthohydrogen in a $^1\Sigma$ state with rotational quantum number 1 and weight 9. Since however these weights apply equally to both phases, $\rho_0/\varpi_0 = 1$. It is important to observe that the resulting low temperature value for i is independent of the relative concentrations of the ortho- and para-species and is therefore unaffected by transitions from one form to the other, provided that such transitions affect both phases equally.

For Na and K we must take $\rho_0/\varpi_0 = 2$. The free atoms have a normal state ^2S of weight 2 and no other states of importance, so that neglecting nuclear spins $\rho_0 = 2$. To see what value ϖ_0 must have we observe that both Na and K crystals are metallic conductors, and that we should apply to them the electron theory of metals developed in Chapter XI. We may think of the atoms in the metal as dissociated into electrons and positive ions (Na^+, K^+); the ions are in a ^1S state of weight unity (nuclear weights being still neglected), and therefore contribute only a factor unity to ϖ_0. The electrons may be considered to form an electron gas obeying the Fermi-Dirac statistics in an enclosure at nearly constant potential. This gas is almost perfectly degenerate at ordinary temperatures. At the absolute zero there exists only one state for the gas—that in which the P electrons occupy the P states of lowest energy, represented by one assembly wave-function. Thus $\varpi_0 = 1$, $\rho_0/\varpi_0 = 2$. The nuclear spin weights cancel if inserted. In thallium vapour the atoms are in a ^2P state for which the lower component is $^2P_{\frac{1}{2}}$ of weight 2. The upper state $^2P_{\frac{3}{2}}$ of weight 4 lies so high above it that it contributes nothing to $f_v(T)$ at the relevant temperatures so that $f_v(T)$ is effectively constant and $\rho_0 = 2$. In the metallic solid we must take $\varpi_0 = 1$ as for the alkalis.

For the monatomic halogen vapours $f_v(T)$ is not strictly constant. The only relevant state of a free halogen is an inverted ^2P term, so that the normal state is $^2P_{\frac{3}{2}}$ of weight 4 while the other term of the doublet, $^2P_{\frac{1}{2}}$ of weight 2, lies slightly higher. If the nuclear spin weight is τ, then

$$f_v(T) = \tau(4 + 2e^{-\Delta P/kT}), \qquad \text{......(655)}$$

where ΔP is the difference of the energies of the two states of the inverted doublet. To determine ϖ_0 convincingly we need the analysis of the molecular halogen crystal given in § 7·31. Anticipating this section we find that a crystal composed of P atoms of a halogen has as one would expect a lowest state represented by τ^P independent wave-functions so that $\varpi_0 = \tau$. It follows that

$$f_v(T)/\varpi_0 = 4 + 2e^{-\Delta P/kT} \qquad \text{......(656)}$$

for any pure halogen assembly. For isotopic mixtures the same result continues to hold, since it holds for each species. The values of the function (656) and its logarithm are given in Table 12 for the relevant temperatures. In Table 11 the mean value 0·7 was used for Cl and 0·6 for Br and I.

TABLE 12.

Effective weights of free halogen atoms.

$$f_v(T)/\varpi_0 = 4 + 2e^{-\Delta P/kT}.$$

Atom	ΔP volts*	Mean temp. °K.	$\Delta P/kT$	$f_v(T)/\varpi_0$	$\log_{10} f_v(T)/\varpi_0$
Cl	0·11	1000	1·28	4·56	0·66
		1700	0·752	4·94	0·69
Br	0·45	1350	3·88	4·04	0·6
I	0·94	1200	9·1	4	0·6

* Electron volts: values from Turner, *Phys. Rev.* vol. 27, p. 397 (1926).

§ 7·3. *Comparison of observed and theoretical diatomic vapour pressure constants.* The calculation of diatomic vapour pressure constants allowing for isotopic mixtures requires us to evaluate i for each molecular species in the mixture using equation (643); it is the term $\log\{v_0/\sigma\varpi_0\}$ which requires delicate study.

For this it is necessary to know how the crystals concerned are constructed—that is whether they should be regarded as built up of atomic, molecular (diatomic) or even of multimolecular units. For example the normal electronic states of the free halogen atoms have weight 4 apart from nuclear factors, but it does not follow and it is not in fact true that the normal state of a halogen crystal of P molecules ($2P$ atoms) has a weight 4^{2P}. Primary subgroupings can and do occur which form states of less weight and of significantly lower energy than the other states derived from the same atomic states. For example two halogen atoms form a halogen molecule in a $^1\Sigma$ electronic state of weight 1, the other 15 states corresponding to the 16 states of the dissociated atoms lie higher and make no contribution at least at low temperatures to the enumeration of the states of the crystal, which may be regarded as built up out of $^1\Sigma$ molecules. Again the normal state of the free NO molecules is a $^2\Pi_{\frac{1}{2}}$ electronic state of weight 2. If all the wave-functions corresponding to these states could be used indiscriminately to form crystal wave-functions, the weight of the lowest state of the crystal of P molecules would be 2^P. This however appears to be incorrect (see § 7·31 (vi)). There is evidence of prior formation of N_2O_2 complexes of weight 2 not 4, whose wave-functions can be used indifferently to form the crystal wave-functions. The lowest state of a crystal of P NO molecules therefore has a weight $2^{\frac{1}{2}P}$.

It is further necessary to know whether or not the molecules can still rotate freely in the crystal at the lowest temperatures at which its specific heats are measured.

It is not yet possible to choose between these and similar alternatives *a priori*, and in fact the correct choice depends on the lowest temperatures to which accurate measurements of specific heats have been carried, from which temperatures they have to be extrapolated to zero. None of the sets of states that we treat as degenerate need be absolutely so, but so long as their energy differences are small compared with kT it is correct to ignore these differences and treat the set as a simple degenerate state with the corresponding extra weight factor. If the observations stop at temperatures sufficiently high, no effect of the ignored separations will be seen in the specific heats and we must include the weight factors in the theory. If however observations are pushed lower ($kT \simeq \Delta\epsilon$), we reach temperatures at which the upper states of the set are gradually emptied and finally may reach still lower temperatures ($kT \ll \Delta\epsilon$) where the upper states can be ignored and the effective states are no longer degenerate or at least not so highly degenerate. During the change over the specific heat will show temporarily exceptionally large values (see §§ 3·7, 21·4) and the extra weight factors removed from ϖ_0 are replaced by an equivalent contribution from the double integral.

These questions must be discussed more deeply in connection with Nernst's Heat Theorem in § 7·5. For the present it is sufficient to have pointed out that the proper specification of ϖ_0 is not absolute but requires us to specify from what temperature the observations of specific heat have been extrapolated to zero, and then to estimate what degeneracy survives in the crystalline wave-functions there. It is not yet always possible to make these specifications *a priori*.

The following table contains all the reliable comparisons of theory and observations for diatomic molecules of which I am aware, followed as before by comments explaining and justifying the adopted values of $\log\{v_0/\sigma\varpi_0\}$. For numerical work, with p in atmospheres and logarithms $\log_{10}$, the diatomic (643) becomes

$$i = 36{\cdot}815 + \tfrac{3}{2}\log_{10} m^* + \log_{10} A + \log_{10}\{v_0/\sigma\varpi_0\}. \qquad \ldots\ldots(657)$$

The general agreement is excellent, but the chosen values of $v_0/\sigma\varpi_0$ for O_2 and NO cannot be unambiguously laid down.

§ 7·31. *Notes on Table* 13. (i) *Hydrogen.* We have already had occasion to point out that hydrogen as ordinarily experimented on must be regarded as a 3 : 1 mixture of effectively distinct gases ortho- and para-hydrogen. The ratio 3 : 1 arises from the ratio of the weights of the nuclear states for the two types. In the vapour $v_0 = 3, 1$ for the ortho- and para-forms respectively, and $\sigma = 2$. In the solid down to 11° K. there are no anomalies in the specific heat and we must assume that the molecules can rotate freely. Anomalies occur at lower temperatures* which correspond to the freezing out of these rotations.† The double integral has been calculated by extrapolating the normal specific heat curve to zero from above 11° K. The weights ϖ_0 are therefore the weights of the lowest states of rotation including the nuclear weights, so that $\varpi_0 = 9, 1$ for the ortho- and para-molecules respectively. We are left with an abnormal term $\tfrac{3}{4}\log 3$ in the vapour pressure constant which would be absorbed into the double integral if less extrapolated specific heats could be used.

(ii) *Other molecules of type X_2 in $^1\Sigma$ states.* All molecules other than H_2 have ceased rotating at comparatively high temperatures and the extrapolations proceed from temperatures at which X_2 is free only to oscillate about a direction of equilibrium, which may be supposed to be unique. Such a unique direction however corresponds to two possible positions of the molecule, for owing to its symmetry it must still be in equilibrium if it is reversed end for end. There are thus two wave-functions for the lowest state of oscillation, but from these just one oscillatory wave-function can be constructed which is symmetrical in the nuclei and one which is antisymmetrical. If the number of possible distinct nuclear orientations allowed by the nuclear spin is ρ, then as in § 5·31

$$\varpi_0 = \tfrac{1}{2}\rho(\rho+1), \quad \tfrac{1}{2}\rho(\rho-1)$$

* Simon, Mendelssohn and Ruhemann, *Naturwiss.* vol. 18, p. 34 (1930).

† Pauling, *Phys. Rev.* vol. 34, p. 430 (1930).

TABLE 13.

Diatomic vapour pressure constants.

Vapour	Electronic normal state	Species and abundance	Molecular weight	$A \times 10^{40}$ gm. cm.^{2}	$v_0/\sigma\varpi_0$*	i calc.	i obs.
H_2	$^1\Sigma$	Para-, $\frac{1}{4}$	2·015	0·463	1/2	−3·72	−3·68 ± 0·03
		Ortho-, $\frac{3}{4}$			1/(2 × 3)		
N_2	$^1\Sigma$	—	28·016	13·8	1/2	−0·18	−0·16 ± 0·03
O_2	$^3\Sigma$	—	32	19·15	3/2	0·53	0·55 ± 0·02
NO	$^2\Pi$	—	30·008	16·4	$\frac{2(1+e^{-170/T})}{\sqrt{2}}$	0·48 (110° K.)	0·55 ± 0·03
						0·70 (∞° K.)	near 110° K.
CO	$^1\Sigma$	—	28·004	15·0	1/1	0·16	−0·07 ± 0·05
HCl	$^1\Sigma$	HCl^{35}, 0·76	35·99	2·656	1/1	−0·42	−0·40 ± 0·03
		HCl^{37}, 0·24	37·99	—			
HBr	$^1\Sigma$	HBr^{79}, 0·54	79·94	3·32	1/1	0·20	0·24 ± 0·04
		HBr^{81}, 0·46	81·94	—			
HI	$^1\Sigma$	—	127·94	4·31	1/1	0·61	0·65 ± 0·05
Cl_2	$^1\Sigma$	$Cl^{35}Cl^{35}$, 0·58	69·97	114	1/2	1·35	1·66 ± 0·08
		$Cl^{35}Cl^{37}$, 0·36	71·96	118			
		$Cl^{37}Cl^{37}$, 0·06	73·96	121			
Br_2	$^1\Sigma$	—	159·70	445	1/2	2·47	2·59 ± 0·10
I_2	$^1\Sigma$	—	255·88	820	1/2	3·03	3·08 ± 0·05

* In $v_0/\sigma\omega_0$ the factors due to nuclear spin have all been cancelled, but all other factors are retained explicitly.

Authorities: For the observed constants, Eucken, *Physikal. Zeit.* vol. 31, p. 361 (1930); for the moments of inertia, H_2, N_2, O_2, Rasetti, *Phys. Rev.* vol. 34, p. 370 (1929); NO, Snow and others, *Proc. Roy. Soc.* A, vol. 124, p. 453 (1929); CO, Snow and Rideal, *Proc. Roy. Soc.* A, p. 125, p. 462 (1929); HCl, Czerny, *Zeit. f. Physik*, vol. 44, p. 252 (1927); HBr, HI, *International Critical Tables*; Cl_2, Elliott, *Proc. Roy. Soc.* A, vol. 127, p. 638 (1930); Br_2, Brown, *Phys. Rev.* vol. 37, p. 1007 (1931); I_2, Mecke, *Zeit. f. Physik*, vol. 42, p. 390 (1927).

according as the nuclei require symmetry or antisymmetry in the spin wave-functions. These same factors occur in v_0. For both ortho- and para-molecules therefore

$$\frac{v_0}{\sigma \varpi_0} = \tfrac{1}{2},$$

and it now makes no difference whether we consider that the ortho- and para-molecules are distinct or not, or whether the complete wave-functions for each molecule have to be symmetrical or antisymmetrical in the nuclei.

(iii) *Molecules of type $X^\alpha X^\beta$ in $^1\Sigma$ states.* Besides the molecules X_2 (ortho- or para-) there are also molecules $X^\alpha X^\beta$ in the isotopic mixture whose nuclei are distinct so that they have no symmetry requirements. For these there must be two oscillatory wave-functions for the lowest state, for they can still be reversed end for end, so that $\varpi_0 = 2\rho_1\rho_2$. For the vapour molecules of this type $v_0 = \rho_1\rho_2$ but $\sigma = 1$. Hence as before

$$\frac{v_0}{\sigma \varpi_0} = \tfrac{1}{2};$$

$v_0/\sigma\varpi_0$ is therefore $\frac{1}{2}$ for every type of molecule in the mixture. The only factors which vary for the different types are the masses and moments of inertia for which mean values have to be taken according to (652).

(iv) *Molecules of type XY in $^1\Sigma$ states.* In general we must expect these molecules to have a dipole moment so that they can only have one unique direction of stable equilibrium—in the direction of the resultant electrostatic field—and cannot be reversed end for end. Hence for every one of these isotopic species $v_0 = \rho_1\rho_2$, $\varpi_0 = \rho_1\rho_2$, $\sigma = 1$ and $v_0/\sigma\varpi_0 = 1$.

(v) *Oxygen.* Oxygen in the vapour is in a $^3\Sigma$ state which can be regarded normally as degenerate and of weight 3; also $\sigma = 2$. The entry in the table shows $\varpi_0 = 1$. This means that we assume that at very low temperatures in the solid supermolecules are formed so that O_2 loses its extra weight 3 due to electronic spins. This might be expected from the analogy of sulphur which forms freely molecules such as S_4. It agrees also with direct measurements of entropy.*

(vi) *Nitric oxide.* We have already seen in § 3·71 that this gas must be considered to have the extra factor $2(1 + e^{-170/T})$ in the partition function for its vapour. If the resulting variable specific heat of the vapour is not corrected for, as it usually is not, then this factor survives in place of v_0, and

* Giauque and Johnston, *J. Amer. Chem. Soc.* vol. 51, p. 2300 (1929). It should be remembered that this comparison of entropies is not an independent check, but merely an equivalent form of the comparison of vapour pressure constants. Solid oxygen at higher temperatures and liquid oxygen are strongly paramagnetic, so that the spins do not lose their weight factor by coupling up between neighbouring molecules until liquid hydrogen temperatures are reached. It is perhaps therefore better to consider this coupling as due to the crystalline field rather than to the formation of O_4 as in the text.

$\sigma = 1$. The factor $\sqrt{2}$ in the denominator is inserted because it has been assumed that the crystal can be regarded as effectively constructed of supermolecules N_2O_2 each of weight 2. This assumption is reasonable but cannot be held to be demanded *a priori* by independent evidence. However it fits the facts.*

§ **7·32.** *Vapour pressures of polyatomic gases.* Similar studies of polyatomic gases (CO_2, NH_3, CH_4) are perfectly possible and have been started by Sterne,† but we shall not include his discussions here. No new questions of principle are involved, merely questions of complication. It is of interest to record the results for the vapour pressure constants. The theoretical formula in (644) and the whole of the elaboration necessary is concerned with the proper evaluation of the factor $v_0/\sigma\varpi_0$ or rather its mean value for different types of non-combining states. Sterne finds

$$i\,(NH_3) = -1{\cdot}55, \quad i\,(CH_4) = -1{\cdot}94.$$

The observed values given by Eucken‡ are

$$i\,(NH_3) = -1{\cdot}50 \pm 0{\cdot}04, \quad i\,(CH_4) = -1{\cdot}97 \pm 0{\cdot}05$$

in excellent agreement.

§ **7·4.** *The reaction isobar in thermodynamics and statistical mechanics.* Let us consider the homogeneous gas reaction $\Sigma_t q_t A_t = 0$. If K_p is the equilibrium constant of this reaction, then it is easy to show by classical thermodynamic arguments similar to those used in § 7·1 that

$$\log K_p = \log\{\Pi_t p_t^{q_t}\},$$
$$= \frac{-(Q_p)_0}{RT} + \frac{\Sigma_t q_t (C_p{}^t)_0}{R}\log T + \int_0^T \frac{dT'}{RT'^2}\int_0^{T'} \{\Sigma_t q_t (C_p{}^t)_1\}\,dT'' + I. \quad \text{......(658)}$$

In this equation q_t is again the number of gram-molecules of the tth species reacting, with a negative sign for those that disappear when the reaction takes place in a specified direction, and p_t is the partial pressure of the tth species. $(Q_p)_0$ is the heat evolved when the reaction goes to completion at the absolute zero and $(C_p{}^t)_0$, $(C_p{}^t)_1$ are the constant and variable parts of the specific heat at constant pressure for the tth species. I is a constant of integration, about which classical thermodynamics has nothing to say; it must not be confused with moments of inertia.

The essential part of the corresponding formula of statistical mechanics is given in equation (532). Converting this to partial pressures we find

$$\log K_p = \log\{\Pi_t p_t^{q_t}\} = \Sigma_t q_t \log\left\{\frac{kT}{V} f_t(T)\right\}. \quad \text{......(659)}$$

* Johnston and Giauque, *J. Amer. Chem. Soc.* vol. 51, p. 3194 (1929).

† Sterne, NH_3, *Phys. Rev.* vol. 39, p. 993 (1932); CH_4, *Phys. Rev.* vol. 42, p. 556 (1932).

‡ Eucken, *Physikal. Zeit.* vol. 31, p. 361 (1930).

If we now expand $\log f_\iota(T)$ in terms of the specific heats as in the preceding sections, we reproduce (658) but also evaluate I in terms of i_ι, finding

$$I = \Sigma_\iota q_\iota\{i_\iota + \log(\varpi_0)_\iota\} = \Sigma_\iota q_\iota j_\iota. \qquad \text{......(660)}$$

It is evident on reference to the vapour pressure formulae that $i_\iota + \log(\varpi_0)_\iota$ depends (as it should) only on the properties of the vapour and makes no reference to the solid states, as the term $\log(\varpi_0)_\iota$ removes the corresponding term from i_ι.

The discussion so far does not apply to mixtures of isotopes; this restriction is removed below. Homonuclear molecules do not however separate into ortho- and para-varieties, since ortho-para transitions can always take place *via* the reaction that is occurring. If we take the high temperature asymptotic form for all rotational partition functions, then it follows from (490) and (491) that $v_0 = \rho^2$, $\sigma = 2$ for all homonuclear, and $v_0 = \rho_1\rho_2$, $\sigma = 1$ for all heteronuclear, molecules.

To include the effect of isotopic mixtures it will be sufficient to consider a simple example as illustration—e.g. the equilibrium* $XX + YY - 2XY = 0$ in which Y is simple but X consists of the two isotopes X^1, X^2, the relative abundance of these atoms being D_1, D_2, $D_1 + D_2 = 1$. With an obvious notation the total partial pressure p_{XX} of the molecules XX of all isotopic types is given by

$$p_{XX} = p_{X^1X^1} + p_{X^1X^2} + p_{X^2X^2}. \qquad \text{......(661)}$$

Similarly

$$p_{XY} = p_{X^1Y} + p_{X^2Y}. \qquad \text{......(662)}$$

We require to evaluate the general equilibrium constant K_p, where

$$K_p = \frac{p_{XX}p_{YY}}{(p_{XY})^2}.$$

Now the equilibrium constants for the similar reactions, in which specified isotopes are concerned, can be immediately written down. We have

$$\frac{p_{X^1X^1}p_{YY}}{(p_{X^1Y})^2} = K_p^{11} = \frac{f_{X^1X^1}f_{YY}}{(f_{X^1Y})^2},$$

$$\frac{p_{X^2X^2}p_{YY}}{(p_{X^2Y})^2} = K_p^{22} = \frac{f_{X^2X^2}f_{YY}}{(f_{X^2Y})^2},$$

$$\frac{p_{X^1X^2}p_{YY}}{p_{X^1Y}p_{X^2Y}} = K_p^{12} = \frac{f_{X^1X^2}f_{YY}}{f_{X^1Y}f_{X^2Y}},$$

where the f's are the partition functions for the species specified by the suffix. On evaluating K_p^{11}, K_p^{22} and $\frac{1}{2}K_p^{12}$ we find at once that they differ only by small factors arising from differences of mass and moments of inertia. If therefore we write

$$K_p^{11} = A(1+\alpha_{11}), \quad K_p^{12} = 2A(1+\alpha_{12}), \quad K_p^{22} = A(1+\alpha_{22}),$$

* It avoids a confusing notation temporarily to drop here the usual molecular notation X_2 for XX.

the α's are small and we find

$$K_p = A\frac{(1+\alpha_{11})(p_{X^1Y})^2 + 2(1+\alpha_{12})p_{X^1Y}p_{X^2Y} + (1+\alpha_{22})(p_{X^2Y})^2}{(p_{X^1Y}+p_{X^2Y})^2}. \quad \ldots(663)$$

Since the α's are small correcting terms we may use approximate values of the p's in their coefficients and as in § 5·31 a sufficient approximation in these coefficients is

$$p_{X^1Y} : p_{X^2Y} = D_1 : D_2. \quad \ldots\ldots(664)$$

Using this ratio we find after a simple reduction that

$$K_p = D_1^2 K_p^{11} + 2D_1 D_2(\tfrac{1}{2}K_p^{12}) + D_2^2 K_p^{22}, \quad \ldots\ldots(665)$$

an equation that may equally well be taken in the form

$$\log K_p = D_1^2 \log K_p^{11} + 2D_1 D_2 \log(\tfrac{1}{2}K_p^{12}) + D_2^2 \log K_p^{22}. \quad \ldots(666)$$

The important part of equation (666) concerns only the constant terms in the K's. Taking the relevant example of (660) in the forms

$$I_{11} = j_{X^1X^1} + j_{YY} - 2j_{X^1Y}, \quad \ldots\ldots(667)$$

$$I_{22} = j_{X^2X^2} + j_{YY} - 2j_{X^2Y}, \quad \ldots\ldots(668)$$

$$I_{12} = j_{X^1X^2} + j_{YY} - j_{X^1Y} - j_{X^2Y}, \quad \ldots\ldots(669)$$

we find that

$$I = D_1^2 I_{11} + 2D_1 D_2(I_{12} - \log 2) + D_2^2 I_{22}, \quad \ldots\ldots(670)$$

$$= [D_1^2 j_{X^1X^1} + 2D_1 D_2(j_{X^1X^2} - \log 2) + D_2^2 j_{X^2X^2}] + j_{YY} - 2[D_1 j_{X^1Y} + D_2 j_{X^2Y}]. \quad \ldots\ldots(671)$$

These equations show at once how I is to be calculated for a mixture of isotopes, owing to the term $-\log\sigma$ ($\sigma = 2$) in I_{11} and I_{22} the term $\log(v_0/\sigma)$ is the same for all the isotopic types except for the nuclear spin factors ρ; but the ρ's cancel completely from (671) because

$$D_1^2 \log \rho_1^2 + 2D_1 D_2 \log \rho_1\rho_2 + D_2^2 \log \rho_2^2 - 2D_1 \log \rho_1 - 2D_2 \log \rho_2 \equiv 0. \quad \ldots\ldots(672)$$

Thus nuclear spins and isotopic mixtures have no effect on the value of I except that the mass and moment of inertia terms must have a mean value taken for the different isotopic species concerned according to (671).

The extension to more complicated cases of isotopic mixtures is immediate and conforms of course to the rule just stated.

Table 14 gives a comparison of theory and observation for those simple gaseous reactions involving only diatomic molecules for which all the relevant data are known.

The agreement is thoroughly satisfactory.

Similar calculations can be made of course for heterogeneous gas reactions in which solids take part in the chemical equation, but as these involve a knowledge of ϖ_0 for the solid phases concerned they are not capable of such unambiguous treatment and we shall not attempt to discuss examples.

TABLE 14.

Comparison of observed and calculated values of I, equation (658) for homogeneous gas reactions.

I calculated by (660) [or (671)] and (657)

Reaction	I calculated	I observed	Authority
$H_2+Cl_2-2HCl=0$	$-1{\cdot}17$	$-1{\cdot}12\pm0{\cdot}2$	(1)
$H_2+Br_2-2HBr=0$	$-1{\cdot}29$	$-1{\cdot}25\pm0{\cdot}45$	(1)
$H_2+I_2-2HI=0$	$-1{\cdot}55$	$-1{\cdot}51\pm0{\cdot}12$	(1)
$2NO-N_2-O_2=0$	$1{\cdot}31$	$0{\cdot}95\pm0{\cdot}3$	(1), (2)
$2Cl-Cl_2=0$	$1{\cdot}53$	$1{\cdot}40\pm0{\cdot}15$	(1)
$2Br-Br_2=0$	$1{\cdot}26$	$1{\cdot}41\pm0{\cdot}05$	(1)
$2I-I_2=0$	$1{\cdot}31$	$1{\cdot}30$	(3)
$H_2{}^1+H_2{}^2-2H^1H^2=0$	$-0{\cdot}63$	$-0{\cdot}58\pm0{\cdot}05$	(4)

(1) Eucken, *Physikal. Zeit.* vol. 30, p. 818 (1929).
(2) These values refer to high temperatures at which $e^{-170/T}\sim 1$.
(3) Cox, *Proc. Camb. Phil. Soc.* vol. 21, p. 541 (1923).
(4) Farkas, *Light and Heavy Hydrogen*, p. 178, Cambridge (1935). Data from Rittenberg, Bleakney and Urey, *J. Chem. Physics*, vol. 2, p. 48 (1934).

§7·5. *Nernst's Heat Theorem, or the third law of thermodynamics.** The relationship (660) between I and the i's is of very great importance. It was first formulated by Nernst in the form

$$I=\Sigma_t q_t i_t \quad \text{(Nernst)}, \qquad \text{......(673)}$$

which if it were generally valid would justify the name *chemical constant* which he gave to i, but which belongs as (660) shows more properly to j. Nernst's formula (673) is true if and only if

$$\Sigma_t q_t \log(\varpi_0)_t = 0. \qquad \text{......(674)}$$

Nernst derived this relationship from his celebrated Heat Theorem or third law of thermodynamics which can be enunciated in various ways. The commonest are:

(*a*) *For any condensed system and any reversible isothermal process*

$$\frac{\partial \Delta A}{\partial T}\to 0 \quad (T\to 0),$$

where ΔA is the maximum work that the reaction can be made to do.

(*b*) *For any condensed system and any reversible isothermal process*

$$\Delta S\to 0 \quad (T\to 0).$$

(*c*) (*Less precisely.*) *At $T=0$ all reactions in condensed systems take place without change of entropy.*

* For an account of this theorem see Nernst, *Die theoretischen und experimentellen Grundlagen des neuen Wärmesatzes* (1918). For a more recent analysis, Lewis and Randall, *Thermodynamics*, esp. chap. XXXI (1924).

The theorem as thus enunciated is a generalization from experimental data on specific heats, heats of combination, and the electromotive force of reversible cells. The body of evidence in its favour is great, but the generalizations as given above were too hastily made. A very careful rediscussion of the law has been given by Lewis and Randall.* They conclude that its original enunciation, as applying to condensed systems other than solids or even rather other than pure crystals, is probably fallacious and that the theorem may be more properly enunciated thus:

(*d*) "*The entropy of each element in some crystalline state can be taken to be zero at the absolute zero of temperature. Every substance then has a finite positive entropy, but at the absolute zero of temperature the entropy may become zero and does so become zero in the case of perfect crystalline substances, including compounds.*"

By this enunciation the behaviour of the entropy of supercooled liquids and solutions is properly left open. It will be observed that Lewis and Randall have been careful so to formulate the theorem that the idea of absolute entropy is not introduced. We can really (if the theorem is true) leave arbitrary the constant S_0 which denotes the entropy of any *element* in a specified crystalline form at $T=0$ and then the constant S_0' for any compound will be the sum of the S_0's for the elements implicated. The content of the theorem is rendered more striking and practical convenience is furthered by putting $S_0=0$ for all elements; so long as this is not taken to imply the existence of an absolute entropy no harm but only good is done.

Equation (673) was derived by Nernst from his theorem by considering the following reversible cycle:

(i) Condense a unit set of reactants to solid form isothermally. The work done and the change of entropy are known in terms of the vapour-pressure equation and involve the i's. [For initial temperatures above a certain limit, a preliminary cooling of the gaseous reactants will be required involving their specific heats. Preliminary expansions will in general also be necessary to adjust pressures to equality with the vapour pressures of the solids.]

(ii) Cool the condensed form to the absolute zero. The entropy change is again known in terms of the specific heats of the solids.

(iii) Allow the reaction to proceed completely at the absolute zero. This may be supposed ideally possible, and by the theorem *there is no change of entropy*.

(iv) Heat up the condensed resultants again as in (ii).

* Lewis and Randall, *loc. cit.*

(v) Evaporate the resultants to gases again as in (i) and adjust pressures and temperatures.

(vi) Let the reaction go back again isothermally in the gas phase, so that the assembly returns to its initial state. The entropy change here will depend on I.

The change of entropy for the cycle must vanish, and it is easily proved that the condition for this is (673).

Nernst's Heat Theorem thus formulated rested on a purely empirical basis which became in fact none too secure as the facts became better known, especially for reactions involving hydrogen.* Its theoretical basis however has now become completely clear when it is regarded as a theorem in statistical mechanics. The precise formulation (*d*) presented as such a theorem can be made to run as follows:

(*d'*) *For any reversible isothermal reaction between perfect crystalline substances*

$$\Delta S/R \to \Sigma_t q_t \log(\varpi_0)_t \quad (T \to 0). \qquad \ldots\ldots(675)$$

If and only if $\Sigma_t q_t \log(\varpi_0)_t = 0$, *Nernst's Heat Theorem holds for this reaction.* We return later to verify the identification (675). Accepting it we see that Nernst's Heat Theorem is a general law of nature (as Nernst believed), if and only if (674) is true without exception.

It is now necessary to proceed with extreme caution. If by $[(\varpi_0)_t]^P$ we mean the weight of the absolutely lowest state of the tth pure crystal of P molecules when all separations no matter how small are regarded as significant, then it is almost certain though not rigorously proved that $(\varpi_0)_t = 1$. With this interpretation Nernst's Heat Theorem is probably true generally. But in order to make this interpretation significant it is necessary to be able to observe the specific heat curve for the crystal down to temperatures for which kT is small compared with the smallest energy separations among the states of the crystal—even those arising from different arrangements of isotopic molecules or different orientations of nuclear spins. Some of these separations are doubtless excessively small and the necessary temperatures probably far less than 0·001° K. This interpretation therefore of Nernst's Heat Theorem though probably true is of little practical importance. As a practical theorem we wish to apply it with quite another interpretation of the $(\varpi_0)_t$, namely when our observations stop at temperatures of the order of a few degrees Kelvin and we ignore all energy separations small compared with say $5k$, or $10k$, in particular all those arising from nuclear spins and isotopic rearrangements. Applied in this practical way Nernst's Heat Theorem is definitely untrue in certain cases, notably for any reaction in which hydrogen is a constituent, but it is then untrue only for rather

* Eucken, *Zeit. f. Physik*, vol. 29, p. 12 (1924).

special reasons. For example, as we shall see in greater detail later, it is untrue for hydrogen only because of the surviving *rotational* weight of the ortho-hydrogen molecules. If the specific heat observations could be carried to a temperature small compared with that at which molecular rotations cease in the solid, or if they could be made on hydrogen properly catalysed to its true equilibrium state instead of on a metastable mixture, this failure of Nernst's theorem would not occur. It fails for reactions containing NO because N_2O_2 apparently survives in the crystal with an electronic weight 2 down to very low temperatures. Apart from such special reasons which appear to be rather exceptional Nernst's Heat Theorem is true in this practical sense. But the mere fact that it does fail sometimes in its natural interpretation means that it must be applied with caution to any reaction not yet studied. It is no disparagement to Nernst's great idea to recognize now its limited or idealized generality. The part that it has played in stimulating a deeper understanding of all these constants and its reaction on the development of the quantum theory itself cannot be overrated.

§7·51. *Absolute entropy.* Another aspect of the same failure is that it is no longer possible to assert that the entropies even of perfect crystals of pure substances can all be assumed to be zero in the idealized state for zero temperature which we reach by extrapolation from accessible temperatures. We may of course assign if we please the value zero to the entropy of any perfect crystal of a pure isotope of a single element in its idealized state at the absolute zero of temperature. But even this will seldom be convenient theoretically on account of nuclear spin. For the purpose of tabulating experimental results some conventional zero must be chosen and the above choice is then often convenient. But its conventional character will no longer be so likely to be overlooked that any importance in the future will be attached to *absolute entropy*, an idea which has caused much confusion and been of very little assistance in the development of the subject.

§7·52. *Entropy limits for idealized states at zero temperatures.* We now return to the detailed verification of the limiting values of S and hence of ΔS specified by (675). It is to be remembered that the whole discussion concerns primarily *idealized* states at zero temperature, by which we mean states reached by extrapolation from say 10° K. which would be actual states if all energy separations, small compared with $10k$, were really identically zero.

From the general formula $S=k\log C$ or from the limits of (586) or (587) as $T\to 0$ the entropy of one gram-molecule of a crystal of a pure species has the limit $R\log \varpi_0$ and of a mixed crystal of several species in proportions D_r, $\Sigma_r D_r=1$, the limit $R\Sigma_r D_r \log(\varpi_0{}^r/D_r)$. It will be convenient to analyse ϖ_0 into the factors $\rho\eta$, where ρ is due to nuclear spins and η to any other surviving weight factor (e.g. rotations of H_2, electronic weight of 2(NO)).

Consider first a pure species with ortho- and para-separations (such as H_2), for which the nuclear spin weight is ρ. For such molecules

$$D_1 = \frac{\frac{1}{2}(\rho-1)}{\rho}, \quad D_2 = \frac{\frac{1}{2}(\rho+1)}{\rho}, \quad \varpi_1 = \tfrac{1}{2}\rho(\rho-1)\eta_1, \quad \varpi_2 = \tfrac{1}{2}\rho(\rho+1)\eta_2,$$

so that

$$S_0(X_2) = R\left[2\log\rho + \frac{\frac{1}{2}(\rho-1)}{\rho}\log\eta_1 + \frac{\frac{1}{2}(\rho+1)}{\rho}\log\eta_2\right]. \quad \ldots(676)$$

If $\eta_1 = \eta_2 = 1$ (as for an oscillating $^1\Sigma$ molecule such as Cl_2^{35}) this result is the same as that obtained directly by treating the lattice as atomic, with nuclear spin weights alone surviving. If $\eta_1 = \eta_2 \neq 1$ the ortho-para separation is irrelevant but we retain the extra entropy $R\log\eta_1$ for the electronic factor, besides the entropy calculated for an atomic lattice as above.

Consider now the more general case of a mixed crystal with ortho- and para-separations (such as Cl_2), for which the *atoms* are present in proportions a_1, a_2 $(a_1 + a_2 = 1)$ with nuclear spin weights ρ_1 and ρ_2. There are then five species of molecule in the crystal, ortho and para $X_2{}^1$, X^2X^1, and ortho and para $X_2{}^2$. These species are present (see § 5·31) in the proportions

$$D_1 = [\tfrac{1}{2}(\rho_1-1)/\rho_1]a_1^2, \quad D_2 = [\tfrac{1}{2}(\rho_1+1)/\rho_1]a_1^2, \quad D_3 = 2a_1a_2,$$
$$D_4 = [\tfrac{1}{2}(\rho_2-1)/\rho_2]a_2^2, \quad D_5 = [\tfrac{1}{2}(\rho_2+1)/\rho_2]a_2^2.$$

Their nuclear spin weights are

$$(\varpi_0)_1 = \tfrac{1}{2}\rho_1(\rho_1-1), \quad (\varpi_0)_2 = \tfrac{1}{2}\rho_1(\rho_1+1), \quad (\varpi_0)_3 = 2\rho_1\rho_2,$$
$$(\varpi_0)_4 = \tfrac{1}{2}\rho_2(\rho_2-1), \quad (\varpi_0)_5 = \tfrac{1}{2}\rho_2(\rho_2+1).$$

The factor 2 in $(\varpi_0)_3$ arises of course from the possibility of turning the unsymmetrical X^1X^2 end for end and not from nuclear spin, but it is part of the symmetry effect and must be included here. Besides these spin and symmetry weights there may be other factors $\eta_1, \ldots, \eta_5$. After some reduction we find that

$$S_0(X_2) = R[2a_1\log\rho_1/a_1 + 2a_2\log\rho_2/a_2] + \sum_1^5 D_s\log\eta_s. \quad \ldots(677)$$

The first set of terms is again exactly the entropy we should derive for an atomic lattice of *two* gram-atoms in which the atoms have only their nuclear spin weights as before. The other terms if they exist are extra. They are zero for all molecules such as Cl_2 which start to solidify in $^1\Sigma$ states, and are oscillating in the crystal.

The case of a crystal of molecules such as XY is very simple. Assuming that only X is a mixture $a_1, a_2\,(a_1+a_2=1)$

$$S_0(XY) = R[a_1 \log \rho_1\rho_Y/a_1 + a_2 \log \rho_2\rho_Y/a_2 + \log \eta]. \quad \text{......(678)}$$

In the absence of symmetry restrictions the η factors will be the same for all molecules in the crystal.

§ **7·53.** *Entropy changes in reactions between idealized states at zero temperature.* The formulae of § 7·52 can be at once employed to calculate ΔS_0. We find that *for any reaction* $\Sigma_t q_t X_t = 0$ *between diatomic molecules* X_t

$$\Delta S_0 = \Sigma_t q_t \log \eta_t. \quad \text{......(679)}$$

If more than one value of η *is concerned for a single constituent, the proper mean value* $\overline{\eta_t}$ *must be taken according to the equation*

$$\log \overline{\eta_t} = \sum_1^s D_r \log(\eta_r)_t. \quad \text{......(680)}$$

Thus neither nuclear spins, isotopic mixtures nor ortho-para separations contribute in themselves anything to ΔS_0, which can only be different from zero if some of the η's are greater than unity. In particular *for any reactions between diatomic molecules all in* $^1\Sigma$ *states in which all the molecules are oscillating about directions of equilibrium in the lattice near the absolute zero*

$$\Delta S_0 = 0.$$

On the contrary whenever the rotational or electronic η factors do occur they are unlikely to cancel and we must expect non-zero values of ΔS_0. Some typical ones are given in Table 15.

TABLE 15.

Some non-zero entropy changes for reactions between idealized crystalline phases at zero temperature.

Reaction	$\Delta S_0/R$
$H_2 + Cl_2 - 2HCl = 0$	$\frac{3}{4} \log 3$
$H_2 + Br_2 - 2HBr = 0$	$\frac{3}{4} \log 3$
$H_2 + I_2 - 2HI = 0$	$\frac{3}{4} \log 3$
$2NO - N_2 - O_2 = 0$	$\log 2$

Similar calculations can be made for polyatomic molecules but they are considerably more complicated when the symmetry properties of systems containing more than two similar nuclei must be taken into account, and we shall not discuss them here. It remains generally true however that nuclear spins and isotopic mixtures never give rise to non-zero values of ΔS_0, which arise only from the somewhat accidental survival of η-factors such as we have already discussed.

§7·54. *Direct comparisons of observed and theoretical entropies.* We have followed in this chapter the conventional method of comparing observed and theoretical vapour pressure and reaction constants i and I. It has become increasingly common of recent years, following the example of G. N. Lewis and his school, to record and to compare the entropies themselves, or rather entropy differences, instead of these constants. There is much to be said for this procedure as entropies are then directly available for the construction of free energies and the application to other thermodynamic problems. We shall therefore close this chapter by recording a few specimen comparisons of this type. The method adopted is to assign by convention the entropy zero to the entropy of the stable crystalline form of the substance ideally extrapolated to the absolute zero, and with this zero to compute from calorimetric observations the entropy of the gas phase at some standard temperature and pressure (reducing this value if

TABLE 16.

Observed and theoretical entropies in calories per mole per degree for various gases (reduced to perfect gases) at one atmosphere pressure and various temperatures. [The probable error in the observed values is ± 0·1 *unit.] All nuclear spin and mixing contributions are omitted.*

Gas	Temp. °K.	ΔS observed	S_0 (η-correction)	$S_0+\Delta S$	S calculated	Authority
H_2	298·1	29·7	$\frac{3}{4}R\log 3$	31·34	31·23	(1)
NO	121·36	43·0	$\frac{1}{2}R\log 2$	43·69	43·75	(2)
O_2	90·13	40·7	0	40·7	40·68	(3)
HCl	188·07	41·3	0	41·3	41·45	(4)
HBr	206·38	45·0	0	45·0	44·92	(5)
HI	237·75	47·9	0	47·9	47·8	(6)
CO	81·61	37·2	—	—	38·3	(7)

(1) Giauque, *J. Amer. Chem. Soc.* vol. 52, p. 4816 (1930). See also Giauque and Johnston, *J. Amer. Chem. Soc.* vol. 50, p. 3221 (1928) and MacGillavry, *Phys. Rev.* vol. 36, p. 1398 (1930).
(2) Johnston and Giauque, *J. Amer. Chem. Soc.* vol. 51, p. 3194 (1929).
(3) Giauque and Johnston, *J. Amer. Chem. Soc.* vol. 51, p. 2300 (1929).
(4) Giauque and Wiebe, *J. Amer. Chem. Soc.* vol. 50, p. 101 (1928).
(5) Giauque and Wiebe, *J. Amer. Chem. Soc.* vol. 50, p. 2193 (1928).
(6) Giauque and Wiebe, *J. Amer. Chem. Soc.* vol. 51, p. 1441 (1929).
(7) Clayton and Giauque, *J. Amer. Chem. Soc.* vol. 54, p. 2610 (1932).

necessary to the corresponding perfect gas value at the same pressure by a suitable equation of state). This entropy difference is perfectly unambiguous once the idealized extrapolation to zero has been properly specified. We then calculate theoretically by equation (592) the entropy of the gas phase. In order to be comparable all contributions of nuclear spin and isotopic mixtures must be omitted in this calculation and a correction made if necessary for any contribution by η-factors omitted in the idealized extrapolation. Table 16 shows a selection of such results based on the work of Giauque and his collaborators. The only discrepancy outstanding is for CO. It should be recorded that the elimination of the spin and mixing terms and the residual η-factors, particularly for H_2, can be done in several ways and the combined terms grouped together in different ways so that the different groups have apparently entirely different meanings. None the less these different groupings are all completely equivalent.

CHAPTER VIII

THE THEORY OF IMPERFECT GASES

§8·1. *General gaseous assemblies with molecules not fully independent.* We have so far only considered assemblies of "isolated" or effectively isolated systems, which almost never interfere with each other. It is only in such assemblies that the energy can be assigned to individual systems rather than to the assembly as a whole, and it is on this partition of the energy among the systems that the foregoing analysis is based. When this independence breaks down between the separate atoms as in a molecule or a crystal, we take the whole complex (molecule or crystal) to be the system. In the worst case the whole assembly must be one system in this sense—the analysis will then apply, but progress is difficult (except by special devices in specially simple cases), unless the wave equation for the whole complex (assembly) can be approximately solved. The essential step is, as always, to evaluate the partition function.

The assemblies which we are now contemplating differ essentially from perfect gases only because there is in the energy of the assembly as a whole a general term W which is a function at least of the positional coordinates of all the constituent systems. The justification for regarding the extra energy W as a *potential energy* depending only on positional coordinates merits close scrutiny. Let us suppose that two simple systems (atoms or molecules) of types α and β have energies E_α, E_β when they are very far apart. These energies will then be functions only of the Hamiltonian coordinates of the systems α and β respectively and each will be independent of the coordinates of the other. The internal energies will be specified by quantum conditions. When however they approach each other this independence must sooner or later cease. The energy of the pair will no longer be $E_\alpha + E_\beta$ but $E_\alpha + E_\beta + E_{\alpha\beta}$, say, where $E_{\alpha\beta}$ is a correcting term, at first small, depending on the coordinates, and perhaps velocities, of both systems. We assume as a first approximation that the effect of α on any β can be expressed by saying that α is surrounded by a constant field of force. This must of course be derived from the mean fields of the moving electric charges in α. The electric charges in β will then have a mean potential energy in the field of α depending on the relative positions of α and β. This, however, is not the whole of the effect, for the field of α may alter the internal energy of β and β that of α. A simple example is polarization in the field of the other body. Such changes of energy are a proper part of $E_{\alpha\beta}$. If α and β approach each other slowly then all the effects must be adiabatic in Ehrenfest's sense, and will depend only on relative coordinates, being independent of velocities. In this case the complete $E_{\alpha\beta}$, derived from all the sources specified, may legitimately be expected to behave like a potential energy depending only on the relative coordinates, whose derivates give the forces. Moreover in actual applications nearly all molecular encounters are slow compared with the velocities of electrons in atoms and molecules, and will therefore be adiabatic in the required sense. This is the assumption as to the nature of $E_{\alpha\beta}$ tacitly made in all discussions of molecular interactions.

The function $E_{\alpha\beta}$ so specified must in general depend on the quantum states of the systems α and β. It is obvious that for atoms and molecules changes of electronic orbits must affect $E_{\alpha\beta}$. It is usual, however, and not unreasonable to assume (at least as an approximation) that $E_{\alpha\beta}$ will be independent of molecular states of rotation and perhaps vibration.* As it is almost universal in gas problems for only one electronic state of atom or molecule to be relevant, the usual approximation should apply. If it does not then, as we shall see, it is only necessary to treat different electronic or vibrational states of atom or molecule as different systems.

We shall proceed throughout on the assumption that classical statistics can be used for the gas. The errors so committed are in general entirely negligible. To the same approximation the entire translatory and potential energy of the gas molecules can be handled classically.

§ **8·2.** *Partition functions for the potential energy of the whole gas.* We can at once proceed to construct a partition function for the potential energy of the whole gas, which in conjunction with previous results will determine all the equilibrium properties.

Let us start by considering a *perfect* gas mixture and examine how to construct a partition function for the whole classical energy of translation of its molecules. The internal energies of the molecules are independent of their translations and will be accounted for as usual by separate partition functions. Let there be N_α, N_β, ... free atoms and molecules of types α, β, ..., internal partition functions $j_\alpha(T)$, $j_\beta(T)$, ..., and masses m_α, m_β, ..., supposed for the moment not to dissociate or combine. If x_α, y_α, z_α, u_α, v_α, w_α are the position and velocity components of a system of type α, then a standard element of phase space for the whole gas has the extension

$$(m_\alpha^{N_\alpha} m_\beta^{N_\beta}\ldots)^3 \prod_{r=1}^{N_\alpha} (dx_\alpha \ldots dw_\alpha)_r \prod_{s=1}^{N_\beta} (dx_\beta \ldots dw_\beta)_s \ldots,$$

and the weight

$$\frac{(m_\alpha^{N_\alpha} m_\beta^{N_\beta}\ldots)^3}{h^{3(N_\alpha+N_\beta+\ldots)}} \prod_{r=1}^{N_\alpha} (dx_\alpha \ldots dw_\alpha)_r \prod_{s=1}^{N_\beta} (dx_\beta \ldots dw_\beta)_s \ldots. \quad \ldots\ldots(681)$$

* If this is not true, then of course $E_{\alpha\beta}$ is a mean value taken in a way not yet specified.

The classical partition function $H(T)$ for the translatory motion is therefore given by

$$H(T)=\frac{(m_\alpha^{N_\alpha}m_\beta^{N_\beta}\dots)^3}{h^{3(N_\alpha+N_\beta+\dots)}}\int\dots\int e^{-\left[\sum_{r=1}^{N_\alpha}\frac{1}{2}m_\alpha(u_\alpha^2+v_\alpha^2+w_\alpha^2)_r+\sum_{s=1}^{N_\beta}\frac{1}{2}m_\beta(u_\beta^2+v_\beta^2+w_\beta^2)_s+\dots\right]/kT}$$
$$\times\prod_{r=1}^{N_\alpha}(dx_\alpha\dots dw_\alpha)_r\prod_{s=1}^{N_\beta}(dx_\beta\dots dw_\beta)_s\dots \qquad \dots\dots(682)$$

It is assumed that there are no external fields of force except local boundary fields. All the space and velocity integrations in (682) are independent and can be carried out one by one, and we find

$$H(T)=\left[\frac{(2\pi m_\alpha kT)^{\frac{3}{2}}V}{h^3}\right]^{N_\alpha}\left[\frac{(2\pi m_\beta kT)^{\frac{3}{2}}V}{h^3}\right]^{N_\beta}\dots,$$
$$=[h_\alpha(T)]^{N_\alpha}[h_\beta(T)]^{N_\beta}\dots \qquad \dots\dots(683)$$

In equation (683) $h_\alpha(T)$, $h_\beta(T)$, ... are the ordinary partition functions for the translatory motion of single molecules of types α, β, ..., and we recover, as we must, the ordinary result. For the complete partition function we have of course to add to $H(T)$ the factor

$$[j_\alpha(T)]^{N_\alpha}[j_\beta(T)]^{N_\beta}\dots \qquad \dots\dots(684)$$

The extension to a classical imperfect gas is immediate. Under the specified conditions $j_\alpha(T)$, $j_\beta(T)$, ... are still separable factors. The function $H(T)$ is altered only by the addition of a term W to the energy, where W, the potential energy of the whole assembly, is a function of all the positional coordinates $(x_\alpha, y_\alpha, z_\alpha)_r$, $(x_\beta, y_\beta, z_\beta)_s$, ..., built up to a first approximation out of the functions $E_{\alpha\beta}$ which we have already specified. The addition of W will not affect the velocity integrations, which can be carried out as before. We therefore find in general

$$H(T)=\left[\frac{(2\pi m_\alpha kT)^{\frac{3}{2}}}{h^3}\right]^{N_\alpha}\left[\frac{(2\pi m_\beta kT)^{\frac{3}{2}}}{h^3}\right]^{N_\beta}\dots\times B(T), \qquad \dots(685)$$

where
$$B(T)=\int\dots\int e^{-W/kT}\prod_{r=1}^{N_\alpha}(dx_\alpha\dots dz_\alpha)_r\prod_{s=1}^{N_\beta}(dx_\beta\dots dz_\beta)_s\dots \qquad \dots\dots(686)$$

The integrals in $B(T)$ are extended over the whole positional phase space accessible to the gas. It is therefore assumed that in no case can $W\to-\infty$, but that ultimately there must be repulsive fields between any two particles so that $W\to+\infty$ when they approach sufficiently closely. No other assumption would be physically consistent with the continued existence of ordinary matter. When $W\equiv 0$, $B(T)=V^{N_\alpha+N_\beta+\dots}$, and $H(T)$ reduces to (683).

From the form of (685) it is clear that classical potential and kinetic energies in a gas can always be handled with separate partition functions. The kinetic energy can be dealt with in the ordinary way, as if the gas were perfect, by ordinary partition functions without the V-factor. The potential energy is accounted for by $B(T)$. It should be observed also that (685) and (686) are perfectly general, so long as all the particles are free to move individually, and apply to classical assemblies in which the imperfections are of any degree and the intermolecular forces of any range, and whether or no there are external fields of force. It is not, of course, true for higher degrees of imperfection that W can be regarded as built up entirely of terms such as $E_{\alpha\beta}$, that is, terms arising from binary encounters. Ternary encounters must next be considered in which three molecules are concerned with a total energy $(E_\alpha+E_\beta+E_\gamma+E_{\beta\gamma}+E_{\gamma\alpha}+E_{\alpha\beta})+E_{\alpha\beta\gamma}$, where $E_{\alpha\beta\gamma}$ depends essentially on the coordinates of all three systems. At higher concentrations encounters of all orders must be successively taken into account. But, assuming W can be constructed, equation (686) continues to give $B(T)$. If there are no external fields of force, W will be a function only of the relative coordinates of the molecules. If there are local boundary fields representing walls of the enclosure, this will still be true in the limit when the whole volume is very large compared with the volume affected by the local fields.

The partition function $B(T)$ refers to what is strictly a single system, the whole gas. It therefore appears by itself, not raised to a high power, in the complex integrals which give the properties of the equilibrium state, resembling in this the partition functions for radiation and for crystals. Some average values derived in the usual way from $B(T)$ may be small; they will none the less be true averages, but will merely be subject to relatively large fluctuations. But any derived average number of molecules which is itself large, like the average values of the theory of the perfect gas, will have its usual validity and its usual insignificant fluctuation which guarantees normality.

In specifying, for example, a statistical state of the assembly in order to derive average values from $B(T)$ it is necessary strictly to specify the positions of *all* molecules. There is also only *one* system, the whole gas. In Chapter II, for example, we proved with the help of a partition function $f(T)$ that the average number $\bar{a}$ of systems satisfying certain conditions represented by part of $f(T)$, $\delta f(T)$ say, was given by

$$\bar{a}=N\delta f(T)/f(T). \qquad \text{......(687)}$$

Here, however, $N=1$; the meaning of (687) is unaltered, but is better appreciated if we say that $\delta f(T)/f(T)$ is the fraction of time during which the gas is in the specified state, or if we like the *frequency ratio* or *probability* of that state. Equation (687) still describes a genuine equilibrium property of the assembly, but only those mean values derived from (687), which have insignificant fluctuations, are physically significant.

The potential energy W will contain terms which depend on the values of the parameters which fix the position of bodies producing external fields. The generalized reaction Y_W of the whole gas arising from W in any configuration is therefore $-\partial W/\partial y$, and

$$\overline{Y_W}=\frac{1}{B(T)}\int\cdots\int\left(-\frac{\partial W}{\partial y}\right)e^{-W/kT}\prod_{r=1}^{N_\alpha}(dx_\alpha\ldots dz_\alpha)_r\prod_{s=1}^{N_\beta}(dx_\beta\ldots dz_\beta)_s\ldots, \quad\ldots\ldots(688)$$

$$=kT\frac{\partial}{\partial y}\log B(T). \quad\ldots\ldots(689)$$

The average value of the potential energy is

$$\overline{W}=kT^2\frac{\partial}{\partial T}\log B(T). \quad\ldots\ldots(690)$$

The former result is, of course, the exact analogue of § 2·8. From it, it follows at once, as in Chapter VI, that the more general assemblies here contemplated obey the laws of thermodynamics, and that $B(T)$ contributes

$$k\log B(T)$$

to the characteristic function Ψ, and

$$k\log B(T)+\overline{W}/T$$

to the entropy.

It is, of course, necessarily true, and becomes particularly obvious in this section, that the partition functions of our theory are identical with the separable factors in the phase integrals of Gibbs. The further developments will therefore hardly differ at all, whether one starts from the canonical ensemble of Gibbs or the conservative dynamical system of Boltzmann. We shall continue for consistency and practical convenience to use the terminology of partition functions.

In applications it is convenient to use an abbreviated notation. We write $d\omega_{\alpha r}$ for $(dx_\alpha dy_\alpha dz_\alpha)_r$ and contract

$$\prod_{r=1}^{N_\alpha}d\omega_{\alpha_r}\prod_{s=1}^{N_\beta}d\omega_{\beta_s}\ldots$$

into

$$\Pi_\alpha(d\omega_\alpha)^{N_\alpha}.$$

Thus

$$B(T)=\int\cdots\int e^{-W/kT}\,\Pi_\alpha(d\omega_\alpha)^{N_\alpha}. \quad\ldots\ldots(691)$$

We shall similarly write $\Sigma_\alpha N_\alpha$ for $N_\alpha+N_\beta+\ldots$.

§ 8·3. *A first approximation for imperfect gases. Short range forces.* We shall now suppose that the field of an α or β, in which $E_{\alpha\beta}$ is sensible, is of strictly limited range, so that there is a certain small volume $v_{\alpha\beta}$ round α in which β (or round β in which α) must lie so that $E_{\alpha\beta}\neq 0$. Otherwise $E_{\alpha\beta}=0$.

For the present we neglect external fields of force. We suppose also that ternary and higher complexes may be neglected, so that W consists only of terms such as $E_{\alpha\beta}$. It can be shown that in so doing we neglect terms in $\log B(T)$ of order at most $N_\alpha(\Sigma_\beta N_\beta v_{\alpha\beta}/V)^2$. We therefore suppose that $\Sigma_\beta N_\beta v_{\alpha\beta}/V$ is small and that terms of higher order than the first in this ratio are to be neglected compared with unity. The terms to be neglected in any step require, however, the most careful scrutiny.*

To evaluate $B(T)$ retaining only binary complexes we consider first an assembly with only one type of molecule. The argument extends at once to any number of types, but the algebra is complicated enough to postpone for the moment. If one α,α pair lies in $v_{\alpha\alpha}$ (shortly v), then $W = E_{\alpha\alpha}$ (more shortly E) and is otherwise zero. Let us first take $W = 0$ everywhere. We obtain the contribution V^N. The rest of $B(T)$ comes from integrating $e^{-W/kT} - 1$ over the whole phase space. This is zero unless at least one v is occupied. Let us put $W = E$ and integrate for the relative coordinates of one pair over v and for the other coordinates over V. We obtain

$$\Sigma \int_v (e^{-E/kT} - 1)\, d\omega \int_{(V)} \cdots \int (d\omega)^{N-1},$$

where Σ is the sum over all pairs, which reduces to

$$V^{N-1} . \tfrac{1}{2} N(N-1) \int_v (e^{-E/kT} - 1)\, d\omega.$$

This contribution is exact so long as only one v is occupied. If two are occupied simultaneously the proper contribution to $B(T) - V^N$ comes from integrating $e^{-(E+E')/kT} - 1$ over v, v' for two sets of relative coordinates, and the others over V. The integrand we have actually used above is, however, $(e^{-E/kT} - 1) + (e^{-E'/kT} - 1)$. It remains, therefore, to integrate in this manner

$$e^{-(E+E')/kT} - 1 - (e^{-E/kT} - 1) - (e^{-E'/kT} - 1) = (e^{-E/kT} - 1)(e^{-E'/kT} - 1).$$

Using this integrand the next contribution is

$$V^{N-2} . (\tfrac{1}{2})^2 \frac{N(N-1)(N-2)(N-3)}{2!} \left\{ \int_v (e^{-E/kT} - 1)\, d\omega \right\}^2 .$$

The numerical coefficient is the number of sets of two pairs. This contribution is again exact so long as there are no sets of three pairs. The correct integrand is then $e^{-(E+E'+E'')/kT} - 1$ taken over v, v', v'' for three sets of relative coordinates and the others over V. The complete integrand we have used so far is

$$\Sigma (e^{-E/kT} - 1)(e^{-E'/kT} - 1) + \Sigma (e^{-E/kT} - 1),$$

* The following presentation is due to Ursell, *Proc. Camb. Phil. Soc.* vol. 23, p. 685 (1927). Many current analyses of this type are completely fallacious, e.g. Jeans, *Dynamical Theory of Gases*, ed. 2, § 218; Fowler, *Proc. Camb. Phil. Soc.* vol. 22, p. 861 (1925).

and there remains

$$(e^{-(E+E'+E'')/kT}-1)-\Sigma\,(e^{-E/kT}-1)\,(e^{-E'/kT}-1)-\Sigma\,(e^{-E/kT}-1)=\Pi\,(e^{-E/kT}-1).$$

Using this integrand the contribution is

$$V^{N-3}\,.\,(\tfrac{1}{2})^3\frac{N!}{3!(N-6)!}\left\{\int_v (e^{-E/kT}-1)\,d\omega\right\}^3.$$

The generality of the method may be established by a simple induction, and the full form of $B(T)$ is therefore

$$B(T)=V^N\sum_{r=0}^{\frac{1}{2}N}\frac{N!}{r!(N-2r)!\,N^r}x^r, \qquad \text{......(692)}$$

where

$$x=\frac{N}{2V}\int_v (e^{-E/kT}-1)\,d\omega, \qquad \text{......(693)}$$

and is therefore small.

It is obvious that the form of $B(T)$, at least for the earlier terms, is approximately given by

$$V^N(1+x)^N=V^N\sum_{r=0}^{N}\frac{N!}{r!(N-r)!}x^r,$$

and one is led to expect (as is in fact true) that

$$B(T)=V^N\{1+x+O(x^2)\}^N, \qquad \text{......(694)}$$

which is an approximation of the desired accuracy. It is *not* sufficient, in order to derive (694), to retain only the first two terms in (692), as is commonly done, and assume that the second is small compared with the first and so on. For the ratio of the first two terms is $(N-1)\,x$, and it is only x which we are entitled to assume small! We proceed, therefore, as follows, using an argument adapted to the generalizations required in § 8·31.

Consider the function $F(x,y)$ defined by the equation

$$F(x,y)=\sum_{r=0}^{\frac{1}{2}N}\frac{N!}{r!(N-2r)!\,N^r}x^r\,y^{N-2r}. \qquad \text{......(695)}$$

By partial differentiation this function obviously satisfies the equation

$$\frac{1}{N}\frac{\partial F}{\partial x}=\left(\frac{1}{N}\frac{\partial}{\partial y}\right)^2 F. \qquad \text{......(696)}$$

Put

$$F=e^{Ng}.$$

Then this equation reduces to

$$\frac{\partial g}{\partial x}=\left(\frac{\partial g}{\partial y}\right)^2+\frac{1}{N}\frac{\partial^2 g}{\partial y^2}, \qquad \text{......(697)}$$

and g is uniquely defined as that solution of (697) which is equal to $\log y$ for

$x=0$. Since N is very large the last term can be neglected. Since also $F=y^N f(x/y^2)$,

$$x\frac{\partial F}{\partial x}+\tfrac{1}{2}y\frac{\partial F}{\partial y}=\tfrac{1}{2}NF,$$

or

$$y\frac{\partial g}{\partial y}=1-2x\frac{\partial g}{\partial x}.$$

Substituting this in (697) we find the approximate equation for g

$$\frac{\partial g}{\partial x}=\frac{1}{y^2}\left(1-2x\frac{\partial g}{\partial x}\right)^2.$$

The actual value of g required is $g(x,1)$; which is that solution of

$$\frac{\partial g}{\partial x}=\left(1-2x\frac{\partial g}{\partial x}\right)^2 \qquad \text{......(698)}$$

which is 0 for $x=0$, terms of order $1/N$ being neglected. This is easily verified directly without the introduction of the auxiliary variable y.

Equation (698) can be solved in finite terms, but of course no physical importance attaches to any terms other than the lowest in x since we have only hitherto included binary complexes. We find formally

$$\frac{dg}{dx}=\frac{1+4x-\sqrt{(1+8x)}}{8x^2},$$

$$g=\log\frac{1-\sqrt{(1+8x)}}{4x}-\frac{1+4x-\sqrt{(1+8x)}}{8x},$$

and the significant terms

$$g=x+O(x^2).$$

Thus

$$\log B(T)=N\{\log V+x+O(x^2)\}, \qquad \text{......(699)}$$

which is equivalent to (684) and contains all the relevant information.

Our result therefore is for a simple gas

$$\log B(T)=N\log\left\{V+\tfrac{1}{2}N\int_v (e^{-E/kT}-1)\,d\omega\right\}, \qquad \text{......(700)}$$

and for the general mixture, to which we devote part of § 8·31,

$$\log B(T)=\Sigma_\alpha N_\alpha \log\left\{V+\tfrac{1}{2}\Sigma_\beta N_\beta\int_{v_{\alpha\beta}} (e^{-E_{\alpha\beta}/kT}-1)\,d\omega_\beta\right\}, \qquad \text{...(701)}$$

$$=\Sigma_\alpha N_\alpha \log V+\Sigma_{\alpha\beta}\frac{N_\alpha N_\beta}{\sigma_{\alpha\beta}V}\int_{v_{\alpha\beta}} (e^{-E_{\alpha\beta}/kT}-1)\,d\omega_\beta. \qquad \text{...(702)}$$

In this equation $\sigma_{\alpha\beta}=1$ $(\alpha\neq\beta)$, $\sigma_{\alpha\alpha}=2$, and $\Sigma_{\alpha\beta}$ means summation over every pair of types α and β. In every case the terms omitted are of order smaller by the factor $\Sigma N_\beta v_{\alpha\beta}/V$ than those retained.

The other partition functions for our assembly are normal, those for the translatory motion being without their V-factors. When therefore we add

$\log B(T)$ to Ψ/k we restore the V-factors to the translatory partition functions and add to Ψ/k the new term

$$\Sigma_{\alpha\beta}\frac{N_\alpha N_\beta}{\sigma_{\alpha\beta}V}\int_{v_{\alpha\beta}}(e^{-E_{\alpha\beta}/kT}-1)\,d\omega_\beta. \qquad \ldots\ldots(703)$$

The integral in (703) is more commonly expressed in polar coordinates, for by ignoring coordinates defining the orientation of molecules we have tacitly assumed that $E_{\alpha\beta}$ is a function of r alone, or may be replaced by a mean value for all orientations. Expressed thus (703) becomes

$$\Sigma_{\alpha\beta}\frac{N_\alpha N_\beta}{\sigma_{\alpha\beta}V}4\pi\int_{v_{\alpha\beta}}(e^{-E_{\alpha\beta}/kT}-1)\,r^2\,dr. \qquad \ldots\ldots(704)$$

The method can easily be extended to systems for which orientations are relevant, and the potential a general function of relative position, but we shall not take up such extensions here.

We have hitherto assumed that $E_{\alpha\beta}$ is due to forces of finite range. It is more usual and convenient in practice to represent molecular forces by forces which fall off like some inverse sth power of the distance. If $s>4$, the integral in (704) will converge when extended over all space. Any such integral, if sufficiently rapidly convergent, can be substituted for the finite integral in (704) and elsewhere, without modification of the argument, for the integral over all space differs negligibly from the integral over a $v_{\alpha\beta}$ of atomic dimensions. We thus find for the extra terms

$$\Sigma_{\alpha\beta}\frac{N_\alpha N_\beta}{\sigma_{\alpha\beta}V}4\pi\int_0^\infty(e^{-E_{\alpha\beta}/kT}-1)\,r^2\,dr. \qquad \ldots\ldots(705)$$

By (690) the average potential energy of the assembly is

$$\overline{W}=\Sigma_{\alpha\beta}\frac{N_\alpha N_\beta}{\sigma_{\alpha\beta}V}4\pi\int_0^\infty E_{\alpha\beta}e^{-E_{\alpha\beta}/kT}\,r^2\,dr. \qquad \ldots\ldots(706)$$

If F_α is the complete partition function for the system of type α without its V-factor, then the complete Ψ/k is given by

$$\Psi/k=\Sigma_\alpha N_\alpha\left\{\log\frac{VF_\alpha}{N_\alpha}+1\right\}+\Sigma_{\alpha\beta}\frac{N_\alpha N_\beta}{\sigma_{\alpha\beta}V}4\pi\int_0^\infty(e^{-E_{\alpha\beta}/kT}-1)\,r^2\,dr. \qquad \ldots\ldots(707)$$

Since $p=T\,\partial\Psi/\partial V$, we find at once

$$p=kT\left[\frac{\Sigma_\alpha N_\alpha}{V}-\frac{1}{V^2}\Sigma_{\alpha\beta}\frac{N_\alpha N_\beta}{\sigma_{\alpha\beta}}4\pi\int_0^\infty(e^{-E_{\alpha\beta}/kT}-1)\,r^2\,dr\right]. \quad \ldots(708)$$

This is the well-known formula of van der Waals, correct to terms in $1/V^2$, for a mixture of imperfect gases.

In concluding this section let us include an external field of force in which the potential energy of the molecule of type α is Ω_α. This must not be so large that the approximations we have made become invalid in any part of the assembly. At the densest point ternary encounters must still be negligible. The condition for this is, of course, that if the densest element of the gas in its average state is treated as a separate assembly, then $\Sigma_\beta N_\beta v_{\alpha\beta}/V$ must still be small there.

The extended form of $\log B(T)$ is easily written down. Wherever before we obtained a factor V by integrating $d\omega_\alpha$ over the whole volume, we now obtain instead the factor

$$A_\alpha(T) = \int_V e^{-\Omega_\alpha/kT}\, d\omega. \qquad \text{......(709)}$$

Wherever before we had to deal with an α,β pair we integrated their relative coordinates over $v_{\alpha\beta}$ and the coordinates of their centre of mass over V, obtaining the factors $V\int_{v_{\alpha\beta}} (e^{-E_{\alpha\beta}/kT}-1)\, d\omega_\beta$. Now, however, we have instead

$$\int_V e^{-(\Omega_\alpha+\Omega_\beta)/kT}\, d\omega \,.\int_{v_{\alpha\beta}} (e^{-E_{\alpha\beta}/kT}-1)\, d\omega_\beta.$$

In place of the old V in this connection we obtain now the factor

$$A_{\alpha\beta}(T) = \int_V e^{-(\Omega_\alpha+\Omega_\beta)/kT}\, d\omega. \qquad \text{......(710)}$$

There are no other alterations and the argument is unaffected. Therefore

$$\log B(T) = \Sigma_\alpha N_\alpha \log\left\{A_\alpha(T) + \tfrac{1}{2}\Sigma_\beta N_\beta \frac{A_{\alpha\beta}(T)}{A_\beta(T)} \int_{v_{\alpha\beta}} (e^{-E_{\alpha\beta}/kT}-1)\, d\omega_\beta\right\}, \qquad \text{......(711)}$$

$$= \Sigma_\alpha N_\alpha \log A_\alpha(T) + \Sigma_{\alpha\beta} \frac{N_\alpha N_\beta}{\sigma_{\alpha\beta}} \frac{A_{\alpha\beta}(T)}{A_\alpha(T)\, A_\beta(T)} \int_{v_{\alpha\beta}} (e^{-E_{\alpha\beta}/kT}-1)\, d\omega_\beta. \qquad \text{......(712)}$$

§ **8·31.** *The general theory of B(T).** In attempting a more general theory than that expounded in § 8·3, it is best to make a fresh start with assemblies of N molecules represented by rigid elastic spheres of one type. For such assemblies $B(T)$ is independent of T and is the volume of $3N$-dimensional space, contained in V^N for which no one of a set of conditions in number $\frac{1}{2}N(N-1)$, of the form

$$(x_r - x_s)^2 + (y_r - y_s)^2 + (z_r - z_s)^2 \geqslant D^2, \qquad \text{......(713)}$$

is violated. Suppose we choose a set of these conditions in number k and calculate the volume of that part of V^N in which they are all violated, and then let B_k denote the sum of these volumes for all possible sets of k conditions. Then

$$B(T) = B_0 - B_1 + B_2 - \ldots + (-)^k B_k + \ldots. \qquad \text{......(714)}$$

* Ursell, *loc. cit.*

This identity follows from the fact that an element of $3N$-space in which s conditions are violated is counted ${}_sC_k$ times in B_k if $0 \leqslant k \leqslant s$ and otherwise not counted; and

$$\sum_{k=0}^{s} (-)^k {}_sC_k = (1-1)^s = 0 \quad (s>0),$$
$$= 1 \quad (s=0).$$

Thus the sum on the right of (714) is precisely that part of $3N$-space in which no conditions are violated, that is, $B(T)$. The approximation of § 8·3 follows immediately.

To evaluate $B(T)$ in general we must expand each B_k. Consider a group of t molecules whose positions can be represented in $3t$-space. At any particular point of this space a certain set of conditions (713) are violated, and there is a surrounding region in which the same conditions are all broken. We can enumerate all the sets of broken conditions (713) in such a way that each set binds the molecules it involves into a single connected group, the connecting links being the broken conditions (713). For any such group there is a symmetry number σ which is defined as the number of permutations of the molecules among themselves which leave the set of defining conditions unaltered. In $3t$-space there is a definite region in which these conditions (and possibly others as well) are violated; by permuting the molecules among themselves we get $t!/\sigma$ such regions, which may of course overlap. We now enumerate the *types* of such sets of conditions in any convenient order, and specify the rth type as follows: the number of molecules involved is t_r with a symmetry number σ_r, and the number of conditions violated is p_r; the extension of $3t_r$-space corresponding to these conditions is η_r. We find it convenient later on to use a symbol ξ_r defined by the equation

$$\xi_r = (-)^{p_r} N^{t_r - 1} \eta_r / (V^{t_r} \sigma_r).$$

Following the definition of B_k we now choose any conditions (713) and calculate the volume of $3N$-space in which they at least are violated. Any such set resolves into a number of sets of the types defined above, the corresponding groups of molecules being disconnected and mutually exclusive. Let there be ν_r sets of the rth type so that

$$k = \Sigma_r \nu_r p_r. \qquad \text{......(715)}$$

The number of sets of conditions for which the ν_r have assigned values is

$$\frac{N!}{(N - \Sigma_r t_r \nu_r)! \, \Pi_r (\nu_r! \, \sigma_r^{\nu_r})}.$$

Hence $$B_k = \Sigma \frac{N! \, V^{N - \Sigma_r t_r \nu_r}}{(N - \Sigma_r t_r \nu_r)! \, \Pi_r (\nu_r! \, \sigma_r^{\nu_r})} \Pi_r \eta_r^{\nu_r},$$

the summation being over all positive integral values of the ν_r satisfying (715). Thus on introducing the ξ_r

$$B(T) = \Sigma_k (-)^k B_k = V^N \Sigma \frac{N! \,\Pi_r \xi_r{}^{\nu_r} N^{-\Sigma_r \nu_r(t_r-1)}}{(N - \Sigma_r t_r \nu_r)! \,\Pi_r \nu_r!}, \quad \text{......(716)}$$

the summation being over all positive integral ν_r. It is obvious that if b is four times the volume of all the molecules (van der Waals' b),

$$\xi_r = O\left\{\left(\frac{b}{V}\right)^{t_r-1}\right\}.$$

We have now to find a means of summing (716). Write

$$F \equiv F(\xi_r, y) = \Sigma \frac{(yN)^{N-\Sigma_r t_r \nu_r} \,\Pi_r (\xi_r N)^{\nu_r} N! \,N^{-N}}{(N - \Sigma_r t_r \nu_r)! \,\Pi_r \nu_r!}.$$

Then
$$\frac{1}{N}\frac{\partial F}{\partial \xi_r} = \left(\frac{1}{N}\frac{\partial}{\partial y}\right)^{t_r} F,$$

and F is defined completely by these equations together with the condition that it is regular in the ξ_r at the origin and takes the value y^N there. Further, if $t_r = t_s$, that is, if the number of molecules concerned in the rth and sth type of connected group is the same, then

$$\frac{\partial F}{\partial \xi_r} = \frac{\partial F}{\partial \xi_s},$$

and F is therefore a function of the sums

$$x_s = (\Sigma_r \xi_r)_{t_r = s}$$

only. We therefore have

$$F = \Sigma \frac{(yN)^{N-\Sigma_s s\nu_s} \,\Pi_s (x_s N)^{\nu_s} N! \,N^{-N}}{(N - \Sigma_s s\nu_s)! \,\Pi_s \nu_s!}, \quad \text{......(717)}$$

together with
$$\frac{1}{N}\frac{\partial F}{\partial x_s} = \left(\frac{1}{N}\frac{\partial}{\partial y}\right)^{s} F. \quad \text{......(718)}$$

Let us now write
$$y = e^z, \quad \delta = \Sigma_s s x_s \frac{\partial}{\partial x_s}.$$

Then from (717)
$$\left(\frac{\partial}{\partial z} + \delta\right)^k F = N^k F, \quad \text{......(719)}$$

and from (718)

$$\frac{1}{N}\frac{\partial F}{\partial x_s} = e^{-sz}\left(\frac{1}{N}\frac{\partial}{\partial z} - \frac{s-1}{N}\right)\cdots\left(\frac{1}{N}\frac{\partial}{\partial z} - \frac{1}{N}\right)\frac{1}{N}\frac{\partial}{\partial z} F.$$

On substituting from (719) we find

$$\frac{1}{N}\frac{\partial F}{\partial x_s} = e^{-sz}\left(1 - \frac{s-1}{N} - \frac{\delta}{N}\right)\cdots\left(1 - \frac{1}{N} - \frac{\delta}{N}\right)\left(1 - \frac{\delta}{N}\right) F.$$

Let us now write, as before, $F = e^{Ng}$

and put $y=1$ or $z=0$. Then after reduction

$$\frac{\partial g}{\partial x_s}=\left(1-\frac{s-1}{N}-\delta g-\frac{\delta}{N}\right)\dots\left(1-\frac{1}{N}-\delta g-\frac{\delta}{N}\right)(1-\delta g),$$

in which the free δ's are to operate on everything to the right of them. This is exact. Making $N\to\infty$ we get

$$\frac{\partial g}{\partial x_s}=(1-\delta g)^s. \qquad \dots\dots(720)$$

If we write h for $1-\delta g$, then in virtue of the definition of δ

$$h=1-\Sigma_s s x_s h^s. \qquad \dots\dots(721)$$

The summation begins for $s=2$. It is easily verified that subject to (721) the partial differential equations

$$\frac{\partial g}{\partial x_s}=h^s \qquad \dots\dots(722)$$

are integrable. The method of evaluating g in series is now clear. We solve (721) for h in series and so form g from (722), putting $g=0$ for $x_s=0$. We then have

$$B(T)=V^N e^{Ng}.$$

The method extends at once to a gas of any number of kinds of molecules; it is sufficient to discuss a mixture of two kinds only, in number N_α and N_β. We must now define the ξ_r so that the volume of $3(t_r+t_r')$-space corresponding to the rth type of group, composed of t_r molecules of the first kind and t_r' of the second, is

$$(-)^{p_r}\xi_r V^{t_r+t_r'}\sigma_r/N_0^{t_r+t_r'-1}.$$

Here N_0 is supposed to be a number of the order of magnitude of N_α and N_β; it may with advantage be taken to be an absolute constant such as Avogadro's number, or else proportional to V. We now find

$$B(T)=V^{N_\alpha+N_\beta}F(\xi_r,1,1), \qquad \dots\dots(723)$$

where

$$F(\xi_r,y,z)=\Sigma\frac{y^{N_\alpha-\Sigma_r t_r\nu_r}z^{N_\beta-\Sigma_r t_r'\nu_r}\Pi_r\xi_r^{\nu_r}}{(N_\alpha-\Sigma_r t_r\nu_r)!\,(N_\beta-\Sigma_r t_r'\nu_r)!\,\Pi_r\nu_r!}N_0^{-\Sigma_r(t_r+t_r'-1)\nu_r}N_\alpha!N_\beta!.$$

From this it follows that

$$\frac{1}{N_0}\frac{\partial F}{\partial \xi_r}=\left(\frac{1}{N_0}\frac{\partial}{\partial y}\right)^{t_r}\left(\frac{1}{N_0}\frac{\partial}{\partial z}\right)^{t_r'}F. \qquad \dots\dots(724)$$

We therefore put

$$x_{r,s}=(\Sigma_q\xi_q)_{t_q=r,\,t_q'=s},$$

so that F is a function of the $x_{r,s}$ only, and $F=e^{N_0g}$. We find eventually, by the same reduction and approximation as before,

$$\frac{\partial g}{\partial x_{r,s}}=\left(\frac{N_\alpha}{N_0}-\delta_\alpha g\right)^r\left(\frac{N_\beta}{N_0}-\delta_\beta g\right)^s=h_\alpha{}^r h_\beta{}^s, \qquad \dots\dots(725)$$

where

$$h_\alpha=\nu_\alpha-\delta_\alpha g=\nu_\alpha-\Sigma_{r,s}r x_{r,s}h_\alpha{}^r h_\beta{}^s, \qquad \dots\dots(726)$$

$$h_\beta=\nu_\beta-\delta_\beta g=\nu_\beta-\Sigma_{r,s}s x_{r,s}h_\alpha{}^r h_\beta{}^s. \qquad \dots\dots(727)$$

We have written ν_α and ν_β here for N_α/N_0 and N_β/N_0, and have to solve these equations for g with the boundary condition $g=0$ for $x_{r,s}=0$. It should be noted that the summations lack the terms $r=1$, $s=0$ and $r=0$, $s=1$. These terms are precisely the h_α and h_β on the left of the equations, which take a fully symmetrical form if transposed. But they must be used in the form in which they are written.

We shall now calculate the terms of the first three orders in h and g, remembering that x_r is $O(b/V)^{r-1}$. For a pure gas we obtain successively from (721) and (722)

$$\begin{aligned} h &= 1, \\ &= 1-2x_2, \\ &= 1-2x_2(1-4x_2)-3x_3, \end{aligned}$$

giving

$$\left.\begin{aligned} g &= x_2, \\ &= x_2-2x_2^2+x_3, \\ &= x_2+(x_3-2x_2^2)+(x_4-6x_2x_3+\tfrac{20}{3}x_2^3). \end{aligned}\right\} \qquad \ldots\ldots(728)$$

For a mixture

$$\begin{aligned} h_\alpha &= \nu_\alpha, & h_\beta &= \nu_\beta, \\ &= \nu_\alpha(1-2x_{2,0}\nu_\alpha-x_{1,1}\nu_\beta), & &= \nu_\beta(1-2x_{0,2}\nu_\beta-x_{1,1}\nu_\alpha), \end{aligned}$$

$$\begin{aligned} h_\alpha = \nu_\alpha &- 2x_{2,0}\nu_\alpha^2(1-4x_{2,0}\nu_\alpha-2x_{1,1}\nu_\beta) \\ &- x_{1,1}\nu_\alpha\nu_\beta(1-2x_{2,0}\nu_\alpha-x_{1,1}[\nu_\alpha+\nu_\beta]-2x_{0,2}\nu_\beta)-3x_{3,0}\nu_\alpha^3-2x_{2,1}\nu_\alpha^2\nu_\beta-x_{1,2}\nu_\alpha\nu_\beta^2, \end{aligned}$$

with a corresponding last formula for h_β. The terms in g of the first two orders are therefore

$$\begin{aligned} g = \{\nu_\alpha^2x_{2,0}+\nu_\alpha\nu_\beta x_{1,1}+\nu_\beta^2x_{0,2}\}+\nu_\alpha^3\{x_{3,0}-2x_{2,0}^2\}+\nu_\beta^3\{x_{0,3}-2x_{0,2}^2\} \\ +\nu_\alpha^2\nu_\beta\{x_{2,1}-2x_{2,0}x_{1,1}-\tfrac{1}{2}x_{1,1}^2\}+\nu_\alpha\nu_\beta^2\{x_{1,2}-2x_{0,2}x_{1,1}-\tfrac{1}{2}x_{1,1}^2\}. \end{aligned} \qquad \ldots\ldots(729)$$

For the terms of the next higher order the coefficients of ν_α^4 and ν_β^4 can be written down from those for a pure gas. The coefficient of $\nu_\alpha^3\nu_\beta$ is

$$x_{3,1}-(4x_{2,0}+x_{1,1})x_{2,1}+\tfrac{1}{3}x_{1,1}^3+2x_{2,0}x_{1,1}^2+8x_{2,0}^2x_{1,1}. \qquad \ldots\ldots(730)$$

The coefficient of $\nu_\alpha\nu_\beta^3$ can be written down from this by symmetry. The coefficient of $\nu_\alpha^2\nu_\beta^2$ is

$$\begin{aligned} x_{2,2}-2x_{2,1}(x_{0,2}+x_{1,1})-2x_{1,2}(x_{2,0}+x_{1,1})+(3x_{2,0}+x_{1,1}+3x_{0,2})x_{1,1}^2 \\ +4x_{2,0}x_{1,1}x_{0,2}. \end{aligned} \qquad \ldots\ldots(731)$$

We shall now calculate the $x_{r,s}$ required for the terms of the first two orders, for hard spherical molecules. For x_2 or $x_{2,0}$ we have a single ξ and $p=1$, $\sigma=2$. Hence

$$x_2 = -\frac{N}{2V^2}\cdot V\cdot\tfrac{4}{3}\pi D^3 = -\tfrac{2}{3}N\pi D^3/V, \qquad \ldots\ldots(732)$$

D being the diameter of the molecule. Similarly,

$$x_{2,0} = -\tfrac{2}{3}N_0\pi D_\alpha{}^3/V, \quad x_{0,2} = -\tfrac{2}{3}N_0\pi D_\beta{}^3/V, \quad x_{1,1} = -\tfrac{4}{3}N_0\pi D_{\alpha\beta}{}^3/V, \quad \text{......(733)}$$

D_α, D_β being the respective diameters, and $D_{\alpha\beta}$ the sum of the radii.

Again, x_3 is the sum of two different ξ's. For one ξ one molecule overlaps two others, but these do not necessarily overlap. Hence $p=2$, $\sigma=2$, and

$$\frac{2\xi V^3}{N^2} = V(\tfrac{4}{3}\pi D^3)^2, \quad \xi = 2x_2{}^2.$$

For the other ξ each pair of molecules must overlap and $p=3$, $\sigma=6$. When the first two have their centres at a distance r less than D, the volume in which the centre of the third molecule must lie is

$$K(r) = 2\int_{\frac{1}{2}r}^{D} \pi(D^2 - x^2)\,dx = 2\pi(\tfrac{2}{3}D^3 - \tfrac{1}{2}D^2 r + \tfrac{1}{24}r^3).$$

Hence

$$\frac{-6\xi V^3}{N^2} = V\int_0^D 4\pi r^2 K(r)\,dr,$$

$$\xi = \frac{-N^2}{6V^2}\,8\pi^2\int_0^D r^2(\tfrac{2}{3}D^3 - \tfrac{1}{2}D^2 r + \tfrac{1}{24}r^3)\,dr,$$

$$= -\tfrac{5}{16}x_2{}^2.$$

Thus
$$x_3 = (2 - \tfrac{5}{16})\,x_2{}^2, \quad \text{......(734)}$$

and for a mixture
$$x_{3,0} = (2 - \tfrac{5}{16})\,x_{2,0}{}^2. \quad \text{......(735)}$$

The next x, $x_{2,1}$, is the sum of three terms. For the first one the β-molecule overlaps each of the α-molecules which do not themselves necessarily overlap. We have $p=2$, $\sigma=2$, and therefore

$$\xi = \tfrac{1}{2}x_{1,1}{}^2.$$

For the second, one of the α-molecules overlaps each of the others and $p=2$, $\sigma=1$,

$$\xi = 2x_{2,0}\,x_{1,1}.$$

For the third each pair of molecules must overlap and $p=3$, $\sigma=2$. When the α-centres are at a distance r, less than D_α, the centre of the β-molecule must lie in a volume

$$K(r) = 2\int_{\frac{1}{2}r}^{D_{\alpha\beta}} \pi(D_{\alpha\beta}{}^2 - x^2)\,dx.$$

Hence
$$\xi = \frac{-N_0^2}{2V^2}\int_0^{D_\alpha} 4\pi r^2 K(r)\,dr,$$

$$= \frac{-4\pi^2 N_0^2 D_\alpha{}^3}{V^2}\,[\tfrac{2}{9}D_{\alpha\beta}{}^3 - \tfrac{1}{8}D_\alpha D_{\alpha\beta}{}^2 + \tfrac{1}{144}D_\alpha{}^3].$$

On substituting in g the second order terms simplify very greatly and we find

$$g = -\left\{\nu_\alpha{}^2 \frac{\frac{2}{3}\pi N_0 D_\alpha{}^3}{V} + \nu_\alpha \nu_\beta \frac{\frac{4}{3}\pi N_0 D_{\alpha\beta}{}^3}{V} + \nu_\beta{}^2 \frac{\frac{2}{3}\pi N_0 D_\beta{}^3}{V}\right\}$$

$$-\tfrac{5}{16}\nu_\alpha{}^3\left(\frac{\frac{2}{3}\pi N_0 D_\alpha{}^3}{V}\right)^2 - 4\pi^2 \nu_\alpha{}^2 \nu_\beta \frac{N_0{}^2 D_\alpha{}^3}{V^2}\{\tfrac{2}{9}D_{\alpha\beta}{}^3 - \tfrac{1}{8}D_{\alpha\beta}{}^2 D_\alpha + \tfrac{1}{144}D_\alpha{}^3\}$$

$$-\tfrac{5}{16}\nu_\beta{}^3\left(\frac{\frac{2}{3}\pi N_0 D_\beta{}^3}{V}\right)^2 - 4\pi^2 \nu_\alpha \nu_\beta{}^2 \frac{N_0{}^2 D_\beta{}^3}{V^2}\{\tfrac{2}{9}D_{\alpha\beta}{}^3 - \tfrac{1}{8}D_{\alpha\beta}{}^2 D_\beta + \tfrac{1}{144}D_\beta{}^3\}. \qquad \text{......(736)}$$

The value of $\log B(T)$ is then, with this g,

$$(N_\alpha + N_\beta)\log V + N_0 g.$$

In terms of N_α and N_β we therefore have

$$\log B(T) = (N_\alpha + N_\beta)\log V - \frac{1}{V}\{N_\alpha{}^2(\tfrac{2}{3}\pi D_\alpha{}^3) + N_\alpha N_\beta(\tfrac{4}{3}\pi D_{\alpha\beta}{}^3) + N_\beta{}^2(\tfrac{2}{3}\pi D_\beta{}^3)\}$$

$$-\frac{1}{V^2}\Big[\tfrac{5}{16}N_\alpha{}^3(\tfrac{2}{3}\pi D_\alpha{}^3)^2 + 4\pi^2 N_\alpha{}^2 N_\beta D_\alpha{}^3\{\tfrac{2}{9}D_{\alpha\beta}{}^3 - \tfrac{1}{8}D_{\alpha\beta}{}^2 D_\alpha + \tfrac{1}{144}D_\alpha{}^3\}$$

$$+\tfrac{5}{16}N_\beta{}^3(\tfrac{2}{3}\pi D_\beta{}^3)^2 + 4\pi^2 N_\alpha N_\beta{}^2 D_\beta{}^3\{\tfrac{2}{9}D_{\alpha\beta}{}^3 - \tfrac{1}{8}D_{\alpha\beta}{}^2 D_\beta + \tfrac{1}{144}D_\beta{}^3\}\Big]. \qquad \text{......(737)}$$

The method of calculating $B(T)$ for a gas of rigid spheres extends at once to any gas, $B(T)$ being given formally by (691). In the integrand $e^{-W/kT}$, W is of course the complete potential energy of the gaseous assembly, which will, speaking roughly, be a sum of terms corresponding to the various groups of molecules engaged in the various elements of $3N$-space in a close encounter. We can make successive approximations to W in any element of $3N$-space by making the groups of molecules which we regard as independent more and more all-inclusive. We shall still obtain for a pure gas in the absence of an external field of force

$$B(T) = V^N e^{Ng}, \qquad \text{......(738)}$$

where g is the same function of the x_r as before, but the x_r are now defined so that $x_r V_r r!/N^{r-1}$ is the integral taken over $3r$-space of a certain quantity which we shall denote by u_r. These quantities u_r must be such that $u_1 = 1$ and that u_r, together with all the contributions from preceding $u_{r'}$ $(r' < r)$, builds up the correct integrand $e^{-E_{12\ldots r}/kT}$, or ϑ_r for short, for the group of r molecules considered by themselves. This u_r is then to be integrated over $3r$-space. It is obvious that

$$\vartheta_2 = u_2 + u_1 u_1, \quad u_2 = \vartheta_2 - 1,$$

$$\vartheta_3 = u_3 + \Sigma\, u_2 u_1 + u_1 u_1 u_1,$$

in which Σ is the sum over all pairs selected from the three molecules. Thus

$$u_3 = \vartheta_3 - \Sigma\,\vartheta_2 + 2.$$

In general we must have $\quad \vartheta_n = \Sigma \Pi\, u_{(i_r)}, \qquad \text{......(739)}$

where the suffixes (i_r) denote any partition of the group of n molecules into (say) s subgroups containing $i_1, i_2, \ldots, i_s$ molecules respectively, and the summation is taken over all possible distinct partitions into any number s of subgroups. If we fix attention on a particular molecule, the nth, we can sum first in (739) all the terms in which the group containing this molecule is the same. This leads to

$$\vartheta_n = u_n + \Sigma\, u_{n-1}\vartheta_1 + \Sigma\, u_{n-2}\vartheta_2 + \ldots, \qquad \ldots\ldots(740)$$

where the u's refer to the various possible subgroups which include the nth molecule and the ϑ's to the complementary subgroups, with $\vartheta_1 = 1$. If, then, we denote by

$$(i_1, i_2, \ldots, i_s)$$

the coefficient of $\vartheta_{i_1}\vartheta_{i_2}\ldots\vartheta_{i_s}$ in u_n we find by equating coefficients in (740), when the nth molecule is in the sth group,

$$0 = (i_1, i_2, \ldots, i_s) + (i_2, i_3, \ldots, i_s) + (i_1, i_3, \ldots, i_s) + \ldots + (i_1, \ldots, i_{s-2}, i_s).$$

In the special case of one group this fails and is replaced by

$$1 = (n).$$

By varying the group in which the nth molecule lies we show at once that, for example,

$$(i_1, i_2, \ldots, i_{s-1}) = (i_2, i_3, \ldots, i_s),$$

so that all the $(s-1)$ group coefficients must be equal. Thence by induction

$$(i_1, i_2, \ldots, i_s) = (-)^{s-1}(s-1)!.$$

Therefore, finally, $$u_n = \Sigma\,(-)^{s-1}(s-1)!\,\vartheta_{i_1}\vartheta_{i_2}\ldots\vartheta_{i_s}, \qquad \ldots\ldots(741)$$

the summation being taken over every distinct partition of the n molecules into subgroups. For example,

$$u_4 = \vartheta_4 - (\Sigma\,\vartheta_3 - \Sigma\,\vartheta_2\vartheta_{2'}) + 2\Sigma\,\vartheta_2 - 6, \qquad \ldots\ldots(742)$$

in which $\Sigma\,\vartheta_3 = e^{-E_{123}/kT} + e^{-E_{234}/kT} + e^{-E_{341}/kT} + e^{-E_{412}/kT}$,

$$\Sigma\,\vartheta_2\vartheta_{2'} = e^{-(E_{12}+E_{34})/kT} + e^{-(E_{13}+E_{42})/kT} + e^{-(E_{14}+E_{23})/kT},$$

$$\Sigma\,\vartheta_2 = e^{-E_{12}/kT} + e^{-E_{23}/kT} + e^{-E_{31}/kT} + e^{-E_{14}/kT} + e^{-E_{24}/kT} + e^{-E_{34}/kT}.$$

When the gas is in a field of force, (741) remains true, but ϑ_1 is no longer 1, and the first term of the series is no longer V^N but $[A(T)]^N$, where

$$A(T) = \iiint e^{-\Omega/kT}\,d\omega,$$

Ω being the potential energy of a single molecule in the external field. It is necessary to redefine x_r using $A(T)$ in place of V. For a mixture we have a V_α and V_β in general different and

$$B(T) = V_\alpha^{N_\alpha} V_\beta^{N_\beta} e^{N_0\sigma},$$

where, in g, $x_{r,s}$ is such that

$$\frac{x_{r,s} V_\alpha{}^r V_\beta{}^s r!\,s!}{N_0{}^{s+r-1}} = \int u_{(r+s)}\, d\omega_{r,s}.$$

The formal connection of the u's with the ϑ's is unchanged by distinctions between molecules of different types, but the ϑ's themselves are changed.

In this extended theory we no longer can say positively *a priori* what the relative orders of x_r and $x_{r,s}$ must be; we can only say that x_r varies as the rth power of the molecular density and that in the general case $x_{r,s}$ occurs in g with a coefficient $(N_\alpha/V_\alpha)^r\,(N_\beta/V_\beta)^s$.

§ **8·4.** *Molecular distribution laws.* The ordinary uniform space distribution law may be derived at once for a gas of any degree of imperfection when there are no external fields, and no long range forces. The frequency ratio for the presence of a selected α molecule in a given element δV is $\delta B(T)/B(T)$, where

$$\delta B(T) = \int_{\delta V} d\omega_\alpha \int_{(V)} \dots \int e^{-W/kT}\, \Pi_\kappa{}'(d\omega_\kappa)^{N_\kappa}, \qquad \dots\dots(743)$$

just one $d\omega_\alpha$ integration over V being omitted. If there are no long range forces or external fields, then W depends only on the relative coordinates of the molecules, and

$$\int_{(V)} \dots \int e^{-W/kT}\, \Pi_\kappa{}'(d\omega_\kappa)^{N_\kappa}$$

must be a constant Q independent of the coordinates of the selected α. Therefore $\delta B(T) = Q\,\delta V$, and obviously $B(T) = QV$. Hence the frequency ratio is $\delta V/V$, and since there are N_α such α's

$$\overline{a_\alpha} = N_\alpha\,\delta V/V, \qquad \dots\dots(744)$$

which is the usual formula. If there are long range forces these *may*, as we shall see, build up the equivalent of external fields from the point of view of any specified element of the assembly. If there are any external fields Ω, then of course (744) is no longer true of the whole assembly but only of an elementary part of it over which the variation of Ω is negligible. In accordance with Gibbs' analysis showing that the laws of thermodynamics include the laws of mechanical equilibrium it is possible to deduce from (701) the distribution law in the field of force and the existence of mechanical equilibrium—the equation $dp = -\nu d\Omega$ being satisfied to the accuracy of the formulae. The investigation, however, is not elegant, and it is better in handling imperfect gases in external fields to apply the laws of statistical mechanics only to *elements* of the gas, and supplement these by the laws of mechanical equilibrium, or general thermodynamic theorems. Since these mechanical laws can be derived from the characteristic function for the whole assembly and lead by themselves to a unique equilibrium state for an assembly at uniform temperature, they must, together with the characteristic functions for the volume elements, be equivalent to the characteristic function for the complete assembly.

By a similar argument we see that the average number $\overline{a_{\alpha\beta}}$ of α,β pairs simultaneously present in selected $d\omega_\alpha$, $d\omega_\beta$ must be given by the formula

$$\overline{a_{\alpha\beta}} = \frac{N_\alpha N_\beta d\omega_\alpha d\omega_\beta}{B(T)} \int_{(V)} \dots \int e^{-W/kT} \Pi_\kappa{}''(d\omega_\kappa)^{N_\kappa}. \quad \text{......(745)}$$

When no effective long range* or external forces are present, the integral in (745) must be a function only of the relative coordinates of the selected $d\omega_\alpha$ and $d\omega_\beta$. We may then write it in the form

$$Qe^{-W_{\alpha\beta}/kT}, \quad \text{......(746)}$$

where $W_{\alpha\beta}$ is defined so that $W_{\alpha\beta}\to 0$ at infinite separation and Q is a constant. *So defined, $W_{\alpha\beta}$ may be called the average potential energy of β in the specified position in the field of α.* It may depend on the average positions of a large number of other molecules and may therefore itself be a function of T. Again,

$$B(T) = \iint Qe^{-W_{\alpha\beta}/kT} d\omega_\alpha d\omega_\beta = QV^2,$$

provided only that $W_{\alpha\beta}\to 0$ rapidly for large separations, as in practice it always does. Thus the average number of pairs is

$$\overline{a_{\alpha\beta}} = N_\alpha N_\beta e^{-W_{\alpha\beta}/kT} d\omega_\alpha d\omega_\beta / V^2. \quad \text{......(747)}$$

After integration with respect to $d\omega_\alpha$ the average number of β's in a selected region near any α anywhere in the assembly is

$$N_\beta e^{-W_{\alpha\beta}/kT} d\omega_\beta / V. \quad \text{......(748)}$$

The factor $1/\sigma_{\alpha\beta}$ is not required in the formulae until we integrate $d\omega_\beta$ over the whole of $v_{\alpha\beta}$ round α. It is not until we do this that we count twice over each α,α pair.

Formulae (747) and (748) are forms of Boltzmann's theorem. It is important to realize, however, the precise meaning of $W_{\alpha\beta}$ for which the theorem is true. In accordance with the discussion in § 6·9 $W_{\alpha\beta}$ may be loosely called the free energy of a β in the field of an α, and it may be that $W_{\alpha\beta} \neq E_{\alpha\beta}$. In the case of short range forces with ternary and higher complexes neglected it is, however, true that $W_{\alpha\beta} = E_{\alpha\beta}$ to the first approximation. For

$$e^{-W_{\alpha\beta}/kT} = e^{-E_{\alpha\beta}/kT} \frac{V^2}{B(T)} \int_{(V)} \dots \int e^{-(W-E_{\alpha\beta})/kT} \Pi_\kappa{}''(d\omega_\kappa)^{N_\kappa},$$

* We shall see that this allows of the consideration of electrostatic forces in ionic media of zero space charge.

and it is obvious that the first approximation to the coefficient of $e^{-E_{\alpha\beta}/kT}$ is unity. Thus for short range forces in assemblies of not too high concentration (747) and (748) take the more familiar forms

$$\overline{a_{\alpha\beta}} = N_\alpha N_\beta e^{-E_{\alpha\beta}/kT} d\omega_\alpha d\omega_\beta / V^2, \qquad \text{......(749)}$$

and

$$N_\beta e^{-E_{\alpha\beta}/kT} d\omega_\beta / V. \qquad \text{......(750)}$$

It is clear, however, that $W_{\alpha\beta} \neq E_{\alpha\beta}$ even for short range forces at higher concentrations as in a liquid, and that the distribution may then be more uniform than is indicated by Boltzmann's law (750).

§ **8·5.** *Generalities on dissociative equilibria in imperfect gases, and the use of the thermodynamic function* Ψ. We can construct a general theory of dissociating imperfect gases as a direct extension of the theory of Chapter V. We will consider for simplicity the theory of § 5·2 which we will here extend to imperfect gases. We there started (for perfect gases) by constructing the generating function (439) which for classical g-functions ($g \equiv \exp$) is equivalent to

$$\Sigma_{a,b} \frac{(\varpi_1{}^1 x_1 z^{\epsilon_1{}^1})^{a_1{}^1} \dots (\varpi_1{}^2 x_2 z^{\epsilon_1{}^2})^{a_1{}^2} \dots (\rho_1 x_1 x_2 z^{\eta_1})^{b_1} \dots}{a_1{}^1! \dots a_1{}^2! \dots b_1! \dots}.$$

This can be partially summed when $\Sigma_r a_r{}^1 = a_1$, $\Sigma_r a_r{}^2 = a_2$, $\Sigma_r b_r = b$, where a_1, a_2, b are fixed to the form

$$\Sigma_{a,b} \frac{\{x_1 V F_1(z)\}^{a_1}}{a_1!} \frac{\{x_2 V F_2(z)\}^{a_2}}{a_2!} \frac{\{x_1 x_2 V F_{12}(z)\}^b}{b!}. \qquad \text{......(751)}$$

When the gases are perfect, (751) sums again, of course, to the familiar form

$$\exp[x_1 V F_1(z) + x_2 V F_2(z) + x_1 x_2 V F_{12}(z)]. \qquad \text{......(752)}$$

When, however, the gases are imperfect the summation to (751) can still be carried out, but the factor $V^{a_1+a_2+b}$ in the terms of (751) must be replaced by $B(z)$, assumed to depend only on a_1, a_2 and b, which can be put in the approximate form

$$V^{a_1+a_2+b}\{1+\alpha_1(a_1,a_2,b)\}^{a_1}\{1+\alpha_2(a_1,a_2,b)\}^{a_2}\{1+\beta(a_1,a_2,b)\}^b. \qquad \text{......(753)}$$

The correcting terms $\alpha_1, \alpha_2, \beta$ are also functions of z and are, of course, the extra terms of (701). They are supposed small, and only their first powers retained. We therefore replace the generating function (752) by Q say, whose approximate form is

$$\Sigma_{a,b} \frac{\{x_1 V F_1(z)(1+\alpha_1)\}^{a_1}}{a_1!} \frac{\{x_2 V F_2(z)(1+\alpha_2)\}^{a_2}}{a_2!} \frac{\{x_1 x_2 V F_{12}(z)(1+\beta)\}^b}{b!}, \qquad \text{......(754)}$$

and operate with Q throughout. Thus, nearly enough,

$$C = \frac{1}{(2\pi i)^3} \iiint \frac{dx_1\, dx_2\, dz}{x_1{}^{X_1+1} x_2{}^{X_2+1} z^{E+1}} Q. \qquad \text{......(755)}$$

The critical point of the integrand is determined as usual by the equations

$$X_1 = x_1 \partial \log Q/\partial x_1, \quad X_2 = x_2 \partial \log Q/\partial x_2, \quad E = z \partial \log Q/\partial z, \quad \text{......(756)}$$

which will have a unique root $\lambda_1, \lambda_2, \vartheta$. The equilibrium state is specified by equations like

$$N_1 = [F_1 \partial \log Q/\partial F_1]_{\lambda_1, \lambda_2, \vartheta}, \quad \text{......(757)}$$

and similar equations. The proof that the assembly obeys the laws of thermodynamics remains valid.

By comparison with the case of a perfect gas the root $\lambda_1, \lambda_2, \vartheta$ is such that the arguments, $\lambda_1 V F_1(\vartheta)$, etc., of Q are all large. We therefore study the asymptotic forms of the function

$$\Sigma_{a,b} \frac{\{A_1(1+\alpha_1)\}^{a_1}}{a_1!} \frac{\{A_2(1+\alpha_2)\}^{a_2}}{a_2!} \frac{\{B(1+\beta)\}^{b}}{b!}$$

for large A, B, the α, β being functions of the a, b which are small compared with unity. To this end we pick out the maximum term of the multiple series by making the first order partial differential coefficients of

$$a_1\left\{\log \frac{A_1(1+\alpha_1)}{a_1} + 1\right\} + a_2\left\{\log \frac{A_2(1+\alpha_2)}{a_2} + 1\right\} + b\left\{\log \frac{B(1+\beta)}{b} + 1\right\}$$

with respect to a_1, a_2, b vanish. It can easily be shown by the usual arguments that the value of the complete multiple series is practically determined by the maximum term, and that the summations may be formally replaced by integrations from $-\infty$ to $+\infty$ on either side of the maximum term. We find after simple reductions that, to the first order in α, β,

$$\log Q \sim A_1\{1+\alpha_1(A_1,A_2,B)\} + A_2\{1+\alpha_2(A_1,A_2,B)\} + B\{1+\beta(A_1,A_2,B)\}. \quad \text{......(758)}$$

We find also that this asymptotic relation can be differentiated. The equilibrium conditions such as (757) are therefore

$$\overline{N_1} = A_1 \partial \log Q/\partial A_1,$$

etc., in which $A_1, \ldots$ are to be replaced by $\lambda_1 V F_1, \ldots$ after differentiation. Thus

$$\overline{N_1} = A_1(1+\alpha_1) + A_1\left(A_1 \frac{\partial \alpha_1}{\partial A_1} + A_2 \frac{\partial \alpha_2}{\partial A_1} + B \frac{\partial \beta}{\partial A_1}\right), \quad \text{......(759)}$$

$$= \lambda_1 V F_1[1+\alpha_1(\overline{N_1},\overline{N_2},\overline{M})]\left[1 + \overline{N_1}\frac{\partial \alpha_1}{\partial \overline{N_1}} + \overline{N_2}\frac{\partial \alpha_2}{\partial \overline{N_1}} + \overline{M}\frac{\partial \beta}{\partial \overline{N_1}}\right] \quad \text{...(760)}$$

to the approximation to which we are working. The law of mass-action is obtained, by eliminating λ_1, λ_2, in the form

$$\log \frac{\overline{N_1}\,\overline{N_2}}{V\overline{M}} = \log \frac{F_1 F_2}{F_{12}} + \left[\frac{\partial}{\partial \overline{N_1}} + \frac{\partial}{\partial \overline{N_2}} - \frac{\partial}{\partial \overline{M}}\right]\{\overline{N_1}\alpha_1 + \overline{N_2}\alpha_2 + \overline{M}\beta\}. \quad \text{......(761)}$$

A similar formula can be given correcting the vapour-pressure equation for imperfection of the vapour phase. We obtain obviously in place of (761)

$$\log\frac{\overline{N}}{V}=\log\frac{F}{\kappa}+\frac{\partial}{\partial\overline{N}}\{\overline{N}\alpha(\overline{N})\}, \qquad \text{......(762)}$$

where κ or $\kappa(T)$ is the partition function for the crystal. At the same time

$$\frac{p}{kT}=\frac{\overline{N}}{V}+\frac{\partial}{\partial V}\{\overline{N}\alpha(\overline{N})\},$$

so that the equilibrium vapour pressure is given by

$$\log p=\log\frac{kTF}{\kappa}+\left[\frac{\partial}{\partial\overline{N}}+\frac{V}{\overline{N}}\frac{\partial}{\partial V}\right]\{\overline{N}\alpha(\overline{N})\}. \qquad \text{......(763)}$$

In the usual first approximation

$$\overline{N}\alpha(\overline{N})=\frac{(\overline{N})^2}{V}2\pi\int_0^\infty\{e^{-E_{\alpha\beta}/kT}-1\}r^2dr,$$

so that
$$\log p=\log\frac{kTF}{\kappa}+\frac{\overline{N}}{V}2\pi\int_0^\infty\{e^{-E_{\alpha\beta}/kT}-1\}r^2dr. \qquad \text{......(764)}$$

The foregoing arguments could easily be generalized, but the general method for imperfect gases in problems of dissociation or of external fields is not particularly convenient. It is of interest to have shown that the complete laws are given as they should be by the properties of C. In practice a more convenient working method can be developed based on the fact that our assemblies are thermodynamic systems, so that thermodynamic theorems may be applied.

Consider an assembly in which the dissociation is fixed, so that it contains N_1, N_2 and M free atoms and molecules. For such an assembly we can at once construct Ψ. By (707) it has the form

$$\Psi/k=N_1\left(\log\frac{VF_1}{N_1}+1\right)+N_2\left(\log\frac{VF_2}{N_2}+1\right)+M\left(\log\frac{VF_{12}}{M}+1\right)$$
$$+N_1\alpha_1+N_2\alpha_2+M\beta, \qquad \text{......(765)}$$

the α, β being the same functions of the N, M as in the foregoing argument. This must hold for any fixed values of N_1, N_2, M, whether or not they happen to agree with the true equilibrium values when dissociation is able to occur. Now suppose that the dissociation, temporarily fixed, is again allowed to proceed in either direction. It is a general thermodynamical principle* that in the final equilibrium state $\Delta\Psi=0$ for any variation of the dissociation (or any other variation consistent with the given temperature and volume).

* See, for example, Planck, *Thermodynamik*, ed. 6, § 151.

Thus the condition of dissociative equilibrium can be at once obtained from the equation

$$\left[\frac{\partial}{\partial N_1}+\frac{\partial}{\partial N_2}-\frac{\partial}{\partial M}\right]\Psi=0, \qquad \ldots\ldots(766)$$

or

$$\log\frac{VF_1}{N_1}+\log\frac{VF_2}{N_2}-\log\frac{VF_{12}}{M}+\left[\frac{\partial}{\partial N_1}+\frac{\partial}{\partial N_2}-\frac{\partial}{\partial M}\right]\{N_1\alpha_1+N_2\alpha_2+M\beta\}=0,$$

which is equation (761) as obtained before.

The foregoing paragraph shows clearly the advantage to be gained in brevity by using thermodynamic arguments at suitable places for these more complicated assemblies. There is the same advantage in discussing imperfect gases in a field of force in this way, as compared with the general statistical method. We apply the general statistical arguments only to construct $\delta\Psi$ for each volume element δV and determine the complete equilibrium by making

$$\Delta(\Sigma\,\delta\Psi)=0$$

for all relevant variations of numbers of particles between the different volume elements.

§ 8·6. *Dissociative equilibria for molecules of finite extension.* The formulae of this chapter take a specially simple form when the constituent systems are of a definite size, without further fields of force. In such a case $E_{\alpha\beta}\to+\infty$ inside $v_{\alpha\beta}$ and is zero elsewhere, and

$$\Psi/k=\Sigma_\alpha N_\alpha\left\{\log\frac{VF_\alpha}{N_\alpha}+1\right\}-\frac{1}{V}\Sigma_{\alpha\beta}\frac{N_\alpha N_\beta v_{\alpha\beta}}{\sigma_{\alpha\beta}}. \qquad \ldots\ldots(767)$$

Equation (767) forms the best starting point for discussing the equilibrium state of an assembly in which one of the systems has a sequence of possible states of definite sizes which differ from state to state.

Suppose there are present in the assembly systems of a certain type in a number of different stationary states of different sizes; these will be initially regarded as distinct systems, specified by different α's. Systems not belonging to this set will be specified by yet other α's. To determine the complete equilibrium state, we construct Ψ as above and vary the N_α among the states of the special systems until $d\Psi=0$ for all such possible variations. A typical variation is to increase N_0 and decrease N_γ by equal amounts. We must therefore have

$$\left[\frac{\partial}{\partial N_0}-\frac{\partial}{\partial N_\gamma}\right]\Psi=0,$$

or, in equilibrium,

$$\frac{\overline{N_\gamma}}{\overline{N_0}}=\frac{F_\gamma\, e^{-\Sigma_\beta \overline{N_\beta} v_{\gamma\beta}/V}}{F_0\, e^{-\Sigma_\beta \overline{N_\beta} v_{0\beta}/V}}. \qquad \ldots\ldots(768)$$

The necessary and sufficient conditions for $d\Psi=0$ for all possible variations of this kind is that (768) should hold for all γ's which specify the different states of the special systems. Now if we were to ignore the differences in size of the various states of the special systems, we should treat them all together and construct a partition function

$$b(T) = \Sigma_\gamma \varpi_\gamma e^{-\epsilon_\gamma/kT}$$

to take account of the distribution of their *internal* energy. Here we temporarily treat each state separately and their separate partition functions are connected by the equations

$$F_\gamma/F_0 = \varpi_\gamma e^{-\epsilon_\gamma/kT}/\varpi_0 e^{-\epsilon_0/kT}. \qquad \text{......(769)}$$

Inserting this ratio in (768) and putting $N = \Sigma_\gamma N_\gamma$ for the total number of special systems, we have*

$$\overline{N_\gamma} = N\varpi_\gamma e^{-\epsilon_\gamma/kT} e^{-\Sigma_\beta \overline{N_\beta} v_{\gamma\beta}/V}/u(T), \qquad \text{......(770)}$$

where $$u(T) = \Sigma_\gamma \varpi_\gamma e^{-\epsilon_\gamma/kT} e^{-\Sigma_\beta \overline{N_\beta} v_{\gamma\beta}/V}. \qquad \text{......(771)}$$

The γ-summation is of course only over all states of the special systems.

In all calculations we have therefore only to replace $b(T)$ and its terms by $u(T)$ and its terms in order to take full account of the excluded volumes. If we use $u(T)$ thus, we may group all the separate states of the special systems together as before. To reconstruct Ψ/k in terms of $u(T)$ we have

$$\frac{VF_\gamma e^{-\Sigma_\beta \overline{N_\beta} v_{\gamma\beta}/V}}{\overline{N_\gamma}} = \frac{Vu(T)\,h(T)}{N} = \frac{VF(T)}{N},$$

where $VF(T)$ is the complete modified partition function. The terms under Σ_α in (767) which belong to the special systems become

$$N\left\{\log\frac{VF(T)}{N} + 1\right\} + \frac{1}{V}\Sigma_\gamma \overline{N_\gamma}(\Sigma_\beta \overline{N_\beta} v_{\gamma\beta}).$$

When every system has been treated in this way we find

$$\Psi/k = \Sigma_r N_r\left\{\log\frac{VF_r(T)}{N_r} + 1\right\} + \frac{1}{V}\Sigma_{\alpha\beta}\frac{\overline{N_\alpha}\,\overline{N_\beta}\,v_{\alpha\beta}}{\sigma_{\alpha\beta}}, \qquad \text{......(772)}$$

where Σ_r is a summation over the separate systems and $\Sigma_{\alpha\beta}$ as before a summation over every pair of states of all the separate systems. It can be verified at once that in the equilibrium state

$$\frac{\partial\Psi}{\partial\overline{N_\gamma}} = 0 \quad (\textit{all } \gamma). \qquad \text{......(773)}$$

Thus the $\overline{N_\gamma}$ of the separate states are only apparent variables in Ψ/k. They do not affect the determination of dissociative equilibrium, which is to be carried out by varying the N_r in (772), without explicit notice of the

* Special cases of (770) and (771) were first given in discussions of high temperature atmospheres by Urey, *Astrophys. J.* vol. 59, p. 1 (1924), and independently by Fermi, *Zeit. f. Physik*, vol. 26, p. 54 (1924).

$\overline{N_\gamma}$. The usual equilibrium laws will be at once obtained in terms of $F_r(T)$ or $\{u(T)\,h(T)\}_r$.

It must be remembered that all the foregoing formulae are necessarily only correct to the first power of $1/V$, so that the exponential correcting factors are largely illusory. At the same time the use of the formulae seems to be justifiable for rough quantitative work, right outside the range in which the corrections are small, in fact for all orders. It will be remembered that states for which the correcting factors are large will *ipso facto* be scarce and therefore affect but little the equilibrium state. A closer examination of this point will, moreover, explain a numerical discrepancy from Urey's work.† Equation (767) may equally well be written in the form

$$\Psi/k = \Sigma_\alpha N_\alpha \left\{ \log \frac{(V - \frac{1}{2}\Sigma_\beta N_\beta v_{\alpha\beta}) F_\alpha}{N_\alpha} + 1 \right\}, \qquad \text{......(774)}$$

which has exactly the same validity so far as terms to the order $1/V$ are concerned. Equation (774), however, is the exact form of Ψ/k for a mixture of gases which obeys exactly van der Waals' equation in the form‡

$$p = kT\Sigma_\alpha \frac{N_\alpha}{V - b_\alpha} \quad (b_\alpha = \tfrac{1}{2}\Sigma_\beta N_\beta v_{\alpha\beta}). \qquad \text{......(775)}$$

If now we work out the equilibrium state by varying the Ψ/k of (774) we find

$$\frac{(V - b_0) F_0}{N_0} e^{-\frac{1}{2}\Sigma_\beta N_\beta \frac{v_{0\beta}}{V - b_\beta}} = \frac{(V - b_\gamma) F_\gamma}{N_\gamma} e^{-\frac{1}{2}\Sigma_\beta N_\beta \frac{v_{\gamma\beta}}{V - b_\beta}}, \qquad \text{......(776)}$$

which to the first order in b/V reduces to (768). If the differences between $V - b_0$ and the $V - b_\gamma$ are ignored, we can cast (776) into Urey's form

$$\frac{F_0}{N_0} e^{-\frac{1}{2}\frac{p}{kT} v_0^*} = \frac{F_\gamma}{N_\gamma} e^{-\frac{1}{2}\frac{p}{kT} v_\gamma^*}, \qquad \text{......(777)}$$

where v_0^* and v_γ^* are mean excluded volumes for the states 0 and γ, and p is the total pressure. This form, as we have seen, is incorrect (by the factor $\frac{1}{2}$ in the exponential) for small values of b/V. This difference is of no importance as the formulae cannot anyhow be exact. What is important is that the approximate agreement of (777) and (768) justifies to some extent the use of the latter for all values of b/V in rough numerical calculations.

It may be noted in conclusion that formulae of exactly the same validity can be obtained for the general assembly in which Ψ is given by (707), so that the excluded volumes are functions of the temperature and may in fact be negative.

§8·7. *Inverse square law forces. Large scale effects.* The only important long range forces ($s \leqslant 4$) which appear to act between actual atoms and molecules are gravitational and electrostatic forces, following the inverse square

† Urey, *loc. cit.* ‡ See equation (848) of Chapter IX, of which this is an obvious generalization.

law ($s=2$). When such forces or external fields are acting, the analysis of § 8·4 must be revised. In equation (743) W must be held to include the long range forces and external fields, and is no longer a function only of the relative coordinates of the systems. Thus

$$\int_{(V)} \cdots \int e^{-W/kT} \Pi_\kappa' (d\omega_\kappa)^{N_\kappa}$$

is no longer necessarily a constant Q independent of the coordinates of the selected α. Instead, we must define a function w by the equation

$$Qe^{-w/kT} = \int_{(V)} \cdots \int e^{-W/kT} \Pi_\kappa' (d\omega_\kappa)^{N_\kappa}, \qquad \text{......(778)}$$

where Q is a constant adjusted so that w takes any convenient value at an assigned point. We then have

$$\overline{a_\alpha} = N_\alpha Q e^{-w/kT} \delta V / B(T), \qquad \text{......(779)}$$

where

$$Q \int_V e^{-w/kT} dV = B(T).$$

Thus (779) reduces at once to

$$\overline{\nu_t} = \overline{\nu_0} e^{-(w_t - w_0)/kT}, \qquad \text{......(780)}$$

where $\overline{\nu_t}$ and $\overline{\nu_0}$ are the average concentrations of the α-molecules in different volume elements δV_t and δV_0. The average potential energy defined by (778) is the potential energy for which alone Boltzmann's theorem (780) is strictly true. The boundary field of an imperfect gas, which is investigated by a special method in § 9·8, is an example of a field such as is considered here —effectively equivalent to an external field though built up from short range forces.

It is desirable to investigate more closely the average potential energy w of this section and the $W_{\alpha\beta}$ of § 8·4, for it does not follow without further investigation that they agree with their values calculated when the rest of the assembly is in its average state. For most purposes of calculation it is almost essential to make this identification owing to the extreme complication of (778). It can easily be seen that the method of § 9·8 for boundary fields is based on this identification.

If we differentiate (778) with respect to the coordinates x, y, z of δV we find

$$\frac{\partial w}{\partial x} \int_{(V)} \cdots \int e^{-W/kT} \Pi_\kappa' (d\omega_\kappa)^{N_\kappa} = \int_{(V)} \cdots \int \frac{\partial W}{\partial x} e^{-W/kT} \Pi_\kappa' (d\omega_\kappa)^{N_\kappa}, \quad \text{...(781)}$$

and two similar equations. In (781) W is of course the total potential energy of the gas phase in any configuration, and therefore $-\partial W/\partial x$, etc. are the force components acting on the selected α-molecule in δV. The function w has therefore been so defined that its partial derivatives are the average values, with α fixed, of the partial derivatives of W, or

$$\frac{\partial w}{\partial x}=\overline{\frac{\partial W}{\partial x}}, \qquad \text{......(782)}$$

and two similar equations. Thus if w is derived by calculating $\partial w/\partial x$ as the average value of the forces acting on the α-molecule—that is, as the force acting on the α-molecule when the rest of the assembly is in its average distribution—such a w is identical with the w of (778). This is the method to be used in § 9·8, which is hereby justified.

In (781) or (782) it is sometimes convenient to distinguish between the parts of $\partial W/\partial x$ which arise from forces of long and short range. If these are distinguished by suffixes l and s so that $W=W_l+W_s$, it may happen that

$$\frac{\partial w_s}{\partial x}=\overline{\frac{\partial W_s}{\partial x}}=0. \qquad \text{......(783)}$$

This will always hold except near a boundary when $W_l=0$, and will continue to hold with the same exception when $W_l\neq 0$ so long as the alterations in $\bar{\nu}_l$ introduced by W_l are insufficient to affect the perfection of the gas laws. When these imperfections begin to matter, W_s must make just such a contribution as to account for the difference between (780), which may be written

$$kT\frac{\partial\bar{\nu}}{\partial s}=-\bar{\nu}\frac{\partial w}{\partial s}, \qquad \text{......(784)}$$

and the laws of hydrostatic equilibrium,

$$\frac{\partial p}{\partial s}=-\bar{\nu}\frac{\partial w_l}{\partial s}. \qquad \text{......(785)}$$

It is an interesting and easy exercise to check the equivalence of (784) and (785) for first order deviations from the perfect gas laws.

Returning to conditions in which (783) is true, we find that

$$\frac{\partial w}{\partial x}\int_{(V)}\cdots\int e^{-W/kT}\,\Pi_\kappa{}'\,(d\omega_\kappa)^{N_\kappa}=\int_{(V)}\cdots\int\frac{\partial W_l}{\partial x}e^{-W/kT}\,\Pi_\kappa{}'\,(d\omega_\kappa)^{N_\kappa}. \qquad \text{...(786)}$$

If we differentiate (786) we find

$$\left[\frac{\partial^2 w}{\partial x^2}-\frac{1}{kT}\left(\frac{\partial w}{\partial x}\right)^2\right]\int_{(V)}\cdots\int e^{-W/kT}\,\Pi_\kappa{}'\,(d\omega_\kappa)^{N_\kappa}$$

$$=\int_{(V)}\cdots\int\left[\frac{\partial^2 W_l}{\partial x^2}-\frac{1}{kT}\left(\frac{\partial W_l}{\partial x}\right)^2\right]e^{-W/kT}\,\Pi_\kappa{}'\,(d\omega_\kappa)^{N_\kappa}, \qquad \text{......(787)}$$

and two similar expressions.* By addition

$$\left[\nabla^2 w - \frac{1}{kT}\Sigma\left(\frac{\partial w}{\partial x}\right)^2\right]\int_{(V)}\cdots\int e^{-W/kT}\,\Pi_\kappa{}'\,(d\omega_\kappa)^{N_\kappa}$$

$$=\int_{(V)}\cdots\int\left[\nabla^2 W_l - \frac{1}{kT}\Sigma\left(\frac{\partial W_l}{\partial x}\right)^2\right]e^{-W/kT}\,\Pi_\kappa{}'\,(d\omega_\kappa)^{N_\kappa}. \qquad \text{......(788)}$$

In (788) we shall take W_l to be a potential energy due to inverse square law forces, and obeying Poisson's equation

$$\nabla^2 W_l = \mu\nu, \qquad \text{......(789)}$$

where ν is the smoothed local concentration of systems in δV in any configuration. The value of μ will depend on the precise mixture of gravitational and electrostatic forces concerned. We find therefore that w satisfies the equation

$$\nabla^2 w - \mu\overline{\nu_\alpha} = -\frac{1}{kT}\Sigma\left\{\overline{\left(\frac{\partial W_l}{\partial x}\right)^2} - \left(\frac{\partial w}{\partial x}\right)^2\right\}. \qquad \text{......(790)}$$

The right-hand side is the mean square fluctuation of the resultant force on the system in δV, divided by $(-kT)$. The mean density $\overline{\nu_\alpha}$ is an average of ν for all configurations in which the α-molecule is fixed at the point concerned (but not counted in ν). The suffix is inserted to distinguish $\overline{\nu_\alpha}$ from $\overline{\nu}$ of equation (780), which is an average for all positions of the α-molecule as well as of the others. In practice, we replace $\overline{\nu_\alpha}$ by $\overline{\nu}$ and omit the fluctuation terms, so that (790) reduces to

$$\nabla^2 w - \mu\overline{\nu} = 0. \qquad \text{......(791)}$$

The equilibrium state of the assembly may then be calculated by the combined use of Boltzmann's and Poisson's equations, (780) and (791), a fertile procedure in frequent use.

There are, however, three points which require critical examination: the smoothing employed in (789), the replacement of $\overline{\nu_\alpha}$ by $\overline{\nu}$, and the neglect of the fluctuation terms on the right-hand side of (790).

In dealing with point charges† smoothing of some sort is essential to mathematical simplicity; if the charges are spread over finite volumes, each defined in position by the coordinates of its centre, the "smoothing" is automatic and inevitable. Thus for point charges we define the smoothed density at any point by some such formula as

$$\nu(x) = \Sigma_\alpha \epsilon_\alpha f(x - x_\alpha), \qquad \text{......(792)}$$

* It is assumed that not only is $\overline{\partial W_s/\partial x} = 0$ but also $\overline{\partial W_s/\partial x . \partial W_l/\partial x} = 0$.

† It is convenient to write in the language of electrostatics, but other fields of force are not excluded.

where x is short for x, y, z, ϵ_α is the charge on an α-molecule, $f(x)$ diminishes rapidly with increasing distance from the origin, and

$$\int_{(V)} f(x)\,d\omega = 1. \qquad \text{......(793)}$$

In the case of charges spread over finite volumes, (792) necessarily holds good, and $\epsilon_\alpha f(x)$ is the actual density in a single particle with its centre at the origin. The possibility of interpenetration is not excluded. The corresponding smoothed function $F^*(x)$, derived from any function of position $F(x)$, is

$$F^*(x) = \int_{(V)} F(x')f(x-x')\,d\omega'. \qquad \text{......(794)}$$

If $F(x_1, x_2, \ldots)$ is a function of many points, the smoothing is to be carried out for all of them, so that

$$F^*(x_1,x_2,\ldots) = \int_{(V)} \ldots \int F(x_1', x_2', \ldots) f(x_1 - x_1') f(x_2 - x_2') \ldots d\omega_1' d\omega_2' \ldots. \qquad \text{......(795)}$$

In the case of charged particles of finite size, this smoothing is again automatic, provided that F is linear in each of the charges concerned—e.g. a potential, or a mutual potential energy.

Equation (789) requires that W_l shall be equal to W^*, the smoothed electrostatic potential energy. It appears to be simplest to replace W by W^* throughout the argument. Thus w would be defined by

$$Qe^{-w/kT} = \int_{(V)} \ldots \int e^{-W^*/kT}\, \Pi_\kappa' (d\omega_\kappa)^{N_\kappa} \qquad \text{......(796)}$$

(compare (778)); and in (780)

$$\bar{\nu}(x) = \Sigma_\alpha N_\alpha \left[\int_{(V)} \ldots \int e^{-W^*/kT}\, \Pi_\kappa' (d\omega_\kappa)^{N_\kappa}\right]_{x_\alpha = x} \Big/ \int_{(V)} \ldots \int e^{-W^*/kT}\, \Pi_\kappa (d\omega_\kappa)^{N_\kappa}. \qquad \text{......(797)}$$

This is not our ordinary equation for the average density; nor is it precisely the smoothed average density, nor even the average smoothed density. In all probability, however, it does not differ considerably from any of these. The difficulties introduced by smoothing are mostly of this kind, and are not likely to be important.

On the other hand, thorough smoothing, when it can be employed, gives the greatest assistance in answering the other two points of criticism. Consider first the relation between $\overline{\nu_\alpha}$ and $\bar{\nu}$. Their difference depends on the difference between the value of ν at the position of the α-molecule and its more normal values. Now the α-molecule induces a considerable excess or deficiency of charge in its immediate neighbourhood; but its effect at some little distance is negligible. If the smoothing function $f(x)$ is appreciable only at very small distances, the effect of the α-molecule's field is an important feature in ν. If, however, $f(x)$ is appreciable also in a considerable volume in which the field of the α-molecule is negligible, the local excess or deficiency has comparatively little effect upon ν. In other words, slight smoothing retains a large part of the difference between $\overline{\nu_\alpha}$ and $\overline{\nu}$, but a more thorough smoothing decreases the divergence.

So long, then, as we are concerned only with smoothed space charges (as in the problem, mentioned below, of a gravitating gas and in the theory of an electron atmosphere, developed in Chapter XI), it is permissible and advantageous to smooth thoroughly, with a function which is effective over a volume large compared with molecular dimensions; $\overline{\nu_\alpha}$ may then be replaced by $\overline{\nu}$. When, however, our whole concern is with concentrations on a molecular scale, as in formula (746) and the Debye-Hückel theory described below, any but the slightest smoothing is impossible, and $\overline{\nu_\alpha}$ and $\overline{\nu}$ may be widely different.

For the validity of the neglect of the fluctuation terms no general conditions have yet been obtained. At present it seems possible to proceed only by verification *a posteriori*.

As an example, consider the equilibrium state of an isothermal gravitating gas of which each small element is effectively perfect.* This is of course a special case of the equilibria of such gaseous masses handled by Emden.† There is no explicit solution for w or for ρ, the mass density at any point, but it is found that if r is the distance from the centre of the gravitating mass

$$\rho \sim \frac{2}{r^2}\left(\frac{kT}{m}\right)\frac{1}{4\pi G},$$

where G is the constant of gravitation. It follows that to the same approximation

$$w = 2kT \log r + const.$$

If we then compare one of the ignored terms $(\partial w/\partial r)^2/kT$ with a term retained such as $\partial^2 w/\partial r^2$, we find that the numerical ratio is 2. If therefore the fluctuation in the resultant force at any point is small compared with the force itself, the neglect of all the terms on the right in (790) is at once justified.

A similar example is the electron atmosphere in equilibrium with a metal at high temperature. If the form of the atmosphere is effectively that of the

* In order to discuss such an assembly completely from the point of view of statistical mechanics it is necessary to idealize the problem so that the mass of gas is contained in a reflecting enclosure so large that molecular impacts on the walls do not effectively alter the position of the centre of gravity of the mass of gas or its total momentum which must be fixed by the conditions of the problem. This is not strictly realizable. Such conditions can be formally accounted for by additional selector variables both for momenta and positional coordinates.

† Emden, *Gaskugeln*; see also for this, Milne, *Trans. Camb. Phil. Soc.* vol. 22, p. 483 (1923).

gap between the parallel plates of an infinite condenser, the problem admits of the exact solution (see (1069))

$$e^{w/2kT} = A \cos \frac{Bx}{\sqrt{(2kT)}},$$

where A and B are constants of which B depends on the electron density at a standard potential. Here again the numerical ratio of the ignored term $(\partial w/\partial x)^2/kT$ to the term retained $\partial^2 w/\partial x^2$ is less than 2, and the same conclusions can be drawn if the same hypothesis is admitted.

The problem of the ignorability of the fluctuation term can be formulated as follows: Let $-\partial\Omega_r/\partial x$ be the x-component of the force at x due to a single molecule in the rth cell of the assembly. Then

$$\frac{\partial W_l}{\partial x} = \Sigma_r a_r \frac{\partial \Omega_r}{\partial x}, \quad \frac{\partial w}{\partial x} = \overline{\frac{\partial W_l}{\partial x}} = \Sigma_r \overline{a_r} \frac{\partial \Omega_r}{\partial x},$$

$$\overline{\left(\frac{\partial W_l}{\partial x}\right)^2} = \Sigma_r \overline{a_r^2} \left(\frac{\partial \Omega_r}{\partial x}\right)^2 + 2\Sigma_{r,s} \overline{a_r a_s} \frac{\partial \Omega_r}{\partial x} \frac{\partial \Omega_s}{\partial x},$$

$$\overline{\left(\frac{\partial W_l}{\partial x}\right)^2} - \left(\frac{\partial w}{\partial x}\right)^2 = \Sigma_r \{\overline{a_r^2} - (\overline{a_r})^2\} \left(\frac{\partial \Omega_r}{\partial x}\right)^2 + 2\Sigma_{r,s} \{\overline{a_r a_s} - \overline{a_r}\,\overline{a_s}\} \frac{\partial \Omega_r}{\partial x} \frac{\partial \Omega_s}{\partial x}.$$

A complete solution requires a knowledge of $\overline{a_r^2} - (\overline{a_r})^2$ and $\overline{a_r a_s} - \overline{a_r}\,\overline{a_s}$ for these complicated assemblies. On general grounds, however, it is quite certain that these quantities will be all positive and very small compared with $(\overline{a_r})^2$ and $\overline{a_r}\,\overline{a_s}$ *provided that the cells need not be taken too small*, that is, provided that smoothing is sufficiently macroscopic. It must be remembered that we are dealing with smoothed functions. The appropriate cell is that volume in which the smoothing function is effective. A thorough smoothing is again what we require, in order to make the cells conveniently large. In evaluating long range large scale effects this is possible, and the above condition can be satisfied. It is not possible to conclude at once that

$$\overline{\left(\frac{\partial W_l}{\partial x}\right)^2} - \left(\frac{\partial w}{\partial x}\right)^2 \ll \left(\frac{\partial w}{\partial x}\right)^2,$$

for the terms $\partial\Omega_r/\partial x$ are not all positive. Their signs depend on the sign of $x - x'$. We can, however, conclude that

$$\overline{\left(\frac{\partial W_l}{\partial x}\right)^2} - \left(\frac{\partial w}{\partial x}\right)^2 \ll \left(\frac{\partial w'}{\partial x}\right)^2,$$

where $\partial w'/\partial x$ is the force component at x, y, z due to all the matter in the assembly on that side of the plane $\xi = x$ which gives the greater value of $\partial w'/\partial x$. It is therefore sufficient to verify *a posteriori* that $\partial^2 w/\partial x^2$ and similar terms are of the same order as $(\partial w'/\partial x)^2/kT$ and similar terms. The neglect of the fluctuation terms is then in general justified.

§ 8·8. *Contributions by intermolecular forces to the free energy of Helmholtz, or the characteristic function.* A similar investigation to that of the last section can be attempted for the local field $W_{\alpha\beta}$ of formula (746). The real purpose of such investigations in equilibrium theory is merely to find a roundabout way of calculating $B(T)$ when a direct attack seems hopeless; for a knowledge of $B(T)$ determines all the equilibrium properties of the assembly. It therefore seems desirable to start with a general study of the relationship of quantities such as $W_{\alpha\beta}$ to $B(T)$ and in particular of how to calculate the corresponding part of the free energy or characteristic function.

The energy $W_{\alpha\beta}$ is defined by the equation

$$e^{-W_{\alpha\beta}/kT}\int_{(V)}\ldots\int \Pi_\kappa''\,(d\omega_\kappa)^{N_\kappa}=\int_{(V)}\ldots\int e^{-W/kT}\,\Pi_\kappa''\,(d\omega_\kappa)^{N_\kappa}. \quad \ldots\ldots(798)$$

On differentiating this equation with respect to x_β we find that

$$\frac{\partial W_{\alpha\beta}}{\partial x_\beta}\int_{(V)}\ldots\int e^{-W/kT}\,\Pi_\kappa''\,(d\omega_\kappa)^{N_\kappa}=\int_{(V)}\ldots\int \frac{\partial W}{\partial x_\beta}e^{-W/kT}\,\Pi_\kappa''\,(d\omega_\kappa)^{N_\kappa}.$$

Thus just as for w, equation (782),

$$\frac{\partial W_{\alpha\beta}}{\partial x_\beta}=\overline{\frac{\partial W}{\partial x_\beta}}. \quad \ldots\ldots(799)$$

This equation states that *the average value of the force acting on the β-molecule at the specified distance from the α-molecule may be derived by differentiating $W_{\alpha\beta}$ defined by* (798).

Let us now suppose that the fields of force which give rise to w or $W_{\alpha\beta}$ are gradually built up from zero by differential additions to the various force centres, *the assembly being at a constant temperature (but not necessarily constant volume) during each successive stage and in the equilibrium state appropriate to the force centres already present.* The only external work done in the rearrangement of the assembly into its equilibrium state at the given temperature in between the "stages" is the work done by the pressure if there is any change of volume. This is of course an ideal process requiring the treatment of individual molecules, but it is a conceivable isothermal reversible thermodynamic process, and the work done on the assembly in the process, including both work of "charging up" and work done by the external pressure, must on general principles be the increase in Helmholtz's free energy due to the establishment of the intermolecular fields. It is analogous to the familiar method of calculating the contribution of such forces to the potential energy, when we suppose that the fields of force are gradually built up from zero, *the various systems of the assembly being fixed in their average final positions.* This is the origin of the familiar $\frac{1}{2}\int\rho W\,dV$ of the theory of attractions. It is sometimes convenient to keep the volume fixed during the charging process; in that case the work of charging up is the total work and equal to the change of Helmholtz's free energy. If on the other hand the pressure is kept fixed and the work done by the external pressure omitted, the work of charging up is equal to the increase in Gibbs' free energy G. A direct statistical proof may be given as follows when the volume is kept constant.

Suppose that every intermolecular energy term is at a fraction σ of its final value. Then the partition function is $B_\sigma(T)$, say, given by

$$B_\sigma(T) = \int_{(V)} \cdots \int e^{-\sigma W/kT} \Pi_\kappa (d\omega_\kappa)^{N_\kappa},$$

and the energy of this force system in its equilibrium state

$$kT^2 \frac{\partial}{\partial T} \log B_\sigma(T) = \frac{\sigma \int_{(V)} \cdots \int W e^{-\sigma W/kT} \Pi_\kappa (d\omega_\kappa)^{N_\kappa}}{\int_{(V)} \cdots \int e^{-\sigma W/kT} \Pi_\kappa (d\omega_\kappa)^{N_\kappa}}.$$

Keeping the distribution laws unaltered and increasing each energy term from σ to $\sigma + d\sigma$ requires an increase in the energy of the force system equal to

$$\frac{d\sigma \int_{(V)} \cdots \int W e^{-\sigma W/kT} \Pi_\kappa (d\omega_\kappa)^{N_\kappa}}{\int_{(V)} \cdots \int e^{-\sigma W/kT} \Pi_\kappa (d\omega_\kappa)^{N_\kappa}}.$$

This increase of energy must be the work necessary to strengthen the force centres from a fraction σ to $\sigma + d\sigma$ of their final values, the assembly being in the equilibrium state corresponding to the fraction σ already present. The total work required to build up the final force system by a reversible isothermal process is therefore

$$\int_0^1 d\sigma \frac{\int_{(V)} \cdots \int W e^{-\sigma W/kT} \Pi_\kappa (d\omega_\kappa)^{N_\kappa}}{\int_{(V)} \cdots \int e^{-\sigma W/kT} \Pi_\kappa (d\omega_\kappa)^{N_\kappa}} = -kT \int_0^1 \frac{d}{d\sigma} \{\log B_\sigma(T)\} d\sigma,$$

$$= -kT \log\{B(T)/V^{\Sigma_\kappa N_\kappa}\}. \quad \ldots(800)$$

But this expression is just $-kT$ times the increase in Ψ/k due to the establishment of the intermolecular fields, and $-T\Psi = F$.

The theorem can obviously be extended to cover the case in which new forces of a given type are established, e.g. electrostatic. We then introduce

$$B_\tau(T) = \int_{(V)} \cdots \int e^{-(W+\tau W_\epsilon)/kT} \Pi_\kappa (d\omega_\kappa)^{N_\kappa}, \quad \ldots\ldots(800\cdot1)$$

and $-kT \log\{B_1(T)/B_0(T)\} = \Delta F = F_\epsilon$ is the work necessary to charge up the ions in a reversible isothermal process, the sizes and other intermolecular forces of the ions retaining their standard values.

The restriction to constant volume can be removed by allowing the complete volume change to take place when the charging process has been completed. The work done by the pressure on the assembly is then $-\int_{V_1}^{V_2} p\,dV$ which is equal to $-kT\log\{B_1(T,V_2)/B_1(T,V_1)\}$. On adding this to the former work term we see that $-kT\log\{B_1(T,V_2)/B_0(T,V_1)\}$ is the total work done on the assembly, so that $-T\Delta\Psi = \Delta F$.

This result can be applied to an electrolyte in the following form. Suppose the electrolyte (volume V) contains N_κ ions of type κ ($\kappa = 1,\ldots,t$) whose final charge is ϵ_κ and charge at any stage of the charging process η_κ. Let $\overline{\psi_\kappa}(\eta_1,\ldots,\eta_t)$ be the average potential at the centre of an ion of type κ due to the charges on all the ions including its own charge. This definition requires us to regard an ion as an electrified sphere of diameter a. This is sufficiently general for our purposes. Then by the preceding theorem

$$dF_\epsilon = \sum_{\kappa=1}^{t} N_\kappa \overline{\psi_\kappa}(\eta_1,\ldots,\eta_t)\,d\eta_\kappa, \qquad \ldots\ldots(801)$$

and

$$F_\epsilon = \sum_{\kappa=1}^{t} N_\kappa \int_0^{\epsilon_\kappa} \overline{\psi_\kappa}(\eta_1,\ldots,\eta_t)\,d\eta_\kappa. \qquad \ldots\ldots(802)$$

This charging process can presumably, since F_ϵ must be a definite function of the final charges, be carried out in any manner, subject to the condition,

$$\sum_{\kappa=1}^{t} N_\kappa \eta_\kappa = 0,$$

that the total charge in the electrolyte is zero. Subject to this condition dF_ϵ must be a perfect differential, from which it follows, the restrictive condition making no difference, that*

$$N_\kappa \frac{\partial \overline{\psi_\kappa}}{\partial \eta_i} = N_i \frac{\partial \overline{\psi_i}}{\partial \eta_\kappa}. \qquad \ldots\ldots(803)$$

Any proposed method of calculating F_ϵ by means of approximate calculations of $\overline{\psi_\kappa}$ must conform to (803).

§**8·81**. *The theory of Debye and Hückel for strong electrolytes. General foundations.* We have seen in § 8·8 that if we can calculate the $\overline{\psi_\kappa}$ we can calculate F_ϵ and thereby obtain the necessary thermodynamic potential to give us all the equilibrium properties of a strong electrolyte. A method of carrying through this calculation has been proposed by Debye and Hückel.†

* Onsager, *Chemical Reviews*, vol. 13, p. 73 (1933).

† Debye and Hückel, *Physikal. Zeit.* vol. 24, pp. 185, 305 (1923). For later references, see Chapter XIII. See also Falkenhagen, *Electrolytes*, Oxford (1934).

It is empirically very successful and has been developed and applied in great detail. We shall therefore study here its position as a deduction from general statistical theory.

In equations (746)–(748) we have shown how to define an energy $W_{\alpha\beta}$ such that the average number of β-systems in a selected volume element near any α-system is

$$N_\beta e^{-W_{\alpha\beta}/kT} d\omega_\alpha / V,$$

so that the average density of the β-systems is

$$N_\beta e^{-W_{\alpha\beta}/kT}/V;$$

$W_{\alpha\beta}$ is a function of r. It necessarily satisfies the equation

$$W_{\alpha\beta} = W_{\beta\alpha}. \qquad \text{......(804)}$$

The first step in the method of Debye and Hückel is to define a function $\overline{\psi_\kappa}(\epsilon_1, \ldots, \epsilon_t, r)$, for short $\overline{\psi_\kappa}(r)$, the average electrostatic potential due to all the other charges, at a distance r from any ion of type κ. The functions $\overline{\psi_\kappa}$ used above are now replaced by $\overline{\psi_\kappa}(\epsilon_1, \ldots, \epsilon_t, 0)$. The definition is

$$\overline{\psi_\kappa}(r) \int_{(V)} \cdots \int e^{-W/kT} \Pi_\kappa' (d\omega_\kappa)^{N_\kappa} = \int_{(V)} \cdots \int \psi_\kappa(r) e^{-W/kT} \Pi_\kappa' (d\omega_\kappa)^{N_\kappa}. \qquad \text{......(805)}$$

Since r is not the coordinate of a system, W is independent of r and therefore

$$\nabla^2 \overline{\psi_\kappa}(r) \int_{(V)} \cdots \int e^{-W/kT} \Pi_\kappa' (d\omega_\kappa)^{N_\kappa} = \int_{(V)} \cdots \int \nabla^2 \psi_\kappa(r) e^{-W/kT} \Pi_\kappa' (d\omega_\kappa)^{N_\kappa},$$

$$= -\frac{4\pi}{D} \int_{(V)} \cdots \int \rho(r) e^{-W/kT} \Pi_\kappa' (d\omega_\kappa)^{N_\kappa}, \qquad \text{......(806)}$$

where D is the dielectric constant of the medium and $\rho(r)$ is the space charge density at r in the given configuration. The precise meaning of D in equation (806) raises difficulties which we shall not examine here. For the moment we may assume D to be the dielectric constant of some medium independent of the distribution laws of the assembly. A deeper discussion is given in § 13·61 and saturation effects are considered in § 12·6. The function $\overline{\psi_\kappa}(r)$ therefore satisfies Poisson's equation

$$\nabla^2 \overline{\psi_\kappa}(r) = -4\pi \overline{\rho_\kappa}/D, \qquad \text{......(806·1)}$$

where

$$\overline{\rho_\kappa} \int_{(V)} \cdots \int e^{-W/kT} \Pi_\kappa' (d\omega_\kappa)^{N_\kappa} = \int_{(V)} \cdots \int \rho(r) e^{-W/kT} \Pi_\kappa' (d\omega_\kappa)^{N_\kappa}. \qquad \text{......(806·2)}$$

Thus $\overline{\rho_\kappa}$ is the average charge density at a distance r from an ion of type κ. Therefore $\overline{\rho_\kappa}$ can be calculated in terms of the $W_{\alpha\beta}$, by means of the equation

$$\overline{\rho_\kappa} = \sum_{\beta=1}^{t} \frac{N_\beta}{V} \epsilon_\beta e^{-W_{\kappa\beta}/kT}. \qquad \text{......(807)}$$

Poisson's equation (806·1) therefore reduces to

$$\nabla^2\overline{\psi_\kappa}+\frac{4\pi}{D}\sum_{\beta=1}^{t}\frac{N_\beta}{V}\epsilon_\beta e^{-W_{\kappa\beta}/kT}=0. \qquad \text{......(807·1)}$$

The essential approximation in Debye and Hückel's theory is now to put

$$W_{\kappa\beta}=\epsilon_\beta\overline{\psi_\kappa}, \qquad \text{......(808)}$$

which yields for $\overline{\psi_\kappa}$ the differential equation

$$\nabla^2\overline{\psi_\kappa}+\frac{4\pi}{D}\sum_{\beta=1}^{t}\frac{N_\beta}{V}\epsilon_\beta e^{-\epsilon_\beta\overline{\psi_\kappa}/kT}=0, \qquad \text{......(808·1)}$$

an equation which has the same form for all κ's. The approximation (808) requires it to be true that

$$\epsilon_\beta\overline{\psi_\kappa}=\epsilon_\kappa\overline{\psi_\beta}. \qquad \text{......(809)}$$

It asserts also that the electrostatic forces are the only forces controlling the distribution of β-systems near κ-systems or rather that up to absolute contact the distribution is uniform when the charges are zero. This is of course untrue as the short range forces are thus neglected, but (808) should be a good approximation at moderate or large distances, so far as the neglect of short range forces is concerned. Besides (809) the $\overline{\psi_\kappa}$ must satisfy the conditions (803).

By this method of approach it is difficult to see what range of validity may be expected for the formulae resulting from (808·1). Certain upper limits for their validity have been laid down in (803) and (809). So soon as these conditions are infringed, the results are illusory if used to such an order of accuracy that the infringing terms are significant. But these conditions are only necessary and the failure might occur earlier, due to the failure of the approximation (808).

One may now attempt to fill this gap by an investigation similar to that of § 8·7. The analysis will not be given in detail since the discussion cannot be made sufficiently rigorous. One may conclude that $W_{\alpha\beta}$, like w, satisfies Poisson's equation, at least on the average for the neighbourhoods of a number of ions if not near a single ion, provided that we may ignore fluctuations and short range forces. The volume elements near any one ion to which this process must be applied are, however, now small on the molecular scale and it is impossible to conclude with certainty by the smoothing arguments used above that the fluctuations are negligible when $(\partial W_{\alpha\beta}/\partial x)^2/kT$ is of the same order as the terms retained. The utmost it is safe to conclude is that $W_{\alpha\beta}$ satisfies Poisson's equation *so long as*

$$\frac{1}{kT}\left(\frac{\partial W_{\alpha\beta}}{\partial x}\right)^2$$

is itself small compared with terms retained in the equation and so long as the effect of short range forces on the electrostatic terms is negligible. This conclusion is made slightly less restrictive by remembering that any non-fluctuating part of $W_{\alpha\beta}$ may be removed before applying this test.

§8·82. *The theory of Debye and Hückel. Explicit formulae.* In illustration of the foregoing argument we shall now derive Debye and Hückel's explicit results for solutions so dilute that the size and shapes of the ions are irrelevant. This investigation and others similar will be essential to us in Chapters XI, XIII and XIV.

Equation (808·1) is soluble in principle as it stands, but requires elaborate treatment, and an explicit recognition of the fact that ions have sizes.* The approximation made by Debye and Hückel which renders (808·1) soluble in finite terms is to assume that for all the important values of r $\epsilon_\beta\overline{\psi_\alpha}/kT$ is small. Since $\overline{\psi_\kappa}$ is spherically symmetrical, the equation then becomes

$$\frac{1}{r^2}\frac{d}{dr}\left(r^2\frac{d\overline{\psi_\alpha}}{dr}\right)=\kappa^2\overline{\psi_\alpha},\quad \kappa^2=\frac{4\pi\epsilon^2}{DkT}\sum_{\beta=1}^{t} z_\beta{}^2\frac{N_\beta}{V}. \qquad \text{......(810)}$$

We have here replaced ϵ_β by $z_\beta\epsilon$ where z_β is the valency, positive or negative, of the ion. The leading term in the expansion of the exponentials in (808·1),

$$-\frac{4\pi\epsilon}{D}\sum_{\beta=1}^{t} z_\beta\frac{N_\beta}{V},$$

vanishes when the average space charge is zero, a condition which may usually be assumed to be satisfied in applications. If it is not fulfilled, the proper solution of $(\nabla^2-\kappa^2)\overline{\psi_\alpha}=const.$ must be added to $\overline{\psi_\alpha}$ so as to satisfy the boundary conditions. The solution of (810) which satisfies the obvious conditions at $r\to 0$ and $r\to\infty$ is

$$\overline{\psi_\alpha}=\frac{z_\alpha\epsilon}{D}\frac{1}{r}e^{-\kappa r}\quad(\kappa>0). \qquad \text{......(811)}$$

The corresponding value of $W_{\alpha\beta}$ is

$$W_{\alpha\beta}=\frac{z_\alpha z_\beta\epsilon^2}{D}\frac{1}{r}e^{-\kappa r}. \qquad \text{......(811·1)}$$

In considering the legitimacy of ignoring a fluctuation term we remember that the portion $z_\alpha z_\beta\epsilon^2/Dr$ of $W_{\alpha\beta}$ is non-fluctuating, being due to the α-ion itself. We may therefore be content to consider

$$W_{\alpha\beta}'=\frac{z_\alpha z_\beta\epsilon^2}{D}\frac{1}{r}(e^{-\kappa r}-1)$$

and to examine whether $(\partial W_{\alpha\beta}'/\partial r)^2/kT$ is or is not small compared with a

* For a more exact treatment, see Chapter XIII.

term retained such as $2(\partial W_{\alpha\beta}'/\partial r)/r$ or $\partial^2 W_{\alpha\beta}'/\partial r^2$ or $\kappa^2 W_{\alpha\beta}'$. The comparison with the first of these is simple. We find a ratio

$$\frac{z_\alpha z_\beta \epsilon^2 \kappa}{2DkT}\left[\frac{1-(1+\kappa r)\,e^{-\kappa r}}{\kappa r}\right].$$

Since for all positive x $\qquad e^x \leqslant 1+x+xe^x,$

it follows that the term in [ ] never exceeds unity, and the omission of the fluctuations may be expected to be legitimate if

$$\frac{z_\alpha z_\beta \epsilon^2 \kappa}{2DkT} \ll 1. \qquad \text{......(811·2)}$$

This condition is equivalent to $W_{\alpha\beta}'/kT$ being small.

We now consider the other necessary tests of the legitimacy of the theory. The value of $\overline{\psi_\alpha}$ obtained does not depend on molecular sizes. To the extent therefore to which it may be used, molecular sizes are unimportant and the identification of $\epsilon_\beta \overline{\psi_\alpha}$ and $W_{\alpha\beta}$ should be legitimate. The values of $\overline{\psi_\alpha}$ moreover satisfy the condition (809) that $\epsilon_\beta \overline{\psi_\alpha} = \epsilon_\alpha \overline{\psi_\beta}$.

Before we apply the test (803) we must extract from $\overline{\psi_\alpha}$ the field of the α-ion itself. The self-energy or energy of charging the ion against its own charge can only be introduced when the ion has a definite non-zero size. It is moreover unimportant to F_ϵ so long as the dielectric constant of the solvent is not altered at constant temperature. The potential at the centre of an α-ion due to the charges of all the other ions is therefore to be derived from

$$\frac{z_\alpha \epsilon}{D}\frac{1}{r}(e^{-\kappa r}-1)$$

and is therefore $\qquad -z_\alpha \epsilon \kappa / D. \qquad$ ......(811·3)

Since $\partial\kappa^2/\partial\epsilon_\alpha$ is proportional to $N_\alpha \epsilon_\alpha$ it is easily verified that the condition (803) is satisfied, and therefore that the proposed solution provides a self-consistent set of $\overline{\psi_\alpha}$ for which the essential values at the centres of the α-ions are given by (811·3).

We can now calculate the electrostatic free energy by integrating (801) or (802) in any convenient manner. If all the charges are reduced to the fraction λ of their final values, then the potentials (811·3) are reduced to the fraction λ^2 since κ varies as ϵ. We find therefore that

$$dF_\epsilon' = \left(-\frac{\kappa\epsilon^2}{D}\sum_{\alpha=1}^{t} N_\alpha z_\alpha^2\right)\lambda^2 d\lambda.$$

The prime to F_ϵ' denotes that the self-energy of the various ions has been omitted from the calculation. Therefore

$$F_\epsilon' = -\frac{\kappa\epsilon^2}{3D}\left(\sum_{\alpha=1}^{t} N_\alpha z_\alpha^2\right), \qquad \text{......(812)}$$

which is the standard result. It satisfies all the tests of consistency and should be valid at least over such a range of concentration that

$$\frac{F_\epsilon'}{\Sigma_\alpha N_\alpha} \ll kT.$$

The standard result (812) can hardly be reached by any argument less deep than the foregoing. At this stage any variation of D is excluded, but it will be observed that all questions of temperature variation of D are irrelevant in forming the contribution to F_ϵ'. But in deducing E_ϵ, the extra internal energy, from F_ϵ' the temperature variation of D when admitted would be relevant and important, and the value of E_ϵ was wrongly given in the earlier work on this theory.

The first attack on the theory of electrostatic effects in gases or solutions was due to Milner.* The various difficulties in the way of a successful calculation were clearly presented by him, but as he did not use Poisson's equation these were not the difficulties which we have encountered here. Another method has been proposed by Kramers.† Both these methods for evaluating $B(T)$ are correct in principle, with difficulties of their own. Both methods confirm the limiting form of Debye's result at great dilutions. But the interest of Debye's result lies in regions where theoretical basis is lacking, and we shall develop the theory further and tentatively use it in later chapters on account of its empirical importance.

* Milner, *Phil. Mag.* vol. 23, p. 551 (1912), vol. 25, p. 743 (1913).
† Kramers, *Proc. Sect. Sci. Amsterdam*, vol. 30, p. 145 (1927).

CHAPTER IX

THE THEORY OF IMPERFECT GASES (*cont.*)

§9·1. *Applications of the theory to simple imperfect gases and binary mixtures.* The number of empirical or semi-empirical equations of state in common use as interpolation formulae or as working digests of tabulated data is very large.* It is only possible here to discuss three of the simplest in relation to the theoretical results of the preceding chapter. For the purposes of this chapter readers who do not wish to follow the theory of Chapter VIII in detail can be content with any one of the three elementary discussions given in §§ 9·6, 9·7, 9·8. Any one of these methods gives for a simple gas, correct to terms in $1/V$,

$$pV = NkT - f(T)/V, \quad f(T) = 2\pi N^2 kT \int_0^\infty r^2(e^{-E/kT} - 1)\, dr. \qquad \text{......(813)}$$

The best known practical equation of state is that of van der Waals, namely,

$$\left(p + \frac{a}{V^2}\right)(V - b) = NkT, \qquad \text{......(814)}$$

where $a \propto N^2$ and $b \propto N$, and are otherwise constants. To the first power of $1/V$ this is equivalent to

$$pV = NkT + \frac{NkTb - a}{V}. \qquad \text{......(815)}$$

It is therefore equivalent, as a first order equation, to the approximation

$$f(T) = 2\pi N^2 kT \int_0^\infty r^2(e^{-E/kT} - 1)\, dr = -NkTb + a. \qquad \text{......(816)}$$

Though obviously incomplete this is useful from its simplicity and a sufficient analogy to the true form. For if the molecules are almost rigid spheres without external fields, then $E = +\infty$ ($r < \sigma$), and $E = 0$ ($r > \sigma$), so that

$$f(T) = -\tfrac{2}{3}\pi N^2 kT\sigma^3.$$

Thus for such a model $b = \tfrac{2}{3}\pi N\sigma^3$ (four times the volume of all the molecules) and $a = 0$. Historically, van der Waals' formula was derived by superposing on this volume effect b the independent effect of weak attractive fields in creating a boundary field (see § 9·8) and thereby diminishing the pressure. In asserting that the effect of this boundary field could be represented by a constant a added as in (815) to the unaltered volume effect, two mistakes are made. One is that the boundary field is calculated ignoring the effects of the intermolecular attractions on the distribution of molecular pairs,

* See, for example, Partington and Shilling, *The specific heats of gases* (1924); Kamerlingh-Onnes and Keesom, "Die Zustandsgleichungen", *Encyk. Math. Wiss.* Bd. V, No. 10; Jeans, *loc. cit.* chaps. VI, VII.

which is equivalent to replacing $(e^{-E/kT}-1)$ by $-E/kT$ in (816). The other is that the effect of the attractions on the volume effect itself is forgotten. When these mistakes are corrected, formulae (813) and (816) are recovered.* The value of (814) as a substitute for (815) rests entirely on its simplicity in applications, but its success is strictly limited.

The equation of state of D. Berthelot has an appreciably greater range of validity since it gives a closer representation of the theoretical first order terms. It is used empirically in either of the forms

$$pV = NkT\left\{1-\frac{p}{NkT}\left(\frac{a'}{NkT^2}-b\right)\right\}, \qquad \text{......(817)}$$

$$\left(p+\frac{a'}{TV^2}\right)(V-b) = NkT, \qquad \text{......(818)}$$

where $a' \propto N^2$ and $b \propto N$ and are otherwise constants. Either of these is equivalent to the first order form

$$pV = NkT + \frac{NkTb - a'/T}{V}, \qquad \text{......(819)}$$

so that they are based on the approximation

$$f(T) = -NkTb + a'/T. \qquad \text{......(820)}$$

If we contemplate a molecular model of an elastic sphere surrounded by an attractive field of force, of potential energy $-P$, we have

$$f(T) = 2\pi N^2 kT\left[-\tfrac{1}{3}\sigma^3 + \int_\sigma^\infty r^2\{e^{P/kT}-1\}\,dr\right]. \qquad \text{......(821)}$$

Thus as before $b = \frac{2}{3}\pi N\sigma^3$, and

$$a' = 2\pi N^2 kT^2\int_\sigma^\infty r^2(e^{P/kT}-1)\,dr. \qquad \text{......(822)}$$

The assumption that a' is independent of T may be expected to be fairly near the truth in suitable regions of temperature. For the extra T-factor makes it possible for $da'/dT=0$ for some T, whereas for a we have always $da/dT<0$. It would perhaps be better still to use a'/T^s instead of a'/T in (817) with the corresponding changes elsewhere, and adjust s to bring the zero of da'/dT into the most important temperature range.

Another equation of state of considerable importance is that of Dieterici. It is used empirically in the forms

$$p(V-b) = NkTe^{-a/NkTV}, \qquad \text{......(823)}$$

$$p(V-b) = NkTe^{-a'/NkT^sV}, \qquad \text{......(824)}$$

in which a, $a' \propto N^2$, $b \propto N$ and s is a constant. The value of s (other than 1)

* Fowler, *Phil. Mag.* vol. 43, p. 785 (1922).

most often used is $\frac{3}{2}$. To the first order in $1/V$ these equations have still the same form

$$pV = NkT + \frac{NkTb - a'/T^{s-1}}{V}, \qquad \text{......(825)}$$

and the same first order validity. They are, however, empirically very much more successful at reproducing observed facts, and this is undoubtedly due to the exponential instead of the additive form of the a-correction. Though we have really only studied first order corrections here, it is not difficult to see that Dieterici's form ought theoretically to be successful over a wider range than van der Waals' or Berthelot's. For we shall show in § 9·8, by a discussion of the boundary field, that approximately

$$pV \doteq NkT \exp\left[\frac{-2\pi N}{V}\int_0^\infty r^2\{e^{-E/kT}-1\}\,dr\right]. \qquad \text{......(826)}$$

The discussion on which this is based is admittedly inadequate, since only first order accuracy was aimed at in the intermolecular distribution law, and there are other approximations. But we may expect qualitative accuracy in the form of these approximations, and the rest of the argument inevitably leads to an equation of the form (826). Dieterici's form is derived by approximating to

$$\frac{2\pi N}{V}\int_0^\infty r^2\{e^{-E/kT}-1\}\,dr$$

with the usual $-b/V + a/NkT^sV$, and replacing $Ve^{-b/V}$ by $V-b$.

For practical use the a, a' and b of these and similar equations are adjusted to give the best fit with the facts over some particular temperature and pressure range, or to reproduce exactly some particular phenomenon such as the critical conditions (see below). It must be remembered that the constants so determined have no direct connection with the interatomic fields of force and cannot be used for anything more than a rough qualitative estimate of these fields or of the sizes of molecules. The mistake of using data from the critical point, for example, for quantitative estimates of molecular diameters has frequently been made. The only correct course is to reduce the observed equation of state to the form*

$$pV = NkT - f(T)/V + O(1/V^2), \qquad \text{......(827)}$$

and thus determine the observational value of $f(T)$, often called the second virial coefficient. Thus determined, $f(T)$ can be directly equated to its theoretical value. We give an account in the next chapter of work of this nature which has succeeded in coordinating into one fairly consistent scheme the requirements of interatomic fields both in gases and in crystals.

* This is the method followed by Kamerlingh-Onnes and Keesom and their collaborators at Leiden, and first correctly applied to the study of atomic fields by Keesom.

§ 9·2. *Critical points and reduced equations of state.* The semi-empirical equations of this section agree in predicting the existence of two types of isothermal separated by a *critical isothermal* for which $T=T_c$. When $T>T_c$, $\partial p/\partial V<0$ for all V. When $T<T_c$, $\partial p/\partial V$ vanishes twice and is positive between these roots. This behaviour can be regarded as a satisfactory description of the observed facts that for any substance there exists a critical temperature above which the substance can exist only in a single phase—the gaseous state, while below there are two possible phases, the gaseous or vapour state and the liquid state, which can co-exist in equilibrium together. Mathematically, the critical isothermal must be determined by the condition that on it the two roots of $\partial p/\partial V=0$ which are real for $T<T_c$ coalesce to form a single double root. This condition is obviously that the critical isothermal is that on which there exists a point, called the *critical point*, at which

$$\partial p/\partial V=\partial^2 p/\partial V^2=0. \qquad \text{......(828)}$$

Combined with the equation of state these equations suffice in general to fix the values of p, V and T for the critical point. These values are usually denoted by p_c, V_c and T_c. At this point the properties of the liquid and vapour phase finally become identical and the two phases fuse into one. The position of the critical point predicted by some of the equations is as follows:

van der Waals— $V_c=3b,\quad p_c=a/27b^2,\quad T_c=8a/27Nkb.$ ......(829)

Dieterici— $V_c=2b,\quad p_c=a/4e^2b^2,\quad T_c=a/4Nkb.$ ......(830)

Dieterici's equation reproduces the position of the critical point with considerable success for many gases. The predicted relation

$$NkT_c/p_cV_c=\tfrac{1}{2}e^2=3\cdot695$$

is particularly successful. The reader should refer to Jeans* for a further discussion. Some typical isothermals for carbon dioxide are shown in Fig. 13, reproduced from Partington and Shilling.†

The equations of state discussed here are alike in possessing only two adjustable constants. In each case, and in all similar cases, these two constants and Nk can be eliminated by introducing instead p_c, V_c and T_c. The equation of state then takes the form

$$\frac{p}{p_c}=f\left(\frac{V}{V_c},\frac{T}{T_c}\right), \qquad \text{......(831)}$$

where f is a function which is the same for all a and b, that is, to this approximation the same for all gases. This can easily be verified directly, or alternatively deduced by a dimensional argument. It is usual to introduce new variables π, v, ϑ, called *reduced variables*, defined by the relations

$$p=\pi p_c,\quad V=vV_c,\quad T=\vartheta T_c.$$

* Jeans, *loc. cit.* † Partington and Shilling, *loc. cit.* p. 37.

The equation of state then takes an absolute form called the *reduced equation of state*. As examples:

van der Waals— $(\pi + 3/v^2)(3v - 1) = 8\vartheta$. ......(832)

Berthelot— $(\pi + 3/v^2\vartheta)(3v - 1) = 8\vartheta$. ......(833)

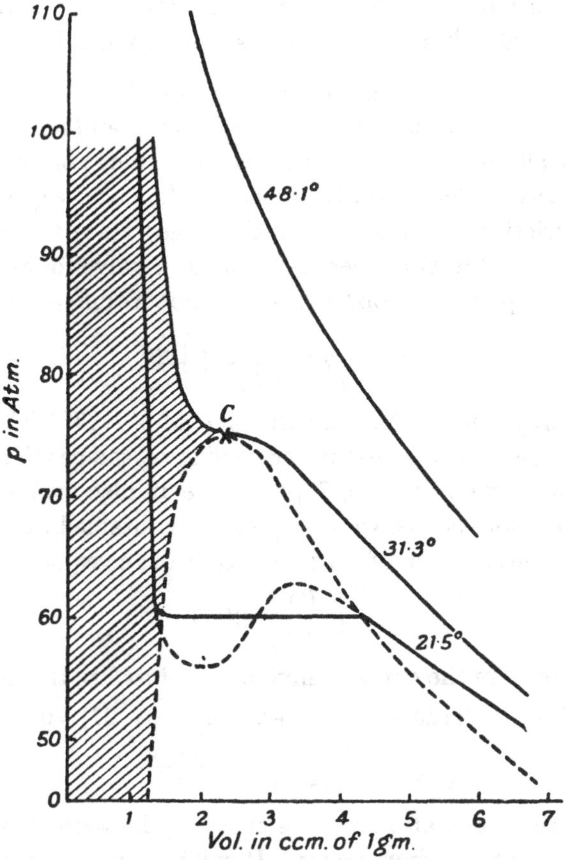

Fig. 13. Isothermals of CO_2. The critical point is at *C*. The shaded portion represents liquid states and the part within the dotted curve liquid-vapour mixtures.

Berthelot modified—An empirical equation used in the reduced forms

$$(\pi + 16/3v^2\vartheta)(v - \tfrac{1}{4}) = \frac{32}{9}\vartheta, \qquad ......(834)$$

$$\pi v = \frac{32}{9}\vartheta\left\{1 + \frac{9\pi}{128\vartheta}\left(1 - \frac{6}{\vartheta^2}\right)\right\}. \qquad ......(835)$$

These of course do not give the critical point 1, 1, 1 as a reduced equation should, but are more successful in the region of moderate deviations from the perfect gas laws.*

Dieterici— $\pi(2v - 1) = \vartheta e^{2-2/(v\vartheta)}$. ......(836)

* See also Henning, *Temperaturmessung*, Braunschweig (1915); Eucken, *Zeit. f. Physik*, vol. 29, p. 1 (1924).

Equal values of the reduced variables are said to be *corresponding* values of the ordinary variables for different gases. The suggestion of these equations that for corresponding temperatures and volumes the pressures have corresponding values for all substances is called *the law of corresponding states.* This law is approximately true over wide ranges of the variables for not too dissimilar molecules, but is by no means true in general.

§ 9·3. *Inversion points in the Thomson-Joule effect.* The thermodynamic theory of the Thomson-Joule effect is well known* and need not be repeated here. Gas is allowed to stream through a valve, porous plug or other throttling device under a steady pressure difference which maintains the flow against frictional resistances. In the steady state there is a temperature difference on the two sides of the plug or valve for an imperfect gas. For a differential pressure drop Δp this temperature difference ΔT is given by

$$\Delta T = \left\{ T\left(\frac{\partial V}{\partial T}\right)_p - V \right\} \frac{\Delta p}{C_p}. \qquad \text{......(837)}$$

Since necessarily $\Delta p < 0$, ΔT has the sign of $V - T(\partial V/\partial T)_p$. The effect may therefore be either a heating or cooling of the gas. The heating and cooling regions in the p, V or p, T planes are divided from one another by *the curve of inversion points*, whose equation is obtained by eliminating one variable from the equation of state by means of the equation

$$T\left(\frac{\partial V}{\partial T}\right)_p - V = 0. \qquad \text{......(838)}$$

In reduced variables the curve of inversion points has the following forms:

van der Waals— $(12\vartheta + \pi - 81)^2 + 216(4\vartheta + \pi - 27) = 0.$ ......(839)

Dieterici— $\pi\vartheta^{s-1} = \{4(s+1) - \vartheta^s\}\, e^{\frac{2s+3}{s+1} - \frac{4}{\vartheta^s}}.$ ......(840)

A diagrammatic presentation of these curves and observed inversion points is shown in Fig. 14 taken from Lewis.† It will be seen at once that Dieterici's equation with $s = \frac{5}{3}$ gives a very faithful representation of the properties of these gases in the neighbourhood of the curve of inversion points. This is a somewhat severe test of any practicable equation of state.

It will be seen on inspection of Fig. 14 that the cooling region is limited in area, the main portion of the p, T plane being the heating region. At the same time for the commoner gases the cooling region practically covers the range of ordinary temperatures and pressures. Hydrogen, helium and neon are exceptions. The limitation of the cooling region is of great importance in liquefaction practice by Linde's process, which is based on the Thomson-

* W. C. McC. Lewis, *A System of Physical Chemistry*, vol. 2, *Thermodynamics*, ed. 2, p. 67; Planck, *Thermodynamik*, ed. 6, § 70; Birtwistle, *Thermodynamics*, chap. VIII.

† W. C. McC. Lewis, *loc. cit.* p. 71.

Joule effect to obtain the reduction of temperature in each cycle. Unless the temperature is low enough for the greater part of the designed pressure drop to lie in the cooling region, the gas will not cool but heat and the liquefaction process cannot be carried out. In the manufacture of liquid hydrogen and helium by Linde's process efficient cooling will not occur unless the hydrogen used has been already cooled with liquid air, and the helium used with liquid hydrogen.

All these aspects of the theory of simple gases have long been fully appreciated, except the precise determination of intermolecular forces to which we devote the following chapter. It therefore seems unnecessary to give further space to them here.

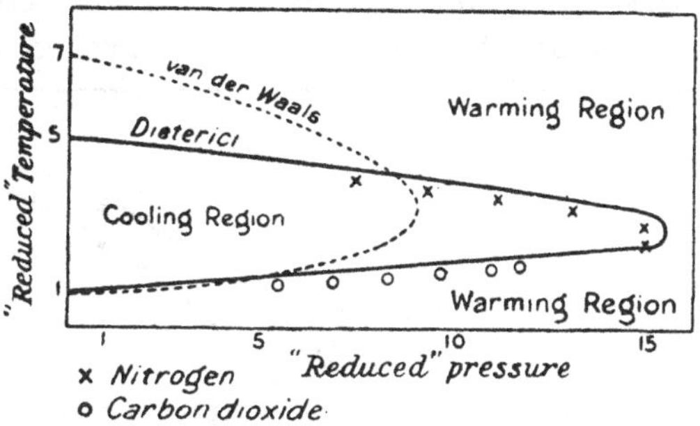

Fig. 14. The curves of inversion points in the Thomson-Joule effect.

§ 9·4. *Binary mixtures.* Except for the variation of their properties with composition, binary (and higher) mixtures of gases which do not react chemically behave like simple gases and no further discussion is required. It does, however, appear desirable to examine the theory of the variation with composition, as curiously erroneous statements have been current.

For a binary mixture equation (708) gives us

$$p=kT\left[\frac{N_1+N_2}{V}-\frac{1}{V^2}\{\tfrac{1}{2}\rho_{11}N_1{}^2+\rho_{12}N_1N_2+\tfrac{1}{2}\rho_{22}N_2{}^2\}\right], \quad \ldots\ldots(841)$$

where

$$\rho_{\alpha\beta}=4\pi\int_0^\infty\{e^{-E_{\alpha\beta}/kT}-1\}r^2dr. \quad \ldots\ldots(842)$$

If we write

$$N_1=xN, \quad N_2=(1-x)N,$$

so that x is the fraction of the first constituent in numbers of molecules, and so approximately the volume fraction at standard temperature and pressure, we have

$$p=kT\left[\frac{N}{V}-\frac{1}{2}\frac{N^2}{V^2}\{\rho_{11}x^2+2\rho_{12}x(1-x)+\rho_{22}(1-x)^2\}\right]. \quad \ldots\ldots(843)$$

It is at once obvious that a linear dependence on x must be the exception rather than the rule. The second virial coefficient must in general be a quadratic function of x, and it is obvious without explicit calculation that the nth virial coefficient must in general be a polynomial of the nth degree in x. The condition for linearity of the second virial coefficient in x is

$$\rho_{11}+\rho_{22}=2\rho_{12}. \qquad \ldots\ldots(844)$$

There is obviously no reason why this should be satisfied in general, but it is clear that it is likely to be true or nearly true when the molecular fields are very closely similar, for the condition states in a sense that the intermolecular forces between molecules 1 and 2 are the mean of those between 1 and 1 and between 2 and 2. This appears to hold for oxygen-nitrogen mixtures with very considerable accuracy,* and was once assumed to hold for helium-neon mixtures for the purpose of deducing the isotherms of pure neon, an assumption shown to be invalid by later work on pure neon.

In an interesting paper, Verschoyle† has established a case of marked

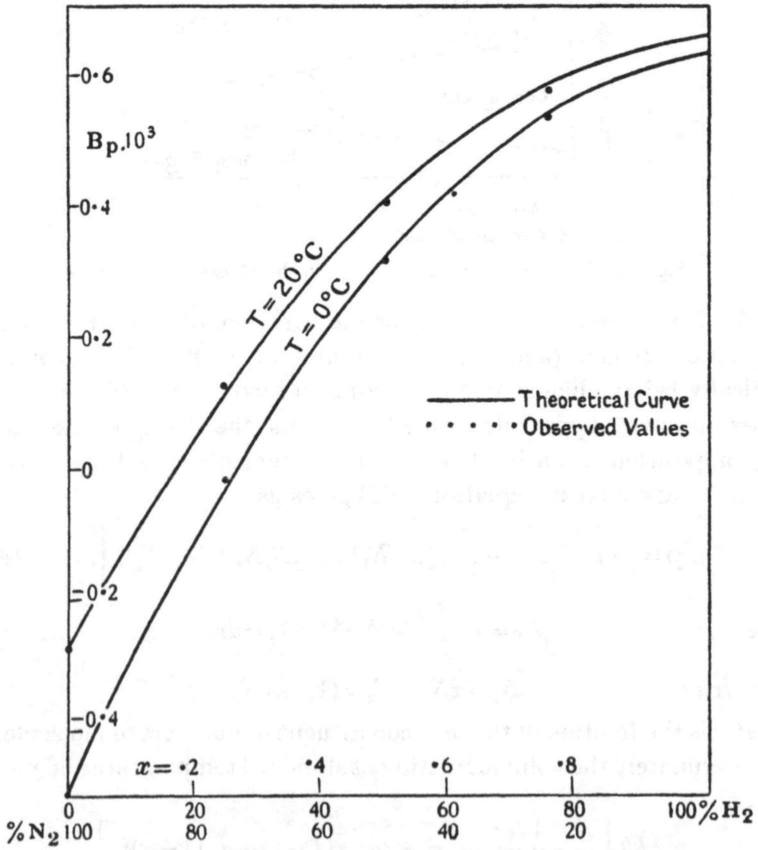

Fig. 15. Variation of the second virial coefficient with composition in hydrogen-nitrogen mixtures.

* Holborn and Otto, *Zeit. f. Physik*, vol. 10, p. 367 (1922), vol. 23, p. 77 (1924).

† Verschoyle, *Proc. Roy. Soc.* A, vol. 111, p. 552 (1926).

failure of (844) by the study of hydrogen-nitrogen mixtures, which has been discussed theoretically by Lennard-Jones and Cook.* Fig. 15 is taken from their paper and shows the variation with composition of the second virial coefficient, proportional to their B_p. The observed points can be fitted reasonably well by a parabola of the theoretical form, though the fit is hardly as good as might have been expected. These mixture curves may be taken as determining ρ_{12} when ρ_{11} and ρ_{22} are already known, and therefore as determining E_{12}, that is, the law of force between unlike molecules. A detailed experimental study of inert gas mixtures in various proportions would provide information of great value in the further development of the theories of Chapter x.

§ **9·5.** *The construction of* Ψ/k. For the logical completion of the elementary methods of § 9·7 and § 9·8 for the study of imperfect gases it is necessary to have a means of deriving the correction to Ψ/k corresponding to the directly calculated correction to the pressure. Such a method is also applicable to the semi-empirical pressure corrections of the preceding sections.

We start from the thermodynamic relation

$$p = T\,\partial\Psi/\partial V. \qquad \text{......(845)}$$

In general a knowledge of the equation of state, p, does not suffice to determine Ψ by integration, for the integration constant, which is unfixed, is an unknown function of N and T. But if we already know completely the exact form Ψ_∞ of Ψ in the limit $V\to\infty$, then (845) is sufficient: and in fact Ψ_∞ is known, for the assembly becomes a perfect gas. Owing to the linear form of (845) the corresponding corrections to p and Ψ must be additive. If we denote the imperfect gas correction to p by p_w and to Ψ by Ψ_w we obtain from (845)

$$\Psi_w = \Psi - \Psi_\infty = -\int_V^\infty \frac{p_w}{T}\,dV. \qquad \text{......(846)}$$

By (813) the general correction is

$$\frac{p_w}{T} = -\frac{1}{2}\frac{N^2k}{V^2}4\pi\int_0^\infty r^2(e^{-E/kT}-1)\,dr.$$

Therefore

$$\frac{\Psi_w}{k} = \frac{1}{2}\frac{N^2}{V}4\pi\int_0^\infty r^2(e^{-E/kT}-1)\,dr, \qquad \text{......(847)}$$

which agrees with (707).

In connection with any semi-empirical equation of state we use this method to determine Ψ_w and so Ψ, and to deduce from Ψ by the usual thermodynamic equations the corrections to C_V and C_p. In this way observed values of p, C_V and C_p can be corrected for deviations from the

* Lennard-Jones and Cook, *Proc. Roy. Soc.* A, vol. 115, p. 334 (1927).

perfect gas laws, and the corresponding p_∞, $(C_V)_\infty$, and $(C_p)_\infty$ determined. For the foregoing empirical equations we have

van der Waals— $$\Psi = Nk\left\{\log\frac{(V-b)F}{N}+1\right\}+\frac{a}{TV}. \qquad \text{......(848)}$$

Berthelot (*V-form*)— $$\Psi = Nk\left\{\log\frac{(V-b)F}{N}+1\right\}+\frac{a'}{T^2V}. \qquad \text{......(849)}$$

Berthelot (*p-form*)— $$\Psi = Nk\left\{\log\frac{(V-b+a'/NkT^2)F}{N}+1\right\}. \qquad \text{......(850)}$$

Dieterici— $$\Psi = \Psi_\infty - Nk\int_{NkT^s V/a'}^{\infty}\left(\frac{e^{-x}}{x-b'}-\frac{1}{x}\right)dx \quad (b' = NkT^s b/a'). \qquad \text{......(851)}$$

The last formula cannot be given in finite terms. Equation (850) is of greater value when we form the other characteristic function Φ in terms of p and T,* based on Gibbs' free energy instead of on Helmholtz's. The relation is

$$\Phi = \Psi - pV/T, \qquad \text{......(852)}$$

and we find

Berthelot (*p-form*)— $$\Phi = Nk\left\{\log\frac{kTF}{p}+\frac{p}{NkT}\left(\frac{a'}{NkT^2}-b\right)\right\}. \qquad \text{......(853)}$$

From Ψ we can at once deduce C_V and so the correction to $(C_V)_\infty$ by the equation

$$C_V = \frac{\partial}{\partial T}\left(T^2\frac{\partial\Psi}{\partial T}\right)_V, \qquad \text{......(854)}$$

and, similarly, from Φ $$C_p = \frac{\partial}{\partial T}\left(T^2\frac{\partial\Phi}{\partial T}\right)_p. \qquad \text{......(855)}$$

§ 9·6. *Alternative methods of calculation. Method* (i). *As a problem in dissociation.* The results of this chapter can be reached in a variety of other ways, some of which will be considered in the following sections. Since we need no longer attempt maximum generality we shall be content to consider the case of a simple imperfect gas of N molecules.

The results for a perfect gas can be extended at once to an imperfect gas (short range forces) by the device† of regarding any pair of molecules within each other's field of force as a *system* to be discussed as a whole. Any single molecule outside the fields of force remains a system as before. The equilibrium state of the assembly can then at once be studied as a problem in the dissociation of perfect gases. We consider only interactions in pairs. Let

* Planck, *Thermodynamik*, ed. 6, § 283 (1922).

† Jeans, *loc. cit.* p. 91. The discussion of dissociation and aggregation there given on a classical basis is inadmissible *in general*. The phenomena are essentially phenomena of the quantum theory. The device is, however, admirably adapted to the discussion of classical systems required here. It is essentially equivalent to Boltzmann's discussion of dissociation, *Vorlesungen über Gastheorie*, II, Abschnitt VI.

$f_1(T)$, equal to $h_1(T)\,j(T)$, as before be the partition function for the free systems of average number $\overline{N_1}$. Let $f_2(T)$ be the partition function for systems which are pairs of molecules in interaction, of average number $\overline{N_2}$. The function $f_2(T)$ will contain the factor $j^2(T)$ dealing with the internal energies. It remains to construct the classical part. The element of phase space for the pair is

$$m^6 dx_1 \dots dw_1 dx_2 \dots dw_2$$

in rectangular Cartesian coordinates and velocities. This can be transformed to coordinates and velocities of the centre of gravity of the pair $x^*, \dots, w^*$ and coordinates and velocities of 1 relative to 2, $\xi, \dots, w$. The Jacobian of the transformation is 1 and the element of phase space is

$$m^6 dx^* \dots dw^* d\xi \dots dw,$$

of effective weight

$$\delta_t = m^6 \frac{dx^* \dots dw^* d\xi \dots dw}{2h^6}, \qquad \text{......(856)}$$

the symmetry number being 2. The corresponding energy is

$$\epsilon_t = \tfrac{1}{2}(2m)\,(\Sigma\, u^{*2}) + \tfrac{1}{2}(\tfrac{1}{2}m)\,(\Sigma\, u^2) + E(\xi,\eta,\zeta). \qquad \text{......(857)}$$

Thus in the classical limit this factor of the partition function is

$$\frac{m^6}{2h^6}\int \dots \int e^{-\epsilon_t/kT} dx^* \dots dw^* d\xi \dots dw, \qquad \text{......(858)}$$

where ϵ_t is given by (857). This reduces at once to

$$h_2(T)\,\frac{(2\pi\frac{1}{2}mkT)^{\frac{3}{2}}}{h^3}\,A(T), \qquad \text{......(859)}$$

where $h_2(T)$ is the ordinary partition function for the motion of the aggregate as a whole, of mass $2m$, and

$$A(T) = \tfrac{1}{2}\int_v e^{-E(\xi,\eta,\zeta)/kT}\, d\xi\, d\eta\, d\zeta. \qquad \text{......(860)}$$

$A(T)$ is obviously the partition function for the potential energy of the pair, and v is the volume in which E is sensible.

The gas here considered of "free" molecules and aggregates obeys the dissociation theory and in particular equation (462). We therefore find

$$\frac{\overline{N_2}}{(\overline{N_1})^2} = \frac{j^2(T)\,h_2(T)\,(2\pi\frac{1}{2}mkT)^{\frac{3}{2}}\,A(T)}{h^3\{j(T)\,h_1(T)\}^2}.$$

To the first approximation the V-factors are just V in both $h_1(T)$ and $h_2(T)$, so that approximately

$$\overline{N_2}/(\overline{N_1})^2 = A(T)/V. \qquad \text{......(861)}$$

In calculating first order corrections v/V and therefore $A(T)/V$ will be small, so that $\overline{N_2}/(\overline{N_1})^2$ will be small and (861) is correct to the order required. $\overline{N_1}$ will differ from N only by first order terms. If we ask for $\overline{(a_2)_t}$, the average number of aggregates with given ranges of relative positional coordinates, we find on adapting (456) that

$$\overline{(a_2)_t} = \tfrac{1}{2}\overline{N_2}\, e^{-E/kT}\,(d\xi\, d\eta\, d\zeta)_t / A(T),$$

and therefore to a sufficient approximation

$$\overline{(a_2)_t} = \tfrac{1}{2}(\overline{N_1})^2\, e^{-E/kT}\,(d\xi\, d\eta\, d\zeta)_t / V. \qquad \text{......(862)}$$

The velocity distribution laws can similarly be shown to be unaltered by the forces.

We can derive at once any of the laws previously established, but it will suffice as an example to calculate the form of Ψ/k. By Theorem 6·31

$$\Psi/k = \overline{N_1}\left(\log\frac{f_1(T)}{\overline{N_1}} + 1\right) + \overline{N_2}\left(\log\frac{f_2(T)}{\overline{N_2}} + 1\right). \qquad \text{......(863)}$$

We require here a more exact evaluation of the V-factor in $f_1(T)$. It is no longer V exactly, because if any one molecule is in the field of another the pair rank as a system and not as free molecules. Thus in $f_1(T)$ the V-factor is approximately apparently

$$V - Nv. \qquad \text{......(864)}$$

This, however, would not be correct, as it would lead to counting the whole of each excluded volume twice over, once for each member of the pair. Thus in $f_1(T)$ the V-factor should be taken to be $V - \tfrac{1}{2}Nv$, and to a sufficient approximation

$$\Psi/k = \overline{N_1}\left\{\log\frac{(1 - \tfrac{1}{2}Nv/V)f_1(T)}{\overline{N_1}} + 1\right\} + \overline{N_2}\left\{\log\frac{f_2(T)}{\overline{N_2}} + 1\right\}.$$

This is most satisfactorily expressed as perfect gas terms plus corrections. Remembering that $\overline{N_1} = N - 2\overline{N_2}$ we find after an easy reduction that

$$\Psi/k = N\left(\log\frac{f_1(T)}{N} + 1\right) + \frac{N^2}{V}\{A(T) - \tfrac{1}{2}v\}, \qquad \text{......(865)}$$

which agrees exactly with (707).

This method is really much simpler to handle than the general method and easily extended to mixtures. It might extend conveniently to the calculation of higher order corrections, though hardly so effectively as the general method of § 8·31.

§ 9·7. *Method* (ii). *The use of the virial of Clausius.* The general method and the method of aggregations evaluate the complete equilibrium laws without the aid of additional theorems. Two other methods of some importance can be used to obtain the equation of state with the aid of the distribution law (750). The first of these to be considered is the use of the virial of Clausius. We give the underlying theorem in a general form due to Milne.*

* Milne, *Phil. Mag.* vol. 50, p. 409 (1925).

The equations of motion of a particle of mass m at x, y, z, moving under a force whose components are X, Y, Z and subject to frictional resistances of the form $-\kappa(\dot{x},\dot{y},\dot{z})$, are

$$m\ddot{x} = X - \kappa\dot{x} \qquad \text{......(866)}$$

and two similar equations. When this is multiplied by $\frac{1}{2}x$ it may be written

$$\frac{1}{4}\frac{d^2}{dt^2}(mx^2) + \frac{1}{4}\frac{d}{dt}(\kappa x^2) - \tfrac{1}{2}m\dot{x}^2 = \tfrac{1}{2}Xx. \qquad \text{......(867)}$$

Hence, on adding the two similar equations,

$$\frac{1}{4}\frac{d^2}{dt^2}(mr^2) + \frac{1}{4}\frac{d}{dt}(\kappa r^2) = \tfrac{1}{2}mv^2 + \tfrac{1}{2}(Xx + Yy + Zz). \qquad \text{......(868)}$$

Now sum this expression over all the systems in an assembly, and integrate over a long time τ. We find

$$\frac{1}{\tau}\left[\frac{1}{4}\frac{d}{dt}(\Sigma\, mr^2) + \tfrac{1}{4}\Sigma\,\kappa r^2\right]_0^\tau = \tfrac{1}{2}\Sigma\,\overline{mv^2} + \tfrac{1}{2}\Sigma\,\overline{Xx + Yy + Zz}. \qquad \text{......(869)}$$

The bars denote time averages from 0 to τ. Now if the state of the assembly is steady, the values of the expression in [ ] must be of the same order at 0 and τ, and will at least display no secular change with τ. Hence the left-hand side of (869) is effectively zero, and we have

$$\tfrac{1}{2}\Sigma\,\overline{mv^2} = -\tfrac{1}{2}\Sigma\,\overline{Xx + Yy + Zz}. \qquad \text{......(870)}$$

This is the theorem of Clausius, who named the expression on the right *the virial*. Provided frictional forces permit of an effectively steady state they do not alter the form of the theorem. We may note also that, provided the forces in the virial include all stresses due to bodies other than the systems to which Σ refers, the theorem is true for any collection of systems not necessarily the whole assembly.

Let us now apply (870) to an assembly consisting of an imperfect gas or to any portion of such assembly enclosed by an imaginary geometrical boundary. In either case the virial is made up of the forces between the molecules and the stresses across the physical or geometrical boundary. This stress per unit area is of course the pressure, and we may insist once again that the pressures on any boundary or across any internal surface are always equal in the absence of surface tension and external fields of force. If dS is a surface element of the boundary and l, m, n the direction cosines of its outward normal, the stress components are $-lpdS$, $-mpdS$, $-npdS$, which contribute to the virial

$$\tfrac{1}{2}p\iint (lx + my + nz)\, dS.$$

By Green's theorem this is equal to

$$\tfrac{1}{2}p\iiint\left(\frac{\partial x}{\partial x} + \frac{\partial y}{\partial y} + \frac{\partial z}{\partial z}\right) dV = \tfrac{3}{2}pV. \qquad \text{......(871)}$$

We have supposed above that the force between two molecules is radial and equal to $-\partial E/\partial r$. Continuing on this basis, if the centres of a pair of molecules are at x, y, z, x', y', z' and the force components are X, Y, Z, X', Y', Z', then

$$X = -\frac{\partial E}{\partial r}\frac{x-x'}{r}, \quad X' = -\frac{\partial E}{\partial r}\frac{x'-x}{r},$$

$$xX + x'X' = -\frac{\partial E}{\partial r}\frac{(x-x')^2}{r}.$$

Thus the contribution of the force between this pair of molecules to the virial is

$$\tfrac{1}{2}r\frac{\partial E}{\partial r},$$

and the total contributions of all intermolecular forces

$$\tfrac{1}{2}\Sigma\, r\frac{\partial E}{\partial r}, \qquad \text{......(872)}$$

summed over all pairs of molecules. Combining (871) and (872) with (870) we obtain finally

$$pV = \tfrac{1}{3}\overline{\Sigma\, mv^2} - \tfrac{1}{3}\Sigma\, r\frac{\partial E}{\partial r}. \qquad \text{......(873)}$$

This is the general form of the equation of state derived from the virial. To interpret it further we need to use distribution laws. For the mean kinetic energy of translation we have

$$\tfrac{1}{3}\overline{\Sigma\, mv^2} = NkT. \qquad \text{......(874)}$$

For the average number of pairs of molecules at a distance apart between r and $r+dr$ we find by (750)

$$\tfrac{1}{2}N^2 e^{-E/kT}\frac{4\pi r^2 dr}{V}. \qquad \text{......(875)}$$

The factor $\frac{1}{2}$ must be introduced when (750) is integrated over all relative directions. Using (874) and (875) we find

$$pV = NkT - \frac{1}{6}\frac{N^2}{V}4\pi\int_0^\infty r^3\frac{\partial E}{\partial r}e^{-E/kT}\,dr.$$

On integrating this by parts in such a way that the conditions of convergence at infinity are satisfied, we find

$$pV = NkT\left[1 - \frac{1}{2}\frac{N}{V}4\pi\int_0^\infty r^2(e^{-E/kT}-1)\,dr\right], \qquad \text{......(876)}$$

which agrees with (708).

The argument must be completed by an appeal to (846) to derive Ψ.

§ 9·8. *Method* (iii). *A direct calculation of the stress per unit area.* It is easily seen that the main conclusions of the calculation of stress in § 5·7 are unaffected by intermolecular forces. The internal stress is necessarily an isotropic pressure, everywhere equal in the absence of external fields to the boundary pressure. It consists of a term arising from the rate of transfer of momentum, which is absolutely unaltered by the forces, together with a new term the average stress per unit area due to the intermolecular forces themselves. To calculate this extra term we have merely to calculate the average force per unit area exerted by all the molecules on one side of a geometrical interface on those on the other This requires a use of the distribution law (750).

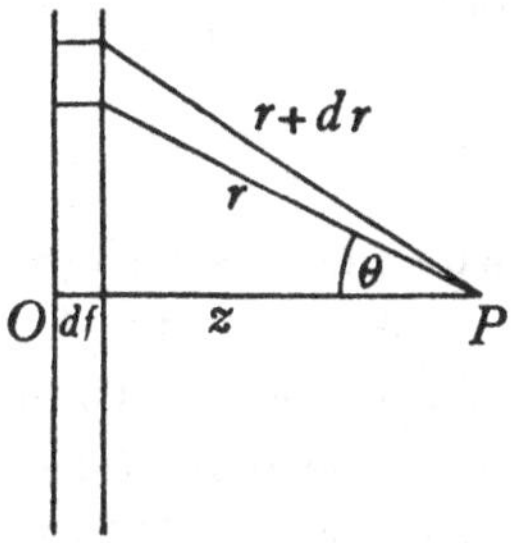

Fig. 16.

Let us consider an infinite plane slab of thickness df and calculate the average force dF exerted by all the molecules in this slab on a molecule at P, distant z from the slab. Our molecules are of course regarded here as point centres of force. The calculation is a generalization of the classical calculation in Laplace's theory of surface tension, generalized so as to apply directly to a molecular structure.* The average number of molecules in the slab per unit area at a distance r from P is†

$$\frac{N}{V}\,df\,e^{-E(r)/kT}.$$

It is safer in this section to show explicitly the argument of E. The average number in the annulus at distances between r and $r+dr$ from P is

$$2\pi r\sin\theta\,.\frac{dr}{\sin\theta}\cdot\frac{N}{V}\,df\,e^{-E(r)/kT},$$

and their resultant repulsion along PO is

$$2\pi\frac{N}{V}\,z\,df\left(-\frac{\partial E}{\partial r}\right)e^{-E(r)/kT}\,dr. \qquad \text{......(877)}$$

To obtain the average repulsion of the whole slab we must integrate (877) for all values of r from z to infinity. This gives

$$dF=-\frac{2\pi NkT}{V}\,z\,df\{e^{-E(z)/kT}-1\}. \qquad \text{......(878)}$$

The average repulsion per unit area on the molecules in a slab of thickness dz is therefore, since there are Ndz/V such molecules,

$$-\frac{2\pi N^2kT}{V^2}\,z\,dz\,df\{e^{-E(z)/kT}-1\}. \qquad \text{......(879)}$$

We now replace z by $\grave{z}+f$, the distance apart of any two slabs, and integrate

* See Rayleigh, *Scientific Papers*, vol. 3, pp. 397, 513 (1890, 1892).

† More generally $E(r)$ must of course be replaced by the mean $W_{\alpha\beta}(r)$ of formula (746).

with respect to both z and f from zero to infinity. We thus obtain the total stress per unit area due to the molecular forces, that is,

$$-\frac{2\pi N^2kT}{V^2}\int_0^\infty\int_0^\infty (z+f)\{e^{-E(z+f)/kT}-1\}\,dz\,df, \qquad \ldots\ldots(880)$$

or

$$-\frac{2\pi N^2kT}{V^2}\int_0^\infty x\{e^{-E(x)/kT}-1\}\,dx\int_0^x dy,$$

or

$$-\frac{2\pi N^2kT}{V^2}\int_0^\infty x^2\{e^{-E(x)/kT}-1\}\,dx, \qquad \ldots\ldots(881)$$

which agrees with (876).

The differential formulae (878) and (879) can also be used in a number of other ways. To obtain the work $dw(f)$ done by the repulsions when a molecule is removed from a distance f from the slab to infinity we have to integrate (878) with respect to z. Thus

$$dw(f)=-\frac{2\pi NkT}{V}\,df\int_f^\infty z\{e^{-E(z)/kT}-1\}\,dz. \qquad \ldots\ldots(882)$$

The work $w(f)$ done when the molecule is removed from a distance f from the plane boundary of a large mass of gas to infinity is

$$-\frac{2\pi NkT}{V}\int_f^\infty df'\int_{f'}^\infty z\{e^{-E(z)/kT}-1\}\,dz,$$

or

$$w(f)=-\frac{2\pi NkT}{V}\left[\int_f^\infty z^2\{e^{-E(z)/kT}-1\}\,dz-f\int_f^\infty z\{e^{-E(z)/kT}-1\}\,dz\right]. \qquad \ldots\ldots(883)$$

The average work done when one molecule is taken from the plane boundary to infinity is $w(0)$.

Again, it is obvious that $dw(-f)=dw(f)$, for the work done in taking the molecule from a distance f on one side of the slab to a distance f on the other is zero. Hence the work $w(-f)$ done by the repulsions when a molecule is removed to infinity from a depth f inside the plane boundary is equal to $w(0)$ together with the work done by the repulsions when the molecule leaves the surface of a finite slab of thickness f. This latter part is of course $w(0)-w(f)$. Thus

$$w(-f)=2w(0)-w(f), \qquad \ldots\ldots(884)$$

$$w(-\infty)=2w(0). \qquad \ldots\ldots(885)$$

The work done when the molecule comes from right inside to the surface is therefore equal to $w(-\infty)-w(0)$ or $w(0)$, which is also the work done when the molecule goes from the surface to infinity. These calculations of work terms are of course all *work done in reversible isothermal processes in which equilibrium conditions are maintained throughout.*

The foregoing formulae take no account of changes of the average density of the molecules near the boundary of the gas. There must in fact be such changes, as a consequence of Boltzmann's theorem. The reversible isothermal work done by the forces on a molecule when it is brought from deep in the gas to a depth f is by the preceding argument $w(-\infty)-w(-f)$, or $w(f)$ as given in (883). Hence by (745) and the following paragraphs $-w(f)$ is just exactly the energy term to be inserted in Boltzmann's theorem, and

$$\nu(f)=\frac{N}{V}e^{w(f)/kT}, \qquad \text{......(886)}$$

where $\nu(f)$ is the equilibrium density at a depth f inside the boundary. A more accurate investigation must proceed by constructing an integral equation for $w(f)$ instead of using (883).*

This calculation of the boundary density and the boundary field can be made to give at once the value of the boundary pressure, and is in fact the correct form of the classical calculations by which van der Waals derived his famous equation. For by (886) the density at the boundary itself is

$$\begin{aligned}\nu(0)&=\frac{N}{V}e^{w(0)/kT},\\ &=\frac{N}{V}\exp\left[-\frac{2\pi N}{V}\int_0^\infty z^2\{e^{-E(z)/kT}-1\}\,dz\right], \qquad \text{......(887)}\end{aligned}$$

and by the usual bombardment argument the pressure is $kT\nu(0)$. Therefore

$$pV=NkT\exp\left[-\frac{2\pi N}{V}\int_0^\infty z^2\{e^{-E(z)/kT}-1\}\,dz\right], \qquad \text{......(888)}$$

which to the order of accuracy agrees with (876).

Further developments of these ideas belong more properly to the theory of surface tension.

* Fowler, *Phil. Mag.* vol. 43, p. 785 (1922).

CHAPTER X

INTERATOMIC FORCES*

§ **10·1.** *Classification of interatomic energies.* The work of the two previous chapters proceeds on the assumption that the forces between molecules and therefore the energy terms such as $E_{\alpha\beta}$ are known, but even now little is known exactly *a priori* about the magnitude of the energies except in the simplest cases of hydrogen (H_2) and helium. It will some day be possible, thanks to the work initiated by Heitler and London,† to derive the interatomic energy and so the forces for any atoms or molecules from their electronic structure. But though no difficulties of principle remain, the day is far distant when such calculations will be a practical possibility. At present we must still rely mainly on indirect methods for such knowledge as we have of intermolecular fields.

The tendency of all molecules to aggregate at low temperatures is sufficient indication of the existence of forces of cohesion between molecules, and the very existence of matter leads of necessity to the conclusion that the forces between molecules become repulsive at short distances. Any adequate representations of an intermolecular energy must therefore satisfy these elementary requirements that its gradient should be positive at great distances and negative at small. Such *a priori* calculations‡ as have yet been completed naturally confirm this general conclusion. The simplest picture of this kind is that molecules consist of hard impenetrable surfaces surrounded by an attractive field. This picture we owe to van der Waals. It leads to the equation of state discussed in Chapter IX and has also had other successes. It is however inadequate to explain the observed compressibility of matter and for this and other reasons must be discarded (apart altogether from requirements of quantum theory). It is convenient, if somewhat artificial, to represent interatomic energies by the superposition of two terms, one yielding an attraction and the other a repulsion, such that the former dominates at large distances. The former we shall refer to as the *van der Waals energy*, and the latter as the *overlap energy*. It is convenient to label the energies in some such way as this, so that they may be more easily differentiated from the energy of the electrostatic repulsions and attractions,

* This chapter was contributed to the first edition by J. E. Lennard-Jones. It has been freely revised for this edition and the calculations repeated with the help of R. A. Buckingham to whom I am greatly indebted.

† Heitler and London, *Zeit. f. Physik*, vol. 44, p. 455 (1927).

‡ Slater and Kirkwood, *Phys. Rev.* vol. 37, p. 682 (1931); Slater, *Phys. Rev.* vol. 32, p. 349 (1928); Kirkwood and Keyes, *Phys. Rev.* vol. 37, p. 832 (1931).

of Coulomb type, between ions with net charges. All interatomic energies are of course ultimately of electrostatic or electromagnetic origin, but it is best to reserve the name electrostatic for the familiar terms ee'/r composing the potential energy of small charged bodies.

These names may be regarded here merely as convenient empirical labels. As we now know from quantum theory the van der Waals energy is due to the polarization of each atom or molecule by the charges of the other. The other energy term which gives dominating repulsions at short distances is non-classical. The overlap energy in general can lead either to strong attractions or strong repulsions at short distances, but in many cases the attractive state is inaccessible owing to Pauli's exclusion principle. When it is not so excluded we have the extra possibility of a state of strong binding—the quantum explanation of chemical combination. In studying intermolecular energies in this chapter we are usually concerned with systems in which all possible chemical bonds have already been formed, systems for which the overlap energy gives uniquely the necessary repulsions.

Where atoms in matter exist permanently in an ionized state, and the work of Arrhenius, of Kossel and others has shown that they often do, the electrostatic energy plays an important part in determining the physical properties of matter in bulk. This energy may be regarded as superimposed on the other energy terms already discussed, though not without modification. Owing to the deformation of the electronic systems of the atoms and ions by the presence of charged ions, or in any other electric field, another energy term comes in which may be called the *energy of primary polarization*. We use this description to distinguish it from the polarization energy of van der Waals' type which might be called the *energy of secondary polarization* since it is due to the polarization of one neutral system by another neutral system. Finally, some molecules are known to possess a permanent electric moment, even in the absence of an electric field. The magnitudes of these moments can often be deduced from a study of the dielectric properties of gases and perhaps liquids (Chapters XII, XXI), and the corresponding intermolecular energy (*the dipole energy*) is then known.

We may thus summarize our classification of interatomic energies under the following headings, arranged in order of simplicity and probably of range:

(1) Electrostatic energy between atoms (or ions) with net charges.

(2) Electrostatic energy between permanent dipoles.

(3) Energy of primary polarization.

(4) van der Waals energy (due to secondary polarization).

(5) Overlap energy.

We shall not, however, discuss the energies in this (the natural) order, as it is more convenient to deal first with the terms of more complex origin but shorter range.

§10·2. *Dependence of the intermolecular forces on orientation.* In general, the forces between molecules depend on their relative orientation as well as on their distance apart, but the mathematical difficulties of dealing with such laws of force in theories of the properties of matter are considerable. Keesom has given a method of deriving the equation of state of a gas of unsymmetrical molecules, though he has only applied it to solid ellipsoids of revolution,* and the expression obtained is not in very good agreement with observation. He has made no attempt to deduce any information of a quantitative character from a comparison of theoretical and observed results. The general method given by Ursell† (see Chapter VIII) is applicable, but has never yet been applied to an unsymmetrical model. Rankine‡ has tried to represent the fields of polyatomic gases by aggregates of overlapping spheres, the sizes of which were fixed by observations of the viscosity of simple gases, but it is difficult to verify these proposed structures from the viscosity of the actual polyatomic gas, as Rankine has tried to do, seeing that no theoretical formula has yet been produced for the viscosity of a gas of unsymmetrical molecules.

We shall therefore confine our attention to those structures which may with reason be regarded as spherically symmetrical; this group naturally includes the inert gases and ions of similar structure. It also includes hydrogen (H_2) which is approximately spherical, and the same analysis can be legitimately applied as a rough approximation to less spherical molecules such as nitrogen (N_2). It is natural that such structures of spherical symmetry should at first have been represented by rigid spheres, for the field, apart from the attractive field, is then completely determined by one parameter—the diameter. This simplicity accounts for its popularity as a molecular model,§ but a conglomeration of such rigid spheres in close packing would be incompressible and fail to possess the observed properties of ordinary solids, and, as we now know *a priori* from quantum theory, rigid spheres (with attractive fields outside them) are not a very good approximation to the actual forces.

* Keesom, *Proc. Sect. Sci. Amsterdam*, vol. 15 (1), p. 240 (1912).

† Ursell, *loc. cit.*

‡ Rankine, *Phil. Mag.* vol. 40, p. 516 (1920); *Proc. Roy. Soc.* A, vol. 98, pp. 360, 369 (1921); *Proc. Phys. Soc. London*, vol. 33, p. 362 (1921); *Phil. Mag.* vol. 42, pp. 601, 615 (1921); *Trans. Far. Soc.* vol. 17, p. 1 (1922).

§ For a discussion of the diameters of such model molecules, determined by a variety of methods, see Jeans, *loc. cit.*; Herzfeld, *Kinetische Theorie der Wärme*, Braunschweig (1925); Grösse und Bau der Moleküle, *Handbuch der Physik*, vol. 22, ed. 2, Berlin (1934).

§10·21. *Form of the overlap energy.* We now know from quantum theory that the overlap energy *at great distances* can be represented accurately by a formula of the type $P(r)\,e^{-r/\rho}$, where $P(r)$ is a polynomial containing positive and negative powers of r. This form can be derived by a perturbation method in the usual way. It would be natural therefore to use some such form to represent the potential energy of the repulsions if such a form were amenable to calculation both for gases and solids. It is however known that this asymptotic form is not necessarily a particularly good approximation to the true form at the smaller distances at which the repulsions matter, and that at the distances at which it is a good approximation the overlap terms are unimportant compared with the van der Waals energy. Such calculations as have been made for closer distances give forms of great complication. In spite therefore of these recent developments of quantum theory there is no compelling theoretical reason to prefer any simple form for the overlap energy to the familiar empirical λr^{-s} which is amenable to calculation for both gases and crystals. As we shall see later the form $Ae^{-r/\rho}$ has been shown by Born and Mayer* to be empirically rather more successful than λr^{-s} in coordinating the properties of crystals. Since, however, it is our main purpose in this chapter to coordinate the properties of both gases and crystals using a single representation of the interatomic energy and since the form of Born and Mayer has not yet been applied and is not easily applicable to the analysis of the second virial coefficient, we have been content here to obtain the best analysis of both gases and crystals that seems possible using the most recent data and the empirical form λr^{-s} for the overlap energy. Such an analysis can achieve as we shall see a considerable modicum of success. The model chosen includes the simple model of the rigid sphere as a special case when $s \to \infty$ with a suitable variation of λ. It must of course be remembered that the form λr^{-s} for the overlap energy obtained by analysis of the data in the manner to be shown here is only a valid representation of this energy term over a restricted range of values of r.

§10·22. *Form of the van der Waals energy.* It is equally convenient to represent the van der Waals energy yielding the attractions at greater distances by a similar function $-\mu r^{-t}$, as Keesom was the first to do empirically, though he superimposed these attractions on the rigid sphere model. Here convenience and theoretical accuracy march hand-in-hand. The leading term in the van der Waals energy must theoretically be of the form $-\mu r^{-6}$. This result is general,† and for a pair of ideal atoms consisting of

* Born and Mayer, *Zeit. f. Physik*, vol. 75, p. 1 (1932).

† Lennard-Jones, *Proc. Phys. Soc. London*, vol. 43, p. 461 (1931), where references to other work will be found.

electrons free to execute linear simple harmonic oscillations about a fixed positive charge it can be easily established as below.*

Let the two "atoms" be at a distance r apart along the z-axis and free to oscillate in this same line, the displacements of their electrons being z_1, z_2. At any instant they form dipoles of moment $z_1\epsilon$, $z_2\epsilon$, with mutual potential energy (when $z_1, z_2 \ll r$) given by

$$-\frac{2\epsilon^2 z_1 z_2}{r^3}.$$

The complete wave equation for the pair is

$$\frac{\partial^2\psi}{\partial z_1^2}+\frac{\partial^2\psi}{\partial z_2^2}+\frac{8\pi^2 m}{h^2}\left(E-\tfrac{1}{2}\kappa z_1^2-\tfrac{1}{2}\kappa z_2^2+\frac{2z_1 z_2\epsilon^2}{r^3}\right)\psi=0, \qquad \text{......(889)}$$

where κ is the constant of the elastic restoring forces. By the substitution

$$\zeta_1=(z_1+z_2)/\sqrt{2}, \quad \zeta_2=(z_1-z_2)/\sqrt{2}$$

(889) is reduced to the equation

$$\frac{\partial^2\psi}{\partial \zeta_1^2}+\frac{\partial^2\psi}{\partial \zeta_2^2}+\frac{8\pi^2 m}{h^2}(E-\tfrac{1}{2}\kappa_1\zeta_1^2-\tfrac{1}{2}\kappa_2\zeta_2^2)\psi=0, \qquad \text{......(890)}$$

where $\kappa_1=\kappa-2\epsilon^2/r^3, \quad \kappa_2=\kappa+2\epsilon^2/r^3.$

The characteristic values of the energy are therefore

$$E=(n_1+\tfrac{1}{2})h\nu_1+(n_2+\tfrac{1}{2})h\nu_2,$$

where $\nu_1=\frac{1}{2\pi}\sqrt{\frac{\kappa_1}{m}}, \quad \nu_2=\frac{1}{2\pi}\sqrt{\frac{\kappa_2}{m}}.$

For the state of lowest energy we find

$$E_0=\tfrac{1}{2}h(\nu_1+\nu_2)=\tfrac{1}{2}h\nu_0\left\{\left(1-\frac{2\epsilon^2}{\kappa r^3}\right)^{\frac{1}{2}}+\left(1+\frac{2\epsilon^2}{\kappa r^3}\right)^{\frac{1}{2}}\right\},$$

where ν_0 is the frequency and $h\nu_0$ the lowest energy of the oscillations when unperturbed. Expanding the square roots we see that the dominant term in the interaction energy is

$$-\frac{h\nu_0\epsilon^4}{4\kappa^2}\frac{1}{r^6}. \qquad \text{......(891)}$$

If the oscillators are isotropic three-dimensional, situated as before, the perturbing potential is $\epsilon^2(x_1x_2+y_1y_2-2z_1z_2)/r^3$ and the final term in the interaction energy

$$-\frac{3h\nu_0\epsilon^4}{8\kappa^2}\frac{1}{r^6}. \qquad \text{......(892)}$$

A rough theoretical method of extending (892) to actual atoms is quoted later.

In view of the foregoing discussion we shall represent the complete intermolecular potential energy (electrostatic terms omitted) by two inverse

* London, *Zeit. f. physikal. Chem.* vol. 11, p. 222 (1930).

power terms, $\lambda r^{-s} - \mu r^{-t}$ $(t=6)$, and the complete specification of the field requires a determination of three more parameters. In order that the field may be effectively attractive at large distances and repulsive at small it is necessary that $s > 6$. Though such a simple function of the distance cannot adequately represent actual intermolecular fields over all distances, it is the most general that has yet yielded to mathematical treatment and as we shall see far from inadequate in applications. The methods of determining the constants which have proved most successful are based on the physical properties of gases and these we now proceed to analyse, taking first the equation of state and then the viscosity. In general where possible we shall complete the mathematical analysis before substituting $t=6$.

§ **10·3.** *The equation of state of gases. The empirical representation of observed isothermals.* In Chapter IX it has been shown that for gases at moderately large dilution

$$pV = NkT\left\{1 + \frac{B}{V} + O\left(\frac{1}{V^2}\right)\right\}, \qquad \text{......(893)}$$

where B is a function of the temperature and of the interatomic energy. Kamerlingh-Onnes* has shown that the observations require in general a similar expansion, and he has expressed the results in the form of an empirical equation of state of the type

$$pV = A_V + \frac{B_V}{V} + \frac{C_V}{V^2} + \frac{D_V}{V^4} + \frac{E_V}{V^6} + \frac{F_V}{V^8}, \qquad \text{......(894)}$$

where the coefficients A, B, ... are functions of the temperature, usually called *first, second, ... virial coefficients*.

There is some divergence in the units used in the observational equation (894). Kamerlingh-Onnes and other workers at Leiden have adopted the international atmosphere as the unit of pressure, the volume being regarded as unity under this unit pressure at 0° C. When this system of units is employed, we shall distinguish the coefficients of (894) by writing them A_V, B_V, ... Workers at Berlin,† on the other hand, have taken the unit of pressure to be equivalent to a column of mercury 1 metre long (under standard conditions) with a corresponding change in the unit of volume. In this case we shall write $\mathfrak{A}_V$, $\mathfrak{B}_V$, ...

Again, it has proved convenient both at Leiden and at Berlin to express the value of pV, not in powers of $1/V$, but in powers of p, so that then we have

$$pV = A_p + B_p p + C_p p^2 + \ldots, \qquad \text{......(895)}$$

and

$$pV = \mathfrak{A}_p + \mathfrak{B}_p p + \mathfrak{C}_p p^2 + \ldots, \qquad \text{......(896)}$$

* Kamerlingh-Onnes, *Comm. Phys. Lab. Leiden*, No. 71, or *Proc. Sect. Sci. Amsterdam*, vol. 4, p. 125 (1902).

† Holborn and Otto, *Zeit. f. Physik*, vol. 23, p. 77 (1924), vol. 30, p. 320 (1924), vol. 33, p. 1 (1925), vol. 38, p. 359 (1926).

respectively. The various coefficients are easily related to each other. Thus we have

$$A_p = A_V = \frac{\mathfrak{A}_p}{(\mathfrak{A}_p)_0 + (\mathfrak{B}_p)_0 l + (\mathfrak{C}_p)_0 l^2}, \qquad \text{......(897)}$$

$$B_p = \frac{B_V}{A_V} = \frac{l\mathfrak{B}_p}{(\mathfrak{A}_p)_0 + (\mathfrak{B}_p)_0 l + (\mathfrak{C}_p)_0 l^2}, \qquad \text{......(898)}$$

where l is the pressure of one atmosphere in the Berlin units, and the suffix 0 refers to the isothermal 0° C. It is thus possible without difficulty to pass from one system of units to another.

For comparison with the theoretical work, the Leiden method of presenting the results is preferable, as the units refer to the normal conditions under which Avogadro's number for the molecular concentration is applicable. We shall regard the Leiden equations, therefore, as the standard experimental equations to which all others are to be converted.

§ **10·31.** *Theoretical expressions for the second virial coefficient.* If we write equation (893) in terms of molecular concentration, thus

$$p = \nu kT(1 + B'\nu), \qquad \text{......(899)}$$

we have*

$$\nu_0 kT = A_V, \quad B'\nu_0 = B_V/A_V, \qquad \text{......(900)}$$

where, by (813) for a spherically symmetrical field,

$$B' = 2\pi \int_0^\infty r^2 \{1 - e^{-E(r)/kT}\}\, dr, \qquad \text{......(901)}$$

or

$$B' = \frac{2\pi}{3kT} \int_0^\infty r^3 \left(-\frac{dE}{dr}\right) e^{-E(r)/kT}\, dr. \qquad \text{......(902)}$$

The condition that B' shall remain finite places a restriction on the molecular models which are possible. For instance, if

$$E(r) = \lambda r^{-s} - \mu r^{-t},$$

the condition requires that $s > 3$, $t > 3$. Equation (900) provides a criterion for any assumed form for $E(r)$.

Although equation (901) gives a formal solution for B' for any field of a spherically symmetrical type, the actual evaluation of the integral has been effected in only three cases, viz. (1) for molecules with a positive interaction energy λr^{-s}, which thus repel according to an inverse $(s+1)$th power law; (2) for molecules behaving like rigid spheres of diameter σ with a negative interaction energy $-\mu r^{-t}$ for $r > \sigma$ which thus attract each other according to an inverse $(t+1)$th power law at greater distances; (3) for molecules with an interaction energy $\lambda r^{-s} - \mu r^{-t}$. The first of these was given by Jeans,† the

* On the Leiden conventions ν_0 is a standard concentration equal to Avogadro's number, $2{\cdot}70 \times 10^{19}$.

† Jeans, *loc. cit.* p. 134.

second by Keesom,* and the third by Lennard-Jones.† For $E(r)=\lambda r^{-s}-\mu r^{-t}$ it is found that

$$B'=\tfrac{2}{3}\pi\left(\frac{\lambda}{\mu}\right)^{3/(s-t)}F(y), \qquad \text{......(903)}$$

where

$$F(y)=y^{3/(s-t)}\left\{\Gamma\left(\frac{s-3}{s}\right)-\sum_{n=1}^{\infty}c_n y^n\right\} \qquad \text{......(904)}$$

and y is a function of temperature given by

$$y=\frac{\mu}{kT}\left(\frac{kT}{\lambda}\right)^{t/s}. \qquad \text{......(905)}$$

The coefficients c_n are given by

$$c_n=\frac{3\Gamma\left(\dfrac{nt+s-3}{s}\right)}{n!(nt-3)}=\frac{3\Gamma\left(\dfrac{nt-3}{s}\right)}{n!\,s}. \qquad \text{......(906)}$$

The formulae for the first two models can be deduced as special cases. Thus for $E(r)=\lambda r^{-s}$ and no negative energy term

$$B'=\tfrac{2}{3}\pi\left(\frac{\lambda}{kT}\right)^{3/s}\Gamma\left(\frac{s-3}{s}\right). \qquad \text{......(907)}$$

For the rigid sphere with a negative energy $E(r)=-\mu r^{-t}$ $(r>\sigma)$,

$$B'=\tfrac{2}{3}\pi\sigma^3\left\{1-\sum_{n=1}^{\infty}\frac{3}{n!(nt-3)}\left(\frac{u}{kT}\right)^n\right\}, \qquad \text{......(908)}$$

where

$$u=\mu/\sigma^t, \qquad \text{......(909)}$$

which is the negative potential energy of two molecules in contact.‡

§ **10·32.** *Comparison of theory and observation.* Equations (903), (907) and (908) express B' in various ways as a function of the temperature, ready for comparison according to equation (900) with the observed value of B_V/A_V. The comparison may be carried out as follows.§ Introduce the new variables X and Y defined by

$$X=(\lambda/\mu)^{3/(s-t)}, \quad Y=(\mu^s/\lambda^t)^{1/(s-t)},$$

* Keesom, *Comm. Phys. Lab. Leiden*, Suppl. 24B, p. 32 (1912).

† Lennard-Jones, *Proc. Roy. Soc.* A, vol. 106, p. 463 (1924). For the case $s=9$, $t=8$ see Zwicky, *Physikal. Zeit.* vol. 22, p. 449 (1921), but he evaluates the integrals only by quadrature.

‡ Keesom, *Proc. Sect. Sci. Amsterdam*, vol. 15 (1), p. 256 (1912); *Comm. Phys. Lab. Leiden*, Suppl. 24B; *Physikal. Zeit.* vol. 22, pp. 129, 643 (1921) has evaluated B' for rigid spheres carrying a permanent electric dipole or a permanent electric quadripole, and has also allowed for polarization of each molecule by the other, as suggested by Debye, *Physikal. Zeit.* vol. 21, p. 178 (1920).

§ The method given here is a modification of that used by Lennard-Jones, *Proc. Roy. Soc.* A, vol. 106, p. 463 (1924), and has the slight advantage that the resulting values of λ and μ vary very smoothly as s is given different values, thus permitting accurate interpolation. Equal weight can also be given to values of B_V/A_V over the whole temperature range. We only consider equation (903) but the method can easily be adapted to the simpler cases.

and rewrite equations (903) and (904) in the form

$$\log X = \log \frac{3}{2\pi} B' - \log F(y), \qquad \text{......(910)}$$

$$\log Y = \frac{s}{s-t} \log y + \log kT. \qquad \text{......(911)}$$

By definition $\log X$ and $\log Y$ are also related to $\log \lambda$ and $\log \mu$ by the equations

$$\log \lambda = \tfrac{1}{3}s \log X + \log Y, \qquad \text{......(912)}$$

$$\log \mu = \tfrac{1}{3}t \log X + \log Y. \qquad \text{......(913)}$$

For a given value of the temperature T, B_V/A_V and therefore B' is taken from the observations, and inserted in (910). Equations (910) and (911) then determine a relation between $\log X$ and $\log Y$ in parametric form (parameter y), since F is a known function, s and t being given. This relation can be shown as a curve in the ($\log X$, $\log Y$) plane. The intersection of two such curves for observed values of B' at two temperatures determines $\log X$ and $\log Y$ and therefore λ and μ uniquely. In practice one takes nine or ten values of B' at regular temperature intervals over the widest possible range of temperature and plots ($\log X$, $\log Y$) curves for each. If the form $\lambda r^{-s} - \mu r^{-t}$ for the chosen values of s and t were an adequate representation of $E(r)$ and if the observed values of B' contained no experimental errors, all the curves would pass through a point; actually most of the intersections can

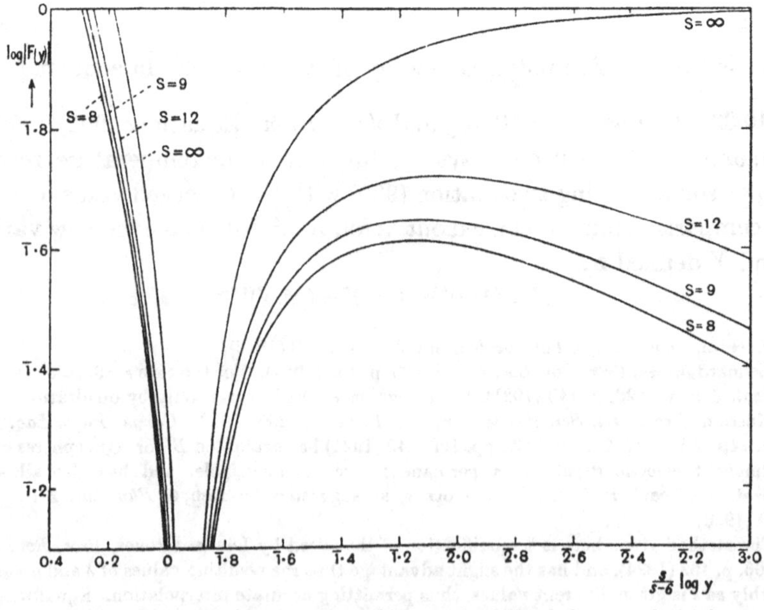

Fig. 17. Theoretical curves from which can be derived the second virial coefficient as a function of T, for $E(r) = \lambda r^{-s} - \mu r^{-6}$.

be made to lie within a fairly well-defined area. By taking the median values of the ordinates and abscissae of all the intersections suitable final values for $\log X$ and $\log Y$ are found, and λ, μ are then derived from (912) and (913).

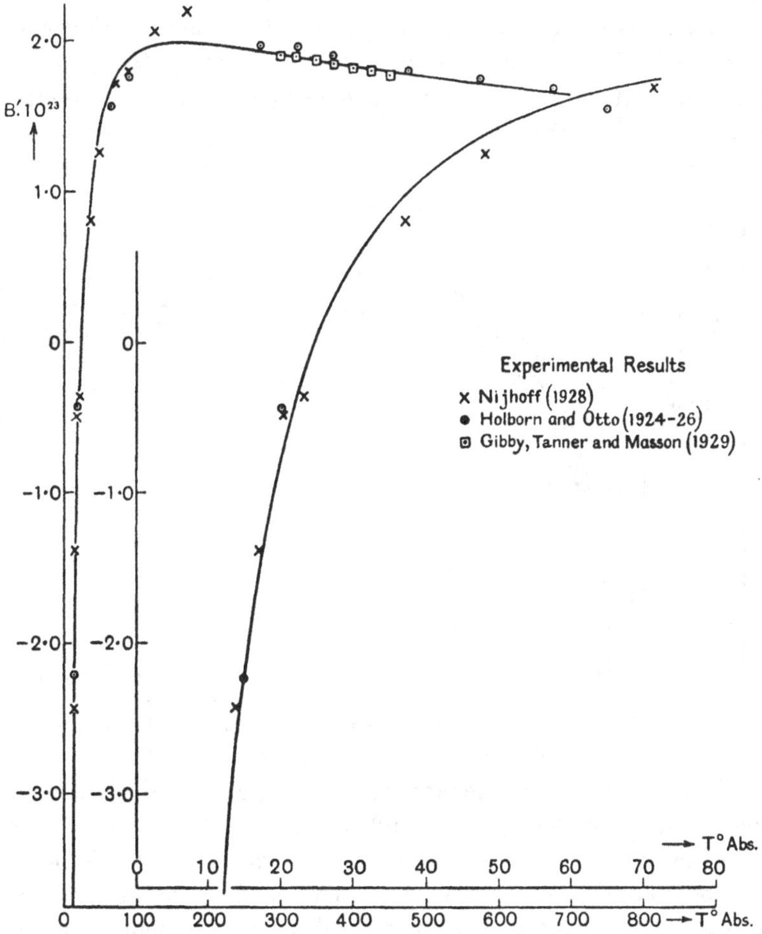

Fig. 17·1. The second virial coefficient as a function of T, with observed values for He, for $E(r) = \lambda r^{-12} - \mu r^{-6}$.

In order to illustrate the differences in the theoretical curves for B' for various values of s and t four curves are given for $\log F(y)$ as a function of $\log y$ in Fig. 17* for the same t and different values of s, and three in Fig. 18 for the same s and different values of t. The curves have been superimposed (with their axes parallel) so that approximately the "Boyle points"

* In this and other figures the left-hand portion of the curves corresponds to negative values of $F(y)$; $\log |F(y)|$ is plotted.

coincide, that is, the points for which $B' = F(y) = 0$. Fig. 17·1 shows a specimen curve for B' itself as a function of T.

It is to be noted that all the curves show a maximum except that for $s = \infty$, which corresponds to the rigid sphere, and this curve tends to an asymptote for infinite y. The physical interpretation of this is that for hard impenetrable molecules there is always an "excluded volume" however high the temperature, whereas for "compressible" molecules (λr^{-s}) there is an ever-growing interpenetration of the molecular fields as the temperature increases, and since the interatomic energies at any given compression are

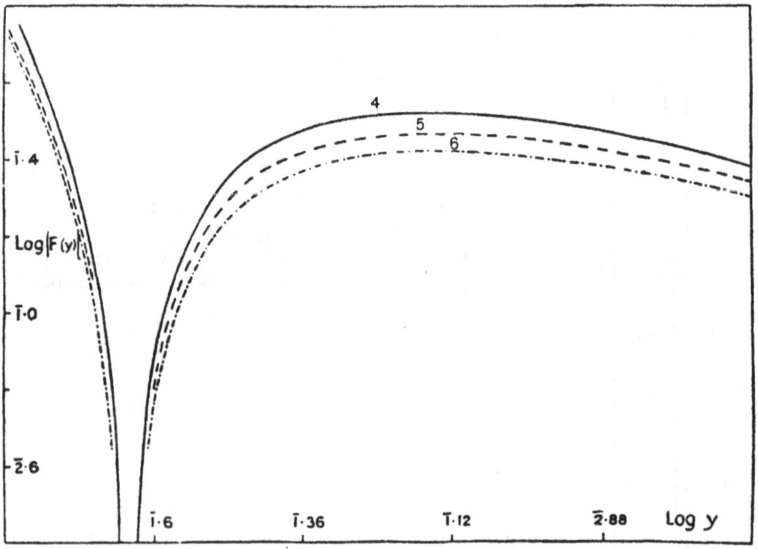

Fig. 18. Theoretical curves from which can be derived the second virial coefficient as a function of T, for $E(r) = \lambda r^{-9} - \mu r^{-t}$ ($t = 4, 5, 6$).

finite, the gas tends to become more and more like an ideal gas. Again, all the curves shown cross the $\log y$ axis and thus have a Boyle point. This is not true, however, of the curve corresponding to equation (907) (for a repulsive field alone). For this reason, the latter curve need not be further considered, as all gases show a reversal in sign of B_V/A_V at low temperatures. The curve for the rigid sphere plus permanent electric dipole, like the curves for all models which comprise hard impenetrable shells, does not possess a maximum. As will be shown presently, the observations of B_V/A_V for helium and neon do exhibit a maximum and thus discount any rigid molecular model. It may be regarded as certain that all other gases would show the same maximum property if the observations could be carried to a high enough temperature.

§ **10·33.** *The observed values of the second virial coefficient: intermolecular energies.* Although observations on the isothermals of gases have been accumulating since the classical experiments of Andrews on carbon dioxide,* a strict comparison of theory and experiment has only been possible since the elaborate analysis of the observational material made by Kamerlingh-Onnes.† His method, or slight modifications of it, has since then been adopted by many subsequent workers, and especially by Holborn and Otto,‡ who have carried out a series of accurate observations on the inert gases and have presented their results in a form comparable with theory. Their experiments cover a wide range of temperature, as is shown in the following table where their results for several gases are summarized. The

TABLE 17.

The observed values of $\log_{10}(B_V/A_V)$.

Temp. °C.	Helium			Neon	Argon	Hydrogen (H_2)	Nitrogen (N_2)
	(1)	(2)	(3)	(2)	(2)	(2)	(2)
−258·0	—	$\bar{4}\cdot7823\,(n)$	—	—	—	—	—
−252·8	—	$\bar{4}\cdot0961\,(n)$	—	—	—	—	—
−216·56	$\bar{4}\cdot6664$	—	—	—	—	—	—
−208·0	—	$\bar{4}\cdot6216$	—	—	—	—	—
−207·9	—	—	—	$\bar{4}\cdot9707\,(n)$	—	$\bar{4}\cdot9129\,(n)$	—
−183·0	—	$\bar{4}\cdot6750$	—	—	—	$\bar{4}\cdot3925\,(n)$	—
−182·75	$\bar{4}\cdot7261$	—	—	—	—	—	—
−182·5	—	—	—	$\bar{4}\cdot5618\,(n)$	—	—	—
−150·0	—	—	—	$\bar{6}\cdot6440$	—	$\bar{4}\cdot1187$	—
−130·0	—	—	—	—	—	—	$\bar{3}\cdot5513\,(n)$
−103·57	$\bar{4}\cdot7350$	—	—	—	—	—	—
−100	—	—	—	$\bar{4}\cdot4588$	$\bar{3}\cdot458\,(n)$	$\bar{4}\cdot6106$	$\bar{3}\cdot3642\,(n)$
− 50	—	—	—	$\bar{4}\cdot6089$	$\bar{3}\cdot227\,(n)$	$\bar{4}\cdot7319$	$\bar{3}\cdot0706\,(n)$
0	$\bar{4}\cdot7094$	$\bar{4}\cdot7231$	$\bar{4}\cdot7233$	$\bar{4}\cdot6770$	$\bar{4}\cdot9939\,(n)$	$\bar{4}\cdot7949$	$\bar{4}\cdot6638\,(n)$
20	$\bar{4}\cdot6970$	—	—	—	—	—	—
50	—	$\bar{4}\cdot7190$	$\bar{4}\cdot7185$	—	$\bar{4}\cdot6920\,(n)$	$\bar{4}\cdot8300$	$\bar{5}\cdot0606\,(n)$
100	—	$\bar{4}\cdot7057$	$\bar{4}\cdot7099$	$\bar{4}\cdot7232$	$\bar{4}\cdot2826\,(n)$	$\bar{4}\cdot8416$	$\bar{4}\cdot4374$
100·35	$\bar{4}\cdot6923$	—	—	—	—	—	—
150	—	—	—	—	$\bar{5}\cdot7159$	—	$\bar{4}\cdot7112$
200	—	$\bar{4}\cdot6934$	$\bar{4}\cdot6931$	$\bar{4}\cdot7648$	$\bar{4}\cdot3191$	$\bar{4}\cdot8452$	$\bar{4}\cdot8353$
300	—	$\bar{4}\cdot6704$	$\bar{4}\cdot6719$	$\bar{4}\cdot7880$	$\bar{4}\cdot6997$	—	$\bar{4}\cdot9642$
400	—	$\bar{4}\cdot6550$	$\bar{4}\cdot6599$	$\bar{4}\cdot7869$	$\bar{4}\cdot8344$	—	$\bar{3}\cdot0206$

For entries marked (n) B' or $F(y)$ is negative.
(1) Kamerlingh-Onnes, *Comm. Phys. Lab. Leiden*, No. 102A (1908).
(2) and (3) Holborn and Otto (*loc. cit.*).

* For a full account of the earlier literature on equations of state, see Kamerlingh-Onnes and Keesom, "Die Zustandsgleichung", *Encyk. der math. Wiss.* vol. 5, p. 615 (1912).

† Kamerlingh-Onnes, *Comm. Phys. Lab. Leiden*, No. 71, or *Proc. Sect. Sci. Amsterdam*, vol. 4, p. 125 (1902).

‡ Holborn and Otto, *Zeit. f. Physik*, vol. 23, p. 77 (1924), vol. 30, p. 320 (1924), vol. 33, p. 1 (1925), vol. 38, p. 359 (1926).

values of B_V/A_V, or rather of $\log(B_V/A_V)$, are given after the conversion of the results to the standard form (894) in the usual units. Three sets of results for helium are given in adjacent columns. one set due to Kamerlingh-Onnes and the other two to Holborn and Otto, deduced from their observations by different methods. The results of Kamerlingh-Onnes are inconsistent with those of Holborn and Otto at low temperatures. The latter are, however, consistent among themselves and, being the more recent observations, have been chosen for comparison with theory.

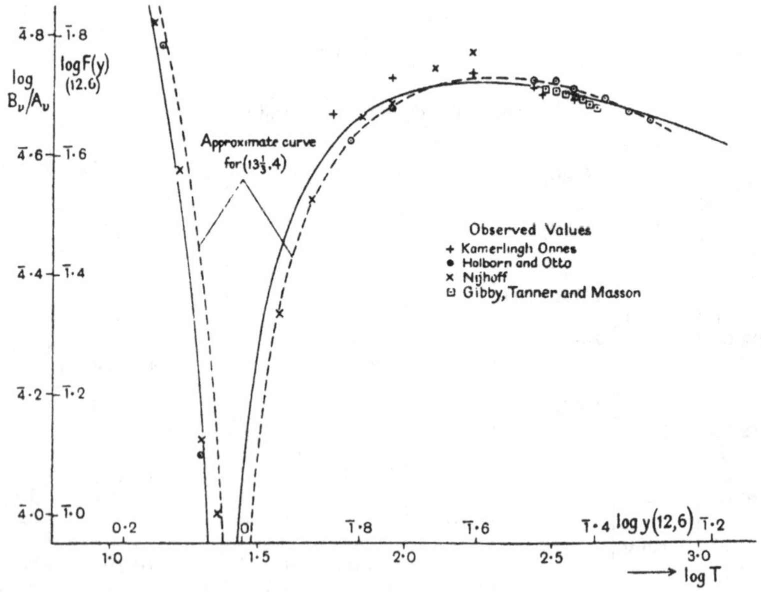

Fig. 19. Showing the determination of the energy constants from the second virial coefficient of He.

Even when by reference to quantum theory we have fixed t at 6 it does not seem possible to determine s, λ and μ uniquely from the equation of state. A good fit can be obtained over a range of values of s (or t for that matter), but when s and t are fixed the equation of state determines λ and μ with some precision. Thus theoretical curves for (s,t) equal to (8,4), (10,4), (10,5), $(13\frac{1}{3},4)$, (9,6), (10,6) and (12,6) can all be made to fit the observations adequately though some are better than others. This is illustrated in Fig. 19 which shows the fit obtainable for helium with (s,t) equal to $(13\frac{1}{3},4)$ and (12,6), which is nearly as perfect as can be expected. To limit this indeterminacy it is necessary to adduce other considerations, either results of quantum theory or properties of crystals.

A rough theoretical formula for the van der Waals energy c/r^6 of two like atoms taking into account only the dipole-dipole term in the interaction has been proposed by Slater and Kirkwood,† namely

$$c = -\tfrac{3}{2}E_0 a_0^{\frac{1}{2}} Z^{*\frac{1}{2}} \alpha^{\frac{3}{2}} = -1{\cdot}24 \times 10^{-23} Z^{*\frac{1}{2}} \alpha^{\frac{3}{2}}, \qquad \ldots\ldots(914)$$

where α is the observed atomic polarizability derived from the refractive index of the gas, Z^* the number of extranuclear electrons, a_0 the radius and E_0 the negative energy of the first Bohr orbit in hydrogen. The experimental values of α for the inert gases and the corresponding values of c are given in Table 18 for the inert gases. If we compare these values of c with

TABLE 18.

Polarizability and van der Waals energy constant (after Slater and Kirkwood) for the inert gases.

Atom	Polarizability $\alpha \times 10^{24}$	van der Waals energy constant $-c \times 10^{60}$
He	0·204	1·62
Ne	0·392	9·62
A	1·63	109·5
Kr	2·465	288
Xe	4·008	731

Observed values of α from C. and M. Cuthbertson, *Proc. Roy. Soc.* A, vol. 84, p. 13 (1910).

the values of μ obtained below from the equation of state, and remember that more accurate calculation is likely still further to increase the polarizability by including for example dipole-quadripole interactions, it is clear that s lies between 8 and 10. If we bring still further evidence into count, and consider the requirements of the compressibility of the alkali halides discussed below, we can still further limit s to values between 8·5 and 9·5. Since, however, our form for $E(r)$ is admittedly not exact, we cannot hope that all applications will agree in indicating a single value of s.

The direct results of the analysis of the equation of state are given in Table 19. This table shows the best values of the energy constants λ and μ for the forms $E(r) = \lambda r^{-s} - \mu r^{-6}$ for integral values of s from 8 to 14. All these values give fair representations of the second virial coefficient.

The intermolecular energy so determined is shown as a function of r for the five gases in Fig. 20. The actual curves are there drawn for $(s,t) = (12,6)$. It is interesting to compare the curve for He with the *a priori* calculations of Slater and Kirkwood‡ which are shown by the dotted curve. The derived energies are in excellent agreement with the calculated ones.

† Slater and Kirkwood, *Phys. Rev.* vol. 37, p. 682 (1931); Kirkwood, *Physikal. Zeit.* vol. 33, p. 57 (1932). Their formula is certainly a better approximation than that of London, but it has been criticized by Hellman, *Acta Physicochim., U.S.S.R.* vol. 2, p. 273 (1935), who suggests that Z^* should denote only the number of electrons in the outermost shell.

‡ Slater and Kirkwood, *loc. cit.*

TABLE 19.

Potential energy constants for gases, calculated from the equation of state.

The intermolecular energy is $E(r) = \lambda r^{-s} - \mu r^{-6}$.

		λ (repulsive)	μ (attractive)
Helium	$S=8$	$0{\cdot}00276 \times 10^{-72}$	$34{\cdot}0 \times 10^{-61}$
	9	$0{\cdot}00476 \times 10^{-80}$	22·6
	10	$0{\cdot}00897 \times 10^{-88}$	17·0
	12	$0{\cdot}0360_5 \times 10^{-104}$	11·7
	14	$0{\cdot}155 \times 10^{-120}$	9·19
Neon	$S=8$	$0{\cdot}0183 \times 10^{-72}$	$20{\cdot}7 \times 10^{-60}$
	9	$0{\cdot}0350 \times 10^{-80}$	$14{\cdot}4_7$
	10	$0{\cdot}0732_5 \times 10^{-88}$	11·4
	12	$0{\cdot}354_5 \times 10^{-104}$	8·32
	14	$1{\cdot}82 \times 10^{-120}$	$6{\cdot}78_5$
Argon	$S=8$	$0{\cdot}314_4 \times 10^{-72}$	$23{\cdot}7 \times 10^{-59}$
	9	$0{\cdot}768 \times 10^{-80}$	17·0
	10	$2{\cdot}04_6 \times 10^{-88}$	13·7
	12	$16{\cdot}2 \times 10^{-104}$	$10{\cdot}3_4$
	14	$136{\cdot}6 \times 10^{-120}$	8·67
Hydrogen H_2	$S=9$	$0{\cdot}051 \times 10^{-80}$	$1{\cdot}78 \times 10^{-59}$
	10	$0{\cdot}116 \times 10^{-88}$	1·42
	12	$0{\cdot}649 \times 10^{-104}$	1·05
Nitrogen N_2	$S=9$	$1{\cdot}32 \times 10^{-80}$	$22{\cdot}7 \times 10^{-59}$
	10	$3{\cdot}93 \times 10^{-88}$	18·5
	12	$37{\cdot}0 \times 10^{-104}$	14·0

Values for He, Ne, A were obtained by the method described in the text.
Values for H_2, N_2 taken from Lennard-Jones, *Proc. Phys. Soc. London*, vol. 43, p. 461 (1931).

§ 10·34. *Gaseous mixtures.* The energy constants just given define the field between two atoms or molecules of the same kind. To extend the work to atoms of different kinds it is necessary to analyse the equation of state for gaseous mixtures. For a binary mixture instead of (899) we have

$$p = kT\{(\nu_1+\nu_2) + \nu_1^2 B'_{11} + 2\nu_1\nu_2 B'_{12} + \nu_2^2 B'_{22}\}, \quad \text{......(915)}$$

where

$$B'_{\alpha\beta} = 2\pi \int_0^\infty r^2\{1 - e^{-E_{\alpha\beta}(r)/kT}\}\, dr. \quad \text{......(916)}$$

The observed isothermals are to be expressed as before by

$$pV = A_V + B_V/V = A_p + B_p p, \quad \text{......(917)}$$

and then by comparison we have

$$A_V = \nu_0 kT, \quad \text{......(918)}$$

$$B_V/A_V = B_p = \nu_0(B'_{11}x_1^2 + 2B'_{12}x_1x_2 + B'_{22}x_2^2), \quad \text{......(919)}$$

or

$$B_{11}x_1^2 + 2B_{12}x_1x_2 + B_{22}x_2^2 = B_V/A_V. \quad \text{......(920)}$$

In (919) ν_0 denotes the concentration of the mixture under standard conditions and

$$B_{\alpha\beta} = \nu_0 B'_{\alpha\beta}, \quad x_1 = \nu_1/\nu, \quad x_2 = \nu_2/\nu, \quad x_1 + x_2 = 1. \quad \text{......(921)}$$

The dependence of B_V/A_V on the relative concentrations has been studied experimentally by Verschoyle* at temperatures of 0° C. and 20° C. for mixtures of hydrogen and nitrogen, and his values were shown in Fig. 15. That figure shows two curves, quadratic in $\nu_1/(\nu_1+\nu_2)$, which are drawn so

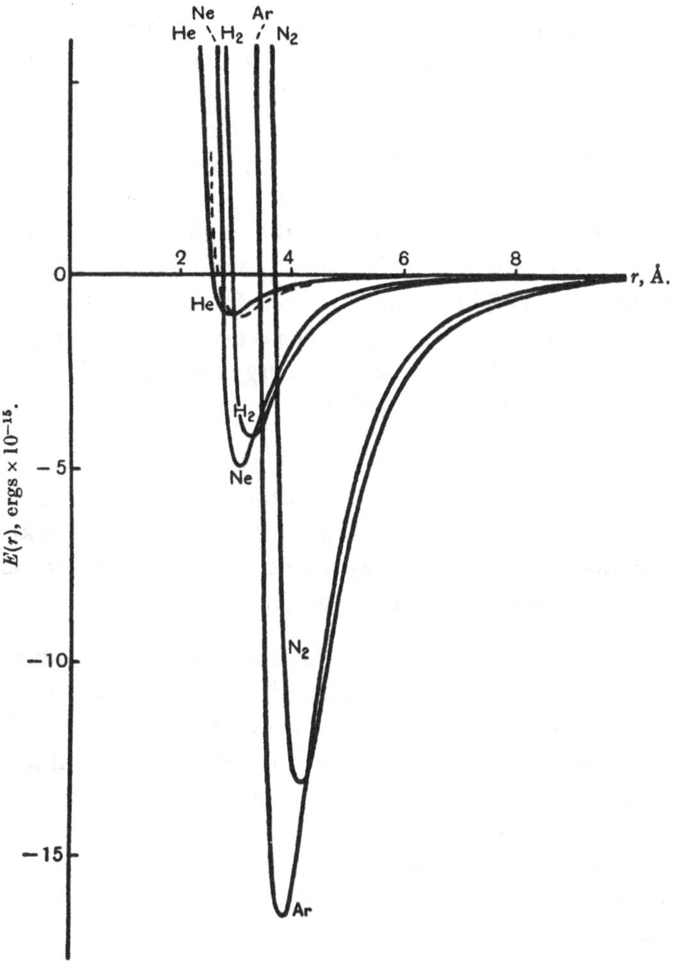

Fig. 20. The potential energies of pairs of inert gas atoms as a function of their distance apart.

that the mean square distances of the observed points from them is a minimum. It is clear that a quadratic function of this kind satisfactorily accounts for the facts, and hence we may deduce from the observations the numerical values of each of the coefficients B_{11}, B_{12} and B_{22} in equation (920).

This process, if continued for a number of temperatures, would determine

* Verschoyle, *Proc. Roy. Soc.* A, vol. 111, p. 552 (1926).

the variation of B_{11}, B_{12} and B_{22} with temperature. Each of these functions could then be dealt with separately, just as the B_V/A_V for a single gas, leading to the evaluation of the four constants of the fields between like and unlike atoms. Unfortunately, there does not appear to be any experimental data so extensive as this for any pair of gases. The work of Verschoyle refers only to two temperatures 0° and 20° C. and is thus scarcely adequate. Tentative calculations have been carried out on the basis of this work, but the results cannot be regarded as final.*

A series of experiments has been carried out by Holborn and Otto† on a mixture of helium and neon of fixed concentration (72·39 per cent. neon and 27·61 per cent. helium) as well as on the single gases, over a temperature range 0° to 400° C. This provides just the minimum information necessary for the determination of B_{11}, B_{12} and B_{22} as functions of temperature and so of the interatomic fields.* An extension of this work to lower temperatures and to mixtures of other concentrations is desirable. The numerical results suggest a simple relation between the energy constants of the fields between like and unlike atoms. It appears that

$$\lambda_{12}^{1/s} = \tfrac{1}{2}(\lambda_{11}^{1/s} + \lambda_{22}^{1/s}), \qquad \ldots\ldots(922)$$

approximately. This means physically that the closest distance of approach of two unlike atoms in a direct encounter with a given relative kinetic energy is equal to the mean of the corresponding distances in the encounter of two pairs of like atoms with the same energy.

§ **10·4.** *The viscosity of a gas.* Another method of determining interatomic energies is to compare the observed and calculated viscosity of the gas at pressures low enough for the gas to be practically perfect. The theory of transport phenomena lies outside the range of this monograph, and we shall only mention in passing such results as have been or might be used to provide information on interatomic energies. The classical theory of transport phenomena is itself so complicated that accurate formulae have only been obtained for certain special spherically symmetrical fields. The quantum theory seriously modifies these results for atoms or molecules so light as helium or hydrogen. At present the conclusion must be drawn that viscosities have scarcely yet been used effectively in accurate determinations of interatomic fields. New investigations are feasible which would enable viscosities to be used again for this purpose and it is much to be hoped that they will be undertaken.

The original investigation by Maxwell‡ applied only to classical atoms

* Lennard-Jones and Cook, *Proc. Roy. Soc.* A, vol. 115, p. 334 (1927).

† Holborn and Otto, *Zeit. f. Physik*, vol. 23, p. 77 (1924).

‡ Maxwell, *Scientific Papers*, vol. 2, p. 26.

repelling as the inverse fifth power of the distance [$E(r)=\lambda r^{-4}$]. This was later generalized by Chapman* and Enskog† to any inverse power law. These authors have also given rigorous calculations for classical atoms behaving on collision like rigid elastic spheres with a weak attractive field surrounding them. Their calculations confirmed the general form of the formula first given by Sutherland for this model.

Another law of interaction for which exact classical calculations can be made is represented by‡

$$E(r)=\lambda r^{-s}-\mu r^{-2}. \qquad \text{......(923)}$$

This special form has little physical significance and cannot continue to represent the interaction at great distances. The corresponding coefficient of viscosity η varies with the temperature according to the form

$$\eta=\eta_0\left(\frac{T}{T_0}\right)^{(s+4)/2s}\frac{1+\sum_{r=1}^{\infty} S_r T_0^{-r(s-2)/s}}{1+\sum_{r=1}^{\infty} S_r T^{-r(s-2)/s}}. \qquad \text{......(924)}$$

The S_r are functions of $\mu/\lambda^{2/s}$ only. When $\mu=0$ the S_r all vanish and the formula reduces to that given by Chapman and Enskog for the model $E(r)=\lambda r^{-s}$, of which the full form is

$$\eta=\epsilon_c\frac{B_s m^{\frac{1}{2}}}{\lambda^{2/s}}T^{(s+4)/2s}, \qquad \text{......(925)}$$

where m is the mass of the molecules, ϵ_c a number differing only trivially from unity and B_s a coefficient depending only on s given by

$$B_s=\frac{5\pi^{\frac{1}{2}}k^{(s+4)/2s}}{4I_2(s-1)\,\Gamma(4-2/s)\,2^{(s-2)/s}}. \qquad \text{......(926)}$$

In this expression k is Boltzmann's constant and $I_2(s-1)$ a function of s which has been tabulated.§

If the attractive field is weak so that $\mu/\lambda^{2/s}$ is small, the first term only of the summation in (924) need be retained. The full form of (924) then reduces, in the notation of (925), to

$$\eta=\frac{B_s m^{\frac{1}{2}}}{\lambda^{2/s}}\frac{T^{\frac{3}{2}}}{T^{(s-2)/s}+S_1}. \qquad \text{......(927)}$$

* Chapman, *Phil. Trans.* A, vol. 211, p. 433 (1912); vol. 216, p. 279 (1915); *Mem. Manchester Lit. and Phil. Soc.* vol. 66, p. 7 (1922).

† Enskog, *Kinetische Theorie der Vorgänge in massig verdünnten Gasen*, Inaug. Diss., Uppsala (1917); also *Arkiv f. Mat. Astro. och Fys.* vol. 16, No. 16 (1921); *Kungl. Svenska Vetenskaps-akademiens Handlingar*, vol. 63, No. 4 (1922).

‡ Lennard-Jones, *Proc. Roy. Soc.* A, vol. 106, p. 441 (1924).

§ Chapman, *loc. cit.* (1922); Lennard-Jones, *loc. cit.*

The well-known formula of Sutherland is obtained from this by letting $s \to \infty$ while $\lambda^{1/s} \to \sigma$, the molecular diameter. We then find

$$\eta = \frac{B_\infty m^{\frac{1}{2}}}{\sigma^2} \frac{T^{\frac{3}{2}}}{T+S_1}, \qquad \text{......(928)}$$

where

$$B_\infty = \frac{5\pi^{\frac{1}{2}} k^{\frac{1}{2}}}{48 I_2(\infty)}. \qquad \text{......(929)}$$

Formulae (927) and (928) apply only to weak attractive fields. The restriction to weak fields has been removed by Hassé and Cook,* for the rigid sphere model surrounded by the field $E(r) = -\mu r^{-4}$, but their results cannot be expressed by an explicit formula for η.

The above survey shows that even these classical results are somewhat incomplete though they have been used with considerable success to analyse observed viscosities and to show that the interatomic fields so derived are in general agreement with those derived from the equation of state. Owing to their incompleteness they can at present only be used for such confirmatory calculations. Moreover, the work of Massey and Mohr† has shown that *all* classical calculations of transport phenomena are inadequate for helium and hydrogen, because for such systems moving at velocities of the order $(kT/m)^{\frac{1}{2}}$ the de Broglie wave length is comparable with the molecular diameter and the quantum theory of collisions must be used.

One may sum up the present situation as follows. For fairly heavy atoms a classical analysis of the viscosity would be adequate, but it can only be carried out accurately for the model used by Hassé and Cook, a model which is known to be inadequate, or for still less adequate models. If nevertheless we make the best analysis possible of the second virial coefficient using this model, the resulting interatomic energies should be in fair agreement. Table 20 shows that this expectation is confirmed.

TABLE 20.

Diameters σ and energy constants μ, $E(r) = -\mu r^{-4}$, for certain gases (rigid sphere model), from comparable analyses of the viscosity and the equation of state.

Gas	Viscosity (Sutherland). Formula (928)		Viscosity (Hassé and Cook)		Equation of state	
	$\sigma \times 10^8$	$4\mu \times 10^{43}$	$\sigma \times 10^8$	$4\mu \times 10^{43}$	$\sigma \times 10^8$	$4\mu \times 10^{43}$
Argon	2·90	2·94	2·98	1·29	3·13	0·704
Nitrogen	3·15	2·96	3·28	1·27	3·38	0·774
Carbon dioxide	3·36	8·17	3·55	3·40	—	—
Air	3·10	2·88	3·22	1·27	—	—

* Hassé and Cook, *Phil. Mag.* vol. 3, p. 977 (1927).

† Massey and Mohr, *Proc. Roy. Soc.* A, vol. 141, p. 434 (1933), vol. 144, p. 188 (1934).

The use of the accurate formula of Hassé and Cook in place of the weak field approximation (928) considerably improves the agreement with the equation of state, and one might expect agreement if accurate calculations could be made for a more suitable model.

We now turn to hydrogen and helium. For helium the experiments of Kamerlingh-Onnes and Weber* show that an excellent representation of the viscosity over the whole range from 15° K. to 457° K. is given by the

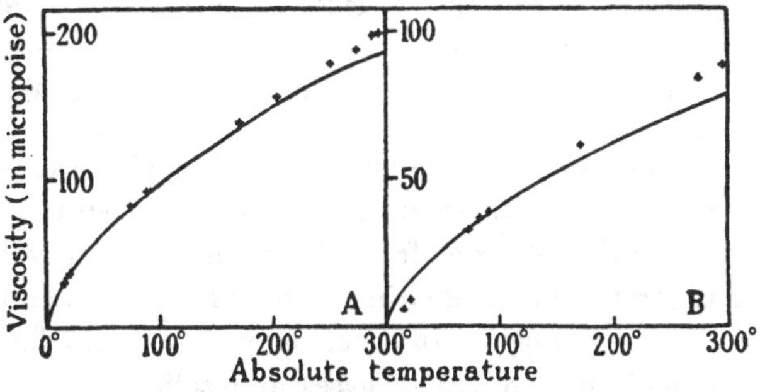

Fig. 21. The observed and theoretical viscosity of *A* helium, *B* hydrogen; rigid sphere model in quantum theory.

classical formula (925) if $s = 13{\cdot}6$, $\eta \propto T^{0{\cdot}647}$. An almost equally good representation of the viscosity of hydrogen (H_2) determined by the same workers is given by (925) if $s = 10$, $\eta \propto T^{0{\cdot}695}$. It has been shown, however, by Massey and Mohr† that, when the viscosity for the rigid sphere model ($s = \infty$) with no attractive field is calculated according to the quantum theory of

TABLE 21.

The viscosity of helium at various temperatures (numerical values in micropoise).

Temp. °K.	Observed η (Onnes and Weber)	Calculated η Quantum theory. Slater's $E(r)$
294·5	199·4	213
203·1	156·4	165
88·8	91·8	98
20·2	35·03	43
15·0	29·46	36

scattering, a temperature variation is obtained very similar to that observed, instead of the classical variation $\eta \propto T^{\frac{1}{2}}$ for this model. In this calculation the proper boundary conditions and the symmetry of the colliding systems must be allowed for. The agreement is shown in Fig 21. The radius of the model has of course been chosen to give the best fit; the actual values chosen

* Kamerlingh-Onnes and Weber, *Comm. Phys. Lab. Leiden*, No. 134 A, B, C (1913).
† Massey and Mohr, *loc. cit.* (1933).

were $\sigma\,(\mathrm{He}) = 2{\cdot}10\,\text{Å}.$; $\sigma\,(\mathrm{H_2}) = 2{\cdot}75\,\text{Å}$. This result at once shows that for these gases no information about interatomic forces can be derived from the viscosity by classical methods. No calculations have been made for any of the fields derived, as in this chapter, from the equation of state, but Massey and Mohr* have calculated the viscosity using Slater's interaction energy

$$E(r) = [7{\cdot}7e^{-2{\cdot}43/a_0} - 0{\cdot}68(a_0/r)^6] \times 10^{-10}. \qquad \ldots\ldots(930)$$

The derived values are somewhat too large especially at low temperatures, but the agreement is quite fair as is shown in Table 21. The agreement could be improved by decreasing the attractive part of Slater's field.

§ **10·5.** *Crystalline salts and similar solids.* The physical properties of certain solids indicate that the main attractions between the constituents are of an entirely different order from those derived from van der Waals energies. The difference in the melting points of rock-salt and solid argon may be quoted as an example. It is necessary to account for this change before the intermolecular forces in gases and solids can be successfully correlated. It may be regarded as established that a solid such as rock-salt consists of an array of closed electronic systems of the inert gas type formed by the transfer of the valency electron from the alkali to the halogen. The structural units differ therefore from the corresponding inert gas atoms first of all by possessing a resultant charge. The electrostatic energy of this charge distribution gives rise to the main part of the cohesion of the crystal and makes it altogether stronger.† X-ray analysis reveals to us the pattern on which the crystals are built. For many substances this information, together with the assumption of the electrostatic energy of the ionic charges and the other energies derived from gas data suitably modified, suffices for a theoretical calculation of their physical properties. The comparison of these calculations with observation provides a stringent test of the assumed interatomic energies.

Long before the nature of the overlap energy leading to the main repulsive terms was properly understood, Born proposed to represent this part of the interionic energy in a salt crystal by the function λr^{-s} and to discuss the equilibrium structure of the crystal using this term and the electrostatic energy, but neglecting in the first instance the heat motion, the van der Waals energy and the zero-point vibrational energy. For such a discussion as we saw in § 4·9 the important quantity is *the potential energy of the crystal*, here most conveniently used in the form of the potential energy *per unit*

* Massey and Mohr, *loc. cit.* (1934).

† For a full account of the older electron theory of valency one may refer to Kossel, *Valenzkräfte und Röntgenspektrum*, ed. 2, Berlin (1914); G. N. Lewis, *Valence* (1916); Sidgwick, *The electron theory of valency*, Oxford (1927).

*cell.** This potential energy may be regarded as made up of two parts (or three if the van der Waals energy is included)

$$\phi = \phi^{(e)} + \phi^{(s)} + \phi^{(t)}; \qquad \text{......(931)}$$

$\phi^{(e)}$, $\phi^{(s)}$ and $\phi^{(t)}$ are respectively the energies per unit cell resulting from the electrostatic, λr^{-s} and μr^{-t} terms.

For a crystal like rock-salt which consists of two interpenetrating face-centred cubic lattices only one parameter r (the closest distance between the ions) can appear in ϕ, and the equilibrium value $r = a$ is determined by

$$[\partial\phi/\partial r]_{r=a} = 0. \qquad \text{......(932)}$$

This gives strictly the value of a for zero temperature, zero-point vibrational energy neglected. Assuming that a is known and $\phi^{(t)}$ negligible, this equation was used by Born to determine λ for given s. Finally s was determined from the measured compressibility κ, which by equations (384) and (385) satisfies the equation

$$\frac{1}{\kappa} = -V\left(\frac{\partial p}{\partial V}\right)_T = \Delta\frac{\partial^2\phi}{\partial\Delta^2}, \qquad \text{......(933)}$$

where Δ is the volume of the unit cell. For the rock-salt type the unit cell may be taken to be a cube of edge $2a$ so that $\Delta = 8a^3$ and then†

$$\frac{1}{\kappa} = \frac{1}{72a}\left[\frac{\partial^2\phi}{\partial r^2}\right]_{r=a}. \qquad \text{......(934)}$$

Equation (934) determines s, but strictly it should be used only with κ_0 the compressibility at zero temperature. Using values of κ for room temperature Born‡ obtained the values of s given in Table 22.

TABLE 22.

Index s of the repulsive energy term (Born).

Salt	$\kappa \times 10^{12}$ obs.	s
NaCl	4·13	7·84
NaBr	5·1	8·61
NaI	6·9	8·45
KCl	5·62	8·86
KBr	6·2	9·78
KI	8·6	9·31

* Great confusion can arise unless care is taken to distinguish between the potential energy of an ion or of a cell in the lattice and the potential energy per ion or per unit cell. Since in calculating the potential energy of the whole system (which is then reduced to ion or cell) every term is reckoned twice over, the potential energy per unit cell is *half* the potential energy of the unit cell in the crystal. In giving numerical formulae it is also essential to specify exactly the unit cell. We shall attempt to be consistent and use only *potential energies of the system per atom or ion or unit cell.*

† Formula (934) is frequently given in the form $\frac{1}{\kappa} = \frac{1}{36a}\left[\frac{\partial^2\phi'}{\partial r^2}\right]_{r=a}$, but ϕ' is then the potential energy of a pair of ions in the lattice; thus $\phi' = 2\phi''$ where ϕ'' is reckoned per pair of ions, and $\phi = 4\phi''$ since there are four pairs of ions for the unit cell chosen in the text.

‡ Born, *Ann. d. Physik*, vol. 61, p. 87 (1919); *Verh. d. Deutsche Physikal. Ges.* vol. 21, pp. 199, 533 (1919).

Assuming an average value $s=9$ for all crystals of this type, Born then derived other properties such as the elastic constants, the total energy, *etc.* Later more accurate calculations* were made with the values of s in Table 22. Our object in this chapter however is to correlate the intermolecular energies for crystals and gases and we shall proceed differently.

In the earlier work it was assumed that all the ions in one crystal, whether like or unlike, repel according to the same law so that for instance the Cl ions in NaCl were assigned a very different overlap energy from those in KCl. It seems more reasonable to assume that two K^+ or two Cl^- ions, of the same electronic structure as argon, possess overlap energy (and van der Waals energy) following a law with the same r-variation as for two argon atoms, that is with the same s but different λ (and μ). The change in λ (and μ) is a secondary result of the ionic charge which changes the scale of the electronic orbits. It will be necessary in later sections to propose rules for relating λ and μ for an ion to λ and μ for the corresponding inert gas. Here it is sufficient to observe that interionic energies in salt crystals may be assumed to be of the form

$$E(r) = \pm\frac{\epsilon^2}{r}+\frac{\lambda}{r^s}-\frac{\mu}{r^t}.$$

The next step is to devise methods for calculating the resulting value of ϕ.

§ **10·6.** *Calculation of $\phi^{(e)}$, the electrostatic potential energy of an ionic crystal.* In the notation of §§ 4·9, 4·91 the electrostatic potential at any point in space $\mathbf{r}$ not occupied by a charge is

$$\phi(\mathbf{r}) = \mathbf{S}_l\Sigma_k\frac{\epsilon_k}{|\mathbf{r}_k^l-\mathbf{r}|}, \qquad \text{......(935)}$$

where ϵ_k is the charge on the kth ion in the basic r unit cell. It follows that the electrostatic potential energy of all the charges in a unit cell is given by

$$\mathbf{S}_l\Sigma'_{k,k'}\frac{\epsilon_k\epsilon_{k'}}{|\mathbf{r}_k^l-\mathbf{r}_{k'}|}, \qquad \text{......(936)}$$

where $\Sigma'_{k,k'}$ denotes as usual that those terms are omitted for which $k=k'$, $l=0$. The *potential energy per unit cell* is therefore given by

$$\phi^{(e)} = \tfrac{1}{2}\mathbf{S}_l\Sigma'_{k,k'}\frac{\epsilon_k\epsilon_{k'}}{|\mathbf{r}_k^l-\mathbf{r}_{k'}|}. \qquad \text{......(937)}$$

The evaluation of $\phi^{(e)}$ with special reference to a crystal was first made by Madelung,† though, unknown to him, a method of evaluation had already been given by Riemann‡ and by Appell.§ Madelung first considered a one-

* Born and Brody, *Zeit. f. Physik*, vol. 11, p. 327 (1922).

† Madelung, *Physikal. Zeit.* vol. 19, p. 524 (1918).

‡ Riemann, *Schwere, Electrizität und Magnetismus*, edited by Hattendorf, ed. 2, para. 23, Hanover (1880).

§ Appell, *Acta Math.* vol. 4, p. 313 (1884), vol. 8, p. 265 (1886); *J. de Math.* (4), vol. 3, p. 5 (1887).

dimensional crystal of lattice constant δ. His method consists in replacing the electric density by a Fourier expansion, and deducing a corresponding Fourier expansion for the potential at a point outside it. (The potential must have the same periodicity as the electric density.) This potential expansion must satisfy Laplace's equation at any point and Gauss' theorem near the line charge. These conditions, together with the fact that the potential must vanish at infinity, determine all the constants in the expression for the potential. For a series of discrete charges ϵ_k ($k=1,2,\ldots,s$) on the unit strip at points x_k, which on the whole are electrically neutral, the potential at a point in space distant ϖ from the line proves to be

$$\phi(x,\varpi)=\frac{2}{\delta}\sum_{n=-\infty}^{\infty}\Sigma_k\,\epsilon_k K_0\left(\frac{2\pi n\varpi}{\delta}\right)e^{2\pi i n(x-x_k)/\delta}, \qquad \ldots\ldots(938)$$

where $K_0(z)$ is the function tabulated by Jahnke and Emde* as $\frac{1}{2}i\pi H_0^{(1)}(iz)$. From the potential of a line charge, that of a plane array of charges and finally that of a space lattice can be deduced.

Improved methods have, however, been given by Ewald† for dealing with three-dimensional lattices. In this case the electric density has a threefold periodicity with respect to the vectors $\boldsymbol{\delta}_1$, $\boldsymbol{\delta}_2$ and $\boldsymbol{\delta}_3$. If p_1, p_2 and p_3 are the lengths of the perpendiculars between consecutive planes containing the pairs of vectors $(\boldsymbol{\delta}_2,\boldsymbol{\delta}_3)$, $(\boldsymbol{\delta}_3,\boldsymbol{\delta}_1)$ and $(\boldsymbol{\delta}_1,\boldsymbol{\delta}_2)$, then the density can be expressed in terms of a threefold Fourier expansion

$$\rho=\mathbf{S}_l\,\rho_l\,e^{i(\mathbf{q}^l\cdot\mathbf{r})}, \qquad \ldots\ldots(939)$$

where $\mathbf{r}$, as usual, is the vector (x,y,z), ρ_l is a coefficient depending on l_1, l_2, l_3, and

$$\mathbf{q}^l=2\pi(l_1\mathbf{b}_1+l_2\mathbf{b}_2+l_3\mathbf{b}_3). \qquad \ldots\ldots(940)$$

In this expression $\mathbf{b}_1$, $\mathbf{b}_2$ and $\mathbf{b}_3$ are vectors in the directions of p_1, p_2 and p_3, with magnitudes $1/p_1$, $1/p_2$ and $1/p_3$.

The potential $\phi(\mathbf{r})$ must have the same periodicity and must be expressible in a series similar to ρ, that is,

$$\phi(\mathbf{r})=\mathbf{S}_l\,c_l\,e^{i(\mathbf{q}^l\cdot\mathbf{r})}. \qquad \ldots\ldots(941)$$

Poisson's equation determines c_l in terms of ρ_l, giving

$$c_l=\frac{4\pi\rho_l}{|\mathbf{q}^l|^2}. \qquad \ldots\ldots(942)$$

If the system is electrically neutral and consists of discrete charges ϵ_k ($k=1,2,\ldots,s$) at positions $\mathbf{r}_k$, then

$$\int_\Delta \rho e^{-i(\mathbf{q}^l\cdot\mathbf{r})}dv=\int_\Delta \mathbf{S}_{l'}\,\rho_{l'}\,e^{i(\mathbf{q}^{l'}-\mathbf{q}^l\cdot\mathbf{r})}dv=\rho_l\Delta, \qquad \ldots\ldots(943)$$

* Jahnke and Emde, *Funktionentafeln*, p. 135, Leipzig (1909). See also Watson, *Bessel's Functions*, p. 78, and Table II, p. 698.

† Ewald, *Ann. d. Physik*, vol. 64, p. 253 (1921); see also Born, *Atomtheorie des festen Zustandes*, ed. 2, p. 723 (1923).

the other terms in $\mathbf{S}_{l'}$ vanishing when the integral is taken over the unit cell of volume Δ. We then get

$$\rho_l = \frac{1}{\Delta}\int_\Delta \rho e^{-i(\mathbf{q}^l.\mathbf{r})}dv = \frac{1}{\Delta}\Sigma_k \epsilon_k e^{-i(\mathbf{q}^l.\mathbf{r}_k)}. \qquad \text{......(944)}$$

The term for which $l_1 = l_2 = l_3 = 0$ vanishes. We thus find

$$\phi(\mathbf{r}) = \frac{4\pi}{\Delta}\mathbf{S}_l{}'\,\Sigma_k \frac{\epsilon_k}{|\mathbf{q}^l|^2}e^{i(\mathbf{q}^l.\mathbf{r}-\mathbf{r}_k)}, \qquad \text{......(945)}$$

$$= \Sigma_k \epsilon_k \psi(\mathbf{r}-\mathbf{r}_k),$$

where
$$\psi(\mathbf{r}) = \frac{4\pi}{\Delta}\mathbf{S}_l{}'\frac{e^{i(\mathbf{q}^l.\mathbf{r})}}{|\mathbf{q}^l|^2}. \qquad \text{......(946)}$$

The solution of the problem depends on the evaluation of this summation and therefore of its transformation into a more rapidly converging form. Since

$$\frac{1}{a} = \frac{1}{a}e^{-a\eta} + \int_0^\eta e^{-a\xi}\,d\xi, \qquad \text{......(947)}$$

we can write
$$\frac{\Delta}{4\pi}\psi(\mathbf{r}) = \mathbf{S}_l{}'\frac{e^{i(\mathbf{q}^l.\mathbf{r})-\eta|\mathbf{q}^l|^2}}{|\mathbf{q}^l|^2} + \int_0^\eta \mathbf{S}_l{}' e^{i(\mathbf{q}^l.\mathbf{r})-\xi|\mathbf{q}^l|^2}\,d\xi. \qquad \text{......(948)}$$

The second integral can be transformed in terms of Gauss' error function

$$G(x) = 1 - F(x) = \frac{2}{\sqrt{\pi}}\int_x^\infty e^{-\alpha^2}d\alpha, \qquad \text{......(949)}$$

which is a known tabulated function; in fact, it can be shown that

$$\psi(\mathbf{r}) = \frac{4\pi}{\Delta}\mathbf{S}_l{}'\frac{e^{i(\mathbf{q}^l.\mathbf{r})-\eta|\mathbf{q}^l|^2}}{|\mathbf{q}^l|^2} + \mathbf{S}_l{}'\frac{G\left(\dfrac{|\mathbf{r}^l-\mathbf{r}|}{2\eta^{\frac{1}{2}}}\right)}{|\mathbf{r}^l-\mathbf{r}|} - \frac{4\pi\eta}{\Delta}. \qquad \text{......(950)}$$

By a suitable choice of η the function ψ is thus divided into two parts, each of which is readily calculable. In practice one or two different values of η are chosen, and from the resulting values of ψ the actual value can be inferred to a high degree of accuracy.

In calculating the potential at a lattice point k, the term $\epsilon_k/(\mathbf{r}-\mathbf{r}_k)$ must be omitted from the summation, and so we have

$$\phi(\mathbf{r}_k) = \Sigma'_{k'}\epsilon_{k'}\psi(\mathbf{r}_k - \mathbf{r}_{k'}) + \epsilon_k\psi^*(0), \qquad \text{......(951)}$$

where
$$\psi^*(0) = \underset{r=0}{\mathrm{Lt}}\left\{\psi(\mathbf{r}) - \frac{1}{r}\right\}. \qquad \text{......(952)}$$

It can be shown that

$$\psi^*(0) = \frac{4\pi}{\Delta}\mathbf{S}_l{}'\frac{e^{-\eta|\mathbf{q}^l|^2}}{|\mathbf{q}^l|^2} + \mathbf{S}_l{}'\frac{G\left(\dfrac{|\mathbf{r}^l|}{2\eta^{\frac{1}{2}}}\right)}{|\mathbf{r}^l|} - \frac{1}{\sqrt{(\pi\eta)}} - \frac{4\pi\eta}{\Delta}. \qquad \text{......(953)}$$

In this way Ewald calculated the potential energy of cubic crystals of the NaCl and CaF_2 types.

The calculations can be extended to other cubic types, though the work may be shortened by improved methods due to Born* and Emersleben† which need not concern us here. It will be sufficient to record the results. It is clear that the electrostatic energy per unit cell or per molecule or unit set of ions can be expressed in the form

$$\phi^{(e)} = -\frac{1}{2}\frac{\epsilon^2}{d}\mathfrak{s}, \qquad \text{......(954)}$$

where ϵ is the electronic charge, d any convenient linear scale parameter and $\mathfrak{s}$ a numerical coefficient. This energy term for the complete crystal is then merely obtained from (954) by multiplying by N, the number of such units in the crystal. The numerical values for $\mathfrak{s}$ given in Table 23 are based on the following specifications. The three tetragonal axes are chosen as axes of reference and d is taken to be the shortest distance between ions of opposite sign resolved along any one of the tetragonal axes. It is better denoted by a_t. The unit group is one molecule.

TABLE 23.

Values of $\mathfrak{s}$ for various types of cubic crystal. The electrostatic energy of the crystal per molecule is

$$\phi^{(e)} = -\frac{1}{2}\frac{\epsilon^2}{a_t}\mathfrak{s}$$

if a_t is the component along the tetragonal axes of the shortest distance between ions of opposite sign.

Type	NaCl	MgO	ZnS	CaF_2	CsCl	Cu_2O‡
$\mathfrak{s}$	$3{\cdot}495{,}11^{7}_{3}$	$13{\cdot}980{,}4^{66}_{56}$	$7{\cdot}565{,}8^{7}_{1}$	$5{\cdot}818{,}^{31}_{25}$	$2{\cdot}035{,}3^{66}_{45}$	$5{\cdot}129$

‡ Corrected by Hund, *Zeit. f. Physik*, vol. 94, p. 11 (1935). The remaining values from Emersleben, *loc. cit.*

§ **10·61.** *The potential energies $\phi^{(s)}$ and $\phi^{(t)}$ per unit cell or per molecule.* The potential energy per unit cell, $\phi^{(s)}$, derived from interionic terms of the form λr^{-s} can be calculated comparatively easily. If an ion situated at $\mathbf{r}_k{}^l$ and an ion at $\mathbf{r}_{k'}$ in the basic unit cell contribute the term

$$\lambda_{k,k'}|\mathbf{r}_k{}^l - \mathbf{r}_{k'}|^{-s},$$

the interaction of the selected ion at $\mathbf{r}_{k'}$ and all other ions at $\mathbf{r}_k{}^l$ contributes $\phi_{k,k'}{}^{(s)}$ say, where

$$\phi_{k,k'}{}^{(s)} = \mathbf{S}_l \lambda_{k,k'}|\mathbf{r}_k{}^l - \mathbf{r}_{k'}|^{-s}, \quad \phi_{k,k'}{}^{(s)} = \phi_{k',k}{}^{(s)}, \qquad \text{......(955)}$$

and $\mathbf{S}_l$ converges if $s > 3$. If $k = k'$ the term $l = 0$ is omitted. With our definition of $\phi^{(s)}$ we have at once

$$\phi^{(s)} = \tfrac{1}{2}\Sigma_k \phi_{k,k}{}^{(s)} + \Sigma_{k \neq k'} \phi_{k,k'}{}^{(s)}, \qquad \text{......(956)}$$

* Born, *Zeit. f. Physik*, vol. 7, p. 124 (1921).
† Emersleben, *Physikal. Zeit.* vol. 24, pp. 73, 97 (1923).

where the second summation is taken over all pairs. For large values of s these summations can be evaluated well enough by direct summation. A better method is to use a formula given by Lennard-Jones and Ingham,* which is analogous in principle to the use of the Euler-Maclaurin sum formula for the Riemann zeta-function. This method has been used to evaluate the expressions (955) appropriate to the three cubic structures *simple*, *body-centred*, and *face-centred*.

(i) *Simple cubic.* We take a unit cell of edge δ containing one atom so that $k=1$. If a is the shortest distance between atoms, then $a=a_l=\delta$. We have

$$\phi^{(s)}=\tfrac{1}{2}\lambda A_s/a^s, \qquad \text{......(957)}$$

where

$$A_s=\mathbf{S}_l'(l_1^2+l_2^2+l_3^2)^{-\frac{1}{2}s}. \qquad \text{......(958)}$$

A_s may conveniently be called a potential energy constant.

(ii) *Body-centred cubic.* We take a unit cell of edge δ containing two atoms ($k=1, 2$) at the cube corners and centres, so that a the shortest distance between atoms is $\sqrt{3}\delta/2$, and $a_l=\tfrac{1}{2}\delta$. Then

$$\phi^{(s)}=\tfrac{1}{2}\lambda_{11}A_{s_{11}}/\delta^{s_{11}}+\lambda_{12}B'_{s_{12}}/\delta^{s_{12}}+\tfrac{1}{2}\lambda_{22}A_{s_{22}}/\delta^{s_{22}}, \qquad \text{......(959)}$$

where A_s has the same meaning as before and

$$B_s'=\mathbf{S}_l\{(l_1+\tfrac{1}{2})^2+(l_2+\tfrac{1}{2})^2+(l_3+\tfrac{1}{2})^2\}^{-\frac{1}{2}s}. \qquad \text{......(960)}$$

For a body-centred cubic crystal consisting of atoms of only one kind so that $\lambda_{11}=\lambda_{12}=\lambda_{22}$ and $s_{11}=s_{12}=s_{22}$, we have

$$\phi^{(s)}=\lambda B_s/a^s, \qquad \text{......(961)}$$

where

$$B_s=\left(\frac{a}{\delta}\right)^s A_s+\left(\frac{a}{\delta}\right)^s B_s'=\left(\frac{\sqrt{3}}{2}\right)^s(A_s+B_s'). \qquad \text{......(962)}$$

(iii) *Face-centred cubic.* For a face-centred cubic lattice of one sort of atom we take a unit cell of edge δ containing four atoms in all, so that a the shortest distance between the atoms is $\delta/\sqrt{2}$, and $a_l=\tfrac{1}{2}\delta$. In this case all the sets of summations can be reduced to a single one and we may write for the potential energy per unit cell of four atoms

$$\phi^{(s)}=2\lambda C_s/a^s, \qquad \text{......(963)}$$

where

$$C_s=2^{\frac{1}{2}s}\mathbf{S}_l'\{(l_2+l_3)^2+(l_3+l_1)^2+(l_1+l_2)^2\}^{-\frac{1}{2}s}. \qquad \text{......(964)}$$

For two interpenetrating face-centred cubic lattices as in NaCl the complete array of atoms is simple cubic and if a is the short distance between neighbours, $a=a_l=\tfrac{1}{2}\delta$, and the potential energy per molecule is

$$\phi^{(s)}=\tfrac{1}{2}\lambda_{11}A''_{s_{11}}/a^{s_{11}}+\lambda_{12}A'_{s_{12}}/a^{s_{12}}+\tfrac{1}{2}\lambda_{22}A''_{s_{22}}/a^{s_{22}}, \qquad \text{......(965)}$$

* Lennard-Jones and Ingham, *Proc. Roy. Soc.* A, vol. 107, p. 636 (1925).

where

$$A_s' = \underset{\substack{l_1+l_2+l_3 \\ odd}}{\mathsf{S}} (l_1^2+l_2^2+l_3^2)^{-\frac{1}{2}s}, \qquad \text{......(966)}$$

$$A_s'' = \underset{\substack{l_1+l_2+l_3 \\ even}}{\mathsf{S}'} (l_1^2+l_2^2+l_3^2)^{-\frac{1}{2}s}, \qquad \text{......(967)}$$

and

$$A_s' + A_s'' = A_s. \qquad \text{......(968)}$$

Exactly similar contributions are made by the van der Waals energies.

The functions A_s, A_s', A_s'', B_s and C_s have been evaluated for integral values of s from 4 to 30, and a selection of values is given in Table 24.

TABLE 24.

Potential energy constants for cubic crystals.

s	A_s	A_s'	A_s''	B_s	C_s
4	$16{\cdot}5323^{28}_{14}$	$10{\cdot}1977^{53}_{36}$	$6{\cdot}33457^{8}_{5}$	$22{\cdot}63872^{3}_{0}$	$25{\cdot}33830^{9}_{2}$
5	$10{\cdot}3775^{32}_{23}$	$7{\cdot}3780^{71}_{60}$	$2{\cdot}99946^{3}_{1}$	$14{\cdot}7585^{13}_{07}$	$16{\cdot}9675^{23}_{16}$
6	$8{\cdot}40192^{7}_{3}$	$6{\cdot}59518^{7}_{2}$	$1{\cdot}806740$	$12{\cdot}2533^{70}_{66}$	$14{\cdot}4539^{23}_{19}$
7	$7{\cdot}4670^{61}_{58}$	$6{\cdot}28624^{7}_{1}$	$1{\cdot}18081^{5}_{3}$	$11{\cdot}05424^{6}_{1}$	$13{\cdot}3593^{90}_{86}$
8	$6{\cdot}94580^{9}_{7}$	$6{\cdot}14568^{9}_{6}$	$0{\cdot}800121$	$10{\cdot}355^{200}_{196}$	$12{\cdot}80193^{9}_{6}$
9	$6{\cdot}6288^{61}_{57}$	$6{\cdot}0767^{64}_{59}$	$0{\cdot}55209^{9}_{7}$	$9{\cdot}8945^{92}_{88}$	$12{\cdot}49254^{9}_{5}$
10	$6{\cdot}4261^{20}_{18}$	$6{\cdot}04139^{4}_{2}$	$0{\cdot}38472^{7}_{6}$	$9{\cdot}564^{403}_{399}$	$12{\cdot}31124^{8}_{4}$
11	$6{\cdot}29229^{7}_{2}$	$6{\cdot}0226^{93}_{87}$	$0{\cdot}26960^{6}_{4}$	$9{\cdot}31326^{5}_{1}$	$12{\cdot}2009^{23}_{18}$
12	$6{\cdot}2021^{50}_{48}$	$6{\cdot}0125^{90}_{87}$	$0{\cdot}18956^{1}_{0}$	$9{\cdot}11418^{5}_{1}$	$12{\cdot}1318^{82}_{78}$
13	$6{\cdot}140^{602}_{598}$	$6{\cdot}0070^{51}_{45}$	$0{\cdot}13355^{3}_{1}$	$8{\cdot}95180^{9}_{6}$	$12{\cdot}08772^{9}_{4}$
14	$6{\cdot}09818^{5}_{4}$	$6{\cdot}00397^{5}_{3}$	$0{\cdot}0942^{12}_{09}$	$8{\cdot}8167^{73}_{69}$	$12{\cdot}05899^{4}_{0}$
15	$6{\cdot}06876^{7}_{3}$	$6{\cdot}00225^{5}_{0}$	$0{\cdot}06651^{3}_{1}$	$8{\cdot}70298^{7}_{3}$	$12{\cdot}04002^{6}_{2}$

With the help of this table it has been shown that atoms, whose interatomic potential energy can be represented by $\lambda r^{-s} - \mu r^{-t}$, will take up the form of a face-centred cubic lattice in preference to a body-centred, and a body-centred in preference to a simple cubic.*

§ **10·7.** *Applications of gas data to crystals of the inert gases.* The apparatus for study of the properties of crystals just prepared can be used quite simply for the inert gases; when these equations of state have been analysed their crystalline forms can be derived at once for any selected form of $E(r)$. The results are given in Table 25 for neon and argon. Corresponding comparisons cannot safely be made for hydrogen and helium for which, especially helium, zero-point vibrational energy is all-important, nor for krypton or xenon for which the equations of state are lacking. The observed crystalline form is face-centred cubic, in agreement with theory for the central forces assumed. From equation (963) for a face-centred cubic lattice it follows that, when

$$\frac{\partial}{\partial a}(\phi^{(s)} + \phi^{(6)}) = 0,$$

$$a_0^{s-6} = \frac{s\lambda C_s}{6\mu C_6}, \quad U_0 = -\tfrac{1}{4}N(\phi^{(s)} + \phi^{(6)}) = \tfrac{1}{2}N\left(1 - \frac{6}{s}\right)\frac{\mu C_6}{a_0^6}. \qquad \text{...(969)}$$

* Lennard-Jones and Ingham, *loc. cit.* When electrostatic energies are also present, the stable form can change from NaCl type to the CsCl type according to the relative importance of the van der Waals energy. Born and Mayer, *Zeit. f. Physik*, vol. 75, p. 1 (1932).

The values for $s = 9$ for neon and for $s = 10+$ for argon are in fair agreement with the observations, but for the correct lattice constant the calculated energies are undoubtedly somewhat too small.

TABLE 25.

Crystal spacing and energies of neon and argon derived from the $E(r)$ of Table 19. (Cubic close packing.)

s	Closest distance of atoms a_0 Å. calc.	a_0 Å. obs.	Heat of evaporation at 0° K. cal./mole U_0 calc.	U_0
	Ne			
8	3·23	3·20	476	Obs.(1) 447
9	$3{\cdot}15_5$	—	512	
10	3·09	—	547	Corrected for zero-point vibrations
12	2·99	—	609	
14	$2{\cdot}91_5$	—	661	590
	A			
8	3·96	3·81	1610	Obs.(2) 1840
9	3·88	—	1734	
10	3·82	—	1848	Corrected for zero-point vibrations
12	$3{\cdot}71_5$	—	2058	
14	3·64	—	2238	2030

The calculated values of the potential energy must be compared with observed heats of evaporation plus the zero-point vibrational energy calculated from equation (353·3). [Θ_D (Ne) = 63, Θ_D (A) = 85.]

(1) Clusius, *Zeit. f. physikal. Chem.* B, vol. 4, p. 1 (1929).

(2) F. Born, *Ann. d. Physik*, vol. 69, p. 473 (1922).

§ 10·71. *Overlap and van der Waals energy constants for atomic ions in crystals.* Before we can proceed further it is necessary to obtain values for λ and μ suitable for the atomic ions in crystals by proper modifications of the values for inert gas atoms. If atoms and ions could really be adequately represented by spherical shells the method of passing from an inert gas atom to the neighbouring ions would be simple. From the refractive indices of crystals or solutions it is possible, as was first shown by Wasastjerna,* to deduce the refractivities of their ions and therefrom (see § 10·72) the relative sizes of atoms and ions of similar structure. The ratios so obtained depend on the average extension of the outer electronic orbits from the nucleus, which depends on the nuclear charge. Now the overlap energy (causing the repulsion) between these closed electronic systems equally depends on the

* Wasastjerna, *Soc. Scient. Fennicae, Comm. Phys. Math.* vol. 1, No. 38 (1923).

extension of the outer orbits, since it is controlled by the degree of overlapping of their wave-functions. Thus when the repulsions are represented by hard impenetrable shells the diameter of the shell may be taken to be directly proportional to the scale of the outer orbits. When the overlap energy is expressed in the form $\lambda_s r^{-s}$ the quantity

$$\sigma_s = [\lambda_s/W]^{1/s} \qquad \text{......(970)}$$

replaces the rigid diameter σ_∞ or σ of the elementary kinetic theory, W being any convenient energy value. This σ_s is then the classical closest distance of approach when two such molecules collide with a kinetic energy of their relative motion equal to W. It therefore seems reasonable to assume that σ_s is proportional to the extension of the outer orbits. We therefore write

$$\sigma_s^+ = (\rho_+/\rho)\,\sigma_s, \qquad \text{......(971)}$$

where ρ_+/ρ is the ratio of the orbital extension of the ion and atom deduced from the refractivity. It follows at once from this that

$$\lambda_s^+ = \left(\frac{\rho_+}{\rho}\right)^s \lambda_s, \qquad \text{......(972)}$$

a result independent of W. Similar formulae are used for negative ions and for ions of higher valency:

$$\lambda_s^- = \left(\frac{\rho_-}{\rho}\right)^s \lambda_s, \quad \lambda_s^{++} = \left(\frac{\rho_{++}}{\rho}\right)^s \lambda_s, \ldots. \qquad \text{......(973)}$$

Combined with (922) these formulae determine all the λ's required in terms of λ_s for the inert gases and ratios of the ρ's derived from the refractivities. Very similar considerations apply to the μ's, but these may be treated less superficially and we return to them in § 10·73.

§ **10·72.** *Use of ionic refractivities.* The ratios of the ρ's can be derived from the refractivities as follows, the method given here being somewhat more refined than that used by Wasastjerna. Kirkwood* has shown that the atomic polarizability is proportional to $(\Sigma_p \overline{r_p^2})^2$, where $\overline{r_p^2}$ is the mean square distance from the nucleus of the pth electron in the atom and the summation extends over all electrons. The greater part of the sum, usually more than 90 per cent. in an atom of rare gas structure, is contributed by the electrons of the outer shell and we may infer that the polarizability of atoms of the same electronic structure varies approximately as the fourth power of the extension of the outer electronic orbits from the nucleus. Since the refractivity R is related to the atomic polarizability by the equation $R = \frac{4}{3}\pi n\alpha$ (see § 12·2), we have

$$\frac{\rho_+}{\rho} = \left(\frac{R_+}{R}\right)^{\frac{1}{4}} = \left(\frac{\alpha_+}{\alpha}\right)^{\frac{1}{4}} \qquad \text{......(974)}$$

* Kirkwood, *Physikal. Zeit.* vol. 33, p. 57 (1932).

if R_+ and R are the mole refractivities for the univalent positive ion and the rare gas atom of the same structure. Similar formulae hold for other ions.

Now the measurement of the refractive index of a crystal does not give directly the refractivities of the ions that compose it, but only their combined refractivity. Moreover, the experiments of Spangenberg† show that the refractivity, especially of negative ions, is not a constant property of the ion but depends on the positive ions with which it is associated. It appears therefore safest to assume that the compact positive ion has a structure not greatly deformed by inclusion in a crystal and therefore to assume for the ratios ρ_+/ρ or ρ_{++}/ρ the values calculated by Slater‡ for free atoms and positive ions from wave-functions of the type $r^{n^*-1}e^{-(Z-s)r/n^*}$. In this expression n^* is an effective quantum number, s a screening constant and Z the nuclear charge. The value of R for the inert gas is known accurately and given in Table 18. The values of R_+ and R_{++} then follow at once from (974). These values are given in Table 26, with values derived in other ways for comparison.

TABLE 26.

Values of ρ_+/ρ and ρ_{++}/ρ from Slater's wave-functions, and mole refractivities for positive ions derived from them.

[The mole refractivity R is $\frac{4}{3}\pi N\alpha$, where N is Avogadro's number.]

Ion	ρ_{ion}/ρ (Slater)	R_{ion}	R_{ion} (Heydweiller)	R_{ion} (Pauling)
Na^+	0·850	0·52	0·65	0·46
K^+	0·877	2·45	2·71	2·12
Rb^+	0·891	3·95	4·10	3·57
Cs^+	0·894	6·50	6·71	6·15
Mg^{++}	0·74	0·30	—	0·24
Ca^{++}	0·77	1·45	1·60	1·19
Sr^{++}	0·80	2·57	2·56	2·18
Ba^{++}	$0{\cdot}80_5$	4·28	5·00	3·94

The values of ρ_{ion}/ρ and R_{ion} are those adopted in the following calculations. The values given by Heydweiller [*Physikal. Zeit.* vol. 26, p. 526 (1925)] are obtained from ions in dilute solution; those by Pauling [*Proc. Roy. Soc.* A, vol. 114, p. 181 (1927)] from calculations; these values are inserted for comparison.

The refractivity R of the negative ion in a given crystal may now be obtained by subtracting from the observed total refractivity the contribution R_+ of the positive ion computed above, and ρ_-/ρ then follows from (974). The values of ρ_-/ρ derived in this way vary from crystal to crystal and in general the smaller the positive ion the greater the deformation of the negative ion and the smaller ρ_-/ρ. The values given in Table 27 were derived

† Spangenberg, *Zeit. f. Kristal.* vol. 57, p. 494 (1923).
‡ Slater, *Phys. Rev.* vol. 36, p. 57 (1930).

in this way from Spangenberg's measurements. At the end of the table values are also given for R_{--} and ρ_{--}/ρ. Measurements of the refractivities of bivalent ions are scarce and no attempt to estimate the variation of ρ_{--}/ρ from crystal to crystal is yet possible. The values given for O^{--} and S^{--} are those originally given by Wasastjerna* and the values for Se^{--} and Te^{--} are extrapolated from them.

We are now in a position to complete the determination of the energy constants λ and μ for all the pairs of ions covered by these tables.

TABLE 27.

Values of ρ_{-}/ρ and R_{-} for negative ions derived from Spangenberg's measurements. Values of ρ_{--}/ρ and R_{--} from Wasastjerna.

Salt	ΣR (Spangenberg)	$\Sigma R - R_+$	R_- adopted	ρ_-/ρ
NaF	2·97	2·45	2·45	1·253
KF	5·00	2·55	2·55	1·265
RbF	6·58	2·63	2·60	1·272
CsF	9·14	2·64	2·60	1·272
NaCl	8·30	7·78	7·78	1·171
KCl	10·5	8·05	8·05	1·181
RbCl	12·0	8·05	8·05	1·181
CsCl	14·5	8·00	8·05	1·181
NaBr	11·1	10·58	10·60	1·141
KBr	13·5	11·05	11·15	1·155
RbBr	15·1	11·15	11·15	1·155
CsBr	17·8	11·30	11·15	1·155
NaI	$16{\cdot}1_5$	15·6	15·60	1·113
KI	18·8	$16{\cdot}3_5$	16·35	1·126
RbI	20·7	$16{\cdot}7_5$	16·75	1·133
CsI	23·2	16·7	16·75	1·133

The values of $\Sigma R - R_+$ are derived as described in the text and smoothed values are entered as R_- and used in the calculations.

Ion	ρ_{--}/ρ	R_{--}	R_{--} (Pauling)
O^{--}	1·45	4·38	9·88
S^{--}	1·38	16·4	26·0
Se^{--}	1·30	17·9	26·8
Te^{--}	1·29	28·6	35·6

§ **10·73**. *The van der Waals energy.* In evaluating constants such as μ_+^+ for the van der Waals energy of a pair of like positive ions we can make direct appeal to Kirkwood's formula (914). According to (914) μ_+^+ will be related to μ for the corresponding pair of atoms by the equation

$$\mu_+^+/\mu = (\alpha_+/\alpha)^{\frac{3}{2}}. \qquad \text{......(975)}$$

* Wasastjerna, *loc. cit.*

It follows from (974) that $\mu_{+}^{+}/\mu = (\rho_{+}/\rho)^{6}$, ...(976)

which is the same as the semi-empirical (972) for the λ's. In the same way

$$\mu_{++}^{++}/\mu = (\rho_{++}/\rho)^{6}, \quad \mu_{-}^{-}/\mu = (\rho_{-}/\rho)^{6}. \qquad \text{......(977)}$$

When the values of μ have not been directly determined, as for Kr and Xe, the theoretical values of c given by Kirkwood's formula are used instead, increased by 20 per cent. to allow for the dipole-quadripole term. The values adopted are shown in Table 29 below.

We require also the van der Waals energy of two unlike atoms or ions. Kirkwood has given the formula

$$c_{12} = 3E_0 a_0^{\frac{3}{2}} \frac{\alpha_1 \alpha_2}{(\alpha_1/Z_1^*)^{\frac{1}{2}} + (\alpha_2/Z_2^*)^{\frac{1}{2}}} \qquad \text{......(978)}$$

as a companion to (914), namely

$$c_{11} = \tfrac{3}{2} E_0 a_0^{\frac{3}{2}} Z_1^{*\frac{1}{2}} \alpha_1^{\frac{3}{2}}, \quad c_{22} = \tfrac{3}{2} E_0 a_0^{\frac{3}{2}} Z_2^{*\frac{1}{2}} \alpha_2^{\frac{3}{2}}.$$

We can then express c_{12} in terms of c_{11} and c_{22} in the form

$$c_{12} = 2\left(\frac{c_{11}^2 c_{22}^2}{Z_1^* Z_2^*}\right)^{\frac{1}{3}} \Big/ \left\{\left(\frac{c_{11}}{Z_1^{*2}}\right)^{\frac{1}{3}} + \left(\frac{c_{22}}{Z_2^{*2}}\right)^{\frac{1}{3}}\right\}, \qquad \text{......(979)}$$

or if $Z_1^* = Z_2^*$, $\qquad c_{12} = 2(c_{11}^2 c_{22}^2)^{\frac{1}{3}}/\{c_{11}^{\frac{1}{3}} + c_{22}^{\frac{1}{3}}\}$. ......(980)

On the basis of these formulae and the values of μ summarized in Table 29 and of ρ_{ion}/ρ in Tables 26 and 27 values of μ_{+}^{+}, μ_{-}^{-}, μ_{+}^{-}, ... for various pairs of ions have been prepared; they are set out in Table 31.

§ **10·74.** *The correlation of the properties of salt crystals and overlap energies. The choice of s.* We have seen that the gas data do not suffice to fix s precisely and it is hardly worth while to carry out sets of calculations such as those of §10·7 for various salts and various values of s with the values of λ and μ of Table 19 appropriately modified according to the rules just laid down. It is sufficient to adopt some crystal property which can be used to fix s sharply before elaborate calculations need be undertaken, and then to complete the study with the selected value of s. The method chosen should be as far as possible independent of the various semi-empirical methods we have had to introduce for fixing λ_{+}^{+}..., μ_{+}^{+}.... The method we shall adopt is due to Hildebrand.†

We start with the thermodynamic formula

$$\left(\frac{\partial E}{\partial V}\right)_T = T\left(\frac{\partial p}{\partial T}\right)_V - p.$$

If κ is the compressibility and α the coefficient of (linear) thermal expansion,

$$\left(\frac{\partial E}{\partial V}\right)_T = \frac{T}{\kappa V}\left(\frac{\partial V}{\partial T}\right)_p - p = \frac{3\alpha T}{\kappa} - p. \qquad \text{......(981)}$$

We may now assume that the change of the vibrational energy with volume

† Hildebrand, *Zeit. f. Physik*, vol. 67, p. 127 (1931).

at any temperature not too high is only a small fraction of the total change in E with volume, so that we may substitute $N\phi$ for E. We may further consider only small pressures of the order of an atmosphere and put $p=0$. We thus have

$$\frac{3\alpha T}{\kappa}=N\frac{\partial\phi}{\partial V}=\frac{\partial\phi}{\partial\Delta},$$

or

$$9a\Delta\alpha T/\kappa=a^2\,\partial\phi/\partial a. \qquad \text{......(982)}$$

For a crystal of the NaCl type the potential energy ϕ *per molecule* is given by

$$\phi=-\tfrac{1}{2}(3{\cdot}495)\frac{z^2\epsilon^2}{a}+\frac{O}{a^s}-\frac{W}{a^6}, \qquad \text{......(983)}$$

where

$$O=\tfrac{1}{2}(\lambda_{+}^{+}+\lambda_{-}^{-})\,A_s''+\lambda_{+}^{-}A_s',\quad W=\tfrac{1}{2}(\mu_{+}^{+}+\mu_{-}^{-})\,A_6''+\mu_{+}^{-}A_6'. \qquad \text{......(984)}$$

The valency of the ions is z and for simplicity it is assumed that s is the same for all ions. For $a=a_0$, its value at zero temperature, $\partial\phi/\partial a_0=0$. If we combine equations (982) and (983), we obtain

$$\frac{9a\Delta\alpha T}{\kappa}=\tfrac{1}{2}(3{\cdot}495)\,z^2\epsilon^2-\frac{sO}{a^{s-1}}+\frac{6W}{a^5}. \qquad \text{......(985)}$$

With this ϕ the volume *per molecule* is Δ. If we eliminate O by means of $\partial\phi/\partial a_0=0$, we find

$$\frac{9a\Delta\alpha T}{\kappa}=\tfrac{1}{2}(3{\cdot}495)\left(1-\left(\frac{a_0}{a}\right)^{s-1}\right)+\frac{6W}{a^5}\left(1-\left(\frac{a_0}{a}\right)^{s-6}\right). \qquad \text{...(986)}$$

The van der Waals energy term, W, is of the nature of a correcting term, and the values of the μ's from Table 31 may be safely used in it as s will not differ much from 9. All the other quantities are observed quantities except s, so that s is thereby determined. The method requires that reliable values of a_0 shall have been obtained by extrapolation to the absolute zero. The results for six alkali halides for which a_0 has been determined by Henglein* are given in Table 28. On the evidence of this table it is clear that $s\sim 9$ and we shall be content for the rest of our calculations of interatomic energies to assume that $s=9$. Similar calculations could be made for body-centred structures.

TABLE 28.

Values of s for alkali halide crystals derived by Hildebrand's method.

[The values of s differ from those given by Hildebrand because the van der Waals energy has been included here.]

Substance	s	Substance	s
NaCl	8·8	KCl	9·1
NaBr	8·7	KBr	8·6
NaI	8·6	KI	8·6

* Henglein, *Zeit. f. Elektrochem.* vol. 31, p. 424 (1925).

TABLE 29.

Adopted values of van der Waals energy constants μ for inert gas atoms with the interaction energy $E(r)=\lambda r^{-9}-\mu r^{-6}$.

$\mu\times10^{60}$

He 2·26	Ne 14·47	A 170·2	Kr 346	Xe 877

TABLE 30.

Adopted values of the overlap energy constants λ for inert gas atoms with the interaction energy $E(r)=\lambda r^{-9}-\mu r^{-6}$. Values of λ from gas data are included for comparison.

Gas	$\lambda\times10^{82}$	
	From crystals	From gases
Ne	4·00	3·50
A	79·5	76·8
Kr	193·6	—
Xe	583	—

TABLE 31.

Adopted values of van der Waals energy constants μ for ion pairs in crystals.

$$\left[E(r)=\pm\frac{\epsilon_1\epsilon_2}{r^2}+\frac{\lambda}{r^9}-\frac{\mu}{r^6}\right]$$

$\mu\times10^{60}$

(i) *Pairs of like ions*

Na^+	5·45	Mg^{++}	2·38	O^{--}	135
K^+	77·5	Ca^{++}	35·5	S^{--}	1175
Rb^+	173	Si^{++}	90·7	Se^{--}	1670
Cs^+	447	Ba^{++}	239	Te^{--}	4045

Ion pair	*Associated positive ion*			
	Na^+	K^+	Rb^+	Cs^+
F^-	55·9	59·2	61·3	61·3
Cl^-	438	462	462	462
Br^-	762	821	821	821
I^-	1665	1785	1850	1850

(ii) *Pairs of unlike ions of opposite sign*

Ion	Na^+	K^+	Rb^+	Cs^+	Mg^{++}	Ca^{++}	Sr^{++}	Ba^{++}
F^-	16·2	67·0	99·9	161	10·5	44·8	69·9	115
Cl^-	42·6	181	262	425	26·7	117	181	298
Br^-	59·7	249	365	590	38·4	164	255	419
I^-	88·4	367	548	885	57·6	245	383	629
O^{--}	23·7	98·2	142	231	14·5	63·6	98·5	162
S^{--}	63·9	273	391	635	38·6	174	266	440
Se^{--}	83·5	346	501	812	51·2	224	347	571
Te^{--}	130·0	537	779	1260	78·9	348	538	886

TABLE 32.

Adopted values of overlap energy constants λ for ion pairs in crystals.

$$\left[E(r) = \pm\frac{\epsilon_1\epsilon_2}{r} + \frac{\lambda}{r^9} - \frac{\mu}{r^6}\right]$$

$\lambda \times 10^{82}$

(i) *Pairs of like ions*

Na^+	0·927	Mg^{++}	0·266	O^{--}	113·5
K^+	24·4	Ca^{++}	7·55	S^{--}	1450
Rb^+	68·7	Sr^{++}	26·0	Se^{--}	2050
Cs^+	212·5	Ba^{++}	82·7	Te^{--}	5190

Ion pair	*Associated positive ion*			
	Na^+	K^+	Rb^+	Cs^+
F^-	30·4	33·2	34·9	34·9
Cl^-	329	355	355	355
Br^-	633	708	708	708
I^-	1525	1695	1790	1790

(ii) *Pairs of unlike ions of opposite sign*

Ion	Na^+	K^+	Rb^+	Cs^+	Mg^{++}	Ca^{++}	Sr^{++}	Ba^{++}
F^-	6·29	28·4	49·4	90·3	4·04	16·9	30·1	55·6
Cl^-	$27{\cdot}9_5$	103	162	$276{\cdot}_5$	18·2	62·1	106	182
Br^-	43·3	154	$237{\cdot}_5$	396	30·1	99·0	159	266
I^-	78·9	$260{\cdot}_5$	406	$657{\cdot}_5$	55·6	170·0	275	433
O^{--}	14·1	54·4	88·5	156	9·06	32·5	55·6	99·0
S^{--}	75·9	$235{\cdot}_5$	358	583	52·3	154	242	385
Se^{--}	97·2	293	441	$714{\cdot}_5$	68·3	192	299	473
Te^{--}	202	561	821	1285	144	385	580	884

§ **10·75.** *The determination of the λ's.* If all the equations of state of the inert gases had been analysed we could now proceed to derive crystal data from the analysed gas data and compare the results with observation. But since gas data are missing for Kr and Xe, this cannot be done for the corresponding crystals. For the sake of a uniform procedure in all cases we shall therefore not adopt any of the λ's of Table 19 but use instead *the crystal spacings of* NaF, KCl, RbBr and CsI *to determine all the* λ's, which can then be used for other crystals, and compared with gas data for Ne and A. Only these four spacings need be assumed *a priori*. For NaF, KCl and RbBr it is only necessary to substitute observed values of a, Δ, α and κ for any one temperature into (985), to put $s=9$ and to calculate W from the μ's of Table 31. The equation then determines O. But for these crystals, in which both ions have the same inert gas structure, we can apply equations (972) and (922) and find

$$O = \tfrac{1}{2}(\lambda_+^+ + \lambda_-^-)\,A_9'' + \lambda_+^- A_9' = \left[\frac{1}{2}\left\{\left(\frac{\rho_+}{\rho}\right)^9 + \left(\frac{\rho_-}{\rho}\right)^9\right\} A_9'' + \left(\frac{\rho_+ + \rho_-}{2\rho}\right)^9 A_9'\right]\lambda, \quad \text{......(987)}$$

so that O fixes λ for the corresponding inert gas and therefore the derived λ_{+}^{+}, λ_{-}^{-}, ... For the body-centred structure of CsI we must use instead of (985) the equation

$$\frac{9\delta\Delta\alpha T}{\kappa}=\delta^2\frac{\partial\phi}{\partial\delta}=2{\cdot}035z^2\epsilon^2-\frac{sO'}{\delta^{s-1}}+\frac{6W'}{\delta^5}, \qquad \text{......(988)}$$

where

$$O'=\tfrac{1}{2}(\lambda_{+}^{+}+\lambda_{-}^{-})A_s+\lambda_{+}^{-}B_s', \quad W'=\tfrac{1}{2}(\mu_{+}^{+}+\mu_{-}^{-})A_6+\mu_{+}^{-}B_6'. \quad \text{...(989)}$$

The rest of the work is unaltered. Values of λ derived in this way are given in Table 30, and a complete set of values of λ_{+}^{+}, λ_{-}^{-}, ... derived from them by means of (972), (922) and the tables of ρ_{+}/ρ, ρ_{-}/ρ, ... are given in Table 32.

§ **10·8.** *Theoretical calculation of crystal properties.* (i) *Lattice constants.* For any alkali halide crystal or similar crystal with bivalent ions which is of the rock-salt type the lattice constant a at any temperature can be calculated at once from (985) and (984) using the tabulated λ's and μ's and $s=9$. The term $9a\Delta\alpha T/\kappa$ is a correcting term in which experimental values may be used for any selected temperature. Similar calculations for the CsCl type of crystal use equations (988) and (989) instead. The equations can be solved for a (or δ) by any convenient method of successive approximation in which the term in W (W') is regarded as a small correction. The convergence of the approximations is rapid.

The observed and calculated values of a and δ are given in Tables 33, 34. For the alkali halides all values refer to 15° C. For the bivalent salts the calculations are much rougher, the thermal expansion term has been neglected and the calculations refer to 0° K.

(ii) *Crystal energies.* We can also calculate the work U_0 required (neglecting zero-point vibrational energy) to disperse completely at zero temperature all the ions contained in (say) one mole of any one of these salts. If $\phi^{(s)}$ is defined by (970) or (976), U_0 is given in terms of ϕ by

$$U_0=-N\phi\,\text{ergs}=-N\phi/J\,\text{calories}.$$

For the NaCl type we have therefore, when N is Avogadro's number,

$$U_0=\frac{N}{J}\left\{\frac{3{\cdot}495z^2\epsilon^2}{2a}-\frac{O}{a^9}+\frac{W}{a^6}\right\}\text{cal./mole.} \qquad \text{......(990)}$$

For the CsCl type

$$U_0=\frac{N}{J}\left\{\frac{2{\cdot}035z^2\epsilon^2}{\delta}-\frac{O'}{\delta^9}+\frac{W'}{\delta^6}\right\}\text{cal./mole.} \qquad \text{......(991)}$$

The values of the crystal energies so calculated are given in Tables 33, 34. They are uniformly slightly larger than those calculated by Born and Mayer* using an exponential repulsive field; the difference may arise because the greater extension of the r^{-9} law field demands too great a

* Born and Mayer, *loc. cit.*

van der Waals term to compensate it. The "observed" energies derived in § 10·9 agree well with Born and Mayer's calculations. Allowing for the outstanding excess for which an origin has just been suggested, the general agreement between gas and crystal data here exhibited is most satisfactory.

TABLE 33.

Properties of alkali halide crystals ($s = 9$).

	$a \times 10^8$ (cm.)		U (kilocal./mole)			$\kappa \times 10^{12}$ (bar.)		
	Calc.	Obs.(1)	Calc.	B. and M.(2)	Obs.(3)	Calc. (1)	Calc. (2)	Obs.(4)
NaF	[2·310]	2·310	228·2	213·4	—	1·54	1·82	2·11
NaCl	2·810	2·814	190·1	183·1	181	3·19	3·85	$4{\cdot}26_3$
NaBr	2·974	2·981	181·4	174·6	176	3·88	4·73	5·08
NaI	3·204	3·231	170·8	163·9	166	5·15	6·36	(7·0)
KF	2·696	2·665	198·4	189·7	—	2·67	3·22	3·31
KCl	[$3{\cdot}138_5$]	$3{\cdot}138_5$	173·1	165·4	163	4·59	5·62	5·63
KBr	3·291	3·293	166·3	159·3	160	5·41	6·68	6·70
KI	3·499	3·526	158·8	150·8	151	6·59	8·25	8·54
RbF	2·872	2·815	187·1	181·6	—	3·34	4·05	(4·3)
RbCl	3·287	3·27	165·7	160·7	159	5·37	6·61	(6·9)
RbBr	[3·427]	3·427	160·6	153·5	157	6·13	7·60	7·94
RbI	3·642	3·663	153·5	145·3	148	7·49	9·42	9·58
CsF	3·072	3·004	176·9	173·7	—	4·22	5·17	(4·6)
CsCl*	4·157	4·110	157·3	152·2	—	5·58	7·06	5·9
CsBr*	4·313	4·287	153·5	146·3	—	6·11	7·81	7·0
CsI*	[4·561]	4·561	147·7	139·1	141	7·06	9·12	9·3

* Body-centred crystals.

(1) Observed values of a taken from Ewald and Hermann, *Strukturbericht* (1931). The values given for CsCl, CsBr, CsI are of δ, not of a. The values in [ ] are assumed.

(2) Born and Mayer, *Zeit. f. Physik*, vol. 75, p. 1 (1932).

(3) See Table 37, except for CsI for which see Mayer, *Zeit. f. Physik*, vol. 61, p. 798 (1930); Mayer and Helmholtz, *Zeit. f. Physik*, vol. 75, p. 19 (1932).

(4) Huggins and Mayer, *J. Chem. Physics*, vol. 1, p. 643 (1933), where full references are given. Values in brackets are estimated.

(iii) *Crystal compressibilities.* We can also derive values for crystal compressibilities which can at once be compared with observation. Starting from the same thermodynamic formula

$$\left(\frac{\partial E}{\partial V}\right)_T = T\left(\frac{\partial p}{\partial T}\right)_V - p,$$

we obtain

$$\left(\frac{\partial^2 E}{\partial V^2}\right)_T = \frac{1}{\kappa V}\left[1 + \frac{T}{\kappa}\left(\frac{\partial \kappa}{\partial T}\right)_V\right] \sim N\frac{\partial^2 \phi}{\partial V^2}$$

using the same approximation as before. It is easily shown that

$$N\frac{\partial^2 \phi}{\partial V^2} = \frac{1}{9N\Delta^2}\left[\delta^2 \frac{\partial^2 \phi}{\partial \delta^2} - 2\delta \frac{\partial \phi}{\partial \delta}\right].$$

On using (982) we thus find that

$$a^2\frac{\partial^2\phi}{\partial a^2}=\delta^2\frac{\partial^2\phi}{\partial\delta^2}=\frac{9\Delta}{\kappa}\left[1+2\alpha T+\frac{T}{\kappa}\left(\frac{\partial\kappa}{\partial T}\right)_V\right]. \qquad \text{......(992)}$$

The terms in [ ] are of the order of small correcting terms and vanish at 0° K. Using experimental values for the terms in [ ] equation (992) enables κ to be calculated, using the calculated values of a or δ. The values for κ for room temperature are given in Table 33. They are uniformly smaller than the observed values. The reason for this is certainly the too rapidly increasing slope of the λr^{-9} term as r diminishes.

TABLE 34.

Properties of bivalent salt crystals ($s=9$).

[Crystals are assumed to be all of NaCl type.]

	$a\times10^8$		U_0 kilo-cal./mole) (calc.)	$\kappa\times10^{12}$	
	Calc.	Obs.†		Calc. (1)	Calc. (2)
MgO	2·12	$2{\cdot}10_4$	976	0·27	0·31
MgS	$2{\cdot}72_5$	$2{\cdot}59_5$	766	0·70	0·82
MgSe	2·82	$2{\cdot}72_5$	743	0·78	0·92
MgTe*	$(3{\cdot}09_5)$	—	—	—	—
CaO	2·38	$2{\cdot}40_1$	876	0·41	0·48
CaS	2·93	$2{\cdot}84_3$	714	0·92	1·08
CaSe	3·01	$2{\cdot}95_6$	700	0·99	1·17
CaTe	$3{\cdot}27_5$	$3{\cdot}17_2$	648	1·33	1·59
SrO	2·52	$2{\cdot}57_5$	829	0·51	0·60
SrS	3·05	$3{\cdot}00_5$	688	1·06	1·25
SrSe	$3{\cdot}12_5$	$3{\cdot}11_5$	675	1·14	1·36
SrTe	3·39	$3{\cdot}32_5$	626	1·53	1·83
BaO	$2{\cdot}68_5$	$2{\cdot}76_5$	780	0·64	0·76
BaS	3·18	$3{\cdot}18_5$	662	1·23	1·46
BaSe	3·25	$3{\cdot}29_5$	652	1·30	1·55
BaTe	$3{\cdot}50_5$	$3{\cdot}49_5$	609	1·70	2·04

* MgTe actually crystallizes in ZnS type with lattice constant 2·75 Å.

† Observed values of a from Ewald and Hermann, *Strukturbericht* (1931).

The following approximate argument shows the effect of this distortion, by comparing these results with those one would expect to get with the exponential form $be^{-r/\rho}$ replacing λr^{-9}. Omitting all refinements we shall use merely the approximate equations

$$\frac{\partial\phi}{\partial a}=0, \quad a^2\frac{\partial^2\phi}{\partial a^2}=\frac{9\Delta}{\kappa}. \qquad \text{......(993)}$$

If we equate $be^{-a/\rho}$ and λa^{-s} and also their gradients $(b/\rho)\,e^{-a/\rho}$ and $s\lambda a^{-s-1}$, we see that we must have $a/\rho\simeq s$. But the second derivatives then differ by containing the extra factors $1/\rho$ ($=s/a$) and $(s+1)/a$ of which the second is too large. For a crystal of rock-salt type we should then have (roughly) by (993), eliminating O,

$$\frac{9\Delta}{\kappa}=\frac{(s-1)(3\cdot495)}{2}\frac{z^2\epsilon^2}{a}+\frac{(s-6)\,6W}{a^6},$$

but if we had replaced $s(s+1)$ in $\partial^2\phi/\partial a^2$ by s^2 to conform better to the exponential form we should have had

$$\frac{9\Delta}{\kappa}=\frac{(s-2)(3\cdot495)}{2}\frac{z^2\epsilon^2}{a}+\frac{(s-7)\,6W}{a^6}. \qquad \text{......(994)}$$

There is a similar change for the CsCl type. Values of κ corrected in this way but using the more accurate formulae (992) and (985) are shown in Table 33 and are in very fair agreement with the observed values. Similar calculations for bivalent salts are recorded in Table 34.

§ **10·81.** *Other properties of crystals.* Besides those properties which we have already discussed, many other crystalline properties can be investigated on the basis of the proposed model and their calculated values compared with observation. It would be beyond the scope of this monograph to undertake further detailed studies of this sort, and we shall therefore be content merely to mention in passing some of the other properties which have been thus investigated.

The elastic properties of an isotropic solid cannot be completely specified by less than two constants of which the compressibility κ may be taken to be one, nor those of the crystals of highest symmetry by less than three. Crystals of lower symmetry may require up to 21 constants. Elastic constants have been studied in general for this model by Born.*

Besides the energy per unit cell, the *surface energy* and the *edge energy* can also be calculated,† thus providing fairly reliable quantitative information of crystal properties which are difficult if not impossible to observe directly. The surface energy σ may be defined as the mutual potential energy Φ_{12} of the parts of a crystal separated by a plane, divided by twice the area F of this separating plane, so that

$$\sigma=-\Phi_{12}/2F. \qquad \text{......(995)}$$

A few results‡ for alkali halide crystals are given in Table 35. The values of σ_{110} for (110) interfaces have also been calculated and it has been shown that $\sigma_{110}/\sigma_{100}>\sqrt{2}$. From this it follows that the 100 planes do not form natural faces of the crystal. This is in accordance with observation.

* Born, *Atomtheorie des festen Zustandes* (1923).

† Lennard-Jones and Taylor, *Proc. Roy. Soc.* A, vol. 109, p. 495 (1925).

‡ The calculations are for old values of the energy constants, and are therefore not strictly comparable with the preceding calculations. The changes of value however are small.

Calculations have not been confined to crystals of NaCl or CsCl type. Crystals such as fluor (CaF_2) have also been studied and it has been shown that energy constants for these ions which give correct results for the simple univalent and bivalent salts will also give the correct lattice constant for fluor.* Similar calculations have been extended to non-cubic crystals of the calcite type.† These calculations can be carried out using the energy constants established by study of simple crystals, provided that in addition a suitable size and configuration is assumed for the carbonate or nitrate ion, that is for the force centres that are to represent it in the model. Some of these calculations are summarized in Table 36; the energy constants therein used have not been revised.

TABLE 35.

Surface energies of certain crystalline salts.

(100 planes, energies in ergs/cm.^{2})

σ_{100}	F	Cl
Na	304	96
K	180	76·6

These and similar investigations are all founded on the approximation that the total configurational energy of the crystal can be built up of interionic or interatomic terms of the form $E(r)$ depending on the distance only between each pair of ionic or atomic centres. It is of course well recognized that this representation cannot always be accurate and that in particular it must fail for that large and important class, the metals. For these it is necessary to return to first principles and to attempt to build up expressions

TABLE 36.

Calculated and observed values of crystal parameters for crystals of calcite type.

[Assumed distances: C to O in CO_3 = 1·08 Å.; N to O in NO_3 = 0·96 Å.]

	$MgCO_3$		$CaCO_3$		$NaNO_3$	
	Calc.	Obs.	Calc.	Obs.	Calc.	Obs.
Rhombohedral angle	102° 24′	103° 21·5′	102° 18′	101° 55′	102° 15′	102° 42·5′
Dist. C → C or N → N nuclei	4·66	4·61	4·96	4·96	5·19	5·15
Crystal energy kilo-cal./mole	771	—	701	—	176	—

* Lennard-Jones and Taylor, *Proc. Roy. Soc.* A, vol. 109, p. 495 (1925).

† Bragg and Chapman, *Proc. Roy. Soc.* A, vol. 106, p. 369 (1924); Chapman, Topping and Morrell, *Proc. Roy. Soc.* A, vol. 111, p. 25 (1926); Lennard-Jones and Dent, *Proc. Roy. Soc.* A, vol. 113, p. 673 (1927); Topping and Chapman, *Proc. Roy. Soc.* A, vol. 113, p. 658 (1927).

for the energy in any configuration by approximate solutions of the appropriate equations of quantum mechanics. The work of Wigner and Seitz on sodium and Seitz on lithium forms the first step in this new study.*

§ **10·9.** *The Born cycle.* There is no direct method of measuring the crystal energies calculated in § 10·8, but Born† has shown how they can be related to observable thermo-chemical quantities. The relations between these quantities are best expressed by a diagram such as that of Fig. 22, first given by Haber.‡

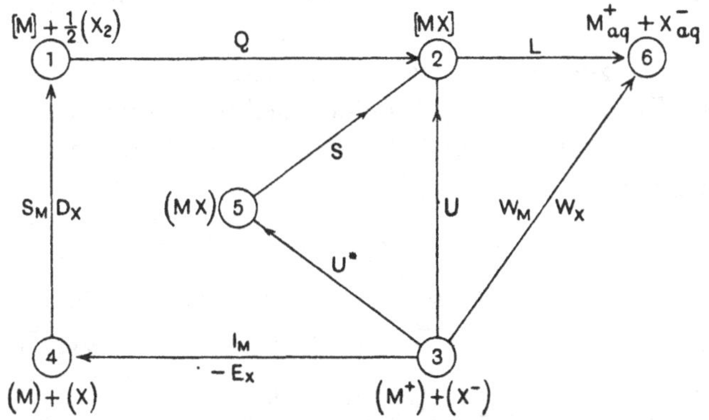

Fig. 22. The Born cycle.

In this figure $[M]$ denotes one gram atom of solid metal, (X_2) one mole of gaseous halogen. Beginning with the state 1, $[M]+\frac{1}{2}(X_2)$, we pass successively to states 2, 3 and 4, in which there is 1 mole of the salt, 1 gram atom each of the ionized metallic vapour and the ionic halogen X^-, and then 1 gram atom each of metallic vapour and atomic halogen respectively. From this state we pass back to the original state, so that on the whole the change of energy is zero. To this simple cycle (the Born cycle) two other subsidiary cycles have been added, that indicated by 5 by Haber§ and that indicated by 6 by Fajans.‖ Heat is given out in the processes indicated by the direction of the arrow in the figure and is then regarded (conventionally) as positive. In the first transformation Q denotes the heat of formation of the salt, in the second U denotes the crystal energy as already considered above. I_M is the work of ionization of 1 gram atom of metal, and E_X is the heat of formation of an halogen ion from the atom and electron (otherwise, the electron affinity). S_M is the heat of evaporation of 1 gram atom of metal

* Wigner and Seitz, *Phys. Rev.* vol. 43, p. 804 (1933), vol. 46, p. 509 (1934); Seitz, *Phys. Rev.* vol. 47, p. 400 (1935). See also Slater, *Rev. Mod. Phys.* vol. 6, p. 209 (1934).

† Born, *Verh. d. Deutsch. Physikal. Ges.* vol. 21, pp. 13, 679 (1919).

‡ Haber, *Verh. d. Deutsch. Physikal. Ges.* vol. 21, p. 750 (1919); cf. also Fajans, *ibid.* pp. 539, 549, 709, 714 (1919).

§ Haber, *loc. cit.*

‖ Fajans, *loc. cit.*

and D_X the heat of dissociation of the diatomic halogen. We then have the relation

$$U = Q + (I_M + S_M) + (D_X - E_X), \qquad \text{......(996)}$$

and all the quantities on the right have been measured. These are summarized in the annexed table. The values of Q are given by Landolt-Börnstein, those of I are known from the atomic spectra. S_M is determined by direct measurement or by observing the vapour-pressure as a function of temperature and using (633) or (523). The heats of dissociation are determined by the use of (658) or (532) and observed equilibria.† Finally, Angerer and Müller,‡ following an idea of Franck's,§ have observed in the halogen gases at high temperature a continuous spectrum with a sharp limit on the long-wave side, which they attribute to X^-, and have thus determined E_X. The values of U determined by this indirect process have been compared with the calculated values in Table 33 above. The agreement is surprisingly good.

TABLE 37.

Crystal energies in kilocals. per mole (observed).‖

	NaCl	NaBr	NaI	KCl	KBr	KI	RbCl	RbBr	RbI
Q	99	90	77	104	97	85	105	99	87
$I_M + S_M$	117+26	117+26	117+26	99+21	99+21	99+21	95+20	95+20	95+20
$D_X - E_X$	27−88	23−80	17−71	27−88	23−80	17−71	27−88	23−80	17−71
U	181	176	166	163	160	151	159	157	148

Before the values of E_X were determined directly, it was usual to eliminate them from equation (996) above, by considering the halogen acids, for which Q_{HX}, E_{HX}, I_H and D_H were known. The values of E so determined differ by 2 or 3 per cent. from those given above.¶

From the subsidiary cycle 2, 3, 5 in the figure we deduce $U = U^* + S$, where U^* is the heat of formation of the salt vapour from the ions of the metal and the halogen, and S is the heat of evaporation from salt crystal to crystal salt vapour. No measurements of either U^* or S have so far been given, so that the relation cannot yet be used to find U.

Fajans†† has used the cycle 2, 6, 3, yielding $U = W_M + W_X - L$, to investigate the magnitude of the heats of hydration W_M and W_X of metallic and

† See Table 11 for sources for this information.

‡ Angerer and Müller, *Physikal. Zeit.* vol. 26, p. 643 (1925).

§ Franck, *Zeit. f. Physik*, vol. 5, p. 428 (1921).

‖ More recent evaluations have been given by Mayer and Helmholtz, *Zeit. f. Physik.* vol. 75, p. 19 (1932); Helmholtz and Mayer, *J. Chem. Physics*, vol. 2, p. 245 (1934). Their values differ only by 2 or 3 in the last figure from the values given here.

¶ Born, *Atomtheorie des festen Zustandes*, p. 751 (1923).

†† Fajans, *Verh. d. Deutsch. Physikal. Ges.* vol. 21, p. 549 (1919).

halogen ions, taking U as known and adding the observed value of L, the heat of solution of the salt in water. He has not been able to determine W_M or W_X uniquely by this cycle alone, but has obtained a series of differences $W_{M_1}-W_{M_2}$, $W_{X_1}-W_{X_2}$, by several methods, the results being consistent among themselves. Thus $W_{Na}-W_K$ is found to be 21,000 calories per mole. The absolute values must be determined by other methods.

§ **10·91.** *Surface forces.* In earlier paragraphs applications of interatomic forces have been made to the calculation of the internal properties of solids. It is equally important that the conditions at a surface should be considered, especially as the action of surfaces is now known to be of great importance in many chemical phenomena. One point of interest is the order of magnitude and extent of the forces outside a crystal, as this has a bearing on theories of adsorption and adhesion. For this purpose the forces outside the (100) plane of a crystal of the rock-salt type have been considered* with the hope that the results might indicate the surface forces in other more complicated cases. This problem is interesting in that it provides an illustration of the forces due to primary polarization as well as of the other forces already considered.

The electrostatic potential of the semi-infinite array of the net positive and negative charges of the ions of valency v in a crystal at a point outside a (100) plane and at a distance of z from it may be shown to be

$$\phi=\frac{2v\epsilon}{a}\sum_{l,m}\frac{(-)^{\frac{1}{2}(l+m)}}{\sqrt{(l^2+m^2)}}\cos\pi\left(\frac{lx}{a}+\frac{my}{a}-\frac{l+m}{2}\right)\frac{e^{-\frac{\pi z}{a}(l^2+m^2)^{\frac{1}{2}}}}{1+e^{-\pi(l^2+m^2)^{\frac{1}{2}}}},$$

summed over all odd values of l and m positive and negative; a is the distance between consecutive planes. The axes of x and y are taken to coincide with the cubic axes. The electrostatic force on a unit charge in a direction normal to the surface is easily deduced to be

$$F_z=-\frac{\partial\phi}{\partial z}=\frac{2\pi v\epsilon}{a^2}\sum_{l,m}\frac{(-1)^{\frac{1}{2}(l+m)}}{1+e^{-\pi\sqrt{(l^2+m^2)}}}e^{-\frac{\pi z}{a}(l^2+m^2)^{\frac{1}{2}}}\cos\pi\left(\frac{lx}{a}+\frac{my}{a}-\frac{l+m}{2}\right). \qquad \text{......(997)}$$

For a fixed value of z, this is a maximum above the lattice points of the crystal (being alternately positive and negative) and is zero at all points midway between them.

Numerical calculations of the value of this force outside KCl are given in Table 38 below. It will be observed that the force is reduced by 1/100 by increasing z/a from 1 to 2, that is, increasing z from 3·14 to 6·28 Å.

* Lennard-Jones and Dent, *Trans. Far. Soc.* vol. 24, p. 92 (1928). The electrostatic forces have also been considered independently by Blüh and Stark (*Zeit. f. Physik*, vol. 43, p. 575 (1927)), though not in the same detail.

These forces apply, of course, only to a charged ion in the neighbourhood of the crystal surface. In the case of a neutral atom, forces of attraction arise from primary and secondary polarization. The former may be calculated in terms of the coefficient of polarizability α. The potential energy of the induced dipole at the centre of the atom in an electric field F is $\frac{1}{2}\alpha F^2$, or, using the potential function ϕ already given above,

$$\Phi = \tfrac{1}{2}\alpha\left\{\left(\frac{\partial\phi}{\partial x}\right)^2 + \left(\frac{\partial\phi}{\partial y}\right)^2 + \left(\frac{\partial\phi}{\partial z}\right)^2\right\}.$$

The force on the dipole perpendicular to the crystal face is then given by

$$F_z = -\frac{\partial\Phi}{\partial z} = -\alpha\left\{\frac{\partial\phi}{\partial x}\frac{\partial^2\phi}{\partial x\,\partial z} + \frac{\partial\phi}{\partial y}\frac{\partial^2\phi}{\partial y\,\partial z} + \frac{\partial\phi}{\partial z}\frac{\partial^2\phi}{\partial z^2}\right\},$$

$$= \frac{64\sqrt{2\pi^3}\epsilon^2 v^2\alpha}{a^5}\frac{e^{-2\sqrt{2\pi}z/a}}{(1+e^{-\sqrt{2\pi}})^2}f(x,y)$$

(very nearly), where $f(x,y)$ represents the variation with respect to x and y. The function $f(x,y)$ is equal to unity over the lattice points and is equal to

TABLE 38.

Adsorptive forces outside KCl.

($a_{KCl} = 3{\cdot}14$ Å.)

z/a	1·0	1·4	2·0	3·0
Force on charge ϵ	$6{\cdot}80\times10^{-5}$	$1{\cdot}14\times10^{-5}$	$7{\cdot}94\times10^{-7}$	$9{\cdot}30\times10^{-9}$
Primary polarization (Argon)	$4{\cdot}74\times10^{-6}$	$1{\cdot}33\times10^{-7}$	$6{\cdot}36\times10^{-10}$	$8{\cdot}85\times10^{-14}$
Secondary polarization (van der Waals attraction) (Argon) ($s=4$)	$1{\cdot}11\times10^{-5}$	$4{\cdot}28\times10^{-6}$	$1{\cdot}75\times10^{-6}$	$6{\cdot}70\times10^{-7}$

zero at such points as $x=\frac{1}{2}a$, $y=\frac{1}{2}a$. The variation above a lattice point is seen from Table 38, where the example of argon near the surface of KCl is considered.

The van der Waals attraction between the electronic systems of the ions in the crystal and an atom outside it can be calculated by a simple extension of the apparatus of § 10·52. It is however sufficiently accurate for this approximate calculation to neglect for example the differences between K^+ and Cl^- and assume that they both attract argon like argon. For argon outside KCl we then have a force normal to the surface which can be put in the form

$$H(z) = \frac{s\mu h^{(s+1)}(z)}{a^{s+1}} \qquad \ldots\ldots(998)$$

when $-\mu r^{-s}$ is the van der Waals energy between the argon atom and any ion in the crystal. In this expression, $h^{(s+1)}(z)$ is a certain summation over all the ions of the crystal, viz.

$$h^{(s+1)}(z) = \sum_{l,m,n} \left(\frac{z}{a}+n\right)\left\{l^2+m^2+\left(\frac{z}{a}+n\right)^2\right\}^{-\frac{1}{2}s+1}, \quad \text{......(999)}$$

the summation extending over all values of l and m, and all positive values of n. These functions have been evaluated for $s=4$, $s=5$ and a series of values of z. The numerical values of $H(z)$ for argon and KCl are given in Table 38.

It will be observed that $H(z)$ falls off very slowly with distance, so much so, in fact, that it becomes the important term at values of $z > 2{\cdot}5a$. It is not difficult to show that $h^{(5)}(z)$ tends asymptotically to z^{-2}, so that at large distances $H(z)$ falls off as the inverse square of the distance. The van der Waals field, usually regarded as a short range force, here becomes one of long range. The reason for this difference in behaviour is that the van der Waals attractions are additive for all ions in the crystal, whereas the forces arising from the charges on the ions of the lattice are alternately of opposite signs and so rapidly neutralize each other.

The results can be applied to estimate the concentration of argon in the neighbourhood of KCl by Boltzmann's formula. We find that, at a distance of 10 Å., the concentration is about two to three times the normal at room temperature.

CHAPTER XI

THE ELECTRON THEORY OF METALS, THERMIONICS AND METALLIC CONDUCTION. SEMI-CONDUCTORS

§11·1. The field of this monograph is a study of the properties of the equilibrium state. In discussing the electron theory of metals and semi-conductors we should strictly therefore exclude such processes as the thermionic emission of electrons and the passage of an electric current, which are processes involving essential departures (even if small ones) from equilibrium conditions. A study of these processes however is our most important means of obtaining information about the electronic nature of the metallic state, and the distribution laws of the equilibrium state can be applied in a direct and significant way to the study of these fundamental phenomena—to exclude them would introduce a lack of balance into our subject matter. Without, therefore, any attempt at a profound discussion of the quantum theory of the free path of an electron in a metal, or of other similar processes, we shall give such an account of the theory of thermionic emission and metallic conduction as can be achieved by simple formal applications and extensions of equilibrium theory. After a short account of the elementary theory of the genuine equilibrium state of a metal we proceed to discuss the phenomena of the emission of electrons by hot bodies and of the associated electron atmospheres in equilibrium under space charge effects. This is followed by a short account of the theory of electronic conduction in metals and semi-conductors. One of the most striking minor triumphs of quantum mechanics has been the interpretation of the distinctive properties of these two types of electronic conductor (and of insulators) in terms of natural characteristics of their equilibrium states.*

§11·2. *Elementary electron theory of a metal.* The elementary theory of a metal is due to Sommerfeld† who inserted the necessary quantum corrections into the classical theory of Drude.‡ To a first approximation one may assume that the long range effects of the electronic charges in the metal are neutralized by the charges of the heavy atomic ions. Thus at first we shall entirely neglect the internal structure of the metal and regard it merely as

* There is now a great choice of literature to which the reader may refer for a more detailed general account of these phenomena: in particular Brillouin, *Die Quantenstatistik* (1931); Nordheim, *Statische und kinetische Theorie des metallischen Zustandes*; Müller-Pouillets, *Lehrbuch der Physik*, vol. 4, part 4 (1934); Sommerfeld and Bethe, *Elektronentheorie der Metalle*, *Handb. d. Physik*, vol. 24 (ed. 2), p. 333 (1933); Bloch, *Handbuch der Experimental Physik*, vol. 11, part 2 (1932).

† Sommerfeld, *Zeit. f. Physik*, vol. 47, p. 1 (1928).

‡ Drude, *Ann. d. Physik*, vol. 1, p. 566, vol. 3, p. 369 (1900).

a home for electrons, where they can move freely in a field of uniform potential energy $-\chi_0$, the potential energy they would have in free space outside the metal being taken as zero. We must now specify how many of the electrons in the metal are to be regarded as "free" to move in this region of uniform potential energy $-\chi_0$. The only strictly logical choice is to take *all* the atomic electrons, but in fact by so doing we gain nothing and do not make a good approximation, as the more tightly an electron is bound the worse is the approximation of motion in a uniform field. It is clear however that since all the atoms are on the same footing we should take at least one electron per atom. We shall therefore proceed by assuming that *either one electron per atom or all the normal valency electrons are effectively free in the metal.* In any event the choice is not important as different choices are largely self compensating. We return to a deeper discussion of the proper specification of a free electron in § 11·5.

To study the equilibrium state of such an assembly of electrons (M in number in a volume V, antisymmetrical statistics, any state of motion of an electron with a weight factor 2 due to spin) we can at once apply the results of §§ 2·4, 2·41, in particular equation (120), which gives*

$$M=\lambda\frac{\partial}{\partial\lambda}Z=\lambda\frac{\partial}{\partial\lambda}2\Sigma_r\log(1+\lambda\vartheta^{\eta_r}),$$

and by equation (222)

$$Z=2\Sigma_r\log(1+\lambda\vartheta^{\eta_r})=2\frac{2\pi(2m)^{\frac{3}{2}}}{h^3}V\int_0^\infty\log(1+\lambda e^{(\chi_0-\eta)/kT})\,\eta^{\frac{1}{2}}d\eta.$$

If we are to apply to this assembly the approximations of § 2·73 we must verify that, in the notation of that section, $\log\mu\gg1$, where $\mu=\lambda e^{\chi_0/kT}$, and μ is evaluated in terms of the electron density n_0 $(=M/V)$ in the metal by means of (227) and (229). Using these formulae we obtain

$$n_0=2\frac{(2\pi mkT)^{\frac{3}{2}}}{h^3}\frac{2}{\sqrt{\pi}}\left\{\tfrac{2}{3}(\log\mu)^{\frac{3}{2}}+\frac{\pi^2}{12}(\log\mu)^{-\frac{1}{2}}\right\}\quad(\mu=\lambda e^{\chi_0/kT}).\qquad\text{......(1000)}$$

We can solve this equation for $\log\mu$ in terms of n_0, still assuming that $\log\mu\gg1$, and obtain

$$\log\mu=\frac{h^2}{8m}\left(\frac{3n_0}{\pi}\right)^{\frac{2}{3}}\frac{1}{kT}\left\{1-\frac{\pi^2}{12}\left[\frac{8m}{h^2}\frac{kT}{(3n_0/\pi)^{\frac{2}{3}}}\right]^2\right\}.\qquad\text{......(1001)}$$

The first approximation to $\log\mu$ is

$$\frac{h^2}{8m}\left(\frac{3n_0}{\pi}\right)^{\frac{2}{3}}\frac{1}{kT}\quad\text{or}\quad 4{\cdot}2\times10^{-11}\frac{n_0^{\frac{2}{3}}}{T}.$$

* Owing to the necessity for the frequent use of $-\epsilon$ for the electronic charge in this chapter, the symbol ϵ is not here used for an energy, and η is generally used instead.

For copper whose density is 8·9 and atomic weight 63 the number of atoms per c.c., and therefore n_0, is about $8{\cdot}5\times10^{22}$. Thus

$$\log\mu = 8\times10^4/T.$$

We see therefore that $\log\mu$ is still 80 at 1000° K. and 40 at 2000° K., amply large enough for the use of the foregoing approximations. For caesium, in which the atomic density has the lowest value known for a metal, n_0 is less by a factor of 10 and $\log\mu$ by a factor of 5, but $\log\mu$ is still large enough for the application of these formulae at any temperature below the melting point of caesium. For any metal therefore we may normally use (1001). We shall see in § 11 that, in special circumstances, for example in Ni, n_0 may have to be taken less than one per atom and m to be an effective mass much greater than the mass of a free electron. The approximation of almost complete degeneracy then no longer holds for all temperatures of importance.

These equations give μ in the form

$$\log\mu = \frac{\eta^*}{kT} - \frac{\pi^2}{12}\frac{kT}{\eta^*}, \qquad \text{......(1002)}$$

where

$$\eta^* = \frac{h^2}{8m}\left(\frac{3n_0}{\pi}\right)^{\frac{2}{3}} \quad (\eta^*/kT \gg 1). \qquad \text{......(1003)}$$

It is easily verified that η^* is the maximum kinetic energy of an electron in the metal at 0° K. when all the n_0 lowest states are filled. For the characteristic energies in a rectangular box of edges a, b, c are

$$\frac{h^2}{8m}\left(\frac{s^2}{a^2}+\frac{t^2}{b^2}+\frac{u^2}{c^2}\right) \quad (s,\ t,\ u,\ \text{positive integers})$$

and the number of these which are less than η^* is

$$\frac{4\pi}{3}abc\left(\frac{2m\eta^*}{h^2}\right)^{\frac{3}{2}}.$$

Each of these states accommodates a pair of electrons of opposite spins and the total number of pairs is $\frac{1}{2}abcn_0$. Equating these numbers we recover (1003).

For $n_0 = 8{\cdot}5\times10^{22}$ the magnitude of η^* in electron volts is 7·05. The value of k is $8{\cdot}60\times10^{-5}$ electron volts per degree. The electron distribution law $n(\eta)$ given in (231) may be written here

$$n(\eta) = \frac{3}{2}\frac{n_0}{(\eta^*)^{\frac{3}{2}}}\frac{\eta^{\frac{1}{2}}d\eta}{1+e^{(\eta-\eta^*)/kT}}. \qquad \text{......(1004)}$$

The Fermi-Dirac factor $1/(1+e^{(\eta-\eta^*)/kT})$ is shown plotted for various values of T in Fig. 23.

Besides equation (1004) we shall have much need for similar distribution laws in which the dependence of the three velocity components is separately shown, and also of a formula for the number of electrons incident on the boundary of the metal per unit area in unit time. Such a formula is provided by equation (217) in which space variations can here be neglected. This formula shows that if $n(u,v,w)\,du\,dv\,dw$ is the number of electrons per c.c. in the given velocity range, then

$$n(u,v,w)=\frac{2(m/h)^3}{1+e^{\frac{1}{2}m(u^2+v^2+w^2)/kT}/\mu}. \qquad \ldots\ldots(1005)$$

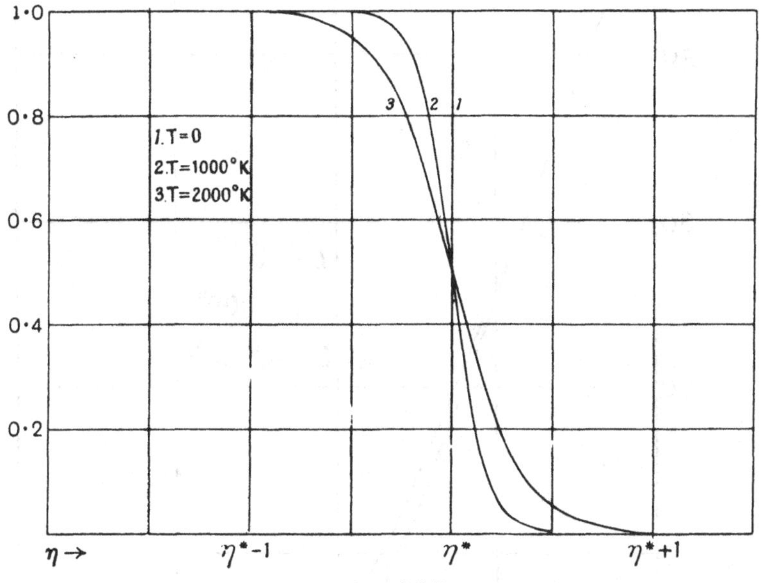

Fig. 23. The function $1/(1+e^{(\eta-\eta^*)/kT})$ for various values of T.

The factor 2 comes from the electron's spin. In terms of the components of momentum p, q, r ($p=mu$, *etc.*)

$$n(p,q,r)=\frac{2/h^3}{1+e^{(p^2+q^2+r^2)/2mkT}/\mu}. \qquad \ldots\ldots(1006)$$

In these equations u, v, w or p, q, r range from $-\infty$ to $+\infty$.

The number of electrons $N(p)\,dp$ incident on unit area in unit time with momentum *normal to the area* in the range p, $p+dp$ is given by

$$N(p)=\frac{p}{m}\int_{-\infty}^{+\infty}\int_{-\infty}^{+\infty} n(p,q,r)\,dq\,dr.$$

The double integral gives the number of suitable electrons per c.c. and p/m their velocity of approach to the surface. On using (1006) and putting $q^2+r^2=\rho^2$ this becomes

$$N(p)=\frac{4\pi p}{h^3 m}\int_0^{\infty}\frac{\rho\,d\rho}{1+e^{(p^2+\rho^2)/2mkT}/\mu}. \qquad \ldots\ldots(1007)$$

The integral can be evaluated and we find

$$N(p) = \frac{4\pi mkT}{h^3}\frac{p}{m}\log(1+\mu e^{-p^2/2mkT}). \qquad \text{......(1008)}$$

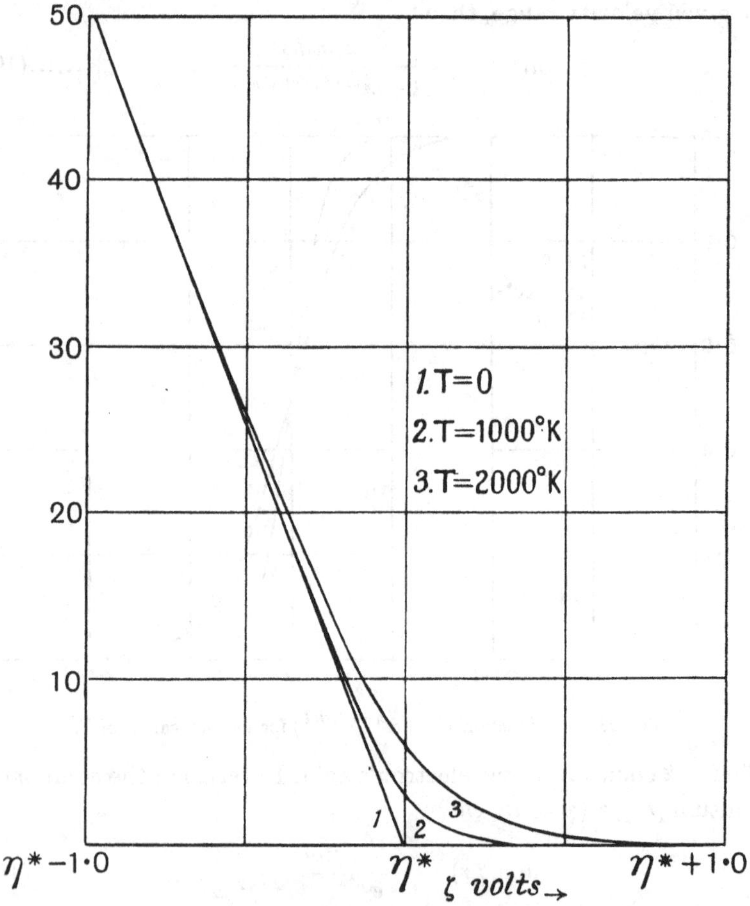

Fig. 24. The number $N(\zeta)$ electrons per unit range of energy incident on the boundary per cm.² per sec. as a function of ζ and T.

If ζ is the *energy* in this normal motion, and $N(\zeta)\,d\zeta$ the number of incident electrons in this given normal energy range, then $2m\zeta = p^2$, $p\,dp = m\,d\zeta$ and

$$N(\zeta) = \frac{4\pi mkT}{h^3}\log(1+\mu e^{-\zeta/kT}). \qquad \text{......(1009)}$$

On inserting the first approximation to μ from (1002) we find

$$N(\zeta) = \frac{4\pi mkT}{h^3}\log(1+e^{(\eta^*-\zeta)/kT}). \qquad \text{......(1010)}$$

There are important approximate forms of (1010), as shown in the following equations:

$$N(\zeta)=\frac{4\pi m}{h^3}\{\eta^*-\zeta+kTe^{-(\eta^*-\zeta)/kT}\} \quad (\eta^*>\zeta,\ \eta^*-\zeta\gg kT), \qquad \text{......(1011)}$$

$$N(\zeta)=\frac{4\pi mkT}{h^3}\left\{\log 2+\frac{1}{2}\frac{\eta^*-\zeta}{kT}\right\} \quad (|\eta^*-\zeta|\ll kT), \qquad \text{......(1012)}$$

$$N(\zeta)=\frac{4\pi mkT}{h^3}e^{-(\zeta-\eta^*)/kT}\{1-\tfrac{1}{2}e^{-(\zeta-\eta^*)/kT}\} \quad (\eta^*<\zeta,\ \zeta-\eta^*\gg kT). \quad \text{...(1013)}$$

The function $N(\zeta)$ is shown for various temperatures in Fig. 24.

§ **11·21.** *Specific heats of metals.* In Chapter IV we were able to give a generally satisfying account of the specific heats of metals (and other solids), in which we entirely neglected any contribution that might be made by the quasi-gas of free electrons which we now explicitly assume a metal to contain. This contribution $\overline{E_\epsilon}$ to the energy per unit volume is, by (226)–(228),

$$\overline{E_\epsilon}=\tfrac{3}{2}kTZ+MkT^2\left(\frac{\partial\log\mu}{\partial T}\right)_\lambda=\tfrac{3}{2}kTZ-M\chi_0. \qquad \text{......(1014)}$$

The value of Z when $\log\mu\gg 1$ is given by

$$Z=2\frac{(2\pi mkT)^{\frac{3}{2}}}{h^3}V\frac{2}{\sqrt{\pi}}\left\{\tfrac{4}{15}(\log\mu)^{\frac{5}{2}}+\frac{\pi^2}{6}(\log\mu)^{\frac{1}{2}}\right\}. \qquad \text{......(1015)}$$

Combining this with (1002) and (1003) we find

$$Z=M\left\{\frac{2}{5}\frac{\eta^*}{kT}+\frac{\pi^2}{6}\frac{kT}{\eta^*}\right\} \qquad \text{......(1016)}$$

and therefore

$$\overline{E_\epsilon}=M\left[-\chi_0+\tfrac{3}{5}\eta^*+\frac{\pi^2}{4}\frac{k^2T^2}{\eta^*}\right]. \qquad \text{......(1017)}$$

For the present we regard n_0 and therefore η^* as a constant independent of T. This is *a priori* a reasonable first assumption for metals of simple univalent and bivalent atoms, for which there is a large energy step between the first (or first two) and the following ionization potentials of the free atom, so that variations in the number of lightly bound electrons per atom seem likely to be of secondary importance. Variations of n_0, and also of χ_0, can be produced by variations of V, but we shall not consider such effects here. This being so the only term in (1017) depending on the temperature is the last and

$$C_V^{\epsilon}=\frac{\partial\overline{E_\epsilon}}{\partial T}=R\frac{\pi^2}{2}\frac{kT}{\eta^*}. \qquad \text{......(1018)}$$

This is to be compared with the classical value $\frac{3}{2}R$.

The ratio of these quantities is only $4{\cdot}1\times 10^{-5}T$ for copper and therefore quite insensible. The fact of primary importance is that the old difficulties of Drude's electronic theory of metals have now been entirely overcome. Properly treated the electron gas is found in general not to contribute sensibly to the specific heat of the metal. We are free to apply the idea of this gas to thermionic, photoelectric, thermoelectric and conductivity problems without upsetting earlier conclusions based on the theory of crystals.

On comparing (1018) with the ordinary lattice specific heat at low temperatures given by Debye's approximation (359), we see that, small as it is, the electronic specific heat should give the dominant term for a metal at sufficiently low temperatures. For a metal, but not for an insulator, the final approach of the specific heat to zero should strictly vary like T not T^3. If we put (1018) in the form

$$C_V^\epsilon/R=\tfrac{1}{2}\pi^2 T/T_\epsilon \quad (T_\epsilon=\eta^*/k), \qquad \text{......(1018·01)}$$

the ratio of the electronic to the lattice specific heat of a metal at low temperatures is

$$\frac{C_V^\epsilon}{C_V}=\frac{5}{24\pi^2}\frac{\Theta^3}{T_\epsilon}\frac{1}{T^2}. \qquad \text{......(1018·1)}$$

For $T_\epsilon \sim 10^5$, $\Theta \sim 2\times 10^2$ this ratio reaches unity between 1° and 2° K. Below this temperature the electronic term is dominant. A dominant term of this very small order of magnitude has actually been measured by Keesom and Koks† for silver. They find that below 3° K. C_V has the value $0{\cdot}00015(5)T$ cal./deg./mole, which agrees exactly with the value calculated from (1018·01) for one free electron per silver atom.

The circumstances in which the electrons can make much bigger contributions can be more conveniently discussed later, and we return to this subject in § 11·55.

The electron pressure may be noted here for reference. According to (237) it is given by

$$p_\epsilon=n_0\left[\tfrac{2}{5}\eta^*+\frac{\pi^2}{6}\frac{k^2T^2}{\eta^*}\right]. \qquad \text{......(1019)}$$

§ **11·22.** *The elementary theory of electron atmospheres.* Let us next consider an assembly of two such enclosures, one of which is in free space and the other inside a metal. This will represent a metal in equilibrium with an atmosphere of evaporated electrons, and since the evaporated electron density n_f will prove to be very small its space charge can initially be neglected, and the number of electrons still in the metal can always be assumed to be normal. [In actual applications any deficiency in the metal would always be replaced by conduction.] We can apply the results of § 2·72 and find for the evaporated electrons

$$\overline{M'}=2\lambda\frac{\partial}{\partial\lambda}\Sigma_r\log(1+\lambda\vartheta^{\eta_r'}), \qquad \text{......(1020)}$$

† Keesom and Koks, *Physica*, vol. 1, p. 770 (1934).

where λ must have the same value as for the metal. Using equations (1000) and (1002)

$$\log\lambda = -\frac{\chi_0-\eta^*}{kT}-\frac{\pi^2}{12}\frac{kT}{\eta^*} = -\frac{\chi}{kT}-\frac{\pi^2}{12}\frac{kT}{\eta^*}. \qquad \text{......(1021)}$$

If χ is positive and of the order of a few electron volts then $\lambda \ll 1$ in spite of the fact that $\log\mu \gg 1$, and the evaporated electrons are effectively classical. That being so, equation (1020) reduces at once to

$$n_f = \frac{\overline{M'}}{V} = 2\frac{(2\pi mkT)^{\frac{3}{2}}}{h^3}\lambda, \qquad \text{......(1022)}$$

which on using (1021) becomes

$$n_f = 2\frac{(2\pi mkT)^{\frac{3}{2}}}{h^3}e^{-\chi/kT-\pi^2kT/12\eta^*} \quad (\chi=\chi_0-\eta^*). \qquad \text{......(1023)}$$

The second term in the exponent can usually be neglected.†

§ **11·23.** *The electron vapour density as a problem in evaporation.* It is worth while to pause for a moment to show that (1023) can also be derived at least formally by the methods of Chapter v. If the metallic electrons are regarded as a quasi-crystal, then by equation (520)

$$\overline{M'} = f(T)/\kappa(T).$$

In this equation $f(T) = 2(2\pi mkT)^{\frac{3}{2}}e^{-\chi/kT}/h^3$ if the energy zero is taken to refer to the state in which all the electrons are condensed into the crystal filling their M lowest states. If $\kappa(T)\simeq\kappa(0)=1$, we at once recover the first approximation to (1023). In that case (523) becomes

$$\log p = -\frac{\chi}{kT}+\tfrac{5}{2}\log T+\log 2\frac{(2\pi m)^{\frac{3}{2}}k^{\frac{5}{2}}}{h^3}, \qquad \text{......(1024)}$$

so that the vapour pressure constant of the electron, i_ϵ, is given by

$$i_\epsilon = \log 2\frac{(2\pi m)^{\frac{3}{2}}k^{\frac{5}{2}}}{h^3}, \qquad \text{......(1025)}$$

which is the normal value for a massive particle of mass m with two orientations.

In the preceding section we have made an explicit (if approximate) evaluation of this $\kappa(T)$. On the analogy of (522), the remaining term in the exponent in (1023) gives this approximation in the form

$$\log\frac{\kappa(T)}{\kappa(0)} = \frac{\pi^2}{12}\frac{kT}{\eta^*}. \qquad \text{......(1026)}$$

† Before the development of the quantum theory of metals formulae substantially equivalent to (1023) were given by O. W. Richardson, *Phys. Rev.* vol. 23, p. 153 (1924); Dushman, *Phys. Rev.* vol. 21, p. 623 (1923); H. A. Wilson, *Phys. Rev.* vol. 24, p. 38 (1924); Schottky, *Zeit. f. Physik*, vol. 34, p. 645 (1925), where other references will be found.

Following up the same analogy we can also express $\kappa(T)$ in the form

$$\log\frac{\kappa(T)}{\kappa(0)}=\int_0^T\frac{dT'}{kT'^2}\int_0^{T'}\sigma(T'')\,dT'', \qquad \text{......(1027)}$$

where $\sigma(T'')$ is *a specific heat per metallic electron*, properly defined. On combining (1026) and (1027) and differentiating twice we see that

$$\sigma=\frac{\pi^2}{6}\frac{k^2T}{\eta^*}. \qquad \text{......(1028)}$$

It will be observed that σ/k so defined does not agree with $C_V{}^\epsilon/R$, nor need it do so, for they are different quantities. $C_V{}^\epsilon$ is the contribution to the total specific heat at constant volume of one gram-molecule of the metal. In setting up the equilibrium atmosphere of electrons however we are not evaporating a fraction of the whole metal, only a fraction of the electrons without change of volume of the metal. It follows from the laws of thermodynamics that the specific heat σ of (1027) is related to $C_V{}^\epsilon$ by the equation

$$\sigma=\left(\frac{\partial C_V{}^\epsilon}{\partial M}\right)_{V,T}. \qquad \text{......(1029)}$$

Since by (1018) $C_V{}^\epsilon \propto M^{\frac{1}{3}}$, $\partial C_V{}^\epsilon/\partial M=\frac{1}{3}C_V{}^\epsilon/M$, so that (1018) and (1028) are consistent with (1029).

Relationships such as (1029) are apt to be overlooked since it is comparatively rarely in statistical mechanics that a quantity such as $C_V{}^\epsilon$ is not proportional to M, the condition that $\partial C_V{}^\epsilon/\partial M \neq C_V{}^\epsilon/M$. There are a variety of other "specific heats" besides this σ which require to be carefully distinguished. *The σ defined by* (1028) *is the increase of the equilibrium specific heat of the metal for the addition of one electron.* As we shall see later this is not the same thing as Kelvin's specific heat of electricity.

§**11·3.** *Thermionic phenomena.* The phenomena of the emission of electricity from hot bodies are well known.† We cannot here describe experimental details. It is sufficient to record that a great range of phenomena lend practical certainty to the view that an incandescent metal emits electrons and to a less extent positive ions at a rate which is extremely sensitive to the temperature. The phenomena actually observed depend in general on these rates of emission, for the system studied experimentally is not (usually) in an equilibrium state. But since there must be an equilibrium state for the corresponding isolated system we are led by these phenomena to believe that when equilibrium is set up between a metal and its surroundings (gas or vacuum), the metal is in equilibrium with a vapour of

† For a general account see O. W. Richardson, *The Emission of Electricity from Hot Bodies*, ed. 2, Longmans (1921); K. T. Compton and Langmuir, *Reviews of Mod. Physics*, vol. 2, p. 123 (1930). The most recent account is Reimann, *Thermionic Phenomena*, Chapman and Hall (1934), from which we have taken the greater part of the numerical data.

electrons and perhaps of positive ions as well. If we have chosen a suitable model this equilibrium so far as it concerns electrons should be adequately described by the formulae of the preceding sections.

It has not proved possible to observe the electron density given by (1023) or the corresponding electron vapour pressure p_ϵ $(=n_f kT)$. The quantity observed is always a current—the maximum current that can be drawn from an incandescent wire per square centimetre per second, by an external voltage large enough to sweep away the electrons as fast as they are emitted, but not large enough to produce a sensible potential gradient near the emitting surface. Such a *saturation current* measures the rate of emission of electrons by the hot solid. About such rates of emission neither thermodynamics nor the equilibrium theory of statistical mechanics have anything to say. We must appeal to some mechanism, which, however, and here the equilibrium theory comes in, must be consistent with the equilibrium state when allowed to act in a normal manner. It must be true, for example, that the rates of emission and return of electrons are equal in the equilibrium state. We now assume that, when the external voltage is applied and the saturation current measured, the rate of emission of electrons is unaltered. Since this rate is determined by the internal state of a conductor, and the voltages in question are not large, this assumption may be accepted.*

§**11·31**. *The saturation current, or rate of emission of electrons.* Now that we possess an adequate picture of the electronic state inside a metal it is possible to calculate directly the rate of emission of electrons under equilibrium conditions. It is possible however to arrive formally at the same result indirectly even more simply by calculating the rate of return of electrons to the metal, and we give this calculation first on account of its historical importance.

The electrons in the vapour have the usual Maxwellian distribution in velocity and position, and therefore by (174) the number of electrons in a volume dV with a velocity between c and $c+dc$ whose direction lies within a solid angle $d\omega$ is

$$n_f dV\left(\frac{m}{2\pi kT}\right)^{\frac{3}{2}} c^2 e^{-mc^2/2kT} dc\, d\omega.$$

The number of electrons N which strike unit area of the metal in unit time is therefore obtained by taking $d\omega = \sin\theta\, d\theta\, d\phi$, $dV = c\cos\theta\,.\,1$, and integrating over all c and over a hemisphere. We find

$$N = \tfrac{1}{4} n_f \frac{2(2kT)^{\frac{1}{2}}}{(\pi m)^{\frac{1}{2}}} = \tfrac{1}{4} n_f \bar{c}. \qquad \text{......(1030)}$$

* The rate is affected by larger applied voltages. See §§ 11·34, 11·35. It should be recorded that the interpretation of the facts here given is still open to doubt (Nottingham, *Phys. Rev.* vol. 49, p. 78 (1936)).

If in the equilibrium state a fraction r of these is reflected again, the number of electrons which return to the metal per unit area in unit time is

$$(1-r)\frac{4\pi mk^2T^2}{h^3}e^{-\chi/kT}, \qquad \text{......(1031)}$$

on using (1030) and (1023), neglecting σ. It follows that the saturation current $-I$ flowing from unit area of the hot body to the collecting electrode is given in electrostatic units by the formula

$$I=(1-r)\frac{4\pi mk^2\epsilon}{h^3}T^2e^{-\chi/kT}. \qquad \text{......(1032)}$$

This is Richardson's emission formula. The energy χ is called *the thermionic work function*. The numerical value of the absolute constant $4\pi mk^2\epsilon/h^3$ is $3{\cdot}60\times10^{11}$ E.S.U., or 120 amperes per square centimetre.

Earlier formulae for I had the constant 60 amp./cm.^{2}, since the weight 2 for the spin of the free electrons was neglected. The reflection coefficient was also neglected. If it may be assumed that r is small, an assumption which is probably correct, for clean metals, then (1032) gives all the information that theory can provide. The investigation however is not strictly complete without a calculation of r, which can be given when a precise form is assumed for the effective potential variations near the surface of the metal.

We shall now give the corresponding direct calculation of the emission, for which purpose we must use (1010). Only those electrons can get out for which $\zeta>\chi_0$. It is assumed that the motion normal to the surface is unaffected by the other velocity components, which is correct when the atomic variations of the potential are neglected. If $D(\zeta)$ is the chance that such an electron will emerge, then the number of electrons emerging from unit area in unit time is

$$\int_{\chi_0}^{\infty}N(\zeta)\,D(\zeta)\,d\zeta.$$

If d is the mean transmission coefficient, this may be written

$$d\int_{\chi_0}^{\infty}N(\zeta)\,d\zeta$$

and approximated to, with the help of (1013), by the expression

$$d\,\frac{4\pi mkT}{h^3}\int_{\chi_0}^{\infty}e^{-(\zeta-\eta^*)/kT}d\zeta=d\,\frac{4\pi mk^2}{h^3}T^2e^{-\chi/kT}. \qquad \text{......(1033)}$$

It is a well-known quantum theorem that transmission coefficients over any barrier are necessarily the same in both directions† for electrons of any energy. Thus the d of (1033) and the $1-r$ of (1031) must be identical and the two formulae agree.

† For example see R. H. Fowler, *Proc. Camb. Phil. Soc.* vol. 25, p. 193 (1929).

§ **11·311.** *Theoretical transmission coefficients.* The actual theoretical values of $D(\zeta)$ and d can now be given. For the abrupt boundary field† shown in Fig. 25,

$$\left.\begin{aligned} D(\zeta) &= \frac{4[(\zeta-\chi_0)\,\zeta]^{\frac{1}{2}}}{[\zeta^{\frac{1}{2}}+(\zeta-\chi_0)^{\frac{1}{2}}]^2} \qquad (\zeta>\chi_0) \\ &= 0 \qquad\qquad\qquad\qquad\quad (\zeta<\chi_0) \end{aligned}\right\}. \qquad \text{......(1034)}$$

From this it follows that, provided $\chi=\chi_0-\eta^*\gg kT$,

$$d = 2\sqrt{\pi}\left(\frac{kT}{\chi_0}\right)^{\frac{1}{2}}; \qquad \text{......(1035)}$$

this has actually a mild temperature dependence, and, for $T\simeq 1500°$ K. and $\chi_0\simeq 4\cdot 5$ volts, is of the order 0·6.

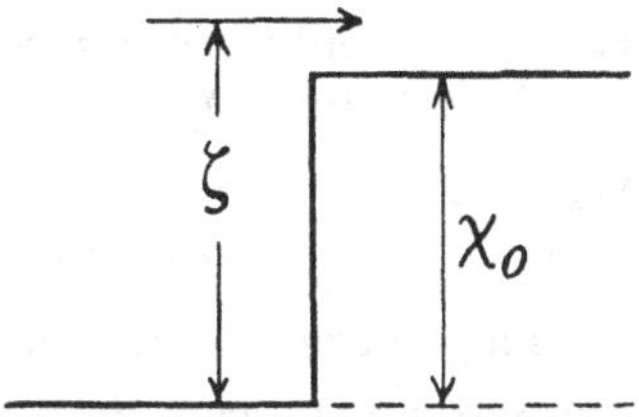

Fig. 25. An abrupt boundary.

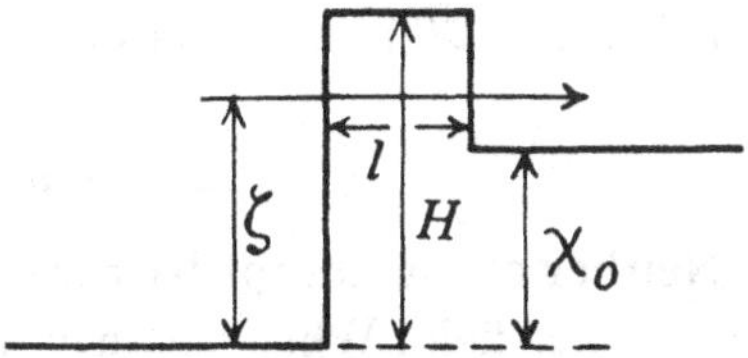

Fig. 26. An abrupt boundary with a potential energy hill.

For the abrupt boundary field shown in Fig. 26 with a potential energy hill of height H ($>\chi_0$) and width l between the inside of the metal and the space outside‡

$$D(\zeta)=0 \quad (\zeta<\chi_0),$$

$$D(\zeta)=\frac{8[(\zeta-\chi_0)\,\zeta]^{\frac{1}{2}}}{\dfrac{H}{\zeta}\dfrac{H-\chi_0}{H-\zeta}\cosh 2\kappa(H-\zeta)^{\frac{1}{2}}\,l+\left[1+\left(\dfrac{\zeta-\chi_0}{\zeta}\right)^{\frac{1}{2}}\right]^2-\left[\left(\dfrac{H-\zeta}{\zeta}\right)^{\frac{1}{2}}-\left(\dfrac{\zeta-\chi_0}{H-\zeta}\right)^{\frac{1}{2}}\right]^2}$$

$$(H>\zeta>\chi_0), \qquad \text{......(1036)}$$

where $\kappa^2=8\pi^2 m/h^2$. Provided that

$$e^{-(H-\chi_0)/kT} \ll e^{-2\kappa l(H-\chi_0)^{\frac{1}{2}}},$$

we find from (1036) that, approximately,

$$d = 8\sqrt{\pi}\,\frac{(kT\chi_0)^{\frac{1}{2}}}{H}\,e^{-2\kappa l(H-\chi_0)^{\frac{1}{2}}}. \qquad \text{......(1037)}$$

This value can be quite small compared with unity, when there is an appreciable barrier of an excess height of the order of 1 or 2 volts and a width of the order of 1–3 Ångstroms.

It will be observed that the important part of $D(\zeta)$ in (1036) when $2\kappa l(H-\zeta)^{\frac{1}{2}}\gg 1$ is

$$(\ldots)\,e^{-2\kappa l(H-\zeta)^{\frac{1}{2}}},$$

† Nordheim, *Zeit. f. Physik*, vol. 46, p. 833 (1928).
‡ Fowler, *Proc. Roy. Soc.* A, vol. 122, p. 36 (1929).

where (...) is a factor of order unity. It can be shown that, for any shape of barrier through which the penetration is slight, $D(\zeta)$ always is of this form with an exponential factor

$$e^{-2\kappa\int_0^{l(\zeta)}(U-\zeta)^{\frac{1}{2}}\,dx} \qquad \text{......(1038)}$$

The meaning of the symbols is explained in Fig. 27.

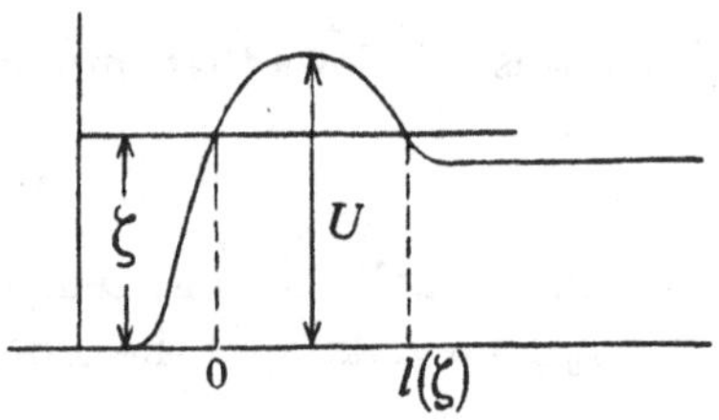

Fig. 27. A gradual boundary with a potential energy hill.

An even better approximation is

$$D(\zeta)=\frac{1}{\left[\cosh\kappa\int_0^{l(\zeta)}(U-\zeta)^{\frac{1}{2}}dx\right]^2}, \qquad \text{......(1039)}$$

which when $D(\zeta)\ll 1$ reduces to

$$D(\zeta)\sim 4e^{-2\kappa\int_0^{l(\zeta)}(U-\zeta)^{\frac{1}{2}}\,dx}. \qquad \text{......(1040)}$$

Neither of the abrupt barriers of Figs. 25, 26 is a good representation of the surface field. When a slowly moving electron is at a distance x from the surface, it will be attracted to the metal by the image force $\epsilon^2/4x^2$, which corresponds to an effective potential energy $-\epsilon^2/4x$. This form of the potential energy must break down very close to the first layer of metal atoms, but if it holds at distances of 10 Ångstroms or more from the surface it means that the top corner of the potential energy curve in Fig. 25 is

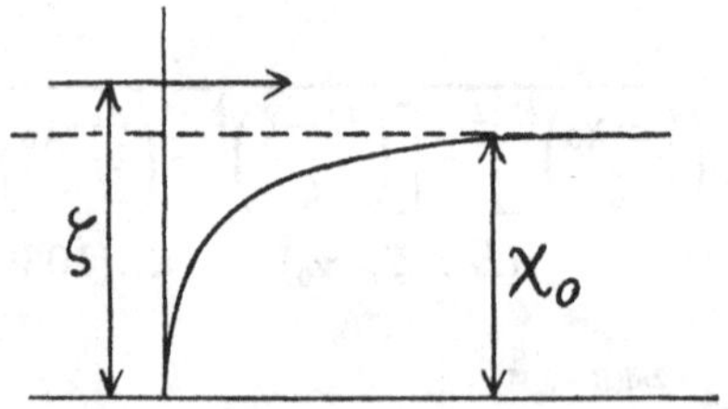

Fig. 28 *a*. The image boundary field.

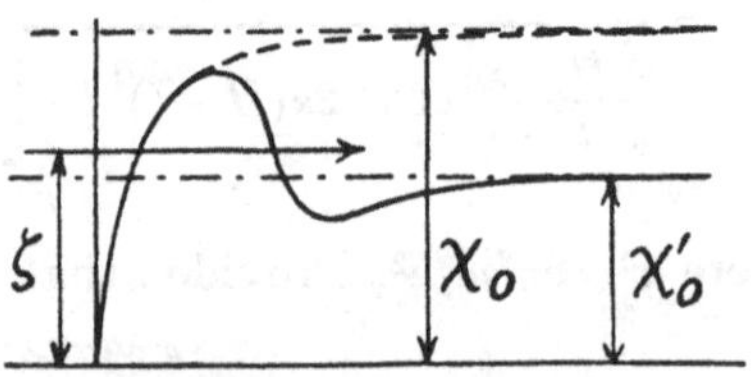

Fig. 28 *b*. The image boundary field with a potential energy hill.

smoothly rounded off. This greatly reduces the electron reflection at the boundary. For the image force boundary shown in Fig. 28 *a* $D(\zeta)$ has been calculated by Nordheim* but cannot be given in a short form. From his result it follows that $D(\zeta)$ tends to a limit not far removed from unity when $\zeta\to\chi_0$ and from this that $d\sim 1$ no matter what the temperature. The probable form of the potential energy for a combination of image field and barrier resulting from the reduction of χ by a surface film is shown in Fig. 28 *b*.

* Nordheim, *Proc. Roy. Soc.* A, vol. 121, p. 626 (1928).

We may summarize this discussion by saying that for the simple model chosen for the metal the saturated thermionic current should be of the form

$$AT^2e^{-\chi/kT}, \qquad \ldots\ldots(1041)$$

where A and χ may be treated as constants and $A \leqslant 120$ amp./cm.² deg.² For clean metals $A \simeq 120$, but for metals covered with surface films A may be very much less.

§ **11·32.** *Comparison of theory and experiment for clean metals, and metals with monomolecular films.* If Richardson's formula for the current, (1041), is compared with the best experiments on carefully cleaned tungsten, regard-

TABLE 39.

Emission data for tungsten.

The entries I in the second column are the observed saturation currents from 0·1825 cm.² hot surface.

Temp. °K.	I milliamperes	$\log_{10}(I/T^2)+10$	b_0
1935·5	0·0934	0·1355	51,890
1986·5	0·1973	0·4378	51,880
2036·0	0·3967	0·7197	51,860
2077·5	0·6784	0·9351	51,880
2086·5	0·7656	0·9838	51,900
2102·0	0·9363	1·0650	51,840
2131·5	1·362	1·2155	51,840
2134·5	1·419	1·2321	51,870
2158·0	1·902	1·3500	51,820
2182·0	2·538	1·4655	51,810
2204·0	3·269	1·5668	51,820
2231·0	4·405	1·6858	51,820
2235·0	4·606	1·7036	51,870
2271·5	6·875	1·8635	51,880
2280·0	7·394	1·8918	51,920
2306·0	9·792	2·0041	51,900
		(4·47 volts =) Mean value =	51,860*

* This value is not the best accepted value for clean tungsten (4·54).

ing A and χ as adjustable constants, an excellent fit is obtained with a value of A about 60. Owing however to the dominance of the exponential term the observations are not capable of fixing A with high accuracy. In fact it is hardly possible by observation to distinguish between (1041) and a formula

$$A'T^{\frac{1}{2}}e^{-b_1/T} \qquad \ldots\ldots(1042)$$

used by Richardson in his earlier work. It is therefore impossible to be confident as yet that there is here a discrepancy between the simple theory and experiment, which demands an elaboration of the theory for its explanation. Similar successes are obtained for comparable values of A on comparing theory and observation for other metals, *when they have been properly cleaned.*

The excellence of the fit obtainable between (1041) and good data is illustrated by the foregoing observations for tungsten given by Dushman. In the analysis it was assumed that $A = 60{\cdot}2$, which was originally supposed to be the theoretically correct value of A, the spin of the electron being then unrecognized, and χ/k was then calculated from the observed current. The values so obtained should be identical and the actual extreme variation in Table 39 is 11 parts in 5000. A fit almost as good could be obtained with $A = 120$.

When we study the values of A and χ for a variety of metals both supposedly clean and with known monomolecular surface films deposited on them we find the values opposite. The values given are those regarded as most reliable by Reimann.*

It will be seen at once that these values are in excellent general agreement with the theory if we may account for the low values of χ and A found for electropositive films in terms of an electrical double layer about one atom thick. Such a layer gives an idealized potential energy curve near the boundary (image force neglected) of the form shown in Fig 29.

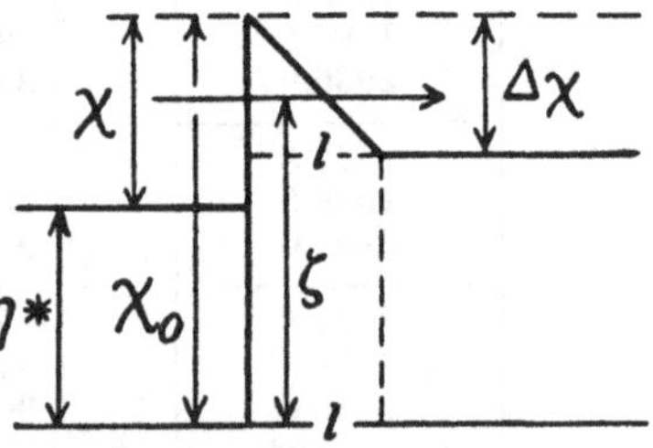

Fig. 29. An abrupt boundary with an ideal double layer (electropositive outwards).

When such a film is present $D(\zeta)$ will by (1040) contain the factor $e^{-2\kappa\int_0^l [\chi_0-\zeta-x\Delta\chi/l]^{\frac{1}{2}}\,dx}$ and d the factor $e^{-\frac{4}{3}\kappa(\Delta\chi)^{\frac{1}{2}}l}$. We shall therefore find a relationship between the value of A and the depression $\Delta\chi$ of the work function below its value for the underlying clean metal of the form

$$\log A = const. - \tfrac{4}{3}\kappa l(\Delta\chi)^{\frac{1}{2}}. \qquad \text{......(1043)}$$

This relationship is in excellent general agreement as to form and order of magnitude with the relationship

$$\log A = \xi - \eta\Delta\chi \quad (\xi,\ \eta\ const.)$$

observed by Richardson and Du Bridge to hold during an outgassing process.† All the abnormally small values of A associated with films in Table 40 can be so interpreted, though the correctness of this is still open to doubt.

* Reimann, *loc. cit.* See also Dushman, *Rev. Mod. Phys.* vol. 2, p. 381 (1930). Full references to original sources are given in Reimann's book.

† Fowler, *Proc. Roy. Soc.* A, vol. 122, p. 36 (1929).

Outstanding discrepancies remain in which A is apparently abnormally large, exceeding 120 by factors which cannot be due to any error in the analysis or the estimate of the available surface. It is probable that these discrepancies are all associated with films, and that in fact we have a temperature-dependent effective work function. For such a film as O on W

TABLE 40.

Thermionic emission constants.

Data as critically selected by Reimann (*loc. cit.*).

Element	χ volts	A amp./cm.^{2} deg.^{2}
	(i) *Clean metals*	
Cs	1·81	162
Ba	2·11	60
Zr	4·12	330
Hf	3·53	14·5
Th	3·38	70
Ta	4·12	60
Mo	4·15	55
W	4·54	60–100
Re	5·1	200
Ni	5·03	1380
Pd	4·99	60
Pt*	5·40	170*
	(ii) *Monomolecular films*	
Th on W	2·63	3·0
Th on Mo	2·58	1·5
Zr on W	3·14	5·0
La, Ce on W	2·71	8·0
Cs on W	1·36	3·2
O on W†	9·2†	5×10^{11}†
Cs on O on W	0·72	0·003
Ba on W	1·56	1·5
Ba on O on W	1·34	0·18

Where a value of 60 is given it usually implies that the best observational value is not very different from 60, differing from it perhaps by a factor less than 2, but that the old theoretical value 60 has been *assumed* in the analysis.

* The values given for Pt even more than for other elements depend on the heat treatment, and it was recently believed that Du Bridge's values $\chi = 6{\cdot}27$, $A = 17{,}000$ (*Phys. Rev.* vol. 31, p. 236 (1928)) represented the best value for the clean metal. It has now been shown by Van Velzer (*Phys. Rev.* vol. 44, p. 831 (1933)) that his wires were probably still contaminated by an oxygen film.

† The significance of the values for O on W is discussed in detail in the text.

there is every reason to believe that the theory does not apply and that χ is not a true work function nor A a transmission coefficient. This interpretation of the χ and A derived from observation can only apply when the whole temperature variation in I/T^2 can be thrown upon the factor $e^{-\chi/kT}$ and none upon A. For such oxygen films we are not dealing with the emission of the same film at different (very high) temperatures; the true χ is much smaller and we have a temperature variable A and χ due to the changing structure of the film. The derived χ (9·22 volts) is also not confirmed as a work function by contact potential measurements (see § 11·38) in agreement with this interpretation.†

It may happen however that the work function, though temperature dependent, is fairly closely of the form $\chi = \chi^* - \alpha kT$ over the whole temperature range. Such an effect lies outside the range of the present model. If modifications can be made to include it, it would satisfactorily account for abnormal values of A, by adding an extra factor e^{α} to A without modifying the agreement between the χ^* derived from the thermionic analysis and the photoelectric threshold $h\nu_0 = \chi(0) = \chi^*$ determined at low temperatures. Du Bridge's supposedly clean Pt, which was, however, probably Pt with an oxygen film, can probably be regarded as an example of this type of χ. The agreement between χ^* and $h\nu_0$ fails at once for any other type of temperature dependence of χ.

§ 11·33. *Magnitude of the potential energy step at a metal surface.* We have seen that this elementary theory demands a fairly definite average negative potential energy χ_0 for an electron inside a metal when its potential energy outside is taken as zero. Direct evidence for such an energy step is most desirable and is provided by the experiments of Davison and Germer‡ and later by Rupp§ on the reflection of electrons by metal crystals. If it is assumed that the velocity and therefore the de Broglie wave length of the beams in the metal are the same as those of the beam in free space, no exact agreement can be obtained between the observed diffracted peaks and those calculated from the crystal spacing. If, however, it is assumed that the beam is accelerated on entry by χ_0 volts, we can obtain excellent agreement for a suitable value of χ_0. The effect of the acceleration appears of course as a refractive index μ (> 1) given by the formula

$$\mu = \sqrt{\frac{X + \chi_0}{X}}, \qquad \text{......(1044)}$$

where X is the energy of the incident electrons before entry. Table 41 shows the analysis of a set of measurements on nickel by Davison and Germer.

Later experiments by Rupp similarly analysed are shown in Table 42. In every case the derived values of η^* are entirely reasonable. We cannot expect that η^* will exactly correspond to the value calculated for 1 or 2 *free* electrons per atom.

These experiments on electron diffraction provide a beautiful confirmation of the whole theory.

† Langmuir and Kingdon, *Phys. Rev.* vol. 34, p. 129 (1929).

‡ Davison and Germer, *Proc. Nat. Acad. Sci.* April (1928).

§ Rupp, *Leipziger Vorträge* (1930).

TABLE 41.

The refraction coefficient for electron waves incident on Ni crystals, chosen to make observed reflections fit the lattice structure and the free wave length of the incident electrons. Experiments of Davison and Germer analysed by Hartree.

X volts energy of incident electrons	Observed refraction coefficient μ	μ^2	$(\mu^2-1)X = \chi_0$	Calculated refraction coefficient with $\chi_0=18$
64	1·14	1·30	19·2	1·132
130	1·07	1·145	18·8	1·067
216	1·04	1·08	17	1·041
328	1·02	1·04	13	·027
449	1·01	1·02	9	1·020
586	1·01	1·02	$11\frac{1}{2}$	1·015

For Ni $\rho=8·9$, $M^*=57$, $\eta^*=11·8$ volts for 2 electrons per atom: $\chi\sim5·0$ volts so that $\chi_0=\chi+\eta^*=16·8$ volts.

TABLE 42.

Observed values of χ_0 (Rupp, electron diffraction) and χ, and derived values of η^* for various metals. Values of χ from thermionic or photoelectric effects, those marked $\sim$ doubtful.

Metal	K	Cu	Ag	Au	Zr	Mo	W	Ni
χ_0 obs.	7·3	13·5	14·0	14·0	10·2	13·5	12·5	16
χ obs.	$\sim$2·2	$\sim$4·4	4·7	4·9	4·1	4·1	4·5	5·0
$\eta^*=\chi_0-\chi$	5·1	9·1	9·3	9·1	6·1	9·4	8·0	11·0

§ **11·34.** *Effects of stronger fields.* (1) *The Schottky effect.* The foregoing formulae for the thermionic emission have been obtained and applied on the assumption that the field applied to collect the electrons has a negligible effect on the emission. As stronger and stronger field strengths are employed this will cease to be true. The current drawn from the emitter ceases to be constant and finally increases with the field strength. This phenomenon was first correctly interpreted in terms of the present model by Schottky.† When a field strength F is applied to the emitter, whose effective potential energy near its surface is given by the image term $-\epsilon^2/4x$ in the absence of a field, the total potential energy for an electron near the boundary will be of the form

$$U=\chi_0-\epsilon^2/4x-\epsilon Fx,$$

illustrated in Fig. 30. The maximum value of U is then depressed by an amount

$$\Delta\chi=\epsilon^{\frac{3}{2}}F^{\frac{1}{2}}. \qquad \text{......(1045)}$$

† Schottky, *Zeit. f. Physik*, vol. 18, p. 63 (1923).

For moderately strong fields the slopes of the potential energy curve near the maximum will be so slow that practically no penetration of the electrons through the barrier will occur; those only will escape whose energies exceed the reduced χ and the transmission coefficient will be unaffected. Thus the first effect of the field is to alter the emission according to the formula

$$I_F = I_0 e^{\epsilon^{\frac{3}{2}} F^{\frac{1}{2}}/kT}, \quad \ldots(1046)$$

where I_0 is the current at zero field strength for the same temperature.

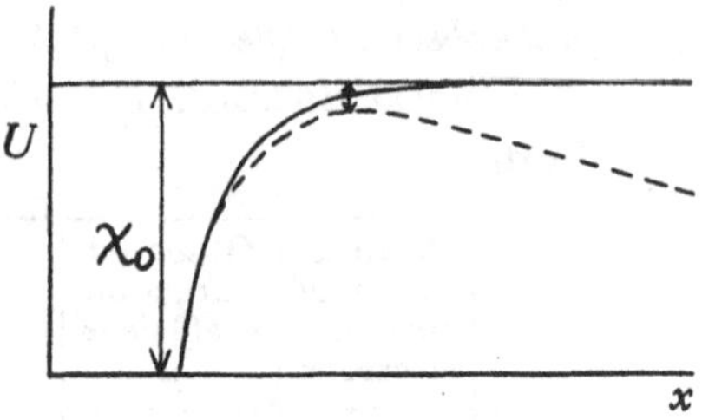

Fig. 30. Change of boundary in the Schottky effect.

An accurate test of this equation is provided by the work of de Bruyne.* According to (1046)

$$T\,\frac{\log I_F - \log I_0}{F^{\frac{1}{2}}}$$

should be constant for all temperatures and field strengths and equal to $\epsilon^{\frac{3}{2}}/k$. de Bruyne's results confirm this with remarkable accuracy up to fields of the order of 10^6 volts/cm. for a wide range of temperatures. Using the mean of his data to evaluate ϵ, one obtains the value $4{\cdot}84 \times 10^{-10}$ in admirable agreement with the accepted value.

§ **11·35.** *Effect of very strong fields.* (2) *Cold emission.* When the field strength at the electrode becomes appreciably greater than 10^6 volts/cm., the slope of the potential energy curve is so great that electrons begin to penetrate through the barrier and a field strength is ultimately reached at which the emerging fraction of the very great number of electrons near the top of the levels normally full swamps the emission from the nearly empty levels near the top of the barrier. The emission becomes nearly independent of the temperature, possibly large even for a cold electrode. A sufficiently accurate formula for the analysis of ordinary observations is obtained by taking $T=0$. We are then concerned with the transmission coefficient for electrons incident on the barrier shown in Fig. 31.† A small correction is made by including the image effect (dotted curve), but this is not important and will be ignored here.

For such a barrier

$$D(\zeta) = \frac{4[\zeta(\chi_0-\zeta)]^{\frac{1}{2}}}{\chi_0} \exp\{-\tfrac{4}{3}\kappa(\chi_0-\zeta)^{\frac{3}{2}}/F\} \quad \left(\kappa^2 = \frac{8\pi^2 m}{h^2}\right). \quad \ldots\ldots(1047)$$

* de Bruyne, *Proc. Roy. Soc.* A, vol. 120, p. 423 (1928).

† Fowler and Nordheim, *Proc. Roy. Soc.* A, vol. 119, p. 173 (1928); Nordheim, *Proc. Roy. Soc.* A, vol. 121, p. 626 (1928).

Using (1011) with $T=0$, we find at once

$$I=\frac{4\pi m\epsilon}{h^3}\int_0^{\eta^*}(\eta^*-\zeta)\,D(\zeta)\,d\zeta,$$

and after simple approximations

$$I=\frac{\epsilon}{2\pi h}\frac{\eta^{*\frac{1}{2}}}{\chi_0\chi^{\frac{1}{2}}}F^2e^{-\frac{4}{3}\kappa\chi^{\frac{3}{2}}/F}, \qquad \text{......(1048)}$$

in which χ is the usual work function. A sufficiently accurate numerical version of this formula, ignoring the differences between χ, χ_0 and η^*, is

$$I=6{\cdot}2\times10^{-6}\frac{F^2}{\chi}e^{-6{\cdot}8\times10^7\chi^{\frac{3}{2}}/F}, \qquad \text{......(1049)}$$

when χ is measured in volts, F in volts/cm. and I in amp./cm.^{2} This formula is in good agreement with the experimental facts† in the nature of its dependence on F. The observed cold emission however actually sets in at field strengths of the order of 10^6 volts/cm. as computed for the cathode surface from the geometry of the apparatus. Equation (1049) however predicts that the emission should set in at 10^7 volts/cm. The observed emission however is never general, but comes from local spots on the cathode. It is clear that these facts are to be explained as due to the roughness of the cathode surface‡ which can make the local field strength at special points greater by just such a factor than the general field strength computed from the geometry. The agreement between theory and experiment is therefore satisfactory.

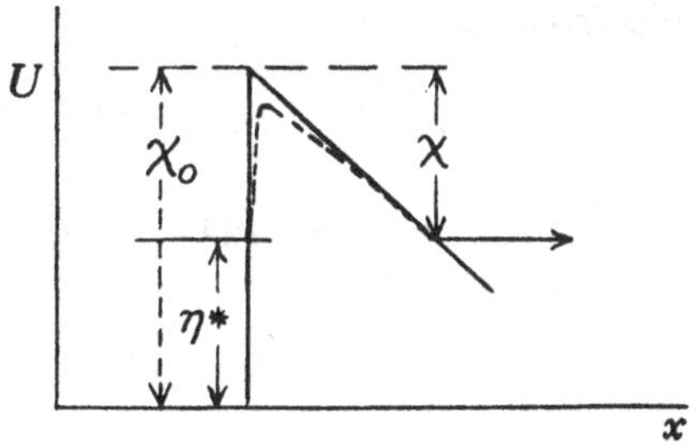

Fig. 31. The boundary effective in the (cold) emission of electrons in very strong fields.

§ **11·36.** *The cooling effect of the evaporation of electrons* has also been investigated experimentally and examined theoretically by Richardson§ and others. If we evaporate N average electrons, then their extra energy in the vapour phase will be

$$N\phi=N\left\{\chi+\tfrac{3}{2}kT-\int_0^T\sigma\,dT\right\}.$$

This, however, is *not* exactly the cooling effect on the metal, for the electrons evaporated are not an average equilibrium group distributed according to

† Millikan and Eyring, *Phys. Rev.* vol. 27, p. 51 (1926); Millikan and Lauritsen, *Proc. Nat. Acad. Sci.* vol. 14, p. 45 (1928); R. H. Fowler, Gossling and Sterne, *Proc. Roy. Soc.* A, vol. **124**, p. 699 (1929).

‡ R. H. Fowler, Gossling and Sterne, *loc. cit.*

§ O. W. Richardson, *loc. cit.* especially chap. v.

Maxwell's law. This can be seen at once if we consider the rates of emission and return *for electrons of given velocity*. The number of electrons of normal velocity between u and $u+du$ which strike unit area per second in the equilibrium state is

$$n_f\left(\frac{m}{2\pi kT}\right)^{\frac{1}{2}} ue^{-mu^2/2kT}\,du.$$

If a fraction $r(u)$ of these is reflected, the number which are re-absorbed per unit area per second is

$$\{1-r(u)\}\,n_f\left(\frac{m}{2\pi kT}\right)^{\frac{1}{2}} ue^{-mu^2/2kT}\,du,$$

which is therefore also the rate of emission. Owing to the extra factor u the rates of emission and return are, as it were, higher for the higher velocities compared with the equilibrium numbers of such electrons present in the gas. The average kinetic energy, in the motion normal to the surface, of the electrons emitted is

$$\tfrac{1}{2}m\int_0^\infty \{1-r(u)\}\,u^3e^{-mu^2/2kT}\,du \bigg/ \int_0^\infty \{1-r(u)\}\,ue^{-mu^2/2kT}\,du.$$

If we may, to a first approximation, ignore the variation of $r(u)$ with u, we have a mean energy

$$\tfrac{1}{2}m\int_0^\infty xe^{-mx/2kT}\,dx \bigg/ \int_0^\infty e^{-mx/2kT}\,dx,$$

or

$$kT.$$

The energy in the motion parallel to the surface has its equilibrium value kT. Thus on these assumptions the cooling effect for the evaporation of N electrons is

$$N(\phi+\tfrac{1}{2}kT)=N\left\{\chi+2kT-\int_0^T \sigma\,dT\right\}. \qquad \text{......(1050)}$$

This is in good agreement with direct experiment. Experiments, however, could hardly fix the coefficient of kT.

§ **11·37**. *The photoelectric effect.* When light falls on a cold metal surface electrons are emitted as soon as the frequency ν of the incident light exceeds a certain threshold frequency ν_0. This is known as the photoelectric effect. The model of a metal here in use accounts at once for the main features of this emission. For when $T\to 0$ there are no electrons in the metal with a kinetic energy greater than η^* and plenty with any energy less than η^*. In order that an electron whose initial energy is η may emerge after absorbing a quantum of energy $h\nu$, it is necessary at the least that

$$h\nu+\eta>\chi_0.$$

The least possible value of ν satisfying this inequality is

$$h\nu=h\nu_0=\chi_0-\eta^*=\chi. \qquad \text{......(1051)}$$

Thus there must be a photoelectric threshold which is equal to the work function. Table 43 compares the observed values of the threshold and the work function where both have been satisfactorily determined. The agreement is excellent.

Experiments on the photoelectric effect are not habitually made at zero temperature, and it is necessary to consider the effect of temperature on the number of available electrons for light of given frequency ν. It is at once evident that there can be no absolutely sharp threshold at temperatures

TABLE 43.

Observed values of the photoelectric threshold $h\nu_0$ for clean metals.

[Compared with observed values of χ.]
Values given in electron volts.

Metal	Cs	Ta	Mo	W	Re	Ni	Pd
$h\nu_0$	~1·9	4·11	4·15	4·54	4·98	5·01	4·97
χ	1·81	4·12	4·15	4·54	5·1	5·03	4·99

Metal	Pt	Ag	Au	Sn β	Sn γ	Sn liq.
$h\nu_0$	[6·30]	4·74	4·90	4·39	4·28	4·17
χ	[6·27] 5·40	—	—	—	—	—

Data from Reimann, *loc. cit.*, where authorities are quoted. The values in [ ] for Pt are from Du Bridge.

other than zero and this is borne out by recent observations now to be discussed.*

In order that an electron may be emitted it is necessary that the energy of its motion normal to the emitting surface should exceed χ_0. Its other velocity components are to a first approximation irrelevant. The rate of emission of such electrons must therefore be expected to be proportional to the intensity of the light, the number of suitable electrons striking unit area of the surface in unit time, the chance that they will pick up the quantum $h\nu$ in the proper velocity component, and the chance that they will then be transmitted through the boundary field.† With the model here in use in which the electrons are free inside the metal the whole excitation takes place in the surface field of the metal. When the boundary field is well

* This was first established beyond reasonable doubt by the work of Mendenhall's laboratory; see Morris, *Phys. Rev.* vol. 37, p. 1263 (1931); Winch, *Phys. Rev.* vol. 37, p. 1269 (1931); Cardwell, *Phys. Rev.* vol. 38, p. 2041 (1931).

† The exact theory which has been given by Mitchell (*Proc. Roy. Soc.* A, vol. 146, p. 442 (1934); *Proc. Camb. Phil. Soc.* vol. 31, p. 416 (1935)) correcting earlier work by Wentzel, *Sommerfeld Festschrift*, *Probleme der modernen Physik*, p. 79 (1928); Tamm and Schubin, *Zeit. f. Physik*, vol. 68, p. 97 (1931); Fröhlich, *Ann. d. Physik*, vol. 7, p. 103 (1930), and others, does not allow *strictly* of such an analysis of the different factors. But this analysis is nearly correct, and allowable for the present discussion. It would take us too far afield to discuss the exact theory further here.

represented by an image field, as for a clean metal, this last chance hardly varies and may be taken to be unity. The chance of absorbing the quantum will vary with ν as in other absorption phenomena, but this variation is not important near ν_0. Thus a good approximation to the photoelectric yield per unit light intensity near the threshold frequency is to take it simply proportional to the number of electrons incident per unit time for which

$$\frac{1}{2m}p^2+h\nu>\chi_0,$$

where p is the initial momentum of the electron normal to the surface.

Equation (1008) gives the number of such electrons as a function of p. The photoelectric yield per unit light intensity is therefore proportional to S, where

$$S=\int_{p^2/2m=\chi_0-h\nu}^{\infty}\frac{4\pi mkT}{h^3}\frac{p}{m}\log(1+\mu e^{-p^2/2mkT})\,dp. \quad \text{......(1052)}$$

On replacing μ by $e^{\eta^*/kT}$ and simplifying the integral, we obtain

$$S=\frac{4\pi mk^2T^2}{h^3}\int_0^{\infty}\log[1+e^{-y+(h\nu-\chi)/kT}]\,dy. \quad \text{......(1053)}$$

This function S should give the photoelectric yield from a clean metal as a function of ν and T for values of ν near the threshold χ/h.

The integral in (1053) cannot be evaluated in finite terms but can be reduced to $\int_0^{u_0}\log(1+u)\,du/u$ which has been tabulated,† or computed from simple expansions. When $(h\nu-\chi)/kT=\delta\leqslant 0$, we find by term by term integration that

$$S=\frac{4\pi mk^2T^2}{h^3}\left[e^{\delta}-\frac{e^{2\delta}}{2^2}+\frac{e^{3\delta}}{3^2}-\dots\right] \quad (\delta\leqslant 0). \quad \text{......(1054)}$$

When $(h\nu-\chi)/kT=\delta\geqslant 0$, we find after a short reduction that

$$S=\frac{4\pi mk^2T^2}{h^3}\left[\frac{\pi^2}{6}+\tfrac{1}{2}\delta^2-\left\{e^{-\delta}-\frac{e^{-2\delta}}{2^2}+\frac{e^{-3\delta}}{3^2}-\dots\right\}\right] \quad (\delta\geqslant 0). \quad \text{......(1055)}$$

We may therefore expect that the photoelectric current I for values of ν near the threshold is given by an equation of the form

$$\frac{I}{T^2}=A\phi\left(\frac{h\nu-\chi}{kT}\right), \quad \text{......(1056)}$$

where A is a constant independent of ν and T and $\phi(\delta)$ is a tabulated function given by the terms in [ ] in (1054) and (1055).

Observations may be analysed by taking logarithms of (1056) and using it in the form

$$\log\frac{I}{T^2}=B+\Phi\left(\frac{h\nu-\chi}{kT}\right), \quad \text{......(1057)}$$

† Pierce's Tables.

where Φ is a known function and B is a constant. For each temperature we can then plot the observations of $\log I/T^2$ against $h\nu/kT$, and can also plot $\Phi(\delta)$ on the same scales. Observations for all temperatures should then be brought to lie upon $\Phi(\delta)$ by suitable shifts of the origin. The shift in the origin of ν determines accurately the true threshold $\chi\,(=h\nu_0)$. An alternative method of analysis has been used by Du Bridge* employing sets of observations for various temperatures and a single frequency. In this method $\Phi(\delta)$ is plotted against $\log \delta$ and the observations are similarly treated. The

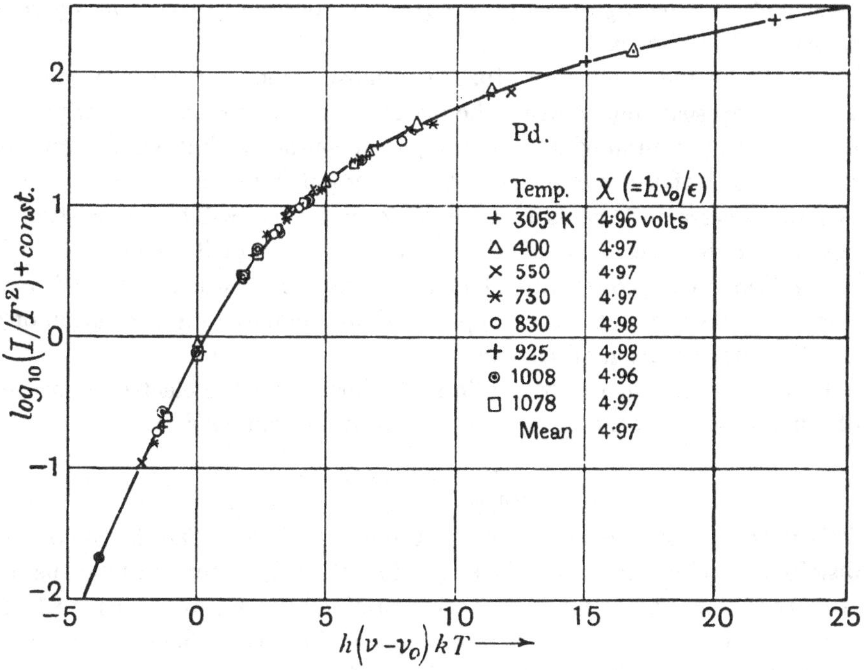

Fig. 32. Analysis of observed photoelectric yields for Pd.

results are in excellent agreement. The function $\Phi(\delta)$ and the analysis of the observations of Du Bridge and Roehr† for Pd are shown in Fig. 32. Some such method as these must always be used in determining the true photoelectric threshold, as empirical extrapolations of the photoelectric yield curve to zero are quite unreliable.‡ In addition to this analysis of the total photoelectric yield Du Bridge § has shown that the same methods may be used to analyse with almost equal success the normal velocity distribution and the total energy distribution of the photoelectrons emitted by a

* Du Bridge, *Phys. Rev.* vol. 43, p. 727 (1933).

† Du Bridge and Roehr, *Phys. Rev.* vol. 39, p. 99 (1932).

‡ For a general account see Du Bridge, *Actualités Scientifiques et Industrielles*, No. 268, *New Theories of the Photoelectric effect* (1935).

§ Du Bridge, *loc. cit.*

clean metal as a function of ν and T. These results provide a beautiful verification of the substantial accuracy of the Fermi-Dirac distribution law for the free electrons of a metal.

§ **11·38.** *Space variations of the electrostatic potential. Contact potentials.* We proceed next to include the effects of electrostatic fields due to the charges or the evaporated electrons themselves. The inclusion of these effects does not alter the form of (1023) but only the physical interpretation of n_f. This formula must remain true in the general case *if* n_f *means the average electron density in the free electron gas immediately outside the effective boundary field of the metal.*

The unaffected validity of (1023) is almost obvious. From the point of view of large scale phenomena, the metal is a conductor at constant potential, and the potential of the electron gas in contact with it will be equal to the potential of the metal, since the potential is a continuous function. Whether electrostatic fields are included or not, there will be no question of potential differences between the metal (just clear of its boundary field) and a thin sheet-like volume element of electron gas in contact with it. The foregoing arguments then apply, with an unaltered interpretation of χ and the interpretation of n_f stated above.

In addition to this result, we find, of course, that the electron densities of different parts of the gas are connected by Boltzmann's equation

$$\frac{n_f}{(n_f)_0} = e^{\epsilon(V-V_0)/kT}. \qquad \text{......(1058)}$$

This is the simplest example of the theorems of § 8·7. The details of the distribution of electrons must be studied by the help of the combined use of (1058) and Poisson's equation, as there examined. To this we return in § 11·4.

We can draw at once some interesting conclusions. Suppose we have an assembly in equilibrium containing two different metals. Then the density of the electron gas in equilibrium with them will not be equal and therefore the immediate neighbourhoods of the two metals cannot be at the same potential. The metals themselves must therefore differ in potential—this is their *contact potential difference*, for it is obviously immaterial to the foregoing argument whether the metals in question are in so-called metallic contact or not. The contact potential difference V_{12} between metals 1 and 2 can now be related at once to quantities already defined. By considering elements of electron gas near the two metals we have by (1023) and (1027)

$$\log\frac{(n_1)_f}{(n_2)_f} = -\frac{\chi_1-\chi_2}{kT} - \int_0^T \frac{dT'}{kT'^2}\int_0^{T'} (\sigma_1-\sigma_2)\,dT'', \qquad \text{......(1059)}$$

and by (1058)

$$\log\frac{(n_1)_f}{(n_2)_f} = \frac{\epsilon V_{12}}{kT}. \qquad \text{......(1060)}$$

[Here V_{12} is the excess of the potential of metal 1 over that of metal 2 in isothermal equilibrium.] We find, therefore,

$$\epsilon V_{12} = \chi_2 - \chi_1 + T\int_0^T \frac{dT'}{T'^2}\int_0^{T'} (\sigma_2 - \sigma_1)\, dT''. \qquad \text{......(1061)}$$

This equation is exact. It shows that as $T \to 0$ the contact potential difference must tend to the difference of the χ's (reckoned in volts), and over a wide range of temperature ϵV_{12} and $\chi_2 - \chi_1$ will be very nearly equal. An equation almost equivalent to (1061) can be derived thermodynamically, and is commonly given in the form

$$\epsilon\left(V_{12} - T\frac{\partial V_{12}}{\partial T}\right) = \chi_2 - \chi_1 + \int_0^T (\sigma_1 - \sigma_2)\, dT'. \qquad \text{......(1062)}$$

This can be obtained by differentiating (1061). The quantity on the right is also called $\phi_2 - \phi_1$, the difference of the average work required at temperature T to evaporate an electron from the two metals. If we further differentiate (1062), we obtain

$$-\epsilon T\frac{\partial^2 V_{12}}{\partial T^2} = \sigma_1 - \sigma_2. \qquad \text{......(1063)}$$

This equation refers only to the equilibrium state with σ's defined by (1029). It must be carefully distinguished from the familiar equation of the same form involving Thomson's specific heats of electricity. The latter refers to a current-carrying state and neither the σ's nor V_{12} have exactly the same meaning. We return to such problems in § 11·74.

Observed contact potentials should satisfy closely the theoretical relationship

$$\epsilon V_{12} = \chi_2 - \chi_1, \qquad \text{......(1064)}$$

the σ-terms being trivial. Contact potentials are difficult to measure satisfactorily under conditions which ensure that the metal surfaces are in the state for which the work function or photoelectric thresholds have been measured. Some recent measurements which more or less satisfy these requirements are collected in Table 43·1.

The experiments on Ni–Fe were complete, and completely substantiate (1064). In the other experiments the χ_1 and χ_2 recorded were not both measured for the actual specimens for which ϵV_{12} was measured and the values are taken from Table 40. From Langmuir and Kingdon experiments too small a value of ϵV_{12} would be derived if there is any reflection of electrons from the surface of the contaminated metal. It is therefore reasonable to hold that these results substantiate (1064) so far as they can be trusted, except for O/W. It is clear that the recorded value of χ for this surface is not in any sense a true work function.

The relationship (1064) can of course be derived for two metals in equilibrium in contact with each other, without appeal to any third vapour phase for the electrons. Provided they can exchange electrons the equilibrium distributions in the two metals must have the same value of λ. By equation (1000)

$$\lambda = \mu e^{-\chi_0/kT} = e^{-(\chi_0-\eta^*)/kT} = e^{-\chi/kT}.$$

This equation refers all energies to a potential energy zero for an electron in free space just outside the metal. It is at once evident that equilibrium is impossible for two metals in contact if both use this same energy zero. The second metal must charge up relative to the first thereby adjusting $(\chi_0)_2$ until $\lambda = e^{-\chi_1/kT} = e^{-(\chi_2-\epsilon V_{12})/kT}$, which is (1064).

Yet another method of proving (1064) is provided by the theory of contacts in § 11·9.

TABLE 43·1.

Measurements of contact potentials, compared with $\chi_2 - \chi_1$.

χ_2	χ_1	$\chi_2-\chi_1$	ϵV_{12} obs.	Authority
Ni, 5·01 ± 0·02	Fe, 4·77 ± 0·02	0·24	0·21 ± 0·01	(1)
W, 4·54	Ba, 2·11	2·43	2·13 ± 0·05	(2)
W, 4·54	Th/W, 2·63	1·91	1·46	(3)
W, 4·54	Cs/W, 1·36	3·18	2·8	(3)
W, 4·54	Cs/O/W, 0·72	3·82	3·1	(3)
W, 4·54	O/W, 9·2	−4·7	−0·8	(3), (4)

(1) Glasoe, *Phys. Rev.* vol. 38, p. 1490 (1931).
(2) Anderson, *Phys. Rev.* vol. 47, p. 958 (1935).
(3) Langmuir and Kingdon, *Phys. Rev.* vol. 34, p. 129 (1929).
(4) A satisfactory rediscussion of this anomaly has been given by Reimann, *Phil. Mag.* vol. 20, p. 594 (1935).

§ 11·4. *Space charge effects. Special electron atmospheres.* It will now be of some interest to examine in detail the equilibrium state of some electron atmospheres in which the space charge effects due to the electrons themselves are important and are included in the calculations. As we have shown in § 8·7 the laws governing the equilibrium distribution are

$$\nabla^2 w = 4\pi\epsilon\bar{\nu}, \quad \bar{\nu}/\bar{\nu}_0 = e^{\epsilon(w-w_0)/kT}. \qquad \text{......(1065)}$$

Not many problems of electron distribution are soluble explicitly in finite terms. The simplest distribution imaginable is a stratification in parallel planes. Such a distribution will be set up between the plates of a parallel plate condenser, and the equations are then soluble exactly. They become

$$d^2w/dx^2 = 4\pi\epsilon\bar{\nu}, \quad \bar{\nu} = \bar{\nu}_0 e^{\epsilon V/kT}, \qquad \text{......(1066)}$$

if we define w to be zero in the plane where $\bar{\nu} = \bar{\nu}_0$. Hence

$$\frac{\epsilon}{kT}\frac{dw}{dx} = \frac{1}{\bar{\nu}}\frac{d\bar{\nu}}{dx}, \quad 4\pi\epsilon\bar{\nu} = \frac{kT}{\epsilon}\frac{d}{dx}\left\{\frac{1}{\bar{\nu}}\frac{d\bar{\nu}}{dx}\right\}.$$

It will be convenient to introduce a new variable y such that

$$\bar{\nu}\,dx = dy.$$

Then

$$d^2\bar{\nu}/dy^2 = 4\pi\epsilon^2/kT,$$

$$\bar{\nu} = \frac{2\pi\epsilon^2}{kT}y^2 + by + c,$$

where b and c are constants of integration. At the symmetrical plane half-way between the two plates (of the same material at the same potential) we shall have by symmetry $\partial w/\partial x = 0$, $\partial\bar{\nu}/\partial y = 0$. If this plane is $x = 0$ and we choose y so that

$$y = \int_0^x \bar{\nu}\,dx,$$

then $\partial\bar{\nu}/\partial y = 0$ for $y = 0$ and $b = 0$. Since $\bar{\nu} > 0$, $c > 0$, and we write

$$\bar{\nu} = \frac{2\pi\epsilon^2}{kT}(y^2 + \lambda^2),$$

so that

$$\frac{dy}{y^2+\lambda^2} = \frac{2\pi\epsilon^2}{kT}dx,$$

$$y = \lambda \tan\frac{2\pi\epsilon^2\lambda}{kT}x,$$

$$\bar{\nu} = \frac{2\pi\epsilon^2\lambda^2}{kT}\sec^2\frac{2\pi\epsilon^2\lambda}{kT}x.$$

We can now determine λ, in terms of the electron density $\bar{\nu}_0$ in the central plane, by the equation

$$\lambda = \left(\frac{kT\bar{\nu}_0}{2\pi\epsilon^2}\right)^{\frac{1}{2}}, \qquad \text{......(1067)}$$

which gives

$$\bar{\nu} = \bar{\nu}_0 \sec^2\left(\left\{\frac{2\pi\epsilon^2\bar{\nu}_0}{kT}\right\}^{\frac{1}{2}} x\right). \qquad \text{......(1068)}$$

For the potential w we have

$$\frac{\epsilon w}{kT} = -2\log\cos\left(\left\{\frac{2\pi\epsilon^2\bar{\nu}_0}{kT}\right\}^{\frac{1}{2}} x\right). \qquad \text{......(1069)}$$

With these equations any desired details may be investigated. We may observe in the first place that if an ideal non-electrical material may be imagined which neither emits, absorbs nor attracts electrons, then, since $\partial w/\partial x = 0$ for $x = 0$ and the electric intensity vanishes, the plane $x = 0$ may be replaced by an ideal non-electrical wall on which the electrons will act merely in virtue of their mechanical momenta. The distribution laws in $\bar{\nu}$ and w will be unaltered by this replacement. The plane of symmetry, or ideal wall, is a locus of equilibrium points which no lines of force cross. The stress per unit area at the surface of the emitter due to electrostatic forces is a negative pressure equal in amount to $2\pi\sigma^2$, where σ is the surface density of the charge on the emitter, or the total atmospheric charge per unit area contained between the emitter and the central plane if the assembly is self-contained and on the whole uncharged. If this distance is a, then

$$\sigma = \epsilon \int_0^a \bar{\nu}\, dx = \epsilon\bar{\nu}_0 \int_0^a \sec^2\left\{\left(\frac{2\pi\epsilon^2\bar{\nu}_0}{kT}\right)^{\frac{1}{2}} x\right\} dx,$$

$$= \left(\frac{kT\bar{\nu}_0}{2\pi}\right)^{\frac{1}{2}} \tan\left\{\left(\frac{2\pi\epsilon^2\bar{\nu}_0}{kT}\right)^{\frac{1}{2}} a\right\}.$$

The electrical negative pressure is

$$kT\bar{\nu}_0 \tan^2\left\{\left(\frac{2\pi\epsilon\bar{\nu}_0}{kT}\right)^{\frac{1}{2}} a\right\},$$

or

$$kT(\bar{\nu}_a - \bar{\nu}_0).$$

At the same time the positive pressure due to the transport of momentum is

$$kT\bar{\nu}_a.$$

The net pressure on the surface of the conductor is therefore $kT\bar{\nu}_0$, which is also of course the pressure at the plane of symmetry or ideal wall, as it must be for mechanical equilibrium.

The distribution of electrons will be unaltered if we suppose the parallel emitters are finite, equal and similarly situated plates connected by non-electrical walls which are normal to them. In that case the pressure at any point on these side walls will be entirely due to the mechanical transport of momentum—there is no electrical stress acting across an element of the wall surface. The pressure at any point will therefore be simply $kT\bar{\nu}_x$ and in particular close to the emitter $kT\bar{\nu}_a$. It is of some importance for thermodynamic arguments to observe that this simple mechanical pressure $kT\bar{\nu}_a$ acts on a realizable wall surface. It is therefore possible to give a simple derivation of the differential form of (1023) by a thermodynamic cycle without ignoring electrostatic effects and without explicit calculation of them.

§ **11·41.** *Further details of electron atmospheres.* If further details of the distribution are required it is natural to follow a classical exposition by von Laue.* We begin by studying the form of the distribution (1068) and (1069) near $x = a$, particularly for large T. Using (1023) without its σ-term for the density at the surface of the metal, we have

$$\bar{\nu}_a = \bar{\nu}_0 \sec^2\left\{\left(\frac{2\pi\epsilon^2\bar{\nu}_0}{kT}\right)^{\frac{1}{2}} a\right\} = A'T^{\frac{3}{2}}e^{-\chi/kT}, \qquad \text{......(1070)}$$

where normally $A' = 4{\cdot}86 \times 10^{15}$. It follows that so long as T is large enough for $e^{-\chi/kT}$ not to swamp A' the right-hand side of (1070) is a large number—

* von Laue, *Jahrb. der Rad. u. Elektronik*, vol. 15, pp. 205, 257 (1928), or later, *Handbuch der Radiologie*, vol. 6, p. 452 (1925). This paper is summarized by Richardson, *loc. cit.*, who gives much additional information. The reader should also refer to Schottky, *Jahrb der Rad. u. Elektronik* (1915), p. 199.

there are plenty of electrons per unit volume. In order to evaluate $\bar{\nu}_0$ for given T we have therefore to solve (1070) for $\bar{\nu}_0$ when the right-hand side is (usually) numerically large. The nature of the root will depend on whether or not, as $\bar{\nu}_0$ increases, $2\pi\epsilon^2\bar{\nu}_0 a^2/kT$ approaches $\pi^2/4$ before $\bar{\nu}_0$ approaches the constant on the right of (1070). Since the former condition is

$$1\cdot04\times 10^{-2}\frac{\bar{\nu}_0 a^2}{T}\to\frac{\pi^2}{4},$$

it is obvious that this happens first unless T is quite small. The first approximation to the root of (1070) is therefore

$$\bar{\nu}_0=\frac{\pi^2}{4a^2}\frac{kT}{2\pi\epsilon^2}, \qquad \text{......(1071)}$$

and the second

$$\bar{\nu}_0=\frac{\pi^2}{4a^2}\frac{kT}{2\pi\epsilon^2}-\alpha, \qquad \text{......(1072)}$$

where

$$\bar{\nu}_0\operatorname{cosec}^2\frac{2a^2\epsilon^2}{kT}\alpha=A'T^{\frac{3}{2}}e^{-\chi/kT},$$

$$\alpha=\frac{\pi}{2a}\left(\frac{k}{2\pi\epsilon^2A'}\right)^{\frac{1}{2}}\frac{kT}{2a^2\epsilon^2}\frac{e^{\chi/2kT}}{T^{\frac{1}{4}}}. \qquad \text{......(1073)}$$

Inserting numerical values for the atomic constants,

$$\bar{\nu}_0=96\frac{\pi^2T}{4a^2}-4\cdot22\times10^{-5}\frac{\pi T^{\frac{3}{4}}}{2a^3}e^{\chi/2kT}.$$

For our immediate purpose the form of w near $x=a$ is more important. By (1069) we have approximately after reduction

$$\frac{\epsilon w}{kT}=-2\log\left\{\tfrac{1}{2}\pi\frac{a'-x}{a}\right\}, \qquad \text{......(1074)}$$

where

$$a'=a\left\{1+\frac{1}{a}\left(\frac{k}{2\pi\epsilon^2A'}\right)^{\frac{1}{2}}\frac{e^{\chi/2kT}}{T^{\frac{1}{4}}}\right\}. \qquad \text{......(1075)}$$

Inserting numerical values

$$a'=a\left\{1+1\cdot40\times10^{-7}\frac{1}{a}\frac{e^{\chi/2kT}}{T^{\frac{1}{4}}}\right\}.$$

Thus for normal values of T and for all practical purposes except for regions in the immediate neighbourhood of the metal surface we may identify a' and a and assert that w behaves as if it had a logarithmic infinity

$$w\sim-\frac{2kT}{\epsilon}\log(a-x) \qquad \text{......(1076)}$$

as $x\to a$. The strict behaviour is that w behaves as if it had a similar logarithmic infinity as x approaches a surface just inside the actual metal by an extremely small distance which tends to zero as $T\to\infty$.

The behaviour of w can be exhibited in its most general aspect by introducing the transformation

$$\psi=\frac{\epsilon w}{kT}+\log\frac{4\pi\epsilon^2\bar{\nu}_0}{kT}. \qquad \text{......(1077)}$$

Then ψ satisfies the differential equation

$$\nabla^2\psi=e^{\psi}, \qquad \text{......(1078)}$$

and the approximate boundary condition

$$\psi\sim-2\log(a'-x)+\log 2. \qquad \text{......(1079)}$$

To the accuracy with which a' can be identified with a both the differential equation and the boundary conditions for ψ are absolute—independent of the temperature, of atomic constants and potentials and of the dimensions of the apparatus. We have only established this by a study of the detailed solution for a parallel plate condenser, but it is clear from the form of the result, which concerns only the immediate neighbourhood of the metal surface, that this boundary condition will continue to hold for all metal surfaces plane or curved, so that all electron atmospheres in enclosures entirely surrounded by metal emitters at a sufficiently high temperature can be studied by solving (1078) for the enclosure subject to the boundary condition

$$[\psi\sim-2\log\delta], \qquad \text{......(1080)}$$

where δ is the normal distance from the boundary.

It does not appear to have been rigorously established that equations (1078) and (1079) suffice to determine ψ uniquely. For physical reasons one may guess that they must do so, and we shall assume it in the rest of the discussion. Some interesting general theorems then follow at once.

Theorem (11·41). *In any enclosure entirely surrounded by hot electrodes all at the same temperature, the electron density at any point not too near the walls is independent of the material of the walls and proportional to the absolute temperature provided this temperature is sufficiently high.*

The electron density depends of course on the size and shape of the enclosure. Equation (1071) provides an example.

Theorem (11·42). *For two similar enclosures at the same temperature, for which Theorem* (11·41) *holds, the electron density at corresponding points is inversely proportional to the square of the linear dimensions.*

Theorem (11·43). *The equilibrium state of the electron atmosphere is characterized by a minimum value of the ratio of the electrostatic energy to the kinetic energy of translation of the electrons.*

The proofs of these theorems are simple and are left to the reader.*

* The reader may refer to von Laue, *loc. cit.*, for further information.

At fairly high temperatures and for not too small a $\bar{\nu}_a$ will be very large compared with $\bar{\nu}_0$. In that case near the emitter there is a large normal electrical stress (tension) and a large mechanical pressure which practically balance since $\bar{\nu}_0$ is trivial. Across a plane, however, normal to the emitting surface there is an electrostatic pressure numerically equal to the tension along the lines of force which therefore just doubles the usual pressure. These considerations will of course continue to hold for surfaces of reasonable curvature, not merely for the plane surfaces of a condenser.

The electron repulsions cause the electron atmosphere to behave near the surface like a surface film of negative surface tension. To examine this more exactly it is necessary to cast (1078) into curvilinear coordinates suitable for the discussion of the immediate neighbourhood of the metal. If the curvature is small it is not difficult to show that equation (1078) takes the form

$$\frac{\partial^2\psi}{\partial\delta^2}+\left(\frac{1}{R_1}+\frac{1}{R_2}\right)\frac{\partial\psi}{\partial\delta}=e^{\psi}, \qquad \text{......(1081)}$$

which is valid in the immediate neighbourhood of the surface if δ is the normal distance from the boundary reckoned positive into the enclosure and R_1 and R_2 are the principal radii of curvature of the boundary, reckoned positive when the centres of curvature lie outside the enclosure. Correct to terms of order $\left(\frac{1}{R_1}+\frac{1}{R_2}\right)$ this equation has a first integral

$$\frac{\partial\psi}{\partial\delta}=-\sqrt{2}\,e^{\frac{1}{2}\psi}-2\left(\frac{1}{R_1}+\frac{1}{R_2}\right)$$

of the correct form. On writing $\frac{1}{2}\psi=\log x$ this equation is easily integrated completely and the required solution is

$$\psi\sim-2\log\left\{\frac{1}{\sqrt{2}}\left(\frac{e^{\left(\frac{1}{R_1}+\frac{1}{R_2}\right)\delta}-1}{\frac{1}{R_1}+\frac{1}{R_2}}\right)\right\}. \qquad \text{......(1082)}$$

With the help of (1077) equation (1082) determines w and so $\bar{\nu}$ in the neighbourhood of any surface of moderate curvature. Equation (1082) reduces to (1079) when $\frac{1}{R_1}+\frac{1}{R_2}\to 0$. For the further development of the properties of the quasi surface film of electrons we refer the reader to von Laue.

In all the foregoing discussion we have ignored the effects of the so-called *image forces* which will alter the distribution laws near the metal surface, so that our conclusions are only valid under conditions and in regions in which these image forces can be neglected. We return to a fuller consideration of this effect in § 11·44.

§ **11·42.** *The emission of positive ions.* In certain cases incandescent solids have been observed to emit positive ions as well as electrons. It is therefore necessary to study the equilibrium theory of an atmosphere of a mixture of positive and negative ions and neutral atoms in equilibrium with the hot solid, which is a simple extension of the work of this chapter.* Theoretically, we must expect to have, for example, an atmosphere of electrons, tungsten ions and neutral tungsten atoms in equilibrium with the solid tungsten, and so on in all similar cases. Actually, the atmosphere in equilibrium with a pure metal will never contain a significant number of ions of that metal at any temperature (below the melting point of the metal) at which experiments can be carried out. The effects actually observed are due to the emission of ions of impurities contained in the metal. When spurious effects due to surface adsorption of gaseous layers have been eliminated the remaining effects are generally due to the emission of singly charged positive ions of the alkali metals. We will present the analysis in such a way that the positive ions are explicitly shown as due to an impurity. A similar analysis can be given when there are no impurities and the positive ions are those of the metal itself. It will be sufficient here to contemplate an atmosphere of electrons, one type of singly charged positive ion, and the corresponding neutral atom together with the impure solid. Metal atoms and ions will be assumed to be absent from the atmosphere. They can be added if required. In considering an atmosphere of ions, electrons and atoms in equilibrium we anticipate in a simple case the general discussion of Chapter XIV. We require here nothing beyond simple reinterpretations of Chapter V.

The formal expression for the number of complexions of this assembly, using classical statistics as a valid approximation for the vapour phase, will be, after (535) generalized,

$$C = \Sigma\, C_{N,P},$$

$$C_{N,P} = \frac{1}{(2\pi i)^3}\iiint \frac{dx\,dy\,dz}{x^{X+1}y^{Y+1}z^{E+1}}\, x^N y^P K_{N,P}(z) \exp\{xf_\epsilon(z) + yf_+(z) + xyf(z)\}. \quad \text{......(1083)}$$

In (1083) $K_{N,P}(z)$ is the partition function for the crystal containing Q permanently non-evaporating metal atoms, N electrons and P (impurity) positive ions. The partition functions $f_\epsilon(z)$, $f_+(z)$ and $f(z)$ refer to electrons, positive ions and neutral (impurity) atoms in the vapour phase. Then we have the usual formulae

$$\overline{N}_{\text{free}} = \lambda f_\epsilon(T), \quad \overline{P}_{\text{free}} = \mu f_+(T), \quad \overline{NP}_{\text{free}} = \lambda\mu f(T), \quad \text{......(1084)}$$

* For a general account of the phenomena, see Richardson, *loc. cit.* The general thermodynamic theory of such atmospheres has been given by von Laue, *Berl. Sitz.* p. 334 (1923).

and from the fact that the electrons and the positive ions are effectively present in the solid $$\lambda = 1/\kappa_1(T), \quad \mu = 1/\kappa_2(T). \qquad \text{......(1085)}$$

We shall expect no alteration in form* for $\kappa_1(T)$ due to the impurity and only slight alterations in the magnitude of σ. We shall therefore still have (1023) for the density of free electrons in the atmosphere in the immediate neighbourhood of the metal. For $\kappa_2(T)$ we shall have the same *form* as for $\kappa_1(T)$, but the magnitude of σ may be entirely different. In fact, at the temperatures concerned it is reasonable to suppose that the specific heat of the solid is "normal" and has the value $3k$ per atom whether metal or impurity. We may go further in fact and assume with sufficient accuracy that $\sigma_2(T) = 3k$ over the whole temperature range for $T > T_0$. In that case

$$\exp\left\{-\int_0^T \frac{dT'}{kT'^2}\int_0^{T'} \sigma\, dT''\right\} \sim const. \times T^{-3},$$

and $$\bar{\nu}_+ = \frac{\bar{P}}{V} = const. \times T^{-\frac{1}{2}} e^{-\chi_2/kT}. \qquad \text{......(1086)}$$

The constant is only determinable if more explicit assumptions can be made about $\kappa_2(T)$. If we may assume, for example, that adding an atom or ion of the impurity to the solid is equivalent to adding a single three-dimensional harmonic oscillator of frequency of ν_0, the weight of all its states being $\kappa_2(0)$, then $$\kappa_2(T) = \kappa_2(0)/(1 - e^{-h\nu_0/kT})^3,$$

$$\bar{\nu}_+ = \frac{\varpi_+}{\kappa_2(0)} \frac{(2\pi m_+ kT)^{\frac{3}{2}}}{h^3} (1 - e^{-h\nu_0/kT})^3 e^{-\chi_2/kT}. \qquad \text{......(1087)}$$

There is little point in expanding these formulae further in the absence of exact measurements to compare them with. We note only that if we measure the current carried by the positive ions, we should expect to find

$$I_+ = BT^{-1} e^{-\chi_2/kT}, \qquad \text{......(1088)}$$

where B is a constant. We may note that the indices of T in (1088) and (1032) add up to 1 and the indices of T in (1086) and (1023) add up to zero. This will always be the case so long as we assume that the neutral atom or positive ion in the metal has 3 practically classical degrees of freedom yielding 6 square terms and the ion and electron in the gas also 3 square terms each or 6 in all. Owing to the fact that we must assume that ionization and recombination are possible in the solid, the product $\lambda\mu$ or $\kappa_1(T)\,\kappa_2(T)$ must be the same as $\kappa(T)$, the corresponding function for the neutral atom in the solid. This secures the relation just mentioned.

* The surface conditions may of course be such that we have an entirely different numerical value of χ.

By eliminating λ and μ between the three equations (1084) we obtain

$$\frac{\overline{N}_{\text{free}}\overline{P}_{\text{free}}}{\overline{N}\overline{P}_{\text{free}}}=\frac{f_\epsilon(T)f_+(T)}{f(T)}. \qquad \text{......(1089)}$$

This is a particular case of the formulae for the ionization equilibrium of a gas at high temperature and is of course independent of the properties of the solid phase. It is mentioned here because some beautiful thermionic measurements by Langmuir* have established the correctness of the theory as applied to the equilibrium between caesium atoms, ions, and electrons. The simple form of (1089), valid in the case to be discussed, is

$$\frac{\bar{\nu}_\epsilon\bar{\nu}_+}{\bar{\nu}_0}=\frac{\varpi_\epsilon\varpi_+}{\varpi_0}\frac{(2\pi mkT)^{\frac{3}{2}}(2\pi m_+kT)^{\frac{3}{2}}}{h^3\{2\pi(m_++m)kT\}^{\frac{3}{2}}}e^{-\chi_0/kT}, \qquad \text{......(1090)}$$

where ϖ_ϵ, ϖ_+ and ϖ_0 are the weights of the normal states of free electrons, positive caesium ions and neutral caesium atoms respectively and χ_0 is the ionization potential. Since $\varpi_\epsilon=\varpi_0=2$, $\varpi_+=1$, $\chi_0=3{\cdot}88$ electron volts and m is negligible compared with m_+, this reduces to

$$\frac{\bar{\nu}_\epsilon\bar{\nu}_+}{\bar{\nu}_0}=\frac{(2\pi mkT)^{\frac{3}{2}}}{h^3}e^{-\chi_0/kT}=K_n, \qquad \text{......(1091)}$$

so that

$$\log_{10}K_n=15{\cdot}385+\tfrac{3}{2}\log_{10}T-\frac{19530}{T}. \qquad \text{......(1092)}$$

Langmuir's test of this equation proceeds as follows. He considers an enclosure in equilibrium with pure tungsten at 1200° K., containing caesium vapour at a measured pressure (ions plus atoms) of (say) 0·001 bar. The number of free electrons in equilibrium with tungsten at this temperature is deduced from (1023) using the known emission constants for pure tungsten. The actual value is $\nu_\epsilon=9{\cdot}25$ per cm.^{3} At this temperature $K_n=5340$, so that $\bar{\nu}_+/\bar{\nu}_0=577$ according to theory. This means that practically all the caesium must be present as ions, which is what is observed, for it is found that above about 1200° K. the positive saturation current flowing to a collecting electrode in given caesium vapour is independent of the temperature of the tungsten, so that presumably above this temperature the tungsten converts every caesium atom that strikes it into an ion and emits only caesium ions at a rate naturally depending only on the caesium vapour density. On the other hand, with thoriated tungsten the equilibrium electron density is $6{\cdot}0\times10^7$ and $\bar{\nu}_+/\bar{\nu}_0=8{\cdot}9\times10^{-5}$. This means that only an insignificant fraction (1 in 11,000) of caesium atoms leaving the thoriated surface is an ion and no positive current should flow. None is observed.

By somewhat different arguments a rough quantitative test of (1092) can be achieved. A pure tungsten filament was raised to 1177° K. in a bulb of

* Langmuir and Kingdon, *Proc. Roy. Soc.* A, vol. 107, p. 61 (1925).

caesium vapour at 70° C. at the vapour pressure of pure caesium for that temperature. The positive and negative ion currents were measured. The electron emission from the tungsten at 1177° K. was $2{\cdot}22 \times 10^{-6}$ ampere per cm.^{2} and the positive ion emission $2{\cdot}06 \times 10^{-6}$. The electron emission is some 10^6 times greater than from pure tungsten at this temperature, so that we are really dealing with a caesiated tungsten surface. This, however, does not alter the arguments. On raising the filament temperature to 1300° K. or more the positive ion current increased to $2{\cdot}43 \times 10^{-3}$ and then remained independent of the filament temperature. This saturation current is therefore a measure of the rate at which caesium atoms and ions strike (and are emitted as ions from) the surface and corresponds to $1{\cdot}52 \times 10^{16}$ atoms or ions per sec. per cm.^{2} At the lower filament temperature (1177° K.) the atoms still strike the filament at the same rate for the conditions in the vapour are unaltered, but the positive ion current is only 1/1180 of its saturation value. This means that of the caesium evaporating 1 in 1180 is an ion. The conditions at the surface of the filament are essentially the same as if it were surrounded by caesium vapour at 1177° K. and at such a concentration as to provide $1{\cdot}52 \times 10^{16}$ impacts per sec. per cm.^{2} This concentration would be

$$\bar{\nu}_0 = 1{\cdot}40 \times 10^{12}.$$

If, then, the filament were in an enclosure at 1177° K. in equilibrium with this concentration of caesium, it would emit electrons and caesium ions at the rates measured $2{\cdot}22 \times 10^{-6}$ and $2{\cdot}06 \times 10^{-6}$ respectively. From these observed currents the corresponding equilibrium concentrations are

$$\bar{\nu}_\epsilon = 2{\cdot}60 \times 10^6, \quad \bar{\nu}_+ = 1{\cdot}19 \times 10^9.$$

From these three values the observed value of the equilibrium constant is

$$K_n = 2210,$$

while the value calculated from (1092) is

$$K_n = 2500.$$

This is excellent agreement. If we express it by examining what temperature makes K_n equal to its observed value we find $T = 1174°$ K. instead of 1177° K., a difference within the uncertainties of the temperature scale.

§ **11·43.** *Space charge effects with positive and negative ions.* The equations so far given for positive ions refer to assemblies of negligible space charge or to the immediate neighbourhood of the emitting surfaces. The general laws for the atmosphere can be studied by an extension of § 11·4. By § 8·7 the average electrostatic density and potential in the atmosphere ρ and w will satisfy the equations

$$\nabla^2 w = -4\pi\rho, \quad \rho = \epsilon\{-(\nu_\epsilon)_0 e^{\epsilon w/kT} + (\nu_+)_0 e^{-\epsilon w/kT}\}. \quad \ldots\ldots(1093)$$

This is the *average* potential. In addition, there will be polarization fields round each positive and negative ion like those considered in § 8·8, which will give rise to small additional effects. These, however, are negligible unless the space charge is zero or very small, when they give rise to the only surviving terms in the electrostatic energy. They are probably never of importance here.

We observe first that $$\nu_\epsilon \nu_+ = (\nu_\epsilon)_0 (\nu_+)_0,$$
and that ν_0 is unaffected by the electrostatic field. This is an example of the general theorem that if the condition of dissociative equilibrium is satisfied at any point of an atmosphere in equilibrium under long-range potentials, then it is in dissociative equilibrium everywhere. The equilibrium constant is independent of position. If we write

$$\alpha = \frac{kT}{2\epsilon} \log \frac{(\nu_+)_0}{(\nu_\epsilon)_0}, \quad \frac{\kappa}{kT} = 8\pi \sqrt{(\nu_\epsilon \nu_+)}, \qquad \text{......(1094)}$$

the equation for w becomes

$$\nabla^2 w = \frac{\epsilon\kappa}{kT} \sinh \frac{\epsilon(w-\alpha)}{kT}. \qquad \text{......(1095)}$$

We naturally only expect to be able to solve this explicitly (if at all) for plane parallel condensers or their equivalent. If then w depends only on x,

$$\frac{d^2 w}{dx^2} = \frac{\epsilon\kappa}{kT} \sinh \frac{\epsilon(w-\alpha)}{kT}.$$

This can be integrated once giving

$$\left(\frac{dw}{dx}\right)^2 = 2\kappa \left[\cosh \frac{\epsilon(w-\alpha)}{kT} - A\right], \qquad \text{......(1096)}$$

where A is a constant of integration. We may notice that κ has a very simple form. It satisfies

$$\kappa = 8\pi \sqrt{(p_\epsilon p_+)} = 8\pi \sqrt{(K_n p_0)}, \qquad \text{......(1097)}$$

where K_n is the equilibrium constant and p_0 the partial pressure of the neutral atoms.

Equation (1096) can be integrated completely in terms of Weierstrass's $\wp$-function. It can be integrated in finite terms with elementary functions when $A = 1$. This case will serve for the general study of the behaviour of V in the neighbourhood of the emitting surfaces, since there the argument of the cosh will in general be large. We then have

$$\frac{dw}{dx} = \pm 2\sqrt{\kappa} \sinh \frac{\epsilon(w-\alpha)}{2kT},$$

which integrates in the form

$$-\log\left\{\pm \tanh \frac{\epsilon(w-\alpha)}{4kT}\right\} = \frac{\epsilon\sqrt{\kappa}}{kT} x,$$

if x is measured from the emitting surface.

§ **11·44.** *Image forces.** Our treatment hitherto has been based on mean potentials and mean densities according to the analysis of § 8·7. But our averaging has been based only on averaging the electrons and ions in the gas phase of the assembly and not on averaging *all* the movable charges in the *assembly*. This latter of course would be the correct procedure and lead to an unexceptionable result for w. But such a procedure seems to be far beyond our resources at present, and a correction is necessary for the polarizing effect of the individual ion on the neighbouring metal surface. If we assume, as seems reasonable, that the metal surface remains in the mean a surface of constant potential, then the polarizing effect is equivalent to the formation of the usual electrical image, and the ion will be attracted to the (plane) surface at a distance δ with a force $\epsilon^2/4\delta^2$, which is in addition to the force arising from the average potential w. Also, unlike w, the image force affects ions of either sign equally. Both are attracted to the metal.

The image force can be derived from a potential energy function $-\epsilon^2/4\delta$. We must suppose, therefore, that the correct atmospheric density law for electrons is

$$\bar{\nu}_\epsilon = (\bar{\nu}_\epsilon)_0\, e^{[\epsilon(w-w_0)+\epsilon^2/4\delta]/kT}, \qquad \text{......(1098)}$$

and for positive ions

$$\bar{\nu}_+ = (\bar{\nu}_+)_0\, e^{[-\epsilon(w-w_0)+\epsilon^2/4\delta]/kT}. \qquad \text{......(1099)}$$

The corresponding equation that w must satisfy is

$$\nabla^2 w = -4\pi\epsilon[-(\bar{\nu}_\epsilon)_0\, e^{\epsilon(w-w_0)/kT} + (\bar{\nu}_+)_0\, e^{-\epsilon(w-w_0)/kT}]\, e^{\epsilon^2/4\delta kT}. \qquad \text{......(1100)}$$

We have not given a rigorous proof of these equations. As we have indicated, this could only come from a proper averaging treatment of all the movable charges, not only of those in the atmosphere. It seems clear, however, that the equations must be of this form, and that the true correction for the inadequate averaging will not be widely different from that proposed. Equation (1100) of course follows logically from (1098) and (1099). The correction cannot hold good indefinitely as $\delta \to 0$. As soon as the specified electron gets within distances of the walls comparable with their departure from an ideal plane conducting surface the polarizing effect will depend on the nature of the surface, the roughness of its microstructure and so on, and finally will reduce to an effect on individual atoms. Thus the apparent infinity in the correcting factor is spurious and the formula suggested cannot hold for values of $\delta < 5 \times 10^{-8}$ cm., or perhaps 10^{-7} cm. Obviously at these distances the discussion fails altogether and we need only pay attention to greater values of δ.

For values of δ of the order of 10^{-4}, 10^{-5} cm. the correction has become quite insensible and our preceding results will hold unaltered. For on inserting numerical values we see that the extra term is

$$4\cdot 15 \times 10^{-4}/\delta T.$$

* Based on von Laue, *loc. cit.* (twice), and Langmuir and Kingdon, *loc. cit.*

At room temperatures ($T=300°$ K.) this is entirely negligible for $\delta > 3\times 10^{-5}$ and at the more usual thermionic temperatures of the order of 1000° K. when $\delta > 10^{-5}$. These are outside limits. Closer investigation shows that marked effects do not even extend so far. Consider for simplicity the case of atmospheres without positive ions. Ignoring the image effect we found in (1076) that

$$w \sim -\frac{2kT}{\epsilon}\log\delta.$$

This was derived from an equation

$$\nabla^2 w = Ae^{\epsilon w/kT}$$

instead of from the equation

$$\nabla^2 w = Ae^{\epsilon w/kT + \epsilon^2/4\delta kT},$$

or, in other words, by neglecting $\epsilon^2/4\delta kT$ compared with $-2\log\delta$. At 300° K. and $\delta = 10^{-6}$ these quantities are 1·4 and 27·6 respectively; for $\delta = 10^{-7}$, 14 and 30. At higher temperatures the main term is unaffected and the image effect proportionally less. We see, therefore, that it is only in the region less than 10^{-6} cm. from the wall that the image force will really alter the solutions already given and at the higher temperatures marked alterations are only caused near $\delta = 10^{-7}$ cm. It must be remembered of course that these neglected terms occur in an exponent and are not simply additive.

When there are positive ions present the image forces can make much more marked qualitative differences, for they lead to the formation of a sheath of positive ions round the emitting surfaces which would be entirely absent were it not for this image effect. The image effect only alters the *ratio* of the concentrations of electrons and positive ions indirectly through its effect on w. The sign of any space charge will be unaltered.

Since all the image effects are confined to thin layers in the immediate neighbourhood of the emitting surfaces, this layer may really be included in the "surface phase" from the point of view of thermodynamic or statistical treatment of volume effects. All our previous arguments are therefore unaffected, if by the "surface of the metal" we mean not so much the actual last fixed metallic atoms as the immediate outside of the surface film at about 10^{-5}, 10^{-6} cm. or so from the last metal atoms. The potential at some such point must then be taken to be the potential of the metal, and the differences of these potentials is the contact potential difference of two different emitters. The question then arises whether the work apparently done against the image forces in the surface layer of the atmosphere is to be included in χ. The answer of course is yes, but caution is required. If we consider two perfectly pure pieces of the same metal, one with a smooth and the other with a rough surface, the work done *in the atmosphere* against the image force would on the average be different for the two pieces. If there were no compensating effect such pieces of metal should have a contact potential difference which there is no evidence for and no reason to expect. There must therefore be some compensation in the surface layer *in the metal*, and it seems necessary to suppose, to avoid spurious contact potential differences, that the compensation is exact or at least that there is exact compensation for all variations due to the mechanical state of the surface.* The same argument for compensation of mechanical states holds for any surface of given composition, whatever impurities are present in or adsorbed on the surface of the pure metal.

Let us summarize this discussion by recalling the complete laws for the equilibrium of the atmosphere which we have obtained.

Let F_ϵ, F_+ be partition functions for the electron and positive ion without their V-factor. Immediately outside the surface layer

$$\bar{\nu}_\epsilon = (\bar{\nu}_\epsilon)_s = \frac{F_\epsilon(T)}{\kappa_1(T)}, \qquad \text{......(1101)}$$

$$\bar{\nu}_+ = (\bar{\nu}_+)_s = \frac{F_+(T)}{\kappa_2(T)}. \qquad \text{......(1102)}$$

The forms of F_ϵ and κ_1 are discussed in § 11·23 and of F_+ and κ_2 in § 11·42. Elsewhere, outside the surface layers

$$\bar{\nu}_\epsilon = (\bar{\nu}_\epsilon)_s \, e^{\epsilon(w-w_s)/kT}, \qquad \text{......(1103)}$$

$$\bar{\nu}_+ = (\bar{\nu}_+)_s \, e^{-\epsilon(w-w_s)/kT}, \qquad \text{......(1104)}$$

where w is an electrostatic potential satisfying

$$\nabla^2 w = 4\pi\epsilon(\bar{\nu}_\epsilon - \bar{\nu}_+). \qquad \text{......(1105)}$$

Inside the surface layer at δ from the metal surface (1105) continues to hold, but (1103) and (1104) are replaced by

$$\bar{\nu}_\epsilon = (\bar{\nu}_\epsilon)_s \, e^{\epsilon(w-w_s)/kT + \epsilon^2/4\delta kT}, \qquad \text{......(1106)}$$

$$\bar{\nu}_+ = (\bar{\nu}_+)_s \, e^{-\epsilon(w-w_s)/kT + \epsilon^2/4\delta kT}. \qquad \text{......(1107)}$$

It will be well to call attention to one last point. Inside the surface layer $\bar{\nu}_\epsilon \bar{\nu}_+$ is no longer constant, but

$$\bar{\nu}_\epsilon \bar{\nu}_+ = (\bar{\nu}_\epsilon)_s (\bar{\nu}_+)_s \, e^{\epsilon^2/2\delta kT}. \qquad \text{......(1108)}$$

At the same time $\bar{\nu}_0$, the concentration of the neutral atoms, is unaffected by the image forces so that

$$\frac{\bar{\nu}_\epsilon \bar{\nu}_+}{\bar{\nu}_0} = \frac{(\bar{\nu}_\epsilon)_s (\bar{\nu}_+)_s}{\bar{\nu}_0} e^{\epsilon^2/2\delta kT} = K_n \, e^{\epsilon^2/2\delta kT}. \qquad \text{......(1109)}$$

It might therefore be thought at first sight that the neutral atoms, ions and electrons are no longer in dissociative equilibrium inside the layer. It seems

* These conclusions may have to be modified if we should include here the active centres of catalytic theories. See for example Constable, *Proc. Roy. Soc.* A, vol. 108, p. 355 (1925), vol. 110, p. 283 (1926).

probable,* however, that this is not the case and that the equilibrium is still complete, the effect of the image forces being merely to decrease the work of dissociation χ by $\epsilon^2/2\delta$. Consider a quasi-Born cycle in which a neutral atom is taken from δ inside the layer to outside the layer (no work required), dissociated there (work χ required), brought back to the layer at δ as two ions (work $\epsilon^2/2\delta$ done) and there allowed to recombine (work χ' done). This is a reversible isothermal cycle, and we must therefore have

$$\chi = \chi' + \epsilon^2/2\delta \qquad \text{......(1110)}$$

in agreement with (1109) and the preceding argument.

§ **11·5.** *The "free" electrons of a metal.* Hitherto we have ignored entirely any periodic structure inside the metal; except for limitations imposed by the surface the electrons have been entirely free. This model must now be generalized before we have an acceptable electronic theory of matter, for the model must at least provide naturally a means of discriminating between conductors and insulators—that is to say we must be able to specify what electrons are more or less "free" and therefore able to conduct and what electrons are "bound" and cannot. This discrimination becomes possible as soon as the periodic variations of potential inside the crystal (metal or insulator) are taken into account.

The motion of an electron in a triply periodic field of force can be studied by either of two methods of approximation, but except in the simplest one-dimensional case the wave equation cannot be solved exactly. We may start with the atoms of the crystal in correct array but at large separations, use atomic wave-functions for the first approximation to the electronic states and enquire how the atomic states are perturbed when the separation is decreased to the separation of the actual crystal. Alternatively we may start with the wave-functions

$$e^{(2\pi i/h)(px+qy+rz)} \quad \left(E = \frac{1}{2m}(p^2+q^2+r^2)\right), \qquad \text{......(1111)}$$

representing completely free electrons with components of momentum p, q, r, and examine how the possible values of p, q, r, and the dependence of the energy E thereon, are modified by small periodic variations in the potential. Both methods naturally lead to the same general results, and one or other may give the better detailed picture according to circumstances —the atomic starting point for the most tightly bound states and the free wave starting point for those least tightly bound.

For our immediate purpose of describing the general nature and distribution of electronic states in a crystal the atomic starting point will prove satisfactory. Consider first a set of N similar nuclei in a regular crystalline

* Langmuir and Kingdon, *loc. cit.*

array expanded to large separations. The nuclei provide for one electron $2N$ similar states for which the three atomic quantum numbers (spin excluded) are specified, one for each direction of spin attached to each nucleus. At large separations these have all the same energy and form in fact one $2N$-ply degenerate state. If now the scale of the nuclear array is reduced this $2N$-ply degenerate state is split by the interactions of the other nuclei into a group of states distributed in energy over a band of energies, whose width increases as the scale of the array diminishes, but remains independent of N at least when N is large. In general the band will contain N distinct energy levels in each of which the electron can have either spin. Such a band can accommodate just $2N$ electrons and no more. The energies of its states will of course be modified by the electronic charges as the later electrons are added, but as always these charges cannot modify the number and general properties of the states. Since the breadth of the band is B say, independent of N, the order of the separation between states of neighbouring energy will be B/N; and this is very small when N is large and for a substantial crystal the set of states in the band forms practically a continuum.

The wave-functions describing the states in the band can be arranged in one or other of two ways. If we consider a finite block of crystal containing N atoms, the wave-functions obeying the appropriate boundary conditions represent stationary electron waves without momentum like the stationary waves on a stretched string with fixed ends. If we wish to arrange the wave-functions so that they represent electrons with a definite momentum—progressive waves like travelling vibrations on a stretched string of great length—we must use boundary conditions which assert that the wave-functions are periodic with period Ga say, where G is a large number and $G^3 = N$. If we use progressive waves we obtain a set of electronic states with the energies

$$E_{l,m,n} = E_0 - \alpha - 2\beta\left(\cos\frac{2\pi l}{G} + \cos\frac{2\pi m}{G} + \cos\frac{2\pi n}{G}\right) \quad (-\tfrac{1}{2}G \leqslant l, m, n < \tfrac{1}{2}G)$$

ranging over the band $E_0 - \alpha \pm 6\beta$. These states and their wave-functions are examined in detail in § 11·52. If on the other hand we use stationary waves we obtain a slightly different set of states with the energies

$$E_{l,m,n} = E_0 - \alpha - 2\beta\left(\cos\frac{\pi l}{G} + \cos\frac{\pi m}{G} + \cos\frac{\pi n}{G}\right) \quad (0 \leqslant l, m, n < G).$$

The energy range is the same, but owing to the degeneracy of the travelling waves in the forward and backward direction the spacing of the energies must be different.

As further electrons are added they will begin to fill up a new band of states derived from another atomic wave-function. It is in fact easy to see that each atomic wave-function corresponding to a definite electron energy is converted by the interaction in the crystal into a band of N states each capable of accommodating two electrons of opposite spin. These energy bands belonging to different atomic wave-functions may or may not overlap; whether they do or not may make an essential difference in the electronic properties of the lattice. The lowest energy band is derived from the wave-function (type $1S$) of the atomic K-electrons. Owing to the very tight binding of the K-electrons the atomic states are scarcely modified even in the actual lattice—the band is narrow and well separated in energy from all other possible electronic states. At any temperature at which the lattice can exist these states are always occupied by $2N$ electrons. The next lowest band will be derived from the wave-functions (types $2S$, $2P$) of the L-electrons. The $2S$-band may or may not overlap with the $2P$-bands, but the $2P$-bands themselves will in general overlap and in cubic crystals the three distinct atomic $2P$ wave-functions must from the symmetry give rise to identical crystalline states, so that there is then one $2P$-band containing N states each of which is represented by six wave-functions (allowing for the two spins) and can accommodate six electrons. The whole set of states may be called the L-bands and contains $8N$ states in all, which will normally be fully occupied. The width of all these bands is at most of the order of a few volts, so that unless the L-electrons are valency electrons the L-bands are well separated in energy from all other possible electronic states. Proceeding in this way we see that the complete set of crystalline electronic states is composed of a series of bands or groups of bands widely separated in energy from the neighbouring bands on either side of them until we come to the bands derived from the atomic wave-functions of the valency electrons themselves. Since band widths here are of the order of a few volts and the separations of the states of atomic valency electrons are of the same order, different valency bands may or may not overlap according to particular properties of the atoms and the lattice. We have spoken throughout of atomic states and an atomic lattice, but all we have said remains equally true of molecular states and a molecular lattice when this is the better representation.

At the absolute zero of temperature the electrons of the lattice will occupy the necessary number of states of lowest energy available. The K-, L-, etc. bands will all be completely full and so on until we come to the highest valency band. This may be a pure band, as in the case of the alkali metals for example, where it is derived from ^2S atomic wave-functions. It then contains $2N$ states (allowing for spin) and only N electrons occupying the lower half of them. In more complicated cases it may be composed of a mixture of states derived from several overlapping bands. It may be completely full of electrons, the next possible electronic state being separated from its highest state by a definite energy gap, or it may be only partly full.

The electrical properties of the crystal will now be entirely different according as the highest band containing electrons at the absolute zero is partly or completely full. If this band is only partly full the highest electrons in the band are free to make transitions under the influence of a small applied electrical field to states of neighbouring energy which are originally unoccupied. By this means a current is set up, and the substance has the electrical properties of a metal. In the absence of an electrical field the symmetry properties of the wave-function and the lattice require that no current can flow at any temperature, as is of course obvious *a priori*. If, however, the highest band is fully occupied, then there are no electrons in the lattice which are in states possessing empty states of neighbouring energy. It can be shown that under the influence of an applied field of ordinary strengths transactions can only be made with extreme rarity* to states separated by even quite a narrow step in energy. The substance in question in this case therefore has the properties of an insulator.

We thus see that "free" electrons—i.e. electrons which are free to convey a current through the lattice under the influence of an applied field no matter how weak—are those and only those which possess empty states of neighbouring energy to which they can make transitions. These states must differ in energy only by negligible amounts and must therefore belong to the same band or group of overlapping bands. Whether or not a given substance possesses free electrons and if so how many is in principle calculable when the arrangements of atoms in the lattice is known. In practice the computations can usually not be carried through with the present mathematical equipment of the quantum theory. We shall pass on to study in greater detail the temperature variation of the number of free electrons in the following sections, and shall then be able to correlate in a satisfactory way the electrical properties of metals and insulators and the intermediate substances called semi-conductors.

§ **11·51**. *Further remarks about the system of energy bands for the electrons in a crystal.* Though we cannot here study the quantum mechanics of an electron in a periodic field, we must be familiar with some of the results of this theory in rather more detail than has been given in the preceeding section.

In the one-dimensional case the electron moves in a potential energy field given by

$$U = U_0 + f(x),$$

* Zener, *Proc. Roy. Soc.* A, vol. 145, p. 523 (1934).

where $f(x)$ is a periodic function of period a satisfying $f(x+a)=f(x)$. Schrödinger's equation is

$$\frac{d^2\psi}{dx^2}+\kappa^2[E-U_0-f(x)]\psi=0 \quad \left(\kappa^2=\frac{8\pi^2 m}{h^2}\right).$$

It can be shown generally that the possible values of E no matter how large are never continuous but are always broken into bands by values of E which are impossible for an electron. For small values of E the width of the bands of possible E's may be small and of impossible E's large in comparison. For large values of E the width of the bands of impossible E's tends to zero, but the impossible values never drop out. The centres of these bands are determined by Bragg's reflection law for the de Broglie wave length λ_B of the electron,

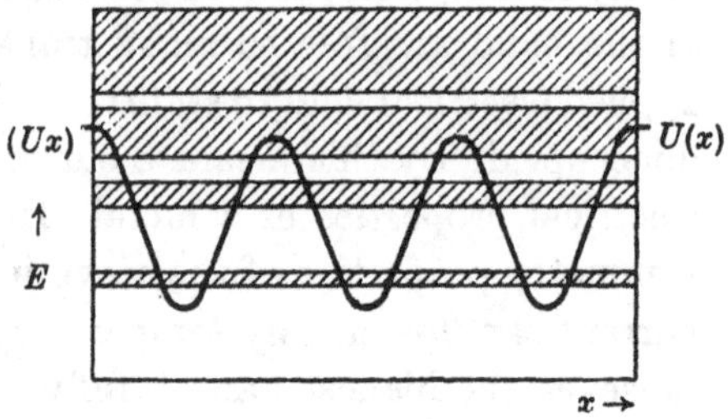

Fig. 33. The shaded strips show the bands of permitted energy levels for an electron in a one-dimensional periodic potential energy field $U(x)$.

$$n\lambda_B=a, \qquad \text{......(1112)}$$

where n is an integer. The arrangement of bands is shown diagrammatically in Fig. 33.

A special model which can be solved exactly has been studied by Kronig and Penney.*

In the actual three-dimensional lattice the arrangement of possible energies is far more complicated. It remains true that there are bands of disallowed energies for electrons moving in any specified direction but it may no longer be true that there are bands of disallowed energies ignoring directions of motion, and in fact for sufficiently great energies the disallowed ranges must always disappear. The possible energies as functions of the momenta in various directions have been studied in great detail by Brillouin.† The bands of disallowed energies are still closely related to the Bragg reflections for the corresponding de Broglie wave lengths, and must be taken account of in any theoretical calculation of the *absolute* value of the electrical conductivity.‡ They play moreover a great part in the structure of metallic alloys. It has been shown recently by Jones§ that Hume-Rothery's rule, that *alloys of the structure of γ-brass always contain very nearly* 21 *valency electrons for every* 13 *atoms in the lattice* (the structure

* Kronig and Penney, *Proc. Roy. Soc.* A, vol. 130, p. 499 (1931).

† Brillouin, *J. de Physique*, vol. 1, p. 377 (1930); *Die Quantenstatistik* (1931); *J. de Physique*, vol. 4, p. 333 (1933), where other references are given.

‡ H. Jones and Zener, *Proc. Roy. Soc.* A, vol. 144, p. 101 (1934).

§ H. Jones, *Proc. Roy. Soc.* A, vol. 144, p. 225 (1934).

being more or less indifferent to what the atoms are so long as the 21/13 ratio is obeyed), can be simply explained in terms of Brillouin's energy discontinuities. It is unfortunately impossible to do justice in a short space to Brillouin's theory and to its very beautiful applications which are just beginning to appear, and we must be content thus to call attention to it.

§ 11·52. *The structure of a simple band.* We shall require some knowledge as to how the energy and momenta depend at least approximately upon quantum numbers for the states of a single band. We shall be content to examine these details for the simplest case and quote more general results when required.

Consider a simple cubic lattice of spacing a, and let the integers g_1, g_2, g_3 characterize the lattice points; and let the potential energy of an electron due to the ion at one lattice point be

$$U_{g_1,g_2,g_3}(x,y,z) = U(x-g_1a, y-g_2a, z-g_3a), \qquad \text{......(1113)}$$

where $U(\alpha,\beta,\gamma)$ is a function of $\alpha^2+\beta^2+\gamma^2$ only. The potential energy of an electron in the crystal is then given by

$$V(x,y,z) = \sum_{g_1,g_2,g_3=-\infty}^{\infty} U_{g_1,g_2,g_3}(x,y,z); \qquad \text{......(1114)}$$

it is assumed that the U's are such that this series converges. Schrödinger's equation for an electron in the crystal is

$$\nabla^2\psi + \kappa^2(E-V)\psi = 0, \qquad \text{......(1115)}$$

and in the isolated atom g_1, g_2, g_3

$$\nabla^2\phi_{g_1,g_2,g_3} + \kappa^2(E-U_{g_1,g_2,g_3})\phi_{g_1,g_2,g_3} = 0.$$

The simplest case, which alone we shall treat here, is that in which ϕ_{g_1,g_2,g_3} is spherically symmetrical representing an atomic S-state, of energy E_0; ϕ_{g_1,g_2,g_3} is then real. We then solve (1115) approximately by putting

$$E = E_0 + \eta, \quad \psi = \sum_{g_1,g_2,g_3} a_{g_1,g_2,g_3}\phi_{g_1,g_2,g_3}$$

and imposing the periodicity conditions on ψ that

$$\psi(x,y,z) = \psi(x+Ga,y,z) = \psi(x,y+Ga,z) = \psi(x,y,z+Ga),$$

where G is a large integer. By the well-known method of perturbation theory the first order equations determining η and the a_{g_1,g_2,g_3} are easily shown to be

$$\sum_{g_1,g_2,g_3} a_{g_1,g_2,g_3} \iiint_{-\infty}^{+\infty} (\eta - U'_{g_1,g_2,g_3})\phi_{g_1,g_2,g_3}\phi_{h_1,h_2,h_3}\,dx\,dy\,dz = 0 \qquad \text{...(1116)}$$

for all integral h_1, h_2, h_3, with

$$U'_{g_1,g_2,g_3} = V - U_{g_1,g_2,g_3}.$$

If the electrons are fairly tightly bound, the atomic wave-functions do not overlap much and we may assume that

$$\iiint \phi_{g_1,g_2,g_3}\phi_{h_1,h_2,h_3}\,dxdydz = 1, 0$$

according as (g_1,g_2,g_3) and (h_1,h_2,h_3) do or do not represent the same atom. Since ϕ and U are both spherically symmetrical, we may also put

$$-\iiint U'_{g_1,g_2,g_3}\phi_{g_1,g_2,g_3}\phi_{h_1,h_2,h_3}\,dxdydz = \alpha, \beta, 0$$

according as (g_1,g_2,g_3) and (h_1,h_2,h_3) represent the same atom, next neighbours, or more distant pairs respectively. Since U' is negative α must be positive, and for fairly tightly bound electrons β will be positive too. Equations (1116) thus reduce to

$$(\eta+\alpha)\,a_{h_1,h_2,h_3} + \beta(a_{h_1+1,h_2,h_3} + a_{h_1-1,h_2,h_3} + a_{h_1,h_2+1,h_3} + a_{h_1,h_2-1,h_3} + a_{h_1,h_2,h_3+1} + a_{h_1,h_2,h_3-1}) = 0. \qquad \text{......(1117)}$$

Since ψ is periodic in the coordinates with period Ga, the a's solving (1117) must be periodic in the h's with period G. The possible solutions are easily shown to be

$$a^{l,m,n}_{h_1,h_2,h_3} = e^{2\pi i(lh_1+mh_2+nh_3)/G}. \qquad \text{......(1118)}$$

Distinct solutions are obtained when and only when l, m, n are a set of integers satisfying the conditions $0 \leqslant l, m, n < G$, or any equivalent inequalities. It is most convenient to assume that G is even (a trivial restriction) and to take

$$-\tfrac{1}{2}G \leqslant l, m, n < \tfrac{1}{2}G.$$

There are just the correct number, G^3, of such distinct solutions. The resulting ψ's are

$$\psi_{l,m,n} = \sum_{h_1,h_2,h_3}^{\infty} e^{2\pi i(lh_1+mh_2+nh_3)/G}\phi_{h_1,h_2,h_3}, \qquad \text{......(1119)}$$

and the corresponding energies are $(\alpha,\beta>0)$

$$E_{l,m,n} = E_0 - \alpha - 2\beta\left(\cos\frac{2\pi l}{G} + \cos\frac{2\pi m}{G} + \cos\frac{2\pi n}{G}\right). \qquad \text{...(1120)}$$

The range of energies covered by the band stretches from $E_0-\alpha-6\beta$ to $E_0-\alpha+6\beta$. For small l, m, n we have approximately

$$E_{l,m,n} = E_0 - \alpha - 6\beta + \frac{4\pi^2\beta}{G^2}(l^2+m^2+n^2). \qquad \text{......(1121)}$$

Near the other extreme we can put $l=\tfrac{1}{2}G-l'$, *etc.* and find for l', m', n' small

$$E_{l,m,n} = E_0 - \alpha + 6\beta - \frac{4\pi^2\beta}{G^2}(l'^2+m'^2+n'^2). \qquad \text{......(1122)}$$

The nature of the wave-function (1119) can best be seen by trying to connect it up with the wave-function (1111) for a free electron. The effect of the triple summation is to repeat the atomic wave-function with a different phase at each lattice point. Thus (1119) is very closely equivalent to a function

$$e^{2\pi i(lx+my+nz)/Ga}\,u(x,y,z), \qquad \text{......(1123)}$$

where $u(x,y,z)$ is a real function with the periodicity of the lattice. But this is of the same form as (1111) except that the free electron waves are modulated by the high frequency factor $u(x,y,z)$.

These formulae at once recall the physical meaning of the quantum numbers l, m, n as momenta. They are from (1123) palpably also *wave numbers*, l being the number of times the electron wave repeats itself in the interval Ga taken along the direction of the x-axis, with similar meanings for m and n. By comparison of (1123) and (1111) we see that

$$l=\frac{Gap}{h}, \quad m=\frac{Gaq}{h}, \quad n=\frac{Gar}{h}, \qquad \text{......(1124)}$$

p, q, r being the components of momentum. We can verify this relationship by using the well-known vector formula for the flux $\mathbf{J}$,

$$\mathbf{J}=\frac{h}{4\pi i m}\iiint(\psi^*\,\mathbf{grad}\,\psi-\psi\,\mathbf{grad}\,\psi^*)\,dx\,dy\,dz. \qquad \text{......(1125)}$$

Equation (1119) gives us $\psi_{l,m,n}$ normalized so that

$$\iiint_0^{Ga}\psi_{l,m,n}\,\psi^*_{l,m,n}\,dx\,dy\,dz=G^3.$$

In order to represent a single electron the expression for ψ must therefore be divided by $G^{\frac{3}{2}}$. Normalizing thus and applying (1125) to the form (1123) we at once recover (1124).†

† These comparisons are only valid when l, m, n or l', m', n' are small compared with G. In the general case the time-dependent wave-function can be put in the form

$$u_{l,m,n}\,e^{-(2\pi i/h)(E_{l,m,n}t-h[lx+my+nz]/Ga)}$$

where $u_{l,m,n}$ is periodic with the period of the lattice, and only mildly dependent on l, m, n—any dependence on l, m, n only enters from the residual differences between (1119) and (1123). If now we build a wave-packet out of such functions using relatively small ranges of l, m, n, the wave-function $\psi(x,t)$ of such a wave-packet can be put in the form

$$\iiint_{l-\Delta l,\dots}^{l+\Delta l,\dots} C_{l,m,n}\,u_{l,m,n}\,e^{-(2\pi i/h)(E_{l,m,n}t-h[lx+my+nz]/Ga)}\,dl\,dm\,dn,$$

which is sufficiently nearly

$$u_{l,m,n}\iiint_{l-\Delta l,\dots}^{l+\Delta l,\dots} C_{l,m,n}\,e^{-(2\pi i/h)(E_{l,m,n}t-h[lx+my+nz]/Ga)}\,dl\,dm\,dn.$$

Identifying the particle velocity with the group velocity of this wave-packet one finds at once by the principle of stationary phase that

$$v_x=\frac{Ga}{h}\frac{\partial E_{l,m,n}}{\partial l}, \quad v_y=\frac{Ga}{h}\frac{\partial E_{l,m,n}}{\partial m}, \quad v_z=\frac{Ga}{h}\frac{\partial E_{l,m,n}}{\partial n}. \qquad \text{......(1125 ft.)}$$

The interpretation of l, m, n just given enables us to write (1121) and (1122) in the form

$$E_{p,q,r}=E_0-\alpha\pm 6\beta\pm\frac{1}{2m^*}(p^2+q^2+r^2), \qquad \text{......(1126)}$$

where
$$m^*=\frac{h^2}{8\pi^2a^2\beta}. \qquad \text{......(1127)}$$

The electrons in these states near the edges of the band behave as free electrons with an effective mass m^* depending on the tightness of their

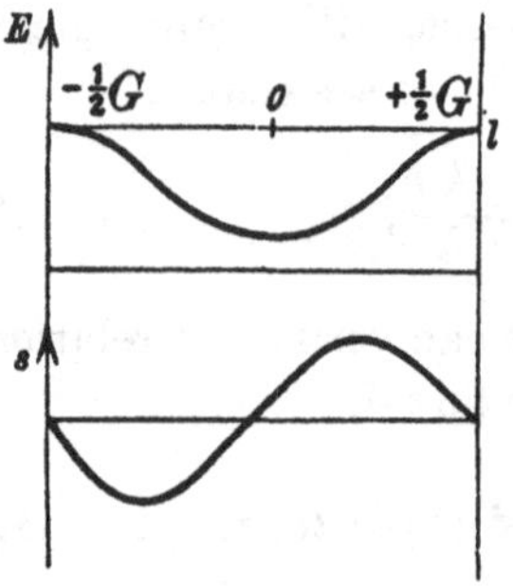

Fig. 34. The energy and momentum of an electron as a function of l for a one-dimensional lattice.

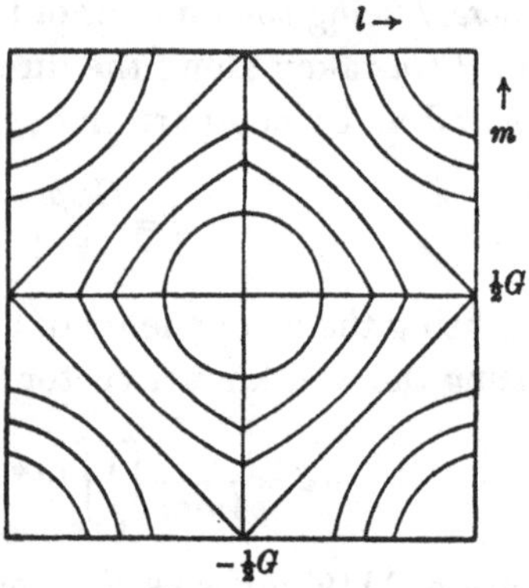

Fig. 35. The dependence of the energy of an electron on l, m for a two-dimensional square lattice.

binding.† Those near the lower edge of the band are normal in that they have an effective *positive* mass. Those near the upper edge of the band are abnormal and have an effectively *negative* mass. This means of course that when a field is applied to them in such a direction as to do work on them and increase their energy, this must show itself by *retarding* instead of accelerating them. They behave in fact as has been shown in detail by Heisenberg and Dirac just as if they were *positive* electrons of normal *positive* mass m^*.

The dependence of the energy on the wave number can be illustrated usefully by the accompanying diagrams. Fig. 34 shows the dependence of the energy and the flux (s) on the wave number for the one-dimensional lattice in which the energy depends on one wave number l only and there is only one cosine term in (1120). Fig. 35 shows curves of constant energy for the two-dimensional lattice as functions of the wave number components l, m. Figs. 36 *a*, *b* show two surfaces of constant energy in three dimensions as functions of the wave number components l, m, n. The first surface is an octahedron with bulging faces, whose vertices lie on the faces of the standard cube whose faces lie at $\pm\frac{1}{2}G$. The second surface is one of greater energy and

† The smaller β the tighter the binding.

represents the energy of the highest occupied levels at the absolute zero when G^3 electrons are accommodated on the $\frac{1}{2}G^3$ lowest levels. Further details will be found in the authorities already quoted.

We can now summarize what we require here of the properties of the states of a simple band, using both the results here given and similar results for more complicated cases. The distinct states of a band for a set of G^3 $(=N)$ atoms in the lattice are in general G^3 in number, each of weight ϖ. [The weight always contains the factor 2 for electron spin and may contain other

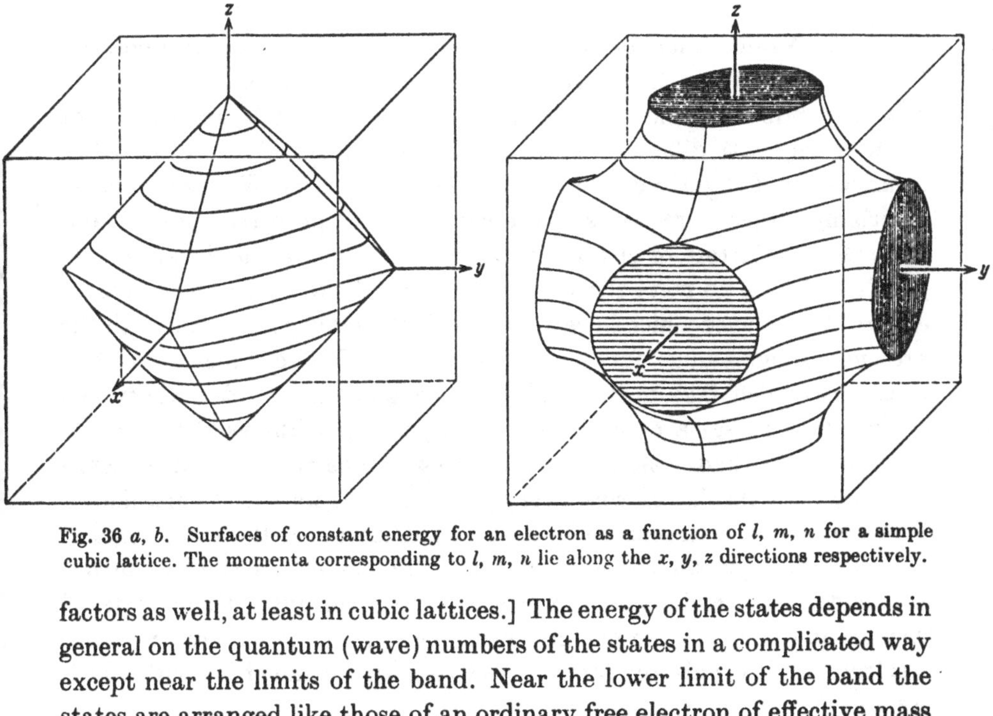

Fig. 36 *a*, *b*. Surfaces of constant energy for an electron as a function of l, m, n for a simple cubic lattice. The momenta corresponding to l, m, n lie along the x, y, z directions respectively.

factors as well, at least in cubic lattices.] The energy of the states depends in general on the quantum (wave) numbers of the states in a complicated way except near the limits of the band. Near the lower limit of the band the states are arranged like those of an ordinary free electron of effective mass m^* different from that of a free electron. Near the upper limit of the band the states are arranged as if they were those of a free electron with negative mass m^*.†

When a band is nearly empty so that the electrons in it are all in the lowest levels, they may be treated as a classical atmosphere of free electrons of mass m^*, ignoring Pauli's exclusion principle when the electrons are too few to get in each other's way. When a band is nearly full and contains only a few vacant states of negative mass, the *vacancies* or holes behave precisely as if they were positive electrons of positive mass m^*. This however holds only

† Rather more generally the equivalent "mass" of the electron need not be a scalar, but can be a symmetrical tensor of rank two [Bronstein, *Physikal. Zeit. d. Sowjetunion*, vol. 2, p. 28 (1932)], but we shall not employ this refinement here.

if the vacancies are few so that there is always an electron in any neighbouring state to allow the hole to move to it. Though electrons in such states must always move *as if* they were positive electrons, it is of course only *holes* in a complete band which have also effectively a positive charge and can therefore be completely described as free positive electrons of positive mass m^*. But when even the upper states of a band are nearly full, the band can be treated as completely full (it is then without importance except for a constant contribution to the total energy), together with a classical atmosphere of positive electrons which are an exact analogue of the few holes.

§ **11·53.** *Metals, insulators and semi-conductors.* We are now in a position to give a qualitative theoretical explanation of how some crystal lattices come to conduct electricity freely and are classed as metals, and others not at all or only with difficulty or at high temperatures and so are classed as insulators or if slightly conducting as semi-conductors. We may start by describing rather more precisely the very different properties of good and bad solid conductors of electricity, which show that a classification into good and bad is no mere trivial one. Good conductors (metals)† have an electrical resistance which always rises with the temperature and with the presence of increasing amounts of impurities. Bad conductors have a resistance which falls very rapidly as the temperature rises (until the temperature passes a definite limit) and generally falls as the impurity content rises. Bad conductors with these properties are called semi-conductors or insulators when very bad. There is no such sharp line of demarcation between semi-conductors and insulators as there is in general between these substances and metals. Semi-conductors can be further subdivided into semi-conductors which carry a current without material change and semi-conductors in which the current is wholly or partly electrolytic. It is only the former, electronic semi-conductors, with which we are concerned in this chapter.

The quantum theory of electrons in a periodic field, of which we have just given an outline, provides at once a theoretical basis for this classification. If the highest band of allowed states, containing any electrons at low temperatures, is only partly full of electrons either because a number of consecutive bands overlap or because there are only enough valency electrons to fill up half the states in a single band, then at all temperatures there are electrons available at the top of the full levels, in number nearly independent of the temperature, which are free to make transitions to neigh-

† Metals as a class of substances have of course many other distinctive properties (such as their ductility when pure) besides their electrical conductivity. Theory has not yet advanced to such a point that these other properties can be correlated with the nature of the electronic bands. For our purposes here a "metal" does not strictly mean more than a metallic conductor.

bouring empty states, and the substance is a good conductor, *a metal.* If on the other hand the highest band containing any electrons at low temperatures is exactly full of electrons and separated from the next higher band by a distinct gap of disallowed energies, there are no electrons at low temperatures free to make any transitions and the substance is an insulator. It has been shown in fact—most explicitly by Zener*—that under the influence of an applied field transitions to the free states of the upper band will only occur with an appreciable frequency when $(\Delta E)^2/F \sim 10^{-8}$, where ΔE is the energy gap in volts and F is the applied field in volts/cm. At higher temperatures the electrons will not be all in the highest normally occupied band, but some will be thermally excited to the higher empty band. The details of this excitation we shall study in the following sections. The general result must be that there will at higher temperatures be an increasing number of free electrons in the almost empty band and therefore also of free holes which function as free positive electrons in the almost full band, both of which can make transitions to neighbouring states and so the substance will conduct more and more freely as the temperature rises. The part played by impurities we shall discuss later. It is already clear that we have the necessary basis for the classification.

It is at present hardly possible to carry through the actual calculations necessary to decide in all cases whether a set of N atoms (or molecules) of a given type when they combine to form a crystal lattice will form a metal or an insulator. Alkali atoms form an atomic lattice and have each a single valency electron in an ns state well separated in energy from the states of the other core electrons. These ns states of N atoms will by themselves form a band capable of accommodating $2N$ electrons and only N electrons are present to go into it. The solid alkalis must therefore be metals. We might go further and try to argue that any atom of odd valency (or molecule of odd residual valency such as NO or TiN) must also form a metal in the solid state. This however is not always correct, as can be seen most clearly by considering the halogens. A halogen atom has an incomplete outer shell containing 7 valency electrons instead of 8. If the atoms could form an atomic lattice the four atomic wave-functions of the outer shell would form a band or bands of states capable of holding $8N$ electrons; only $7N$ would be present and a metal would result. But in fact this is not what occurs: the halogen atoms from energy considerations prefer first to form molecules of two atoms rather than crystal lattices, and the lattices are built up from these molecules, not from atoms. The molecular electronic states are all fully occupied by two electrons each and can form bands of crystal states which are all completely full (or completely empty) so that the solid is an

* Zener, *Proc. Roy. Soc.* A, vol. 145, p. 523 (1934).

insulator. In the same way, while TiN forms a typical metal, NO does not. The molecules first polymerize to N_2O_2 in which all the molecular electronic states are fully occupied, and give rise to fully occupied lattice bands.

The converse of this is also true. Atoms such as the alkaline earths with electronic states all full in the free state need not necessarily give rise to insulators, for if there are other atomic states not far removed in energy from the ground state, as there are for the alkaline earths, the corresponding bands may overlap and the lattice have the properties of a metal. It is at present therefore still necessary to discuss the electrical and other properties of solids on the basis of explicit assumptions as to the nature and arrangement of the electronic bands, having shown that the theory provides a natural place for all the types that it is necessary to use.

§ 11·54. *The function of impurities in supplying "free" electrons.* We have seen in § 11·52 that the periodic field of a perfectly regular lattice offers no obstacle whatever to the movement through it of an electron of suitable velocity components. The whole resistance of a metal to the passage of an electric current must arise from irregularities in the lattice which will give an electron in any state of motion a finite mean free path. Such irregularities will always arise from the thermal agitation of the lattice, and are provided also by any foreign atoms (impurities) that may be present. This is the only part of any importance played by the impurities present in a metal—they shorten the mean free path, and thereby increase the resistance in agreement with observation.

In an insulator or semi-conductor they play an additional part which is far more vital. Suitable impurity atoms embedded in the lattice will provide electronic states which do not fuse with the band states of the lattice, but remain isolated and to a first approximation without effect on the usual bands. If the impurity is a suitable one it may then function in one or other of two ways. It may provide an isolated electron level higher than the top of the normally occupied band, and therefore more ready to supply an electron to the empty band above where it can conduct. If the impurity level lies below the bottom of this band, thermal excitation is required to ionize it and we shall obtain a strongly temperature-dependent number of electrons available to conduct. This is the typical case of an *impurity semi-conductor*. If the impurity level lies above the bottom of the same band, then it will be always ionized at all temperatures and we shall have a number of conducting electrons independent of the temperature and proportional to the impurity content. Such a rather surprising substance might be called an *impurity metal* and it is probable that such substances actually exist.

The second type of impurity might provide a possible localized home for an electron above the top of the normally occupied band. Such a home, normally empty, could accommodate an electron from the lower band. Electrons from the lower band could be more easily transferred by thermal excitation to these low lying localized levels than to the states of the empty upper band. As a result we shall have a number of effective free positive electrons in the lower band, which will again give a strongly temperature-dependent conductivity, this time by positive electrons. Both types of impurity might well be present together.

§11·541. We have now in our possession a sufficient description of the electronic levels in a crystal lattice to apply them to explain many of its equilibrium properties. We start in the next section by applying the probable structure of the electronic bands in nickel to explain in detail its excess specific heat over and above the contribution of the lattice vibrations (Chapter IV) and the ferromagnetic contribution (§ 12·9). This explanation extends no doubt to other metals. We follow this in §§ 11·6 sqq. by developing the quantitative theory of the distribution of free electrons, negative and positive, in semi-conductors of various types, and apply the results to explain their electrical properties. In §§ 11·7 sqq. we give the formal theory of electrical and thermal conductivities for both metals and semi-conductors but *only formally where a knowledge of the free path is required*, making no attempt to calculate the free path *a priori*. Such applications of quantum mechanics lie beyond the scope of this monograph.* We conclude (§§ 11·9 sqq.) with a discussion of electrical contacts, and a theory of rectification.

§11·55. *The electronic specific heat of nickel and other metals.* The observed value of C_V for nickel differs in several ways from any value which could be associated with lattice vibrations distributed according to any reasonable distribution law $g(\nu)\,d\nu$. At very low temperatures, 0–20° K., for which the important frequencies lie in the region in which $g(\nu) \sim \alpha\nu^2$ and $C_V^{\text{lat}} \sim \beta T^3$, Keesom and Clark† have found the values of C_V shown in Fig. 37·1. These values correspond to a normal lattice contribution $C_V^{\text{lat}} = 464{\cdot}5(T/\Theta_D)^3$, equation (359), with $\Theta_D = 413°$, a value derived from the elastic constants of nickel, together with an extra term which as Fig. 37·2 shows is accurately represented over the whole range of its importance by the expression

$$0{\cdot}001744T \text{ cal./deg./mole.}$$

At considerably higher temperatures, but below the Curie point, there is

* For the theory of the free path and the resulting complete theory of metallic (and other) conduction see Sommerfeld and Bethe, *loc. cit.*, where full reference will be found to other writers, who have developed the theory, notably Bloch, Peierls, Bethe and Nordheim.

† Keesom and Clark, *Physica*, vol. 2, p. 513 (1935).

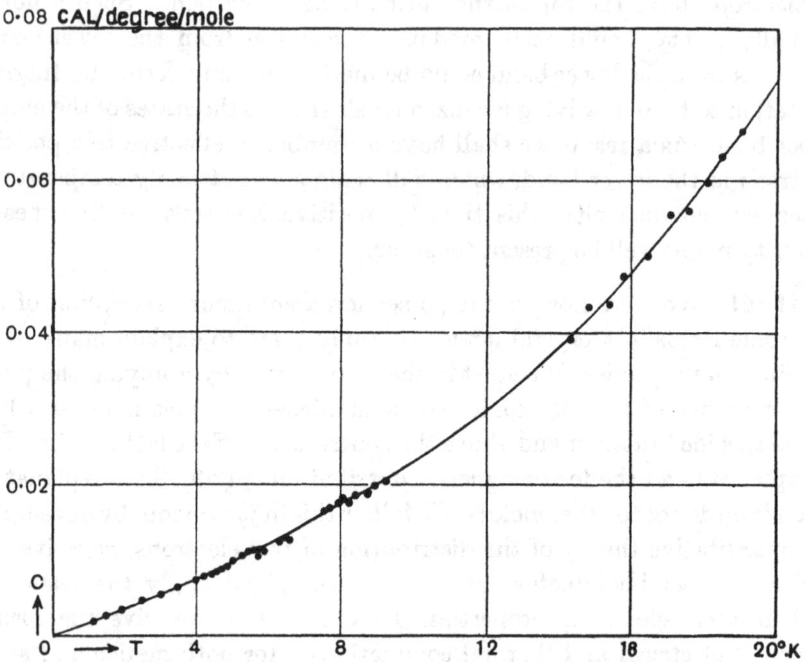

Fig. 37·1. The specific heat of Ni at very low temperatures.

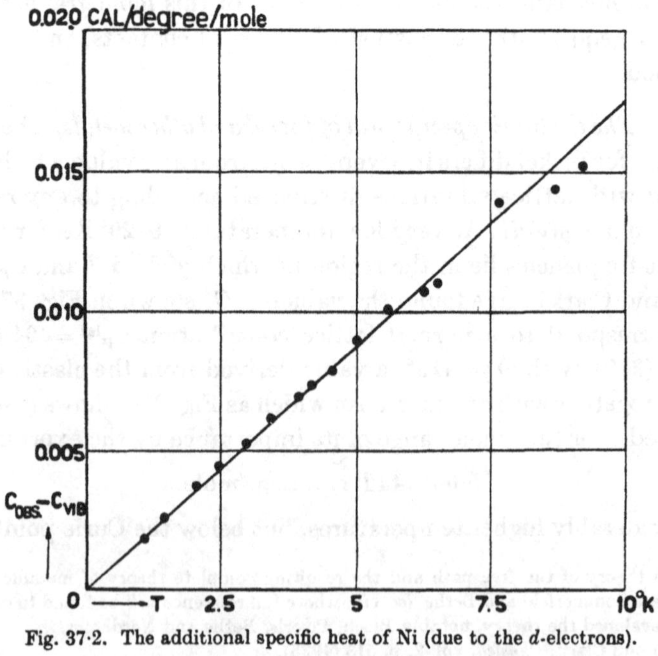

Fig. 37·2. The additional specific heat of Ni (due to the *d*-electrons).

a large excess specific heat associated with the disappearance of the ferromagnetism as the temperature rises, which we need not consider further here. Above the Curie point (631° K.) contributions from this source should entirely disappear, but for nickel the value of C_V does not fall above the Curie point to a value comparable with 6 cal./deg./mole which is the value for the lattice vibrations in this region of temperature. The *minimum* value for C_p observed above the Curie point ranges from 7·3 to 7·9 according to various investigators.* Taking even the lowest value and reducing it to C_V by equation (390) we obtain the value 7·0, compared with a similar corrected value for copper of 6·1 at this temperature. [Some excess about the value 6 may be expected to arise from the anharmonic terms in the strain energy

TABLE 43·2.

Observed values of C_V at high temperatures for various metals.

Temp. ° C.	Pd	Pt	Cu	Ag	Au
500	6·59	6·38	6·2	6·0	6·0
900	7·07	—	—	6·13	—
1000	7·15	6·65	6·5	—	6·12
1300	7·25	—	—	—	—
1500	7·23	—	—	—	—
1600	—	6·8	—	—	—

of the lattice.] Considering all the values it is clear that C_V exceeds C_V^{lat} above the Curie point by 0·9–1·3 cal./deg./mole and this excess appears to increase with the temperature.

Similar contributions appear at high temperatures for other transition metals as is shown by Table 43·2.

Both these excesses in C_V for Ni and their absence for the noble metals can be simply explained as an electronic specific heat if the probable structure of the electronic bands in these metals is taken into account. It has been shown by Mott,† that in the nickel lattice there are two bands of energy levels of comparable energy derived from the 3*d* and 4*s* electronic states of the free atom. The states in these bands are arranged as shown in Fig. 38. In Ni neither band is completely full, but in Cu and the other noble metals only the 4*s*-band is partially full. In Ni there appears to be 0·6 electron per atom in the 4*s*-band and 0·6 electron per atom too few in the 3*d*-band to fill it. The actual number 0·6 is that number which gives the metal its correct saturated magnetization, at low temperatures, since the

* Mott, "Discussion on supraconductivity and low temperatures", *Proc. Roy. Soc.* A, vol. 152, p. 42 (1935).

† Mott, *Proc. Phys. Soc. London*, vol. 47, p. 571 (1935).

magnetization is due to the incomplete d-band only. The states of the s-band appear to correspond to electrons which are almost perfectly free and are therefore spaced as if the electron had its natural rest mass m_0, while in the states of the d-band the electrons are rather tightly bound, the overlap of the atomic wave-functions is small, and the states are distributed as if the electron had a large effective mass m^*. The nearly free electrons of the s-band will, we know, make a negligible contribution to the specific heat and it remains to examine the contribution by the holes in the d-band. Since there are many less than one per atom and the states in the d-band have probably an extra weight factor, it is certainly safe to use the approximation (1126) for the energies of all the states that matter.

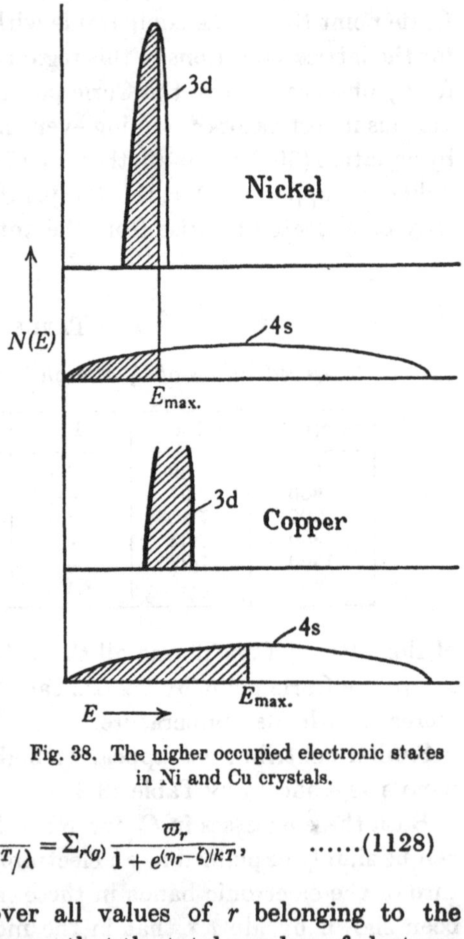

Fig. 38. The higher occupied electronic states in Ni and Cu crystals.

The complete formulation of the theory for two such overlapping bands is as follows. By equation (124) the average number $\overline{M_g}$ of electrons in any group of levels whatever is given by

$$\overline{M_g} = \Sigma_{r(g)} \frac{\varpi_r}{1 + e^{\eta_r/kT}/\lambda} = \Sigma_{r(g)} \frac{\varpi_r}{1 + e^{(\eta_r - \zeta)/kT}}, \qquad \text{......(1128)}$$

where $\Sigma_{r(g)}$ denotes summation over all values of r belonging to the group; λ may have to be fixed either so that the total number of electrons is correct, or by equality with its value for some other group of electrons in common equilibrium, according to the conditions of the problem. In the present problem there are two groups, the d-band containing $10N$ states, and $(10-x)N$ electrons, where N is the number of atoms in volume V, and the s-band containing $2N$ states and xN electrons. In the s-band

$$\overline{M_s} = \Sigma_{r(s)} \frac{\varpi_r}{1 + e^{(\eta_r - \zeta)/kT}}.$$

It is convenient to take the energy zero at the bottom of the s-band and sufficient to use the approximation (1126) for η_r. Then, by following the derivation of (222), we see at once that

$$\overline{M_s}=2\,\frac{2\pi(2m)^{\frac{3}{2}}V}{h^3}\int_0^\infty \frac{\eta^{\frac{1}{2}}\,d\eta}{1+e^{(\eta-\zeta)/kT}}. \qquad \text{......(1128·1)}$$

We see further that the average energy of these electrons is

$$\overline{E_s}=2\,\frac{2\pi(2m)^{\frac{3}{2}}V}{h^3}\int_0^\infty \frac{\eta^{\frac{3}{2}}\,d\eta}{1+e^{(\eta-\zeta)/kT}}. \qquad \text{......(1128·2)}$$

It is convenient to express the distribution in the d-band in terms of positive electrons or holes. The average number of these is

$$\overline{M_d^+}=\Sigma_{r(d)}\,\varpi_r-\Sigma_{r(d)}\frac{\varpi_r}{1+e^{(\eta_r-\zeta)/kT}}=\Sigma_{r(d)}\frac{\varpi_r}{1+e^{(\zeta-\eta_r)/kT}}.$$

The holes are all near the top of the band and we can therefore apply (1126) which takes the form

$$\eta_r=\Delta E-\frac{1}{2m^*}(p^2+q^2+r^2),$$

where ΔE is the energy by which the top of the d-band exceeds the bottom of the s-band. Then

$$\overline{M_d^+}=\varpi^*\,\frac{2\pi(2m^*)^{\frac{3}{2}}V}{h^3}\int_0^\infty \frac{\eta^{\frac{1}{2}}\,d\eta}{1+e^{\{\eta-(\Delta E-\zeta)\}/kT}}. \qquad \text{......(1128·3)}$$

The average energy of the electrons in this band is

$$\overline{E_d}=\Sigma_{r(d)}\frac{\varpi_r\,\eta_r}{1+e^{(\eta_r-\zeta)/kT}}=\textit{const.}-\Sigma_{r(d)}\frac{\varpi_r\,\eta_r}{1+e^{(\zeta-\eta_r)/kT}}.$$

This can easily be reduced to

$$\overline{E_d}=\textit{const.}-\overline{M_d^+}\Delta E+\varpi^*\,\frac{2\pi(2m^*)^{\frac{3}{2}}V}{h^3}\int_0^\infty \frac{\eta^{\frac{3}{2}}\,d\eta}{1+e^{\{\eta-(\Delta E-\zeta)\}/kT}}. \qquad \text{......(1128·4)}$$

Finally ζ is fixed as a function of T by the condition

$$\overline{M_s}=\overline{M_d^+}.$$

At zero temperature $\zeta\to\zeta_0$ and both bands are completely degenerate. We may assume also that $x\to x_0\sim 0{\cdot}6$. We then find that

$$x_0N=2\,\frac{2\pi(2m)^{\frac{3}{2}}V}{h^3}\,\tfrac{2}{3}\zeta_0^{\frac{3}{2}}=\varpi^*\,\frac{2\pi(2m^*)^{\frac{3}{2}}V}{h^3}\,\tfrac{2}{3}(\Delta E-\zeta_0)^{\frac{3}{2}}$$

by the usual evaluation of the integrals in (1128·1) and (1128·3) as $T\to 0$. These equations determine ζ_0 and ΔE; ζ_0 is large and of the order of 5 volts but $\Delta E-\zeta_0$ may be much smaller owing to ϖ^* and m^*.

At higher temperatures ζ must vary but the s-band will remain almost completely degenerate so that we continue to have

$$xN\sim 2\,\frac{2\pi(2m)^{\frac{3}{2}}V}{h^3}\,\tfrac{2}{3}\zeta^{\frac{3}{2}}.$$

The variations of ζ necessary to preserve the equality $\overline{M_s} = \overline{M_d^+}$ will be on the scale of $\Delta E - \zeta_0$ and therefore quite small compared with ζ_0. We may therefore neglect them in a first approximation; it follows that $x \sim x_0$ at all relevant temperatures and the d-band may be discussed as if it were a separate set of electrons of constant number, $\Delta E - \zeta$ varying so that this constancy is maintained.

Under these conditions C_V^{ϵ} is to be obtained by differentiating (1128·4). At low temperatures the usual approximations for degenerate assemblies will hold, and we have by (1018), in the present notation,

$$C_V^{\epsilon}/R = x_0[\tfrac{1}{2}\pi^2 T/T_{\epsilon}], \qquad \text{......(1128·5)}$$

where

$$T_{\epsilon} = \frac{\Delta E - \zeta_0}{k} = \frac{h^2}{2m^*k}\left(\frac{3x_0 n}{4\varpi^*\pi}\right)^{\frac{2}{3}}. \qquad \text{......(1128·6)}$$

This will fit Keesom and Clark's measurements if $T_{\epsilon} = 3340$. At higher temperatures C_V^{ϵ} must be evaluated from (1128·3) and (1128·4) by numerical integration. This has been done by Mott† whose results are shown in Fig. 39. It is easily seen that C_V^{ϵ} must be a function of T/T_{ϵ} only, where T_{ϵ} is given by (1128·6). We may now compare the calculated contribution with the observed values above the Curie point, remembering that for the ferromagnetic d-band at low temperatures ϖ^* may be 2 or 3, only states of one spin being allowed, but that above the Curie point all spin states may be used indifferently so that ϖ^* increases by a factor 2. We must therefore use a value $3340/2^{\frac{2}{3}} = 2110$ for T_{ϵ}. At 630° K., T/T_{ϵ} has the value 0·30 and the specific heat per electron is about 1·0k, or (0·6) R = 1·2 cal./deg./mole. At 1300° K. this will rise to 1·6 cal./deg./mole. These values correspond to those observed within the error of the experiments.

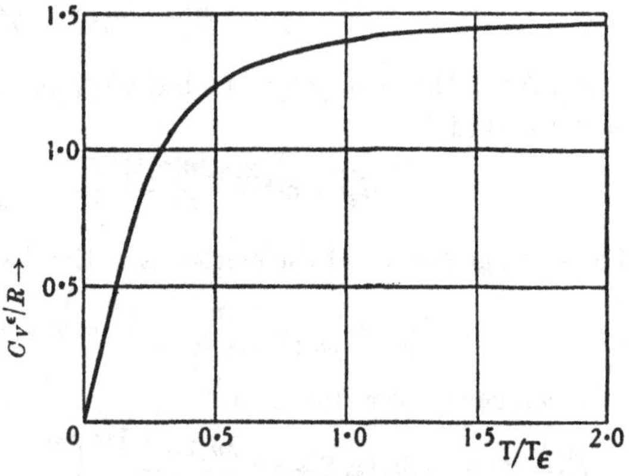

Fig. 39. The specific heat of a degenerate electron gas as a function of T.

It remains to examine what value of β is required for the d-band to give this value of T_{ϵ}. On combining (1128·6) and (1127), taking $\varpi^* = 3$ and remembering that $na^3 = 1$, we see that

$$\beta = \left(\frac{4\pi}{x_0}\right)^{\frac{2}{3}} \frac{kT_{\epsilon}}{4\pi^2} \sim \frac{kT_{\epsilon}}{5·2}.$$

† Mott, *Proc. Roy. Soc.* A, vol. 152, p. 42 (1935).

When $T_\epsilon = 3340$, $\beta \sim 5{\cdot}5 \times 10^{-2}$ electron volt and the width of the d-band would be $\sim 0{\cdot}3$ electron volt, which is an acceptable value. It seems likely that further studies of this type will prove extremely fruitful.†

Before leaving this subject it may be convenient to derive a more general formula given by Bethe‡ which can be applied to calculate the specific heat contributed by any assembly of a constant number M of nearly degenerate electrons in which the number of states in the energy range η, $\eta + d\eta$ is $N(\eta)\,d\eta$. It follows at once from the foregoing formulae that

$$M = \int_0^\infty \frac{N(\eta)\,d\eta}{1+e^{(\eta-\zeta)/kT}}, \quad \overline{E_\epsilon} = \int_0^\infty \frac{\eta N(\eta)\,d\eta}{1+e^{(\eta-\zeta)/kT}}.$$

These integrals may be evaluated asymptotically by the methods described in § 11·71, by converting them to the form (1171·1) by integration by parts. It follows at once that, if $S(\zeta) = \int_0^\zeta N(\eta)\,d\eta$, $T(\zeta) = \int_0^\zeta \eta N(\eta)\,d\eta$,

$$M = S(\zeta) + \frac{\pi^2 k^2 T^2}{6} N'(\zeta), \qquad \text{......(1128·7)}$$

$$\overline{E_\epsilon} = T(\zeta) + \frac{\pi^2 k^2 T^2}{6} \{\zeta N'(\zeta) + N(\zeta)\}. \qquad \text{......(1128·8)}$$

On differentiating (1128·7) and (1128·8) and eliminating $d\zeta/dT$, we find

$$C_V{}^\epsilon = \frac{d\overline{E_\epsilon}}{dT} = \frac{\pi^2}{3} k^2 T N(\zeta). \qquad \text{......(1128·9)}$$

In this approximation ζ may be replaced by η^*, its limit when $T \to 0$. It is easily verified that this formula reduces to (1018) for an assembly of degenerate free electrons. It follows also from (1128·9) that in the nearly degenerate region the contribution of two overlapping bands is always equal to the sum of the contributions of the two bands calculated independently.

§ **11·6.** *Electron distributions in semi-conductors.* We shall consider a standard case in which we are concerned with two bands of levels 1, 2, of which the former is empty and the latter full at the absolute zero. Band 2 when full will be supposed to contain N^* electrons for a volume V of the lattice. The states of the bands have weights ϖ_1, ϖ_2 each respectively, so that $N^*/\varpi_2 = N$, the number of atoms in volume V contributing states to the band. The effective masses of electrons in states near the limits of the bands are m_1 and m_2 respectively, and the free paths of ordinary electrons in the upper band or positive electrons in the lower band l_1 and l_2. The energy

† See a forthcoming book by Mott and Jones, Oxford Press. A more exact study by Slater (in course of publication) has fully confirmed these elementary calculations.

‡ Sommerfeld and Bethe, *loc. cit.* p. 430.

zero will be taken for convenience at the top of band 2; the energy interval from there to the bottom of band 1 is ΔE_1.

The equilibrium concentration of free electrons and free holes (functioning as positive electrons) for such a semi-conductor which we shall call *intrinsic* can be simply determined as an ordinary dissociative equilibrium:

Free Electron + Free Hole $\rightleftarrows$ *Bound Electron.*

Suppose there are $\overline{N_1}$ free electrons and therefore in this case also $\overline{N_1}$ free holes, in a volume V. Then by (459)

$$\frac{(\overline{N_1})^2}{N^* - \overline{N_1}} = \frac{f_1(T) f_2(T)}{g(T)}, \qquad \text{......(1129)}$$

where $f_1(T)$ is the partition function for the free electron, a classical particle of mass m_1 with states of weight ϖ_1 in a volume V, $f_2(T)$ the same for a mass m_2 with states of weight ϖ_2, and $g(T)$ the partition function for a bound electron in band 2 which is practically full. We can therefore assimilate the exact problem to the classical one by treating all the N^* states of the band 2 as equivalent and taking $g(T) = N^*$. Since $\overline{N_1} \ll N^*$ we have therefore

$$\overline{N_1}^2 = f_1(T) f_2(T). \qquad \text{......(1130)}$$

With the specified band structure and energy zero

$$f_1(T) = \varpi_1 \frac{(2\pi m_1 kT)^{\frac{3}{2}}}{h^3} V e^{-\Delta E_1/kT}, \quad f_2(T) = \varpi_2 \frac{(2\pi m_2 kT)^{\frac{3}{2}}}{h^3} V,$$

so that

$$\frac{N_1}{V} = (\varpi_1 \varpi_2)^{\frac{1}{2}} \frac{\{2\pi (m_1 m_2)^{\frac{1}{2}} kT\}^{\frac{3}{2}}}{h^3} e^{-\frac{1}{2}\Delta E_1/kT}. \qquad \text{......(1131)}$$

It may happen however that the free electrons in band 1 are supplied primarily by localized impurities, when band 2 can be ignored. The dissociative equilibrium is then

Free Electron + Bound Hole $\rightleftarrows$ *Bound Electron.*

If N_0 is the number of impurity levels in a volume V and if the energy zero is now taken in the impurity levels at a depth $\Delta E'$ below the bottom of band 1, the equation of dissociative equilibrium is

$$\frac{(\overline{N_1})^2}{N_0 - \overline{N_1}} = \frac{f_1(T) N_0}{N_0},$$

or when $\overline{N_1} \ll N_0$

$$\frac{\overline{N_1}}{V} = \left(\frac{N_0}{V}\right)^{\frac{1}{2}} \varpi_1^{\frac{1}{2}} \frac{(2\pi m_1 kT)^{\frac{3}{4}}}{h^{\frac{3}{2}}} e^{-\frac{1}{2}\Delta E'/kT}. \qquad \text{......(1132)}$$

These results, due to A. H. Wilson,† will be generalized in a more systematic way in the next section.

† A. H. Wilson, *Proc. Roy. Soc.* A, vol. 133, p. 458 (1931), vol. 134, p. 277 (1932).

§ **11·61**. *General formulae for semi-conductors.* For a general discussion it is best to start with (1128). We can at once distinguish various cases.

(i) *Intrinsic semi-conductors. Energy zero at the top of band* 2. If the band 1 is nearly empty $e^{\eta_r/kT} \gg \lambda$ throughout the band 1. Thus

$$\overline{N_1} = \lambda \Sigma_{(b_1)} \varpi_r e^{-\eta_r/kT} = \lambda f_1(T),$$

or

$$\overline{N_1} = \lambda \varpi_1 \frac{(2\pi m_1 kT)^{\frac{3}{2}}}{h^3} V e^{-\Delta E_1/kT}. \qquad \text{......(1133)}$$

Band 2 is practically full so that $e^{\eta_r/kT} \ll \lambda$ throughout the band. Thus, on expanding (1128),

$$N^* - \overline{N_1} = \Sigma_{(b_2)} \varpi_r (1 - e^{\eta_r/kT}/\lambda) = N^* - f_2(T)/\lambda,$$

or

$$\overline{N_1} = \frac{\varpi_2}{\lambda} \frac{(2\pi m_2 kT)^{\frac{3}{2}}}{h^3} V. \qquad \text{......(1134)}$$

Combining (1133) and (1134)

$$\lambda = \left(\frac{\varpi_2}{\varpi_1}\right)^{\frac{1}{2}} \left(\frac{m_2}{m_1}\right)^{\frac{3}{4}} e^{\frac{1}{2}\Delta E_1/kT}. \qquad \text{......(1135)}$$

Our previous result (1131) follows at once on combining (1134) and (1135).

(ii) *Normal impurity semi-conductors.* N_0 *impurity levels in volume V. Energy zero at the top of band* 2. *Electrons in band* 1 *from both band* 2 *and the impurities.* Denote the average numbers of electrons in band 1 and holes in band 2 by $\overline{N_1}$ and $\overline{N_2}$ respectively. Then $\overline{N_1}$ is still given by (1133) and $\overline{N_2}$ by (1134). The number of electrons in the impurity levels is $N_0 - \overline{N_1} + \overline{N_2}$, so that

$$N_0 - \overline{N_1} + \overline{N_2} = N_0 \frac{1}{1 + e^{\Delta E_2/kT}/\lambda}$$

if it is assumed that the impurity levels lie at a height ΔE_2 ($< \Delta E_1$) above the top of band 2. Thus

$$\overline{N_1} - \overline{N_2} = \frac{N_0}{\lambda e^{-\Delta E_2/kT} + 1}; \qquad \text{......(1136)}$$

on combining (1133), (1134) for $\overline{N_2}$ and (1136) the equation for λ is seen to be

$$\lambda \varpi_1 \frac{(2\pi m_1 kT)^{\frac{3}{2}}}{h^3} e^{-\Delta E_1/kT} = \frac{N_0/V}{\lambda e^{-\Delta E_2/kT} + 1} + \frac{\varpi_2}{\lambda} \frac{(2\pi m_2 kT)^{\frac{3}{2}}}{h^3}. \qquad \text{......(1137)}$$

For any given values of the lattice and impurity constants ΔE_1, ΔE_2, N_0, m_1, m_2, ϖ_1 and ϖ_2 the values of λ as a function of T can easily be computed from this formula. For small values of T the lower band states are unimportant, $\overline{N_2} \simeq 0$, $N_1 \ll N_0$ and $\lambda e^{-\Delta E_2/kT} \gg 1$. Then

$$\lambda = \left(\frac{N_0}{V}\right)^{\frac{1}{2}} \varpi_1^{-\frac{1}{2}} \frac{h^{\frac{3}{2}}}{(2\pi m_1 kT)^{\frac{3}{4}}} e^{\frac{1}{2}(\Delta E_1 + \Delta E_2)/kT}. \qquad \text{......(1138)}$$

Equations (1138) and (1133) reproduce (1132). For larger values of T the impurity contribution is swamped by the electrons from band 2 and we recover the formulae for the intrinsic semi-conductor.

(iii) *Abnormal impurity semi-conductors. N_0' impurity levels in volume V. Energy zero at the top of band 2. Holes in band 2 due to supply of electrons to the impurities and to band* 1. This case differs from (ii) only in that the number of electrons on the impurities is $\overline{N_2}-\overline{N_1}$ so that

$$\overline{N_2}-\overline{N_1}=\frac{N_0'}{1+e^{\Delta E_2'/kT}/\lambda}. \qquad \text{......(1139)}$$

The equation for λ is now

$$\lambda\varpi_1\frac{(2\pi m_1 kT)^{\frac{3}{2}}}{h^3}e^{-\Delta E_1/kT}+\frac{N_0'/V}{1+e^{\Delta E_2'/kT}/\lambda}=\frac{\varpi_2}{\lambda}\frac{(2\pi m_2 kT)^{\frac{3}{2}}}{h^3}. \qquad \text{......(1140)}$$

For small values of T the upper band states are unimportant, $\overline{N_1}\simeq 0$, $\overline{N_2}\ll N_0'$ and $e^{\Delta E_2'/kT}\gg\lambda$. Then

$$\lambda=\left(\frac{V}{N_0'}\right)^{\frac{1}{2}}\varpi_2^{\frac{1}{2}}\frac{(2\pi m_2 kT)^{\frac{3}{4}}}{h^{\frac{3}{2}}}e^{\frac{1}{2}\Delta E_2'/kT}. \qquad \text{......(1141)}$$

Equations (1141) and (1134) for $\overline{N_2}$ give

$$\frac{\overline{N_2}}{V}=\left(\frac{N_0'}{V}\right)^{\frac{1}{2}}\varpi_2^{\frac{1}{2}}\frac{(2\pi m_2 kT)^{\frac{3}{4}}}{h^{\frac{3}{2}}}e^{-\frac{1}{2}\Delta E_2'/kT}. \qquad \text{......(1142)}$$

For larger values of T the states of the upper band take charge and we recover the formulae for an intrinsic semi-conductor.

(iv) *Semi-conductor with impurities of both types. N_0 and N_0' impurity levels in volume V. Energy zero at the top of band 2. Impurity electron donors, N_0, at a height ΔE_2. Impurity electron acceptors, N_0', at a height $\Delta E_2'$.* It is necessary to suppose that

$$\Delta E_2<\Delta E_2'<\Delta E_1$$

or else the electron acceptors will take up electrons from the donors at low temperatures and we shall fall back on one or other of cases (ii) or (iii) according as donors or acceptors are the more abundant. The levels are shown in Fig. 40. The equations for $\overline{N_1}$ and $\overline{N_2}$ remain unaltered. The equation for λ is now

$$\lambda\varpi_1\frac{(2\pi m_1 kT)^{\frac{3}{2}}}{h^3}e^{-\Delta E_1/kT}+\frac{N_0'/V}{1+e^{\Delta E_2'/kT}/\lambda}=\frac{N_0/V}{1+\lambda e^{-\Delta E_2/kT}}+\frac{\varpi_2}{\lambda}\frac{(2\pi m_2 kT)^{\frac{3}{2}}}{h^3}. \qquad \text{......(1143)}$$

Great variety of behaviour is now possible. For example, for small T the second and third terms must dominate the equation and therefore in that range

$$\lambda=\left(\frac{N_0}{N_0'}\right)^{\frac{1}{2}}e^{\frac{1}{2}(\Delta E_2'+\Delta E_2)/kT} \qquad \text{......(1144)}$$

On comparing the number of free electrons and free holes for such λ's, we see that if $\Delta E_2' > \Delta E_1 - \Delta E_2$, $\overline{N_2} \ll \overline{N_1}$ and the properties will be those of a normal impurity semi-conductor. If, however, $N_0' > N_0$ by a considerable factor, it can be shown that we pass as the temperature rises to a range in which there are many more free holes than free electrons when the semi-conductor will be abnormal. Finally, as always, for higher temperatures we must reach a stage when the numbers of free electrons and free holes are large and approximately equal. The interest of this example is that it reproduces the complicated observed behaviour of cuprite* which is normal at low temperatures and abnormal at moderate ones.

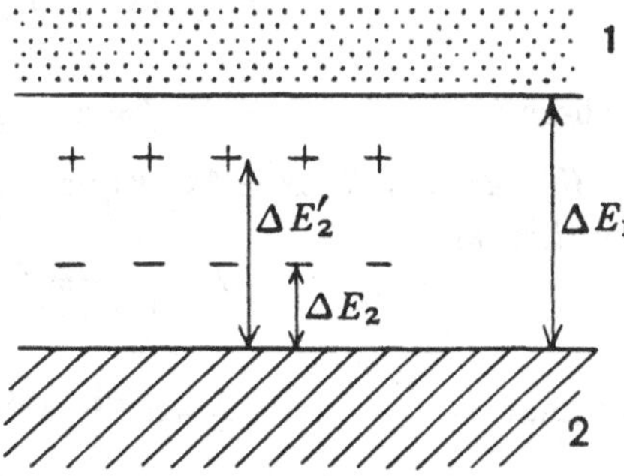

Fig. 40. Electron levels in an impurity semi-conductor with both electron donors (−) and electron acceptors (+).

§ **11·62**. *Work functions and contact potentials for semi-conductors.* In order to study contact equilibria for metals and semi-conductors we have only to equate values of λ, taking care that a consistent energy zero is used. It is best for this purpose to change the definition and take as zero the energy of an electron at rest outside one of the substances. The energy outside any other then differs by the contact potential energy. To study the thermionic work function for a semi-conductor we can combine (1022) and (1030), neglecting the transmission coefficient, into the emission formula

$$I = A\lambda T^2, \qquad \text{......(1145)}$$

and insert the proper value of λ. If E_x is the energy interval from the top of band 2 to the energy of an electron at rest outside the semi-conductor, this energy being assumed to lie in band 1, at a level to which our various approximations apply, then the formulae of the preceding section are altered only by the addition of the factor $e^{-E_x/kT}$ to every formula for λ. We consider a number of special cases.

(i) *Intrinsic semi-conductor*. The thermionic emission formula is

$$I = A\left(\frac{\varpi_2}{\varpi_1}\right)^{\frac{1}{2}}\left(\frac{m_2}{m_1}\right)^{\frac{3}{4}} T^2 e^{-\chi/kT}, \qquad \text{......(1146)}$$

where

$$\chi = E_x - \tfrac{1}{2}\Delta E_1. \qquad \text{......(1147)}$$

* Schottky and Waibel, *Physikal. Zeit.* vol. 34, p. 858 (1933).

(ii) *Normal impurity semi-conductor.* (*a*) *Low* T, $\overline{N_1} \gg \overline{N_2}$. Equation (1138) corrected by $e^{-E_x/kT}$ gives λ. Then

$$I = AT^{\frac{3}{4}} e^{-\chi'/kT} \varpi_1^{-\frac{1}{2}} \frac{h^{\frac{3}{2}}}{(2\pi m_1 k)^{\frac{3}{4}}} \left(\frac{N_0}{V}\right)^{\frac{1}{2}}, \qquad \text{......(1148)}$$

where

$$\chi' = E_x - \tfrac{1}{2}(\Delta E_1 + \Delta E_2). \qquad \text{......(1149)}$$

(*b*) *Medium* T, $\overline{N_1} \simeq N_0$. Equation (1137) corrected by $e^{-E_x/kT}$ gives for λ in this range

$$\lambda = \frac{N_0}{V} \frac{1}{\varpi_1} e^{(\Delta E_1 - E_x)/kT} \frac{h^3}{(2\pi m_1 kT)^{\frac{3}{2}}}, \qquad \text{......(1150)}$$

so that

$$I = AT^{\frac{1}{2}} e^{-\chi''/kT} \frac{N_0}{V} \frac{1}{\varpi_1} \frac{h^3}{(2\pi m_1 k)^{\frac{3}{2}}}, \qquad \text{......(1151)}$$

where

$$\chi'' = E_x - \Delta E_1.$$

This is actually the classical emission formula for a conductor containing a fixed number N_0 of classical electrons of mass m_1 in a volume V. At higher temperatures we pass over in the forms of case (i).

Similar formulae can be given for abnormal semi-conductors which we shall not stay to detail.

It should be remembered that all these formulae are only valid so long as the rate of emission of electrons is very slow compared with the rate of readjustment of the internal equilibrium. The process creating the free electrons and free holes must be able to maintain the normal equilibrium supply. This is an essential assumption and there is no simple means of investigating its validity—nor any *a priori* reason why it should be true.

Volta contact potentials, determined by equating λ's for two substances, are best studied by taking some standard comparison substance—the most suitable is an ideal metal with a simple Fermi-Dirac distribution of electrons, and we shall neglect for this ideal standard all terms but those of the highest order. We therefore take the standard to be an enclosure in which the potential energy of an electron is $-W_i$ relative to our chosen zero, that is zero for an electron at rest outside the standard substance. Then for such a metal

$$\lambda = e^{-(W_i - \eta^*)/kT} = e^{-\chi_0/kT} \qquad \left(\eta^* = \frac{h^2}{8m_0}\left(\frac{3N_0}{\pi V}\right)^{\frac{2}{3}}\right); \qquad \text{......(1152)}$$

the suffix 0 refers to the standard, m_0 is the ordinary rest mass of the electron and χ_0 the work function.

We define the Volta contact potential V_0^x of any substance x relative to our standard metal to be the amount by which the potential of x must exceed the standard when they are in equilibrium together. To avoid confusion of signs let $|\epsilon|$ denote the numerical charge on the electron, the actual charge on an ordinary electron being $-|\epsilon|$. Then all the energy levels of electrons and holes alike in substance x have been raised by $-|\epsilon|\, V_0{}^x$ relative to those of the metal. Expressions for λ for the substance x must therefore be multiplied by

$$e^{-|\epsilon| V_0{}^x/kT}$$

Using the values of λ already given, we obtain the following results:

(i) *Intrinsic semi-conductor—standard metal.*

$$-|\epsilon|\, V_0{}^x = \chi_x - \chi_0 - kT \log\left\{\left(\frac{\varpi_2}{\varpi_1}\right)^{\frac{1}{2}}\left(\frac{m_2}{m_1}\right)^{\frac{3}{4}}\right\}. \qquad \text{......(1153)}$$

(ii) *Normal impurity semi-conductor—standard metal.*

$$-|\epsilon|\, V_0{}^x = \chi_x - \chi_0 - kT \log\left\{\left(\frac{N_0}{V}\right)^{\frac{1}{2}} \frac{1}{\varpi_1{}^{\frac{1}{2}}} \frac{h^{\frac{3}{2}}}{(2\pi m_1 kT)^{\frac{3}{4}}}\right\}. \qquad \text{......(1154)}$$

(iii) *Abnormal impurity semi-conductor—standard metal.*

$$-|\epsilon|\, V_0{}^x = \chi_x - \chi_0 + kT \log\left\{\left(\frac{N_0}{V}\right)^{\frac{1}{2}} \frac{1}{\varpi_2{}^{\frac{1}{2}}} \frac{h^{\frac{3}{2}}}{(2\pi m_2 kT)^{\frac{3}{4}}}\right\}. \qquad \text{......(1155)}$$

The first approximations in all cases show that as usual the contact potential difference of a semi-conductor and a metal is equal to the difference of the work functions. On referring to the formulae for λ we see that this means that the top of the full levels of the Fermi-distribution of electrons in the metal must lie at a level just half way between the two sets of levels, one full and one empty, which control the electron distribution—for an intrinsic semi-conductor half way between the two bands and so on.

The second approximations show that the contact potential may be more highly temperature-dependent for impurity semi-conductors than for a metal, and one can have still more important variations of contact potential as λ for these semi-conductors changes over from one type of formula to another. It is perhaps worth recording that for two examples of the same impurity semi-conductor with different amounts of impurity we have

$$-|\epsilon|\, V_x{}^y = -|\epsilon|(V_0{}^y - V_0{}^x) = -\tfrac{1}{2}kT \log \frac{N_y}{N_x}. \qquad \text{......(1156)}$$

This is of classical form and strictly non-metallic. It might amount to as much as 0·1 volt.

For a semi-conductor the effective photoelectric threshold obtained by extrapolating to zero a curve of current against exciting frequency can differ at low temperatures from the thermionic work function. The few excited electrons in band 1 will give a tail to the photoelectric yield curve which is too faint to be detected, and the curve will appear to stop at a frequency which is just sufficient to eject electrons from band 2 out of the metal—this frequency is given by $h\nu = E_x \neq E_x - \frac{1}{2}\Delta E_1$.

§ 11·7. *Formal theory of electrical and thermal conduction by the electrons in a metal.* In order to proceed any further with the theory of the electrical properties of metals or semi-conductors, it is necessary to develop at least formally the theory of states of steady flow which are not equilibrium states. Just such a development has been given by Sommerfeld† by generalizing to Fermi-Dirac statistics the classical theories of Lorentz‡ and Bohr,§ but making no attempt to calculate the quantum mean free path l which occurs in the formulae. We shall also confine attention to bands or sets of states in which the kinetic energy is sufficiently accurately given by $(p^2+q^2+r^2)/2m$, where p, q, r are the components of the momentum. We shall however finally have to extend the calculations to include two sets of such states, one of them electrons and the other holes, both of which may contribute to the conductivity.

The number dn of electrons per unit volume in the metal, or in an isolated band of states, with (group) velocity components in the range

$$u,\ u+du;\quad v,\ v+dv;\quad w,\ w+dw\quad (u^2+v^2+w^2=\mathsf{U}^2)$$

satisfies, by (1005) slightly generalized,

$$dn=\varpi\left(\frac{m}{h}\right)^3 f_0(\mathsf{U})\,du\,dv\,dw=f_0\,d\omega, \qquad \text{......(1157)}$$

where m is the (effective) mass of the electron and

$$f_0=\frac{1}{1+e^{\frac{1}{2}m\mathsf{U}^2/kT}/\lambda}=\frac{1}{1+e^{\eta/kT}/\lambda}. \qquad \text{......(1158)}$$

In (1158) we have taken the energy zero at the bottom of the Fermi distribution of free electrons or at the lower edge of the isolated band.‖ When we have an ideal metal with nearly free electrons $m=m_0$ and $\lambda=e^{\zeta/kT}$, where $\zeta\sim\eta^*$, the energy of the highest occupied level at zero temperature. When the group is effectively classical dn reduces to

$$dn=\varpi\left(\frac{m}{h}\right)^3\lambda e^{-\frac{1}{2}m\mathsf{U}^2/kT}\,du\,dv\,dw,$$

$$=n\left(\frac{m}{2\pi kT}\right)^{\frac{3}{2}}e^{-\frac{1}{2}m\mathsf{U}^2/kT}\,du\,dv\,dw, \qquad \text{......(1159)}$$

where n is the total electron density in the group. For generality we shall work throughout with (1158). Similar formulae apply to a group of classical holes (free positive electrons), which we shall discuss later.

Suppose now that these effectively free electrons of charge $-|\epsilon|$ are

† Sommerfeld, *Zeit. f. Physik.* vol. 47, p. 1 (1928); Sommerfeld and Bethe, *loc. cit.*

‡ Lorentz, *The Theory of Electrons*, Teubner (1905).

§ Bohr, *Studier over Metallernes Elektrontheori* (1911).

‖ In collating these formulae with § 11·6 we must allow for the different choice of energy zero.

subjected to electric fields F_x and F_y and a magnetic field H_z (for short H) referred to right-handed axes. Then the electrons (classical or quantal) move as if subject to forces

$$-|\epsilon|(F_x+vH), \quad -|\epsilon|(F_y-uH).$$

The distribution function no longer remains f_0, for the accelerating forces above cause transitions to new states of motion thereby altering f_0 to a value $f(u,v,w,x,y,z)$ until collisions with the lattice, which must tend always to restore f_0, can balance the effect of the accelerating forces. It is clear from the symmetry that we can write

$$f=f_0+u\phi(\mathsf{U})+v\psi(\mathsf{U}),$$

where $u\phi$ and $v\psi$ are quantities of the first order small compared with f and f_0.

When such a distribution has been established we have a state of steady flow. The electrical currents J_x, J_y per cm.² parallel to the x- and y-axes are clearly given by

$$J_x=-|\epsilon|\int ufd\omega, \quad J_y=-|\epsilon|\int vfd\omega.$$

From the symmetry these reduce at once to

$$J_x=-\tfrac{1}{3}|\epsilon|\int \mathsf{U}^2\phi(\mathsf{U})\,d\omega, \quad J_y=-\tfrac{1}{3}|\epsilon|\int \mathsf{U}^2\psi(\mathsf{U})\,d\omega. \quad \text{......(1160)}$$

The thermal currents W_x, W_y in ergs per cm.² per second are

$$W_x=\tfrac{1}{2}m\int u\mathsf{U}^2fd\omega, \quad W_y=\tfrac{1}{2}m\int v\mathsf{U}^2fd\omega$$

which reduce to

$$W_x=\tfrac{1}{6}m\int \mathsf{U}^4\phi(\mathsf{U})\,d\omega, \quad W_y=\tfrac{1}{6}m\int \mathsf{U}^4\psi(\mathsf{U})\,d\omega. \quad \text{......(1161)}$$

Inspection of (1160) and (1161) shows that the crux of the matter is the calculation of $\phi(\mathsf{U})$ and $\psi(\mathsf{U})$. On appealing to Boltzmann's integro-differential equation, discussed in detail in Chapter XVII, we see that in a steady state, in which $\partial f/\partial t=0$, we must have

$$-\left\{\dot{u}\frac{\partial f}{\partial u}+\dot{v}\frac{\partial f}{\partial v}+\dot{w}\frac{\partial f}{\partial w}+u\frac{\partial f}{\partial x}+v\frac{\partial f}{\partial y}+w\frac{\partial f}{\partial z}\right\}+\left[\frac{\partial f}{\partial t}\right]_{\text{coll}}=0.$$

Since the electrons are effectively free

$$\dot{u}=-|\epsilon|(F_x+vH)/m, \quad \dot{v}=-|\epsilon|(F_y-uH)/m$$

and Boltzmann's equation reduces to

$$-\frac{|\epsilon|}{m}\left\{(F_x+vH)\frac{\partial f}{\partial u}+(F_y-uH)\frac{\partial f}{\partial v}\right\}+u\frac{\partial f}{\partial x}+v\frac{\partial f}{\partial y}-\left[\frac{\partial f}{\partial t}\right]_{\text{coll}}=0. \quad \text{......(1162)}$$

Now the collision terms themselves must be terms of the first order proportional to the deviation of f from f_0. Moreover, if l is the mean free path, U/l is the number of effective collisions per second made by the electron with the lattice irregularities. The contribution to the collision term for a single U must strictly of course depend on the deviations of f from f_0 for all values of u, v, w, but if l is properly interpreted a formally correct expression must be obtained by putting*

$$\left[\frac{\partial f}{\partial t}\right]_{\text{coll}} = -\frac{\mathsf{U}}{l}\{u\phi(\mathsf{U})+v\psi(\mathsf{U})\}. \qquad \text{......(1163)}$$

The minus sign must be taken because the collisions must always tend to reduce the abnormality in f. Combining (1162) and (1163), we find

$$-\frac{|\epsilon|}{m}\left\{(F_x+vH)\frac{\partial f}{\partial u}+(F_y-uH)\frac{\partial f}{\partial v}\right\}+u\frac{\partial f}{\partial x}+v\frac{\partial f}{\partial y} = -\frac{\mathsf{U}}{l}\{u\phi(\mathsf{U})+v\psi(\mathsf{U})\}. \qquad \text{......(1164)}$$

We can now replace f by f_0 in all terms except those which have H as a factor, since they are all first order terms. Replacing f by f_0 in the H-terms eliminates H entirely so that f must be retained here. We thus find

$$-|\epsilon|\left\{F_x u\frac{\partial f_0}{\partial \eta}+F_y v\frac{\partial f_0}{\partial \eta}\right\}+u\frac{\partial f_0}{\partial x}+v\frac{\partial f_0}{\partial y}-\frac{|\epsilon|}{m}H\{v\phi(\mathsf{U})-u\psi(\mathsf{U})\}$$
$$= -\frac{\mathsf{U}}{l}\{u\phi(\mathsf{U})+v\psi(\mathsf{U})\}, \qquad \text{......(1165)}$$

terms in uv, u^2v and uv^2 cancelling automatically. The coefficients of u and v must vanish separately so that

$$\frac{\mathsf{U}}{l}\phi+\frac{|\epsilon|H}{m}\psi = -\left\{-|\epsilon|\,F_x\frac{\partial f_0}{\partial \eta}+\frac{\partial f_0}{\partial x}\right\}, \qquad \text{......(1166)}$$

$$-\frac{|\epsilon|H}{m}\phi+\frac{\mathsf{U}}{l}\psi = -\left\{-|\epsilon|\,F_y\frac{\partial f_0}{\partial \eta}+\frac{\partial f_0}{\partial y}\right\}. \qquad \text{......(1167)}$$

On solving these equations we find

$$\phi(\mathsf{U}) = \frac{-l/\mathsf{U}}{1+|\epsilon|^2H^2l^2/m^2\mathsf{U}^2}\left[\left(-|\epsilon|F_x\frac{\partial f_0}{\partial \eta}+\frac{\partial f_0}{\partial x}\right)-\frac{l}{\mathsf{U}}\frac{|\epsilon|H}{m}\left(-|\epsilon|F_y\frac{\partial f_0}{\partial \eta}+\frac{\partial f_0}{\partial y}\right)\right], \qquad \text{......(1168)}$$

$$\psi(\mathsf{U}) = \frac{-l/\mathsf{U}}{1+|\epsilon|^2H^2l^2/m^2\mathsf{U}^2}\left[\left(-|\epsilon|F_y\frac{\partial f_0}{\partial \eta}+\frac{\partial f_0}{\partial y}\right)+\frac{l}{\mathsf{U}}\frac{|\epsilon|H}{m}\left(-|\epsilon|F_x\frac{\partial f_0}{\partial \eta}+\frac{\partial f_0}{\partial x}\right)\right] \qquad \text{......(1169)}$$

* The form (1163) is only strictly correct if the collisions of the electrons with the lattice are elastic and merely *scatter* the electrons. None the less the conclusions we shall draw are true independently of this restriction. It would take us too far afield to attempt to remove it. See Sommerfeld and Bethe, *loc. cit.*, sect. D.

On substituting these equations in (1160) and (1161) we obtain expressions for the electrical and thermal currents, which can be evaluated as soon as the special conditions of the particular problem have been fixed. It must be remembered that l may be a function of η.

In the evaluations which are to follow it is clear that integrals of the type $\int p(\mathsf{U})(\partial f_0/\partial \eta)\,d\omega$ will play an important part. They can all be simply expressed in terms of the functions

$$K_s = -\frac{8\pi\varpi m}{3h^3}\int_0^\infty l(\eta)\,\eta^s \frac{\partial f_0}{\partial \eta}\,d\eta, \qquad \text{......(1170)}$$

which we proceed to evaluate (i) for a simple Fermi-Dirac distribution almost completely degenerate and (ii) for a classical group of electrons (or holes).

§**11·71.** *The free path integrals.* (i) *Degenerate case.* Equation (1158) shows us that in this case

$$-\frac{\partial f_0}{\partial \eta} = \frac{1}{kT}\frac{1}{(1+e^{(\eta-\zeta)/kT})(1+e^{(\zeta-\eta)/kT})}. \qquad \text{......(1171)}$$

Thus $-\partial f_0/\partial \eta$ is only sensible when $\eta \simeq \zeta$. It is therefore possible to evaluate asymptotically any integral of the form

$$K(q) = -\int_0^\infty q(\eta)\frac{\partial f_0}{\partial \eta}\,d\eta \qquad \text{......(1171·1)}$$

by putting $\eta = \zeta + x$ and $q(\eta) = q(\zeta) + xq'(\zeta) + \frac{1}{2!}x^2 q''(\zeta) + \ldots.$ The typical term is therefore

$$\frac{q^{(i)}(\zeta)}{i!\,kT}\int_{-\zeta}^\infty \frac{x^i\,dx}{(1+e^{x/kT})(1+e^{-x/kT})},$$

which is only trivially altered if the range of integration is changed to $-\infty, \infty$. These integrals can be converted by integration by parts to the same form as the integral in § 2·73 and can be evaluated by those or similar methods. The integrals all vanish for i odd, and we shall require only those for $i = 0, 2$ of which the first is elementary. The values are

$$\frac{1}{kT}\int_{-\infty}^\infty \frac{dx}{(1+e^{x/kT})(1+e^{-x/kT})} = 1, \qquad \text{......(1172)}$$

$$\frac{1}{kT}\int_{-\infty}^\infty \frac{x^2\,dx}{(1+e^{x/kT})(1+e^{-x/kT})} = \frac{\pi^2 k^2 T^2}{3}, \qquad \text{......(1173)}$$

so that

$$K(q) = q(\zeta) + \frac{\pi^2 k^2 T^2}{6} q''(\zeta). \qquad \text{......(1174)}$$

It follows from this that

$$K_s = \frac{8\pi\varpi m}{3h^3}\left[l(\zeta)\,\zeta^s + \frac{\pi^2 k^2 T^2}{6}\frac{d^2}{d\zeta^2}\{l(\zeta)\,\zeta^s\}\right]. \qquad \text{......(1175)}$$

§ **11·72.** *The free path integrals.* (ii) *Classical group of electrons.* In the classical case

$$-\frac{\partial f_0}{\partial \eta}=\frac{\lambda}{kT}e^{-\eta/kT}$$

and the important energies are all fairly small. We may therefore usually neglect variations of l with η. Then it follows at once that

$$K_s=\frac{8\pi m\varpi}{3h^3}l\lambda(kT)^s s!. \qquad \text{......(1176)}$$

This can be more usefully expressed in terms of n, the number of free electrons per unit volume in the group. On comparing (1159) and the preceding equation, we find that K_s reduces to

$$K_s=\frac{4}{3}\frac{ln}{(2\pi m)^{\frac{1}{2}}}(kT)^{s-\frac{1}{2}}s!, \qquad \text{......(1177)}$$

since
$$\log\lambda=\log n-\log\frac{(2\pi mkT)^{\frac{3}{2}}\varpi}{h^3}. \qquad \text{......(1177·1)}$$

§ **11·73.** *Thermal and electrical currents for zero magnetic field.* In this important special case we can put $H=0$, $F_y=0$ and $\psi(\mathsf{U})=0$ in (1168), (1160) and (1161). We may then put

$$\frac{\partial f_0}{\partial x}=-\frac{\partial f_0}{\partial \eta}\left\{\frac{kT}{\lambda}\frac{\partial\lambda}{\partial x}+\frac{\eta}{T}\frac{\partial T}{\partial x}\right\}. \qquad \text{......(1178)}$$

From these equations and the value of $d\omega$ in (1157) it follows at once that

$$J_x=\frac{8\pi|\epsilon|\varpi m}{3h^3}\left[\left(|\epsilon|F_x+\frac{kT}{\lambda}\frac{\partial\lambda}{\partial x}\right)\int_0^\infty l(\eta)\,\eta\left(-\frac{\partial f_0}{\partial\eta}\right)d\eta+\frac{1}{T}\frac{\partial T}{\partial x}\int_0^\infty l(\eta)\,\eta^2\left(-\frac{\partial f_0}{\partial\eta}\right)d\eta\right],$$

$$W_x=-\frac{8\pi\varpi m}{3h^3}\left[\left(|\epsilon|F_x+\frac{kT}{\lambda}\frac{\partial\lambda}{\partial x}\right)\int_0^\infty l(\eta)\,\eta^2\left(-\frac{\partial f_0}{\partial\eta}\right)d\eta+\frac{1}{T}\frac{\partial T}{\partial x}\int_0^\infty l(\eta)\,\eta^3\left(-\frac{\partial f_0}{\partial\eta}\right)d\eta\right].$$

In terms of the free path integrals (1170) these equations reduce to

$$J_x=K_1\left(|\epsilon|^2F_x+|\epsilon|\frac{kT}{\lambda}\frac{\partial\lambda}{\partial x}\right)+K_2|\epsilon|\frac{1}{T}\frac{\partial T}{\partial x}, \qquad \text{......(1179)}$$

$$W_x=-K_2\left(|\epsilon|F_x+\frac{kT}{\lambda}\frac{\partial\lambda}{\partial x}\right)-K_3\frac{1}{T}\frac{\partial T}{\partial x}. \qquad \text{......(1180)}$$

(i) *The electrical conductivity.* In a uniform block of metal at constant temperature the terms in $\partial/\partial x$ vanish and

$$J_x=K_1|\epsilon|^2F_x.$$

The specific conductivity κ is defined by the equation $J=\kappa F$, J being the current per square centimetre flowing under potential gradient F. Thus

$$\kappa=|\epsilon|^2 K_1. \qquad \ldots\ldots(1181)$$

For the simple degenerate electron gas it follows by (1175) that

$$\kappa=\frac{8\pi\varpi m|\epsilon|^2}{3h^3}\,l(\zeta)\,\zeta. \qquad \text{(F-D)}\ldots\ldots(1182)$$

On using $\zeta\sim\eta^*$ and the formula for η^* which in the present notation is

$$\eta^*=\tfrac{1}{2}m\mathsf{U}^{*2}=\frac{h^2}{8m}\left(\frac{6n}{\pi\varpi}\right)^{\frac{2}{3}},$$

we can reduce (1182) to the form

$$\kappa=\frac{|\epsilon|^2 nl}{m\mathsf{U}^*}. \qquad \text{(F-D)}\ldots\ldots(1183)$$

For a classical group of electrons it follows by (1177) that

$$\kappa=\frac{4}{3}\frac{|\epsilon|^2 ln}{(2\pi mkT)^{\frac{1}{2}}}, \qquad \text{(C)}\ldots\ldots(1184)$$

or with

$$\bar{\mathsf{U}}=2\sqrt{\frac{2kT}{\pi m}},$$

$$\kappa=\frac{1}{3}\frac{|\epsilon|^2 ln\bar{\mathsf{U}}}{kT}=\frac{8}{3\pi}\frac{|\epsilon|^2 ln}{m\bar{\mathsf{U}}}. \qquad \text{(C)}\ldots\ldots(1185)$$

The formula (1183) will reproduce the observed conductivities of the best metallic conductors if l is of the order of 100 times the lattice constant at ordinary temperatures and if further the whole of the temperature variation in the conductivity is given by the variation in l. The studies of Bloch, Peierls, Bethe and others have in fact shown that l must be of this order and that l and therefore κ vary like T^{-1} at high temperatures and probably like T^{-5} at low as in fact they are observed to do.† A comparison of the observed and calculated temperature variation of the conductivity is shown in Table 44 for two noble metals.

(ii) *The thermal conductivity* is naturally measured in a uniform metal with a temperature gradient carrying no electrical current. The electron distribution must therefore adjust itself so that $J_x=0$, while $F_x\neq 0$. Eliminating F_x from (1180) by means of this condition we find

$$W_x=-\left(K_3-\frac{K_2^2}{K_1}\right)\frac{1}{T}\frac{\partial T}{\partial x}.$$

If the thermal conductivity θ is defined by $W_x=-\theta\,\partial T/\partial x$, we see that

$$\theta=\frac{K_3K_1-K_2^2}{TK_1}. \qquad \ldots\ldots(1186)$$

† The low temperature form is still doubtful. See Sommerfeld and Bethe, *loc. cit.* §§ 37, 40, 41.

TABLE 44.

Comparison of the observed and calculated variations of the conductivity with temperature for pure copper and gold.

[The theory is only completely evaluated for high and low temperatures and has been semi-empirically interpolated between.]
From Sommerfeld and Bethe, *loc. cit.* p. 532.

Au $[\Theta_D = 173°]$			Cu $[\Theta_D = 315°]$		
Temp. °K.	(κ_0/κ) calc.	(κ_0/κ) obs.	Temp. °K.	(κ_0/κ) calc.	(κ_0/κ) obs.
273·2	1	1	273·2	1	1
87·43	0·2645	0·2551	195·2	0·662	0·658
78·86	0·2276	0·2187	90·2	0·1847	0·1804
57·8	0·1356	0·1314	81·2	0·1451	0·141
20·4	0·00604	0·0058	20·4	0·0059	0·0051
18·9	0·00346	0·0035	4·2	$< 10^{-5}$	$< 10^{-5}$
14·3	0·00117	0·00137			
12·1	0·00051	0·00048			
11·1	0·00033	0·00030			
4·2	3×10^{-6}	3×10^{-6}			

[In the observed values the residual resistance due to strains and impurities has been eliminated.]

TABLE 44·1.

Comparison of observed values of $\theta/\kappa T$ with the value given by (1188). *This theoretical value is*

$$\frac{\theta}{\kappa T} = 2{\cdot}72 \times 10^{-13} \text{ E.S.U.}, = 2{\cdot}45 \times 10^{8} \text{ E.M.U.}, = 2{\cdot}45 \times 10^{-8} \text{ watt-ohm/(deg)}^2.$$

[Observed values of $10^8\theta/\kappa T$ in watt-ohm/(deg.)^2.]
From *International Critical Tables*, vol. 5, p. 218.

Metal	Temperature °K.				
	103	173	273	301	373
Al	1·50	1·81	2·09	2·19	2·27
Ag	2·04	2·29	2·33	2·36	2·37
Cd	2·39	2·43	2·40	2·43	2·44
Cu	1·85	2·17	2·30	2·29	2·32
Fe	3·10	2·98	2·97	2·76	2·85
Ni	2·92	2·59	2·59	2·40	2·44
Pb	2·55	2·54	2·53	2·46	2·51
Sn	2·48	2·51	2·49	2·53	2·49
Zn	2·20	2·39	2·45	2·31	2·33

On substituting for the K's from (1175), retaining only the terms of highest surviving order, we find

$$\theta = \frac{8\pi^3}{9} \frac{\varpi m k^2 l(\zeta)\,\zeta T}{h^3}. \qquad \text{(F-D)......(1187)}$$

On comparing (1182) and (1187) we obtain the Law of Wiedermann-Franz in the form

$$\frac{\theta}{\kappa T} = \frac{\pi^2}{3} \frac{k^2}{|\epsilon|^2}. \qquad \text{(F-D)......(1188)}$$

Using classical values for the K's from (1177) we find

$$\theta = \frac{8}{3}\frac{lnk^2T}{(2\pi mkT)^{\frac{1}{2}}}, \qquad \text{(C)......(1189)}$$

so that

$$\frac{\theta}{\kappa T} = 2\frac{k^2}{|\epsilon|^2}. \qquad \text{(C)......(1190)}$$

It has of course long been recognized that $\theta/\kappa T$ is very nearly an absolute constant for all metals. A selection of experimental values is shown in Table 44·1 which are in excellent agreement with (1188), particularly at temperatures which are not too low.

§ **11·74.** *The reversible thermal effects, associated with the passage of a current.* When a current flows the rate of evolution of heat in a given element of the conductor is no longer given solely by the difference of the values of W for the inflow and outflow. Other terms arise from the work done on the current carrying electrons by the field F, work which is continually converted into heat by inelastic collisions with the lattice. These have hitherto been ignored, but they must be brought in here or the energy account cannot be balanced. If therefore we introduce the quantity Q defined as the rate of evolution of heat energy in unit volume of the conductor, it follows from general considerations of energy that

$$Q = JF - \partial W/\partial x. \qquad \text{......(1191)}$$

From (1179) it follows, on dropping the suffix x and rearranging, that

$$F = \frac{J}{|\epsilon|^2 K_1} - \frac{kT}{|\epsilon|\lambda}\frac{\partial\lambda}{\partial x} - \frac{K_2}{|\epsilon| K_1}\frac{1}{T}\frac{\partial T}{\partial x}.$$

On substituting for F in (1191) and (1180) and thence for W in (1191) we find after reduction that

$$Q = \frac{J^2}{\kappa} + \frac{J}{|\epsilon|}kT\frac{\partial}{\partial x}\left(\frac{K_2}{kTK_1} - \log\lambda\right) + \frac{\partial}{\partial x}\left(\theta\frac{\partial T}{\partial x}\right). \qquad \text{......(1192)}$$

The first term is the Joule heat, the last is the heat contributed by thermal conduction. The middle term contains all the reversible effects that change sign with J which we wish to study. The form of the coefficient of J—namely $T\,\partial/\partial x(...)$—is vital, since this form gives the reversible effects their thermodynamic relationships. The necessary and sufficient condition for the appearance of this form in (1192) is the equality of the two coefficients K_2 in (1179) and (1180). This equality holds good for the most general possible formulation of the problem of metallic conduction, as has been shown by Uehling.* The equality of these coefficients is a direct consequence of the general quantal theorem that the probability of a process and

* Uehling, *Phys. Rev.* vol. 39, p. 821 (1931).

its reverse are equal, which is itself merely a direct interpretation of the Hermitian character of all matrices representing physical quantities.

Before giving the statistical theory of the reversible thermoelectric effects we shall define them in the manner usual to large scale physics and quote their thermodynamic relationships. These thermodynamic relationships are deduced on the assumption that the reversible effects can be separated from the irreversible first and last terms in (1192). This postulate has rightly often been questioned, since it is not possible by any idealization to make both the first and third terms depending on J^2 and J^0 small compared with the reversible term proportional to J. None the less it is now certain that the thermodynamic results must be true for any quantal model of a metal.

The effects are as follows: (i) When a current of density J flows in a uniform conductor with a temperature gradient $\partial T/\partial x$ in the direction of J, there is an absorption of heat by the conductor from the surroundings at a rate

$$\sigma^{(t)} J \frac{\partial T}{\partial x} \qquad \text{......(1193)}$$

per unit volume of the conductor. The coefficient $\sigma^{(t)}$ is *Thomson's specific heat of electricity*.

(ii) When a current passes from a conductor 1 of one material to conductor 2 of another material, all at constant temperature, there is an absorption of heat at a rate

$$J\Pi_{1\to 2}(T) \qquad \text{......(1194)}$$

per unit area of the junction. This is called *the Peltier effect*.

(iii) When a circuit of two different metals 1 and 2 has the two junctions maintained at different temperatures T_0, T an electromotive force acts in it. This is called *the Seebeck effect* and we shall denote the thermal electromotive force in such a circuit, in the limit of vanishing current, by Θ. If the direction of Θ is taken positive when it drives a current from metal 1 to metal 2 over the junction at temperature T, then it follows from the first law of thermodynamics that

$$\Theta = \Pi_{1\to 2}(T) - \Pi_{1\to 2}(T_0) + \int_{T_0}^{T} (\sigma_1^{(t)} - \sigma_2^{(t)})\, dT. \qquad \text{......(1195)}$$

On applying the second law one finds also

$$\frac{\Pi_{1\to 2}(T)}{T} - \frac{\Pi_{1\to 2}(T_0)}{T_0} + \int_{T_0}^{T} \frac{\sigma_1^{(t)} - \sigma_2^{(t)}}{T}\, dT = 0. \qquad \text{......(1196)}$$

By differentiation equivalent relationships can be obtained in the forms

$$\Pi_{1\to 2} = T\frac{d\Theta}{dT}, \qquad \text{......(1197)}$$

$$\sigma_1^{(t)} - \sigma_2^{(t)} = -T\frac{d}{dT}\frac{\Pi_{1\to 2}(T)}{T} = -T\frac{d^2\Theta}{dT^2}. \qquad \text{......(1198)}$$

Since the signs of these effects are apt to be confusing we display the definitions diagrammatically below. All these formulae and definitions are well known.

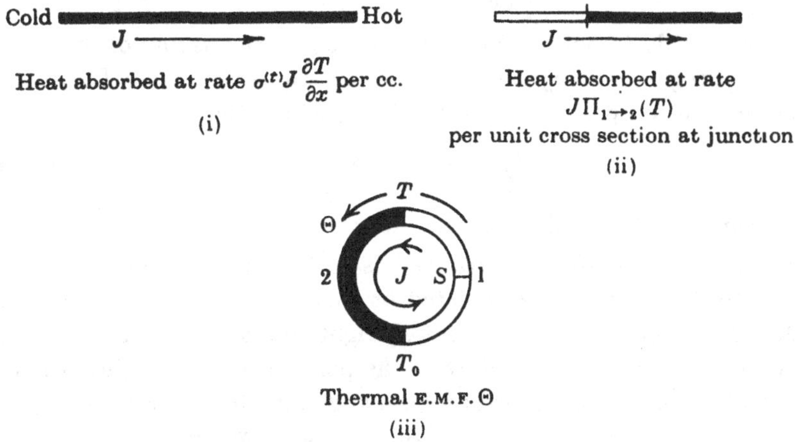

Fig. 41. Conventions for the thermoelectric effects.

§11·75. *Thermoelectric effects. Statistical theory.* (i) *Thomson's specific heat of electricity.* On referring to (1192) it is at once evident that heat is reversibly *absorbed* per unit volume at the rate

$$-\frac{J}{|\epsilon|} kT \frac{\partial}{\partial x}\left(\frac{K_2}{kTK_1} - \log \lambda\right).$$

On comparing this with (1193) we see that

$$\sigma^{(t)} = -\frac{kT}{|\epsilon|}\frac{\partial}{\partial T}\left(\frac{K_2}{kTK_1} - \log \lambda\right). \qquad \text{......(1199)}$$

By the definition $\sigma^{(t)} > 0$ means that heat is absorbed when J and $\partial T/\partial x$ are both positive in the same direction. Since J and the ordinary electron current are of opposite signs, $\sigma^{(t)} > 0$ means that heat is *evolved* when the electrons go to places of higher temperature and $\sigma^{(t)} < 0$ means that heat is then *absorbed*. Thus for a metal whose electron levels are reasonably well represented by those of a degenerate gas we should expect to find $\sigma^{(t)} < 0$. Both signs are found in common metals, but for the alkalis, which we must expect to be the metals best typified by the simple model, the observed values of σ are actually negative.

(ii) *The Peltier effect.* If we assume that the transition from the electron states of metal 1 to those of metal 2 is made continuously so that (1192) can apply, it follows that the rate of absorption of heat per unit cross-section in the transition region is

$$-\frac{J}{|\epsilon|} kT \int_1^2 \frac{\partial}{\partial x}\left[\frac{K_2}{kTK_1} - \log \lambda\right] dx.$$

It follows from (1194) that

$$\Pi_{1\to 2} = -\frac{kT}{|\epsilon|}\left[\left(\frac{K_2}{kTK_1} - \log\lambda\right)_2 - \left(\frac{K_2}{kTK_1} - \log\lambda\right)_1\right]. \quad \text{...(1200)}$$

Since the result does not depend in any way on the nature of the assumed transition region we may assume that it will continue to hold for an abrupt junction. On comparing (1200) and (1199) we see that

$$\frac{d}{dT}\left(\frac{\Pi_{1\to 2}}{T}\right) = \frac{\sigma_2^{(t)} - \sigma_1^{(t)}}{T}, \quad \text{......(1201)}$$

in agreement with the thermodynamic (1198).

(iii) *The thermal electromotive force* is to be calculated on open circuit, that is to say it must be calculated by putting $J_x = 0$ in (1179) and integrating the effective field gradient F_x right round the circuit back to its starting point. Suppose we start at the point S in Fig. 41 (iii), where the temperature may be supposed uniform and equal to T. From (1179)

$$F_x = -\frac{k}{|\epsilon|}\left(T\frac{\partial \log\lambda}{\partial x} + \frac{K_2}{kTK_1}\frac{\partial T}{\partial x}\right).$$

We require Θ which is $-\oint F_x dx$. In the first place

$$\oint T\frac{\partial \log\lambda}{\partial x}dx = [T\log\lambda] - \oint \log\lambda\frac{\partial T}{\partial x}dx.$$

The integrated part is a one valued function and its contribution therefore vanishes. Thus

$$\Theta = -\oint F_x dx = \frac{k}{|\epsilon|}\oint\left(\frac{K_2}{kTK_1} - \log\lambda\right)dT. \quad \text{......(1202)}$$

It follows at once that

$$\Theta = -\frac{k}{|\epsilon|}\int_{T_0}^{T}\left[\left(\frac{K_2}{kTK_1} - \log\lambda\right)_2 - \left(\frac{K_2}{kTK_1} - \log\lambda\right)_1\right]dT. \quad \text{......(1203)}$$

On referring to (1200) we see that

$$\Theta = \int_{T_0}^{T}\frac{\Pi_{1\to 2}(T)}{T}dT, \quad \text{......(1204)}$$

in agreement with the thermodynamic (1197).

§11·76. ***Explicit formulae for*** $\sigma^{(t)}$, $\Pi_{1\to 2}$ ***and*** Θ. (i) ***The degenerate Fermi-Dirac gas of electrons.*** The phenomena all depend on the function

$$\frac{K_2}{kTK_1} - \log\lambda = \frac{1}{kT}\left(\frac{K_2}{K_1} - \zeta\right).$$

On using (1175) we see that the highest order terms cancel and that the highest order surviving terms are

$$\frac{\pi^2 kT}{3}\left(\frac{l'(\zeta)}{l(\zeta)} + \frac{1}{\zeta}\right). \quad \text{(F-D)......(1205)}$$

In this expression ζ can be replaced by η^*. To the first approximation all the thermoelectric effects vanish and their actual values should be very small, in complete agreement with the facts. The resulting formulae are

$$\sigma^{(t)} = -\frac{\pi^2 k^2 T}{3|\epsilon|}\left(\frac{l'(\eta^*)}{l(\eta^*)}+\frac{1}{\eta^*}\right). \qquad \text{(F-D)......(1206)}$$

$$\Pi_{1\to 2} = -\frac{\pi^2 k^2 T^2}{3|\epsilon|}\left[\left(\frac{l'(\eta^*)}{l(\eta^*)}+\frac{1}{\eta^*}\right)_2 - \left(\frac{l'(\eta^*)}{l(\eta^*)}+\frac{1}{\eta^*}\right)_1\right]. \qquad \text{(F-D)......(1207)}$$

If l'/l depends on the temperature as in fact it may well do, then we can get nothing simpler for Θ than the integral resulting from substituting (1207) for $\Pi_{1\to 2}$ in (1204). If l'/l is independent of T, as for example theory requires at high temperatures, then this integral can be evaluated and gives

$$\Theta = -\frac{\pi^2 k^2}{6|\epsilon|}\left[\left(\frac{l'(\eta^*)}{l(\eta^*)}+\frac{1}{\eta^*}\right)_2 - \left(\frac{l'(\eta^*)}{l(\eta^*)}+\frac{1}{\eta^*}\right)_1\right](T^2 - T_0^2). \qquad \text{(F-D)......(1208)}$$

(ii) *A classical group of electrons.* In this case it follows from (1177) and (1177·1) that

$$\frac{K_2}{kTK_1} - \log\lambda = 2 + \log\frac{(2\pi mkT)^{\frac{3}{2}}\varpi}{nh^3}.$$

The resulting formulae are

$$\sigma^{(t)} = -\frac{3}{2}\frac{k}{|\epsilon|}\left[1 - \frac{2}{3}\frac{T\,\partial\log n}{\partial T}\right]. \qquad \text{(C)......(1209)}$$

$$\Pi_{1\to 2} = -\frac{kT}{|\epsilon|}\left[\left(\log\frac{m^{\frac{3}{2}}\varpi}{n}\right)_2 - \left(\log\frac{m^{\frac{3}{2}}\varpi}{n}\right)_1\right]. \qquad \text{(C)......(1210)}$$

For quasi-classical groups of electrons in semi-conductors we do *not* have $(m^{\frac{3}{2}}\varpi)_2 = (m^{\frac{3}{2}}\varpi)_1$. But in the special case in which this condition is satisfied, e.g. for two different samples of the same impurity semi-conductor,

$$\Pi_{1\to 2} = \frac{kT}{|\epsilon|}\log\frac{n_2}{n_1}. \qquad \text{(C)......(1211)}$$

With the same simplification

$$\Theta = \frac{k}{|\epsilon|}\int_{T_0}^{T}\log\frac{n_2}{n_1}\,dT. \qquad \text{(C)......(1212)}$$

§ **11·77.** *General discussion of thermoelectric effects.* Owing to their thermodynamic relationships it is in general sufficient to discuss the values of $\sigma^{(t)}$. When the theory agrees satisfactorily with observed values of $\sigma^{(t)}$, agreement over $\Pi_{1\to 2}$ and Θ may be presumed.

On comparing (1206) and (1209) we see that

$$\frac{\sigma^{(t)}[\text{F-D}]}{\sigma^{(t)}[\text{C}]} \simeq \frac{kT}{\eta^*} \simeq 10^{-5}T.$$

All the effects therefore are very small compared with the values that classical theory would lead one to expect. This is in agreement with observation.

Again all the effects tend to zero as $T \to 0$, since $\sigma^{(t)} \infty T$, $\Pi_{1\to 2} \infty T^2$. In fact all the effects tend to zero as $T \to 0$ in such a way that all the entropy terms in (1196) tend to zero as T and $T_0 \to 0$. This may be held to be a manifestation of Nernst's third law of thermodynamics. It is not conformed to by the classical values. The general type of variation with T appears to be in good agreement with experiment for fairly high temperatures.

In order to compare precisely the values of $\sigma^{(t)}$ derived from the theory with observed values one must evaluate $l'(\eta^*)/l(\eta^*)$. This is not yet possible with certainty except in the simplest case. It then appears however that for fairly high temperatures and not too tightly bound electrons†

$$\eta^* l'/l = 2, \qquad \text{......(1213)}$$

in which case

$$\sigma^{(t)} = -\frac{\pi^2 k^2 T}{|\epsilon|\,\eta^*}. \qquad \text{......(1214)}$$

If the electrostatic value of $|\epsilon|$ is inserted in (1214), then $\sigma^{(t)}/T$ is given in absolute electrostatic units. On multiplying by 300 it is converted to volts/(deg.)^2.

This approximation should hold best for the alkalis other than lithium owing to their tightly bound cores and very loosely bound valency electrons which will be as nearly as possible free in the metal. For other metals, even the noble metals and lithium, the binding of the valency electrons is far greater, and the electrons of the cores are better able to interfere with the valency electron. The foregoing approximations are thus of dubious applicability.

It has recently been pointed out by Sommerfeld‡ by analysis of measurements by Bidwell§ that these expectations are in fact realized. The Thomson coefficient for lead is known to be extremely small at ordinary temperatures. We may therefore measure Θ or $\Pi_{1\to 2}$ for a circuit in which metal 1 is lead and 2 is the alkali and neglect $\sigma_1^{(t)}$. Proceeding thus we obtain the values of Table 45.

Considering the refinement and smallness of the effect this is excellent agreement. The positive values of $\sigma^{(t)}$ for the noble metals and Li for example can be ascribed to a breakdown of the theory, perhaps to the failure of (1213) for more tightly bound electrons. Such a breakdown is to be expected.

† Sommerfeld and Bethe, *loc. cit.* § 36.
‡ Sommerfeld, *Phys. Rev.* vol. 45, p. 65 (1934), or *Naturwiss.* (1934), p. 49.
§ Bidwell, *Phys. Rev.* vol. 23, p. 357 (1924).

TABLE 45.

Observed and calculated values of $\sigma^{(t)}/T$ in microvolts/(deg.)^2 for the alkali metals. Observed values from the thermoelectric power against Pb *assuming* $\sigma_{Pb}{}^{(t)}=0$. *Calculated values from* (1214).

Metal	Li	Na	K	Rb	Cs
$n\times 10^{-22}$	4·66	2·67	1·30	1·08	0·86
$\eta^*\times 10^{12}$ ergs	7·4	5·1	3·2	2·8	2·4
$\sigma^{(t)}/T$ calc.	−0·016	−0·023	−0·036	−0·041	−0·048
$\sigma^{(t)}/T$ obs.	+0·40	−0·0282	−0·0275	−0·069	−0·062

If we take $T_0=0$ in (1195) it may be combined with (1197) into the form

$$\Theta - T\frac{d\Theta}{dT} = \int_0^T (\sigma_1{}^{(t)} - \sigma_2{}^{(t)})\, dT'. \qquad \text{......(1215)}$$

It is important to compare this with (1063), namely

$$-|\epsilon|\left(V_{12} - T\frac{dV_{12}}{dT}\right) = \chi_2 - \chi_1 + \int_0^T (\sigma_1 - \sigma_2)\, dT'. \qquad \text{......(1216)}$$

It is at once apparent that there is a complete analogy between the temperature-variable part of $-|\epsilon|V_{12}$ and Θ, between σ and $\sigma^{(t)}$. All the other thermodynamic relationships are similarly congruent. None the less on comparing the values of σ and $\sigma^{(t)}$ in (1028) and (1206) we see that these quantities are fundamentally distinct. They do not refer to the same regime. V_{12} and σ refer to the strict equilibrium state when no current can flow. Θ and $\sigma^{(t)}$ refer to a state of zero current, it is true, but to such a state defined as the limit of current carrying steady states when the current tends to zero. These states need not be and in fact are not the same, a possibility pointed out long ago by O. W. Richardson.† It is however only for the quantum model used here that this distinction becomes obvious. None the less Θ can be properly regarded as the temperature-variable part of a contact potential difference and the value of Θ for a circuit as made up of contributions from all the junctions that it contains. In the simple case of a two metal circuit with junctions at T_0 and T we have here

$$\Theta = \int_{T_0}^T \frac{\Pi_{1\to 2}(T')}{T'}\, dT'.$$

This however can be generalized to junctions of all sorts not merely to metallic ones.

The distinction between σ and $\sigma^{(t)}$ suggests that in all these problems careful definition is required of the quantity measured before calculation is

† O. W. Richardson, *The emission of electricity by hot bodies*, ed. 2 (1921). See also Schottky, *Zeit. f. Physik*, vol. 34, p. 645 (1925).

attempted. Equations (1179) and (1180) are always correct but give different results according to the conditions imposed by the problem. The various possible σ's have been carefully defined by Frank and Sommerfeld.†

It remains to mention one further effect found by Bridgman,‡ that there may be a Peltier effect between differently orientated crystals of the same material at the same temperature. This cannot arise from a difference in η^* which must be independent of direction, but can be due to a directional dependence of the mean free path. It is thus in principle possible for such an effect to occur, but in view of (1213) it is likely to be very small for the simple model here discussed and it is likely that it should be referred rather to the effective mass of the electron which can also be a function of direction in a more elaborate theory.

§11·78. *The electrical properties of a simple semi-conductor.* (i) *An intrinsic semi-conductor* with only a small number of electrons in the upper band 1, for which the contribution by the free holes in the lower band 2 may be neglected, has a conductivity given by the classical formula (1184) in which n $(=N_1/V)$ is given by (1130). We see at once that for such a substance

$$\kappa \propto l T e^{-\frac{1}{2}\Delta E_1/kT}.$$

The actual value of the constant of proportionality is of no importance here. Since at ordinary temperatures $l \propto T^{-1}$, we find for such a substance

$$\kappa = \kappa_0 e^{-\frac{1}{2}\Delta E_1/kT}. \qquad \text{......(1217)}$$

The Thomson coefficient is given by (1209) for the same value of n. Thus

$$\sigma^{(l)} = \frac{k}{|\epsilon|}\frac{\frac{1}{2}\Delta E_1}{kT}. \qquad \text{......(1218)}$$

This is very large compared with even the classical value of $\sigma^{(l)}$ for a fixed number of electrons, and *a fortiori* compared with the ordinary values for metals. The Peltier coefficient for such a substance against a normal metal can be calculated

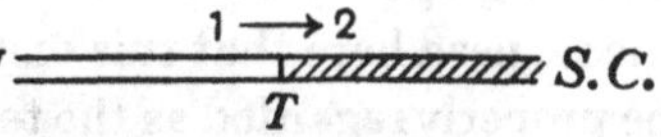

Fig. 42. The Peltier effect for a metal-semi-conductor junction.

from (1200) if we insert the classical value of $\left(\frac{K_2}{kTK_1} - \log\lambda\right)$ for the semi-conductor and neglect the contribution of the metal. We thus find

$$\Pi_{1\to 2} = -\frac{kT}{|\epsilon|}\left(2 + \log\frac{(2\pi m_1 kT)^{\frac{3}{2}}\varpi_1}{n_1 h^3}\right), \qquad \text{......(1219)}$$

or on using (1130)

$$\Pi_{1\to 2} = -\frac{kT}{|\epsilon|}\left\{2 + \log\left(\frac{\varpi_1}{\varpi_2}\right)^{\frac{1}{2}}\left(\frac{m_1}{m_2}\right)^{\frac{3}{4}} + \frac{\frac{1}{2}\Delta E_1}{kT}\right\}. \qquad \text{......(1220)}$$

† Frank and Sommerfeld, *Rev. Mod. Phys.* vol. 3, p. 1 (1931).
‡ Bridgman, *Proc. Nat. Acad. Sci.* vol. 14, p. 943 (1928).

The thermoelectric power of the circuit in Fig. 43 is given to the same approximation by

$$\Theta = -\frac{k}{|\epsilon|}\left[\left\{2+\log\left(\frac{\varpi_1}{\varpi_2}\right)^{\frac{1}{2}}\left(\frac{m_1}{m_2}\right)^{\frac{3}{4}}\right\}(T-T_0)+\frac{\frac{1}{2}\Delta E_1}{k}\log\frac{T}{T_0}\right], \quad \text{......(1221)}$$

in which the $\log T/T_0$ term is overwhelmingly the more important. It will be observed that, if $T > T_0$, $\Theta < 0$ so that *the normal direction of positive current in this circuit is from the semi-conductor to the metal at the hotter junction*. The magnitude of Θ may be large. For example if $\frac{1}{2}\Delta E_1/|\epsilon|$ is 0·8 volt the electromotive force for $T_0 = 273°$ K. is approximately $0{\cdot}8\,\delta T/T_0$ or nearly 3 millivolts per degree. The electromotive force in such a circuit whose junctions are at 273° K. and 373° K. respectively would be so large as 0·25 volt.

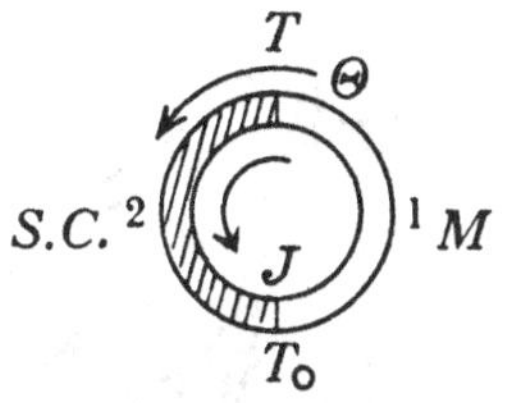

Fig. 43. The thermoelectric power of a metal-semi-conductor circuit.

(ii) *A normal impurity semi-conductor* has a conductivity for which by (1131)

$$\kappa \propto n_1 l T^{-\frac{1}{2}} \propto n_0^{\frac{1}{2}} T^{-\frac{3}{4}} e^{-\frac{1}{2}\Delta E_1'/kT}.$$

We may therefore write

$$\kappa = \kappa_0' n_0^{\frac{1}{2}} T^{-\frac{3}{4}} e^{-\frac{1}{2}\Delta E_1'/kT}. \quad \text{......(1222)}$$

The power of T is unimportant. The temperature variation is similar to (1217) and the effect of impurity content (n_0) is explicitly shown. At higher temperatures the conducting electrons will primarily be supplied by the more numerous electrons of the lower band and the conductivity will tend to the value corresponding to (1217). A beautiful example of such behaviour in cuprite (Cu_2O) of varying oxygen impurity content has been given by Joffé* whose results are shown in Fig. 44.

Formula (1219) continues to hold for the Peltier effect against a normal metal. On inserting the value of n_1 from (1131), we find

$$\Pi_{1\to 2} = -\frac{kT}{|\epsilon|}\left(2+\tfrac{1}{2}\log\frac{(2\pi m_1 kT)^{\frac{3}{2}}\varpi_1}{n_0 h^3}+\frac{\frac{1}{2}\Delta E_1'}{kT}\right), \quad \text{......(1223)}$$

in which of course n_0 is the impurity concentration. There is a corresponding formula for Θ in which the dominant term is again

$$-\frac{k}{|\epsilon|}\left(\frac{\frac{1}{2}\Delta E_1'}{k}\log\frac{T}{T_0}\right). \quad \text{......(1224)}$$

The dominant term in (1220) or (1223), namely $-\frac{1}{2}\Delta E_1/|\epsilon|$, has a simple physical interpretation. This term implies that (to this approximation) the heat absorbed at the junction is $+\frac{1}{2}\Delta E_1$ for every *electron* that passes from the metal 1 to the semi-conductor 2. On referring to § 11·61 we see that the electron levels of a metal and an intrinsic semi-conductor in contact in

* Joffé, *Actualités Sci. et Indus.*, No. 202, "Semi-conducteurs électronique", p. 35 (1935).

equilibrium with each other must be arranged as shown in Fig. 45. When an electron passes from 1 to 2 it is in general an electron from the level x as lower electrons cannot get through. To replace it from the general supply of electrons and so maintain the equilibrium distribution in 1, an electron

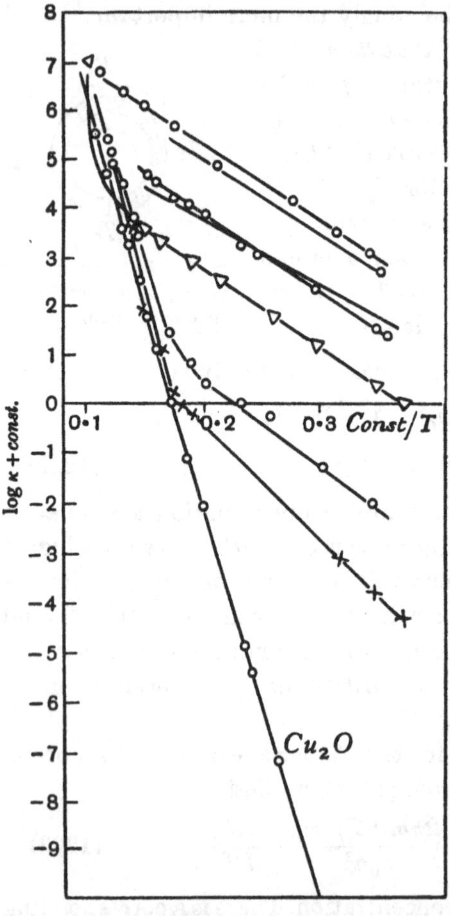

Fig. 44. The conductivity of cuprite Cu_2O with various additional amounts of excess oxygen; the amount of excess oxygen increases upwards.

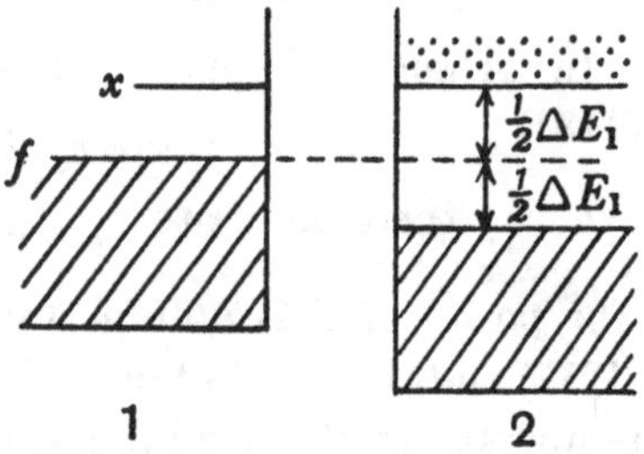

Fig. 45. Explanation of the metal-semi-conductor Peltier effect.

must be taken from the top of the nearly full levels at f since electrons are normally supplied by conduction to the boundary at this level. This requires an absorption of energy $\frac{1}{2}\Delta E_1$.

Some general comment on these results is perhaps called for at this stage. Many impurity semi-conductors do not conform to the equation

$$\kappa \propto e^{-\frac{1}{2}\Delta E/kT}$$

at all accurately even when allowance is made for the proper $T^{-\frac{3}{4}}$ factor.

This fact has been taken to imply that the theory is entirely wrong; all that it implies however is that ΔE has not the same value for all the contributing impurities. Finally it is well to bear in mind that for a metal the number of electrons in the conducting band is (practically) an absolute constant and independent of T and it is meaningless to invoke large changes of n to explain metallic anomalies, changes which necessarily occur for a semi-conductor. The effect of temperature must mimic a change of n in a more subtle way by slightly shifting the emphasis among the conducting electrons from one region of the band of states to another in which the density of possible states and their distribution with respect to Brillouin's zones and discontinuities varies in an important manner. Such effects cannot be incorporated in the elementary theory discussed here, nor is there any need to attempt to do so.

(iii) *An abnormal impurity semi-conductor* conducts by holes (or positive electrons) at the top of the lower band 2. Its conductivity therefore still obeys the formula (1222) since only $|\epsilon|^2$ is concerned in the exact formula. But *all the formulae for* $\sigma^{(t)}$, $\Pi_{1\to 2}$ *and* Θ *contain the factor* $-|\epsilon|$ *for electrons and* $+|\epsilon|$ *for holes and therefore are repeated with altered sign.*

A simple verification that the signs of all these effects do in fact change can be obtained by approximate calculation of the Peltier heat as above. The state of affairs is now that shown in Fig. 46, and when electrons flow from metal 1 to semi-conductor 2 they must go through at level x, thereby creating new holes in the metal distribution at that level. These are filled by the conduction electrons supplied at level f and there is therefore an *evolution* of energy $\frac{1}{2}\Delta E_1'$ per electron passing.

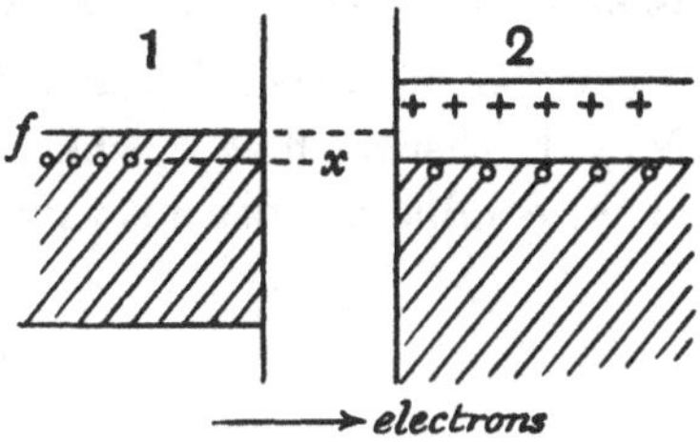

Fig. 46. Diagram showing the origin of the Peltier heat at a metal-semi-conductor contact for an abnormal impurity semi-conductor.

Substances are known, e.g. cuprite, molybdenite, with the properties of these semi-conductors—in particular with a thermoelectric power of this large order against a normal metal, and of either sign. We have however in these calculations considered only conduction by free electrons *or* by free holes. In general semi-conductors will possess both and it is necessary to extend the theory as in § 11·8 to give an adequate treatment of the effects.

§ **11·79.** *Transverse effects. The isothermal Hall effect.* We have so far considered only the effects that occur in linear circuits when from the symmetry the electrons tend only to move along the direction of flow of the current. We have assumed in short that there is no magnetic field of any importance acting. When, however, there is a transverse magnetic field we must consider two electric currents given by (1160) and two thermal currents (1161), one of each along and one across the conductor, and use the complete expressions (1168) and (1169) for $\phi(\mathsf{U})$ and $\psi(\mathsf{U})$. The new effects that now come in may be grouped together as *the transverse effects*. There are of course also modifications of the old effects which depend on H^2 which we have not space to discuss.

These transverse effects are rather numerous. We shall enumerate them but shall only complete the calculations for the Hall effect. The coefficients of the other effects can easily be constructed from the given formulae.* The transverse effects are the following:

(1) *The isothermal Hall effect* is the production of a transverse potential gradient F_y normal to J_x and H when no transverse electric current is allowed to flow and the temperature is maintained uniform. The conditions are

$$J_y = \frac{\partial T}{\partial x} = \frac{\partial T}{\partial y} = 0 \quad (W_y \neq 0,\ W_x \neq 0),$$

$$F_y = R_i J_x H. \qquad \text{......(1225)}$$

R_i is the isothermal Hall coefficient.

(2) *The isothermal Nernst effect* is the production of F_y normal to W_x and H by the thermal flow W_x when no electric current flows and the temperature of the conductor does not vary transversely. The conditions are

$$J_x = J_y = \frac{\partial T}{\partial y} = 0,$$

$$F_y = Q_i \frac{\partial T}{\partial x} H. \qquad \text{......(1226)}$$

Q_i is the isothermal Nernst coefficient.

(3) *The (adiabatic) Ettinghausen effect* is the production of a transverse temperature gradient $\partial T/\partial y$ normal to J_x and H when no transverse thermal or electrical flow is allowed in an otherwise isothermal conductor. The conditions are

$$W_y = J_y = \frac{\partial T}{\partial x} = 0,$$

$$\frac{\partial T}{\partial y} = P_a J_x H. \qquad \text{......(1227)}$$

P_a is the (adiabatic) Ettinghausen coefficient.

* They are given in detail by Frank and Sommerfeld, *Rev. Mod. Phys.* vol. 3, p. 1 (1931).

(4) *The adiabatic Hall effect* is the production, under the same conditions as (3), of a transverse potential gradient. The conditions are

$$W_y = J_y = \frac{\partial T}{\partial x} = 0 \quad (\partial T/\partial y \neq 0),$$

$$F_y = R_a J_x H. \qquad \text{......(1228)}$$

R_a is the adiabatic Hall coefficient.

(5) *The (adiabatic) Righi-Leduc effect* is the production of a transverse temperature gradient $\partial T/\partial y$ normal to W_x and H by the thermal flow W_x when no electric current flows and there is no transverse flow of heat. The conditions are

$$J_x = J_y = W_y = 0,$$

$$\frac{\partial T}{\partial y} = S_a \frac{\partial T}{\partial x} H. \qquad \text{......(1229)}$$

S_a is the (adiabatic) Righi-Leduc coefficient.

(6) *The adiabatic Nernst effect* is the production of a transverse potential gradient under the same conditions as in (5). The conditions are

$$J_x = J_y = W_y = 0,$$

$$F_y = Q_a \frac{\partial T}{\partial x} H. \qquad \text{......(1230)}$$

Q_a is the adiabatic Nernst coefficient.

The description adiabatic is applied to those effects for which a condition is $W_y = 0$; isothermal to those for which $\partial T/\partial y = 0$. The Ettinghausen and Righi-Leduc effects are essentially adiabatic and the adjective is usually omitted from their descriptions.

Between the various coefficients there exists one equation based on energy considerations, which should be of general validity—the relation of Bridgman and Lorentz that

$$Q_i = \theta P_a / T. \qquad \text{......(1231)}$$

Returning to complete the calculations for the isothermal Hall effect in a uniform conductor we find that the conditions require that f_0 should be independent of x and that, by (1160), we should make

$$\int \mathsf{U}^2 \psi(\mathsf{U})\, d\omega = 0,$$

where, by (1169) simplified to the conditions of this problem,

$$\psi(\mathsf{U}) = -\frac{l}{\mathsf{U}}\left[-|\epsilon| F_y \frac{\partial f_0}{\partial \eta} + \frac{l}{\mathsf{U}} \frac{|\epsilon| H}{m}\left(-|\epsilon| F_x \frac{\partial f_0}{\partial \eta}\right)\right].$$

It follows on dropping constant factors, since $d\omega = \varpi(m/h)^3\, du\, dv\, dw$, that

$$\int \mathsf{U}^4 \psi(\mathsf{U})\, d\mathsf{U} = 0,$$

or $$\int_0^\infty \frac{\partial f_0}{\partial \eta} d\eta \left[|\epsilon| F_y l(\eta) \frac{2\eta}{m} + \frac{|\epsilon|^2 H F_x}{m} l^2(\eta) \left(\frac{2\eta}{m}\right)^{\frac{1}{2}} \right] = 0. \quad \text{......(1232)}$$

We find at once that for the Fermi-Dirac case

$$\frac{F_y}{HF_x} = -\frac{|\epsilon|}{m} \frac{l(\zeta)}{(2\zeta/m)^{\frac{1}{2}}}. \qquad \text{(F-D)......(1233)}$$

Since $$F_x = J_x/\kappa, \quad \kappa = \frac{8\pi\varpi m|\epsilon|^2}{3h^3} l(\zeta)\,\zeta,$$

$$R_i = \frac{F_y}{HJ_x} = -\frac{3h^3}{4\pi\varpi|\epsilon|(2m\zeta)^{\frac{3}{2}}}.$$

Since $\zeta \sim \eta^*$ and $\eta^* = (h^2/8m)(6n/\pi\varpi)^{\frac{2}{3}}$,

$$R_i = -\frac{1}{|\epsilon| n}. \qquad \text{(F-D)†......(1234)}$$

For a classical group of electrons, starting again from (1232), we find

$$\frac{F_y}{HF_x} = -\frac{|\epsilon|}{2m} l \sqrt{\left(\frac{\pi m}{2kT}\right)}. \qquad \text{(C)}$$

Since $$\kappa = \frac{4}{3} \frac{|\epsilon|^2 ln}{(2\pi m kT)^{\frac{1}{2}}},$$

we find $$R_i = -\frac{3\pi}{8} \frac{1}{|\epsilon| n}. \qquad \text{(C)......(1235)}$$

If the current is carried by a group of holes, then the sign of the effect is reversed, the numerical value in (1235) being unaltered.

Equation (1234) shows that the Hall coefficient for a metal (so far as it can be represented by this approximation) should be independent of the temperature, of negative sign, and of order of magnitude 2×10^{-13} in absolute electrostatic units, or 6×10^{-3} in electromagnetic. This is of the correct order for most ordinary metals, but for some metals it is of the wrong sign. It must be supposed that such metals conduct mainly in a band of states nearly full of electrons,‡ but we shall not attempt to elaborate such a theory. Equation (1235) applied to semi-conductors shows that

$$R_i = R_0 T^{-s} e^{\frac{1}{2}\Delta E_1/kT}, \qquad \text{......(1236)}$$

where $s = \frac{3}{2}, \frac{3}{4}$. The temperature dependence is very large and is controlled

† This formula contains n directly, and therefore it matters essentially how many of the atomic electrons in a metal we should really regard as "free". The exact formula for a metal of atoms with one valency electron and lattice constant a is $R_i = -a^3/|\epsilon|$ which is of course the same as (1234). But in this form the effect of the inner electrons is seen at once to be zero, and the equations (F-D) (1234) and (C) (1235), though superficially alike, are in origin quite distinct.

‡ Peierls, *Zeit. f. Physik*, vol. 53, p. 255 (1929).

by the exponential factor. Variations of R_i of this type are well known for semi-conductors and R_0 may be of either sign. R_i may further change sign, an effect which we shall examine in the next section. A knowledge of R_i for a semi-conductor is of great importance since by (1235) R_i determines n.

§**11·8.** *Composite semi-conductors. General theory.* In the foregoing electronic theory of metallic conduction we have considered only a single group of carriers which may be a Fermi-Dirac gas of electrons or a sparse group of electrons effectively classical or a similar sparse group of holes. The first case represents a first approximation to a metal and the other cases represent semi-conductors under special limiting conditions. In general however it is not possible to discuss semi-conductors so simply for they must be held to contain a group of free electrons and a group of free holes, both of which collaborate in carrying the current. We can start by treating the free electrons and the free holes as independent. We shall see later that this independence cannot be complete and that excitation and recombination must be allowed for.

The necessary formal generalizations are easily made. We shall study in particular the thermoelectric power and the Hall effect as the signs of these quantities provide important information about the nature of the semi-conductor.

Let us distinguish the two sets of carriers by suffices 1, 2. The equations of the preceding sections apply to each set separately if we attach the proper suffix to ϕ, ψ, l, $|\epsilon|$, m, f_0, and $d\omega$, except that the equations for the total current must include both groups thus:

$$J_x = \tfrac{1}{3}\epsilon_1 \int \mathsf{U}^2 \phi_1(\mathsf{U})\, d\omega_1 + \tfrac{1}{3}\epsilon_2 \int \mathsf{U}^2 \phi_2(\mathsf{U})\, d\omega_2, \qquad \text{......(1237)}$$

$$J_y = \tfrac{1}{3}\epsilon_1 \int \mathsf{U}^2 \psi_1(\mathsf{U})\, d\omega_1 + \tfrac{1}{3}\epsilon_2 \int \mathsf{U}^2 \psi_2(\mathsf{U})\, d\omega_2. \qquad \text{......(1238)}$$

In these equations the suffixed ϵ is the algebraic charge on the electron, and we shall ultimately put $\epsilon_1 = -|\epsilon|$, $\epsilon_2 = |\epsilon|$. The only points requiring special attention are to decide on the proper subsidiary conditions in any given problem, and on what correction if any to apply for the interaction of the groups.

§**11·81.** *The thermoelectric power of a semi-conductor against a metal.* On referring to §11·75 (iii) we see that Θ is to be calculated by putting $J_x = 0$ in (1237) or the equivalent equation for the rest of the circuit and calculating $-\oint F_x dx$ with the value of F_x so obtained. The temperature gradient in the x-direction is assumed to be so small that the equilibrium values of n_1 and

n_2 are everywhere established. The primary equation in the semi-conductor obtained by extending (1179) is therefore

$$(K_1)_1\left(\epsilon_1{}^2F_x-\epsilon_1\frac{kT}{\lambda_1}\frac{\partial\lambda_1}{\partial x}\right)-(K_2)_1\epsilon_1\frac{1}{T}\frac{\partial T}{\partial x}$$
$$+(K_1)_2\left(\epsilon_2{}^2F_x-\epsilon_2\frac{kT}{\lambda_2}\frac{\partial\lambda_2}{\partial x}\right)-(K_2)_2\epsilon_2\frac{1}{T}\frac{\partial T}{\partial x}=0. \quad \text{......(1239)}$$

In deriving (1179) we took the energy zero at the bottom of the electron levels in their own band. Hence in forming this extension we must redefine λ in this way for each group, when it will be given by (1177·1). Putting in also the classical values for the K's from (1177), we find

$$\left(\frac{l_1n_1}{m_1^{\frac{1}{2}}}\epsilon_1{}^2+\frac{l_2n_2}{m_2^{\frac{1}{2}}}\epsilon_2{}^2\right)F_x-kT\left(\frac{l_1\epsilon_1}{m_1^{\frac{1}{2}}}\frac{\partial n_1}{\partial x}+\frac{l_2\epsilon_2}{m_2^{\frac{1}{2}}}\frac{\partial n_2}{\partial x}\right)$$
$$-\tfrac{1}{2}k\frac{\partial T}{\partial x}\left(\frac{l_1n_1\epsilon_1}{m_1^{\frac{1}{2}}}+\frac{l_2n_2\epsilon_2}{m_2^{\frac{1}{2}}}\right)=0.$$

Now since the electrons in each group are effectively classical

$$\frac{1}{m_1^{\frac{1}{2}}}:\frac{1}{m_2^{\frac{1}{2}}}=\overline{\mathsf{U}}_1:\overline{\mathsf{U}}_2,\qquad \frac{1}{\overline{\mathsf{U}}}\frac{\partial\overline{\mathsf{U}}}{\partial x}=\frac{1}{2}\frac{1}{T}\frac{\partial T}{\partial x}.$$

The last equation can therefore be rewritten in the classical form

$$(l_1n_1\overline{\mathsf{U}}_1\epsilon_1{}^2+l_2n_2\overline{\mathsf{U}}_2\epsilon_2{}^2)F_x=kT\left(l_1\epsilon_1\frac{\partial}{\partial x}(n_1\overline{\mathsf{U}}_1)+l_2\epsilon_2\frac{\partial}{\partial x}(n_2\overline{\mathsf{U}}_2)\right). \quad \text{......(1240)}$$

If we ignore the smaller contributions from the rest of the circuit we shall get a formula which should give the correct leading terms, but which need not, and in fact does not, reduce to (1221) when $n_2=0$, since we cannot here evaluate $-\oint F_x dx$ using the integration by parts that leads to (1202). This will only be possible when the correct interactions between the groups are included. With these approximations and the sign conventions of Fig. 43,

$$\Theta=-\oint F_x dx=\left[\int_{T_0}^{T}F_x dx\right]_{\text{semi-conductor}}$$

Confining attention to a small temperature range we can write

$$\frac{\partial\Theta}{\partial T}=F_x\frac{\partial x}{\partial T},$$

so that

$$\frac{\partial\Theta}{\partial T}=kT\frac{l_1\epsilon_1\frac{\partial}{\partial T}(n_1\overline{\mathsf{U}}_1)+l_2\epsilon_2\frac{\partial}{\partial T}(n_2\overline{\mathsf{U}}_2)}{l_1n_1\overline{\mathsf{U}}_1\epsilon_1{}^2+l_2n_2\overline{\mathsf{U}}_2\epsilon_2{}^2}. \quad \text{......(1241)}$$

When $n_2=0$, n_1 has the value for an intrinsic semi-conductor, and $\epsilon_1=-|\epsilon|$. Equation (1241) then agrees with (1221) omitting the terms in ϖ_1/ϖ_2, m_1/m_2.

This omission is a measure of the inaccuracy in the present discussion; it will usually be unimportant.

(i) *Intrinsic semi-conductor.* $n_1 = n_2$ given by (1130); $-\epsilon_1 = \epsilon_2 = |\epsilon|$. The approximate extended formula (1241) now gives us

$$-\frac{d\Theta}{dT} = \frac{(l_1\overline{U_1} - l_2\overline{U_2})(2k + \frac{1}{2}\Delta E_1/T)}{(l_1\overline{U_1} + l_2\overline{U_2})|\epsilon|}. \qquad \text{......(1242)}$$

From the nature of the electronic states in the lattice it seems that we must conclude that $l_1\overline{U_1} > l_2\overline{U_2}$. It follows that for an intrinsic semi-conductor we shall always have $d\Theta/dT < 0$, so that $\Theta < 0$ with the sign conventions of Fig. 43. This means that the *positive* current flows from the semi-conductor to the metal at the hotter junction, which may be regarded as the normal sign of the effect. The value of Θ may be quite large. If for example $\frac{1}{2}\Delta E_1/|\epsilon| = 0{\cdot}8$ volt, the electromotive power of the semi-conductor against a normal metal is of the order of 3 millivolts per degree. Such large values appear actually to occur.

(ii) *For impurity semi-conductors* we have merely to use (1241) with the values of n_1 and n_2 as functions of T determined in § 11·61. We can write (1241) in the form

$$-\frac{d\Theta}{dT} = \frac{k}{|\epsilon|}\,\frac{\frac{1}{2}(l_1 n_1\overline{U_1} - l_2 n_2\overline{U_2}) + T\left(l_1\overline{U_1}\frac{\partial n_1}{\partial T} - l_2\overline{U_2}\frac{\partial n_2}{\partial T}\right)}{l_1 n_1\overline{U_1} + l_2 n_2\overline{U_2}}.$$

We must suppose as before that $l_1 \geqslant l_2$, $\overline{U_1} > \overline{U_2}$. For a *normal* impurity semi-conductor we have necessarily $n_1 > n_2$ and $\partial n_1/\partial T > \partial n_2/\partial T$. The latter equation follows because n_1 is supplied both from the impurities and from the lower band and therefore must increase faster than n_2. For such a substance $d\Theta/dT < 0$ and the sign of the effect must be normal.

For an abnormal impurity semi-conductor we have at low temperatures $n_1 \simeq 0$, $\partial n_1/\partial T \ll \partial n_2/\partial T$ so that the sign of $d\Theta/dT$ must start by being positive and the effect is abnormal at low temperatures. At higher temperatures we continue to have $\partial n_2/\partial T > \partial n_1/\partial T$ tending to an equality as the temperature rises. On the other hand $n_2 > n_1$ but $l_1\overline{U_1} > l_2\overline{U_2}$ and the sign of the effect is no longer definite. In fact since $n_2 \sim n_1$ as T increases, the *thermoelectric power must vanish and change sign at higher temperatures.*

It will be observed that equation (1237) does not in general make both partial currents vanish. In the steady state of zero total current there is therefore actually a flow of electrons in one direction in the semi-conductor and an equal flow of holes in the same direction. At any particular place therefore, n_1 and n_2 will be increasing at equal rates. They will increase or diminish until the increased or diminished rate of recombination exactly balances the contribution made by the flow. The assumption of a normal n_1 and n_2 (with a modified velocity distribution law) is therefore not exactly correct as we have already pointed out from another aspect.

§ **11·82.** *The isothermal Hall coefficient for a semi-conductor.* On referring to the definition of the Hall coefficient we see that we have to make $J_y = 0$ in (1238) when the conditions are uniform in the x-direction and in particular the temperature is uniform everywhere. We may therefore interpret the condition $\partial T/\partial x = 0$ in the more general sense that it requires us to impose the conditions $\partial f_1/\partial x = \partial f_2/\partial x = 0$ on both groups of carriers. The conductor could well be a closed circuit when a value of $\partial f_i/\partial x \neq 0$ is impossible. With these conditions, neglecting H, we obtain from (1237) the additive relation

$$\kappa = \kappa_1 + \kappa_2 \qquad \left(\kappa_i = \frac{|\epsilon|^2 l_i n_i \overline{\mathsf{U}_i}}{3kT}\right),$$

which is *a priori* obvious. The conditions across the flow are however less simple. The condition of no space charge requires

$$\frac{\partial n_1}{\partial y} = \frac{\partial n_2}{\partial y}, \qquad \text{......(1243)}$$

assuming that no change with y occurs in the number of bound electrons; one more condition however is required, for we cannot assert by a uniformity argument that $\partial n_1/\partial y = \partial n_2/\partial y = 0$. This condition obviously is that the equal rates of increase of electrons and holes by transport to any neighbourhood must be exactly balanced by the increased rate of recombination due to the presence of a slight excess of electrons and holes. These effects must be proportional in a first approximation to $\partial n_1/\partial y$ since the deviation from the equilibrium value of n_1 at any point will vary in this way. Hence the partial transverse currents themselves must be proportional to $\partial n_1/\partial y$ and we can write (1238) in the form

$$\int \mathsf{U}^2 \psi_1(\mathsf{U})\, d\omega_1 = A\frac{\partial n_1}{\partial y}, \quad \int \mathsf{U}^2 \psi_2(\mathsf{U})\, d\omega_2 = A\frac{\partial n_2}{\partial y}.$$

On inserting the proper values of ψ from (1169) the equations easily reduce to

$$\left.\begin{aligned} \frac{-l_1\epsilon_1 n_1 \overline{\mathsf{U}_1}}{kT} F_y + l_1\overline{\mathsf{U}_1}\frac{\partial n_1}{\partial y} + \frac{l_1^2\epsilon_1^2 n_1 H F_x}{m_1 kT} &= A\frac{\partial n_1}{\partial y}, \\ \frac{-l_2\epsilon_2 n_2 \overline{\mathsf{U}_2}}{kT} F_y + l_2\overline{\mathsf{U}_2}\frac{\partial n_2}{\partial y} + \frac{l_2^2\epsilon_2^2 n_2 H F_x}{m_2 kT} &= A\frac{\partial n_2}{\partial y}. \end{aligned}\right\} \qquad \text{......(1244)}$$

The elimination of $\partial n_1/\partial y$ $(=\partial n_2/\partial y)$ in the general case is not very illuminating. We record only the two limiting cases:

(*a*) $A \ll l\overline{\mathsf{U}}$. Then

$$R_i\kappa = \frac{F_y}{HF_x} = \left(\frac{l_1\epsilon_1{}^2 n_1}{m_1\overline{\mathsf{U}}_1} - \frac{l_2\epsilon_2{}^2 n_2}{m_2\overline{\mathsf{U}}_2}\right) \Big/ (\epsilon_1 n_1 - \epsilon_2 n_2), \qquad \ldots\ldots(1245)$$

which reduces to

$$R_i\kappa = -\frac{3\pi}{8}\frac{\kappa_1 - \kappa_2}{|\epsilon|(n_1+n_2)}. \qquad \ldots\ldots(1246)$$

(*b*) $A \gg l\overline{\mathsf{U}}$. Then

$$R_i\kappa = -|\epsilon|\left(\frac{l_1{}^2 n_1}{m_1} - \frac{l_2{}^2 n_2}{m_2}\right) \Big/ (l_1\overline{\mathsf{U}_1}n_1 + l_2\overline{\mathsf{U}_2}n_2). \qquad \ldots\ldots(1247)$$

The correct formula must lie somewhere in between: The important factors in (1245) and (1247) can also be put in the respective forms

$$l_1\overline{\mathsf{U}_1}n_1 - l_2\overline{\mathsf{U}}_2 n_2, \quad l_1{}^2\overline{\mathsf{U}_1{}^2}n_1 - l_2{}^2\overline{\mathsf{U}_2{}^2}n_2.$$

The results can now be applied very simply to the various models. We see at once that an intrinsic semi-conductor and a normal impurity semi-conductor will both have a normal sign for their Hall coefficient at all temperatures. The abnormal impurity semi-conductor will have a Hall coefficient of abnormal sign at low temperatures, which will vanish and change sign as the temperature rises. The signs of the thermoelectric power and the Hall effect both behave in the same way, but they need not alter at the same temperature. Over a limited temperature range one might have any combination of signs. The theory suggests however that, for any substance properly represented by these models, if the sign of either effect becomes abnormal so will the sign of the other effect at sufficiently low temperatures. In fact the signs of all the effects that depend on the first power of ϵ will all become abnormal together at sufficiently low temperatures, and normal at sufficiently high ones. For any substance which does not conform a still more complicated model must be used.

§ **11·9.** *Metallic contacts and the nature of strongly rectifying contacts.* We have so far not considered the conditions under which electrons can be transferred from one conductor to another across an ordinary mechanical contact. We shall discuss this problem in this section, but with no attempt at a refined theory, for which the time is hardly ripe.

A contact between two ideal metals may be pictured as a potential hill separating two regions of constant potential. The shape of the hill may be determined roughly as follows: When the two metals are brought together and equilibrium is set up the equilibrium contact potential difference is established between them, determined by the equality of the parameter λ for both metals (§ 11·38). It is sufficiently accurate here to describe this condition by saying that in equilibrium the total energy of the highest electron levels occupied at low temperatures is the same in both metals.

The hill would therefore have the form shown in Fig. 47*b* if it were not modified by the image effect. The actual form as modified will be more like that shown in Fig. 47*c*.

From the point of view of electron exchanges this equilibrium condition must be such that equal numbers of electrons pass in each direction through the barrier in any given time. By equation (1010) the number of electrons $N_1(\zeta)\,d\zeta$ incident on the barrier in metal 1 per unit area per unit time in a given range ζ, $\zeta+d\zeta$ of energy in their motion normal to the surface is

$$N_1(\zeta)\,d\zeta=\frac{4\pi mkT}{h^3}\log\{1+e^{(\eta_1{}^*-\zeta)/kT}\}\,d\zeta. \qquad \text{......(1248)}$$

In this formula the energy zero $\zeta=0$ is at the bottom of the free electron levels in metal 1. There is a similar formula for metal 2, with a similar energy

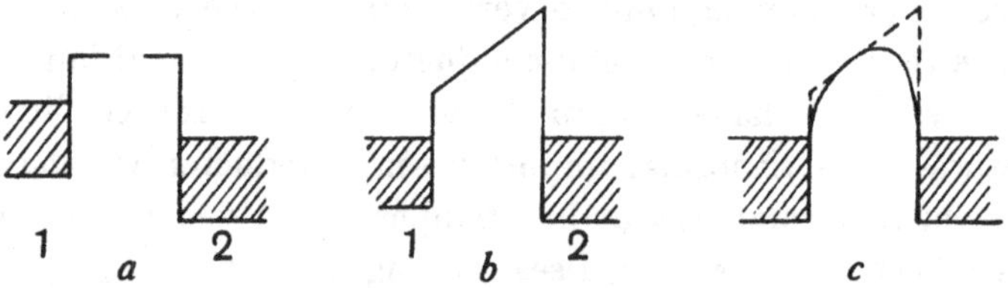

a *b* *c*

Fig. 47. Barriers at metal contacts: (*a*) separated ideal metals; (*b*) ditto adjusted to equilibrium; (*c*) barrier modified by image effect.

zero. Reduced to the same energy zero as for metal 1 the formula for $N_2(\zeta)\,d\zeta$ becomes

$$N_2(\zeta)\,d\zeta=\frac{4\pi mkT}{h^3}\log\{1+e^{(\eta_1{}^*-\zeta)/kT}\}\,d\zeta \quad (\zeta>-(\eta_2{}^*-\eta_1{}^*)).$$

Now there is a general theorem† that for any hill

$$D_{1\to2}(\zeta)=D_{2\to1}(\zeta) \quad (\zeta>0),$$

where $D_{1\to2}(\zeta)$ and $D_{2\to1}(\zeta)$ are the fractions of the incident electrons of energy ζ that penetrate the barrier in the directions $1\to2$ and $2\to1$ respectively. Further it is obvious that $D_{2\to1}(\zeta)=0$ $(\zeta<0)$. There is therefore the exact balance required in the electrons passing in the two directions.‡

When a current flows from one metal to the other the distribution laws are altered and the balance can be upset sufficiently to carry any reasonable current without an appreciable potential jump across the contact. But if the contact is a bad one an appreciable potential difference may be necessary across it before a significant current will pass. A potential difference across

† Fowler, *Proc. Camb. Phil. Soc.* vol. 25, p. 193 (1929).

‡ Strictly speaking in calculating the number of electrons which pass from metal 1 to metal 2 and from 2 to 1 it is necessary to take account of the number of empty levels in the receiving metal into which the entering electron can go. This introduces correcting factors $1-\alpha n_2(\zeta)$ for the flow $1\to2$ and $1-\alpha n_1(\zeta)$ for the flow $2\to1$. Since the flows are themselves proportional to $n_1(\zeta)$ and $n_2(\zeta)$ the extra terms exactly balance and it is not necessary to include them.

the contact imposes a lack of balance in the electron exchange which can be of a higher order than the small terms due to the disturbance of the distribution functions in each metal.

It is evident that the current flowing across the contact due to change of distribution functions without an extra contact potential difference will be symmetrical in the current so that there will then be no rectifying effect. It might however be thought at first sight that rectification could occur at a bad contact between two different metals, since the shape of the potential barrier will no longer remain the same, and its transparency can therefore differ for the two signs of the potential difference imposed; this is illustrated in Fig. 48.

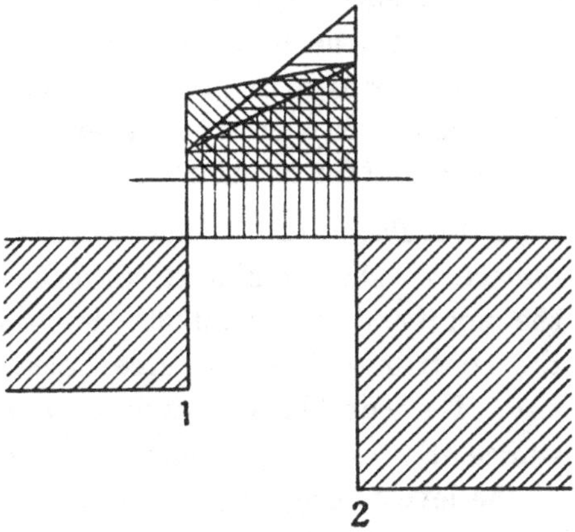

Fig. 48.

It is found however on numerical investigation that though a contact with a rather low transparency factor (a bad contact) can give an unsymmetrical current-voltage curve and therefore can rectify an alternating current to some extent, such a contact has only a slight rectification which varies with the voltage in a manner entirely different to the large and important rectifying effects which are actually observed for suitable contacts.

It is now realized that all contacts which rectify strongly are contacts between a semi-conductor and a metal. More than this, a good rectifying contact cannot be made by just pressing together a metal and a good semi-conductor. It is essential that the surface layer of the semi-conductor should be a very bad conductor, almost an insulator. This has been established by showing that if the surface layer is etched away rectification ceases. A typical rectifying contact therefore consists of a metal separated by a very thin, almost insulating, layer from a good semi-conductor. The best known and best understood example is the copper oxide rectifier. This is made by oxidizing one face of a block of metallic copper to an appreciable depth. The oxidation converts the copper to Cu_2O which is almost pure next to the mother copper but oxygen rich elsewhere. The pure Cu_2O forms the almost insulating layer and the $Cu_2O + O$ a good impurity semi-conductor to back it up. The depth of the essential almost insulating layer is estimated at

$\eta 10^{-6}$ cm. by capacity measurements, η being the dielectric constant, or at 10^{-5} cm. by the depth of etching necessary to remove the effect. It is probable that η is about 10. Such plates are used pressed together in a pile, and possess another contact between Cu_2O+O and metallic copper, where the back of the next plate is pressed against the oxidized surface. This is a contact between a metal and a good semi-conductor, and does not rectify appreciably. The direction of easy flow of the current is for electrons to pass from the metal to the semi-conductor at the intimate contact where the insulating layer is present. It remains to examine whether these effects can be accounted for by the theory.

§ **11·91.** *The current-voltage relationship for an idealized metal-semi-conductor contact.* We shall confine attention to the exchange of free electrons between the metal and semi-conductor ignoring the contribution of the free holes if any. The equilibrium state is given sufficiently nearly by adjusting the top of the levels occupied in the metal at low temperatures midway between the bottom of the empty band and the top of the full band in the semi-conductor as shown in Fig. 49. There may also be a potential hill between these levels which belong to the interiors of the two substances.

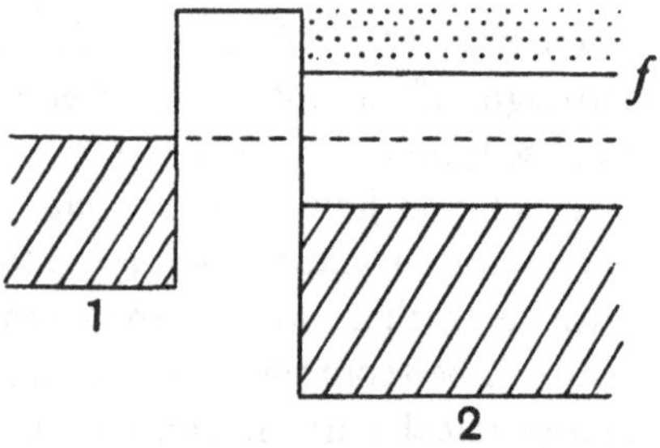

Fig. 49. A metal-semi-conductor contact in equilibrium.

The electrons that can penetrate the barrier in either direction lie above f and are distributed classically. The number in the metal incident on the barrier per unit area per unit time in the range $\zeta, \zeta+d\zeta$ for the energy of their motion normal to the barrier is by (1248), in this high energy region,

$$N_1(\zeta)\,d\zeta = \frac{4\pi mkT}{h^3}\,e^{-(\zeta-\eta^*)/kT}.$$

Since exactly the same number pass in the reverse direction, we must also have for the semi-conductor

$$N_2(\zeta)\,d\zeta = \frac{4\pi mkT}{h^3}\,e^{-(\zeta-\eta^*)/kT} \quad (\zeta > \eta^* + \tfrac{1}{2}\Delta E_1).$$

It is easily verified directly from the properties of the semi-conductor that this result holds to the present order of approximation. The total current passing $1 \rightarrow 2$ per unit area is therefore

$$-\frac{4\pi m|\epsilon|kT}{h^3}\int_{\eta^*+\frac{1}{2}\Delta E_1}^{\infty} D(\zeta)\,e^{-(\zeta-\eta^*)/kT}\,d\zeta,$$

where $D(\zeta)$ is the transparency factor of the barrier. It is convenient to change the energy zero to the level f in the semi-conductor, so that ζ then becomes the kinetic energy of the electrons in the semi-conductor. The current density $1 \rightarrow 2$ is then

$$-\frac{4\pi m|\epsilon|kT}{h^3} e^{-\frac{1}{2}\Delta E_1/kT} \int_0^\infty D(\zeta)\, e^{-\zeta/kT}\, d\zeta. \qquad \text{......(1249)}$$

The same current flows in the reverse direction.

Now suppose that a potential difference V is applied across the contact in such a manner that the electron levels in the metal are *raised* relative to those of the semi-conductor by the amount $|\epsilon|V$. The transparency factor for given ζ will now be altered to some extent by change of shape of the barrier. Let us therefore denote it by $D(\zeta, V)$. Then the current density flowing $1 \rightarrow 2$ becomes

$$-\frac{4\pi m|\epsilon|kT}{h^3} e^{-\frac{1}{2}\Delta E_1/kT} \int_0^\infty D(\zeta, V)\, e^{-(\zeta - |\epsilon|V)/kT}\, d\zeta,$$

and the reverse current $2 \rightarrow 1$

$$-\frac{4\pi m|\epsilon|kT}{h^3} e^{-\frac{1}{2}\Delta E_1/kT} \int_0^\infty D(\zeta, V)\, e^{-\zeta/kT}\, d\zeta.$$

The extra factor $e^{|\epsilon|V/kT}$ in the former arises because for given ζ we now select the electron in the metal from a different level in the Maxwell distribution, so that the supply at all levels is multiplied by this constant factor. The net current density flowing $1 \rightarrow 2$ is therefore

$$-\frac{4\pi m|\epsilon|k^2T^2}{h^3} e^{-\frac{1}{2}\Delta E_1/kT} (e^{|\epsilon|V/kT} - 1)\, \overline{D}(V), \qquad \text{......(1250)}$$

where
$$\overline{D}(V) = \int_0^\infty D(xkT, V)\, e^{-x}\, dx. \qquad \text{......(1251)}$$

By considering a simple square hump it is at once clear that $\overline{D}(V)$ decreases when V increases and if the hump has a low transparency this increase will be very marked. The formulae (1250) and (1251) hold for values of V of either sign, and for a range of the order of $\frac{1}{2}\Delta E_1$.

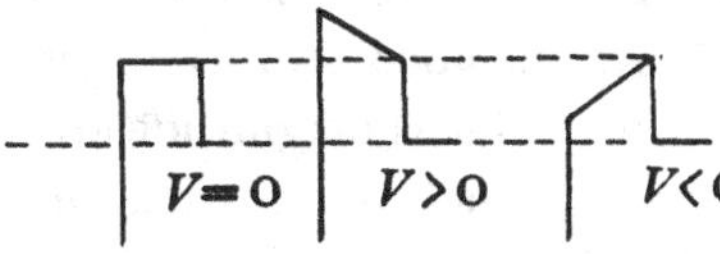

Fig. 50. Illustrating the change of transparency factor.

§11·92. *A barrier of high transparency.* If the barrier between the metal and semi-conductor is unimportant so that $D(\zeta) \simeq 1$, then $\overline{D} \simeq 1$ and the current-voltage relationship for the contact is

$$J(V) = -\frac{4\pi m|\epsilon|k^2T^2}{h^3} e^{-\frac{1}{2}\Delta E_1/kT} (e^{|\epsilon|V/kT} - 1),$$

$$= J_s(e^{|\epsilon|V/kT} - 1). \qquad \text{......(1252)}$$

This has exactly the right general form to represent a rectifying contact. When $V > 0$ the current increases rapidly with V. When $V < 0$ the current tends to saturate at the fixed value J_s no matter how large V may be. The formulae are however only valid if V is of the order $\frac{1}{2}\Delta E_1$ or less. For larger values of V there is still strong rectification, but not given by (1252).

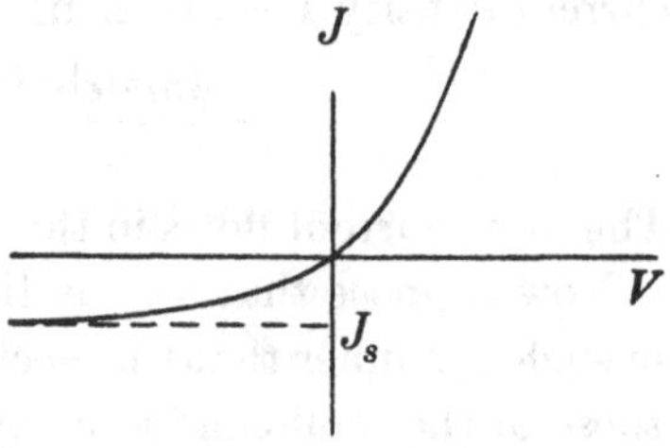

Fig. 51. The current-voltage relationship for an ideal rectifying contact.

The formulae can only apply to actually observed rectifications if the value of J_s is of the correct order. If the J_s of these formulae were very much larger than the actual current densities rectified, then the voltage required to carry the actual currents would be small ($|\epsilon| V \ll kT$) and there would be no rectification. On examining (1252) we see that J_s is the thermionic emission from a metal of work function $\frac{1}{2}\Delta E_1$ at temperature T. Therefore

$$J_s = 120T^2 e^{-\frac{1}{2}\Delta E_1/kT} \text{ amp./cm.}^2. \qquad \text{......(1253)}$$

At room temperature (300° K.) this gives a value $10^{7-8\cdot4\Delta E_1}$ amp./cm.². Thus a value of ΔE_1 equal to 1 or slightly greater will give a value of J_s in reasonable accord with the facts. The rectifying power rapidly weakens as the temperature is raised in agreement with observation.

If we may interpret observed rectifications (at least approximately) in this very simple way, the essential nature of the very highly insulating layer between the metal and the good semi-conductor is at once accounted for. A value of $\Delta E_1 = 1$ represents an almost perfect insulator at ordinary temperatures. Direct contact between a metal and a good semi-conductor ($\Delta E_1 \simeq 0\cdot4$) would not rectify at all. The rectifying contact is between the metal and the very bad semi-conductor which presumably shades gradually off into the good semi-conductor at perhaps 10^{-5} cm. away. When current densities of this magnitude are required to flow in a semi-conductor, that is, current densities requiring all the electrons to move in one direction with velocities comparable to their velocities of thermal agitation, the ordinary conductivity theory no longer applies. None the less one can use the usual classical formula for κ and (1130) for n to estimate the potential gradient in the semi-conductor required to carry the current. One finds that, if a voltage V is applied between the metal and the good semi-conductor with 10^{-5} cm. of bad semi-conductor between, then somewhat the larger part of V must be used up in the semi-conductor. This is consistent with the good rectification observed over a range of about 5 volts, only about $\frac{1}{2}$–1 volt being required across the actual contact according to this version of the theory. In putting forward this explanation however we have ignored the "contact" between the bad and the good semi-conductor which might rectify in the reverse direction. We shall discuss this point more closely in § 11·94.

§ **11·93.** *A barrier of low transparency.* If D is small for small values of ζ, there are two cases to consider. In equation (1251) $D(0)$ may be so small that the main contribution to $\overline{D}$ does not come from values of x nearly zero, but from values of x much higher, corresponding perhaps to values of ζ near the top of the barrier. In this case all rectification is at once lost. Rectification occurs only because the effective electron distributions are unsymmetrically limited on the two sides (by the forbidden band of the semi-conductor). If the limit is imposed by the barrier it is necessarily imposed symmetrically. We shall therefore assume that $D(0)$ though small is still large enough for the dominant contributions to $\overline{D}$ to come from the neighbourhood of $\zeta = 0$.

For a hump of the shape shown in Fig. 49 the important factor in $D(\zeta, 0)$ is

$$\exp[-2\kappa l(H-\zeta)^{\frac{1}{2}}] \quad (\kappa^2 = 8\pi^2 m/h^2),$$

where l is the breadth and H the height of the hump. This is the appropriate value of the more general transparency factor

$$\exp\left[-2\kappa \int_0^l (U-\zeta)^{\frac{1}{2}}\, dx\right].$$

The transparency factor in $D(\zeta, V)$ is therefore

$$\exp\left[-2\kappa \int_0^l (H + |\epsilon| Vx/l - \zeta)^{\frac{1}{2}}\, dx\right].$$

This contains approximately the extra factor

$$\exp\left[-\frac{\kappa l |\epsilon| V}{2(H-\zeta)^{\frac{1}{2}}}\right],$$

and the dominant terms in $\overline{D}$ are of the form

$$\overline{D_0} \exp\left[-\frac{\kappa l |\epsilon| V}{2H^{\frac{1}{2}}}\right].$$

The current-voltage relationship is now

$$J = J_s(e^{|\epsilon| V/kT} - 1)\, e^{-\alpha V}. \qquad \text{......(1254)}$$

Provided $\alpha \ll |\epsilon|/kT$ this is still a strongly rectified current, though less completely rectified than (1252), since the reverse current does not really saturate but eventually increases without limit. A formula such as (1254) gives an even better representation of observed currents than (1252), using rather small values of α.

§ **11·94.** *General discussion of rectification.* In the experimental study of rectification the current-voltage relationship observed is the current as a function of the voltage applied between two conductors, one the metal and the other the good semi-conductor. We have seen that the theory demands that there should be at most a trivial rectification at any contact between metals, or at a good contact between a metal and a good semi-conductor, or at any really bad contact, where the usual current carrying electrons in the semi-conductor can no longer penetrate through the potential barrier. There is a strong rectification of just the type observed at a contact between a metal and a very bad semi-conductor, provided that the contact is a good or fairly good one. All this is in excellent apparent agreement with the observations, but the difficulty remains that this explanation really requires us to consider two contacts: (1) metal/bad semi-conductor, and (2) bad semi-conductor/good semi-conductor. It is easy to see that if we use the same simple models that we have used so far there will be a rectifying effect at the second contact exactly similar and in a reverse direction to the rectifying effect at the second contact, and it is not immediately possible to assert that the whole rectification will not be thereby wiped out. Many solutions have been proposed to deal with this difficulty none of them really convincing, and a much more elaborate discussion may be required of the current carrying mechanism in a semi-conductor when the current is nearly saturated. It remains perhaps safe to believe that the general explanation of rectification here offered by the theory of a single good or fairly good contact is essentially correct and provides the required explanation of observed contact rectification, but that the precise manner in which this simple theory is to be applied to the observations is as yet undetermined.*

* In thus concluding this chapter it is proper to call the attention of the reader to the fact that many of the views adopted here both on semi-conductors and thermionic emission are still matters of controversy. Reference should be made to Joffé, *Actualités Sci. et Indus.*, No. 202, "Semi-conducteurs électronique" (1935) and Schottky, *Handbuch d. exper. Physik*, vol. 13, part 2.

CHAPTER XII

ELECTRIC AND MAGNETIC SUSCEPTIBILITIES. FERROMAGNETISM

§ **12·1**. *The electromagnetic theory of susceptibilities.* The theory of the dielectric constant of a gas (or other material body) or of its (dia-) paramagnetic susceptibility falls necessarily into two parts. In the first, which belongs to electromagnetic theory, we trace the connection between the external field applied to the medium, and its electric or magnetic polarization, determining incidentally how the polarization of the surrounding matter affects the field acting on any particular element. In the second, which alone is statistical, we derive the connection between the electric (magnetic) force and the polarization when the material medium is built up of molecules of given type or types, distributed in equilibrium in the field of force acting on them.

It is, strictly, beyond the scope of this monograph to consider the details of the electromagnetic part of the theory, nor should it be necessary to do so. None the less this classical part of the theory* is not so easily freed from misconceptions and fallacies, as is evident from recent discussions.† An analysis by Darwin‡ has made the conflict particularly clear. It seems worth while therefore to start here with a short account of the electromagnetic part of the theory, largely based on Darwin's work.§

The primary definitions of electric (magnetic) force **E** (**H**) refer to a vacuum; they are vectors which *in vacuo* satisfy the relations

$$\int_C \mathbf{E}\,.\,\mathbf{ds} = 0, \quad \int_C \mathbf{H}\,.\,\mathbf{ds} = 0, \qquad \text{......(1255)}$$

$$\int_S \mathbf{E}\,.\,\mathbf{d\sigma} = 4\pi \int_V \rho\, d\omega, \quad \int_S \mathbf{H}\,.\,\mathbf{d\sigma} = 0; \qquad \text{......(1256)}$$

C is any closed circuit, S is any closed surface enclosing the volume V, and ρ is the electric charge density. The last integral vanishes since we experience no accumulations of magnetism.

It is well recognized to be convenient to extend these equations to circuits C and surfaces S which are not necessarily *in vacuo*.

To extend (1255) to the interior of matter we surround the path C by a narrow pipe and remove the matter entirely from this pipe. In this slightly

* For an exposition see Lorentz, *The theory of electrons*, chap. IV, or Livens, *The theory of electricity*, § 237, ed. 1.

† For example, Hartree, *Proc. Camb. Phil. Soc.* vol. 25, p. 97 (1929), vol. 27, p. 143 (1932); *Nature*, vol. 132, p. 929 (1933); Kronig, *Proc. Sect. Sci. Amsterdam*, vol. 35, p. 7 (1933).

‡ Darwin, (I) *Proc. Roy. Soc.* A, vol. 146, p. 17 (1934); (II) vol. 151, p. 572 (1935).

§ See also Debye, *Marx's Handbuch der Radiologie*, vol. 6, p 597 (1925).

modified world equations (1255) remain true, provided we use the forces $\mathbf{E}'$, $\mathbf{H}'$ *measured in this world.* It is only the components of the forces along the pipe that are relevant to (1255). It is easy to show that for a sufficiently small pipe these components are independent of the size and shape of the pipe, and are themselves the components of vectors $\mathbf{E}'$ and $\mathbf{H}'$ which may be determined by measurement in a pipe in a suitably chosen direction. These vectors may be conveniently referred to as the electric (magnetic) *pipe-force* respectively. In virtue of (1255) they are the gradients of potentials and are frequently referred to simply as electric (magnetic) force as if they were such forces *in vacuo.*

To extend (1256) to the interior of matter we make a different modification by enclosing S between two parallel surfaces and removing all the matter between them, so that S lies in a thin crack-shaped cavity. Equations (1256) remain true for forces $\mathbf{E}''$, $\mathbf{H}''$ measured in this modified world. Only the components of these forces normal to the crack-cavity are relevant to (1256), and it is again easy to show that for a sufficiently thin crack these normal components are the components of vectors $\mathbf{E}''$, $\mathbf{H}''$, which can be measured in suitably orientated cracks. These vectors may be called the electric (magnetic) *crack-forces.* Since they are the vectors that conserve equations (1256), they are more commonly called the electric (magnetic) induction, and denoted by the symbols $\mathbf{D}(\mathbf{B})$.

The need to distinguish so carefully between the pipe-force and the crack-force is due of course to the polarization of the material, and the difference between the two forces to the different portions of polarized material removed in the two cases. In an isotropic medium, which is all we need consider at present, the vectors $\mathbf{E}'$, $\mathbf{D}$ and $\mathbf{P}$ (electric) are all parallel as also are $\mathbf{H}'$, $\mathbf{B}$ and $\mathbf{P}$ (magnetic), this last being usually denoted by $\mathbf{I}$ and called the (intensity of) magnetization.

By filling up the distant parts of the pipe until only a crack-shaped cavity is left, much thinner than the width of the pipe, about the point at which $\mathbf{D}(\mathbf{B})$ is to be determined, it is easy to establish the fundamental relation between $\mathbf{E}'$, $\mathbf{D}$ and $\mathbf{P}$ or between $\mathbf{H}'$, $\mathbf{B}$ and $\mathbf{I}$. The material put back is polarized to a dipole strength $\mathbf{P}(\mathbf{I})$ per unit volume and the force exerted by these dipoles in the remaining cavity must be the difference between the pipe- and crack-forces. It follows at once that

$$\mathbf{D}=\mathbf{E}'+4\pi\mathbf{P}, \quad \mathbf{B}=\mathbf{H}'+4\pi\mathbf{I}. \qquad \text{......(1257)}$$

These equations are usually summarized by the introduction of coefficients called the *susceptibility* and the *dielectric constant* (*permeability*). The susceptibilities κ are defined by

$$\mathbf{P}=\kappa_e\mathbf{E}', \quad \mathbf{I}=\kappa_m\mathbf{H}'; \qquad \text{......(1258)}$$

the dielectric constant (permeability) by

$$\mathbf{D}=\delta_e\mathbf{E}', \quad \mathbf{B}=\delta_m\mathbf{H}' \qquad (\delta=1+4\pi\kappa). \qquad \text{......(1259)}$$

The susceptibility thus defined is the volume susceptibility—the dipole strength per unit volume induced by unit pipe-force strength. It must be noted that it is the pipe-force in (1258) and (1259) and never E_{ext} or H_{ext} though sometimes their difference is insignificant. The difference is due of course to surface charges on the dielectric, and in changing from E_{ext} to E' for example we have already made a great allowance for the effect of the polarization of the medium on the force acting at any point of it. This important difference is illustrated in detail by the calculations of § 12·12.

These definitions are quite general for isotropic media and easily generalized for crystalline ones. Experiments determine directly δ and therefore κ. We have still to examine how κ is to be correlated with quantities that can be deduced from atomic theory.

§ **12·11.** *The polarizing force* **F**. If our medium is composed of atoms of known properties, we can at once calculate its polarization if we know the average polarizing force **F**, electric or magnetic, acting on each atom. This force however is not necessarily either the crack-force or the pipe-force, for the force required is the force acting on the atoms in the actual material when no matter has been removed. Before we can determine P in terms of the pipe-force—that is to say, calculate κ—we must determine the polarizing force in terms of the pipe-force and the polarization. This can be done by starting with a pipe-shaped cavity parallel to the vectors **E**′ and **P**, in which therefore **E**′ is the total force acting, and restoring the polarized matter which has been removed from this pipe. The effects will be proportional to **P** so that we shall have

$$\mathbf{F}=\mathbf{E}'+\nu\mathbf{P}. \qquad \text{......(1260)}$$

The contribution of the restored matter can be considered in two parts: (i) the contribution from the matter lying inside a small sphere round the point in question and (ii) the contribution of the rest, so that $\nu=\nu_1+\nu_2$.

The factor ν_2 is easily calculated. The polarized matter in the distant parts of the pipe exerts outside itself the same forces as a charge density P_n over its surface, where P_n is the normal component of **P**. This is zero on the walls of the pipe, P over the distant ends whose effect at the origin is negligible and $P\cos\theta$ over the surface of the small sphere. It is easily shown that this surface charge leads to a force $\frac{4}{3}\pi\mathbf{P}$ inside the sphere in the same direction as **P**. Thus $\nu_2=\frac{4}{3}\pi$.

The essential difficulty of the problem lies in the evaluation of ν_1, the contribution of the matter in the immediate neighbourhood of the point or atom at which **F** is required. There is in fact no general solution, the correct value of ν_1 depending on the nature of the polarized matter itself. The difficulty is of the nature of a change in the order of two limiting processes. Different results are thus obtained which properly correspond to different physical properties of the material. The principal results are as follows:

Lorentz's Lemma. If the material consists of isolated polarizable atoms, in disordered array as in a gas, or in a regular array with cubic or tetrahedral symmetry about the selected atom, then*

$$\nu_1 = 0.$$

The potential Ω at x, y, z due to a dipole at $(l,m,n)a$ of strength α pointing in the direction λ, μ, ν is given by

$$\Omega = -\frac{\alpha[\lambda(la-x)+\mu(ma-y)+\nu(na-z)]}{\{(la-x)^2+(ma-y)^2+(na-z)^2\}^{\frac{3}{2}}}. \qquad \text{......(1261)}$$

By a theorem due to Gauss the average value of the potential arising from any distribution of poles taken over any sphere which contains none of the poles within or on itself is equal to the value of the potential at the centre of the sphere. The same theorem is true for the average value over the sphere of the derivatives of Ω, the electric (magnetic) force, which can always be regarded as the potential of a more elaborate set of poles. Thus if we are concerned with the average value of the potential or electric (magnetic) force due to a set of dipoles taken over an atom with spherical symmetry, so that any weighting factors are functions of r only, then the average value required is the value at the centre $x=y=z=0$. If the atoms are small compared with the average distance of the other dipoles, this central value is of course the correct value without any other restriction. But even when the atom is not small, this central value will generally be good enough so that we have only to consider the average value of

$$\Omega_0 = -\alpha(\lambda l+\mu m+\nu n)/a^2, \qquad \text{......(1262)}$$

or of its derivatives such as

$$-\frac{\partial\Omega_0}{\partial x} = \frac{\alpha\lambda(3l^2-1)}{a^3}. \qquad \text{......(1263)}$$

Since for completely irregular distributions or for lattices with tetrahedral or cubic symmetry about the origin $\overline{l}=\overline{m}=\overline{n}=0$, $3\overline{l^2}=1$, ... for any given values of a, α, λ, μ, ν, the average values of Ω_0, $\partial\Omega_0/\partial x$, ... all vanish. The lemma is therefore established.

Lorentz's lemma depends on the possibility of drawing a sphere outside which the matter may be treated as a continuous medium, and inside which it consists of effectively discrete atoms. If the polarizable elements of the matter are not really discrete atoms but effectively free electrons among

* This is a convenient title for the lemma, but it is far from correct to say that Lorentz enunciated it in a form so complete. See his *Theory of Electrons* (1916), p. 138, and Notes 54, 55.

positive ions as in a metal or the Kennelly-Heaviside layer we might still hope to apply the lemma for a *periodic* electric force, but the proof breaks down and the result in fact is wrong. An argument, which is however quite unreliable, can be constructed suggesting this failure: we can equally legitimately regard the polarizable material as smoothed out into positive and negative continua before the sphere is drawn. The polarization P inside the sphere then produces the effect of a charge $-P\cos\theta$ over the surface of the sphere which exactly neutralizes the $P\cos\theta$ above. It will be seen that we have now inverted the order of the two limit operations of excluding the small sphere and smoothing out the actual matter into a continuum. Thus in such a case, so one argues, $\nu_1 = -\nu_2$ and $\nu = 0$. It appears therefore that we may have one or other of the two formulae

$$\mathbf{F} = \mathbf{E}' + \tfrac{4}{3}\pi\mathbf{P}, \qquad \text{......(1264)}$$

$$\mathbf{F} = \mathbf{E}' \qquad \text{......(1265)}$$

in extreme cases, and therefore, presumably, intermediate forms as well.

The last remarks make no claim to rigour—they suffice merely to show that a much deeper discussion is required, and that the result must be expected to depend on the exact nature of the atoms or ions of the material.

The second part of the problem will be concerned with evaluating P as a function of F from atomic and statistical theory. Without anticipating details we may record here that the approximation

$$P = \gamma F, \qquad \text{......(1266)}$$

where γ is a constant, is usually sufficiently good. In that case when F is given by (1264)

$$\gamma = \frac{3}{4\pi}\frac{\delta - 1}{\delta + 2} \quad \text{(Lorentz's formula)}, \qquad \text{......(1267)}$$

and when F is given by (1265)

$$\gamma = \frac{1}{4\pi}(\delta - 1) \quad \text{(Sellmeyer's formula)}. \qquad \text{......(1268)}$$

Both formulae may be either electric or magnetic.

§**12·12**. *An idealized dielectric.* We shall gain further insight into the problem by considering a dielectric composed of n "atoms" per unit volume each of which is a rigid sphere of radius a uniformly polarized to such a degree that each atom has a dipole moment p. All these dipoles are parallel to each other and to the pipe-force E'. We shall consider an infinite slab of such a dielectric bounded by plane faces normal to the z-axis at $z = 0$ and $z = b$; E' is parallel to the z-axis. The smoothed out

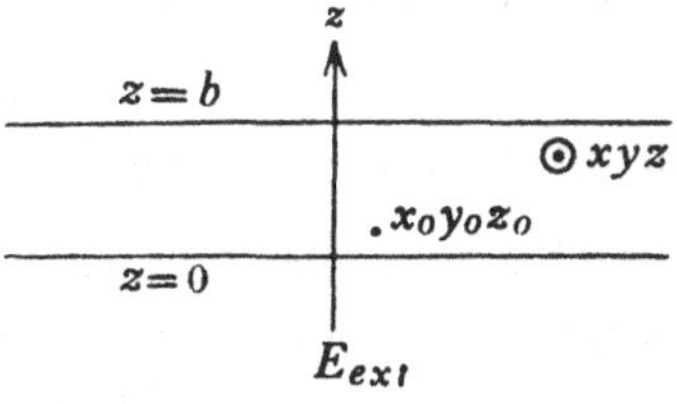

Fig. 52. A slab of dielectric.

polarization of the dielectric is of course given by $P = np$. The dielectric is subject to an external force E_{ext} parallel to the z-axis to which its polarization is due. The actual electric force at any point among these atoms fluctuates wildly and we require certain average values for F which will naturally depend on the way the average is taken.

Let us start by calculating the z-component G_z of the force acting at a point x_0, y_0, z_0 due to the slab of dipole atoms. If this point is inside an atom, then the force due to this atom itself is to be excluded. In any event therefore the point x_0, y_0, z_0 may be regarded as surrounded by a small sphere containing no atoms.

An "atom" whose centre is at x, y, z exerts a force at x_0, y_0, z_0 whose z-component is

$$\frac{3(z_0 - z)^2 - r^2}{r^5} p.$$

Hence
$$G_z(x_0, y_0, z_0) = \Sigma_{x,y,z} \frac{3(z_0 - z)^2 - r^2}{r^5} p.$$

This summation must be turned into an integration. For the more distant parts of the slab this presents no difficulty. The region of integration is bounded by the plane faces of the slab and a small sphere round x_0, y_0, z_0 of radius A. The smoothing implied by integration is illegitimate near the sphere A, but Lorentz's lemma applies here and tells us that no contribution arises from the atoms between the sphere A and a larger sphere B at least when we take the average value of G_z over all points x_0, y_0, z_0. Thus

$$\overline{G_z}(x_0, y_0, z_0) = \int_{\text{slab}} np \frac{\partial}{\partial z}\left(\frac{z_0 - z}{r^3}\right) d\omega, \qquad \text{......(1269)}$$

over the specified range of integration. Evaluation presents no difficulty, and we find that if x_0, y_0, z_0 is inside the slab

$$\overline{G_z}(x_0, y_0, z_0) = -\tfrac{8}{3}\pi np, \qquad \text{......(1270)}$$

while outside the slab $\overline{G_z}(x_0, y_0, z_0) = 0$. Other components must by symmetry average to zero. To this we must add the externally applied field (due to distant bodies) so that the average force at any point x_0, y_0, z_0 inside the slab is

$$E_{\text{ext}} - \tfrac{8}{3}\pi P.$$

But since by their definitions and ordinary properties

$$E' + 4\pi P = D_{\text{int}} = E_{\text{ext}},$$

the average internal field is
$$E' + \tfrac{4}{3}\pi P,$$

agreeing with our previous result. It is to be remembered that if x_0, y_0, z_0 happens to lie inside any atom, the internal field of that atom has been deliberately omitted. This omission is correct if we require the average field tending to polarize a particular discrete atom.

Suppose however we do a similar calculation for the potential at x_0, y_0, z_0 due to the dipoles in the slab. Then

$$V(x_0,y_0,z_0)=\int_{\text{slab}} np\frac{\partial}{\partial z}\left(\frac{1}{r}\right)d\omega. \qquad \text{......(1271)}$$

There is now no convergence difficulty near $r=0$ but we still have to use Lorentz's lemma to justify the smoothing for small values of r. Thus the value we find is again an average value. Evaluation of the integral is easy and we obtain

$$\left.\begin{aligned}\overline{V}(x_0,y_0,z_0) &= 4\pi np(z_0-\tfrac{1}{2}b) & (0<z_0<b)\\ &= 2\pi npb & (z_0>b)\\ &= -2\pi npb & (z_0<0)\end{aligned}\right\}. \qquad \text{......(1272)}$$

Thus, calculating in this way, we find that there is on the average a force $-4\pi P$ inside the slab due to its own dipoles, so that the average force inside the slab which can be derived from the potential is just the pipe-force E'.

This relationship between the potential and the pipe-force is of course essential to satisfy (1255) and it is important to examine the difference between these two averages. We have shown in fact that the average force *along any path* doing work for example on a test electron moving along that path is E', while the average force *at any point* (if within an atom omitting the internal field of that atom) is on the other hand $E'+\frac{4}{3}\pi P$. The necessary reconciliation is brought about by recalling that *on the average* any path of length l will contain a length $\frac{4}{3}\pi a^3 nl$ inside atoms since the atoms occupy the fraction $\frac{4}{3}\pi a^3 n$ of all space. Inside any one of the atoms there is a constant force $-p/a^3$ along the z-axis due to its own polarization, so that an average force $-\frac{4}{3}\pi P$ due to the insides of the atoms has been omitted if we use the point average force along a path. If the atoms are very small a straight path may be drawn to miss them nearly all; but if the average path is fairly taken it will cut a few and their effects will just be large enough to counterbalance their rarity. Moreover the average force along a path specially selected to avoid all the atoms will still be E' because such a path *ex hypothesi* must spend too much of its length in the equatorial planes and too little near the axes of the dipoles for the point average force to apply. There is no reason to expect equality between a space average and the average along a specially selected path in that space.

We conclude therefore that in such a dielectric the average polarizing force F acting on an atom of the dielectric is $E'+\frac{4}{3}\pi P$, but that the average force acting on an electron moving freely along any path is E'. Whether this average strictly applies to the actual path of an electron requires still further investigation.

§ **12·13.** *The magnetic case. A similar model.* The results of § 12·12 apply equally to the magnetic case so long as the atoms may be represented by magnetic dipoles. The average polarizing force acting on any atom of the medium is $H' + \frac{4}{3}\pi I$ and the average magnetic force along a path is the pipe-force H'. Magnetic dipoles however are very far from an adequate representation of actual magnetized atoms in which the magnetic elements are almost certainly to be regarded rather as small electric currents due to circulating electrons or, if it is preferred, to atomic states with angular momentum. We shall therefore consider an alternative model in which the uniformly polarized spheres are replaced by spherical current sheets so arranged that they give precisely the same external fields.* This is achieved if the spheres carry currents in planes normal to the z-axis of strength

$$(p/\tfrac{4}{3}\pi a^3)\,dh$$

in any element of surface of height dh. With any such model the result is unaltered that the average polarizing force acting on a single atom is $H' + \frac{4}{3}\pi I$ so long as Lorentz's lemma applies; but for the average force along a path the result is entirely different. There is now no reversed force inside each atom, but instead a force in the same direction as P. This is possible because the magnetic potential of a current is not one-valued. Now if a test pole is taken from a great distance away near the axis of a small current circuit to a great distance away in the other direction of the axis without threading the circuit no work is done, but if it threads the circuit the work done is $4\pi i$ per unit pole strength, where i is the current. Since the average length of any path which lies inside the atomic spheres is $\frac{4}{3}\pi a^3 nl$ as before, the total current strength threaded on the average on such a path in the direction of the field is pnl, and the extra work done $4\pi pnl$. Thus on a path which penetrates the atoms to an average amount there is an average extra force $4\pi P$ acting over and above the average force on a similar path which just goes round all the atoms. The same extra force will act to deflect a moving electron so long as it freely penetrates the atoms. On a path which just goes round all the atoms the forces are exactly the same as those for atoms with rigid dipoles, and on such a path for such atoms the average force is the same as for a straight path which penetrates the atoms where necessary to do so, since now the magnetic potential of each atom is one-valued. We have already proved that in this case the average force along the path is the pipe-force, H'. Hence for magnetic atoms with current circuits freely penetrable by magnetic poles or moving electrons the average force along the path is the crack-force $B = H' + 4\pi P$. This result presumably applies to the deflection of the very fast electrons of the cosmic rays penetrating magnetized iron.

* Darwin, *loc. cit.* (II), has given a much more elegant and general discussion leading to the same result which is independent of the distribution of currents in the model atom. The currents need only be assumed to vanish outside some definite atomic radius.

§ **12·14.** *An alternative treatment.* In view of the subtlety of the problem it is perhaps worth while to give an independent discussion, by considering a sphere of the material in question small compared with the wave length of the electromagnetic field (if this field is time variable) but large enough to contain many atoms, and calculating the electric dipole moment excited in it by an external electric field of strength $\mathbf{F}\sin\nu t$. This field penetrates right through the atoms of the sphere without modification being due to distant bodies. When a sphere of dielectric constant δ_e is placed in the presence of an electrostatic field of strength $\mathbf{F}$, the dipole moment induced is

$$\frac{\delta_e-1}{\delta_e+2}a^3\mathbf{F},$$

and the field of this dipole represents the sole external effect of the sphere. A similar result holds for a small sphere (of permeability unity) in a time variable field so long as retardation effects can be neglected. The external field of the sphere is that of the dipole

$$\frac{\delta_e-1}{\delta_e+2}a^3\mathbf{F}\sin\nu t. \qquad \text{......(1273)}$$

We can however also calculate the effective dipole moment directly. By equating this to (1273) we evaluate the dielectric constant in terms of atomic properties and thereby can decide between the formulae (1264) and (1265), though that analysis of the internal field is irrelevant to this method of attack.

Let us consider first in this manner material composed of discrete atoms, each of which may be thought of as containing a number of virtual oscillating electrons. In this way quantum requirements can be taken account of, and the reactions of the atoms to the field treated classically. Let us start by assuming that the virtual electrons of any one type oscillate with the same amplitude and phase in each atom. It is necessary to show that each electron is then unaffected by the displacements of all the others; that being so it is affected only by the external field and the initial assumption is correct. But this independence follows at once from the theorem that in a spherical hollow in a sphere of uniformly polarized material the polarization exerts no force. The smoothing and averaging necessary to apply this theorem to the actual material under discussion have been shown to be legitimate by the arguments of the preceding sections.

Let $\boldsymbol{\xi}_r$, ν_r, ϵ_r be the displacement, frequency and effective charge of the virtual electron of type r. Then

$$m\ddot{\boldsymbol{\xi}}_r=-m\nu_r^2\boldsymbol{\xi}_r+\epsilon_r\mathbf{F}\sin\nu t, \qquad \text{......(1274)}$$

and therefore

$$\boldsymbol{\xi}_r=\frac{\epsilon_r\mathbf{F}\sin\nu t}{m(\nu_r^2-\nu^2)}. \qquad \text{......(1274·1)}$$

If the sphere contains n atoms per unit volume and we sum (1274) over the various types of virtual electron and over all atoms, we see that the sphere has a dipole moment

$$\tfrac{4}{3}\pi a^3 n\,\Sigma_r \frac{\epsilon_r^2 \mathbf{F}\sin\nu t}{m(\nu_r^2-\nu^2)}.$$

On comparing this with (1273) we see that

$$\frac{3}{4\pi}\frac{\delta_e-1}{\delta_e+2}=n\,\Sigma_r\frac{\epsilon_r^2}{m(\nu_r^2-\nu^2)}, \qquad \text{......(1275)}$$

agreeing with Lorentz's formula (1267).

Now consider material consisting of N free electrons ($N=\tfrac{4}{3}\pi a^3 n$) immersed in a medium with a uniform continuous positive charge, just neutralizing the charges of the electrons. The total charge of the positive medium is $N|\epsilon|$ so that the potential at an internal point is

$$N|\epsilon|\frac{3a^2-r^2}{2a^3}. \qquad \text{......(1276)}$$

Electrons may emerge temporarily from the sphere into a region where the potential is strictly not given by (1276), but for convenience (1276) may still be used without affecting our discussion. The equations of motion of the fth electron are

$$m\ddot{\mathbf{x}}_f=-\frac{N\epsilon^2}{a^3}\mathbf{x}_f-\frac{\partial}{\partial\mathbf{x}_f}\Sigma_{g\neq f}\frac{\epsilon^2}{|\mathbf{x}_g-\mathbf{x}_f|}+\epsilon\mathbf{F}\sin\nu t,$$

so that

$$m\,\Sigma_f\,\ddot{\mathbf{x}}_f=-\frac{N\epsilon^2}{a^3}\Sigma_f\,\mathbf{x}_f+N\epsilon\mathbf{F}\sin\nu t. \qquad \text{......(1277)}$$

The interactions of the electrons have now disappeared, as they should, because a collision between a pair of electrons does not alter their resultant acceleration, nor therefore the radiation they scatter. From (1277) we find at once that the dipole moment of the sphere, $\epsilon\,\Sigma_f\,\mathbf{x}_f$, is given by

$$\epsilon\,\Sigma_f\,\mathbf{x}_f=\frac{N\epsilon^2\mathbf{F}\sin\nu t}{-m\nu^2+N\epsilon^2/a^3},$$

to be equated to (1273). Therefore

$$\frac{\delta_e-1}{\delta_e+2}a^3=\frac{N\epsilon^2}{-m\nu^2+N\epsilon^2/a^3}.$$

From this it follows at once that

$$\frac{\delta_e-1}{4\pi}=-\frac{3N\epsilon^2}{4\pi m\nu^2 a^3}=-\frac{n\epsilon^2}{m\nu^2}. \qquad \text{......(1278)}$$

This result agrees with Sellmeyer's formula (1268). The term on the right agrees of course with the corresponding term in (1275) when we put $\nu_r=0$.

We have still not solved the really important problem which is that of a material consisting of free electrons and positive ions, which can be taken to be protons without affecting the illustrative value of the discussion. The only convincing treatment is that of Darwin,† but this is too long to give here, especially since we shall not be concerned with such media in the rest of this chapter. Darwin shows that it is sufficient to discuss the problem with fixed protons taking averages over all the electrons with special care of the contributions made by electrons when in close collision with one of the protons. It turns out that *if the energy of the electron in the collision is positive so that the electron is really free of the proton*, the average contribution made by the colliding electron when it is in a small sphere round the proton is to a sufficient approximation the same as if the central proton were absent. Thus only distant protons matter and these may be thought of as smoothed out into a continuum. The whole effect of close collisions being negligible we fall back on the electron-continuum case just discussed and find a result in accord with Sellmeyer's formula. The importance of Darwin's discussion is that it shows that the distinction between free and bound electrons is a vital one, and that it is only when the colliding electron is really free that the effect of the adjacent proton cuts out.

Darwin has extended his discussion to a material which contains both discrete atoms and free electrons, and proved the following theorem:

Darwin's theorem. Form the expression for $3(\delta_e'-1)/(\delta_e'+2)$ *by summation over all types of atom or other system to which Lorentz's formula* (1267) *applies when they are present by themselves. Form the expression for* $\delta_e''-1$ *by summation over the free electrons as if they were present by themselves. Then the dielectric constant* δ_e *for the material as a whole is given by*

$$\delta_e=\delta_e'+\delta_e''-1. \qquad \text{......(1279)}$$

A somewhat similar theorem can be proved for the magnetic case in which the magnetism is due partly to free and partly to bound electrons.‡

§ **12·2.** *The classical theory of the dielectric constant of a gas or liquid of discrete atoms or molecules.*§ We now take up in earnest the statistical part of our problem. According to the foregoing discussion Lorentz's formula will apply to any assembly of discrete molecules. The formula $P=\gamma F$ may be put in the form

$$P=n\theta^*F, \qquad \text{......(1280)}$$

where θ^* is the average dipole moment induced in each of the molecules (n per unit volume) of the substance by the polarizing force F. The parameter θ^* is in general a function of F. Equation (1267) may then be re-written in the form

$$\frac{\delta-1}{\delta+2}=\frac{4\pi}{3}n_r\theta_r^*. \qquad \text{......(1281)}$$

† Darwin, *loc. cit.* (I). ‡ Darwin, *loc. cit.* (II).

§ For the rest of this chapter reference should be made for more detailed information to Van Vleck, *Electric and Magnetic Susceptibilities*, Oxford (1932).

This may be extended at once to molecular mixtures in the form

$$\frac{\delta-1}{\delta+2}=\frac{4\pi}{3}\Sigma n_r\theta_r^*. \qquad \text{......(1282)}$$

The quantity $(\delta-1)/(\delta+2)$ which is thus seen to be additive is called the *refractivity*. We have already used (1282) in Chapter x. This formula applies also to polarization by electromagnetic radiation fields; δ is then usually replaced by μ^2, the square of the refractive index, and the θ_r^* depend on the frequency ν. An example of this for a particular type of molecule was given in (1275). We shall be mainly concerned with static fields here.

In calculating the static θ^* the results will depend on the particular assumptions made as to the structure of the molecules. It was first shown by Debye† that the dielectric properties of many gases and liquids are satisfactorily accounted for, if we assume that the molecules are rigid dipoles of approximately constant electric moment and in addition have ordinary polarizable isotropic electronic structures. On this assumption the classical theory of θ^* proceeds as follows.

Consider as the model molecule a rigid solid of revolution without axial spin, of moment of inertia A free to turn about its centre of mass, with an electric doublet of strength α directed along its axis of symmetry. Let the body be subject to a polarizing force F. Then the Hamiltonian function for the motion of the rigid body is

$$H=\frac{1}{2A}\left(p_\theta{}^2+\frac{1}{\sin^2\theta}p_\phi{}^2\right)-\alpha F\cos\theta.$$

The angle θ is so measured that $\theta=0$ when the positive direction of the dipole points along the field of force F. The partition function for the rotations and orientations of such molecules will therefore be

$$R(T)=\frac{1}{h^2}\iiiint\exp\left[-\frac{1}{kT}\left\{\frac{1}{2A}\left(p_\theta{}^2+\frac{1}{\sin^2\theta}p_\phi{}^2\right)-\alpha F\cos\theta\right\}\right]dp_\theta dp_\phi d\theta d\phi,$$

$$=\frac{2\pi AkT}{h^2}2\pi\int_0^\pi\exp\left(\frac{\alpha F\cos\theta}{kT}\right)\sin\theta\,d\theta, \qquad \text{......(1283)}$$

$$=\frac{8\pi^2AkT}{h^2}\frac{\sinh(\alpha F/kT)}{\alpha F/kT}. \qquad \text{......(1284)}$$

The fraction of molecules with their coordinate θ between θ and $\theta+d\theta$ will therefore be

$$\frac{\exp\left(\frac{\alpha F\cos\theta}{kT}\right)\sin\theta\,d\theta}{\int_0^\pi\exp\left(\frac{\alpha F\cos\theta}{kT}\right)\sin\theta\,d\theta}. \qquad \text{......(1285)}$$

† Debye, *Phys. Zeit.* vol. 13, p. 97 (1912).

The average polarization $\theta^* F$ is the resolved part of α along the direction of the field averaged over all molecules distributed according to (1285), and is therefore given by

$$\theta^* = \frac{\dfrac{\alpha}{F}\displaystyle\int_0^\pi \exp\left(\frac{\alpha F\cos\theta}{kT}\right)\cos\theta\sin\theta\, d\theta}{\displaystyle\int_0^\pi \exp\left(\frac{\alpha F\cos\theta}{kT}\right)\sin\theta\, d\theta},$$

$$= \frac{kT}{F^2}\left[\frac{\alpha F}{kT}\coth\frac{\alpha F}{kT} - 1\right]. \qquad \text{......(1286)†}$$

This formula can of course be obtained by differentiating (1284) as we show in detail in § 12·3. In all or nearly all practical cases $\alpha F/kT$ is very small, so that this reduces to

$$\theta^* = \frac{1}{3}\frac{\alpha^2}{kT}. \qquad \text{......(1287)}$$

The coefficient $\frac{1}{3}$ arises as the average value of $\cos^2\theta$ over a sphere.

Two restrictions in the foregoing account are easily removed. We have first to introduce a term for the polarizability of the electronic structure of the molecule. This obviously might depend on the orientation θ, ϕ, and we should then have the Hamiltonian function

$$H = \frac{1}{2A}\left(p_\theta^2 + \frac{1}{\sin^2\theta}p_\phi^2\right) - \alpha F\cos\theta - \tfrac{1}{2}\beta(\theta,\phi)\, F^2,$$

and the polarization $\alpha\cos\theta + \beta(\theta,\phi)\, F$ instead of $\alpha\cos\theta$. We therefore find for the average polarization

$$\theta^* = \frac{\dfrac{1}{F}\displaystyle\int_0^\pi\int_0^{2\pi} \exp\left(\frac{\alpha F\cos\theta + \frac{1}{2}\beta F^2}{kT}\right)(\alpha\cos\theta + \beta F)\sin\theta\, d\theta\, d\phi}{\displaystyle\int_0^\pi\int_0^{2\pi} \exp\left(\frac{\alpha F\cos\theta + \frac{1}{2}\beta F^2}{kT}\right)\sin\theta\, d\theta\, d\phi}.$$

The leading terms reduce to

$$\theta^* = \frac{1}{3}\frac{\alpha^2}{kT} + \overline{\beta}, \qquad \text{......(1288)}$$

where
$$\overline{\beta} = \frac{1}{4\pi}\int_0^\pi\int_0^{2\pi} \beta\sin\theta\, d\theta\, d\phi.$$

Secondly it is obviously unnecessary to restrict ourselves to a rigid body with an axis of symmetry without axial spin. Any rigid body with moments of inertia A, B, C, spinning in any manner, may be considered, which contains an electric doublet fixed in it and is also of polarizable structure. We shall equally obtain (1288).‡ As an example of such more general calculations consider a non-polarizable general rigid body, with a dipole of strength

† This formula is more familiar in the theory of paramagnetism and is there due to Langevin, *J. de Physique*, ser. 4, vol. 4, p. 678 (1905). See also § 12·5.

‡ Van Vleck, *Phys. Rev.* vol. 30, p. 31 (1927) especially § 6; *Electric and Magnetic Susceptibilities*, p. 37 (1932).

α fixed in any direction relative to the principal dynamical axes, subject to a polarizing force F along the z-axis. For such a system the partition function is

$$R(T)=\frac{1}{h^3}\int^{(6)}\cdots\int \exp\left[-\frac{1}{kT}\{\epsilon-\alpha F\cos(\widehat{\alpha z})\}\right]dp_\theta dp_\phi dp_\psi d\theta d\phi d\psi,$$

where ϵ is given as a function of the variables of integration by the expression (194·1). The p-integrations are unaffected and can be carried out as before, giving

$$R(T)=\frac{(8\pi^3ABCk^3T^3)^{\frac{1}{2}}}{h^3}\int_0^\pi\int_0^{2\pi}\int_0^{2\pi}\exp\left\{\frac{\alpha F\cos(\widehat{\alpha z})}{kT}\right\}\sin\theta\, d\theta\, d\phi\, d\psi.$$

Now take new axes and angles θ', ϕ', ψ' so that $(\widehat{\alpha z})=\theta'$. Then

$$\sin\theta\, d\theta\, d\phi\, d\psi=\sin\theta'\, d\theta'\, d\phi'\, d\psi'$$

and we obtain, apart from a constant factor, exactly the same integrals as before, and the same value for θ^*.

The general classical result is therefore

$$\frac{\delta_e-1}{\delta_e+2}=\tfrac{4}{3}\pi n\left(\frac{1}{3}\frac{\alpha^2}{kT}+\bar{\beta}\right), \qquad \text{......(1289)}$$

so long as $\alpha F/kT$ is small. The contributions of different molecular species on the right are additive.

§ 12·21. *Comparison with experiment.* The formulae of § 12·2 were compared with experiment with marked success by Debye,† and more recently excellent new comparisons have been rendered possible by the observations of Zahn.‡ It is obvious that the classical theory cannot be correct at very low temperatures, but the quantum theory agrees, as we shall see, with (1289) at temperatures covering all observations.

In observations on gases δ_e is very nearly unity and therefore very accurate observations of δ_e are required to establish the value of δ_e-1 with satisfactory precision. The necessary accuracy can be obtained by the use of the modern technique of oscillating circuits. The variation of capacity of a condenser containing the gas under observation as dielectric is detected and measured by its effect on the electrical beats of two high frequency circuits slightly out of tune. In analysis of the observations δ_e+2 can be taken to be equal to 3. Referring to (1289) we see that

$$(\delta_e-1)\,T/n=4\pi\left(\frac{1}{3}\frac{\alpha^2}{k}+\bar{\beta}T\right),$$

† Debye, *loc. cit.*

‡ Zahn, *Phys. Rev.* vol. 24, p. 400 (1924); *ibid.* vol. 27, p. 329 (1926); Smyth and Zahn, *J. Amer. Chem. Soc.* vol. 47, p. 2501 (1925); K. Compton, *Science*, vol. 63, p. 53 (1926). See also Van Vleck, *Electric and Magnetic Susceptibilities*, chap. III.

or in terms of gas pressure

$$(\delta_e - 1)\, T^2 = \frac{4\pi p}{k}\left(\frac{1}{3}\frac{\alpha^2}{k} + \overline{\beta} T\right). \qquad \text{......(1290)}$$

We therefore plot $(\delta_e - 1)\, T^2$ or better $(\delta_e - 1)\, vT$, where v is the volume per mole, against the temperature and expect to find a straight line whose slope determines $\overline{\beta}$ and whose intercept on the axis of $(\delta_e - 1)\, vT$ the value of α. The principal interest centres in α.

Fig. 53, taken from K. Compton,* shows typical examples of such plots, which are excellent straight lines, and Table 46 shows the values of dipole moments so determined for a variety of gases.

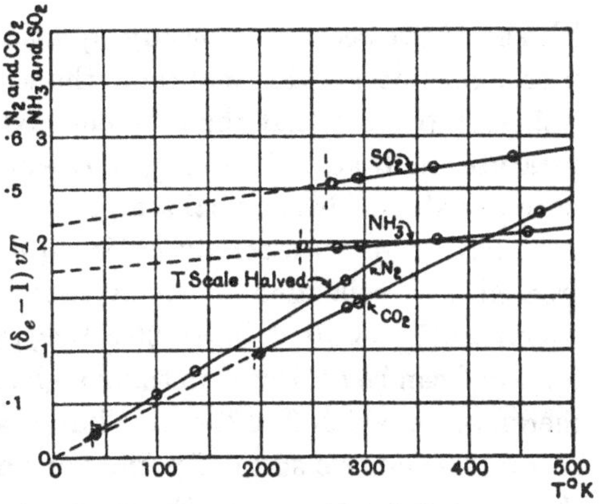

Fig. 53. The analysis of dielectric constants by Debye's equation.

The large value for water vapour confirms and in fact demands the triangular model for this molecule which we have already accepted in Chapter III. The zero value for carbon dioxide confirms the straight model adopted in that chapter.

TABLE 46.

Dipole moments of gaseous molecules from Debye's equation.

Gas	$\alpha \times 10^{18}$	Gas	$\alpha \times 10^{18}$	Gas	$\alpha \times 10^{18}$	Gas	$\alpha \times 10^{18}$
HCl	1·03	H_2O	1·84	CH_4	0·00	C_6H_6†	0·00
HBr	0·79	H_2S	1·10	CH_3Cl	1·89	C_6H_5F†	1·4
HI	0·38	C_2H_2	0·00	CH_2Cl_2	1·59	C_6H_5Cl†	$1{\cdot}5_5$
NH_3	1·48	C_2H_4	0·00	$CHCl_3$	0·95	C_6H_5Br†	1·5
PH_3	0·55	C_2H_6	0·00	CCl_4	0·00	C_6H_5I†	1·3
AsH_3	0·15	N_2	0·00	NO	>0	$C_6H_5NO_2$†	3·9
SO_2	1·61	CO_2	0·00	CO	0·12	CH_3OH	1·7

A summary of the best work to 1931 is given by Van Vleck, *loc. cit.*, from whom most of the above values have been selected. An elaborate collection to 1926 was given by Blüh, *Physikal. Zeit.* vol. 27, p. 226 (1926) supplementing Debye, *Handbuch der Radiologie*, vol. 6 (1925). Values for PH_3 and AsH_3 from H. E. Watson, *Proc. Roy. Soc.* A, vol. 117, p. 43 (1927).

† Values measured in solution in non-polar solvents.

§ **12·22.** *Applications to liquids.* Formula (1289) can also be applied with considerable success to many liquids, but in this case in its complete form, for δ_e is no longer nearly unity. If we solve (1289) for δ_e for a polar liquid we find

$$\delta_e = 1 + \frac{3T_c}{T - T_c}\left(1 + \frac{3kT\overline{\beta}}{\alpha^2}\right), \qquad \text{......(1291)}$$

* K. Compton, *loc. cit.*

where
$$T_c = \frac{4\pi n\alpha^2}{9k(1-\frac{4}{3}\pi n\overline{\beta})}. \qquad \text{......(1292)}$$

If the liquid is non-polar ($\alpha = 0$), this reduces to
$$\delta_e = 1 + 4\pi n\overline{\beta}/(1-\tfrac{4}{3}\pi n\overline{\beta}). \qquad \text{......(1293)}$$

We shall have occasion to study equations of the type of (1291) much more thoroughly later in this chapter in the sections on ferromagnetism. Here it is sufficient to note that there should exist for polar liquids a critical temperature and that, as $T \to T_c$, $\delta_e \to \infty$, while for higher temperatures δ_e falls to normal values. This critical temperature is of real physical importance, for it is quite high for strongly polar liquids. For example if water could be treated as a collection of freely rotating dipoles—steam molecules—for which $\alpha = 1{\cdot}84 \times 10^{-18}$, $n = 6{\cdot}06 \times 10^{23}/18$, we should find $T_c = 1200$. (The term in $\overline{\beta}$ can be neglected in this rough analysis.) At ordinary temperatures therefore for which $T < T_c$ such a liquid would be in the electrical analogue of a ferromagnetic state of permanent natural polarization. It is probable that in substances such as Rochelle salt the water of crystallization should be regarded as just such a liquid, but no ordinary polar liquid has such properties. Liquids polar and non-polar may be divided fairly sharply into two classes. Class (1) contains the non-polar liquids for all of which $1-\frac{4}{3}\pi n\overline{\beta} > 0$ and those polar liquids whose dipoles are rather weak so that T_c calculated from (1292) is small compared with the temperatures of the liquid range. For these liquids equation (1293) or (1291) fits the observed dielectric constants fairly well, with values of α and $\overline{\beta}$ in reasonable agreement with the values found from the corresponding vapour phase. We may conclude that such liquids are mainly unassociated, and that the dipoles (when present) rotate freely and play no very important part in the liquid structure. Class (2) contains the liquids with strong dipoles for which T_c is comparable with or larger than the temperatures of the liquid range. From the fact that they are not permanently polarized it follows that the dipoles do not rotate freely and the molecules must in the main be linked more or less rigidly to neighbouring molecules by their dipoles. We may conclude that the liquids are strongly associated and that the dipoles play a dominant part in determining the liquid structure. These conclusions fit in with other evidence and have been used recently in an attempt to elucidate the structure of water itself.*

It is perhaps of some interest to show by a general argument why polar liquids must become associated and so less strongly polar until $T_c < T$; they are then no longer "ferromagnetic"—their normal state is unpolarized but with a very large dielectric constant. If we start with the approximation of freely rotating dipoles for which $T_c > T$, then we have the ferromagnetic

* Bernal and Fowler, *J. Chem. Physics*, vol. 1, p. 515 (1933).

case and (see §§ 12·9 *sqq.*) almost all the dipoles will normally be aligned in the same direction. This means that the state of minimum free energy is to this extent an organized one, in spite of the loss of entropy due to the elimination of orientational disarrangement. But now in general in a *liquid* a further diminution of free energy may be possible by a process of association, in which neighbouring dipoles, instead of being all aligned parallel, tie up to each other in blocks and chains, perhaps closed, forming in fact definite dipole bonds. In the water of crystallization of a Rochelle salt on the other hand such bonds cannot form for the water molecules are in fixed positions. Such a process will introduce little or no entropy change by change of orientational disorder since the orientations remain ordered, and it must therefore occur so long as the total energy is diminished more by the formation of the dipole bonds than it is increased by a loss of the negative energy of polarization. In general one must expect that such bonds can form in strongly polar liquids but not in Rochelle salt, and that when they form they will greatly diminish the number and the strength of the dipoles available for orientation. Thus T_c will fall until the normal state of the liquid calculated by (1291) and (1292) for the remaining available dipoles is no longer a permanently polarized one, but merely one of large δ_e —how large these general arguments naturally do not permit one to say. A Rochelle salt on the other hand which at temperature T contains sufficient orientatable water to make $T_c > T$ cannot escape in this way and should show real "ferromagnetic" polarization.

§ **12·3.** *Thermodynamic relations and the approach to the quantum theory of dielectric polarization.*† We have so far presented the theory of dielectrics in the most familiar way, but there is much to be gained by a more general thermodynamic treatment. We recall that $R(T)$ is a partition function and as such equivalent to a thermodynamic potential, and all relevant molar equilibrium properties, for example P, should be derivable from $R(T)$ by differentiation. Comparing (1284) and (1286) we see at once that for molecules carrying permanent dipoles

$$P = n\theta^* F = nkT\frac{\partial}{\partial F}\log R(T). \qquad \text{......(1293·1)}$$

This equation is not merely true in this particular case, but is perfectly general. It is simple to give a general proof of (1293·1) similar to the proof of (233) for $\overline{X}$. For if the n localized systems in unit volume are distributed among a set of possible states whose unperturbed energies are ϵ_u and average dipole moments, in the direction of F, $\alpha_u + \beta_u F$, then the perturbation energy is $-\alpha_u F - \frac{1}{2}\beta_u F^2$ and

$$P = \Sigma_u \overline{a_u}(\alpha_u + \beta_u F) = \lambda\Sigma_u(\alpha_u + \beta_u F)\,\varpi_u e^{-(\epsilon_u - \alpha_u F - \frac{1}{2}\beta_u F^2)/kT}.$$

† This section is based on Debye, *Physikal. Zeit.* vol. 27, p. 67 (1926).

Since $\lambda = n/f(T), \quad f(T) = \Sigma_u \varpi_u e^{-(\epsilon_u - \alpha_u F - \frac{1}{2}\beta_u F^2)/kT},$

$$P = nkT \frac{\partial}{\partial F} \log f(T), \qquad \text{......(1294)}$$

which is the desired relation, since only the orientational part of $f(T)$ is relevant.

It is next necessary to examine in detail the form of (1294) in the most general case required, namely with

$$f(T) = \Sigma_u \varpi_u e^{-(\epsilon_u - \alpha_u F - \frac{1}{2}\beta_u F^2)/kT}. \qquad \text{......(1295)}$$

We may suppose that the terms in F have had the effect of breaking up a set of degenerate states of the same energy into less degenerate sets of different energies, and write

$$f(T) = \Sigma_j e^{-\epsilon_j/kT} (\Sigma_s \varpi_s e^{(\alpha_s F + \frac{1}{2}\beta_s F^2)/kT}). \qquad \text{......(1296)}$$

We may suppose further that $\alpha_s F + \frac{1}{2}\beta_s F^2$ is so small that only the lowest order terms need be retained so that

$$f(T) = \Sigma_j e^{-\epsilon_j/kT} \left(\Sigma_s \varpi_s + (F/kT) \Sigma_s \alpha_s \varpi_s + \tfrac{1}{2} F^2 \left\{ \Sigma_s \varpi_s \alpha_s^2/k^2T^2 + \Sigma_s \varpi_s \beta_s/kT \right\} \right).$$

In this equation we may replace $\Sigma_s \varpi_s$ by ϖ_j the weight of the jth degenerate state, and $\Sigma_j \varpi_j e^{-\epsilon_j/kT}$ by $f_0(T)$ the partition function for zero external field. We may also suppose that

$$\Sigma_s \alpha_s \varpi_s = 0. \qquad \text{......(1297)}$$

If (1297) were not true, there would be a polarization effect independent of field strength—no such effect is known, or predicted either by the classical or quantum theory of the effect. Then

$$f(T) = f_0(T) + \tfrac{1}{2} F^2 \Sigma_j e^{-\epsilon_j/kT} \left\{ \Sigma_s \varpi_s \alpha_s^2/k^2T^2 + \Sigma_s \varpi_s \beta_s/kT \right\}, \qquad \text{......(1298)}$$

$$P = \frac{nF}{f_0(T)} \Sigma_j e^{-\epsilon_j/kT} \left\{ \Sigma_s \varpi_s \alpha_s^2/kT + \Sigma_s \varpi_s \beta_s \right\}. \qquad \text{......(1299)}$$

It is easy to recover the classical formula from (1299), for classically $\alpha_s = \alpha \cos\theta$ and $\Sigma_s \varpi_s \cos^2\theta = \frac{1}{3}\varpi_j$ so that

$$P = \frac{nF}{f_0(T)} \Sigma_j \varpi_j e^{-\epsilon_j/kT} \left\{ \frac{1}{3}\frac{\alpha^2}{kT} + \overline{\beta} \right\} = nF \left\{ \frac{1}{3}\frac{\alpha^2}{kT} + \overline{\beta} \right\},$$

as before, α and $\overline{\beta}$ being assumed independent of j.

In the quantum theory the classical "states" that have appeared in the foregoing calculations are of course regrouped into the true quantum states. We must not expect all the details of the classical calculation to be repeated, and in fact we shall find that in the quantum theory it may be that $\alpha_s = 0$ and the whole effect comes from β. All we can demand is that for large T the quantum theory shall agree with (1289). Being a second and not a first order effect the older quantum theory never succeeded in giving a correct account satisfying the limiting condition.

§ **12·31**. *Applications of quantum mechanics.* (i) *The rigid rotator without axial spin.** It is never possible in this monograph to give detailed quantum solutions of problems whose results we require for statistical applications. It would be necessary to develop the solutions at far too great a length and we must content ourselves with quoting results.

The unperturbed energies of the rigid rotator without axial spin are (§ 2·22)

$$\epsilon_j = \frac{h^2}{8\pi^2 A} j(j+1) \quad (j=0, 1, 2, \ldots), \qquad \ldots\ldots(1300)$$

and the weights of these states are $2j+1$. We have now to find the values of ϵ_j when perturbed by an external field F, in which the rigid dipole α has a potential energy $-\alpha F \mu$. The first order energy changes are linear combinations of the integrals

$$\int_{-1}^{+1} \int_0^{2\pi} \psi_{j,s} \psi_{j,s'} \mu \, d\mu \, d\phi,$$

where the $\psi_{j,s}$'s are the spherical harmonics of order j. These all vanish when $s \neq s'$ owing to the ϕ integration and, when $s=s'$,

$$\int_{-1}^{+1} \mu \{P_j^s(\mu)\}^2 \, d\mu$$

always vanishes because $\{P_j^s(\mu)\}^2$ is an even function of μ. All the integrals therefore vanish and *there are no first order changes in the energy*. This is in agreement with observation on the infra-red bands of HCl. These show no linear Stark effect.†

The quadratic terms have also been evaluated.‡ It is found that for the possible values $0 \leqslant s \leqslant j$

$$\beta_s = \frac{8\pi^2 A \alpha^2}{h^2} \frac{3s^2 - j(j+1)}{j(j+1)(2j-1)(2j+3)} \quad (j \neq 0), \qquad \ldots\ldots(1301)$$

$$\beta_0 = \frac{8\pi^2 A \alpha^2}{3h^2} \quad (j=s=0). \qquad \ldots\ldots(1302)$$

[In the limit of large quantum numbers this goes over into the form

$$\beta_s \sim \frac{4\pi^2 A \alpha^2}{h^2 j^2} \left(\frac{3s^2}{2j^2} - \frac{1}{2} \right), \qquad \ldots\ldots(1303)$$

which is the classical result for a dipole rotating with total angular momentum $jh/2\pi$ and a resolved momentum $sh/2\pi$ about the direction of the field.]

* Correct versions have been given by Mensing and Pauli, *Phys. Zeit.* vol. 27, p. 509 (1926); Kronig, *Proc. Nat. Acad. Sci.* vol. 12, pp. 488, 608 (1926), in abstract by Van Vleck, *Nature*, Aug. 14, 1926, and Manneback, *Phys. Zeit.* vol. 28, p. 72 (1927).

† Barker, *Astrophys. J.* vol. 58, p. 201 (1923).

‡ Kronig or Mensing and Pauli, *loc. cit.* Both authors calculate the average electric moment parallel to F rather than the energy term, but the analysis is of course equivalent.

The degeneracy has not been completely resolved. There are two states of equal energy when $s \neq 0$ and only one when $s = 0$. Therefore $\varpi_s = 2$, $s \neq 0$, and $\varpi_0 = 1$, from which it follows that

$$\Sigma_s \varpi_s \beta_s = 0 \quad (j \neq 0), \qquad \text{......(1304)}$$

since

$$\sum_{s=-j}^{s=+j} \{3s^2 - j(j+1)\} = 0. \qquad \text{......(1305)}$$

It follows at once that

$$\theta^* = \frac{3}{4\pi}\frac{\delta_e - 1}{\delta_e + 2}\frac{1}{n} = \frac{8\pi^2 A\alpha^2}{3h^2}\frac{1}{f_0(T)}. \qquad \text{......(1306)}$$

In this account we have of course treated *rigid* dipoles. There will still be an extra approximately constant term arising from the distortion of the electronic structures, and we therefore write the complete formula for *rigid polarizable dipoles without axial spin*

$$\theta^* = \frac{3}{4\pi}\frac{\delta_e - 1}{\delta_e + 2}\frac{1}{n} = \frac{8\pi^2 A\alpha^2}{3h^2}\frac{1}{f_0(T)} + \bar{\beta}. \qquad \text{......(1307)}$$

The value of $f_0(T)$ or

$$\sum_{j=0}^{\infty} (2j+1)\, e^{-\sigma j(j+1)} \quad \left(\sigma = \frac{h^2}{8\pi^2 AkT}\right)$$

has already been given in § 3·3. For sufficiently large T

$$f_0(T) \sim \frac{1}{\sigma} = \frac{8\pi^2 AkT}{h^2},$$

with a numerical error about $\frac{1}{3}\sigma$ of this. Thus on substituting for $f_0(T)$ in (1307) we recover the classical value. The error in the classical value for the temperatures of Zahn's and similar experiments is at most of the order of 0·5 per cent. and so insensible.

The quantum theory of gaseous dielectric polarization contains the remarkable feature that only molecules in their quantum states $j = 0$ contribute to the polarization. This has a strict classical analogy as has been pointed out by Pauli.† If we group together all orbits of equal classical moment of momentum $jh/2\pi$ and average classically *by integration* over all orientations, the average value of s^2/j^2 is $\frac{1}{3}$ and the average value of β_s is still zero. Thus classically all *rotating* molecules make no contribution. The sole contribution comes from those few molecules which execute small vibrations about a position of rest.

§ **12·32.** *The isotropic character of the dielectric constant.* We have so far considered that the direction of the electric field itself serves to define a definite origin of spherical polar coordinates for the molecules. The electrical effects are however of the second order and small at that, and a magnetic field may have an overriding orientational effect. Whatever the relative directions

† Pauli, *Zeit. f. Physik*, vol. 6, p. 319 (1921).

of the electric and magnetic fields the first order electric effect vanishes ($\alpha_s = 0$) as before. To see this it is only necessary to verify that

$$\int_{-1}^{+1}\int_{0}^{2\pi} \cos\gamma\,\psi_{j,s}\psi_{j,s'}\,d\mu\,d\phi = 0 \quad (all\ s,\ s'),$$

where $$\cos\gamma = \cos\theta\cos\chi + \sin\theta\sin\chi\cos\phi$$

and χ is the angle between the electric and magnetic fields. The quadratic effect β_s will continue to be given by (1301) when the directions of the electric and magnetic fields coincide. When they do not a different value might perhaps be expected, but if they are at right angles the same values of β_s and β_0 are found as before. Thus the response to the electric field is exactly the same whether the molecules are orientated along or across the field, and therefore, as easily follows, *in whatever direction they are orientated or even if their precessional axes are not orientated at all but distributed uniformly in all directions*. This result of course only holds neglecting terms depending on H the strength of the orientating magnetic field. It implies in accordance with observation that there is no magnetic double refraction for long waves independent of the strength of the magnetic field. Similar arguments show that such double refraction must be absent for all wave lengths.

The model treated in § 12·31 represents adequately diatomic molecules with $^1\Sigma$ normal states, for example the halogen hydrides and N_2, and the conclusions of the theory apply directly to these gases. The results are however far more general as we shall see in § 12·4.

§ **12·33.** *Applications of quantum mechanics.* (ii) *The symmetrical top model.** For other molecules a more complicated model is essential and the necessary analysis can be carried through for the symmetrical rigid top. We may assume, in accordance with the symmetry, that the dipole moment lies along the axis of symmetry, and shall indicate the calculations for the case in which the electric field itself removes the degeneracy. The system is then characterized by three quantum numbers j, r, s, of which s defines the (trivial) precession of the resultant angular momentum about the field. The unperturbed energies (neglecting this precession) are

$$\epsilon_{j,r} = \frac{h^2}{8\pi^2}\left\{\frac{1}{A}j(j+1) + \left(\frac{1}{C} - \frac{1}{A}\right)r^2 + const.\right\}, \qquad \text{......(1308)}$$

the subsidiary quantum numbers r and s are subject to the restrictions $|r| \leqslant j$, $|s| \leqslant j$. It is then found that

$$\alpha_{r,s} = \frac{rs}{j(j+1)}\alpha, \qquad \text{......(1309)}$$

$$\beta_{r,s} = -\frac{8\pi^2 A\alpha^2}{h^2}\left\{\frac{(j^2-r^2)(j^2-s^2)}{j^3(2j-1)(2j+1)} - \frac{\{(j+1)^2-r^2\}\{(j+1)^2-s^2\}}{(j+1)^3(2j+1)(2j+3)}\right\}. \qquad \text{...(1310)}$$

† Kronig, *loc. cit.* (2).

The degeneracy is fully resolved and $\varpi_s = 1$. Substituting in (1299), remembering that Σ_j and ϵ_j are now $\Sigma_{j,r}$ and $\epsilon_{j,r}$, we verify that $\Sigma_s \varpi_s \alpha_s = 0$, and find

$$\theta^* = \frac{\alpha^2}{f_0(T)} \Sigma_{j,r} e^{-\epsilon_{j,r}/kT} \left\{ \frac{1}{kT} \frac{r^2(2j+1)}{3j(j+1)} + \frac{8\pi^2 A}{h^2} \frac{r^2(2j+1)}{3j^2(j+1)^2} \right\}. \quad \text{......(1311)}$$

We notice that the polarization is contributed quite differently for this model and the former for which $r=0$. There are linear terms as well as quadratic—there should be a linear Stark effect in all states except those for which $r=0$. We notice also that when $r=0$ (1311) reduces to (1306), the only contribution arising when $j=0$ also, and then only from the quadratic terms.

To find the limiting form of θ^* for large T we replace the sums by integrals in the usual manner. With the abbreviations

$$\sigma = \frac{h^2}{8\pi^2 AkT}, \quad \tau = \frac{h^2}{8\pi^2 kT}\left(\frac{1}{C} - \frac{1}{A}\right),$$

the sum in the numerator is

$$\frac{1}{3kT}\int_0^\infty e^{-\sigma j(j+1)} \left\{ \frac{2j+1}{j(j+1)} + \frac{1}{\sigma} \frac{2j+1}{j^2(j+1)^2} \right\} dj \int_{-j}^{+j} e^{-\tau r^2} r^2 dr,$$

and in the denominator

$$\int_0^\infty e^{-\sigma j(j+1)} (2j+1)\, dj \int_{-j}^{+j} e^{-\tau r^2} dr.$$

These can be simplified by integration by parts to the forms

$$\frac{1}{3kT} \frac{2}{\sigma} \int_0^\infty e^{-\sigma j(j+1) - \tau j^2} \frac{j}{j+1} dj$$

and

$$\frac{2}{\sigma} \int_0^\infty e^{-\sigma j(j+1) - \tau j^2} dj.$$

Since σ and τ are small, the important parts of the former of these integrals arises for large j and $j/(j+1) \sim 1$. Their ratio is therefore $1/3kT$, and we find again the classical result

$$\theta^* = \frac{\alpha^2}{3kT},$$

to which a constant deformation term may be added.

§ **12·4.** ***General theory of a complex molecule carrying a permanent dipole.*** † In view of §§ 12·31, 12·33 which lead to the same Langevin-Debye formula for different models by entirely different routes, it is natural to suppose that this result is of great generality for ordinary temperatures, as it is in the classical theory. This has been proved by Van Vleck to be a simple consequence of the summation rules, and the spectroscopic stability rules of quantum mechanics, that hold quite generally for those molecular systems which when unperturbed have orientational degeneracy. We can hardly

† This section describes the results of Van Vleck, *loc. cit.*

digress here sufficiently far to give an adequate account of Van Vleck's proof, though it is in fact quite simple, and must be content with a precise statement of his result and a discussion of its consequences.

In order that the Langevin-Debye formula should hold for sufficiently large T it is only necessary to assume that the dipole carrier (atom or molecule) has a "permanent" electric (or magnetic) dipole moment which is the same for the whole group of normal states, and that these normal states have energy steps from one to the next among themselves small compared with kT, or in other words that the precessional frequencies of *all* the moments of momenta in the normal state are small compared with kT/h. At the same time this group of normal states must be separated from all other (excited) states by energy steps large compared with kT. These conditions are usually strictly complied with by all simple atoms and molecules. When they are complied with it does not matter whether the normal atom is originally strictly degenerate or already perturbed by a field of any strength weak enough for *second* order perturbations to be neglected. It is unnecessary to specify precisely the degree of complication of the normal group of states. The group may be built up by the composition of any number of moments of momenta with *slow* rates of precession, so that the proof applies equally to rigid molecules, and to molecules in which the orbital and electronic moments of momenta play an important part. Under these conditions formula (1288) for θ^* will still be true, α and $\overline{\beta}$ being independent of the direction of the field even when another field in some other direction is already orientating the molecules. In short α and $\overline{\beta}$ are independent of the degeneracy and of the manner in which it is removed We obtain the same value for θ^* for any type of spatial quantization or if there is no spatial quantization at all. All the molecules in Table 46 are covered by Van Vleck's theorem, and the dipole moments may be legitimately deduced in the manner of § 12·21.

The general proof includes a proof that both terms in the dielectric constant of a gas are independent of the effect of a magnetic field H so far as first order terms in H are concerned. This is in agreement with observation. It has been shown that θ^* for He, O_2 and air changes by less than 10, 0·4, 0·4 per cent. respectively in fields up to 8000 gauss, and for NO and HCl by less than 8 per cent. and 1 per cent. up to 4800 gauss.† There might be an effect depending on H^2 observable only in very large fields, but this has not been investigated either theoretically or experimentally.

These generalities also hold in general for the refractive index just as for the dielectric constant.

† Weatherby and Wolf, *Phys. Rev.* vol. 27, p. 769 (1926); Mott-Smith and Daily, *Phys. Rev.* vol. 28, p. 976 (1926). To reach even these accuracy limits requires of course extreme accuracy of measurement of δ_e: 1 in 500,000 for He, O_2; 1 in 100,000 for NO, HCl.

In Van Vleck's proof and the foregoing discussion it is assumed that a relation equivalent to (1297) holds in general, even when there is an already existing magnetic field. It is conceivable that cases might exist of a kind of magnetoelectric directive effect in which the equivalent of (1297) fails. There would then be a state of electric polarization produced by a magnetic field and *vice versa*. The energy term concerned is still really quadratic in the perturbations. It is of type HF, but linear therefore in H and F. The foregoing theory would continue to hold in such cases, but only for the additional polarization produced by the electric field itself. The effect could only be found with molecules which possess permanent non-zero electric and magnetic moments which are more or less rigidly bound together and not at right angles to each other, so that the orientation of one produces a net orientation of the other.* There is no known example of such a gas, though the effect might be expected for NO.†

A sufficient formal example of the nature of the effect is provided by the classical rigid rotator of § 12·2 with both an electric doublet of strength α and a magnetic doublet of strength μ along its axis of symmetry, in parallel fields F and H. Then the rotational partition function is an obvious extension of (1284), namely

$$R(T) = \frac{8\pi^2 AkT}{h^2}\,\frac{\sinh\{(\alpha F + \mu H)/kT\}}{(\alpha F + \mu H)/kT}. \qquad \text{......(1312)}$$

Using (1294) and retaining only the largest terms, the polarization P is given by

$$P = \frac{n\alpha\mu H}{3kT}. \qquad \text{......(1313)}$$

A simple extension of this result to a model in which the electric and magnetic dipoles are fixed in any directions in the rotator making an angle χ with each other leads to

$$P = \frac{n\alpha\mu\cos\chi}{3kT}H. \qquad \text{......(1314)}$$

The explanation of the absence of this effect in NO is interesting. An explanation by assuming that $\cos\chi = 0$ is impossible, for the dipole moment must lie along the figure axis, and it is certain from the evidence of the band spectrum that the orbital moment of momentum and therefore the magnetic moment has an axial component. The classical explanation would be that NO molecules are of two kinds, right- and left-handed, of equal energies and therefore of equal concentrations, so that for half of them the axial com-

* When the magnetic moment is entirely due to the spins of the electrons, a magnetic field orientating the magnetic moment might be practically without effect on the molecular framework carrying the electric moment. This explanation might hold for the normal states of atoms and molecules whenever these are Σ-terms, but does not apply to NO whose normal state is a Π-term.

† Debye, *Zeit. f. Physik*, vol. 36, p. 300 (1926); Huber, *Physikal. Zeit.* vol. 27, p. 619 (1926).

ponent of μ is parallel to α and for the other half anti-parallel. The quantum version of this is that the states of any NO molecule are in fact all double forming a Λ-type doublet, but that each of these states is formed by superposition of equal amounts of right- and left-handed wave-functions. The component of electronic angular momentum in the $\overrightarrow{\mathrm{NO}}$ direction is therefore constant in magnitude but indeterminate in sign and $\overline{\cos\chi}=0$.†

§ **12·5.** *Para- and dia-magnetism of gases and ions in solution.* The foregoing theory can be transferred bodily from electric to magnetic effects, and becomes the statistical theory of para- and dia-magnetism. The relationships between the polarization of the medium, here the magnetization I, the magnetic pipe-force H', the induction (crack-force) B, the magnetizing force F, the magnetic dipole strength per molecule κ^*, the volume susceptibility κ_m and the permeability δ_m are exactly the same as those between P, E', D, F, θ^*, κ_e and δ_e respectively. We have in fact the standard relationships

$$I=n\kappa^*(H'+\tfrac{4}{3}\pi I),$$
$$B=4\pi I+H'=\delta_m H' \quad (\delta_m=1+4\pi\kappa_m).$$

These lead at once to

$$\frac{\delta_m-1}{\delta_m+2}=\tfrac{4}{3}\pi n\kappa^*, \quad n\kappa^*=\frac{\kappa_m}{1+\frac{4}{3}\pi\kappa_m}.$$

We shall drop the magnetic suffix m whenever no confusion can arise.

In magnetic problems it is usual to work in terms of susceptibilities, and it is almost always true except in ferromagnetics that $\frac{4}{3}\pi\kappa\ll 1$. We shall make this assumption except where it is explicitly stated otherwise. It is then allowable to ignore the differences between B, H' and F and to identify them all with the magnetic force H *in vacuo* outside the medium.

The calculation of κ^* is identical in form with the calculation of θ^*. We can for example construct a partition function for the rotations and orientations of any molecule including magnetic effects in the form

$$f(T)=\Sigma\,\varpi_u e^{-(\epsilon_u-\mu_u H-\frac{1}{2}\beta_u{}^2H^2)/kT} \qquad \text{......(1315)}$$

and find at once

$$I=nkT\frac{\partial}{\partial H}\log f(T). \qquad \text{......(1316)}$$

Any rigid body model with fixed magnetic moment μ treated classically leads at once to an extra factor

$$\frac{\sinh(\mu H/kT)}{\mu H/kT} \qquad \text{......(1317)}$$

in the partition function $f(T)$ and therefore to Langevin's formula

$$\kappa^*=\frac{kT}{H^2}\left[\frac{\mu H}{kT}\coth\frac{\mu H}{kT}-1\right]. \qquad \text{......(1318)}$$

† For further details and references see Van Vleck, *Electric and Magnetic Susceptibilities*, p. 280.

For the common case of $\mu H/kT$ small we have

$$\kappa^* = \frac{1}{3}\frac{\mu^2}{kT}. \qquad \text{......(1319)}$$

This is the *paramagnetic susceptibility*.

The additional constant term $\bar{\beta}$ of the dielectric theory reappears here too. It is usually of negative sign and represents the diamagnetic effect of the magnetic field on the electronic orbits in the atoms or molecules. In certain cases, however, notably among the rare earths, it contains an important positive part which may even contribute the greater part of the paramagnetic effect. The complete value of κ^* is therefore

$$\kappa^* = \frac{1}{3}\frac{\mu^2}{kT} + \bar{\beta}. \qquad \text{......(1320)}$$

Paramagnetic moments are always associated with mechanical moment of momentum. When large they always arise from unbalanced electronic orbits or unbalanced spins of the electrons themselves. The proton and many other atomic nuclei possess magnetic moments which are of the order of one thousandth of those arising from electrons. The rotating nuclei of a polar molecule can also give rise to a magnetic moment but this again is very small. These small terms and the diamagnetic part of $\bar{\beta}$ in (1320) can always be ignored compared with the electronic paramagnetic terms when these do not vanish identically, except for a degenerate gas of free electrons (§ 12·7).

For a magnetically polar molecule the symmetrical top model with its magnetic moment along the axis of figure is thus the only not entirely inadequate model which can be discussed in detail with any simplicity. The analysis of § 12·33 would then apply with obvious changes and lead to Langevin's formula. The general theory of Van Vleck described in § 12·4 shows however that Langevin's formula, or better (1320), applies generally, and that too without change in μ or $\bar{\beta}$ whether or not there is spatial quantization.†

Atomic and molecular magnetic dipoles are measured in terms of Bohr's magneton. It is well known that any central orbit of area S described in time τ by an electron has a mechanical moment of momentum $2mS/\tau$ and is equivalent to a magnet of moment $\epsilon S/\tau c$. The ratio of magnetic to mechanical momentum has therefore the standard value $\epsilon/2mc$, a ratio which is

† Doubt had been thrown on this independence of spatial quantization by the experiments of Glaser, *Ann. d. Physik*, vol. 75, p. 1059 (1924). Later work (Lehrer, *Zeit. f. Physik*, vol. 37, p. 155 (1926); Hammer, *Proc. Nat. Acad. Sci.* vol. 12, p. 597 (1926)) had shown that Glaser's effect was spurious, being probably due to inadequate drying of the gases used, and that the susceptibility was accurately proportional to the pressure at low pressures for air, O_2, CO_2, N_2 and H_2.

preserved by quantum mechanics. For a system containing electronic moment of momentum M, purely orbital in origin, we have

$$M=\frac{h}{2\pi}\{j(j+1)\}^{\frac{1}{2}},\quad \mu=\frac{\epsilon h}{4\pi mc}\{j(j+1)\}^{\frac{1}{2}}, \qquad \text{......(1321)}$$

where j is the angular momentum quantum number. The value of $\epsilon h/4\pi mc$, which was equal to μ for the normal one-quantum orbit of hydrogen in the older quantum theory, is known as Bohr's magneton. Its numerical value is $9{\cdot}23\times10^{-21}$ electrostatic units. The empirical unit, Weiss's magneton, is smaller by the factor 4·967. It is easily seen that exact multiples of Bohr's magneton are not to be expected in observations.

The ratio of magnetic to mechanical momentum for the spin of an electron has on the other hand the double value ϵ/mc. If the resultant electronic momentum is contributed partly by orbital momentum and partly by spin, we have

$$\mu=\frac{\epsilon h}{4\pi mc}g\{j(j+1)\}^{\frac{1}{2}}. \qquad \text{......(1322)}$$

The factor g is known as Landé's splitting factor. In a molecule there is in addition the angular momentum of the rotation of the molecule as a whole which does not contribute to μ.

§ **12·51**. *Paramagnetism of gases*. There are only two known gases which are normally paramagnetic, O_2 and NO. For these theory and experiment are in exact agreement. The normal state of O_2 is a term $^3\Sigma_1$, so that its residual electronic moment of momentum is due purely to electronic spin and its magnetic moment should be given by (1322) with $g=2$, $j=1$. The conditions of Van Vleck's general theorem (§ 12·4) are well satisfied, and we obtain an atomic susceptibility given by (1319) or (1320) with μ equal to $2\sqrt{2}$ Bohr magnetons, irrespective of the precise nature of the coupling of this momentum vector to the molecular frame. The observed value is 2·84 Bohr magnetons.

The theory of NO is not so simple. The normal electronic state is of type $^2\Pi_{\frac{1}{2}}$. The other state of the doublet is $^2\Pi_{\frac{3}{2}}$ with a separation of 122 wave numbers. At room temperatures this separation is comparable with kT/hc ($=200$). We have therefore to take account of the equilibrium distribution between the two states of the doublet and also to make special calculations since the usual condition that all precessional frequencies are small compared with kT/h is not fulfilled. Van Vleck's result† is that $\kappa^*=(\mu_{\text{eff}})^2/3kT$, where

$$\mu_{\text{eff}}=2\sqrt{\left\{\frac{1-e^{-x}+xe^{-x}}{x(1+e^{-x})}\right\}}\quad\left(x=\frac{h\Delta\nu}{kT}\right).$$

† Van Vleck, *Electric and Magnetic Susceptibilities*, p. 270.

The excellent agreement between theory and experiment is shown in Fig. 54.

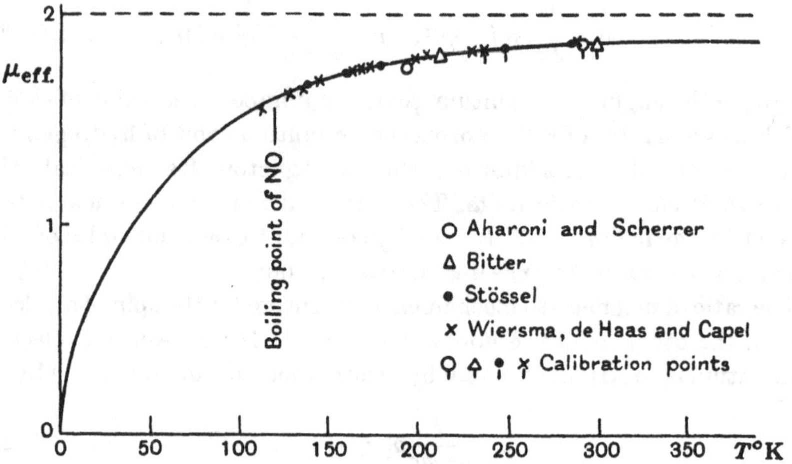

Fig. 54. The effective average magnetic dipole moment of NO.

§12·52. *Paramagnetism of atomic ions.* Similar considerations* apply to the para- and dia-magnetism of atoms or atomic ions. Generally speaking these must be studied in liquid solutions, or even in solids, rather than in gases, but the theory of the effect is the same, as it does not matter whether or not there is spatial quantization so long as the ion is rotationally free to respond to the orientating field. The theory of paramagnetism for an atomic ion is simpler than for a molecule, since the magnetic moment is always directly proportional to the total mechanical momentum and the independence of spatial quantization can be established very simply. If the magnetic field lies along the z-axis the atomic paramagnetic susceptibility is easily shown by the argument yielding (1299) to be proportional to the average value of M_z^2/kT, where M_z is the z-component of the total angular momentum M. When there is spatial quantization

$$M_z = sh/2\pi \quad (-j \leqslant s \leqslant j).$$

Thus the average value of M_z^2 is

$$\frac{1}{2j+1}\sum_{s=-j}^{s=+j}\frac{s^2h^2}{4\pi^2} = \tfrac{1}{3}j(j+1)\frac{h^2}{4\pi^2} = \tfrac{1}{3}M^2; \qquad \text{......(1323)}$$

for the square of the total angular momentum is $j(j+1)$. This result holds whether j has integral or half-integral values. The result that $\overline{M_z^2} = \tfrac{1}{3}M^2$ with spatial quantization means of course that $\overline{M_z^2}$ has the same value as with random orientations, which is Van Vleck's theorem in this special case. In comparing the theory with experiments and determining atomic magnetic

* Van Vleck, *loc. cit.*

moments it must be remembered that M is given by (1321) and that the magnetic moment associated with M is gM not M Bohr magnetons.

It is now possible to make a fairly satisfactory comparison between observed and theoretical values of μ for atomic ions in solution or in solids for the rare earth and iron groups of elements. Data for the palladium and platinum groups are still somewhat fragmentary and will not be discussed. Table 47 gives the data for the rare earth group. The values of μ in Bohr magnetons given in column 6 are calculated from (1322) with the values of

TABLE 47.

Observed and calculated magneton numbers for the atomic ions of the rare earth group in solutions.

Ion	No. of electrons in 4_3 orbits	Theoretical normal term	j	g	Magnetons calc. by (1322)	Magnetons calc. corrected by Van Vleck	Magnetons (observed)	
							Cabrera	St Meyer
La^{+++}	0	^1S	0	—	0·00	0·00	0	0
Ce^{+++}	1	^2F	$\frac{5}{2}$	$\frac{6}{7}$	2·54	2·56	2·39	2·77‡
Pr^{+++}	2	^3H	4	$\frac{4}{5}$	3·58	3·62	3·60	3·47
Nd^{+++}	3	^4J	$\frac{9}{2}$	$\frac{8}{11}$	3·62	3·68	3·62	3·51
Il^{+++}	4	^5J	4	$\frac{3}{5}$	2·68	2·83	—	—
Sm^{+++}	5	^6H	$\frac{5}{2}$	$\frac{2}{7}$	0·84	1·55–1·65	1·54?	1·32
Eu^{+++}	6	^7F	0	—	0·00	3·40–3·51	3·61	3·12
Gd^{+++}	7	^8S	$\frac{7}{2}$	2	7·9	7·9	8·2	8·1
Tb^{+++}	8	^7F	6	$\frac{3}{2}$	9·7	9·7	9·6	9·0
Ds^{+++}	9	^6H	$\frac{15}{2}$	$\frac{4}{3}$	10·6	10·6	10·5	10·6
Ho^{+++}	10	^5J	8	$\frac{5}{4}$	10·6	10·6	10·5	10·4
Er^{+++}	11	^4J	$\frac{15}{2}$	$\frac{6}{5}$	9·6	9·6	9·5	9·4
Tu^{+++}	12	^3H	6	$\frac{7}{6}$	7·6	7·6	7·2	7·5
Yb^{+++}	13	^2F	$\frac{7}{2}$	$\frac{8}{7}$	4·5	4·5	4·4	4·6
Lu^{+++}	14	^1S	0	—	0·00	0·00	0	0

‡ For Pr^{++++}.

The alternative values in Van Vleck's column for Sm^{+++} and Eu^{+++} are obtained by using two different estimates of the screening constant for 4_3 orbits in calculating the multiplet intervals—the smaller value to a screening constant 33 and the larger to 34.

j and g for the normal term of the corresponding ions specified in columns 1–5. The normal terms are those derived from general spectroscopic laws for the specified numbers of 4_3 electrons and have not been checked by spectroscopic evidence. They are not however open to doubt.* It will be seen that there is excellent agreement except for Sm^{+++} and Eu^{+++}. The cause of the discrepancy here has been established by Van Vleck† who has shown that for these ions the term in $\bar{\beta}$ makes an important contribution, and the lowest state of the normal multiplet is not well separated from the next higher one.

* Hund, *Linienspektren und periodische System der Elemente* (1927).
† Van Vleck, *loc. cit.* p. 245.

Van Vleck's corrected values are given in column 7 and are in excellent agreement with the observations. We conclude that the rare earth ions in solution have their magnetic elements (the incomplete group of 4_3 electrons) so deeply buried in their structure that they behave magnetically in solutions (and sometimes even at ordinary temperatures in crystals) just as if they were undistorted free ions, in almost exact agreement with statistical theory.

Since the susceptibility of Sm^{+++} and Eu^{+++} arises from more than one state of the multiplet and largely from the $\overline{\beta}$ term, the magneton numbers for these ions are markedly temperature-dependent. These susceptibilities are shown as functions of the temperature in Figs. 55 *a*, *b*. The normal temperature variation would be given by a rectangular hyperbola, with the axes for asymptotes.

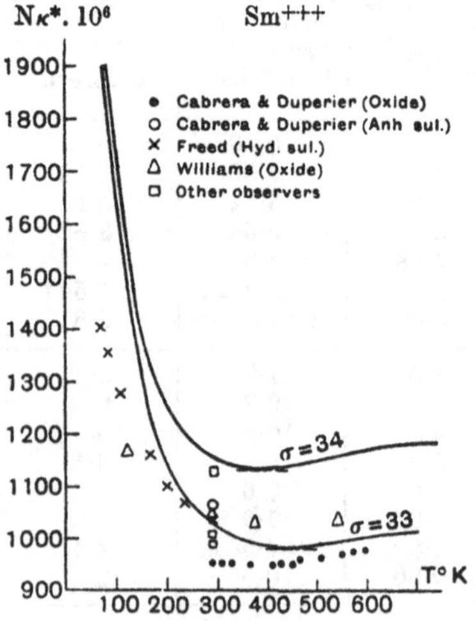

Fig. 55 *a*.

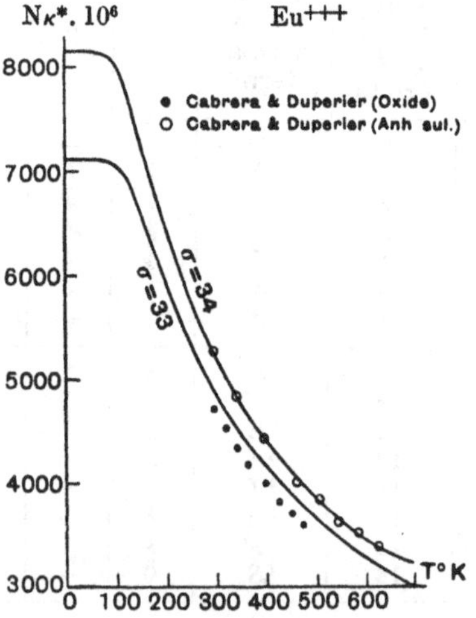

Fig. 55 *b*.

The observed and theoretical susceptibilities per mole ($N\kappa^*$) for Sm^{+++} and Eu^{+++} in various salts ; σ is the screening constant adopted in the calculations.

Similar calculations can be made for the iron transition group. It is at once seen that for the ions of this group there is no agreement between the observed magnetons and those calculated either for the lowest term or for the complete multiplet, except for Mn^{++} and Fe^{+++} where the multiplet is an *S*-term. This suggests that the orientatable element in these ions is only the electron spin and that the orbital angular momentum is "quenched" by the interaction with the field of the surrounding atoms which has not spherical symmetry. The incomplete group of 3_2 orbits is here the outermost group in the atom and is not well screened from neighbouring ions as is the 4_3 group in the rare earths by the completed 5_0 and 5_1 groups. This interpretation is confirmed by comparing the observed values with the last calculated column headed "spin only", with which the agreement is on the whole very good.† It can be shown that the orbital magnetic moment will actually be quenched in this way, leaving the spin moment unaffected, provided that the asymmetrical external fields are sufficiently strong.‡

TABLE 48.

Observed and calculated magneton numbers for atomic ions of the iron transition group.

Atomic ion	No. of electrons in 3_2 orbits	Theoretical normal terms	j	g	Magnetons calculated for: Lowest term only	Magnetons calculated for: Actual multiplet	Magnetons calculated for: Spin only	Magnetons observed in: Solutions	Magnetons observed in: Salts
$K^{+}\ldots V^{5+}$	0	^1S	0	—	0·00	0·00	0·00	0·00	0·00
Sc^{++}						2·57		—	—
Ti^{+++}	1	^2D	$\frac{3}{2}$	$\frac{4}{5}$	1·55	2·18	1·73	—	—
V^{++++}						1·78		1·75	1·79
Ti^{++}	2	^3F	2	$\frac{2}{3}$	1·63	3·36	2·83	—	—
V^{+++}						2·73		2·76–2·85	—
V^{++}						3·60		3·81–3·86	—
Cr^{+++}	3	^4F	$\frac{3}{2}$	$\frac{2}{5}$	0·77	2·97	3·87	3·68–3·86	3·82
Mn^{++++}						2·47		4·00	—
Cr^{++}	4	^5D	0	—	0·00	4·25	4·90	4·80	—
Mn^{+++}						3·80		—	5·05
Mn^{++}	5	^6S	$\frac{5}{2}$	2	5·92	5·92	5·92	5·2–5·96	5·85
Fe^{+++}						5·92		5·94	5·4–6·0
Fe^{++}	6	^5D	4	$\frac{3}{2}$	6·70	6·54	4·90	5·33	5·0–5·5
Co^{++}	7	^4F	$\frac{9}{2}$	$\frac{4}{3}$	6·64	6·56	3·87	4·6–5·0	4·4–5·2
Ni^{++}	8	^3F	4	$\frac{5}{4}$	5·59	5·56	2·83	3·23	2·9–3·4
Cu^{++}	9	^2D	$\frac{5}{2}$	$\frac{6}{5}$	3·55	3·53	1·73	1·8–2·0	1·8–2·2
Cu^{+}, Zn^{++}	10	^1S	0	—	0·00	0·00	0·00	0·00	0·00

§ **12·6.** *Saturation effects.* Throughout both the electrical and the magnetic discussion we have so far assumed that the approximation of replacing (1286) by (1287), or the equivalent, can be made—that is that $\alpha F/kT$ or $\mu F/kT$ is sufficiently small. So long as this is true θ^* or κ^* is independent of F and the relation between P (I) and F a linear one. For strong fields and low temperatures this approximation cannot be made. Higher order terms in the distortion effect of the field become significant, but, what is more important, a saturation effect sets in when the dipoles still available for orientation are beginning to become appreciably scarcer. The electrical saturation effect must be taken into account in any proper theory of the nature of a polar liquid in the immediate neighbourhood of a dissolved ion. If the induced polarization of the atoms can be neglected, the Langevin

† Sommerfeld, *Physikal. Zeit.* vol. 24, p. 360 (1923); Bose, *Zeit. f. Physik*, vol. 43, p. 864 (1927); Stoner, *Phil. Mag.* vol. 8, p. 250 (1929).

‡ Van Vleck, *loc. cit.* p. 287.

formula (1286) remains correct for this case, but we shall not attempt to discuss the phenomenon further here, as it is hardly yet susceptible to direct experiment.

The phenomenon of paramagnetic saturation at low temperatures and great field strengths is more important and has been extensively studied for ions of the rare earth and iron groups. The necessary theory is now very simple. An atomic ion has no variable rotational energy but merely so many different possible orientations in the magnetic field. In amplification of § 12·51 we see therefore that it has an orientational partition function, in the field H,

$$O(T) = \sum_{s=-j}^{j} e^{sg\mu_B H/kT}, \quad \text{......(1324)}$$

where μ_B is one Bohr magneton. Thus

$$O(T) = \frac{\sinh\{(j+\frac{1}{2})g\mu_B H/kT\}}{\sinh\{\frac{1}{2}g\mu_B H/kT\}}. \quad \text{......(1325)}$$

It then follows at once that

$$I = n\kappa^* H = nkT\frac{\partial}{\partial H}\log O(T),$$

$$= ng\mu_B\left[(j+\tfrac{1}{2})\coth\frac{(j+\frac{1}{2})g\mu_B H}{kT} - \tfrac{1}{2}\coth\frac{\frac{1}{2}g\mu_B H}{kT}\right]. \quad \text{...(1326)}$$

When $H \to \infty$, I tends to its saturation value

$$I_\infty = ng\mu_B j.$$

For small values of H we have the result

$$I = nH\left[\frac{1}{3}\frac{g^2\mu_B{}^2 j(j+1)}{kT}\right],$$

which we have already proved and used in § 12·52. The saturated value of I corresponds classically to a dipole $g\mu_B j$ per atom while the initial value of I corresponds to $g\mu_B[j(j+1)]^{\frac{1}{2}}$ per atom. This quantal distinction arises from the non-commuting properties of the components of angular momentum. The function defined by $I = I_\infty B_j(g\mu_B H/kT)$ is often called Brillouin's function.†

A particularly important special case arises when $j = \frac{1}{2}$ (g is then 2). We have then

$$I = n\kappa^* H = n\mu_B \tanh\frac{\mu_B H}{kT}, \quad \text{......(1327)}$$

$$[I_\infty = n\mu_B;\ I \sim n\mu_B{}^2 H/kT]. \quad \text{......(1328)}$$

When $g\mu_B \to 0$, $j \to \infty$ in such a way that $jg\mu_B \to \alpha$, we recover the classical Langevin function of equation (1318).

† Brillouin, *J. de Physique*, vol. 8, p. 74 (1927).

Fig. 56 shows a number of these functions drawn for the same value of the initial slope, that is of $\mu_{\text{eff}} = g\mu_B[j(j+1)]^{\frac{1}{2}}$ and different values of j; they are shown as functions of $H\mu_{\text{eff}}/kT$. The curve for $j=\frac{7}{2}$ should fit the observations for the Gd^{+++} ion. The crosses in the figure show Woltjer and Onnes'* observations on hydrated gadolinium sulphate $Gd_2(SO_4)_3 . 8H_2O$ and are in perfect agreement with the theory.

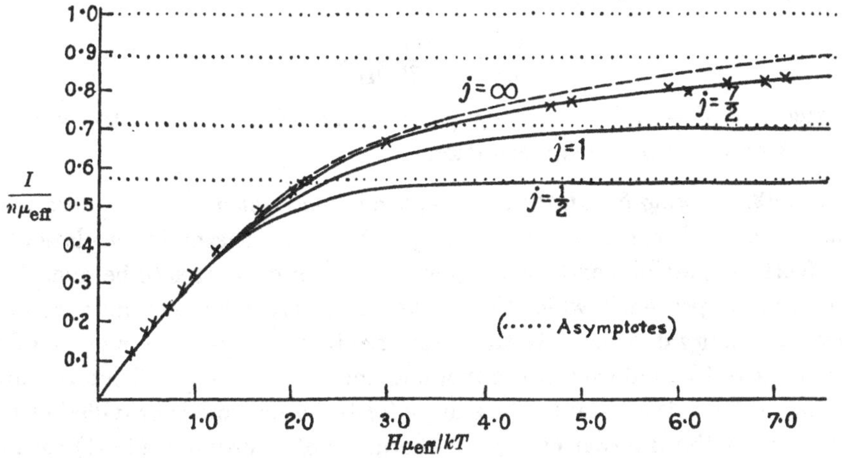

Fig. 56. Curves showing magnetization as a function of $H\mu_{\text{eff}}/kT$ (Brillouin's functions). The approach to saturation is clearly shown. The × are observations for gadolinium sulphate, for which the theory gives the curve marked $j=\frac{7}{2}$ in excellent agreement.

§ **12·61.** *Entropy of magnetization.* We consider the thermodynamics of polarizable media in greater detail in § 12·91 as an introduction to ferromagnetism, and to some extent also in § 12·3. In this section we shall merely supplement these results by giving such formulae as are required to understand the theory of cooling by demagnetization, a process which has been shown to be of great importance in the attainment of very low temperatures.†

It is simplest to start by considering the entropy contribution made to an assembly by the orientations only of N atoms or ions such as those of the preceding section. By (594) this entropy contribution for $N\,(=nV)$ atoms in a given field H is

$$S(H,T) = Nk\left[\log O(T) + T\frac{\partial \log O(T)}{\partial T}\right], \qquad \text{......(1329)}$$

$O(T)$ being given by (1325). The explicit formula is

$$S(H,T) = Nk\log\frac{\sinh\{(j+\frac{1}{2})\,g\mu_B H/kT\}}{\sinh\{\frac{1}{2}g\mu_B H/kT\}} - \frac{HIV}{T}, \qquad \text{......(1330)}$$

* Woltjer and Onnes, *Comm. Phys. Lab. Leiden*, No. 167c (1923).

† Giauque, *J. Amer. Chem. Soc.* vol. 49, p. 1864 (1927); Debye, *Ann. d. Physik*, vol. 81, p. 1154 (1926).

where I is the intensity of magnetization as given by (1326). When $T \to 0$ for fixed H, we have $S(H,T) \to S(H, 0) = 0$. When $H \to 0$ for fixed T, we have $S(H,T) \to S(0, T) = Nk \log(2j+1)$; results which are otherwise evident. The energy contribution of the magnetization is given by

$$E = NkT^2 \frac{\partial}{\partial T} \log O(T) = -HIV. \qquad \text{......(1331)}$$

It is finally easy to show that

$$\frac{\partial S}{\partial H} = -\frac{HV}{T} \frac{\partial I}{\partial H} < 0,$$

so that $S(0, T) - S(H,T)$ is always positive, the orientational entropy being always diminished by magnetization.

§ **12·62.** *Cooling by adiabatic demagnetization*. Let us now suppose that such an assembly, magnetized at temperature T, is thermally insulated (as perfectly as possible) and the magnetic field removed. It is to be remarked that this is a process in which the entropy of the assembly remains constant, but the energy does not. As the magnetization changes with change of H currents will be induced in the magnetizing coils tending to keep the flux through the coils constant. The energy for these currents is provided at the expense of the internal energy of the assembly. Formula (1331) cannot therefore be used, but we can proceed as follows. Let $S_A(T)$ be the entropy of the assembly apart from the orientational term here considered. Then since during demagnetization the total entropy is constant and since $S(0, T)$ is independent of T,

$$S_A(T) - S_A(T') = S(0, T) - S(H,T). \qquad \text{......(1332)}$$

The right-hand side of (1332) can reach values as high as 3·2 cal./deg. for $H = 20{,}000$ gauss, $T = 1{\cdot}3°$ K. and an assembly containing one gram-ion of gadolinium. Since at these low temperatures specific heats normally become small, compared with 6 cal./deg./gram-atom the cooling on demagnetization can reach values of the order of the total initial temperature. Such cooling experiments have been successfully carried out by Giauque,* Simon,† and de Haas.‡

The formulae here used would allow the cooling on demagnetization to be greater than the initial temperature, so that negative temperatures would be reached contrary to the laws of thermodynamics. This cannot actually happen even for the ideal assembly here discussed because at sufficiently low temperatures it would necessarily become ferromagnetic and therefore fail to demagnetize when the field is removed. Actual assemblies fail to follow

* Giauque and MacDougall, *Phys. Rev.* vol. 43, p. 768 (1933); vol. 44, p. 235 (1933).
† Kürti and Simon, *Proc. Roy. Soc.* A, vol. 149, p. 152 (1935).
‡ de Haas, Wiersma and Kramers, *Physica*, vol. 13, p. 175 (1933).

these formulae even earlier for another reason, namely because the field to which the magnetic elements of the ions are subjected must contain some fraction which has the symmetry of the crystal and therefore less than spherical symmetry. Such a field will split the states of the ions so that they are no longer degenerate when $H=0$. This splitting will make itself felt as soon as kT becomes of the same order of magnitude as the energy differences of the splitting and result in a large increase in the specific heat. These aspects of the effect are discussed further in Chapter XXI.

§ **12·7.** *The para- and dia-magnetism of a degenerate gas of free electrons. The paramagnetism of the alkalis.** Since electrons carry a magnetic dipole, being systems with $j=\frac{1}{2}$, $g=2$, a gas of free electrons must be paramagnetic. When the density of the electrons is so low that classical statistics may be used, the results of the preceding sections may be applied at once as if the electrons were localized systems. When however the gas is degenerate and the Fermi-Dirac statistics must be used, a new investigation is necessary.

Let us consider N electrons in a volume V $(N/V=n)$ and field H and adapt (224) to this more general problem by grouping separately the electrons orientated down the field and against the field. For such an assembly, the factor 2 in (224) drops out and

$$Z=Z_{+}+Z_{-},$$

$$Z_{\pm}=\frac{(2\pi mkT)^{\frac{3}{2}}V}{h^3}\frac{2}{\sqrt{\pi}}\int_0^\infty x^{\frac{1}{2}}\log(1+\mu_{\pm}e^{-x})\,dx \quad (\mu_{\pm}=\lambda e^{\pm\mu_B H/kT}). \qquad \text{......(1333)}$$

The magnetization is determined by the usual equation

$$IV=kT\,\partial Z/\partial H. \qquad \text{......(1334)}$$

The interest of the calculation lies in the form of I for a nearly degenerate assembly. In this case equation (229) may be used for $Z_{\pm}$ from which we find

$$Z=\frac{(2\pi mkT)^{\frac{3}{2}}V}{h^3}\frac{2}{\sqrt{\pi}}\left\{\tfrac{4}{15}[(\log\mu_{+})^{\frac{5}{2}}+(\log\mu_{-})^{\frac{5}{2}}]+\frac{\pi^2}{6}[(\log\mu_{+})^{\frac{1}{2}}+(\log\mu_{-})^{\frac{1}{2}}]\right\}, \qquad \text{......(1335)}$$

or, correct to terms in H^2,

$$Z=\frac{(2\pi mkT)^{\frac{3}{2}}V}{h^3}\frac{2}{\sqrt{\pi}}\left[\tfrac{8}{15}(\log\lambda)^{\frac{5}{2}}+\frac{\pi^2}{3}(\log\lambda)^{\frac{1}{2}}+\frac{\mu_B{}^2H^2}{k^2T^2}(\log\lambda)^{\frac{1}{2}}\right]. \qquad \text{......(1336)}$$

By (1334), the volume susceptibility κ is given by

$$\kappa=\frac{I}{H}=2\frac{(2\pi mkT)^{\frac{3}{2}}}{h^3}\frac{2}{\sqrt{\pi}}\frac{\mu_B{}^2}{kT}(\log\lambda)^{\frac{1}{2}}, \qquad \text{......(1337)}$$

* Pauli, *Zeit. f. Physik*, vol. 41, p. 81 (1927); Frenkel, *Zeit. f. Physik*, vol. 49, p. 31 (1928).

where λ is given by (1001). It is easily verified from (1336) that λ is unaffected by H to the desired order. Thus

$$\kappa = \frac{4\pi m \mu_B{}^2}{h^2}\left(\frac{3n}{\pi}\right)^{\frac{1}{3}}\left\{1 - \frac{8\pi^2}{3}\frac{m^2k^2T^2}{h^4(3n/\pi)^{\frac{4}{3}}}\right\}. \qquad \ldots\ldots(1338)$$

The temperature-dependent term is negligible at all ordinary temperatures. The remaining small constant paramagnetic susceptibility is of the order of the small susceptibilities observed for the alkalis which are paramagnetic, but temperature-independent. It is however also of the order of the usual diamagnetic susceptibilities, and it is necessary before comparing (1338) with experiment to examine whether a gas of free electrons has also any diamagnetic susceptibility due to modification of the orbits in the magnetic field.

§ **12·71.** *Absence of diamagnetism for a gas of free electrons in classical theory.* In classical theory the energy of an electron is not affected by a magnetic field (apart from its energy of orientation if it is thought to carry a magnetic dipole) and therefore Maxwell's and Boltzmann's distribution laws continue to apply to its translatory motion unaltered. If therefore one considers the current carried by the electrons in any direction past any fixed point in the gas, this current is always exactly zero whether or not a magnetic field is acting. The gas of classical free electrons can therefore have no diamagnetic susceptibility, since no current is generated by the magnetic field. It is however easy to cast doubt on this argument by considering the paths of the actual electrons, which (in the plane normal to H) are all circles in the same sense about the field, each one making a diamagnetic contribution (see Fig. 57). The contradiction is removed when it is remembered that the electron gas must have a boundary and that near the boundary the average movement of the electrons must by repeated reflection be a creep round the boundary in the opposite sense. These few large orbits exactly balance the effect of the many smaller orbits in which electrons circulate in the ordinary direction.*

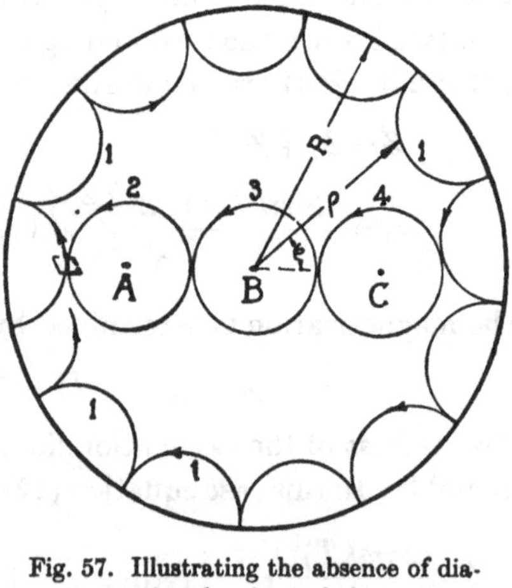

Fig. 57. Illustrating the absence of diamagnetism for classical electrons.

* Bohr, *Studier over Metallernes Elektrontheori*, Dissertation, Copenhagen (1911).

§ 12·72. *Diamagnetism of a gas of free electrons in the quantum theory.* In the quantum theory however this balance is no longer exact and diamagnetism results. The reason for this is that in the quantum theory the wave-functions are modified by the boundary, for example they must tend to zero at a rigid wall, and further that even apart from this boundary effect the possible momenta normal to the field which can vary continuously in the absence of a magnetic field are quantized in the presence of the field, facts which destroy the perfect classical balance. This was first pointed out by Landau.* We shall follow Darwin's† exposition based on a simple soluble model in which the walls of the enclosure are represented by a special field of force. The result for a free electron gas is obtained by making this field tend to zero.

Consider an enclosure bounded by rigid plane walls a distance a apart normal to the axis of z along which the magnetic field H acts, and by the potential energy $\frac{1}{2}B(x^2+y^2)$ at right angles to H. We shall ultimately represent free electrons by letting $B \to 0$. Schrödinger's equation for the electrons in such an enclosure may be shown to be

$$-\frac{h^2}{8\pi^2 m}\left(\frac{\partial^2\psi}{\partial r^2}+\frac{1}{r}\frac{\partial\psi}{\partial r}+\frac{1}{r^2}\frac{\partial^2\psi}{\partial\theta^2}+\frac{\partial^2\psi}{\partial z^2}\right)-\frac{ih\omega}{2\pi}\frac{\partial\psi}{\partial\theta}+\tfrac{1}{2}(B+m\omega^2)r^2\psi=E\psi, \qquad \text{......(1339)}$$

where ω $(=|\epsilon|H/2mc)$ is the Larmor precession. The equation separates in these variables and the characteristic values, which are all we require, are easily shown to be

$$E_{l,n,s}=\frac{h\omega}{2\pi}[l+b(2n+|l|+1)]+\frac{h^2}{8ma^2}s^2 \quad \left(b=\left(1+\frac{B}{m\omega^2}\right)^{\frac{1}{2}}\right)$$

$$(-\infty<l<\infty,\ 0\leqslant n<\infty,\ 1\leqslant s<\infty). \qquad \text{......(1340)}$$

To determine the equilibrium properties of this assembly we have to construct and evaluate

$$Z=2\sum_{s=1}^{\infty}\sum_{n=0}^{\infty}\sum_{l=-\infty}^{\infty}\log(1+\lambda e^{-E_{l,n,s}/kT}), \qquad \text{......(1341)}$$

or if λ is small then merely the partition function

$$F(T)=2\sum_{s=1}^{\infty}\sum_{n=0}^{\infty}\sum_{l=-\infty}^{\infty}e^{-E_{l,n,s}/kT}. \qquad \text{......(1342)}$$

When λ is small the s-summation merely gives a factor independent of H and may be omitted. We have then to evaluate

$$\sum_{n=0}^{\infty}\sum_{l=-\infty}^{\infty}e^{-\alpha[l+b(2n+|l|+1)]} \quad \left(\alpha=\frac{h\omega}{2\pi kT}=\frac{\mu_B H}{kT}\right), \qquad \text{...(1343)}$$

which sums at once to $\dfrac{e^{\alpha}}{(e^{(b+1)\alpha}-1)(1-e^{-(b-1)\alpha})}$.

* Landau, *Zeit. f. Physik*, vol. 64, p. 629 (1930).
† Darwin, *Proc. Camb. Phil. Soc.* vol. 27, p. 86 (1931).

Since $$\kappa_d = \frac{NkT}{VH}\frac{\partial \log F(T)}{\partial H} = \frac{n\mu_B}{H}\frac{\partial \log F(T)}{\partial \alpha},$$

we find on letting $B \to 0, b \to 1$

$$\kappa_d = \frac{n\mu_B}{H}\left[-\coth\frac{\mu_B H}{kT} + \frac{kT}{\mu_B H}\right]. \qquad \text{......(1344)}$$

To this diamagnetic κ must be added the paramagnetic effect of the spins to obtain the complete result. By (1327) this is

$$\frac{n\mu_B}{H}\tanh\frac{\mu_B H}{kT}.$$

Thus the complete result for classical statistics is

$$\kappa = \frac{n\mu_B}{H}\left[\frac{kT}{\mu_B H} - \frac{2}{\sinh(2\mu_B H/kT)}\right]. \qquad \text{......(1345)}$$

When the saturation can be neglected, this reduces to

$$\kappa = \tfrac{2}{3}n\mu_B{}^2/kT, \qquad \text{......(1346)}$$

which is two-thirds of the value uncorrected for orbital diamagnetism. It is clear that the (paramagnetic) orientational energy could be incorporated from the start by replacing (1341) by

$$Z = \sum_{s=1}^{\infty}\sum_{n=0}^{\infty}\sum_{l=-\infty}^{\infty} \log\{1+\lambda\exp[-\alpha(l+1+b[2n+|l|+1]) - h^2s^2/8ma^2kT]\}$$
$$+\log\{1+\lambda\exp[-\alpha(l-1+b[2n+|l|+1]) - h^2s^2/8ma^2kT]\}. \qquad \text{......(1347)}$$

For a degenerate gas it is necessary to evaluate (1341) or (1347) for λ large. Greater accuracy is required than replacing the sums by integrals, and we shall therefore use the formula

$$\sum_{x=\nu_1}^{\nu_2} F(x) = \int_{\nu_1-\frac{1}{2}}^{\nu_2+\frac{1}{2}} F(x)\,dx - \frac{1}{24}\Big[F'(x)\Big]_{\nu_1-\frac{1}{2}}^{\nu_2+\frac{1}{2}}. \qquad \text{......(1348)}$$

Since a comparison of (1344) and (1345) shows that there is no universal relationship between the diamagnetic and paramagnetic parts of the susceptibility, we shall be content to evaluate (1341) and (1347) for almost complete degeneracy. In that case if $\lambda = e^{\beta}$, β large, the summand may be taken to be

$$\beta - \{\alpha l + b(2n+|l|+1)\} - \gamma s^2$$

to be summed over all those values of l, n, s for which it is positive. It is convenient to sum in the order s, l, n. The s-summation is independent of H and merely provides a weighting factor. It is therefore sufficient to use the ordinary integral approximation. For the l and n summations (1348), or its equivalent, is necessary. We then find without difficulty that

$$Z = \frac{16}{105}\frac{1}{\gamma^{\frac{1}{2}}}\frac{\beta^{\frac{7}{2}}}{\alpha^2(b^2-1)} + \frac{1}{\gamma^{\frac{1}{2}}}\left[-\frac{2}{9}\frac{1}{b^2-1} - \frac{1}{9}\right]\beta^{\frac{3}{2}}. \qquad \text{......(1349)}$$

The first term is independent of H. The second contains in $1/(b^2-1)$ a term proportional to H^2.

The interpretation of the absolute meaning of this result would be slightly tedious. This can be avoided by comparing it with the similar deduction from (1347) which contains the paramagnetic term. On summing this in the same way we see that the only difference is the loss of a factor 2 and the substitution of two partial sums in which the β of (1349) is replaced by $\beta+\alpha$ and $\beta-\alpha$ respectively. Thus

$$Z_1=\frac{8}{105}\frac{1}{\gamma^{\frac{1}{2}}}\frac{(\beta+\alpha)^{\frac{7}{2}}+(\beta-\alpha)^{\frac{7}{2}}}{\alpha^2(b^2-1)}+\frac{1}{2\gamma^{\frac{1}{2}}}\left[-\frac{2}{9}\frac{1}{b^2-1}-\frac{1}{9}\right]\{(\beta+\alpha)^{\frac{3}{2}}+(\beta-\alpha)^{\frac{3}{2}}\}. \qquad \text{......(1350)}$$

The H^2 terms in (1349) and (1350) are respectively

$$-\frac{2}{9}\frac{1}{\gamma^{\frac{1}{2}}}\frac{\beta^{\frac{3}{2}}}{(b^2-1)}, \quad \left(\tfrac{2}{3}-\tfrac{2}{9}\right)\frac{1}{\gamma^{\frac{1}{2}}}\frac{\beta^{\frac{3}{2}}}{(b^2-1)}.$$

Thus the diamagnetic term is again $-\frac{1}{3}$ of the paramagnetic term as in the classical limit. The true value of the paramagnetic susceptibility of a degenerate electron gas is therefore

$$\kappa=\frac{8\pi m{\mu_B}^2}{3h^2}\left(\frac{3n}{\pi}\right)^{\frac{1}{3}}. \qquad \text{......(1351)}$$

Such comparison with observation as is possible for the alkalis is shown in Table 49. The order of magnitude of the variation with atomic number is roughly correct. It must be remembered that the large positive ions of Rb and Cs will make a considerable diamagnetic contribution which has been ignored in these calculations.

TABLE 49.

Observed and calculated molar susceptibilities for the alkali metals.

	Na	K	Rb	Cs
$10^7\kappa$ calc.	4·4	3·5	3·3	3·0
$10^7\kappa$ obs.	5·8	5·1	0·6	−0·5

§ **12·8.** *Dissociative equilibria in magnetic fields.* We take up again a question left over from § 5·8, as to possible effects of magnetic forces on dissociative equilibria, when some or all of the systems concerned possess permanent magnetic moments. A typical example is the dissociation of the halogens, for halogen atoms must be paramagnetic with a normal state $^2P_{\frac{3}{2}}$ $(g=\frac{4}{3})$, while the halogen molecules are known to be diamagnetic. We shall be content to discuss only the simple typical case of the reaction $X_2 \rightleftharpoons 2X$, where X is an atom. We shall retain only terms linear in H in the energies used in the partition functions for the atoms, which give the whole paramagnetic effect in this case, and ignore all magnetic effects on the molecules. The atoms may be assumed to be orientated by the field and to be internally in their lowest quantum state. The magnetic states will be assumed to be non-degenerate, of weight unity (§ 14·2).

The partition function (1296) must be generalized to include space-variable magnetic fields. Let us start by considering the various magnetic states separately. Then the atom in the sth state will have a potential energy $-\mu_s H$ in the field and therefore a partition function

$$f_s(T) = F(T)\, V_s(T), \qquad \text{......(1352)}$$

where $$F(T) = \frac{(2\pi mkT)^{\frac{3}{2}}}{h^3}, \quad V_s(T) = \iiint_V e^{\mu_s H/kT} dq_1\, dq_2\, dq_3. \qquad \text{......(1353)}$$

The partition function for the atoms as a whole will therefore be

$$f(T) = \Sigma_s f_s(T) = F(T)\, V'(T) \quad [V'(T) = \Sigma_s V_s(T)]. \qquad \text{......(1354)}$$

The partition function $VG(T)$ for the molecules is spatially constant and need not be further specified. The numbers of atoms of any magnetic type in a selected volume element will be given by the usual formula of type (560).

In the absence of the magnetic field the dissociative equilibrium is fixed by the equation

$$\frac{\overline{X_2}/V}{(\overline{X}/V)^2} = \frac{G(T)}{j^2 F^2(T)}, \qquad \text{......(1355)}$$

where j is the number of magnetic states. When the magnetic field is acting

$$\frac{\overline{X_2}}{(\overline{X})^2} = \frac{G(T)}{F^2(T)} \times \frac{V}{[V'(T)]^2}. \qquad \text{......(1356)}$$

In any selected volume element δV, let $\overline{x_s}$ be the average number of atoms in the sth magnetic state and $\overline{x_2}$ the number of molecules. Then

$$\frac{\overline{x_s}}{\delta V} = \frac{\overline{X_s}}{V_s(T)} e^{\mu_s H/kT} = \frac{\overline{X}}{V'(T)} e^{\mu_s H/kT},$$

and if $$\overline{x} = \Sigma_s \overline{x_s},$$

$$\frac{\overline{x_2}/\delta V}{(\overline{x}/\delta V)^2} = \frac{\overline{X_2}/V}{\{\overline{X}/V'(T)\}^2} \frac{1}{(\Sigma_s e^{\mu_s H/kT})^2},$$

$$= \frac{G(T)}{F^2(T)} \frac{1}{(\Sigma_s e^{\mu_s H/kT})^2}. \qquad \text{......(1357)}$$

There is nothing to the discredit of this equation, but unlike the corresponding equation (563) it yields a space-variable equilibrium constant. Its form however suggests that the fundamental reaction should be regarded as

$$X_2 \rightleftarrows X_s + X_{-s}$$

for the various values of s and not crudely as $X_2 \rightleftarrows 2X$. If this is right, then magnetic and mechanical moments parallel to the field are conserved in recombination or dissociation, and for each fundamental constituent of the reaction we have a space-constant equilibrium constant as before. Equation (1357) is of course unaltered by these considerations as to mechanisms of interaction, but the fundamental equations of the dissociative equilibria are

$$\frac{\overline{x_2}/\delta V}{\overline{x_s}/\delta V\,.\,\overline{x_{-s}}/\delta V}=\frac{G(T)}{F^2(T)}\quad (all\ s). \qquad \text{......(1358)}$$

It seems likely that the reaction *must* be of this form. It is difficult to see how *all* processes of dissociation can generate angular momentum parallel to the field, but this is necessary if the dissociation (and recombination) are to be of the more general type.

§ **12·9.** *Ferromagnetism.* In bodies of the type hitherto discussed the magnetic polarization (intensity of magnetization I) is always a very small fraction of the applied magnetic field, so that differences between B, H, H' and the effective field F can be entirely neglected. In sharp distinction the essential property of a ferromagnetic at low temperatures is that a very large magnetization can be produced by a small field H, and that moreover this magnetization can continue to exist when H is removed, provided the temperature is less than a critical value. The relationship between H and I at constant T starting from an apparently unmagnetized state is somewhat as shown in Fig. 58.

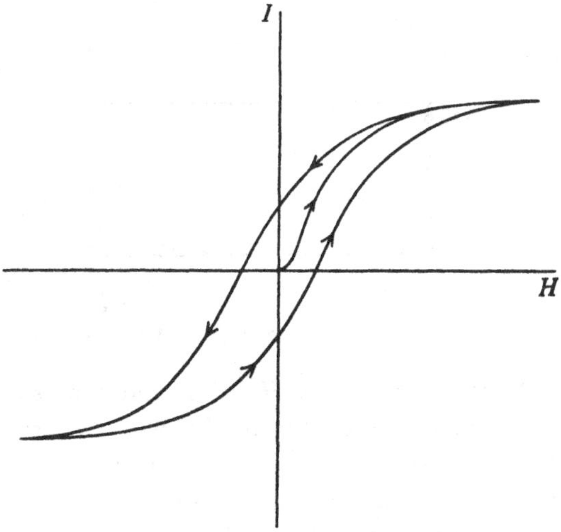

Fig. 58. The hysteresis-loop of an ordinary specimen of a ferromagnetic.

The persistence of the magnetization when H is removed is called *remanence* and the whole I-H curve the *hysteresis-loop*. It indicates a non-conservative system; there is an irreversible conversion of magnetic energy into heat in each cycle specified by the area of the hysteresis-loop, $\oint I\,dH$.

The shape and size of the hysteresis-loop is however *not* a primary property of a ferromagnetic. It is highly structure-sensitive in Zwicky's sense; the property of acquiring rapidly, in small fields H, a large magnetization which then is unaffected by further increase of the field is within wide limits structure-insensitive, and must be regarded as fundamental. If we examine a single crystal of iron or nickel, the better the crystal and the purer the metal the smaller the area of the hysteresis-loop. From such experiments one concludes that the ideal I-H curve which the theory of ferromagnetics must begin by explaining is a discontinuous but reversible curve of the type shown in Fig. 59. The discontinuity $2I_0$ $(+I_0 \to -I_0)$ at $H=0$ must be a function of T which vanishes when $T \geqslant T_c$. T_c is a critical temperature called *the temperature of the Curie point.* As H increases at constant temperature I increases, but slowly, so that very strong fields are required to increase I appreciably beyond I_0 except near the Curie point. I never exceeds I_∞ which represents complete saturation of the magnetization. As $T \to 0$, $I_0(T) \to I_\infty$.

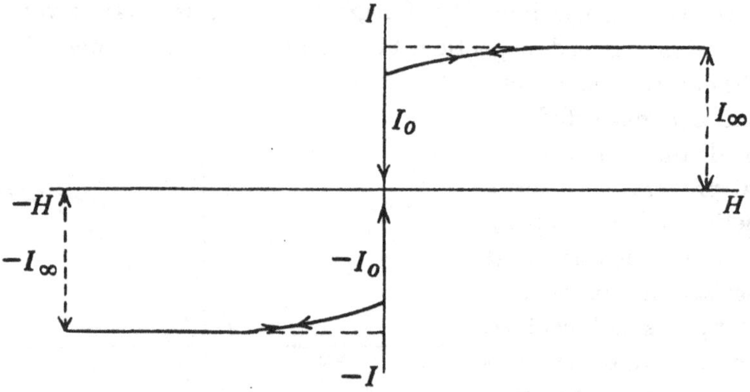

Fig. 59. The ideal I-H curve for a ferromagnetic.

For ferromagnetics H, H', B and F must be carefully distinguished. The quantity which must be used however depends on the nature of the problem in hand. In constructing a proper theory of a ferromagnetic one must consider not individual atoms but the whole magnetic material or at least a substantial part of it. If one considers the whole substance, then the interactions of the elementary magnets it contains must be included in constructing its energy values and it is only the field H due to external bodies which may be thought of as acting on the magnetic substance as a whole. It is also true when we discuss the thermodynamics of magnetization that it is the external field H which must be considered to act on the body. The internal interactions affect only the energy of the body as a function of the magnetization. If one considers a substantial block of the magnetic substance for which the rest of the magnetic substance is an external body, then it is F that must be used. In certain cases F may, as we shall see, be simplified to H'. Finally when one starts to construct a theory of ferromagnetism as Weiss did, by generalizing the theory of the susceptibility of a gas, the effective field acting on the individual magnetic elements is of course F.

A satisfactory formal theory of ferromagnetism was first given by Weiss using the theory of Langevin for paramagnetics expounded in the previous sections. In the exact theory of § 12·5 we have

$$I = n\mu L(x) \quad (x = \mu F/kT),$$

F being as usual the effective magnetizing force acting on the molecular magnetic dipoles. Thus strictly $F = H' + \frac{4}{3}\pi I$ if the magnetism arises from bound atomic electrons. If we take these formulae as exact, we can get nothing like ferromagnetism at ordinary temperatures. But if we suppose that we have somehow overlooked some important term in the energy, which in fact (see § 12·91) need not be magnetic, we may write in place of $\mu(H' + \frac{4}{3}\pi I)$

$$\mu(H + \nu I). \qquad \text{......(1359)}$$

There is here no sense in a distinction between H and H'. I is then given by

$$I = n\mu L(x) = n\mu L\left\{\frac{\mu(H + \nu I)}{kT}\right\}. \qquad \text{......(1360)}$$

We have now to determine I as a function of H from (1360). To do so we write $y = I/n\mu$ and plot on the same diagram the two functions

$$y = L(x), \quad y = \frac{kT}{n\mu^2\nu}x - \frac{H}{n\mu\nu}. \qquad \text{......(1361)}$$

The intersections determine possible values of I. The form of $L(x)$ is shown

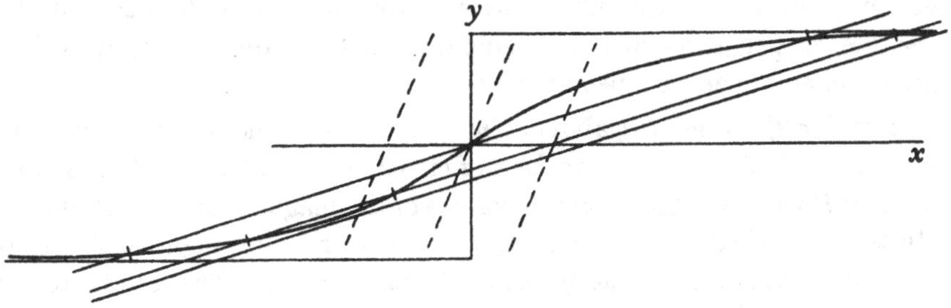

Fig. 60. Determination of values of I satisfying (1360).

in Fig. 60. It is monotonic, of steadily decreasing slope as $x \to \infty$, the maximum slope being $\frac{1}{3}$ at $x = 0$.

(i) If the slope of the straight line is greater than that of $L(x)$ at $x = 0$, i.e. if

$$T > \frac{1}{3}\frac{n\mu^2\nu}{k} = T_c, \qquad \text{......(1362)}$$

there can only be one intersection for any value of H. I is a monotone function of H, vanishing with H and tending to saturation ($\pm$) as $H \to \infty(\pm)$. This is ordinary paramagnetism corrected for what are effectively extra terms in F. A more exact calculation shows that in this case, for not too large values of H,

$$I = \frac{1}{3}\frac{n\mu^2}{k(T-T_c)}H. \qquad \text{......(1363)}$$

This is the modified law of Curie, well known to be obeyed by ferromagnetics when $T > T_c$. It applies also, as a better approximation than (1319), to many paramagnetics not known to be ferromagnetic, with T_c, that is ν, of either sign. The excellent agreement of (1363) with the observations for nickel is shown in Table 50.

TABLE 50.

Paramagnetic susceptibility of Ni, $T > T_c$.

Theory: $\kappa = \dfrac{I}{H} = \dfrac{0{\cdot}00555}{T-645}$.

Temp. °K.	$10^6\kappa$ obs.*	$10^6\kappa$ calc.
699	103·4	103·1
734	63·6	62·4
782	40·7	40·4
829	29·8	30·2
888	22·8	22·9
989	16·2	16·2
1095	12·4	12·3

* Weiss and Foex, *Archives d. Sci. phys. et nat., Genève* (4), vol. 31, pp. 4, 89 (1911) using polycrystalline material.

This agreement is obtained by regarding the number n of contributing electron spins as an adjustable constant. According to (1363) the number must be 0·3 per atom in disagreement with the value 0·6 derived from saturation and from the specific heat.

(ii) If $T < T_c$ as defined above, then the curves cut in three points for small values of H or one point for sufficiently large values of H. For small values of H we have three possible values of the magnetization, but of these only the numerically greatest root, which has the same sign as H, will be found to represent a thermodynamically stable state. [The middle root is unstable and physically meaningless. The smaller extreme root is metastable and might be expected to represent a state of remanent magnetization. But the ideal theory cannot be pushed so far and no such identification can be made.]

One can see at once from the figure that, except when T is nearly equal to T_c, I is very insensitive to H. This accords with the facts. By approximate solutions we easily find (I_∞ is the same as $n\mu$)

$$I = I_\infty \sqrt{\frac{5}{3}}\left(1-\frac{T}{T_c}\right)^{\frac{1}{2}} \quad (T \to T_c - 0), \qquad \text{......(1364)}$$

$$I = I_\infty\left(1-\frac{1}{3}\frac{T}{T_c}\right) \quad (T \to 0). \qquad \text{......(1365)}$$

Equation (1364) accords with the facts, but equation (1365) does not. The fall of I from I_∞ as T increases is more gradual. This failure is corrected in Heisenberg's version which substitutes the function $\tanh x$ for $L(x)$, or better still in Bloch's (§ 12·942).

We thus see that Weiss's theory with $L(x)$ or better $\tanh x$ for its polarization function gives an excellent general account of primary ferromagnetic phenomena, provided that we may use a suitable value of ν. The necessary values of ν are

$$\text{Ni } 14{,}000; \quad \text{Fe } 12{,}000; \quad \text{Co } 8{,}600,$$

whereas the theoretical value is $\frac{4}{3}\pi$ if we calculate ν purely from the magnetic forces. But what Weiss's theory really asserts is that the equilibrium must be calculated *as if* there were a large energy term per unit volume, $-I(H+\nu I)$, of no matter what origin, depending on the magnetization in this way. This will be made clearer in the next section, on considering the thermodynamics of magnetic phenomena, which we include here because there is no convenient treatment in the usual textbooks.

§ **12·91.** *The thermodynamics of magnetic phenomena.** For any system we have the relation

$$dE = T\,dS + \delta A,$$

where dE and dS are the increases of internal energy and entropy and δA the work done *on* the body during any specified small change of state. The special problems of the thermodynamics of polarizable media are concerned entirely with the calculation of δA.

The correct form of δA can only be specified when the nature of the external arrangements for producing the magnetic field have been laid down, owing to the different reactions of the field producing agencies to a change of magnetization in the specimen under discussion.

It is desirable for us to discuss the energy as a function of the magnetization I in a constant magnetic field. To do so it is necessary to suppose that the magnetic field is produced by external electric currents maintained by batteries or dynamos. These conditions are also those in which experiments are carried out.

We will suppose to start with that H is uniform over the specimen, which is isotropic so that $\mathbf{H}$ is always parallel to $\mathbf{I}$, and the specimen itself of volume V is represented outside itself by a magnet of total magnetization VI. Suppose that in time δt the total magnetization increases by $V\delta I$. This will induce an electromotive force in the circuit $\alpha V\,\delta I/\delta t$, where α is a geometrical constant, in such a direction as to keep the flux through the circuit constant, that is in the opposite direction to the current i which produces the field H. To keep H constant the batteries or dynamos must do more work and this

* A more profound study has recently been given by Guggenheim, *Proc. Roy. Soc.* A, vol. 155, pp. 49, 70 (1936).

work is clearly the work done on the specimen. The amount of this work is (current) × (E.M.F.) × δt. But the current is proportional to the magnetic field, being equal to H/α. Hence the work done on the specimen is $VH\delta I$. Change of E so long as I is constant can involve no (magnetic) work, as there is no change of external field of the body, no change of flux through the circuit and therefore no back E.M.F.

There may also be a work term associated with a volume change. Thus in this simple case

$$\delta A = -p\delta V + VH\delta I. \qquad \text{......(1366)}$$

It is easy to see that this can be generalized to

$$\delta A = -p\delta V + \int (\mathbf{H} \,.\, \delta\mathbf{I})\, dV, \qquad \text{......(1367)}$$

when **H** and **I** are not necessarily uniform or parallel. It is to be remembered that in these formulae H is the field produced by external bodies and contains no term arising from I.

As an example of a different work term for different conditions, consider a small body in a field H due to rigidly fixed permanent magnetic poles. There is a force on the body $VI\partial H/\partial x$ in the direction of x increasing if I and H are both directed along the axis of x. This force will do work $VI\delta H$ on the body if the body is allowed to move through a distance δx. This work might naturally appear as kinetic energy, but if it is turned into internal energy we have a state of affairs in which

$$\delta A = -p\delta V + VI\delta H. \qquad \text{......(1368)}$$

The work term (1366) corresponds to the conditions we wish to examine. It is often permissible to ignore volume changes, and it is then more convenient to use the total magnetization VI as a single variable, which in this section only will be denoted by Σ. Then

$$dE = T\,dS + H\,d\Sigma. \qquad \text{......(1369)}$$

This function $E(S,\Sigma)$ is a thermodynamic potential, which will give us all that we require. But it is not always so convenient a function as

$$f(T,H) = E - ST - H\Sigma \quad (df = -S\,dT - \Sigma\,dH), \qquad \text{......(1370)}$$

or Planck's characteristic function $\Psi = -f/T$, which is the function constructed by the statistical method, with the properties

$$T\frac{\partial\Psi}{\partial H} = \Sigma, \quad T^2\frac{\partial\Psi}{\partial T} = E - \Sigma H, \quad C_V = \frac{d}{dT}\left(T^2\frac{\partial\Psi}{\partial T}\right). \qquad \text{......(1371)}$$

Strictly $\Psi = \Psi(V,T,H)$, and we have also $p = T\,\partial\Psi/\partial V$ unless we are ignoring volume changes. This specific heat is strictly $C_{V,H}$, defining the heat required to raise the temperature at constant volume and constant magnetic field.

We have tacitly assumed, in asserting that (1369) defines a thermodynamic potential, that there is a one-one relationship between H and Σ for given T. This precludes hysteresis, but is correct for our idealized system.

By writing (1369) in the form $dS = dE/T - Hd\Sigma/T$ and asserting that $S(T,\Sigma)$ exists, we find at once that

$$\left(\frac{\partial E}{\partial \Sigma}\right)_T = -T^2 \frac{\partial}{\partial T}\left(\frac{H}{T}\right)_\Sigma \quad (H = H(\Sigma, T)). \qquad \text{......(1372)}$$

If Weiss's theory is correct, even if only in the very general sense that

$$\frac{\Sigma}{\Sigma_\infty} = L\left\{\frac{\mu(H + \nu\Sigma/V)}{kT}\right\}, \qquad \text{......(1373)}$$

where $L(x)$ may be Langevin's function itself, or any other reasonable function of the same general form, then (1373) can be inverted and written in the form

$$H = \frac{kT}{\mu}\Lambda\left(\frac{\Sigma}{\Sigma_\infty}\right) - \frac{\nu\Sigma}{V}. \qquad \text{......(1374)}$$

This equation may be taken to be the complete formal expression of Weiss's theory and represents the facts with considerable accuracy for constant ν.* This being so, it follows from (1372) that

$$\frac{\partial E}{\partial \Sigma} = -\frac{\nu\Sigma}{V},$$

and therefore

$$E = E(T, 0) - \tfrac{1}{2}\nu\Sigma^2/V. \qquad \text{......(1375)}$$

Conversely, if we merely have this dependence of E on Σ for no matter what reason, then, by reversing the argument,

$$\frac{H + \nu\Sigma/V}{kT} = \Lambda'\left(\frac{\Sigma}{\Sigma_\infty}\right),$$

and we shall certainly have ferromagnetism so long as we have a suitable form for Λ'. The form of Λ' is of course not determined by thermodynamics.

It will be useful to record here one more formula. Since by definition

$$C_V dT = T(dS)_{V,H},$$

$$C_V = \left(\frac{\partial E(T,0)}{\partial T}\right)_{\Sigma,V} - \left(H + \frac{\nu\Sigma}{V}\right)\left(\frac{d\Sigma}{dT}\right)_V. \qquad \text{......(1376)}$$

When $H = 0$ this may of course be obtained by operating with d/dT on (1375).

From the thermodynamic theory interesting magnetocaloric effects (and similar electrocaloric ones) can be deduced. We have not space to discuss these, for which reference can be made to Debye's article† or the original papers.‡

* Weiss, *J. de Physique*, vol. 2, p. 170 (1921).

† Debye, *loc. cit.*, p. 437.

‡ Weiss, *Archives d. Sci. phys. et nat., Genève*, vol. 45, p. 329 (1918); *J. de Physique*, vol. 6, p. 161 (1921).

§ **12·92.** *Heisenberg's theory of ferromagnetism.** We have now to try and find a model which will give us an E with the correct dependence on $\Sigma\,(=VI)$. Such a model, as we have seen in the preceding section will probably lead us to all the successes of Weiss's theory. We shall of course achieve all our thermodynamic results by constructing and summing the model's partition function.

Since magnetization is a property of the whole system, it is useless to consider an assembly of many practically independent systems. Interactions are the essence of the effect. We consider a single system composed of a large number of "carriers" (atoms) arranged in a regular array as in a crystal. Each carrier carries a single orientating subsystem which is the magnetic element. We have to determine at least roughly the energies and weights of the accessible states of such a system including the interactions of the subsystems, and so form the partition function. The system is subject also to the external field H.

The first question that arises is: What is the magnetic element? Fortunately the correct choice is the simplest possible—the spin of the electron itself. Orbital angular momentum of an electron in an atom, subject to interactions with neither spherical nor axial symmetry, cannot remain quantized at all, and has a mean value zero when the interactions are strong as they must be here. But the coupling of the spins to the unsymmetrical fields of their neighbours is weak, so that the resultant spin (and component spin parallel to H) of the electrons remains well quantized even in the lattice. It will clearly only be necessary to consider electrons in the incomplete shells of the atoms in the lattice. A completed shell has neither resultant orbital angular momentum nor electron spin and is so tightly coupled up in general that the neighbours will hardly disturb it.

This necessary choice of the magnetic element explains in passing the so-called gyromagnetic anomaly in the Richardson-Einstein-de Haas effect. If a delicately suspended ferromagnet is suddenly magnetized, it is found to acquire a definite amount of angular momentum, and the theory of the effect obviously requires that

$$\frac{\text{change of magnetization}}{\text{acquired angular momentum}}=\frac{\text{magnetic moment of carrier}}{\text{angular momentum of carrier}}.$$

It is regularly observed that this ratio is *twice* that to be expected from orbital contributions, but agrees exactly with an electron spin origin for the whole magnetization, since the g-factor of electron spin is 2.†

* Heisenberg, *Zeit. f. Physik*, vol. 49, p. 619 (1928).

† Refined experiments by Barnett [*Rev. Mod. Phys.* vol. 7, p. 129 (1935)] on the reciprocal effect—production of magnetization by spinning the specimen—indicate a ratio somewhat less than 2 so that a small percentage of the magnetization may properly belong to orientatable orbital momentum.

Experiments by Sucksmith [*Proc. Roy. Soc.* A, vol. 135, p. 276 (1932)] have succeeded in detecting the Richardson effect in paramagnetic rare earth salts. Values of g other than 2 are here to be expected and are found.

Since orbital angular momentum is in general totally quenched, it can play no essential primary role in ferromagnetism. We can therefore simplify our model by taking as carriers atoms in S states with orientatable electron spin.

The calculation of the states of such a system of interacting carriers differs only in complication from the calculation by Heitler and London of the states of two interacting hydrogen atoms. No new principles enter. It is the large non-classical exchange energies which provide the energy term depending on the magnetization required for Weiss's theory, whose molecular field thus receives a rational explanation.

We consider therefore a system of $2N$ atoms regularly arranged in a space lattice of volume V, simplifying the problem by assuming that only interactions between nearest neighbours in the lattice need be taken into account, and that the exchange integral for each pair of nearest neighbours is the same. Each atom has z nearest neighbours and contains y electrons whose spins are parallel and orientate as a whole.

§ **12·93.** *The formulation of Heisenberg's partition function.* We can first classify the states of the system by the total component of electron spin parallel to the magnetic field H,

$$m\frac{h}{2\pi} \quad (-Ny \leqslant m \leqslant Ny),$$

remembering that all such accessible states must be antisymmetrical in the electrons. The total number of such states, f_m, is the number of ways in which the spins of the individual atoms can be arranged to give a resultant m. We divide up the $2N$ atoms into $y+1$ groups according to their contribution to m $(\frac{1}{2}y, \frac{1}{2}y-1, \ldots, -\frac{1}{2}y+1, -\frac{1}{2}y)$, the groups containing $a, b, c, \ldots l$ atoms respectively, and the possible arrangements are all those satisfying

$$a+b+\ldots+l=2N, \quad (\tfrac{1}{2}y)a+(\tfrac{1}{2}y-1)b+\ldots+(-\tfrac{1}{2}y)l=m.$$

The total number of such arrangements is the coefficient of ξ^m in

$$[\xi^{\frac{1}{2}y}+\xi^{\frac{1}{2}y-1}+\ldots+\xi^{-\frac{1}{2}y}]^{2N}. \qquad \ldots\ldots(1377)$$

We can leave f_m in this general form. When $y=1$ it obviously reduces to

$$\frac{2N!}{(N-m)!\,(N+m)!}. \qquad \ldots\ldots(1378)$$

Before proceeding it is necessary to assure ourselves that to every one of the arrangements enumerated in (1377) or (1378) based on specified atomic states of the given multiplicity there corresponds just one wave-function for the assembly antisymmetrical in all the electrons. This is a special case of the general theorems proposed in § 5·11. It is therefore only necessary to assure ourselves that to each of the f_m ways of arranging the spins there corresponds exactly one wave-function for the whole system symmetrical or antisymmetrical, according as y is even or odd, in the $2N$ atoms. This however follows immediately, for when we select any one of the f_m arrangements we specify the spin component for each atom and therefore specify completely each atomic wave-function. These are all distinct and from them can be formed just one wave-function for the $2N$ atoms with the required symmetry.

We can now begin to construct the partition function. The f_m states with resultant spin $mh/2\pi$ have an energy in the external field H equal to

$$-2Hm\mu_B, \qquad \text{......(1379)}$$

where $\mu_B = \epsilon h/(4\pi\mu c)$ is Bohr's magneton. Let their energies apart from the magnetic field be E_j, $j = 1, \ldots, f_m$. Then the partition function for the system is

$$K = \sum_{m=-Ny}^{Ny} \exp\left[\frac{2m}{kT}\mu_B H\right] \sum_{j=1}^{f_m} \exp\left[-\frac{E_j}{kT}\right]. \qquad \text{......(1380)}$$

Various methods of approximation have to be adopted to calculate the E_j and the sum involving them. At very low temperatures only the smallest E_j are relevant. They can be best investigated by the methods of Bloch and Slater of which some account is given in § 12·942. Such investigations however tell us nothing about the Curie point which is the most interesting region. Here the only success has been achieved by Heisenberg's guess that it is legitimate to approximate to the energies E_j for given m by regarding them as a Gaussian distribution of energies about a mean value $\overline{E_m}$, or even more roughly still to approximate by writing

$$\sum_{j=1}^{f_m} \exp\left(-\frac{E_j}{kT}\right) = f_m \exp\left(-\frac{\overline{E_m}}{kT}\right),$$

where $\overline{E_m}$ is the arithmetic mean of the E_j. We will discuss first this roughest approximation, since for $y = 1$ we can give an elementary proof of the formula for $\overline{E_m}$.

If we refer to Heitler and London's theory of the interaction of two hydrogen atoms—a calculation which is typical of any two atoms, each with a single unpaired electron in an S-state—we find that for $m = 1$ there is one state and the energy $\overline{E_m}$ is $J_E - J$ while for $m = 0$ there are two states and $\overline{E_m}$ is equal to J_E. J_E is an integral representing the classical electrostatic interaction energy of the two systems, and J is an integral representing the exchange energy. Thus for parallel spins we get an exchange energy $-J$ in the mean energy which we do not get when the spins are antiparallel. It is reasonable to suppose that, as we are only considering interactions of neighbours, we may generalize this result to our system of $2N$ atoms. We shall therefore find a term $-J$ in the mean energy for given m whenever a neighbouring pair of spins are parallel.

The total number of ways in which we can arrange two spins as neighbours both parallel to H is the number of ways of choosing two neighbouring atoms $(2Nz/2!)$ times the number of ways of putting the other $2N-2$ spins $N+m-2$ parallel to H and $N-m$ antiparallel in the other atoms, that is, in all,

$$\frac{(2N-2)!}{(N-m)!\,(N+m-2)!}Nz.$$

The total number of ways of arranging two spins as neighbours both antiparallel to H is similarly

$$\frac{(2N-2)!}{(N-m-2)!\,(N+m)!}Nz.$$

The total energy contribution by the exchange effects is therefore

$$-J\left\{\frac{(2N-2)!}{(N-m-2)!(N+m)!}+\frac{(2N-2)!}{(N-m)!\,(N+m-2)!}\right\}Nz,$$

and the number of states in the set is $(2N)!/\{(N-m)!\,(N+m)!\}$. Thus we get a contribution to $\overline{E_m}$ of

$$-zJ\frac{N^2+m^2}{2N}, \qquad \text{......(1381)}$$

neglecting the difference between N and $N-1$. To this must be added a classical term J_E independent of m. When $y>1$ the result is only altered* by having zy for z and Ny for N, so that

$$\overline{E_m}=J_E-zyJ\frac{N^2y^2+m^2}{2Ny}, \qquad \text{......(1382)}$$

though (1382) cannot be proved in this simple way.

As a result of these elementary calculations we find the approximate partition function

$$K=\sum_{m=-Ny}^{Ny} f_m \exp\left[-\frac{J_E}{kT}+\alpha m+\beta\frac{m^2+N^2y^2}{2Ny}\right], \qquad \text{......(1383)}$$

where
$$\alpha=\frac{2\mu_B H}{kT}, \quad \beta=\frac{zyJ}{kT}$$

for shortness. This elementary account is due to Heisenberg. We see at once from (1381) that if J is large and positive we have the correct type of dependence of $\overline{E_m}$ on m demanded by (1375) and can proceed hopefully with the evaluation of K.

Heisenberg's original deeper discussion of K is as follows. For the proposed quantum model not only is m a "good" quantum number, but also s, the resultant spin of the state. We can therefore still further subdivide the states into states of given s and m, $|m|\leqslant s$. Every state of given s $(\geqslant |m|)$

* Heisenberg, *Probleme der modernen Physik*, p. 114 (1928).

belongs to the set of f_m states of given m. Thus if ϕ_s is the number of states of given s, then

$$f_{|m|}=\phi_{|m|}+\phi_{|m|+1}+\dots+\phi_{Ny}, \qquad \text{......(1384)}$$

$$\phi_s=f_s-f_{s+1}. \qquad \text{......(1385)}$$

We can therefore write down a more detailed form of K, namely, in the absence of a magnetic field H,

$$K=\sum_{s=0}^{Ny}(2s+1)\sum_{j=1}^{\phi_s}e^{-E_j^{(s)}/kT}. \qquad \text{......(1386)}$$

The factor $(2s+1)$ occurs because each state of spin s has $(2s+1)$ orientations and is to that degree degenerate in the absence of H. When $H\neq 0$ the orientational degeneracy is removed and we have

$$K=\sum_{s=0}^{Ny}\sum_{m=-s}^{s}\exp\left(\frac{2m\mu_B}{kT}H\right)\sum_{j=1}^{\phi_s}\exp\left(-\frac{E_j^{(s)}}{kT}\right). \qquad \text{......(1387)}$$

As we cannot evaluate the $E_j^{(s)}$ this exact form is still of little use. For the study of the region of the Curie point Heisenberg* has approximated to it by writing $E_j^{(s)}=\overline{E}_s+\Delta E_s$ and assuming that the ΔE_s are densely distributed in a Gaussian error curve. The $\overline{E}_s$ and $\overline{(\Delta E_s)^2}$ can be computed from the secular perturbation equations; we can only quote the results here:

$$\overline{E}_s=-zy\frac{s^2+y^2N^2}{2yN}J+J_E \quad (y\geqslant 1). \qquad \text{......(1388)}$$

$$\overline{(\Delta E_s)^2}=\Delta_s^2=z\frac{(y^2N^2-s^2)(3y^2N^2-s^2)}{4yN^3}J^2 \quad (y=1). \text{......(1389)}$$

For $y>1$ the exact value of Δ_s^2 is unknown. We shall use the general symbol Δ_s as far as possible. It is now possible without trouble to reduce the partition function to the form

$$K=\sum_{m=-yN}^{yN}e^{\alpha m}\sum_{s=|m|}^{yN}\phi_s\exp\left[\beta\frac{s^2+y^2N^2}{2yN}-\frac{J_E}{kT}+\frac{1}{2}\frac{\Delta_s^2}{k^2T^2}\right]. \qquad \text{......(1390)}$$

From here a rather tedious argument leads us to the result

$$K=O(1)\sum_{m=-yN}^{yN}f_m\exp\left[\alpha m+\beta\frac{m^2+y^2N^2}{2yN}-\frac{J_E}{kT}+\frac{1}{2}\frac{\Delta_m^2}{k^2T^2}\right], \qquad \text{......(1391)}$$

where $O(1)$ is a factor leading to a negligible term in $\log K$. This is the standard form of K in Heisenberg's theory. It is however unnecessary to refer to s at all to arrive at (1391). We can obtain it directly from (1380) in place of (1383) by taking into account a Gaussian spread in the f_m values of E_j.

The factors in K or $K(T,V,H)$, which are independent of m, should contain all the main terms which give cohesion, rigidity and elastic properties in general to the crystal. This they palpably fail to do, for the model and the calculations are far too restricted. We can properly replace them by semi-

* Heisenberg, *loc. cit.*; Van Vleck, *loc. cit.* p. 322.

empirical terms later on, and for the present omit them altogether. We are thus left with a magnetic factor K_m of the complete partition function given by

$$K_m = K_m(T,V,H) = \sum_{m=-yN}^{yN} f_m \exp\left[\alpha m + \frac{\beta m^2}{2yN} + \frac{1}{2}\frac{\Delta_m{}^2}{k^2T^2}\right]. \quad \text{......(1392)}$$

To proceed further we need an asymptotic form for f_m, which we can obtain from (1377) and Cauchy's theorem by the usual method of steepest descents. For

$$f_m = \frac{1}{2\pi i}\int_\gamma \frac{d\xi}{\xi^{m+1}}[P(\xi)]^{2N} \quad (P(\xi) = \xi^{\frac{1}{2}y} + \xi^{\frac{1}{2}y-1} + \ldots + \xi^{-\frac{1}{2}y}),$$

where γ is any contour going once counterclockwise round $\xi = 0$. If ξ_m is fixed as the positive real root of

$$\frac{d}{d\xi}\frac{[P(\xi)]^{2N}}{\xi^m} = 0, \quad \frac{\xi_m P'(\xi_m)}{P(\xi_m)} = \frac{m}{2N}, \quad \text{......(1393)}$$

then effectively

$$f_m = [P(\xi_m)]^{2N}/\xi_m{}^m,$$

and we can write

$$K_m = \sum_{m=-yN}^{yN} \exp\left[2N\log P(\xi_m) - m\log\xi_m + \alpha m + \beta\frac{m^2}{2yN} + \frac{1}{2}\frac{\Delta_m{}^2}{k^2T^2}\right], \quad \text{......(1394)}$$

ξ_m being determined as a function of m by (1393).

§ **12·931.** *The final evaluation of K_m. Existence of ferromagnetism.* We have lastly to evaluate K_m by searching for the maximum term and therefore for the maximum exponent in (1394). In differentiating with respect to m, ξ_m must be allowed to vary, but from the definition of ξ_m the coefficient of $d\xi_m/dm$ vanishes. Thus m_0, the value of m for any stationary term, is a root of the simultaneous equations

$$\log\xi_{m_0} = \alpha + \beta\frac{m_0}{yN} + \frac{\Delta_{m_0}\Delta'_{m_0}}{k^2T^2}, \quad \frac{\xi_{m_0}P'(\xi_{m_0})}{P(\xi_{m_0})} = \frac{m_0}{2N}.$$

We can write these equations more simply with the notation

$$\zeta = \frac{m_0}{Ny}, \quad \xi_{m_0} = e^{2x};$$

ζ is the fraction of absolute saturation achieved at the root m_0. Then stationary values are given by

$$2x = \alpha + \beta\zeta + \frac{\Delta(\zeta)\Delta'(\zeta)}{k^2T^2}, \quad \text{......(1395)}$$

$$\zeta = \frac{1}{y}\frac{ye^{xy} + (y-2)e^{x(y-2)} + \ldots + (-y)e^{-xy}}{e^{xy} + e^{x(y-2)} + \ldots + e^{-xy}}. \quad \text{......(1396)}$$

The right-hand side of (1396) is Brillouin's function. When $y = 1$, $\zeta = \tanh x$. Equations (1395) and (1396) on comparison with (1361) can already be seen to be about to yield us ferromagnetism, but logically we have still to complete the evaluation of K_m by finding its maximum term.

To be explicit we now take the form (1389) for $\Delta_s{}^2$, from which it follows that

$$\frac{\Delta(\zeta)\,\Delta'(\zeta)}{k^2T^2}=\frac{\beta^2}{zy}\{\tfrac{1}{2}\zeta^3-\zeta\}.$$

Equation (1395) then becomes

$$2x=\alpha+\beta\left(1-\frac{\beta}{yz}\right)\zeta+\frac{\beta^2}{2yz}\zeta^3. \qquad \text{......(1397)}$$

The slope of the Brillouin function at the origin is

$$\left(\frac{d\zeta}{dx}\right)_{\zeta=0}=\frac{1}{y}\frac{y^2+(y-2)^2+\ldots+(-y)^2}{y+1}=\frac{y+2}{3}. \qquad \text{......(1398)}$$

If the slope of the curve (1397) at $\zeta=0$ is less than $\frac{1}{3}(y+2)$, we have non-zero roots for $\alpha=0$ and roots large compared with α when $\alpha\neq0$. If the slope is greater, then we have only a zero or small root. This critical condition is

$$\tfrac{1}{2}\beta\left(1-\frac{\beta}{yz}\right)>\frac{3}{y+2}. \qquad \text{......(1399)}$$

The equation

$$\beta_c\left(1-\frac{\beta_c}{yz}\right)=\frac{6}{y+2}$$

determines a critical temperature T_c in terms of J, which will of course represent the Curie point.

To prove that when it exists the large root of the same sign as α gives the absolute maximum of the terms of K_m we can proceed as follows, confining ourselves to $y=1$ and omitting the terms in β^2 to avoid complications of little significance. To this approximation

$$K_m=\sum_{m=-\nu N}^{\nu N}\exp[2N\log 2N-(N+m)\log(N+m)-(N-m)\log(N-m)$$
$$+\alpha m+\beta m^2/2N],$$

and stationary terms occur where

$$\alpha+\beta\zeta=\log\frac{1+\zeta}{1-\zeta}, \qquad \text{......(1400)}$$

at which points

$$Q=\frac{d^2}{dm^2}[\text{exponent}]=-\frac{2N}{N^2-m^2}+\frac{\beta}{N}.$$

When $\beta<2$, $Q<0$ (*all m*), so that the single stationary value is a maximum. When $\beta>2$, $Q>0$ for $m=0$ and $Q<0$ when $m^2\to N^2$; Q can vanish for just one value of m^2 between 0 and N^2. For values of α for which there are three stationary values, roots of (1400), the roots of Q for $\pm$ values of m must occur between the middle and both extreme roots. For the extreme roots therefore $Q<0$, and the stationary values are maxima. For greater values of α for which there is only one root of (1400), since $d\zeta/d\alpha>0$ the corresponding value of m has increased, $Q<0$, and the single stationary value is still a maximum.

It remains to consider which of the extreme roots, when they exist, gives the absolute maximum. For $\alpha=0$ the maxima are identical by symmetry. We examine therefore the variation of the exponent at the maximum as α varies. Thus

$$\frac{d(\text{exponent})}{d\alpha}=m_0+\frac{\partial(\text{exponent})}{\partial m_0}\frac{\partial m_0}{\partial x}=m_0,$$

and the exponent increases with α at the root for which $m_0>0$. The absolute maximum occurs at that root which has the same sign as α.

Returning to the general case, we can now apply the familiar argument, easily rendered rigorous, that since N is very large K_m is effectively equal to its maximum term. Therefore

$$\log K_m=\log K_m(V,T,H)$$
$$=2N\left[\log\{e^{xy}+e^{x(y-2)}+\ldots+e^{-xy}\}-y\zeta x+\tfrac{1}{2}\alpha y\zeta+\tfrac{1}{4}\beta\left(1-\frac{\beta}{yz}\right)y\zeta^2+\frac{\beta^2}{16z}\zeta^4\right], \quad \ldots\ldots(1401)$$

where x and ζ are determined by the proper root of

$$2x=\alpha+\beta\left(1-\frac{\beta}{yz}\right)\zeta+\frac{\beta^2}{2y}\zeta^3 \quad \left[\frac{\partial\log K_m}{\partial\zeta}=0\right], \quad \ldots\ldots(1402)$$

$$\zeta=\frac{1}{y}\frac{ye^{xy}+(y-2)\,e^{x(y-2)}+\ldots+(-y)\,e^{-xy}}{e^{xy}+\ldots+e^{-xy}} \quad \left[\frac{\partial\log K_m}{\partial x}=0\right]. \quad \ldots\ldots(1403)$$

We recall that K_m depends on T through the denominators of α and β, on H through α and on V through J and therefore through β.

This is the magnetic factor in the partition function. Since in general

$$IV=kT\left(\frac{\partial}{\partial H}\log K_m\right)_{V,T},$$

we find, in virtue of (1402) and (1403), that

$$I=\frac{kT}{V}\frac{\partial\log K_m}{\partial\alpha}\frac{\partial\alpha}{\partial H} \quad \left(\alpha=\frac{2\mu_B H}{kT}\right),$$
$$=2\mu_B m_0/V. \quad \ldots\ldots(1404)$$

The intensity of magnetization is therefore m_0/V double-magnetons, where m_0 corresponds to the proper root of (1402) and (1403). Saturated intensity I_∞ is Ny/V double-magnetons so that

$$I=I_\infty\zeta. \quad \ldots\ldots(1405)$$

§ **12·94.** *Phenomena of the Curie point. Its existence, and the behaviour of the specific heat.* We have already seen that the condition for the ferromagnetic state is

$$\beta\left(1-\frac{\beta}{yz}\right)>\frac{6}{y+2} \quad \left(\beta=\frac{zyJ}{kT}\right). \quad \ldots\ldots(1406)$$

This condition requires that β should lie between β_1 and β_2, where

$$\beta_1, \beta_2 = \frac{yz}{2}\left[1 \pm \sqrt{\left\{1 - \frac{24}{y(y+2)z}\right\}}\right];$$

these β's are real only if $\qquad y(y+2)z > 24. \qquad$ ......(1407)

We shall see that probable values of y and z are

$$\text{Ni: } y \sim 1,\ z=12; \quad \text{Co: } y \sim 2,\ z=12; \quad \text{Fe: } y \sim 3,\ z=8;$$

for these actual values (1407) is always amply satisfied. Roughly speaking β must lie between yz and $6/(y+2)$. The upper limit yz to β, a lower limit to T, corresponds to no physical feature and is an inadequacy of the theory. What really happens is that as T decreases the Gaussian approximation becomes inadequate and the states of low energy, not the states of mean energy, take charge. The essential condition, equivalent to neglecting the Gaussian spread altogether, is

$$\beta > \frac{6}{y+2}, \qquad T < T_c = \frac{y(y+2)z}{6}\frac{J}{k}. \qquad \text{......(1408)}$$

From this equation, an assumed y and an observed T_c J can be calculated. J must be positive for ferromagnetism and, for $T_c \cong 10^3$, $J \cong 10^{-1} - 10^{-2}$ electron volt. The ortho-para separations in helium and other atoms are of the order of 10^{-1} volt, which admits the proposed explanation.

The extra part of the internal energy that depends on the magnetization must, by general theory, be given by

$$\overline{E_m} = kT^2\left(\frac{\partial \log K_m}{\partial T}\right)_{V,H} = kT^2\left[\frac{\partial \log K_m}{\partial \alpha}\frac{\partial \alpha}{\partial T} + \frac{\partial \log K_m}{\partial \beta}\frac{\partial \beta}{\partial T}\right],$$

$$= 2NkT^2\left[-\frac{1}{2}\frac{\alpha y\zeta}{T} - \frac{1}{4}\frac{\beta}{T}\left(1 - \frac{2\beta}{yz}\right)y\zeta^2 - \frac{\beta^2}{8zT}\zeta^4\right]. \qquad \text{......(1409)}$$

Let us consider one mole ($2Nk = R$) in zero field ($\alpha = 0$). Then above the Curie point $\zeta = 0$ and therefore $\overline{E_m} = 0$. Below the Curie point the extra part of the specific heat, $C_V{}^m$, is given by

$$C_V{}^m = \frac{d\overline{E_m}}{dT} = RT^2\frac{d}{dT}\left[-\frac{1}{4}\frac{\beta}{T}\left(1 - \frac{2\beta}{yz}\right)y\zeta^2 - \frac{\beta^2}{8zT}\zeta^4\right]. \qquad \text{......(1410)}$$

This expression does not tend to zero as $T \to T_c$ from below and $C_V{}^m$ is therefore discontinuous. The value of the discontinuity $\Delta C_V{}^m$ affords an important comparison between theory and experiment. Since $\zeta^2 \to 0$ the theoretical value of the discontinuity is

$$\Delta C_V{}^m = C_V{}^m(T_c - 0) - C_V{}^m(T_c + 0) = RT^2\left[-\frac{1}{4}\frac{\beta}{T}\left(1 - \frac{2\beta}{yz}\right)y\frac{d\zeta^2}{dT}\right]_{T \to T_c}. \qquad \text{......(1411)}$$

Since $T \sim T_c$ it is sufficiently accurate for our purposes to take

$$\beta\left(1-\frac{2\beta}{yz}\right) \sim \beta\left(1-\frac{\beta}{yz}\right) \sim \frac{6}{y+2}. \qquad \text{......(1412)}$$

To determine $d\zeta^2/dT$ we return to equations (1402) and (1403) for small ζ and x, from which we can derive that

$$\frac{d\zeta^2}{dT} \sim \frac{2(y+2)}{3T_c} \frac{A}{Ay - B/(y+2)}$$

with $A = y^2 + (y-2)^2 + \ldots + (-y)^2$, $B = y^4 + (y-2)^4 + \ldots + (-y)^4$.

We thus find

$$\Delta C_V{}^m = \frac{R}{1 - \dfrac{B}{y(y+2)A}}. \qquad \text{......(1413)}$$

This formula *overestimates* $\Delta C_V{}^m$, since we have everywhere made the approximation (1412) in the numerator.

It is at once clear that the theory gives results of the right order, but the

TABLE 51.

The discontinuity $\Delta C_V{}^m$ cal./deg./mole in the specific heat at the Curie point.

Substance	$\Delta C_V{}^m$ theory	$\Delta C_V{}^m$ Honda*	$\Delta C_V{}^m$ Weiss†	y from saturation
Fe	4·4 ($y=3$)	4·6	6·8	2
Co	4·0 ($y=2$)	2·8	—	1·5
Ni	3·0 ($y=1$)	2·1	1·7	0·6
Fe_3O_4 (magnetite)	4·4 ($y=3$)	—	6·1	—
—	5·0 ($y=\infty$)	—	—	—

* Honda, *Zeit. f. Physik*, vol. 63, p. 147 (1930).

† Weiss, *Archives d. Sci. phys. et nat., Genève*, vol. 42, p. 378 (1917), vol. 43, pp. 22, 113, 199 (1917).

Some magnetic minerals appear to have even larger discontinuities.

agreement is not close and in fact no exact agreement at all is possible for Fe, since on the present version of the theory the observed values are greater than the greatest possible theoretical value. The values of y in the last column are derived from measurements of the saturated magnetization at low temperatures. They do not agree with the most natural integral values given in the second column, or with the values which would be derived from the paramagnetic susceptibility above the Curie point. We have seen in §§ 11·541, 11·55 how such fractional values may be expected to arise, and the present version of the theory is not supple enough to cater for them. If we bear in mind these refinements which must be incorporated in the construction and evaluation of a more accurate K_m, we may rest well satisfied with the result of the test of the rough theory shown in Table 51.

One obvious improvement of the theory is to remove the restriction that all the y electrons supplied by one atom orientate as a whole; this restriction is almost automatically removed when we consider electron states as distributed over bands, but in the present theory it can be approximately removed as follows, when we can see that the removal will lead to greater values of $\Delta C_V{}^m$. For consider the extreme case when the interactions of the y-electrons in a single atom are identical with their interactions with the y-electrons in a neighbouring atom. This is the same as the preceding model but with $y'=1$, $z'=zy+y-1$, while one mole of such a system supplies $2Ny$ orientating subsystems and therefore the effects are all y times as great. It follows that for this model $\Delta C_V{}^m = 3y$ cal./deg./mole.

We note in passing that it is found that the behaviour of the specific heat near the Curie point is practically independent of any magnetization of the system in bulk. This must be accepted as definite proof that ferromagnetics even when apparently unmagnetized consist of microelements magnetized to their natural saturation for the temperature in question, but in directions so orientated that the bulk-magnetization is neutralized. It is convenient to introduce a word to describe this characteristic state of affairs. We shall say that a ferromagnetic below the Curie point, unmagnetized in bulk, is *micromagnetized*, and that when we magnetize it in bulk it becomes *holomagnetized*; this process consists primarily in removing the randomness of the directions of magnetization of the microelements. We discuss its features in detail in later sections.

§12·941. *Change of size at the Curie point.* To study the change of size at the Curie point we must use a more complete partition function $K(T,V,H)$, which we take in the form

$$\log K(T,V,H) = \log K_m(T,V,H) - F(V)/kT,$$

where $F(V)$ is the energy of the crystal at low temperatures as a function of V. It is legitimate to neglect the effect of the thermal vibration of the lattice in the first approximation.

Since $p = kT\partial \log K/\partial V$, and to a sufficient approximation $p=0$, and since K_m depends on V only through J and therefore β, we find

$$2N\frac{\partial \beta}{\partial V}\left\{\tfrac{1}{4}y\left(1-\frac{2\beta}{yz}\right)\zeta^2+\frac{1}{8}\frac{\beta}{z}\zeta^4\right\}-\frac{1}{kT}\frac{\partial F}{\partial V}=0.$$

When $T > T_c$ the size is given by $(\partial F/\partial V)_0 = 0$. Therefore

$$\frac{\partial F}{\partial V} = \delta V\left(\frac{\partial^2 F}{\partial V^2}\right)_0 = \frac{\delta V}{V_0}\frac{1}{\kappa_0},$$

where κ_0 is the volume compressibility. We thus find

$$\frac{\delta V}{V_0} = 2Nyz\kappa_0\frac{\partial J}{\partial V_0}\left\{\tfrac{1}{4}y\left(1-\frac{2\beta}{yz}\right)\zeta^2+\frac{1}{8}\frac{\beta}{z}\zeta^4\right\}. \qquad \text{......(1414)}$$

This change of volume as T passes through T_c is of course to be superposed on the ordinary thermal expansion here neglected. It appears therefore as an anomalous variation in the coefficient α of thermal expansion, in the region just below the Curie point. Figs. 61, 62 show the effect on α for nickel and the integrated effect on $\delta l/l$ for iron. The anomaly must be estimated by extrapolating the curves in a normal manner from above the Curie point and taking the difference between these extrapolated and the observed values. The data are as follows:

Fe*. $T_c = 1041°$K., $\kappa_0 = 0{\cdot}6 \times 10^{-12}$, $2N/V_0 = 8{\cdot}6 \times 10^{22}$, $y = 3$, $z = 8$.

At 991° K. $\zeta = 0{\cdot}4$, $\delta V/V_0 = 3\delta l/l = 3{\cdot}3 \times 10^{-4}$.

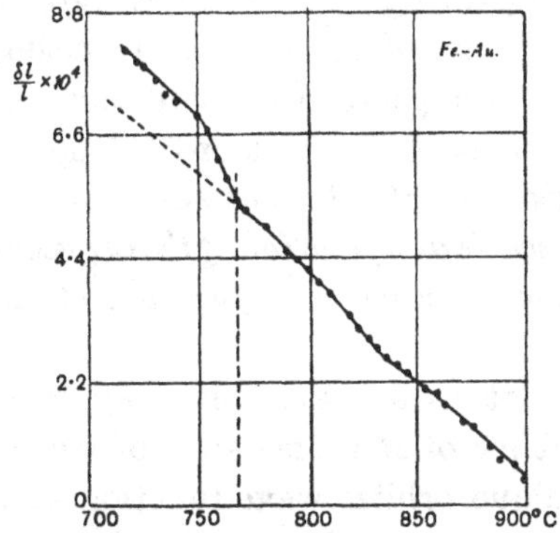

Fig. 61. The differential linear expansion of Fe and Au showing the anomaly in Fe near the Curie point.

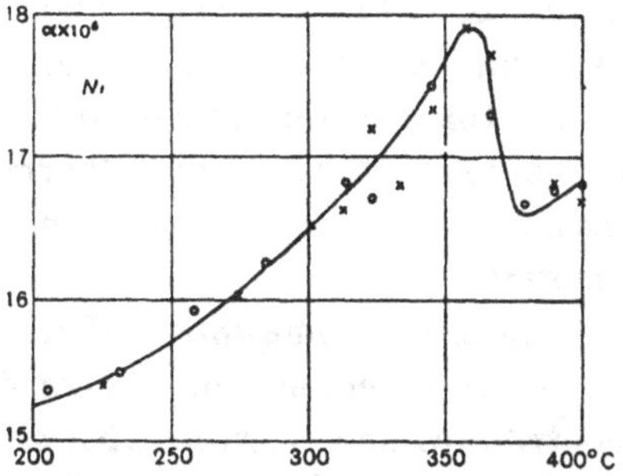

Fig. 62. The linear thermal expansion coefficient for Ni, showing its variation near the Curie point.

From (1414) neglecting the terms in β we thus obtain

$$zV_0\frac{\partial J}{\partial V_0} = 1{\cdot}7 \times 10^{-14}, \quad z\frac{\partial J}{\partial a} = 2{\cdot}1 \times 10^{-6},$$

where a is the distance between nearest neighbours. From T_c we find also $zJ = 5{\cdot}7 \times 10^{-14}$ and therefore

$$\frac{a}{J}\frac{\partial J}{\partial a} = 0{\cdot}9.$$

Ni†. $T_c = 647°$ K., $\kappa_0 = 0{\cdot}6 \times 10^{-12}$, $2N/V_0 = 9{\cdot}4 \times 10^{22}$, $y = 1$, $z = 12$.

At 493° K. $\zeta = 0{\cdot}7$, $\partial V/V_0 = 3\delta l/l = -2{\cdot}7 \times 10^{-4}$.

From these data

$$z\frac{\partial J}{\partial a} = -4{\cdot}7 \times 10^{-6}, \quad zJ = 1{\cdot}8 \times 10^{-13},$$

$$\frac{a}{J}\frac{\partial J}{\partial a} = -0{\cdot}6.$$

* Benedicks, *J. Iron and Steel Institute*, vol. 89, p. 407 (1914).

† Colby, *Phys. Rev.* vol. 30, p. 506 (1910).

Assuming the correctness of the theoretical background, these may be called observed values of $\partial J/\partial a$. We can make only a rough theoretical estimate thus: The form of J for the interaction of two normal hydrogen atoms* at a distance a is

$$J = \sum_{n=-1}^{3} A_n \left(\frac{a}{a_0}\right)^n e^{-2a/a_0}, \qquad \text{......(1415)}$$

where the A_n's are all of the same order and a_0 is the radius of Bohr's first orbit in hydrogen. If we may use an expression of the same form here, with a/a_0 between 1 and 2, then $(a/J)(\partial J/\partial a)$ will be of order unity but may be of either sign. Thus the theory is perfectly consistent with the observed values.

§12·942. *Better calculations of $K(V,T,H)$ especially for low temperatures.*† It is possible to improve on Heisenberg's guess of $\overline{E_m}$ and the resulting value of $K(V,T,H)$, particularly for low temperatures, by discussing systematically the energies and wave-functions of the electrons belonging to an atomic lattice, rather in the manner of §11·51. Here however we are concerned with *the energy of the crystal as a whole as a function of the arrangement of all the electron spins, not of a single electron as a function of its momenta.*

A notation somewhat different from that used previously may conveniently be employed. We consider a lattice of N atoms each of which contributes one electron with a given unique orbital wave-function and freely orientatable spin. Let the number of spins pointing against the field H (negative spins) be denoted by r. The total spin resolved along the field is as usual denoted by $mh/2\pi$, so that $2m = N - 2r$.

The atoms are labelled with numbers 1, 2, ..., N, and those with negative spins occur at $f_1, f_2, \ldots, f_r$, where $1 \leqslant f_1 < f_2 < \ldots < f_r \leqslant N$. This is a definite state of the assembly as a whole represented by one wave-function antisymmetrical in all the electrons, which we shall call $\psi(f_1, f_2, \ldots, f_r)$. We shall assume that these ψ's are all orthogonal and normalized; there are 2^N of them in all for all r and

$$\frac{N!}{r!\,(N-r)!}$$

for a given value of r. If we neglect all the spin and exchange interactions betweent he electrons of different atoms, all these ψ's represent states of the same energy, for they are all solutions, to this approximation, of Schrödinger's equation

$$\{H(1) + H(2) + \ldots + H(N) - E_0\}\Psi = 0.$$

* Heitler and London, *loc. cit.*

† Bloch, *Zeit. f. Physik*, vol. 61, p. 206 (1930). See also Sommerfeld and Bethe, *loc. cit.*; Nordheim, *loc. cit.*; Epstein, *Phys. Rev.* vol. 41, p. 91 (1932); the last investigation however contains oversights pointed out by Bethe.

This equation has next to be completed by the interaction terms and becomes

$$\{H(1)+H(2)+\ldots+H(N)+\sum_{s<t} V_{st}-E\}\Psi=0, \qquad \ldots\ldots(1416)$$

to be solved approximately by E and Ψ of the form

$$E=E_0+\eta, \quad \Psi=\sum_{(f)} a(f_1,\ldots,f_r)\,\Psi(f_1,\ldots,f_r). \qquad \ldots\ldots(1417)$$

The a's and η are determined by the usual first order equations

$$-\eta a(f_1,\ldots,f_r)+\int \psi^*(f_1,\ldots,f_r)\,(\sum_{s<t} V_{st})\,[\sum_{(f')} a(f_1',\ldots,f_r')\,\psi(f_1',\ldots,f_r')]\,d\omega=0. \qquad \ldots\ldots(1418)$$

These equations simplify greatly. If $r \neq r'$, the terms of the interaction matrix vanish, so that we may confine attention at any time to a constant value of r. Since each term V_{st} depends on the coordinates of only two electrons, the only terms in the matrix which give a non-zero result are those for which $(f_1',\ldots,f_r')=(f_1,\ldots,f_r)$ and those for which $(f_1',\ldots,f_r')$ differs from $(f_1,\ldots,f_r)$ by the interchange of a single pair of opposite spins. [Interchanges of like spins do not give a new value of ψ.] The secular equations (1418) therefore reduce to sets of equations of constant r of the form

$$(A-\eta)\,a(f_1,\ldots,f_r)+\sum_{(f')}[a(f_1,\ldots,f_r)-a(f_1',\ldots,f_r')]\,J_{f,f'}, \qquad \ldots\ldots(1419)$$

in which $J_{f,f'}$ is the exchange integral for that pair of atoms for which a negative and positive spin have been interchanged between the sets (f) and (f'). A represents an interaction energy (partly Coulombian) which is independent of the number and arrangement of negative spins, and can therefore be absorbed in E_0 and disregarded for the rest of the present investigation. Since the exchange integral decreases rapidly with the distance apart of the pair of atoms, it is sufficient to retain exchanges between nearest neighbours, and to assume that $J_{f,f'}$ has the same value for every pair of nearest neighbours. Equations (1419) then reduce to

$$-\eta a(f_1,\ldots,f_r)+J\sum_{(nb)}[a(f_1,\ldots,f_r)-a(f_1',\ldots,f_r')], \qquad \ldots\ldots(1420)$$

the summation being for all (f') over the nearest neighbours of (f). Particular examples of these equations can easily be written down explicitly. For a linear chain and $r=1$, for example, we have

$$-\eta a(f_1)+J[2a(f_1)-a(f_1+1)-a(f_1-1)]=0. \qquad \ldots\ldots(1421)$$

For a linear chain and $r=2$, and provided that $f_2>f_1+1$,

$$-\eta a(f_1,f_2)+J[4a(f_1,f_2)-a(f_1+1,f_2)-a(f_1-1,f_2)$$
$$-a(f_1,f_2+1)-a(f_1,f_2-1)]=0, \qquad \ldots\ldots(1422)$$

but if $f_2=f_1+1$,

$$-\eta a(f_1,f_1+1)+J[2a(f_1,f_1+1)-a(f_1-1,f_1+1)-a(f_1,f_1+2)]=0. \qquad \ldots\ldots(1423)$$

Suitable boundary or periodicity conditions must also be imposed.

Equations (1421) admit the solution

$$a(f_1)=e^{i\kappa_1 f_1} \qquad \text{......(1424)}$$

for any value of κ_1, the corresponding value of η being given by

$$\eta=4J\sin^2\tfrac{1}{2}\kappa_1. \qquad \text{......(1425)}$$

Equations (1422) admit the solution

$$a(f_1,f_2)=e^{i(\kappa_1 f_1+\kappa_2 f_2)} \qquad \text{......(1426)}$$

for any κ_1, κ_2, the corresponding value of η being given by

$$\eta=4J(\sin^2\tfrac{1}{2}\kappa_1+\sin^2\tfrac{1}{2}\kappa_2). \qquad \text{......(1427)}$$

For r negative spins the main equations analogous to (1422) again admit the solution

$$a(f_1,\ldots,f_r)=e^{i(\kappa_1 f_1+\ldots+\kappa_2 f_2)} \qquad \text{......(1428)}$$

for any values of κ_1, ..., κ_2 with

$$\eta=4J\left(\sum_{\nu=1}^{r}\sin^2\tfrac{1}{2}\kappa_\nu\right). \qquad \text{......(1429)}$$

If we now impose a periodicity condition on $a(f_1,\ldots,f_r)$ that it shall repeat itself when any f is increased by N, then

$$\kappa_\nu=2\pi k_\nu/N \quad (k_\nu=0,1,\ldots,N-1). \qquad \text{......(1430)}$$

In order to complete the solution it is necessary to make the $a(f_1,\ldots f_r)$ satisfy also the conditions (1423). This can be done for $r=2$ by taking suitable linear combinations of the four possible solutions

$$e^{i(\kappa_1 f_1+\kappa_2 f_2)},\quad e^{-i(\kappa_1 f_1+\kappa_2 f_2)},\quad e^{i(\kappa_2 f_1+\kappa_1 f_2)},\quad e^{-i(\kappa_2 f_1+\kappa_1 f_2)},$$

all of which correspond to the same value of η, and similarly for general values of r. When this is done it is found that the periodicity conditions are no longer exactly (1430). We can now say that the exact energy is given by the equation

$$\eta=4J\sum_{\nu=1}^{r}\sin^2\frac{\pi k_\nu}{N}, \qquad \text{......(1431)}$$

in which the k_ν are not strictly integers, but determined by certain complicated phase relations. It can be shown however that the smaller k_ν are very close to the simple integral values given by (1430).

The number of distinct states apparently given by (1428) and (1430) for given r is $(N+r-1)!/r!(N-1)!$. The correct number is smaller, being of course $N!/r!(N-r)!$. It has been shown however by Bethe that (1431) may be used to give the correct numbers and energies of the states for small spin wave numbers k_ν and values of r which are not too large.

The generalization to lattices in two and three dimensions is easy. We shall consider only a simple cubical array ($N=G^3$) in which the atoms are labelled with the numbers (f,g,h). The nearest neighbours to (f,g,h) are $(f\pm 1,g,h)$, $(f,g\pm 1,h)$ and $(f,g,h\pm 1)$. In place of (1424) we now have

$$a(f_1,g_1,h_1)=e^{i(\kappa_1 f_1+\lambda_1 g_1+\mu_1 h_1)},$$

with a corresponding energy

$$\eta=4J(\sin^2\tfrac{1}{2}\kappa_1+\sin^2\tfrac{1}{2}\lambda_1+\sin^2\tfrac{1}{2}\mu_1). \qquad \text{......(1432)}$$

The periodicity conditions are now

$$\kappa_\nu=2\pi k_\nu/G,\quad \lambda_\nu=2\pi l_\nu/G,\quad \mu_\nu=2\pi m_\nu/G, \qquad \text{......(1433)}$$

where the k_ν, l_ν, m_ν may be assumed to be nearly integers when small. The general energy value replacing (1431) is

$$\eta=4J\sum_{\nu=1}^{r}\left(\sin^2\frac{\pi k_\nu}{G}+\sin^2\frac{\pi l_\nu}{G}+\sin^2\frac{\pi m_\nu}{G}\right). \qquad \text{......(1434)}$$

For small values of k_ν, l_ν, m_ν for which alone (1432) is reliable we may write

$$\eta=B\sum_{\nu=1}^{r}(k_\nu^2+l_\nu^2+m_\nu^2)=B\sum_{\nu=1}^{r}L_\nu^2 \quad (B=4\pi^2 J/G^2). \qquad \text{......(1435)}$$

This approximation will be valid for the least spin energies when $J>0$. Strictly speaking our approximations can hold only for small r. When $r>\frac{1}{2}N$ and especially when r is nearly equal to N we obtain an exactly similar set of states in terms of the few positive spins, but these can be omitted.*

We can now form the partition function for the orientational energy in a field H to the approximation which uses (1435) for η, the total energy of a state being $\eta-2m\mu_B H$, and all possible distinct sets of integers k_ν, l_ν, m_ν $(0\leqslant k_\nu, l_\nu, m_\nu\leqslant G-1)$ for the number of states. This enumeration is therefore the same as if the spin waves were particles with momentum components (k_ν,l_ν,m_ν) obeying the Einstein-Bose statistics. We have at once

$$K_m=e^{N\mu_B H/kT}\sum_{r=0}^{\frac{1}{2}N}\sum_{k_\nu,l_\nu,m_\nu=0}^{G-1}e^{-\left(2r\mu_B H+B\sum_{\nu=1}^{r}L_\nu^2\right)\Big/kT}$$

If we assume that $B>0$ the terms for large r will be unimportant at low temperatures; it is then easy to show by re-expansion that K_m is effectively given by

$$K_m=e^{N\mu_B H/kT}\prod_{k,l,m=0}^{G-1}\frac{1}{1-e^{-(2\mu_B H+BL^2)/kT}}; \qquad \text{......(1436)}$$

$$\log K_m=\frac{N\mu_B H}{kT}-\sum_{k,l,m=0}^{G-1}\log\{1-e^{-(2\mu_B H+BL^2)/kT}\}.$$

* Epstein, *loc. cit.* has discussed in detail the effect of including these complementary terms for $H=0$, when they make a considerable difference to the apparent form of the results. Since however the physically important result is for small non-zero H or $\underset{H\to 0}{\mathrm{Lt}}$, these refinements will not be considered here.

This summation can be replaced by an integral and if $J/kT \gg 1$ the upper limit of this integral can be taken to be infinity. Thus

$$\log K_m = \frac{N\mu_B H}{kT} - 4\pi \int_0^\infty \log\{1 - e^{-(2\mu_B H + BL^2)/kT}\} L^2 dL. \quad \ldots\ldots(1437)$$

This integral, of a type already familiar to us, can be evaluated as a series in a convenient form. If we substitute $BL^2/kT = x$ and use (1435), we find easily

$$\log K_m = \frac{N\mu_B H}{kT} - \frac{N(kT)^{\frac{3}{2}}}{2\pi^2 J^{\frac{3}{2}}} \int_0^\infty \log\{1 - e^{-x-2\mu_B H/kT}\} x^{\frac{1}{2}} dx.$$

If we expand the logarithm and integrate term by term, we find

$$\log K_m = \frac{N\mu_B H}{kT} + \frac{N}{4}\left(\frac{kT}{\pi J}\right)^{\frac{3}{2}} \sum_{t=1}^{\infty} \frac{e^{-t(2\mu_B H/kT)}}{t^{\frac{5}{2}}}. \quad \ldots\ldots(1438)$$

It follows at once that

$$I = \frac{kT}{V}\frac{\partial \log K_m}{\partial H} = \frac{N\mu_B}{V}\left\{1 - \frac{1}{2}\left(\frac{kT}{\pi J}\right)^{\frac{3}{2}} \sum_{t=1}^{\infty} \frac{e^{-t(2\mu_B H/kT)}}{t^{\frac{3}{2}}}\right\}. \quad \ldots(1439)$$

Similar results can be obtained for other three-dimensional lattices.

When $H \to 0$ $$I \to \frac{N\mu_B}{V}\left\{1 - 0\cdot1325\left(\frac{kT}{J}\right)^{\frac{3}{2}}\right\}, \quad \ldots\ldots(1440)$$

or $$\zeta \to 1 - 0\cdot1325(kT/J)^{\frac{3}{2}}.$$

This represents the ferromagnetic state since we have already assumed that $J \gg kT$. Thus a simple three-dimensional cubic array (and similarly any other cubic arrays) give ferromagnetism at sufficiently low temperatures provided $J > 0$. If we attempt to do the same calculations for linear or square arrays, the only essential difference is that we have the factors L^0 and L respectively in the integrand of (1437) in place of L^2. This leads to factors $t^{\frac{1}{2}}$ and t in place of $t^{\frac{3}{2}}$ in the denominators of the series terms in (1439). The series then does not converge for $H = 0$ and I does not remain comparable with $N\mu_B$ when $H \to 0$. This means that there are a large number of negative spins in the equilibrium state in this limit and our approximations

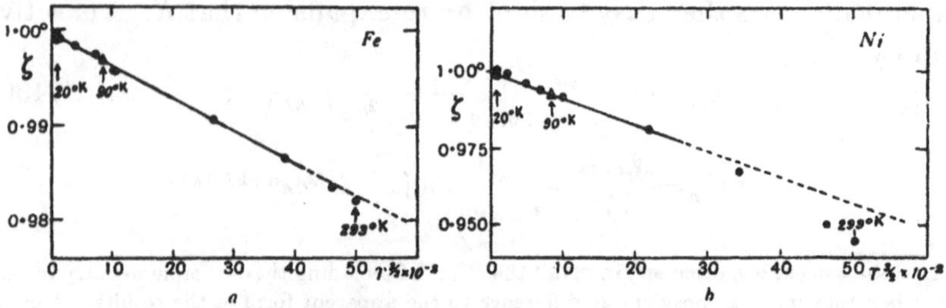

Fig. 63. The observed and theoretical variations of ζ with T at low temperatures for (*a*) Fe, (*b*) Ni.

do not apply. This breakdown must obviously be interpreted as meaning that such arrays, while strongly paramagnetic, are not ferromagnetic. They are however only of theoretical interest.

Figs. 63 compare observations of the magnetization as a function of T with the various approximate forms of the theory here developed. Figs. 63 *a*, *b* compare the $T^{\frac{3}{2}}$-law for very low temperatures with the observations* for Fe and Ni. It will be seen that the observations conform accurately to the theory. By comparing them also with the form $\zeta = (1 - \alpha T^2)$ Weiss has shown that the $T^{\frac{3}{2}}$-law gives an appreciably better fit. Fig. 63 *c* compares the general trend of the observed values of ζ with the theoretical values given by equations (1402) and (1403) for $y = 1$ which reduce, omitting β^2/z, to

$$2x = \beta\zeta, \quad \zeta = \tanh x.$$

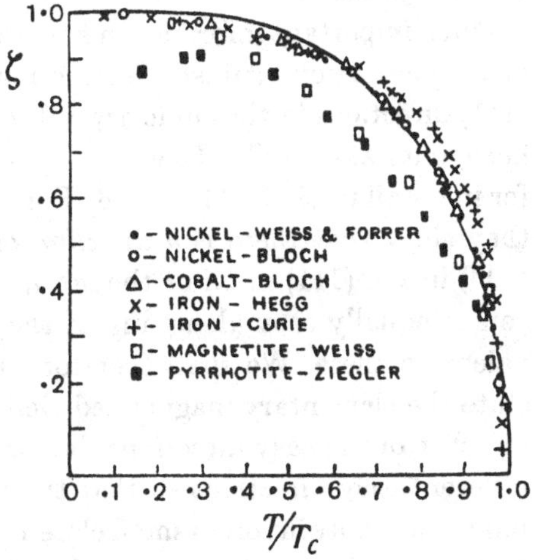

Fig. 63 *c*. The observed variation of ζ as a function of T/T_c and the theoretical variation for the equation $\zeta = \tanh \frac{1}{2}\beta\zeta$.

The theoretical curve for $y = 2$ lies slightly lower. We know that no function of this type can give the correct form near $T = 0$, but for medium values and values near the Curie point the general fit with the observations is excellent using $y = 1$ or 2.

§ 12·95. *Phenomena of holomagnetization. The "block" structure of a ferromagnetic specimen.* As we saw in § 12·94 the phenomena of the Curie point lead us to the view that ferromagnetics, even single crystals of a ferromagnetic, are always naturally micromagnetized to an intensity depending on the temperature, but practically independent of magnetic fields or of holomagnetization. This magnetization would be adequately described by the Weiss-Heisenberg theory already given if the calculations of K_m could be completed. In the following sections we shall suppose that we are in possession of such an idealized completed theory, which however ignores all the magnetic energy terms themselves except HIV. There are however such energy terms arising from the interactions of the magnetic dipoles (*s-s* interactions) and from the interactions of the spins and their own atomic orbits (*l-s* interactions). These are all of the nature of second order corrections and only their form can be predicted accurately. But with the

* Weiss, *Comptes Rendus*, vol. 198, p. 1893 (1934).

help of terms of the proper form and the expected order of magnitude the complicated phenomena associated with the passage from the micro- to the holo-magnetized state *in single crystals* can be satisfactorily explained, together with finer details of the holomagnetized state itself. It is essential to discuss observational material on single crystals. The effects in polycrystalline specimens can easily be understood in terms of the properties of single crystals.

Much important information has been obtained by comparing the curves of magnetization against effective magnetic field for the (100), (110) and (111) directions in the cubic crystals of Fe and Ni and along and across the hexagonal axis in Co. These curves are shown, and their form accounted for in detail in §§ 12·963–12·966. For the present it is sufficient to note that they show that there is *a direction of easy magnetization* in each crystal, (100) in Fe, (111) in Ni or the hexagonal* axis in Co. Magnetization is of course equally easy along any of the directions crystallographically equivalent to these. We shall therefore assume that in the micromagnetized state the elementary magnetized elements or blocks are magnetized in one or other of the easy directions, but with random arrangements among the different easy directions so that the resultant magnetization is zero. This model probably involves small closed circuits of magnetic flux in the crystal. Holomagnetization in the easy direction will then only involve changes of magnetization of the blocks from one easy direction to another, which might ideally be supposed to be able to take place reversibly in vanishingly small applied magnetic fields, requiring in the ideal limit, reached perhaps in the ideal perfect crystal, no expenditure of energy and involving no hysteresis loss. This idealized model, which we shall adopt, would not be permissible if it had to be strictly interpreted. For directions of easy magnetization are necessarily separated by harder directions on any path and weak fields could not be sufficient to pull the magnetization round. A truer version of the theory has been given by Bloch† who points out that the individual blocks are not permanent units whose magnetism must be turned as a whole but that the holomagnetization really grows by the blocks magnetized in the required direction eating up their wrongly orientated neighbours.

Though these autophagous blocks give the more correct picture, the model of distinct permanent blocks, with magnetization freely switchable from one easy direction to another, is a sufficiently good approximation to use in the analysis at the present stage. We shall use this model neglecting the effects of the fields of the surrounding blocks on each individual block,

* Below 250° C. Between this temperature and the transition point, the basal plane is an easy plane of magnetization.

† Bloch, *Zeit. f. Physik*, vol. 74, p. 295 (1932).

except in so far as this is taken care of *in the mean* by classical theory in specifying the effective magnetic force F; as we consider here only a part of the magnetic substance F is no longer equal to H. As shown in § 12·11 we have $F = H' + \frac{4}{3}\pi I$, or more accurately $\mathbf{F} = \mathbf{H}' + \frac{4}{3}\pi\mathbf{I}$, when $\mathbf{H}'$ and $\mathbf{I}$ are not necessarily parallel, $\mathbf{I}$ being here the natural magnetization $\mathbf{I}_s$ of a microelement depending only on the temperature. But we can ignore the term $\frac{4}{3}\pi\mathbf{I}_s$ since it leads only to energy terms independent to the present approximation of all the variables, such as degree or direction of holomagnetization, in which we shall be interested. We shall assume further that $\mathbf{H}'$ is constant in spite of the block structure, when the specimen is so cut that $\mathbf{H}'$ would be constant in genuinely uniform isotropic material. In that case we have

$$\mathbf{H}' = \mathbf{H} - \gamma\mathbf{I}_h, \qquad \text{......(1441)}$$

where γ is the demagnetizing coefficient depending on the geometry of the specimen and $\mathbf{I}_h$ is the holomagnetization.

§ **12·951.** *The "dipole" and "quadripole" energy of magnetization.* Let us consider next how the energy of a microelement or a large scale specimen holomagnetized to saturation will depend on the direction cosines l, m, n of magnetization in a cubic crystal. For simplicity we can consider only absolute saturation corresponding strictly to zero temperature. Since this extra energy must have the symmetry of the cube and be of even order, it can only be of the form

$$A'(l^2+m^2+n^2)+B'(l^4+m^4+n^4)+C'(l^2m^2+m^2n^2+n^2l^2)$$

plus higher order terms. Since $l^2+m^2+n^2=1$, this reduces to irrelevant constants and a term

$$B(l^4+m^4+n^4). \qquad \text{......(1442)}$$

To make sure of the *origin* of this term is a quantum problem, which we cannot discuss here. It was first proposed as a quadripole effect by Mahajani* and is generally referred to as *the quadripole energy*. It is now held to arise from the (l-s) interactions.†

In the absence of an external field the only energy terms depending on direction are just (1442), which must be a minimum subject to $l^2+m^2+n^2=1$ for the directions of natural (i.e. easy) magnetization. By the usual methods we find therefore that if $B>0$ the minimum value of (1442) occurs in the direction (111) and the maximum in (100) (and equivalent directions in each case), while if $B<0$ the minimum is along (100) and the maximum along (111). These directions will agree with the facts if

$$\text{Ni}, B>0; \quad \text{Fe}, B<0.$$

* Mahajani, *Phil. Trans.* A, vol. 228, p. 63 (1929).

† Powell, *Proc. Roy. Soc.* A, vol. 130, p. 812 (1930); Bloch and Gentile, *Zeit. f. Physik*, vol. 70, p. 395 (1931).

In the close packed hexagonal lattice of Co the quadratic terms must be of the form

$$A_1'l^2+A_2'(m^2+n^2) \quad (A_1' \neq A_2'), \qquad \ldots\ldots(1443)$$

the quadripole terms must also from the symmetry depend only on l^2 and m^2+n^2. They therefore reduce effectively to $A''l^4$. The terms of importance may therefore be taken to be of the form

$$Al^2+A'l^4, \qquad \ldots\ldots(1444)$$

where l is the cosine of the angle between the direction of magnetization and the hexagonal axis. If as turns out to be the case A and A' have the same sign, this is a minimum for $l=1$ when $A<0$, which agrees with the facts, since the easy directions are along the hexagonal axis for Co. Since the magnetic energy of a set of magnetic dipoles with this symmetry must be of the form (1443), these terms are commonly known as *the dipole energy*.

§ **12·952.** *Corrected partition functions for microelements, and holo-magnetized systems, or for collections of microelements.* To a microelement naturally magnetized to saturation, or to a large scale specimen holo-magnetized by a suitable external field, we may apply Heisenberg's theory, equations (1401)–(1404). For a microelement the effective field F must be used. This will differ from the pipe-force H' by terms due to the local magnetization. So far as these terms are due to the magnetization of the microelement in question they do not affect the effective external field acting on the element. We shall therefore in the following sections which are largely concerned with microelements use H' as the best available first approximation to the true F.

The magnetization ζ determined by (1401)–(1404) is effectively independent of H' and therefore of α. For present purposes we can neglect the terms in β^2/z. We then have the simplified equations

$$\log K_m(V,T,0)=2N[\log\{e^{xy}+\ldots+e^{-xy}\}-y\zeta x+\tfrac{1}{4}\beta y\zeta^2], \quad \ldots\ldots(1445)$$

where x and ζ are determined by $\partial K_m/\partial x=\partial K_m/\partial\zeta=0$, or

$$2x=\beta\zeta, \quad \zeta=\frac{1}{y}\left[\frac{ye^{xy}+\ldots+(-y)e^{-xy}}{e^{xy}+\ldots+e^{-xy}}\right]=B_{\frac{1}{2}y}(2x). \quad \ldots\ldots(1446)$$

This $K_m(V,T,0)$ is an approximate evaluation of a partition function, with all the properties of these functions.

Thus evaluated *all* magnetic energy terms have been systematically omitted, and they can now be inserted. If λ, μ, ν are the direction cosines of the magnetic pipe-force H', then there is a directional energy term

$$-H'IV(\lambda l+\mu m+\nu n) \qquad \ldots\ldots(1447)$$

due to the orientation of the "magnet" of strength I per unit volume in the magnetic field. There is also an energy term

$$VB(I)(l^4+m^4+n^4) \quad [VA(I)l^2+VA'(I)l^4] \qquad \ldots\ldots(1448)$$

depending on the direction of magnetization relative to the axes of the crystal, taken to coincide with the axes of coordinates.

These terms should be added to the energy terms in the original partition function before the approximate summation, but since the whole effective contribution comes from states indistinguishable from the equilibrium state, and since the extra energy terms are small compared with the main (exchange) energies, we can insert I_s (the equilibrium value) for I in (1447) and (1448) and treat these terms as extra constant factors during the evaluation. We thus get the corrected (partial) partition function

$$\log K_m(V,T,\mathbf{H}',l,m,n) = \log K_m(V,T,0) + \frac{H'I_sV}{kT}(\lambda l + \mu m + \nu n) - \frac{VB(I_s)}{kT}(l^4 + m^4 + n^4). \qquad \text{......(1449)}$$

This is still not a true partition function, for while V, T and $\mathbf{H}'$ can be specified by the external conditions l, m, n should be deducible from them. Strictly (1449) multiplied by $d\omega$ specifies the contribution to the true partition function made by those states for which I_s lies in $d\omega$ with a mean direction (l,m,n). The complete partition function is

$$K_m{}^*(V,T,\mathbf{H}') = \int K_m(V,T,\mathbf{H}',l,m,n)\,d\omega \qquad \text{......(1450)}$$

and the equilibrium values of (l,m,n) should be determined in the usual way by equations of the type

$$\bar{l}K_m{}^* = \int lK_m(V,T,\mathbf{H}',l,m,n)\,d\omega. \qquad \text{......(1451)}$$

But since even a microelement is a system very large on the atomic scale $H'I_sV/kT$ $(H' \neq 0)$ and $VB(I_s)/kT$ will both be very large indeed, so that the whole contribution to the integrals in (1450) and (1451) comes from values of (l,m,n) which make

$$H'I_sV(\lambda l + \mu m + \nu n) - VB(I_s)(l^4 + m^4 + n^4) \qquad \text{......(1452)}$$

a maximum, subject to

$$l^2 + m^2 + n^2 = 1. \qquad \text{......(1453)}$$

It is therefore only necessary to maximize (1449) as it stands for (l,m,n) subject to $l^2+m^2+n^2=1$. The permissible values of (l,m,n) are therefore roots of the equations

$$H'I_s\lambda - 4Bl^3 + \Lambda l = 0, \qquad \text{......(1454)}$$

$$H'I_s\mu - 4Bm^3 + \Lambda m = 0, \qquad \text{......(1455)}$$

$$H'I_s\nu - 4Bn^3 + \Lambda n = 0, \qquad \text{......(1456)}$$

where Λ is the undetermined multiplier. We thus obtain all the properties that we require for microelements or completely holomagnetized crystals.

Incompletely holomagnetized crystals are collections of many microelements each of which can be regarded as a supermolecule whose partition function for its internal states is approximately $K_m(V,T,\mathbf{H}',l,m,n)$, when the direction of its magnetization is specified. Let the various permissible roots or groups of roots of equations (1454)–(1456) be labelled with a suffix i, and let N_i be the number of elements whose direction of magnetization is given by the ith root. Then $\Sigma_i N_i = N$, the total number of microelements. It will be convenient not to enumerate all the roots singly, but to group together all the roots, ϖ_i in number, which yield equal values of K_m, which we shall denote by $K_m^{(i)}$. Then the partition function for the N microelements is

$$\mathscr{K}_m = \frac{N!\,\varpi_1^{N_1}\varpi_2^{N_2}\dots}{N_1!\,N_2!\dots}(K_m^{(1)})^{N_1}(K_m^{(2)})^{N_2}\dots. \qquad \dots\dots(1457)$$

This can be used exactly as an ordinary partition function if the N_i's are fixed. But if, as is generally the case, the N_i's may have any values subject to $\Sigma_i N_i = N$ and their average values are the quantities to be determined, then the complete partition function is

$$\mathscr{K}_m^* = \sum_{(N_i)} \mathscr{K}_m = (\varpi_1 K_m^{(1)} + \varpi_2 K_m^{(2)} + \dots)^N. \qquad \dots\dots(1458)$$

The average values $\overline{N_i}$ are determined exactly as in § 2·3. We find at once

$$\overline{N_i} = \varpi_i K_m^{(i)}/(\Sigma_j \varpi_j K_m^{(j)}). \qquad \dots\dots(1459)$$

Now for moderate values of H' the magnetic terms that differentiate the $K_m^{(i)}$ are all large compared with kT even for a microelement. It follows therefore from equation (1459) that all the $\overline{N_i}$'s are effectively zero except those corresponding to the greatest value of the magnetic terms. If there are several values of (l_i,m_i,n_i) corresponding to this greatest value, then the $\overline{N_i}$ for each such set taken separately will be equal. This conclusion holds *a fortiori* for larger values of H'.

For $H' = 0$ on the other hand the magnetic terms correspond to the various easy directions of magnetization and are all equal. Equal numbers $\overline{N_i}$ of microelements will therefore according to (1459) be found orientated in each easy direction with zero resultant holomagnetization. This is in full agreement with the requirements already laid down for our model in § 12·95. But the equations we have so far set up do not enable us in practice to trace the transition from $H' = 0$ to moderate values of H'. For very small values of H' neglected magnetic energies arising from interblock interactions are comparable with the *differences* between the magnetic energies in the various $K_m^{(i)}$, and the whole argument breaks down. We shall see that this transition corresponds to the early part of the magnetization curve below the "knee", a part which is ideally almost vertical. From the point of view here adopted we may fairly regard the holomagnetization in this region as indeterminate. But the properties of such states can be discussed by *specifying* the holomagnetization and determining the $\overline{N_i}$ subject to this extra condition; all the magnetic energy terms can be neglected.

To formulate this new condition, let γ_i be the cosine of the angle between (l_i, m_i, n_i) and the direction of holomagnetization I_h. Then

$$\Sigma_i \gamma_i N_i = N I_h / I_s.$$

It is assumed for simplicity that the microelements are all of the same size V. In place of (1458) we now have

$$\mathscr{H}_m{}^* = \sum_{(N_i)} \mathscr{H}_m,$$

subject to $$\Sigma_i N_i = N, \quad \Sigma_i \gamma_i N_i = N I_h / I_s.$$

By the usual processes we therefore find

$$\mathscr{H}_m{}^* = \frac{1}{2\pi i} \int \frac{dx}{x^{NI_h/I_s+1}} (\varpi_1 x^{\gamma_1} K_m{}^{(1)} + \varpi_2 x^{\gamma_2} K_m{}^{(2)} + \ldots)^N, \quad \ldots\ldots(1460)$$

and $$\overline{N_i} = N \varpi_i \xi^{\gamma_i} K_m{}^{(i)} / (\Sigma_j \varpi_j \xi^{\gamma_j} K_m{}^{(j)}), \quad \ldots\ldots(1461)$$

where ξ is determined as the root of the equation

$$\frac{\varpi_1 \gamma_1 x^{\gamma_1} K_m{}^{(1)} + \varpi_2 \gamma_2 x^{\gamma_2} K_m{}^{(2)} + \ldots}{\varpi_1 x^{\gamma_1} K_m{}^{(1)} + \varpi_2 x^{\gamma_2} K_m{}^{(2)} + \ldots} = \frac{I_h}{I_s}. \quad \ldots\ldots(1462)$$

Since to our approximation the $K_m{}^{(i)}$'s are all equal, these equations simplify to

$$\frac{\overline{N_i}}{N} = \frac{\varpi_i \xi^{\gamma_i}}{\Sigma_j \varpi_j \xi^{\gamma_j}}, \quad \ldots\ldots(1463)$$

where ξ is the root of

$$\frac{\varpi_1 \gamma_1 x^{\gamma_1} + \varpi_2 \gamma_2 x^{\gamma_2} + \ldots}{\varpi_1 x^{\gamma_1} + \varpi_2 x^{\gamma_2} + \ldots} = \frac{I_h}{I_s} = \delta. \quad \ldots\ldots(1464)$$

Similar simple considerations apply to non-cubic crystals such as Co. in which there are only two directions of easy magnetization, $l = \pm 1$. In place of (1452) and (1453) we have to make

$$H' I_s V(\lambda l + \mu m + \nu n) - VA(I_s) l^2 - VA'(I_s) l^4 \quad \ldots\ldots(1465)$$

a maximum subject to $l^2 + m^2 + n^2 = 1$. This gives the equations

$$H' I_s \lambda - (2A + \Lambda) l - 4A' l^3 = 0, \quad \ldots\ldots(1466)$$

$$H' I_s \mu - \Lambda m = 0, \quad \ldots\ldots(1467)$$

$$H' I_s \nu - \Lambda n = 0. \quad \ldots\ldots(1468)$$

We emphasize once again that the H' of all these formulae should strictly be the effective magnetic force F, but that the difference is here unimportant and that H' is related to H the external field applied to the specimen by (1441).

§ **12·96.** *The deviation effect in non-cubic crystals, in particular Co.* When a single crystal of a ferromagnetic is holomagnetized to saturation by a sufficient field it is found that except for particular directions of the field the direction of magnetization and the direction of the field do not coincide. This phenomenon is called *the deviation effect* and is simply explained by the existence of the dipole and quadripole energy terms of the preceding section. The more complicated effect in cubic crystals is similarly treated in the next section.

To the required approximation everything is symmetrical about the hexagonal axis. We can therefore suppose that $\nu = n = 0$, and confine the vectors $\mathbf{H}'$ and $\mathbf{I}_s$ to the (x,y) plane. Eliminating Λ from (1466) and (1467) we find

$$-\frac{2A}{I_s H'} - \frac{4A'}{I_s H'} l^2 = \frac{\mu}{m} - \frac{\lambda}{l}. \qquad \text{......(1469)}$$

If $\mathbf{H}'$ makes an angle ϕ with the hexagonal axis and $\mathbf{I}_s$ an angle ψ, the deviation $\Delta = \phi - \psi$ is given by

$$\sin\Delta = -\frac{A + A'}{I_s H'} \sin 2\psi + \frac{1}{2}\frac{A'}{I_s H'} \sin 4\psi. \qquad \text{......(1470)}$$

If $A < 0$ then $\Delta > 0$ when ψ is in the first quadrant, so that $\mathbf{H}'$ deviates more than $\mathbf{I}_s$ from the hexagonal axis, which is, for $A < 0$, the direction of easy magnetization; $\mathbf{H}$ as it were drags $\mathbf{I}_s$ reluctantly away from the easy direction. Formula (1470) gives an excellent representation of the observations* if

$$-A = 4{\cdot}14 \times 10^6,$$
$$-A' = 1{\cdot}01 \times 10^6.$$

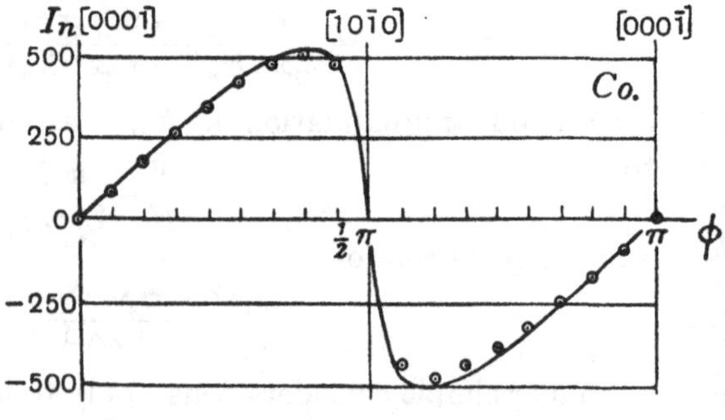

Fig. 64. The normal component of magnetization for a single crystal of Co in a field H' of 9016 gauss, as a function of the inclination (ϕ) of H' to the hexagonal axis of the crystal.

The theoretical curve for Δ is compared with Kaya's† observations in Fig. 64. The curve shows $I_n = I_s \sin\Delta$ or the component of I_s normal to H' for a field H' of 9016 gauss.

§ **12·961.** *The deviation effect in the cubic crystals of Fe and Ni.* In cubic crystals there is no significant dipole term, but a significant quadripole energy. In moderate fields sufficient to holomagnetize the crystals, the directions of $\mathbf{H}'$ and $\mathbf{I}_s$ again do not in general coincide. According to the discussion of § 12·952 the directions of natural magnetization in zero field (easy directions) are those that make $-B(l^4 + m^4 + n^4)$ a maximum, that is the 6 directions (± 100), (0 ± 10), (00 ± 1) when $B < 0$ (as for Fe), or the 8 directions $(\pm 1 \pm 1 \pm 1)$ when $B > 0$ (as for Ni).

* Gans and Czerlinsky, *Ann. d. Physik*, vol. 16, p. 625 (1933).

† Kaya, *Sci. Reports Tôkohu*, vol. 17, p. 1165 (1928).

We have to solve the equations (1454), (1455) and (1456) for the root which makes (1452) a maximum, for sufficiently large values of H'. The general solution would be complicated and we confine attention to special cases, supposing that the observations are made on specimens so cut that the demagnetizing terms have the proper symmetry.

Case (i). $\mathbf{H}'$ *lies in a cubic face so that* $\nu=0$, *the specimen being cut so that this plane is a plane of symmetry.*

If $\nu=0$ there are *a priori* three possible values of n, but the non-zero values are irrelevant to our problem, since the plane $\nu=0$ is a plane of symmetry. From the remaining equations

$$\frac{4Bl^3-H'I_s\lambda}{l}=\frac{4Bm^3-H'I_s\mu}{m}\quad(l^2+m^2=1),$$

which leads to $$\sin\Delta=\sin(\phi-\psi)=-\frac{B}{H'I_s}\sin 4\psi. \qquad \text{......(1471)}$$

The values of B required for different specimens are somewhat variable, but for any one specimen excellent agreement is found between (1471) and the

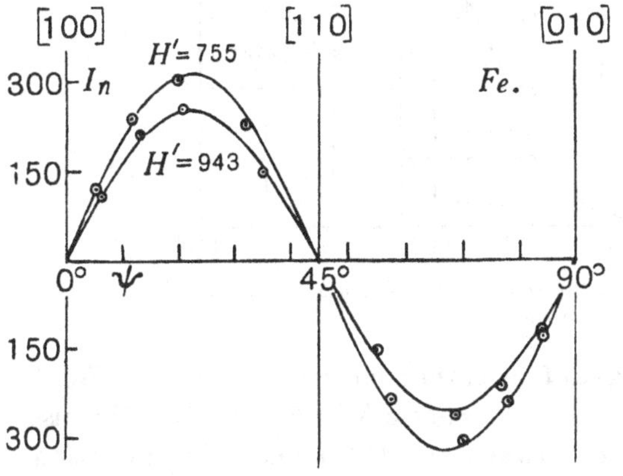

Fig. 65 *a*. The component of magnetization normal to H' in the (001) plane of a single crystal of Fe as a function of the direction ψ of the resultant magnetization.

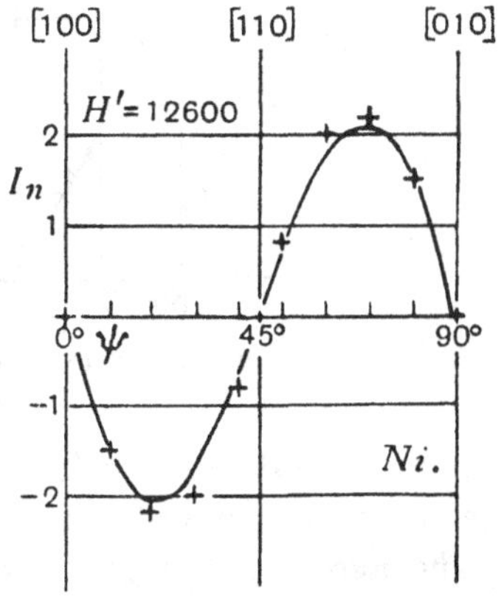

Fig. 65 *b*. The component of magnetization normal to H' in the (001) plane of a single crystal of Ni as a function of the direction ψ of the resultant magnetization.

observations. Figs. 65 *a*, *b** show the agreement for $I_n=I_s\sin\Delta$ for specimens of Fe and Ni with (Fe) $B=-2{\cdot}39\times10^5$, (Ni) $B=2{\cdot}9\times10^4$. The vector $\mathbf{H}'$ is again always farther than $\mathbf{I}_s$ from the nearest easy direction of magnetization (cube edge, Fe, $B<0$; cube diagonal, Ni, $B>0$).

* Gans and Czerlinsky, *loc. cit.*; observation of Honda and Kaya, *Sci. Reports Tôhoku*, vol. 15, p. 721 (1926). Similar results are given for Fe by the observations of Webster, *Proc. Roy. Soc.* A, vol. 107, p. 504 (1925).

Case (ii). $\mu = \nu$, $\mathbf{H}'$ *being normal to a face diagonal. The specimen is so cut that the plane* (01 − 1) *is a plane of symmetry. This plane contains the* (100), (111) *and* (011) *directions.*

Since $\mu = \nu$, $m = n$ is a possible root and the only relevant one from the symmetry. Thus

$$l, m, n = \cos\psi, \quad \frac{1}{\sqrt{2}}\sin\psi, \quad \frac{1}{\sqrt{2}}\sin\psi,$$

$$\lambda, \mu, \nu = \cos\phi, \quad \frac{1}{\sqrt{2}}\sin\phi, \quad \frac{1}{\sqrt{2}}\sin\phi,$$

where ϕ and ψ are again the angles between $\mathbf{H}'$ and $\mathbf{I}_s$ and the (100) axis. Equations (1454)–(1456) then lead to

$$\sin\Delta = \sin(\phi - \psi) = -\frac{B}{H'I_s}\sin 2\psi\{3\cos^2\psi - 1\}. \quad \text{......(1472)}$$

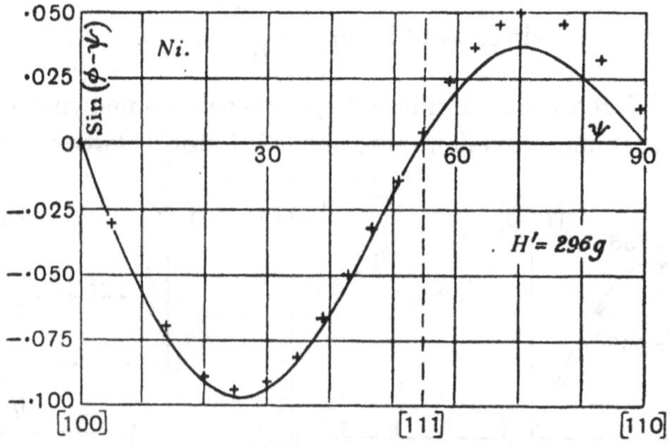

Fig. 66. The observed and theoretical values of sin Δ for the (1 − 10) plane of a single crystal of Ni.

Here again $\mathbf{H}'$ is always farther than $\mathbf{I}_s$ from the nearest easy direction. Fig. 66 shows* the observed and theoretical values of $\sin\Delta$ for Ni for $H' = 296$ gauss. The agreement is good except between the (111) and (110) directions where it is only fair. By including the next possible term in the magnetization energy of the form $B'l^2m^2n^2$ Gans and Czerlinsky† have obtained good agreement over the whole range. The extra term leaves the (001) plane deviations unaffected.

Case (iii). *There is a plane of symmetry normal to the trigonal* (111) *axis; if H' lies in this plane,* $\lambda + \mu + \nu = 0$. In a specimen cut so that this plane is a plane of symmetry one would expect $\mathbf{I}_s$ to obey the same symmetry condition which requires $l + m + n = 0$, but this does not satisfy the equations (1454)–(1456) since in general Σl^3 does not vanish with Σl. The observations

* Powell, *Proc. Roy. Soc.* A, vol. 130, p. 167 (1931).

† Gans and Czerlinsky, *loc. cit.*

appear to show that the symmetry condition keeps control; subject to this extra condition we should obtain the equilibrium state by maximizing

$$H'I_s(\lambda l+\mu m+\nu n)-B(l^4+m^4+n^4) \quad (\lambda+\mu+\nu=0)$$

subject to $\quad l^2+m^2+n^2=1, \quad l+m+n=0.$

The conditions are $\quad -4Bl^3+HI_s\lambda+\Lambda l+M=0$

and two similar equations. Eliminating Λ and M from these, we find

$$-4B\begin{vmatrix} l^3 & m^3 & n^3 \\ l & m & n \\ 1 & 1 & 1 \end{vmatrix}+H'I_s\begin{vmatrix} \lambda & \mu & \nu \\ l & m & n \\ 1 & 1 & 1 \end{vmatrix}=0.$$

The coefficient of B has $(l+m+n)$ as a factor and so vanishes. Therefore

$$\begin{vmatrix} \lambda & \mu & \nu \\ l & m & n \\ 1 & 1 & 1 \end{vmatrix}=0, \quad l+m+n=0, \quad \lambda+\mu+\nu=0,$$

so that $l/\lambda=m/\mu=n/\nu$ or $\mathbf{I}_s$ is parallel to $\mathbf{H}'$. This agrees with the facts except for very small fields and even then the observed deviations are small compared with those of cases (i) and (ii).

§ **12·962.** *The two stages of magnetization.* As we have already mentioned, magnetization curves of single crystals show clearly two distinct stages.

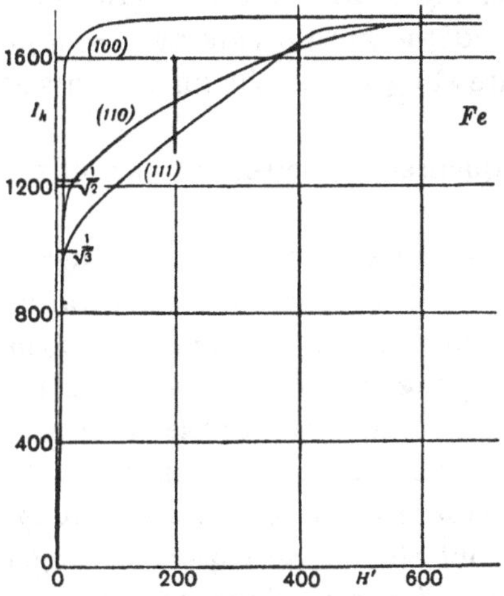

Fig. 67. Observed magnetization curves for a single crystal of Fe.

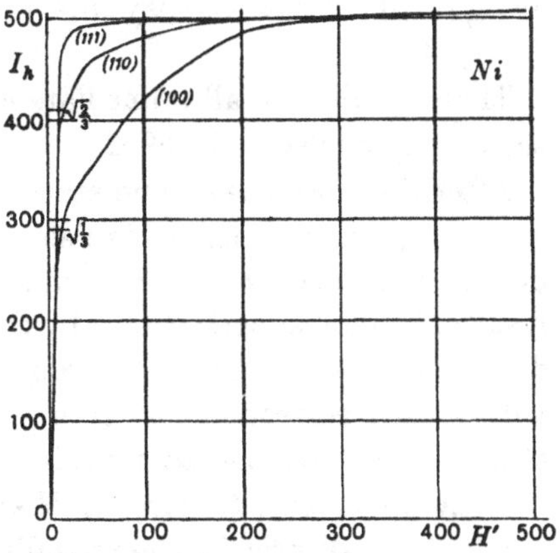

Fig. 68. Observed magnetization curves for a single crystal of Ni.

These stages are here shown in Figs. 67, 68 and 69 for Fe, Ni and Co respectively.* Stage one is a rapid (ideally infinitely rapid) increase of magnetiza-

* Fe, Ni, Webster, *Proc. Phys. Soc. London*, vol. 42, p. 431 (1930); Co, Gans and Czerlinsky, *Ann. d. Physik*, vol. 16, p. 625 (1933). The observed points for cobalt are from Kaya, *Sci. Rep. Tôhoku*, vol. 17, p. 1165 (1928).

tion in very weak fields, up to a certain maximum value. If the direction of holomagnetization is one of the easy directions for a microelement, then this maximum value is complete saturation. It should correspond according to our theory to a state in which all the microelements are orientated in groups along the directions of easy magnetization nearest to the direction of holomagnetization, the groups being so arranged that the resultant holomagnetization is in the desired direction. The maximum values attainable in this way in the important special cases are as detailed below, in excellent agreement with the observations. At the proper point a conspicuous bend or knee occurs in the magnetization curve.

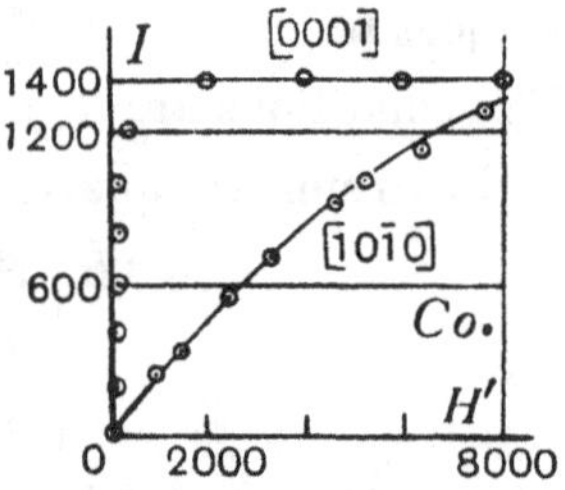

Fig. 69. Theoretical magnetization curves and observed values of the magnetization for a single crystal of Co.

Co. *Easy directions: Hexagonal axis* ($\pm$). For holomagnetization normal to the hexagonal axis the knee occurs at $I_h = 0$.

Fe. *Easy directions* (100), *etc.* For holomagnetization along (111) the nearest easy directions are (100), (010) and (001) and the knee occurs at $I_h = I_s/\sqrt{3}$; along (110) the nearest easy directions are (100) and (010) and the knee occurs at $I_h = I_s/\sqrt{2}$.

Ni. *Easy directions* (111). For holomagnetization along (100) the nearest easy directions are $(1 \pm 1 \pm 1)$ and the knee occurs at $I_h = I_s/\sqrt{3}$; along (110) the nearest easy directions are (11 ± 1) and the knee occurs at $I_h = I_s\sqrt{\tfrac{2}{3}}$.

These results are all immediate consequences of the second part of the theory expounded in § 12·952.

After this knee the second stage of holomagnetization sets in; a stage in which a gradually increasing field H' pulls the direction of magnetization of each microelement away from an easy direction to the direction of the resultant holomagnetization. In the foregoing sections in discussing the deviation effect we have not troubled to consider whether the stationary values used correspond to true maxima. This is not necessary for the strongish fields concerned, for it is obvious that the correct roots must make the deviation tend to zero as $H' \to \infty$. But this investigation no longer applies in weaker fields when the direction of $\mathbf{I}_s$ is deviating gradually from an easy direction. The formulae giving Δ as a function of ψ are unique, but the root of ψ as a function of ϕ must be chosen differently. In the following sections the complicated phenomena that can occur will be sufficiently illustrated by the discussion of important special cases. It will be assumed throughout that all the microelements are subject only to the effective magnetic force F which need not be distinguished from the pipe-force H'.

§ **12·963.** *Magnetization of Co normal to the hexagonal axis.* There is no stage one magnetization. Stage two starts from $I_h = 0$ with equal numbers of microblocks orientated $\pm$ along the hexagonal axis. Subject to the magnetic force H' normal to this axis, the magnetization of each will deviate through an angle ψ, such that $I_h/I_s = \sin\psi$. Applying equation (1470) with $\phi = \frac{1}{2}\pi$, we find

$$\cos\psi = -\frac{A+A'}{I_s H'}\sin 2\psi + \frac{1}{2}\frac{A'}{I_s H'}\sin 4\psi,$$

and since ψ must vanish with H, the $\cos\psi$ factor is irrelevant and the required root is given by

$$\frac{I_h}{I_s} = \sin\psi, \quad I_s H' = -2A\sin\psi - 4A'\sin^3\psi. \qquad \text{......(1473)}$$

The good general agreement between (1473) and the observations is shown in Fig. 69. Complete saturation should be achieved when

$$H' = -(2A + 4A')/I_s \sim 8600\,\text{g}.$$

§ **12·964.** *The magnetization curves of Fe.* ($B < 0$.)

Case (i). *Magnetization along the* (110) *direction in the* (001) *plane.* At the end of stage one, half the microelements are magnetized along (100) and the other half along (010), in a negligible external field. The microelements are now all subjected to a force H' along (110) which will affect them all similarly. Their magnetizations will all be deflected towards H' through the angle ψ, where ψ is the proper root of the equation (1471), which here takes the form

$$\sin(\tfrac{1}{4}\pi - \psi) = \frac{-B}{I_s H'}\sin 4\psi \quad (\psi = 0,\ H' = 0), \qquad \text{......(1474)}$$

and at the same time $\qquad \cos(\frac{1}{4}\pi - \psi) = I_h/I_s. \qquad$ ......(1475)

The (I_h, H') relationship is simplified by writing $\frac{1}{4}\pi - \psi = \epsilon$. Then

$$\frac{I_h}{I_s} = \cos\epsilon, \quad \sin\epsilon = \frac{-B}{I_s H'}\sin 4\epsilon,$$

from which the proper root is given by

$$\frac{I_h}{I_s}\left(2\frac{I_h^2}{I_s^2} - 1\right) = \frac{I_s H'}{-4B}. \qquad \text{......(1476)}$$

The factor $\sin\epsilon = 0$ gives only a minimum of (1452). Complete saturation is reached for $H' = -4B/I_s$, and the magnetization curve joins the line $I_h = I_s$ at an acute angle, $(\partial I_h/\partial H)_s = \frac{1}{8}I_s/H_s$. This curve is in excellent agreement with the measurements of Honda, Masumoto and Kaya, shown in Fig. 69·1.*

* Bitter, *Phys. Rev.* vol. 39, p. 337 (1932); Honda, etc., *Sci. Reports Tôhoku*, vol. 17, p. 111 (1928).

Case (ii). *Magnetization along the* (111) *direction*. At the end of stage one magnetization one third of the microelements are magnetized along (100), one third along (010) and one third along (001). The microelements are now all subjected to a force H' along (111), which again affects them all alike; their magnetization is deflected towards H' through the angle ψ in the relevant plane of type (011). Equation (1472) applies and here becomes

$$\sin(\alpha-\psi)=\frac{-B}{H'I_s}\sin 2\psi\{3\cos^2\psi-1\}\quad\left(\cos\alpha=\frac{1}{\sqrt{3}}\right),$$

with

$$\cos(\alpha-\psi)=I_h/I_s.$$

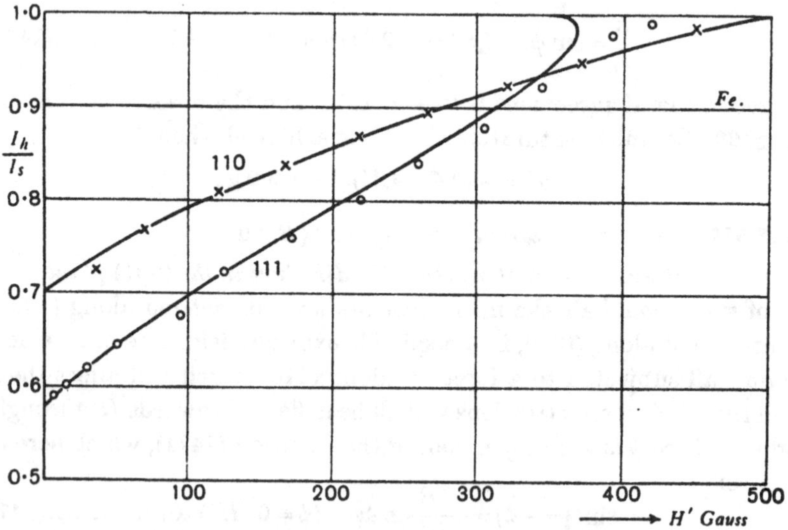

Fig. 69·1. Theoretical magnetization curves and observed magnetizations for the (110) and (111) directions in a single crystal of Fe as a function of H'.

Writing $\alpha-\psi=\epsilon$, these equations become

$$\sin\epsilon\left[\cos\epsilon-\cos(4\alpha-3\epsilon)-\frac{2H'I_s}{-3B}\right]=0\quad\left(\cos\epsilon=\frac{I_h}{I_s}\right).\quad\text{......(1477)}$$

For values of $H'<-8B/3I_s$ the factor in [ ] has one root which is the required maximum of (1452), and $\sin\epsilon=0$ gives a minimum. For values of H such that $-8B/3I_s<H'<-(2{\cdot}96)\,B/I_s$ the factor in [ ] gives two roots of which the lesser value of $\cos\epsilon$ gives a maximum and the other a minimum, while $\sin\epsilon=0$ gives a maximum. For $H'>-(2{\cdot}96)\,B/I_s$, $\sin\epsilon=0$ is the only root. The complete curve is shown in Fig. 69·1. If we assume that equilibrium corresponds always to the greater maximum of (1452), the magnetization curve is as shown by the continuously drawn curve in the figure, the final step being discontinuous at about

$$H'=-(2{\cdot}9)\,B/I_s=H_s'.$$

This curve is also in excellent agreement with the measurements, though naturally these do not reproduce the discontinuity of the idealized theory. This part of the curve is much altered by including the term $B'l^2m^2n^2$.

§ **12·965.** *The magnetization curves for Ni.* $(B>0.)$

Case (i). *Magnetization along the* (100) *direction.* At the end of stage one one quarter of the microelements will be orientated along each of the four easy directions $(1\pm1\pm1)$ in a negligible external field. The microelements are now all subjected to H' along (100), which affects them all similarly by deflecting their magnetizations towards H' in the (110) or similar plane through an angle ϵ. Equation (1472) applies with $\phi=0$ in the form

$$\sin(-\psi)=\frac{-B}{H'I_s}\sin 2\psi\{3\cos^2\psi-1\}. \qquad \text{......(1478)}$$

The factor $\sin\psi$ is irrelevant and $\cos\psi=1/\sqrt{3}$ when $H'=0$. Moreover $\cos\psi=I_h/I_s$. Therefore

$$\frac{H'I_s}{B}=2\frac{I_h}{I_s}\left\{3\frac{I_h^2}{I_s^2}-1\right\} \qquad \text{....(1479)}$$

The deviation vanishes and saturation is reached when

$$H'=H_s'=4B/I_s,$$

and at this point $(\partial I/\partial H)_s\neq 0$. This theoretical magnetization curve is in fair agreement with the observations as shown in Fig. 69·2.

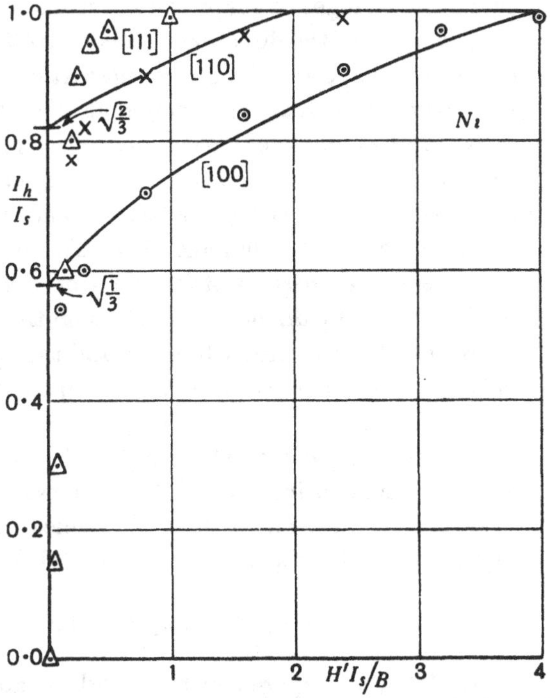

Fig. 69·2. Theoretical magnetization curves and observed magnetizations for the (111), (110) and (100) directions in a single crystal of Ni as a function of H'.

Case (ii). *Magnetization along the* (110) *direction.* At the end of stage one magnetization one half of the microelements will be orientated along each of the two easy directions (11 ± 1) in a negligible external field. The microelements are now all subjected to H' along (110) and equation (1472) again applies but now with $\phi=\frac{1}{2}\pi$ and $\psi=\frac{1}{2}\pi-\epsilon$, where ϵ is the angle between $\mathbf{I}_s$ and (110), so that $\cos\epsilon=I_h/I_s$. Thus

$$\sin\epsilon=\frac{-B}{H'I_s}\sin 2\epsilon\{3\sin^2\epsilon-1\} \quad (H'=0,\ \cos\epsilon=\sqrt{\tfrac{2}{3}}),$$

and the sin ϵ factor is irrelevant. Therefore

$$\frac{H'I_s}{B}=2\frac{I_h}{I_s}\left\{3\frac{I_h^2}{I_s^2}-2\right\}. \qquad \text{......(1480)}$$

Saturation is reached when $H'=H_s'=2B/I_s$ and at that point $(\partial I/\partial H)_s \neq 0$. There is again fair agreement with the observations as shown in Fig. 69·2.

§ **12·97.** *Magnetostriction in cubic crystals as a function of magnetization.* During stage one magnetization the microelements are merely turning round. This could cause no change of shape of the body in bulk if a microelement (or other body magnetized to saturation) were of true cubic structure. But if the unmagnetized state is a perfect cube the magnetized state need not be, and in fact is not, and the change of shape will depend on how the microelements are arranged for given I_h. These arrangements can be derived at once from the formulae of § 12·952. During stage two magnetization other changes of shape can occur due to the distortion accompanying the diversion of the directions of magnetization from easy directions. Such changes of shape are called *magnetostriction*. We shall show in the following sections that the magnetostriction at all stages of holomagnetization in Fe or Ni can be completely accounted for in terms of our general theory and two empirical constants which may be taken to be the proportional changes of length parallel to the magnetization when the specimen is holomagnetized to saturation along (100) and (111). To complete the theory we should be able to compute these constants in terms of the elastic constants of the metal and the magnetic dipole energy of a slightly distorted cubic lattice. Though there have been several attempts* and though the principles on which the calculation must proceed are agreed upon, no satisfactory calculation has yet been made.

Suppose that a microelement or other saturated crystal is magnetized with direction cosines $\alpha_1, \alpha_2, \alpha_3$ and measured along a line with direction cosines $\beta_1, \beta_2, \beta_3$. These direction cosines refer to axes coinciding with the crystallographic axes of the cubic crystal. Then†

$$\left(\frac{\delta l}{l}\right)_{\alpha,\beta}=c+\kappa_1\sum_{i\neq j}\alpha_i^2\beta_j^2-\kappa_2\sum_{i\neq j}\alpha_i\beta_i\alpha_j\beta_j. \qquad \text{......(1481)}$$

In (1481) κ_1 and κ_2 are constants and the additive term c, which may be a function of the β's, occurs because we do not know what to call the standard length of the specimen. This standard must be fixed later by a suitable convention, and we need not consider it further for the moment. We can only measure *changes of length in a given direction for changes of the direction*

* Akulov, *Zeit. f. Physik*, vol. 52, p. 389 (1928), vol. 59, p. 254 (1930); Becker, *Zeit. f. Physik*, vol. 62, p. 253 (1930), vol. 64, p. 660 (1930); Powell, *Proc. Camb. Phil. Soc.* vol. 27, p. 561 (1931).

† Heisenberg, *Zeit. f. Physik*, vol. 69, p. 287 (1931).

of magnetization. The formula (1481) is dictated entirely by the requirements of cubic symmetry. In the first place the formula (apart from c) must be even in the α's and the β's separately, for reversal of either of these directions is irrelevant. It must be unaltered by interchange of any pair of suffixes in the α's and β's simultaneously and by simultaneous change of sign of any pair α_i, β_i. The terms in κ_1 and κ_2 are then the only possible ones of less than fourth order in the α's.

It is important to see exactly how this formula may be used during stage one magnetization. The direction of measurement (β) then remains unaltered for the microelements or the whole body while the direction of magnetization (α) changes, so that c is constant. But owing to the symmetry of the κ-terms in α's and β's we can equally well use the formula for a microelement with the α's fixed and the β's changed if we still keep c constant. For convenience therefore we may proceed thus as if c were an absolute constant. The total result for the whole body will be obtained by summing over the contributions of each microelement.

§ **12·971.** *Magnetostriction curves for Fe crystals.*

Case (i). *Magnetization along* (100). Since (100) is an easy direction, the whole holomagnetization is stage one and the magnetostriction throughout can be expressed in terms of the shape of the natural microelements.

For a microelement

$$\left(\frac{\delta l}{l}\right)_{(100),(100)}=c \quad \text{(Longitudinal effect),} \qquad \text{......(1482)}$$

$$\left(\frac{\delta l}{l}\right)_{(0\beta_2\beta_3),(100)}=\left(\frac{\delta l}{l}\right)_{(100),(0\beta_2\beta_3)}=c+\kappa_1 \quad \text{(Transverse effect).} \qquad \text{......(1483)}$$

If N_1, N_2, N_3 are the numbers of microelements in the conglomerate body orientated along, normal to and against $\mathbf{H}'$, acting along (100), then for the whole specimen measured parallel to $\mathbf{H}'$

$$\frac{\delta l}{l}=\frac{N_1}{N}c+\frac{N_2}{N}(c+\kappa_1)+\frac{N_3}{N}c=c+\kappa_1\frac{N_2}{N}. \qquad \text{......(1484)}$$

We can now lay down our standard shape for the specimen by making $\delta l/l=0$ for the state of no holomagnetization when by § 12·95 $N_2/N=\frac{2}{3}$. Thus

$$c+\tfrac{2}{3}\kappa_1=0. \qquad \text{......(1485)}$$

For complete holomagnetization $N_2=0$

$$\left(\frac{\delta l}{l}\right)_{\text{sat}}=c=1{\cdot}95\times10^{-5}. \qquad \text{......(1486)}$$

This numerical value fits Webster's observations. Kaya's require $1{\cdot}68\times10^{-5}$. For partial holomagnetization

$$\frac{\delta l}{l}=c\left(1-\frac{3}{2}\frac{N_2}{N}\right).$$

By (1463) and (1464)
$$\frac{N_2}{N}=\frac{4}{\xi+4+1/\xi}, \qquad \text{......(1487)}$$

where
$$\frac{\xi-1/\xi}{\xi+4+1/\xi}=\frac{I_h}{I_s}=\delta. \qquad \text{......(1488)}$$

A simple reduction leads to

$$\frac{N_2}{N}=\tfrac{2}{3}\{2-\sqrt{(3\delta^2+1)}\},$$

so that
$$\frac{\delta l}{l}=c\left(-1+\sqrt{\left\{1+3\frac{I_h^2}{I_s^2}\right\}}\right). \qquad \text{......(1489)}$$

This formula is compared with the observations in Fig. 70.

Case (ii). *Magnetization along* (111). In stage one magnetization it follows from § 12·592 that the microelements merely turn end for end which does not change their shape. Thus we have $\delta l/l\equiv 0$ up to $I_h=I_s/\sqrt{3}$.

In stage two magnetization the direction of the magnetization, and therewith the shape of any microelement, changes. But all are similarly situated with respect to (111) so that the change of $\delta l/l$ in bulk is the same as for any microelement.

If ϵ is the angle between the direction of magnetization and (111), so that $\cos\epsilon=I_h/I_s$, then

$$\alpha_1=\sqrt{\tfrac{1}{3}}\cos\epsilon+\sqrt{\tfrac{2}{3}}\sin\epsilon,\quad \alpha_2=\alpha_3=\sqrt{\tfrac{1}{2}}\,(1-\alpha_1^2)^{\frac{1}{2}},$$

and

$$\left[\left(\frac{\delta l}{l}\right)_{(111),(111)}\right]_{\text{Bulk}}=\left(\frac{\delta l}{l}\right)_{(\alpha_1\alpha_2\alpha_3),(111)}=-\tfrac{2}{3}\kappa_2\{\sqrt{2}\,\alpha_1(1-\alpha_1^2)^{\frac{1}{2}}+\tfrac{1}{4}(1-\alpha_1^2)\}. \qquad \text{......(1490)}$$

This reduces to

$$\left[\left(\frac{\delta l}{l}\right)_{(111),(111)}\right]_{\text{Bulk}}=-\tfrac{1}{3}\kappa_2\left(3\frac{I_h^2}{I_s^2}-1\right). \qquad \text{......(1491)}$$

For saturation we have $\alpha_1=1/\sqrt{3}$, $\delta l/l=-\tfrac{2}{3}\kappa_2=-1{\cdot}7\times10^{-5}$ to fit Webster's observations, or $-1{\cdot}29\times10^{-5}$ to fit Kaya's. These results are again compared with observation in Fig. 70.

Case (iii). *Magnetization along* (110). In stage one magnetization N_1 microelements have their magnetization directed along (100) and (010), N_2 along (00 ± 1) and N_3 along (-100) and $(0-10)$. Formula (1481) (with $\kappa_1=-\tfrac{3}{2}c$) gives

$$\left(\frac{\delta l}{l}\right)_{\left({\pm100 \atop 0\pm10}\right),(110)}=\tfrac{1}{4}c,\quad \left(\frac{\delta l}{l}\right)_{(00\pm1),(110)}=-\tfrac{1}{2}c. \qquad \text{......(1492)}$$

Thus $$\left[\left(\frac{\delta l}{l}\right)_{(110),(110)}\right]_{\text{Bulk}}=\frac{c}{4N}(N_1+N_3-2N_2). \qquad ......(1493)$$

The formula of § 12·952 gives us at once

$$\frac{N_1+N_3-2N_2}{N}=\frac{\xi^{1/\sqrt{2}}+\xi^{-1/\sqrt{2}}-2}{\xi^{1/\sqrt{2}}+\xi^{-1/\sqrt{2}}+1},$$

where $$\frac{\frac{1}{\sqrt{2}}\xi^{1/\sqrt{2}}-\frac{1}{\sqrt{2}}\xi^{-1/\sqrt{2}}}{\xi^{1/\sqrt{2}}+\xi^{-1/\sqrt{2}}+1}=\frac{I_h}{I_s}=\delta.$$

We find easily

$$\left[\left(\frac{\delta l}{l}\right)_{(110),(110)}\right]_{\text{Bulk}}=\frac{c}{4}\left\{2-\sqrt{\left(4-6\frac{I_h^2}{I_s^2}\right)}\right\} \quad \left(I_h\leqslant\frac{I_s}{\sqrt{2}}\right). \qquad ...(1494)$$

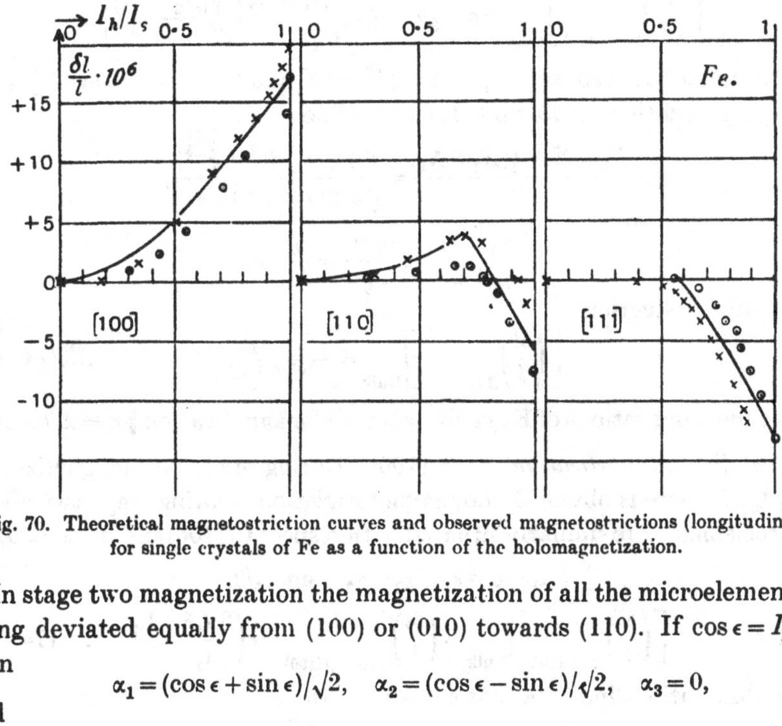

Fig. 70. Theoretical magnetostriction curves and observed magnetostrictions (longitudinal) for single crystals of Fe as a function of the holomagnetization.

In stage two magnetization the magnetization of all the microelements is being deviated equally from (100) or (010) towards (110). If $\cos\epsilon=I_h/I_s'$, then

$$\alpha_1=(\cos\epsilon+\sin\epsilon)/\sqrt{2},\quad \alpha_2=(\cos\epsilon-\sin\epsilon)/\sqrt{2},\quad \alpha_3=0,$$

and

$$\left[\left(\frac{\delta l}{l}\right)_{(110),(110)}\right]_{\text{Bulk}}=\left(\frac{\delta l}{l}\right)_{(\alpha_1\alpha_2\alpha_3),(110)}=c(1-\tfrac{3}{4}\{\alpha_1^2+\alpha_2^2\})-\kappa_2\alpha_1\alpha_2,$$

$$=\tfrac{1}{4}c-\tfrac{1}{2}\kappa_2\left\{2\frac{I_h^2}{I_s^2}-1\right\}. \qquad ......(1495)$$

All these results are in excellent agreement with Webster's or Kaya's measurements for suitable empirical values of c and κ_2, as the curves of Fig. 70 show.

§ **12·972.** *Magnetostriction curves for Ni crystals.* Very similar considerations apply to Ni crystals and it will be sufficient to summarize the results.

Case (i). *Magnetization along* (111). Stage one magnetization proceeds to saturation. There are four classes of microelement, N_1 along (111), N_2 along $(11-1)$, $(1-11)$ or (-111), N_3 along $(-1-1-1)$ and N_4 along $(-1-11)$, $(-11-1)$ and $(1-1-1)$. We find at once

$$\left(\frac{\delta l}{l}\right)_{(111),(111)}=\left(\frac{\delta l}{l}\right)_{(-1-1-1),(111)}=c+\tfrac{2}{3}\kappa_1-\tfrac{2}{3}\kappa_2,$$

$$\left(\frac{\delta l}{l}\right)_{(11-1),(111)}=\left(\frac{\delta l}{l}\right)_{(-1-11),(111)}=c+\tfrac{2}{3}\kappa_1+\tfrac{2}{9}\kappa_2,$$

$$\left[\left(\frac{\delta l}{l}\right)_{(111),(111)}\right]_{\text{Bulk}}=c+\tfrac{2}{3}\kappa_1-\tfrac{2}{3}\kappa_2\left\{\frac{N_1+N_3-\frac{1}{3}(N_2+N_4)}{N}\right\}.$$

For the unmagnetized state $N_1=N_3=\frac{1}{3}N_2=\frac{1}{3}N_4$ and by convention $c+\frac{2}{3}\kappa_1=0$. Applying equation (1463) and (1464), we find

$$\frac{N_1+N_3-\frac{1}{3}(N_2+N_4)}{N}=\frac{\xi+\xi^{-1}-(\xi^{\frac{1}{3}}+\xi^{-\frac{1}{3}})}{\xi+3\xi^{\frac{1}{3}}+3\xi^{-\frac{1}{3}}+\xi^{-1}},$$

where

$$\frac{\xi+\xi^{\frac{1}{3}}-\xi^{-\frac{1}{3}}-\xi^{-1}}{\xi+3\xi^{\frac{1}{3}}+3\xi^{-\frac{1}{3}}+\xi^{-1}}=\frac{I_h}{I_s}=\delta.$$

Thus after reduction

$$\left[\left(\frac{\delta l}{l}\right)_{(111),(111)}\right]_{\text{Bulk}}=-\tfrac{2}{3}\kappa_2\frac{I_h^2}{I_s^2}. \qquad \text{......(1496)}$$

To fit the observations of Kaya for saturated magnetization $\frac{2}{3}\kappa_2=2\cdot7\times10^{-5}$.

Case (ii). *Magnetization along* (100). During stage one magnetization, $I_h\leqslant I_s/\sqrt{3}$, there is obviously no magnetostriction. During stage two all the microelements are similarly situated with respect to (100) and if $\cos\epsilon=I_h/I_s$, then

$$\alpha_1=\cos\epsilon,\quad \alpha_2=\alpha_3=\sin\epsilon/\sqrt{2},$$

and

$$\left[\left(\frac{\delta l}{l}\right)_{(100),(100)}\right]_{\text{Bulk}}=\left(\frac{\delta l}{l}\right)_{(\alpha_1\alpha_2\alpha_3),(100)}=c\left(\frac{3}{2}\frac{I_h^2}{I_s^2}-\frac{1}{2}\right). \qquad \text{......(1497)}$$

The observations indicate that $c=-5\cdot1\times10^{-5}$.

Case (iii). *Magnetization along* (110). During stage one magnetization, $I_h\leqslant I_s\sqrt{\frac{2}{3}}$, there are three classes of microelement to consider: N_1 along (11 ± 1), N_2 along $\pm(1-1\pm1)$ and N_3 along $-(11\pm1)$. For these microelements we have

$$\left(\frac{\delta l}{l}\right)_{\pm(11\pm1),(110)}=-\tfrac{1}{3}\kappa_2,\quad \left(\frac{\delta l}{l}\right)_{\pm(1-1\pm1),(110)}=\tfrac{1}{3}\kappa_2,$$

$$\left[\left(\frac{\delta l}{l}\right)_{(110),(110)}\right]_{\text{Bulk}}=-\tfrac{1}{3}\kappa_2\left[\frac{N_1+N_3-N_2}{N}\right].$$

Using again the methods of § 12·952, we find

$$\left[\left(\frac{\delta l}{l}\right)_{(110),(110)}\right]_{\text{Bulk}} = -\tfrac{1}{2}\kappa_2 \frac{I_h^2}{I_s^2} \quad (I_h \leqslant \sqrt{\tfrac{2}{3}} I_s). \quad \text{......(1498)}$$

During stage two magnetization, if as usual $\cos\epsilon = I_h/I_s$, then

$$\alpha_1 = \alpha_2 = \frac{1}{\sqrt{2}}\cos\epsilon, \quad \alpha_3 = \sin\epsilon,$$

and
$$\left[\left(\frac{\delta l}{l}\right)_{(110),(110)}\right]_{\text{Bulk}} = \left(\frac{\delta l}{l}\right)_{(\alpha_1\alpha_2\alpha_3),(110)},$$

$$\doteq c\left(\frac{3}{4}\frac{I_h^2}{I_s^2} - \tfrac{1}{2}\right) - \tfrac{1}{2}\kappa_2 \frac{I_h^2}{I_s^2} \quad (I_h \geqslant I_s\sqrt{\tfrac{2}{3}}). \text{......(1499)}$$

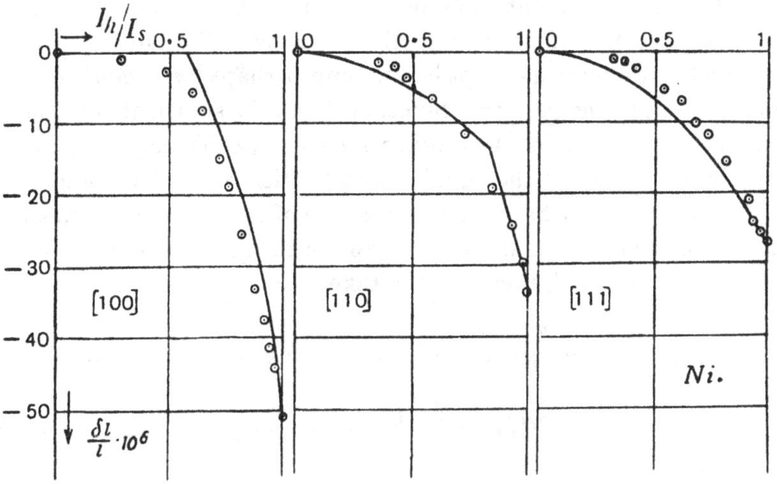

Fig. 71. Theoretical magnetostriction curves and observed magnetostrictions for single crystals of Ni as a function of the holomagnetization.

The general agreement between these formulae and Kaya's measurements is fair, and is shown in Fig. 71. The observed values given are very uncertain for the smaller magnetizations, since the magnetostriction and the magnetization were not simultaneously observed.

CHAPTER XIII†

APPLICATIONS TO LIQUIDS AND SOLUTIONS

§ **13·1.** *Introduction.* It is at present far beyond the power of statistical mechanics to deduce rigorously the properties of a liquid from a given molecular model. Most liquids of our common experience are composed of very complex molecules. Even if we possessed detailed experimental knowledge of the liquid states of the inert gases we could hardly yet hope to give a satisfactory theoretical interpretation. It is not that the method is not sufficiently clear. It ought to be possible for such simple liquids to give a satisfactory account of their properties as follows. For liquids with heavy molecules (i.e. all liquids except helium and perhaps hydrogen‡) a classical approximation to the partition function should be sufficient. We may then formulate the problem in one or other of two ways. We may adopt as a first rough approximation to the liquid a model which regards it as a quasi-solid in which each molecule has a natural position of equilibrium differing merely from a molecule in a solid in that its position of equilibrium wanders about. If we adopt this model, then we may take

$$-\frac{F}{kT}=\frac{\Psi}{k}=N\log J(T)+\log B^*(T), \qquad \text{......(1500)}$$

$$B^*(T)=\int\!\dots\!\int e^{-W/kT}\,\Pi_\kappa\,(d\omega_\kappa), \qquad \text{......(1501)}$$

in which W is the complete potential energy of the liquid in any configuration,§ and $J(T)$ is the ordinary partition function for the translational kinetic energy and internal and rotational energies of the molecules—frequently much the same as in a gas. In calculating $B^*(T)$ after this model the molecules must not be allowed to change places, so that if we allowed $W\to 0$ the limit of $B^*(T)$ would have to be $V^N/N!$. This model of course cannot validly be used in this limit.

If we adopt the model of an extremely imperfect gas, then we can proceed

† This chapter has been completely recast by Mr E. A. Guggenheim.

‡ The hydrogen rich liquids such as water, ammonia and methane in which each molecule contains only one atom other than hydrogen probably form a partial exception and require a quantum treatment of their rotational energy content. Bernal and Tamm, *Nature*, vol. 135, p. 229 (1935).

§ More strictly W is only that part of the potential energy of the liquid which depends on the configuration of the centres of mass of every molecule since (1501) refers only to such coordinates. Potential energy arising from internal vibrations and rotations of the molecules (if any) must be regarded as absorbed in $J(T)$.

exactly as in Chapter VIII and use Theorem 6·31 for systems occupying a common region of space. We may then take

$$-\frac{F}{kT}=\frac{\Psi}{k}=N\left(\log\frac{J(T)}{N}+1\right)+\log B(T), \qquad \text{......(1502)}$$

$$B(T)=\int_{(V)}\cdots\int e^{-W/kT}\,\Pi_\kappa(d\omega_\kappa), \qquad \text{......(1503)}$$

where the integration in $B(T)$ is now to be calculated without restriction for all configurations. We shall easily see later that these formulations lead to identical results.

A statistical theory of liquids is still lacking, not because of a failure of theory but because of the almost prohibitive difficulty of calculating $B(T)$ or $B^*(T)$. The utmost that we can hope to do at present is to calculate quantitatively the changes in $B(T)$ or $B^*(T)$ when certain changes are made in the force system W or in the nature of the systems composing the assembly—that is to calculate the properties of solutions given an empirical knowledge of the properties of the pure components, and even this only tentatively. We shall give here merely a sketch of how a theory of solutions can be approached by application of the general theorems of statistical mechanics.

§ **13·11.** *A pure liquid.* Let us start by recalling how (1503) is evaluated for a gas. The volume V of a gas is determined entirely by that of its container, and for a perfect gas $W=0$ when all the molecules are anywhere inside V, and $W=\infty$ when any molecule is outside V, so that $B(T)=V^N$. For an imperfect gas the condition "W small" replaces $W=0$, so that integration over the whole configuration space still means integration over a hypervolume V^N of an integrand not exactly unity. This leads when developed to the formulae of Chapter VIII.

For a liquid however the conditions are essentially different, for the volume V is not imposed by the containing vessel but by the intermolecular forces. The volume V, provided it exceeds a certain lower limit, determines merely how many molecules are in a vapour phase. We shall assume that these are of negligible number and neglect them. For a simple liquid phase W must have a pronounced minimum for certain configurations in which all the N molecules are packed together in a volume V, the actual volume of the liquid. The large heat of evaporation and small compressibility of a liquid show that W will be much greater (effectively infinite) for all configurations differing sensibly from V in that volume. Since obviously this minimum value $W_{\min}$ must be proportional to N, we may write

$$W_{\min}=Nu(T), \qquad \text{......(1504)}$$

where $w(T)$ is independent of N. We thus obtain

$$B(T) = e^{-Nw/kT} \int \dots \int \Pi_\kappa (d\omega_\kappa),$$

where the integration is to be extended over all configurations sufficiently nearly consistent with the condition $W = W_{\min} = Nw$. The extent of these configurations can be estimated formally thus.† Consider the molecules forming the liquid to be in one definite configuration satisfying $W = W_{\min}$. Then there are obviously $N!$ configurations of identical energy which can be constructed by interchanging the molecules. When we are using (1502) and (1503) these interchanges are to be included. Further, round each such configuration we can specify a region of neighbouring configurations in which W differs inappreciably from $W_{\min}$ by assigning to each molecule a region v over which this molecule can be displaced more or less independently of the other molecules without significantly breaking the condition $W = W_{\min}$. Thus

$$B(T) = e^{-Nw/kT} N! \, v^N, \qquad \text{......(1505)}$$

$$-\frac{F}{kT} = \frac{\Psi}{k} = N\left\{\log vJ(T) - \frac{w}{kT}\right\}. \qquad \text{......(1506)}$$

Both v and w will depend on T in a manner which these considerations provide no means of evaluating. We do not therefore gain from (1506) any insight into the properties of a pure liquid, but we can easily extend (1506) to dilute solutions where it becomes significant.

§ **13·12.** *A pure liquid as a quasi-crystal.* Before proceeding with this extension it is of interest to examine the other approach provided by (1500) and (1501). It is obvious that with the same approximations

$$B^*(T) = v^N e^{-Nw/kT},$$

so that (1506) is unaffected. We can however with the quasi-crystal picture make more explicit assumptions and so reach explicit formulae for v, w, F and Ψ as functions of T, which have been used successfully by Mott‡ to explain the electrical properties of liquid metals.

The proposed model of a liquid is one in which it is assumed that the molecules vibrate about positions of equilibrium as in a solid with a characteristic frequency or spectrum of frequencies. The difference from a solid is merely that the positions of equilibrium are themselves free to wander slowly about, wandering which is without effect on $J(T)$ or $B^*(T)$ to a first

† Guggenheim, *Proc. Roy. Soc.* A, vol. 135, p. 181 (1932).
‡ Mott, *Proc. Roy. Soc.* A, vol. 146, p. 465 (1934).

approximation. For simplicity we shall assume that the vibrations have a common frequency ν_L. Thus

$$W = W_{\min} + \sum_{1}^{3N} \tfrac{1}{2}m(2\pi\nu_L)^2 x_r^2,$$

$$B^*(T) = e^{-Nw/kT}\left[\int_{-\infty}^{\infty} e^{-m(2\pi\nu_L x_r)^2/2kT}\,dx_r\right]^{3N},$$

$$v = \left[\frac{2\pi kT}{m(2\pi\nu_L)^2}\right]^{\frac{3}{2}}. \qquad \text{......(1507)}$$

For this simple model we shall also have

$$J(T) = \left(\frac{2\pi mkT}{h^2}\right)^{\frac{3}{2}},$$

so that

$$-\frac{F}{kT} = \frac{\Psi}{k} = N\left\{\log\left(\frac{kT}{h\nu_L}\right)^3 - \frac{w}{kT}\right\}. \qquad \text{......(1508)}$$

It is of course unnecessary to have used the classical approximation here. The assumption that the liquid has a minimum configurational potential energy of Nw' and is otherwise equivalent to a degenerate oscillator of $3N$ freedoms each of frequency ν_L leads at once by (88) and (598) to

$$-\frac{F}{kT} = \frac{\Psi}{k} = N\left\{-3\log(1 - e^{-h\nu_L/kT}) - \frac{w}{kT}\right\}, \qquad \text{......(1509)}$$

if w is taken to include both w' and the zero-point energy of oscillation $\frac{3}{2}h\nu_L$. Equation (1509) reduces to (1508) when $h\nu_L \ll kT$. In the whole formulation up to this point the volume occupied by the liquid is invariable and there is no dependence on pressure.

§ **13·13.** *An elementary (thermodynamic) theory of melting.* Mott has used (1508) to give an elementary theory of melting, applicable to the melting of simple atomic substances such as the metals. We may suppose that in the regularly ordered configuration of the solid crystal the assembly contains less potential energy than in the irregular configurations of the liquid, but owing to its regularity and rigidity a natural frequency ν_S which is greater than ν_L. We may therefore write

$$F_S = N\left\{-kT\log\left(\frac{kT}{h\nu_S}\right)^3 + w_S\right\}, \qquad \text{......(1510)}$$

$$F_L = N\left\{-kT\log\left(\frac{kT}{h\nu_L}\right)^3 + w_L\right\}, \qquad \text{......(1511)}$$

where $w_L > w_S$, $\nu_L < \nu_S$. The condition determining the melting point is that liquid and solid can be in equilibrium together, that is $\delta F = 0$ for a change from one form to the other at constant temperature, or

$$3\log\frac{\nu_S}{\nu_L} = \frac{w_L - w_S}{kT}. \qquad \text{......(1512)}$$

For the proposed model there is no change of specific heat on melting; for both phases $C_V = 3R$. By calculating the energy or entropy of the phases it is easy to show that the latent heat of melting Λ defined as $E_L - E_S$ or $T(S_L - S_S)$ is given by

$$\Lambda = N(w_L - w_S). \qquad \text{......(1513)}$$

It follows at once that if Λ refers to one gram-molecule

$$\frac{\nu_L}{\nu_S} = e^{-\frac{1}{3}\Lambda/RT}. \qquad \text{......(1514)}$$

Mott* has applied this relationship to explain the increase of electrical resistance on melting for a normal metal. In the absence of large volume charges on melting or complicated rearrangements of the electron levels the

TABLE 52.

Showing Λ and T_M for various metals, and the observed and calculated values of κ_S/κ_L at the melting point.

Metal	Λ	T_M° K.	κ_S/κ_L calc.	κ_S/κ_L obs.	$h\nu_S/k$
Li	830	459	1·84 (1·57)	1·68	510
Na	630	370	1·77 (1·58)	1·45	200
K	570	335	1·75 (1·67)	1·55	126
Rb	520	311	1·76	1·61	85
Cs	500	299	1·75	1·66	68
Cu	2750	1356	1·97	2·07	310
Ag	2630	1233	2·0	1·9	215
Au	3180	1336	2·22	2·28	175
Al	1910	933	2·0 (1·8)	1·64	400
Cd	1500	594	2·3	2·0	168
Pb	1120	590	1·90	2·07	90
Sn	[1700]	505	[3]	2·1	—
Tl	1470	580	2·2	2·0	96
Zn	1700	692	2·3	2·1	235
Hg	560	234	2·23	3·2–4·9	97
Bi	2600	544	5·0	0·43	—
Ga	1320	303	4·5	0·58	—
Sb	4660	903	5·7	0·67	—

The values of Λ are in calories per mole. The values in [ ] are uncertain. The values in ( ) have been corrected for the fact that $\kappa \propto \nu^2$ strictly only when $T \gg h\nu/k$, a condition which is not well satisfied for these metals; compare cols. 3 and 6.

conductivity of a metal varies as ν^2, the square of the characteristic temperature.† Thus at the melting point, $T = T_M$,

$$\frac{\kappa_L}{\kappa_S} = e^{-\frac{2}{3}\Lambda/RT_M}. \qquad \text{......(1515)}$$

The right-hand side of (1515) depends on calorimetric quantities only. The excellent agreement between the (electrically) observed and calculated values of κ_S/κ_L is shown in Table 52. For the last group the theory ob-

* Mott, *loc. cit.*

† See, for example, Sommerfeld and Bethe, *Handb. d. Physik*, vol. 24 (2), p. 507 (1933).

viously fails, but failure is to be expected here due to factors omitted in using $\kappa \propto \nu^2$.

The substantial success of (1515) suggests that the quasi-crystal model is at least a fair representation of a simple liquid near its melting point. This thermodynamic theory of melting however is far from a complete molecular theory, for it makes no attempt to explain the *sharpness* of the transition from liquid to solid. This sharpness which is also characteristic of other transitions such as those from one crystalline form to another is the most striking characteristic of the process and must be a natural deduction from any complete theory. It is not possible to carry this discussion any further here.

§ 13·2. *Ideal solutions.* We now extend the previous results to a mixture of N_α molecules of type α and N_β of type β.† Adopting the quasi-solid starting point, the free energy and characteristic function are now of the form

$$-\frac{F}{kT}=\frac{\Psi}{k}=N_\alpha \log J_\alpha + N_\beta \log J_\beta + \log B^*(T).$$

J_α and J_β have meanings similar to that of J above. $B^*(T)$ is given by

$$B^*(T)=\int\!\dots\!\int e^{-W/kT}(d\omega_\alpha)^{N_\alpha}(d\omega_\beta)^{N_\beta},$$

in which W is the configurational potential energy and the integral is extended over all distinct configurations. As in § 5·61 however the number of such distinct configurations‡ is now no longer 1 but $(N_\alpha+N_\beta)!/N_\alpha!\,N_\beta!$. A simple rough evaluation of $B^*(T)$ is now possible in the special case in which there are no long range electrostatic forces and N_β/N_α is so small that one may neglect its square. When these two conditions are fulfilled we call the mixture *an ideal solution of* β *(the solute) in* α *(the solvent)*. Under these conditions two molecules β will be within reach of each other so seldom that such configurations can be neglected. W will therefore still have a pronounced minimum of the form

$$W_{\min}=N_\alpha w_\alpha(T)+N_\beta w_\beta(T), \qquad \text{......(1516)}$$

for as compared with (1504) there are merely a number N_β of centres of distortion which make equal contributions to $W_{\min}$, distinct from the contribution of the normal α molecule. It follows at once that we may write $B^*(T)$ in the form

$$B^*(T)=e^{-(N_\alpha w_\alpha+N_\beta w_\beta)/kT}\,\frac{(N_\alpha+N_\beta)!}{N_\alpha!\,N_\beta!}\,(v_\alpha)^{N_\alpha}(v_\beta)^{N_\beta}. \qquad \text{......(1517)}$$

† Guggenheim, *loc. cit.*

‡ If the two sorts of molecules are markedly dissimilar in size and shape this argument can hardly be retained, but the result (1518) remains a fair approximation.

It follows that

$$-\frac{F}{kT}=\frac{\Psi}{k}=N_\alpha\left\{\log\frac{J_\alpha v_\alpha(N_\alpha+N_\beta)}{N_\alpha}-\frac{w_\alpha}{kT}\right\}+N_\beta\left\{\log\frac{J_\beta v_\beta(N_\alpha+N_\beta)}{N_\beta}-\frac{w_\beta}{kT}\right\}. \quad ......(1518)$$

In (1518) J_α, J_β, v_α, v_β, w_α, w_β are in general all functions of T.

§**13·21.** *Partial vapour pressures of an ideal solution.* If we add to our assembly a vapour phase, which is a perfect gas mixture, then the free energy F_a of the complete assembly will be

$$F_a=F+F',$$

where F is given by (1518) and F', after (597), by

$$-\frac{F'}{kT}=N_\alpha'\left\{\log\frac{V'J_\alpha'}{N_\alpha'}+1\right\}+N_\beta'\left\{\log\frac{V'J_\beta'}{N_\beta'}+1\right\}. \quad ......(1519)$$

Primes will be consistently used to denote the vapour phase. The condition of equilibrium is $\delta F_a=0$ subject to $\delta T=0$ and $\delta(V+V')=0$, but since for a given number of molecules δV is negligible compared with $\delta V'$ this reduces to $\delta V'=0$. The condition that F_a is unaffected by a transference of either molecule from one phase to the other is satisfied provided that

$$\frac{\partial F}{\partial N_\alpha}=\frac{\partial F'}{\partial N_\alpha'},\quad \frac{\partial F}{\partial N_\beta}=\frac{\partial F'}{\partial N_\beta'}. \quad ......(1520)$$

Combining (1518), (1519) and (1520), we find

$$\log\frac{J_\alpha v_\alpha(N_\alpha+N_\beta)}{N_\alpha}-\frac{w_\alpha}{kT}=\log\frac{J_\alpha'V'}{N_\alpha'}, \quad ......(1521)$$

$$\log\frac{J_\beta v_\beta(N_\alpha+N_\beta)}{N_\beta}-\frac{w_\beta}{kT}=\log\frac{J_\beta'V'}{N_\beta'}. \quad ......(1522)$$

It follows that the partial vapour pressure of the solvent is given by

$$p_\alpha'=\frac{N_\alpha'kT}{V'}=\frac{N_\alpha}{N_\alpha+N_\beta}\frac{J_\alpha'kT}{J_\alpha v_\alpha}e^{w_\alpha/kT}, \quad ......(1523)$$

an equation which can be cast into the form

$$p_\alpha'=\frac{N_\alpha}{N_\alpha+N_\beta}(p_\alpha')_0, \quad ......(1524)$$

$(p_\alpha')_0$ being the vapour pressure of the pure solvent. This is *Raoult's Law.* Similarly

$$p_\beta'=\frac{N_\beta}{N_\alpha+N_\beta}\frac{J_\beta'kT}{J_\beta v_\beta}e^{w_\beta/kT}, \quad ......(1525)$$

so that at constant temperature

$$p_\beta'\propto N_\beta/(N_\alpha+N_\beta), \quad ......(1526)$$

which is *Henry's Law.*

§ **13·22.** *Lowering of the freezing point.* A familiar purely thermodynamic argument shows that *any* dilute solution is in equilibrium with the pure solid solvent at a slightly lower temperature, $T_M{}^0 - \Delta T_M$, than the freezing point $T_M{}^0$ of the pure liquid and that ΔT_M is given by*

$$\frac{\Lambda}{R(T_M{}^0)^2}\Delta T_M = \log\frac{(p_\alpha{}')_0}{p_\alpha{}'}. \qquad \ldots\ldots(1527)$$

Using Raoult's law this becomes

$$\frac{\Lambda}{R(T_M{}^0)^2}\Delta T_M = \log\frac{N_\alpha + N_\beta}{N_\alpha}. \qquad \ldots\ldots(1528)$$

The natural statistical proof of (1528) as a property of ideal solutions is as follows: For the pure solid phase by (598)

$$-\frac{F_S}{kT} = N_\alpha{}^S\{\log\kappa(T)\}.$$

Combining this with (1518) and the condition of equilibrium, we see that

$$\log\kappa(T_M) = \log\frac{J_\alpha(T_M)\,v_\alpha(T_M)\,(N_\alpha + N_\beta)}{N_\alpha} - \frac{w_\alpha(T_M)}{kT_M}.$$

In the same notation

$$\log\kappa(T_M{}^0) = \log J_\alpha(T_M{}^0)\,v_\alpha(T_M{}^0) - w_\alpha(T_M{}^0)/kT_M{}^0.$$

Since $T_M - T_M{}^0$ is small, these equations can be combined in the form

$$\log\frac{N_\alpha + N_\beta}{N_\alpha} = \frac{\Delta T_M}{kT_M{}^2}\left[kT_M{}^2\frac{\partial}{\partial T_M}\log\frac{J(T_M)\,v(T_M)\,e^{-w_\alpha(T_M)/kT_M}}{\kappa(T_M)}\right].$$

But by the nature of a partition function the term in [ ] is the excess energy content of the liquid per molecule over the energy content of the solid, so that (1528) is established.

§ **13·23.** *Effects of pressure.* Hitherto we have ignored the pressure entirely as a variable affecting the liquid, which has thereby been taken to be incompressible, with a mean molecular volume V/N ($= v(T)$) depending on the temperature alone.† For such a liquid, in sharp contrast to a gas, it is not possible to use V as a variable defining its state. If the liquid is actually compressible V becomes again a legitimate variable but a very inconvenient one, since formulae become indeterminate instead of particularly simple

* The symbol Λ for the latent heat of melting per gram-molecule, or mole, has already been introduced in § 13·13. In the notation introduced in the next section it would be written Λ^*.

† Throughout the rest of this chapter we attempt to conform to a notation not too remote from that customary in physical chemistry. It is however necessary systematically to distinguish *molecular* from *molar* quantities. Since we are primarily concerned with the former, it is not feasible to complicate the notation by distinguishing them by a special affix and we therefore (when necessary) distinguish molar quantities by an asterisk: e.g. $v(T)$ the volume per molecule, $v^*(T)$ the volume per mole (gram-molecule). Thus $v^* = \mathsf{N}v$, where N is Avogadro's number, which will be denoted throughout this chapter by this special N.

when the compressibility tends to zero. It is necessary therefore to use p together with T, N_α and N_β to define the state of the assembly. The functions w_α, w_β, v_α and v_β are strictly functions of p as well as of T, but cannot be explicitly calculated. Owing to this mathematical inconvenience of p as a fundamental variable, progress is only possible by special devices.

The free energy F satisfies the equation $\partial F/\partial V = -p$. Let us denote the value of F (and other quantities) for zero pressure by attaching a suffix or affix zero. Then

$$F = F_0 + \int_V^{V_0} p\,dV, \qquad \text{......(1529)}$$

$$-\frac{F_0}{kT} = \frac{\Psi_0}{k} = N_\alpha\left\{\log\frac{J_\alpha v_\alpha^0(N_\alpha+N_\beta)}{N_\alpha} - \frac{w_\alpha^0}{kT}\right\} + N_\beta\left\{\log\frac{J_\beta v_\beta^0(N_\alpha+N_\beta)}{N_\beta} - \frac{w_\beta^0}{kT}\right\}.$$

We may safely assume for ordinary pressures that

$$V = V_0(1-\kappa p), \qquad \text{......(1530)}$$

where κ is independent of p. Moreover for an ideal solution we must have

$$V = N_\alpha v_\alpha + N_\beta v_\beta$$

for the same reasons that require (1516). The v's are functions of T and p only and represent the average volume required by one molecule of each type. To conform, as nearly as may be, to thermodynamic usage we shall use *partial molar volumes* v_α^*, v_β^*, and the relation becomes

$$V = (N_\alpha v_\alpha^* + N_\beta v_\beta^*)/\mathbf{N}. \qquad \text{......(1531)}$$

Relations (1530) and (1531) only hold simultaneously if

$$v_\alpha^* = v_\alpha^{0*}(1-\kappa_\alpha p), \quad v_\beta^* = v_\beta^{0*}(1-\kappa_\beta p), \qquad \text{......(1532)}$$

where the κ's are independent of p and of the composition. We have therefore assumed that

$$V = \frac{N_\alpha}{\mathbf{N}} v_\alpha^{0*}(1-\kappa_\alpha p) + \frac{N_\beta}{\mathbf{N}} v_\beta^{0*}(1-\kappa_\beta p); \qquad \text{......(1533)}$$

it follows from (1533), (1529) and the formula for F_0 that

$$-F = T\Psi = N_\alpha\left\{kT\log\frac{J_\alpha v_\alpha^0(N_\alpha+N_\beta)}{N_\alpha} - w_\alpha^0 - \frac{1}{2}\frac{v_\alpha^{0*}}{\mathbf{N}}\kappa_\alpha p^2\right\}$$
$$+ N_\beta\left\{kT\log\frac{J_\beta v_\beta^0(N_\alpha+N_\beta)}{N_\beta} - w_\beta^0 - \frac{1}{2}\frac{v_\beta^{0*}}{\mathbf{N}}\kappa_\beta p^2\right\}. \qquad \text{......(1534)}$$

For use with the variables T, p the thermodynamic potential† is Gibbs'

† As usages differ, especially among physical chemists, it is well to insist here that in this monograph the name *thermodynamic potential* is used for any one of the functions E, H, F or G (or any other function) expressed in the proper variables such that the other thermodynamic variables may be obtained from it by immediate partial differentiation. The name *characteristic function* is used only for Ψ and Φ, Planck's variants of F and G. The four thermodynamic potentials named above are of course the energy, heat function, Helmholtz's free energy (or work function) and Gibbs' free energy respectively.

free energy G or Planck's second characteristic function Φ. Since $G = F + pV$,

$$-G = T\Phi = N_\alpha\left\{kT\log\frac{J_\alpha v_\alpha{}^0(N_\alpha+N_\beta)}{N_\alpha} - w_\alpha{}^0 - p\frac{v_\alpha{}^{0*}}{\mathsf{N}}(1-\tfrac{1}{2}\kappa_\alpha p)\right\}$$

$$+N_\beta\left\{kT\log\frac{J_\beta v_\beta{}^0(N_\alpha+N_\beta)}{N_\beta} - w_\beta{}^0 - p\frac{v_\beta{}^{0*}}{\mathsf{N}}(1-\tfrac{1}{2}\kappa_\beta p)\right\}. \quad \text{......(1535)}$$

Gibbs' partial or chemical potentials per mole, μ_α, μ_β, which must be written $\mu_\alpha{}^*$, $\mu_\beta{}^*$ in the notation of this chapter, are defined by

$$\mu_\alpha{}^* = \mathsf{N}\left(\frac{\partial G}{\partial N_\alpha}\right)_{T,p,N_\beta}, \quad \mu_\beta{}^* = \mathsf{N}\left(\frac{\partial G}{\partial N_\beta}\right)_{T,p,N_\alpha} \quad \text{......(1536)}$$

For our ideal solutions they have therefore the forms

$$\mu_\alpha{}^* = RT\log\frac{N_\alpha}{J_\alpha v_\alpha{}^0(N_\alpha+N_\beta)} + \mathsf{N}w_\alpha{}^0 + pv_\alpha{}^{0*}(1-\tfrac{1}{2}\kappa_\alpha p), \quad \text{......(1537)}$$

$$\mu_\beta{}^* = RT\log\frac{N_\beta}{J_\beta v_\beta{}^0(N_\alpha+N_\beta)} + \mathsf{N}w_\beta{}^0 + pv_\beta{}^{0*}(1-\tfrac{1}{2}\kappa_\beta p). \quad \text{......(1538)}$$

From these formulae we can deduce all the equilibrium properties of such solutions by purely thermodynamic reasoning. We need not therefore elaborate the details, and shall confine ourselves to a discussion of osmotic pressure by way of illustration.

§ **13·24.** *Osmotic pressure.* The *osmotic pressure* of a solution is defined to be the extra pressure that must be applied to the solution to keep it in equilibrium for the solvent molecules with the pure solvent at a given external pressure. The definition implies that solvent molecules (but not solute molecules) can pass freely to and fro between the solution and the pure solvent, for instance through a semi-permeable membrane. The condition of equilibrium is therefore

$$\mu_\alpha{}^*(T,p+P,N_\alpha,N_\beta) = \mu_\alpha{}^*(T,p,N_\alpha,0),$$

P being the osmotic pressure. We find at once

$$Pv_\alpha{}^{0*}\{1-\kappa_\alpha(p+\tfrac{1}{2}P)\} = RT\log\frac{N_\alpha+N_\beta}{N_\alpha}. \quad \text{......(1539)}$$

Neglecting compressibility this simplifies to

$$Pv_\alpha{}^{0*} = RT\log\frac{N_\alpha+N_\beta}{N_\alpha}. \quad \text{......(1540)}$$

§ **13·25.** *Heat of dilution.* By definition an ideal solution is one in which the interaction between solute molecules is negligible. Any two such solutions therefore, having the same solvent, temperature and pressure, will mix without absorption or evolution of heat. This may be rigorously established by a thermodynamic calculation of the total heat H since in *any* process at constant temperature and pressure, involving no work other than that done by the pressure, the increase in H is equal to the heat taken in. But by its definition and (1535)

$$H = -T^2\left(\frac{\partial G/T}{\partial T}\right)_{p,N_\alpha,N_\beta} = T^2\left(\frac{\partial \Phi}{\partial T}\right)_{p,N_\alpha,N_\beta} = N_\alpha H_\alpha + N_\beta H_\beta, \quad \text{......(1541)}$$

where H_α and H_β are independent of N_α and N_β. By considering more general ideal solutions we see at once that the total heat terms are all strictly additive. Any two such solutions therefore will mix without absorption or evolution of heat. In particular this applies to the mixing of one such solution with pure solvent. In other words *the heat of (further) dilution of an ideal solution is zero.*

§ **13·3.** *Further applications of the formulae for ideal solutions.* (1) *Perfect solutions.* In evaluating $B(T)$ for ideal solutions we assumed that among the $(N_\alpha + N_\beta)!/(N_\alpha! N_\beta!)$ permutations of the molecules among themselves those leading to configurations with two or more β molecules in contact were negligible in number. It is not really necessary to be so strict, for all that it is necessary to assume is that such configurations have a negligible effect on $B(T)$. The formulae for ideal solutions which must hold when $N_\beta \ll N_\alpha$ may also hold when this condition is poorly satisfied and the solutions are not so very dilute. In the extreme case in which β-β configurations occupy the same volume and possess the same energy as α-β or α-α configurations the laws of ideal solutions will hold for all concentrations.† It is well known that such solutions exist. They are known as *perfect solutions.* In perfect solutions it is no longer necessary to draw any distinction between solvent and solute. Both species obey Raoult's law in the form

$$p_\alpha' = (p_\alpha')_0 N_\alpha/(N_\alpha + N_\beta), \quad p_\beta' = (p_\beta')_0 N_\beta/(N_\alpha + N_\beta). \quad \text{...(1542)}$$

§ **13·31.** (2) *Several solute species.* The theory extends at once to solutions of mixed solutes and to dissociative equilibrium between the solute species, provided that the resulting ionization makes a negligible contribution to $B(T)$. The simple theory of ideal solutions will thus account satisfactorily for the properties of weak electrolytes, in which there is some small degree of dissociation into ions as shown by the conductivity. For strong electrolytes in which the dissociation into ions is almost complete the simple theory is inadequate.

§ **13·32.** (3) *Ideal solutions at extreme dilution.* We shall next consider an ideal solution of general solute species in a given solvent which we may take to be water. The necessary generalization of (1537) is

$$\mu^*_{H_2O} = \mu^{0*}_{H_2O} + p v^{0*}_{H_2O}(1 - \tfrac{1}{2}\kappa_{H_2O} p) + RT \log \frac{N_{H_2O}}{N_{H_2O} + \Sigma_\beta' N_\beta}, \quad \text{...(1543)}$$

† The necessary condition for a perfect solution can be formulated slightly more generally still. See § 13·4.

where $\mu^{0*}_{H_2O}$ is the chemical potential of H_2O in the pure solvent at zero pressure, a function only of T, and the prime in Σ_β' denotes that the solvent is excluded from the summation. We now neglect the compressibility and suppose that

$$\Sigma_\beta' N_\beta \ll N_{H_2O}, \quad \Sigma' N_\beta v_\beta \ll N_{H_2O} v_{H_2O}.$$

On expanding the logarithm and retaining only the first order term we obtain

$$\mu^*_{H_2O} = \mu^{0*}_{H_2O}(T) + pv^*_{H_2O} - RT(\Sigma_\beta' N_\beta)/N_{H_2O}.$$

If c_β^* denotes the concentration of the βth solute species in moles/c.c., this may be written (v_{H_2O} and $v^0_{H_2O}$ need not be distinguished)

$$\mu^*_{H_2O} = \mu^{0*}_{H_2O}(T) + p\,v^*_{H_2O} - RTv^*_{H_2O}(\Sigma_\beta' c_\beta^*). \quad \text{......(1544)}$$

Equation (1544) forms a convenient starting point for the study of the equilibria of solutions at extreme dilution. Equations (1528) and (1539) take the corresponding forms

$$\Delta T_M = \frac{RT_M{}^2 v^*_{H_2O}}{\Lambda^*}(\Sigma_\beta' c_\beta^*), \quad P = RT(\Sigma_\beta' c_\beta^*). \quad \text{......(1545)}$$

These formulae are due to van 't Hoff. The formula for P is of perfect gas type.

The chemical potential of the solute species can be similarly simplified. The same approximations lead easily to

$$\mu_\beta^* = \mu_\beta^{0*}(T) + p\,v_\beta^* + RT \log c_\beta^*. \quad \text{......(1546)}$$

For a given solvent, μ_β^{0*} is independent of the composition of the solution.

§ **13·4.** *S-regular solutions.* The difficulties in evaluating $B(T)$ are so great that only for ideal or perfect solutions have accurate formulae been so far obtained, and then only accurate in form—no evaluation in terms of intermolecular forces has been possible. The method of treatment can however be extended somewhat further still when certain simplifications can be made.† Let us consider a mixture of N_α α-molecules and N_β β-molecules, and assume that both types are approximately spherical and of equal radius. Let us assume further that they "pack" in the same way so that each α- and β-molecule has on the average the same number r of neighbours. If the molecules are strictly close-packed, $r = 12$, but there is no need here to assign any particular value to r provided its value is the same for both types of molecule. We shall assume next that the two liquids mix completely in all proportions without volume change. Finally we shall assume that in any configuration of the proper total volume the potential energy W may be regarded as the sum of contributions from each pair of molecules in direct contact. The wider definition of Hildebrand is not yet susceptible of statistical analysis. To avoid confusion we refer to solutions obeying the restricted definition as *s-regular* (strictly regular).

† Regular solutions were first defined by Hildebrand (*loc. cit.* p. 535) using a wider definition than that here given which is due to Guggenheim (*Proc. Roy. Soc.* A, vol. 148, p. 305 (1935)).

Let us denote the potential energy of normal configurations of the pure α-liquid of N_α molecules by $N_\alpha w_\alpha$ (as usual) and that of the pure β-liquid by $N_\beta w_\beta$. Then according to the stated assumptions the liquid formed by mixture of these two pure liquids, with their molecules in an arbitrary configuration, will have a potential energy of the form

$$W = (N_\alpha - X)\, w_\alpha + (N_\beta - X)\, w_\beta + 2Xw_{\alpha\beta}. \qquad \text{......(1547)}$$

There are $\frac{1}{2}r(N_\alpha + N_\beta)$ pairs of close neighbours in the mixture in all. If rX of these are (α,β) pairs, then the numbers of α,α and β,β pairs must both be reduced by $\frac{1}{2}rX$ below these maxima. Since the energy W is made up of the sum of contributions of each pair of close neighbours the result (1547) follows at once, $2w_\alpha/r$, $2w_\beta/r$ and $2w_{\alpha\beta}/r$ being the contribution of each α,α, β,β and α,β pair respectively. We can give another interpretation of $w_{\alpha\beta}$ as follows: If we start with the pure liquids and interchange an interior α-molecule with an interior β, the total increase of potential energy will be

$$2\{2w_{\alpha\beta} - w_\alpha - w_\beta\}.$$

We shall denote this quantity by 2λ and write

$$W = N_\alpha w_\alpha + N_\beta w_\beta + X\lambda. \qquad \text{......(1548)}$$

We can then put $B'(T)$ in the form

$$B'(T) = \exp\left(-\frac{N_\alpha w_\alpha + N_\beta w_\beta}{kT}\right)\int \ldots \int e^{-X\lambda/kT} (d\omega_\alpha)^{N_\alpha} (d\omega_\beta)^{N_\beta}. \qquad \text{...(1549)}$$

Defining a mean value $\bar{X}$ by the equation

$$e^{-\bar{X}\lambda/kT} \int \ldots \int (d\omega_\alpha)^{N_\alpha} (d\omega_\beta)^{N_\beta} = \int \ldots \int e^{-X\lambda/kT} (d\omega_\alpha)^{N_\alpha} (d\omega_\beta)^{N_\beta},$$

and evaluating the remaining integral as in § 13·2, we find

$$B'(T) = \exp\left(-\frac{N_\alpha w_\alpha + N_\beta w_\beta + \bar{X}\lambda}{kT}\right) \frac{(N_\alpha + N_\beta)!}{N_\alpha!\, N_\beta!}\, v_\alpha^{N_\alpha} v_\beta^{N_\beta}. \qquad \text{......(1550)}$$

The formulae of § 13·2 for $-F/kT$ and Ψ/k are affected by the addition of the term $-\bar{X}\lambda/kT$ to the right-hand side. The same addition must be made to $-G/kT$ and Φ/k.

While w_α, w_β, $w_{\alpha\beta}$ depend on the temperature in an unknown manner, the differential quantity $\lambda = 2w_{\alpha\beta} - w_\alpha - w_\beta$ will only vary with the temperature in so far as the change in tightness of packing may affect $w_{\alpha\beta}$ differently from w_α and w_β. We may reasonably expect this differential effect to be small and we shall therefore treat λ as independent of the temperature. The corresponding changes in other thermodynamic functions follow at once from the changes in F or G. (An extra $-\lambda\partial\bar{X}/\partial T$ in S, and $\lambda(\bar{X} - T\partial\bar{X}/\partial T)$ in E and H.) The thermodynamics of *s*-regular solutions reduces therefore to evaluating $\bar{X}$.

Without evaluating $\bar{X}$ we can at once draw the conclusion that, if $\lambda=0$, *s-regular solutions are perfect*; the necessary condition is

$$w_\alpha + w_\beta = 2w_{\alpha\beta}. \qquad \text{......(1551)}$$

To evaluate $\bar{X}$ various assumptions have been made* all equivalent to putting

$$(\bar{X})^2 = (N_\alpha - \bar{X})(N_\beta - \bar{X}). \qquad \text{......(1552)}$$

The physical meaning of (1552) is that there is completely random mixing, or that the change of entropy on mixing is the same as for an ideal solution. Solving (1552) for $\bar{X}$, we obtain

$$\bar{X} = N_\alpha N_\beta/(N_\alpha + N_\beta). \qquad \text{......(1553)}$$

It is easy to see however that (1552) is incorrect.† By its definition $\bar{X}$ must be a function of N_α, N_β and λ/kT and so for given N_α, N_β a function of λ/kT. In particular when $\lambda/kT=\infty$ $\bar{X}=0$, when $\lambda/kT=0$ $\bar{X}$ satisfies (1553) and when $\lambda/kT=-\infty$ $\bar{X}$ is equal to the lesser of N_α and N_β. It seems obvious that $\bar{X}$ increases steadily as λ/kT decreases from $+\infty$ to $-\infty$. For a given non-zero value of λ $\bar{X}$ cannot be independent of T.

An equation for $\bar{X}$ much nearer the truth than (1552) is

$$(\bar{X})^2 = (N_\alpha - \bar{X})(N_\beta - \bar{X})\, e^{-2\lambda/rkT}. \qquad \text{......(1554)}$$

Since $2\lambda/r$ is the work required to change an α,α pair and a β,β pair into two α,β pairs and $\bar{X}$, $N_\alpha-\bar{X}$ and $N_\beta-\bar{X}$ are proportional to the numbers of α,β, α,α and β,β pairs respectively, this is an equation of quasi-dissociative equilibrium of the correct form. On solving for $\bar{X}$ we obtain

$$\bar{X} = \frac{-(N_\alpha+N_\beta)+\{(N_\alpha+N_\beta)^2+4N_\alpha N_\beta(e^{2\lambda/rkT}-1)\}^{\frac{1}{2}}}{2(e^{2\lambda/rkT}-1)}. \qquad \text{......(1555)}$$

If $2\lambda/rkT \gg 1$, α,β pairs in contact must be rare and the liquid must be expected to separate into two phases a dilute solution of β in α and another of α in β. If on the contrary $-2\lambda/rkT \gg 1$ there will practically be chemical combination between α and β. We can therefore expect the mixture to behave as a single phase without any complications due to compound formation only when $e^{2\lambda/rkT}=O(1)$. It is therefore of interest to examine the value of $\bar{X}$ when

$$|e^{2\lambda/rkT}-1| < 1. \qquad \text{......(1556)}$$

On expanding (1555) when (1556) is satisfied we find approximately

$$\bar{X} = \frac{N_\alpha N_\beta}{N_\alpha+N_\beta}\left[1-\frac{N_\alpha N_\beta}{(N_\alpha+N_\beta)^2}(e^{2\lambda/rkT}-1)\right]. \qquad \text{......(1557)}$$

This shows that (1553) is a first approximation for $\bar{X}$ but is valid only when $|2\lambda/rkT| \ll 1$ unless N_β/N_α is small. For a sufficiently dilute solution (1557)

* Hildebrand, *J. Amer. Chem. Soc.* vol. 51, p. 69 (1929); Scatchard, *Chem. Rev.* vol. 8, p. 321 (1931).

† Guggenheim, *loc. cit.*

is valid no matter what the value of λ and (1553) is a valid first approximation to it. Formula (1555) and not (1553) should be used for $\bar{X}$ in all thermodynamic functions for s-regular solutions, but (1555) may be approximated to as indicated above.

§ **13·5.** *Strong electrolytes.* When certain substances such as common salt are dissolved in water, the solution has a comparatively high conductivity, showing that charged ions must be present, and the effects of the solute on the colligative properties of the solution (vapour pressure of solvent, freezing point, osmotic pressure) is much larger than predicted by the laws of ideal solutions if the solute is present mainly as molecules of NaCl. We owe to Arrhenius the suggestion that for such substances, called *strong electrolytes*, the solute is composed largely of the independent systems Na^+ and Cl^-. Study of the optical properties of such solutions also leads to the conclusion that at least in dilute and moderately concentrated solutions there are very few NaCl molecules, and in many cases the properties of the solution can be accurately accounted for on the assumption that no undissociated molecules at all are present.*

It is of interest to examine whether this complete dissociation of salts is to be expected theoretically.† The equilibrium point of the simple dissociation $\alpha\beta \rightarrow \alpha + \beta$ is given by

$$\frac{N_\alpha N_\beta}{V N_{\alpha\beta}} = \frac{f_\alpha f_\beta}{f_{\alpha\beta}},$$

where the f's are the usual partition functions, modified so far as may be necessary by the effects of the surrounding solvent. If we start with the partition functions for free α-, β-, and $\alpha\beta$-systems at ordinary temperatures (α, β monatomic), we may use the formulae of Chapters II and III and so obtain as in (463)

$$\frac{f_\alpha f_\beta}{f_{\alpha\beta}} = \left(\frac{2\pi(m_\alpha m_\beta/m_{\alpha\beta})\,kT}{h^2}\right)^{\frac{3}{2}} \frac{h^2}{8\pi^2 AkT}\frac{1}{q(T)}e^{-\chi/kT},$$

where $q(T)$ is the vibrational partition function for $\alpha\beta$. This lies between 1 at low temperatures and $h\nu/kT$, ν the vibration frequency of the molecule, at high, and can be assumed to be 1 with sufficient accuracy here. If a is the distance apart of Na^+ and Cl^- in the molecule, then $A = m_\alpha m_\beta a^2/m_{\alpha\beta}$. We therefore have

$$\frac{N_\alpha N_\beta}{V N_{\alpha\beta}} = \left(\frac{2\pi(m_\alpha m_\beta/m_{\alpha\beta})\,kT}{h^2}\right)^{\frac{1}{2}} \frac{e^{-\chi/kT}}{4\pi a^2}. \qquad \text{......(1558)}$$

* Bjerrum, *Proc. 7th Int. Cong. Pure and Applied Chem.* London (1909); *Zeit. f. Electrochem.* vol. 24, p. 321 (1918).

† Cf. Guggenheim, *Report of Chem. Sect. Brit. Ass., Centenary Meeting*, p. 58 (1931).

The results of Chapter x show that the law of force between Na^+ and Cl^- may be taken to be

$$\frac{\epsilon^2}{r^2}-\frac{2\cdot 5}{(10^8 r)^{10}},$$

which gives $a=2\cdot 4\times 10^{-8}$ cm., and $\chi=5\cdot 3$ electron-volts. Using these values, and reducing the concentrations to *moles per litre*,* $c_\alpha^\dagger$, $c_\beta^\dagger$, $c_{\alpha\beta}^\dagger$, we find for $T=300^\circ$ K.

$$K^\dagger=\frac{c_\alpha^\dagger c_\beta^\dagger}{c_{\alpha\beta}^\dagger}=10^2 e^{-\chi/kT}\simeq 10^{-88}. \qquad \ldots\ldots(1559)$$

Thus *in vacuo* NaCl is (naturally) completely undissociated.

We must now introduce the modifications due to the solvent. If we continue to regard the free ions as *unhydrated*—that is as free systems not specifically associated with any water molecule, whose motion on the average has the characteristics of a free particle in a gas except for the extra frequency of collision—we have yet to remember that the energy of dissociation in the solvent will be less than χ. If we regard the solvent merely as a continuous medium of dielectric constant D, the energy of separation would be reduced to χ/D. We thus obtain in water ($D\sim 80$) $K^\dagger\sim 7$, but this procedure must overestimate the real correction to χ in a molecular solvent. In the very strong fields near an ion a saturation effect sets in, which will reduce considerably the effective value of D. The correct value of $K^\dagger$ for NaCl must be very much less than 7 (as calculated on this basis), and a very considerable degree of incompleteness in the dissociation is predicted for ordinary concentrations, in distinct disagreement with the actual facts. The error can only have entered by ignoring *hydration*, the intimate effect of the solvent molecules on the free ions.

Similar calculations can be made for other ion pairs such as

$$(MgCl)^+ \rightleftarrows Mg^{++}+Cl^-, \quad K^\dagger=0\cdot 2,$$

and

$$Na(ClO_4) \rightleftarrows Na^+ + (ClO_4)^-, \quad K^\dagger\sim 10^4.$$

In the latter the much greater value of $K^\dagger$ arises from the great number of rotational states for the free ion $(ClO_4)^-$, which are practically as numerous as those of the original molecule so that we have put $R_{(ClO_4)^-}/R_{Na(ClO_4)}\sim 1$.

Comparison of the results for NaCl, $(MgCl)^+$ and $Na(ClO_4)$ indicates one way in which our neglect of hydration has led us to gross underestimates of $K^\dagger$. We may have greatly underestimated the partition functions for the free ions by ignoring the possibilities provided by the neighbouring water molecules for new states of motion not available to a free ion *in vacuo*. An estimate of this effect may be made by adopting the model of a liquid given

* Concentrations are so commonly given in moles (gram-molecules or gram-ions) per litre solution instead of per c.c. that in this chapter a special notation seems desirable.

in § 13·12, and taking Na^+, Cl^- and NaCl in the solution as units in the quasi-crystal structure.

Consider a quasi-crystalline mixed liquid containing N_0 water molecules and N_α, N_β, N_γ molecules of three other systems which will ultimately be taken to be Na^+, Cl^- and NaCl in dissociative equilibrium. Let $f_0, f_\alpha, f_\beta, f_\gamma$ be the effective partition functions for their respective states of motion associated with given positions in the quasi-lattice. Then the arguments giving (1518) give here

$$-\frac{F}{kT}=\frac{\Psi}{k}=N_0\log\frac{(N_0+N_\alpha+N_\beta+N_\gamma)f_0}{N_0}+N_\alpha\log\frac{(N_0+N_\alpha+N_\beta+N_\gamma)f_\alpha}{N_\alpha}$$
$$+N_\beta\log\frac{(N_0+N_\alpha+N_\beta+N_\gamma)f_\beta}{N_\beta}+N_\gamma\log\frac{(N_0+N_\alpha+N_\beta+N_\gamma)f_\gamma}{N_\gamma}. \quad \text{......(1560)}$$

The condition of dissociative equilibrium is

$$\left[\frac{\partial}{\partial N_\alpha}+\frac{\partial}{\partial N_\beta}-\frac{\partial}{\partial N_\gamma}\right]\Psi=0,$$

so that, when $N_\alpha+N_\beta+N_\gamma \ll N_0$,

$$\frac{N_\alpha N_\beta}{N_0 N_\gamma}=\frac{f_\alpha f_\beta}{f_\gamma}. \quad \text{......(1561)}$$

The complete partition functions for α, β, γ are now $N_0 f_\alpha$, $N_0 f_\beta$ and $N_0 f_\gamma$. It is also easy to deduce these forms directly from the classical partition function

$$\frac{1}{h^3}\int\cdots\int\exp\left[-\frac{1}{kT}\left\{\frac{1}{2m}(p_1{}^2+p_2{}^2+p_3{}^2)+W(q_1,q_2,q_3)\right\}\right]dp_1\ldots dq_3, \quad \text{......(1562)}$$

in which the potential energy W has a similar well-marked minimum associated with each water molecule, or quasi-lattice point. The integral thus breaks up into N_0 equal parts one for each water molecule, and reduces at once to $N_0 f_\alpha$.

If w_α is the minimum of $W(q_1,q_2,q_3)$ in (1562), it is obvious that $N_0 f_\alpha$ cannot exceed

$$\frac{V(2\pi m_\alpha kT)^{\frac{3}{2}}}{h^3}e^{-w_\alpha/kT}. \quad \text{......(1563)}$$

It is easy to verify however that for reasonable types of field (1563) can reach a value nearly of this order. If w_α and w_β are negative, so that $-w_\alpha$, $-w_\beta$ are "heats of hydration at the absolute zero", the calculated values of $K^\dagger$ will be increased by the factor $e^{-(w_\alpha+w_\beta)/kT}$. This factor, previously omitted, may be expected to be of such an order as largely to cancel the over correction for D in the earlier estimate, but it cannot of itself lead to large values of $K^\dagger$ unless there is also a large heat of hydration and a correspondingly large heat of solution for the salt. Large heats of solution are rare, and therefore a still more careful analysis is required.

In formulating and evaluating (1562) we have thought of any water molecule as providing locally a field W in which the ion can move, but as itself unaffected by the attachment of the ion. It is just here that the extra effect enters. The states of motion of the ion relative to its attached water molecule have already been accounted for in (1563). [The mass m_α should more strictly be replaced by $m_\alpha m_{H_2O}/(m_\alpha + m_{H_2O})$ for relative oscillations.] An extra factor in f_α is provided by the change from f_0 for the water molecule to f_0' for the (H_2O-ion) complex. If we regard the water molecule as rigid and assume that both its translations and rotations are approximately simple harmonic oscillations of frequencies ν_0 and ν_1 respectively, we may approximate to f_0 with the expression

$$\left(\frac{kT}{h\nu_0}\right)^3 \left(\frac{kT}{h\nu_1}\right)^3.$$

If it were held by the same field, the complex would have a translational frequency ν_0' given by

$$\nu_0' = \left(\frac{m_{H_2O}}{m_\alpha + m_{H_2O}}\right)^{\frac{1}{2}} \nu_0.$$

One of its moments of inertia and therefore one of its rotational freedoms is unaltered. The other two moments of inertia will increase from about 2×10^{-40} to 10^{-38}, by a factor of about 50. It follows that $f_0' \simeq 10^2 f_0$. Since this increase presumably applies to both ions, $K^\dagger$ is increased by this effect by the factor 10^4. Even if considerable allowance must be made for the stronger fields holding the complex than those holding the H_2O molecule, $K^\dagger$ remains large enough for dissociation to be practically complete up to normal solutions (1 mole per litre) of such salts.

In conclusion we may summarize this discussion by saying that the observed degree of (almost complete) dissociation of such a salt as NaCl in aqueous solution cannot be accounted for if the ions Na^+ and Cl^- are effectively free systems, even if the dielectric constant of the medium has its full value down to molecular distances. When the effects of the interaction of the ions with individual molecules of solvent (hydration) are taken into account the high degree of dissociation follows naturally out of general theory, the effect of the factor due to the great increase of possible states of motion for the hydrated ion being even more important than the energy factor in determining the degree of dissociation.

§ **13·51**. *Generalities concerning hydration of ions.* We have seen in the preceding section that in order that a simple salt may be highly dissociated its ions must be intimately associated with at least one water molecule, and therefore in this general sense hydrated. To form a more precise picture of hydration, and in particular to determine it quantitatively, various methods have been used. Methods based on the transport of water by ions in non-aqueous solvents are really irrelevant as then the water is held by the ion against the attraction of the molecules of the less polar solvent not against the attraction of other water molecules. The number of attached water molecules per ion thus comes out too high. Methods based on mobilities in aqueous solution assume the validity of Stokes' law for the terminal velocity of a sphere in a viscous liquid to extend much further than can be justified. A simpler and theoretically sounder method is provided by a study of the densities of ionic solutions. In sufficiently dilute solution the partial molar volumes become constant and additive. They can be determined experimentally.

Now it is clear* from a variety of evidence that water has an irregular 4-coordinated structure, of open type, very similar to the regular 4-coordinated structure of ice. Each water molecule is generally surrounded by four others arranged more or less at the corners of a regular tetrahedron. On the other hand when H_2O molecules occur coordinated round ions in crystals they are closely packed round the ion and the number of H_2O molecules coordinated to a given ion is just that number for which room is allowed by the relative sizes of the ion and the water molecule. Suppose now that an ion is introduced into liquid water. If it has a sufficient attraction for H_2O molecules it will cause some of these to pack tightly round it. The contraction associated with this packing can be calculated by comparing the volume occupied by each water molecule round the ion in a crystal with the molecular volume of pure water. In many cases this contraction is greater than the volume of the unhydrated ion and the partial ionic volume will then be negative.

A quantitative study of partial molar volumes shows that all monatomic positive ions except the large Rb^+ and Cs^+ are completely hydrated in the sense that they are surrounded in water by a tightly packed layer of as many water molecules as their size allows, just as they are in crystals. The exceptions Rb^+ and Cs^+ are so large that their surface field is insufficiently strong to coordinate the water. Monatomic negative ions are in general large and except for F^- appear to be unhydrated. Polyatomic negative ions are *a fortiori* unhydrated, while all multivalent monatomic ions are hydrated.

According to this more developed view of the structure of liquid water the computation of the degree of dissociation of a strong electrolyte becomes still more complicated than appears in § 13·5. The tentative calculations of that section remain however sufficient to show that hydration is an important factor in enhancing the dissociation for monatomic ions, especially

* Bernal and Fowler, *J. Chem. Physics*, vol. 1, p. 515 (1933).

multivalent ones. As the calculations of that section also show, such enhancement is not required for polyatomic ions and, as we see here, it probably does not occur for them.

§ **13·52.** *The "anomalies" of strong electrolytes. The osmotic coefficient.* We have just seen that strong electrolytes, in water at least, are almost completely dissociated at low concentrations, certainly up to tenth molar ($c^{\dagger} \sim 0{\cdot}1$). At such concentrations to regard the solute as 100 per cent. dissociated into ions is at least as good an approximation as to assign any other constitution to it. It is therefore of great interest and importance to discuss theoretically the behaviour of solutions of such completely dissociated electrolytes at low concentrations—so low that the solutions are *extremely dilute* in the sense of § 13·32. If they were *ideal* they would then obey the equation (1544) here repeated,

$$\mu^*_{H_2O} = \mu^{0*}_{H_2O}(T) + pv^*_{H_2O} - RTv^*_{H_2O}(\Sigma_\beta' c_\beta^*).$$

If on the other hand they do not obey these laws, the deviation may be taken account of by a coefficient‡ g defined by

$$\mu^*_{H_2O} = \mu^{0*}_{H_2O}(T) + pv^*_{H_2O} - gRTv^*_{H_2O}(\Sigma_\beta' c_\beta^*) \qquad \text{.......(1564)}$$

For a solution obeying (1564) equations (1545) become

$$\Delta T_M = \frac{RT_M^2 v^*_{H_2O}}{\Lambda^*} g(\Sigma_\beta' c_\beta^*), \quad P = RTg(\Sigma_\beta' c_\beta^*). \qquad \text{......(1565)}$$

Thus g is equal to the ratio of the observed lowering of the freezing point (or observed osmotic pressure) to the values calculated for the same solution assumed to be ideal; it is called the *osmotic coefficient*. It might equally well be called the freezing point coefficient.

In aqueous solutions of non-electrolytes g does not usually deviate appreciably from unity at concentrations much below molar ($c^{\dagger} \sim 1$). In solutions of strong electrolytes, however, g deviates considerably from unity for concentrations as low as thousandth molar ($c^{\dagger} \sim 0{\cdot}001$). These deviations were formerly known as *the anomalies of strong electrolytes*. They are clearly due to the effects of the long range electrostatic forces hitherto neglected, which we must now include in the manner foreshadowed in §§ 8·8–8·82.

§ **13·6.** *Applications and developments of the theory of Debye and Hückel.* Having already discussed in Chapter VIII the meaning of the average electrostatic potential $\overline{\psi_\alpha}$ round an α-ion of charge ϵ_α, we can now simplify the notation and denote this field by ψ_α and the charge on the ion by $z_\alpha \epsilon$, so that z_α is the *valency*. Under certain conditions, already fully specified,

‡ Bjerrum, *Zeit. f. Elektrochem.* vol. 24, p. 325 (1918).

ψ_α will satisfy the Poisson-Boltzmann equation (808·1) which since ψ_α is spherically symmetrical reduces to

$$\frac{1}{r^2}\frac{\partial}{\partial r}\left(r^2\frac{\partial\psi_\alpha}{\partial r}\right)=-\frac{4\pi\epsilon}{D}\sum_{\beta=1}^{t} z_\beta\frac{N_\beta}{V}e^{-z_\beta\epsilon\psi_\alpha/kT}. \qquad \text{......(1566)}$$

If all the exponents in (1566) are small for all important values of r, it takes the approximate form, first discussed by Debye and Hückel,

$$\frac{1}{r^2}\frac{\partial}{\partial r}\left(r^2\frac{\partial\psi_\alpha}{\partial r}\right)=\kappa^2\psi_\alpha \quad \left(\kappa^2=\frac{4\pi\epsilon^2}{DkT}\sum_{\beta=1}^{t} z_\beta{}^2\frac{N_\beta}{V}\right), \qquad \text{......(1567)}$$

already solved for the simplest case in § 8·82. A slightly more general solution will first be developed here. We shall now not entirely neglect the sizes or short range forces between the ions, but shall assume that they may be represented by assigning the same diameter to them all or by using a common *mean* diameter a. In view of other uncertainties and approximations a more refined treatment is hardly justified. Then (1567) holds for $r>a$, while for $r<a$ there can be no other ion present and the electric intensity for values of r just less than a must be $z_\alpha\epsilon/r^2D$. We have also

$$\psi_\alpha=\frac{1}{r}(Be^{-\kappa r}+Ce^{\kappa r}) \quad (r\geqslant a),$$

where B and C are constants. To satisfy the boundary condition for $r\to\infty$ we must have $C=0$, and to satisfy the boundary condition for the continuity of electric induction at $r=a$ we must have

$$\frac{z_\alpha\epsilon}{a^2D}=\frac{1}{a^2}Be^{-\kappa a}(1+\kappa a),$$

from which it follows that $\quad B=\dfrac{z_\alpha\epsilon}{D}\dfrac{e^{\kappa a}}{1+\kappa a}.$

The boundary value of ψ_α is therefore

$$\psi_\alpha(a)=\frac{z_\alpha\epsilon}{aD}\frac{1}{1+\kappa a}, \qquad \text{......(1568)}$$

and the average potential due to the rest of the ions at the surface, and so inside the interior of a specified α, is

$$\psi_\alpha(a)-\frac{z_\alpha\epsilon}{aD}=\frac{-\kappa z_\alpha\epsilon}{D}\frac{1}{1+\kappa a}. \qquad \text{......(1569)}$$

The sufficient condition that the fluctuations should be negligible is still, as in § 8·82, that

$$\frac{z_\alpha z_\beta\epsilon^2\kappa}{DkT}\ll 1. \qquad \text{......(1570)}$$

Since κ vanishes with the concentration, this condition is certainly satisfied for solutions sufficiently dilute. To study numerical values we write

$$z_\beta^2 \frac{N_\beta}{V} = 6{\cdot}06 \times 10^{20} z_\beta^2 c_\beta^\dagger.$$

If we define *the ionic strength* I of the solution by the equation

$$I = \tfrac{1}{2}\Sigma_\beta z_\beta^2 c_\beta^\dagger,$$

then

$$\kappa = 0{\cdot}324 \times 10^8\, I^{\frac{1}{2}}$$

for water at 0° C. With the same values of κ, D and T (1570) reduces to $z_\alpha z_\beta I^{\frac{1}{2}} \ll 1$, and *the condition of smallness may be taken to be satisfied for uni-univalent solutions hundredth molar* ($c^\dagger \sim 0{\cdot}01$) *or less.*

We conclude that there is a range of fairly great dilution in which the use of the Poisson-Boltzmann equation (1566) may give an accurate account of the ionic distribution laws, by providing an adequate method of approximating to $B(T)$. The use however of Debye and Hückel's approximation (1567) in place of (1566) is not thereby justified and it remains to examine in § 13·7 whether a more accurate account can be given by avoiding this approximation, which at this stage has been introduced merely for mathematical convenience. As was pointed out in § 8·81, ψ_α must satisfy the relation (803) which in the notation of this section requires that

$$c_\alpha^\dagger \frac{\partial \psi_\alpha}{\partial z_\beta} = c_\beta^\dagger \frac{\partial \psi_\beta}{\partial z_\alpha} \qquad \qquad \text{......(1571)}$$

near the centre of the ion in question. From formula (1569) and the definition of κ in (1567) it is easily verified that (1571) is satisfied. Since moreover we must have $W_{\alpha\beta} = W_{\beta\alpha}$ it is necessary that $\psi_\alpha/z_\alpha = \psi_\beta/z_\beta$ and this condition is also satisfied. Thus the solutions we have obtained are internally self-consistent. We have emphasized this point because when attempts are made to solve (1566) more accurately they lead as we shall see to solutions lacking this self consistency and it is therefore by no means certain that such solutions are any more valuable than that given in this section.

§ **13·61.** *Thermodynamic functions in the theory of Debye and Hückel.* We have already seen in § 8·8 how to calculate the contribution to F made by intermolecular forces in general and interionic electrostatic forces in particular. Formula (812) in § 8·82 gives a limiting form for F_ϵ the contribution to F when the charges on the ions are built up from zero to their actual values in a reversible isothermal process at constant volume. In obtaining this result we have had to assume (implicitly) that the dielectric constant was independent of the charges and of the pressure since the pressure will change during the charging process in order to preserve the constant volume. It seems likely that this assumption, that the isothermal D is constant, will be most nearly fulfilled at constant pressure. The more exact meaning of D is discussed at the end of this section.

Let us assume that the solution of discharged ions is ideal, and that in the actual solution at the same temperature and pressure

$$F=F_\iota+F_\epsilon,\quad G=G_\iota+G_\epsilon,\quad V=V_\iota+V_\epsilon, \qquad \text{......(1572)}$$

F_ι, G_ι and V_ι being the values of these functions in the discharged ideal solution. The work to be done on the solution in charging up at constant pressure and temperature is then equal to W_v+W_ϵ, where W_ϵ is the electrostatic work and W_v the hydrostatic work to be done by the external pressure. Then

$$F_\epsilon=W_v+W_\epsilon,\quad W_v=-pV_\epsilon,\quad G_\epsilon=W_\epsilon. \qquad \text{......(1573)}$$

We have already calculated the simplest case of W_ϵ in § 8·82. We now see that if this calculation is emended by taking account of variations of V in κ—they are in fact negligible—it will give with greater reliability G_ϵ. The calculation may now be extended by using (1569) in place of (811·3). We thus find that, ignoring the V changes,

$$G_\epsilon=-\frac{\epsilon^2\kappa}{D}\Sigma_\beta N_\beta z_\beta{}^2\int_0^1\frac{\lambda^2\,d\lambda}{1+\lambda\kappa a}, \qquad \text{......(1574)}$$

$$=-\frac{\epsilon^2\kappa}{3D}\Sigma_\beta N_\beta z_\beta{}^2\tau(\kappa a),$$

where

$$\tau(x)=\frac{3}{x^3}[\log(1+x)-x+\tfrac{1}{2}x^2]. \qquad \text{......(1575)}$$

For high dilutions $\tau(x)\sim 1$ and (1574) reduces to

$$G_\epsilon=-\frac{\epsilon^2\kappa}{3D}\Sigma_\beta N_\beta z_\beta{}^2. \qquad \text{......(1576)}$$

The function $\tau(x)$ is tabulated in Table 53.

TABLE 53.

Values of the Debye-Hückel functions $\tau(x)$ and $\sigma(x)$ for even values of x^2.

x^2	x	$\tau(x)$	$\sigma(x)$	x^2	x	$\tau(x)$	$\sigma(x)$
0·000	0·0000	1·000	1·000	0·030	0·1732	0·886	0·786
0·001	0·0316	0·976	0·954	0·040	0·2000	0·870	0·759
0·002	0·0447	0·967	0·936	0·050	0·2236	0·857	0·738
0·003	0·0557	0·960	0·922	0·060	0·2449	0·846	0·717
0·004	0·0633	0·954	0·912	0·070	0·2646	0·835	0·700
0·005	0·0707	0·949	0·902	0·080	0·2828	0·827	0·685
0·006	0·0775	0·945	0·893	0·090	0·3000	0·819	0·671
0·007	0·0837	0·941	0·886	0·100	0·3162	0·811	0·659
0·008	0·0894	0·937	0·879	0·110	0·3317	0·803	0·646
0·009	0·0947	0·934	0·871	0·120	0·3464	0·796	0·636
0·010	0·1000	0·931	0·866	0·150	0·3873	0·778	0·607
0·020	0·1414	0·905	0·818	0·200	0·4472	0·752	0·569

We can now justify the neglect of V-variations in κ by showing that $V_\epsilon \ll V_\iota$, by means of the relation $V_\epsilon = \partial G_\epsilon/\partial p$. It will be sufficient in investigating orders of magnitude to use (1576). Then since $\kappa \propto (V_\iota D)^{-\frac{1}{2}}$

$$V_\epsilon = -\frac{\epsilon^2\kappa}{3D}(\Sigma_\beta N_\beta z_\beta^2)\left(-\frac{1}{2}\frac{1}{V_\iota}\frac{\partial V_\iota}{\partial p} - \frac{3}{2}\frac{1}{D}\frac{\partial D}{\partial p}\right).$$

Since for the applicability of the theory we have $\epsilon^2\kappa/DkT \ll 1$, it follows that

$$\left|\frac{V_\epsilon}{V_\iota}\right| \ll \tfrac{1}{3}(\Sigma_\beta c_\beta z_\beta^2)\, RT\left[-\frac{1}{2}\frac{1}{V_\iota}\frac{\partial V_\iota}{\partial p} - \frac{3}{2}\frac{1}{D}\frac{\partial D}{\partial p}\right].$$

Even for a solution as concentrated as one molar $c^* = 10^{-3}$, $c^\dagger = 1$, $\frac{1}{3}(\Sigma_\beta c_\beta z_\beta^2) RT$ will be at most of the order of 10 atmospheres, while the compressibility $-\partial V/V\partial p$ of a liquid is of the order 10^{-4} and $\partial D/D\,\partial p$ at most comparable. It follows that even for a one molar solution

$$|V_\epsilon/V_\iota| \ll 10^{-3},$$

and for more dilute solutions it will be still smaller, thus justifying our neglect of the variation in V in the charging process.

The contribution F_ϵ of the electrostatic forces to Helmholtz's free energy is given by

$$F_\epsilon = G_\epsilon - pV_\epsilon = G_\epsilon - p\,\partial G_\epsilon/\partial p. \qquad \text{......(1577)}$$

Their contribution to the partial potential $\mu^*{}_{H_2O}$ of the solvent, according to (1536), is given by

$$\mu_\epsilon^*{}_{H_2O} = \mathsf{N}\frac{\partial G_\epsilon}{\partial \kappa}\frac{\partial \kappa}{\partial V}\frac{\partial V}{\partial N_{H_2O}},$$

$$= \frac{\epsilon^2\kappa}{6D}(\Sigma_\beta{}' c_\beta^* z_\beta^2)\, v^*{}_{H_2O}\,\sigma(\kappa a), \qquad \text{......(1578)}$$

where

$$\sigma(x) = \frac{3}{x^3}\left\{1 + x - \frac{1}{1+x} - 2\log(1+x)\right\}. \qquad \text{......(1579)}$$

Values of the function $\sigma(x)$ are also given in Table 53. From (1578) we obtain at once a formula for g, the osmotic coefficient defined in (1564) Comparing this definition with the corresponding equation for an ideal solution, we find by subtraction

$$(1-g)\,v^*{}_{H_2O}\,RT\,\Sigma_\beta{}' c_\beta^* = \mu^*{}_{H_2O} - \mu_\iota^*{}_{H_2O} = \mu_\epsilon^*{}_{H_2O}.$$

Hence, according to (1578)

$$1 - g = \frac{\epsilon^2\kappa}{6DkT}\frac{\Sigma_\beta{}' c_\beta^* z_\beta^2}{\Sigma_\beta{}' c_\beta^*}\,\sigma(\kappa a). \qquad \text{......(1580)}$$

Formulae (1574) and (1576) for G_ϵ contain D, the dielectric constant of the medium, which has been introduced as a parameter entirely external to the distribution laws of the assembly. In fact D is temperature (and pressure) dependent, showing thereby that it is strictly a property derived from the distribution laws—of course from the orientations of the molecular dipoles. In forming other thermodynamic functions from G this variation of D with T and p must of course be taken into account. Though to do this is essential to preserve the ordinary thermodynamic relationships and must be a better approximation to the truth than ignoring the variation of D altogether, we have no right to assume that the terms so derived are exactly correct. D has entered our equations as a coefficient in certain energy terms and strictly speaking all energy terms (before averaging begins) must be functions only of configurations. In calculating the two free energies it is necessary to form this complete configurational energy W and evaluate $\iiint e^{-W/kT}\,d\Omega$ integrated over all possible configurations in a single stage. In allowing a temperature variable D to enter our W we have already approximated illegitimately by averaging in two stages, first over the dipole orientations and then over the ionic configurations. A series of successive partial averagings cannot give exactly the true result. Nevertheless this successive partial averaging is all that we can attempt to carry through, and the results given by it, though inaccurate, should be of the correct form and order of magnitude.

§ **13·62.** *Heat of dilution in the theory of Debye and Hückel.* The heat of complete dilution L of a given solution is defined as the heat *absorbed* when a large (effectively infinite) quantity of the pure solvent is added to the solution at constant temperature and pressure. If H is the heat function for the original solution, H_0 for the pure solvent added and H_∞ for the resultant mixture, then the definition states that

$$L = H_\infty - (H + H_0). \qquad \text{......(1581)}$$

For an ideal solution, on the other hand, the heat functions are additive (see § 13·25) so that

$$L^\iota = H_\infty{}^\iota - (H^\iota + H_0{}^\iota) = 0.$$

Moreover $H_\infty = H_\infty{}^\iota$, $H = H^\iota + H_\epsilon$, $H_0 = H_0{}^\iota$, so that

$$L = -H_\epsilon = -T^2\frac{\partial \Phi_\epsilon}{\partial T} = T^2\frac{\partial}{\partial T}\left(\frac{G_\epsilon}{T}\right). \qquad \text{......(1582)}$$

The strict thermodynamic meaning of this $\partial/\partial T$ is $[\partial/\partial T]_{p,N_2,\ldots}$. Hence*

$$L = -H_\epsilon = T^2\left[\left(\frac{\partial}{\partial T}\right)_{D,V,a} + \frac{\partial D}{\partial T}\left(\frac{\partial}{\partial D}\right)_{T,V,a} + \frac{\partial V}{\partial T}\left(\frac{\partial}{\partial V}\right)_{T,D,a} + \frac{\partial a}{\partial T}\left(\frac{\partial}{\partial a}\right)_{T,D,V}\right]\frac{G_\epsilon}{T}. \qquad \text{......(1583)}$$

* The original formula of Debye and Hückel contained only the $\partial/\partial T$ term and is incorrect. The need to include $\partial D/\partial T$ was pointed out by Bjerrum, *Zeit. f. physikal. Chem.* vol. 119, p. 145 (1926) and $\partial V/\partial T$ by Scatchard, *J. Amer. Chem. Soc.* vol. 53, p. 2037 (1931) and independently by Gatty, *Phil. Mag.* vol. 11, p. 1082 (1931). That there should be a $\partial a/\partial T$ term was suggested by Gross and Halpern, *Physikal. Zeit.* vol. 26, p. 403 (1925), and by Bjerrum, *Trans. Far. Soc.* vol. 23, p. 445 (1927). The diameter a is quasi-empirical, and must certainly be expected to be temperature-dependent.

As nothing is known of $\partial a/\partial T$ we can only evaluate (1583) for dilutions so great that G_ϵ is given by (1576). Then

$$L = -H_\epsilon = -\frac{\epsilon^2\kappa}{2D}(\Sigma_\beta' N_\beta z_\beta^2)\left\{1+\frac{T}{D}\frac{\partial D}{\partial T}+\frac{1}{3}\frac{T}{V}\frac{\partial V}{\partial T}\right\} \quad \text{.......(1584)}$$

Inserting numerical values for aqueous solutions at 25° C. we obtain for the heat of complete dilution in calories per mole of electrolyte whose molecule yields q_+ cations of valency z_+ and q_- anions of valency z_-,

$$L = -H_\epsilon = A\{\tfrac{1}{2}(q_+ + q_-)|z_+z_-|\}^{\frac{3}{2}} c^{\dagger\frac{1}{2}}, \quad \text{......(1584·1)}$$

where

$$A = 1375\left(1+\frac{T}{D}\frac{\partial D}{\partial T}+\frac{1}{3}\frac{T}{V}\frac{\partial V}{\partial T}\right).$$

Since for water at room temperatures $(T/D)\,\partial D/\partial T$ is roughly $-1{\cdot}4$, the greater part of this term is cancelled by the unit term. Thus A is very sensitive

TABLE 54.

Values of D, $\partial D/\partial T$ and $1+(T/D)\,\partial D/\partial T$ for water at various temperatures.

12·5° C.			25° C.				40° C.			Authorities
D	$\frac{\partial D}{\partial T}$	$1+\frac{T}{D}\frac{\partial D}{\partial T}$	D	$\frac{\partial D}{\partial T}$	$1+\frac{T}{D}\frac{\partial D}{\partial T}$	$-A$	D	$\frac{\partial D}{\partial T}$	$1+\frac{T}{D}\frac{\partial D}{\partial T}$	
3·16	−0·379	−0·300	78·54	−0·361	−0·371	477	73·28	−0·341	−0·457	Wyman (1)
3·29	−0·393	−0·346	78·57	−0·362	−0·374	480	73·41	−0·326	−0·390	Drake, etc. (2)
3·34	−0·379	−0·298	78·77	−0·353	−0·337	430	73·71	−0·322	−0·367	Drude (3)
2·77	−0·368	−0·269	78·26	−0·349	−0·329	417	73·18	−0·330	−0·411	Drude (3)
2·81	−0·404	−0·393	77·84	−0·385	−0·474	617	72·24	−0·349	−0·512	Kockel (4)
9·42	−0·368	−0·323	75·40	−0·289	−0·142	160	71·48	−0·248	−0·086	Cuthbertson, etc. (5

(1) Wyman, *Phys. Rev.* vol. 35, p. 623 (1930).
(2) Drake, Pierce and Dow, *Phys. Rev.* vol. 35, p. 613 (1930).
(3) Drude, *Wied. Ann.* vol. 59, p. 48 (1896). The first entry gives the figures derived by an interpolation formula; the second set were derived graphically.
(4) Kockel, *Ann. d. Physik*, vol. 77, p. 430 (1925).
(5) Cuthbertson and Maass, *J. Amer. Chem. Soc.* vol. 52, p. 483 (1930).

to variations in the value assigned to $\partial D/\partial T$ which is not particularly well determined experimentally, so that A cannot be fixed with an accuracy better than about 15 per cent. The term in $\partial V/\partial T$ is well determined but considerably less than the uncertainty in $1+(T/D)\,\partial D/\partial T$, and will therefore be neglected. In Table 54 relevant values by various authors are given, arranged roughly in order of reliability.

At concentrations above tenth molar the heats of dilution for various electrolytes in water are highly specific being of either sign. The only accurate measurements at high dilutions are those of Lange and Robinson* and their collaborators. These are in good general agreement with the theory. They

* Lange and Robinson, *Chem. Rev.* vol. 9, p. 89 (1931).

show for example that at high dilutions $H_\epsilon > 0$, so that on further dilution heat is evolved as required by (1584·1). They show secondly a variation approximately proportional to the square root of the concentration. Thirdly it is found that the limiting slope of the H_ϵ, $\sqrt{c^\dagger}$ curves is determined primarily by the valency type of the electrolyte; being least for uni-univalent electrolytes, considerably greater for uni-bivalent and greater still for bi-bivalent electrolytes. All this is in agreement with (1584·1). The determination of the absolute value of this limiting slope is experimentally of extreme difficulty owing to the great dilutions at which measurements are required. Thus no clear cut comparison can be made between the theory and experiment on this point even for uni-univalent electrolytes. It is fair to say, however, that within the uncertainty of the experimental data there is no inconsistency between the absolute value of this limiting slope in theory and experiment.

§ **13·63.** *Comparison of theory and experiment for the osmotic coefficient.* If we insert numerical values for water at 0° C. and introduce the ionic strength I $(=\frac{1}{2}\Sigma_\beta' c_\beta^\dagger z_\beta^2)$, we obtain* for g

$$1-g=0{\cdot}374 I^{\frac{1}{2}} \frac{\Sigma_\beta' c_\beta^\dagger z_\beta^2}{\Sigma_\beta' c_\beta^\dagger} \sigma(\tfrac{1}{3} a I^{\frac{1}{2}}), \qquad \text{......(1585)}$$

when a is measured in Ångström units. When all ions present have the same numerical valency z, the valency factor simplifies to z^2. When all the positive ions are of one valency z_+ and the negative ions of another z_-, the valency factor reduces to $|z_+ z_-|$. The formula for g is then

$$1-g=0{\cdot}374|z_+ z_-| I^{\frac{1}{2}} \sigma(\tfrac{1}{3} a I^{\frac{1}{2}}) \qquad \text{......(1586)}$$

and as $I \to 0$, $\sigma \to 1$. If then $1-g$ is plotted against the square root of the ionic strength for solutions of single electrolytes we should obtain curves with a slope near the origin equal to $0{\cdot}374|z_+ z_-|$, and the slope should decrease as the concentration increases.

Experimental values for g are obtained by taking the ratio of the freezing point depression of the solution to its value for an ideal solution of the same concentration. As the dilution increases the freezing point depression decreases approximately in proportion to the concentration and so the accuracy of g varies inversely as the concentration. The relative accuracy of $1-g$ decreases still more rapidly. Instead of comparing observed and theoretical values of g or $1-g$ it is therefore better to compare observed and theoretical values of the lowering of the freezing point over a whole range of ionic strengths for an arbitrarily chosen value of a. The results of such a comparison may be summarized as follows: In practically all cases, where

* The coefficient in σ is more exactly 1/3·08 but the round value is accurate enough for most purposes, especially as the adjustable parameter a is semi-empirical.

reliable freezing point data are available for dilute aqueous solutions of strong electrolytes, there is agreement within the accuracy of the measurements between the observed and calculated freezing point lowering *if the value assigned to a is suitably adjusted.* For example the data of Hovorka and Rodebush* agree with the theory with an accuracy of 0·0001° at all ionic strengths up to 0·01 if the values in Table 55 are assigned to a. These values of a are of the right order of magnitude. The best data for some other electrolytes are fitted by assigning to a values that are much smaller than the possible closest distance of approach of the ions—for instance 0·4 Å. for KNO_3† and 0·0 Å. for KIO_3.‡ The true value for the closest distance of approach of two ions in a solution should be either very nearly the same as the mean of the ionic diameters determined in crystals, or greater if the ions are permanently hydrated. The true values cannot be less. It must be reluctantly admitted that the parameter a is not a real mean ionic diameter, but rather a parameter correcting for a whole variety of theoretical imperfections. This is especially so for solutions containing small ions.

TABLE 55.

Values of a, fitting the Debye-Hückel values of g to Rodebush's measurements.

Electrolyte	KCl	$CsNO_3$	K_2SO_4	$Ba(NO_3)_2$	$MgSO_4$	$CuSO_4$	$La_2(SO_4)_3$
a, Å.	3·8	3·0	3·0	2·1	3·0	2·2	3·0

§ **13·7.** *More accurate solutions of the Poisson-Boltzmann equation.* The impossibly small diameters of the preceding section can in part at least be ascribed to the inaccuracy of the approximation by which (1566) is replaced by (1567), an inaccuracy which enters so soon as $z\epsilon\psi$ is comparable with kT. More accurate methods have been developed but they are necessarily rather elaborate; for the sake of simplicity we shall consider only a solution containing N molecules of a symmetrical electrolyte whose ions have charges $\pm z\epsilon$, and are all of the same size. This restriction is not serious since we shall find reason to believe that the results of the extended calculations can only be valid (if at all) in this case. We shall also only consider explicitly the first stage of the extension.

The Poisson-Boltzmann equation for this simple assembly reduces to

$$\frac{1}{r^2}\frac{d}{dr}\left(r^2\frac{d\psi_\alpha}{dr}\right)=\frac{8\pi\epsilon}{D}z\frac{N}{V}\sinh\frac{z\epsilon\psi_\alpha}{kT}. \qquad \text{......(1587)}$$

* Hovorka and Rodebush, *J. Amer. Chem. Soc.* vol. 47, p. 1614 (1925). The values of a actually given in this paper are computed incorrectly owing to a misprint in Debye and Hückel's table of values of σ. (Private communication to E. A. G.)

† Adams, *J. Amer. Chem. Soc.* vol. 37, p. 481 (1915).

‡ Hall and Harkins, *J. Amer. Chem. Soc.* vol. 38, p. 2658 (1916).

The solution of this equation is required for which $\psi_\alpha \to 0$, $d\psi_\alpha/dr \to 0$ $(r \to \infty)$ and $d\psi_\alpha/dr = -z\epsilon/Da^2$ $(r = a)$. It is convenient to introduce the notation

$$\kappa = +\left(\frac{8\pi\epsilon^2}{DkT} z^2 \frac{N}{V}\right)^{\frac{1}{2}}, \quad \rho = \kappa r, \quad x = \kappa a, \quad y = \frac{z\epsilon\psi_\alpha}{kT}, \quad b = \frac{z^2\epsilon^2}{DkTa}. \quad \text{......(1588)}$$

Then

$$\frac{1}{\rho^2}\frac{d}{d\rho}\left(\rho^2 \frac{dy}{d\rho}\right) = \sinh y, \quad \text{......(1589)}$$

with the boundary conditions

$$y \to 0, \quad dy/d\rho \to 0 \qquad (\rho \to \infty),$$

$$\frac{dy}{d\rho} = -\frac{b}{x} \qquad (\rho = x).$$

Debye and Hückel's approximation consists in putting $\sinh y = y$.

There is no difficulty in principle in obtaining the desired solution without this approximation. It has been achieved in an elementary way by Müller,* who used the approximation $\sinh y = y$ only for sufficiently large values of ρ. For such values of ρ the solution of Debye and Hückel

$$y(\rho) = Ae^{-\rho} \quad \text{......(1590)}$$

is valid, but for smaller values of ρ $y(\rho)$ deviates from (1590) and the value of A cannot be determined from the boundary condition at $\rho = x$. Müller therefore continued (1590) inwards by graphical integration to $\rho = x$, determined A to fit the boundary condition, and so finally found $y(x)$ as a function of b and x.

An analytical determination of the solution of (1589) has also been given by Gronwall.† The complete solution of the problem given by Gronwall, La Mer and Sandved‡ is too lengthy to develop here and we shall be content to describe Gronwall's first step beyond Debye and Hückel's approximation. Equation (1589) may be written

$$\frac{d}{d\rho}\left(\rho^2 \frac{dy}{d\rho}\right) - \rho^2 y = \rho^2 \phi(y), \quad \text{......(1591)}$$

where

$$\phi(y) = \sinh y - y = \sum_{n=1}^{\infty} y^{2n+1}/(2n+1)!. \quad \text{......(1592)}$$

We now transform the differential equation into an integral equation by means of the Green's function of the left-hand side,§ obtaining

$$y(\rho) = \frac{bx}{1+x}\frac{e^{x-\rho}}{\rho} - \frac{1}{2\rho}\int_x^\infty \left\{e^{-|t-\rho|} - \frac{1-x}{1+x}e^{2x-t-\rho}\right\} \phi\{y(t)\}\, t\, dt. \quad \text{......(1593)}$$

This form is exact. It may be verified by direct differentiation and attention to the boundary conditions.

* Müller, *Physikal. Zeit.* vol. 28, p. 324 (1927); vol. 29, p. 78 (1928).
† Gronwall, *Proc. Nat. Acad. Sci.* vol. 13, p. 198 (1927).
‡ Gronwall, La Mer and Sandved, *Physikal. Zeit.* vol. 29, p. 358 (1928).
§ See Courant and Hilbert, *Methoden der mathematischen Physik*, vol. 1, pp. 273–275 (1924).

The Debye-Hückel solution neglects ϕ entirely. A next approximation is obtained by substituting this crude solution in ϕ. As $\kappa \to 0$ and so $x \to 0$ this term is of the order x^2, and the error at this stage can be shown to be of order $x^3 \log x$. We have therefore

$$y(x) \simeq \frac{b}{1+x} - \frac{1}{1+x}\int_x^\infty e^{x-t}\phi\left(\frac{bx}{1+x}\frac{e^{x-t}}{t}\right)t\,dt, \qquad \text{......(1594)}$$

$$= \frac{b}{1+x} - \frac{x^2}{1+x}\sum_{n=1}^{\infty}\frac{\{b/(1+x)\}^{2n+1}}{(2n+1)!}\int_0^\infty e^{-2(n+1)x\alpha}\frac{d\alpha}{(1+\alpha)^{2n}},$$

$$\simeq b(1-x+x^2) - x^2\sum_{n=1}^{\infty}\frac{b^{2n+1}}{(2n+1)!\,(2n-1)}.$$

We thus find on reverting to the original notation that

$$\psi_\alpha(a) \simeq \frac{z\epsilon}{Da} - \frac{z\epsilon\kappa}{D}\left(1 - \kappa a + \kappa a\sum_{n=1}^{\infty}\frac{(z^2\epsilon^2/DkTa)^{2n}}{(2n+1)!\,(2n-1)}\right). \qquad \text{......(1595)}$$

The more complete discussion of Gronwall, La Mer and Sandved shows that (1595) may be replaced by

$$\frac{z\epsilon\psi_\alpha}{kT} = \frac{z^2\epsilon^2/DkTa}{1+\kappa a} + \sum_{s=1}^{\infty}\left(\frac{z^2\epsilon^2}{DkTa}\right)^{2s+1} y_{2s+1}(\kappa a), \qquad \text{......(1595·1)}$$

where the $y_{2s+1}(\kappa a)$ are determinable functions of κa.

The equation for $\psi_\beta(a)$ is, in this symmetrical case, derived from (1595) by changing the sign of z. The potential due to the distribution about the central ion is given by the second term correct to terms in $\kappa^2 a^2$. Using (1595) for $\psi_\alpha(a)$ and charging up the ions as in § 8·82, we find that

$$G_\epsilon = -\frac{2Nz^2\epsilon^2\kappa}{3D}\left(1 - \tfrac{3}{4}\kappa a + \tfrac{3}{2}\kappa a\sum_{n=1}^{\infty}\frac{(z^2\epsilon^2/DkTa)^{2n}}{(2n+2)!\,(2n-1)}\right). \qquad \text{......(1596)}$$

The more elaborate work of Gronwall, La Mer and Sandved is correct to higher powers of κa.

If the solution either of Müller or of Gronwall, La Mer and Sandved is applied to the experimental data more reasonable values are obtained for a than those of § 13·63. In fact Gronwall has shown that, if the values of a derived from the exact theory (equation (1591)) are denoted by a_G and those of the preceding section by a_H, a_G remains positive even if $a_H \to -\infty$. We shall not however go further into the details of these solutions of the Poisson-Boltzmann equation because there is considerable doubt whether their validity extends any further than those of the equation of Debye and Hückel. As we have already seen in § 8·8, the conditions of self-consistency

$$\epsilon z_\alpha \psi_\beta = \epsilon z_\beta \psi_\alpha = W_{\alpha\beta}, \quad N_\beta\frac{\partial\psi_\beta}{\partial z_\alpha} = N_\alpha\frac{\partial\psi_\alpha}{\partial z_\beta}$$

must be satisfied by the ψ's in the general case. Owing to the special symmetry of (1595) the condition $\epsilon z_\alpha\psi_\beta = \epsilon z_\beta\psi_\alpha$ is here automatically satisfied.

Owing to the restriction of (1595) to two sets of equivalent ions only the order of charging cannot be changed (preserving neutrality) in this special case, so that the second condition does not operate. For this special case therefore the only condition that can fail is the equality of $\epsilon z_\alpha \psi_\beta$ to $W_{\alpha\beta}$ and it is possible that, though this condition cannot be exactly fulfilled, its failure is not serious. It is apparent at once in the general case that the ψ_α do not depend on the N_α and z_α solely through the combination $\Sigma_\beta N_\beta z_\beta^2$ which occurs in κ, but through other combinations $\Sigma_\beta N_\beta z_\beta^{2+r}$ as well. The conditions of self-consistency then inevitably fail, and far more completely than in the symmetrical case. It must be regretfully admitted that this most promising method of evaluating $B(T)$ cannot be carried *logically* beyond the first crudest approximation. The extended theory of Gronwall, La Mer and Sandved at least for symmetrical electrolytes does not appear to be so radically involved in these inner failures, and as a semi-empirical extension can apparently be used with confidence so long as the Poisson-Boltzmann equation itself is valid.

§ **13·71.** *Bjerrum's treatment of ion association.* We have seen that attempts to improve on Debye and Hückel's theory by solving more accurately the Poisson-Boltzmann equation are in general doomed to failure in spite of their empirical success. The Debye-Hückel theory itself is valid when both

$$\frac{z_\alpha z_\beta \epsilon^2 \kappa}{DkT} \ll 1, \quad \frac{z_\alpha z_\beta \epsilon^2}{DkTa} \ll 1.$$

The second condition is independent of the concentration, and it is this restriction to large ions which we have still to find a general means of avoiding.

An entirely different method of avoiding this restriction has been proposed by Bjerrum,* at first sight less elegant and more arbitrary than that of Gronwall, but free from inner contradictions. The principle and application of Bjerrum's method is simple so long as all ions have the same numerical valency and we shall confine attention to this case.

It is clear that the radius

$$q = \left| \frac{z_\alpha z_\beta \epsilon^2}{2DkT} \right| \qquad \text{......(1597)}$$

is a critical radius in any theory, for if $a \gg q$ then Debye and Hückel's theory applies in its original form. We imagine therefore the centre of every ion surrounded by a sphere of radius q. For each ion there are then the following alternatives: There may be (i) no other ion, (ii) one other ion of opposite charge, (iii) one other ion of the same charge, or (iv) more than one other ion, inside the spherical shell $r = q$. The relative frequency of these arrangements

* Bjerrum, *Kgl. Danske Vid. Selsk., Math.-fys. Medd.* vol. 7, No. 9 (1926).

can easily be evaluated approximately, for in calculating distributions in which one ion (or a few ions) are within a certain small region immediately round the central ion the screening effect of the other (distant) ions may be ignored. The potential due to the central ion at a distance r can therefore be given the simple value $z\epsilon/Dr$. The average number of ions of the opposite or the same charge within the sphere $r=q$ is therefore

$$\frac{N}{V}\int_a^q 4\pi r^2 \exp\left(\pm\frac{z^2\epsilon^2}{DrkT}\right)dr,$$

the + sign referring to ions of opposite and the − to ions of the same charge. For sufficiently small values of N this number is small compared with unity even for the + sign and still smaller for the negative sign. We may then say that we have effectively a fraction α ($\alpha \ll 1$) of ions with an ion of the opposite sign within the sphere $r=q$, and a negligible fraction with an ion of the same sign. The fraction with two ions within $r=q$ can easily be seen to be of order at most α^2 and may therefore also be assumed to be negligible. The remaining fraction $1-\alpha$ has no other ion within $r=q$. Of the four possible arrangements only (i) and (ii) are effectively present.

The novel idea in Bjerrum's treatment is to deal with these two classes of ions separately. The fraction α is called *associated ion pairs* and the fraction $1-\alpha$ free or unassociated ions. Bjerrum's approximation is now to ignore the effect of the electrostatic field of an associated pair on the remaining ions and to apply Debye and Hückel's theory of the free ions, assuming of course that they have an effective diameter q, for if they come closer than q they cease to count as free.

For sufficiently small concentrations we have shown that

$$\alpha=\frac{N}{V}\int_a^q 4\pi r^2 \exp(z^2\epsilon^2/DrkT)\,dr. \qquad \text{......(1598)}$$

To extend the theory, at least roughly, to values of α for which the condition $\alpha \ll 1$ fails, we can calculate α by treating the distribution as one of dissociative equilibrium between the associated ion pairs and the free ions, in the manner of § 9·6. The condition of dissociative equilibrium takes the form

$$\frac{(1-\alpha)^2 N}{\alpha}=\frac{f_1 f_2}{f_{12}},$$

where f_1, f_2 and f_{12} are the partition functions for the free ions and the associated pair respectively. Only the configurational factors can contribute to this ratio and it is easily shown that

$$\frac{f_1 f_2}{f_{12}}=\frac{V}{4\pi\int_a^q r^2 \exp(z^2\epsilon^2/DrkT)\,dr},$$

if the electrostatic forces between the distant ions are ignored. To this approximation therefore

$$\frac{(1-\alpha)^2 N/V}{\alpha} = K_\infty = \left[4\pi\int_a^q r^2 \exp(z^2\epsilon^2/DrkT)\,dr\right]^{-1}, \quad \text{......(1599)}$$

and to the next approximation

$$\frac{(1-\alpha)^2 N/V}{\alpha} = K_c, \quad \text{......(1600)}$$

where K_c/K_∞ may be derived thermodynamically from Debye and Hückel's value for G_ϵ. G_ϵ, and therefore the ratio so calculated, is a function of $\kappa = \kappa(\alpha)$ which must be given the value

$$\kappa(\alpha) = \left(\frac{8\pi N(1-\alpha)\,z^2\epsilon^2}{DkTV}\right)^{\frac{1}{2}}. \quad \text{......(1601)}$$

Thus K_c and α can only be evaluated together, by a process of successive approximation. The necessary calculations have been carried through by Bjerrum for aqueous solutions of uni-univalent electrolytes at 18° C. and the values obtained for α are given in Table 56.

TABLE 56.

Degree of association α of uni-univalent ions in water at 18° C.

$a \times 10^8$ cm.	2·82	2·35	1·76	1·01	0·70	0·47
q/a	2·5	3	4	7	10	15
$c^\dagger$ in moles per litre						
0·0001	—	—	—	—	0·001	0·027
0·0002	—	—	—	—	0·002	0·049
0·0005	—	—	—	0·002	0·006	0·106
0·001	—	0·001	0·001	0·004	0·011	0·177
0·002	0·002	0·002	0·003	0·007	0·021	0·274
0·005	0·002	0·004	0·007	0·016	0·048	0·418
0·01	0·005	0·008	0·012	0·030	0·083	0·529
0·02	0·008	0·013	0·022	0·053	0·137	0·632
0·05	0·017	0·028	0·046	0·105	0·240	0·741
0·1	0·029	0·048	0·072	0·163	0·336	0·804
0·2	0·048	0·079	0·121	0·240	0·437	0·854

The entries of this table can easily be transformed to apply to other valencies and other solvents by making use of the fact that the osmotic coefficient g depends only on the ratios $q:a:c^{-\frac{1}{3}}$ or is a function only of cz^6/D^3T^3 and of aDT/z^2. The method of computing g according to this theory should be sufficiently obvious and it is left to the reader.

Values of $g = g\{\kappa(\alpha)\}$ are derived from (1580) using the corrections necessary to allow for α. By comparing the values of g obtained from freezing point measurements with calculated values much more reasonable values of a are obtained than from the simple theory, especially for bi-bivalent electrolytes, and for uni-univalent electrolytes in solvents of lower dielectric constant such as the alcohols.

§ **13·72.** *Fuoss's discussion of the choice of q.* Before leaving this subject it is proper to enquire why precisely the distance q should have been taken arbitrarily as the critical radius distinguishing between free and associated ions. The answer is that the exact value of q is unimportant, and that the final result is not greatly different if we replace q by e.g. $\frac{1}{2}q$ or $2q$. A revised form of the theory due to Fuoss* makes this point clearer by virtually eliminating the arbitrary choice of q.

In Fuoss's treatment *every* ion is paired with some one ion of the opposite sign according to the following convention. *A positive ion and a negative ion, the centre of which lies at a distance between r and r + dr from the positive ion, are defined to be an ion pair provided that no other unpaired negative ion lies within a sphere of radius r drawn round the positive ion.* Let the centre of a particular positive ion be taken as origin and let $G(r)\,dr$ be the probability that this ion forms a pair with one of the N negative ions at a distance between r and $r+dr$, there being no unpaired negative ion nearer than a distance r. We may then assert that $G(r)$ will be proportional to N; to $4\pi r^2\,dr/V$; to the Boltzmann factor $\exp(z^2\epsilon^2/DrkT)$ (this neglects screening); and finally to the probability $f(r)$ that an unpaired negative ion is not already present in the volume $4\pi r^3/3$. The probability that the first unpaired negative ion is present in the spherical shell $x < r \leqslant x + dx$ is by definition $G(x)\,dx$ and therefore the probability $f(r)$ that no such ion is anywhere present in the sphere of volume $4\pi r^3/3$ is

$$f(r) = 1 - \int_a^r G(x)\,dx.$$

Combining these statements we see that we have asserted that

$$G(r)\,dr = \frac{4\pi N}{V}\,r^2\,dr\,e^{2q/r}\left\{1 - \int_a^r G(x)\,dx\right\}. \qquad \text{......(1602)}$$

For small r the probability $G(r)\,dr$ must reduce to the ordinary average number of ions of opposite sign, that is to

$$G(r) \sim \frac{4\pi N}{V}\,r^2 e^{2q/r}.$$

With this boundary condition equation (1602) is easily solved giving

$$G(r) = \frac{4\pi N}{V}\,r^2 \exp\left\{\frac{2q}{r} - \frac{4\pi N}{V}\int_a^r x^2 e^{2q/x}\,dx\right\}. \qquad \text{......(1603)}$$

* Fuoss, *Trans. Far. Soc.* vol. 30, p. 967 (1934).

It is easily verified that

$$\int_a^{V^{\frac{1}{3}}} G(r)\,dr \simeq \int_a^{\infty} G(r)\,dr = 1.$$

We now study the dependence of $G(r)$ on r. We see that $G(r)$ has stationary values when

$$\frac{2}{r} - \frac{2q}{r^2} - \frac{4\pi N}{V} r^2 e^{2q/r} = 0. \qquad \text{......(1604)}$$

There is always one maximum value, and there is also a minimum if $q > a$. If the concentration is fairly small so that $q \ll (V/N)^{\frac{1}{3}}$ and if also $q > a$, then the minimum occurs at

$$r = q\left(1 + \frac{4\pi N}{V}\frac{e^2}{2} q^3 + \ldots\right)$$

or at $r \simeq q$, and the maximum at

$$r = (V/2\pi N)^{\frac{1}{3}}\{1 - 2q(V/2\pi N)^{-\frac{1}{3}} + \ldots\}$$

or at $r \simeq (V/2\pi N)^{\frac{1}{3}} = \rho$. These give

$$G_{\min} \simeq G(q) \simeq \frac{4\pi N}{V} e^2 q^2;$$

$$G_{\max} \simeq G(\rho) \simeq \left(\frac{16\pi N}{e^2 V}\right)^{\frac{1}{3}};$$

while for ions in contact $\quad G(a) \simeq \dfrac{4\pi N}{V} a^2 e^{2q/a}.$

It is thus approximately true that $G(a) \propto N/V$ while $G(\rho) \propto (N/V)^{\frac{1}{3}}$ so that at low concentrations $G(\rho) \gg G(a)$, while at high the order will be reversed. For given solvent, temperature, valency, and ionic diameter there will be a concentration for which these two probabilities are equal. This concentration $c_0^{\dagger}$ has the value

$$c_0^{\dagger} = \frac{97}{(10^8 a)^3} e^{-3q/a}.$$

If for example we take $T = 300°$ K.,

$$a = 4{\cdot}6 \times 10^{-8} \text{ cm.},$$

then

$$\log_{10} C_0^{\dagger} = -78{\cdot}7z^2/D.$$

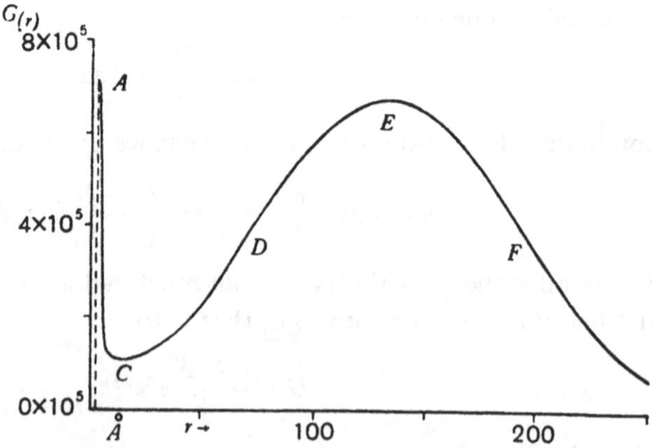

Fig. 72. Showing the distribution function $G(r)$ as a function of r.

Thus for aqueous solutions of a bi-bivalent electrolyte the critical concentration is about ten thousandth molar. An example of $G(r)$ is shown in Fig. 72 for this case. The various characteristic distances in Ångström units are $a=4{\cdot}6$, $(V/2N)^{\frac{1}{3}}=202$, $1/\kappa=153$, $q=14$, $\rho=138$, and their variations as $c^\dagger$ changes can be derived at once from the foregoing formulae. The maximum at E and the inflexions at D and F move however only at rates proportional to $c^{\dagger\frac{1}{3}}$ both vertically and horizontally while A and C move upwards at rates proportional to $c^\dagger$. It follows that over a very wide range of values of $c^\dagger$ there are very few ion pairs at C in the neighbourhood of $r\simeq q$ compared with those round A or E or both. It is therefore legitimate to divide the ion pairs up into these two classes by an arbitrary convention provided that the dividing radius is chosen near q as Bjerrum has done. It is clearly then the correct first approximation to treat the two distinct classes as contributing to the electrical energy according to the Debye-Hückel theory for the distant pairs while giving zero for the associated pairs. There is therefore no ground for the assertion of Gronwall, La Mer and Sandved* that Bjerrum's assumptions are arbitrary.

§ **13·8.** *Specific interactions of ions.* The most valuable contribution of the theory of Debye and Hückel consists of the limiting laws for infinite dilution, (1576) and the formulae derivable from them. The formulae involving the mean ionic diameter a may be fairly satisfactory if a is not too small. When a is small they fail, and the variations, such as those of Bjerrum and of Gronwall, which attempt to correct for this failure are all cumbersome and none of them exact. Moreover the treatment of the ions as rigid spheres is far too crude to take account of the specific properties of various ions of the same valency, for the ions differ not merely in size but also in shape and polarizability, all of which factors have to be represented by a change of a in this theory and its extensions. It is an obvious suggestion for improvement that in a solution containing several kinds of ions independent values of a should be assigned to each ion pair. It would be necessary however to ensure that the partial potentials of the various ions satisfy the relationships

$$\frac{\partial \mu_\alpha^{\epsilon}}{\partial N_\beta}=\frac{\partial \mu_\beta^{\epsilon}}{\partial N_\alpha}$$

and any simplification of the calculation would be liable to fail to preserve these necessary relationships.

Anything like a complete calculation being at present apparently out of the question, an attempt has been made by Guggenheim‡ to solve the simpler problem: *Given the equilibrium properties of dilute solutions of each single electrolyte, to calculate the specific equilibrium properties of a solution containing several such electrolytes.* A standard of comparison is required to which we can refer the properties of any actual solution of electrolytes.

* Gronwall, La Mer and Sandved, *loc. cit.*

‡ Guggenheim, *Proc. 18th Scandinavian Scientific Congress*, Copenhagen (1929); *Phil. Mag.* vol. 19, p. 588 (1935).

Among a variety of possible standards the most convenient seems to be an imaginary electrolyte containing a mixture of ions of the same concentrations and valencies as those of the actual electrolyte, which accurately obeys the formulae of Debye and Hückel for a definite value of the parameter a. The value we shall actually choose for a is $3{\cdot}08 \times 10^{-8}$ cm.

Let us now consider two actual ions, α and β, in a given relative configuration. Their mutual potential energy $W_{\alpha\beta}$ can be regarded as the sum of the two terms

$$W_{\alpha\beta} = W^{\sigma}{}_{\alpha\beta} + W^{\pi}{}_{\alpha\beta}, \qquad \text{......(1605)}$$

where $W^{\sigma}{}_{\alpha\beta}$ is the value for the two standard ions of the corresponding valencies and $W^{\pi}{}_{\alpha\beta}$ measures the specific (peculiar) deviations of the interaction of the given ion pairs from that of the corresponding standard ion pair. For all except small distances $W^{\sigma}{}_{\alpha\beta}$ is effectively the Coulomb energy, while $W^{\pi}{}_{\alpha\beta}$ is negligible except at very small distances. We shall now make the assumption, strictly illegitimate, that the contribution of $W^{\pi}{}_{\alpha\beta}$ to the thermodynamic functions can be added as a correction additional to the electrostatic terms arising from $W^{\sigma}{}_{\alpha\beta}$. It may be expected that this approximation will be reasonably accurate so long as the contribution is small compared with the electrostatic terms. We then write G_{ϵ} as before for the contribution of the interionic potentials to G and may break this up into

$$G_{\epsilon} = G_{\epsilon}{}^{\sigma} + G_{\epsilon}{}^{\pi}, \qquad \text{......(1606)}$$

$G_{\epsilon}{}^{\sigma}$ being given by (1574) for the chosen a while $G_{\epsilon}{}^{\pi}$ is an additive term vanishing when $W^{\pi}{}_{\alpha\beta} = 0$. Now this additive term arises from effectively short range forces and may therefore be evaluated as a contribution to $\log B(T)$ by the methods of Chapter VIII for imperfect gases. The arguments there used must be generalized so that the standard value of $B(T)$ is not V^N for N ions but contains the interionic energies $W^{\sigma}{}_{\alpha\beta}$. Whereas in Chapter VIII we found an extra contribution to Helmholtz's free energy of

$$\frac{kT}{V} \Sigma_{\alpha,\beta} \frac{N_\alpha N_\beta}{\sigma_{\alpha\beta}} \int (1 - e^{-W_{\alpha\beta}/kT})\, d\omega,$$

the integrand being effectively non-zero over a short range only, so here the extra contribution will be, to the same approximation,*

$$\frac{kT}{V} \Sigma_{\alpha,\beta} \frac{N_\alpha N_\beta}{\sigma_{\alpha\beta}} \int e^{-W^{\sigma}{}_{\alpha\beta}/kT} (1 - e^{-W^{\pi}{}_{\alpha\beta}/kT})\, d\omega; \qquad \text{......(1607)}$$

in (1607) $\Sigma_{\alpha,\beta}$ denotes summation over every type of ion pair, and the symmetry number $\sigma_{\alpha\beta}$ is 1 when $\alpha \neq \beta$ and 2 when $\alpha = \beta$. The expression (1607) is strictly $F_{\epsilon}{}^{\pi}$, calculated by establishing the $W^{\pi}{}_{\alpha\beta}$ at constant temperature and volume. For reasons similar to those given in § 13·61 it is

* The factor V in the denominator is here inexact. It will be a more complicated expression of the order and dimensions of V, depending on the temperature. We shall see however that the distinction need not be preserved.

however probably more accurate to take it as equal to G_ϵ^π. We thus obtain the following expression

$$G_\epsilon = G_\epsilon^\sigma + G_\epsilon^\pi,$$
$$= -\frac{\epsilon^2\kappa}{3D}\Sigma_\beta N_\beta z_\beta^2 \tau(\kappa a) + \frac{kT}{V}\Sigma_{\alpha,\beta}\frac{N_\alpha N_\beta}{\sigma_{\alpha\beta}}\int e^{-W^\sigma_{\alpha\beta}/kT}(1-e^{-W^\pi_{\alpha\beta}/kT})\,d\omega. \qquad \text{......(1608)}$$

Not all the specific terms are equally important; the term arising from pairs of ions of the same sign for which $W^\sigma_{\alpha\beta}>0$ will be small compared with those arising from pairs of ions of opposite sign for which $W^\sigma_{\alpha\beta}<0$.† We may therefore simplify G_ϵ to the expression

$$G_\epsilon = -\frac{\epsilon^2\kappa}{3D}\Sigma_\beta N_\beta z_\beta^2\tau(\kappa a) + \frac{kT}{V}\Sigma_{R,X} N_R N_X v_{R,X}, \qquad \text{......(1609)}$$

where R is any cation and X any anion, and $v_{R,X}$ is a coefficient depending only on the temperature, but specific to the pair of ions in question. These $v_{R,X}$'s may each be determined by the properties of a single electrolyte, and the equilibrium properties of a general mixed electrolyte will then be derivable from (1609). For example the specific contribution $\mu^{\pi *}_{H_2O}$ to the partial potential of the solvent water will be

$$\mu^{\pi *}_{H_2O} = \mathsf{N}\frac{\partial G_\epsilon^\pi}{\partial N_{H_2O}} = -v^*_{H_2O}\,kT\,\Sigma_{R,X}\frac{N_R N_X}{V^2}v_{R,X},$$
$$= -\mathsf{N}RT\,v^*_{H_2O}\,\Sigma_{R,X}\,c_R^*\,c_X^*\,v_{R,X}. \qquad \text{......(1610)}$$

For the sake of brevity we shall discuss the implications of this formula for the case in which all the electrolytes in the solution are of the same valency type, each molecule dissolved yielding q_+ cations of valency z_+ and q_- anions of valency z_- so that $q_+z_+ + q_-z_- = 0$. Let c^* be the total concentration of electrolyte in moles per unit volume so that

$$c^* = \frac{1}{q_+}\Sigma_R c_R^* = \frac{1}{q_-}\Sigma_X c_X^*,$$

and let the composition of the solution be defined by the fractions $r_1, \ldots, r_\sigma$ for the cations and $x_1, \ldots, x_\tau$ for the anions, such that

$$c_{R_1}^* = r_1\Sigma_R c_R^* = r_1 q_+ c^*, \quad c_{X_1}^* = x_1\Sigma_X c_X^* = x_1 q_- c^*,$$

etc. Then

$$\mu^{\pi *}_{H_2O} = -\mathsf{N}RT\,v^*_{H_2O}\,q_+q_-c^{*2}\,\Sigma_{R,X}\,r_R x_X v_{R,X}. \qquad \text{......(1611)}$$

The corresponding contribution to $1-g$ as in § 13·61 is

$$\mu^{\pi *}_{H_2O}/v^*_{H_2O}\,RT\,\Sigma_\beta c_\beta^*,$$

or

$$-\frac{q_+q_-}{q_++q_-}\,c^*\,\Sigma_{R,X}\,r_R x_X\,\mathsf{N}v_{R,X};$$

† Cf. Brönsted, *J. Amer. Chem. Soc.* vol. 44, p. 877 (1922). His formulation of the "principle of specific interaction" states that: "Ions are uniformly influenced by ions of their own sign and specifically influenced only by ions of the opposite signs."

this may be shortened by writing

$$\bar{q} = \frac{2q_+q_-}{q_+ + q_-}, \quad l^*_{R,X} = \tfrac{1}{2}\mathrm{N}v_{R,X}.$$

Then the complete expression, in the notation of § 13·63 is

$$1 - g = 0{\cdot}374|z_+z_-|\, I^{\frac{1}{2}}\, \sigma(I^{\frac{1}{2}}) - \bar{q}\, c^* \, \Sigma_{R,X}\, r_R x_X l^*_{R,X}. \qquad \text{......(1612)}$$

For single electrolytes this reduces to

$$1 - g = 0{\cdot}374|z_+z_-|\, I^{\frac{1}{2}}\, \sigma(I^{\frac{1}{2}}) - \bar{q}\, c^* l^*_{R,X}. \qquad \text{......(1613)}$$

There is excellent agreement between the best freezing point data and (1613) for single electrolytes of various valency types. The values of the $l^*_{R,X}$ which best fit the data are given in Table 57. With these values the disagreement between theory and observation is probably less than the experimental error for all ionic strengths up to tenth molar. Such data as are available for mixed electrolytes appear to be in good agreement with the theory.†

TABLE 57.

Values of $l^*_{R,X}$ fitting freezing point data for simple electrolytes

$$1 - g = 0{\cdot}374|z_+z_-|\, I^{\frac{1}{2}}\, \sigma(I^{\frac{1}{2}}) - \bar{q}\, c^* l^*_{R,X}.$$

Electrolyte	$\bar{q}$	$l^*_{R,X}$	$\bar{q}l^*_{R,X}$	Authority
HCl	1	+0·275	+0·275	(4)
LiCl	1	−0·223	−0·223	(8)
NaCl	1	+0·135	+0·135	(3), (8)
KCl	1	+0·065	+0·065	(8)
...	—	+0·083	+0·083	(1), (5)
TlCl	1	+0·40	+0·40	(4)
$LiNO_3$	1	+0·26	+0·26	(7)
$NaNO_3$	1	0·00	0·00	(7)
KNO_3	1	−0·24	−0·24	(7)
...	—	−0·29	−0·29	(1)
$CsNO_3$	1	0·00	0·00	(5)
$NaIO_3$	1	−0·40	−0·40	(2)
KIO_3	1	−0·40	−0·40	(2)
Na_2SO_4	$\frac{4}{3}$	−0·45	−0·60	(6)
K_2SO_4	$\frac{4}{3}$	0·00	0·00	(2), (5)
$Ba(NO_3)_2$	$\frac{4}{3}$	−0·41	−0·55	(6), (5)
$MgSO_4$	1	0·00	0·00	(2), (5)
$CuSO_4$	1	−1·7	−1·7	(5)
$La(NO_3)_3$	$\frac{3}{2}$	+2·6	+3·9	(2)
$La_2(SO_4)_3$	$\frac{12}{5}$	0·00	0·00	(5)

(1) Adams, *J. Amer. Chem. Soc.* vol. 37, p. 481 (1915).
(2) Hall and Harkins, *ibid.* vol. 38, p. 2658 (1916).
(3) Harkins and Roberts, *ibid.* vol. 38, p. 2676 (1916).
(4) Randall and Vanselow, *ibid.* vol. 46, p. 2418 (1924).
(5) Hovorka and Rodebush, *ibid.* vol. 47, p. 1614 (1925).
(6) Randall and Scott, *ibid.* vol. 49, p. 647 (1927).
(7) Scatchard, Jones and Prentiss, *ibid.* vol. 54, p. 2690 (1932).
(8) Scatchard and Prentiss, *ibid.* vol. 54, p. 2696 (1932), vol. 55, p. 4355 (1933).

† For further details, see: Brönsted, *J. Amer. Chem. Soc.* vol. 44, p. 877 (1922); Guggenheim, *Phil. Mag.* vol. 19, p. 588 (1935).

CHAPTER XIV

ASSEMBLIES OF ATOMS, ATOMIC IONS AND ELECTRONS

§ **14·1.** *Introductory.* We have already had occasion to consider incidentally examples of gaseous assemblies in which atoms, ions and electrons are present in dissociative equilibrium. We shall have occasion in the following chapters to discuss systematically assemblies, especially at high and very high temperatures, which are composed entirely of atomic ions and electrons. These discussions of course have reference to the state of matter in stars. It will be necessary to give a general survey of the present state of atomic theory, at least on the formal descriptive side, so that we may be able to write down at will the partition function, or an effective approximate form of it, for the internal energy of any atomic ion. We have then to reformulate the general theory of dissociative equilibrium in terms of atomic ions and electrons instead of atoms and molecules and introduce correcting terms for the electric charges of the particles. This would be a simple matter were it not for the outstanding difficulty of the convergence of the partition functions which must be disposed of in some way during the process. We are not in this chapter concerned with assemblies of electrons or atoms dense enough to require the statistics of Fermi-Dirac or Einstein-Bose. All the formulae used are based on classical statistics. The more general case will be considered later in Chapter XVI.

We shall assume that no molecules are present in the assembly; they can easily be included when required. Let

M_0^z be the (average) number of neutral atoms of atomic number Z in a volume V;

M_r^z be the (average) number of such atoms r-times ionized;

N be the (average) number of free electrons.

In general atoms r-times ionized must be defined to mean nuclei accompanied by $Z-r$ electrons, each of which has insufficient energy to effect an escape. These $Z-r$ electrons combine together to form the stationary states of the ion in which the state of each electron can be described by four quantum numbers. The most convenient state of conventionally zero energy is that state of the assembly in which the only constituents are electrons and bare nuclei at rest at infinite separation. The bare nucleus may usually be assumed to be structureless. This is of course untrue, but except when we discuss the break-up and re-formation of nuclei their structure is irrelevant*

* This is of course only correct because we are not concerned here with formation of molecules. We have seen in the study of H_2 in § 3·4, and again in § 7·31, that the nuclear weights are not only relevant but important.

and it is a legitimate simplification to regard them as structureless massive charged points, for which the standard weight $dp_1 \ldots dq_3/h^3$ is assigned to the element of phase space $dp_1 \ldots dq_3$. The bare nucleus will therefore have a partition function $VF_z^z(T)$ of the usual form (172·1). The electron has a spin with two possible orientations in a magnetic field. We shall use m for the mass of the electron and m^z for the bare nucleus Z. It will seldom be necessary to distinguish between m^z and $m^z + rm$ $(r \leqslant z)$. The partition function for the free electron will therefore be $VG(T)$, where

$$G(T) = \frac{2(2\pi mkT)^{\frac{3}{2}}}{h^3}, \qquad \ldots\ldots(1614)$$

which we have already used in Chapter XI.

Consider next the normal state of each atomic ion. Let the successive ionization energies of the atom Z be $\chi_0^z, \chi_1^z, \ldots, \chi_{z-1}^z$ and the weights of the normal states ϖ_r^z. These χ's are all to be defined with reference to a series of normal states or states of least energy so that χ_r^z is the work required to remove one electron from an atom, which has already lost r electrons and is then at rest in the state of lowest energy possible for its remnant of $(Z-r)$ electrons, and leave it again in its state of lowest possible energy for the remnant of $(Z-r-1)$ electrons, the atom and the extracted electron being at rest at infinite separation. For the r-times ionized atom in its normal state the partition function $VF_r^z(T)$ is therefore given by

$$F_r^z(T) = \frac{(2\pi m^z kT)^{\frac{3}{2}}}{h^3} \varpi_r^z e^{(\chi_r^z + \ldots + \chi_{z-1}^z)/kT}. \qquad \ldots\ldots(1615)$$

Each ion possesses in addition a set of stationary excited states of greater energy content. If every excited state could be treated formally as a constituent of a perfect gas this would cause ϖ_r^z to be replaced by $b_r^z(T)$, where

$$b_r^z(T) = \sum_{s=0} (\varpi_r^z)_s e^{-\{\chi_r^z - (\chi_r^z)_s\}/kT}. \qquad \ldots\ldots(1616)$$

The state $s=0$ is the normal state and we continue to write ϖ_r^z and χ_r^z instead of $(\varpi_r^z)_0$ and $(\chi_r^z)_0$. The energy of excitation is $\chi_r^z - (\chi_r^z)_s$, so that $(\chi_r^z)_s$ is the energy corresponding to the sth spectral *term* (suitably ordered) of the r-times ionized atom Z. In fact $b(T)$ does not converge. The alterations necessary will be discussed at length later in §§ 14·4, 15·4. It is evident from (1616) that an application of statistical mechanics to such assemblies requires a working knowledge of or approximation to $(\varpi_r^z)_s$ and $(\chi_r^z)_s$ for all r, s and z.

§ **14·2**. *General features of atomic structure.* We shall start by summarizing the general features of atomic structure which will form the basis of our discussion. Following Bohr it is well known that the electrons in the atom can be classified first of all according to their principal quantum number n (1,2,3,...) and azimuthal quantum number k $(0,1,2,...,n-1)$ as n_k electrons. The broad outlines of the periodic table of the elements can be at once accounted for in this way if it is remembered that just so many electrons and no more can be packed into each of these groups or subgroups. It was early recognized that the maximum number of electrons in any group with principal quantum number n must be $2n^2$ to fit the periodic table. The numbers of electrons in the n_k subgroups must be $2(2k+1)$, as was first shown by Stoner and Pauli. We recall of course that

$$2\sum_0^{n-1}(2k+1)=2n^2.$$

These numbers of electrons follow of course from the requirements of antisymmetry, that no two electrons shall have the same wave-function, a requirement usually known as Pauli's principle, especially in this connection. When any group or subgroup of electrons is full it forms a symmetrical structure without mechanical or magnetic moment and interacts with other electrons at least to a first approximation like a simple central field of force.

We have started by introducing the closed groups in the atom on account of their descriptive importance, but logically they are complex and are arrived at at a later stage. We turn next therefore to describe the states of a hydrogen-like atom with just one electron, observing that owing to the symmetry of closed groups any atomic ion containing only closed groups behaves qualitatively exactly like a bare nucleus in forming states for the next electron. There are only quantitative energy differences in the states due to the different effective central field.

The states of the hydrogen-like atom containing one spinning electron are described by four quantum numbers n, k, j and s.* The principal quantum number n takes the values (1,2,3,...). The azimuthal quantum number k takes the values $(0,1,2,...,n-1)$. The angular momentum of the spin of the electron is $\frac{1}{2}h/2\pi$. This momentum compounds with the "orbital" momentum k to give the total angular momentum j of the atom. The j values of the atom are therefore

$$k=0,\quad j=\tfrac{1}{2};$$

$$k\geqslant 1,\quad j=k\pm\tfrac{1}{2}.$$

The total moment of momentum of the atom is represented as usual by $(h/2\pi)\{j(j+1)\}^{\frac{1}{2}}$. If the atom (or electron and orbit) as a whole is orientated by an external magnetic field, then the possible components of the momentum along the field are $sh/2\pi$, where s takes the $(2j+1)$ values

$$-j\leqslant s\leqslant +j.$$

* This is usually denoted by m, which we avoid to save confusion with the mass of the electron.

The complete set of states therefore corresponds to the following sets of values of the four quantum numbers n, k, j, s.

TABLE 58.

The quantum numbers of the states of a one-electron atom.

n	k	j	s	Totals		
1	0	$\frac{1}{2}$	$\frac{1}{2}, -\frac{1}{2}$	2		2
2	0	$\frac{1}{2}$	$\frac{1}{2}, -\frac{1}{2}$	2		8
	1	$\frac{3}{2}$	$\frac{3}{2}, \frac{1}{2}, -\frac{1}{2}, -\frac{3}{2}$	4	6	
		$\frac{1}{2}$	$\frac{1}{2}, -\frac{1}{2}$	2		
3	0	$\frac{1}{2}$	$\frac{1}{2}, -\frac{1}{2}$	2		18
	1	$\frac{3}{2}$	$\frac{3}{2}, \frac{1}{2}, -\frac{1}{2}, -\frac{3}{2}$	4	6	
		$\frac{1}{2}$	$\frac{1}{2}, -\frac{1}{2}$	2		
	2	$\frac{5}{2}$	$\frac{5}{2}, \frac{3}{2}, \frac{1}{2}, -\frac{1}{2}, -\frac{3}{2}, -\frac{5}{2}$	6	10	
		$\frac{3}{2}$	$\frac{3}{2}, \frac{1}{2}, -\frac{1}{2}, -\frac{3}{2}$	4		
4	0	$\frac{1}{2}$	$\frac{1}{2}, -\frac{1}{2}$	2		32
	1	$\frac{3}{2}$	$\frac{3}{2}, \frac{1}{2}, -\frac{1}{2}, -\frac{3}{2}$	4	6	
		$\frac{1}{2}$	$\frac{1}{2}, -\frac{1}{2}$	2		
	2	$\frac{5}{2}$	$\frac{5}{2}, \frac{3}{2}, \frac{1}{2}, -\frac{1}{2}, -\frac{3}{2}, -\frac{5}{2}$	6	10	
		$\frac{3}{2}$	$\frac{3}{2}, \frac{1}{2}, -\frac{1}{2}, -\frac{3}{2}$	4		
	3	$\frac{7}{2}$	$\frac{7}{2}, \frac{5}{2}, \frac{3}{2}, \frac{1}{2}, -\frac{1}{2}, -\frac{3}{2}, -\frac{5}{2}, -\frac{7}{2}$	8	14	
		$\frac{5}{2}$	$\frac{5}{2}, \frac{3}{2}, \frac{1}{2}, -\frac{1}{2}, -\frac{3}{2}, -\frac{5}{2}$	6		

We see at once that they verify the formulae $2n^2$ and $2(2k+1) = 2k+2+2k$ already given, and, assuming Pauli's principle, that the general structure of the periodic table must follow. We note in addition that structurally the spectrum of hydrogen is exactly analogous to that of an alkali; the only difference arises from the accidental coincidence of certain terms in the hydrogen spectrum for different values of k which do not coincide for an alkali owing to the energy variation with k in a non-Coulomb central field. To an approximation which is in general ample for our applications the terms of a true one-electron spectrum depend only on n and are given by the Balmer formula

$$\chi_n = RhcZ^2/n^2, \qquad \text{......(1617)}$$

where R is Rydberg's constant. The typical one-electron spectrum is thus (in agreement with observation) what is called a *doublet* spectrum. It consists of sets of terms labelled ^2S, ^2P, ^2D, ^2F, ^2G, ... corresponding to $k = 0, 1, 2, 3, 4, \ldots$, of which the ^2S terms are *single* and all the others *double*. Different terms with the same label correspond to different values of n, but their structure is always independent of n. All the terms are completely resolved by a magnetic field so that no degeneracy remains. The weight unity is therefore attached to each magnetic state, and we shall find that this agrees with the limiting principle. The weight of all ^2S terms is therefore 2, of the ^2P terms $4+2=6$, of the ^2D $6+4=10$ and so on, or in general the weight of each separate term is $2j+1$. At the moment this is only justified for strictly hydrogen-like atoms, but we shall see directly that on the same basis all closed configurations have also weight unity and the same set of weights therefore apply to the states of all atoms and ions constructed of closed groups plus one extra electron.

When we come to atoms with more than one extra electron the same principles can still be applied. Subject always to exclusions by Pauli's principle, if we have two electrons in n, k, j, s and n', k', j', s' orbits, originally thought of as independent of each other, we obtain thus one possible atomic state. Subsequent introduction of the mutual perturbations may alter the energy but cannot touch the existence of the state. For given n, k and n', k', for example, the variations of j, s and j', s' are independent and the total number of states found should be

$$2^2(2k+1)(2k'+1). \qquad \text{......(1618)}$$

This is in fact correct. If we allow k and k' also to vary, the total number of states for given n and n' should be

$$2^2 n^2 n'^2, \qquad \text{......(1619)}$$

which is again correct. In practice of course the actual states of an atom with two electrons do not present themselves in this way. The atom as it were constructs itself by compounding the four momentum vectors of the two orbits and the two electrons in a certain order of tightness of binding. The four vectors have possible components in a specified direction ranging between $\pm k$, $\pm k'$, $\pm \frac{1}{2}$, $\pm \frac{1}{2}$, respectively. It can be shown without much difficulty that the order of compounding is without effect on the total number of terms and total weight or number of magnetic states. The composition commonest in actual two-electron spectra is based on the formation of the vector orbital momentum l from k and k',

$$|k-k'| \leqslant l \leqslant k+k' \quad (l=k \text{ if } k'=0), \qquad \text{......(1620)}$$

and the electronic momentum r from $\frac{1}{2}$ and $\frac{1}{2}$,

$$r=1,\ 0. \qquad \text{......(1621)}$$

Then l and r combine to give the j-values specifying the total moment of momentum of the atomic system (or rather its maximum resolved part), namely

$$|l-r| \leqslant j \leqslant l+r \quad (j=l \text{ if } r=0). \qquad \text{......(1622)}$$

In the particular case of two electrons we therefore get

$$j=l, \qquad \text{......(1623)}$$

$$j=l+1,\ l,\ l-1, \qquad \text{......(1624)}$$

that is sets of singlets and triplets containing four terms in all for given l of total weight $4(2l+1)$. If $k'=0$, $l=k$, and this agrees with (1618). If k, $k' \neq 0$, then we may suppose $k' \leqslant k$ (otherwise we interchange them), and the possible l values are $(k+k')$, $(k+k'-1)$, ..., $(k-k')$. The total number of terms is $4(2k'+1)$ and the total weight $4\{\sum_{k-k'}^{k+k'}(2l+1)\}$ or $4(2k+1)(2k'+1)$ in agreement with (1618). Subject still to exclusions by Pauli's principle the argument can be extended to any number of electrons extra to the closed groups. The total number of states formed by q such electrons in $(n_1)_{k_1}$, ..., $(n_q)_{k_q}$ orbits is always

$$2^q(2k_1+1)\dots(2k_q+1), \qquad \dots\dots(1625)$$

or in n_1, ..., n_q orbits for any k's

$$2^q n_1{}^2 \dots n_q{}^2. \qquad \dots\dots(1626)$$

Let us now try to put the maximum number $2(2k+1)$ of electrons into any one subgroup of azimuthal quantum number k, remembering Pauli's principle. For all these electrons n and k are the same and therefore one at least of j and s must differ for any pair of electrons. Since there are exactly $2(2k+1)$ different pairs of possible values of j and s, there is one way and one way only in which an antisymmetrical wave-function can be constructed. There must be one electron in each orbit, and since the values of s are symmetrical about zero the resultant j for the atom is zero, and we have a single state of weight unity. This is the theoretical basis of our previous assertions about the properties of closed groups and subgroups.

There are of course always reductions in the number of states given by (1625) and (1626) whenever two electrons have the same (n,k) or the same n. This is of particular importance for the smallest possible values of n and k, when it describes the fact that certain otherwise expected spectral terms do not occur. For larger values of n it can usually be ignored, for what will be required is the asymptotic form of (1626) for large n, and it is easy to see that (1626) remains *asymptotically* true in spite of Pauli's principle. It will be sufficient to consider a simple case of two electrons in states of the same n, and to exclude *all* states of the same k which of course is a gross overestimate. The total number of states by (1626) would be $4n^4$. By the other estimate the actual number is at least

$$4\sum_{0}^{n-1}{}_{k,k'}(2k+1)(2k'+1) \quad (k \neq k'),$$

or

$$4n^4 - \sum_{0}^{n-1}(2k+1)^2,$$

which is asymptotically still $4n^4$.

In assigning these total weights we have been counting together all states constructed out of so many orbits of given (n,k)'s or given n's. It is often permissible to group the orbits of higher quantum numbers in this way because the differences of the energies of the various states in the group are not significant. For states of lower quantum numbers this will not always be true, though it is often even then legitimate to group together all terms formed out of orbits of given (n,k). We should therefore complete these rules by formulating the corresponding rules for the weights of single states (e.g. *one* of the P or D terms of an alkali spectrum) and for a group of multiple terms (e.g. the *pair* of P or D terms of an alkali spectrum). The necessary analysis has already been implicitly given.

The terms of any atomic spectrum can be conveniently classified into multiple terms and the multiple terms into sets of sequences to which the labels S, P, D, F, G, ... are attached. There is just one such set for a one-electron spectrum; in complex cases there may be many more. The labels $S, P, D, F, G, \ldots$ still correspond to the values $0, 1, 2, \ldots$ of l. The number of components of any multiple term has a maximum value ρ (equal to $2r+1$). The number of components in the multiple S, P, D, F, G, ... terms is always the *lesser* of the two numbers $(2l+1,\ 2r+1)$. A set of terms of maximum multiplicity* ρ is labelled ${}^{\rho}S$, ${}^{\rho}P$, ${}^{\rho}D$, ${}^{\rho}F$, ... In no case can $\rho-1$ be greater than the number of electrons forming the incomplete group of orbits being compounded together. The individual terms of a multiple term are distinguished by their j values derived from (1622) and are labelled

$$ {}^{\rho}S_j,\ {}^{\rho}P_j,\ {}^{\rho}D_j,\ {}^{\rho}F_j,\ \ldots. $$

A number n can be prefixed to this symbol to specify the current number of the term in the sequence, or sets of numbers n_k can be prefixed to specify the quantum numbers of the group of orbits out of which it is constructed. In any case any such term is degenerate and splits into $2j+1$ magnetic states in an external field, so that its weight is $2j+1$.

One other type of grouping is sometimes employed. We may group together all the terms arising from the addition of q electrons in given n_k orbits to an atomic core which is not composed of closed configurations but has some of its electrons in an incomplete group of given $n'_{k'}$ orbits. Remembering the effects of Pauli's principle the set of orbits composing the core will give rise to a countable number of states ϖ_c, which is the weight of the core if its energy differences are insignificant. The result of this is of course that the total number of states of the final system is no longer given by (1625) and (1626) but is larger by the extra factor ϖ_c.

* For this the symbol r (or R or $2R$) is more often used, but we have used r for the associated quantum number.

We collect together the leading results:

(1) The weight of a single term ${}^{\rho}S_j$, ${}^{\rho}P_j$, ${}^{\rho}D_j$, ... in any spectrum of any atom is

$$2j+1. \qquad \text{......(1627)}$$

(2) The total weight of any multiple term ${}^{\rho}S$, ${}^{\rho}P$, ${}^{\rho}D$, ... $(l=0, 1, ...)$ in any spectrum of any atom is

$$(2r+1)(2l+1) \quad (2r+1=\rho). \qquad \text{......(1628)}$$

(3) The total weight of all terms arising from q outer electrons in given n_k orbits attached to a core of total weight ϖ_c is

$$2^q(2k_1+1)...(2k_q+1)\varpi_c. \qquad \text{......(1629)}$$

If the core is a bare nucleus or consists of closed groups of electrons, then $\varpi_c=1$. This formula is subject to reductions when any of the q outer electrons are in orbits of the same n and k.

(4) The total weight of all terms arising from q outer electrons in orbits of given principal quantum numbers n attached to a core of total weight ϖ_c is

$$2^q n_1^2 ... n_q^2 \varpi_c. \qquad \text{......(1630)}$$

If the core is a bare nucleus or consists of closed groups of electrons, then $\varpi_c=1$. This formula is also subject to reductions when any of the q outer electrons are in orbits of the same n.

There are no exceptions to these rules.

In conclusion we shall find it useful in applications to have a table of the weights of the lowest states for a number of atoms. The weight which is of most value in this connection is the sum of the weights of all terms in which all the electrons are in orbits of the normal (least possible) values of n and k. In cases of doubt as to which orbit is normal after $Z=18$ (e.g. between 3_2 and 4_0) the most useful value refers to states of ions of large core charge. For these there is no doubt which is the normal orbit for the effect of the smaller n overwhelms that of the larger k. In these calculations full account has been taken of Pauli's principle. The weight is the weight of the group of normal terms for the atom named or for any atomic ion with greater nuclear charge and the same (stated) number of electrons. After atomic number 18 the normal state of the atom and singly charged ion with the stated number of electrons may be different, as one or two of the 3_2 orbits may be initially replaced by 4_0 orbits. The atomic symbols are therefore inserted purely as a descriptive reminder, and it is *not* implied that the weights necessarily apply to the normal state of a neutral atom with Z electrons, but only to the normal state of ions of nuclear charge Z and the stated number of electrons provided Z is large enough. In the range of this table it is probably sufficient that Z should exceed the number of electrons by two or more.

TABLE 59.

Weights of the group of normal states for various atomic ions.

Atom	No. of electrons in groups: Closed	No. of electrons in groups: Unclosed* [type]	Weight	Atom	No. of electrons in groups: Closed	No. of electrons in groups: Unclosed* [type]	Weight
1 H	0	1 $[1_0]$	2	19 K	18	1 $[3_2]$	10
2 He	2	0	1	20 Ca	18	2 $[3_2]$	45
3 Li	2	1 $[2_0]$	2	21 Sc	18	3 $[3_2]$	120
4 Be	4	0	1	22 Ti	18	4 $[3_2]$	210
5 B	4	1 $[2_1]$	6	23 V	18	5 $[3_2]$	252
6 C	4	2 $[2_1]$	15	24 Cr	18	6 $[3_2]$	210
7 N	4	3 $[2_1]$	20	25 Mn	18	7 $[3_2]$	120
8 O	4	4 $[2_1]$	15	26 Fe	18	8 $[3_2]$	45
9 F	4	5 $[2_1]$	6	27 Co	18	9 $[3_2]$	10
10 Ne	10	0	1	28 Ni	28	0	1
11 Na	10	1 $[3_0]$	2	29 Cu	28	1 $[4_0]$	2
12 Mg	12	0	1	30 Zn	30	0	1
13 Al	12	1 $[3_1]$	6	31 Ga	30	1 $[4_1]$	6
14 Si	12	2 $[3_1]$	15	32 Ge	30	2 $[4_1]$	15
15 P	12	3 $[3_1]$	20	33 As	30	3 $[4_1]$	20
16 S	12	4 $[3_1]$	15	34 Se	30	4 $[4_1]$	15
17 Cl	12	5 $[3_1]$	6	35 Br	30	5 $[4_1]$	6
18 A	18	0	1	36 Kr	36	0	1

* In the older theory all these suffixes would have been increased by unity.

We have still to consider the form of the term values χ. Except for hydrogen-like atoms exact formulae cannot be given. But in the simpler one- and two-electron spectra most sequences of terms can be put approximately in Rydberg's form

$$\chi_n = \frac{RhcC^2}{(n-\alpha)^2}, \qquad \text{......(1631)}$$

where C is the core charge or charge on the rest of the atom other than a single outer electron, and α is a constant, provided we consider terms of one sequence only, in which only the principal quantum number n of a *single* series electron is allowed to vary. It is frequently important to use exact values of the earlier larger χ's. When these are required observed values must be taken. For the later smaller terms (1631) will usually suffice, or sometimes even rougher approximations such as zero.

The only important quantities remaining to be specified before we can handle assemblies of ions and electrons are therefore the ionization energies χ_r^z and to a less extent $(\chi_r^z)_s$ for small s. As we have said, these must in general be taken from observation, but this can only be done directly when the corresponding spectrum has been fully analysed. Thus χ_0^z and χ_1^z (sometimes χ_2^z, χ_3^z and even χ_r^z for several more values of r) have usually been determined in this way, but much work will be required before the higher values are thus determined. It must be remembered that these are the *successive* ionization energies of the atom. The removal of (say) the qth electron often corresponds to a process well known and accurately observed in X-ray spectroscopy, but the energy values derived from X-ray spectroscopy are valueless to us. For these energies are the energies required to remove certain electrons from an intact atom or molecule, while we require

TABLE 60.

Successive ionization energies for oxygen ($Z = 8$).

Ionization energy			Accuracy
Symbol	ν/R	Electron volts	
χ_7^8	64	865	O
χ_6^8	54	730	B
χ_5^8	10·2	140	A
χ_4^8	7·8	105	C
χ_3^8	5·7	77	C
χ_2^8	4·0	55	O
χ_1^8	2·6	35	O
χ_0^8	1·0	13½	O

TABLE 61.

Successive ionization energies for iron ($Z = 26$).

Ionization energy			Accuracy
Symbol	ν/R	Electron volts	
χ_{25}^{26}	676	9150	O
χ_{24}^{26}	645	8730	A
χ_{23}^{26}	149	2010	A
χ_{22}^{26}	141	1910	A
χ_{21}^{26}	131	1770	A
χ_{20}^{26}	123	1660	A
χ_{19}^{26}	116	1570	A
χ_{18}^{26}	104	1410	B
χ_{17}^{26}	97	1310	B
χ_{16}^{26}	90	1220	B
χ_{15}^{26}	35·7	480	A
χ_{14}^{26}	32·2	435	A
χ_{13}^{26}	29	390	B
χ_{12}^{26}	26	350	C
χ_{11}^{26}	22	300	C
χ_{10}^{26}	21	280	C
χ_9^{26}	—	250	C
χ_8^{26}	16	220	C
χ_7^{26}	11	150	C
χ_6^{26}, χ_5^{26}, χ_4^{26}, χ_3^{26}, χ_2^{26}, χ_1 (bracketed)	—	Average 80	D
χ_0^{26}	0·60	8·15	O

to remove the same electron when all the outer more loosely bound electrons are already gone. This often requires twice as much energy—in certain cases it can even require five times as much.

TABLE 62.

Successive ionization energies for silver ($Z = 47$).

Ionization energy			Accuracy	Ionization energy			Accuracy
Symbol	ν/R	Electron volts		Symbol	ν/R	Electron volts	
χ_{46}^{47}	2210	30,000	O	χ_{21}^{47}	62	850	D
χ_{45}^{47}	2150	29,000	B	χ_{20}^{47}, χ_{19}^{47}	—	Average 800	D
χ_{44}^{47}	516	6,980	A	χ_{18}^{47}	36	500	B
χ_{43}^{47}	500	6,770	A	χ_{17}^{47}	—	450	C
χ_{42}^{47}	480	6,500	B	χ_{16}^{47}	24	350	D
χ_{41}^{47}	460	6,200	C	χ_{15}^{47}, χ_{14}^{47}, χ_{13}^{47}, χ_{12}^{47}, χ_{11}^{47}	—	Average 300	D
χ_{40}^{47}	440	6,000	C	χ_{10}^{47}	11	150	C
χ_{39}^{47}	420	5,700	C	χ_{9}^{47}, χ_{8}^{47}, χ_{7}^{47}, χ_{6}^{47}, χ_{5}^{47}, χ_{4}^{47}, χ_{3}^{47}, χ_{2}^{47}, χ_{1}^{47}	—	Average 80	D
χ_{38}^{47}	—	5,500	C	χ_{0}^{47}	0·56	$7\frac{1}{2}$	O
χ_{37}^{47}	390	5,300	C				
χ_{36}^{47}	170	2,300	A				
χ_{35}^{47}	160	2,160	B				
χ_{34}^{47}	154	2,100	C				
χ_{33}^{47}	150	2,000	C				
χ_{32}^{47}	140	1,900	C				
χ_{31}^{47}	135	1,800	C				
χ_{30}^{47}	—	1,700	C				
χ_{29}^{47}	120	1,600	C				
χ_{28}^{47}	105	1,400	C				
χ_{27}^{47}, χ_{26}^{47}, χ_{25}^{47}, χ_{24}^{47}, χ_{23}^{47}, χ_{22}^{47}	—	Average 1,200	D				

In these tables O denotes observed or equally certain values.
A ,, theoretical estimates with error probably less than 3 per cent.
B ,, ,, ,, ,, ,, ,, 10 ,,
C ,, ,, ,, ,, ,, ,, 30 ,,
D ,, estimates quite possibly in error by more than 30 per cent.

We must fall back therefore on theoretical asymptotic formulae and extrapolations by their means of known results. It has been shown by Hartree* that the majority of the χ_r^z can be fixed with reasonable security in this way. Hartree has constructed tables for oxygen, iron and silver as representative atoms, and others can be constructed by his methods. But most calculations of highly ionized assemblies such as stellar interiors can be carried through for representative atoms or simple mixtures and need not employ large varieties of atoms.

* Hartree, *Proc. Camb. Phil. Soc.* vol. 22, p. 464 (1924).

In constructing and using a table of successive ionization energies we must assume a definite order in which the electrons are to be removed (or to return) which is the same as the order of tightness of binding. In accordance with the arguments of the earlier part of this section we assume the order

$$2(1_0),\ 2(2_0),\ 6(2_1),\ 2(3_0),\ 6(3_1),\ 10(3_2),\ 2(4_0),\ 6(4_1),\ 10(4_2),\ 2(5_0).$$

There are of course the well-known temporary departures from this order already mentioned, and the two 5_0 orbits do not follow the 4_2 orbits but the 4_3 orbits for the heaviest elements. We shall not usually make calculations explicitly for these. We give below tables taken from Hartree's paper. They were calculated in 1924 and could be made more accurate if revised now in the light of later evidence. But they are amply accurate enough for the purpose for which they are required. Those for oxygen and iron have been so revised in parts.

§ **14·3.** *The partition function for a single electron, bound or free, in the presence of a nucleus.* It remains to show that the weights assigned in the preceding section are consistent with the limiting principle so that the partition function

$$\Sigma_s \varpi_s e^{-\chi_s/kT}$$

passes over continuously into the classical form

$$\frac{2^{q'}}{h^{3q'}}\int e^{-\chi/kT}\,dp_1\ldots dq_{3q'}$$

for large quantum numbers. Since the χ's concerned are all small and tend to zero for large quantum numbers like $1/n^2$, this reduces to showing that

$$\varpi_s \sim \frac{2^{q'}}{h^{3q'}}\int dp_1\ldots dq_{3q'}, \qquad \text{......(1632)}$$

when the integral is extended over the proper region of phase space. The factor $2^{q'}$ allows for the two orientations of each electron. In the rest of the calculation the electron can be treated as structureless. We shall start with a detailed analysis of the limiting form of the partition function for a single electron bound or free in the presence of a fixed nucleus of charge $Z\epsilon$, which we require in full later in the chapter.*

Consider for simplicity a volume V in the form of a sphere of radius A with the nucleus fixed at its centre. The classical partition function for a single movable electron is, in polar coordinates,

$$f(T)=\frac{2}{h^3}\iiiint e^{-\chi/kT}\,p^2dp\,d\Omega_p\,r^2dr\,d\Omega_r, \qquad \text{......(1633)}$$

where

$$\chi=\frac{1}{2m}p^2-\frac{Z\epsilon^2}{r}. \qquad \text{......(1634)}$$

* Planck, *Ann. d. Physik*, vol. 75, p. 673 (1924).

The elements of solid angle $d\Omega_p$ and $d\Omega_r$ define the directions of the momentum and position vectors respectively. Thus $f(T)$ can be written

$$f(T)=2\frac{8\pi^2(2m)^{\frac{3}{2}}}{h^3}\iint e^{-\chi/kT}(\chi r^2+Z\epsilon^2 r)^{\frac{1}{2}}\,r\,dr\,d\chi. \quad \text{......(1635)}$$

We will suppose that A is so large that a χ' can be chosen so that

$$Z\epsilon^2/A \ll \chi' \ll kT. \quad \text{......(1636)}$$

This requirement is usually satisfied in practice with an ample margin. Then the contributions to $f(T)$ can be divided into three parts:

(1) $f_1(T)$. $\infty>\chi>0$. Electron Free. Classical.

(2) $f_2(T)$. $0>\chi>-\chi'$. Electron Bound. Effectively Classical.*

(3) $f_3(T)$. $-\chi'>\chi>-\infty$. Electron Bound. Quantized.

In (3) the integral form of the partition function must of course be replaced by the usual sum over the possible stationary states.

Case (1). In $f_1(T)$ we have $\chi \gg Z\epsilon^2/r$ over practically the whole of the effective domain of integration. Hence we replace the factor $(\chi r^2+Z\epsilon^2 r)^{\frac{1}{2}}$ by $\chi^{\frac{1}{2}}r$ and find

$$f_1(T)=2\frac{(2\pi mkT)^{\frac{3}{2}}V}{h^3},$$

the usual formula for the partition function of a free electron. Corrections for the neglect of $Z\epsilon^2/r$ in $f_1(T)$ will be made later by applying Debye and Hückel's theory of ionized media.

Case (2). In $f_2(T)$ we have effectively $\chi/kT=0$. Putting $\alpha=-\chi$ we find

$$f_2(T)=2\frac{8\pi^2(2m)^{\frac{3}{2}}}{h^3}\int_0^{\chi'}d\alpha\int_0^{a}(Z\epsilon^2 r-\alpha r^2)^{\frac{1}{2}}\,r\,dr, \quad \text{......(1637)}$$

where a is the *smaller* of $Z\epsilon^2/\alpha$ and A.

Case (3). In $f_3(T)$ we have to replace the integral by the quantized sum

$$f_3(T)=\sum_1^{n'} 2n^2 e^{\chi_n/kT} \quad (\chi_n=RhcZ^2/n^2), \quad \text{......(1638)}$$

and

$$n'=\{RhcZ^2/\chi'\}^{\frac{1}{2}}. \quad \text{......(1639)}$$

We now return to evaluate $f_2(T)$, putting $r=(Z\epsilon^2/\alpha)\sin^2\phi$. Then

$$f_2(T)=2\frac{16\pi^2(2m)^{\frac{3}{2}}}{h^3}(Z\epsilon^2)^3\int_0^{\chi'}\frac{d\alpha}{\alpha^{\frac{5}{2}}}\int_0^{a}\sin^4\phi\cos^2\phi\,d\phi,$$

where a is $\arcsin(\alpha A/Z\epsilon^2)^{\frac{1}{2}}$ if this is real or else $\frac{1}{2}\pi$. The double integral therefore divides into

$$\int_0^{Z\epsilon^2/A}\frac{d\alpha}{\alpha^{\frac{5}{2}}}\int_0^{\arcsin(\alpha A/Z\epsilon^2)^{\frac{1}{2}}}\sin^4\phi\cos^2\phi\,d\phi+\int_{Z\epsilon^2/A}^{\chi'}\frac{\pi}{32}\frac{d\alpha}{\alpha^{\frac{5}{2}}},$$

or

$$\frac{2}{(Z\epsilon^2/A)^{\frac{3}{2}}}\int_0^{\frac{1}{2}\pi}\frac{\cos\theta\,d\theta}{\sin^4\theta}\int_0^{\theta}\sin^4\phi\cos^2\phi\,d\phi+\frac{\pi}{32}\int_{Z\epsilon^2/A}^{\chi'}\frac{d\alpha}{\alpha^{\frac{5}{2}}}.$$

* If the limiting principle is satisfied, as we shall shortly verify.

The repeated integral can be evaluated by integration by parts and is found to have the value $\frac{1}{9}-\frac{1}{96}\pi$. Thus

$$f_2(T)=2\frac{16\pi^2(2m)^{\frac{3}{2}}(Z\epsilon^2)^3}{h^3}\left[\frac{2}{9}\frac{A^{\frac{3}{2}}}{(Z\epsilon^2)^{\frac{3}{2}}}-\frac{\pi}{48}\frac{1}{(\chi')^{\frac{3}{2}}}\right],$$

$$=2Z^3(Rhc)^{\frac{3}{2}}\left[\frac{32}{9\pi}\frac{A^{\frac{3}{2}}}{(Z\epsilon^2)^{\frac{3}{2}}}-\frac{1}{3}\frac{1}{(\chi')^{\frac{3}{2}}}\right], \qquad \text{......(1640)}$$

$$=2\left\{\frac{32}{9\pi}\frac{(ZRhcA)^{\frac{3}{2}}}{\epsilon^3}-\tfrac{1}{3}n'^3\right\}. \qquad \text{......(1641)}$$

We can now see at once that the limiting principle is obeyed for an atom with a single excited electron. For the contribution to the phase integral, corresponding, according to (1639), to energies χ' between $n'\pm\frac{1}{2}$, say, is by (1641) the difference of the values of $f_2(T)$ for $n'\pm\frac{1}{2}$ or

$$\tfrac{2}{3}\{(n'+\tfrac{1}{2})^3-(n'-\tfrac{1}{2})^3\}$$

which is asymptotically $2n'^2$. This result can be extended at once to the case of a number of electrons each independently in specified orbits of large quantum number $n_1, \ldots, n_q$. The foregoing analysis applies formally to each electron if Z denotes the proper effective nuclear charge. Since Z disappears from the result, its actual specification is unnecessary. The corresponding phase space is asymptotically $2^q n_1^2 \ldots n_q^2$.

§ **14·4.** *The approximate characteristic function. The method of excluded volumes.* A direct and accurate evaluation of Ψ for the assemblies contemplated in this chapter would be an affair of some difficulty. There are two methods possible for approximations due in essentials to Urey and Planck which are subject to quite different adverse criticisms. The fact that they confirm each other qualitatively and even roughly quantitatively can be regarded as some justification for a belief that the resulting formulae for Ψ/k are a fair approximation to the truth.

In the theory of Urey and Fermi* we treat the various atoms and atomic ions as possessing an actual volume from which they entirely exclude other systems, as in van der Waals' elementary theory of an imperfect gas. We use the formulae of § 8·6 and assume that the ionic volume in an excited state has a radius of the order of the diameter of the central orbit described (after Bohr's theory) by the most highly excited electron. The resulting excluded volumes are therefore really fictitious. Physical reality can only be ascribed to them by the somewhat doubtful argument that they represent that region of space which must be empty for the ion in question to exist in that state at all. To the expression for Ψ/k so obtained we add a correction Ψ_ϵ/k for the till then neglected electrostatic charges.

* Urey, *Astrophys. J.* vol. 49, p. 1 (1924); Fermi, *Zeit. f. Physik*, vol. 26, p. 54 (1924).

In the theory of Planck we proceed *initially* more logically by trying to generalize the calculations of the preceding section into a simplified calculation of $B(T)$. But the simplifications which have to be made are rather severe, and it is satisfactory that the form of the result is checked by the other method using an entirely different type of approximation.

The Ψ/k for the theory of Urey and Fermi, omitting the electrostatic term, has already been given in equation (772). Let us denote the atomic ion of general type (r,s,z) by the suffixes α or β for shortness, and use the suffix ϵ for quantities characteristic of the electron. Σ_β will then mean a summation over all atomic types, and $\Sigma_{\alpha,\beta}$ a summation over all pairs of atomic types; free electrons are excluded from either summation. Let the average excluded volume of the (r,s,z)-ion for interaction with an electron be $(v_r^z)_{s,\epsilon}$, and for interaction with an atomic ion β, $(v_r^z)_{s,\beta}$. Then in the notation of this chapter we have

$$\Psi/k = N\left(\log\frac{VG}{N}+1\right)+\Sigma_{r,z}M_r^z\left(\log\frac{VF_r^z}{M_r^z}+1\right)+\frac{1}{V}\Sigma_{\alpha,\beta}\frac{M_\alpha M_\beta v_{\alpha\beta}}{\sigma_{\alpha\beta}}, \quad \text{...(1642)}$$

in which

$$F_r^z = \frac{(2\pi m^z kT)^{\frac{3}{2}}}{h^3}u_r^z(T)\,e^{(\chi_r^z+\ldots+\chi^z_{z-1})/kT}, \quad G = 2\,\frac{(2\pi mkT)^{\frac{3}{2}}}{h^3}, \quad \text{...(1643)}$$

$$u_r^z(T) = \Sigma_s(\varpi_r^z)_s\exp[-\{\chi_r^z-(\chi_r^z)_s\}/kT-\{N(v_r^z)_{s,\epsilon}+\Sigma_\beta M_\beta(v_r^z)_{s,\beta}\}/V]. \quad \text{......(1644)}$$

In transforming Ψ/k we have left unmodified the partition function for the free electron. The excluded volume corrections, when not small, are only qualitatively correct. To determine $(M_r^z)_s$ we have

$$\frac{(M_r^z)_s}{(\varpi_r^z)_s\exp[-\{\chi_r^z-(\chi_r^z)_s\}/kT-\{N(v_r^z)_{s,\epsilon}+\Sigma_\beta M_\beta(v_r^z)_{s,\beta}\}/V]} = \frac{M_r^z}{u_r^z(T)}. \quad \text{......(1645)}$$

To the Ψ/k of (1642) we must add the contributions of the radiation in the enclosure, and of the electrostatic potentials. The former is properly additive, the latter is not, and if the excluded volumes had a genuine physical existence, the electrostatic and excluded volume effects would interact and ought to be introduced together. Since however we can only aim at qualitative correctness here, we shall be content with the rough approximation of adding a separate electrostatic term. If we could adopt the approximations of the theory of Debye and Hückel, the arguments of § 13·61 would give us here

$$\Psi_\epsilon/k = \frac{2\sqrt{\pi}\,\epsilon^3}{3V^{\frac{1}{2}}(kT)^{\frac{1}{2}}}\{N+\Sigma_{r,z}r^2M_r^z\}^{\frac{3}{2}}. \quad \text{......(1646)}$$

In calculating this we have assumed that any ion of type (r,s,z) can be treated as a point charge of charge $r\epsilon$.

The expression (1646) for the electrostatic correction is of course far from accurate. Since the ions are all small and the positive ions of high valency none of the conditions of validity of Debye and Hückel's theory are satisfied. Moreover we have ignored the correction of § 2·64 necessary to distinguish between free and bound electrons. It is in fact true that the Poisson-Boltzmann equation (1566) can be altered to take account of § 2·64, and Gronwall's method can then be applied to it.* The solution is more difficult than that of § 13·7, and raises interesting points in mathematics and of physical interpretation. No such correction of the Debye-Hückel theory can however produce a formula applicable to stellar conditions even for the most massive star, since the fundamental internal inconsistencies of the theory for unsymmetrical ions are left uncorrected.

Formula (1646) owing to its compact and explicit form remains however the most convenient way of estimating rapidly the effect of electrostatic forces on the compressibility and other properties of a sample of stellar material. In using it it is essential to have some check on the probable size and sign of its errors. Such controls have been provided in particular cases by the investigations of Eddington† who concludes that (1646) *overestimates* the electrostatic correction by a factor of the order of *three* for the conditions typical of stars of mass about equal to that of the sun.

In the approximations of Urey and Fermi the excluded volumes are treated as spherical and the radius of the interacting system β is neglected compared with the radius of the (r,s,z)-system itself. This introduces no error at the moderate or low temperatures which they themselves discuss where highly excited orbits are rare. The radius of the (r,s,z)-system is assumed to be the semi-diameter of the orbit of the most highly excited electron or more strictly the aphelion distance when the orbit is not closed. As a result we obtain from (1644) for hydrogen

$$u_0^1(T) = \sum_{s=1}^{\infty} 2s^2 \exp\left\{-\frac{\chi_0^1}{kT}\left(1-\frac{1}{s^2}\right) - \alpha s^6\right\}, \qquad \text{......(1647)}$$

where
$$\alpha = \frac{4\pi}{3}\frac{N+\Sigma_\beta M_\beta}{V}a^3 = \frac{4\pi}{3}\frac{p_\epsilon + p_a}{kT}a^3. \qquad \text{......(1648)}$$

In (1648) a is the radius of the 1-quantum orbit in hydrogen, $5{\cdot}34\times10^{-9}$ cm., and χ_0^1 in (1647) its ionization energy, 13·54 volts. More generally at higher temperatures a better approximation to (1644) is obviously provided by replacing s^6 by $(s_0^2+s^2)^3$, where as_0^2 is the average radius of all the interacting systems. This improvement will not affect orders of magnitude and we shall not investigate it in detail.

* Unpublished work by Gaunt. A summary is given in *Monthly Not. R.A.S.* vol. 88, p. 369 (1928).
† Eddington, *Monthly Not. R.A.S.* vol. 86, p. 1 (1925), vol. 88, p. 352 (1928).

To formulate this theory quantitatively for other atoms with one excited electron we obviously replace αs^6 by $\alpha n_s^6/(r+1)^3$, where n_s is the effective quantum number and $(r+1)$ the core charge. Since the excluded volumes are only of importance for states of great excitation, it will be sufficiently accurate to group all states of given principal quantum number together with the approximate excluded volume $\alpha s^6/(r+1)^3$ and the weight $2s^2\varpi_c$. For an atom with any number of excited electrons we must presumably replace n_s by $(n_s)_{\max}$, denoting thereby the greatest effective quantum number among the excited electrons in the sth state. In general we shall be able to arrange the states into series in which the quantum defect is roughly constant and n_s increases by unity from term to term. These quantum defects will vary considerably with the k of the greatest orbit and with variations in any of the quantum numbers of the other excited orbits, but as a first approximation it will be legitimate to ignore these variations and group together all the terms in $u(T)$ which have a given principal quantum number for the greatest orbit. We can then write

$$u_r^z(T) = \Sigma_s \{\Sigma_t' (\varpi_r^s)_t e^{-\{\chi_r^z - (\chi_r^s)_t\}/kT}\} e^{-\alpha n_s^6/(r+1)^3}, \quad \text{......(1649)}$$

where Σ' is summed over all states in which the principal quantum numbers of every electron are less than or equal to s and one at least is equal to s.

To elucidate (1649) further we must group together the terms that belong to given numbers of highly excited electrons; let there be q of these and $z-r$ electrons in all. Then for these states we can ignore the variations of $(\chi_r^s)_t$ and take $\chi_r^s - (\chi_r^s)_t$ to be effectively

$$\chi_r^s + \ldots + \chi^s_{r+q-1},$$

the energy required to remove entirely the q highly excited electrons from their normal orbits. These states contribute to $u_r^z(T)$

$$e^{-(\chi_r^s + \ldots + \chi^s_{r+q-1})/kT} \Sigma_s \{\Sigma_t' (\varpi_r^s)_t\} e^{-\alpha n_s^6/(r+1)^3}.$$

If we ignore reductions in the ϖ's for equivalent orbits, which do not affect the terms of highest order, we find from the rules that, summed over all states with the q principal quantum numbers of the excited electrons less than or equal to s,

$$\Sigma_t (\varpi_r^z)_t = \varpi^s_{r+q} 2^q \sum_{t_1, t_2 \ldots \leqslant s} t_1^2 t_2^2 \ldots t_q^2,$$

$$\sim \varpi^s_{r+q} \frac{2^q s^{3q}}{3^q}.$$

Therefore, by differentiation,

$$\Sigma_t' (\varpi_r^z)_t \sim \frac{3q\varpi^s_{r+q} 2^q}{3^q} s^{3q-1}.$$

These states therefore contribute to $u_r^z(T)$

$$\frac{3q\varpi^s_{r+q} 2^q}{3^q} e^{-(\chi_r^s + \ldots + \chi^s_{r+q-1})/kT} \Sigma_s s^{3q-1} e^{-\alpha n_s^6/(r+1)^3}.$$

In this summation we shall for this approximation omit the quantum defect and replace the sum by the integral

$$\int_0^\infty s^{3q-1}\, e^{-\alpha s^6/(r+1)^3}\, ds,$$

or

$$\tfrac{1}{6}\Gamma(\tfrac{1}{2}q)\left\{\frac{(r+1)^3}{\alpha}\right\}^{\frac{1}{2}q}.$$

The contribution to $u_r^z(T)$ is therefore

$$\Gamma(1+\tfrac{1}{2}q)\,\varpi^z_{r+q}\,2^q\left\{\frac{(r+1)^3}{9\alpha}\right\}^{\frac{1}{2}q} e^{-(\chi_r^z+\ldots+\chi^z_{r+q-1})/kT}. \quad \ldots\ldots(1650)$$

There is a similar contribution for every possible value of q. The complete result may be written

$$u_r^z(T)=\varpi_r^z+\sum_{q=1}^{q=z-r}\Gamma(1+\tfrac{1}{2}q)\,\varpi^z_{r+q}\,2^q\left\{\frac{(r+1)^3}{9\alpha}\right\}^{\frac{1}{2}q} e^{-(\chi_r^z+\ldots+\chi^z_{r+q-1})/kT}. \quad \ldots\ldots(1651)$$

It will often be found that one term in (1651) is dominant for given values of the density and temperature. In such a case nearly all the atoms M_r^z present will have just q excited electrons and the rest $z-r-q$ in normal orbits. If there is at most a single excited electron

$$u_r^z(T)=\varpi_r^z+\pi^{\frac{1}{2}}\varpi^z_{r+1}\left\{\frac{(r+1)^3}{9\alpha}\right\}^{\frac{1}{2}} e^{-\chi_r^z/kT}. \quad \ldots\ldots(1652)$$

This is the simplest generalization of the theory of Urey and Fermi. These authors do not give these approximate summations, whose accuracy is ample for most applications.

It is naturally possible to calculate average values derived from $u_r^z(T)$ by the same process. The average energy content of any atom is given by

$$kT^2\frac{\partial}{\partial T}\log u_r^z(T),$$

in which of course only the exponents $\{\chi_r^z-(\chi_r^z)_s\}/kT$ are to be differentiated. We find

$$kT^2\frac{\partial u_r^z(T)}{\partial T}=\sum_{q=1}^{q=z-r}\{\chi_r^z+\ldots+\chi^z_{r+q-1}\}\,\Gamma(1+\tfrac{1}{2}q)\,\varpi^z_{r+q}\,2^q\left\{\frac{(r+1)^3}{9\alpha}\right\}^{\frac{1}{2}q}$$
$$\times\, e^{-(\chi_r^z+\ldots+\chi^z_{r+q-1})/kT}. \quad \ldots\ldots(1653)$$

When the term $q=q^*$ is predominant in (1651) and so also in (1653) we have

$$kT^2\frac{\partial}{\partial T}\log u_r^z(T)=\chi_r^z+\ldots+\chi^z_{r+q^*-1}. \quad \ldots\ldots(1654)$$

The energy content is the same as if these q^* electrons were *free and at rest relative to the ion.*

Other expressions which occur in the general formulae are

$$\frac{1}{V}\Sigma_\beta M_\beta v_{\beta,\epsilon}, \quad \frac{1}{V^2}\left[N\Sigma_\beta M_\beta v_{\beta,\epsilon}+\Sigma_{\alpha,\beta}\frac{M_\alpha M_\beta v_{\alpha,\beta}}{\sigma_{\alpha\beta}}\right].$$

These may be evaluated as

$$-\Sigma_{r,z}M_r^z\frac{\partial}{\partial N}\log u_r^z(T), \quad \tfrac{1}{2}\Sigma_{r,z}M_r^z\left(\frac{\partial}{\partial V}-\frac{N}{V}\frac{\partial}{\partial N}\right)\log u_r^z(T)$$

respectively.

The accuracy required in these formulae is reached by retaining only the dominant term $q=q^*$ in (1651). To this approximation

$$\frac{\partial}{\partial N}\log u_r^z(T)=-\tfrac{1}{2}q^*\frac{\partial}{\partial N}\log\alpha=-\frac{\tfrac{1}{2}q^*}{N+\Sigma_\beta M_\beta}, \qquad \text{......(1655)}$$

$$\left\{\frac{\partial}{\partial V}-\frac{N}{V}\frac{\partial}{\partial N}\right\}\log u_r^z(T)=\frac{\tfrac{1}{2}q^*}{V}\frac{2N+\Sigma_\beta M_\beta}{N+\Sigma_\beta M_\beta}. \qquad \text{......(1656)}$$

In general q^*, the average number of bound but highly excited electrons, will vary with r and z and must be written $(q_r^z)^*$. Then

$$\frac{1}{V}\Sigma_\beta M_\beta v_{\beta,\epsilon}=\frac{1}{2}\frac{\Sigma_{r,z}(q_r^z)^* M_r^z}{N+\Sigma_\beta M_\beta}, \qquad \text{......(1657)}$$

$$\frac{1}{V^2}\left[N\Sigma_\beta M_\beta v_{\beta,\epsilon}+\Sigma_{\alpha,\beta}\frac{M_\alpha M_\beta v_{\alpha,\beta}}{\sigma_{\alpha\beta}}\right]=\frac{\Sigma_{r,z}(q_r^z)^*M_r^z}{2V}\frac{N+\tfrac{1}{2}\Sigma_\beta M_\beta}{N+\Sigma_\beta M_\beta}. \qquad \text{......(1658)}$$

§ **14·5.** *Planck's theory.* In the foregoing theory there is an arbitrary element, the choice of excluded volume. This is physically and theoretically unsatisfactory. Instead of attempting to patch up the theory of imperfect gases in this way so as to apply to assemblies of atoms, ions and electrons, Planck has attempted to make a direct simplified calculation of Gibbs' phase integral, and so $B(T)$, for such an assembly by generalizing the calculation of § 14·3.

For an assembly consisting of one electron and one fixed nucleus ($Z\epsilon$) contained in a sphere of radius A about the nucleus, the calculations already given are almost complete. For the electron free we have

$$f(T)=f_1(T)=2\frac{(2\pi mkT)^{\frac{3}{2}}V}{h^3}.$$

For the electron bound

$$\begin{aligned}f'(T)&=f_2(T)+f_3(T),\\ &=\sum_{n=1}^{n'}2n^2e^{\chi_n/kT}+2\left\{\frac{32}{9\pi}\frac{(ZRhcA)^{\frac{3}{2}}}{\epsilon^3}-\tfrac{1}{3}n'^3\right\}.\end{aligned}$$

Now by familiar arguments we have approximately

$$\sum_{n=1}^{n'}2n^2e^{\chi_n/kT}=2\{e^{\chi_1/kT}+\tfrac{1}{3}n'^3\}.$$

Hence
$$f'(T) = 2e^{\chi_1/kT} + \frac{64}{9\pi}\frac{(ZRhcA)^{\frac{3}{2}}}{\epsilon^3}. \qquad \text{......(1659)}$$

In constructing partition functions hitherto we have found it convenient to take the normal state of lowest energy as standard. With this convention
$$f'(T) = 2 + \frac{64}{9\pi}\frac{(ZRhcA)^{\frac{3}{2}}}{\epsilon^3}e^{-\chi_1/kT}. \qquad \text{......(1660)}$$

In all such approximations it may be necessary to include a finite number of terms besides the first term for the normal state before the terms of high excitation are lumped together with the remainder. Thus the complete form of $f'(T)$ would be
$$f'(T) = \sum_1^s \varpi_s e^{-(\chi_1-\chi_s)/kT} + \frac{64}{9\pi}\frac{(ZRhcA)^{\frac{3}{2}}}{\epsilon^3}e^{-\chi_1/kT}. \qquad \text{......(1661)}$$
In general either the first term ($s=1$) or the remainder is dominant (the latter for large values of T). When however the remainder is negligible, it is often necessary in discussing fine points, such as the appearance in absorption of lines arising from the first few excited states, to include the first few terms in the former part of $f'(T)$. Only the first term contributes sensibly to the numerical value of $f'(T)$ itself, but the other terms control the distribution of the small fraction of atoms in the early excited states on which the phenomena in question depend.

When the assembly of volume V contains a number of fixed nuclei and one electron, $f(T)$ is still given by (1633) with
$$\chi = \frac{1}{2m}p^2 - \Sigma_\alpha \frac{Z\epsilon^2}{r_\alpha}.$$
It is however now impossible to carry out the exact integration, and the most reasonable simplifying assumption is that when $\chi < 0$ all terms in Σ_α are negligible except the largest. This means that we treat as free all states of the electron in which its energy is greater than the energy of escape from the nearest nucleus when all nuclei have the same charge, and is in a sense equivalent to ignoring molecule formation. This simplification should be reliable so long as the nuclei are not too close together. Each nucleus then makes a contribution to the phase integral like (1635) and therefore like (1660), which may be described by saying that each nucleus has this partition function for a bound electron, if A is so chosen that on the average the bound electron is nearer to the selected nucleus than to any other. If there are M nuclei we must therefore take
$$M\,.\,\tfrac{4}{3}\pi A^3 = V.$$
The essential part of the condition (1636) is then that
$$Z\epsilon^2\left(\frac{M}{V}\right)^{\frac{1}{3}} \ll kT. \qquad \text{......(1662)}$$

The method can now be generalized for a number of nuclei of different positive charges. The condition that the electron should be bound to nucleus 1 rather than to nucleus 2 is now naturally

$$Z_1/r_1 > Z_2/r_2.$$

This means that to each nucleus $(Z\epsilon)$ we must attach an average volume proportional to Z^3. If we now define a radius A by the equation

$$\tfrac{4}{3}\pi A^3 \Sigma_z Z^3 M^z = V, \qquad \ldots\ldots(1663)$$

then the actual radius A_1 for use with a nucleus $(Z_1\epsilon)$ will be given by

$$A_1{}^3 = Z_1{}^3 A^3 = \frac{3}{4\pi}\frac{Z_1{}^3 V}{\Sigma_z Z^3 M^z}. \qquad \ldots\ldots(1664)$$

The partition function for an electron bound to a nucleus $(Z_1\epsilon)$ may therefore be written

$$2 + e^{-\chi_1/kT}\left[\frac{64}{9\pi}\left(\frac{3}{4\pi}\right)^{\frac{1}{2}}\frac{(Rhc)^{\frac{3}{2}}}{\epsilon^3}\right]\frac{Z_1{}^3 V^{\frac{1}{2}}}{(\Sigma_z Z^3 M^z)^{\frac{1}{2}}}. \qquad \ldots\ldots(1665)$$

The permanent constant in [ ] will be written Q; it has the numerical value

$$Q = 1{\cdot}017 \times 10^{12}\,\text{cm.}^{-\frac{3}{2}}.$$

Let us now allow the nuclei their natural freedom of movement. Their mean velocities are very slow compared with the velocities of the electrons, and to a first rough approximation the rule for fixing the particular nucleus to which an electron with negative energy is bound is not affected. It is of course strictly the relative kinetic energy of electron and nucleus which should be used in the binding rule, and we replace this by the kinetic energy of the electron (relative to the centre of mass of the assembly as a whole). A closer approximation here would be of interest. To the approximation at present proposed the foregoing formulae can be regarded as independent of the motion of the nuclei. The calculations of the phase integral of the assembly for the nuclei therefore take their ordinary form and yield the ordinary partition functions for massive particles; the electrostatic forces between the nuclei themselves lead to no complications (being repulsive), and are (or should be) allowed for in the term (1646).

Let us next suppose that there is more than one electron in the assembly, but that no nucleus can catch more than one electron. We can ignore also the repulsive force between the electrons. Then the foregoing analysis suffices to determine the partition function of any nucleus which has caught an electron, and the whole value of $B(T)$ can best be evaluated by the usual combinatory rules of § 5·2. In our notation therefore

$$u^{z_1}{}_{z_1-1}(T) = 2 + Q\,\frac{Z_1{}^3 V^{\frac{1}{2}}}{(\Sigma_z Z^3 M^z)^{\frac{1}{2}}}\exp(-\chi^{z_1}{}_{z_1-1}/kT), \qquad \ldots\ldots(1666)$$

and using (1666) we can apply all the usual formulae.

To extend these arguments to the capture of more than one electron is not difficult, provided rather rough approximations are sufficient. We shall obviously approximate fairly closely to the holding power of an atom r-times ionized by assuming that it holds excited electrons like a point charge $(r+1)\epsilon$. This approximation will be very good when all the electrons except one are in normal or nearly normal orbits. In general it must *underestimate* the efficiency of the ion at holding its last electron. An *overestimate* of the efficiency can be obtained by assuming that the ion holds like a point charge $(r+q)\epsilon$, where q is the total number of its electrons in highly excited orbits.

In order to estimate the value of $u_r^z(T)$ we again consider separately the parts arising for various specified numbers q of highly excited electrons. The weight of the remaining core of the atom with electrons in normal orbits is ϖ^z_{r+q}. This core now catches q electrons in succession into highly excited orbits acting on one assumption like a point charge $(r+1)\epsilon$ and on the other like $(r+q)\epsilon$. The contribution to $u_r^z(T)$, estimated according to the foregoing version of Planck's theory, will therefore be the continued product of contributions for each electron in order with an extra factor ϖ^z_{r+q} for the weight of the core. According therefore to the holding power assumed we find the following limiting approximations for $u_r^z(T)$:

$$u_r^z(T) = \varpi_r^z + \sum_{q=1}^{q=z-r} \varpi^z_{r+q}(r+1)^{3q} Q^q \left\{\frac{V}{\Sigma_{r,z}(r+1)^3 M_r^z}\right\}^{\frac{1}{2}q} \times e^{-(\chi_r^z + \dots + \chi^z_{r+q-1})/kT}, \qquad \dots\dots(1667)$$

or

$$u_r^z(T) = \varpi_r^z + \sum_{q=1}^{q=z-r} \varpi^z_{r+q}(r+q)^{3q} Q^q \left\{\frac{V}{\Sigma_{r,z,q}(r+q)^3 (M_r^z)_q}\right\}^{\frac{1}{2}q} \times e^{-(\chi_r^z + \dots + \chi^z_{r+q-1})/kT}. \qquad \dots\dots(1668)$$

In (1668) $(M_r^z)_q$ denotes the average number of atomic ions of atomic number Z, r-times ionized, with q highly excited electrons. These equations give upper and lower limits for $u_r^z(T)$. A closer approximation than either can probably be obtained by considering an $(r+q)$-times ionized atom and letting it catch q electrons in succession into excited orbits. If we assume that the number of orbits so obtained is not altered by later captures and that at each stage the ion captures like a point ion of the new net charge, we can replace $(r+q)^q$ in (1668) by $(r+1)\dots(r+q)$ and obtain

$$u_r^z(T) = \varpi_r^z + \sum_{q=1}^{q=z-r} \varpi^z_{r+q}(r+1)^3 \dots (r+q)^3 Q^q \times \left\{\frac{V}{\Sigma_{r,z,q}(r+1)^{3/q}\dots(r+q)^{3/q}(M_r^z)_q}\right\}^{\frac{1}{2}q} e^{-(\chi_r^z + \dots + \chi^z_{r+q-1})/kT}. \qquad \dots\dots(1669)$$

If no other considerations entered one should prefer (1669) to (1667) or (1668) in applications. The clustering of the *free* electrons round the positive ion will however decrease its holding power for highly excited electrons, and terms of large q are mainly important for large N/V when this shielding is largest. When as here this shielding is not directly allowed for, formula (1667) is to be preferred. It is undoubtedly at this point, that is in the correct enumeration of bound states, that the present theory is weakest, and a better method of enumeration is greatly to be desired.

These $u_r^z(T)$ may be used for a direct construction of Ψ, which is formally the Ψ for a mixture of perfect gases with radiation and the electrostatic terms added. There are in this theory as here developed no excluded volumes. The residual atomic cores of electrons in normal orbits do possess volumes which can be taken account of in the usual way. But the effect of these is usually extremely small.

In applications the most important combination is $u^z_{r+1}(T)/u_r^z(T)$. It sometimes happens that one term of (1667) is dominant for a given density and temperature, and then the dominant terms of $u^z_{r+1}(T)$, $u_r^z(T)$ generally correspond to equal numbers of electrons in normal orbits, and a difference of one highly excited electron. In such a case

$$\frac{u^z_{r+1}(T)}{u_r^z(T)} = \frac{e^{\chi_r^z/kT}}{(r+1)^3 Q}\left\{\frac{V}{\Sigma_{r,z}(r+1)^3 M_r^z}\right\}^{-\frac{1}{2}}. \quad \text{......(1670)}$$

If s is the average number of free electrons per atom, then it is defined by

$$\Sigma_{r,z} r M_r^z = s\Sigma_{r,z} M_r^z = N.$$

It will then be permissible as a very rough approximation to write

$$\Sigma_{r,z}(r+1)^3 M_r^z = (s+1)^3 \Sigma_{r,z} M_r^z = \frac{(s+1)^3}{s} N,$$

and so

$$\frac{u^z_{r+1}(T)}{u_r^z(T)} = \frac{\{(s+1)^3/s\}^{\frac{1}{2}} e^{\chi_r^z/kT}}{(r+1)^3 Q(V/N)^{\frac{1}{2}}}. \quad \text{......(1671)}$$

When we now form the equations of dissociative equilibrium the exponential factors vanish. The ratios M^z_{r+1}/M_r^z are controlled entirely by the $(r+1)^3$ factor. The χ_r^z only control the numbers of electrons in normal orbits.

As a check on the somewhat speculative approximations which have been necessary at certain stages in this chapter, it is well to compare the formulae for $u_r^z(T)$ given by the two theories in the simplest case of a single excited electron. When the permanent constants are given their numerical values we find:

For the theory of Urey and Fermi,

$$u_r^z(T) = \varpi_r^z + 7{\cdot}40 \times 10^{11} \varpi^z_{r+1}(r+1)^{\frac{3}{2}} \left(\frac{s}{s+1}\right)^{\frac{1}{2}} \left(\frac{V}{N}\right)^{\frac{1}{2}} e^{-\chi_r^z/kT}.$$

For the theory of Planck,

$$u_r^z(T) = \varpi_r^z + 1{\cdot}017 \times 10^{12} \varpi^z_{r+1} \frac{(r+1)^3}{s+1} \left(\frac{V}{N}\right)^{\frac{1}{2}} e^{-\chi_r^z/kT}.$$

These formulae are in substantial agreement.

The following chapters consist largely of applications of these formulae.

CHAPTER XV

ATMOSPHERIC PROBLEMS*

§15·1. *Scope of Chapters* XV *and* XVI. In this chapter and the following we shall set out to apply our general theorems to special problems of the properties of matter in a gaseous state which is nearly perfect. We shall of course discuss only such problems as arise out of the study of equilibrium states of such matter or as can be treated at once by application of the properties of the equilibrium state and the laws of mechanisms detailed in Chapters XVII and XIX. Problems essentially requiring the theory of transport phenomena or of radiative equilibrium are therefore excluded. The problems that present themselves are of two classes: (1) atmospheric problems, this chapter, and (2) problems of the interior of a gaseous star, Chapter XVI. The equilibrium and quasi-equilibrium properties of atmospheres—extensive assemblies of perfect gas constituents in a strong external field of force—consist only of the properties of assemblies which can be treated as isothermal. We have already derived Dalton's law for the distribution of the various constituents in an atmosphere of perfect gases. Until recently the only other problem discussed was the rate of escape of molecules into space from the boundary of the atmosphere of a planet or a star.† Thanks however largely to the work of Saha and Milne a number of other interesting atmospheric problems have been proposed and solved at least to a first approximation. The contents of this chapter consist therefore of discussions of the following problems: (1) The equilibrium of an isothermal *ionized* atmosphere and the permanent electrical fields and charges existing in it. (2) The behaviour of the absorption spectra formed by stellar reversing layers. (3) The normal escape of molecules from the atmosphere of the earth or a star. (4) The formation by radiation pressure of tenuous high-level atmospheres (chromospheres); (4) in outline only. Of these the first alone refers to a true equilibrium state. The second and third can properly be treated to a first approximation by using equilibrium properties. In the fourth equilibrium properties play a smaller part. It is however justifiable to give a sketch of them for the sake of a systematic account of this whole group of problems. An account of the third is rendered the more desirable because great advances in treatment have been made since Jeans' account was written, and the theory is now probably in a final form. When all is said however it is obvious that these researches merely deal with special features

* In the revision of Chapters XV and XVI I am greatly indebted to Mr S. Chandrasekhar, who has drafted all the new matter, and revised the old.

† Jeans, *loc. cit.* chap. XV.

of a stellar atmosphere. A proper theory embracing them all, based as it must be on a study of the assembly subject to the flux of radiation from below, would lie outside our range. Though no such theory has yet been completed, great advances have recently been made in the study of the absorption spectra formed by stellar reversing layers, and we shall summarize some of the results of the more complete theory which cannot itself be included.

In an atmospheric problem the main field of force may be regarded as external to and independent of the assembly discussed. In a star, considered as a whole, this is not so, and the main field arises from the gravitation of the matter of the assembly itself. In Chapter XVI therefore the start corresponding to that of Chapter XV would be a discussion of the isothermal gravitating gas sphere of ionized material. In the absence of electrical forces there is nothing to add to Emden.* We could make some comment following Rosseland† on the effect of electrical forces on the radial distribution of different elements. The applicability of these remarks to an actual (non-isothermal) star is however doubtful. It is obviously outside our province to attempt any discussion of the large scale interior constitution of a star, which depends on many other factors besides the properties of the equilibrium state of a given body of matter. For this the reader will naturally turn to Eddington.‡ But whatever the large scale structure of a star, the small scale structure is essentially that of matter in the most complete thermodynamic equilibrium.§ The equilibrium properties of stellar material are important, and the main part of Chapter XVI is therefore devoted to an attempt to calculate as accurately as possible the equilibrium state of matter at stellar temperatures and pressures.

§ **15·2**. *The equilibrium of an ionized atmosphere.*‖ We start by discussing an atmosphere of a single primary constituent, say Ca, in equilibrium with its ionization products Ca^+ and electrons. We later consider extensions to more complicated atmospheres, but these are not easy to make exactly. By (198) and (565) for the sth constituent

$$\overline{n_s} = (\overline{n_s})_0\, e^{-(m_s\phi+\epsilon_s\psi)/kT}. \qquad \ldots\ldots(1672)$$

It is necessary to include an electrostatic potential ψ as well as the gravitational ϕ, since Dalton's law entails a separation of the charges. By (791) for each constituent

$$\nabla^2(m_s\phi+\epsilon_s\psi) = 4\pi m_s G\{\Sigma_s m_s \overline{n_s}\} - 4\pi\epsilon_s\{\Sigma_s \epsilon_s \overline{n_s}\},$$

* Emden, *Gaskugeln*.

† Rosseland, *Monthly Not. R.A.S.* vol. 84, p. 729 (1924).

‡ Eddington, *The internal constitution of the stars* (1926); Jeans, *Astronomy and Cosmogony* (1928).

§ Eddington, *loc. cit.* p. 21.

‖ Milne, *Proc. Camb. Phil. Soc.* vol. 22, p. 493 (1925).

where G is the constant of gravitation. As these hold for all m_s and ϵ_s we must have

$$\nabla^2\phi = 4\pi G\Sigma_s m_s \overline{n_s}, \qquad \text{......(1673)}$$

$$\nabla^2\psi = -4\pi\Sigma_s \epsilon_s \overline{n_s}. \qquad \text{......(1674)}$$

These are the complete equations. Owing however to the smallness of Gm_s^2 compared with ϵ_s^2 we may usually neglect altogether the gravitational field of the atmosphere itself and replace (1673) by $\nabla^2\phi = 0$, so that $\phi = gz$ if the curvature of the atmosphere may be neglected, and otherwise $\phi = -GM/r$, where M is the mass of the central body and r the distance from its centre.

By (566) the ionization constant is constant in space, so that the condition of ionization equilibrium affects only the constants $(\overline{n_s})_0$. The whole equilibrium problem may therefore be solved without reference to this condition, provided the constants of integration $(\overline{n_s})_0$ are adjusted to satisfy it. If the suffixes 1, 2, 3 refer respectively to the neutral atom, the ion and the electron, we have therefore, curvature neglected,

$$\overline{n_1} = (\overline{n_1})_0\, e^{-m_1 gz/kT}, \qquad \text{......(1675)}$$

$$\overline{n_2} = (\overline{n_2})_0\, e^{-(m_2 gz + \epsilon\psi)/kT}, \qquad \text{......(1676)}$$

$$\overline{n_3} = (\overline{n_3})_0\, e^{-(m_3 gz - \epsilon\psi)/kT}, \qquad \text{......(1677)}$$

coupled with

$$\frac{d^2\psi}{dz^2} = -4\pi\epsilon(\overline{n_2} - \overline{n_3}). \qquad \text{......(1678)}$$

For an atmosphere stratified in planes ψ will be a function of z alone. We recall also that $m_1 = m_2 + m_3$ and m_3 is very small.

Let us write

$$\epsilon\psi' = -\tfrac{1}{2}(m_2 - m_3)\, g + kTf, \qquad \text{......(1679)}$$

denoting differentiations with respect to z by primes, and differentiate logarithmically (1676) and (1677). Then

$$\left.\begin{aligned} \frac{\overline{n_2}'}{\overline{n_2}} &= -\frac{\tfrac{1}{2}(m_2 + m_3)\, g}{kT} - f, \\ \frac{\overline{n_3}'}{\overline{n_3}} &= -\frac{\tfrac{1}{2}(m_2 + m_3)\, g}{kT} + f, \end{aligned}\right\} \qquad \text{......(1680)}$$

$$f' = \frac{4\pi\epsilon^2}{kT}(\overline{n_3} - \overline{n_2}). \qquad \text{......(1681)}$$

We can reintegrate (1680) in the form

$$\left.\begin{aligned} \overline{n_2} &= (\overline{n_2})_0 \exp\left(-\alpha z - \int_{z_0}^{z} f\, dz\right), \\ \overline{n_3} &= (\overline{n_3})_0 \exp\left(-\alpha z + \int_{z_0}^{z} f\, dz\right), \end{aligned}\right. \quad \left(\alpha = \frac{\tfrac{1}{2}(m_2 + m_3)\, g}{kT}\right). \qquad \text{...(1682)}$$

From equations (1681) and (1682) we can eliminate $\overline{n_2}$ and $\overline{n_3}$, and from the result eliminate $\int f dz$ by differentiation. We find

$$\left\{\frac{f''+\alpha f'}{f}\right\}^2 - f'^2 = 4\left(\frac{4\pi\epsilon^2}{kT}\right)^2 (\overline{n_2})_0 (\overline{n_3})_0 e^{-2\alpha z}. \qquad \ldots\ldots(1683)$$

With the help of this equation we may determine the behaviour of f and so the general characteristics of the atmosphere.

§ **15·21.** *Form of f for large positive z, at the outside boundary of the atmosphere.* If the resultant charge on the whole body and its atmosphere is zero, then at the outer boundary $\psi' = 0$ and

$$f \to \frac{\frac{1}{2}(m_2 - m_3) g}{kT} = f_\infty.$$

We search for an approximate solution of (1683) by replacing f by f_∞ and writing $f' = e^{-\alpha z}\xi$. Then ξ satisfies

$$\xi'^2 - f_\infty^2 \xi^2 = 4\left(\frac{4\pi\epsilon^2}{kT}\right)^2 (\overline{n_2})_0 (\overline{n_3})_0 f_\infty^2. \qquad \ldots\ldots(1684)$$

Therefore
$$\xi = \frac{4\pi\epsilon^2}{kT} \sqrt{(\overline{n_2}\,\overline{n_3})_0}\, \{\gamma \exp f_\infty z - \gamma^{-1} \exp(-f_\infty z)\},$$

of which the second term is negligible compared with the first; γ is a constant of integration. Returning to f' and integrating, we find

$$f = f_\infty - \frac{4\pi\epsilon^2}{kT} \sqrt{(\overline{n_2}\,\overline{n_3})_0}\, \gamma \frac{e^{-(\alpha - f_\infty)z}}{\alpha - f_\infty},$$

$$= \frac{\frac{1}{2}(m_2 - m_3) g}{kT} - \frac{4\pi\epsilon^2}{m_3 g} \sqrt{(\overline{n_2}\,\overline{n_3})_0}\, \gamma e^{-m_3 g z/kT}. \qquad \ldots\ldots(1685)$$

It is easy to show that the error in this equation is $O(e^{-2m_3 g z/kT})$. From this it follows that

$$\int^z f dz = \frac{\frac{1}{2}(m_2 - m_3) g z}{kT} + \frac{4\pi\epsilon^2 kT}{m_3^2 g^2} \sqrt{(\overline{n_2}\,\overline{n_3})_0}\, \gamma e^{-m_3 g z/kT} + \delta,$$

where δ is a constant of integration. Inserting this in (1682) we find

$$\overline{n_2} = \frac{(\overline{n_2})_0 e^{-m_3 g z/kT}}{D \exp\left[\dfrac{4\pi\epsilon^2 kT}{m_3^2 g^2} \sqrt{(\overline{n_2}\,\overline{n_3})_0}\, \gamma e^{-m_3 g z/kT}\right]}, \qquad \ldots\ldots(1686)$$

$$\overline{n_3} = (\overline{n_3})_0 e^{-m_3 g z/kT} D \exp\left[\frac{4\pi\epsilon^2 kT}{m_3^2 g^2} \sqrt{(\overline{n_2}\,\overline{n_3})_0}\, \gamma e^{-m_3 g z/kT}\right], \ldots(1687)$$

and in order to satisfy (1681) we must have $D = \gamma \sqrt{(\overline{n_2}/\overline{n_3})_0}$.

We can draw interesting deductions. Since the extra exponentials tend to unity as $z \to +\infty$, we have *ultimately*

$$\left.\begin{aligned}\overline{n_2} &\sim \sqrt{(\overline{n_2}\,\overline{n_3})_0}\,\gamma e^{-m_2 g z/kT},\\ \overline{n_3} &\sim \sqrt{(\overline{n_2}\,\overline{n_3})_0}\,\gamma e^{-m_3 g z/kT},\end{aligned}\right\} \quad (z \text{ very large}), \qquad \text{......(1688)}$$

which is Dalton's law, when the electrostatic forces have become trivial. The "when" is however instructive, for this occurs when

$$\frac{4\pi\epsilon^2 kT}{m_3{}^2 g^2}\sqrt{(\overline{n_2}\,\overline{n_3})_0}\,\gamma e^{-m_3 g z/kT}$$

is small compared with unity. On using (1688) we see that this condition is

$$\frac{4\pi\epsilon^2 kT\overline{n_3}}{m_3{}^2 g^2}, \quad \frac{4\pi\epsilon^2 p_3}{m_3{}^2 g^2} \ll 1. \qquad \text{......(1689)}$$

For electrons on the sun $4\pi\epsilon^2/m_3{}^2 g^2$ is $4{\cdot}85 \times 10^{27}$. Thus the partial pressure of the electrons must be well below 10^{-28} dyne/cm.^{2} or 10^{-34} atmosphere before Dalton's law becomes effective. We have then reached a region so tenuous that the density there is far below the probable density in interstellar space! In no region which can be considered as belonging to the atmosphere of the sun or of any particular star do we even approach the condition (1689).

§ **15·22.** *The form of f for large negative z, at the base of the atmosphere.* On referring to equation (1683) we see that the right-hand side tends to infinity as $z \to -\infty$, and therefore if f has a limit at all that limit must be zero. There are no physically possible alternatives. For large negative z the equation therefore approximates to

$$f'' + \alpha f' = \frac{8\pi\epsilon^2}{kT}\sqrt{(\overline{n_2}\,\overline{n_3})_0}\,e^{-\alpha z} f. \qquad \text{......(1690)}$$

This equation must be solved asymptotically as $z \to -\infty$, α and the coefficient of $e^{-\alpha z} f$ being numerically small. By the substitution* $f = \alpha e^{-\eta}$ $(\eta \to +\infty)$ it is easy to show that

$$\eta' = -\left[\frac{8\pi\epsilon^2\sqrt{(\overline{n_2}\,\overline{n_3})_0}}{kT}\right]^{\frac{1}{2}} e^{-\frac{1}{2}\alpha z} + \tfrac{1}{4}\alpha + O(e^{\frac{1}{2}\alpha z}).$$

From this it follows that to a sufficient approximation

$$f = \alpha \exp\left[-\frac{2}{\alpha}\left\{\frac{8\pi\epsilon^2\sqrt{(\overline{n_2}\,\overline{n_3})_0}}{kT}\right\}^{\frac{1}{2}} e^{-\frac{1}{2}\alpha z} + O(\alpha z)\right]. \qquad \text{......(1691)}$$

From this form it follows at once that $\int_{-\infty} f\,dz$ converges, and approximately

$$\int_{-\infty}^{z} f\,dz = \alpha\left[\frac{8\pi\epsilon^2\sqrt{(\overline{n_2}\,\overline{n_3})_0}}{kT}\right]^{-\frac{1}{2}} e^{\frac{1}{2}\alpha z} \exp\left[-\frac{2}{\alpha}\left\{\frac{8\pi\epsilon^2\sqrt{(\overline{n_2}\,\overline{n_3})_0}}{kT}\right\}^{\frac{1}{2}} e^{-\frac{1}{2}\alpha z}\right].$$

* We insert α to take care of the dimensional factor in f.

We can now suppose that $-\infty$ is chosen for z_0 in (1682). Then since $f' \to 0$ $(\overline{n_2})_0 = (\overline{n_3})_0$. For sufficiently large negative z we shall therefore find approximately

$$\overline{n_2} = \overline{n_3} = (\overline{n_2})_0 \, e^{-\alpha z}. \qquad \text{......(1692)}$$

The meaning of this distribution law is that the ions and electrons are distributed as if they were both of mass equal to their mean mass—with no tendency to separate out and consequently no electric field. This fusion we have shown to occur at the base of the atmosphere, or in the interior of the star, but the question of interest is where for this purpose does the interior start. It starts obviously as soon as $\int_{-\infty} f dz$ is small compared with unity, that is to say as soon as

$$\frac{2}{\alpha}\left[\frac{8\pi\epsilon^2\sqrt{(\overline{n_2}\,\overline{n_3})_0}}{kT}\right]^{\frac{1}{2}} e^{-\frac{1}{2}\alpha z}, \quad \frac{8\epsilon}{(m_2+m_3)\,g}\,[2\pi p_3]^{\frac{1}{2}} \gg 1.$$

For calcium on the sun this reduces numerically to

$$5{\cdot}4 \times 10^9 \sqrt{p_3} \gg 1,$$

which is satisfied as soon as p_3 the electron pressure is greater than say 10^{-16} dyne/cm.^{2}, or 10^{-22} atmosphere. Even this figure corresponds to a density less than that of interstellar space, so that so far as separation of electrons and ions is concerned, the interior of the star, in which separation is impossible, may be taken to include the whole of the star and any atmosphere that can properly be regarded as private to it. All we have to do in any problem of an ionized atmosphere in equilibrium is to use (1692) instead of Dalton's law. To produce this there is a constant electric field acting outwards of intensity $\frac{1}{2}(m_2 - m_3)\,g/\epsilon$, but the separation of charge necessary to produce this is altogether trivial.

§ **15·23.** *Further observations.* If the charge on the star has a surface density σ, then, as $z \to +\infty$,

$$-\epsilon\psi' \to 4\pi\sigma.$$

This alters slightly the form of the solution as $z \to +\infty$, but does not affect the conditions as $z \to -\infty$, since these are independent of f_∞. Hence the conclusions of § 15·22 are valid whatever the charge on the star.

Again we have shown that $f \to 0$ as $z \to -\infty$ and that

$$f \to \tfrac{1}{2}(m_2 - m_3)\,g/kT,$$

a positive quantity as $z \to +\infty$. From the form of the relation for f', namely

$$f' = A\left[e^{\int_{-\infty}^{z} f dz} - e^{-\int_{-\infty}^{z} f dz}\right] \quad (A > 0),$$

it follows that if f starts positive for large negative z then $f' > 0$, f increases and f' can never vanish. Since f has to be positive for large positive z, it follows that f and f' must both always be positive and f steadily increases.

For the uncharged star this implies that the resultant charge down to any level is always negative since as $f>0$, $\overline{n_3}>\overline{n_2}$. But this negative charge is excessively minute. Applying Gauss's theorem to (1679) we see that the excess charge σ per unit area is given by

$$-4\pi\sigma=-\tfrac{1}{2}(m_2-m_3)\,g/\epsilon.$$

The excess number of electrons per unit area is therefore

$$(m_2-m_3)\,g/8\pi\epsilon^2,$$

which for ionized calcium on the sun is 0·3!

§ **15·24.** *Multivalent ions and their compensating free electrons.* The theory for an atmosphere of a single set of positive ions of charge $+v_2\epsilon$ and the corresponding electrons is very similar. In place of the equations (1676)–(1678) we have

$$\overline{n_2}=(\overline{n_2})_0\,e^{-(m_2gz+v_2\epsilon\psi)/kT}, \qquad \text{......(1693)}$$

$$\overline{n_3}=(\overline{n_3})_0\,e^{-(m_3gz-\epsilon\psi)/kT}, \qquad \text{......(1694)}$$

$$\psi''=-4\pi\epsilon(v_2\overline{n_2}-\overline{n_3}). \qquad \text{......(1695)}$$

We now define f by the equation

$$\epsilon\psi'=-\frac{(m_2-m_3)g}{v_2+1}+kTf,$$

and replace the last equations by

$$\overline{n_2}=(\overline{n_2})_0\exp\left(-\alpha z-v_2\int_{z_0}^{z}f\,dz\right),$$

$$\overline{n_3}=(\overline{n_3})_0\exp\left(-\alpha z+\int_{z_0}^{z}f\,dz\right), \qquad \left(\alpha=\frac{m_2+v_2m_3}{v_2+1}\frac{g}{kT}\right),$$

$$f'=\frac{4\pi\epsilon^2}{kT}(\overline{n_3}-v_2\overline{n_2}),$$

the equation satisfied by f being now

$$\left\{\frac{f''+\alpha f'}{f}-f'\right\}\left\{\frac{f''+\alpha f'}{f}+v_2f'\right\}^{v_2}=\left[\frac{4\pi\epsilon^2}{kT}(v_2+1)\,e^{-\alpha z}\right]^{v_2+1}(\overline{n_3})_0^{\,v_2}\,v_2(\overline{n_2})_0. \qquad \text{......(1696)}$$

The conclusions which can be drawn from this equation correspond exactly to those drawn from (1683). The ions and electrons never separate out appreciably in the atmosphere proper, and for large negative z, $f\to 0$, $f'\to 0$. If then z_0 is taken as $-\infty$, we find $(\overline{n_3})_0=v_2(\overline{n_2})_0$ and the distribution laws

$$\overline{n_2}=(\overline{n_2})_0\,e^{-\alpha z}, \qquad \text{......(1697)}$$

$$\overline{n_3}=(\overline{n_2})_0\,v_2e^{-\alpha z}. \qquad \text{......(1698)}$$

There are just v_2 times as many electrons as ions, and both possess the exponential factor of a neutral particle of mass $(m_2+v_2m_3)/(v_2+1)$.

§ 15·25. *Atmospheres with more than one positive ion.* This case is unfortunately much more complicated, and it is of course more typical of actual atmospheres. It is sufficient for illustration to consider two positive ions and electrons. The neutral atoms required by the dissociative equilibrium are also present, but need not be explicitly considered. The equations then are

$$\overline{n_1} = (\overline{n_1})_0 \, e^{-(m_1 gz + v_1 \epsilon\psi)/kT},$$

$$\overline{n_2} = (\overline{n_2})_0 \, e^{-(m_2 gz + v_2 \epsilon\psi)/kT},$$

$$\overline{n_3} = (\overline{n_3})_0 \, e^{-(m_3 gz - \epsilon\psi)/kT},$$

$$\psi'' = -4\pi\epsilon(\overline{n_1} v_1 + \overline{n_2} v_2 - \overline{n_3}).$$

If it happens that

$$\frac{m_1 - m_3}{v_1 + 1} = \frac{m_2 - m_3}{v_2 + 1}, \qquad \text{......(1699)}$$

the substitution

$$\epsilon\psi' = -\frac{m_1 - m_3}{v_1 + 1} g + kTf$$

reduces the leading terms in all three exponentials to equality. We are then led to an equation for f similar to but more complicated than (1696), but from which the same conclusions can be drawn. This equality will never be satisfied by the main constituents of an atmosphere, even approximately.

When (1699) is not satisfied a formal approximation to the solution of the set of equations proposed appears to be difficult. This however is not a serious drawback, for in view of the extremely small separation of electric charges which we have found in soluble cases we may safely assume that there is no separation in the more general case, so that

$$\overline{n_1} v_1 + \overline{n_2} v_2 - \overline{n_3} = (\overline{n_1})_0 \, v_1 + (\overline{n_2})_0 \, v_2 - (\overline{n_3})_0 = 0. \qquad \text{......(1700)}$$

Starting from this proper simplification Eddington* has attempted to obtain further information as follows. If we write out equation (1700) in full it takes the form

$$(\overline{n_1})_0 \, v_1 e^{-(m_1 gz + v_1 \epsilon\psi)/kT} + (\overline{n_2})_0 \, v_2 e^{-(m_2 gz + v_2 \epsilon\psi)/kT} = \{(\overline{n_1})_0 \, v_1 + (\overline{n_2})_0 \, v_2\} \, e^{-(m_3 gz - \epsilon\psi)/kT}. \qquad \text{......(1701)}$$

Strictly speaking equation (1701) determines ψ as a function of z and the problem is solved. It is however difficult to extract much information from it as it stands. Differentiating it with respect to z we find after rearrangement that

$$\epsilon\psi' = -g\left\{\frac{v_1(v_1+1)\,\overline{n_1}\mu_1 + v_2(v_2+1)\,\overline{n_2}\mu_2}{v_1(v_1+1)\,\overline{n_1} + v_2(v_2+1)\,\overline{n_2}} - m_3\right\}, \qquad \text{......(1702)}$$

in which we have written for shortness

$$\mu_1 = \frac{m_1 + v_1 m_3}{v_1 + 1}, \quad \mu_2 = \frac{m_2 + v_2 m_3}{v_2 + 1}.$$

* Eddington, *The internal constitution of the stars* (1926), § 192.

It will be observed that the coefficient of g is such that this last equation might be cast in the form
$$\epsilon\psi' = -g(\bar{\mu} - m_3),\qquad\qquad\ldots\ldots(1703)$$
where $\bar{\mu}$ is a suitably weighted mean value of the μ's for the different species. An equation of this form holds for any number of species. Eddington has now suggested that one may (roughly) integrate (1703) or rather its generalization for a spherical body of gas as if $\bar{\mu}$ were constant through the gas. If this is done one obtains formulae for the distribution of the various species of which the formula
$$\frac{\overline{n_2}}{\overline{n_3}} = \frac{(\overline{n_2})_0}{(\overline{n_3})_0} e^{-gz(v_2+1)(\mu_2-\bar{\mu})/kT}\qquad\qquad\ldots\ldots(1704)$$
is typical. Such formulae may be taken to show that those species for which $\mu < \bar{\mu}$ at any level will tend to concentrate into the higher levels and those for which $\mu > \bar{\mu}$ to concentrate into the lower levels; but the effect of this concentration itself will be to modify $\bar{\mu}$ in such a way as everywhere greatly to cut down the space rate of concentration. The degrees of concentration given by Eddington on the assumption of μ constant must therefore be overestimated, but by exactly how much has not been determined.

Some reliable information may be obtained in a simple way as follows. Let us suppose that in a certain region the conditions are dominated by the first constituent. Then approximately
$$\epsilon\psi = -\left(\frac{m_1 - m_3}{v_1 + 1}\right) gz,$$
$$\left.\begin{aligned}\overline{n_1} &= (\overline{n_1})_0 e^{-\alpha_1 z},\\ \overline{n_3} &= (\overline{n_3})_0 e^{-\alpha_1 z},\end{aligned}\right\}\quad \alpha_1 = \frac{g}{kT}\frac{m_1 + v_1 m_3}{v_1 + 1} = \frac{\mu_1 g}{kT},$$
$$\overline{n_2} = (\overline{n_2})_0 e^{-\alpha_1' z},\qquad\qquad\ldots\ldots(1704\cdot1)$$
where
$$\alpha_1' = \frac{(v_1+1)m_2 - v_2(m_1 - m_3)}{v_1 + 1}\frac{g}{kT}.\qquad\qquad\ldots\ldots(1704\cdot2)$$
We find that $\alpha_1' < \alpha_1$ if
$$\frac{m_2 - m_3}{v_2 + 1} < \frac{m_1 - m_3}{v_1 + 1}\quad\text{or}\quad \mu_2 < \mu_1,$$
and we shall certainly find a region of control by the first constituent if we can go deep enough into the atmosphere $(z \to -\infty)$ without other disturbances. Generalizing we might say that in an isothermal atmosphere the deepest levels will be controlled by the constituent for which
$$\mu_1 \quad\text{or}\quad (m_1 - m_3)/(v_1 + 1)$$
is greatest, but this only holds if the effectively undisturbed isothermal region is extensive enough to differentiate between the various constituents. In the region of control by $\overline{n_1}$ the rate of space variation of $\overline{n_2}$ given by α_1' may even be reversed so that n_2 increases upwards $(\alpha_1' < 0)$.

As we rise in the atmosphere ($z \to +\infty$) this state of things must reverse and control must pass over to the second constituent. We shall then have

$$\epsilon\psi = -\left(\frac{m_2 - m_3}{\nu_2 + 1}\right) gz,$$

$$\left.\begin{aligned} \overline{n_2} &= (\overline{n_2})_0\, e^{-\alpha_2 z}, \\ \overline{n_3} &= (\overline{n_3})_0\, e^{-\alpha_2 z}, \end{aligned}\right\} \quad \alpha_2 = \frac{m_2 + \nu_2 m_3}{\nu_2 + 1}\,\frac{g}{kT},$$

$$\overline{n_1} = (\overline{n_1})_0\, e^{-\alpha_2' z},$$

where $$\alpha_2' = \frac{(\nu_2 + 1)\, m_1 - \nu_1(m_2 - m_3)}{\nu_2 + 1}\,\frac{g}{kT}.$$

We can verify that $\alpha_2' > \alpha_2$ as it should be for this region.

A special case of particular interest is that in which the constituents 1 and 2 are the single and double ions of the same neutral atom 0. Then $m_1 = m_0 - m_3$, $m_2 = m_0 - 2m_3$, $\nu_1 = 1$, $\nu_2 = 2$. Applying these formulae we find that in the deep region

$$\overline{n_0} = (\overline{n_0})_0\, e^{-m_0 gz/kT}, \quad \overline{n_1} = (\overline{n_1})_0\, e^{-\frac{1}{2} m_0 gz/kT},$$

$$\overline{n_2} = (\overline{n_2})_0, \qquad \overline{n_3} = (\overline{n_3})_0\, e^{-\frac{1}{2} m_0 gz/kT},$$

so that the concentration of double ions does not alter. In the high region

$$\overline{n_0} = (\overline{n_0})_0\, e^{-m_0 gz/kT}, \quad \overline{n_1} = (\overline{n_1})_0\, e^{-\frac{2}{3} m_0 gz/kT},$$

$$\overline{n_2} = (\overline{n_2})_0\, e^{-\frac{1}{3} m_0 gz/kT}, \quad \overline{n_3} = (\overline{n_3})_0\, e^{-\frac{1}{3} m_0 gz/kT}.$$

Here we leave this application. Our main conclusion is that large scale electrical effects profoundly alter the vertical distribution of the ions and electrons in an atmosphere, in general refusing to let them separate. In effect they alter g for ions and electrons in the way we have attempted to calculate, but the actual fields required to do this are very small, and apart from this modification of g the resulting fields and charges can always be neglected. Certain similar applications to stellar interiors will be made in the next chapter. We should in conclusion record the warning that an actual atmosphere is submitted to a strong one-sided flow of radiation from the photosphere of the star, and that selective action of this radiation may seriously modify any conclusions drawn when gravity is the only external force acting. The atmosphere cannot then be an assembly in statistical equilibrium, and the type of effect which may enter is discussed in § 15·5.

§ 15·3. *Stellar absorption spectra.* The atmosphere of a star, when we have reached the deeper levels of the preceding sections, still of extreme tenuity, will consist of a mixture of atoms, ions and electrons in a state in which each constituent behaves approximately like a perfect gas. The mixture as a whole is electrically neutral as the separation of electrons and ions in the gravitational field is trivial. We shall verify later that the deviations from the perfect gas laws required by the theory developed in Chapter XIV are insignificant for the regions to be explored in this section. We shall regard this atmosphere to a first approximation as an isothermal homogeneous slab in statistical equilibrium, but subject to a flow of radiation of all wave lengths, corresponding to a higher temperature from the lower levels (*photosphere*) of the star. Such a slab forms an idealized *reversing layer*, forming by specific absorption dark lines in the continuous spectrum. Our object in these sections is to show how the theorems of statistical mechanics can be applied in a general way to explain the behaviour of these *absorption lines*, particularly in regard to the march of their intensities as we pass through the series of spectral types, and to deduce at least rough information as to the temperatures and pressures in the reversing layers of most stars. We shall not enter into great detail, since a full account is given in Miss Payne's *Stellar Atmospheres** to which the reader should refer. These sections will therefore be confined to a summary, with supplements to her account.

The first successful quantitative application of the theory of statistical (or rather in his case thermodynamic) equilibrium to stellar reversing layers is due to Saha.† An atom absorbs a different optical spectrum for each stage of ionization, and in fact a different set of lines for each stationary state belonging to each stage, and therefore the relative intensities of the absorption lines of its successive spectra in the spectrum of any star must give some indication of the relative numbers of atoms in the various stages of ionization in the reversing layer, and therefore of the temperature and pressure.

The early applications of this idea may be divided into two main groups. The first is typified by comparisons between the spectra of the normal solar reversing layer and sun spots, and between spectra of giants and dwarfs of the same spectral type. It was shown that the intensity differences are largely explained by changes in the degree of ionization resulting either from temperature differences (sun and spot) or from pressure differences (giant and dwarf).‡ A similar successful comparison may be drawn between the spectra of the reversing layer and chromosphere (flash spectrum).‡ In the second group attention is devoted to the general march of the intensity of a line, or group of lines, through the sequence of stellar spectral types, and an attempt is made to deduce the temperature scale from the positions in this sequence of the first and last ("marginal") appearances of the line. At such a point the fraction of atoms in the reversing layer capable of absorbing the

* Miss Payne, *Stellar Atmospheres*, Harvard Monographs, No. 1 (1925). More recent accounts to which reference should be made are Miss Payne, *Stars of High Luminosity*, Harvard Monographs, No. 5 (1931); Russell, *loc. cit.* in § 15·34.

† Saha, *Phil. Mag.* vol. 40, pp. 472, 809 (1920); *Proc. Roy. Soc.* A, vol. 99, p. 136 (1921).

‡ Saha, *loc. cit.* (1) and (2); Russell, *Astrophys. J.* vol. 55, p. 119 (1922).

line must be very small, and if the pressure is known the temperature can be calculated.*

But the precision of the early calculations was questionable owing to the difficulty of formulating the conditions for the marginal appearance of a line.† Firstly we do not know *a priori* how small the "very small" fraction of atoms must be at marginal appearance. Secondly the point of marginal appearance will depend on the relative abundance of the element giving the line—other things being equal lines due to a more abundant element will persist to a smaller fractional concentration. A similar difficulty may arise from different atomic absorption coefficients for different lines. Large uncertainties may arise from all these causes.

However, the general qualitative adequacy of the theory carried the conviction that imperfections of this kind were imperfections of our knowledge and not of the theory; it is reasonable both *a priori* and on the evidence to conclude that in the main the intensities of absorption lines vary in the same sense as the numbers of atoms capable of absorbing them. The next step was to formulate this conclusion explicitly: *Other things being equal, the intensity of a given absorption line in a stellar spectrum varies always in the same sense as the concentration of atoms in the reversing layer capable of absorbing the line.* The foregoing difficulties are then in the main avoided, if we devote attention in the first instance to the place in the stellar sequence at which a given line attains its *maximum intensity*. This will be attained at the maximum concentration of atoms capable of absorbing the line, and the conditions therefore only involve the temperature and pressure. We do not now require to know the relative abundance of the various elements, the efficiency of any atomic state as an absorber, or the absolute number of effective atoms required to form a line. The temperature at which, for a given pressure, a given line of known series relationships attains its maximum is simply deducible from the properties of the equilibrium state. A number of such calculations will be given in the following pages. Consequently in the first instance each observed maximum of a line in the stellar sequence connects the temperature and pressure of the reversing layer at that point of the sequence. This appears to be the most satisfactory way to apply Saha's theory quantitatively to fix stellar temperatures and pressures.‡ We therefore summarize the results of such calculations. But we can then attempt to refine them in various ways, and show that each observed maximum can probably be made to determine directly a temperature in the

* Saha, *loc. cit.* (3), and *Zeit. f. Physik*, vol. 6, p. 40 (1921); H. H. Plaskett, "The Spectra of three *O*-type stars", *Pub. Dom. Astrophys. Obs.* vol. 1, No. 30 (1922).

† These difficulties were discussed by Milne, *The Observatory*, vol. 46, p. 113 (1923).

‡ Such calculations were made by R. H. Fowler and Milne, *Monthly Not. R.A.S.* vol. 83, p. 403 (1923), vol. 84, p. 499 (1924); R. H. Fowler, *Monthly Not. R.A.S.* vol. 85, p. 970 (1925).

stellar sequence, without assuming a pressure. The pressure is itself determined by the theory, being that of the layer in which the absorption line is formed. We find that this refined theory, largely due to Milne and Russell, requires the height of this layer to vary in a definite way from line to line. Instead of assuming a pressure, we assume that a certain function of the properties of the atom and its abundance and gravity on the surface of the star may be assumed roughly constant, the value of the constant depending on whether we are following the intensities along the giant or the dwarf sequence. There is reasonable ground for belief that this assumption is true on the average, so that the derived temperature scale should be reliable. The final step, not here reached, will be to regard each observed maximum in the known temperature scale as fixing the value of this function and to deduce therefrom abundance factors or absolute values of atomic absorption coefficients. For this step we require a quantitative theory of the formation of an absorption line. Such a theory is now available from the work of Milne* but to discuss it lies beyond the range of this monograph.

Lines absorbed by the neutral atom in its normal state, which we shall call *normal lines*, will be shown theoretically always to decrease in intensity as we traverse the spectral types from M to higher temperatures. This is almost obvious without calculation, for the fraction in the normal state (initially practically unity) can only decrease as the temperature increases. An exception to this absence of maximum could occur if in atmospheres of very low temperature the atoms in question were all removed by condensation or chemical combination. Such considerations are however not usually of importance and will not be referred to again. It is easy to see that all other lines should have a maximum somewhere. Consider first a line absorbed by some excited state of a neutral atom, which we shall refer to as a *subordinate line*. The fraction of atoms in this excited state is the product of two factors: (1) The fraction of the atoms not yet ionized; (2) the fraction of these neutral atoms in the proper state. The first factor decreases steadily from 1 to 0 as T increases, while the second increases steadily from a very small value. This must lead somewhere to a maximum in the product. Similarly for the normal lines of any ion, we start at low temperatures with atmospheres in which there are no such ions, and pass through a stage at which almost all are once ionized, to a final stage in which all are in still higher stages of ionization. Again we find a maximum.†

In the first stage of the discussion we find it convenient to calculate

* Milne, *loc. cit.* in § 15·34.

† In identifying this maximum with the observed maximum we assume that the average abundance of this atom changes only very slowly (if at all) along the stellar sequence. There is every *a priori* consideration in favour of this assumption which is of course essential. Otherwise, for example, the maximum of the Balmer lines in A-type stars might be due to a maximum in the absolute abundance of hydrogen there, a barren and unsatisfying conclusion.

pressures for maxima at a given temperature. The general result of the comparison with observation is that observed and theoretical maxima can be fitted together into a consistent scheme provided that the pressure in the layers which reverse strongly ordinary subordinate lines is of the order of 10^{-5} atmosphere. Similarly in layers which reverse strongly normal lines of atoms and ions and nearly normal lines of the metallic transition elements the pressure must be of the order 10^{-9}–10^{-10} atmosphere. At the second more refined stage we see that this result is expected and that on the whole all observations fit into a logically consistent whole, leaving outstanding a number of interesting minor discrepancies.

The observational material on the positions of maxima in the sequence of giant stars is mainly due to Miss Payne and Menzel.* For dwarfs there is not the same sequence of material, and we have mainly cross comparisons with the giant sequence at spectral types F and G.

§ **15·31.** *The statistical theory of absorption lines.* To carry out the calculations suggested in § 15·3 we use equations (1645)† and the method of § 8·6 to determine the $(M_r^z)_s/V$, and then differentiate with respect to T to determine maximum values. But the general formulae simplify greatly here. In the first place χ_r^z/χ_{r+1}^z is never greater than about 0·7 for any atomic ions with which we have to deal in reversing layers and is usually nearer 0·5. The successive stages of ionization are thus well separated, and it is found that we never have to consider more than two consecutive stages of any atom at any one time for a subordinate line or three at most for a normal line. Secondly, excluded volumes and the electrostatic energy can be ignored entirely, except of course for the convergency factors due to the former in $u_r^z(T)$. We shall find further that in all calculations of maxima $u(T)$ reduces to its first or at most its first two or three terms, for which the excluded volumes can be ignored, and that the highly excited states make no effective contribution. We shall therefore start the calculations with these simplifications, which are easily justified *a posteriori*.

It is convenient to express the laws of dissociative equilibrium in terms of p_ϵ the partial pressure of the free electrons, and $x_0, x_1, \ldots, x_s$ the fractions of any atom present in the neutral, singly ionized, or s-times ionized states. Then the equations of dissociation (479), generalized as in § 8·6, reduce to

$$\frac{x_{r+1}}{x_r} p_\epsilon = \frac{2(2\pi m)^{\frac{3}{2}} (kT)^{\frac{5}{2}}}{h^3} \frac{u_{r+1}(T)}{u_r(T)} e^{-\chi_r/kT}, \qquad \ldots\ldots(1705)$$

with $$\Sigma_r x_r = 1.$$

* Miss Payne, Harvard Circular, No. 258 (1924): *loc. cit.* chap. VIII; Menzel, Harvard Circular, Nos. 252, 256 (1924).

† It is here a matter of indifference whether we use Urey's or Planck's form for $u_r^z(T)$, since the states of high excitation prove negligible.

We need not here retain the affix z. Let $(n_r)_s$ be the fraction of all the atoms of one type which are r-times ionized and in their sth state. Then (1645) reduces to

$$(n_r)_s = x_r(\varpi_r)_s e^{-\{\chi_r-(\chi_r)_s\}/kT}/u_r(T).$$

For discussing the *maxima* of the states of the neutral atom we may ignore x_r for $r \geqslant 2$ owing to the rapidly increasing χ's. Inserting numerical values in (1705), we find*

$$x_0 = \frac{u_0(T)\, e^{\chi_0/kT}}{u_0(T)\, e^{\chi_0/kT} + 0\cdot664 u_1(T)\, T^{\frac{5}{2}}/p_\epsilon}, \qquad \text{......(1706)}$$

$$(n_0)_s = \frac{(\varpi_0)_s\, e^{(\chi_0)_s/kT}}{u_0(T)\, e^{\chi_0/kT} + 0\cdot664 u_1(T)\, T^{\frac{5}{2}}/p_\epsilon}. \qquad \text{......(1707)}$$

The maximum value of $(n_0)_s$ for given pressure occurs where $\partial(n_0)_s/\partial T = 0$, or where

$$p_\epsilon = \frac{0\cdot664 u_1(T)}{u_0(T)} \left\{ \frac{(\chi_0)_s + \frac{5}{2}kT + kT^2 u_1'/u_1}{\chi_0 - (\chi_0)_s - kT^2 u_0'/u_0} \right\} T^{\frac{5}{2}} e^{-\chi_0/kT}. \qquad \text{......(1708)}$$

This gives the pressure in dynes/cm.^{2} which can be converted to atmospheres with sufficient accuracy by multiplication by 10^{-6}. We have also

$$(x_0)_{\max} = \frac{(\chi_0)_s + \frac{5}{2}kT + kT^2 u_1'/u_1}{\chi_0 + \frac{5}{2}kT + kT^2(u_1'/u_1 - u_0'/u_0)}, \qquad \text{......(1709)}$$

$$\{(n_0)_s\}_{\max} = (x_0)_{\max} \frac{(\varpi_0)_s\, e^{-\{\chi_0-(\chi_0)_s\}/kT}}{u_0(T)}. \qquad \text{......(1710)}$$

The same formulae hold for the maxima of the subordinate lines of the r-times ionized atom, when we replace the suffix 0 by r and ignore all stages of ionization except r and $r+1$. A simple verification, which we omit, is required first to show that for the relevant temperatures $x_r + x_{r+1} \simeq 1$.

We conclude that any subordinate line should have a maximum given by these formulae. For any normal line of a neutral atom $(\chi_0)_s = \chi_0$ and $(n_0)_s$ has no maximum for $kT^2 u_0'/u_0$ reduces to zero, and (1708) merely gives $T = 0$. Formula (1709) reduces to $(x_0)_{\max} = 1$ in this case.

We must next examine the numerical values of $u_0(T)$, $u_1(T)$ and their differential coefficients. From (1708) it follows that near the maximum $p_\epsilon/T^{\frac{5}{2}}$ and $e^{-\chi_0/kT}$ must be approximately equal, and we shall find that the actual orders of both are about 10^{-8}. For the approximate form of $u_r(T)$ we take (1651), in which α is given by (1648). We may take $p_\epsilon = p_a$ approximately. The numerical value of $[(r+1)^3/9\alpha]^{\frac{1}{3}}$ is therefore of the order 2×10^5 at most in a reversing layer. It follows at once from (1651) that we have at most one highly excited electron, and that the states of high excitation of

* The numerical factor here is twice that (0·332) used in the work of R. H. Fowler and Milne, because we have here corrected for the weight 2 of the free electron. All the weights used in the two papers (*loc. cit.*) are wrong, though the results are hardly affected. Effectively correct weights were used by R. H. Fowler, *loc. cit.*

even this one electron make a negligible contribution to $u_r(T)$ (less than 1 per cent.). In reversing layer problems therefore it is only the first few terms of $u_r(T)$ which need be considered, and usually only the first (constant) term and the term to which the subordinate line under discussion belongs. We verify in the same way that the contributions of highly excited terms to $kT^2u_r'(T)/u_r(T)$ are equally negligible in the equations for the maxima.

Returning now to the early terms, suppose that for simplicity

$$u_0(T) = \varpi_0 + (\varpi_0)_1 e^{-\{\chi_0 - (\chi_0)_1\}/kT}. \qquad \text{......(1711)}$$

Then

$$\frac{kT^2u_0'(T)}{u_0(T)} = \{\chi_0 - (\chi_0)_1\}\left[1 - \frac{\varpi_0}{u_0(T)}\right]. \qquad \text{......(1712)}$$

Unless $\chi_0 - (\chi_0)_1$, the energy of excitation, is small compared with kT, the effect of the second term in $u_0(T)$ is negligible both in (1711) and (1712). Even if $\chi_0 - (\chi_0)_1$ is small, $u_0'(T)$ will still be of no importance in the formula for p_ϵ unless this second state of small energy of excitation is also the state in which the atoms must be in order to absorb the line in question. In that special case $(\chi_0)_s = (\chi_0)_1$ and the denominator of p_ϵ becomes

$$\{\chi_0 - (\chi_0)_1\}\,\varpi_0,$$

instead of $\{\chi_0 - (\chi_0)_1\}\,u_0(T)$, which is the value when $u_0'(T)$ is ignored. When the excited states are nearly normal this correction may be considerable. No effect of this sort can be produced by $kT^2u_1'(T)/u_1(T)$, which can always be neglected. We can therefore replace (1708) by

$$p_\epsilon = \frac{0{\cdot}664u_1(T)}{\varpi_0}\,\frac{(\chi_0)_s + \frac{5}{2}kT}{\chi_0 - (\chi_0)_s}\,T^{\frac{5}{2}}e^{-\chi_0/kT}, \qquad \text{......(1713)}$$

which in this form is always valid if ϖ_0 is taken to mean the sum of the weights of just those states which differ in energy from the normal state by an amount which is very small compared with $\chi_0 - (\chi_0)_s$. In calculating $(x_0)_{\max}$ we can always ignore $u_0'(T)$ and $u_1'(T)$ so that

$$(x_0)_{\max} = \frac{(\chi_0)_s + \frac{5}{2}kT}{\chi_0 + \frac{5}{2}kT}. \qquad \text{......(1714)}$$

To discuss a normal line of the ionized atom we must retain three stages. Then

$$\frac{x_2}{x_1} = \frac{0{\cdot}664}{p_\epsilon}\,\frac{u_2(T)}{u_1(T)}\,T^{\frac{5}{2}}e^{-\chi_1/kT},$$

$$\frac{x_1}{x_0} = \frac{0{\cdot}664}{p_\epsilon}\,\frac{u_1(T)}{u_0(T)}\,T^{\frac{5}{2}}e^{-\chi_0/kT},$$

$$x_0 + x_1 + x_2 = 1, \quad n_1 = \varpi_1 x_1/u_1(T).$$

If we solve these equations for n_1 we find that

$$n_1 = \frac{\varpi_1}{u_1(T) + u_0(T)\dfrac{p_\epsilon}{0{\cdot}664}T^{-\frac{5}{2}}e^{\chi_0/kT} + u_2(T)\dfrac{0{\cdot}664}{p_\epsilon}T^{\frac{5}{2}}e^{-\chi_1/kT}}. \qquad \text{......(1715)}$$

Thus the maximum of n_1 will occur when

$$kT^2u_1'(T)-\frac{p_\epsilon}{0{\cdot}664}\{\chi_0+\tfrac{5}{2}kT-kT^2u_0'/u_0\}u_0(T)\,T^{-\frac{5}{2}}e^{\chi_0/kT}$$
$$+\frac{0{\cdot}664}{p_\epsilon}\{\chi_1+\tfrac{5}{2}kT+kT^2u_2'/u_2\}u_2(T)\,T^{\frac{5}{2}}e^{-\chi_1/kT}=0. \quad \text{......(1716)}$$

If we can ignore the term in $u_1'(T)$, we then find

$$p_\epsilon=0{\cdot}664\left\{\frac{\chi_1+\frac{5}{2}kT+kT^2u_2'/u_2}{\chi_0+\frac{5}{2}kT-kT^2u_0'/u_0}\cdot\frac{u_2(T)}{u_0(T)}\right\}^{\frac{1}{2}}T^{\frac{5}{2}}e^{-\frac{1}{2}(\chi_0+\chi_1)/kT}. \quad \text{......(1717)}$$

The order of the two terms retained from (1716) is $(u_0u_2)^{\frac{1}{2}}e^{-\frac{1}{2}(\chi_1-\chi_0)/kT}$ at the maximum, and this maximum occurs at a lower temperature than that at which the maxima of any subordinate lines of the ionized atom occur. The conclusions as to the negligibility of the terms of high excitation hold therefore *a fortiori*. The order of $kT^2u_1'(T)$ will therefore be

$$\{\chi_1-(\chi_1)_1\}e^{-\{\chi_1-(\chi_1)_1\}/kT},$$

which may be important in (1716) if there is an excitation potential of the ionized atom comparable with $\frac{1}{2}(\chi_1-\chi_0)$. When this is the case (1716) must be solved as a quadratic for p_ϵ. At the maximum, if $u_1'(T)$ can be ignored

$$(n_1)_{\max}=\frac{\varpi_1}{u_1(T)+O\{(u_0u_2)^{\frac{1}{2}}e^{-\frac{1}{2}(\chi_1-\chi_0)/kT}\}}, \quad \text{......(1718)}$$

which is approximately equal to $\varpi_1/u_1(T)$ and therefore nearly unity.

If, as can happen, the term in $u_1'(T)$ is of larger order than the other two terms at the suggested maximum (1717), then the third term in (1716) can be ignored and we have

$$p_\epsilon=\frac{0{\cdot}664kT^2u_1'(T)}{u_0(T)\{\chi_0+\frac{5}{2}kT-kT^2u_0'/u_0\}}T^{\frac{5}{2}}e^{-\chi_0/kT}. \quad \text{......(1719)}$$

The maximum is due in this case to the switch over of the majority of the atoms from the normal state to the first excited state as the temperature rises, and in the determination of the maximum we can ignore second stage ionization entirely. If for simplicity we write here

$$u_1(T)=\varpi_1+(\varpi_1)_1e^{-\{\chi_1-(\chi_1)_1\}/kT},$$

then
$$p_\epsilon=\frac{0{\cdot}664(\varpi_1)_1\{\chi_1-(\chi_1)_1\}}{u_0(T)\{\chi_0+\frac{5}{2}kT-kT^2u_0'/u_0\}}T^{\frac{5}{2}}e^{-\{\chi_0+\chi_1-(\chi_1)_1\}/kT}. \quad \text{...(1720)}$$

In this case
$$(n_1)_{\max}=\frac{\varpi_1}{u_1(T)+O\{(\varpi_1)_1e^{-\{\chi_1-(\chi_1)_1\}/kT}\}}, \quad \text{......(1721)}$$

which again differs only a little from $\varpi_1/u_1(T)$ or unity.

In certain cases in calculating maxima of subordinate lines of an ionized element of very small energy of excitation it may be necessary to use the apparatus of these paragraphs taking account of *three* successive stages of ionization. But no general discussion of the negligibility of the various $u(T)$ and $u'(T)$ can expect to cover all cases, and when doubt arises it is always a simple matter to write down an explicit 2- or 3-term formula for the whole relevant part of $u(T)$ and use the exact values of $u(T)$ and $u'(T)$.

§ **15·32.** *Numerical calculations.* We present in this section numerical applications to a selection of the available observational material. To do more would take up too much space, and the selection is wide enough to justify the theory. It is obvious from a cursory inspection of the rest of the material that it will fit into the same scheme. The observed positions of the maxima are taken from Menzel or Miss Payne, and the spectral data from Backer and Goudsmit.*

Throughout the calculations we group together all the lines of a multiplet, ignoring the energy differences in the initial and in the final state. This is legitimate, and our discussion therefore refers to the maximum of the multiplet as a whole rather than the maxima of its component lines, but these are indistinguishable. The only noteworthy inaccuracy in this simplification is that the fraction of atoms at maximum is that capable of absorbing *some* line of the multiplet. The fraction capable of absorbing one given line is naturally less, and in some applications this point may need attention. We tabulate for each element or each group of lines of each element the data† used in the calculations. This is followed by the observed position of the maximum, and then by a few calculations for a series of maximum temperatures of the electron pressure, the fraction of atoms in the required stage of ionization, the fraction in the required stationary state, and the product of this last fraction by the electron pressure.

(i) V. $1\,^6D-{}^6F$ (part); $\lambda\lambda$ 4395, 4390, 4385, 4379; $\chi_0=6{\cdot}30$, $(\chi_0)_1=6{\cdot}02$;

$$u_0(T)=28+30e^{-0{\cdot}28/kT}+20e^{-1{\cdot}05/kT},\quad u_1(T)=25+35e^{-0{\cdot}32/kT}.$$

For the least blended lines no maximum is observed. It lies below *M* 3.

T_{max}	p_ϵ atmos.	$(x_0)_{max}$	$\{(n_0)_1\}_{max}$	$(n_0)_1 p_\epsilon$
2500	$1{\cdot}07\times10^{-9}$	0·96	$2{\cdot}17\times10^{-1}$	$2{\cdot}32\times10^{-10}$
3000	$2{\cdot}44\times10^{-7}$	0·96	$2{\cdot}54\times10^{-1}$	$6{\cdot}2\times10^{-8}$

(ii) Ba^+. $1\,^2S-1\,^2P$; $\lambda\lambda$ 4934, 4554; $\chi_0=5{\cdot}18$, $\chi_1=9{\cdot}95$;

$$\tfrac{1}{2}(\chi_1-\chi_0)=2{\cdot}385>1{\cdot}512=\chi_1-(\chi_1)_1;$$

$$u_0(T)=1+5e^{-1{\cdot}41/kT},\quad u_1(T)=2+10e^{-1{\cdot}512/kT},\quad u_2(T)=1.$$

* Backer and Goudsmit, *Atomic Energy States*, New York (1932).

† Energies in electron volts.

[Variations of $u_0(T)$ may be ignored.] The maximum must lie at $M\,3$ or just below. $(n_1)_{max}=1$.

T_{max}	p_ϵ atmos.	$(n_1)\,p_\epsilon$
2500	$1{\cdot}66\times10^{-11}$	$1{\cdot}66\times10^{-11}$
3000	$4{\cdot}6\ \times10^{-9}$	$4{\cdot}6\ \times10^{-9}$
3500	$2{\cdot}72\times10^{-7}$	$2{\cdot}72\times10^{-7}$

(iii) Cr. $1\,^5S-{}^5P$; $\lambda\lambda$ 4497; $\chi_0=6{\cdot}75$, $(\chi_0)_1=5{\cdot}81$;

$$u_0(T)=7+5e^{-0{\cdot}94/kT}+25e^{-1{\cdot}00/kT},\quad u_1(T)=6+30e^{-0{\cdot}40/kT}+20e^{-1{\cdot}20/kT}.$$

The observed maximum lies at $M\,1$ (Menzel).

T_{max}	p_ϵ atmos.	$(x_0)_{max}$	$\{(n_0)_1\}_{max}$	$(n_0)_1 p_\epsilon$
2500	$5{\cdot}0\ \times10^{-11}$	0·87	$7{\cdot}8\ \times10^{-3}$	$3{\cdot}9\ \times10^{-13}$
3000	$1{\cdot}74\times10^{-8}$	0·87	$1{\cdot}60\times10^{-2}$	$2{\cdot}78\times10^{-10}$
3500	$1{\cdot}21\times10^{-6}$	0·87	$2{\cdot}66\times10^{-2}$	$3{\cdot}2\ \times10^{-8}$

(iv) Sr^+. $1\,^2S-1\,^2P$; $\lambda\lambda$ 4216, 4078; $\chi_0=5{\cdot}665$, $\chi_1=10{\cdot}97$;

$$\tfrac{1}{2}(\chi_1-\chi_0)=2{\cdot}652>1{\cdot}815=\chi_1-(\chi_1)_1;$$

$$u_0(T)=1,\quad u_1(T)=2+10e^{-1{\cdot}815/kT},\quad u_2(T)=1.$$

The observed maximum lies at $M\,0$ (Menzel) or $K\,2$ (Payne), the latter is probably the more reliable. $(n_1)_{max}=1$.

T_{max}	p_ϵ atmos.	$(n_1)\,p_\epsilon$
3000	$2{\cdot}42\times10^{-10}$	$2{\cdot}42\times10^{-10}$
3500	$2{\cdot}02\times10^{-8}$	$2{\cdot}02\times10^{-8}$
4000	$6{\cdot}8\ \times10^{-7}$	$6{\cdot}8\ \times10^{-7}$

(v) Mg. $1\,^3P-1\,^3S$; $\lambda\lambda$ 5184, 5173, 5167; $\chi_0=7{\cdot}61$, $(\chi_0)_1=4{\cdot}92$;

$$u_0(T)=1,\quad u_1(T)=2;\quad (\varpi_0)_1=9.$$

The observed maximum lies at $K\,2$–5 but is poorly determined.

T_{max}	p_ϵ atmos.	$(x_0)_{max}$	$\{(n_0)_1\}_{max}$	$(n_0)_1 p_\epsilon$
3500	$2{\cdot}42\times10^{-8}$	0·68	$8{\cdot}1\ \times10^{-4}$	$1{\cdot}95\times10^{-11}$
4000	$7{\cdot}2\ \times10^{-7}$	0·68	$2{\cdot}47\times10^{-3}$	$1{\cdot}78\times10^{-9}$
4500	$1{\cdot}14\times10^{-5}$	0·69	$5{\cdot}7\ \times10^{-3}$	$6{\cdot}4\ \times10^{-8}$
5000	$1{\cdot}08\times10^{-4}$	0·69	$1{\cdot}19\times10^{-2}$	$1{\cdot}28\times10^{-6}$

(vi) Fe. $1\,^3F-{}^3X$; $X=D, F, G$; some 16 lines in all; $\chi_0=8{\cdot}15$, $(\chi_0)_2=6{\cdot}59$;

$$u_0(T)=25+35e^{-0{\cdot}95/kT}+21e^{-1{\cdot}56/kT},\quad u_1(T)=30+28e^{-0{\cdot}33/kT}+\dots.$$

The observed maximum is well determined at $K\,2$.

T_{max}	p_ϵ atmos.	$(x_0)_{max}$	$\{(n_0)_2\}_{max}$	$(n_0)_2 p_\epsilon$
3500	$4{\cdot}0\ \times10^{-9}$	0·82	$3{\cdot}7\ \times10^{-3}$	$1{\cdot}47\times10^{-11}$
4000	$1{\cdot}67\times10^{-7}$	0·83	$6{\cdot}8\ \times10^{-3}$	$1{\cdot}13\times10^{-9}$
4500	$3{\cdot}3\ \times10^{-6}$	0·83	$1{\cdot}10\times10^{-2}$	$3{\cdot}6\ \times10^{-8}$

(vii) Ca^+. $1\,^2S-1\,^2P$; $\lambda\lambda$ 3968, 3934; $\chi_0=6{\cdot}08$, $\chi_1=11{\cdot}82$;

$$\tfrac{1}{2}(\chi_0-\chi_1)=2{\cdot}87>1{\cdot}69=\chi_1-(\chi_1)_1;$$

$$u_0(T)=1,\quad u_1(T)=2+10e^{-1\cdot 69/kT},\quad u_2(T)=1.$$

There are observations of a maximum at $K\,0$ (Menzel), but it is possible that there is no maximum before $M\,0$ (Payne). $(n_1)_{\max}=1$.

$T_{\max}$	p_ϵ atmos.	$(n_1)\,p_\epsilon$
3000	$6{\cdot}8\times10^{-11}$	$6{\cdot}8\times10^{-11}$
3500	$7{\cdot}4\times10^{-9}$	$7{\cdot}4\times10^{-9}$
4000	$2{\cdot}56\times10^{-7}$	$2{\cdot}56\times10^{-7}$
4500	$3{\cdot}9\times10^{-6}$	$3{\cdot}9\times10^{-6}$

(viii) Zn. $1\,^3P-1\,^3S$; $\lambda\lambda$ 4810, 4722 (4680); $\chi_0=9{\cdot}34$, $(\chi_0)_1=5{\cdot}33$; $u_0(T)=1$, $u_1(T)=2$; $(\varpi_0)_1=9$. The observed maximum is well determined at $G\,0$.

$T_{\max}$	p_ϵ atmos.	$(x_0)_{\max}$	$\{(n_0)_1\}_{\max}$	$(n_0)_1p_\epsilon$
5000	$1{\cdot}38\times10^{-6}$	0·61	$4{\cdot}9\times10^{-4}$	$6{\cdot}8\times10^{-10}$
6000	$8{\cdot}4\times10^{-5}$	0·62	$2{\cdot}52\times10^{-3}$	$2{\cdot}12\times10^{-7}$

There are well-determined maxima of many lines of Fe^+ and Ti^+ at $F\,5$, some rising to $F\,0$ and some falling to $G\,0$, but χ_1 for these elements is not yet well determined. The values 16·5 and 13·6 volts are given by Backer and Goudsmit for Fe^+ and Ti^+ respectively. The following calculation is carried out for 15 volts and represents a mean value for the Fe^+ and Ti^+ lines rather than an accurate calculation for any single line of either element.

(ix) Fe^+. $2\,^4P-1\,^4F$, $^4D'$; $\lambda\lambda$ 4173·3, 4179, 4417; approximately $\chi_1=15$, $(\chi_1)_s=12{\cdot}3$; $u_1(T)=60$, $u_2(T)=37$; $(\varpi_1)_s=12$. There is a well-observed maximum at $F\,5$.

$T_{\max}$	p_ϵ atmos.	$(x_1)_{\max}$	$\{(n_1)_1\}_{\max}$	$(n_1)_1p_\epsilon$
6000	$1{\cdot}37\times10^{-9}$	0·83	$8{\cdot}9\times10^{-4}$	$1{\cdot}22\times10^{-12}$
7000	$1{\cdot}30\times10^{-7}$	0·84	$1{\cdot}89\times10^{-3}$	$2{\cdot}46\times10^{-10}$

(x) Mg^+. $1\,^2D-1\,^2F$; $\lambda\lambda$ 4481; $\chi_1=15{\cdot}00$, $(\chi_1)_1=6{\cdot}15$; $u_1(T)=2$, $u_2(T)=1$; $(\varpi_1)_1=10$. There is a well-observed maximum at $A\,2$.

$T_{\max}$	p_ϵ atmos.	$(x_1)_{\max}$	$\{(n_1)_1\}_{\max}$	$(n_1)_1p_\epsilon$
9,000	$9{\cdot}0\times10^{-6}$	0·48	$2{\cdot}58\times10^{-5}$	$2{\cdot}30\times10^{-10}$
10,000	$8{\cdot}2\times10^{-5}$	0·48	$8{\cdot}2\times10^{-5}$	$6{\cdot}8\times10^{-9}$
11,000	$5{\cdot}3\times10^{-4}$	0·49	$2{\cdot}12\times10^{-4}$	$1{\cdot}12\times10^{-7}$

(xi) H. Balmer series; $\lambda\lambda$ 4861, etc.; $\chi_0 = 13{\cdot}54$, $(\chi_0)_1 = 3{\cdot}385$; $u_0(T) = 2$, $u_1(T) = 1$; $(\varpi_0)_1 = 8$. The maximum is at A 0, by definition of the type.

T_{max}	p_ϵ atmos.	$(x_0)_{max}$	$\{(n_0)_1\}_{max}$	$(n_0)_1 p_\epsilon$
9,000	$3{\cdot}3 \times 10^{-5}$	0·34	$2{\cdot}76 \times 10^{-6}$	$9{\cdot}2 \times 10^{-11}$
10,000	$2{\cdot}62 \times 10^{-4}$	0·35	$1{\cdot}05 \times 10^{-5}$	$2{\cdot}75 \times 10^{-9}$
11,000	$1{\cdot}44 \times 10^{-3}$	—	—	—

(xii) Si^+. $1\,^2D - 1\,^2F$; $\lambda\lambda$ 4131, 4128; $\chi_1 = 16{\cdot}27$, $(\chi_1)_1 = 6{\cdot}47$; $u_1(T) = 6$, $u_2(T) = 1$; $(\varpi_1)_1 = 10$. The observed maximum is well defined at A 0.

T_{max}	p_ϵ atmos.	$(x_1)_{max}$	$\{(n_1)_1\}_{max}$	$(n_1)_1 p_\epsilon$
10,000	$5{\cdot}9 \times 10^{-6}$	0·47	$8{\cdot}8 \times 10^{-6}$	$5{\cdot}2 \times 10^{-11}$
11,000	$4{\cdot}3 \times 10^{-5}$	0·47	$2{\cdot}51 \times 10^{-5}$	$1{\cdot}08 \times 10^{-9}$

(xiii) C^+. $1\,^2D - 1\,^2F$; $\lambda\lambda$ 4267; $\chi_1 = 24{\cdot}28$, $(\chi_1)_1 = 6{\cdot}31$; $u_1(T) = 6$, $u_2(T) = 1$; $(\varpi_1)_1 = 10$. The observed maximum is well defined at B 3.

T_{max}	p_ϵ atmos.	$(x_1)_{max}$	$\{(n_1)_1\}_{max}$	$(n_1)_1 p_\epsilon$
14,000	$1{\cdot}96 \times 10^{-6}$	0·34	$1{\cdot}88 \times 10^{-7}$	$3{\cdot}7 \times 10^{-13}$
16,000	$3{\cdot}7 \times 10^{-5}$	0·35	$1{\cdot}25 \times 10^{-6}$	$4{\cdot}6 \times 10^{-11}$
18,000	$3{\cdot}7 \times 10^{-4}$	0·36	$5{\cdot}5 \times 10^{-6}$	$2{\cdot}02 \times 10^{-9}$

(xiv) He. Sharp and diffuse series of par- and ortho-helium; $\lambda\lambda$ 4712, 4471, 4922, etc.; $\chi_0 = 24{\cdot}47$, $(\chi_0)_1 = 3{\cdot}48$ (mean); $u_0(T) = 1$, $u_2(T) = 2$; $(\varpi_0)_1 = 3$ (par-He) or 9 (ortho-He). The observed maximum is well defined at B 2–3.

T_{max}	p_ϵ atmos.	$(x_0)_{max}$	$\{(n_0)_1\}_{max}$ [par-He]	$(n_0)_1 p_\epsilon$
16,000	$2{\cdot}68 \times 10^{-4}$	0·248	$1{\cdot}76 \times 10^{-7}$	$4{\cdot}7 \times 10^{-11}$
18,000	$2{\cdot}76 \times 10^{-3}$	0·259	$1{\cdot}00 \times 10^{-6}$	$2{\cdot}76 \times 10^{-9}$

(xv) N^+. $1\,^1P' - 1\,^1P$; λ 3995; $\chi_1 = 29{\cdot}48$, $(\chi_1)_s = 11{\cdot}06$; $u_1(T) = 9$, $u_2(T) = 6$; $(\varpi_1)_s = 3$. The observed maximum is at B 3–5.

T_{max}	p_ϵ atmos.	$(x_1)_{max}$	$\{(n_1)_s\}_{max}$	$(n_1)_s p_\epsilon$
18,000	$8{\cdot}5 \times 10^{-5}$	0·45	$1{\cdot}03 \times 10^{-6}$	$8{\cdot}7 \times 10^{-11}$
20,000	$1{\cdot}35 \times 10^{-3}$	0·45	$3{\cdot}4 \times 10^{-6}$	$4{\cdot}6 \times 10^{-9}$

(xvi) Si^{++}. $1\,^3S - 2\,^3P$; $\lambda\lambda$ 4552, 4567, 4574; $\chi_2 = 33{\cdot}35$, $(\chi_2)_s = 14{\cdot}40$; $u_2(T) = 1$, $u_3(T) = 2$; $(\varpi_2)_s = 3$. There is a well-defined maximum at B 1–2.

T_{max}	p_ϵ atmos.	$(x_2)_{max}$	$\{(n_2)_s\}_{max}$	$(n_2)_s p_\epsilon$
18,000	$2{\cdot}48 \times 10^{-5}$	0·49	$7{\cdot}1 \times 10^{-6}$	$1{\cdot}77 \times 10^{-10}$
20,000	$2{\cdot}82 \times 10^{-4}$	0·50	$2{\cdot}44 \times 10^{-5}$	$6{\cdot}9 \times 10^{-9}$

(xvii) O^+. $1\,^2P - 1\,^2D$; $\lambda\lambda$ 4417, 4415; $\chi_1 = 35{\cdot}00$, $(\chi_1)_s = 11{\cdot}53$; $u_1(T) = 4 + 10e^{-3{\cdot}34/kT} + 6e^{-5{\cdot}04/kT}$, $u_2(T) = 9$; $(\varpi_1)_s = 6$. There is a well-defined maximum at $B1$. [A number of other O^+ lines with $(\chi_1)_s$ varying down to 8·5 volts have maxima at $B1$ or perhaps at slightly higher temperatures towards $B0$.]

T_{max}	p_ϵ atmos.	$(x_1)_{max}$	$\{(n_1)_s\}_{max}$	$(n_1)_s p_\epsilon$
18,000	$4{\cdot}7\times10^{-6}$	0·40	$1{\cdot}13\times10^{-7}$	$5{\cdot}3\times10^{-13}$
20,000	$5{\cdot}7\times10^{-5}$	0·40	$5{\cdot}0\times10^{-7}$	$2{\cdot}84\times10^{-11}$
22,000	$4{\cdot}5\times10^{-4}$	0·41	$1{\cdot}65\times10^{-6}$	$8{\cdot}2\times10^{-10}$

(xviii) Si^{+++}. $2\,^2S-2\,^2P$; λλ 4089, 4116; $\chi_3=44{\cdot}94$, $(\chi_3)_s=20{\cdot}98$; $u_3(T)=2$, $u_4(T)=1$; $(\varpi_3)_s=2$. The observed maximum occurs in *O*-type stars at Plaskett's *O* 9.

T_{max}	p_ϵ atmos.	$(x_3)_{max}$	$\{(n_3)_s\}_{max}$	$(n_3)_s p_\epsilon$
24,000	$1{\cdot}14\times10^{-5}$	0·52	$4{\cdot}7\times10^{-6}$	$5{\cdot}4\times10^{-11}$
26,000	$7{\cdot}4\times10^{-5}$	0·52	$1{\cdot}16\times10^{-5}$	$8{\cdot}7\times10^{-10}$

(xix) C^{++}. $3\,^3S-3\,^3P$; λλ 4649, 4651, 4653; $\chi_2=46{\cdot}35$, $(\chi_2)_s=18{\cdot}05$; $u_2(T)=1$, $u_3(T)=2$; $(\varpi_2)_s=3$. There is a well-defined maximum near that of Si^{+++} at *O* 9.

T_{max}	p_ϵ atmos.	$(x_2)_{max}$	$\{(n_2)_s\}_{max}$	$(n_2)_s p_\epsilon$
24,000	$1{\cdot}73\times10^{-5}$	0·45	$1{\cdot}52\times10^{-6}$	$2{\cdot}62\times10^{-11}$
26,000	$1{\cdot}20\times10^{-4}$	0·46	$4{\cdot}3\times10^{-6}$	$5{\cdot}3\times10^{-10}$

(xx) N^{++}. $1\,^2S-2\,^2P_2$; λ 4098; $\chi_2=47{\cdot}19$, $(\chi_2)_s=19{\cdot}20$; $u_2(T)=6$, $u_3(T)=1$; $(\varpi_2)_s=2$. The maximum has not been reached among the *O*-type stars investigated.

T_{max}	p_ϵ atmos.	$(x_2)_{max}$	$\{(n_2)_s\}_{max}$	$(n_2)_s p_\epsilon$
26,000	$7{\cdot}3\times10^{-6}$	0·47	$5{\cdot}7\times10^{-7}$	$4{\cdot}2\times10^{-12}$
28,000	$4{\cdot}1\times10^{-5}$	0·47	$1{\cdot}49\times10^{-6}$	$6{\cdot}1\times10^{-11}$
30,000	$1{\cdot}82\times10^{-4}$	0·48	$3{\cdot}1\times10^{-6}$	$5{\cdot}7\times10^{-10}$

(xxi) He^+. "4686 series"; λ 4686; $\chi_1=54{\cdot}16$, $(\chi_1)_s=6{\cdot}02$; $u_1(T)=2$, $u_2(T)=1$; $(\varpi_1)_s=18$.

"Pickering series"; λλ 5412, etc.; $(\chi_1)_{s'}=3{\cdot}39$; $(\varpi_1)_{s'}=32$.

The lines seem nearly to have reached their maximum at the end of Plaskett's sequence of *O*-stars. The observed maximum may be taken to occur at *O* 5. The calculations are for the 4686 series.

T_{max}	p_ϵ atmos.	$(x_1)_{max}$	$\{(n_1)_s\}_{max}$ ["4686"]	$(n_1)_s p_\epsilon$
30,000	$1{\cdot}06\times10^{-5}$	0·206	$1{\cdot}45\times10^{-8}$	$1{\cdot}54\times10^{-13}$
36,000	$5{\cdot}9\times10^{-4}$	0·222	$3{\cdot}5\times10^{-7}$	$2{\cdot}03\times10^{-10}$
40,000	$4{\cdot}7\times10^{-3}$	0·233	$1{\cdot}76\times10^{-6}$	$8{\cdot}3\times10^{-9}$

In addition to these maxima the observations show no maximum for any normal lines of any neutral atom in the sequence of giants, in full accord with the theory. To illustrate the way in which the concentration of atoms in a given state in a reversing layer varies with the temperature according to statistical theory, using merely the equation (1707), we reproduce a figure* showing typical curves. These are only roughly accurate and for applications are now superseded by the curves given by Russell (§ 15·34).

§ 15·33. *Discussion and theoretical developments. A first approximation.* It will be evident from a survey of these tables that there is excellent general agreement between theory and observation as to the positions of the maxima in the sequence of giants, on certain conditions. We have to suppose that the partial pressure of the electrons in the layer in which the lines originate is of the order 10^{-9}–10^{-10} atmosphere for the normal lines of the ionized atoms Ca^+, Sr^+ and Ba^+, and the nearly normal lines of very small energy of excitation such as those of V, Cr (and Ti); we suppose further that it increases steadily as the energy of excitation increases (compared with the energy of ionization) until it reaches 10^{-5} atmosphere for subordinate lines of average excitation energy such as those of Mg, Mg^+ and Si^{+++}, and finally perhaps 10^{-3} atmosphere for the lines of exceptionally great excitation energy such as those of He and He^+. The elementary theory of the maxima was based on a given value of p_ϵ. This is confirmed by the more complete data so long as we confine ourselves to lines of similar depth in the various atoms. The variation of the necessary p_ϵ with variation of depth, which is the same thing as variation with $(n_s)_{max}$, suggests the next stage in a more refined investigation.

Consider any slab of atmosphere in equilibrium at a mean temperature T. The concentration of any atomic species of mass m and charge $v\epsilon$ is given approximately by $$\nu = \nu_0 e^{-mgz/kT - v\epsilon\psi/kT}.$$

We have already seen that the electrostatic term ψ may profoundly modify the equilibrium distribution of ions. We must be content to assume that

$$\nu = \nu_0 e^{-m'gz/kT}, \qquad \qquad \text{......(1722)}$$

where m' is used for the effective atomic mass. Consider the variation of homologous slabs of these atmospheres along the stellar sequence. Homologous slabs will be determined by equal changes in the exponent in (1722), and therefore the thickness of a given slab will be fixed by an equation like

$$\left[\frac{m'gz}{kT}\right]_0^1 = A,$$

where A is some constant such as 1 or 10, and 0 and 1 refer to the bottom and

* Fowler and Milne, *loc. cit.* (1).

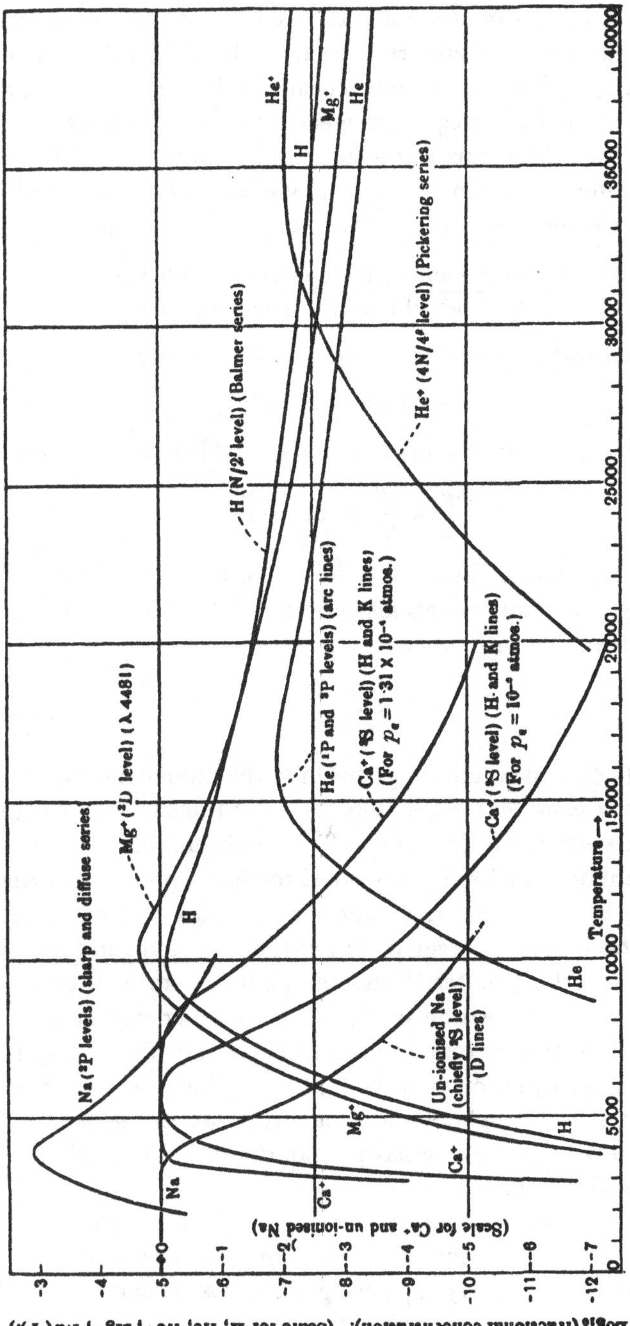

Fig. 73. Showing typical curves of atomic fractional concentrations, calculated without allowing for further successive ionizations, and with inaccurate weights.

top of the slab. Thus the thickness of homologous slabs varies as $T/m'g$ from star to star, and atom to atom. The number of any atomic species per cm.^{2} of the stellar surface therefore also varies like $T/m'g$. [In an atmosphere perfectly mixed by large scale convections, such as the lower atmosphere of the earth, m' will have a common mean value for all atoms.]

If p_ϵ is the partial pressure of the electrons at any level in the slab, then the concentration of electrons ν_ϵ is given by $\nu_\epsilon = p_\epsilon/kT$. If ω is an abundance factor for the element in question, defined by the equation

$$\omega = \frac{\text{Number of nuclei of given atomic number per cm.}^3}{\text{Number of free electrons per cm.}^3},$$

then the concentration ν of these atomic nuclei satisfies

$$\nu = \omega\nu_\epsilon = \omega p_\epsilon/kT,$$

and the number of these nuclei per cm.^{2} of the stellar surface in a given slab will vary as

$$\frac{\omega p_\epsilon}{kT} \times \frac{T}{m'g}, \quad \text{or} \quad \frac{\omega p_\epsilon}{m'g}.$$

If $(n_r)_s$ is the fraction of these nuclei which provide atoms r-times ionized in state s, then the number of atoms per cm.^{2} of the stellar surface in a state to absorb a given line will vary as

$$\frac{\omega p_\epsilon (n_r)_s}{m'g}. \qquad \text{......(1723)}$$

In (1723) ω, $(n_r)_s$, and m' are functions of Z the atomic number.

We must now consider more closely the conditions under which a strong absorption line is formed. There must be enough suitable atoms per cm.^{2} to form it. It cannot therefore be formed at too low a pressure, or too high in the reversing layer. But at the same time it must be formed as high as possible in the reversing layer in order that the temperature difference between the absorbing material and the photosphere may be as great as possible.* The actual position of the effective layer will be fixed by the balancing of these two factors. Viewed from the outside we may regard an absorption line as a particular wave length at which the absorption coefficient of the stellar atmosphere is abnormally great, so that we can only see into the star down to an abnormally small depth, a depth which is that of the effective slab for this line.

In our elementary calculations we start by determining maxima by $d(n_r^z)_s/dT = 0$, that is for a given value of $\omega p_\epsilon/m'g$. This means that for the selected value of (effectively) p_ϵ/g we get the best absorption at the temperature so determined. It remains to examine how to choose or fix the

* See § 15·34 for a quantitative formulation.

proper value of p_ϵ for use in this way. The *maximum* concentration of suitable atoms per cm.[2] in the slab of the chosen p_ϵ varies as

$$\frac{\omega\{(n_r^z)_s\}_{\max} p_\epsilon}{m'g}.$$

The value of p_ϵ for use here must be large enough to give enough suitable atoms but no larger. Now it is well known from general evidence that to produce a strong absorption line one needs about the same number per cm.[2] of any sort of atom in the suitable state—or more generally that the number which is required will vary as $1/(\alpha_r^z)_s$, where α is the atomic absorption coefficient for selective absorption of this line by atoms in the suitable state. The argument therefore suggests that

$$\frac{\omega\{(n_r^z)_s\}_{\max} p_\epsilon}{m'g} = \frac{A}{(\alpha_r^z)_s}, \qquad \text{......(1724)}$$

where A is a constant, which this elementary reasoning cannot fix. There is considerable evidence that $(\alpha_r^z)_s$ is of the order 10^8 and not subject to much variation from state to state or atom to atom for the lines of stellar importance.

In equation (1724) there are several factors which it is legitimate to ignore in a first approximation, for the argument is not sufficiently refined to take account of their probable variations. We have already made some comment on m', or rather $m'g$. This really represents the force of gravity on the atom as modified by electrostatic fields and radiation pressure, in particular for the pressure due to the selective absorption forming the line. Gravity itself increases along the giant sequence approximately as T^4, but it is very doubtful if this variation will remain effective in $m'g$. Again ω must decrease as T and therefore ν_ϵ increases, but this is a slow variation, and it will vary by unknown amounts from atom to atom. It seems reasonable to conclude from this rough argument *that* $\{(n_r^z)_s\}_{\max} p_\epsilon$ *calculated in the preceding section should have a roughly constant value for all maxima of absorption lines in stellar spectra*. This conclusion is tentative. Argument and conclusion must both be replaced by a more exact theory.

This conclusion, it will be seen, at once accounts for the variations of p_ϵ required by the calculations on which we commented before. We see now that if the constant value of $\{(n_r^z)_s\}_{\max} p_\epsilon$ is say 2×10^{-10} (p_ϵ in atmospheres), the theoretical maxima fit coherently with the observations, and determine with some precision and convincingness a stellar temperature scale. We give below the resulting temperatures derived by interpolation from the tables. It will be observed that the equation

$$\{(n_r^z)_s\}_{\max} p_\epsilon = 2 \times 10^{-10}$$

determines $T_{\max}$ directly, without explicit assumption as to p_ϵ, as the root of the equation (for subordinate lines)

$$2 \times 10^{-10} = \frac{0{\cdot}664(\varpi_r{}^z)_s\{(\chi_r{}^z)_s + \frac{5}{2}kT\}^2 u^z{}_{r+1}(T)}{\{\chi_r{}^z - (\chi_r{}^z)_s\}\{\chi_r{}^z + \frac{5}{2}kT\}\{u_r{}^z(T)\}^2} T^{\frac{5}{2}} e^{-\{2\chi_r{}^z - (\chi_r{}^z)_s\}/kT}. \quad \text{......(1725)}$$

There are of course slight modifications when u_r' and u'_{r+1} are included. We derive maxima from (1725) by assuming that $m'g/\omega\alpha$ is roughly constant from atom to atom and star to star, and have fixed its mean value by a rough average over all the stellar material.

TABLE 63.

The stellar temperature scale derived from absorption line maxima.

Atom	Temperature of maximum	Spectral type of observed maximum
V	2,500	Below *M* 3
Ba+	2,700	*M* 3 or below
Cr	2,900	*M* 1
Sr+	3,000	*M* 0 [*K* 5]
Mg	3,700	*K* 2–5
Fe	3,800	*K* 2
Ca+	3,100	*K* 0 [? *M* 0]
Zn	4,800	*G* 0
Fe+	7,000	*F* 5
Mg+	9,000	*A* 2
H	9,250	*A* 0 [*A* 2]
Si+	10,500	*A* 0
N+	18,400	*B* 3–5
C+	16,800	*B* 3
He	16,800	*B* 2–3
Si++	18,000	*B* 1–2
O+	21,100	*B* 0–1
Si+++	25,000	*O* 9
C++	25,400	*O* 9
N++	29,100	Not observed
He+	36,000	Not observed

This scale is eminently satisfactory, and there are few anomalies in the individual entries. When the uncertainties in the material are remembered, there remains only one glaring inconsistency, the value for N^+. It must be admitted that for whatever cause, this maximum does not fit in with this approximate theory. The approximate theory also cannot account for the Ca^+ maximum for type *K* 0, but the more refined theory which we shall discuss in the next section makes important modifications in this case.

§ **15·34.** *The formation of absorption lines. A second approximation.* In the previous sections we have shown in a general way how statistical theory enables us to give a rough explanation of the observed linear sequence of stellar spectra. A more detailed and logical explanation can only be obtained by combining statistical theory with a proper theory of the actual formation of the absorption lines themselves. We need something more exact to replace the quantity "average concentration of atoms in a state fit to absorb the line" which underlay our first approximation. We must also define more precisely what we mean by the *intensity* of an absorption line. The complete theory, which is largely due to Milne,* involves a solution of the problem of radiative transfer in stellar atmospheres and lies outside the scope of this monograph. We must content ourselves with a brief account of general principles.

The theory of the formation of an absorption line originated with Schuster's† classical discussion of the following problem: *A layer of gas is placed in front of a bright background which radiates monochromatic light of frequency ν with an outward flux equal to πG_ν for each cm.[2] of radiating surface. The layer of gas contains N atoms (or molecules) per cm.[2] of the radiating surface and each atom has a scattering coefficient‡ α_ν. The flux emerging from the layer of gas is πF_ν per cm.[2] surface. The problem is to determine F_ν/G_ν.* It can be shown that

$$r_\nu = \frac{F_\nu}{G_\nu} = \frac{1}{1+\frac{3}{4}N\alpha_\nu}. \qquad \text{......(1726)}$$

If now the material above the radiating surface is characterized by different values of α_ν for different frequencies, the Schuster-Milne formula gives the residual intensity r_ν as a function of ν. This formula has obvious applications to the theory of the contours of absorption lines, but it must be remembered that in using it we assume that there is something in a real star corresponding roughly to Schuster's radiating surface. In practice the "photosphere" lying at the base of the reversing layer is taken to correspond to this radiating surface; but this is merely a matter of words unless the photosphere is more sharply defined.

If α_ν is a sensitive function of ν in the neighbourhood of a characteristic frequency ν_0 and otherwise zero, then r_ν is unity except near ν_0 and as ν passes through ν_0 r_ν will pass through a sharp minimum. We thus obtain a fair representation of a line contour. Hence a comparison of the observed contour with the Schuster-Milne contour combined with a quantum determination of α_ν provides an astrophysical determination of N the number of

* Milne, *Monthly Not. R.A.S.* vol. 89, pp. 2, 17, 157 (1928); McCrea, *Monthly Not. R.A.S.* vol. 91, p. 836 (1931); Milne and Chandrasekhar, *Monthly Not. R.A.S.* vol. 92, p. 150 (1932); Milne, *Phil. Trans.* A, vol. 228, p. 421 (1929).

† Schuster, *Astrophys. J.* vol. 21, p. 1 (1905). The formula (1726) is due to Milne (*loc. cit.*). Schuster gave a less accurate form of it.

‡ A scattering coefficient is defined as follows: The flux of radiation in a beam is I_ν. It falls on a slab of thickness δx containing n atoms per c.c., and emerges as a flux $I_\nu - \delta I_\nu$, the rest of the light being deflected (scattered) unaltered in frequency into other directions. If $\delta I_\nu = I_\nu n\alpha_\nu \delta x$, α_ν is called the atomic scattering coefficient for radiation of frequency ν.

scattering atoms above the photosphere. Such comparisons and determinations of N have been made by Stewart and by Unsöld.*

The idealization of stellar atmospheres to Schuster's conditions forms the basis of an acceptable theory, but the theory must provide at the same time for fixing the precise level of the photosphere, for otherwise "the number of atoms above the photosphere" has no precise meaning and Stewart's and Unsöld's determinations of N are useless. In actual stellar atmospheres the upper layers contributing to the line absorption continuously shade off into deeper layers contributing to the general opacity for light of all frequencies. To give a precise meaning therefore to these determinations of N we must answer the following question: *What is the height of the transparent atmosphere (i.e. one without general opacity) equivalent to the infinite column of the real atmosphere in which the atoms at greater depths are lost more and more in the general fog (opacity)?* This question has been answered by Milne. If

$$\tau_\nu = \int_x^\infty \kappa_\nu \rho \, dx,$$

where κ_ν is the coefficient of absorption† of the atmosphere per unit mass averaged over the frequencies of the absorption line in question, τ_ν is called the optical thickness of the atmosphere down to depth x. Then if x is chosen so that

$$\tau_\nu = \frac{4}{3}\frac{r_\nu}{r_\nu + 1} \qquad \text{......(1727)}$$

x is the level of the photosphere to be used with the Schuster-Milne formula, and the values of N determined by Stewart's and Unsöld's methods mean numbers of atoms above the depth x. The physical meaning of τ_ν is that it is the optical thickness (or fogginess) that the atmosphere must have so that the atoms down to this level (x) when fully viewed against a bright background would produce the observed line contour. In the real atmosphere the atoms below $\tau_\nu(x)$ are contributing some effect, but the atoms down to τ_ν are not fully viewed, and if fully viewed would produce the same effect as all the atoms down to $\tau_\nu = \infty$ more and more obscured as the depth increases by the general fog.

The relation (1727) enables us at once to define the intensity of an absorption line more precisely. Consider a given absorption line in a number of spectra and select in each that point on the line contour where the intensity

* Stewart, *Astrophys. J.* vol. 59, p. 30 (1924); Unsöld, *Zeit. f. Physik*, vol. 44, p. 793 (1927), vol. 46, p. 765 (1928).

† The coefficient of absorption is defined just as the scattering coefficient so that the flux in a beam (of mixed frequencies) falling on a slab of scattering material of density ρ is reduced by $\delta I_\nu = I_\nu \kappa_\nu \rho \, \delta x$ in passing through the slab, where κ_ν is the average absorption coefficient of the material for the frequencies in the beam. The difference is that the radiation so removed is not re-emitted without change of wave length but converted into heat energy and re-emitted as temperature radiation in equilibrium with the absorber at its actual temperature.

ratio r has a previously assigned value, say $r=\frac{1}{3}$. Determine the corresponding frequency $\nu_{\frac{1}{3}}$. Then $|\nu_{\frac{1}{3}}-\nu_0|$, where ν_0 is the centre of the line, measures the width of the line for the prescribed value of r, $r=\frac{1}{3}$. In general this width will vary from spectrum to spectrum. But by Schuster's formula the selective optical thickness, $N\alpha_\nu$, will be constant. Since $\alpha_\nu \propto (\nu-\nu_0)^{-2}$ we find

$$N \propto (\nu_{\frac{1}{3}}-\nu_0)^2. \qquad \text{......(1728)}$$

Consequently a study of line widths for given r provides information about the number of atoms N above a certain optical depth τ_ν given by (1728). In particular the line will be widest for given r when N down to given τ_ν attains its maximum. We may assert that *the intensity of an absorption line, defined as the line width for a prescribed ratio r, is proportional to the square root of the number of atoms in a column of the atmosphere which has a given optical thickness* τ $(=\frac{4}{3}r/(r+1))$ *in the continuous background.*

This formulation explains at once, what was at first thought to be a paradox, why the lines of neutral atoms are enhanced as we pass from dwarfs to giants of the same spectral type. Other things being supposed equal the consequence of a smaller surface gravity on the giant is a general reduction in the mean pressure in the reversing layer, causing increased ionization and a reduction in the concentration of neutral atoms. But at the same time the reduction in the pressure makes the stellar atmosphere more generally transparent, so that to traverse a column of assigned optical thickness we have to descend to much greater actual depths. Hence if the effect of the increased transparency more than compensates for the lower mean concentration of neutral atoms, the observed increase in intensity will be accounted for. Actual calculations show that this increase must in fact take place.

We have shown that according to this more accurate theory we have to calculate the number of atoms in a given state down to a given τ. It is a comparatively simple matter to do this, given the electron pressure p_ϵ at the depth corresponding to τ. We shall illustrate the calculation by a simple example.

Let ρ be the atmospheric density at any level, n the atomic concentration of a particular atom of mass m and β the proportion by mass of these particular atoms to the whole atmosphere at this level.* Then

$$mn=\beta\rho. \qquad \text{......(1729)}$$

Let y be the depth measured inwards from a suitable level of reference. Neglecting radiation pressure, the equation of hydrostatic equilibrium is

$$dp/dy=g\rho. \qquad \text{......(1730)}$$

* The m used here may require to be modified to allow for the effects discussed in §§ 15·2–15·25. But it is unnecessary to take explicit notice of any such effect.

Let x_0, x_1, x_2, ... be, as before, the fractions of these atoms which are neutral, once ionized, etc., N their total number down to a depth h, and N_r the number r-times ionized. Then

$$N=\int_0^h n\,dy, \quad N_r=\int_0^h n x_r\,dy. \qquad \text{......(1731)}$$

Neglecting any variation of β with y it follows that

$$N=\frac{\beta(p-p_0)}{mg}, \quad N_r=\frac{\beta}{mg}\int_{p_0}^{p} x_r\,dp. \qquad \text{......(1732)}$$

We can choose the reference level so that effectively $p_0=0$. Let p_ϵ be the electron pressure at depth h. The typical ionization formula (1705) can be written

$$\frac{x_{r+1}}{x_r}p_\epsilon=K_{r+1}, \qquad \text{......(1733)}$$

where K_{r+1} is a function only of T. The general method of procedure is now to express p, x_0, x_1, ... in terms of p_ϵ, K_1, K_2, ... and evaluate the integrals (1732) taking T and therefore the K_r's as constants. The expressions for p, x_0, x_1, ... differ according to the conditions of the problem. To take an example: if the atoms in question are just becoming once ionized in the presence of an excess of atoms already once ionized, we have $x_1=x$, $x_0=1-x$, $p_\epsilon=\frac{1}{2}p$. From (1732) and (1733) we then deduce

$$N_0=\frac{2\beta}{mg}\left[p_\epsilon-K_1\log\left(1+\frac{p_\epsilon}{K_1}\right)\right], \qquad \text{......(1734)}$$

$$N_1=\frac{2\beta}{mg}K_1\log\left(1+\frac{p_\epsilon}{K_1}\right). \qquad \text{......(1735)}$$

In these equations p_ϵ is the electron pressure at depth h. To determine it we have to calculate the optical thickness τ down to the same depth. For this purpose we need the coefficient of general absorption in the continuous spectrum. Under stellar atmospheric conditions this is of the form*

$$\kappa_\nu=\gamma(\nu)\,p_\epsilon/T^{\frac{7}{2}}, \qquad \text{......(1736)}$$

where $\gamma(\nu)$ is a slowly varying function of the frequency ν. Hence τ is given by

$$\tau=\int_0^h \kappa_\nu \rho\,dy=\frac{1}{g}\int_0^h \kappa_\nu\,dp,$$

and in this example

$$\tau=\frac{const.\,p_\epsilon^{\,2}}{g\ \ T^{\frac{7}{2}}}. \qquad \text{......(1737)}$$

In general we have an expression of the form

$$(N_r)_s=\frac{\beta}{mg}\phi(T,p_\epsilon), \qquad \text{......(1738)}$$

$(N_r)_s$ being the number of atoms per cm.^{2} of photosphere r-times ionized

* Gaunt, *Phil. Trans.* A, vol. 229, p. 163 (1930).

and in their sth state, and associated with (1738) an expression for the optical thickness of the form

$$\tau = \frac{\gamma(\nu)}{g}\,\psi(T, p_\epsilon). \qquad \ldots\ldots(1739)$$

In the proposed method of analysis we fix attention on points of the contour of an absorption line with a specified value of r which are therefore associated with a given value of τ. The method of maxima therefore consists in making (1738) a maximum for variations of T and p_ϵ subject to (1739). The maximum is therefore determined by

$$\frac{\partial\phi}{\partial T}\frac{\partial\psi}{\partial p_\epsilon} - \frac{\partial\phi}{\partial p_\epsilon}\frac{\partial\psi}{\partial T} = 0. \qquad \ldots\ldots(1740)$$

$T = T_{\max}$ (and the associated p_ϵ) are therefore determined as the roots of (1740) and (1739).

To examine the effect of variations of g we differentiate (1738) keeping T constant and treating p_ϵ as a function of g given by (1739). We thus determine the sign of $d(N_r)_s/dg$. For the simple example given above in (1734), (1735) and (1737) it is easily proved that

$$\frac{dN_0}{dg} < 0, \quad \frac{dN_1}{dg} < 0, \quad \frac{d}{dg}\left(\frac{N_1}{N_0}\right) < 0,$$

thus showing that the lines of the neutral and the once ionized atoms must both be enhanced as we pass from dwarfs to giants *of the same surface temperature*, and the once ionized lines enhanced more than those of the neutral atom, results which are nearly enough equivalent to asserting that there is a similar enhancement on passage from dwarfs to giants *of the same spectral type*.

The above theory has been applied in great detail and with great success by Russell* whose final results are summarized in Figs. 74 and in Table 64.

TABLE 64.

Revised stellar temperature scale (after Russell) derived from more accurate analysis of observed maxima.

Element	Calculated temperature of maximum	Observed spectral type of maximum	Calculated spectral type of maximum
Si	5,300	*G*0	*G*2
H (Hγ)	7,500	*A*0	*A*7
O	7,400	—	*A*8
He	16,200	*B*3	*B*4
O^+	21,000	*B*1	*B*2
Si^{++}	20,000	*B*1	*B*3
Si^{+++}	26,000	*B*0	*B*0
He^+	36,000	*O*	*O*7

* Russell, *Astrophys. J.* vol. 78, p. 239 (1933); or *Mt Wilson Contribution*, No. 477.

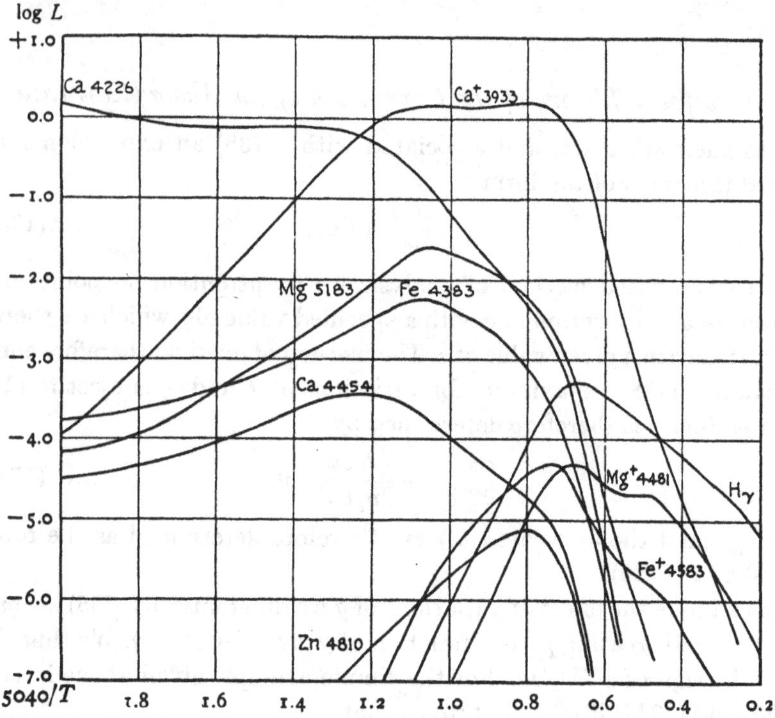

Fig. 74 *a*. Behaviour of the absorption lines of metals and hydrogen for stars of the main sequence. The ordinates in all three figures give the relative strengths of the absorption lines on a logarithmic scale.

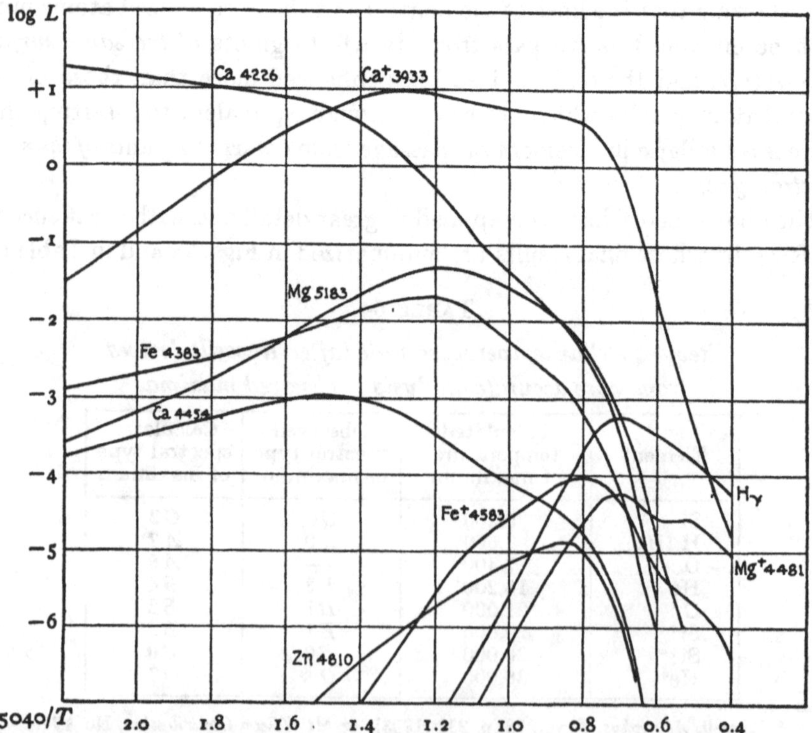

Fig. 74 *b*. Behaviour of the absorption lines of metals and hydrogen for the sequence of giant stars.

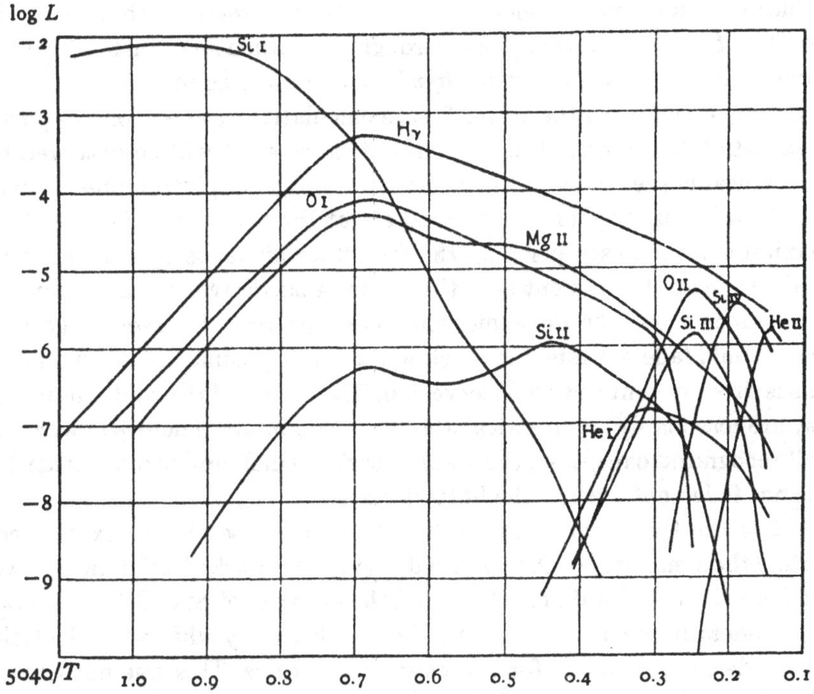

Fig. 74c. Behaviour of the absorption lines in high temperature stars of the main sequence.

One of the most important results of Russell's analysis is the explanation of the position of the maximum of the H and K lines of Ca^+ in spectral type K. The maximum lies in type K because in giant stars the diminished pressure and enhanced transparency of the stellar atmosphere at low temperatures throws this maximum forward to the point at which first stage ionization is completed. Finally we may notice on comparing Tables 64 and 63 that Russell's results preserve the original results as a good first approximation which can in any event only be improved upon where detailed information about line contours is available.

§ 15·35. *The points of marginal appearance. Decay of lines past their maxima.* When the stellar temperature scale is known, with or without the help of the positions of the maxima of absorption lines, it is possible to draw some information about the factor $\omega\alpha$ (§ 15·33) or $\beta\alpha_\nu$ (§ 15·34) from the temperatures of the points of marginal appearance of an absorption line. When a line is just visible we may suppose that the concentration of suitable atoms required to form it must reach a certain minimum value proportional to $1/\alpha$; that is $\alpha(n_r)_s$ has some fairly definite value. Since we know p_ϵ and T we can derive from this a corresponding value of $\omega\alpha$. Since the values of α are all much the same, we can thus derive values of ω or rather relative values of ω for different elements—that is their relative abundance. The result is of course admittedly only a rough approximation. Such calculations have been begun by Miss Payne (*q.v.*) with interesting results.

In connection with these and future calculations there is one important observation to be made from the statistical side. It will be observed that the typical curves of the concentration of atoms in given states shown in Fig. 73 are markedly unsymmetrical about the maximum. The rise to the maximum is much steeper than the fall off for higher temperatures. So far as the suggested argument goes the value of $\omega\alpha$ derived from first and last appearance should be the same, and these theoretical curves suggest that lines should fade out much more slowly than they come in. For many lines this is directly contrary to observation. The observed rise and fall in intensity are reasonably symmetrical about the maximum. There is however one well-recognized exception, the Balmer series, which unlike most other lines has not faded out even in the hottest stars.

While the theory can perhaps hardly claim to be able to explain completely the symmetry of the observed curves, the marked difference between the behaviour of the Balmer lines and the lines say of Si or Si^+ and metallic lines generally is accounted for by the simple theory when the calculations are made so as to allow for successive ionizations. This was not done for Fig. 73, but has been included by Russell in preparing Figs. 74. Any line of the spectra H I, He II, etc. is a line of the last spectrum the atom can emit, for it has no further electrons to lose. Spectra such as Si I, Fe I, Si II, etc. belong on the other hand to atoms that can lose a regular sequence of further electrons at not too widely spaced ionization potentials. Spectra such as Si IV, Mg II and probably Na I will be effectively the last spectra of these atoms, for the next electron is separated by a big step in ionization potential, and will not become effectively removable at the temperatures of stellar reversing layers. The numerically different behaviour of "last" and "not last" spectra can at once be established.

The behaviour of any state of a "last" spectrum will continue to be given for temperatures far beyond $T_{\max}$ by equation (1707), for no further stage of ionization is reached. The behaviour of a state of a "not last" spectrum is determined on the contrary by (1705) and (1706) as before, but with

$$x_0 + x_1 + \ldots + x_r + x_{r+1} = 1, \qquad \ldots\ldots(1741)$$

if the $(r+1)$th state is the last ever relevant. This leads to (1707) near the maximum of a line of the neutral atom where (1741) reduces to $x_0 + x_1 = 1$. But at higher temperatures we have successively $x_1 + x_2 = 1, \ldots\, x_r + x_{r+1} = 1$, approximately, and $(n_0)_s$ will be substantially diminished. The physical reason for this decrease is easily seen. So long as the commonest ion has merely to catch one electron to get into the required state, the chance of doing so diminishes as the temperature rises, it is true, but not excessively rapidly. But if it has to catch two or more electrons, its chance of being in the proper state diminishes more or less like the square or higher power of the former chance. In the case specified we have

$$\frac{x_r}{x_0} p_\epsilon{}^r = (0{\cdot}664)^r\, T^{\frac{5}{2}r} e^{-(\chi_0+\chi_1+\ldots+\chi_{r-1})/kT} \frac{u_r(T)}{u_0(T)},$$

and so

$$(n_0)_s = \frac{x_r}{u_r(T)} \frac{(\varpi_0)_s\, e^{-\{\chi_0-(\chi_0)_s\}/kT}}{\left(\dfrac{0{\cdot}664}{p_\epsilon}\right)^r T^{\frac{5}{2}r} e^{-(\chi_0+\chi_1+\ldots+\chi_{r-1})/kT}}. \quad \ldots\ldots(1742)$$

At high temperatures, far past the maximum, this can be combined with $x_r + x_{r+1} = 1$ and the usual relation (1705) between x_r and x_{r+1}. We thus find

$$(n_0)_s = \frac{(\varpi_0)_s\, e^{-\{\chi_0-(\chi_0)_s\}/Tk}}{u_r(T) + \dfrac{0{\cdot}664}{p_\epsilon} T^{\frac{5}{2}} e^{-\chi_r/kT} u_{r+1}(T)} \times \frac{1}{\left(\dfrac{0{\cdot}664}{p_\epsilon}\right)^r T^{\frac{5}{2}r} e^{-(\chi_0+\chi_1+\ldots+\chi_{r-1})/kT}}. \quad \ldots\ldots(1743)$$

This equation holds through the region in which $x_r + x_{r+1} = 1$ sufficiently nearly. It holds also for spectra other than the first if we replace the suffixes 0 by t and the *index* r by $r-t$.

Simple calculation for an idealized case will best show the effect of the extra term in (1743). We will compare the spectrum H I with an idealized Si I of the same term values and ionization potential, 13·54 volts, followed by successive ionization potentials at 20, 31·5 and 45 volts, and carry through the calculations for a temperature of 25,000° K. and $p_\epsilon = 4 \times 10^{-5}$ atmosphere, roughly the maximum of the Si IV lines. For simplicity we will ignore differences of weight and take $(\varpi_0)_s$ and $u_r(T)$ all unity. Dropping factors the same for both atoms we find

$$\text{H I} \quad (n_0)_s \propto \frac{1}{1+\dfrac{0{\cdot}664}{p_\epsilon} T^{\frac{5}{2}} e^{-13{\cdot}54/kT}} = 10^{-6{\cdot}48},$$

$$\text{Si I} \quad (n_0)_s \propto \frac{1}{1+\dfrac{0{\cdot}664}{p_\epsilon} T^{\frac{5}{2}} e^{-45/kT}} \times \frac{1}{\left(\dfrac{0{\cdot}664}{p_\epsilon}\right)^3 T^{\frac{15}{2}} e^{-65/kT}} = 10^{-14{\cdot}89}.$$

There is thus an extra $10^{-8{\cdot}4}$ at 25,000° K. reducing the lines of the idealized Si I compared with those of H I. The fraction of H atoms capable of absorbing the Balmer lines at maximum is about 4×10^{-6}, which is only reduced by an extra factor of $10^{-2{\cdot}3}$ at 25,000° K. The reduction factor of similar Si I atoms is therefore $10^{-10{\cdot}7}$.

§ **15·4.** *The escape of molecules from the boundary of an isothermal atmosphere.* The rate of escape of molecules from a planetary atmosphere is a problem which has exercised theoretical physicists for many years. The earlier calculations, which started with Johnstone Stoney, have been sufficiently described by Jeans,* but they all contain an unsatisfactory feature in the arbitrary choice of a ceiling to the atmosphere. The rate of flow of molecules past the ceiling, possessing more than the velocity of escape, is then evaluated and taken to be the rate of loss of molecules from the atmosphere. A numerical result can only be reached at all by this method for the molecular distribution law typical of an isothermal gas in a constant field. The calculations have however been greatly improved and cast into a final form by Milne† and Lennard-Jones,‡ whose work entirely supercedes the earlier discussions.

The first advance on the old calculations was made by Milne by introducing the idea of the "cone of escape". For simplicity he regarded all the other molecules as fixed except those whose escape is in question, and for a given position in the atmosphere calculated the solid angle above the moving molecule, not screened by other molecules. The restriction of fixing the other molecules is not very serious, and he also evaluates in some detail the atmospheric distribution laws under the assumption $T \propto r^{-n}$, where r is the distance from the centre of the star or planet. He then shows that the number of molecules moving past a given level into the cone of escape has a fairly sharp maximum at a particular level in the atmosphere—the level of escape—and takes this maximum value as equal to the rate of escape.

In spite of the advance made in this calculation, which defines properly the hitherto undefined level of escape or ceiling, there are still not entirely satisfactory features, for obviously the number of escaping molecules crossing a given level must continually increase as the level rises, starting from zero at the surface of the planet. Though the maximum Milne calculates may and in fact does represent fairly accurately the escaping molecules integrated for all levels, we may reasonably ask for a still better theory, and the remedy lies in a more accurate enumeration of the molecules starting off at a given level into the cone of escape. Incidentally we can remove the restriction that the other molecules are fixed. These are the final improvements due to Lennard-Jones. We shall give some account of his work here, which is carried out for isothermal atmospheres. An extension to Milne's general case has not been made. It would be laborious and lie outside our scope. For the isothermal case Milne's approximate theory is satisfactorily confirmed, so that Milne's results are probably reliable in general. We make

* Jeans, *Dynamical Theory of Gases*, chap. xv.

† Milne, *Trans. Camb. Phil. Soc.* vol. 22, p. 483 (1923).

‡ J. E. Lennard-Jones, *Trans. Camb. Phil. Soc.* vol. 22, p. 535 (1923).

these applications the occasion for presenting a general calculation of the mean free path and the general method of enumerating molecules by the volume element in which they last suffered a collision—matter which we have not found place for elsewhere, but which is of great importance in many applications of statistical mechanics which lie just beyond our scope.

All calculations are of course made on the assumption that the extremely slow rates of escape are without sensible effect on the equilibrium distribution laws.

§ **15·41**. *The free paths of molecules in a uniform or non-uniform gas.* In accurate calculations of the type that follow the important quantity is not so much the average distance between consecutive collisions (free path) travelled by all molecules as the free path for a molecule of given velocity, or what is the same thing, the chance that in time dt a molecule of velocity c will suffer a collision. This idea was introduced by Tait.* The molecules are supposed to be rigid elastic spheres of diameter σ.

Consider a molecule of velocity c. The chance of collision with a second molecule of velocity c' in an element of time dt is equal to the number of molecules of this type contained in a cylinder of cross-section $\pi\sigma^2$ and length Vdt, where V is the relative velocity. If the direction of c' with respect to c be described by the usual Eulerian angles θ, ϕ, then the number of molecules in this cylinder having velocities between c' and $c'+dc'$ and moving in $d\omega$ about θ, ϕ is

$$\nu\left(\frac{m}{2\pi kT}\right)^{\frac{3}{2}} e^{-\frac{1}{2}mc'^2/kT}c'^2 dc' \sin\theta\, d\theta\, d\phi \times \pi\sigma^2 V\, dt. \qquad \text{......(1744)}$$

The relative velocity V is of course given by

$$V^2 = c^2 + c'^2 - 2cc'\cos\theta. \qquad \text{......(1745)}$$

If we denote by $\Theta(c)\,dt$ the average number of collisions in time dt (the chance of a collision in dt) with any other molecule, then

$$\Theta(c) = \pi\nu\sigma^2\left(\frac{m}{2\pi kT}\right)^{\frac{3}{2}} \int_0^\infty \int_0^\pi \int_0^{2\pi} e^{-\frac{1}{2}mc'^2/kT} c'^2 V dc' \sin\theta\, d\theta\, d\phi. \qquad \text{......(1746)}$$

The ϕ-integration is immediate; the θ-integration can be carried out by transforming to the variable V (c and c' fixed). By (1745)

$$\sin\theta\, d\theta = V dV/cc'.$$

The limits of integration for V are $|c-c'|$ and $c+c'$. We are thus left with

$$\Theta(c) = \tfrac{4}{3}\pi^2\nu\sigma^2\left(\frac{m}{2\pi kT}\right)^{\frac{3}{2}}\left[\int_c^\infty e^{-\frac{1}{2}mc'^2/kT}c'(c^2+3c'^2)\,dc' + \frac{1}{c}\int_0^c e^{-\frac{1}{2}mc'^2/kT}c'^2(c'^2+3c^2)\,dc'\right]. \qquad \text{...(1747)}$$

* Tait, *Edin. Trans.* vol. 33, p. 74 (1886).

The first integral can, the second cannot, be evaluated in finite terms. We find, after simple reductions,

$$\Theta(c)=\frac{2\sqrt{\pi}\,\nu\sigma^2kT}{mc}\,\psi\left\{c\sqrt{\left(\frac{m}{2kT}\right)}\right\}, \qquad ......(1748)$$

where $$\psi(x)=xe^{-x^2}+(2x^2+1)\int_0^x e^{-y^2}dy. \qquad ......(1749)$$

The average number of collisions in a small path length dl is therefore

$$\Theta(c)\,dl/c.$$

This is therefore the chance of the free path terminating in dl. The chance of continuing unaffected is therefore

$$1-\frac{\Theta(c)}{c}\,dl.$$

Let $f(l)$ be the fraction of c-molecules projected in a given direction from a given origin that describe without a collision a path greater than l. The fraction of these molecules that survive a further distance dl without a collision is $f(l+dl)$, and therefore by the last argument

$$f(l+dl)=f(l)\left(1-\frac{\Theta(c)}{c}\,dl\right),$$

$$\frac{1}{f}\frac{\partial f}{\partial l}=-\frac{\Theta(c)}{c}. \qquad ......(1750)$$

It follows generally that

$$f(l)=e^{-\int_0^l \frac{\Theta(c)}{c}dl}; \qquad ......(1751)$$

in this equation at the worst both c and ν in $\Theta(c)/c$ may depend on l and the direction of projection and ν also on the origin. If we ignore changes of c and write

$$\theta(c)=\Theta(c)/\nu c,$$

a function of c only, then

$$f(l)=e^{-\theta(c)\int_0^l \nu(x_0,y_0,z_0,l,\theta,\phi)\,dl} \qquad ......(1752)$$

It follows that the fraction of c-molecules with a path length between l and $l+dl$ before a collision is

$$\theta(c)\,\nu\,dl\,e^{-\theta(c)\int_0^l \nu\,dl} \qquad ......(1753)$$

When ν is constant (1752) reduces to

$$f(l)=e^{-\nu l\theta(c)}=e^{-\Theta(c)\,l/c}=e^{-l/\lambda(c)}, \qquad ......(1754)$$

defining *the mean free path* $\lambda(c)$. The fraction of paths between l and $l+dl$ reduces to

$$\frac{dl}{\lambda(c)}\,e^{-l/\lambda(c)}. \qquad ......(1755)$$

These are well-known formulae. The name mean free path is justified by the equation

$$\int_0^\infty \frac{l\,dl}{\lambda(c)} e^{-l/\lambda(c)} = \lambda(c). \qquad \text{......(1756)}$$

Similar arguments hold in the general case using (1751). The mean free path of the molecules may then be defined by putting

$$\int_0^{\lambda(c)} \frac{\Theta(c)}{c}\,dl = 1,$$

or

$$\int_0^{\lambda(c)} \frac{\psi\left\{c\sqrt{\left(\frac{m}{2kT}\right)}\right\}}{c^2}\nu\,dl = \frac{m}{2\sqrt{\pi}\sigma^2 kT}. \qquad \text{......(1757)}$$

If c is large compared with the average molecular velocity, so that the argument of ψ is large,

$$\frac{\psi(x)}{x^2} \to 2\int_0^\infty e^{-y^2}\,dy = \sqrt{\pi},$$

and formula (1757) becomes

$$\int_0^{\lambda(\infty)} \nu\,ds = \frac{1}{\pi\sigma^2}. \qquad \text{......(1758)}$$

This equation gives the length of a cylinder of base $\pi\sigma^2$ which contains on the average just one molecule, and the free path is that which we get by regarding all the other molecules as fixed. For uniform density we have the well-known result

$$\lambda(\infty) = 1/\pi\nu\sigma^2. \qquad \text{......(1759)}$$

The integral in (1751) or (1752) may, and in atmospheric problems does, converge when $l \to \infty$. In that case the definition of $\lambda(c)$ by (1754) or (1757) fails—it would make $\lambda(c)$ infinite, and the integral in (1756) diverges. This means that as the path increases $f(l) \to f(\infty)$ so that a non-zero fraction of the original molecules survives to execute infinite free paths. The fraction which so survives is

$$e^{-\int_0^\infty \frac{\Theta(c)}{c}dl}, \quad \text{or} \quad e^{-\theta(c)\int_0^\infty \nu\,dl},$$

if variations of c may be neglected.

§ **15·42.** *The number of molecules with specified velocities which cross or strike a given area in given time.* We have already had occasion in § 11·2 to use such a formula, and it will be considered in greater detail in § 17·8. The formula (1968) there obtained is

$$\nu\left(\frac{m}{2\pi kT}\right)^{\frac{3}{2}} c^3 e^{-\frac{1}{2}mc^2/kT}\cos\theta\,dc\,d\omega\,dS. \qquad \text{......(1760)}$$

This is the number of c-molecules which strike an area dS moving within a solid angle $d\omega$ making an angle θ with the normal to dS. The number which cross a geometrical interface is the same, and similarly obtained. The molecular density at the wall or interface is ν.

Since this is also the number reflected from such an element of surface in the equilibrium state, the formula (1760) must hold for all equilibrium states, whatever the fields of force in the gas. It applies therefore in atmospheric problems for atmospheres *in equilibrium*, even when the free paths are very long. It is however of some interest to obtain this formula in a more special way, using the idea of the free path, so as to see what contributions the various elements of the gas make to the actual number.

The molecules in (1760) must have had their origin, that is their last collision, somewhere in a cylinder of infinite length on the base dS with generators parallel to the molecular velocity. If the molecules move in a field of force, the "cylinder" becomes a tube of flow enclosed by the family of free trajectories passing through the perimeter of dS. In accounting for the molecules which cross dS we shall *associate all the molecules with the volume element in which they last collided*, a classification which is important in many applications. Molecules which cross dS with velocity c in $d\omega$ start from their last collision as c_0-molecules moving in $d\omega_0$.

When the gas is in equilibrium the number of collisions per unit volume *after* which one molecule has specified velocities must be equal to the number *before* which one molecule had the same specified velocities. Hence in a length dl of such a cylinder or tube there will be

$$\nu_0\left(\frac{m}{2\pi kT}\right)^{\frac{3}{2}} e^{-\frac{1}{2}mc_0^2/kT}c_0^2 dc_0 d\omega_0 \Theta(c_0)\, dl \cos\theta_0 dS_0$$

such collisions per unit time, since $\Theta(c_0)\,dt$ is the fraction of c_0-molecules that suffer a collision in time dt. This can be written

$$\nu_0^2\left(\frac{m}{2\pi kT}\right)^{\frac{3}{2}} e^{-\frac{1}{2}mc_0^2/kT}c_0^3\theta(c_0)\cos\theta_0 dc_0 d\omega_0 dl dS_0. \quad \text{......(1761)}$$

This is the number of proper molecules which are shot off per unit time from an element at distance l along the tube from dS so as to cross dS if undisturbed. Not all of these reach dS. The number which succeed is reduced by collisions to the fraction

$$e^{-\int_0^l \theta(c_0)\nu_0 dl}$$

The total number of c-molecules crossing dS in the specified direction in unit time is therefore

$$\left(\frac{m}{2\pi kT}\right)^{\frac{3}{2}}\int_0^\infty \nu_0^2 e^{-\frac{1}{2}mc_0^2/kT}c_0^3\theta(c_0)\cos\theta_0 d\omega_0 dS_0 dc_0 e^{-\int_0^l \theta(c_0)\nu_0 dl} dl. \quad \text{...(1762)}$$

Such formulae can sometimes be used in non-equilibrium states provided Maxwell's law remains valid. To see how in equilibrium (1762) reduces to

(1760) we consider the simplest case of a uniform field of force at right angles to dS. Then $dS_0 = dS, \quad \tfrac{1}{2}mc_0{}^2 = \tfrac{1}{2}mc^2 + mgz_0$,

$$c_0{}^2 dc_0 d\omega_0 = du_0 dv_0 dw_0 = J\left(\frac{u_0, v_0, w_0}{u, v, w}\right) c^2 dc\, d\omega,$$

$$= \frac{\partial w_0}{\partial w} c^2 dc\, d\omega = \frac{w}{w_0} c^2 dc\, d\omega.$$

Also $$\nu_0 e^{-mgz_0/kT} = \nu.$$

Hence (1762) reduces to

$$\nu\left(\frac{m}{2\pi kT}\right)^{\frac{3}{2}} c^3 e^{-\frac{1}{2}mc^2/kT} \cos\theta\, dc\, d\omega\, dS \int_0^\infty \nu_0 \theta(c_0)\, e^{-\int_0^l \theta(c_0)\nu_0\, dl'} dl,$$

which is (1760). This reduction holds equally in the general case.

§ **15·43.** *Free paths in an upper atmosphere.* We will now apply these results to an upper atmosphere where the free paths are very long, so that the change of ν along the path cannot always be neglected. When account is taken of the decrease of gravity with height we have, at a distance r from the centre of the planet,

$$\nu = \nu_0 e^{-\frac{mg}{kT}\frac{a(r-a)}{r}}, \qquad \text{......(1763)}$$

where ν_0 is the density at the base of the isothermal part of the atmosphere, g is the value of gravity and a is the radius of the planet (both strictly for the base of the isothermal atmosphere). As is well known this formula leads to a finite atmospheric density at infinity. The "atmosphere" would then fill space and there would be no problem of escape. It is not however consistent with obvious physical facts to regard such an extensive atmosphere as belonging in any sense to a particular body. The atmosphere has effectively ended at distances comparable with the planet's radius at most. The chief difficulty in (1763) arises from neglecting the mass of the atmosphere itself and treating g as constant. Milne* has shown that when this is taken into account an appropriate formula applicable at all distances is

$$\nu = \nu_0\left(\frac{r_0}{r}\right)^2 e^{-q_0(1-r_0/r)} \quad (q_0 = mgr_0/kT - 2), \qquad \text{......(1764)}$$

where r_0 refers to any convenient level in the atmosphere. This gives an atmosphere of zero density at infinity though of infinite mass. The mass of the atmosphere does not really seriously affect the distribution law in any important region or the rate of escape, but the use of (1764) avoids troublesome difficulties in the calculations.

We shall now investigate in such an atmosphere the free paths of c-molecules, with the simplification that we shall ignore the effect on the free

* Milne, *loc. cit.* equation (25).

path of the change of c in the gravitational field and the curvature of the path. The first of these steps is justified because for the molecules that will interest us (escaping molecules) c must be large compared with $(2kT/m)^{\frac{1}{2}}$ in any region in which collisions matter. In the integrand of (1751) ψ/c^2 is only slowly variable and we may use (1752). The second is justified because the length of the exact hyperbolic path is sufficiently represented by its asymptote.

If the c-molecule starts at radius r at θ to the vertical and proceeds to R after a path s, then the exponent in (1752) is

$$-\theta(c)\int_0^l \nu(R)\,ds,$$

where with sufficient accuracy $R=r+s\cos\theta$. This neglects terms due to the curvature of the layers of the atmosphere of order $(s\sin\theta/R)^2$. Using (1764) this reduces to

$$-\frac{\nu_0 r_0 \theta(c)}{q_0 e^{q_0}\cos\theta}\{e^{q_0 r_0/r}-1\}. \qquad \text{......(1765)}$$

The fraction of c-molecules projected from a level r at an angle θ with the vertical which could escape to infinity without a further collision is given by taking the exponential of (1765).

§ **15·44.** *The loss of molecules from a simple isothermal atmosphere.* By a trivial adaptation* of (1761) we see that the number of c-molecules shot off in unit time from a spherical shell of radius r and thickness dr at an angle between θ and $\theta+d\theta$ with the vertical is

$$2\pi\nu^2\left(\frac{m}{2\pi kT}\right)^{\frac{3}{2}} e^{-\frac{1}{2}mc^2/kT}c^3\theta(c)\sin\theta\,d\theta\,dc\times 4\pi r^2\,dr. \qquad \text{......(1766)}$$

We have hitherto been studying only the conditions for avoiding further collisions. We must now introduce the further condition for escape that $c>c_g$, where

$$c_g = 2ga^2/r. \qquad \text{......(1767)}$$

We find therefore at once that dL, the total number of escaping molecules produced by the level r, is

$$dL = 8\pi^2\nu^2 r^2\,dr\left(\frac{m}{2\pi kT}\right)^{\frac{3}{2}}\int_{c_g}^{\infty} e^{-\frac{1}{2}mc^2/kT}c^3\theta(c)\,dc$$
$$\times\int_0^{\frac{1}{2}\pi}\exp\left[-\frac{\nu_0 r_0\theta(c)}{q_0 e^{q_0}\cos\theta}\{e^{q_0 r_0/r}-1\}\right]\sin\theta\,d\theta. \qquad \text{......(1768)}$$

To find L, the total rate of escape, dL must be integrated with respect to r from the surface of the planet to infinity.

There is no need to give the further work in detail since no further question of principle arises, and the calculations are somewhat laborious. Molecules

* In (1761) the volume element was $\cos\theta_0\,dl\,dS_0$, which here becomes $4\pi r^2\,dr$.

of course theoretically escape from all levels, but owing to the exponential factor in (1768) only effectively when the argument does not much exceed unity. In order to obtain a concrete picture we may usefully simplify (1768) by replacing the exponential by zero when its argument is greater than unity and unity when its argument is less than unity. In fact since only orders of magnitude are of interest this simplification may be used in computations. This is the method used by Lennard-Jones (*q.v.*).

To obtain an idea of the level at which escape effectively begins we quote the results obtained by this simplified method of calculation for the outer helium* atmosphere of the earth, for which $g=981$, $r_0=6{\cdot}39\times10^8$, $T=219°$ K., $k/m=2{\cdot}08\times10^7$. We have therefore $q_0=135{\cdot}6$. We may take further $\sigma=2{\cdot}0\times10^{-8}$, $\nu_0=7{\cdot}46\times10^{12}$ (base of the stratosphere assumed to lie at 20 km. height). We then find that the exponent falls to unity for the most favourable conditions $c=\infty$, $\theta=0$ at a critical height r_c given by

$$r_c=1{\cdot}085r_0.$$

The critical level is about 540 km. above the base of the stratosphere. Above this height we can define a rapidly expanding cone of escape in such a way that the exponent becomes less than unity inside a cone of given semi-vertical angle about the vertical for $c=\infty$ as soon as r reaches the values given in the table below.

TABLE 65.

Variation of cone of escape with height. (*Earth's helium.*)

Semi-vertical angle of cone	$r_c/r_0=1{\cdot}085$ $(r-r_c)/r_0$	$r-r_0$ km.
0°	0	540
25°	0·00086	545
45°	0·0030	559
65°	0·0075	588
85°	0·0212	676

Inside the cone of escape for given r and θ the velocity must exceed a certain lower limit to reduce the exponent to unity. The variation of this with angle is shown in Fig. 75. The variation of finite free path with direction (defined as the distance required to reduce the undeflected beam to $1/e$ of its initial strength) at a point in the upper atmosphere is shown in Fig. 76.

In terms of the critical level r_c so defined it can be shown by treatment of (1768) that with sufficient accuracy

$$L=\frac{2\pi\sigma^2\nu_0^{\,2}r_0^{\,4}e^{-2q_0}e^{q_0r_0/r_c}}{r_c}\left(\frac{2\pi kT}{m}\right)^{\frac{1}{2}}. \qquad \text{......(1769)}$$

* Chapman and Milne, *Quart. J. Roy. Met. Soc.* vol. 46, p. 357 (1920). This paper is the source of all the meteorological data in the discussion here.

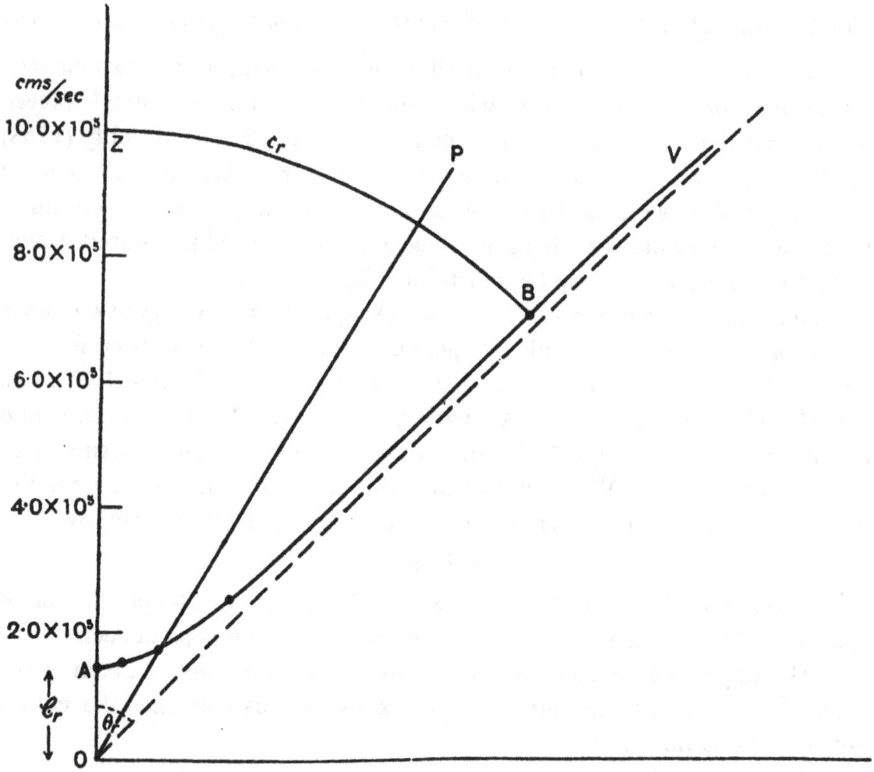

Fig. 75. Showing the variation of the velocity of escape with angle for a level in the earth's atmosphere at which the least velocity of escape $\mathscr{C}_r$ is $1{\cdot}6 \times 10^5$ cm./sec.

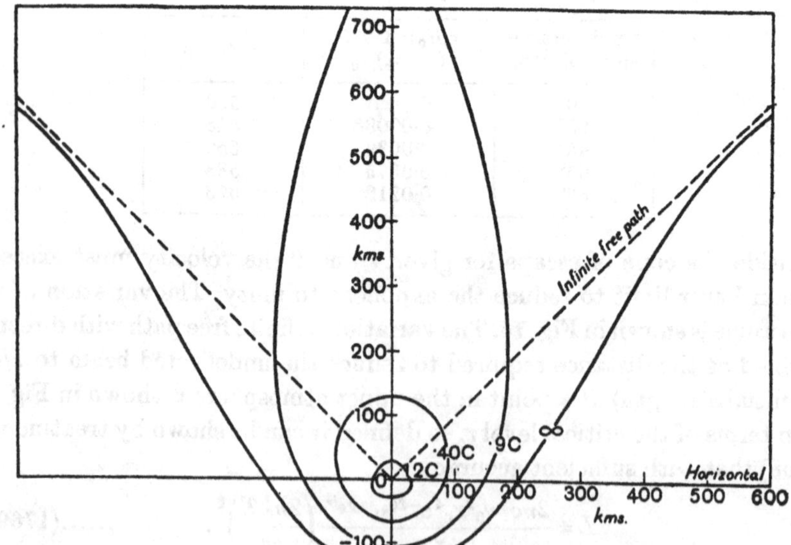

Fig. 76. Showing the variation of free paths with direction in the earth's upper atmosphere.

The critical level r_c as we have defined it makes the exponent in (1768) unity for $\cos\theta = 1$, $c = \infty$, $\theta(c) = \pi\sigma^2$ (see (1748), etc.). Hence

$$e^{q_0 r_0/r_c} - 1 = \frac{q_0 e^{q_0}}{\pi \nu_0 \sigma^2 r_0}, \qquad \text{......(1770)}$$

an equation in which, owing to the size of q_0, 1 may be neglected. The use of this reduces (1769) to

$$L = \frac{2\nu_0 q_0 r_0{}^3 e^{-q_0}}{r_c}\left(\frac{2\pi kT}{m}\right)^{\frac{1}{2}} = \frac{2\nu_0 g r_0{}^4 e^{-mgr_0/kT}}{r_c}\left(\frac{2\pi m}{kT}\right)^{\frac{1}{2}}. \qquad \text{......(1771)}$$

The equation (1771) will hold so long as r_c is free to be fixed by (1770), that is until r_c falls to the solid surface of the planet. After that, when r_c is so fixed, the equation (1769) can be shown to hold. Equation (1771) makes L proportional to ν_0 and independent of σ. The independence of σ is interesting and might have been foreseen. For the rate of production of possible escaping molecules is proportional to σ^2, and the rate at which these are thwarted by collisions is inversely proportional to σ^2.

Table 66 gives various calculations of the rate of escape by these formulae for various gases from various bodies. The rates are in all cases slower than

TABLE 66.

(1) *Rate of escape of helium from Mars.*

Isothermal atmosphere.

Surviving molecular density at the planet's surface	Years elapsed; 173° K.		Years elapsed; 273° K.	
	Helium atmosphere	Mixed atmosphere	Helium atmosphere	Mixed atmosphere
10^{15}	0	0	0	0
10^{12}	$2{\cdot}97 \times 10^{9}$	$2{\cdot}97 \times 10^{9}$	$1{\cdot}58 \times 10^{4}$	$1{\cdot}58 \times 10^{4}$
10^{9}	$5{\cdot}94 \times 10^{9}$	$5{\cdot}94 \times 10^{9}$	$3{\cdot}16 \times 10^{4}$	$3{\cdot}16 \times 10^{4}$
10^{6}	$9{\cdot}77 \times 10^{11}$	$8{\cdot}91 \times 10^{9}$	$2{\cdot}08 \times 10^{6}$	$4{\cdot}74 \times 10^{4}$
10^{5}	$9{\cdot}77 \times 10^{12}$	$1{\cdot}19 \times 10^{10}$	$2{\cdot}08 \times 10^{7}$	$6{\cdot}32 \times 10^{4}$
10^{4}	$9{\cdot}77 \times 10^{13}$	$1{\cdot}49 \times 10^{10}$	$2{\cdot}08 \times 10^{8}$	$7{\cdot}90 \times 10^{4}$
10^{3}	$9{\cdot}77 \times 10^{14}$	$1{\cdot}79 \times 10^{10}$	$2{\cdot}08 \times 10^{9}$	$9{\cdot}48 \times 10^{4}$

(2) *Earth's hydrogen.*

	219° K.		300° K.	
$1{\cdot}89 \times 10^{13}$	0	0	0	0
$1{\cdot}89 \times 10^{8}$	2×10^{24}	2×10^{24}	$2{\cdot}68 \times 10^{16}$	$2{\cdot}68 \times 10^{16}$

those given by Jeans, so that the interest lies rather in the cases which are regarded by him as critical. The table also includes calculations for mixed atmospheres for which the original paper should be consulted. The much greater rate of loss of the lightest constituent from a mixed atmosphere in the final stages is due to the effect of collisions with the other constituents in keeping up the supply of faster moving molecules. The calculations for mixed atmospheres give of course a better representation of the facts for the later stages, but the later stages are not of much astronomical importance.

Applications of these formulae may also be made to determine the limits of the electrostatic potential of a star set by the possibilities of escape in this way of electrons or positive nuclei. For these reference should be made to Milne's paper.*

§ 15·5. *Problems of the calcium chromosphere.* Observations at eclipses of the spectrum of the high level gases surrounding the sun show that the H and K lines of ionized calcium are emitted up to very great heights above the apparent limb. Heights of 14,000 km. above the limb were recorded in 1905 by Mitchell,† and still greater heights have been recorded in more recent work. This means of course that calcium ions must exist continually (presumably more or less in an equilibrium state) up to such heights, and the problem at once arises as to how they are supported. A simple calculation shows at once that no such extensive atmosphere can possibly exist in ordinary statistical equilibrium. Beyond the apparent limb of the sun we should expect the atmosphere to consist of matter which in equilibrium would have the properties of a perfect gas, and therefore we should expect the atmospheric density law to hold—that is

$$\nu = \nu_0 e^{-mgz/kT},$$

and g has its solar value $2{\cdot}73 \times 10^4$. Even if we assume a temperature of 6000° K. (the actual temperature must be less), we find that this means a density ratio at the top and bottom of a 10,000 km. layer of 10^{800}. This would be somewhat reduced by the electrostatic fields as in § 15·2. If the atmosphere were a simple mixture of calcium ions and electrons the factor would be 10^{400}, but in no case can we suppose that the index of the factor is reduced much below this order. If therefore we suppose that we have a calcium atmosphere in ordinary equilibrium the density ratio must fall off at this prodigious rate. The density at the sun's limb may not be exactly known, but it is certain that it is not excessively large. The pressure is certainly not of order much greater than one atmosphere, and if this were all due to calcium atoms, the density would be 10^{18}. A reduction of this to 10^{-382} leaves us practically no calcium atoms at all, certainly not enough to be visible. In fact the thickness of the effective calcium atmosphere would be of the order of 100 km. at most, rather than 10,000 km. or more.

* Milne, *loc. cit.*

† Mitchell, *Astrophys. J.* vol. 38, p. 407 (1913).

This has long been recognized, and the consequence admitted that the calcium atoms seen at eclipses which constitute the permanent calcium chromosphere cannot be matter in approximately thermodynamic equilibrium. They cannot therefore be matter supported by the ordinary material pressure gradient of which the underlying mechanism is molecular collisions, and the only other possible mechanism of support is radiation pressure. This idea has been developed in quantitative form by Milne* in a series of important papers, with the result that much new light has been thrown on the nature of such atmospheres supported by radiation and many allied questions, and the observed facts roughly but satisfactorily explained. We shall give an account here of so much of these researches as does not require a study of the general laws of the flux of radiation, and summarize the rest. We must naturally make free use of the laws of radiative processes which are collected in Chapter XIX, to which we shall refer forward here.

If the pressure arising from the reaction of the atom to the flow of radiation from the photosphere is to balance the force of gravity, the pressure must, on the atomic scale, be very large indeed. An atom of course interacts with light of all frequencies, but only strongly with light of the frequency of its own absorption lines (§ 19·2) or of a frequency great enough to ionize it (§ 19·3). For Ca^+ near the sun there will be practically no photo-ionization since the absorption edge lies too far in the ultra-violet, so that the calcium chromosphere must be supported (at least mainly) by the radiation pressure due to the formation of the H and K absorption lines themselves.

The H and K lines of Ca^+ form a close doublet $1\,^2S - 1\,^2P$. It will obviously be legitimate in a first survey to imagine that the two lines are fused to form a single one of their united strength. The states of the atom concerned are then to be thought of as two only, of weights 2 and 6. We shall find that any Ca^+ atom is only in the excited ^2P state for a very short fraction of all time, so that multiple absorptions to still higher states will be comparatively infrequent and may be disregarded. From the $1\,^2P$ state the excited Ca^+ atom can return either to the normal $1\,^2S$ state emitting H and K or to the intermediate metastable $1\,^2D$ state emitting the lines $\lambda\lambda$ 8498, 8542, 8662, which may also be regarded as fused, and the ^2D state as a single state of weight 10. It follows that this group known to astrophysicists as X should be visible in the sun's chromosphere wherever H and K are visible and with closely connected intensities. Owing to photographic difficulties in the deep red and infra-red region of these lines this has not yet been established observationally, but the X lines have been traced to great heights by Curtis

* Milne, *Monthly Not. R.A.S.* vol. 84, p. 354 (1924), vol. 85, p. 111 (1924), vol. 86, pp. 8, 578 (1925, 1926).

and Burns.* The existence of this metastable ^2D state is a complication to the analysis, which however may be avoided at first by the following considerations. Since the state is metastable and the densities in the region to be discussed turn out to be extremely low, a Ca^+ atom in the ^2D state will remain there permanently until it reabsorbs one of the X lines and returns to $1\,^2P$. Consequently for present purposes the ^2D state acts like an extra normal state. If the wave lengths of H and K and X were approximately equal we could even regard the $1\,^2S$ and $1\,^2D$ states as fused into a state of weight 12. Though this cannot be done, it is clear that a first discussion omitting all reference to the $1\,^2D$ state will be correct in essentials, liable merely to subsequent numerical correction.

We propose therefore the simplified problem of the formation of a high level chromosphere of Ca^+ atoms (and electrons) supported by the pressure of the absorbed H and K radiation, the atoms being idealized systems of two stationary states only, of weights 2 and 6. The problem is not an equilibrium one, but the state of the matter in this atmosphere must be determined by enumerating and balancing all the atomic events, using the atomic formulae of the equilibrium theory of § 19·2. The only atomic events of importance are absorption and emission of H and K, satisfying Einstein's laws.

Consider the state of an atom at the upper boundary of the chromosphere —that is so high that there are too few atoms above it to alter effectively the outflowing H and K radiation or to return radiation to the atom from above. We may suppose that the sun's continuous photospheric radiation $I(\nu)$ is that of a black body at temperature T (about 6000° K.).† From the point of view of the Ca^+ atom this may to a first approximation be regarded as uniformly distributed over the hemisphere below it and reduced by absorption below to a fraction r of the photospheric value. Such an atom must on the whole be in equilibrium so that the average rate of absorption of upward momentum from the radiation must exactly balance the rate of increase of downward momentum due to gravity. Spontaneouse mission of radiation by the atom is isotropic and so contributes nothing on the average to the momentum. Stimulated emissions are directed, but for the frequencies in question too rare to need inclusion.

By (2015) the chance of absorption by a normal atom in time dt is $B_1{}^2 I(\nu)\,d\nu dt$ in isotropic ν-radiation. The chance is reduced here to

$$\tfrac{1}{2} r B_1{}^2 I(\nu)\,d\nu dt.$$

* Eclipse of 1928, Jan. 24. Curtis and Burns, *Pub. Allegheny Obs.* vol. 6, p. 95 (1925). More recently Davidson, *Monthly Not. R.A.S.* vol. 88, p. 30 (1927), has recorded preliminary comparisons of H, K and X by observations without an eclipse.

† Fabry and Buisson, *Comptes Rendus*, vol. 175, p. 156 (1922); H. H. Plaskett, *Pub. Dom. Astrophys. Obs.* vol. 2, p. 253 (1923).

The average time τ' that a normal atom remains in the normal state before absorbing* is therefore given by

$$1/\tau' = \tfrac{1}{2} r B_1{}^2 I(\nu)\,d\nu. \qquad \ldots\ldots(1772)$$

Neglecting stimulated emissions, the chance of emitting in time dt and the average life τ in the excited state are given by $A_2{}^1$ and $\tau = 1/A_2{}^1$. Using (2020) and Planck's law for $I(\nu)$ we find

$$\frac{\tau}{\tau'} = \frac{\tfrac{1}{2} r B_1{}^2 I(\nu)\,d\nu}{A_2{}^1} = \frac{\varpi_2}{\varpi_1}\frac{\tfrac{1}{2} r}{e^{h\nu/kT}-1}. \qquad \ldots\ldots(1773)$$

Now the average life in both states together, during which the atom absorbs just one quantum, is $\tau+\tau'$, and the upward momentum absorbed with this quantum is $h\nu\cos\theta/c$ which averaged over the hemisphere has the value $\tfrac{1}{2}h\nu/c$. It follows that for equilibrium

$$mg(\tau+\tau') = \tfrac{1}{2}h\nu/c. \qquad \ldots\ldots(1774)$$

For the numerical values $T = 6000^\circ$ K., $\lambda = 3950$, $h\nu/kT = 6{\cdot}05$, $m = 40\times 1{\cdot}65\times 10^{-24}$ and $g = 2{\cdot}73\times 10^4$, $\varpi_2/\varpi_1 = 3$, we find

$$\tau/\tau' = r\times 3{\cdot}54\times 10^{-3},$$
$$\tau+\tau' \simeq \tau' = 4{\cdot}6\times 10^{-5},$$
$$\tau = r\times 1{\cdot}62\times 10^{-7}.$$

The Ca^+ atoms here considered on the upper boundary of the chromosphere will be almost ideally undisturbed and their absorption lines will have their natural narrow width. The observed H and K lines of the sun and many stars are broad, but it can only be the centre of the line with which we are concerned in the upper chromosphere. The observed value of r for the centre of the H and K lines averaged over the sun's disc is about 0·11. We thus find

$$\tau = 1{\cdot}8\times 10^{-8}, \quad A_2{}^1 = 5{\cdot}5\times 10^7.$$

The simplicity of the argument and the data leading to this result are noticeable, and give this determination of τ great weight.†

A number of interesting deductions may be drawn from these formulae. Equations (1773) and (1774) may be used to derive a formula for r, namely

$$r = \frac{\varpi_1}{\varpi_2}\frac{4mgc\tau}{h\nu}(e^{h\nu/kT}-1). \qquad \ldots\ldots(1775)$$

* The argument from the chance in time dt to the fraction not having absorbed after time t and the mean life τ' is that of § 15·41.

† An important contribution to the theory has been made by Unsöld, *Zeit. f. Physik*, vol. 44, p. 793 (1927). He points out that we now know $A_2{}^1$ and τ theoretically with some certainty, for the strength of the combined H and K absorption coefficient is almost exactly the same as if each atom carried one classical electron vibrating with this frequency. The true value of τ is slightly less than the value derived by Milne. The atoms are more absorptive. The discrepancy may be accounted for by the proportion of doubly ionized calcium atoms which must be supported in equilibrium by the radiation pressure of the H and K lines alone.

If $F(\nu)$ is the residual central intensity in the line, so that $F(\nu)=rI(\nu)$, this can be expressed very simply in the form

$$F(\nu)=\frac{\varpi_1}{\varpi_2}\frac{8mg\nu^2\tau}{c}. \qquad \ldots\ldots(1776)$$

If the right-hand side of equation (1775) is greater than unity there can be no absorption line, and the type of high level chromospheric equilibrium here contemplated cannot occur. This will happen if the atom is too heavy, if g is too large, if ν/T is too large, or if τ is too large. Using the value of τ deduced from the sun for Ca^+, the critical photospheric temperature below which a calcium chromosphere cannot be formed is 4400° K. for a star with the same surface value of gravity as the sun.

If the right-hand side of (1775) is less than unity a calcium chromosphere may still be impossible, for the photospheric ν-radiation will be reduced below $I(\nu)$ in the lower absorbing layers (reversing layer). But so long as the r of (1775) is less than r_1, the reduction ratio when the ν-radiation is free of the pressure-supported layers, a chromosphere will be formed at once, for equilibrium is impossible for Ca^+ atoms at the top of the reversing layer, and they will be driven out by the radiation pressure until a screen is formed sufficiently deep to cut down the finally emergent ν-radiation to the fraction r.

Values of r for various stars have been given by Milne as follows:

TABLE 67.

Residual intensities in the centre of the H and K lines.

Star	Mass [Sun's mass = 1]	g	T	r
Capella	4·2	$0{\cdot}067\times10^4$	5,500	0·010
Sirius	2·43	$1{\cdot}73\times10^4$	11,000	0·0043
Plaskett's massive *O*-star	86	$0{\cdot}56\times10^4$	24,000	0·00019

This table shows that for hot stars the residual intensity in the centre of the line must be very small indeed if a calcium chromosphere in equilibrium is formed. It is probable that the formation in these cases is never complete for two reasons. Firstly there must always be a loss of Ca^+ atoms from the chromosphere for a variety of causes of which the most important may be photoionization. For stars with a hot photosphere this will be no longer negligible. Secondly there will be a loss by radiative ejection from the star into space which will be continually going on especially during the attempted formation of a chromosphere of a hot star.

The simple arguments so far presented essentially fix only *the upper boundary conditions* for the calcium chromosphere. They apply only to the normal absorption lines of an atom (or ion) whose ionization potential is so high that photoionization hardly occurs. All that we have said of Ca^+ applies also to Sr^+ and Ba^+ but probably to no other atoms whose radiations we can observe in stellar spectra. For example the D-lines of sodium are found in the chromosphere only to 1000 km. They could be supported like Ca^+ by radiation pressure if the ionization potential (5·1 volts) were probably not too low. Hydrogen and helium extend to considerable heights in the chromosphere but mainly by absorption of light in the visible region from excited not from normal states. For helium certainly there is a metastable state concerned, the lowest state of the triplet system ^3S, which is almost certainly the reason for the occurrence of the lines of the triplet system much higher in the chromosphere than the singlets. In general however the discussion of the support of such atoms is much more complicated.

In all the arguments above we may suppose that the calcium ions are accompanied by an equal number of free electrons, since the electrostatic fields will prevent any separation of the charges as in § 15·2.

We have yet to show that the density distribution in space of the calcium chromosphere set up in this way is such that the chromosphere can extend to the great heights at which it is observed. For this the reader must refer to the original papers.* The result of this investigation is that the pressure of radiation at any given level is proportional to $F(\nu)$ defined above and to $(n_1-n_2)/(n_1+n_2)$. The ratio n_2/n_1 decreases outwards and

$$(n_1-n_2)/(n_1+n_2)$$

increases outwards and at great heights tends to a limit fixed by the boundary conditions already discussed. The density ρ of the chromosphere at height x is given by

$$\rho=\frac{8\pi kT_0h\nu^3\Delta\nu}{mg^2c^3(x+x_0)^2}\frac{\varpi_1}{\varpi_1+\varpi_2}. \qquad \ldots\ldots(1777)$$

In this formula T_0 is the "temperature" of the chromosphere assumed uniform, as measured by its pressure, that is by the mean kinetic energy of its constituent systems, $\Delta\nu$ is the effective width of the absorption line of residual intensity $F(\nu)$ and x_0 a constant fixed by the theory, equal to the height of the equivalent homogeneous atmosphere. Heights are measured from an assumed level at which the intensity of ν-radiation has its photospheric value. Curvature of the layers and change of g with height have been neglected.†

Milne discusses also the case in which at the upper boundary the radiation pressure supports a fraction $1-\mu$ of the weight of the atom and shows that

* Milne, *loc. cit.* (2).

† An extension of the theory to a spherical sun has been made by P. A. Taylor, *Monthly Not. R.A.S.* vol. 87, p. 605 (1927).

the distribution law changes over rapidly to the exponential type as if the atoms had an effective mass μm. Such chromospheres he calls "partially supported", and those with $\mu=0$ above "fully supported". In the case of partial support, remembering the electrostatic effects (equation (1692)), we have

$$\rho=\rho_0 e^{-\mu m g x/2kT_0}. \qquad \text{......(1778)}$$

The theory therefore gives just the type of extended atmosphere required. For calcium on the sun the density theoretically decreases in a fully supported chromosphere by 1/23 in 14,000 km., compared with 10^{-800} when radiation pressure is neglected or 10^{-8} and 10^{-80} for $\mu=0\cdot01$ or $0\cdot1$. Values of μ as great as 0·01 must be impossible on the sun.

§ **15·51.** *Further developments in the theory of the chromosphere.* Milne's theory outlined in the preceding section has removed the outstanding difficulty caused by the great extension of the calcium chromosphere. There remain however very serious difficulties over the finer details of the chromospheric structure. The recent work of Pannekock and Minnaert and of Menzel* has shown that to a rough approximation the density distribution in the chromosphere mimics that of an isothermal atmosphere at a temperature so high that the calcium atoms would have random velocities of the order of 15 km./sec. Menzel and McCrea† further have given evidence based on Doppler broadening to show that these velocities are real.

At first sight these facts seem to be in such marked disagreement with Milne's theory of support in monochromatic radiative equilibrium, that entirely different theories have been put forward, none of which however could be regarded as logical deductions from the laws of atomic physics and statistical mechanics. A theory which preserves the satisfactory features of Milne's theory and at the same time removes its outstanding difficulties has however recently been proposed by Chandrasekhar. This theory has not yet reached a final form but it promises to prove completely satisfactory.

Chandrasekhar starts from the observational result that the emergent flux of radiation in the chromospheric lines is not constant over the sun's surface but oscillates considerably about a mean value equal to the residual intensity in the line used by Milne. Spectroheliograms taken in calcium light indicate that the emergent flux varies from place to place by a factor of 2 or 3, the average distance on the sun's surface between neighbouring maxima being about 10,000 km. If in conformity with Milne's ideas we assume that the mean emergent flux corresponds to full support, then calcium ions over a hot spot will be accelerated upwards and over a cold

* Pannekock and Minnaert, *Verhand. d. Kon. Akad. v. Weten. Amsterdam*, vol. 13, No. 5 (1928); Menzel, *Lick Obs. Publ.* vol. 17, p. 1 (1931).

† McCrea, *Monthly Not. R.A.S.* vol. 89, pp. 483, 718 (1929).

spot accelerated downwards, and from a consideration merely of dimensions it follows that the velocity fluctuations so engendered will be of the order $\sqrt{(\lambda g)}$, where λ is the linear scale of the flux fluctuations over the surface. If $\lambda = 5000$ km., then $\sqrt{(\lambda g)} = 36{\cdot}8$ km./sec. This is of the correct order of magnitude. Following out these ideas Chandrasekhar has shown, with suitable approximations, that the density distribution, though not strictly an exponential function of the height, is not widely divergent from the exponential law suggested by the observations, and that the atoms should actually possess the observed velocities.

We are still, however, far from a complete theory of the chromosphere. The hydrogen chromosphere for example still presents an unsolved problem. For further information reference should be made to a recent discussion by McCrea.*

* McCrea, *Monthly Not. R.A.S.* vol. 95, p. 80 (1935).

CHAPTER XVI

APPLICATIONS TO STELLAR INTERIORS*

§ **16·1.** *Rosseland's theorem.* We shall start this chapter with some comment on an application by Rosseland† of the theorems of §§ 15·2–15·25 to the far interior of a star. He there discusses the effect of electrical fields on the relative distribution of different elements in the interior of a star, as we have done for atmospheres in the sections quoted.

It will be sufficient to consider the assembly of § 15·25, but we must now use the general equations (1672), (1673) and (1674), since both gravitational and electrical fields arise from the matter of the assembly itself. The point made by Rosseland is that, if (1699) is satisfied, then the substitution

$$\psi = -\frac{m_1 - m_3}{\nu_1 + 1}\phi = -\frac{m_2 - m_3}{\nu_2 + 1}\phi \qquad \text{......(1779)}$$

reduces all the exponential factors in the three distribution laws to the common value

$$E = e^{-\frac{\phi}{kT}\left(\frac{m_1+\nu_1 m_3}{\nu_1+1}\right)} = e^{-\frac{\phi}{kT}\left(\frac{m_2+\nu_2 m_3}{\nu_2+1}\right)} \qquad \text{......(1780)}$$

The equation for ϕ remains

$$\nabla^2\phi = 4\pi G E\{m_1(\overline{n_1})_0 + m_2(\overline{n_2})_0 + m_3(\overline{n_3})_0\},$$

and the equation for ψ becomes

$$\nabla^2\psi = \frac{4\pi\epsilon^2(\nu_1+1)}{m_1 - m_3} E\{\nu_1(\overline{n_1})_0 + \nu_2(\overline{n_2})_0 - (\overline{n_3})_0\}.$$

These two equations are consistent and the relation (1779) is a possible solution of the problem if the values of $(\overline{n_1})_0$, $(\overline{n_2})_0$, $(\overline{n_3})_0$ are suitably fixed, that is if the ratio of the mass density and charge density at a particular point has the correct value. The analysis of §§ 15·2–15·25 shows however that in that simple case even if the initial condition is not satisfied the exact solution approaches the corresponding particular one with extreme rapidity, and one may safely assume that it does so here too. In the interior of a star therefore whose material consists of any number of mixed ions, if the conditions (1699) equivalent to (1779) are satisfied for each pair, the gravitational and electrostatic potentials will be everywhere in the ratio (1779) and the constituents will be everywhere mixed in the same proportions. The space variation common to all components is given by (1780).

Interesting conclusions could be drawn if this uniform constitution could be established even in the restricted isothermal case. It is therefore neces-

* I have to thank Dr Hartree for the numerical calculations in this chapter.

† Rosseland, *Monthly Not. R.A.S.* vol. 84, p. 720 (1924).

sary to examine (1699) with some care to see how closely it could be satisfied in a star. Since atoms are largely reduced to nuclei and free electrons, and since the masses and charges of nuclei are very nearly proportional, it seems at first sight as if (1699) would be very nearly satisfied for stellar material, and Rosseland's theorem might apply. But the satisfaction is not exact. For oxygen, iron and silver nuclei, for example, the values of $(m_1 - m_3)/(v_1 + 1)$ are 1·78, 2·08, 2·25. In order to assert that the electrical forces really maintain the isothermal stellar material at a constant constitution, an almost exact equality would be required. We may conclude that this effect will very much reduce the rate at which the relative concentrations of the heavier elements increase towards the centre, but on the stellar scale the concentration to the centre *in an isothermal star* would still be overwhelmingly great.

As we have said, further consideration of large scale effects is beyond our scope, and we pass on to the study of the properties of matter in equilibrium at stellar temperatures and densities.

§ **16·2.** *Stellar material.* In the radiative theory of the steady states of a star the important quantity that must be provided by statistical theory is the equation of state of the stellar material. The material consists of atomic ions and electrons in dissociative equilibrium, and the equations of Chapter XIV permit this equilibrium state to be calculated so long as the density is not too high.

To a first approximation stellar material behaves like a perfect gas of that number of constituents which is required by the dissociative equilibrium. It is therefore customary to write its equation of state as the perfect gas equation in terms of the temperature, density and *the mean molecular weight* μ. In the first instance this is calculated from the number of atomic ions and free electrons, but any corrections to the pressure, for example for electrostatic effects, can be incorporated as corrections to the mean molecular weight. It is this corrected mean molecular weight μ^* which is required in stellar theory.

In addition to the mean molecular weight, the pulsation theory of Cepheid variables requires a knowledge of the ratio γ of the specific heats of the material together with the radiation it contains—or in other words of its adiabatic curves in p, V coordinates.

Both μ^* and γ can be derived at once from the formulae of statistical theory given in Chapter XIV. The calculations however are somewhat intricate and have not been completed on any reliable theory† except for

† Fairly extensive calculations were made by R. H. Fowler and Guggenheim, *Monthly Not. R.A.S.* vol. 85, p. 939 (1925), using an earlier less accurate version of the theory. Their results have been used by Eddington, *loc. cit.* especially § 180, to check the more elementary approximations. The alterations made in their results for μ^* by the more exact theory will therefore be noted here.

μ^*, for which some results are given in this chapter. Finally stellar theory requires a calculation of the mean absorption coefficient of stellar material in the state specified for the radiation passing through it, which is practically black body radiation of its own temperature. This calculation leads us far afield into atomic theory and we shall not attempt to give it here.

We will now summarize the formulae to be used, with the notation of Chapter XIV:

$$\frac{\Psi}{k}=N\left\{\log\frac{VG}{N}+1\right\}+\Sigma_{r,z}M_r^z\left\{\log\frac{VF_r^z}{M_r^z}+1\right\}+\frac{8\pi^5k^3VT^3}{45c^3h^3}+\frac{\Psi_\epsilon}{k}. \quad\text{......(1781)}$$

The electrostatic contribution Ψ_ϵ/k is not accurately known. Debye and Hückel's theory (§§ 13·61, 14·4) gives

$$\frac{\Psi_\epsilon}{k}=\frac{2\sqrt{\pi}\,\epsilon^3}{3V^{\frac{1}{2}}(kT)^{\frac{3}{2}}}\{N+\Sigma_{r,z}r^2M_r^z\}^{\frac{3}{2}}. \quad\text{......(1782)}$$

We have also, repeating former equations,

$$G(T)=2\,\frac{(2\pi mkT)^{\frac{3}{2}}}{h^3}, \quad\text{......(1783)}$$

$$F_r^z(T)=\frac{(2\pi m^z kT)^{\frac{3}{2}}}{h^3}\,u_r^z(T)\,e^{\{\chi_r^z+\ldots+\chi^z_{z-1}\}/kT}, \quad\text{......(1784)}$$

$$u_r^z(T)=\varpi_r^z+\sum_{q=1}^{q=z-r}\varpi^z_{r+q}(r+1)^{3q}Q^q\left\{\frac{V}{\Sigma_{p,z}(p+1)^3M_p^z}\right\}^{\frac{1}{2}q}$$
$$\times e^{-\{\chi_r^z+\ldots+\chi^z_{r+q-1}\}/kT}, \quad\text{......(1785)}$$

$$Q=1{\cdot}017\times10^{12},\quad N=\Sigma_{r,z}rM_r^z. \quad\text{......(1786)}$$

We define s^z, the average number of free electrons per atom of type Z, by the equation

$$s^zM^z=s^z\Sigma_rM_r^z=\Sigma_r rM_r^z, \quad\text{......(1787)}$$

so that

$$N=\Sigma_z s^zM^z. \quad\text{......(1788)}$$

The pressure p is given by $p=T\,\partial\Psi/\partial V$, ...(1789)

the material (as opposed to the radiation) pressure by (1789) in which the term arising from the radiative term σVT^3 in (1781) is omitted. The density ρ is given by

$$\rho=\Sigma_z M^zm^z/V. \quad\text{......(1790)}$$

If the dependence of F_r^z on V is ignored, so that the gas mixture is perfect, we should have

$$p=\frac{R\rho T}{\mu},\quad \mu=\frac{\Sigma_z M^zm_*^z}{N+\Sigma_z M^z}, \quad\text{......(1791)}$$

where R is the gas constant per mole and m_*^z the chemical atomic weight. As it is we have

$$p=\frac{R\rho T}{\mu^*} \quad\text{......(1792)}$$

defining μ^*.

The equations of dissociative equilibrium are obtained by making $d\Psi = 0$ for all variation of the type

$$\delta N = \delta M^z_{r+1} = -\delta M_r^z = \delta\alpha. \qquad \text{......(1793)}$$

If T, ρ and the ratios of the various M^z's are then given (matter of given chemical constitution, temperature and density), the equations can be solved completely.

The actual procedure is rather to assume values for N/V and T, and solve by successive approximations. It must be remembered that Planck's theory is only shown to give a valid approximation to $B(T)$ so long as in all cases the private catching region of each ion is so large that on its boundary the potential energy of an electron due to the central nucleus is small compared with kT. This condition is (1662) in the simple case there considered. It easily extends to

$$\epsilon^2\left\{\frac{\Sigma_{r,z}(r+1)^3 M_r^z}{V}\right\}^{\frac{1}{3}} \ll kT,$$

which may be taken with sufficient accuracy to be

$$\frac{s+1}{s^{\frac{1}{3}}}\epsilon^2\left(\frac{N}{V}\right)^{\frac{1}{3}} \ll kT. \qquad \text{......(1794)}$$

It must be remembered also that Ψ_ϵ is not accurately given by (1782). We have to conclude at present that as soon as the terms arising from Ψ_ϵ become important in evaluating μ^* the present theory ceases to be reliable, and, in view of Eddington's discussion,† that the electrostatic correction is overestimated by (1782).

For simplicity in the actual calculations the equations of dissociative equilibrium have been taken in their perfect gas form, ignoring the effect of the variations on the $u_r^z(T)$. It is easily verified that the omitted factors never differ much from unity; the greatest value they take is about $10^{0\cdot 1}$. Therefore their omission can hardly affect seriously or give an undue bias to the calculated values of μ^*. It is not worth while including them until we have a proper theory of the electrostatic effects.

Calculations have so far been made only for stars composed of iron which are given below in Table 68, and are shown plotted in Fig. 77.

The first six columns in Table 68 need no comment. The seventh gives the μ^* of (1792). When we apply (1789) to (1781) we find

$$p = \frac{kT}{V}\left(N + \Sigma_{r,z} M_r^z + \Sigma_{r,z} M_r^z V \frac{\partial}{\partial V}\log u_r^z\right).$$

Now the various terms in u_r^z correspond to different numbers of excited electrons, and therefore $V\partial \log u_r^z/\partial V$ takes the simple form

$$\frac{\Sigma_q \frac{1}{2}q(M_r^z)_q}{\Sigma_q (M_r^z)_q} = \tfrac{1}{2}\overline{q_r^z}.$$

* Eddington, *loc. cit.* § 14·4.

TABLE 68.

Equilibrium states of matter in stars of iron ($m_* = 55{\cdot}84$).

(Electrostatic corrections are omitted except where specially recorded.)

T° K.	ρ gm./c.c.	Number of electrons: Free per c.c. $=N/V$	Free per atom $=s$	Highly excited per atom $=\bar{q}$	Mean molecular weight μ	μ^*	μ^* with electrostatic correction for same N/V and T	μ' assuming no excited electrons same N/V and T	μ^* Fowler and Guggenheim same ρ and T
$3{\cdot}41 \times 10^6$	$1{\cdot}286 \times 10^{-5}$	$3{\cdot}32 \times 10^{18}$	23·96	0·04	2·237	2·235	—	2·234	—
	$1{\cdot}291 \times 10^{-4}$	$3{\cdot}32 \times 10^{19}$	23·87	0·13	2·245	2·239	—	2·234	—
	$1{\cdot}307 \times 10^{-3}$	$3{\cdot}32 \times 10^{20}$	23·58	0·40	2·272	2·253	—	2·235	2·33
	$1{\cdot}385 \times 10^{-2}$	$3{\cdot}32 \times 10^{21}$	22·25	1·59	2·402	2·322	—	2·251	2·36
$6{\cdot}82 \times 10^6$	$1{\cdot}244 \times 10^{-4}$	$3{\cdot}32 \times 10^{19}$	24·78	0·04	2·166	2·164	—	2·169	—
	$1{\cdot}282 \times 10^{-3}$	$3{\cdot}32 \times 10^{20}$	24·04	0·13	2·230	2·224	—	2·219	—
	$1{\cdot}307 \times 10^{-2}$	$3{\cdot}32 \times 10^{21}$	23·57	0·45	2·273	2·252	—	2·232	2·32
	$1{\cdot}398 \times 10^{-1}$	$3{\cdot}32 \times 10^{22}$	22·05	1·93	2·422	2·325	—	2·235	2·33
$1{\cdot}364 \times 10^7$	$1{\cdot}188 \times 10^{-3}$	$3{\cdot}32 \times 10^{20}$	25·93	0·06	2·073	2·071	2·075	2·069	—
	$1{\cdot}196 \times 10^{-2}$	$3{\cdot}32 \times 10^{21}$	25·77	0·17	2·085	2·079	2·092	2·073	—
	$1{\cdot}235 \times 10^{-1}$	$3{\cdot}32 \times 10^{22}$	24·95	0·60	2·151	2·127	2·161	2·105	2·28
	1·383	$3{\cdot}32 \times 10^{23}$	22·28	2·42	2·398	2·280	2·353	2·185	2·30
	$1{\cdot}702 \times 10$	$3{\cdot}32 \times 10^{24}$	18·18	5·97	2·911	2·519	2·729	2·239	—
$2{\cdot}728 \times 10^7$	$1{\cdot}195 \times 10^{-1}$	$3{\cdot}32 \times 10^{22}$	25·80	0·20	2·083	2·076	—	2·068	—
	1·222	$3{\cdot}32 \times 10^{23}$	25·22	0·74	2·130	2·100	—	2·072	2·18
	$1{\cdot}364 \times 10$	$3{\cdot}32 \times 10^{24}$	22·59	3·16	2·367	2·218	—	2·098	2·22
	$1{\cdot}689 \times 10^2$	$3{\cdot}32 \times 10^{25}$	18·25	7·04	2·901	2·452	—	2·193	2·26
$5{\cdot}456 \times 10^7$	1·196	$3{\cdot}32 \times 10^{23}$	25·77	0·23	2·086	2·077	—	2·068	—
	$1{\cdot}227 \times 10$	$3{\cdot}32 \times 10^{24}$	25·11	0·87	2·138	2·103	—	2·070	—
	$1{\cdot}384 \times 10^2$	$3{\cdot}32 \times 10^{25}$	22·27	3·60	2·400	2·227	—	2·084	—
	$1{\cdot}707 \times 10^3$	$3{\cdot}32 \times 10^{26}$	18·05	7·45	2·931	2·452	—	2·194	—

$\mu^* = m^*/(s + 1 + \frac{1}{2}\bar{q})$.

It follows that $$p = \frac{kT}{V}\{N + \Sigma_{r,z} M_r^z(1 + \tfrac{1}{2}\overline{q_r^z})\},$$

and therefore that

$$\mu^* = \frac{\Sigma_z M^z m_*^z}{N + \Sigma_{r,z} M_r^z(1 + \frac{1}{2}\overline{q_r^z})} = \frac{\Sigma_z M^z m_*^z}{N + \Sigma_z M^z(1 + \frac{1}{2}\overline{q^z})}, \quad \text{......(1795)}$$

when electrostatic effects are neglected.

On inspection of these results we see that the values of μ^* seem to differ in a rather irregular way (but not seriously) from the old results of Fowler

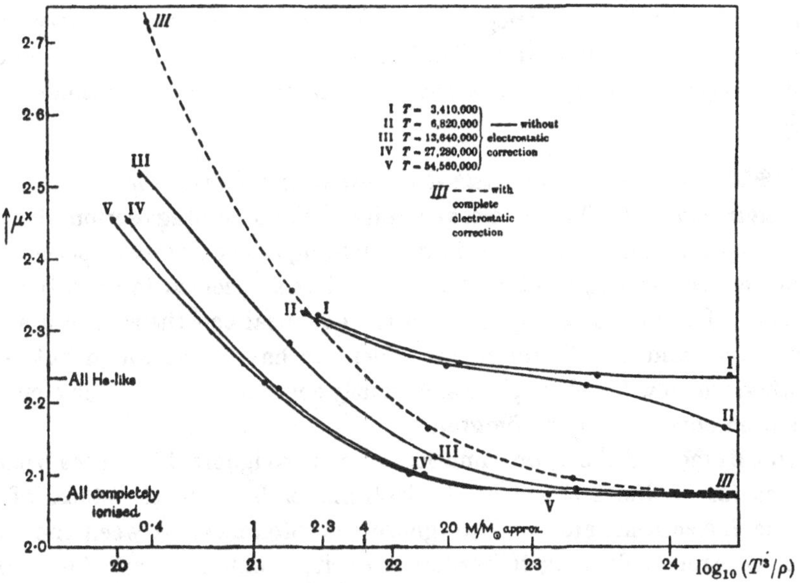

Fig. 77. The mean molecular weight μ^* of stellar material (iron).

and Guggenheim except for the highest densities, where the new values are higher as they should be. The irregular differences arise from a number of partially compensating changes in the data and the theory used, and are not significant. It is satisfactory to record that the old values are not far wrong at least for the more massive stars.

The control calculation with the estimated electrostatic correction has only been made for one temperature, but this is sufficient, for its importance can be seen to depend almost entirely on the mass of the star, that is on T^3/ρ, and hardly at all on T or ρ separately for given T^3/ρ. The following comments therefore apply sufficiently nearly to all temperatures. The estimated correction becomes sensible ($1\frac{1}{2}$ per cent. in μ^*) at a mass about twenty times the sun's mass, and has increased to $3\frac{1}{2}$ per cent. for a mass about 1·7 times the sun's mass. The average number of highly excited electrons per atom is here about 2·4, calculated of course neglecting the screening power of free electrons. Comparing μ^* and μ' we see that we have reached a mass at which the number of highly excited electrons has an appreciable effect on μ^*. For the reasons given we cannot claim that our theory gives a reliable value for $\bar{q}$ (the true value being less), so that we have here reached the limit of safety; the next entries show that $\bar{q}$ and the electrostatic correction are beginning to increase rapidly. The electrostatic correction has however been overestimated by a factor of the order of three. We may conclude that the values of μ^* as calculated without electrostatic corrections are sufficiently reliable down to masses twice the mass of the sun. By a cancellation of neglected effects the values given for μ^* without electrostatic correction are probably nearly correct down to masses equal to the mass of the sun. Below that mass none of our calculations can be relied upon.

§ **16·21.** *An approximate method for evaluating the mean molecular weight of stellar material.* The numerical results of the preceding section illustrate the general principles involved in determining as accurately as possible the mean molecular weight of stellar material at a specified temperature and density. But in actual applications the chemical constitution of a star is not known and it is therefore convenient to have some rough but ready method which will yield rapidly sufficiently accurate results. Such a method has been developed by Strömgren.†

In this method the approximation made is to ignore differences between the states of first and second K-electrons, or between the states of any of the L-electrons, etc., and to ignore the differences between the states of L-electrons belonging to normal or excited configurations. Any atomic configuration is then specified sufficiently by the numbers $(n_K, n_L, n_M, \ldots)$ specifying the number of its K-, L-, M-, ... electrons. The energy of such a state with a nucleus of atomic number z may be taken to be

$$-n_K \chi_K{}^z - n_L \chi_L{}^z - n_M \chi_M{}^z - \ldots,$$

where the χ's are constants representing the mean ionization potentials of the various shells, and the weight of the state by

$$\varpi^K{}_{n_K} \varpi^L{}_{n_L} \varpi^M{}_{n_M} \ldots.$$

When Pauli's exclusion principle is allowed for it is easy to verify that

$$\varpi^K{}_{n_K} = {}_2C_{n_K}, \quad \varpi^L{}_{n_L} = {}_8C_{n_L}, \quad \varpi^M{}_{n_M} = {}_{18}C_{n_M}, \quad \ldots \qquad (1796)$$

Similar formulae could be used to replace the weights given in Table 59.

We next observe that the equations of dissociative equilibrium can be cast into the following form. We can consider the equilibrium between

† B. Strömgren, *Zeit. f. Astrophysik*, vol. 4, p. 119 (1932), vol. 7, p. 222 (1933).

atoms in the single configuration $(n_K, n_L, n_M, \ldots)$ for short (n) and free electrons and the corresponding bare nucleus which is here denoted by the configuration $(0,0,0,\ldots)$ or (0). It is obvious that the equation of dissociative equilibrium then takes the form

$$\frac{N^{n_K+n_L+n_M+\cdots}M_{(0)}{}^z}{M_{(n)}{}^z} = \frac{[VG(T)]^{n_K+n_L+n_M+\cdots}[VF_z{}^z(T)]}{VF_{(n)}{}^z(T)}. \quad \ldots\ldots(1797)$$

This equation embodies a general principle of considerable value. It shows that dissociative equilibria can always be handled when convenient as between single states or single groups of states of all the systems concerned instead of as between the whole corpus of states.

Now in the present problem $VF_z{}^z(T)$ and $VF_{(n)}{}^z(T)$ both contain as common factor the partition function for free particles of mass m^z in a volume V. Dropping these factors

$$\frac{F_{(n)}{}^z(T)}{F_z{}^z(T)} = {}_2C_{n_K} e^{n_K\chi_K{}^z/kT} \times {}_8C_{n_L} e^{n_L\chi_L{}^z/kT} \times {}_{18}C_{n_M} e^{n_M\chi_M{}^z/kT} \times \ldots . \quad \ldots\ldots(1798)$$

It follows that

$$\frac{M_{(n)}{}^z}{M_{(0)}{}^z} = \left[\frac{N}{VG(T)}\right]^{n_K+n_L+n_M+\cdots} {}_2C_{n_K} e^{n_K\chi_K{}^z/kT} \times {}_8C_{n_L} e^{n_L\chi_L{}^z/kT} \times {}_{18}C_{n_M} e^{n_M\chi_M{}^z/kT} \times \ldots . \quad \ldots\ldots(1799)$$

The evaluation of the total number of bound electrons by this method will be sufficiently illustrated by the calculation of $\overline{n_L}$. Let

$$\frac{Ne^{\chi_L{}^z/kT}}{VG(T)} = x, \quad \ldots\ldots(1800)$$

so that $M_{(n)}{}^z/M_{(0)}{}^z$ depends on n_L only through the factor ${}_8C_{n_L}x^{n_L}$. It follows that

$$\overline{n_L} = \frac{\sum_{r=0}^{\infty} r\, {}_8C_r x^r}{\sum_{r=0}^{\infty} {}_8C_r x^r} = \frac{8x(1+x)^7}{(1+x)^8} = \frac{8}{1+1/x}. \quad \ldots\ldots(1801)$$

The other shells give similar contributions. Quite generally we see that the number of bound electrons of principal quantum number n round a nucleus of atomic number z is

$$\frac{2n^2}{1+\dfrac{VG(T)}{N}e^{-\chi_n{}^z/kT}}. \quad \ldots\ldots(1802)$$

It has been convenient to follow Strömgren in this derivation in order to show the nature of the approximations involved. We now see by inspection of (1802) that it is, as it must be, of the standard electronic form

$$\frac{\varpi_r}{1+e^{\epsilon_r/kT}/\lambda},$$

in which λ has been evaluated from the density N/V of the classical free electrons.

Strömgren has used (1802) to compute the electrons retained by the elements O, Na, Mg, Si, K, Ca and Fe for various values of T and $VG(T)/N$, considering only K-, L- and M-electrons. Electrons if present in higher orbits have important excluded volumes which cut down their numbers and they can be entirely neglected without serious error. He assumed that in the stellar material O, (Na + Mg), Si, (K + Ca), and Fe are present by weight in the ratio 8 : 4 : 1 : 1 : 2 ("Russell's Mixture"), and calculated the number of free particles per unit mass of the mixture for various values of T and $VG(T)/N$. The following Table 69 summarizes his results. The unit mass chosen is the mass of one hydrogen atom.

TABLE 69.

The average number $\bar{q}$ of free particles per unit mass (the mass of one hydrogen atom) of Russell's mixture.

$\log_{10} T$	$\log_{10}\{VG(T)/N\}$							
	3	4	5	6	7	8	9	10
6·4	—	—	—	0·46	0·49	0·50	0·51	0·52
6·6	—	—	0·48	0·51	0·52	0·53	0·53	0·54
6·8	—	0·48	0·51	0·53	0·53	0·53	0·54	0·54
7·0	0·44	0·50	0·52	0·53	0·54	0·54	0·54	0·54
7·2	0·46	0·51	0·53	0·54	0·54	0·54	0·54	0·54
7·4	0·47	0·51	0·53	0·54	0·54	0·54	0·54	0·54
7·6	0·47	0·51	0·53	0·54	0·54	0·54	0·54	0·54

The value 0·54 is the limiting value for complete ionization for all the atoms of the mixture.

The table can be used to determine μ^* for matter which contains any proportion of hydrogen and Russell's mixture. If the proportions are x parts hydrogen and $1-x$ parts mixture by weight, then

$$\mu^* = \frac{1}{2x + \bar{q}(1-x)}. \qquad \text{......(1803)}$$

In stellar applications μ^* is usually determined "astrophysically" from Eddington's mass-luminosity-effective temperature relation and (1803) is then used to determine x. Strömgren and Eddington† have both made such calculations and found that according to this analysis there should be about 30 per cent. hydrogen by weight in the material of the sun, and a larger percentage in more massive stars. Further discussion of these developments lies outside the range of this monograph.

† Eddington, *Monthly Not. R.A.S.* vol. 92, p. 471 (1932).

§ **16·3.** *Matter of great density.* We have not yet investigated whether matter of the densities and temperatures appearing in Table 68, or of still greater densities, can really exist and still more behave approximately like a gas of mixed dissociating constituents approximately perfect. We have omitted all mention of excluded volumes in the foregoing discussion. It is these, and the fields of force arising from permanent electronic structures, which will make the mixture depart from perfection—in particular reduce largely its very high compressibility. The highly excited electrons and the electrostatic corrections do not have any such effect, and can be ignored in this connection.

A complete formulation of the problem might have started by assigning to every possible type of electronic structure with its electrons in their lowest or at least in tightly bound orbits a definite field of force—with sufficient accuracy an excluded volume, such that no part of any pair of such excluded volumes may overlap. The characteristic function for such an assembly can be written down to a first approximation, using § 8·6. It is easy to see however that such elaboration is unnecessary, for the large volumes of the more complex electronic structures will make them occur on the average still less often than in the assembly for which the calculations have been made, where their excluded volumes are all zero. If therefore we take the simplified assembly for which the calculations have been made, estimate reliable excluded volumes for the dominant tightly bound electronic structures, and show that such excluded volumes are insignificant, the existence of matter at such densities with the properties of a perfect gas will be established, and the foregoing procedure fully justified.

On examining the details of the calculations it appears that the most important surviving electronic structure contains two electrons only in their normal orbits—that is helium-like iron. Apart from the excess nuclear charges which contribute only to the electrostatic correction (the complete electrostatic correction *increases* the compressibility), we may be certain that the helium-like iron ions will interact as if the radius of their excluded volumes bore the same ratio to the radius of the excluded volume of two ordinary helium atoms as do the radii of the corresponding Bohr orbits of the two electrons.* Allowing for screening, the ratio of the radii of the orbits of helium-like iron and normal helium is about $1{\cdot}7/25{\cdot}7$. The excluded volumes for helium-like iron are therefore smaller than those of helium by a factor $(1{\cdot}7/25{\cdot}7)^3$ or, say, $1/(15)^3$.

For a gas of M helium-like iron ions in a volume V and N $(=24M)$

* The radius of an orbit of the old quantum theory is a reliable guide to the position of maximum density of electricity (the important region) in quantum mechanics.

free electrons we shall have to a first approximation an equation of state

$$\frac{pV}{kT}=N+M+\frac{1}{V}\{NMv_{ec}+\tfrac{1}{2}M^2v_{cc}\}.$$

The electrons are practically without volume. The excluded volume for a free electron and a core is v_{ec} and for two cores v_{cc}. It is perhaps legitimate to suppose that $v_{ec}=\frac{1}{8}v_{cc}$ on the average. The ratio of the correcting term to the main term is therefore nearly enough Mv_{ec}/V. In the same way the ratio of the correcting term to the main term for normal helium is (say) $\frac{1}{2}M'v'_{cc}/V$. The excluded volumes affect the pressure of ordinary helium gas by 1 per cent. at a density of about 0·0043 gm./c.c. The 1 per cent. correction will therefore be reached in a gas of helium-like iron plus electrons at a density of

$$0\cdot0043\times\tfrac{56}{4}\times(15)^3\times4=840.$$

In two important respects we have probably underestimated the density for a 1 per cent. correction in this calculation. The greater temperature of the stellar material must diminish rather than increase corresponding excluded volumes, and it is likely that the volume excluded to free electrons may not be comparable with $\frac{1}{8}v_{cc}$. We may conclude that stellar material will in general reach densities of the order of 1000 gm./c.c. or more before we find any departures from the compressibility of a perfect gas due to those properties of finite extension of the constituent systems which make the ordinary gases of our experience imperfect.

Whether we can go further than this—to densities of 10,000 or 100,000 with the same compressibility—the theory we have developed is unable to say. Such matter can only be discussed when we have a proper method of including electrostatic effects and use the correct Fermi-Dirac statistics for the free electrons. We shall not attempt to introduce the electrostatic corrections more accurately, but neglecting them shall apply the Fermi-Dirac statistics to stellar conditions. The effect of Pauli's exclusion principle undoubtedly dominates the properties of matter at very high densities such as are met with in white dwarf stars and controls their ultimate fate as "black dwarfs". We shall take this opportunity therefore of deriving the exact equation of state of matter of great density (degenerate matter) allowing for the relativistic variation of mass with velocity.

§ **16·31.** *Assemblies of electrons with relativistic energies.* At extremely great densities the equation of state of matter is entirely controlled by the electron pressure; the contributions of the heavy nuclei can be entirely ignored. We have already discussed electron assemblies obeying Fermi-Dirac statistics in the region of degeneracy in § 2·73 and again in Chapter XI, when we used freely the formulae for such assemblies to build up the electron theory of metals. In § 11·21 equation (1019) we gave the formula for the electron pressure which we shall require here for stellar applications, namely

$$p_\epsilon = \tfrac{2}{5} n_0 \eta^* = \frac{\pi}{60} \frac{h^2}{m} \left(\frac{3n_0}{\pi}\right)^{\frac{5}{3}}. \qquad \text{......(1804)}$$

This formula was derived on the basis of Schrödinger's equation, and therefore assumes a non-relativistic relationship $E = p_*^2/2m$ between momentum p_* and kinetic energy E. For present purposes therefore, where it will be necessary to consider also the more general relativistic relationship

$$E = mc^2 \left\{ \left(1 + \frac{p_*^2}{m^2c^2}\right)^{\frac{1}{2}} - 1 \right\}, \qquad \text{......(1805)}$$

it will be convenient to obtain (1804) *de novo* by a slightly different arrangement of the arguments.

Consider an assembly of N electrons in a volume V. The amount of six-dimensional phase space available for an electron whose momentum lies between p_* and $p_* + dp_*$ in absolute value is then $4\pi V p_*^2 dp_*$ and the number of states available to such electrons is accordingly

$$2 \frac{V}{h^3} 4\pi p_*^2 dp_*. \qquad \text{......(1806)}$$

The generating function Z can therefore be put in the form

$$Z = \frac{8\pi V}{h^3} \int_0^\infty p_*^2 \log(1 + \lambda e^{-E/kT})\, dp_*, \qquad \text{......(1807)}$$

which is easily reduced to (224) when $E = p_*^2/2m$. On using the relativistic (1805) this can be put in the forms

$$Z = \frac{8\pi V}{c^3h^3} \int_0^\infty (E + mc^2) \{E(E + 2mc^2)\}^{\frac{1}{2}} \log(1 + \lambda e^{-E/kT})\, dE; \qquad \text{......(1808)}$$

$$Z = \frac{8\pi V}{c^3h^3} \int_{mc^2}^\infty (E'^2 - m^2c^4)^{\frac{1}{2}} E' \log(1 + \lambda e^{-E'/kT})\, dE'. \qquad \text{......(1808·1)}$$

In (1808·1) E' is the total energy including the rest energy mc^2 and in consequence λ has been differently defined.

From these formulae, with the help of the standard relationships

$$N = \lambda \frac{\partial Z}{\partial \lambda}, \quad \bar{E} = kT^2 \partial Z/\partial T, \quad p = kT \partial Z/\partial V,$$

all the details of statistical equilibrium with relativistic energies can be derived. For example using (1807) we have the formulae

$$N = \frac{8\pi V}{h^3} \int_0^\infty \frac{p_*^2 dp_*}{1 + e^{E/kT}/\lambda}, \qquad \text{......(1809)}$$

$$\frac{pV}{kT} = Z, \quad p = \frac{8\pi}{3h^3} \int_0^\infty \frac{p_*^3 (\partial E/\partial p_*)\, dp_*}{1 + e^{E/kT}/\lambda}. \qquad \text{......(1809·1)}$$

§ **16·32.** *Classical assemblies of relativistic electrons.*† We pause for a moment to obtain the limiting forms taken by these formulae when the assembly is still classical and the exclusion principle can be ignored but not the relativistic variation of mass with velocity. This will occur at temperatures sufficiently high. We then can simplify (1808·1) to

$$Z=\frac{8\pi V\lambda}{c^3h^3}\int_{mc^2}^{\infty}(E'^2-m^2c^4)^{\frac{1}{2}}E'e^{-E'/kT}dE'$$

and by the substitution $E'=mc^2\cosh\mu$ to

$$Z=\frac{8\pi Vm^3c^3\lambda}{h^3}\int_0^{\infty}e^{-(mc^2/kT)\cosh\mu}\sinh^2\mu\cosh\mu\,d\mu.$$

This can be expressed in terms of the Bessel function $K_n(x)$ in the form‡

$$Z=\frac{8\pi Vm^3c^3\lambda}{4h^3}\left\{K_3\left(\frac{mc^2}{kT}\right)-K_1\left(\frac{mc^2}{kT}\right)\right\}. \qquad \text{......(1810)}$$

The case of interest to us is when $kT\gg mc^2$; in the other limiting case $mc^2\gg kT$ the relativistic effects can be ignored and the formulae reduce to those of the ordinary classical assembly. Either from (1810) or more directly by neglecting mc^2 in (1808) we then obtain

$$Z=\frac{16\pi Vk^3T^3}{c^3h^3}\lambda. \qquad \text{......(1811)}$$

The coefficient of λ in (1810) is the relativistic form of the ordinary partition function and (1811) shows its limiting form for exceedingly high temperatures.

§ **16·33.** *The equation of state of completely degenerate matter.* When $T\to 0$ the factor $1/(1+e^{E/kT}/\lambda)$ tends to 0 or 1; its limit is unity when $p_*<p_*^0$ and otherwise zero, p_*^0 being fixed by the condition that (1809) then gives the correct value of N. The limiting formulae are therefore §

$$\frac{N}{V}=\frac{8\pi}{3h^3}(p_*^0)^3, \qquad \text{......(1812)}$$

$$p=\frac{8\pi}{3h^3}\int_0^{p_*^0}p_*^3\frac{\partial E}{\partial p_*}dp_*, \qquad \text{......(1813)}$$

$$\bar{E}=\frac{8\pi V}{h^3}\int_0^{p_*^0}Ep_*^2dp_*. \qquad \text{......(1814)}$$

† See Jüttner, *Ann. d. Physik*, vol. 34, p. 856 (1911); *Zeit. f. Physik*, vol. 47, p. 542 (1928).

‡ See Watson, *Bessel Functions*, p. 181, Cambridge (1922).

§ These formulae can be derived in an entirely elementary way from (1806) and the condition that when $T\to 0$ the lowest N states are all full and the rest empty. The pressure is then given as the rate of transport of momentum across any interface in the assembly which can be calculated as one-third of the integral of (the number of electrons with momentum between p_* and $p_*\times dp_*$) $\times(p_*)\times$ (the velocity of these electrons $(=\partial E/\partial p_*)$).

From (1813) and (1814) it follows that

$$p = \frac{8\pi}{3h^3}(p_*^{\,0})^3 E(p_*^{\,0}) - \frac{\bar{E}}{V}. \qquad \text{......(1815)}$$

If the relativistic mass-variation with velocity is neglected so that $E = p_*^{\,2}/2m$, (1815) becomes

$$p = \frac{8\pi}{15mh^3}(p_*^{\,0})^5.$$

Combining this with (1812) we reobtain (1804). In stellar applications to the theory of white dwarfs one assumes that the matter is sufficiently dense and cool for (1804) to be a valid approximation. The equation of state of the material is then

$$p = K_1 \rho^{\frac{5}{3}}, \qquad \text{......(1816)}$$

where

$$K_1 = \frac{\pi}{60}\left(\frac{3}{\pi}\right)^{\frac{2}{3}} \frac{h^2}{m(\mu^* m_H)^{\frac{5}{3}}} = \frac{9{\cdot}89 \times 10^{12}}{(\mu^*)^{\frac{5}{3}}}. \qquad \text{......(1817)}$$

In (1817) μ^* is again the mean molecular weight of the stellar material on the chemical scale and m_H the mass in grams of a conventional hydrogen atom of atomic weight unity. We have used the relation $\rho = (N/V)\mu^* m_H$ which agrees with μ^* defined in (1795) only if the electrons are much more numerous than the nuclei. In other cases it defines μ^*.

For matter in this completely condensed state a simple estimate shows that the negative potential energy is very roughly equal to† $8{\cdot}9 \times 10^{13}\rho^{\frac{1}{3}}$, while from (1814) the total kinetic energy of the assembly is roughly $4{\cdot}2 \times 10^{12}\rho^{\frac{2}{3}}$. Comparing these two estimates we see that for matter of the density occurring in the white dwarfs, about 10^5 gm./c.c., the total kinetic energy is about twice the negative potential energy. This fact removes a difficulty, originally pointed out by Eddington, that on classical statistics dense matter would ultimately contain far less energy than the same matter expanded in the form of atoms at rest at great separations. But if the dense matter behaves according to the theory here given, a sample extracted from a star could reconstruct itself as an expanded gas at a very high temperature.

§ **16·34** *Relativistic degeneracy.* We will now consider the exact form of the degenerate equation of state allowing for the variation of mass with velocity. On substituting (1805) in (1813) we have

$$p = \frac{8\pi}{3mh^3}\int_0^{p_*^{\,0}} \frac{p_*^{\,4}\,dp_*}{(1 + p_*^{\,2}/m^2c^2)^{\frac{1}{2}}}. \qquad \text{......(1818)}$$

† R. H. Fowler, *Monthly Not. R.A.S.* vol. 87, p. 114 (1926). The index of ρ must be $\frac{1}{3}$ on almost any view of dense matter. The actual proportionality constant here given is uncertain and likely to be overestimated.

This can be integrated in the form

$$p=\frac{\pi m^4c^5}{3h^3}f(x)\quad\left(x=\frac{p_*{}^0}{mc}\right), \qquad \text{......(1819)}$$

where

$$f(x)=x(2x^2-3)(x^2+1)^{\frac{1}{2}}+3\operatorname{arc\,sinh} x. \qquad \text{......(1820)}$$

The parameter x is fixed by the equations

$$\frac{N}{V}=\frac{8\pi m^3c^3}{3h^3}x^3,\quad \rho=\frac{N}{V}\mu^* m_H, \qquad \text{......(1821)}$$

which yield

$$\rho=9{\cdot}885\times10^5\mu^*x^3. \qquad \text{......(1822)}$$

The function $f(x)$ has been tabulated by Chandrasekhar.†

From (1820) it follows that as $x\to0$

$$f(x)\sim\tfrac{8}{5}x^5, \qquad \text{......(1823)}$$

while as $x\to\infty$

$$f(x)\sim2x^4. \qquad \text{......(1824)}$$

Hence as $x\to\infty$

$$p\sim\frac{2\pi m^4c^5}{3h^3}x^4,$$

or

$$p\sim K_2\rho^{\frac{4}{3}}, \qquad \text{......(1825)}$$

where

$$K_2=\left(\frac{3}{\pi}\right)^{\frac{1}{3}}\frac{hc}{8(\mu^*m_H)^{\frac{4}{3}}}=\frac{1{\cdot}228\times10^{15}}{\mu^{*\frac{4}{3}}}. \qquad \text{......(1826)}$$

Comparing (1817) and (1825) we see that these limiting forms of the exact equation of state yield the same value for the pressure for a density ρ' given by

$$\rho'=\left(\frac{K_2}{K_1}\right)^3=1{\cdot}9\times10^6\mu^*. \qquad \text{......(1827)}$$

We infer that for densities greater than 10^6 the relativistic effects become important.

Finally it is worth noticing that the setting in of relativistic degeneracy does not revive any difficulties of the nature of that discussed in § 16·33. This is clear on comparing (1826), which also gives the kinetic energy, with the estimated negative potential energy $8{\cdot}9\times10^{13}\rho^{\frac{4}{3}}$. The total kinetic energy is more than ten times the negative potential energy.

The equations of state of degenerate matter here discussed have been applied to the theory of stellar structures particularly by Milne‡ and Chandrasekhar,§ but such applications lie outside the scope of this monograph.‖

† Chandrasekhar, *Monthly Not. R.A.S.* vol. 95, p. 225 (1935). Such calculations were first made by Stoner, *Monthly Not. R.A.S.* vol. 91, p. 444 (1931), but less accurately and less extensively.

‡ Milne, *Monthly Not. R.A.S.* vol. 91, p. 4 (1931), vol. 92, p. 610 (1932).

§ Chandrasekhar, *Monthly Not. R.A.S.* vol. 95, p. 321 (1935).

‖ It has recently been contended by Eddington, *Monthly Not. R.A.S.* vol. 95, p. 194 (1935), that (1805) is invalid and does not apply to an electron in any stationary state in an enclosure, but only to electrons represented by progressive waves. If his contentions are correct then the formulae derived from (1805) are meaningless and we must always use formulae derived from $E=p_*{}^2/2m$ (i.e. Schrödinger's equation) for electrons represented by standing waves in an enclosure.

§ **16·4.** *Assemblies of electrons and positive electrons.* It is now certain that a quantum of radiation of sufficiently high frequency on interaction with matter, or even two quanta interacting with each other, can be converted into a pair of positive and negative electrons, and conversely that a pair of positive and negative electrons can be annihilated, emitting in the process one or two quanta of radiation. The minimum energy required to create a pair is $2mc^2$ and such effects will therefore only occur with a frequency to make the pairs significant for the equilibrium properties of the assembly at temperatures of the order $2mc^2/k$, that is 10^{10} degrees. But at such temperatures, if they anywhere exist, the number of ordinary electrons is no longer fixed, but only the excess of the number of ordinary electrons over the number of positives. The modified equilibria so established can be calculated at once by a simple extension of the usual method.

Let ϵ_r be the energies of the states of an electron and $2mc^2+\eta_r$ those for a positive electron. Let N be the number of electrons when there are no positives, and the energy zero be taken to be the state in which there are no positives and the N electrons are all in their lowest state. The electrons and positive electrons each obey the Fermi-Dirac statistics. We therefore form the generating function

$$\Pi_r (1+xz^{\epsilon_r})\, \Pi_r \left(1+\frac{1}{x} z^{2mc^2+\eta_r}\right),$$

and can see at once that the total number of complexions is given by the coefficient of $x^N z^E$. Thus

$$C=\frac{1}{(2\pi i)^2}\iint \frac{dx\,dz}{x^{N+1}z^{E+1}}\, \Pi_r (1+xz^{\epsilon_r})\, \Pi_r \left(1+\frac{1}{x} z^{2mc^2+\eta_r}\right). \quad \text{...(1828)}$$

The necessary extensions of the usual theorems to integrands of this type which have an essential singularity at the origin are easily made. It then follows that, if $\overline{N_-}$ and $\overline{N_+}$ are the average numbers of ordinary and positive electrons present in the assembly,

$$\overline{N_-}=\lambda\frac{\partial}{\partial\lambda}\Sigma_r \log(1+\lambda e^{-\epsilon_r/kT}), \quad \text{......(1829)}$$

$$\overline{N_+}=-\lambda\frac{\partial}{\partial\lambda}\Sigma_r \log(1+e^{-(2mc^2+\eta_r)/kT}/\lambda), \quad \text{......(1830)}$$

λ being determined by the condition that

$$\overline{N_-}-\overline{N_+}=N. \quad \text{......(1831)}$$

From these equations any desired formulae can be obtained. We shall be content to discuss here the classical limit since the temperature is so high before N_+ is sensible that degeneracy is unlikely to have set in. In that case, if we ignore the variation of mass with velocity,

$$\overline{N_-}=\lambda\frac{2(2\pi mkT)^{\frac{3}{2}}V}{h^3},\quad \overline{N_+}=\frac{1}{\lambda}\frac{2(2\pi mkT)^{\frac{3}{2}}V}{h^3}e^{-2mc^2/kT},\quad \text{......(1832)}$$

so that the number of pairs is determined by the "reaction isochore"

$$\frac{\overline{N_+}\,\overline{N_-}}{V^2}=\frac{4(2\pi mkT)^3}{h^3}e^{-2mc^2/kT}.\qquad \text{......(1833)}$$

The pressure is given by the obvious equation $p=(\overline{N_+}+\overline{N_-})\,kT/V$ which can easily be reduced to

$$p=kT\left\{\frac{N^2}{V^2}+\frac{16(2\pi mkT)^3}{h^3}e^{-2mc^2/kT}\right\}^{\frac{1}{2}}.\qquad \text{......(1834)}$$

The pressure is always greater than, and on these approximations tends at high temperatures asymptotically to, the value

$$p_{\pm}=\frac{4(2\pi m)^{\frac{3}{2}}k^{\frac{5}{2}}}{h^3}T^{\frac{5}{2}}e^{-mc^2/kT}.\qquad \text{......(1835)}$$

At these high temperatures however the neglect of the variation of mass with velocity is hardly justified, and the limiting form will be better represented by taking the partition function from (1811) and so using $16\pi Vk^3T^3/c^3h^3$ in place of $2(2\pi mkT)^{\frac{3}{2}}V/h^3$. The argument is otherwise unaltered. We then find

$$\frac{\overline{N_+}\,\overline{N_-}}{V^2}=\frac{256\pi^2k^6T^6}{c^6h^6}e^{-2mc^2/kT},\qquad \text{......(1836)}$$

$$p_{\pm}=\frac{32\pi k^4T^4}{c^3h^3}e^{-mc^2/kT}.\qquad \text{......(1837)}$$

The pressure $p_{\pm}$ may be regarded as the unavoidable material pressure in an assembly due to the creation of pairs. It is interesting to compare it with the ordinary radiation pressure at various temperatures. The maximum value of the ratio $p_{\pm}/p_R$ according to (1835) occurs at a temperature of about 4×10^9 and is then only 10^{-4}. The value of $p_{\pm}$ according to (1837) can be larger. Inserting numerical values into (1837) and (332) we have

$$p_{\pm}=4{\cdot}66\times10^{-15}T^4e^{-5{\cdot}91\times10^9/T},\quad p_R=2{\cdot}53\times10^{-11}T^4.$$

Thus $p_{\pm}/p_R$ tends to the limit $1{\cdot}8\times10^{-4}$ as $T\to\infty$, and is otherwise always smaller.

Though these pressures are always trivial compared with the radiation pressure, the actual electron densities that they represent are still very high. When $T\sim mc^2/k\sim6\times10^9$ the material pair density $mp_{\pm}/kT$ reaches values between 10^3 and 10^4. Interesting speculations based on this result have been made by Heitler.*

* Heitler, *Proc. Camb. Phil. Soc.* vol. 31, p. 243 (1935).

§16·5. *Assemblies in which nuclear transformations can occur and reach equilibrium.* It is now recognized that, provided the necessary energy conditions are satisfied, very large numbers of nuclear transformations can occur, so much so that we may conclude that, given time, nuclei could attain relative concentrations in equilibrium. It is still doubtful whether nuclei should be regarded as composed solely of protons and neutrons and whether these are independent particles or should be regarded as capable of transformation into each other with the emission of positive or negative electrons. But these doubts concern mainly the mechanisms by which equilibrium is reached. Whatever the ultimate constituents of nuclei may be, an analysis into protons and neutrons or protons and electrons will equally enable us to study conveniently the statistical equilibrium of an assembly of nuclei. We shall adopt here an analysis into protons and neutrons as likely to be closer to the truth than any other.

Discussions of such assemblies with transmutation obviously provide a basis for an equilibrium theory of the abundance of the elements and calculations of this kind have been made in some detail by Sterne.* We shall give a brief sketch of this work without intending to imply that the actual abundances of the elements in terrestrial or stellar matter must correspond to equilibrium abundances under some particular conditions. Indeed, since we believe for good reasons that hydrogen is the dominant constituent of stellar matter, it is palpable that such matter is not now and probably never has been in an equilibrium state for nuclear transformations, though it may be on its way thither. On the other hand we shall be able to draw some conclusions which may have significance for the abundances observed in the earth's crust and in meteorites.

Let us consider then an assembly of protons, neutrons and various composite nuclei. The electrons necessary to neutralize the assembly are present but need not be referred to explicitly. The protons and neutrons obey separately the Fermi-Dirac statistics; nuclei of odd mass number (odd total number of protons and neutrons) obey the same statistics, and nuclei of even mass number the Einstein-Bose. We take the state of zero energy of the assembly to be that in which all the nuclei are completely dissociated into protons and neutrons at rest in their lowest states. The energy of formation Q_s of a composite nucleus of rest mass m_s, that is its energy in its lowest state relative to the defined energy zero, is then given by

$$Q_s = c^2[m_s - Z_s m_p - N_s m_n].$$

In this formulation c is the velocity of light, m_p and m_n are the rest masses of the proton and the neutron, which it is sufficiently accurate to equate, Z_s and N_s are the numbers of protons and neutrons in the nucleus respec-

* Sterne, *Monthly Not. R.A.S.* vol. 93, p. 736 (1933).

tively. It then follows by the usual arguments that, if $\overline{X_p}$, $\overline{X_n}$ and $\overline{X_s}$ are the average numbers of free protons, free neutrons and nuclei of type s in the assembly,

$$\overline{X_p}=\eta\frac{\partial}{\partial\eta}\Sigma_r\log(1+\eta e^{-\epsilon_r/kT}),$$

$$\overline{X_n}=\zeta\frac{\partial}{\partial\zeta}\Sigma_r\log(1+\zeta e^{-\epsilon_r'/kT}),$$

$$\overline{X_s}=\mu_s\frac{\partial}{\partial\mu_s}\Sigma_r\pm\log(1\pm\mu_s e^{-\epsilon_r/kT}),$$

where
$$\mu_s=\eta^{Z_s}\zeta^{N_s}e^{-Q_s/kT}. \qquad \text{......(1838)}$$

The upper signs refer to nuclei of odd, the lower to even mass number, and the ϵ_r enumerate the kinetic energies of the states of motion of the various systems, the least possible value being zero for each system. Using these equations and given total numbers of protons and neutrons, the equilibrium concentrations can be calculated in any given case. We shall not however give any such calculations, since the most interesting consequences of these equations are qualitative.

Equation (1838) may be written in the form

$$T\log\mu_s=Z_s(T\log\eta)+N_s(T\log\zeta)+\frac{c^2}{R}[(Z_s+N_s)\,m_p{}^*-m_s{}^*],$$

$$=Z_s(T\log\eta)+N_s(T\log\zeta)+\frac{c^2}{10^4R}(Z_s+N_s)\,(f_p-f_s).$$
$$\text{......(1839)}$$

In the form (1839) f_p and f_s are Aston's packing fractions. They are so defined that for any atom† of mass $m_s{}^*$ on the scale $O^{16}=16$,

$$m_s{}^*=(Z_s+N_s)\,(1+10^{-4}f_s).$$

We observe further that each pair of variables $\overline{X_p}$ and η, $\overline{X_n}$ and ζ, $\overline{X_s}$ and μ_s increase together and that for even nuclei, when the lower signs are taken, $\overline{X_s}\to\infty$ as $\mu_s\to 1$, so that $\mu_s<1$.

We can gain valuable information by using $T\log\eta$ $(=x)$ and $T\log\zeta$ $(=y)$ as cartesian coordinates and drawing the "abundance lines"

$$Z_sx+N_sy+\frac{c^2}{10^4R}(Z_s+N_s)\,(f_p-f_s)=0 \qquad \text{......(1840)}$$

for all even nuclei. Since $\mu_s<1$ only points below and to the left of all these abundance lines represent possible states of the assembly. Moreover when the representative point (x, y) moves towards the abundance lines, the assembly must consist almost solely of that even element whose abundance line is first approached by the representative point. Fig. 78 shows that this

† In an expression such as $(Z_s+N_s)\,m_p{}^*-m_s{}^*$ we need not distinguish between nuclear and atomic masses.

line can belong to light, heavy or medium elements in various regions of the diagram. In the central region that abundance line is first approached for which the perpendicular distance from the origin

$$\frac{c^2}{10^4 R} \frac{Z_s + N_s}{(Z_s^2 + N_s^2)^{\frac{1}{2}}} (f_p - f_s) \qquad \text{......(1841)}$$

is greatest. Since $(Z_s + N_s)/(Z_s^2 + N_s^2)^{\frac{1}{2}}$ is never greatly different from $\sqrt{2}$ for nuclei of even mass number, this perpendicular is greatest for that nucleus for which $f_p - f_s$ is greatest or for which the *packing fraction* f_s reaches its greatest negative value. This occurs for iron or among the nuclei of the iron group. It is possible therefore to find conditions—very extreme conditions it is true—for which the nuclear equilibrium is one in which the nuclei of the iron group are in great excess as they are in fact in iron meteorites and perhaps the earth's core.

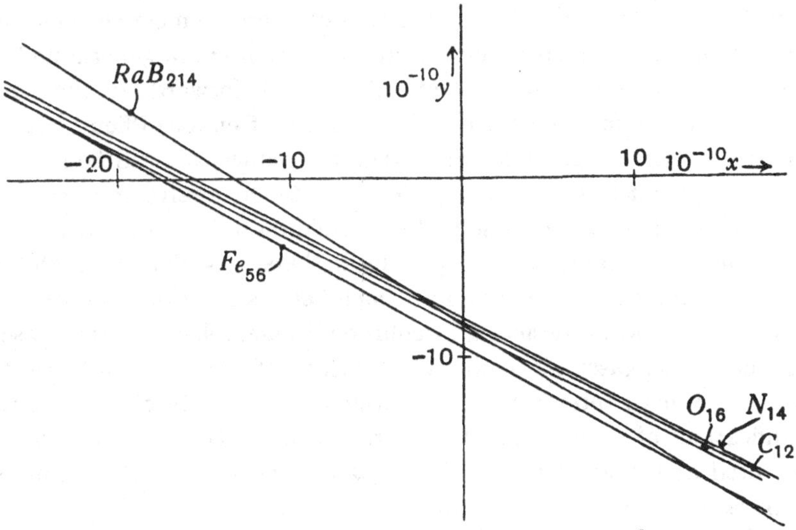

Fig. 78. Abundance lines for elements of even mass number.

No such effect can occur under conditions for which classical statistics are sufficiently accurate. For then the concentration $\overline{X_s}$ is roughly proportional to μ_s and that nucleus is dominant for which

$$Z_s x + N_s y + \frac{c^2}{10^4 R}(Z_s + N_s)(f_p - f_s)$$

has the least negative value. Roughly speaking this is that nucleus for which

$$\frac{c^2}{10^4 R}(Z_s + N_s)(f_p - f_s)$$

has the greatest value or that nucleus which has the greatest mass defect—nuclei from the neighbourhood of tin. We know of no matter for which a maximum abundance is realized in this region.

CHAPTER XVII

MECHANISMS OF INTERACTION. COLLISION PROCESSES

§17·1. *The nature of the equilibrium state.* We have pointed out in the introductory chapter that the Equilibrium Theory of Statistical Mechanics is essentially of a thermodynamic nature. Its laws are independent of all mechanisms of interaction. It has merely to be supposed that the necessary interactions can occur. It is only when we begin to discuss the far more difficult theory of (non-equilibrium) steady states of flow that the actual laws of the mechanisms acting become relevant, or can be deduced from the experimental facts. It was also pointed out that the laws of equilibrium, being thus in some sense universal, must be conformed to by the laws of any mechanism. The actual details of the laws of interaction between molecular systems and between such systems and radiation are in general no longer obscure, being given by theoretically straightforward applications of quantum mechanics. But such applications are often too difficult or tedious to carry through in practice. Statistical mechanics can still help towards the development of the theory of steady states by analysing carefully the restrictions that its laws impose on the laws of interaction, so as to leave vague for further calculation as little as possible in these laws. This is valuable help, and this chapter gives an account of such help as can thus be given to the study of the laws of interaction by collision in assemblies of perfect gases, or between perfect gases and solid walls. The laws of interaction with radiation are treated in Chapter XIX. The treatment is primarily classical, but the modifications of form required by Fermi-Dirac or Einstein-Bose statistics are noted at the appropriate places. Quantum mechanics introduces no modification of principle into this field.

In the classical kinetic theory the laws of interaction of structureless molecules are all that interest us; we visualize molecular encounters as purely conservative collisions either of elastic spheres or point centres of force. This conception has of course long proved fruitful in the study of transport phenomena initiated by the work of Maxwell. By way of introduction to this chapter we shall present the classical theorems of Maxwell and Boltzmann, including the latter's famous H-theorem, from this slightly unusual angle.

In the equilibrium state of a classical perfect gas the positions and velocities of each set of molecules satisfy Maxwell's law. The distribution function for any one set is, let us say,

$$f(u,v,w)\,du\,dv\,dw \qquad \text{......(1842)}$$

per unit volume, and the function f must be the same *whatever the mechanism by which it is set up*. This is the position at which we are left by the general mechanismless equilibrium theory. It remains to examine whether the usual mechanism of classical conservative collisions is a possible mechanism which will preserve (1842). Now if a certain type of collision between molecules 1 and 2 with velocities u_1, v_1, w_1 and u_2, v_2, w_2 transforms these into u_1^*, v_1^*, w_1^* and u_2^*, v_2^*, w_2^*, then the number of such collisions per unit volume per unit time is

$$f_1 f_2 du_1 \ldots dw_2 V p\, dp\, d\psi. \qquad \ldots\ldots(1843)$$

In this expression V is the relative velocity of the two molecules before the encounter begins, and p and ψ are polar coordinates, in a plane through one molecule normal to the direction of V, which define the position of the initial asymptote of the necessary relative orbit. The symbol f_1 is short for $f(u_1,v_1,w_1)$, etc. These collisions all *destroy* molecules of velocities u_1, v_1, w_1 and u_2, v_2, w_2 and *create* molecules of velocities u_1^*, v_1^*, w_1^* and u_2^*, v_2^*, w_2^*. Conversely, since any conservative orbit can also be travelled in the reverse sense there are a set of reverse collisions, in number

$$f_1^* f_2^* du_1^* \ldots dw_2^* V^* p^* dp^* d\psi^*, \qquad \ldots\ldots(1844)$$

which *destroy* molecules of velocities u_1^*, v_1^*, w_1^* and u_2^*, v_2^*, w_2^* and *create* molecules of velocities u_1, v_1, w_1 and u_2, v_2, w_2. The relation between (1843) and (1844) can be simplified. By applying Liouville's theorem to the element of phase space of the conservative system formed by the two molecules we find that

$$du_1 \ldots dw_2 V p\, dp\, d\psi = du_1^* \ldots dw_2^* V^* p^* dp^* d\psi^*. \qquad \ldots(1845)\dagger$$

The proposed collision mechanism will therefore certainly conserve the equilibrium state if this requires that

$$f_1 f_2 = f_1^* f_2^*, \qquad \ldots\ldots(1846)$$

a relation equivalent to Maxwell's law. Since equation (1846) is a sufficient condition for the preservation of equilibrium, the classical collision mechanism is therefore a possible mechanism which can act and preserve the equilibrium state, a conclusion which was of course sufficiently obvious without calculation.

§**17·11.** *Detailed balancing.* It is important to observe that equation (1846) has far wider implications than we have just drawn, for it asserts that there is a *detailed balancing* of all the individual types of collision which can be specified by the velocity exchanges as above. According to (1846) exactly the same number of collisions of any one type and of the corre-

† Since the systems are conservative we have also $V = V^*$. If the relative orbit is described under a central force and so is plane and symmetrical about the apse $d\psi = d\psi^*$, $p = p^*$, $dp = dp^*$. Thus in this simple case $du_1 \ldots dw_2 = du_1^* \ldots dw_2^*$.

sponding reverse type must occur per unit volume per second. This condition of detailed balancing is naturally sufficient for the *preservation of equilibrium*, but asserts (at least at first sight) far more than is asserted by the mere requirement that equilibrium is preserved. It does not necessarily follow without further investigation that detailed balancing is also *necessary* and therefore equivalent to the demand for the preservation of equilibrium. On certain assumptions however it is possible to prove this equivalence for the classical collision mechanism and the proof, given in § 17·2, constitutes Boltzmann's H-theorem. Thus for the particular mechanism of classical conservative collisions under central forces the two requirements of

(*a*) *Detailed balancing,*

(*b*) *Preservation of equilibrium*

are equivalent.

For other mechanisms it is not always possible to prove this equivalence, and in fact examples are known for which condition (*b*) is definitely less restrictive than condition (*a*). The study of particular mechanisms in this way has in the past been a study of considerable importance by which much light was thrown on the laws of particular mechanisms before their fundamental properties were properly understood through the development of quantum mechanics. *We now know that all mechanisms of interaction conform to the requirement of detailed balancing.* This is guaranteed by the Hermitian character of all interaction matrices in quantum mechanics. The frequency of transition from a state r to a state s is governed by a factor of the form $|V_{rs}|^2$, where V is a certain matrix and V_{rs} the term in its rth row and sth column. The frequency of the reverse transition differs only by containing the factor $|V_{sr}|^2$ and for Hermitian matrices these factors are always equal.

In spite of the fact that we now know that all mechanisms conform to detailed balancing and know so much more about mechanisms in general thanks to quantum mechanics, it is still not without value to study particular mechanisms from the statistical point of view without appeal to quantum mechanics. One can in this way better than in any other obtain a deep insight into the nature of the equilibrium state. We shall analyse particular mechanisms keeping both requirements in mind, in order to see when requirements (*a*) and (*b*) are or are not equivalent from the statistical point of view.

§ 17·12. *Unit mechanisms.* In classical kinetic theory molecules are structureless. In the extensions now required their electronic structure becomes relevant and in general more than one mechanism may be causing a specified change of state in any system. The question then presents itself whether or not we can discuss separately the effects of the separate mechanisms. The arguments that follow are those of the classical radiation theory. Suppose we have a system X which can undergo a specified change by interaction with other systems Y or Z. Suppose we can effectively change the concentrations $[Y]$ and $[Z]$ independently of each other, of $[X]$ and of the temperature. This will be the case if Y and Z are separable in the equilibrium state from each other and from X. They will be so separable for example if they refer to different atoms, but not if they refer to different states of the same atom. Interactions between X and Y and X and Z will then certainly occur with unconnected frequencies. This still applies if Y means the temperature radiation in the assembly, with the proviso that then the "concentration"—the density of the radiation—cannot be varied in equilibrium independently of the temperature. Let P_Y be the frequency with which (X, Y) interactions occur per unit volume at unit concentrations, converting X from a state 1 to a state 2, and Q_Y be the corresponding frequency of the reverse process, for an assembly in full statistical equilibrium. We consider now only an assembly of effectively perfect gases, and suppose that the systems are distributed entirely independently of each other, in accordance with the ideas of classical statistical mechanics. Then we see that for equilibrium we must have

$$\begin{aligned}[X_1]\{P_{\text{rad}}+[Y]P_Y+[Z]P_Z+\dots+[Y]^2P_{YY}+\dots+[Y][Z]P_{YZ}+\dots\\ +[Y]P_{Y,\text{rad}}+\dots\}\\ =[X_2]\{Q_{\text{rad}}+[Y]Q_Y+[Z]Q_Z+\dots+[Y]^2Q_{YY}+\dots+[Y][Z]Q_{YZ}+\dots\\ +[Y]Q_{Y,\text{rad}}+\dots\}.\end{aligned} \quad \dots\dots(1847)$$

We have here allowed for all sorts of interactions, radiative, simple and multiple collision processes, and mixed radiative and collision processes. Now so long as $[Y]$, $[Z]$, ... are independent variables, and there is a unique equilibrium relationship between $[X_1]$ and $[X_2]$, this equation has important consequences. If the ratio $[X_1]/[X_2]$ is independent of $[Y]$, $[Z]$, ... and so is a function of the temperature only, then

$$\begin{aligned}\frac{P_{\text{rad}}}{Q_{\text{rad}}}=\frac{P_Y+P_{Y,\text{rad}}}{Q_Y+Q_{Y,\text{rad}}}=\frac{P_Z+P_{Z,\text{rad}}}{Q_Z+Q_{Z,\text{rad}}}=\dots\\ =\frac{P_{YY}}{Q_{YY}}=\dots=\frac{P_{YZ}}{Q_{YZ}}=\dots=\frac{[X_2]}{[X_1]}.\end{aligned} \quad \dots\dots(1848)$$

The effects of plain interactions by collision between X and Y and interaction by collision accompanied by the absorption or emission of radiation cannot be separated, and such terms really occur in every fraction. Apart from this however we see that the ratio of the frequencies of each separate process and its reverse process must be equal to a definite function of the temperature $[X_1]/[X_2]$. The process and its reverse are inseparable and form together what we may call a *unit mechanism*. We can go still further than this later. For the present we can be content to note the independence of the different processes, whose laws therefore can be separately analysed.

If the change from X_1 to X_2 is one of dissociation, the arguments are the same, though the form is a little different. Suppose for example that X_1 dissociates into X_2 and Y. Then $[X_2][Y]/[X_1]$ is a function of the temperature only, and we have

$$\frac{0}{Q_{\text{rad}}} = \frac{0}{Q_Z + Q_{Z,\text{rad}}} = \ldots = \frac{P_{\text{rad}}}{Q_Y + Q_{Y,\text{rad}}} = \frac{P_Y + P_{Y,\text{rad}}}{Q_{YY} + Q_{YY,\text{rad}}} = \ldots$$
$$= \frac{P_Z + P_{Z,\text{rad}}}{Q_{YZ} + Q_{YZ,\text{rad}}} = \ldots = \frac{[X_2][Y]}{[X_1]}. \quad \ldots\ldots(1849)$$

Thus, naturally enough, no process in which no Y is concerned can make up an X_1 from an X_2. The Q's of all such processes must vanish. More generally every process must pair off with a reverse process in which one more Y is concerned—the Y in fact which is to be caught by the X_2 to make it into an X_1. The pairing off is therefore such that it is possible as a dynamical reversal of the original process, and the laws for such pairs can in general be discussed separately.

No difference is made in these conclusions by the requirements of Einstein-Bose or Fermi-Dirac statistics. The number of transitions by X's from state 1 to state 2 under the influence of radiation for example is then no longer $[X_1]P_{\text{rad}}$ but depends on the number of systems already in the state 2 and in particular must vanish in the Fermi-Dirac statistics when the state 2 is full. The number of transitions must be now assumed to be

$$[X_1]P_{\text{rad}}\{A_2 \pm [X_2]\},$$

where A_2 is a constant defining the number of distinct states in the group 2; the + sign refers to the Einstein-Bose and the − to Fermi-Dirac statistics. The number of reverse transitions is similarly now $[X_2]Q_{\text{rad}}\{A_1 \pm [X_1]\}$. The same extra factors occur for all types of transition, and equations (1848) are replaced by

$$\frac{P_{\text{rad}}}{Q_{\text{rad}}} = \frac{P_Y + P_{Y,\text{rad}}}{Q_Y + Q_{Y,\text{rad}}} = \frac{P_Z + P_{Z,\text{rad}}}{Q_Z + Q_{Z,\text{rad}}} = \ldots = \frac{P_{YY}}{Q_{YY}} = \ldots = \frac{P_{YZ}}{Q_{YZ}} = \frac{[X_2]/\{A_2 \pm [X_2]\}}{[X_1]/\{A_1 \pm [X_1]\}}. \quad \ldots\ldots(1850)$$

From these equations all the same conclusions can be drawn.

It remains impossible by these general arguments to separate the effects of, for example, two different states of the same atom, and these must be considered together in any further analysis. We have however shown that it follows merely from the assumption of a unique equilibrium state that the laws of interaction between any molecule and radiation or any molecule and any other group of atoms or molecules must lead separately to the laws of statistical equilibrium. The precise *forms* (1848)–(1850) in which these conclusions are embodied are those for perfect gases. There are analogous forms for the interaction of solids and gases which we shall also discuss. The conclusions can be extended in slightly different forms to imperfect gases, but these we shall not discuss here.

§ **17·2.** *The numbers of classical collisions of given type in a gas.* The general form of the number of collisions of a given type per unit volume per unit time has already been given in (1842). To derive it we recall that $f_1 du_1 dv_1 dw_1$, which we shall here contract to $f_1 do_1$, is the number of molecules of the first type in unit volume at any time, and near each of them there is a volume $Vpdpd\psi$ in which the centre of the molecule of the second type must lie in order to effect contact within the specified time. The number of such molecules is $f_2 do_2$ per unit volume; hence the formula. For elastic spheres of diameter σ, if θ is the angle between V and the line of centres at collision, we have $p = \sigma \sin\theta$ and $dp = \sigma\cos\theta d\theta$. The element $pdpd\psi$ can then be conveniently replaced by

$$\sigma^2 \cos\theta d\Omega,$$

where $d\Omega$ is an element of solid angle about a direction specifying the line of centres at the instant of collision. The number of collisions then becomes

$$f_1 f_2 \sigma^2 V \cos\theta do_1 do_2 d\Omega. \qquad \text{......(1851)}$$

This is the number of collisions which destroy a pair of molecules u_1, v_1, w_1 and u_2, v_2, w_2 and create a pair u_1^*, v_1^*, w_1^* and u_2^*, v_2^*, w_2^*. Since $do_1 do_2 = do_1^* do_2^*$ and $d\Omega = d\Omega^*$, the corresponding number of reverse collisions creating a pair u_1, v_1, w_1 and u_2, v_2, w_2 and destroying u_1^*, v_1^*, w_1^* and u_2^*, v_2^*, w_2^* is

$$f_1^* f_2^* \sigma^2 V \cos\theta do_1 do_2 d\Omega. \qquad \text{......(1852)}$$

In these collisions the velocities after collision are given in terms of the velocities before and the direction cosines of the line of centres by the equations

$$\left.\begin{aligned} u_1^* &= u_1 + lV\cos\theta, \quad & u_2^* &= u_2 - lV\cos\theta, \\ v_1^* &= v_1 + mV\cos\theta, \quad & v_2^* &= v_2 - mV\cos\theta, \\ w_1^* &= w_1 + nV\cos\theta, \quad & w_2^* &= w_2 - nV\cos\theta, \\ V^2 &= (u_2-u_1)^2 + (v_2-v_1)^2 + (w_2-w_1)^2, \\ V\cos\theta &= l(u_2-u_1) + m(v_2-v_1) + n(w_2-w_1). \end{aligned}\right\} \qquad \text{......(1853)}$$

We shall confine attention in this section to properties of the equilibrium state deducible from (1851) when f has Maxwell's form

$$f = \nu\left(\frac{m}{2\pi kT}\right)^{\frac{3}{2}} e^{-\frac{1}{2}m(u^2+v^2+w^2)/kT}, \qquad \text{......(1854)}$$

and ν is the molecular density. The number of collisions per unit volume per unit time in which the relative velocity lies between V and $V+dV$ is an important quantity. If the velocity of the centre of gravity of the pair is $\mathsf{U}, \mathsf{V}, \mathsf{W}$, and the velocity of 2 relative to 1 is ξ, η, ζ, then

$$\mathsf{U}=\tfrac{1}{2}(u_1+u_2), \quad \xi=u_2-u_1, \text{ etc.}$$

If we transform (1851) to these new variables, we find

$$do_1\,do_2=d\mathsf{U}\,d\mathsf{V}\,d\mathsf{W}\,d\xi\,d\eta\,d\zeta,$$

$$\tfrac{1}{2}m(u_1{}^2+v_1{}^2+w_1{}^2)+\tfrac{1}{2}m(u_2{}^2+v_2{}^2+w_2{}^2)=m(\mathsf{U}^2+\mathsf{V}^2+\mathsf{W}^2)+\tfrac{1}{4}mV^2,$$

and the number of collisions of the specified type is

$$\nu^2\left(\frac{m}{2\pi kT}\right)^3 e^{-\{m(\mathsf{U}^2+\mathsf{V}^2+\mathsf{W}^2)+\frac{1}{4}mV^2\}/kT}\sigma^2 V\cos\theta\,d\mathsf{U}\ldots d\zeta\,d\Omega. \quad \ldots\ldots(1855)$$

If we integrate this for all values of the motion of the centre of gravity of the system, which is always unimportant in questions of collisions, we find

$$\nu^2\left(\frac{m}{4\pi kT}\right)^{\frac{3}{2}} e^{-\frac{1}{4}mV^2/kT}\sigma^2 V\cos\theta\,d\xi\,d\eta\,d\zeta\,d\Omega. \quad \ldots\ldots(1856)$$

If we express the relative velocity in spherical polar coordinates V, θ, ϕ, with the direction of the line of centres for axis, we find

$$\nu^2\left(\frac{m}{4\pi kT}\right)^{\frac{3}{2}} e^{-\frac{1}{4}mV^2/kT}\sigma^2 V^3\sin\theta\cos\theta\,dV\,d\theta\,d\phi\,d\Omega. \quad \ldots\ldots(1857)$$

For a given direction of the line of centres θ can range from 0 to $\frac{1}{2}\pi$, and ϕ from 0 to 2π, and

$$\int_0^{\frac{1}{2}\pi}\int_0^{2\pi}\sin\theta\cos\theta\,d\theta\,d\phi=\pi.$$

The direction of the line of centres can then range over the whole sphere, but every collision is then counted twice over, since every collision is counted separately with the molecules interchanged. Dividing by 2 to correct for this, we find the total number of collisions per unit volume per unit time, with relative velocity between V and $V+dV$, to be

$$2\pi^2\nu^2\sigma^2\left(\frac{m}{4\pi kT}\right)^{\frac{3}{2}} e^{-\frac{1}{4}mV^2/kT}V^3\,dV. \quad \ldots\ldots(1858)$$

This is the important result. If we finally integrate over all relative velocities we obtain for the total number of collisions per unit volume per second the well-known formula

$$\nu^2\sigma^2\left(\frac{4\pi kT}{m}\right)^{\frac{1}{2}}. \quad \ldots\ldots(1859)$$

We shall require also similar formulae for collisions of unlike molecules. Suppose these have masses m_1 and m_2, diameters σ_1 and σ_2. The distance apart of the centres on collision is now σ_{12}, say, where

$$\sigma_{12}=\tfrac{1}{2}(\sigma_1+\sigma_2).$$

If σ is replaced by σ_{12} the form of (1851) is then still valid. The reduction to the motion of the centre of gravity and the relative motion now however requires the equations

$$(m_1+m_2)\mathsf{U}=m_1u_1+m_2u_2, \quad \xi=u_2-u_1, \text{ etc.}$$

In these variables $\quad do_1 do_2 = d\mathsf{U}\,d\mathsf{V}\,d\mathsf{W}\,d\xi\,d\eta\,d\zeta,$

$$\tfrac{1}{2}m_1(u_1^2+v_1^2+w_1^2)+\tfrac{1}{2}m_2(u_2^2+v_2^2+w_2^2)=\tfrac{1}{2}(m_1+m_2)(\mathsf{U}^2+\mathsf{V}^2+\mathsf{W}^2) + \frac{1}{2}\frac{m_1m_2}{m_1+m_2}V^2.$$

Formula (1856) is therefore replaced by

$$\nu_1\nu_2\left(\frac{m_1m_2}{2\pi(m_1+m_2)kT}\right)^{\frac{3}{2}} e^{-\frac{1}{2}\frac{m_1m_2}{m_1+m_2}V^2/kT}\sigma_{12}{}^2V\cos\theta\,d\xi\,d\eta\,d\zeta\,d\Omega, \quad \text{......(1860)}$$

and formula (1858) by

$$4\pi^2\nu_1\nu_2\sigma_{12}{}^2\left(\frac{m_1m_2}{2\pi(m_1+m_2)kT}\right)^{\frac{3}{2}} e^{-\frac{1}{2}\frac{m_1m_2}{m_1+m_2}V^2/kT}V^3dV. \quad \text{......(1861)}$$

Since the molecules are now distinct, collisions are not counted twice over as before. The total number of collisions is now

$$2\nu_1\nu_2\sigma_{12}{}^2\left\{\frac{2\pi(m_1+m_2)kT}{m_1m_2}\right\}^{\frac{1}{2}}. \quad \text{......(1862)}$$

It is frequently convenient in applications to rewrite (1854), (1858) and (1861) in terms of kinetic energy, or kinetic energy of the relative motion. If we write

$$\tfrac{1}{2}m(u^2+v^2+w^2)=\eta,$$

then Maxwell's law for the number of molecules per unit volume with kinetic energy between η and $\eta+d\eta$ becomes

$$\frac{2\pi\nu}{(\pi kT)^{\frac{3}{2}}}e^{-\eta/kT}\eta^{\frac{1}{2}}\,d\eta. \quad \text{......(1863)}$$

This will frequently be written $\mu(\eta)\,d\eta$, and the molecules with this kinetic energy will be called η-molecules. Similarly, if

$$\tfrac{1}{4}mV^2=\eta,$$

the number of collisions per unit volume per unit time in which the kinetic energy of the relative motion lies between η and $\eta+d\eta$ is

$$\frac{2\nu^2\sigma^2}{(kT)^{\frac{3}{2}}}\left(\frac{\pi}{m}\right)^{\frac{1}{2}}e^{-\eta/kT}\eta\,d\eta. \quad \text{......(1864)}$$

If $$\frac{1}{2}\frac{m_1m_2}{m_1+m_2}V^2=\eta,$$

the number of collisions between molecules of different types per unit volume per unit time in which the kinetic energy of the relative motion lies between η and $\eta+d\eta$ is

$$\frac{2\nu_1\nu_2\sigma_{12}{}^2}{(kT)^{\frac{3}{2}}}\left(\frac{2\pi(m_1+m_2)}{m_1m_2}\right)^{\frac{1}{2}}e^{-\eta/kT}\eta\,d\eta. \quad \text{......(1865)}$$

§17·3. *Boltzmann's H-theorem for a simple gas.* We have examined in §17·11 the logical position of this famous theorem in the equilibrium theory as here developed. We may enunciate it most satisfactorily thus:

Theorem 17·3. *Boltzmann's H-theorem. In an assembly of perfect gases, in which the only interactions between the systems are conservative collisions with central forces, the preservation of an equilibrium state requires detailed balancing.*

Since detailed balancing ($f_1 f_2 = f_1^* f_2^*$), which is equivalent to Maxwell's law, obviously preserves the equilibrium state, the theorem tells us that detailed balancing, Maxwell's law and the preservation of an equilibrium state are in the case of this mechanism all equivalent. The idea of Boltzmann's proof is of course to construct a function H (practically the entropy) whose constancy requires detailed balancing. We start with the simplest case of a gas of a single kind of hard elastic spherical molecule of diameter σ in the absence of external fields.

For such a gas f can be changed only by molecular collisions,† so that we can combine (1851) and (1852) and integrate over all the velocities of the second molecule to give its time variation in the form

$$\frac{df_1}{dt} = \sigma^2 \iint (f_1^* f_2^* - f_1 f_2)\, V \cos\theta\, do_2\, d\Omega. \qquad \text{......(1866)}$$

The integrand is considered to be expressed as a function of $u_1, \ldots, w_2$ by means of equations (1853).

It will be observed that it is essential to the argument that the same molecular distribution laws of the equilibrium state, whatever they may be, should hold for specified volume elements near selected molecules as for the gas as a whole. This is true according to the general equilibrium theory. In discussions which seek to avoid the use of the general theory it becomes a necessary explicit assumption.

Consider now the function

$$H = \int f_1 \log f_1\, do_1, \qquad \text{......(1867)}$$

an expression depending only on the number of molecules and the form of f, and therefore constant in the equilibrium state. Then

$$\frac{dH}{dt} = \int (1 + \log f_1) \frac{df_1}{dt}\, do_1,$$

$$= \sigma^2 \iiint (1 + \log f_1)(f_1^* f_2^* - f_1 f_2)\, V \cos\theta\, do_1\, do_2\, d\Omega. \qquad \text{......(1868)}$$

† Except by the boundary fields or collisions with the walls. For the effect of these see §17·8. They are there shown not to affect the argument.

But it is equally true that

$$H=\int f_2 \log f_2 do_2,$$

and that
$$\frac{df_2}{dt}=\sigma^2 \iint (f_1{}^* f_2{}^* - f_1 f_2)\, V\cos\theta\, do_1\, d\Omega;$$

combining these we find an expression similar to (1868) with $\log f_2$ in place of $\log f_1$, and combining this with (1868) we find

$$\frac{dH}{dt}=\tfrac{1}{2}\sigma^2 \iiint (2+\log f_1 f_2)(f_1{}^* f_2{}^* - f_1 f_2)\, V\cos\theta\, do_1\, do_2\, d\Omega. \quad \text{......(1869)}$$

But by precisely similar arguments we can start with

$$H=\int f_1{}^* \log f_1{}^* do_1{}^*$$

and find

$$\frac{dH}{dt}=\tfrac{1}{2}\sigma^2 \iiint (2+\log f_1{}^* f_2{}^*)(f_1 f_2 - f_1{}^* f_2{}^*)\, V^*\cos\theta^*\, do_1{}^*\, do_2{}^*\, d\Omega^*. \quad \text{....(1870)}$$

The variables of integration may now be changed from the starred variables back to the old variables, and

$$V^*\cos\theta^*\, do_1{}^*\, do_2{}^*\, d\Omega^* = V\cos\theta\, do_1\, do_2\, d\Omega.$$

If we make this change and combine (1869) and (1870) we find

$$\frac{dH}{dt}=-\tfrac{1}{4}\sigma^2 \iiint (\log f_1 f_2 - \log f_1{}^* f_2{}^*)(f_1 f_2 - f_1{}^* f_2{}^*)\, V\cos\theta\, do_1\, do_2\, d\Omega, \quad \text{......(1871)}$$

$$\leqslant 0, \quad \text{......(1872)}$$

because the integrand is never negative. The equilibrium state is possible if and only if

$$f_1 f_2 = f_1{}^* f_2{}^*,$$

that is if there is detailed balancing.

We include for completeness the familar proof that

$$\log f_1 + \log f_2 = \textit{const.}\ \text{(coll.)}$$

implies that f is Maxwell's distribution function. Let $\chi(u,v,w)$ be any function of the velocities of a molecule such that $\chi_1+\chi_2$ is constant in a collision. Then

$$\log f = \chi$$

is a solution of the equation for the distribution function, and $\log f - \chi$ satisfies the same equation. The most general solution is therefore

$$\log f = \alpha_1 \chi' + \alpha_2 \chi'' + \alpha_3 \chi''' + \dots.$$

where the χ's are all the functions of u, v, w such that $\chi_1+\chi_2$ is constant in a collision. We know of five such functions, energy, three components of momentum, and mass. There are no others possible, for the four constant relations involving velocities give four independent relations between the six u_1^*, ..., w_2^* and the old u_1, ..., w_2. Two relations must be left unfixed in this way to depend essentially on the direction cosines of the line of centres at impact. Thus

$$\log f = \alpha_1 + \alpha_2 mu + \alpha_3 mv + \alpha_4 mw + \alpha_5 \tfrac{1}{2}m(u^2+v^2+w^2),$$

$$f = Ae^{-jm\{(u-u_0)^2+(v-v_0)^2+(w-w_0)^2\}},$$

where A, j, u_0, v_0, w_0 are constants, which is of the required form.

It will be observed that the detailed collision relations hardly enter into this proof. It is only necessary that there should be some relations giving the u^*, ..., in terms of the u, ..., which conserve momentum, energy and the extension of the element of phase space of the system.

§ **17·31.** *Extensions of Boltzmann's H-theorem.* The extension to the case of a number of different types of molecules is very simple. In place of equation (1866), when classical collisions are the only mechanisms acting, we find a set of equations of the form

$$\frac{df_r}{dt} = \Sigma_s \sigma_{rs}^2 \iint (f_r^* f_s^* - f_r f_s)\, V \cos\theta\, do_s\, d\Omega. \qquad \text{......(1873)}$$

If we now define the function H by the equation

$$H = \Sigma_s \int f_s \log f_s\, do_s$$

and apply the same analysis, we find

$$\frac{dH}{dt} = -\tfrac{1}{4}\Sigma_r \Sigma_s \sigma_{rs}^2 \iiint (\log f_r f_s - \log f_r^* f_s^*)(f_r f_s - f_r^* f_s^*) V \cos\theta\, do_r\, do_s\, d\Omega.$$
$$\text{......(1874)}$$

From this it follows that H is constant if and only if

$$f_r f_s = f_r^* f_s^* \quad (\textit{all } r,\ s),$$

that is to say if we have detailed balancing for every choice of a pair of molecules in collision. From the equation for a pair of similar molecules Maxwell's law follows as before. From the dissimilar pairs it follows by the same argument that the mass motion and temperature of each species of gas must be the same.

Let us finally extend the theorem to general classical encounters under central forces between molecules subjected to an external field of force. The distribution function now depends on the position in the gas and we may write it explicitly for a volume element $d\omega$ $(= dx\,dy\,dz)$,

$$f(u,v,w,x,y,z,t)\, d\omega,$$

so that $\iint f\, do\, d\omega$ extended to a unit volume is the molecular density there.

We will assume that force components X_r, Y_r, Z_r per unit mass act on the molecule of the rth kind, so that its equations of motion are

$$\dot{x}_r = u_r, \quad \dot{u}_r = X_r, \text{ etc.}$$

If we now consider all the molecules of the rth type in the element $do\,d\omega$ at u, v, w, x, y, z, t, we see that, *apart from collisions*, after a time dt these molecules will lie in an equal cell of the same space and velocity ranges as before but centred about the point

$$u_r + X_r dt, \quad v_r + Y_r dt, \quad w_r + Z_r dt, \quad x_r + u_r dt, \quad y_r + v_r dt, \quad z_r + w_r dt, \quad t + dt.$$

Moreover since the motion is determinate these molecules are the only ones (apart from collisions) which at time $t + dt$ have the u, v, w, x, y, z so specified. The number of such molecules is by definition

$$f_r(u_r + X_r dt, \ldots, z_r + w_r dt, t + dt)\, do_r\, d\omega,$$

and we have just seen that this differs from $f_r do_r d\omega$ only by the collision term. Thus proceeding to the limit $dt \to 0$,

$$\frac{\partial f_r}{\partial t} = -\left(X_r \frac{\partial}{\partial u_r} + Y_r \frac{\partial}{\partial v_r} + Z_r \frac{\partial}{\partial w_r} + u_r \frac{\partial}{\partial x_r} + v_r \frac{\partial}{\partial y_r} + w_r \frac{\partial}{\partial z_r}\right) f_r + \left[\frac{\partial f_r}{\partial t}\right]_{\text{coll}} \qquad \ldots\ldots(1875)$$

For general classical encounters under central forces we have

$$\left[\frac{\partial f_r}{\partial t}\right]_{\text{coll}} = \Sigma_s \iiint (f_r^* f_s^* - f_r f_s)\, Vp\, dp\, d\psi\, do_s, \qquad \ldots\ldots(1876)$$

where the starred velocities are given in terms of the initial velocities (and p and ψ) by the detailed laws derived from the central orbit. Equations (1875) and (1876) together form a very important integro-differential equation for f which should be satisfied for all states (not merely equilibrium ones) of a gas of molecules undergoing classical encounters. It is known as *Boltzmann's equation*.

§**17·32.** *The equations of mean values.* To derive the general form of the H-theorem we must introduce the equations of mean values. These equations form the starting point for all accurate investigations of transport phenomena in gases,† but these lie outside the range of this monograph and we shall formulate them only to lead up to the H-theorem.

Let $\phi(u,v,w,x,y,z,t)$ be any function of the arguments specified, such that all the integrals in the following arguments are absolutely convergent. If we multiply every term in Boltzmann's equation for f_r by $\phi_r do_r$ and

† A general account in Jeans, *loc. cit.* chaps. VIII–XIV. The main recent detailed investigations are Chapman, *Phil. Trans.* vol. 211, p. 433 (1911), vol. 216, p. 279 (1916), vol. 217, p. 115 (1917); Enskog, *Inaug. Diss. Upsala* (1917), *Arkiv för Matematik*, vol. 16, No. 16 (1921), *Kungl. Svenska Akad.* vol. 63, No. 4 (1922); J. E. Jones, *Phil. Trans.* vol. 223, p. 1 (1922).

integrate for all velocities of the molecules in any specified volume element, we obtain

$$\int \phi_r \frac{\partial f_r}{\partial t} do_r + \Sigma_x \int u_r \phi_r \frac{\partial f_r}{\partial x_r} do_r + \Sigma_x X_r \int \phi_r \frac{\partial f_r}{\partial u_r} do_r$$

$$= \Sigma_s \iiiint \phi_r (f_r{}^* f_s{}^* - f_r f_s)\, Vp\,dp\,d\psi\,do_r\,do_s,$$

$$= \Sigma_s \Delta_{sr}(\phi_r),$$

say. In this notation $\Delta_{sr}(\phi_r)$ denotes the rate of change in the average value of ϕ_r in the given volume element for molecules of type r produced by collisions with molecules of type s. We can express the last equation more simply in terms of mean values if we remember that

$$\int f_r \psi_r do_r = N_r \overline{\psi_r},$$

where $\overline{\psi_r}$ is the mean value of ψ_r and N_r the total number of molecules of type r in the given volume element. Then

$$\int \phi_r \frac{\partial f_r}{\partial t} do_r = \frac{\partial}{\partial t}(N_r \overline{\phi_r}) - N_r \overline{\frac{\partial \phi_r}{\partial t}},$$

$$\int u_r \phi_r \frac{\partial f_r}{\partial x_r} do_r = \frac{\partial}{\partial x_r}(N_r \overline{u_r \phi_r}) - N_r \overline{u_r \frac{\partial \phi_r}{\partial x_r}},$$

$$\int \phi_r \frac{\partial f_r}{\partial u_r} do_r = -\int f_r \frac{\partial \phi_r}{\partial u_r} do_r = -N_r \left(\overline{\frac{\partial \phi_r}{\partial u_r}}\right).$$

Inserting all these expressions we obtain

$$\frac{\partial}{\partial t}(N_r \overline{\phi_r}) + \Sigma_x \frac{\partial}{\partial x_r}(N_r \overline{u_r \phi_r}) = N_r \left\{ \overline{\frac{\partial \phi_r}{\partial t}} + \Sigma_x \overline{u_r \frac{\partial \phi_r}{\partial x_r}} + \Sigma_x X_r \overline{\frac{\partial \phi_r}{\partial u_r}} \right\} + \Sigma_s \Delta_{sr}(\phi_r). \quad \text{......(1877)}$$

These are the equations of mean values. Putting

$$\phi_r = \log f_r$$

we obtain for the right-hand side of (1877)

$$\int f_r \left[\frac{\partial}{\partial t} + \Sigma_x u_r \frac{\partial}{\partial x_r} + \Sigma_x X_r \frac{\partial}{\partial u_r} \right] \log f_r do_r + \Sigma_s \Delta_{sr}(\log f_r),$$

or

$$\int \left[\frac{\partial}{\partial t} + \Sigma_x u_r \frac{\partial}{\partial x_r} + \Sigma_x X_r \frac{\partial}{\partial u_r} \right] f_r do_r + \Sigma_s \Delta_{sr}(\log f_r). \quad \text{......(1878)}$$

On referring to Boltzmann's equation, (1875), and (1876) we see that (1877) can be written

$$\int \left\{ \Sigma_s \iiint (f_r{}^* f_s{}^* - f_r f_s)\, Vp\,dp\,d\psi\,do_s \right\} do_r + \Sigma_s \Delta_{sr}(\log f_r),$$

and the first terms obviously vanish. The equation of mean values reduces here to

$$\frac{\partial}{\partial t}(N_r\overline{\log f_r}) + \Sigma_x \frac{\partial}{\partial x_r}(N_r\overline{u_r \log f_r}) = \Sigma_s \Delta_{sr}(\log f_r), \quad \text{......(1879)}$$

which is to hold for every volume element. Written at length this becomes

$$\frac{\partial}{\partial t}\int f_r \log f_r do_r + \Sigma_x \frac{\partial}{\partial x_r}\int u_r f_r \log f_r do_r$$
$$= \Sigma_s \iiiint \log f_r (f_r^* f_s^* - f_r f_s)\, Vp\,dp\,d\psi\,do_r\,do_s. \quad \text{......(1880)}$$

When there are space variations the function H is naturally defined by

$$H = \Sigma_r \iint f_r \log f_r do_r d\omega.$$

Multiplying (1880) by $d\omega$, integrating over the whole gas and summing for all types of molecule, we find

$$\frac{dH}{dt} + \Sigma_r \Sigma_x \int \frac{\partial}{\partial x_r}\int u_r f_r \log f_r do_r d\omega$$
$$= \Sigma_r \Sigma_s \iiiiint \log f_r (f_r^* f_s^* - f_r f_s)\, Vp\,dp\,d\psi\,do_r\,do_s\,d\omega,$$

and by the usual repetitions

$$= -\tfrac{1}{4}\Sigma_r \Sigma_s \iiiiint (\log f_r f_s - \log f_r^* f_s^*)(f_r f_s - f_r^* f_s^*)\, Vp\,dp\,d\psi\,do_r\,do_s\,d\omega.$$

The last terms on the left must vanish on integration, for the integration includes the whole gas and therefore extends to regions in which the molecular density vanishes. In a volume element in which the molecular density vanishes f_r must also vanish and therewith $\int u_r f_r \log f_r do_r$. We therefore end by obtaining

$$\frac{dH}{dt} = -\tfrac{1}{4}\Sigma_r \Sigma_s \iiiiint (\log f_r f_s - \log f_r^* f_s^*)(f_r f_s - f_r^* f_s^*)\, Vp\,dp\,d\psi\,do_r\,do_s\,d\omega.$$
$$\text{......(1881)}$$

This is the general H-theorem and requires as before detailed balancing for the equilibrium state.

§ **17·33.** *The general form of Maxwell's law for equilibrium states.* It is not without interest to complete the argument and see how the laws of the most general equilibrium state follow from $f_r f_s = f_r^* f_s^*$ and Boltzmann's equation (1875) which in this case reduces to

$$\Sigma_x u_r \frac{\partial f_r}{\partial x_r} + \Sigma_x X_r \frac{\partial f_r}{\partial u_r} = 0. \quad \text{......(1882)}$$

The form of f_r for detailed balancing in each volume element is of course still

$$f_r = \nu_r \left(\frac{jm_r}{\pi}\right)^{\frac{3}{2}} e^{-jm_r\{\Sigma(u_r-u_0)^2\}},$$

but so far as detailed balancing is concerned ν_r, j, u_0, v_0, w_0 may all be functions of x, y, z. We may substitute this value of f_r in (1882) and require the equation to be satisfied for all u_r, v_r, w_r. On equating coefficients we find

$$\frac{\partial j}{\partial x} = \frac{\partial j}{\partial y} = \frac{\partial j}{\partial z} = 0,$$

$$\frac{\partial u_0}{\partial x} = \frac{\partial v_0}{\partial y} = \frac{\partial w_0}{\partial z} = 0,$$

$$\frac{\partial v_0}{\partial x} + \frac{\partial u_0}{\partial y} = \frac{\partial w_0}{\partial y} + \frac{\partial v_0}{\partial z} = \frac{\partial u_0}{\partial z} + \frac{\partial w_0}{\partial x} = 0.$$

The first three of these equations require that the temperature shall be constant throughout the assembly. The next six are the equations which characterize a rigid body. They specify that, though the mass motion may vary from point to point in the assembly in the equilibrium state, the variation is such that the assembly moves as a rigid body, with a certain uniform rotation superposed on a translation. These equations being satisfied we are left with

$$\Sigma_x u_r \left\{\frac{1}{\nu_r}\frac{\partial \nu_r}{\partial x} - 2jm_r X_r\right\} + 2jm_r \Sigma_x X_r u_0 \equiv 0.$$

This can only be satisfied if X_r, Y_r, Z_r are derived from a potential function χ_r, so that

$$\nu_r = (\nu_r)_0 e^{-2jm_r\chi_r} \quad ((\nu_r)_0 \textit{ constant}),$$

and finally

$$\Sigma_x X_r u_0 = 0.$$

The last equation demands that the lines of force and the lines of flow must be everywhere at right angles.

The complete specification of the general equilibrium state is therefore that *the gas moves as a rigid body, with temperature everywhere the same, in a field of force of potential χ_r, with a molecular distribution function*

$$f_r = (\nu_r)_0 \left(\frac{m_r}{2\pi kT}\right)^{\frac{3}{2}} e^{-\{\frac{1}{2}m_r\Sigma(u_r-u_0)^2 + m_r\chi_r\}/kT},$$

in which $(\nu_r)_0$ is constant and u_0, v_0, w_0 the velocity components of a rigid body motion such that the lines of flow are everywhere normal to the gradient of χ_r.

§17·34. *Quantum reformulation of the effects of gas collisions on distribution laws.** The analysis of §§17·2, 17·3 needs only slight emendations to yield the number of collisions of a given type per unit volume per unit time

* Started by Jordan, *Zeit. f. Physik*, vol. 41, p. 711 (1927); Ornstein and Kramers, *Zeit. f. Physik*, vol. 42, p. 481 (1927); Bothe, *Zeit. f. Physik*, vol. 46, p. 327 (1928). More complete versions by Nordheim, *Proc. Roy. Soc.* A, vol. 119, p. 689 (1928); Nordheim and Kikuchi, *Zeit. f. Physik*, vol. 60, p. 652 (1930); Uehling and Uhlenbeck, *Phys. Rev.* vol. 43, p. 552 (1933).

and the form of Boltzmann's H-theorem strictly according to quantum mechanics. We shall not aim at maximum generality, but merely at a sufficient account to show how the reformulation can be carried through. In unit volume of the gas there are $f_2 do_2$ molecules of type 2, each of which may be regarded as a scattering centre for an incident stream of molecules of type 1 approaching with relative velocity V. Quantal scattering, like classical, leaves the motion of the centre of mass of the pair unaffected, and provides a definite chance of deflecting the relative velocity through an angle θ into a solid angle $d\omega$. The stream of molecules incident on each scattering centre has a flux density of molecules per cm.^{2} per sec. given by

$$Vf_1 do_1.$$

Quantum theory enables us to calculate explicitly, when the law of interaction of the molecules is given, a target area $\sigma(V,\theta)\,d\omega$, which for symmetrical molecules depends only on V and θ, such that each scattering centre deflects molecules into $d\omega$ from an incident beam of unit flux density at a rate

$$\sigma(V,\theta)\,d\omega.$$

In our problem therefore each such scattering centre produces a pair of molecules with their relative velocity V in $d\omega$ at a rate

$$Vf_1 do_1 \sigma(V,\theta)\,d\omega.$$

The rate of production of such pairs [velocity components u_1^*, v_1^*, w_1^*; u_2^*, v_2^*, w_2^*] by such collisions per cm.^{3} of the gas is therefore

$$f_1 f_2 do_1 do_2 V\sigma(V,\theta)\,d\omega. \qquad \text{......(1883)}$$

The required formula is however not yet complete. We have taken no account of the statistics satisfied by the molecules so that (1883) is only correct if there are present in the assembly no molecules with the velocity components of the product pair. To allow for these interferences† we have to multiply (1883) by the factor

$$(1 \pm f_1^*/A_1^*)(1 \pm f_2^*/A_2^*),$$

taking the upper signs for the Einstein-Bose and the lower for the Fermi-Dirac statistics. The A's are such that $A_1^* do_1^*$, $A_2^* do_2^*$ give the number of possible quantum states in the velocity ranges do_1^*, do_2^*, so that these factors vanish with the lower signs when the f's have such values that all the possible states are full. With the definition of f here in use

$$A_1^* = A_1 = \varpi_1 m_1^3/h^3, \quad \text{and} \quad A_2^* = A_2 = \varpi_2 m_2^3/h^3,$$

but it is unnecessary to use these actual forms. The complete quantum formula therefore for the number of collisions per unit volume per unit time

† It is impossible to claim that the insertion of the factor can be seen *a priori* to be correct for the Einstein-Bose statistics. For the Fermi-Dirac however it can be seen at once to be at least plausible.

which convert a pair of molecules $(u_1,v_1,w_1;u_2,v_2,w_2)$ into a pair $(u_1^*,v_1^*,w_1^*;$ $u_2^*,v_2^*,w_2^*)$ by deflecting the relative velocity through an angle θ into a solid angle $d\omega$ is

$$f_1 f_2(1 \pm f_1^*/A_1^*)(1 \pm f_2^*/A_2^*)\, V\sigma(V,\theta)\, do_1 do_2 d\omega. \quad \text{......(1884)}$$

Little progress can be made with this formula on the lines of § 17·2 unless the statistical interference factors can be neglected so that (1883) with classical statistics and quantal interactions is sufficiently accurate. When (1883) may be used the equilibrium value of f_1 is still of Maxwell's form (1854). If the velocity components are expressed in terms of the components of the velocity of the centre of mass and the relative velocity so that

$$\mathsf{U} = (m_1u_1 + m_2u_2)/(m_1+m_2), \quad \xi = u_2 - u_1, \ etc.,$$

the number of collisions of the specified type is

$$\nu_1\nu_2\left(\frac{(m_1m_2)^{\frac{1}{2}}}{2\pi kT}\right)^3 e^{-\{\frac{1}{2}(m_1+m_2)(\mathsf{U}^2+\mathsf{V}^2+\mathsf{W}^2)+\frac{1}{2}m_1m_2V^2/(m_1+m_2)\}/kT}\, V\sigma(V,\theta)\, d\mathsf{U} \ldots d\zeta\, d\omega,$$
$$\text{......(1885)}$$

which differs only from (1855), generalized to unlike molecules, in replacing $\sigma^2\cos\theta$ by $\sigma(V,\theta)$. On integrating over all velocities of the centre of mass of the pair we find as before

$$\nu_1\nu_2\left(\frac{m_1m_2}{2\pi(m_1+m_2)\,kT}\right)^{\frac{3}{2}} e^{-\frac{1}{2}m_1m_2V^2/(m_1+m_2)kT}\, V\sigma(V,\theta)\, d\xi\, d\eta\, d\zeta\, d\omega.$$
$$\text{......(1886)}$$

This is the number of collisions in the gas per cm.^{3} per sec. in which the molecules have relative velocity V in a given direction, which is deflected by the collision through a given angle θ into a given solid angle $d\omega$. We find the total number of collisions at relative velocity between V and $V+dV$ by integrating independently over all directions of V and all directions $d\omega$ in any order. We thus find

$$4\pi\nu_1\nu_2\left(\frac{m_1m_2}{2\pi(m_1+m_2)\,kT}\right)^{\frac{3}{2}} e^{-\frac{1}{2}m_1m_2V^2/(m_1+m_2)kT}\, V^3 S(V)\, dV, \quad \text{...(1887)}$$

where
$$S(V) = \int \sigma(V,\theta)\, d\omega \quad \text{......(1888)}$$

integrated over the complete solid angle 4π. It will be seen that these formulae do not differ in form in any essential way from those of § 17·2.

The revised form of Boltzmann's H-theorem is more important. Each collision enumerated in (1884) removes one molecule from the set $f_1 do_1$. The final velocities are fixed in terms of the initial velocities and the direction of $d\omega$, so that to find the total rate of removal of molecules from the group

$f_1 do_1$ we have to integrate (1884) with respect to do_2 and $d\omega$. These collisions therefore contribute to df_1/dt the term

$$-\iint f_1 f_2 (1 \pm f_1^*/A_1^*)(1 \pm f_2^*/A_2^*)\, V\sigma(V,\theta)\, do_2 d\omega. \quad \text{...(1889)}$$

We must now account for the reverse collisions which create molecules in the group $f_1 do_1$. The general collisions of this type in which a pair of molecules with components $(u_1^*, v_1^*, w_1^*; u_2^*, v_2^*, w_2^*)$ becomes a pair $(u_1, v_1, w_1; u_2, v_2, w_2)$ by deflection of the relative velocity through an angle θ into the solid angle $d\omega^*$ occur to a number

$$f_1^* f_2^* (1 \pm f_1/A_1)(1 \pm f_2/A_2)\, V^* \sigma^*(V^*, \theta)\, do_1^* do_2^* d\omega^* \quad \text{...(1890)}$$

per unit volume per unit time. In this expression it is possible to replace $V^* \sigma^* do_1^* do_2^* d\omega^*$ by $V \sigma\, do_1 do_2 d\omega$, on account of the reversibility condition, a fact that is most easily seen as follows by changing over to the velocity of the centre of mass and the relative velocity. In these coordinates

$$do_1^* do_2^* = \mathsf{C}^{*2} d\mathsf{C}^* d\Omega^* V^{*2} dV^* d\omega$$

and

$$do_1 do_2 = \mathsf{C}^2 d\mathsf{C}\, d\Omega\, V^2 dV\, d\omega^*.$$

The solid angle elements occur in this manner because the element of solid angle $d\omega$ which contains the relative velocity before the reverse collision is the same as that which contains the relative velocity after the direct collision, and similarly for $d\omega^*$. Moreover

$$V = V^*, \quad \mathsf{C} = \mathsf{C}^*, \quad d\Omega = d\Omega^*$$

since none of these quantities are affected by the collision. The reversibility condition therefore reduces to $\sigma = \sigma^*$, a relationship which expresses the Hermitian character of the interaction matrix, and is therefore an immediate consequence of quantum mechanics. We are now in a position to combine (1889) and (1890), thereby obtaining

$$\frac{df_1}{dt} = \iint [f_1^* f_2^* (1 \pm f_1/A_1)(1 \pm f_2/A_2) - f_1 f_2 (1 \pm f_1^*/A_1^*)(1 \pm f_2^*/A_2^*)] \times V\sigma(V,\theta)\, do_2 d\omega, \quad \text{......(1891)}$$

in which the integrand is expressed in terms of $u_1, v_1, w_1; u_2, v_2, w_2$ and the direction cosines of $d\omega$.

Let us now restrict attention to a simple gas, so that f_1 and f_2 represent the same distribution law, and consider the function

$$H = \int [f_1 \log f_1 - (A_1 + f_1) \log(1 + f_1/A_1)]\, do_1, \quad \text{......(1892)}$$

which is the analogue of (1867). The lower signs (Fermi-Dirac statistics) can be included in the argument by changing the sign of A_1. Then, due to collisions,

$$\frac{dH}{dt} = \int \log \frac{f_1}{1 + f_1/A_1} \frac{df_1}{dt}\, do_1$$

which reduces on using (1891) to

$$\frac{dH}{dt}=\iiint \log\frac{f_1}{1+f_1/A_1}\left[f_1{}^*f_2{}^*\left(1+\frac{f_1}{A_1}\right)\left(1+\frac{f_2}{A_2}\right)-f_1f_2\left(1+\frac{f_1{}^*}{A_1{}^*}\right)\left(1+\frac{f_2{}^*}{A_2{}^*}\right)\right]$$
$$\times V\sigma(V,\theta)\,do_1do_2d\omega. \quad \text{......(1893)}$$

But it is equally true for a simple gas that

$$\frac{dH}{dt}=\int \log\frac{f_2}{1+f_2/A_2}\frac{df_2}{dt}do_2,$$

$$\frac{df_2}{dt}=\iint\left[f_1{}^*f_2{}^*\left(1+\frac{f_1}{A_1}\right)\left(1+\frac{f_2}{A_2}\right)-f_1f_2\left(1+\frac{f_1{}^*}{A_1{}^*}\right)\left(1+\frac{f_2{}^*}{A_2{}^*}\right)\right]V\sigma(V,\theta)\,do_1d\omega. \quad \text{......(1894)}$$

Equations (1893) and (1894) can be combined in the form

$$\frac{dH}{dt}=\frac{1}{2}\iiint \log\frac{f_1f_2}{(1+f_1/A_1)(1+f_2/A_2)}\left[f_1{}^*f_2{}^*\left(1+\frac{f_1}{A_1}\right)\left(1+\frac{f_2}{A_2}\right)\right.$$
$$\left.-f_1f_2\left(1+\frac{f_1{}^*}{A_1{}^*}\right)\left(1+\frac{f_2{}^*}{A_2{}^*}\right)\right]\times V\sigma(V,\theta)\,do_1do_2d\omega. \quad \text{......(1895)}$$

We can also start with

$$H=\int[f_1{}^*\log f_1{}^*-(A_1{}^*+f_1{}^*)\log(1+f_1{}^*/A_1{}^*)]\,do_1{}^*,$$

and by the same devices as in § 17·3 reach the final result

$$\frac{dH}{dt}=-\frac{1}{4}\iiint \log\frac{f_1f_2(1+f_1{}^*/A_1{}^*)(1+f_2{}^*/A_2{}^*)}{f_1{}^*f_2{}^*(1+f_1/A_1)(1+f_2/A_2)}$$
$$\times\left[f_1f_2\left(1+\frac{f_1{}^*}{A_1{}^*}\right)\left(1+\frac{f_2{}^*}{A_2{}^*}\right)-f_1{}^*f_2{}^*\left(1+\frac{f_1}{A_1}\right)\left(1+\frac{f_2}{A_2}\right)\right]$$
$$\times V\sigma(V,\theta)\,do_1do_2d\omega. \quad \text{......(1896)}$$

From (1896) it follows as before that

$$\frac{dH}{dt}\leqslant 0,$$

and therefore that equilibrium is possible if and only if

$$\frac{f_1f_2}{(1+f_1/A_1)(1+f_2/A_2)}=\frac{f_1{}^*f_2{}^*}{(1+f_1{}^*/A_1{}^*)(1+f_2{}^*/A_2{}^*)} \quad \text{......(1897)}$$

when the velocity components $u_1, v_1, w_1; u_2, v_2, w_2; u_1{}^*, v_1{}^*, w_1{}^*; u_2{}^*, v_2{}^*, w_2{}^*$ are connected by the conservation conditions holding in a collision. It follows at once that the equilibrium state demands detailed balancing and that for a gas with no mass motion

$$\frac{f_1}{1+f_1/A_1}=e^{-\frac{1}{2}mc^2/kT}$$

which is equivalent to the usual formulae of Einstein-Bose or Fermi-Dirac statistics.

We conclude by observing that a strict analogue of Boltzmann's integro-differential equation (1875) remains true in quantum mechanics if the space-variation of the external fields of force remains so slow that the considerations of Darwin* apply, showing that the wave-packet representing a particle undisturbed by collisions moves in nearly uniform external fields like a classical particle. The distribution function f therefore satisfies the equation

$$\frac{\partial f_1}{\partial t}+X\frac{\partial f_1}{\partial u_1}+Y\frac{\partial f_1}{\partial v_1}+Z\frac{\partial f_1}{\partial w_1}+u_1\frac{\partial f_1}{\partial x}+v_1\frac{\partial f_1}{\partial y}+w_1\frac{\partial f_1}{dz}=\left[\frac{df_1}{dt}\right]_{\text{coll}},$$

where the right-hand side is given by (1891).

§ 17·4. *Collisions of the first and second kind.*† The next mechanism which we shall examine is that of inelastic impacts between electrons and atoms. It is well known that such collisions occur freely as soon as the energy of the electron exceeds a certain minimum energy, which is that required to excite the atom hit from its normal stationary state to the next state of greater energy. No such process can occur and preserve equilibrium unless it is accompanied by a corresponding reverse process, the two together forming a unit mechanism. The associated process is here *a superelastic collision*, called by Klein and Rosseland a collision of the second kind. This balance of inelastic and superelastic collisions is of course now clearly understood as an example of quantal reversibility (§ 17·11). The analysis we shall use here is purely statistical; its importance was first made clear by Klein and Rosseland.

We start as usual with a classical formulation. The frequency of occurrence of any type of collision can always be expressed in terms of *a target area*, which must be a function solely of the atom hit and of the relative velocity and line of impact of the electron hitting it. This target area which determines the chance of a successful collision may in an individual collision depend on the orientation of the atom or molecule struck relative to the line of impact, and on other variables. The average target is however usually all that matters, and is obtained by integration over all these other variables, which are then no longer effective.

The electrons are so light that in a first treatment the velocities of the atoms can be neglected compared with the velocities of the electrons, and the atoms treated as fixed (or of infinite mass). If ν_1 is the density of the type of atom or molecule under discussion, then the number of collisions per unit volume per unit time between these atoms and η-electrons in which the

* Darwin, *Proc. Roy. Soc.* A, vol. 117, p. 258 (1927). For a more complete proof see Nordheim and Kikuchi, *loc. cit.*

† Klein and Rosseland, *Zeit. f. Physik*, vol. 4, p. 46 (1921).

line of impact lies at a distance between p and $p+dp$ from the atomic nucleus (or centre of mass of the molecule) is by (1863)

$$\nu_1 \, . \, 2\pi p\, dp \, . \left(\frac{2\eta}{m}\right)^{\frac{1}{2}} \mu(\eta)\, d\eta. \qquad \text{......(1898)}$$

If $\Sigma_1^{\,2}(p,\eta)$ is the probability that such a collision will excite the atom struck from state 1 to state 2, the number of such successful collisions is

$$\nu_1 \, . \, 2\pi p\, dp \, . \, \Sigma_1^{\,2}(p,\eta) \left(\frac{2\eta}{m}\right)^{\frac{1}{2}} \mu(\eta)\, d\eta.$$

If we now define a function $S_1^{\,2}(\eta)$ by the equation

$$S_1^{\,2}(\eta) = 2\pi \int_0^\infty \Sigma_1^{\,2}(p,\eta)\, p\, dp,$$

we find that the total number of successful collisions of the first kind by η-electrons per unit volume per unit time in the equilibrium state is

$$\nu_1 S_1^{\,2}(\eta) \left(\frac{2\eta}{m}\right)^{\frac{1}{2}} \mu(\eta)\, d\eta. \qquad \text{......(1899)}$$

The foregoing analysis in terms of an impact parameter p is not valid in quantum mechanics unless p is large compared with the de Broglie wave length of the electron. When this condition is not fulfilled we start as in § 17·34 with an incident stream of electrons of flux density $(2\eta/m)^{\frac{1}{2}} \mu(\eta)\, d\eta$ electrons per unit area per unit time impinging on ν_1 centres each of which possesses a total target area $S_1^{\,2}(\eta)$ for excitation from state 1 to state 2. The result (1899) remains unaffected.

This expression is only relevant when $\eta \geqslant \eta_{12}$, where η_{12} is the extra energy of the atom in state 2 over that in state 1. When $\eta < \eta_{12}$, $S_1^{\,2}(\eta) = 0$.

By the general discussion of § 17·1 the reverse process can only be one which occurs with a frequency proportional to ν_2 and to the electron density. It must therefore be some sort of collision. It is hardly possible to suppose that any other sort of collision can be concerned except those which are the direct reverse of the inelastic impacts. These are collisions in which an η-electron interacts with an excited atom and removes its superfluous energy, leaving it in its normal state. The energy removed is carried off by the electron as extra kinetic energy of translation, so that the collision may be termed superelastic. The total number of successful superelastic collisions by η-electrons per unit volume per second can be written

$$\nu_2 S_2^{\,1}(\eta) \left(\frac{2\eta}{m}\right)^{\frac{1}{2}} \mu(\eta)\, d\eta, \qquad \text{......(1900)}$$

where ν_2 is the density of the atoms in the state 2. It is possible that $S_2^{\,1}(\eta) > 0$ for any value of η.

It must be noted that $S_1^{\,2}(\eta)$ and $S_2^{\,1}(\eta)$ have a purely atomic significance;

they may be called the mean effective target areas of the atom for η-electrons. It is important to observe that these areas by definition cannot depend on any statistical parameter such as temperature or density.*

By considering the conditions for the preservation of the $\mu(\eta)$-law for the velocity distribution of the electrons a fundamental relation can be established between $S_1{}^2(\eta)$ and $S_2{}^1(\eta)$. The concentrations ν_1 and ν_2 are connected by the relation (see (197))

$$\nu_2/\nu_1 = (\varpi_2/\varpi_1)\, e^{-\eta_{12}/kT}. \qquad \text{......(1901)}$$

The number η-electrons destroyed by inelastic collisions is

$$\nu_1 S_1{}^2(\eta) \left(\frac{2\eta}{m}\right)^{\frac{1}{2}} \mu(\eta)\, d\eta \quad (\eta \geqslant \eta_{12}),$$

$$0 \qquad (\eta < \eta_{12}).$$

The number of η-electrons created by inelastic collisions is

$$\nu_1 S_1{}^2(\eta + \eta_{12}) \left\{\frac{2(\eta+\eta_{12})}{m}\right\}^{\frac{1}{2}} \mu(\eta+\eta_{12})\, d\eta \quad (\eta > 0).$$

The number of η-electrons destroyed by superelastic collisions is

$$\nu_2 S_2{}^1(\eta) \left(\frac{2\eta}{m}\right)^{\frac{1}{2}} \mu(\eta)\, d\eta \quad (\eta > 0).$$

The number of η-electrons created by superelastic collisions is

$$\nu_2 S_2{}^1(\eta - \eta_{12}) \left(\frac{2(\eta-\eta_{12})}{m}\right)^{\frac{1}{2}} \mu(\eta-\eta_{12})\, d\eta \quad (\eta \geqslant \eta_{12}),$$

$$0 \qquad (\eta < \eta_{12}).$$

Putting in the equilibrium values of $\mu(\eta)$ and ν_2/ν_1 we see at once that the action will balance and leave the equilibrium undisturbed if and only if

$$e^{-\eta_{12}/kT} F_1{}^2(\eta) - F_1{}^2(\eta - \eta_{12}) = 0,$$

where $$F_1{}^2(\eta) = \alpha[\varpi_1(\eta+\eta_{12})\, S_1{}^2(\eta+\eta_{12}) - \varpi_2 \eta S_2{}^1(\eta)] \quad (\eta \geqslant 0),$$
$$= 0 \qquad (\eta < 0),$$

and α is a constant. But by a step by step argument this condition can be shown only to be satisfied if

$$F_1{}^2(\eta) = 0 \quad (\textit{all } \eta).$$

Hence $$\varpi_1(\eta+\eta_{12})\, S_1{}^2(\eta+\eta_{12}) = \varpi_2 \eta S_2{}^1(\eta). \qquad \text{......(1902)}$$

This is Klein and Rosseland's result. It guarantees the preservation of the

* It should of course be remembered that if we experiment with directed electron streams and atoms orientated by magnetic fields (e.g. Skinner and Appleyard, *Proc. Roy. Soc.* A, vol. 117, p. 224 (1927)) the target areas for a given switch may well be different from those for unorientated collisions. It is obvious that this is in no way contradictory to the statements in the text.

equilibrium distribution law of electron velocities for all T. It also guarantees the preservation of the distribution law of atomic states (1901). For the rate of destruction of atoms in state 1 by this mechanism is

$$\nu_1 \int_{\eta_{12}}^{\infty} S_1{}^2(\eta) \left(\frac{2\eta}{m}\right)^{\frac{1}{2}} \mu(\eta)\, d\eta,$$

and the rate of creation

$$\nu_2 \int_0^{\infty} S_2{}^1(\eta) \left(\frac{2\eta}{m}\right)^{\frac{1}{2}} \mu(\eta)\, d\eta.$$

These rates will balance when

$$\varpi_1 \int_0^{\infty} (\eta + \eta_{12})\, S_1{}^2(\eta + \eta_{12})\, e^{-\eta/kT}\, d\eta = \varpi_2 \int_0^{\infty} \eta S_2{}^1(\eta)\, e^{-\eta/kT}\, d\eta, \quad \text{......(1903)}$$

which is satisfied for all values of T in virtue of (1902).

It should be observed in passing that conversely (1902) can be deduced directly from the fact that (1903) must hold for all T. Equation (1903) is of the general form

$$\int_0^{\infty} f(\eta)\, e^{-\eta/kT}\, d\eta = 0 \quad (all\ T), \qquad \text{......(1904)}$$

and there exists the following

Lemma. If $f(\eta)$ is a continuous function of η for $\eta \geqslant 0$ and satisfies the general conditions of Fourier's integral theorem, and if

$$\int_0^{\infty} f(\eta)\, e^{-\eta/kT}\, d\eta = 0 \quad (all\ T),$$

then

$$f(\eta) = 0 \quad (\eta \geqslant 0).$$

This lemma is a direct corollary of Fourier's integral theorem. There is no physical reason to question the applicability of the conditions of the lemma to the function $f(\eta)$ derived from the atomic target areas in (1903).

The relation (1902) just obtained on the hypothesis of the preservation of the equilibrium state can be obtained at once on the hypothesis of detailed balancing by equating (1899), with η replaced by $\eta + \eta_{12}$, to (1900). Thus for this simple mechanism of interaction between electrons and atoms with only two stationary states the hypotheses of preservation and detailed balancing are again equivalent.

One might expect to be able to deduce from the preservation hypothesis that a relation of the form (1902) holds for the frequencies of every possible switch by collision in an atom with n levels. It appears however that no such deduction can be drawn. An atom with n possible levels has $\frac{1}{2}n(n-1)$ possible switches, all of which we must regard as forming a single mechanism.

A detailed presentation of the case $n = 3$ is instructive. There are three switches (to and fro) to consider. *Sufficiently slow* η-electrons are created only by collisions of the first kind at a rate

$$\nu_1 S_1^2(\eta+\eta_{12})\left\{\frac{2(\eta+\eta_{12})}{m}\right\}^{\frac{1}{2}}\mu(\eta+\eta_{12})\,d\eta$$
$$+\nu_1 S_1^3(\eta+\eta_{13})\left\{\frac{2(\eta+\eta_{13})}{m}\right\}^{\frac{1}{2}}\mu(\eta+\eta_{13})\,d\eta$$
$$+\nu_2 S_2^3(\eta+\eta_{23})\left\{\frac{2(\eta+\eta_{23})}{m}\right\}^{\frac{1}{2}}\mu(\eta+\eta_{23})\,d\eta. \quad\text{......(1905)}$$

They are destroyed only by collisions of the second kind at a rate

$$\{\nu_2 S_2^1(\eta)+\nu_3 S_3^1(\eta)+\nu_3 S_3^2(\eta)\}\left(\frac{2\eta}{m}\right)^{\frac{1}{2}}\mu(\eta)\,d\eta. \quad\text{......(1906)}$$

The condition for the preservation of the equilibrium state can be written

$$e^{-\eta_{12}/kT}F_1^2(\eta)+e^{-\eta_{13}/kT}\{F_1^3(\eta)+F_2^3(\eta)\}=0 \quad (\textit{all } T), \quad\text{......(1907)}$$

an equation valid only when η is less than the least of η_{12}, η_{13} and η_{23}. When η is unrestricted the first terms in (1905) and (1906) continue to give the numbers of η-electrons created by the switch $1\to 2$ and destroyed by the switch $2\to 1$. There are now in addition

$$\nu_1 S_1^2(\eta)\left(\frac{2\eta}{m}\right)^{\frac{1}{2}}\mu(\eta)\,d\eta$$

η-electrons *destroyed* by the switch $1\to 2$ and

$$\nu_2 S_2^1(\eta-\eta_{12})\left\{\frac{2(\eta-\eta_{12})}{m}\right\}^{\frac{1}{2}}\mu(\eta-\eta_{12})\,d\eta$$

η-electrons *created* by the switch $2\to 1$. There are similar terms for the other two switches. The complete form of (1907) is therefore

$$-\{F_1^2(\eta-\eta_{12})+F_1^3(\eta-\eta_{13})\}+e^{-\eta_{12}/kT}\{F_1^2(\eta)-F_2^3(\eta-\eta_{23})\}$$
$$+e^{-\eta_{13}/kT}\{F_1^3(\eta)+F_2^3(\eta)\}=0 \quad (\textit{all } T, \eta). \quad\text{......(1908)}$$

This is equivalent to the three equations

$$\begin{aligned} F_1^2(\eta-\eta_{12})+F_1^3(\eta-\eta_{13})&=0,\\ F_1^2(\eta)\qquad -F_2^3(\eta-\eta_{23})&=0,\\ F_1^3(\eta)\qquad +F_2^3(\eta)\qquad&=0,\end{aligned}$$

of which only two are independent. They reduce to

$$F_1^2(\eta)=F_2^3(\eta-\eta_{23})=-F_1^3(\eta-\eta_{23}). \quad\text{......(1909)}$$

We can infer that equation (1902), $F_1^2(\eta)=0$, still holds for $\eta<\eta_{23}$, but we can infer nothing more as to the vanishing of the F's. The assumption of only two stationary states is equivalent to making η_{23} infinite. If we were to exclude all switches by collision of types $2\to 3$ and $3\to 2$, then $F_2^3(\eta)=0$, and we can infer that $F_1^2(\eta)=0$ and $F_1^3(\eta)=0$ for all η. This would be the expected generalization of (1902). But it can only be made when by some means or other we can rule out the possibility of the cyclic process $1 \to 2$, $2 \to 3$, $3 \to 1$. Such processes seem quite natural,* and are not incompatible with the hypothesis of preservation. In actual fact all the F's must vanish and detailed balancing must hold to satisfy the reversibility requirements of quantum mechanics.

It remains to determine the conditions under which the distribution of atoms between the three states is preserved. For state 1 this can be written

$$e^{-\eta_{12}/kT}\int_0^\infty F_1^{\,2}(\eta)\,e^{-\eta/kT}\,d\eta + e^{-\eta_{13}/kT}\int_0^\infty F_1^{\,3}(\eta)\,e^{-\eta/kT}\,d\eta = 0,$$

and finally reduced to the form

$$\int_0^\infty \{F_1^{\,2}(\eta) + F_1^{\,3}(\eta - \eta_{23})\}\,e^{-\eta/kT}\,d\eta = 0 \quad (all\ T). \quad \ldots\ldots(1910)$$

Condition (1910) is equivalent to one of the equations (1909) and gives us nothing new. In the same way for states 2 and 3 we get

$$\int_0^\infty \{-F_1^{\,2}(\eta) + F_2^{\,3}(\eta - \eta_{23})\}\,e^{-\eta/kT}d\eta = 0 \quad (all\ T),$$

$$\int_0^\infty \{F_1^{\,3}(\eta) + F_2^{\,3}(\eta)\}\,e^{-\eta/kT}d\eta = 0 \quad (all\ T),$$

which are also equivalent to components of (1909). Equations (1909) are therefore necessary and sufficient for the preservation of the equilibrium state. In the general case we shall obtain on the preservation hypothesis just $n-1$ necessary and sufficient relations between $\frac{1}{2}n(n-1)$ functions F.

§ **17·41.** *Inelastic and superelastic collisions between heavy systems.* Owing to the small mass and high average velocity of the electron compared with an atom it was legitimate to ignore the momentum of the electron and the velocity of the atom in the preceding section. When both interacting systems are of comparable mass the calculations must be revised. The action must now depend only on the relative velocity V of the interacting systems. There are now obviously two sorts of interaction possible—(*a*) An atom of type 1 may play the part of an electron and excite an atom of type 0 with expenditure of energy $(\eta_0)_{12}$. (*b*) An excited atom of type 1 may excite an atom of type 0 with the expenditure of its energy of excitation returning itself to its normal state. Such a collision may be either inelastic or superelastic according as energy is taken up from or surrendered to the relative motion of the systems.

* In all atoms this particular cycle might be ruled out by selection rules, but other cycles such as $1 \to 2$, $2 \to 3$, $3 \to 4$, $4 \to 1$ will always be possible.

Case (a). Ordinary inelastic collisions between heavy systems. If we replace the $\pi\sigma_{12}^2$ of (1865) by the more general $S_1^2(\eta)$ for the effective target area for excitation, then the number of successful collisions per unit volume per second will be

$$\frac{4\nu_0^{(1)}\nu_1}{(kT)^{\frac{3}{2}}}\left(\frac{m_0+m_1}{2\pi m_0 m_1}\right)^{\frac{1}{2}} e^{-\eta/kT}\eta S_1^2(\eta)\,d\eta. \qquad \text{......(1911)}$$

We use $\nu_0^{(1)}$ to denote the concentration of systems 0 in state 1. We recall that

$$\eta = \frac{1}{2}\frac{m_0 m_1}{m_0+m_1}V^2,$$

the kinetic energy of the relative motion. This is the number of collisions per unit volume which destroy members of $\nu_0^{(1)}$, create members of $\nu_0^{(2)}$, destroy pairs of molecules of relative velocity V and create other pairs for which the energy of the relative motion is

$$\eta' = \eta - \eta_{12}, \qquad \text{......(1912)}$$

and the relative velocity

$$V'^2 = V^2 - 2\frac{m_0+m_1}{m_0 m_1}\eta_{12}.$$

The number of reverse processes creating $\nu_0^{(1)}$ out of $\nu_0^{(2)}$ by collisions with relative energy η' will be

$$\frac{4\nu_0^{(2)}\nu_1}{(kT)^{\frac{3}{2}}}\left(\frac{m_0+m_1}{2\pi m_0 m_1}\right)^{\frac{1}{2}} e^{-\eta'/kT}\eta' S_2^1(\eta')\,d\eta'. \qquad \text{......(1913)}$$

On the hypothesis of detailed balancing if η and η' are connected by (1912) the expressions (1911) and (1913) must be equal. This gives on reduction

$$\varpi_1(\eta+\eta_{12})\,S_1^2(\eta+\eta_{12}) = \varpi_2\eta S_2^1(\eta), \qquad \text{......(1914)}$$

which is (naturally) the same relation as (1902). Here again if there are only two states of the system of type 0 concerned and no other complications it can be shown that, since S may not depend on T, (1914) is necessary and sufficient on the hypothesis of preservation only. Relations similar to (1914) may be expected to hold when the systems of type 0 have a number of stationary states.

We observe finally that

$$S_1^2(\eta) = 0 \quad (\eta < \eta_{12}), \qquad \text{......(1915)}$$

that is when the energy of the motion relative to the centre of gravity is less than η_{12}. If momentum as well as energy is to be conserved, it is of course only the energy of this relative motion which is available for excitation. It is moreover easy to see that simple conditions for the preservation of the laws of equilibrium are impossible unless they conserve momentum as well as energy.

Case (*b*). *Transference of excitation.* This case is very similar, but the direct and reverse processes are now practically identical. We start with the system of type 0 normal ($\nu_0^{(1)}$), and the system of type 1 excited ($\nu_1^{(2)}$), and end with the system of type 0 excited ($\nu_0^{(2)}$), and type 1 normal ($\nu_1^{(1)}$), and conversely. The equation of relative energy is now

$$\eta' + (\eta_0)_{12} = \eta + (\eta_1)_{12}, \qquad \text{......(1916)}$$

and the condition of detailed balancing as before

$$(\varpi_0)_1 (\varpi_1)_2 \eta S_1^{\,2}(\eta) = (\varpi_0)_2 (\varpi_1)_1 \eta' S_2^{\,1}(\eta'), \qquad \text{......(1917)}$$

subject to (1916). One or other of the S's will be zero for a range of relative energies less than a definite limit depending on the relative sizes of $(\eta_0)_{12}$ and $(\eta_1)_{12}$. As before these relations follow from the preservation hypothesis only, if there are no added complications.

§ **17·5.** *Practical applications of the theory of inelastic collisions.* Some direct consequences of the hypotheses of preservation and detailed balancing have been developed in the foregoing sections. Quantitative applications require exact knowledge of one of a pair of S's or target areas, but the mere qualitative knowledge that one S is not zero implies that the other S is also different from zero, and this leads to the recognition of the existence of processes which might otherwise have been overlooked. In the hands of Franck and others the theory, mainly used qualitatively, has been of first-class importance in the interpretation of a variety of phenomena, some of which we shall mention briefly.

The most striking of such qualitative confirmations of the theory is provided by the work of Cario* on the excitation of special lines in the spectra of thallium and silver by illuminating a mixture of the vapours of mercury and thallium or silver with the resonance radiation (λ 2536) of mercury. It was possible to show with certainty that the thallium or silver lines were excited when and only when excited mercury atoms were present in the mixed vapour, and the energy relations were such that this may be regarded as a complete proof of the process (*b*) of excitation discussed in § 17·41. It was shown further that the thallium lines must be emitted by abnormally swiftly moving thallium atoms, in virtue of their very small absorption in the thallium vapour. This confirms in a general way the laws of conservation of momentum in the process of excitation. For some energy is left over when the excitation is transferred and can only be used up in increasing the relative kinetic energy, and so on the average the actual kinetic energy of both atoms. These studies of excitation have also been extended to other atoms such as Cd, Na, In, Sb and As.

* Cario, *Zeit. f. Physik*, vol. 10, p. 185 (1922); Cairo and Franck, *Zeit. f. Physik*, vol. 17, p. 202 (1923); Donat, *Zeit. f. Physik*, vol. 29, p. 345 (1924).

In addition to this it has been shown by Franck, and Cario and Franck, in other papers that excited mercury atoms can probably dissociate H_2. These authors have also accounted for many of the effects of pressure, in particular the effects of foreign inert gases on the resonance spectra of mercury, sodium and the iodine molecule. The strength of the resonance radiation emitted in these gases is greatly reduced by too high a pressure or by the presence of an inert gas. The energy of excitation is used up in superelastic collisions instead of in re-emission of the resonance line. The same authors have also given a theory of the red sensitization of photographic plates by a red-absorbent dye. All these phenomena are intimately connected with the foregoing principles.* The same ideas are of importance in a discussion of the mechanism of excitation underlying so-called unimolecular gas reactions.

§ **17·6.** *The process of ionization by electronic impacts.* We shall again start with fixed atoms, equivalent to atoms of infinite mass. We know that processes exist in which an η-electron (or α-particle) can knock another electron out of an atom provided that $\eta > \eta_0$, where η_0 is the energy necessary to remove this electron. This process of ionization can certainly occur when η only just exceeds η_0, so that the process can occur without the emission of any radiation, nor need radiation be absorbed. The rest of the energy $\eta - \eta_0$ can be distributed in any manner between the two electrons after the collision. Coupled with this process of ionization there must be some process of capture which for all values of the temperature and concentrations will preserve the laws of dissociative equilibrium. The only possible process of capture is, after § 17·1, a collision of three bodies, two electrons and one ionized atom in which the two electrons so interact that one of them is bound by the atom and the other is thrown off with the kinetic energy of both before collision and the ionization energy η_0 thrown in. If the superfluous energy is not to be radiated it can only be carried off by at least one other material body!

We again start classically. The number of impacts by η-electrons on atoms with their line of impact between p and dp is given by (1898). Let us suppose that the only possible result (other than trivial elastic collisions) is ionization (provided $\eta > \eta_0$), after which there will be two free electrons, a new ζ-electron and the old one, now an $(\eta - \eta_0 - \zeta)$-electron. It is necessary that $\eta - \eta_0 - \zeta \geqslant 0$. Let $\Sigma_1^2(p,\zeta,\eta)\,d\zeta$ be the probability that such an encounter gives rise to a new ζ-electron. If we introduce the function

$$S_1^2(\zeta,\eta) = 2\pi \int_0^\infty p\Sigma_1^2(p,\zeta,\eta)\,dp, \qquad \text{......(1918)}$$

* Franck, *Zeit. f. Physik.* vol. 9, p. 259 (1922); Franck and Cairo, *Zeit. f. Physik,* vol. 11, p. 161 (1922).

then we find that the total number of such collisions is

$$\nu_1 . \left(\frac{2\eta}{m}\right)^{\frac{1}{2}} \mu(\eta)\,d\eta\,.\,S_1^{\,2}(\zeta,\eta)\,d\zeta. \qquad \text{......(1919)}$$

The total rate of production of ions in this way must be obtained by integrating with respect to ζ from 0 to $\eta-\eta_0$, and then with respect to η from η_0 to infinity. There is a simple physical meaning for

$$\int_0^{\eta-\eta_0} S_1^{\,2}(\zeta,\eta)\,d\zeta.$$

It is the mean effective collision area for ionization by η-electrons. Equation (1919) obviously continues to hold in quantum mechanics when we can distinguish between the incident and ejected particle.

The reverse three-body process can be similarly formulated. Consider first of all an ionized atom in process of being hit by a ζ-electron on a line of impact between p_1 and p_1+dp_1. There are

$$\nu_2\,.\,2\pi p_1 dp_1 . \left(\frac{2\zeta}{m}\right)^{\frac{1}{2}} \mu(\zeta)\,d\zeta$$

of these occurrences per unit volume per second. The other electron is to enter into collision on a path on which (if undisturbed) it would reach its apse at a time between τ and $\tau+d\tau$ seconds after the first electron reached its apse. With every collision with a ζ-electron there is therefore associated a volume

$$\left(\frac{2\xi}{m}\right)^{\frac{1}{2}} d\tau\,.\,2\pi p_2 dp_2$$

in which a ξ-electron may lie, whose passage through its apse on a p_2-line of impact would occur in the specified interval. The number of ξ-electrons in such a volume is

$$d\tau\,.\,2\pi p_2 dp_2 . \left(\frac{2\xi}{m}\right)^{\frac{1}{2}} \mu(\xi)\,d\xi.$$

The number of three-body collisions in which ζ-electrons and ξ-electrons on p_1- and p_2-lines of impact and time difference τ are concerned is therefore

$$\nu_2 . \left(\frac{2\zeta}{m}\right)^{\frac{1}{2}} \mu(\zeta)\,d\zeta . \left(\frac{2\xi}{m}\right)^{\frac{1}{2}} \mu(\xi)\,d\xi\,.\,2\pi p_1 dp_1\,.\,2\pi p_2 dp_2\,.\,d\tau. \qquad \text{......(1920)}$$

This number has been so specified that it disregards the relative directions of the line of impact of the electrons and the atomic orientation. Averaged over all such directions and orientations there will be a certain definite probability

$$\Sigma_2^{\,1}(p_1,p_2,\tau,\zeta,\xi)$$

that any one of these collisions will lead to a successful interaction in which the ζ-electron is bound and the ξ-electron thrown off as a $(\xi+\zeta+\eta_0)$-electron. Writing

$$4\pi^2\int_0^\infty p_1 dp_1\int_0^\infty p_2 dp_2\int_{-\infty}^{+\infty}\Sigma_2{}^1(p_1,p_2,\tau,\zeta,\xi)\,d\tau = S_2{}^1(\zeta,\xi), \quad ...(1921)$$

which will then be a purely atomic function of dimensions $[L]^4[T]$, we can express the number of successful triple collisions in which a ζ-electron is bound by the interaction of a ξ-electron in the form

$$\nu_2\cdot\left(\frac{2\zeta}{m}\right)^{\frac{1}{2}}\mu(\zeta)\,d\zeta\,.\left(\frac{2\xi}{m}\right)^{\frac{1}{2}}\mu(\xi)\,d\xi\,.\,S_2{}^1(\zeta,\xi). \quad ......(1922)$$

These collisions each create a neutral atom and a $(\xi+\zeta+\eta_0)$-electron at the expense of an ion, a ξ-electron and a ζ-electron. Equation (1922) continues to hold in quantum mechanics when we can distinguish between the particle bound and the particle rejected in the collision.

The appropriate law of dissociative equilibrium is, after (479) and (1643),

$$\frac{\nu_2\,\nu_\epsilon}{\nu_1}=\frac{(2\pi mkT)^{\frac{3}{2}}}{h^3}\frac{2\varpi_2}{\varpi_1}e^{-\eta_0/kT}. \quad ......(1923)$$

If the suggested processes form a possible mechanism they must preserve (1923) subject only to a purely atomic relation between $S_1{}^2$ and $S_2{}^1$, which turns out to be very simple. On the hypothesis of detailed balancing we must equate at once, subject to (1923), the numbers (1919) and (1922), with the proper relation between η, ζ and ξ. This means that we must write $\eta_0+\zeta+\xi$ for η in (1919) and identify the ζ's in the two expressions. We then find after a simple reduction that the relation between $S_1{}^2$ and $S_2{}^1$ must be

$$(\eta_0+\zeta+\xi)\,S_1{}^2(\zeta,\eta_0+\zeta+\xi)=\frac{16\pi m\varpi_2}{\varpi_1 h^3}\,\xi\zeta S_2{}^1(\zeta,\xi) \quad (all\ \zeta,\ \xi), \quad ...(1924)$$

$$S_1{}^2(\zeta,\eta)=0 \quad (\eta<\eta_0+\zeta).$$

It is however not without interest to see once again how far one can get on the simple preservation hypothesis, supposing that this process of ionization is the only process of the type that is acting.

Consider first the preservation of the electron distribution law for ζ-electrons, $\zeta<\eta_0$. By the process of ionization new ζ-electrons are produced by η-electrons at the rate (1919) and the old η-electrons are converted to ζ-electrons when these new ones are $(\eta-\eta_0-\zeta)$-electrons. The combined rate of production of ζ-electrons in ionization by η-electrons is therefore

$$\nu_1\left(\frac{2\eta}{m}\right)^{\frac{1}{2}}\mu(\eta)\,d\eta\,\{S_1{}^2(\zeta,\eta)+S_1{}^2(\eta-\eta_0-\zeta,\eta)\}\,d\zeta \quad (\eta>\eta_0+\zeta),$$

and otherwise zero. The total rate of creation of ζ-electrons is therefore

$$\nu_1 d\zeta \int_{\eta_0+\zeta}^{\infty} \left(\frac{2\eta}{m}\right)^{\frac{1}{2}} \{S_1{}^2(\zeta,\eta) + S_1{}^2(\eta-\eta_0-\zeta,\eta)\}\,\mu(\eta)\,d\eta \quad \text{...(1925)}$$

per unit volume by the ionization process. This holds for all ζ, but when $\zeta<\eta_0$ there is no destruction of ζ-electrons by this process.

Taking now the capture process we see at once that it creates no ζ-electrons ($\zeta<\eta_0$), for the resulting free electron has always an energy greater than η_0. It destroys each time both a ζ- and a ξ-electron. The number of ζ-electrons destroyed by binding (ζ-type) is therefore

$$\nu_2 \left(\frac{2\zeta}{m}\right)^{\frac{1}{2}} \mu(\zeta)\,d\zeta \int_0^{\infty} \left(\frac{2\xi}{m}\right)^{\frac{1}{2}} S_2{}^1(\zeta,\xi)\,\mu(\xi)\,d\xi.$$

The corresponding number destroyed by conversion into fast electrons (ξ-type) is

$$\nu_2 \left(\frac{2\zeta}{m}\right)^{\frac{1}{2}} \mu(\zeta)\,d\zeta \int_0^{\infty} \left(\frac{2\xi}{m}\right)^{\frac{1}{2}} S_2{}^1(\xi,\zeta)\,\mu(\xi)\,d\xi.$$

The total rate of destruction of ζ-electrons per unit volume is therefore

$$\nu_2 \left(\frac{2\zeta}{m}\right)^{\frac{1}{2}} \mu(\zeta)\,d\zeta \int_0^{\infty} \left(\frac{2\xi}{m}\right)^{\frac{1}{2}} \{S_2{}^1(\zeta,\xi) + S_2{}^1(\xi,\zeta)\}\,\mu(\xi)\,d\xi \quad \text{...(1926)}$$

for all values of ζ. For preservation of the distribution laws for $\zeta<\eta_0$ we must therefore equate (1925) and (1926). When we use the laws of the equilibrium state the resulting equation reduces to

$$\int_0^{\infty} e^{-\alpha/kT}(\eta_0+\zeta+\alpha)\{S_1{}^2(\zeta,\eta_0+\zeta+\alpha) + S_1{}^2(\alpha,\eta_0+\zeta+\alpha)\}\,d\alpha$$
$$= \frac{16\pi m\varpi_2}{\varpi_1 h^3} \int_0^{\infty} e^{-\alpha/kT}\alpha\zeta\{S_2{}^1(\zeta,\alpha) + S_2{}^1(\alpha,\zeta)\}\,d\alpha \quad (\zeta<\eta_0). \quad \text{......(1927)}$$

This holds for all T, and it follows at once from the Lemma that we must have

$$(\eta_0+\zeta+\alpha)\{S_1{}^2(\zeta,\eta_0+\zeta+\alpha) + S_1{}^2(\alpha,\eta_0+\zeta+\alpha)\}$$
$$= \frac{16\pi m\varpi_2}{\varpi_1 h^3}\,\alpha\zeta\{S_2{}^1(\zeta,\alpha) + S_2{}^1(\alpha,\zeta)\} \quad \text{......(1928)}$$

for all α and all $\zeta<\eta_0$.

It is necessary in addition that the rates of production and destruction of atomic ions should balance. The condition is easily obtained. It is

$$\nu_1 \int_{\eta_0}^{\infty} \left(\frac{2\eta}{m}\right)^{\frac{1}{2}} \mu(\eta)\,d\eta \int_0^{\eta-\eta_0} S_1{}^2(\zeta,\eta)\,d\zeta$$
$$= \nu_2 \int_0^{\infty}\int_0^{\infty} \left(\frac{2\zeta}{m}\right)^{\frac{1}{2}} \left(\frac{2\xi}{m}\right)^{\frac{1}{2}} \mu(\zeta)\,\mu(\xi)\,S_2{}^1(\zeta,\xi)\,d\zeta\,d\xi, \quad \text{......(1929)}$$

which reduces with the aid of the lemma, and the laws of the equilibrium state, to

$$(\eta_0+\alpha)\int_0^{\alpha} S_1{}^2(\zeta,\eta_0+\alpha)\,d\zeta = \frac{16\pi m\varpi_2}{\varpi_1 h^3}\int_0^{\alpha} \zeta(\alpha-\zeta)\,S_2{}^1(\zeta,\alpha-\zeta)\,d\zeta \quad (\textit{all } \alpha).$$
$$\text{......(1930)}$$

This condition however is not entirely new. For if we write $\alpha-\zeta$ for α in (1928) that equation becomes

$$(\eta_0+\alpha)\{S_1{}^2(\zeta,\eta_0+\alpha)+S_1{}^2(\alpha-\zeta,\eta_0+\alpha)\}$$
$$=\frac{16\pi m\varpi_2}{\varpi_1 h^3}\zeta(\alpha-\zeta)\{S_2{}^1(\zeta,\alpha-\zeta)+S_2{}^1(\alpha-\zeta,\zeta)\}.$$

If both sides of this equation are integrated with respect to ζ from 0 to α, it reduces to (1930), which will therefore be satisfied in virtue of (1928) when (1928) has been established for unlimited ζ.

It remains to consider the electron balance for $\zeta>\eta_0$. Expressions (1925) and (1926) still give the rates of creation by ionization and destruction by capture. There is now in addition a destruction of ζ-electrons by ionization at a rate

$$\nu_1\left(\frac{2\zeta}{m}\right)^{\frac{1}{2}}\mu(\zeta)\,d\zeta\int_0^{\zeta-\eta_0}S_1{}^2(\alpha,\zeta)\,d\alpha, \qquad \ldots\ldots(1931)$$

equal of course to the rate of formation of ions by ζ-electrons. There is also creation by capture. The first argument of $S_2{}^1(\alpha,\xi)$ refers to the electron that is bound. The energy of the electron left free on binding an α-electron in interplay with a ξ-electron is $\xi+\alpha+\eta_0$. If therefore $\xi+\alpha+\eta_0=\zeta$, we have to sum over all triple encounters of the type

$$\alpha,\ \zeta-\alpha-\eta_0 \quad (0<\alpha<\zeta-\eta_0),$$

and shall so find that ζ-electrons are created at a rate

$$\nu_2 d\zeta\int_0^{\zeta-\eta_0}\left(\frac{2\alpha}{m}\right)^{\frac{1}{2}}\frac{2(\zeta-\alpha-\eta_0)}{m}\mu(\alpha)\,\mu(\zeta-\alpha-\eta_0)\,S_2{}^1(\alpha,\zeta-\alpha-\eta_0)\,d\alpha.\ \ldots(1932)$$

The electron balance equation now contains four terms, and equates (1925) plus (1932) to (1926) plus (1931). But on account of (1930) the extra terms (1931) and (1932) balance by themselves. For they will so balance if

$$\zeta\int_0^{\zeta-\eta_0}S_1{}^2(\alpha,\zeta)\,d\alpha=\frac{16\pi m\varpi_2}{\varpi_1 h^3}\int_0^{\zeta-\eta_0}\alpha(\zeta-\alpha-\eta_0)\,S_2{}^1(\alpha,\zeta-\alpha-\eta_0)\,d\alpha,$$

which is (1930).

The hypothesis of preservation therefore demands only the necessary and sufficient condition* (1928) for all α and ζ; the hypothesis of detailed balancing requires the somewhat more restrictive (1924). But we have carried through the analysis quasi-classically to this extent that we have all the time assumed that the two particles (electrons) concerned are distinguishable. When the two particles are both electrons, which is the case here discussed, this is not correct, and the necessity for using antisymmetrical wave-functions in the interaction calculation brings in all the effects

* This result was first given completely correctly by R. Becker, *Zeit. f. Physik*, vol. 18, p. 325 (1923), eq. (34).

of electron exchanges. It is impossible to propose any formula such as (1919) for the number of collisions in which an incident η-electron knocks a ζ-electron out of an atom and *itself* becomes an $(\eta-\eta_0-\zeta)$-electron. We can only propose a formula for the number of collisions in which an η-electron is incident on an atom and ζ- and $(\eta-\eta_0-\zeta)$-electrons emerge. If we change the notation and consider collisions in which an $(\eta_0+\alpha+\zeta)$-electron produces ζ- and α-electrons, then the number of such collisions could be written

$$\nu_1 \frac{2(\eta_0+\alpha+\zeta)}{m} \mu(\eta_0+\alpha+\zeta)\,[S_1{}^2(\zeta,\eta_0+\alpha+\zeta)+S_1{}^2(\alpha,\eta_0+\alpha+\zeta)]\,d\zeta,$$

the terms in [ ] having no separate meaning. In the same way it is not possible to distinguish in recombination collisions between the bound and rejected electron, and only the expression $[S_2{}^1(\zeta,\alpha)+S_2{}^1(\alpha,\zeta)]$ retains a meaning. When allowance is made therefore for the indistinguishability of the electrons, we shall arrive at (1928) on the basis either of detailed balancing or of preservation. The really correct form of the relationship between the target areas for ionization and recombination is obtained by casting (1928) into the form

$$(\eta_0+\zeta+\alpha)\,\mathbf{S}_1{}^2(\zeta,\eta_0+\zeta+\alpha)=\frac{16\pi m\varpi_2}{\varpi_1 h^3}\alpha\zeta\,\mathbf{S}_2{}^1(\zeta,\alpha); \quad \text{......(1933)}$$

$\mathbf{S}_1{}^2$ is the target area for ionization by an $(\eta_0+\zeta+\alpha)$-electron with the production of a pair of α- and ζ-electrons; $\mathbf{S}_2{}^1$ is the "target" for recombination under simultaneous bombardment by uniform isotropic streams of α- and ζ-electrons.

After these illustrations we shall confine further analysis to the hypothesis of detailed balancing, since we know that this condition is in fact fulfilled. The examples given suffice to show that detailed balancing is not a necessary consequence merely of the existence of an equilibrium state.

§17·61. *Approximate form of the target areas* $S_1{}^2(\zeta,\eta)$ *and* $S_2{}^1(\zeta,\eta)$. Values of $S_1{}^2(\zeta,\eta)$ and $S_2{}^1(\zeta,\eta)$ can in principle be derived directly from quantum mechanics, and many such calculations have in fact been made. It is however very difficult to make such calculations except to rough approximations, and seldom that the results can be cast into a closed analytical form. It is therefore perhaps still worth while to put on record the result of the classical theory of Thomson* and Bohr† for $S_1{}^2(\zeta,\eta)$ and the derived result for $S_2{}^1(\zeta,\eta)$; the classical theory is in fair accord with the observations on the ionization produced by β-rays, and with β-ray ranges.

If for the moment p denotes the distance of the line of impact of the

* J. J. Thomson, *Phil. Mag.* vol. 23, p. 449 (1912).
† Bohr, *Phil. Mag.* vol. 24, p. 10 (1913), vol. 30, p. 581 (1915).

η-electron from the electron, which is to be knocked out of the atom and become a ζ-electron, there is (classically) a precise relation between this p and ζ, namely*

$$p^2=\frac{\epsilon^4}{\eta^2}\left(\frac{\eta}{\zeta+\eta_0}-1\right), \qquad \text{......(1934)}$$

$$-2\pi p\,dp=\frac{\pi\epsilon^4}{\eta}\frac{d\zeta}{(\zeta+\eta_0)^2}. \qquad \text{......(1935)}$$

This is the effective target area of the present theory for each electron of ionization potential η_0. In the notation of the preceding section $\Sigma_1{}^2(p,\zeta,\eta)\,d\zeta=1$ for this area and is otherwise zero. Thus

$$S_1{}^2(\zeta,\eta)=\frac{\pi\epsilon^4}{\eta}\frac{1}{(\zeta+\eta_0)^2} \qquad \text{......(1936)}$$

for each electron. For the process of capture we deduce

$$S_2{}^1(\zeta,\alpha)=\frac{\varpi_1 h^3}{16\pi m\varpi_2}\frac{1}{\alpha\zeta}\frac{\pi\epsilon^4}{(\zeta+\eta_0)^2}. \qquad \text{......(1937)}$$

This makes the mean effective "time-(area)^2" for capture tend to infinity as $\alpha\to 0$ or $\zeta\to 0$. This means that very slow electrons are very good at being caught or at helping the capture of others.

§ 17·7. *General frequency relations for 2- and 3-body encounters leading to dissociation and recombination.* We shall now reformulate the results of the preceding few sections accurately for bodies of comparable masses. We have already sufficiently analysed a 2-body encounter; we have now to extend this analysis to 3-body encounters.

The nature of any encounter can only be a function of the *relative* motion of the systems, and the motion of the centre of gravity of the systems will be unaltered by any interaction in which momentum is conserved. For an encounter of 3-bodies we therefore write

$$\left.\begin{aligned}(m_0+m_1+m_2)\,\mathsf{U}&=m_0u_0+m_1u_1+m_2u_2\\ \xi_1&=u_1-u_0\\ \xi_2&=u_2-u_0\end{aligned}\right\},\ \textit{etc.} \qquad \text{......(1938)}$$

These equations can be solved at once for u_0, u_1, u_2, and we find

$$\left|\frac{\partial(\mathsf{U},\xi_1,\xi_2)}{\partial(u_0,u_1,u_2)}\right|=1, \qquad \text{......(1939)}$$

$$m_0u_0{}^2+m_1u_1{}^2+m_2u_2{}^2=(m_0+m_1+m_2)\,\mathsf{U}^2$$
$$+\frac{1}{m_0+m_1+m_2}\{m_1(m_0+m_2)\,\xi_1{}^2-2m_1m_2\xi_1\xi_2+m_2(m_0+m_1)\,\xi_2{}^2\}. \qquad \text{......(1940)}$$

* See, for example, R. H. Fowler, *Proc. Camb. Phil. Soc.* vol. 21, p. 521 (1923).

We shall use similar equations for the other components, and θ for the angle between V_1, (ξ_1,η_1,ζ_1) and V_2, (ξ_2,η_2,ζ_2).

Now the number of triple combinations in which systems of types 0, 1, 2, masses m_0, m_1, m_2 and velocity components u_0, u_1, u_2, *etc.*, lie simultaneously in volume elements $d\omega_0$, $d\omega_1$, $d\omega_2$ is by the usual formula

$$\nu_0\nu_1\nu_2\frac{(m_0m_1m_2)^{\frac{3}{2}}}{(2\pi kT)^{\frac{9}{2}}}e^{-\eta/kT}\,do_0\,do_1\,do_2\,d\omega_0\,d\omega_1\,d\omega_2, \quad \text{......(1941)}$$

where

$$\eta=\tfrac{1}{2}m_0(u_0^2+v_0^2+w_0^2)+\tfrac{1}{2}m_1(u_1^2+v_1^2+w_1^2)+\tfrac{1}{2}m_2(u_2^2+v_2^2+w_2^2).$$

Using the substitutions of the preceding paragraph this reduces to

$$\nu_0\nu_1\nu_2\frac{(m_0m_1m_2)^{\frac{3}{2}}}{(2\pi kT)^{\frac{9}{2}}}\exp\Bigg[-\frac{1}{2kT}\Bigg\{(m_0+m_1+m_2)(\mathsf{U}^2+\mathsf{V}^2+\mathsf{W}^2)$$
$$+\frac{1}{m_0+m_1+m_2}[m_1(m_0+m_2)V_1^2-2m_1m_2V_1V_2\cos\theta$$
$$+m_2(m_0+m_1)V_2^2]\Bigg\}\Bigg]d\omega_0\,d\omega_1\,d\omega_2\,d\mathsf{U}\ldots d\zeta_2. \quad \text{......(1942)}$$

General 3-body collisions may be classified according to the position of the lines of impact of bodies 1 and 2 on body 0 (asymptotes of orbits) and the time interval between the instants at which the undisturbed relative orbits of (1,0) and (2,0) would bring these pairs closest together. To obtain the number of triple collisions per unit volume per unit time in which the lines of impact of 1 and 2 on 0 lie between p_1 and p_1+dp_1 and p_2 and p_2+dp_2 with a time interval between τ and $\tau+d\tau$ we take

$$d\omega_0=1,\quad d\omega_1=2\pi p_1\,dp_1V_1,\quad d\omega_2=2\pi p_2\,dp_2V_2\,d\tau. \quad \text{......(1943)}$$

We next change (1942) into spherical polar coordinates for the relative velocities, using the direction of V_1 as the polar axis from which to specify the angles defining the direction of V_2. Then

$$d\xi_1\ldots d\zeta_2=V_1^2dV_1\sin\psi\,d\psi\,d\chi V_2^2dV_2\sin\theta\,d\theta\,d\phi. \quad \text{......(1944)}$$

We are only interested in the relative configurations of the orbits, specified by p_1, p_2, τ, V_1, V_2, and θ. The other variables may be eliminated by integration. We then find that the number of collisions so specified is

$$8\pi^2\nu_0\nu_1\nu_2\frac{\{m_0m_1m_2/(m_0+m_1+m_2)\}^{\frac{3}{2}}}{(2\pi kT)^3}$$
$$\times\exp\left[-\frac{m_1(m_0+m_2)V_1^2-2m_1m_2V_1V_2\cos\theta+m_2(m_0+m_1)V_2^2}{2(m_0+m_1+m_2)kT}\right]$$
$$\times 2\pi p_1\,dp_1\,2\pi p_2\,dp_2\,d\tau V_1^3V_2^3\sin\theta\,dV_1\,dV_2\,d\theta. \quad \text{......(1945)}$$

The process of capture is a triple encounter of this type resulting in a radiationless union of the bodies 0 and 2. The relative motion of (0,2) and 1 is to take off the superfluous energy. If $\Sigma_2^1(p_1,p_2,\tau;V_2,\theta,V_1)$ is the probability of this event, and

$$S_2^1(V_2,\theta,V_1)=4\pi^2\int_0^\infty p_1\int_0^\infty p_2\int_{-\infty}^{+\infty}\Sigma_2^1\,dp_1\,dp_2\,d\tau, \quad \text{......(1946)}$$

then the rate of captures per unit volume of the (V_2,θ,V_1) type is

$$8\pi^2\nu_0\nu_1\nu_2\frac{\{m_0m_1m_2/(m_0+m_1+m_2)\}^{\frac{3}{2}}}{(2\pi kT)^3}S_2^1(V_2,\theta,V_1)\,e^{-\eta/kT}V_1^3V_2^3\sin\theta\,dV_1\,dV_2\,d\theta. \quad \text{......(1947)}$$

In this formula η is the energy of the relative motion before the event, which is given explicitly in [ ] in (1945). The energy of the relative motion of the 2-body system (0,2) and 1 after the event is $\eta+\eta_0$, where η_0 is the energy of dissociation of (0,2).

By analogy with (1861) the number of 2-body collisions of type (p,V) between (0,2) and 1 per unit volume per second is

$$4\pi\nu_{02}\nu_1\left\{\frac{(m_0+m_2)\,m_1}{2\pi kT(m_0+m_1+m_2)}\right\}^{\frac{3}{2}}\exp\left[-\frac{m_1(m_0+m_2)\,V^2}{2(m_0+m_1+m_2)\,kT}\right]2\pi p\,dp\,V^3dV. \quad \text{......(1948)}$$

The suffix 02 refers to the body (0,2). These collisions are effective if they result in the dissociation of (0,2) into 0 and 2. There are then three bodies moving with certain relative velocities and the type of collision depends on the distribution of the available energy. Let

$$\frac{m_0m_2}{m_0+m_2}\Sigma_1^2(p;V_2,\theta,V)\,V_2\sin\theta\,dV_2\,d\theta \quad \text{......(1949)}$$

be the probability that a body 1 with relative velocity V will so break up (0,2) that 2 is thrown off with velocity between V_2 and V_2+dV_2 relative to 0, and in such a direction that the angle between V_2 and the velocity V_1 of the body 1 relative to the body 0 after collision lies between θ and $\theta+d\theta$. If V_2 and θ are arbitrarily specified, then V_1 is determined by the conservation of energy, which is

$$\frac{1}{2}\frac{m_1(m_0+m_2)}{m_0+m_1+m_2}V^2$$
$$=\eta_0+\frac{1}{2}\frac{m_1(m_0+m_2)\,V_1^2-2m_1m_2V_1V_2\cos\theta+m_2(m_0+m_1)\,V_2^2}{m_0+m_1+m_2}. \quad \text{......(1950)}$$

We now write as usual

$$S_1^2(V_2,\theta,V)=2\pi\int_0^\infty p\Sigma_1^2\,dp. \quad \text{......(1951)}$$

The number of dissociations of type (V_2,θ,V) per unit volume per unit time is therefore

$$4\pi\nu_{02}\nu_1\left\{\frac{(m_0+m_2)m_1}{2\pi kT(m_0+m_1+m_2)}\right\}^{\frac{3}{2}}\frac{m_0m_2}{m_0+m_2}S_1^{\,2}(V_2,\theta,V)$$
$$\times\exp\left[-\frac{m_1(m_0+m_2)V^2}{2(m_0+m_1+m_2)kT}\right]V^3V_2\sin\theta\, dV\, dV_2\, d\theta. \quad \text{......(1952)}$$

Let us now assume that there is detailed balancing. Then we must assert here that if the V's are related by (1950) the expressions (1952) and (1947) must be equal. The resulting relation between $S_2^{\,1}$ and $S_1^{\,2}$ refers only to the process $(0,2)\rightleftarrows 0+2$, each body being in a unique internal state before and after the interaction. To see that the process preserves the equilibrium laws with a purely atomic relation between $S_2^{\,1}$ and $S_1^{\,2}$, we must consider the equilibrium laws for just such a reaction. For bodies with more than one internal state, including in this strictly speaking both internal oscillations and rotations, the equilibria of the separate states must be discussed with separate coefficients S. For the bodies here considered the law of dissociative equilibrium, after (479), takes the form

$$\frac{\nu_0\nu_2}{\nu_{02}}=\left\{\frac{m_0m_2}{m_0+m_2}\right\}^{\frac{3}{2}}\frac{(2\pi kT)^{\frac{3}{2}}}{h^3}\frac{\varpi_0\varpi_2}{\varpi_{02}}e^{-\eta_0/kT}. \quad \text{......(1953)}$$

Inserting this value of $\nu_0\nu_2/\nu_{02}$ in the equation balancing (1952) and (1947) we find

$$V^3S_1^{\,2}(V_2,\theta,V)\,dV=\left(\frac{m_0m_2}{m_0+m_2}\right)^2\frac{2\pi\varpi_0\varpi_2}{\varpi_{02}h^3}V_1^{\,3}V_2^{\,2}S_2^{\,1}(V_2,\theta,V_1)\,dV_1. \quad \text{......(1954)}$$

On differentiating (1950) we find

$$V\,dV=V_1\,dV_1\left(1-\frac{m_2}{m_0+m_2}\frac{V_2}{V_1}\cos\theta\right), \quad \text{......(1955)}$$

and, using this in (1954),

$$V^2S_1^{\,2}(V_2,\theta,V)\left\{1-\frac{m_2}{m_0+m_2}\frac{V_2}{V_1}\cos\theta\right\}$$
$$=\left(\frac{m_0m_2}{m_0+m_2}\right)^2\frac{2\pi\varpi_0\varpi_2}{\varpi_{02}h^3}V_1^{\,2}V_2^{\,2}S_2^{\,1}(V_2,\theta,V_1). \quad \text{......(1956)}$$

An important special process of this type is *ionization*, in which m_2 is negligibly small compared with m_0. In that case when the body 1 is of atomic (not electronic) mass classical dynamics requires that $V_2<2V$, while V_1 and V are only slightly different. Thus V_2/V_1 is of the order unity at most, $\varpi_2=2$ and the $\{\ \}$ reduces to unity. Thus (1956) becomes with sufficient accuracy

$$V^2S_1^{\,2}(V_2,\theta,V)=\frac{4\pi\varpi_0m_2^{\,2}}{\varpi_{02}h^3}V_1^{\,2}V_2^{\,2}S_2^{\,1}(V_2,\theta,V_1), \quad \text{......(1957)}$$

with the energy relation

$$\frac{1}{2}\frac{m_0 m_1}{m_0+m_1}V^2=\eta_0+\frac{1}{2}\frac{m_0 m_1}{m_0+m_1}V_1^2+\tfrac{1}{2}m_2V_2^2. \qquad \text{......(1958)}$$

These equations are the result of ignoring the momentum of the electron in the equations of conservation.

Now that θ no longer appears explicitly in (1957) and (1958), it is possible and often convenient to reformulate these relations more in accord with the relations of the earlier sections, where both the bodies 1 and 2 were electrons. We recall that $\Sigma_2^1(p_1,p_2,\tau;V_2,\theta,V_1)$ is the fraction of all $(p_1,p_2,\tau;V_2,\theta,V_1)$-collisions that result in capture. The derived function $S_2^1(V_2,\theta,V_1)$ is the "target" ($[L]^4[T]$) for (V_2,θ,V_1)-collisions to be successful, and so leads to the number of successful (V_2,θ,V_1)-collisions when these are distributed at random in p_1, p_2 and τ. If $S_2^{1*}(V_2,V_1)$ is the mean value of this with respect to θ, so that

$$S_2^{1*}(V_2,V_1)=\frac{1}{2}\int_0^\pi S_2^1(V_2,\theta,V_1)\sin\theta\,d\theta, \qquad \text{......(1959)}$$

then S_2^{1*} is the $[L]^4[T]$ "target" for successful (V_1,V_2)-collisions distributed at random in p_1, p_2, τ and θ.

In a similar way $\Sigma_1^2(p;V_2,\theta,V)\,m_2V_2\sin\theta\,dV_2\,d\theta$ is that fraction of (p,V)-collisions which result in (V_2,θ,V_1)-relative motions after causing dissociation. Then

$$S_1^2(V_2,\theta,V)\,m_2V_2\sin\theta\,dV_2\,d\theta$$

is the $[L]^2$ target for V-collisions distributed at random in p to have the result specified. If finally

$$S_1^{2*}(V_2,V)=\int_0^\pi S_1^2(V_2,\theta,V)\sin\theta\,d\theta, \qquad \text{......(1960)}$$

then $S_1^{2*}m_2V_2\,dV_2$ is the target for V-collisions distributed at random in p to have (V_2,V_1) relative velocities after dissociation. The complete ionization target for V-collisions distributed at random in p is

$$\int_0^{(V_2)_{\max}} S_1^{2*}(V_2,V)\,m_2V_2\,dV_2. \qquad \text{......(1961)}$$

The relation between S_1^{2*} and S_2^{1*} must be

$$V^2S_1^{2*}(V_2,V)=\frac{8\pi\varpi_0 m_2^{\,2}}{\varpi_{02}h^3}V_1^2V_2^2S_2^{1*}(V_2,V_1) \qquad \text{......(1962)}$$

with the energy relation (1958). Expressed in terms of energy (1962) is identical with the formula (1924) for electron impact. It is easy to see that all the formulae of this paragraph which contain S's but not Σ's retain their validity in quantum mechanics.

It is beyond the range of this monograph to discuss the detailed forms of these target areas.

§ 17·71. *The laws of detailed balancing for general collisions.* It is perhaps worth while in conclusion to consider a very general formulation of the laws of detailed balancing for collisions given by Dirac,* which brings out the main features better than the discussion of § 17·7.

We start by observing that in its ordinary form Maxwell's law for the density-in-velocity or density-in-momentum of systems per unit volume is invariant for a transformation from any set of axes to another moving relatively to the former with constant velocity. For relative to the old axes the density-in-momentum is

$$f = Ae^{-(p_1^2+p_2^2+p_3^2)/2mkT},$$

and if the transformation is $p_1' = p_1 + \delta$, then relative to the second set the density-in-momentum is

$$f' = Ae^{-\{(p_1'-\delta)^2+p_2'^2+p_3'^2\}/2mkT}$$

which is unaltered ($f=f'$).

We then consider a general encounter between n material systems in which the rth system has initially a momentum in a region $(dp_1dp_2dp_3)_r$, and in which as a result of the encounter n' material systems leave the scene of action, the rth having a momentum in a region $(dp_1'dp_2'dp_3')_r$. A material system may be of course any molecule, atom, or ion in any specified stationary state, or a free electron, but not here a quantum of radiation. The n' systems must have the same material constituents, but recombined in any manner whatever. The velocity of the centre of gravity of the systems both before and after is $\mathbf{C}$. We then transform to a frame of reference in which the centre of gravity is at rest, and use a zero suffix to distinguish quantities measured in this frame, which we call the normal frame. If we assume for the present that the momenta before and after the encounter are all independent, the number of such encounters per unit volume per second will be of the form

$$\left[\prod_1^n f_r(dp_1dp_2dp_3)_r\right]_0 . \phi . \prod_1^{n'} (dp_1'dp_2'dp_3')_r, \qquad \ldots\ldots(1963)$$

where ϕ is an atomic probability coefficient which must be independent of T and $\mathbf{C}$ and depend only on the momenta of the systems in the normal frame of reference both before and after the encounter. It is unnecessary here to analyse ϕ further. It must be supposed to contain the velocity factors for the speeds with which the various systems approach the scene of action. In the same way the corresponding number of reverse encounters is

$$\left[\prod_1^n f_r'(dp_1'dp_2'dp_3')_r\right]_0 . \phi' . \prod_1^n (dp_1dp_2dp_3)_r. \qquad \ldots\ldots(1964)$$

* Dirac, *Proc. Roy. Soc.* A, vol. 106, p. 581 (1924).

By the principle of detailed balancing we may equate these two expressions, and obtain

$$[f_1 \cdots f_n]_0 \phi = [f_1' \cdots f_{n'}']_0 \phi'. \qquad \text{......(1965)}$$

Our provisional assumption that all the momenta are independent is not true. There are at least seven relations (energy, and zero momenta in the normal frame before and after the encounter), and there may be more in special cases. We can allow for this most simply by making ϕ and ϕ' zero except when these necessary conditions are fulfilled. There will then be fewer differentials in (1963), but on account of the complete reversibility the same differentials will drop out of (1964), and (1965) will remain generally true. If we now transform back to the frame in which the whole assembly is at rest, owing to the invariant property of Maxwell's law, we find the general collision relation

$$f_1 f_2 \cdots f_n \phi = f_1' f_2' \cdots f_n' \phi'. \qquad \text{......(1966)}$$

Now if $F_1, \ldots, F_n$ are the partition functions for the internal energies of the corresponding systems, and $\nu_1, \ldots, \nu_n$ their total concentrations, then, in the equilibrium state,

$$f_1 f_2 \cdots f_n = \frac{\varpi_1 \cdots \varpi_n}{F_1 \cdots F_n} \frac{\nu_1 \cdots \nu_n}{\prod_1^n (2\pi m_r kT)^{\frac{3}{2}}} e^{-\Sigma\left\{\epsilon_r + \frac{1}{2m_r}(p_1^2+p_2^2+p_3^2)_r\right\}/kT},$$

with a corresponding expression for $f_1' f_2' \ldots f_n'$. There is also the equation of dissociative equilibrium

$$\frac{\nu_1 \ldots \nu_n}{\nu_1' \ldots \nu_n'} = \frac{F_1 \ldots F_n}{F_1' \ldots F_n'} \prod_1^n \frac{(2\pi m_r kT)^{\frac{3}{2}}}{h^3} \Big/ \prod_1^{n'} \frac{(2\pi m_r' kT)^{\frac{3}{2}}}{h^3},$$

and the energy equation

$$\sum_1^n \left\{\epsilon_r + \frac{1}{2m_r}(p_1^2+p_2^2+p_3^2)_r\right\} = \sum_1^{n'} \left\{\epsilon_r' + \frac{1}{2m_r'}(p_1'^2+p_2'^2+p_3'^2)_r\right\}.$$

Combining these equations we find the relation

$$\frac{\phi'}{\phi} = \frac{\varpi_1 \ldots \varpi_n}{\varpi_1' \ldots \varpi_{n'}'} (h^3)^{n'-n}. \qquad \text{......(1967)}$$

The relationships of the preceding sections are special cases of (1967) in which the coefficients S have been defined somewhat differently from ϕ and ϕ'.*

§ **17·8.** *Laws of interaction of gaseous molecules with solid walls.* We can apply the general arguments of § 17·12 to show that collisions of molecules with the walls of the containing vessel must separately be capable of preserving the equilibrium state. For the equilibrium state of the gas is

* Dirac, *loc. cit.*, presents the argument somewhat differently and derives the equations of dissociative equilibrium, *etc.*, from the principle of detailed balancing by substantially the reverse of the foregoing argument. The principle of the invariance of f which must be then known *a priori* can, he shows, be deduced from the properties of the Lorentz transformation.

independent of the shape of the enclosure containing the gas, and by varying the shape of the enclosure the relative importance of the *surface* can be varied independently of all other parameters. Collisions with the walls vary in frequency as the surface multiplied by the molecular density and therefore differently from collisions between molecules ([density]^2) or radiative effects (density).

We have already made use of this principle in discussing the relation between the emission and absorption of electrons at a metal surface in § 11·31. It will therefore only be necessary to generalize the analysis of that section here. If the gaseous phase is practically perfect, and contains a set of systems (atoms, molecules or electrons) at concentration ν, then the number of such systems with velocities between c and $c+dc$ which strike unit area of any solid surface per unit time in a direction within a solid angle $d\Omega$ at an angle θ with the normal to the surface is

$$\nu\left(\frac{m}{2\pi kT}\right)^{\frac{3}{2}} c^3 e^{-mc^2/2kT} \cos\theta \, dc \, d\Omega. \qquad \text{......(1968)}$$

This is the number of such systems which is destroyed in unit time by unit area of wall. By the same reasoning an equal number *moving in the reverse direction* must be thrown off by the wall in unit time in order to preserve equilibrium.

As we are no longer dealing with single atoms and molecules we can no longer argue that elementary processes must be independent of the temperature.

In very high vacua collisions with walls are all important in controlling the equilibrium state, and a series of researches notably by Knudsen*, Millikan† and Langmuir‡ have been undertaken to elucidate the properties of the equilibrium state and steady non-equilibrium states under such conditions. These investigations have thrown much light on the nature of the mechanism which we call "collision with a wall". The conclusion is that in general the greater part of the molecules striking a solid surface actually condense on the surface (stick to it) for a limited period and are then thrown off again. Their direction and velocity of ejection will then naturally have no connection with their direction and velocity of incidence. If this idea of complete lack of correlation is correct, then the molecules evaporating must be thrown off by the wall at a rate given by (1968). This is a conclusion of considerable importance in the researches quoted. It embodies what is known as *Lambert's law of diffuse reflection*. It contains

* Knudsen, *Ann. d. Physik*, vol. 28, pp. 75, 999 (1909), and a series of other papers up to vol. 34, p. 593 (1911).

† Millikan, *Phys. Rev.* vol. 21, p. 217, vol. 22, p. 1 (1923).

‡ Langmuir, *Trans. Far. Soc.* vol. 17, pp. 607, 621 (1921).

however more information than Lambert's law which refers only to distribution with angle. When we study steady states in which we have no longer a temperature equilibrium, the T in (1968) ceases to have a definite meaning. The general form of the molecular emission law may be expected to hold good, but T becomes a parameter determined by the temperature of the incident molecules and the temperature and other properties of the wall. A great part of Knudsen's researches deals with particular cases of this type.

In other cases we have evidence, from the rate of transfer of momentum to the walls during steady states of flow, that an appreciable fraction of the incident molecules do not condense, or at least do not communicate momentum to the wall on impact. In such cases it is usual (and probably adequate) to describe the interaction by means of an accommodation coefficient f which is such that the interaction proceeds as if the fraction f of all the incident molecules condenses on impact and $1-f$ is reflected according to the laws of reflection of light ("specular reflection"). It is easy to see that specular reflection also conserves (1968). It is usually assumed that f is independent of c and θ.

There is good evidence for condensation and consequent uncorrelated re-emission obeying (1968). Beside this perfectly diffuse reflection the only simple types of reflection which preserve (1968) are perfectly specular reflection and reflection by direct reversal of path. The latter is physically unacceptable. The former has undoubtedly been used to supplement perfectly diffuse reflection on the ground of its simplicity. A more correct analysis would doubtless fuse both the perfectly diffuse and perfectly specular reflections together into a single law with a varying correlation between the direction of incidence and all possible directions of re-emission. It would present no difficulty to formulate such laws satisfying (1968), but at present they do not appear to be of interest.

CHAPTER XVIII

CHEMICAL KINETICS IN GASEOUS SYSTEMS

§ **18·1.** *General nature of reactions in gaseous assemblies.** When gases which undergo a chemical reaction are mixed, it is natural to look to the collisions between the reacting molecules for the source of the rearrangements that occur. The ideas and the formulae of the preceding chapter should therefore enable a satisfactory account to be given of those gaseous reactions which do not depend observably on radiation,† and proceed sufficiently slowly for the calculation of collisions by the equilibrium theory to be applicable. Classical statistics can always be used. We shall see that this expectation is in general fulfilled, but there is at least one quite exceptional example in explaining which the theory is severely strained. It is possible that some other considerations may enter.

Let us start by defining more closely what we mean by "sufficiently slow" for the reactive collisions not to upset the numbers of collisions calculated on the equilibrium theory. All reactions of course proceed to their equilibrium point, at which all the considerations of the equilibrium theory must apply. But in chemical kinetics we are concerned with the speed of unbalanced reactions proceeding primarily in one direction, and it is these which we try to record by observation. In order to calculate such speeds from the equilibrium theory we have to assume that certain types of collisions are effective, and that these types occur (in spite of the one-sided reaction) with a frequency corresponding to that which would be deduced from the properties of an equilibrium state. It will be best to examine the requirements of this condition in the various special cases discussed.

In the examination of a gaseous reaction the first point to be established is that it *is* a reaction between gases (homogeneous reaction) and not a reaction occurring primarily between gas molecules condensed on the walls of the containing vessel—catalysed by the walls—(a heterogeneous reaction). We shall only discuss homogeneous gas reactions here: heterogeneous reactions are more naturally discussed as part of the kinetic theory of surfaces. Though catalysis by the walls dominates the great majority of apparently homogeneous reactions, quite a considerable number of simple homogeneous reactions are now known. A selection of those that are known

* For a general account see Hinshelwood, *The Kinetics of Chemical Change in Gaseous Systems* ed. 1 (1926), ed. 3 (1933). This chapter is based almost entirely on Hinshelwood's book. I do not however agree with his conclusion that in some cases all collisions with sufficient energy lead to reaction nor with the derivation and use of some of his formulae.

† The best known example of a photosensitive reaction, or photochemical change, in gases is the reaction $H_2 + Cl_2 \rightarrow 2HCl$ under the influence of visible light.

to be simple and homogeneous, for which the velocity of a perfectly definite molecular event can be and has been observed, are described in detail by Hinshelwood.*

Having established the homogeneity of a reaction, its velocity is studied as a function of the temperature and the concentrations. The dependence on the concentration generally classifies the reactions into different orders of which we need only consider two here, namely:

First order or unimolecular reactions, with a velocity proportional to the concentration so that

$$-\frac{d\nu}{dt}=\kappa\nu, \qquad \text{......(1969)}$$

where κ depends only on the temperature.

Second order or bimolecular reactions, with a velocity proportional to the square or product of the concentrations so that

$$-\frac{d\nu}{dt}=\kappa\nu^2 \quad \text{or} \quad -\frac{d\nu_1}{dt}=-\frac{d\nu_2}{dt}=\kappa\nu_1\nu_2, \qquad \text{......(1970)}$$

and κ again depends only on the temperature. This function κ is called *the velocity constant* of the reaction. These equations of course only hold before the products of the reaction interfere in any way such as by beginning the reverse reaction. But κ is usually measured† under conditions in which (1969) and (1970) are sufficiently exhaustive. We shall not discuss the methods by which the κ of these equations is determined in practice.

When we compare the observed rates of reaction expressed in numbers of molecules per second with the numbers of collisions per molecule per second we find that at most one collision in 10^8 can lead to a reaction. Gaseous reactions cannot ordinarily be quantitatively studied if the initial concentrations of reactants fall to half value in a time as short as one second, and the usual time to half value is of the order of at least a minute. But by (1859) the number of collisions per molecule per second is

$$\nu\sigma^2\left(\frac{4\pi kT}{m}\right)^{\frac{1}{2}}.$$

Inserting numerical values with $\sigma=3\times10^{-8}$, $k=1{\cdot}37\times10^{-16}$, $T=273$, $\nu=2{\cdot}7\times10^{19}$, $m=1{\cdot}65\times10^{-24}A$, where A is the chemical molecular weight, this reduces to

$$\frac{1{\cdot}3\times10^{10}}{\sqrt{A}}.$$

This is to be compared with a number of reactive collisions at most of the order one per second. Thus reactive collisions are completely exceptional.

* Hinshelwood, *loc. cit.* In his first edition he described *all* the then known examples. The difference in presentation between the editions of 1926 and 1933 is a record of the great development of the subject during that period.

† See Hinshelwood, *loc. cit.* ed. 3, pp. 68, 77, for a more general case.

The clue to this behaviour is provided by the strikingly large temperature variation of the velocity constant κ. Whereas the total number of collisions varies as $\sqrt{T}$, that is hardly at all over wide ranges of temperature, the velocity constant ordinarily doubles itself for a rise of temperature of about 10° C. This suggests at once that the effective collisions are selected not from all collisions, but only from all collisions with more than a certain large minimum of distributable energy.

The actual form of κ which can be successfully compared with observation is suggested by equilibrium considerations. Consider for definiteness a bimolecular reaction between unlike molecules of types 1 and 2 which combine to form two other molecules of types 3 and 4. There is an equilibrium point which is given by

$$\frac{\nu_1\nu_2}{\nu_3\nu_4}=\frac{f_1(T)f_2(T)}{f_3(T)f_4(T)}=\mathrm{K}, \qquad \text{......(1971)}$$

where K is the equilibrium constant. For *all* concentrations we know from the velocity measurements that the rate of destruction of systems 1 and 2, with creation of systems 3 and 4, is $\kappa\nu_1\nu_2$. By the arguments of § 17·12 the rate of creation of systems 1 and 2 and destruction of 3 and 4 must be of the form $\kappa'\nu_3\nu_4$, and at the equilibrium point these are equal, so that

$$\mathrm{K}=\kappa'/\kappa. \qquad \text{......(1972)}$$

Now it follows from the definition of K, by (480), (481), that

$$\frac{d\log \mathrm{K}}{dT}=\frac{q}{kT^2}=\frac{Q}{RT^2}, \qquad \text{......(1973)}$$

where q is the excess of the average energy content of systems 1 and 2 at temperature T over that of 3 and 4, so that Q is the "heat of the reaction". Therefore

$$\frac{d\log\kappa'}{dT}-\frac{d\log\kappa}{dT}=\frac{q}{kT^2}.$$

The form of this equation suggests putting

$$\frac{d\log\kappa}{dT}=\frac{\zeta}{kT^2}, \quad \frac{d\log\kappa'}{dT}=\frac{\zeta'}{kT^2},$$

where

$$\zeta-\zeta'=-q$$

and the known approximate constancy of q suggests that perhaps ζ and ζ' are also roughly constant. In that case we find after integration

$$\kappa=Ae^{-\zeta/kT}. \qquad \text{......(1974)}$$

This is the well-known empirical equation of Arrhenius for the velocity constant of a homogeneous gaseous reaction. As so far presented it is purely tentative, but by plotting $\log\kappa$ against $1/T$ for observations over a sufficient range of temperatures, it is found that (1974) with A and ζ constant gives an entirely adequate representation of the facts. The energy ζ is called *the heat of activation of the reaction*, for, as we shall see, it is closely related to the necessary minimum disposable energy present in a possibly effective collision. We shall show in the following sections how the collision mechanism gives an entirely adequate account of the phenomena we have described.

§ **18·2.** *Simple theory of bimolecular reactions.* Equation (1865) gives us the number of collisions per unit volume per unit time in which the kinetic energy of the relative motion lies between η and $\eta+d\eta$ for simple unlike molecules. Let us suppose that $\sigma(\eta)$ is the fraction of these collisions that lead to reaction. Then

$$\kappa=\frac{2\sigma_{12}^2}{(kT)^{\frac{3}{2}}}\left\{\frac{2\pi(m_1+m_2)}{m_1m_2}\right\}^{\frac{1}{2}}\int_0^\infty \sigma(\eta)\,e^{-\eta/kT}\eta\,d\eta. \qquad ......(1975)$$

In order to mimic (1974) the simplest assumption is that

$$\sigma(\eta)=0\ \ (\eta\leqslant\zeta),\quad \sigma(\eta)=\alpha\ \ (\eta>\zeta),$$

where α is a constant less than or equal to unity. On this assumption

$$\kappa=2\alpha\sigma_{12}^2\left\{\frac{2\pi(m_1+m_2)}{m_1m_2}kT\right\}^{\frac{1}{2}}e^{-\zeta/kT}\left(\frac{\zeta}{kT}+1\right). \qquad ......(1976)$$

This is very nearly of the prescribed form. It is to be observed that the T-variation of κ is so dominated by the exponential term when (as in all actual examples) ζ/kT is fairly large, that the experiments cannot possibly distinguish between

$$Ae^{-\zeta/kT} \quad \text{and} \quad A'T^se^{-\zeta/kT}$$

for any moderate value of s. The same difficulty has been already encountered in Richardson's thermionic formulae. This simple assumption therefore yields, for fairly large values of ζ/kT, the formula for the velocity constant

$$\kappa=A'T^{-\frac{1}{2}}e^{-\zeta/kT}, \qquad ......(1977)$$

in which

$$A'=2\alpha\sigma_{12}^2\left\{\frac{2\pi k(m_1+m_2)}{m_1m_2}\right\}^{\frac{1}{2}}\frac{\zeta}{k}. \qquad ......(1978)$$

The theory thus gives us at once a satisfactory form for κ. We have no *a priori* knowledge of ζ/k, and it must therefore be determined from Arrhenius' equation. When this has been done, the theory gives us κ completely in terms of a coefficient α which must be less than unity, and a "molecular diameter" σ_{12}. For like molecules $m_1=m_2$, and the first factor 2 drops out from any formula for the number of collisions but not for the velocity constant, since two molecules react for each successful collision. Equations (1977) and (1978) therefore hold for like or unlike molecules.

The formula (1978) contains the product of two coefficients α and $\pi\sigma_{12}^2$, neither of which (it may be argued) is precisely defined alone. There is, however, some virtue in not amalgamating them. One can suppose ideally that $\pi\sigma_{12}^2$ has been determined (though possibly as a function of the temperature) by viscosity measurements, when it will have the physical meaning of the effective target area for momentum exchange in an encounter, or from general ideas of molecular magnitudes at any relative velocity. The effective target area for reaction at sufficiently high relative velocity, $\alpha\pi\sigma_{12}^2$, is then a distinct quantity, which as we shall shortly see is probably appreciably smaller than $\pi\sigma_{12}^2$.

To proceed further we must consider an actual example, and choose the decomposition of HI discussed by Hinshelwood. This is primarily homogeneous and bimolecular from 550° K. to 780° K., and its temperature variation satisfies Arrhenius's equation accurately with $\zeta/k = 22{,}000$ ($Q = 44{,}000$ calories). With $\sigma = 2 \times 10^{-8}$, $m_1 = m_2 = 128 \times 1{\cdot}65 \times 10^{-24}$, and $\sqrt{T} = 25$ this gives

$$A' = 5{\cdot}0 \times 10^{-8}\alpha, \quad \kappa = 2 \times 10^{-9}\alpha e^{-22{,}000/T}.$$

From the definition of κ
$$\frac{1}{\nu}\frac{d\nu}{dt} = -\kappa\nu.$$

Therefore the fraction of molecules reacting in one second at a concentration of one gram-molecule per litre ($\nu = 6{\cdot}06 \times 10^{20}$) and a temperature 556° T is

$$8 \times 10^{-6}\alpha.$$

The observed value is $3{\cdot}52 \times 10^{-7}$. The observed value is thus obtained* if $\alpha = 1/23$.

This is entirely satisfactory so far as it goes. It only remains for us to verify that this fraction of reactive collisions is small enough for the calculation of collisions with energy more than ζ to be substantially unaffected, i.e. to verify that the reaction is "sufficiently slow". A certain proportion of collisions with relative energy more than ζ will concern at least one molecule whose last collision was also one of the same class. This is the phenomenon *of the persistence of velocities*.† In default of an exact theory of transport phenomena, correction for persistence of velocities was successful in removing the greater part of the numerical error in the simpler theory of these phenomena. In order to be certain that the equilibrium calculations are adequate it is sufficient to assure oneself that allowance for persistence of velocities is unimportant here. The proportion of collisions affected can be fairly high, but even if it is nearly unity only the fraction α at most will be removed by the reaction and the effect on numerical values cannot possibly reach 10 per cent. At the same time the numerical values might begin to be seriously affected if α were larger. The interpretation given by Hinshelwood to his result $\alpha = 1$ is therefore hardly acceptable. It is necessary for α to be

* Hinshelwood, *loc. cit.*, concludes that approximately $\alpha = 1$, but he has used an inaccurate formula for the number of collisions.

† Jeans, *loc. cit.* pp. 260, 275, 312.

small for the simple theory to apply at all. But of course the results of a more exact theory are not likely to be widely different, and will only differ in a numerical factor.

Summing up, we may say that the simple theory of homogeneous bimolecular reactions, namely that *they are reactions by collision which can only occur in a fairly small fraction of collisions in which the relative kinetic energy of the molecules exceeds a certain lower limit determined by Arrhenius's equation*, gives a most satisfying account of the observed facts. We have discussed it only with reference to the reaction

(1) $2HI \rightarrow H_2 + I_2$,

but the following other reactions have been shown by Hinshelwood to fit fairly well into the same theory:

(2) $H_2 + I_2 \rightarrow 2HI$,
(3) $2O_3 \rightarrow 3O_2$,
(4) $2N_2O \rightarrow 2N_2 + O_2$,
(5) $2Cl_2O \rightarrow 2Cl_2 + O_2$,

with other more complicated examples in which the homogeneous bimolecular reaction must be disentangled from other simultaneous effects.

It is at the same time to be remarked that there is not much margin in the above theory as it stands. A different form for $\sigma(\eta)$ might at once require $\alpha > 1$. For example, if $\sigma(\eta) = \alpha\zeta/\eta$ the number of effective collisions is practically unchanged, but if in conformity with some of the collision targets of the preceding chapter we have, say,

$$\sigma(\eta) = \frac{\alpha\zeta^2}{\eta}\left(\frac{1}{\zeta} - \frac{1}{\eta}\right),$$

we find the number of collisions smaller by loss of the factor ζ/kT. The number of collisions is also smaller if calculated as in § 18·21 below. Effective collisions in which the relative kinetic energy of translation exceeds ζ may therefore be somewhat scarce, and it seems to my mind better to recognize even here that there are other sources of energy as yet unexplored in this connection. We proceed therefore in § 18·3 to develop formulae for the number of collisions with given transferable energy in which the transferable energy from all sources is taken into account.

§ **18·21.** *Bimolecular reactions, using only the head-on component of the relative velocity.* It has frequently been proposed that in calculations of the number of collisions with sufficient energy only those collisions should be included in which the energy of the motion along the line of centres at impact exceeds the critical value. Though there is no particular merit in such a formulation from the point of view of quantum mechanics, it is perhaps worth while to put the correct formulae for this proposal on record since incorrect ones are frequently in use.*

In formula (1860) we have an expression for the number of collisions per unit volume per second in which the line of centres at impact makes an angle θ with the relative velocity whose components lie in the ranges ξ, $\xi+d\xi$; η, $\eta+d\eta$; ζ, $\zeta+d\zeta$. We can rearrange this in the form

$$\nu_1\nu_2\left(\frac{m_1m_2}{2\pi(m_1+m_2)kT}\right)^{\frac{3}{2}} e^{-\frac{1}{2}m_1m_2(\xi^2+\eta^2+\zeta^2)/(m_1+m_2)kT}\xi\, d\xi\, d\eta\, d\zeta\, .\, \sigma_{12}{}^2 d\Omega. \qquad \text{......(1979)}$$

We can then integrate (1979) with respect to η and ζ to obtain all collisions with a relative velocity along the line of centres between ξ and $\xi+d\xi$ occurring on the selected element of surface $\sigma_{12}{}^2 d\Omega$ and after that integrate over the whole surface of the sphere of radius σ_{12}. We thus obtain

$$2\nu_1\nu_2\sigma_{12}{}^2\left(\frac{2\pi(m_1+m_2)}{m_1m_2kT}\right)^{\frac{1}{2}} e^{-\eta/kT}\, d\eta \qquad \text{......(1980)}$$

for the number of collisions per unit volume per unit time in which the energy of the component of the relative motion along the line of centres at impact lies between η and $\eta+d\eta$. The number of such collisions in which this energy exceeds ζ is

$$2\nu_1\nu_2\sigma_{12}{}^2\left(\frac{2\pi(m_1+m_2)\,kT}{m_1m_2}\right)^{\frac{1}{2}} e^{-\zeta/kT}. \qquad \text{......(1981)}$$

By comparison with (1976) we see that values of κ based on this use of head-on components only are smaller by lack of the factor $(\zeta/kT+1)$. The loss of this factor would already lead to known difficulties in the application of the theory to the decomposition of HI.

§ **18·22.** *The reverse reactions.* In any discussion of a reaction mechanism it must never be overlooked that the reverse reaction must be able to occur and form a unit mechanism which will preserve by itself the equilibrium state. For a simple reaction such as $X+Y \rightleftarrows Z+W$ or $X+Y \rightleftarrows 2Z$ the reaction is bimolecular in both directions and no new features are found in the reverse direction. But other reactions may not be so simple. If for example† the bimolecular decomposition of ozone occurs in a single step correctly represented by $2O_3 \rightarrow 3O_2$, then the reverse process must be a termolecular collision between three oxygen molecules. It has been suggested that the reaction proceeds by the stages (1) $O_3 \rightleftarrows O_2+O$ in equilibrium, the reaction being rapid, followed by (2) $O_3+O \rightarrow 2O_2$, the slow

* Hinshelwood, *loc. cit.* ed. 3, p. 123, argues as if the velocity component of each molecule along the line of centres contributes each one square term to the available energy, apparently forgetting that it is only the single component of the *relative* velocity that can be effective.

† Hinshelwood, *loc. cit.* p. 80.

reaction whose rate is measured. Since the concentrations of O_3 and O in the presence of excess of O_2 must be proportional, the measured reaction will have a rate proportional to the square of the concentration of O_3. If this mechanism is the true one, it is the first (fast) reaction which contains the termolecular encounter. The ozone molecule cannot just break up into O_2+O but will require a collision to dissociate it, and the true form of this reaction is probably $O_3+O_2 \rightleftarrows 2O_2+O$. The function of the inert O_2 could probably be played by any other molecule.

Since the direct occurrence of termolecular reactions can thus be inferred, it is of special interest to examine whether any such reactions can be experimentally studied, and their termolecular character established by direct experiment. Few such reactions are known, but

$$2NO+O_2 \rightarrow 2NO_2$$

has been shown by Bodenstein* to be such a reaction, by verifying that its velocity obeys the equation

$$\frac{d[NO_2]}{dt}=\kappa[NO]^2[O_2].$$

This velocity constant κ is almost independent of the temperature, and actually falls slightly as the temperature rises. Such behaviour is reasonable if the activation energy is practically zero.

§ **18·3.** *The transferable energy in collisions, including internal energy of the molecules.* It is hardly possible to obtain simple formulae if we consider the rotations and internal vibrations of the molecules as quantized. It will however be sufficiently accurate for present purposes to assume that the molecules are equivalent to molecules whose Hamiltonian function contains s square terms (kinetic or potential) associated with such low frequencies that classical mechanics applies. Then the classical partition function for the internal energy of such a molecule is

$$\int \ldots \int e^{-\eta/kT}\, dp_1 \ldots dq_t,$$

where $p_1, \ldots, p_t, q_1, \ldots, q_t$ are a complete set of Hamiltonian coordinates and

$$\eta = \sum_1^s \alpha_t \mu_t^2,$$

μ being either a p or q; all the p's must occur. The integration is extended over all values from $-\infty$ to $+\infty$ for any μ, and with respect to the other geometrical variables over their proper ranges. To find the fraction with

* Bodenstein and Lindner, *Zeit. f. physikal. Chem.* vol. 100, p. 68 (1922).

energy between η_1 and $\eta_1+d\eta_1$ we have to take out that part of the complete integration which corresponds to

$$\eta_1 \leqslant \sum_1^s \alpha_i \mu_i^2 \leqslant \eta_1 + d\eta_1.$$

The number of molecules per unit volume with this internal energy is therefore

$$\nu_1 \frac{\underset{[\eta_1 \leqslant \Sigma \alpha_i \mu_i^2 \leqslant \eta_1 + d\eta_1]}{\int \dots \int} e^{-\eta/kT} d\mu_1 \dots d\mu_s}{\underset{[-\infty \leqslant \mu_i \leqslant +\infty]}{\int \dots \int} e^{-\eta/kT} d\mu_1 \dots d\mu_s}. \qquad \dots\dots(1982)$$

By the well-known procedure of Dirichlet* this can be reduced to

$$\nu_1 \frac{\eta_1^{\frac{1}{2}s-1} e^{-\eta_1/kT} d\eta_1}{\int_0^\infty \eta^{\frac{1}{2}s-1} e^{-\eta/kT} d\eta}, \qquad \dots\dots(1983)$$

or

$$\frac{\nu_1}{\Gamma(\frac{1}{2}s)} \left(\frac{\eta_1}{kT}\right)^{\frac{1}{2}s-1} e^{-\eta_1/kT} \frac{d\eta_1}{kT}. \qquad \dots\dots(1984)$$

The number of molecules with internal energy greater than η_0 is

$$\frac{\nu_1}{\Gamma(\frac{1}{2}s)} \frac{1}{(kT)^{\frac{1}{2}s}} \int_{\eta_0}^\infty \eta^{\frac{1}{2}s-1} e^{-\eta/kT} d\eta, \qquad \dots\dots(1985)$$

which, for η_0/kT large, is approximately†

$$\frac{\nu_1}{\Gamma(\frac{1}{2}s)} \left(\frac{\eta_0}{kT}\right)^{\frac{1}{2}s-1} e^{-\eta_0/kT}. \qquad \dots\dots(1986)$$

Let us now combine (1984) with (1865). We find that the number of collisions per unit volume per second between a molecule of type 1 with energy between η_1 and $\eta_1+d\eta_1$ and a molecule of type 2 with energy between η_2 and $\eta_2+d\eta_2$ and relative kinetic energy ξ is

$$\frac{2\nu_1\nu_2\sigma_{12}^2}{\Gamma(\frac{1}{2}s_1)\Gamma(\frac{1}{2}s_2)} \left\{\frac{2\pi(m_1+m_2)}{m_1 m_2}\right\}^{\frac{1}{2}} \frac{e^{-(\eta_1+\eta_2+\xi)/kT}}{(kT)^{\frac{1}{2}s_1+\frac{1}{2}s_2+\frac{3}{2}}} \eta_1^{\frac{1}{2}s_1-1} \eta_2^{\frac{1}{2}s_2-1} \xi \, d\eta_1 d\eta_2 d\xi. \qquad \dots\dots(1987)$$

So far as energy considerations go it is conceivable that any collision in which $\eta_1+\eta_2+\xi \geqslant \eta_0$ may have a non-zero probability of producing an active molecule or causing a reaction. The total number of such collisions is

* Whittaker and Watson, *Modern Analysis*, ed. 3, p. 258.

† It must be emphasized that s in these equations is the number of square terms in the energy, not the number of degrees of freedom, so that any harmonic oscillation contributes 2 to s, and any free rotation 1.

therefore obtained by integrating (1987) over all η_1, η_2 and ξ satisfying $\eta_1+\eta_2+\xi\geqslant\eta_0$. This number reduces by simple substitutions* to

$$\frac{2\nu_1\nu_2\sigma_{12}^2}{\Gamma(\frac{1}{2}s_1+\frac{1}{2}s_2+2)}\left\{\frac{2\pi(m_1+m_2)}{m_1m_2}\right\}^{\frac{1}{2}}\frac{1}{(kT)^{\frac{1}{2}s_1+\frac{1}{2}s_2+\frac{3}{2}}}\int_{\eta_0}^{\infty}e^{-\lambda/kT}\lambda^{\frac{1}{2}s_1+\frac{1}{2}s_2+1}\,d\lambda, \quad\text{...(1988)}$$

which, for η_0/kT large, is approximately

$$\frac{2\nu_1\nu_2\sigma_{12}^2}{\Gamma(\frac{1}{2}s_1+\frac{1}{2}s_2+2)}\left\{\frac{2\pi(m_1+m_2)}{m_1m_2}kT\right\}^{\frac{1}{2}}\left(\frac{\eta_0}{kT}\right)^{\frac{1}{2}s_1+\frac{1}{2}s_2+1}e^{-\eta_0/kT}. \quad\text{......(1989)}$$

If the two molecules are of the same species, then of course $m_1=m_2$ and $\sigma_{12}=\sigma$ and the factor 2 must be removed, for every collision will as usual be found to have been counted twice over in all the foregoing formulae.

The fraction of all collisions with "enough" available energy is easily seen to be

$$\frac{(\eta_0/kT)^{\frac{1}{2}s_1+\frac{1}{2}s_2+1}e^{-\eta_0/kT}}{\Gamma(\frac{1}{2}s_1+\frac{1}{2}s_2+2)}, \quad\text{......(1990)}$$

which may be very large indeed compared with the fraction $e^{-\eta_0/kT}$, often used in error in this connection.

If it is denied (as is perhaps natural in certain applications) that the internal energy of the second molecule is ever available to serve towards the activation energy of the first, then the formulae (1988) and (1989) will still apply if we put $s_2=0$. In fact more generally we can in these formulae always use s_1 and s_2 for the number of square terms in the internal energy *whose energy content is available for redistribution in the collision.*

The total number of activations can of course only be expressed in terms of a probability coefficient $\sigma(\eta_1,\eta_2,\xi)$. Thus on multiplying (1987) by $\sigma(\eta_1,\eta_2,\xi)$ and integrating, we find

$$\kappa=\frac{2\sigma_{12}^2}{\Gamma(\frac{1}{2}s_1)\Gamma(\frac{1}{2}s_2)}\left\{\frac{2\pi(m_1+m_2)}{m_1m_2}\right\}^{\frac{1}{2}}\frac{1}{(kT)^{\frac{1}{2}s_1+\frac{1}{2}s_2+\frac{3}{2}}}$$
$$\times\iiint_{\eta_1+\eta_2+\xi\geqslant\eta_0}e^{-(\eta_1+\eta_2+\xi)/kT}\sigma(\eta_1,\eta_2,\xi)\,\eta_1^{\frac{1}{2}s_1-1}\eta_2^{\frac{1}{2}s_2-1}\xi\,d\eta_1\,d\eta_2\,d\xi; \quad\text{......(1991)}$$

the preceding formulae arise from putting $\sigma=1$ when $\eta_1+\eta_2+\xi\geqslant\eta_0$.

When internal energies are taken into account a more accurate investigation of the precise use of Arrhenius's equation is necessary. In practice the observations are used to plot $\log\kappa$ against $1/T$ and determine a slope, which defines the activation energy ζ of the equation

$$\frac{d\log\kappa}{dT}=\frac{\zeta}{kT^2}.$$

* Whittaker and Watson, *loc. cit.*

The theoretical κ derived by putting simple forms for σ in (1991) is however generally of the form

$$\kappa = B\left(\frac{\eta_0}{kT}\right)^t e^{-\eta_0/kT},$$

so that

$$\frac{d\log\kappa}{dT} = \frac{\eta_0}{kT^2} - \frac{t}{T}.$$

Thus the η_0 of the theory and the ζ determined by observation are connected by the equation

$$\eta_0 = \zeta + tkT. \qquad \text{......(1992)}$$

The apparent constancy of the experimental ζ in no way prevents ζ being really of the theoretical form (1992), for the experiments could not detect these variations, tkT being small compared with η_0. In calculating whether there are sufficient energetic collisions to give the observed rate of reaction with a small efficiency α we must use the η_0 of (1992) with the observed ζ and a mean value of T. We thus retain the correct temperature variation of κ.

If we now examine how these considerations affect the typical homogeneous bimolecular reaction between simple molecules we see that for a molecule such as HI we must have at least $s = 2$, or for a triatomic molecule $s = 3$ at least. Thus for the reaction $2HI \rightarrow H_2 + I_2$ ($s = 2$ say) we have to replace the factor

$$\frac{\zeta}{kT} e^{-\zeta/kT}$$

of (1976) by the factor

$$\frac{1}{6}\left(\frac{\eta_0}{kT}\right)^3 e^{-\eta_0/kT}$$

of (1989), with $\eta_0 = \zeta + 3kT$, ζ having its observed value. There are therefore substantially more collisions with enough energy than the simple theory indicates, the extra factor being about 13, which leaves an ample margin.

§ **18·4.** *Homogeneous unimolecular reactions.* There are well-known difficulties in the theory of those unimolecular reactions which are apparently insensitive to radiation. The simplest assumption which will account in any way for the facts is to suppose that the reaction is not elementary but consists of bimolecular activation and deactivation processes, which by themselves would keep up a normal equilibrium between activated and inert molecules, while superposed on this there is a definite chance for the spontaneous disintegration of the activated molecules. Provided this disintegration is slow compared with the activation process, the equilibrium concentration of activated molecules will be unaffected. This concentration will therefore be proportional to the total concentration and the reaction will apparently obey the unimolecular law. If this theory is correct, then there should come a concentration for any unimolecular reaction below which its rate begins to fall below the expected unimolecular rate, for the underlying bimolecular process is then beginning to work too slowly. This form of collision theory for unimolecular reactions is usually associated with the name of Lindemann.*

The state of affairs according to this theory is formally very simple. Let x and y be the concentrations of activated and inert molecules at any time. Let the total number of activations and deactivations in time dt be $Z'ydt$ and $Zxdt$ respectively. Let $Bxdt$ be the number of disintegrations of activated molecules. Then the differential equations which control the process are

$$\left.\begin{aligned} \frac{dx}{dt} &= Z'y-(Z+B)x, \\ \frac{dy}{dt} &= -Z'y+Zx. \end{aligned}\right\} \qquad \text{......(1993)}$$

On Lindemann's theory B is a molecular constant and Z and Z' are themselves proportional to $x+y$ or perhaps rather are of the form $Ax+By+Cz$, where z is the concentration of any diluents plus the gaseous products of reaction.

If we solve equations (1993), assuming that Z and Z' are constants, the general solution takes the form

$$x=L_1e^{-\lambda_1 t}+L_2e^{-\lambda_2 t}, \quad y=M_1e^{-\lambda_1 t}+M_2e^{-\lambda_2 t}, \qquad \text{......(1994)}$$

where λ_1 and λ_2 are the roots of the equation

$$D^2-D(Z+Z'+B)+BZ'=0. \qquad \text{......(1995)}$$

Using the initial condition that the equilibrium is undisturbed by disintegration, which is $(Z'y)_0=(Zx)_0$, we find that the coefficients in (1994) satisfy

$$\frac{L_1}{Z'-\lambda_1}=\frac{M_1}{Z}=\frac{M_2}{-Z\lambda_1/\lambda_2}=\frac{L_2}{-(Z'-\lambda_2)\lambda_1/\lambda_2}.$$

This is exact and the rate is of course not that of a unimolecular reaction. If now we suppose that B/Z is small the values of λ_1 and λ_2 reduce approximately to

$$\lambda_1=BZ'/(Z+Z'), \quad \lambda_2=Z+Z'.$$

Since $(x/y)_0$ is in general rather small, Z'/Z will also be rather small and at any rate less than unity. Then $L_2/L_1=O(B/Z)$ and $M_2/M_1=O(BZ'/Z^2)$, which is still smaller. Therefore the second terms in (1994) are negligible even initially and *a fortiori* at all later times owing to their much more powerful exponential factor. They are still negligible initially and so always if $(x/y)_0$ is altered by terms of order B/Z. The solution thus reduces to

$$x=X_0e^{-BZ't/(Z+Z')}, \quad y=Y_0e^{-BZ't/(Z+Z')}, \qquad \text{......(1996)}$$

which is of unimolecular form.

* Lindemann, *Trans. Far. Soc.* vol. 17, p. 599 (1921).

A more exact treatment of equations (1993) or more general forms can be given by treating them as of the form

$$\left.\begin{aligned}\frac{dx}{dt}&=f(t)\,y-\{g(t)+B\}\,x,\\ \frac{dy}{dt}&=-f(t)\,y+g(t)\,x,\end{aligned}\right\} \qquad \text{......(1997)}$$

the coefficients $f(t)$ and $g(t)$ being slowly varying functions of the time. The precise variation of $f(t)$ and $g(t)$ is only assignable *a posteriori*, but this does not affect the argument. We can then apply the general theory of such equations, which is equivalent to the theory of the asymptotic forms of solutions of such equations for large values of a parameter.* It follows from this theory that the first approximation to the solutions of (1997) is of the form

$$x=L_1e^{-\int\lambda_1dt}+L_2e^{-\int\lambda_2dt},\quad y=M_1e^{-\int\lambda_1dt}+M_2e^{-\int\lambda_2dt},$$

where λ_1 and λ_2 are functions of t which are the roots of (1995) with Z and Z' replaced by $f(t)$ and $g(t)$. Since $\lambda_1=BZ'/(Z+Z')=Bx/(x+y)$, λ_1 is independent of the concentrations, and therefore of t, when B/Z is small. The L_1, L_2, M_1, M_2 are constants to this approximation. We obtain the same unimolecular form as before, with an accuracy dependent on the slowness of the variation of $f(t)$ and $g(t)$. The fundamental condition for the validity of the unimolecular forms is that B/Z should be small, or $Bx/Z'y$ small, that is that the number of disintegrations in time dt should be small compared with the total number of activations or deactivations in the same time.

From equations (1993)

$$\frac{1}{\nu}\frac{d\nu}{dt}=\frac{1}{x+y}\frac{d(x+y)}{dt}=\frac{Bx}{x+y}=\frac{BZ'}{Z+Z'}=\kappa.$$

Thus in this theory κ is B times the fraction of activated molecules. If these are molecules with more internal energy than η_0, then by (1986)

$$\kappa=\frac{B}{\Gamma(\frac{1}{2}s)}\left(\frac{\eta_0}{kT}\right)^{\frac{1}{2}s-1}e^{-\eta_0/kT}. \qquad \text{......(1998)}$$

This will therefore fit the observed form of Arrhenius's equation if

$$\eta_0=\zeta+(\tfrac{1}{2}s-1)\,kT. \qquad \text{......(1999)}$$

For the correctness of the theory there is also the over-riding condition that Bx or $\kappa\nu$ is small compared with Zx or the number of activating collisions. The maximum number of such collisions is given by (1989). and the actual number should be a small fraction of this.

* Schlesinger, *Math. Ann.* vol. 63, p. 277 (1907); Birkhoff, *Trans. Amer. Math. Soc.* vol. 9, p. 219 (1908); Fowler and Lock, *Proc. Lond. Math. Soc.* vol. 20, p. 127 (1922).

If we use the exact form (1985) and define ζ as usual by $\kappa = Ae^{-\zeta/kT}$, a simple physical meaning can be given to ζ. It is easy to show by differentiation that

$$\zeta = kT^2 \frac{d}{dT} \log \kappa = \frac{\int_{\eta_0}^{\infty} \eta^{\frac{1}{2}s} e^{-\eta/kT}\, d\eta}{\int_{\eta_0}^{\infty} \eta^{\frac{1}{2}s-1} e^{-\eta/kT}\, d\eta} - \tfrac{1}{2}skT.$$

Thus ζ is equal to "the average energy of the activated molecules" less "the average energy of all the molecules".

§ **18·41.** *Numerical discussion of some unimolecular reactions.* We will now discuss shortly, in the order of the severity of their demands on the theory, five reactions known to be homogeneous and (apparently) unimolecular.*

(1) *Decomposition of gaseous diethyl ether.* This reaction proceeds at a convenient rate between 700–860° K. and is homogeneous and unimolecular down to pressures of 200 mm. Hg. Below that pressure of the reactant the reaction proceeds more slowly and approximates to a bimolecular type. The final result of the reaction is roughly

$$C_2H_5 . O . C_2H_5 \rightarrow CO + 2CH_4 + \tfrac{1}{2}C_2H_4,$$

though of course this does not represent the primary process which is probably the formation of CO and two unstable hydrocarbons. Sufficient admixture of H_2, for example a partial pressure of 300 mm. Hg at 800° K., will preserve the unimolecular rate unaltered down to a pressure of 40 mm. Hg of the reactant. He, N_2 and the reaction products have no marked effects. The velocity constant observed is

$$\log \kappa = 26{\cdot}47 - 53{,}000/RT. \qquad \ldots\ldots(2000)$$

(2) *Decomposition of gaseous dimethyl ether.* This is very similar. The final result of the reaction is

$$CH_3 . O . CH_3 \rightarrow CH_4 + H_2 + CO,$$

proceeding at a convenient rate in the range 700–825° K. unimolecularly down to a pressure of 400 mm. Hg. A pressure of 400 mm. Hg of admixed H_2 at 775° K. will preserve the unimolecular rate to a pressure of 30 mm. Hg of the reactant. N_2, He, CO and CO_2 have no such effect. The velocity constant observed is

$$\log \kappa = 30{\cdot}36 - 58{,}500/RT. \qquad \ldots\ldots(2001)$$

(3) *Decomposition of gaseous propionic aldehyde.* The main feature of the reaction is

$$C_2H_5CHO \rightarrow CO + [\text{various hydrocarbons}];$$

* These reactions are the first five to be proved homogeneous and unimolecular. A large number are now known, but these five form a representative sample. Details are taken from Hinshelwood, *loc. cit.*

it is of convenient speed for 725–875° K., and unimolecular down to a pressure of about 80 mm. Hg. No effect of admixed gases has been detected. The velocity constant observed is

$$\log \kappa = 28{\cdot}56 - 55{,}000/RT. \qquad \ldots\ldots(2002)$$

These three reactions can be discussed together.

The molecules concerned are all fairly complicated and have a rather large number of degrees of freedom which might have their classical energies. We shall find in all these cases that Lindemann's theory gives a completely satisfactory explanation of the observations even if we only take into account the internal energy of the molecule to be activated and put $s_2 = 0$. By way of making the calculations precise we shall take $\sigma_{12} = 10^{-7}$, $s_2 = 0$, and determine for what value of s_1 the number of collisions given by (1989) falls to 100 times the value of $\kappa\nu$ for the least value of ν for which the reaction remains unimolecular. Since we deal with collisions of like molecules $m_1 = m_2$ and the factor 2 falls out.

We find that this condition is fulfilled:

For $C_2H_5.O.C_2H_5$, $\nu = 2{\cdot}4 \times 10^{18}$, with s_1 between 6 and 7.
For $CH_3.O.CH_3$, $\nu = 5 \times 10^{18}$, with s_1 between 10 and 11.
For C_2H_5CHO, $\nu = 1 \times 10^{18}$, with s_1 about 8.

These values of s_1 are all acceptable. The molecules may be thought of as consisting of at least three loosely bound structural units. The first of these gives a rigid framework to which the others are fitted. Ignoring torsional oscillations each of the others has 3 freedoms in this framework yielding 6 square terms, and the rotations of the whole complex another 3. Values of s_1 up to 15 at least are thus to be expected. There is thus an ample margin for an activation rate slow compared with the number of sufficiently energetic collisions and a disintegration rate slow compared with the rate of activation, even when we do not admit that any part of the internal energy of the other molecule is available.

In the case of the two reactions maintained by a sufficient pressure of H_2, the activating collisions must be supplied by collisions between H_2 and the reacting molecule. There is no theoretical difficulty in this. The difficulty is rather to understand why all diluent molecules do not act in the same way. This point lies deeper than the simple collision theory. It is likely that it is connected with the slow rate of transfer between kinetic and vibrational energy found in the study of the specific heats of simple gases other than H_2 (Chapter III).

We now pass to two entirely different cases.

(4) *Decomposition of gaseous acetone.* This reaction is homogeneous and unimolecular and of convenient speed for the temperatures 780–900° K.

and shows no signs of deviation from the unimolecular law down to pressures of 100 mm. Hg. No diluents tested have been found to have any effect. The nature of the reaction is

$$CH_3 . CO . CH_3 \rightarrow CO + [\text{hydrocarbons}],$$

and the velocity constant observed is

$$\log \kappa = 34{\cdot}95 - 68{,}500/RT. \qquad \text{......(2003)}$$

On carrying out the same calculations as before, we find if we take $\nu = 10^{18}$ and $s_1 = 15$ that there are only just about twice as many possible activations as disintegrations. The unimolecular law could not possibly be maintained on this margin. Even if we assume $s_1 = 24$ (24 square terms) we have only a marginal factor of about 30. There might well be 15 relevant square terms in the acetone molecule, or even so many as 24, but this is hardly enough, and there is no evidence that the reaction does not remain unimolecular to still lower pressures.

We have hitherto confined attention to the internal energy of the molecule which is to be activated in the collision. But this is arbitrary, for there is no *a priori* reason why some or indeed all of the energy in certain coordinates of the other molecule may not in certain circumstances be available for activation. The consequences of such an assumption of availability are considered in the next section in the light of the theory of detailed balancing. Such an assumption makes a large difference, for the fraction of activated molecules in equilibrium depends only on s_1 and is unaltered, but the number of possible activating collisions is largely increased. If we take $s_1 = s_2 = 15$ there is a marginal factor of $3{\cdot}6 \times 10^4$ which is probably ample. A comparatively small value of s_2 will increase the previous margin substantially.

(5) *Decomposition of nitrogen pentoxide.* This, the best known and most exhaustively investigated homogeneous unimolecular reaction, shows no signs of deviation from the unimolecular law down to pressures of 0·06 mm. Hg but at about this pressure it has recently been shown that a distinct falling off in the rate sets in. The convenient temperature range is from 273–340° K. No effect of any diluent in maintaining the rate at pressures below 0·06 mm. Hg has been recorded. Admixtures of numerous gases have been shown to be without effect on the rate of the reaction, the only positive effects being obtained with various organic vapours which are themselves attacked by N_2O_5 and may therefore be disregarded here. The result of the reaction is

$$N_2O_5 \rightarrow N_2O_4 + \tfrac{1}{2}O_2,$$

but of course this does not represent the actual mechanism which probably may be described by the following sequence of processes: (i) $N_2O_5 \rightarrow N_2O_3 + O_2$, unimolecular and slow determining the measured rate; (ii) $N_2O_3 \rightarrow NO + NO_2$, very fast; (iii) $NO + N_2O_5 \rightarrow 3NO_2$ (or $NO_2 + N_2O_4$), also fast. Thus each primary decomposition of N_2O_5 consumes two molecules. The NO_2 and N_2O_4 molecules produced may be assumed to come to dissociative equilibrium. The velocity constant observed is

$$\log \kappa = 31{\cdot}45 - 24{,}700/RT. \qquad \text{......(2004)}$$

For $\nu = 2{\cdot}13 \times 10^{15}$ (0·06 mm. Hg at 0° C.), $\sigma_{12} = 10^{-7}$ and $s_1 = 16$ we find a maximum rate of activation 10 times *less* than the observed rate of disintegration. It is possible to suppose that at the temperatures concerned the energy content of N_2O_5 is even greater than is represented by 16 square terms, but even an increase to 24 square terms only increases the maximum activation rate to twice the observed rate of disintegration. To retain Lindemann's form of the theory it is therefore essential to include in the available energy the energy of the other molecule. If we take $s_1 = 16$, $s_2 = 14$, we find an activation rate 10^3 times the observed rate of disintegration. It is feasible to suppose that either s_1 or s_2 could be greater so that the margin is probably ample. But the conditions for obtaining such a margin are extreme. It is impossible to obtain it, practically speaking, unless the greater part of the internal energy in both molecules is available for the activation energy of one, and such activations occur in something more than one per thousand of all sufficiently energetic collisions.†

§18·5. *The requirements of detailed balancing.* In view of the extreme form of the theory required to account for the decomposition of N_2O_5 it is desirable to examine the consequences of the assumption we have been driven to, that all the energy in a large number of freedoms in both molecules is available for the activation of one, and is actually so used in a fraction of all sufficiently energetic collisions which may be as large as 1/1000 or even possibly 1/100. Obviously not *all* "collisions with enough energy" can be activations, for the class must include all deactivations as well.

When $\eta_1 + \eta_2 + \xi > \eta_0$, a collision with the initial conditions η_1, η_2, ξ *can* yield a pair of molecules with energies between $\eta_1^*, \eta_1^* + d\eta_1^*$ ($\eta_1^* > \eta_0$) and $\eta_2^*, \eta_2^* + d\eta_2^*$, the rest of the energy being absorbed in ξ^*. The effective target for such an exchange, in conformity with the notation of the preceding chapter, will be taken to be

$$S(\eta_1, \eta_2, \xi; \eta_1^*, \eta_2^*)\, d\eta_1^* d\eta_2^*.$$

Then in the equilibrium state (or other state not seriously disturbed from

† The conditions are extreme, but not so extreme as was at one time thought; see for example this section of the first edition of this book.

this) the number of collisions per unit volume and unit time which convert η_1, η_2, ξ into $\eta_1^*, \eta_2^*, \xi^*$ is proportional to

$$\eta_1^{\frac{1}{2}s_1-1}\eta_2^{\frac{1}{2}s_2-1}\xi e^{-(\eta_1+\eta_2+\xi)/kT}\,d\eta_1 d\eta_2 d\xi \times S(\eta_1,\eta_2,\xi;\eta_1^*,\eta_2^*)\,d\eta_1^* d\eta_2^*. \qquad \text{......(2005)}$$

By the same argument the number of collisions which convert $\eta_1^*, \eta_2^*, \xi^*$ into η_1, η_2, ξ is

$$(\eta_1^*)^{\frac{1}{2}s_1-1}(\eta_2^*)^{\frac{1}{2}s_2-1}\xi^* e^{-(\eta_1^*+\eta_2^*+\xi^*)/kT}\,d\eta_1^* d\eta_2^* d\xi^* \times S(\eta_1^*,\eta_2^*,\xi^*;\eta_1,\eta_2)\,d\eta_1 d\eta_2. \qquad \text{......(2006)}$$

By the principle of detailed balancing these must be equal. Moreover, since $\eta_1+\eta_2+\xi=\eta_1^*+\eta_2^*+\xi^*$, for given η's, $d\xi=d\xi^*$. Hence

$$(\eta_1^*)^{\frac{1}{2}s_1-1}(\eta_2^*)^{\frac{1}{2}s_2-1}\xi^* S(\eta_1^*,\eta_2^*,\xi^*;\eta_1,\eta_2) = \eta_1^{\frac{1}{2}s_1-1}\eta_2^{\frac{1}{2}s_2-1}\xi S(\eta_1,\eta_2,\xi;\eta_1^*,\eta_2^*). \qquad \text{......(2007)}$$

If now it is to be possible for the number of activations to be comparable with the number of collisions with enough energy, we must have

$$\iint S(\eta_1,\eta_2,\xi;\eta_1^*,\eta_2^*)\,d\eta_1^* d\eta_2^*$$

of the same order as $\pi\sigma_{12}^2$ (equal to $\alpha\pi\sigma_{12}^2$ say), when $\eta_1+\eta_2+\xi \geqslant \eta_0$ and $\eta_1^* > \eta_0$. For the sake of investigating orders of magnitude we will take S constant over the ranges of η_1^* and η_2^* in which it has a non-zero value. Then, very roughly,

$$S(\eta_1,\eta_2,\xi)\int_{\eta_0}^{\eta_1+\eta_2+\xi} d\eta_1^* \int_0^{\eta_1+\eta_2+\xi-\eta_1^*} d\eta_2^* = \alpha\pi\sigma_{12}^2,$$

or

$$S(\eta_1,\eta_2,\xi) = 2\alpha\pi\sigma_{12}^2/(\eta_1+\eta_2+\xi-\eta_0)^2. \qquad \text{......(2008)}$$

There is nothing unacceptable in (2008). Then by (2007)

$$S(\eta_1^*,\eta_2^*,\xi^*;\eta_1,\eta_2) = \frac{2\alpha\pi\sigma_{12}^2}{(\eta_1+\eta_2+\xi-\eta_0)^2}\left(\frac{\eta_1}{\eta_1^*}\right)^{\frac{1}{2}s_1-1}\left(\frac{\eta_2}{\eta_2^*}\right)^{\frac{1}{2}s_2-1}\frac{\xi}{\xi^*}. \qquad \text{......(2009)}$$

To find the corresponding total collision area for deactivations we have to integrate with respect to η_1 and η_2, the ranges being respectively 0 to η_0 and 0 to η_0 or $\eta_1^*+\eta_2^*+\xi^*-\eta_1$ whichever is the lesser. It is not necessary however to carry out this integration explicitly. We can see at once from (2009) that in certain circumstances the deactivation target must be very large compared with the activation target $\alpha\pi\sigma_{12}^2$. Such a molecular property is not entirely impossible and dismissible on *a priori* grounds. Formula (2009) means that a very slow molecule or a molecule with exceptionally little energy finds it exceptionally easy to bring about deactivation. In the present state of molecular theory all we can do is to bear such possibilities constantly in mind†.

† A valuable critique of the extreme collision theory here presented will be found in Tolman, Yost and Dickinson, *Proc. Nat. Acad. Sci.* vol. 13, p. 188 (1927).

§ **18·51.** *The reverse process for unimolecular reactions.* We have accepted as a mechanism for unimolecular reactions a mechanism of which one stage consists of the spontaneous explosion of an activated molecule into two or more, probably two parts. This stage is closely analogous to the radioactive disintegration of an atomic nucleus by the expulsion of an α-particle. If this mechanism is correct, then the reverse process consists of a collision between two particles leading to a transition which unites them into one. There is no difficulty in supposing that such processes can occur to the necessary frequency. One is tempted to argue that they can practically never occur because the relative kinetic energy will never be adjusted exactly to fit any possible value of the quantized internal energy of the united system, and no third body is concerned in the process to carry away the excess energy. This objection however overlooks the fact that, just because the combined system is unstable and able to explode, its levels cannot be perfectly sharp, and the breadths of its levels will exactly enable the reverse recombinations to occur at the proper rate. It is of course also possible that the process of explosion involves the absorption of a quantum of radiation lying within a wide range of frequencies which could be quite low. In that case the reverse process would involve the emission of a quantum in the same band, and no difficulty would arise even with sharp levels for the united system. No such explanation is necessary however, and any such explanation is perhaps ruled out by the recorded insensitivity of the reactions to radiation, though it is doubtful if the low frequency region has been sufficiently studied for this conclusion to be certain.

§ **18·6.** We have confined the foregoing account of homogeneous gaseous reactions to the simplest collision mechanism. Interesting and important questions are raised by following up the subsequent history of the products of reaction which may often contain excessive amounts of energy, especially when the reaction is thermodynamically exothermic with a large heat of activation. In these cases we shall expect to find the formation of reaction chains and explosion waves. It would take us too far afield to follow up these possibilities which have been studied by numerous investigators and expounded in a notable work by Semenoff.* Nor have we in this chapter discussed photo-sensitive gaseous reactions. Some account of these is given in the next chapter. It should be mentioned that the homogeneous unimolecular gaseous reactions of this chapter have frequently been considered to be due primarily to a radiative process on account of their unimolecular character. It is however almost impossible to maintain such a theory in the face of the abnormal molecular absorption coefficients which it requires,†

* Semenoff, *Chain Reactions*, Oxford Press (1934).

† Christiansen and Kramers, *Zeit. f. physikal. Chem.* vol. 104, p. 451 (1923).

which would have to manifest themselves in the absorption of external radiation. When the attempt is made to modify such a theory by claiming that these abnormal coefficients are only typical of the interaction of two molecules in resonance with one another,* we are really abandoning the radiative theory altogether and again groping after a theory of molecular interaction, of the nature of quantum resonance, which may just as properly be regarded as a collision process. This may be the direction in which an improved theory will one day be found.†

* G. N. Lewis and Smith, *J. Amer. Chem. Soc.* vol. 47, p. 1508 (1925); G. N. Lewis, *Proc. Nat. Acad. Sci.* vol. 13, p. 623 (1927).

† The quantum theory of gas reactions has recently made great advances, the basis of which is described in outline in § 21·8.

CHAPTER XIX

MECHANISMS OF INTERACTION. RADIATIVE PROCESSES

§ **19·1.** *The nature of radiative processes.* The analysis of § 17·12 was arranged to cover both collisions and radiative processes. We recapitulate the conclusions as to the latter. If X_1 and X_2 denote two different states of the same system such that the equilibrium ratio of the concentrations $[X_1]$ and $[X_2]$ is a function of the temperature only, and if P_{rad} and Q_{rad} are the frequencies with which an X_1 is converted to an X_2 and an X_2 to an X_1 by interaction with equilibrium radiation alone, then

$$\frac{P_{\mathrm{rad}}}{Q_{\mathrm{rad}}}=\frac{[X_2]}{[X_1]}. \qquad \text{......(2010)}$$

If the change from X_1 to X_2 is one of dissociation, so that an expression such as $[X_2][Y]/[X_1]$ is a function of the temperature only, then

$$\frac{P_{\mathrm{rad}}}{Q_Y+Q_{Y,\,\mathrm{rad}}}=\frac{[X_2][Y]}{[X_1]}. \qquad \text{......(2011)}$$

Here Q_Y means the frequency of interaction of an X_2 and a Y to form an X_1 *without* any radiative action, and $Q_{Y,\,\mathrm{rad}}$ that of the similar interaction *with* radiative action.

It is now possible to analyse this result a little further and see that in general

$$Q_Y=0.$$

In general there will be a difference of energy between the free X_2+Y and the combined X_1. Consequently no radiationless interaction between X_2 and Y is possible, resulting in X_1 and conserving energy. In general therefore $Q_Y=0$, and we have

$$\frac{P_{\mathrm{rad}}}{Q_{Y,\,\mathrm{rad}}}=\frac{[X_2][Y]}{[X_1]}. \qquad \text{......(2012)}$$

In this interaction of course the radiation can adjust the energy balance. We proceed to examine in detail the consequences of (2010) and (2012). We shall content ourselves with classical statistics; extensions to quantum statistics can easily be made.

§ **19·2.** *Interaction of radiation with the stationary states of fixed atoms.* This problem was the first of such problems to be discussed in this way—by Einstein* in a classical paper. Consider first an atom with only two stationary states 1 and 2, of negative energies χ_1 and χ_2, $\chi_1>\chi_2$, and weights ϖ_1 and ϖ_2. The atom can proceed from state 1 to state 2 with absorption of radiation and from state 2 to state 1 with emission of radiation. This

* Einstein, *Physikal. Zeit.* vol. 18, p. 121 (1917).

radiation will be monochromatic, of frequency ν given by $h\nu=\chi_1-\chi_2$, *Bohr's frequency condition*; we need not at this stage consider the finite sharpness of the upper state. It follows from the properties of the partition function (Boltzmann's law) that the equilibrium ratio of the numbers of atoms in the two states is

$$\frac{\varpi_1}{\varpi_2 e^{-(\chi_1-\chi_2)/kT}}. \qquad \ldots\ldots(2013)$$

We have now to specify the dependence of the rates of emission and absorption of the atoms in either state on the intensity of the radiation of frequency ν. These specifications are of course immediate consequences of the quantum theory, but may be introduced *a priori* as natural assumptions. Let $I(\nu)\,d\nu$ be the intensity of radiation of frequency between ν and $\nu+d\nu$, that is the quantity of energy in radiation of this type which crosses unit area normal to its path per unit solid angle per second. It is connected with the density of this radiation $\rho(\nu)\,d\nu$ by the equation

$$I(\nu)=\frac{c}{4\pi}\rho(\nu). \qquad \ldots\ldots(2014)$$

We may now assume that the chance of absorption of one quantum by an atom in state 1 in time dt is

$$B_1{}^2 I(\nu)\,d\nu dt, \qquad \ldots\ldots(2015)$$

and that the chance of emission of one quantum by an atom in state 2 in time dt is

$$\{A_2{}^1+B_2{}^1 I(\nu)\,d\nu\}\,dt, \qquad \ldots\ldots(2016)$$

the A's and B's being atomic constants. They are commonly referred to as Einstein's A's and B's or Einstein's coefficients. The form of (2015) is the obvious one to choose, that of (2016) is more obscure and the second term might *a priori* be overlooked. It is necessary, as we shall see, for the mechanism to preserve Planck's law for $\rho(\nu)$, and it can be seen to be necessary *a priori* by a deeper consideration of classical* or quantal† radiation theory.

In equilibrium the absorption and emission of quanta and the numbers of switches from $1\to 2$ and $2\to 1$ must be equal. Using therefore (2013), Bohr's frequency condition, (2015) and (2016) this necessary and sufficient equality is equivalent to

$$\varpi_1 B_1{}^2 I(\nu)\,d\nu=\varpi_2 e^{-h\nu/kT}\{A_2{}^1+B_2{}^1 I(\nu)\,d\nu\}. \qquad \ldots\ldots(2017)$$

Solving for $I(\nu)$ we find

$$I(\nu)\,d\nu=\frac{A_2{}^1}{(\varpi_1/\varpi_2)B_1{}^2 e^{h\nu/kT}-B_2{}^1}. \qquad \ldots\ldots(2018)$$

On comparing this with Planck's law,

$$I(\nu)\,d\nu=\frac{2h\nu^3\,d\nu}{c^2}\,\frac{1}{e^{h\nu/kT}-1}, \qquad \ldots\ldots(2019)$$

* Van Vleck, *Phys. Rev.* vol. 24, pp. 330, 347 (1924).

† Dirac, *Quantum Mechanics*, ed. 2, § 67.

we see that the form is correct and equilibrium will be preserved if

$$\varpi_1 B_1{}^2 = \varpi_2 B_2{}^1 = \varpi_2 A_2{}^1 \frac{c^2}{2h\nu^3 d\nu}. \qquad ......(2020)$$

These relations are extremely important.

Some comment is called for on other arrangements of the foregoing argument. It has been arranged here purely to derive the relations (2020) and to verify that our assumptions formulate a possible unit mechanism. We may note too that there is here no difference between preservation of equilibrium and detailed balancing.

In the first place the forms (2015) and (2016), though the only forms consistent with classical or quantal radiation theory, are not only the forms which satisfy (2019). If we replace them by $f_a(I(\nu)\,d\nu)\,dt$ and $f_e(I(\nu)\,d\nu)\,dt$ respectively, the functions f_a and f_e have only to satisfy the necessary and sufficient relation

$$f_e(x) = \frac{2h\nu^3 d\nu/c^2 + x}{x} f_a(x). \qquad ......(2021)$$

In the next place we have actually assumed more results of the equilibrium theory and atomic theory than are strictly necessary for the proof. Einstein, in his original presentation, assumed the forms (2015) and (2016), Wien's displacement law (a theorem of pure thermodynamics) and Boltzmann's law, and *deduced* from these premises Bohr's frequency condition and Planck's law. Eddington* has recast the discussion and taken as his premises Wien's law and Bohr's frequency condition, and *deduced* Boltzmann's law and Planck's law. These theorems however are of interest from points of view different from that adopted in this monograph.

The discussion extends at once to atoms with any number of stationary states. For each pair of stationary states which are connected by a radiative transition there is a set of relations identical with (2020). Moreover this is true whether we work on the hypothesis of preservation or of detailed balancing. The difficulty of § 17·4 does not arise here, for in general we may assume that every possible transition affects radiation of a different frequency.

§ **19·21.** *Numerical values.* We can proceed to estimate numerical values without specific reference to quantum mechanics. The connection between $B_1{}^2$ and the mass absorption coefficient is obtained as follows. Of the energy of radiation

$$I(\nu)\,d\nu\,d\omega \cos\theta\,dS\,dt$$

incident in time dt within a solid angle $d\omega$ on a slab of matter of area dS and thickness dx at an angle θ with the normal to the slab, the fraction

$$k_\nu \rho\,dx \sec\theta$$

* Eddington, *Phil. Mag.* vol. 50, p. 803 (1925).

will be absorbed, ρ being the mass density of the atomic distribution. This expression defines the mass absorption coefficient k_ν. In terms of B_1^2 the energy absorbed is

$$\overline{a_1}\, dx\, dS \times B_1^2 I(\nu)\, d\nu\, dt \frac{d\omega}{4\pi} \times h\nu,$$

where $\overline{a_1}$ is the number of atoms in the state 1 per unit volume. Equating the two amounts we find

$$k_\nu = \frac{B_1^2 h\nu \overline{a_1}}{4\pi\rho}. \qquad \text{......(2022)}$$

We can write $\rho = ma$, where m is the atomic mass and a the total concentration. Then

$$k_\nu = \frac{B_1^2 h\nu \overline{a_1}}{4\pi m a}. \qquad \text{......(2023)}$$

Unless $\overline{a_1}/a$ is small the mass absorption coefficients are extremely large. The direct experimental evidence for the Hg line λ 2536 ($1\,{}^1S$–$2\,{}^3P$), the Ca^+ lines H, K, $\lambda\lambda$ 3968, 3933 ($1\,{}^2S$–$2\,{}^2P$) of the chromosphere and the Na lines D ($1\,{}^2S$–$2\,{}^2P$), all lines for which in the conditions of observation $\overline{a_1}/a = 1$, is that k_ν is of the order 10^9. Einstein's absorption coefficient is then of the order 0·1 to 0·5.

In order to evaluate A_2^1 numerically we have to know the value of $d\nu$. The meaning of this is of course that B_1^2 is a mean absorption coefficient integrated through the line, no line of any set of atoms (even if all at rest) being of mathematically zero breadth. The total absorption is therefore the mean absorption coefficient B_1^2 multiplied by the mean breadth $d\nu$, and it is $B_1^2 d\nu$ or more strictly $\int_{\nu_0-\infty}^{\nu_0+\infty} \alpha_\nu d\nu$ which is correlated with the emission coefficient A_2^1. If we estimate on observational evidence the ordinary line breadth at 10^{-10} cm., then for $\lambda = 3000$, $\delta\nu = 3 \times 10^9$, $B_1^2 d\nu = 1{\cdot}5 \times 10^9$. The numerical order of A_2^1 is then 2×10^7. There is direct experimental confirmation of such a value, for we observe that $1/A_2^1$ is the mean life of the atom in the excited state 2 before it spontaneously radiates and returns to the state 1. This mean life τ is therefore approximately

$$\tau = 5 \times 10^{-8} \text{ sec.}$$

Mean lives of this order or rather shorter are determined by experiments such as Wien's on the light emitted by streams of positive rays and by the theory of the chromosphere. This is also the mean time of radiation suggested by classical radiation theory and confirmed by quantum mechanics.

The stimulated emissions are often of negligible importance numerically. The ratio of stimulated to spontaneous emissions is

$$\frac{B_2^1 I(\nu)\, d\nu}{A_2^1} = \frac{c^2 I(\nu)}{2h\nu^3} = \frac{1}{e^{h\nu/kT} - 1}.$$

For the region of the spectrum near λ_{max} $h\nu/kT$ is about 5, so this fraction is negligible and *a fortiori* for all higher frequencies. For very low frequencies the fraction approximates to $kT/h\nu$ and the stimulated emissions become important.

The A's and B's and the natural widths of spectral lines can now all be calculated in principle by quantum mechanics for all systems. Complication prevents the actual execution of the calculation for any but the simplest spectra, but where the calculations have been carried out, as for hydrogen, helium and the alkalis,* the results are in full accord with the experimental evidence.

§ 19·3. *The photoelectric liberation of electrons from fixed atoms.*† Besides the line emission and absorption spectrum of an atom, associated with transitions between stationary states, there is also a continuous emission and absorption spectrum associated with capture and loss of electrons. We will again consider the atoms as fixed, and at first for simplicity as possessing a single stationary state of negative energy χ and weight ϖ_0. The process to be analysed consists of absorption of radiation of frequency ν such that $h\nu > \chi$ with the ejection of an electron of velocity v or energy η such that

$$\eta = \tfrac{1}{2}mv^2 = h\nu - \chi, \qquad \text{......(2024)}$$

which is Einstein's law of the photoelectric effect. The relic of the atom is an ion in its normal (at present sole) stationary state of weight ϖ_1. If y is the concentration of free electrons and x_0, x_1 the concentrations of (single state) atoms and ions, the equilibrium state is characterized by

$$\frac{x_1 y}{x_0} = \frac{(2\pi mkT)^{\frac{3}{2}}}{h^3} \frac{2\varpi_1}{\varpi_0} e^{-\chi/kT}. \qquad \text{......(2025)}$$

To obtain this result we use (479) and (1643), the $u_r^z(T)$ being here all unity. The mechanism is controlled by (2012) and the reverse process is the encounter of an ion and a free electron resulting in capture with emission of radiation.

Let $\psi(\nu) I(\nu) d\nu dt$ be the chance that a neutral atom will in time dt, under the influence of isotropic ν-radiation of intensity $I(\nu) d\nu$, become ionized by absorption of a quantum $h\nu$. The total number of ν-quanta absorbed in time dt will therefore be (per unit volume)

$$x_0 \psi(\nu) I(\nu) d\nu dt. \qquad \text{......(2026)}$$

* The first calculations of this type were given by Schrödinger, *Ann. d. Physik*, vol. 80, p. 437 (1926); Heisenberg, *Zeit. f. Physik*, vol. 39, p. 499 (1926); Sugiura, *J. de Physique*, vol. 8, p. 113 (1927), and *Phil. Mag.* vol. 4, p. 495 (1927).

† Milne, *Phil. Mag.* vol. 47, p. 209 (1924). This is the most complete original account. See also Kramers, *Phil. Mag.* vol. 46, p. 836 (1923), and R. Becker, *Zeit. f Physik*, vol. 18, p. 325 (1923).

By (1898) the number of collisions in time dt between η-electrons and ions in which the line of impact lies between p and $p+dp$ is

$$x_1 \,.\, 2\pi p\, dp \,.\, \left(\frac{2\eta}{m}\right)^{\frac{1}{2}} \mu(\eta)\, d\eta\, dt.$$

Let the probability that such an encounter results in capture with emission of ν-radiation be

$$f(p,\eta) + I(\nu)\, g(p,\eta).$$

We shall find that stimulated captures are necessary here as for line emission and absorption in order to conserve Planck's law. Then the total number of captures with emission of ν-radiation in time dt is

$$x_1 \,.\, 2\pi p\, dp \,.\, \{f(p,\eta) + I(\nu)\, g(p,\eta)\} \left(\frac{2\eta}{m}\right)^{\frac{1}{2}} \mu(\eta)\, d\eta\, dt.$$

In order to conserve energy η and ν are connected by (2024) so that we have also

$$d\eta = h\, d\nu.$$

Inserting the equilibrium value of $\mu(\eta)$ from (1863) and writing

$$F(\eta) = 2\pi \int_0^\infty p f(p,\eta)\, dp, \quad G(\eta) = 2\pi \int_0^\infty p g(p,\eta)\, dp,$$

we find the total number of captures of η-electrons in time dt with emission of ν-radiation to be

$$\frac{2\pi x_1 y}{(\pi k T)^{\frac{3}{2}}} \{F(\eta) + I(\nu)\, G(\eta)\} \left(\frac{2}{m}\right)^{\frac{1}{2}} \eta e^{-\eta/kT}\, d\eta\, dt. \qquad \text{......(2027)}$$

Either for preservation or detailed balancing (2026) and (2027) must be equated. On using (2025), (2024) and its resulting differential relation this gives

$$\eta F(\eta) = \frac{2h\nu^3}{c^2} \eta G(\eta) = \frac{h^3 \nu^3}{4\pi m c^2} \frac{\varpi_0}{2\varpi_1} \psi(\nu), \qquad \text{......(2028)}$$

or

$$\psi(\nu) = \frac{4\pi m c^2}{h^3 \nu^3} \frac{2\varpi_1}{\varpi_0} \eta F(\eta) \quad (\eta = h\nu - \chi). \qquad \text{......(2029)}$$

Of course

$$\psi(\nu) = 0 \quad (\nu < \chi/h).$$

These relations (2028) are independent of the classical formulation adopted in this paragraph and remain true in quantum mechanics. The ratio of stimulated to spontaneous captures is again $1/(e^{h\nu/kT} - 1)$, and is still less important than for line emission, since in this process we are never concerned with frequencies less than χ/h.

§ **19·31.** *Extensions to complicated atoms.* If we consider more complicated atoms with more than one stationary state, then we can always write the equilibrium relation between single states of the neutral atom and ion in the form*

$$\frac{(x_1)_t y}{(x_0)_s} = \frac{(2\pi m k T)^{\frac{3}{2}}}{h^3} \frac{2(\varpi_1)_t}{(\varpi_0)_s} e^{-\chi_s^t/kT}, \qquad \text{......(2030)}$$

* First pointed out explicitly by Milne, *Phil. Mag.* vol. 50, p. 547 (1925).

where $\chi_s{}^t$ is the ionization energy required to remove an electron from the neutral atom in its sth state and leave the ion in its tth state. This can be effected by radiation of frequency ν ejecting an η-electron, provided now that

$$h\nu = \eta + \chi_s{}^t.$$

We introduce exactly the same coefficients $F(\eta)$, $G(\eta)$ and $\psi(\nu)$ as before for each such photoelectric process. On the principle of detailed balancing it follows at once that the relations (2028) hold for each set of these coefficients. On the preservation hypothesis matters are rather complicated and it is hardly of sufficient interest to discuss them in detail. The balancing of ν-quanta and η-electrons involves in general more than one set of coefficients. Only the atomic balance involves a single set (and then only when the ion is assumed to have only a single state), and the condition of atomic balance for all T does not imply so much as (2028). We may be content to expect that, as in § 17·4, preservation could be shown to demand less restrictive conditions than detailed balancing.

It is obvious that the foregoing argument applies to any two consecutive stages of ionization.

§ **19·32.** *Free-free transitions.* In addition to radiative captures transitions from one free orbit to another are possible with emission or absorption of radiation. The greater part of "white" X-radiation is of this nature. The probability of these transitions can in principle be calculated by quantum mechanics, but the calculations are very complicated. In the meantime the laws for such processes can easily be formulated, but do not give much significant information, since the process and its reverse are essentially the same, with change of sign of ν.

By (1898) and (1863) the number of encounters per atom per unit time with (η,p)-electrons is

$$2\pi p\,dp\,.\left(\frac{2}{m}\right)^{\frac{1}{2}}\frac{2\pi y}{(\pi kT)^{\frac{3}{2}}}e^{-\eta/kT}\eta\,d\eta.$$

The chance of a switch to an (η',p')-electron with absorption of ν-radiation, $h\nu = \eta' - \eta$, is naturally taken to be

$$I(\nu)\,\sigma(\eta,p;\eta',p')\,d\eta'\,.\,2\pi p'dp'.$$

If

$$S_a(\eta,\eta') = 4\pi^2\int_0^\infty\int_0^\infty pp'\sigma(\eta,p;\eta',p')\,dp\,dp',$$

then the number of switches of η-electrons to η'-electrons per atom per unit time with absorption of ν-radiation is

$$\left(\frac{2}{m}\right)^{\frac{1}{2}}\frac{2\pi y}{(\pi kT)^{\frac{3}{2}}}e^{-\eta/kT}I(\nu)\,\eta S_a(\eta,\eta')\,d\eta\,d\eta', \qquad \text{......(2031)}$$

a result which remains true in quantum mechanics. In exactly the same way we must then take the number of switches of η'-electrons to η-electrons per atom per unit time with emission of ν-radiation to be

$$\left(\frac{2}{m}\right)^{\frac{1}{2}}\frac{2\pi y}{(\pi kT)^{\frac{3}{2}}}e^{-\eta'/kT}\{1+\alpha I(\nu)\}\eta' S_e(\eta',\eta)\,d\eta\,d\eta'. \qquad \text{......(2032)}$$

On the principle of detailed balancing we must equate (2031) and (2032) obtaining

$$e^{h\nu/kT}\eta S_a(\eta,\eta')\,I(\nu)=\eta' S_e(\eta',\eta)\{1+\alpha I(\nu)\}. \qquad \text{......(2033)}$$

If $I(\nu)$ is to satisfy Planck's law, then we must have

$$\alpha=c^2/2h\nu^3, \qquad \text{......(2034)}$$

$$\eta S_a=\frac{c^2}{2h\nu^3}\eta' S_e. \qquad \text{......(2035)}$$

The analogous process to this in Chapter XVII would be merely a simple 3-body encounter with energy exchanges between the 3 bodies, from which we should have derived little of importance.

§ **19·33.** *Numerical values of continuous absorption coefficients.* Little is known experimentally or from direct astrophysical evidence as to the numerical values of the mass absorption coefficient which is derived from the $\psi(\nu)$ of (2029) by the equation analogous to (2023),

$$k_\nu=\frac{h\nu}{4\pi m}\,\frac{\Sigma\,\psi^{(s)}(\nu)\,\overline{a_s}}{a}. \qquad \text{......(2036)}$$

We do know that the continuous absorption grades off continuously into the massed line absorption at the series limit and that there is no infinity at the limit itself. Hence $\underset{\eta\to 0}{\mathrm{Lt}}\,\eta F(\eta)$ must be finite and non-zero. This means that the chance of capture of a very slow electron must ultimately vary like $1/\eta$ or $1/v^2$. It does not imply however that $\psi(\nu)\propto\nu^{-3}$ and $k_\nu\propto\nu^{-2}$, and in fact these relations do not seem to be true. The X-ray evidence is that $k_\nu\propto\nu^{-2\cdot 6}-\nu^{-3}$. Direct calculations of $\psi(\nu)$ and so k_ν for atoms with one electron, which probably apply to the absorption of X-rays and roughly to the optical absorption by any simple atom or ion, have now been made.*

§ **19·4.** *General processes involving emission and absorption of radiation.* We have considered hitherto only simple radiative processes in which a single quantum is absorbed or emitted. In order to generalize these considerations on the basis of detailed balancing to apply to processes involving two or more quanta such as the Compton effect (scattering of radiation by free electrons) a convenient way is to generalize the method of § 17·71 to include radiation,† at first a single quantum only. To define the encounters

* Oppenheimer, *Zeit. f. Physik*, vol. 41, p. 268 (1927); Sugiura, *loc. cit.*

† Dirac, *loc. cit.* (2). First discussed from the present point of view by Pauli, *Zeit. f. Physik*, vol. 18, p. 272 (1923).

we must now specify in addition the solid angle $d\Omega$ within which the radiation is directed, and also, when continuous ranges of frequencies are concerned, the frequency range $d\nu$. These specifications are first required in the normal frame of reference $(d\Omega_0, d\nu_0)$, the momentum of the radiation being taken into account if it is significant. As before the probability coefficients ϕ and ϕ' are to be independent of **C** and therefore also of $I(\nu)$, though they may of course depend on the direction of the radiation. We find as before that the only simple assumption is that, in the normal frame, the frequency of the absorption process is proportional to $I_0(\nu_0)$ and the frequency of the emission process to $\{1+\alpha I_0(\nu_0)\}$. The equation of detailed balancing, see (1965), now reduces in the normal frame to

$$I_0(\nu_0)\,[f_1 f_2 \dots f_n]_0 \phi = \{1+\alpha I_0(\nu_0)\}\,[f_1' f_2' \dots f_{n'}']_0 \phi'.$$

Since $I(\nu)/\nu^3$ is invariant under a Lorentz transformation* this reduces, in the original frame in which the assembly as a whole is at rest, to

$$\frac{1}{\alpha} f_1 f_2 \dots f_n \phi = \left\{\frac{\nu^3}{\alpha \nu_0^3 I(\nu)} + 1\right\} f_1' f_2' \dots f_{n'}' \phi'. \qquad \dots\dots(2037)$$

It is of course only this form for the ratios of the emission and absorption

* See, for example, Einstein, *Physikal. Zeit.* vol. 18, p. 121 (1917).

The proof is simple and may be repeated here for reference. In the original frame K, in which the assembly as a whole is at rest, the radiation is isotropic in all frequencies and of intensity in a given range and direction

$$I(\nu)\, d\nu\, d\omega.$$

Consider a system at rest in a frame K' moving with velocity v in the frame K along the x-axis. A given bundle of radiation of intensity $I(\nu)\, d\nu\, d\omega$ in K will belong to an interval $d\nu'$ and solid angle $d\omega'$ and be of intensity

$$I'(\nu', \theta')\, d\nu'\, d\omega'$$

in K', where θ' is the angle between the x'-axis and $d\omega'$.. Obviously in K', I' is not isotropic but depends on θ', not on ϕ'.

Between these two expressions for the intensity or energy density of a given beam there must obviously be the same relation as between the squares of the corresponding frequencies, so that

$$\frac{I'(\nu', \theta')\, d\nu'\, d\omega'}{I(\nu)\, d\nu\, d\omega} = \left(\frac{\nu'}{\nu}\right)^2.$$

The transformation equations (2050) give us at once Doppler's law and the aberration in the form

$$\nu' = \frac{\nu\{1-(v/c)\cos\theta\}}{(1-v^2/c^2)^{\frac{1}{2}}}, \quad \cos\theta' = \frac{\cos\theta - v/c}{1-(v/c)\cos\theta}.$$

It follows that

$$\frac{d\nu'}{d\nu} = \frac{\nu'}{\nu}, \quad \frac{d\omega'}{d\omega} = \frac{d\cos\theta'}{d\cos\theta} = \left(\frac{\nu}{\nu'}\right)^2.$$

Therefore (to all orders)

$$\frac{I'(\nu', \theta')}{I(\nu)} = \left(\frac{\nu'}{\nu}\right)^3.$$

The full expression for $I'(\nu', \theta')$ in terms of ν' and θ' correct to v/c is

$$I'(\nu', \theta') = \left\{I(\nu') + \frac{v}{c}\nu' \cos\theta' \frac{\partial I}{\partial \nu}\right\}\left(1 - 3\frac{v}{c}\cos\theta'\right). \qquad \dots\dots(2038)$$

An alternative method is to apply the general theorem used by Dirac (see footnote to § 17·71, p. 696) that for any set of systems (here quanta) the law of density-in-momentum is invariant for a Lorentz transformation. If the density-in-momentum of ν-quanta per unit volume in K is

$$\mu(\nu, \theta, \phi)\, \nu^2 d\nu\, d\omega,$$

coefficients which it is necessary and sufficient to postulate. Writing Planck's law in the form

$$\frac{2h}{c^2}\frac{\nu^3}{I(\nu)}+1=e^{h\nu/kT},$$

we see that, if

$$\alpha=\frac{c^2}{2h\nu_0{}^3}, \qquad \text{......(2039)}$$

the equation (2037) of detailed balancing reduces to

$$\frac{2h\nu_0{}^3}{c^2}f_1f_2\ldots f_n\phi=e^{h\nu/kT}f_1'f_2'\ldots f_{n'}'\phi'. \qquad \text{......(2040)}$$

Only in this form can the relation be satisfied with coefficients ϕ and ϕ' independent of T and **C**. It will be observed that ν_0, the frequency of the necessary radiation, is measured in the normal frame and so is a constant of the process independent of **C**. The rest of the argument proceeds exactly as in § 17·71 except that the energy relation is

$$h\nu+\sum_1^n\left\{\epsilon_r+\frac{1}{2m_r}(p_1{}^2+p_2{}^2+p_3{}^2)_r\right\}=\sum_1^{n'}\left\{\epsilon_r'+\frac{1}{2m_r'}(p_1'^2+p_2'^2+p_3'^2)_r\right\}.$$

The final result of the introduction of the equations of equilibrium is now

$$\frac{\phi'}{\phi}=\frac{2h\nu_0{}^3}{c^2}\frac{\varpi_1\ldots\varpi_n}{\varpi_1'\ldots\varpi_{n'}'}(h^3)^{n'-n}. \qquad \text{......(2041)}$$

We have considered above a process and its reverse in which a single quantum is absorbed and emitted respectively. It is easy to generalize the argument to cover processes in which any number of quanta are absorbed and emitted with the corresponding emissions and absorptions in the reverse process. In the normal frame the equation of detailed balancing must obviously take the form

$$\Pi_s\{I_0(\nu_0{}^{(s)})\}\,\Pi_t\{1+\alpha_tI_0(\nu_0{}^{(t)})\}[f_1f_2\cdots f_n]_0\phi$$
$$=\Pi_s\{1+\alpha_sI_0(\nu_0{}^{(s)})\}\,\Pi_t\{I_0(\nu_0{}^{(t)})\}[f_1'f_2'\cdots f_{n'}']_0\phi', \qquad \text{......(2042)}$$

then $\mu(\nu,\theta,\phi)$ is invariant. But since we are reckoning in quanta $\mu(\nu,\theta,\phi)\,\nu^2d\nu d\omega$ and $\{I(\nu)/\nu\}\,d\nu d\omega$ are proportional. Therefore $I(\nu)/\nu^3$ is invariant.

We therefore give a proof of this general invariance theorem, which is of course only a rearrangement of the proof above. Let the four space-time momenta of the particle of rest mass m_0 be m_1, m_2, m_3 and m_4 connected by the relation

$$m_4{}^2-(m_1{}^2+m_2{}^2+m_3{}^2)=m_0{}^2c^2.$$

The density-in-momentum per unit volume of a set of particles is $\mu(m_1,m_2,m_3)\,dm_1dm_2dm_3$. This is the number per unit volume of a specified group of particles and therefore transforms according to the same law as the reciprocal of a volume moving with the velocity of the particles, that is, the same law as m_4. Hence

$$\frac{\mu(m_1,m_2,m_3)\,dm_1dm_2dm_3}{m_4}$$

is invariant. But by the equations of transformation

$$J\left(\frac{m_1',m_2',m_3'}{m_1,m_2,m_3}\right)=\frac{m_4'}{m_4}.$$

Therefore $dm_1dm_2dm_3/m_4$ is invariant, so that $\mu(m_1,m_2,m_3)$ is invariant, which is the theorem.

which reduces in the original frame to

$$\Pi_s\left(\frac{1}{\alpha_s}\right)\Pi_t\left\{\frac{(\nu^{(t)})^3}{\alpha_t(\nu_0^{(t)})^3 I_0(\nu^{(t)})}+1\right\}f_1f_2\dots f_n\phi$$
$$=\Pi_s\left\{\frac{(\nu^{(s)})^3}{\alpha_s(\nu_0^{(s)})^3 I_0(\nu^{(s)})}+1\right\}\Pi_t\left(\frac{1}{\alpha_t}\right)f_1'f_2'\dots f_{n'}'\phi'. \quad\dots\dots(2043)$$

Provided that all the "coefficients of stimulation" α satisfy

$$\alpha_s=\frac{c^2}{2h(\nu_0^{(s)})^3},\quad \alpha_t=\frac{c^2}{2h(\nu_0^{(t)})^3}, \quad\dots\dots(2044)$$

equation (2043) reduces, on using Planck's law, to

$$\Pi_s\frac{2h(\nu_0^{(s)})^3}{c^2}e^{h(\Sigma_t\nu^{(t)})/kT}f_1f_2\dots f_n\phi$$
$$=\Pi_t\frac{2h(\nu_0^{(t)})^3}{c^2}e^{h(\Sigma_s\nu^{(s)})/kT}f_1'f_2'\dots f_{n'}'\phi'. \quad\dots\dots(2045)$$

The energy equation is

$$\Sigma_s h\nu^{(s)}+\sum_1^n\left\{\epsilon_r+\frac{1}{2m_r}(p_1^2+p_2^2+p_3^2)\right\}$$
$$=\Sigma_t h\nu^{(t)}+\sum_1^{n'}\left\{\epsilon_r'+\frac{1}{2m_r'}(p_1'^2+p_2'^2+p_3'^2)\right\},$$

and the final relation between the ϕ's,

$$\frac{\phi'}{\phi}=\frac{\Pi_s\dfrac{2h(\nu_0^{(s)})^3}{c^2}}{\Pi_t\dfrac{2h(\nu_0^{(t)})^3}{c^2}}\frac{\varpi_1\dots\varpi_n}{\varpi_1'\dots\varpi_{n'}'}(h^3)^{n'-n}. \quad\dots\dots(2046)$$

It is easily seen that in no other way can we eliminate T and **C** from the equation of detailed balancing and so obtain a statistically acceptable relation between the ϕ's. It is not necessary to take the exact form chosen for the various emission and absorption coefficients as functions of $I_0(\nu_0)$. Only the ratio of each pair is relevant and determinate. If however the form chosen is accepted, then we must accept also the following general law:

Any atomic process which results in the emission of one or more quanta of radiation is stimulated by external incident radiation of the same frequency as that of any of the emitted quanta, the ratio of the stimulated to the spontaneous emission being proportional to the intensity of the incident radiation divided by the cube of its frequency and independent of the nature of the process concerned, the direction of the stimulated radiation being the same as that of the incident radiation.

§ **19·41.** *The Debye-Compton effect.** Compton's process of the scattering of radiation by free electrons is an example of a process to which the foregoing general theory will apply. In this case there is one material system,

* A. H. Compton, *Phys. Rev.* vol. 21, p. 483 (1923); Debye, *Physikal. Zeit.* vol. 24, p. 161 (1923).

$n'=n=1$, one quantum of frequency ν is absorbed and one quantum of frequency ν' is emitted, the difference between ν and ν' being of course controlled by the laws of conservation of energy and momentum. The weights ϖ_1 and ϖ_1' are equal. Consequently the scattering process will preserve the equilibrium state (Maxwell's law for the electrons and Planck's law for the radiation) if

$$\frac{\phi'}{\phi}=\left(\frac{\nu_0}{\nu_0'}\right)^3. \qquad \text{......(2047)}$$

Here ν_0 and ν_0' are the frequencies of the incident and scattered radiation respectively in the normal frame. By (2042) the chance of one electron scattering ν_0-radiation into ν_0'-radiation must be

$$\left\{I_0(\nu_0)+\frac{c^2}{2h\nu_0'^3}I_0(\nu_0)\,I_0(\nu_0')\right\}\phi, \qquad \text{......(2048)}$$

which is the result first given by Pauli. Reduced to the ordinary frame, the chance of the elementary scattering process must be

$$\left\{\frac{I(\nu)}{\nu^3}+\frac{c^2}{2h}\frac{I(\nu)}{\nu^3}\frac{I(\nu')}{\nu'^3}\right\}\phi\nu_0^3. \qquad \text{......(2049)}$$

Further details of the interaction are not necessary for formulating its laws, but it is perhaps desirable to summarize the details here for reference. In the original frame we have initially a quantum with energy $E=h\nu$, and (vector) momentum $\mathbf{\Gamma}=h\nu/c$ and an electron with (vector) momentum

$$\mathbf{G}=m_0\mathbf{v}(1-\beta^2)^{-\frac{1}{2}} \quad (\beta=|\mathbf{v}|/c),$$

and energy

$$U=m_0c^2\left\{1+\frac{\mathbf{G}^2}{m_0^2c^2}\right\}^{\frac{1}{2}}=\frac{m_0c^2}{(1-\beta^2)^{\frac{1}{2}}}.$$

The quantum and the electron "collide" and go off with a new frequency ν' and new velocity $\mathbf{v}'$ in new directions; energy and momentum are conserved, so that

$$\mathbf{G}+\mathbf{\Gamma}=\mathbf{G}'+\mathbf{\Gamma}', \quad E+U=E'+U'.$$

We are not concerned here with the distribution of scattering with angle. If the electron is initially at rest in the ordinary frame we derive at once Compton's formula for the change of frequency of the radiation scattered at an angle θ with the direction of the primary beam,

$$\nu\nu'(1-\cos\theta)=\frac{m_0c^2}{h}(\nu-\nu'),$$

or

$$\lambda'-\lambda=\frac{h}{m_0c}(1-\cos\theta).$$

In order to understand the nature of the interaction we reduce to the normal frame. The transformation equations are of course typified by

$$(\Gamma_x)_0 = \frac{\Gamma_x - \mathbf{C}E/c^2}{(1-\mathrm{B}^2)^{\frac{1}{2}}}, \quad (\Gamma_y)_0 = \Gamma_y, \quad (\Gamma_z)_0 = \Gamma_z, \quad E_0 = \frac{E - \mathbf{C}\Gamma_x}{(1-\mathrm{B}^2)^{\frac{1}{2}}} \quad (\mathrm{B} = \mathbf{C}/c), \qquad \text{......(2050)}$$

with similar equations for $\mathbf{G}$ and U. In the normal frame there is no resultant momentum, so that

$$\mathbf{G}_0 + \mathbf{\Gamma}_0 = \mathbf{G}_0' + \mathbf{\Gamma}_0' = 0.$$

To reduce to the normal frame we choose the x-axis in the direction of $\mathbf{G} + \mathbf{\Gamma}$ and take

$$\mathbf{C} = c^2 |\mathbf{G} + \mathbf{\Gamma}| / (E + U).$$

When the total momentum is zero

$$h\nu_0 = m_0 c^2 \beta_0 / (1 - \beta_0^2)^{\frac{1}{2}},$$

and from the energy relation $(E_0 + U_0 = W_0)$

$$\beta_0 = \frac{(W_0/m_0c^2)^2 - 1}{(W_0/m_0c^2)^2 + 1}.$$

It follows that β_0, and therefore v_0, and therefore ν_0 are fixed by W_0, so that in the normal frame

$$\nu_0 = \nu_0', \quad v_0 = v_0'.$$

The nature of the interaction in the normal frame is thus simply that the quantum and the electron make a head-on collision and rebound with directly reversed momenta. The chance of the scattering process in the normal frame must be

$$\left\{ \frac{I(\nu_0)}{\nu_0^3} + \frac{c^2}{2h} \left(\frac{I(\nu_0)}{\nu_0^3} \right)^2 \right\} \phi \nu_0^3.$$

§ **19·42.** *Further inferences from Dirac's discussion.* The discussion of § 19·4 proceeds throughout on the basis of detailed balancing, and the requirement that the atomic laws in the normal frame shall be independent of T and $\mathbf{C}$. Certain of the conclusions implicitly drawn there deserve further emphasis.

(1) Every quantum of radiation concerned in every process must possess *both energy and momentum.* For every one must transform according to the usual Lorentz transformation in order that the atomic laws may be independent of $\mathbf{C}$. Thus all emitted quanta are emitted in a definite direction.

(2) All processes, radiative or not, contemplated in § 19·4 preserve Maxwell's law for the density-in-momentum of material systems and Planck's law for the radiation. Thus in particular if we generalize the discussions of § 19·2 and § 19·3 and allow the atoms to move (thereby exchanging momentum as well as energy with the quanta), the atoms will take up (as they must) Maxwell's distribution law.

We have arrived at these important results on the basis of detailed balancing and the use of very general arguments. On account of their importance it is worth while to pause here and consider the more specialized but more direct arguments by which Einstein in his original investigation was led to the conclusion that in the process of line absorption and emission every quantum must be directed (have momentum), and that then Maxwell's velocity distribution law for the atoms would be preserved. Einstein considered only the mean square atomic velocity. We shall give the discussion as completed by Milne.* From the present point of view this discussion may be regarded as a study of the process of line absorption and emission by atoms free to move on the assumption of the preservation hypothesis. It is satisfactory to confirm all the conclusions on this narrower hypothesis in the simple case of an atom with one pair of stationary states.

§ **19·5.** *Extensions of Einstein's argument to free atoms and the conditions for the preservation of Maxwell's law by line absorption and emission.* In a frame in which an atom is at rest the radiation is as we have seen not isotropic. The atom meets more radiation from ahead than behind, so that absorption tends to slow it up while the emission (in this frame) is isotropic and without mean effect. There is therefore on the average a deceleration to the first order proportional to v, so that

$$\frac{dv}{dt} = -\lambda v.$$

The actual value of λ will be investigated later. We might at first suppose that each atom, apart from atomic collisions, would ultimately come to rest, but this is not so. Superposed on the steady deceleration there will be in any given time interval τ a net gain of velocity u, arising from fluctuations in the directions of absorption and emission. The mean value of u by definition is zero, and obviously to the first order in v independent of the slight anisotropy of the radiative field—that is independent of v. It is then clear that a necessary condition for the equilibrium state is that

$$\overline{(ve^{-\lambda\tau} + u)^2} = \overline{v^2}, \qquad \text{......(2051)}$$

which means that $\overline{v^2}$ is unaltered after time τ by the two radiative effects. The further development of these ideas belongs more properly to Chapter XX, but may be conveniently taken up at once. Since $\overline{u} = 0$ and the individual values of u are independent of v, $\overline{uv} = 0$ and the equation (2051) reduces to

$$\overline{u^2} = \overline{v^2}(1 - e^{-2\lambda\tau}), \qquad \text{......(2052)}$$

or for small τ

$$\overline{v^2} = \overline{u^2}/2\lambda\tau. \qquad \text{......(2053)}$$

* Milne, *Proc. Camb. Phil. Soc.* vol. 23, p. 465 (1926). Compare the discussion of *displacements* in § 20·62.

The direct calculation of $\overline{u^2}$ is also simple, but is temporarily postponed. In the foregoing, velocities may be interpreted everywhere as velocity components in any given direction.

When u^2 and λ have been calculated directly from the properties of the assumed mechanism $\overline{v^2}$ is determined by (2053), and Einstein showed that it had its proper equilibrium value. This verification is not however, as Einstein pointed out, a complete solution to the problem. But further detailed investigation of this particular mechanism is rendered unnecessary by the following general theorem.*

Theorem 19·5. *If the centre of mass of the atoms moving with velocity (or velocity component) v moves according to the equation*

$$\frac{dv}{dt} = -\lambda v, \qquad \text{......(2054)}$$

and if in addition in any small interval τ each atom acquires a velocity (or velocity component) increment u, independent of v, such that $\overline{u} = 0$, then in the steady state the velocity (or velocity component) distribution function $f(v)\,dv$ is given by

$$f(v) = ce^{-\frac{1}{2}v^2/\overline{v^2}}, \qquad \text{......(2055)}$$

where $\overline{v^2}$ is given by (2053) *and c is a constant.*

This theorem is of somewhat wide importance since it applies under the conditions stated, which need not correspond to statistical equilibrium. In the present application it justifies confining a discussion of the particular mechanism to the values of λ and $\overline{u^2}$. We take its proof next, confining the discussion to one dimension, or one velocity component.

Of the atoms moving with velocity v at a given instant let the fraction $\phi(v,w)\,dw$ acquire increments of velocity between w and $w+dw$ during the succeeding interval τ. Consider the atoms at the end of this interval which are moving with velocities between v' and $v'+dv'$. Those of them which were originally v-atoms have had an increment w, where

$$w = v' - v, \quad dw = dv';$$

they therefore form the fraction $\phi(v,v'-v)\,dv'$ of the v-atoms, which were originally a fraction $f(v)\,dv$ of the whole. Hence if $F(v')\,dv'$ is the new distribution function,

$$F(v')\,dv' = dv' \int_{-\infty}^{+\infty} f(v)\,\phi(v,v'-v)\,dv. \qquad \text{......(2056)}$$

The condition for the preservation of equilibrium (or more generally a steady state) is $F \equiv f$ or

$$f(v') = \int_{-\infty}^{+\infty} f(v)\,\phi(v,v'-v)\,dv. \qquad \text{......(2057)}$$

* Milne, *loc. cit.*, whose account we follow. See also Fokker, *Ann. d. Physik*, vol. 43, p. 810 (1914); Planck, *Berl. Sitz.* p. 324 (1917).

The increment w is given by

$$w=(ve^{-\lambda\tau}-v)+u,$$

or for τ small

$$w=-\lambda\tau v+u.$$

The function $\phi(v,-\lambda\tau v+u)\,du$ is by definition the fraction of v-atoms which acquire by fluctuations increments between u and $u+du$ in time τ. By hypothesis this fraction is independent of v, and we therefore write

$$\phi(v,-\lambda\tau v+u)=\psi(u). \qquad \text{......(2058)}$$

The function $\psi(u)$ satisfies

$$\psi(u)=\psi(-u), \quad \int_{-\infty}^{+\infty}\psi(u)\,du=1, \quad \int_{-\infty}^{+\infty}u\psi(u)=\overline{u}=0,$$

and we shall write

$$\int_{-\infty}^{+\infty}u^2\psi(u)\,du=\overline{u^2}.$$

In order to make use of (2058), and this is the point of Milne's proof, we make the substitution

$$v=(v'-x)/(1-\lambda\tau).$$

Then equation (2057) becomes

$$f(v')=\int_{-\infty}^{+\infty}f\left(\frac{v'-x}{1-\lambda\tau}\right)\phi\left(\frac{v'-x}{1-\lambda\tau},\,-\lambda\tau\frac{v'-x}{1-\lambda\tau}+x\right)\frac{dx}{1-\lambda\tau},$$

and after using (2058) this becomes

$$f(v')=\frac{1}{1-\lambda\tau}\int_{-\infty}^{+\infty}f\left(\frac{v'-x}{1-\lambda\tau}\right)\psi(x)\,dx. \qquad \text{......(2059)}$$

We can now for small τ expand f in powers of x and integrate term by term. We find after rearrangement

$$-f(v')=\frac{1}{\lambda\tau}\left[f\left(\frac{v'}{1-\lambda\tau}\right)-f(v')+\frac{\frac{1}{2}\overline{u^2}}{(1-\lambda\tau)^2}f''\left(\frac{v'}{1-\lambda\tau}\right)+\dots\right].$$

It is necessary to assume that $\overline{u^4}, \overline{u^6}, \dots$ are of a higher order in τ than u^2. Letting now $\tau\to 0$ we find that

$$\tfrac{1}{2}f''(v')+[f(v')+v'f'(v')]\ \underset{\tau\to 0}{\mathrm{Lt}}\ \frac{\lambda\tau}{\overline{u^2}}=0. \qquad \text{......(2060)}$$

Let us now put

$$\underset{\tau\to 0}{\mathrm{Lt}}\ \frac{\lambda\tau}{\overline{u^2}}=\mu.$$

Then the complete solution of the differential equation (2060) is

$$f(v')=e^{-\mu v'^2}\left[A\int_0^{v'}e^{\mu q^2}\,dq+B\right].$$

Since

$$\int_{-\infty}^{+\infty}f(v')\,dv'=1$$

and the A-contribution does not converge, we must have

$$A=0, \quad B=(\mu/\pi)^{\frac{1}{2}}.$$

Then $$\overline{v'^2} = \left(\frac{\mu}{\pi}\right)^{\frac{1}{2}} \int_{-\infty}^{+\infty} v^2 e^{-\mu v^2} dv = \frac{1}{2\mu},$$

and finally $$f(v) = \left(\frac{1}{2\pi\overline{v^2}}\right)^{\frac{1}{2}} e^{-\frac{1}{2}v^2/\overline{v^2}}, \quad \left(\overline{v^2} = \tfrac{1}{2}\operatorname*{Lt}_{\tau=0} \frac{\overline{u^2}}{\lambda\tau}\right). \qquad \text{......(2061)}$$

§ **19·51.** *Corollaries and extensions of Theorem* 19·5. A more exact treatment of the theorem, not confined to small τ, is of considerable interest, at least mathematical. The ψ defined in (2058), though independent of v, depends on τ, as of course does $\overline{u^2}$. When τ is no longer restricted to be small we must everywhere replace $1-\lambda\tau$ by the exact $e^{-\lambda\tau}$. Equation (2059) then takes the exact form

$$f(v') = e^{\lambda\tau} \int_{-\infty}^{+\infty} f[(v'-x)\, e^{\lambda\tau}]\, \psi(x,\tau)\, dx. \qquad \text{......(2062)}$$

This equation must hold for all v' and τ and the right-hand side must be independent of τ. We can show that these conditions serve to determine not only $f(v')$ but also $\psi(x,\tau)$, that is both the steady distribution law and the law of diffusion of random velocity among a subgroup of atoms, moving initially all with the same velocity.

By the study of small τ we have already shown that

$$f(v) = \left(\frac{\mu}{\pi}\right)^{\frac{1}{2}} e^{-\mu v^2}.$$

Inserting this in (2062) and changing the variable of integration we find

$$e^{-\mu v'^2} = \int_{-\infty}^{+\infty} e^{-\mu v^2} \psi(v' - ve^{-\lambda\tau}, \tau)\, dv.$$

If we differentiate this with respect to τ and simplify by integration by parts we find

$$\int_{-\infty}^{+\infty} \left\{\frac{\frac{1}{2}\lambda e^{-2\lambda\tau}}{\mu} \psi_{xx} - \psi_\tau\right\} e^{-\mu v^2} dv = 0,$$

in which the variable x has the value $v' - ve^{-\lambda\tau}$ after differentiation. This equation thus reduces for fixed τ to the form

$$\int_{-\infty}^{+\infty} g(v''-v)\, e^{-\mu v^2} dv = 0 \quad \textit{(all } v''\textit{)}$$

which requires that $g=0$.* We thus have

$$\frac{\frac{1}{2}\lambda e^{-2\lambda\tau}}{\mu} \psi_{xx} = \psi_\tau, \qquad \text{......(2063)}$$

* A simple proof of this theorem can be given under sufficiently wide conditions by connecting it with Fourier's integral theorem. We assume that g is differentiable and that $\int_{-\infty}^{+\infty} |g|\, dx$ exists. Then from the hypothesis of the theorem to be proved

$$0 = \int_{-\infty}^{+\infty} dx \int_0^{\infty} e^{ixt} g(x-v)\, e^{-v^2} dv = \int_{-\infty}^{+\infty} d\xi \int_0^{\infty} e^{i(\xi+v)t} g(\xi)\, e^{-v^2} dv,$$
$$= \left(\int_0^{\infty} e^{-v^2+ivt} dv\right)\left(\int_{-\infty}^{+\infty} g(\xi)\, e^{i\xi t} d\xi\right).$$

which (on the proper time scale) is equivalent to the standard equation of diffusion. It is reducible to the ordinary form by the change of variable

$$t = 1 - e^{-2\lambda\tau}.$$

Its solution for a "point source", that is for an initial concentration of all the particles at the origin of x, is, when the constant is adjusted so that

$$\int_{-\infty}^{+\infty} \psi(x,\tau)\,dx = 1,$$

$$\psi(x,\tau) = \left(\frac{\mu}{\pi}\right)^{\frac{1}{2}} \frac{e^{\lambda\tau}}{(e^{2\lambda\tau}-1)^{\frac{1}{2}}} e^{-\frac{\mu e^{2\lambda\tau} x^2}{e^{2\lambda\tau}-1}}. \qquad \text{......(2064)}$$

This function exhibits the growth of the distribution law for the random velocities. Its initial value is zero except for $x = 0$, and as $\tau \to \infty$

$$\psi(x,\tau) \sim \left(\frac{\mu}{\pi}\right)^{\frac{1}{2}} e^{-\mu x^2}.$$

Thus the subgroup of atoms moving initially with a given velocity ultimately acquires a Maxwellian distribution, which is naturally independent of its initial velocity.

§ **19·52.** *Einstein's calculation of* λ *and* $\overline{u^2}$. We proceed next to the calculation of λ. By (2015) the number of absorptions of isotropic radiation per atom in state 1 per unit time is

$$B_1{}^2 I(\nu)\,d\nu.$$

This can be generalized to radiation in a given solid angle, thus applying to anisotropic cases. The number of absorptions is then

$$B_1{}^2 I'(\nu',\theta')\,d\nu' d\omega'/4\pi,$$

measured in the frame K' moving with the atom. The coefficient $B_1{}^2$ cannot depend on orientation unless there is a field orientating the atoms. The number of absorptions per atom in unit time averaged over all the atoms is therefore

$$\frac{\varpi_1 e^{\chi_1/kT}}{f(T)} B_1{}^2 I'(\nu',\theta')\,d\nu' d\omega'/4\pi. \qquad \text{......(2065)}$$

We recall that χ_1 and χ_2 are negative energies and $f(T) = \varpi_1 e^{\chi_1/kT} + \varpi_2 e^{\chi_2/kT}$. Every such absorption conveys to the atoms X'-momentum to the amount

$$\frac{h\nu'}{c} \cos\theta'.$$

The first integral factor is equal to $\frac{1}{2}\sqrt{\pi}\, e^{-\frac{1}{4}t^2}$ and never vanishes. Therefore

$$\int_{-\infty}^{+\infty} g(\xi) \begin{matrix}\cos\\ \sin\end{matrix} (\xi t)\,d\xi = 0 \quad (\textit{all } t).$$

But by Fourier's integral theorem (Hobson, *Functions of a Real Variable*, vol. 2, p. 727)

$$g(x) = \frac{1}{\pi}\int_0^{\infty} dv \int_{-\infty}^{+\infty} g(t) \cos\{v(t-x)\}\,dt = 0,$$

since the inner integral vanishes.

The stimulated emissions must also be anisotropic, for they must take place always in the direction of the incident radiation. We have as yet not considered this point, but it is required by the classical analogy discussed by Van Vleck.* More forcibly, unless the stimulated emissions are in the same direction as the incident radiation, they will not concern light of the same frequency in all frames of reference, and the whole argument of § 19·4 applied to this process must break down. Due to the radiation incident in $d\omega'$ there are

$$\frac{\varpi_2 e^{\chi_2/kT}}{f(T)} B_2^{\,1} I'(\nu',\theta')\,d\nu'd\omega'/4\pi$$

such emissions per atom per unit time, each contributing to the atom the recoil momentum

$$-\frac{h\nu'}{c}\cos\theta'.$$

In the frame K' the natural emissions must be isotropic and contribute nothing to λ.

If now we collect these results and recall that $\varpi_1 B_1^{\,2} = \varpi_2 B_2^{\,1}$ we find that the total average momentum contributed by the radiation to each atom per unit time is

$$\frac{h\nu'}{4\pi c f(T)} \varpi_1 B_1^{\,2}\{e^{\chi_1/kT} - e^{\chi_2/kT}\}d\nu' \int I'(\nu',\theta')\cos\theta'\,d\omega'.$$

Using (2038) and carrying out the integration we find that this rate of transfer of momentum is

$$-v\frac{h\nu'}{c^2 f(T)}\left\{I(\nu') - \tfrac{1}{3}\nu'\frac{\partial I(\nu')}{\partial \nu'}\right\}\varpi_1 B_1^{\,2} d\nu'\,(e^{\chi_1/kT} - e^{\chi_2/kT}).$$

The coefficient of $-v$ is by definition $m\lambda$, and the formula now contains only the isotropic $I(\nu')$. Using Planck's law and the relation $h\nu' = \chi_1 - \chi_2$ we reduce this to

$$\lambda = \frac{h^2\nu'^2}{c^2}\,\frac{\varpi_1 e^{\chi_1/kT}}{f(T)}\,\frac{B_1^{\,2} I(\nu')\,d\nu'}{3mkT}. \qquad \text{......(2066)}$$

We may observe that in this calculation we have for simplicity used a frame K' and neglected the change of mass of the atom on absorbing $h\nu$. This is sufficiently accurate here, but the exact method is of course that of § 19·4.

We have finally to calculate $\overline{u^2}$, the fluctuational increment of velocity in time τ, to an approximation which ignores the slight anisotropy and treats the atom as at rest. A single absorption or emission at an angle θ_r exerts an impulse $(h\nu'/c)\cos\theta_r$ on the atom in the X-direction. Thus for any atom

$$mu = (h\nu'/c)\,\Sigma_r \cos\theta_r,$$

* Van Vleck, *loc. cit.* in § 19·2.

and averaged over all atoms $\bar{u}=0$, and

$$m^2\overline{u^2}=\frac{h^2\nu'^2}{c^2}\overline{(\Sigma_r \cos^2\theta_r+2\Sigma_r\cos\theta_r\cos\theta_s)}.$$

The average value of the product term is obviously zero. The average value of the square terms is $\frac{1}{3}n$, where n is the total number of emissions and absorptions in time τ, that is double the number of the absorptions. Thus by (2065)

$$n=\tau\frac{2\varpi_1 e^{\chi_1/kT}}{f(T)}B_1{}^2I(\nu')\,d\nu',$$

and finally

$$\overline{u^2}=\frac{h^2\nu'^2}{c^2}\frac{\varpi_1 e^{\chi_1/kT}}{f(T)}\frac{2B_1{}^2I(\nu')\,d\nu'}{3m^2}\tau. \qquad \text{......(2067)}$$

Comparing (2066) and (2067) we find

$$\frac{\overline{u^2}}{2\lambda\tau}=\frac{kT}{m},$$

so that by (2053)

$$\tfrac{1}{2}m\overline{v^2}=\tfrac{1}{2}kT,$$

as required for preservation of the equilibrium state.

If we did not assume Planck's law for $I(\nu)$ in reducing (2066) and (2067), we should eventually find

$$\tfrac{1}{2}m\overline{v^2}=\frac{m\overline{u^2}}{4\lambda\tau}=\frac{\tfrac{1}{2}h\nu'}{(1-e^{-h\nu'/kT})\left(3-\dfrac{\nu'}{I}\dfrac{\partial I}{\partial\nu'}\right)}. \qquad \text{......(2068)}$$

§ **19·6.** *The emission and absorption of solids.* A solid body, or for that matter a liquid or a sufficiently dense gas, can generally be regarded as emitting light of all wave lengths (e.g. incandescent filaments, electric arcs, the sun). In an assembly in equilibrium this is a surface effect, and therefore must preserve equilibrium independently of all volume effects.

In unit time the energy of ν-radiation which strikes a unit surface within a solid angle $d\omega$ at an angle θ with the normal to the surface is

$$I(\nu)\cos\theta\,d\nu\,d\omega.$$

In order to preserve equilibrium (isotropic radiation) the same amount of ν-radiation must be sent back in the reverse direction. This must hold for all ν and all T. The surfaces of condensed bodies are usually described by an absorption coefficient $k(\nu,\theta)$, which is the fraction of ν-radiation incident at an angle θ that they absorb. The rest of the radiation, the fraction $1-k(\nu,\theta)$, is reflected in some manner, but with a definite phase relation to the incident light. The existence of a phase relation is the defining property of reflection, and forces us to distinguish between the reflected radiation and the radiation re-emitted after absorption for which there is no such phase relation. It is this point which leads (at least on classical theories) to a discussion somewhat more complicated than that required for the impact of particles on walls. If there is absorption there must of course be emission as well, and the emission coefficient $\epsilon(\nu,\theta)$ is so defined that the ν-radiation emitted in unit time by unit area into a solid angle $d\omega$ at any angle θ with the normal to the surface is

$$\epsilon(\nu,\theta)\cos\theta\, d\nu\, d\omega.$$

The radiation reflected from *all* incident beams into the same solid angle may be defined to be

$$r(\nu,\theta)\cos\theta\, d\nu\, d\omega.$$

The hypothesis of the preservation of equilibrium requires that

$$I(\nu)=\epsilon(\nu,\theta)+r(\nu,\theta). \qquad \text{......(2069)}$$

According to the hypothesis of detailed balancing we are entitled to assert separately the equality of the absorbed and re-emitted fractions and of the unabsorbed and total reflected fraction, i.e. that

$$r(\nu,\theta)=\{1-k(\nu,\theta)\}\, I(\nu), \qquad \text{......(2070)}$$

$$\epsilon(\nu,\theta)=k(\nu,\theta)\, I(\nu). \qquad \text{......(2071)}$$

This is a familiar result of the classical radiation theory. It will be true on the hypothesis of preservation alone if for example the reflection is perfectly specular or perfectly diffuse, or obeying any mixture of these two laws, or if the surface is such that Helmholtz's reciprocal theorem on definite beams of light holds.* It does not seem that Helmholtz's theorem can extend to the most general condition of reflective scattering contemplated here, but it is undoubtedly true for most reflections actually found in practice.

A body for which $k(\nu,\theta)=1$ is usually called *a black body*, or, since

$$\epsilon(\nu,\theta)=I(\nu),$$

a full radiator. A body for which $k(\nu,\theta)$ is a constant (less than unity) is called *a grey body*. It emits radiation distributed according to Planck's law, but at less than the normal rate.

The surface coefficients actually depend on the state of polarization of the incident beam. It is necessary to imagine $I(\nu)$ broken up into two plane polarized components, one in the plane of incidence and the other at right angles to it. The theory is easily extended to cover this distinction.

§ 19·7. *Photoelectric emission of solids.* The emission of electrons by cold solids under the influence of radiation has already been discussed in § 11·37. The energy of the electron and the frequency of the incident light are connected by Einstein's equation $\eta=h\nu-\chi$, where χ is the photoelectric

* See Lorentz, *Problems of Modern Physics*, Note 17 (1927).

threshold or work function of the metal. The converse of this effect is the emission of ν-radiation when an electron of energy $h\nu-\chi$ strikes the solid. Such effects have never been recorded and are perhaps too scarce and too difficult ever to observe. That radiation is emitted in stopping the electrons striking a metal target (X-ray emission) is familiar enough. The radiation of continuous frequency so obtained is probably almost entirely atomic in origin, arising from free-free transitions in the field of a single atom; the characteristic radiations are obviously atomic. The interest of the converse photoelectric effect would lie in $h\nu$ exceeding (by a few volts) the energy of the impinging electron.

We shall not give the details in this case, but it follows at once by arguments now familiar and the use of (1023) that if ϕ is the chance that a ν-quantum striking the solid shall produce a photoelectron, and

$$\phi'\left(1+\frac{c^2}{2h\nu^3}I(\nu)\right)$$

is the chance that an η-electron ($\eta=h\nu-\chi$) striking the solid shall be absorbed with the production of a ν-quantum of radiation, then

$$\phi=\frac{mc^2}{h^2\nu^2}\eta\phi'. \qquad \text{......(2072)}$$

In the region of $\lambda\,3000$ ϕ may perhaps be numerically comparable with unity, but ϕ' is then of the order 10^{-5}–10^{-6}. The converse effect is thus very small. It must also be remembered that the calculations leading to (2072) refer to the equilibrium state, and that a solid is no longer (like an atom in a given state) a definite system, but itself depends on the temperature. The solid may be capable at high temperatures of capturing impinging electrons in this way with the calculated frequency, but it does not follow that the same solid cold, bombarded by electrons, will be able to achieve the same fraction of captures. There may not be the same vacant orbits for the electrons to occupy.

§ **19·8.** *Photochemical reactions and chemi-luminescence.* It is a familiar fact that certain gaseous reactions are initiated by the absorption of external radiation—usually light of the visible region or nearby. It is however difficult to find simple unambiguous examples, in which we can be perfectly sure of what is happening, so that we shall be content to discuss such mechanisms and their converses in general terms without attempting precise applications of any theoretical results.

In the photochemical mechanism the primary action is the absorption of a single quantum of ν-radiation. There is no known case of dependence on a power of $I(\nu)$ higher than the first, so that only a single quantum can normally be concerned. The result of this absorption may be either

(1) An activated molecule.

(2) Immediate dissociation.

In either case we may expect (and find) a threshold frequency ν_0 below which radiation is ineffective. We may observe further that, whether $\nu > \nu_0$ or $\nu < \nu_0$, ν-radiation must be ineffective unless it is absorbed by the molecule and therefore that only the lines and continuous stretches (if any) of the absorption spectrum of the molecule can be effective photochemically. This point is well brought out by the practice of sensitizing photographic plates to red light by staining them with red-absorbent dyes. The plates are normally insensitive to red light not because $\nu_{red} < \nu_0$ but because they do not absorb the red light. Once absorbed the energy of the red light is transferable by other mechanisms, and the photographic reaction occurs.

In simple cases we shall expect to find with either mechanism (1) or (2) that just one molecule is transformed for each quantum of effective light absorbed. This is called *Einstein's law of photochemical equivalence*. There is however no reason to expect its universal validity, owing to secondary effects such as the formation of reaction chains carried on by the products of the primary photochemical act. Einstein's law is found to hold in many cases, but there are marked exceptions, such as the reaction

$$H_2 + Cl_2 \rightarrow 2HCl,$$

in which the number of reacting pairs may be many million times the number of light quanta absorbed.

The converse reactions are either

(1) Emission of ν-radiation and deactivation.

(2) Recombination of two free atoms or molecules with emission of ν-radiation to get rid of the excess energy.

Experiments will certainly be able to distinguish between (1) and (2) and no doubt will do so soon, but owing to subsidiary effects this is never easy. In the meantime we may be certain that examples of both types of mechanism occur in nature. A quantum of visible light represents a heat of activation per gram-molecule of 50,000 calories, more or less, which is just of the order of the heats of activation with which we are familiar in reactions proceeding by collision. On the other hand, examples of the converse mechanism (2) are known, though not in the gaseous state. This is the phenomenon of chemi-luminescence.*

* See, for example, Noddack, *Handbuch d. Physik*, vol. 23, p. 631, art. *Photochemie* (1926).

CHAPTER XX

FLUCTUATIONS

§20·1. We have hitherto, in calculating average values and asserting that they represent properties of the assembly characterizing the equilibrium state, generally ignored the fact that at any moment the actual assembly will always deviate more or less from the average state. From the definition of average value the average of these deviations must of course vanish, but the average numerical value of the deviation will not vanish, and will be a measure of the closeness with which the assembly conforms to the theoretical equilibrium state. It was necessary for the logical development to point out at once, as we did in Chapter II, that in general these average deviations are insignificant compared with the average values themselves. This at once justifies our treatment of the equilibrium state and the possession of all its average values as a normal property of the assembly in the sense familiarized by Jeans. We have also had occasion to prove certain theorems, particularly in Chapter XIX in connection with Einstein's discussion of the interaction of radiation and atoms, which are closely concerned with fluctuation problems and will be applied further.

It is therefore our primary object here to develop systematic methods of calculating such average deviations, (*a*) for the sake of the logical development of the whole subject, and (*b*) for direct application where possible to physical observations such as Brownian movement and opalescence near the critical point. In taking up these calculations we find at once that the average value of the numerical deviation is awkward to handle and is naturally replaced by a calculation of the average of the square of the deviation. This is moreover the quantity ordinarily required in applications, and we have usually spoken of this quantity hitherto as the *fluctuation*. If P is any quantity whose average value is $\overline{P}$, then the fluctuation in this sense is $\overline{(P-\overline{P})^2}$. In devising general methods of calculating these fluctuations, we find it almost equally simple to calculate $\overline{(P-\overline{P})^n}$ for any positive integral value of n. We shall therefore speak of all such quantities in this chapter as fluctuations, and refer to particular ones as fluctuations of order n, the "fluctuations" of other chapters being here fluctuations of the second order. Fluctuations of orders other than the second are not of much immediate physical importance, but the results are elegant and perhaps not without intrinsic interest. They generalize in many cases the corresponding results due to Gibbs.*

* Gibbs, *Elementary Principles in Statistical Mechanics*, chap. VII. The results for non-dissociating assemblies were given by Darwin and Fowler, *Proc. Camb. Phil. Soc.* vol. 21, p. 391 (1922), but the discussion there given for dissociating assemblies is inadequate and is here revised.

§20·11. *General formulae for $\overline{a_r^n}$, $\overline{a_r^n a_s^m}$, etc., and $\overline{E_A^n}$.* Before starting to discuss fluctuations it is convenient to assemble certain formulae for average values which are immediate generalizations of those of §2·4. We can obtain full generality by considering first an assembly of two types of permanent quantized systems, obeying one or other of the possible forms of statistics (unspecified). The results extend at once to any number of types of systems and the form for the classical limit can be deduced. Dissociation is for the present excluded and considered separately later.

In the notation of §2·4

$$C=\frac{1}{(2\pi i)^3}\iiint\frac{dx\,dy\,dz}{x^{M+1}y^{N+1}z^{E+1}}\,\Pi_t\,g(xz^{\epsilon_t})\,\Pi_t\,g_1(yz^{\eta_t}),$$

$$C\overline{a_r}=\frac{1}{(2\pi i)^3}\iiint\frac{dx\,dy\,dz}{x^{M+1}y^{N+1}z^{E+1}}\left[x\frac{\partial}{\partial x}g(xz^{\epsilon_r})\right]\Pi_{t\neq r}\,g(xz^{\epsilon_t})\,\Pi_t\,g_1(yz^{\eta_t}).$$

The last equation is easily generalized by use of the same arguments and we find

$$C\overline{a_r^n}=\frac{1}{(2\pi i)^3}\iiint\frac{dx\,dy\,dz}{x^{M+1}y^{N+1}z^{E+1}}\left[\left(x\frac{\partial}{\partial x}\right)^n g(xz^{\epsilon_r})\right]\Pi_{t\neq r}\,g(xz^{\epsilon_t})\,\Pi_t\,g_1(yz^{\eta_t}).\quad\text{......(2073)}$$

A further generalization on the same lines is at once possible to

$$C\overline{a_r^n a_s^m}=\frac{1}{(2\pi i)^3}\iiint\frac{dx\,dy\,dz}{x^{M+1}y^{N+1}z^{E+1}}\left[\left(x\frac{\partial}{\partial x}\right)^n g(xz^{\epsilon_r})\right]$$
$$\times\left[\left(x\frac{\partial}{\partial x}\right)^m g(xz^{\epsilon_s})\right]\Pi_{t\neq r,s}\,g(xz^{\epsilon_t})\,\Pi_t\,g_1(yz^{\eta_t}).\quad\text{......(2074)}$$

It is clear that any number of such factors may be introduced.

No integral has yet been given for $C\overline{E_A}$, since $\overline{E_A}$ was derived from $\Sigma_r\,\overline{a_r}\epsilon_r$. We can easily construct one, and in fact by considering $\overline{(\Sigma_r\,a_r\epsilon_r)^n}$ construct integrals for $C\overline{E_A^n}$. The general term in this expression is of the form

$$\frac{n!}{s!\,s'!\,s''!\dots}\,\epsilon_r^{\,s}\epsilon_{r'}^{\,s'}\epsilon_{r''}^{\,s''}\,\overline{a_r^{\,s}a_{r'}^{\,s'}a_{r''}^{\,s''}}\dots$$

corresponding to a general term in the integrand of (2074) of the form

$$\frac{n!}{s!\,s'!\,s''!\dots}\,\epsilon_r^{\,s}\epsilon_{r'}^{\,s'}\epsilon_{r''}^{\,s''}\dots\left[\left(x\frac{\partial}{\partial x}\right)^s g(xz^{\epsilon_r})\right]\left[\left(x\frac{\partial}{\partial x}\right)^{s'} g(xz^{\epsilon_{r'}})\right]\left[\left(x\frac{\partial}{\partial x}\right)^{s''} g(xz^{\epsilon_{r''}})\right]$$
$$\times\dots\times\Pi_{t\neq r,r',r'',\dots}\,g(xz^{\epsilon_t}).$$

The ϵ_r-factors can be absorbed by replacing the operators $x\partial/\partial x$ by $z\partial/\partial z$. By an application of Leibnitz's theorem it is then easily seen that the last expression reduces to

$$\left(z\frac{\partial}{\partial z}\right)^n \Pi_t\,g(xz^{\epsilon_t}).$$

It follows that

$$C\overline{E_A^n}=\frac{1}{(2\pi i)^3}\iiint\frac{dx\,dy\,dz}{x^{M+1}y^{N+1}z^{E+1}}\left[\left(z\frac{\partial}{\partial z}\right)^n \Pi_t\,g(xz^{\epsilon_t})\right]\Pi_t\,g_1(yz^{\eta_t}).\quad\text{......(2075)}$$

§ **20·2.** *General fluctuations in the partition of energy.* It would be possible to evaluate the dominant terms in $\overline{E_A{}^n}$ directly from (2075) by the usual application of steepest descents, but we really require the dominant term in $\overline{(E_A-\overline{E_A})^n}$, and in passing to this by expansion and use of $\overline{E_A{}^n}$ a number of leading terms cancel and a more or less complete expansion of $\overline{E_A{}^n}$ must be used. We avoid this difficulty by first constructing an exact integral for $C\overline{(E_A-\overline{E_A})^n}$, which can be simply evaluated in the usual way.

A change of notation is expedient. Put $x=e^u$, $y=e^v$, $z=e^w$,

$$g(xz^{\epsilon_t})=g(e^{u+\epsilon_t w})=\exp\{G(u+\epsilon_t w)\}, \qquad \text{......(2076)}$$

with a similar expression for G_1. Then

$$C\overline{E_A{}^n}=\frac{1}{(2\pi i)^3}\iiint_{\gamma'} du\,dv\,dw\exp[-Mu-Nv-Ew+\Sigma_t G_1(v+\eta_t w)]$$
$$\times\left(\frac{\partial}{\partial w}\right)^n\exp[\Sigma_t G(u+\epsilon_t w)], \qquad \text{......(2077)}$$

where the contours γ' are now straight lines from $\log\lambda-i\pi$ to $\log\lambda+i\pi$, $\log\mu-i\pi$ to $\log\mu+i\pi$, $\log\vartheta-i\pi$ to $\log\vartheta+i\pi$ respectively. These contours are of course those required for the application of the method of steepest descents. To form the integral for $C\overline{(E_A-\overline{E_A})^n}$ we replace E by $\overline{E_A}+\overline{E_B}$ and again use Leibnitz's theorem; thus

$$C\overline{(E_A-\overline{E_A})^n}=C[\overline{E_A{}^n}-{}_nC_1\overline{E_A{}^{n-1}}\overline{E_A}+{}_nC_2\overline{E_A{}^{n-2}}(\overline{E_A})^2-\ldots(-)^n(\overline{E_A})^n],$$
$$=\frac{1}{(2\pi i)^3}\iiint_{\gamma'} du\,dv\,dw\exp[-Mu-Nv-Ew+\Sigma_t G_1(v+\eta_t w)]$$
$$\times\left[\left(\frac{\partial}{\partial w}\right)^n-{}_nC_1\overline{E_A}\left(\frac{\partial}{\partial w}\right)^{n-1}+{}_nC_2(\overline{E_A})^2\left(\frac{\partial}{\partial w}\right)^{n-2}-\ldots(-)^n(\overline{E_A})^n\right]$$
$$\times\exp[\Sigma_t G(u+\epsilon_t w)]$$
$$=\frac{1}{(2\pi i)^3}\iiint_{\gamma'} du\,dv\,dw\exp[-Mu-Nv-\overline{E_B}w+\Sigma_t G_1(v+\eta_t w)]$$
$$\times\left(\frac{\partial}{\partial w}\right)^n\exp[\Sigma_t G(u+\epsilon_t w)-\overline{E_A}w]. \qquad \text{......(2078)}$$

This is the desired integral. We know that effectively the whole value of this integral is provided by the immediate neighbourhood of the saddle point at $\log\lambda$, $\log\mu$, $\log\vartheta$. We therefore write $u=\log\lambda+i\alpha$, *etc.*, and G for $G(\log\lambda+\epsilon_t\log\vartheta)$, and recall that in this notation

$$\overline{E_A}=\left[\frac{\partial}{\partial w}\Sigma_t G(u+\epsilon_t w)\right]_{\log\lambda,\log\mu,\log\vartheta}=\Sigma_t\epsilon_t G'.$$

Then $\Sigma_t G(u+\epsilon_t w)-\overline{E_A}w-Mu=\Sigma_t G-\overline{E_A}\log\vartheta-M\log\lambda$

$$-\tfrac{1}{2}\Sigma_t(\alpha+\epsilon_t\gamma)^2G''-\tfrac{1}{6}i\Sigma_t(\alpha+\epsilon_t\gamma)^3G'''+\ldots.$$

There is a similar expression for $\Sigma_t G_1(v+\epsilon_t w) - \overline{E_B}w - Nv$. The integrations are now with respect to α, β, γ and range effectively from $-\infty$ to $+\infty$, while other terms such as $\Sigma_t(\alpha+\epsilon_t\gamma)^3 G'''$ are still small. Thus

$$C\overline{(E_A-\overline{E_A})^n} = \frac{(i)^{-n}}{(2\pi)^3} e^{\Sigma_t G - \overline{E_A}\log\vartheta - M\log\lambda + \Sigma_t G_1 - \overline{E_B}\log\vartheta - N\log\mu}$$
$$\times \iiint_{-\infty}^{+\infty} d\alpha\, d\beta\, d\gamma\, e^{-\frac{1}{2}\Sigma_t(\beta+\eta_t\gamma)^2 G_1''}\{1+O(\Sigma_t(\beta+\eta_t\gamma)^3 G_1''')\}$$
$$\times \left(\frac{\partial}{\partial\gamma}\right)^n [e^{-\frac{1}{2}\Sigma_t(\alpha+\epsilon_t\gamma)^2 G''}\{1+O(\Sigma_t(\alpha+\epsilon_t\gamma)^3 G''')\}],$$

an expression in which the O-terms may be differentiated. In the special case $n=0$ we obtain C, and the O-terms cannot contribute to the leading term in the integral which is

$$\iiint_{-\infty}^{+\infty} d\alpha\, d\beta\, d\gamma\, e^{-\frac{1}{2}\Sigma_t(\beta+\eta_t\gamma)^2 G_1'' - \frac{1}{2}\Sigma_t(\alpha+\epsilon_t\gamma)^2 G''} = J \quad \text{......(2079)}$$

say. Thus

$$\overline{(E_A-\overline{E_A})^n} = \frac{(i)^{-n}}{J} \iiint_{-\infty}^{+\infty} d\alpha\, d\beta\, d\gamma\, e^{-\frac{1}{2}\Sigma_t(\beta+\eta_t\gamma)^2 G_1''}\{1+O(\Sigma_t(\beta+\eta_t\gamma)^3 G_1''')\}$$
$$\times \left(\frac{\partial}{\partial\gamma}\right)^n [e^{-\frac{1}{2}\Sigma_t(\alpha+\epsilon_t\gamma)^2 G''}\{1+O(\Sigma_t(\alpha+\epsilon_t\gamma)^3 G''')\}]. \quad \text{......(2080)}$$

The order of the O-terms may be expressed more clearly by writing them in the forms $O[N(\beta+\eta\gamma)^3]$ and $O[M(\alpha+\epsilon\gamma)^3]$ respectively.

The further approximation to (2080) depends on the parity of n. It is clear that in any case the first differentiation of the bracket $\{1+O[M(\alpha+\epsilon\gamma)^3]\}$ does not alter the order of the integrand. Instead of the leading term 1, we then have a term of order $M(\alpha+\epsilon\gamma)^2$. But every time we differentiate $e^{-\frac{1}{2}\Sigma_t(\alpha+\epsilon_t\gamma)^2 G''}$ we multiply the integrand by $\Sigma_t(\alpha+\epsilon_t\gamma)\epsilon_t G''$, which is of order $M(\alpha+\epsilon\gamma)$ and increases its order after differentiation by $\sqrt{M}$. Thus the highest order term arises from differentiating the exponential n times and the O-terms are both of lower order, *provided that this highest order term does not happen to vanish identically on integration.* It does not vanish for n even but does so vanish for n odd, when further consideration is required to which we return. If now we put $n=2v$, we find for the required asymptotic form of $\overline{(E_A-\overline{E_A})^{2v}}$

$$\overline{(E_A-\overline{E_A})^{2v}} = \frac{(-)^v}{J} \iiint_{-\infty}^{+\infty} d\alpha\, d\beta\, d\gamma\, e^{-\frac{1}{2}\Sigma_t(\beta+\eta_t\gamma)^2 G_1''} \left(\frac{\partial}{\partial\gamma}\right)^n e^{-\frac{1}{2}\Sigma_t(\alpha+\epsilon_t\gamma)^2 G''}.$$
$$\text{......(2081)}$$

To evaluate this integral its integrand may be written for shortness in the form

$$e^{-\frac{1}{2}(B\beta^2+2U\beta\gamma+C_2\gamma^2)} \left(\frac{\partial}{\partial\gamma}\right)^n e^{-\frac{1}{2}(A\alpha^2+2V\gamma\alpha+C_1\gamma^2)}$$

and the α- and β-integrations may be carried out immediately, the order of the operations $\int d\alpha$ and $\left(\frac{\partial}{\partial\gamma}\right)^n$ being inverted. We are then left with

$$\overline{(E_A-\overline{E_A})^{2v}} = \frac{(-)^v}{J}\frac{2\pi}{(AB)^{\frac{1}{2}}}\int_{-\infty}^{+\infty} e^{-\frac{1}{2}(C_2-U^2/B)\gamma^2}\left(\frac{\partial}{\partial\gamma}\right)^n e^{-\frac{1}{2}(C_1-V^2/A)\gamma^2}\,d\gamma. \quad \text{......(2082)}$$

Let us now write

$$I_v=(-)^v\int_{-\infty}^{+\infty}\phi_2\phi_1^{(2v)}\,d\gamma=(-)^{v+r}\int_{-\infty}^{+\infty}\phi_2^{(r)}\phi_1^{(2v-r)}\,d\gamma;$$

the latter form follows by repeated integration by parts, and ϕ_1, ϕ_2 are short for $e^{-\frac{1}{2}(C_2-U^2/B)\gamma^2}$, $e^{-\frac{1}{2}(C_1-V^2/A)\gamma^2}$ respectively. Then since

$$\phi_1'=-\gamma(C_2-U^2/B)\,\phi_1,\quad \phi_2'=-\gamma(C_1-V^2/A)\,\phi_2,$$

we find that

$$\frac{C_1+C_2-U^2/B-V^2/A}{(C_1-V^2/A)(C_2-U^2/B)}I_v=(-)^v\int_{-\infty}^{+\infty}\gamma(\phi_2\phi_1^{(2v-1)}+\phi_1\phi_2^{(2v-1)})\,d\gamma.$$

On integration by parts the right-hand side becomes

$$2I_{v-1}+(-)^{v-1}\int_{-\infty}^{+\infty}\gamma(\phi_2'\phi_1^{(2v-2)}+\phi_1'\phi_2^{(2v-2)})\,d\gamma,$$

and by continued repetition

$$2vI_{v-1}+\int_{-\infty}^{+\infty}\gamma(\phi_2^{(v)}\phi_1^{(v-1)}+\phi_1^{(v)}\phi_2^{(v-1)})\,d\gamma$$

or

$$(2v-1)\,I_{v-1}.$$

Thus

$$I_v=(2v-1)\dots 3.1\left[\frac{(C_1-V^2/A)(C_2-U^2/B)}{C_1+C_2-V^2/A-U^2/B}\right]^v I_0, \quad \text{......(2083)}$$

and it is easily verified that

$$\frac{2\pi}{(AB)^{\frac{1}{2}}}I_0=J.$$

Thus

$$\overline{(E_A-\overline{E_A})^{2v}}=(2v-1)\dots 3.1\left[\frac{(C_1-V^2/A)(C_2-U^2/B)}{C_1+C_2-V^2/A-U^2/B}\right]^v,$$

$$=(2v-1)\dots 3.1\left[\overline{(E_A-\overline{E_A})^2}\right]^v, \quad \text{......(2084)}$$

$$\overline{(E_A-\overline{E_A})^2}=\frac{(C_1-V^2/A)(C_2-U^2/B)}{C_1+C_2-V^2/A-U^2/B}. \quad \text{......(2085)}$$

This result must now be interpreted. The physical variables M, N, $\overline{E_A}$ and $\overline{E_B}$ are given in terms of λ, μ, ϑ and the G's by familiar equations which in the present notation take the form

$$M=\Sigma_t G',\quad N=\Sigma_t G_1',\quad \overline{E_A}=\Sigma_t\epsilon_t G',\quad \overline{E_B}=\Sigma_t\eta_t G_1'.$$

We regard these variables as functions of λ, μ and ϑ. Then

$$C_1 = \Sigma_l \epsilon_l^2 G'' = \vartheta \partial \overline{E_A}/\partial\vartheta, \quad C_2 = \Sigma_l \eta_l^2 G_1'' = \vartheta \partial \overline{E_B}/\partial\vartheta; \quad \text{......(2086)}$$

$$A = \Sigma_l G'' = \lambda \partial M/\partial\lambda, \qquad B = \Sigma_l G_1'' = \mu \partial N/\partial\mu; \quad \text{......(2087)}$$

$$V = \Sigma_l \epsilon_l G'' = \vartheta \partial M/\partial\vartheta = \lambda \partial \overline{E_A}/\partial\lambda; \quad \text{......(2088)}$$

$$U = \Sigma_l \eta_l G_1'' = \vartheta \partial N/\partial\vartheta = \mu \partial \overline{E_B}/\partial\mu. \quad \text{......(2089)}$$

We can now show that

$$C_1 - \frac{V^2}{A} = \vartheta \frac{\partial \overline{E_A}}{\partial\vartheta} - \lambda \frac{\partial \overline{E_A}}{\partial\lambda} \vartheta \frac{\partial M}{\partial\vartheta} \Big/ \lambda \frac{\partial M}{\partial\lambda} = \vartheta \left(\frac{\partial \overline{E_A}}{\partial\vartheta}\right)_M, \quad \text{......(2090)}$$

and similarly that
$$C_2 - \frac{U^2}{B} = \vartheta \left(\frac{\partial \overline{E_B}}{\partial\vartheta}\right)_N. \quad \text{......(2091)}$$

The first equality in (2090) follows at once from the relations (2086)–(2089). For the second we note that when M is constant λ varies and therefore

$$\vartheta \left(\frac{\partial \overline{E_A}}{\partial\vartheta}\right)_M = \vartheta \frac{\partial \overline{E_A}}{\partial\vartheta} + \vartheta \frac{\partial \overline{E_A}}{\partial\lambda} \left(\frac{\partial\lambda}{\partial\vartheta}\right)_M, \quad \text{......(2092)}$$

and since
$$0 = \frac{\partial M}{\partial\lambda} d\lambda + \frac{\partial M}{\partial\vartheta} d\vartheta,$$

$$\left(\frac{\partial\lambda}{\partial\vartheta}\right)_M = -\frac{\partial M}{\partial\vartheta} \Big/ \frac{\partial M}{\partial\lambda}. \quad \text{......(2093)}$$

Combining (2092) and (2093) we obtain the second equality in (2090). We thus obtain finally the relation

$$\overline{(E_A - \overline{E_A})^2} = \frac{\vartheta(\partial \overline{E_A}/\partial\vartheta)_M \, \vartheta(\partial \overline{E_B}/\partial\vartheta)_N}{\vartheta(\partial \overline{E}/\partial\vartheta)_{M,N}} = \vartheta\left(\frac{\partial \overline{E_A}}{\partial\vartheta}\right)_M \left[1 - \frac{\vartheta(\partial \overline{E_A}/\partial\vartheta)_M}{\vartheta(\partial \overline{E}/\partial\vartheta)_{M,N}}\right]. \quad \text{......(2094)}$$

Fluctuations of the energy of all even orders are thus completely determined.

We now return to the odd orders. To an equivalent approximation these all vanish, and they are actually of order lower by $\sqrt{M}$ or $\sqrt{N}$ than the corresponding even orders. We must retain the exact terms in $\Sigma_l (\beta + \eta_l \gamma)^3 G_1'''$ and $\Sigma_l (\alpha + \epsilon_l \gamma)^3 G'''$ in (2080), the terms $\Sigma_l (\alpha + \epsilon_l \gamma)^4 G^{\mathrm{iv}}$ *etc.* and higher terms being negligible. We then find after simple reductions

$$\overline{(E_A - \overline{E_A})^{2v-1}} = P_v + Q_v,$$

$$P_v = \frac{(-)^v}{6J} \iiint_{-\infty}^{+\infty} d\alpha\, d\beta\, d\gamma\, e^{-\frac{1}{2}(B\beta^2 + 2U\beta\gamma + C_2\gamma^2)} [\Sigma_l (\beta + \eta_l \gamma)^3 G_1'''] \times \left(\frac{\partial}{\partial\gamma}\right)^{2v-1} e^{-\frac{1}{2}(A\alpha^2 + 2V\gamma\alpha + C_1\gamma^2)},$$

$$Q_v = \frac{(-)^{v-1}}{6J} \iiint_{-\infty}^{+\infty} d\alpha\, d\beta\, d\gamma\, e^{-\frac{1}{2}(A\alpha^2 + 2V\gamma\alpha + C_1\gamma^2)} [\Sigma_l (\alpha + \epsilon_l \gamma)^3 G'''] \times \left(\frac{\partial}{\partial\gamma}\right)^{2v-1} e^{-\frac{1}{2}(B\beta^2 + 2U\beta\gamma + C_2\gamma^2)}.$$

It is sufficient to consider one of these expressions. Carrying through the α- and β-integrations we find for P_v

$$P_v = \frac{(-)^v}{6J}\frac{2\pi}{(AB)^{\frac{1}{2}}}\int_{-\infty}^{+\infty} d\gamma\, e^{-\frac{1}{2}(C_2-U^2/B)\gamma^2}$$
$$\times\left[\gamma^3\Sigma_l\left(\eta_l-\frac{U}{B}\right)^3 G_1''' + \frac{3\gamma}{B}\Sigma_l\left(\eta_l-\frac{U}{B}\right)G_1'''\right]\left(\frac{\partial}{\partial\gamma}\right)^{2v-1} e^{-\frac{1}{2}(C_1-V^2/A)\gamma^2}.$$

By using the identities

$$\gamma e^{-\frac{1}{2}C_2'\gamma^2} = -\frac{1}{C_2'}\frac{\partial}{\partial\gamma}(e^{-\frac{1}{2}C_2'\gamma^2}),\quad \gamma^3 e^{-\frac{1}{2}C_2'\gamma^2} = \left[-\frac{1}{C_2'^3}\left(\frac{\partial}{\partial\gamma}\right)^3 - \frac{3}{C_2'^2}\frac{\partial}{\partial\gamma}\right]e^{-\frac{1}{2}C_2'\gamma^2},$$

we can reduce this expression to integrals already evaluated, and find

$$P_v = \tfrac{1}{6}(2v-1)\dots 3\,.\,1\left[\frac{(C_2-U^2/B)(C_1-V^2/A)}{C_1+C_2-U^2/B-V^2/A}\right]^v$$
$$\times\left[\left\{3-\frac{(2v+1)(C_1-V^2/A)}{C_1+C_2-U^2/B-V^2/A}\right\}\frac{\Sigma_l(\eta_l-U/B)^3G_1'''}{(C_2-U^2/B)^2} + 3\frac{\Sigma_l(\eta_l-U/B)G_1'''}{B(C_2-U^2/B)}\right]. \quad \text{......(2095)}$$

The corresponding expression for Q_v is

$$Q_v = -\tfrac{1}{6}(2v-1)\dots 3\,.\,1\left[\frac{(C_2-U^2/B)(C_1-V^2/A)}{C_1+C_2-U^2/B-V^2/A}\right]^v$$
$$\times\left[\left\{3-\frac{(2v+1)(C_2-U^2/B)}{C_1+C_2-U^2/B-V^2/A}\right\}\frac{\Sigma_l(\epsilon_l-V/A)^3G'''}{(C_1-V^2/A)^2} + 3\frac{\Sigma_l(\epsilon_l-V/A)G'''}{A(C_1-V^2/A)}\right]. \quad \text{......(2096)}$$

These expressions cannot be substantially simplified. We observe however that it is easily proved that

$$\Sigma_l(\epsilon_l-V/A)^3G'' = \left[\left(\vartheta\frac{\partial}{\partial\vartheta}\right)^2\overline{E_A}\right]_M, \quad \text{......(2097)}$$

$$\Sigma_l(\epsilon_l-V/A)G'''/A = \lambda\frac{\partial}{\partial\lambda}\left(\frac{\partial\overline{E_A}}{\partial M}\right)_\vartheta, \quad \text{......(2098)}$$

with the help of which the formal evaluation of $\overline{(E_A-\overline{E_A})^{2v-1}}$ may be completed. The order of $\overline{(E_A-\overline{E_A})^{2v-1}}$ in M or N is $v-1$, that is $\frac{1}{2}(2v-1)-\frac{1}{2}$, while that of $\overline{(E_A-\overline{E_A})^{2v}}$ is $\frac{1}{2}(2v)$.

An important special case of these formulae occurs when $N \gg M$, so that the M systems of the first type A are a small part of a very much larger assembly. This corresponds to the physical condition of being in a thermostat or bath formed by the B's. The formulae then simplify. We derive at once from (2084) and (2094)

$$\overline{(E_A-\overline{E_A})^{2v}} = (2v-1)\dots 3\,.1[\vartheta(\partial\overline{E_A}/\partial\vartheta)_M]^v. \quad \text{......(2099)}$$

From (2095)–(2098) we find also

$$\overline{(E_A-\overline{E_A})^{2v-1}}=(2v-1)\ldots 3\,.\,1[\vartheta(\partial\overline{E_A}/\partial\vartheta)_M]^{v-1}$$

$$\times\left\{\frac{\frac{1}{3}(v-1)\left[\left(\vartheta\dfrac{\partial}{\partial\vartheta}\right)^2\overline{E_A}\right]_M}{\vartheta(\partial\overline{E_A}/\partial\vartheta)_M}-\tfrac{1}{2}\lambda\frac{\partial}{\partial\lambda}\left(\frac{\partial\overline{E_A}}{\partial M}\right)_\vartheta\right\}. \quad \ldots\ldots(2100)$$

Formula (2099) (and also (2094)) is the same for all statistics. Formula (2100) simplifies further for classical statistics when $(\partial\overline{E_A}/\partial M)_\vartheta$ is independent of λ and the second term vanishes. Then

$$\overline{(E_A-\overline{E_A})^{2v-1}}=\tfrac{1}{3}(v-1)\,(2v-1)\ldots 3\,.\,1\left[\vartheta\frac{\partial\overline{E_A}}{\partial\vartheta}\right]_M^{v-2}\left[\left(\vartheta\frac{\partial}{\partial\vartheta}\right)^2\overline{E_A}\right]_M. \quad \ldots .(2101)$$

Formulae (2099) and (2101) give the dominant terms of the formulae given by Gibbs,* which alone are relevant when M is large.

In their final forms these fluctuations are obviously independent of the assumption that the "rest" of the assembly is composed of only one sort of system. It is easy to carry through these calculations, for the extra variables of integration introduced by the extra systems can be eliminated by integration before we start on the processes of this section. The only factor in the formulae depending on systems other than the A's, $\vartheta[\partial E/\partial\vartheta]_{M,N}$ is merely replaced by

$$\vartheta\left[\frac{\partial E}{\partial\vartheta}\right]_{M,N,\ldots}.$$

We observe in conclusion that integrals for more complicated fluctuations can be set up in a similar way. It is no use setting up an integral, for $\overline{C(E_A-\overline{E_A})^m(E_B-\overline{E_B})^n}$ for example, in an assembly of only two types of system since in such an assembly $E_A-\overline{E_A}\equiv-(E_B-\overline{E_B})$. We therefore formulate it for an assembly of three types when it takes the form

$$\overline{C(E_A-\overline{E_A})^m(E_B-\overline{E_B})^n}$$

$$=\frac{1}{(2\pi i)^4}\iiiint_{\gamma'}dudvdwdx\left\{\left(\frac{\partial}{\partial x}\right)^m\exp[\Sigma_t G(u+\epsilon_t x)-\overline{E_A}x-Mu]\right\}$$

$$\times\left\{\left(\frac{\partial}{\partial x}\right)^n\exp[\Sigma_t G_1(v+\eta_t x)-\overline{E_B}x-Nv]\right\}\exp[\Sigma_t G_2(w+\zeta_t x)-\overline{E_C}x-Ow].$$

$$\ldots\ldots(2102)$$

Such integrals cannot be evaluated for general values of m and n by the devices of the preceding section but for small values of m and n may be evaluated directly without difficulty.

* Gibbs, *loc. cit.* p. 78. Gibbs' formulae all refer to $N \gg M$.

§ **20·21.** *General fluctuations of $\overline{a_r}$.* The determination of the general fluctuations of a_r can be carried out in the same way as for E_A. From (2073) we can at once derive the equation

$$C\overline{(a_r-\overline{a_r})^n}=\frac{1}{(2\pi i)^3}\iiint_{\gamma'} du\,dv\,dw$$
$$\times\exp[-(M-\overline{a_r})\,u-Nv-Ew+\Sigma_{t\neq r}G(u+\epsilon_t w)+\Sigma_t G_1(v+\eta_t w)]$$
$$\times\left(\frac{\partial}{\partial u}\right)^n\exp[G(u+\epsilon_r w)-\overline{a_r}u]. \quad\text{......(2103)}$$

In this discussion it is convenient to allow the energy values to be degenerate, so that the substitution (2076) is replaced by

$$[g(xz^{\epsilon_t})]^{\varpi_t}=\exp[G(u+\epsilon_t w)], \quad\text{......(2104)}$$

and we have
$$\overline{a_r}=G'(\log\lambda+\epsilon_r\log\vartheta). \quad\text{......(2105)}$$

Since effectively the whole of the integral is again contributed by the immediate neighbourhood of the saddle point, we can obtain at once by the same arguments the analogue of (2080), namely

$$\overline{(a_r-\overline{a_r})^n}=\frac{(i)^{-n}}{J}\iiint_{-\infty}^{+\infty}d\alpha\,d\beta\,d\gamma\,e^{-\frac{1}{2}\Sigma_{t\neq r}(\alpha+\epsilon_t\gamma)^2G''-\frac{1}{2}\Sigma_t(\beta+\eta_t\gamma)^2G_1''}$$
$$\times\{1+O[\Sigma_{t\neq r}(\alpha+\epsilon_t\gamma)^3G'''+\Sigma_t(\beta+\eta_t\gamma)^3G_1''']\}$$
$$\times\left(\frac{\partial}{\partial\alpha}\right)^n[e^{-\frac{1}{2}(\alpha+\epsilon_r\gamma)^2G''}\{1+O((\alpha+\epsilon_r\gamma)^3G''')\}]. \quad\text{......(2106)}$$

The evaluation of this integral is not so straightforward as those of the preceding sections, as a transformation of variables must be made at once. We shall be content to carry through the calculations only for the case n even, equal to $2v$ say, when the O-terms can be neglected as before. Fluctuations of odd order vanish as before to this approximation, and the actual order of their largest surviving terms is lower than that corresponding to the fluctuations of even order by a factor of order $\sqrt{(\overline{a_r})}$.

For fluctuations of even order, $2v$, we have therefore to evaluate

$$\overline{(a_r-\overline{a_r})^{2v}}$$
$$=\frac{(-)^v}{J}\iiint_{-\infty}^{+\infty}d\alpha\,d\beta\,d\gamma\,e^{-\frac{1}{2}\Sigma_{t\neq r}(\alpha+\epsilon_t\gamma)^2G''-\frac{1}{2}\Sigma_t(\beta+\eta_t\gamma)^2G_1''}\left(\frac{\partial}{\partial\alpha}\right)^{2v}e^{-\frac{1}{2}(\alpha+\epsilon_r\gamma)^2G''}. \quad\text{......(2107)}$$

The G'' in the last factor is of course short for $G''(\log\lambda+\epsilon_r\log\vartheta)$. To evaluate (2107) or any similar integral other than the specially simple ones of the preceding sections, where the order of $\int d\alpha$ and $\partial/\partial\gamma$ can be inverted, it is best to start by transforming α to α' so as to reduce the differentiated terms to the form

$$\left(\frac{\partial}{\partial\alpha'}\right)^{2v} e^{-\frac{1}{2}\alpha'^2G''}.$$

We can then integrate with respect to β and γ. All the operations are independent of v and the form of the result must be

$$C\overline{(a_r-\overline{a_r})^{2v}} = (-)^v\, Q\int_{-\infty}^{+\infty} e^{-\frac{1}{2}P\alpha'^2}\left(\frac{\partial}{\partial\alpha'}\right)^{2v} e^{-\frac{1}{2}G''\alpha'^2}\,d\alpha',$$

where P and Q are independent of v. It follows at once that

$$\overline{(a_r-\overline{a_r})^{2v}} = (2v-1)\dots 3\,.\,1\left(\frac{PG''}{P+G''}\right)^v, \qquad \text{......(2108)}$$

$$= (2v-1)\dots 3\,.\,1\left[\overline{(a_r-\overline{a_r})^2}\right]^v. \qquad \text{......(2109)}$$

This familiar form therefore holds for all such fluctuations, but the explicit value can only be obtained by employing sufficient detail to evaluate P.

We will now complete the argument for this particular fluctuation, which is most easily done by direct evaluation of $\overline{(a_r-\overline{a_r})^2}$ as follows.

Put $\alpha' = \alpha - \epsilon_r\gamma$. Then

$$\overline{(a_r-\overline{a_r})^2} = -\frac{1}{J}\iiint_{-\infty}^{+\infty} d\alpha'\,d\beta\,d\gamma\, e^{-\frac{1}{2}\Sigma_{t\neq r}\{\alpha'+(\epsilon_t-\epsilon_r)\gamma\}^2G''-\frac{1}{2}\Sigma_t(\beta+\eta_t\gamma)^2G_1''}\left(\frac{\partial}{\partial\alpha'}\right)^2 e^{-\frac{1}{2}\alpha'^2G''},$$

$$= \frac{1}{J}\iiint_{-\infty}^{+\infty} d\alpha'\,d\beta\,d\gamma\,[G''-\alpha'^2(G'')^2]\, e^{-\frac{1}{2}\Sigma_t\{\alpha'+(\epsilon_t-\epsilon_r)\gamma\}^2G''-\frac{1}{2}\Sigma_t(\beta+\eta_t\gamma)^2G_1''}.$$

We recall that

$$J = \iiint_{-\infty}^{+\infty} d\alpha\,d\beta\,d\gamma\, e^{-\frac{1}{2}\Sigma_t(\alpha+\epsilon_t\gamma)^2G''-\frac{1}{2}\Sigma_t(\beta+\eta_t\gamma)^2G_1''},$$

$$= \iiint_{-\infty}^{+\infty} d\alpha'\,d\beta\,d\gamma\, e^{-\frac{1}{2}\Sigma_t\{\alpha'+(\epsilon_t-\epsilon_r)\gamma\}^2G''-\frac{1}{2}\Sigma_t(\beta+\eta_t\gamma)^2G_1''},$$

and can therefore see at once that

$$\overline{(a_r-\overline{a_r})^2} = G'' + 2(G'')^2\frac{\partial}{\partial(\Sigma_t G'')}\log J, \qquad \text{......(2110)}$$

if J is expressed as a function of the coefficients of the quadratic in α', β, γ in the last form for J. It is easily verified that

$$J = \frac{(2\pi)^{\frac{3}{2}}}{\left[\begin{array}{r}(\Sigma_t G'')(\Sigma_t G_1'')\{\Sigma_t(\epsilon_t-\epsilon_r)^2 G''+\Sigma_t\eta_t^2G_1''\}-(\Sigma_t G'')(\Sigma_t\eta_t G_1'')^2\\ -(\Sigma_t G_1'')\{\Sigma_t(\epsilon_t-\epsilon_r)G''\}^2\end{array}\right]^{\frac{1}{2}}}.$$

It follows at once that

$$\overline{(a_r-\overline{a_r})^2} = G'' - (G'')^2\frac{(\Sigma_t G_1'')\{\Sigma_t(\epsilon_t-\epsilon_r)^2G''+\Sigma_t\eta_t^2G_1''\}-(\Sigma_t\eta_t G_1'')^2}{\left[\begin{array}{l}(\Sigma_t G'')(\Sigma_t G_1'')\{\Sigma_t(\epsilon_t-\epsilon_r)^2G''+\Sigma_t\eta_t^2G_1''\}\\ \quad-(\Sigma_t G'')(\Sigma_t\eta_t G_1'')^2-(\Sigma_t G_1'')\{\Sigma_t(\epsilon_t-\epsilon_r)G''\}^2\end{array}\right]}.$$

$$\text{......(2111)}$$

Equations (2086)–(2093) suffice to show that

$$\Sigma_t(\epsilon_t-\epsilon_r)^2 G''+\Sigma_t\eta_t^2 G_1''-(\Sigma_t\eta_t G_1'')^2/(\Sigma_t G_1'')-\{\Sigma_t(\epsilon_t-\epsilon_r)G''\}^2/(\Sigma_t G'')$$
$$=\vartheta[\partial E/\partial\vartheta]_{M,N}.$$

We can therefore cast (2111) into the simpler form

$$\overline{(a_r-\overline{a_r})^2}=G''-\frac{(G'')^2}{\Sigma_t G''}-\frac{(G'')^2\{\Sigma_t(\epsilon_t-\epsilon_r)G''\}^2}{(\Sigma_t G'')^2\,\vartheta[\partial E/\partial\vartheta]_{M,N}}. \quad \text{......(2112)}$$

Since $G''=\lambda\partial\overline{a_r}/\partial\lambda$, further use of equations (2086)–(2089) enables us to put (2112) in the form

$$\overline{(a_r-\overline{a_r})^2}=\lambda\frac{\partial\overline{a_r}}{\partial\lambda}-\frac{(\lambda\partial\overline{a_r}/\partial\lambda)^2}{\lambda\partial M/\partial\lambda}-\frac{\left[\epsilon_r-\dfrac{\lambda\partial\overline{E_A}/\partial\lambda}{\lambda\partial M/\partial\lambda}\right]^2\left(\dfrac{\lambda\partial\overline{a_r}}{\partial\lambda}\right)^2}{\vartheta[\partial E/\partial\vartheta]_{M,N}}. \quad \text{......(2113)}$$

It is easy to show that

$$\vartheta\left[\frac{\partial\overline{a_r}}{\partial\vartheta}\right]_M=\lambda\frac{\partial\overline{a_r}}{\partial\lambda}\left[\epsilon_r-\frac{\lambda\partial E_A/\partial\lambda}{\lambda\partial M/\partial\lambda}\right],$$

so that (2113) can be cast into the final form

$$\overline{(a_r-\overline{a_r})^2}=\lambda\frac{\partial\overline{a_r}}{\partial\lambda}-\frac{(\lambda\partial\overline{a_r}/\partial\lambda)^2}{\lambda\partial M/\partial\lambda}-\frac{(\vartheta[\partial\overline{a_r}/\partial\vartheta]_M)^2}{\vartheta[\partial E/\partial\vartheta]_{M,N}}. \quad \text{......(2114)}$$

Equations (2114) and (2109) together determine all the fluctuations of a_r of even order.

Under bath conditions when E is very large compared with any quantity connected with systems of the first type, the last term can be neglected and (2114) reduces to

$$\overline{(a_r-\overline{a_r})^2}=\lambda\frac{\partial\overline{a_r}}{\partial\lambda}\left[1-\frac{\lambda\partial\overline{a_r}/\partial\lambda}{\lambda\partial M/\partial\lambda}\right]. \quad \text{......(2115)}$$

This is valid under bath conditions for all statistics. The result commonly given,* though not in this general form, is

$$\overline{(a_r-\overline{a_r})^2}=\lambda\frac{\partial\overline{a_r}}{\partial\lambda} \quad \text{......(2116)}$$

which, as we now see, is valid under bath conditions if also $\overline{a_r}$ is small compared with M. For the Fermi-Dirac statistics,

$$\overline{a_r}=\frac{\varpi_r}{e^{\epsilon_r/kT}/\lambda+1},$$

and equation (2116) gives

$$\overline{(a_r-\overline{a_r})^2}=\overline{a_r}-(\overline{a_r})^2/\varpi_r. \quad \text{......(2117)}$$

* See, for example, Einstein, *Berl. Sitz.* (1924), p. 261; (1925), p. 3; Pauli, *Zeit. f. Physik*, vol. 41, p. 81 (1927).

This result was first given by Pauli. The fluctuation vanishes when $T\to 0$, since the assembly is then tight packed. For the Einstein-Bose statistics,

$$\overline{a_r}=\frac{\varpi_r}{e^{\epsilon_r/kT}/\lambda-1},$$

$$\overline{(a_r-\overline{a_r})^2}=\overline{a_r}+(\overline{a_r})^2/\varpi_r. \qquad \text{......(2118)}$$

Equation (2118) was first shown by Einstein to yield the correct fluctuation of radiation when applied to a collection of light quanta (§ 20·41).

Returning for a moment to the general formula (2114) let us consider the form it takes for classical statistics, which will hold also effectively for many ordinary assemblies. Classically $G\equiv\exp$, so that

$$G'=G''=\overline{a_r},\quad \Sigma_t G'=\Sigma_t G''=M.$$

Equations (2113) and (2114) then reduce to the forms

$$\overline{(a_r-\overline{a_r})^2}=\overline{a_r}-\frac{(\overline{a_r})^2}{M}-\frac{(\overline{a_r})^2(\epsilon_r-\overline{E_A}/M)^2}{\vartheta[\partial E/\partial\vartheta]_{M,N}}, \qquad \text{......(2119)}$$

$$\overline{(a_r-\overline{a_r})^2}=\overline{a_r}-\frac{(\overline{a_r})^2}{M}-\frac{(\vartheta[\partial\overline{a_r}/\partial\vartheta]_M)^2}{\vartheta[\partial E/\partial\vartheta]_{M,N}}. \qquad \text{......(2120)}$$

By obvious extensions we can construct integrals for, and so evaluate, all such expressions as

$$\overline{(a_r-\overline{a_r})^m(a_s-\overline{a_s})^n},\quad \overline{(a_r-\overline{a_r})^m(b_s-\overline{b_s})^n}$$

for given values of m and n, the general reduction formula of the type (2109) being naturally no longer available. The analogues of (2103) in these cases are

$$C\overline{(a_r-\overline{a_r})^m(a_s-\overline{a_s})^n}=\frac{1}{(2\pi i)^3}\iiint_{\gamma'}du\,dv\,dw$$
$$\times\exp[-(M-\overline{a_r}-\overline{a_s})u-Nv-Ew+\Sigma_{t\neq r,s}G(u+\epsilon_t w)+\Sigma_t G_1(v+\eta_t w)]$$
$$\times\left(\frac{\partial}{\partial u}\right)^m\exp[G(u+\epsilon_r w)-\overline{a_r}u]\left(\frac{\partial}{\partial u}\right)^n\exp[G(u+\epsilon_s w)-\overline{a_s}u];\qquad \text{......(2121)}$$

$$C\overline{(a_r-\overline{a_r})^m(b_s-\overline{b_s})^n}=\frac{1}{(2\pi i)^3}\iiint_{\gamma'}du\,dv\,dw$$
$$\times\exp[-(M-\overline{a_r})u-(N-\overline{b_s})v-Ew+\Sigma_{t\neq r}G(u+\epsilon_t w)+\Sigma_{t\neq s}G_1(v+\eta_t w)]$$
$$\times\left(\frac{\partial}{\partial u}\right)^m\exp[G(u+\epsilon_r w)-\overline{a_r}u]\left(\frac{\partial}{\partial v}\right)^n\exp[G_1(v+\eta_s w)-\overline{b_s}v]. \qquad \text{......(2122)}$$

We shall content ourselves with quoting the results required in calculating second order fluctuations in the external reactions of the assembly. These are

$$\overline{(a_r-\overline{a_r})(a_s-\overline{a_s})} = -\frac{(\lambda\,\partial\overline{a_r}/\partial\lambda)(\lambda\,\partial\overline{a_s}/\partial\lambda)}{\lambda\,\partial M/\partial\lambda} - \frac{\vartheta[\partial\overline{a_r}/\partial\vartheta]_M\,\vartheta[\partial\overline{a_s}/\partial\vartheta]_M}{\vartheta[\partial E/\partial\vartheta]_{M,N}}, \quad \ldots(2123)$$

$$\overline{(a_r-\overline{a_r})(b_s-\overline{b_s})} = -\frac{\vartheta[\partial\overline{a_r}/\partial\vartheta]_M\,\vartheta[\partial\overline{b_s}/\partial\vartheta]_N}{\vartheta[\partial E/\partial\vartheta]_{M,N}}. \quad \ldots\ldots(2124)$$

These results are valid in any statistics. For classical statistics the former can be simplified by replacing $\lambda\,\partial\overline{a_r}/\partial\lambda$ by $\overline{a_r}$, *etc.*

§ **20·3.** *Fluctuations in the external reactions.* The generalized reaction of the systems of type A on an external body is given by

$$Y = \Sigma_r\left(-\frac{\partial\epsilon_r}{\partial y}\right)a_r, \quad \ldots\ldots(2125)$$

where y is the corresponding coordinate defining the position of the external body. The average reaction is

$$\overline{Y} = \Sigma_r\left(-\frac{\partial\epsilon_r}{\partial y}\right)\overline{a_r} = \Sigma_r\left(-\frac{\partial\epsilon_r}{\partial y}\right)G'(\log\lambda+\epsilon_r\log\vartheta). \quad \ldots\ldots(2126)$$

The second order fluctuation of Y is

$$\overline{(Y-\overline{Y})^2} = \Sigma_r\left(\frac{\partial\epsilon_r}{\partial y}\right)^2\overline{(a_r-\overline{a_r})^2} + 2\Sigma_{r\neq s}\frac{\partial\epsilon_r}{\partial y}\frac{\partial\epsilon_s}{\partial y}\overline{(a_r-\overline{a_r})(a_s-\overline{a_s})}.$$

Applying formulae (2114) and (2123) we find that

$$\overline{(Y-\overline{Y})^2} = \lambda\frac{\partial}{\partial\lambda}\Sigma_r\left(\frac{\partial\epsilon_r}{\partial y}\right)^2\overline{a_r} - \frac{(\lambda\,\partial\overline{Y}/\partial\lambda)^2}{\lambda\,\partial M/\partial\lambda} - \frac{(\vartheta[\partial\overline{Y}/\partial\vartheta]_M)^2}{\vartheta[\partial E/\partial\vartheta]_{M,N}}. \quad \ldots\ldots(2127)$$

This expression can be cast into alternative forms. If we differentiate (2125) and (2126) with respect to y keeping λ, ϑ constant, we find first

$$\frac{\partial\overline{Y}}{\partial y} = \overline{\frac{\partial Y}{\partial y}} + \Sigma\left(-\frac{\partial\epsilon_r}{\partial y}\right)\frac{\partial\overline{a_r}}{\partial y}, \quad \frac{\partial\overline{a_r}}{\partial y} = \log\vartheta\,\frac{\partial\epsilon_r}{\partial y}\lambda\frac{\partial\overline{a_r}}{\partial\lambda}.$$

It follows that

$$\overline{(Y-\overline{Y})^2} = \frac{\partial\overline{Y}/\partial y - \overline{\partial Y/\partial y}}{\log 1/\vartheta} - \frac{(\lambda\,\partial\overline{Y}/\partial\lambda)^2}{\lambda\,\partial M/\partial\lambda} - \frac{(\vartheta[\partial\overline{Y}/\partial\vartheta]_M)^2}{\vartheta[\partial E/\partial\vartheta]_{M,N}}. \quad \ldots\ldots(2128)$$

If however we differentiate keeping M and not λ constant, we find

$$\left(\frac{\partial\overline{Y}}{\partial y}\right)_M = \overline{\left(\frac{\partial Y}{\partial y}\right)_M} + \Sigma\left(-\frac{\partial\epsilon_r}{\partial y}\right)\left(\frac{\partial\overline{a_r}}{\partial y}\right)_M, \quad \left(\frac{\partial\overline{a_r}}{\partial y}\right)_M = \log\vartheta\left(\frac{\partial\epsilon_r}{\partial y} + \frac{\lambda\,\partial\overline{Y}/\partial\lambda}{\lambda\,\partial M/\partial\lambda}\right)\lambda\frac{\partial\overline{a_r}}{\partial\lambda}.$$

It then follows that

$$\frac{(\partial\overline{Y}/\partial y)_M - \overline{(\partial Y/\partial y)_M}}{\log 1/\vartheta} = \lambda\frac{\partial}{\partial\lambda}\Sigma_r\left(\frac{\partial\epsilon_r}{\partial y}\right)^2\overline{a_r} - \frac{(\lambda\,\partial\overline{Y}/\partial\lambda)^2}{\lambda\,\partial M/\partial\lambda},$$

and therefore that

$$\overline{(Y-\overline{Y})^2}=\frac{(\partial\overline{Y}/\partial y)_M-\overline{(\partial Y/\partial y)_M}}{\log 1/\vartheta}-\frac{(\vartheta[\partial\overline{Y}/\partial\vartheta]_M)^2}{\vartheta[\partial E/\partial\vartheta]_{M,N}} \quad \text{......(2129)}$$

Gibbs* gives a formula for $\overline{(Y-\overline{Y})^2}$ equivalent to (2129) without the last term, which is of course negligible under his condition that N is large compared with M. Equation (2129) holds for all statistics.

With the help of (2124) it is easy to show that (2129) is formally unaltered if Y refers to the total reaction due to all groups of systems instead of to the partial reaction of a single group.

In (2129) all the terms except $\overline{\partial Y/\partial y}$ or $\Sigma_r \overline{a_r}(-\partial^2\epsilon_r/\partial y^2)$ have an obvious interpretation. This term lies deeper and is compared by Gibbs to an elasticity.

In illustration we shall apply these results to the limiting case in which the reaction is a pressure, and the mechanics and statistics are classical. We can then use the ordinary partition function $f(\vartheta)$. It is sufficient to consider the reaction with a part of the wall only, which may be taken to be plane and represented by a movable piston in a cylinder of cross-section A, which completes the enclosure. We cannot progress without *some* definite assumption as to the field of force near the wall. We shall suppose its potential is D/d^s, where d is the distance of the molecule from the wall, and D and s are constants. If D is small and $s > 4$, this will adequately represent an intense local field. If y is the length of the cylinder and x_1, x_2, x_3 rectangular cartesian coordinates, x_1 along the cylinder, then

$$\epsilon_r=\frac{1}{2m}(p_1{}^2+p_2{}^2+p_3{}^2)_r+[D/(y-x_1)^s]_r,$$

$$f(\vartheta)=\frac{(2\pi m)^{\frac{3}{2}}A}{h^3(\log 1/\vartheta)^{\frac{3}{2}}}\int_0^y \exp\left\{-\log 1/\vartheta\,\frac{D}{(y-x_1)^s}\right\}dx_1. \quad \text{......(2130)}$$

When the field is sufficiently local, or D nearly zero, we have the usual result,

$$f(\vartheta)=\frac{(2\pi m)^{\frac{3}{2}}Ay}{h^3(\log 1/\vartheta)^{\frac{3}{2}}}. \quad \text{......(2131)}$$

By direct differentiation of (2130) we show further that to the same approximation

$$\frac{1}{f}\frac{\partial f}{\partial y}=\frac{1}{y}. \quad \text{......(2132)}$$

It follows at once from the usual formula

$$\overline{Y}=\frac{M}{\log 1/\vartheta}\frac{\partial}{\partial y}\log f(\vartheta)$$

that

$$Y=M/(y\log 1/\vartheta),\quad \overline{Y}y=PV=MkT. \quad \text{......(2133)}$$

We see then that we arrive at (2133) whatever the law of force. But when

* Gibbs, *loc. cit.* p. 81.

we come to calculate $\overline{\partial Y/\partial y}$ we find that it depends essentially on the form of the law. Thus

$$\overline{\frac{\partial Y}{\partial y}} = \Sigma_r\left(-\frac{\partial^2\epsilon_r}{\partial y^2}\right)\overline{a_r} = -\frac{M}{y}\int_0^v \frac{Ds(s+1)}{(y-x_1)^{s+2}}\exp\left\{-\log 1/\vartheta\frac{D}{(y-x_1)^s}\right\}dx_1,$$

$$= -\frac{(s+1)M}{y}\int_{D/y^s}^{\infty}\left(\frac{q}{D}\right)^{1/s} e^{-q\log 1/\vartheta}\,dq.$$

Now D/y^s is the potential of the wall field at the other end of the cylinder, and is indistinguishable from zero. Therefore

$$\overline{\frac{\partial Y}{\partial y}} = -\frac{(s+1)M}{yD^{1/s}}\frac{\Gamma(1+1/s)}{(\log 1/\vartheta)^{1+1/s}}. \qquad \text{......(2134)}$$

This depends essentially on D and s and moreover must tend to infinity if $D\to 0$ or $s\to\infty$. The largeness of $\overline{\partial Y/\partial y}$ and therewith the fluctuation $\overline{(Y-\overline{Y})^2}$ is however then to be expected, for in the limit the reaction of the boundary with a single molecule must itself become infinite, although $\overline{Y}$ retains its usual value. In spite of this, however, if we calculate roughly the order of $\overline{(Y-\overline{Y})^2}/(\overline{Y})^2$ for reasonable values of D and s, we find that it is still negligibly small. Equation (2134) and the usual form for E for a monatomic gaseous assembly lead to

$$\overline{(Y-\overline{Y})^2} = -\frac{5}{2}\frac{M}{y^2(\log 1/\vartheta)^2} + \frac{sM}{yD^{1/s}}\frac{\Gamma(2+1/s)}{(\log 1/\vartheta)^{2+1/s}},$$

$$\frac{\overline{(Y-\overline{Y})^2}}{(\overline{Y})^2} = \frac{\overline{(p-\overline{p})^2}}{(\overline{p})^2} = \frac{1}{M}\left\{-\tfrac{5}{2} + s\Gamma(2+1/s)\frac{V}{A(D/kT)^{1/s}}\right\}.$$

A reasonable assumption as to the wall field is to take $s=4$ and suppose that $(D/kT)/(y-x_1)^s$ is small, 10^{-2} say, for $y-x_1=10^{-7}$ cm. Thus $D/kT=10^{-30}$. With $V=1$, $A=1$ and $M=2{\cdot}7\times 10^{19}$ we have

$$\overline{(Y-\overline{Y})^2}/(\overline{Y})^2 = 4{\cdot}7\times 10^{-12}.$$

Other second order fluctuations involving reactions, such as

$$\overline{(Y-\overline{Y})(Z-\overline{Z})},\quad \overline{(Y-\overline{Y})(E_A-\overline{E_A})},$$

both mentioned by Gibbs, may be similarly calculated. We find for any statistics

$$\overline{(Y-\overline{Y})(Z-\overline{Z})} = \left[\overline{\frac{\partial Y}{\partial z}} - \overline{\frac{\partial Y}{\partial z}}\right]_M \Big/ \log 1/\vartheta - \frac{\vartheta[\partial\overline{Y}/\partial\vartheta]_M\,\vartheta[\partial\overline{Z}/\partial\vartheta]_M}{\vartheta[\partial\overline{E}/\partial\vartheta]_{M,N}}, \qquad \text{......(2135)}$$

$$\overline{(Y-\overline{Y})(E_A-\overline{E_A})} = \vartheta\left[\frac{\partial\overline{Y}}{\partial\vartheta}\right]_M\left(1-\frac{\vartheta[\partial\overline{E_A}/\partial\vartheta]_M}{\vartheta[\partial\overline{E}/\partial\vartheta]_{M,N}}\right). \qquad \text{......(2136)}$$

Of course, since the forces have a potential,

$$\overline{\frac{\partial Y}{\partial z}} = \overline{\frac{\partial Z}{\partial y}},\quad \frac{\partial\overline{Y}}{\partial z} = \frac{\partial\overline{Z}}{\partial y}.$$

§ 20·31. *Fluctuations of dissociation in dissociating assemblies.* We have so far considered only assemblies in which the numbers of component systems are fixed, but the calculations can easily be extended in their general form to dissociating assemblies. We shall not however give the most general form of these calculations, and, since actual dissociating assemblies can almost always be regarded as classical, we shall be content to restrict the discussion to classical dissociating assemblies, starting with the simple assembly of § 5·2. If we transform equation (440) by writing X, Y for X_1, X_2,

$$x_1 = e^u, \quad x_2 = e^v, \quad z = e^w$$

and remember that classically $g \equiv \exp$, we have at once

$$C = \frac{1}{(2\pi i)^3} \iiint_{\gamma'} du\,dv\,dw$$
$$\times \exp[-Xu - Yv - Ew + e^u F_1(w) + e^v F_2(w) + e^{u+v} F_{12}(w)], \quad \ldots\ldots(2137)$$

where $F(w) = \Sigma_r \varpi_r e^{w\epsilon_r}$ is the ordinary partition function in the present notation. On examining equation (441) we see that the extra factor necessary to provide an integral for $C\overline{M_1}$ is obtained by operating on the factor $\exp(e^u F_1)$ of the integrand with the operator $\partial/\partial u$. It follows at once by obvious repetitions of the operation that

$$C\overline{M_1^n} = \frac{1}{(2\pi i)^3} \iiint_{\gamma'} du\,dv\,dw$$
$$\times \exp[-Xu - Yv - Ew + e^v F_2 + e^{u+v} F_{12}] \left(\frac{\partial}{\partial u}\right)^n \exp(e^u F_1). \quad \ldots\ldots(2138)$$

Therefore by a familiar argument

$$C\overline{(M_1 - \overline{M_1})^n} = \frac{1}{(2\pi i)^3} \iiint_{\gamma'} du\,dv\,dw$$
$$\times \exp[-(X - \overline{M_1})\,u - Yv - Ew + e^v F_2 + e^{u+v} F_{12}] \left(\frac{\partial}{\partial u}\right)^n \exp(e^u F_1 - \overline{M_1} u).$$
$$\ldots\ldots(2139)$$

We can now evaluate (2139) by the methods already developed. We shall not consider further fluctuations of odd order. For fluctuations of even order, $2v$, it is easy to reduce (2139) to the form

$$\overline{(M_1 - \overline{M_1})^{2v}} = \frac{(-)^v}{J} \iiint_{-\infty}^{+\infty} d\alpha\,d\beta\,d\gamma \left(\frac{\partial}{\partial \alpha}\right)^{2v} \exp\left[-\frac{1}{2}\left\{\overline{M_1}\alpha^2 + 2\overline{E_1}\alpha\gamma + \vartheta \frac{\partial \overline{E_1}}{\partial \vartheta}\gamma^2\right\}\right]$$
$$\times \exp\left[-\frac{1}{2}\left\{\overline{M_2}\beta^2 + 2\overline{E_2}\beta\gamma + \vartheta \frac{\partial \overline{E_2}}{\partial \vartheta}\gamma^2 + \overline{N}(\alpha+\beta)^2 + 2\overline{E_{12}}(\alpha+\beta)\gamma + \vartheta \frac{\partial \overline{E_{12}}}{\partial \vartheta}\gamma^2\right\}\right].$$
$$\ldots\ldots(2140)$$

The various symbols have their usual meanings, and $\partial/\partial\vartheta$ means differentiation with λ and μ constant. The integral for J is

$$J=\iiint_{-\infty}^{+\infty} d\alpha\, d\beta\, d\gamma$$

$$\times \exp\left[-\frac{1}{2}\left\{X\alpha^2+Y\beta^2+\vartheta\frac{\partial E}{\partial\vartheta}\gamma^2+2(\overline{E_2}+\overline{E_{12}})\,\beta\gamma+2(\overline{E_1}+\overline{E_{12}})\,\gamma\alpha+2\overline{N}\alpha\beta\right\}\right], \quad \text{......(2141)}$$

$$=\frac{(2\pi)^{\frac{3}{2}}}{\left\{XY\vartheta\frac{\partial E}{\partial\vartheta}+2\overline{N}(\overline{E_1}+\overline{E_{12}})(\overline{E_2}+\overline{E_{12}})-X(\overline{E_2}+\overline{E_{12}})^2-Y(\overline{E_1}+\overline{E_{12}})^2-\vartheta\frac{\partial E}{\partial\vartheta}\overline{N}^2\right\}^{\frac{1}{2}}} \quad \text{......(2142)}$$

It follows at once from the form of (2140) that

$$\overline{(M_1-\overline{M_1})^{2v}}=(2v-1)\ldots 3\,.\,1\left[\overline{(M_1-\overline{M_1})^2}\right]^v. \quad \text{......(2143)}$$

It is therefore again only necessary to consider $v=1$. The extra factor in $\overline{(M_1-\overline{M_1})^2}$ beyond the integrand of J is

$$\overline{M_1}-(\overline{M_1}\alpha+\overline{E_1}\gamma)^2.$$

It follows that

$$\overline{(M_1-\overline{M_1})^2}=\overline{M_1}+2\left(\overline{M_1}^2\frac{\partial}{\partial X}+\overline{M_1}\overline{E_1}\frac{\partial}{\partial\overline{E_1}}+\overline{E_1}^2\frac{\partial}{\partial(\vartheta\,\partial E/\partial\vartheta)}\right)\log J,$$

from which it follows by a straightforward reduction that

$$\overline{(M_1-\overline{M_1})^2}=\frac{\overline{M_1}\,\overline{M_2}\,\overline{N}}{XY-\overline{N}^2}$$

$$-\frac{\left\{\overline{E_1}-\overline{M_1}\dfrac{Y(\overline{E_1}+\overline{E_{12}})-\overline{N}(\overline{E_2}+\overline{E_{12}})}{XY-\overline{N}^2}\right\}^2}{\vartheta\dfrac{\partial E}{\partial\vartheta}-\dfrac{X(\overline{E_2}+\overline{E_{12}})^2+Y(\overline{E_1}+\overline{E_{12}})^2-2\overline{N}(\overline{E_1}+\overline{E_{12}})(\overline{E_2}+\overline{E_{12}})}{XY-\overline{N}^2}}. \quad \text{...(2144)}$$

This is rather complicated and there is no very simple general form. The last denominator can be cast into the form

$$\vartheta\left[\frac{\partial E}{\partial\vartheta}\right]_{\overline{M},\overline{N}}+\frac{\overline{M_1}\,\overline{M_2}\,\overline{N}}{XY-\overline{N}^2}\left(\vartheta\frac{\partial}{\partial\vartheta}\log\kappa\right)^2,$$

where κ is the equilibrium constant F_1F_2/F_{12}; and the numerator is

$$\left(\frac{\overline{M_1}\,\overline{M_2}\,\overline{N}}{XY-\overline{N}^2}\right)^2\left(\vartheta\frac{\partial}{\partial\vartheta}\log\kappa\right)^2.$$

We can therefore write (2144) in the form

$$\overline{(M_1-\overline{M_1})^2}=\frac{\overline{M_1}\,\overline{M_2}\,\overline{N}}{XY-\overline{N}^2}\left[1-\frac{\dfrac{\overline{M_1}\,\overline{M_2}\,\overline{N}}{XY-\overline{N}^2}\left(\vartheta\dfrac{\partial}{\partial\vartheta}\log\kappa\right)^2}{\vartheta\left[\dfrac{\partial E}{\partial\vartheta}\right]_{\overline{M},\overline{N}}+\dfrac{\overline{M_1}\,\overline{M_2}\,\overline{N}}{XY-\overline{N}^2}\left(\vartheta\dfrac{\partial}{\partial\vartheta}\log\kappa\right)^2}\right]. \quad ...(2145)$$

This shows the type of formula we should expect in more complex examples. We note as a check that, if there is no "molecule", $\overline{N}=0$ and the fluctuation vanishes. We may note also that the argument is unaffected by the presence of other sets of systems in the assembly, so long as these do not combine with or are not formed out of systems of the types already under discussion. The only residual effect of such other systems is to increase $\vartheta[\partial E/\partial\vartheta]_{\overline{M},\overline{N}}$, which will continue to refer to the whole assembly, though it may not retain this simple form. When the whole assembly is large (bath conditions), we may therefore simplify (2145) by the omission of the last term, and write

$$\overline{(M_1-\overline{M_1})^2}=\frac{\overline{M_1}\,\overline{M_2}\,\overline{N}}{XY-\overline{N}^2}. \quad ......(2146)$$

From the symmetry of (2145) or (2146) it follows that $\overline{(M_2-\overline{M_2})^2}$ has the same value, as also does $\overline{(N-\overline{N})^2}$, since in this case $N-\overline{N}=-(M_1-\overline{M_1})$.

§ **20·32.** *Fluctuations of energy in dissociating assemblies.* The fluctuations of the energy can be studied in a similar way. By following out the usual steps we find easily that

$$C\overline{(E_1-\overline{E_1})^n}=\frac{1}{(2\pi i)^3}\iiint_{\gamma'} du\,dv\,dw$$
$$\times\exp[-Xu-Yv-(E-\overline{E_1})\,w+e^vF_2+e^{u+v}F_{12}]\left(\frac{\partial}{\partial w}\right)^n\exp[e^uF_1(w)-\overline{E_1}w]. \quad ......(2147)$$

The further study of (2147) also follows the usual course. The form of (2143) is preserved. In the explicit calculations the extra factor in the integrand of $\overline{(E_1-\overline{E_1})^2}$ is

$$\vartheta\frac{\partial\overline{E_1}}{\partial\vartheta}-\left(\vartheta\frac{\partial\overline{E_1}}{\partial\vartheta}\gamma+\overline{E_1}\alpha\right)^2.$$

After the usual reductions we find that

$$\overline{(E_1-\overline{E_1})^2}=\vartheta\frac{\partial\overline{E_1}}{\partial\vartheta}-\frac{Y\overline{E_1}^2}{XY-\overline{N}^2}$$
$$-\frac{\left\{\vartheta\dfrac{\partial\overline{E_1}}{\partial\vartheta}-\overline{E_1}\dfrac{Y(\overline{E_1}+\overline{E_{12}})-\overline{N}(\overline{E_2}+\overline{E_{12}})}{XY-\overline{N}^2}\right\}^2}{\vartheta\dfrac{\partial E}{\partial\vartheta}-\dfrac{X(\overline{E_2}+\overline{E_{12}})^2+Y(\overline{E_1}+\overline{E_{12}})^2-2\overline{N}(\overline{E_1}+\overline{E_{12}})(\overline{E_2}+\overline{E_{12}})}{XY-\overline{N}^2}}.$$

To reduce this to a more intelligible form we use

$$\vartheta\left[\frac{\partial \overline{E_1}}{\partial \vartheta}\right]_{\overline{M_1}} = \vartheta\frac{\partial \overline{E_1}}{\partial \vartheta} - \frac{(\overline{E_1})^2}{\overline{M_1}}.$$

Then

$$\overline{(E_1-\overline{E_1})^2} = \vartheta\left[\frac{\partial \overline{E_1}}{\partial \vartheta}\right]_{\overline{M_1}} + \frac{\overline{M_1}\,\overline{M_2}\,\overline{N}}{XY-\overline{N}^2}\left(\frac{\overline{E_1}}{\overline{M_1}}\right)^2$$

$$-\frac{\left\{\vartheta\left[\dfrac{\partial \overline{E_1}}{\partial \vartheta}\right]_{\overline{M_1}} + \dfrac{\overline{E_1}}{\overline{M_1}}\dfrac{\overline{M_1}\,\overline{M_2}\,\overline{N}}{XY-\overline{N}^2}\,\vartheta\dfrac{\partial}{\partial \vartheta}\log\kappa\right\}^2}{\vartheta\left[\dfrac{\partial \overline{E}}{\partial \vartheta}\right]_{\overline{M},\overline{N}} + \dfrac{\overline{M_1}\,\overline{M_2}\,\overline{N}}{XY-\overline{N}^2}\left(\vartheta\dfrac{\partial}{\partial \vartheta}\log\kappa\right)^2}. \quad \text{......(2148)}$$

In this form the relationships of the new formula to preceding ones are obvious. If $\overline{N}=0$ it reduces to (2094). If $\vartheta[\partial E/\partial\vartheta]_{\overline{M},\overline{N}}$ is large, so that the whole assembly is large compared with the systems under discussion, then, using (2146),

$$\overline{(E_1-\overline{E_1})^2} = \vartheta\left[\frac{\partial \overline{E_1}}{\partial \vartheta}\right]_{\overline{M_1}} + \left(\frac{\overline{E_1}}{\overline{M_1}}\right)^2\overline{(M_1-\overline{M_1})^2}. \quad \text{......(2149)}$$

Equation (2149) states that, under bath conditions, the energy fluctuation is equal to the energy fluctuation for fixed dissociation plus the fluctuation of the energy resulting from the fluctuation in the dissociation. This additive result is not however true in general, since for less extensive assemblies the fluctuations in $\overline{M_1}$ and in the energy content for fixed $\overline{M_1}$ are not independent.

§20·33. *Other fluctuations.* It should now be sufficiently clear how to construct exact integral expressions for any fluctuation such as

$$\overline{(M_r-\overline{M_r})^m(N_s-\overline{N_s})^n}, \quad \overline{(M_r-\overline{M_r})^m(E_s-\overline{E_s})^n}, \ \textit{etc.},$$

for the most general gaseous dissociating assembly. We shall not write down these integrals, still less attempt to evaluate them here, as they are obviously complicated. It is sufficient to have established a direct method by which they can be calculated if required.

It remains however to adapt § 20·21 to dissociating assemblies. When a molecule AB is included, classical statistics are used, and the notation is slightly altered to conform to Chapter v, equation (2103) becomes

$$C\overline{(a_r{}^1-\overline{a_r{}^1})^n} = \frac{1}{(2\pi i)^3}\iiint_{\gamma'} du\,dv\,dw\left(\frac{\partial}{\partial u}\right)^n \exp\{\varpi_r{}^1 e^{u+\epsilon_r{}^1 w} - \overline{a_r{}^1}u\}$$

$$\times\exp[-(X-\overline{a_r{}^1})\,u - Yv - Ew + e^u\{F_1(w) - \varpi_r{}^1 e^{\epsilon_r{}^1 w}\} + e^v F_2(w) + e^{u+v}F_{12}(w)].$$

When we carry through the usual calculations we find that (2109) continues to hold, and finally that an explicit form for $\overline{(a_r^1-\overline{a_r^1})^2}$ is

$$\overline{(a_r^1-\overline{a_r^1})^2}=\overline{a_r^1}-\frac{(\overline{a_r^1})^2\,Y}{XY-\overline{N}^2}-\frac{(\overline{a_r^1})^2\left\{\epsilon_r^1-\frac{\overline{E_1}}{\overline{M_1}}+\frac{\overline{M_2}\,\overline{N}}{XY-\overline{N}^2}\,\vartheta\frac{\partial}{\partial\vartheta}\log\kappa\right\}^2}{\vartheta\left[\frac{\partial \overline{E}}{\partial\vartheta}\right]_{\overline{M},\overline{N}}+\frac{\overline{M_1}\,\overline{M_2}\,\overline{N}}{XY-\overline{N}^2}\left(\vartheta\frac{\partial}{\partial\vartheta}\log\kappa\right)^2}. \quad \text{......(2150)}$$

This reduces to (2119) when $\overline{N}=0$.

§20·4. *Formal consequences of general fluctuation theorems.* An interesting consequence of the general form, e.g. (2108), of our fluctuation theorems may be noted here. Retaining only terms of the highest order we may say that if

$$\overline{(P-\overline{P})^2}=\mu,$$

then

$$\overline{(P-\overline{P})^{2n-1}}=O(\mu^{n-1})\{1+O(1/E)\},$$

$$\overline{(P-\overline{P})^{2n}}=1\,.\,3\ldots(2n-1)\,\mu^n\{1+O(1/\mu)+O(1/E)\}. \quad \text{......(2151)}$$

We may pass at once from these equations to a general distribution law in P for examples of the assembly with the accuracy of (2151) by means of a theorem due to Pólya,* which we quote here.

Theorem 20·4. *The distribution function $f(x)$ is continuous. The moments t_m of $f(x)$,*

$$t_m=\int_{-\infty}^{+\infty}x^m\,df(x)\quad(m=0,1,2,\ldots),$$

are assumed to satisfy the condition that

$$\overline{\mathop{\mathrm{Lt}}_{m\to 0}}\frac{\sqrt[2m]{(t_{2m})}}{m}$$

is finite. If a sequence of distribution functions $f_1(x),\ldots,f_n(x),\ldots$ satisfies the infinite set of limiting conditions

$$\mathop{\mathrm{Lt}}_{n\to\infty}\int_{-\infty}^{+\infty}x^m\,df_n(x)=t_m\quad(m=0,1,2,\ldots),$$

then

$$\mathop{\mathrm{Lt}}_{n\to\infty}f_n(x)=f(x)$$

uniformly in any interval.

In equations (2151) let us substitute

$$P=\overline{P}+\gamma\sqrt{\mu}.$$

* Pólya, *Math. Zeit.* vol. 8, p. 171 (1920). The integrals must be taken as Stieltjes' integrals.

Then we have proved that

$$\overline{\gamma^{2n}} = 1\,.\,3\,.\ldots.(2n-1)\left\{1 + O\left(\frac{1}{E}\right) + O\left(\frac{1}{\mu}\right)\right\},$$

$$\overline{\gamma^{2n-1}} = O\left(\frac{1}{\sqrt{\mu}}\right) + O\left(\frac{1}{E}\right).$$

Now let E, μ tend to infinity in fixed ratios, which means taking larger and larger assemblies. The γ-moments tend to the moments of the distribution function

$$f(\gamma)\,d\gamma = \frac{1}{(2\pi)^{\frac{1}{2}}} e^{-\frac{1}{2}\gamma^2}\,d\gamma. \qquad \ldots\ldots(2152)$$

Pólya's theorem applies, and it follows that (2152) is the actual limit for infinite assemblies of the distribution function of examples of the assembly, uniformly in any fixed interval of γ. In terms of P the distribution function therefore tends to the limit

$$f(P)\,dP = \frac{1}{(2\pi\mu)^{\frac{1}{2}}} e^{-\frac{1}{2}(P-\overline{P})^2/\mu}\,dP, \qquad \ldots\ldots(2153)$$

uniformly in any fixed interval $\overline{P} \pm z\sqrt{\mu}$.

This theorem provides us with the simplest means of completing the proof that the possession of $\overline{P}$ is a normal property of a sufficiently large assembly. It will be a normal property provided only

$$\int_{\overline{P}-p}^{\overline{P}+p} f(P)\,dP = 1 - \epsilon(p),$$

where ϵ can be made very small while $p/\overline{P}$ is still itself very small. But this equation is obviously equivalent to

$$\frac{1}{\sqrt{\mu}}\int_p^\infty e^{-\frac{1}{2}x^2/\mu}\,dx = \epsilon(p),$$

of which the left-hand side is $O\left(\frac{1}{p}e^{-p^2/2\mu}\right)$. This can be made small compared with unity for small values of $p/\overline{P}$ provided only $(\overline{P})^2/\mu$ is very large, which is the general result of this chapter.

The elegant formula (2153), which we have just obtained from the purely analytical side, really gives no information not already contained in Einstein's quasi-thermodynamical formulae (627) and (628). Taking these in the form (627) and comparing with (2153) we see that these two equations are attempting to assert the same relation and that they succeed if

$$-(\Sigma\,\Delta S)/k = \tfrac{1}{2}(P-\overline{P})^2/\mu.$$

In this chapter we are only working accurately for comparatively small displacements from the true equilibrium. If we further confine our attention to a small part of a large assembly, then this equation is equivalent to

$$\frac{1}{k}\frac{\partial^2 S}{\partial \overline{P}^2} = -\frac{1}{\mu}, \qquad \text{......(2154)}$$

or

$$\overline{(P-\overline{P})^2} = -k\Big/\frac{\partial^2 S}{\partial \overline{P}^2}. \qquad \text{......(2155)}$$

It is easy to verify that (2154) is correct in the simple cases to which alone *all* the calculations apply. For example, if $P = E_A$, then in bath conditions $\mu = kT^2 \partial \overline{E_A}/\partial T$ and $\partial S/\partial \overline{E_A} = 1/T$, which verifies (2154). Similarly if $P = a_r$ in a gaseous assembly, then S as a function of $\overline{a_r}$ contains $\overline{a_r}\{\log[f(T)/\overline{a_r}]+1\}$, $\mu = \overline{a_r}$, and (2154) is again verified. A similar verification holds for $\overline{M_1}$.

§ **20·41.** *Special cases.* The special formulae that are of primary importance among those of this chapter refer to bath conditions and are

(G) $$\overline{(E-\overline{E})^2} = \vartheta\frac{\partial \overline{E}}{\partial \vartheta} = kT^2\frac{\partial \overline{E}}{\partial T}, \qquad \text{......(2156)}$$

(G) $$\overline{(a_r-\overline{a_r})^2} = \lambda\frac{\partial \overline{a_r}}{\partial \lambda} - \frac{(\lambda\,\partial \overline{a_r}/\partial\lambda)^2}{\lambda\,\partial M/\partial \lambda}, \qquad \text{......(2157)}$$

(C) $$\overline{(a_r-\overline{a_r})^2} = \overline{a_r}(1-\overline{a_r}/M), \qquad \text{......(2158)}$$

(C) $$\overline{(M_1-\overline{M_1})^2} = \frac{\overline{M_1}\,\overline{M_2}\,\overline{N}}{XY-\overline{N}^2}, \qquad \text{......(2159)}$$

and in particular (2156). Those of the formulae true for all statistics are denoted by (G), those limited to classical statistics by (C). The form taken by (2156) for special systems should be noted.

(i) A gas of N structureless classical atoms

$$\overline{(E-\overline{E})^2} = \tfrac{3}{2}Nk^2T^2 = (\overline{E})^2/\tfrac{3}{2}N. \qquad \text{......(2160)}$$

(ii) A set of N Planck's oscillators of frequency ν

$$\overline{(E-\overline{E})^2} = \frac{Nh^2\nu^2 e^{h\nu/kT}}{(e^{h\nu/kT}-1)^2} = \frac{(\overline{E})^2}{N} + h\nu\overline{E}. \qquad \text{......(2161)}$$

(iii) Temperature ν-radiation in unit volume. Regarding this as the energy of the correct number of Planck oscillators, so that

$$\rho(\nu)\,d\nu = N(\nu)\,d\nu\frac{h\nu}{e^{h\nu/kT}-1} \quad \left(N(\nu)\,d\nu = \frac{8\pi\nu^2 d\nu}{c^3}\right),$$

we find

$$\overline{\{\rho(\nu)\,d\nu - \overline{\rho(\nu)}\,d\nu\}^2} = \frac{8\pi h^2\nu^4 e^{h\nu/kT}\,d\nu}{(e^{h\nu/kT}-1)^2},$$

$$= h\nu\overline{\rho(\nu)}\,d\nu + \frac{c^3\{\overline{\rho(\nu)}\}^2\,d\nu}{8\pi\nu^2}. \qquad \text{......(2162)}$$

Formula (2162) has played an important part in the history of the theory of radiation. It will be observed that $h\nu\overline{\rho(\nu)}\,d\nu = h^2\nu^2\overline{n(\nu)}\,d\nu$, where $\overline{n(\nu)}\,d\nu$ is the average number of ν-quanta per unit volume. On referring to (2158) we see that this term gives just the fluctuation we should get if the extreme light-quantum view of radiation as "classical" particles could be adopted. On the other hand this term cannot be interpreted on the classical wave theory, for all fluctuations by interference must depend on the square of the energy density. For the second term the reverse is true; classical light quanta cannot account for it, while the classical wave theory accounts for it naturally by interference. These critical remarks of course apply not to the result itself as a property of the equilibrium state, but only to the mechanisms proposed for its maintenance.

A satisfactory alternative derivation of (2162) can be given by regarding ν-radiation as light quanta each of energy $h\nu$ obeying Einstein-Bose statistics. It follows then from (2118) that

$$\overline{\{\rho(\nu)\,d\nu - \overline{\rho(\nu)}\,d\nu\}^2} = h^2\nu^2\overline{\{n(\nu)\,d\nu - \overline{n(\nu)}\,d\nu\}^2} = h^2\nu^2\left[\overline{n(\nu)}\,d\nu + \frac{\{\overline{n(\nu)}\}^2\,d\nu}{N(\nu)}\right],$$
$$= h\nu\overline{\rho(\nu)}\,d\nu + c^3\{\overline{\rho(\nu)}\}^2\,d\nu/8\pi\nu^2,$$

as before.

§20·5. *The scattering of light by liquids and gases. Opalescence near the critical point.* When a beam of light is passed through a homogeneous gas or liquid, a certain small proportion is scattered out of the beam by the molecules of the gas or liquid in each element of the path. The amount scattered is very small for a gas or a liquid not near its critical point, but still of measurable intensity, e.g. the blue of a clear sky. As a liquid nears its critical point the intensity of the scattered light increases, and at the critical point itself the liquid glows strongly with a peculiar shine known as the phenomenon of *critical opalescence*. Owing to its more striking character attention is often concentrated on the phenomenon of critical opalescence, but all the phenomena of the scattering of light by liquids and gases when the wave length of the light is long compared with the average distance apart of the scattering agents (molecules) form a single whole and may be discussed together.*

It can be shown† that the scattering of such light must depend on the irregular spacing of the scattering centres, and that it is therefore a problem of the fluctuations in the distribution of molecules in given small volume elements. If the molecules were regularly spaced at their average spacing as in a crystal there would be no scattering at all. Any scattering actually observed in crystals can be traced to imperfections or foreign bodies. This

* Perrin, *La théorie du rayonnement et les quanta* (1912); "Les preuves de la réalité moléculaire"

† For a simple account see Lorentz, *The Problems of Modern Physics* (1927), § 21.

absence of scattering persists until the wave length falls to molecular dimensions, and we have a true scattering but only in definite directions, more commonly known as X-ray reflection or diffraction.

From the statistical point of view therefore the phenomena are best approached as an example of molecular fluctuations in the manner of von Smoluchowski,* using the general formula (628). We consider a fluid which in its normal equilibrium state contains n_0 molecules in a volume v_0 at a pressure p_0. The actual volume at any given moment occupied by this body of molecules is v, only small values of $(v-v_0)/v_0$ being important. The rest of the fluid is a large volume, the whole being isothermal. Then by general thermodynamical theorems $\Sigma\,\Delta A$ is the maximum work which the assembly can do in returning to its normal equilibrium state; since for any body in an isothermal expansion $dA=-p\,dv$,

$$\Sigma\,\Delta A=\int_v^{v_0}(p-p_0)\,dv.$$

The fraction of examples of the assembly in which we shall find these n_0 molecules at a volume between v and $v+dv$ is therefore

$$W(v)\,dv=\mu e^{-\frac{1}{kT}\int_v^{v_0}(p-p_0)\,dv}\,dv,$$

where μ is a constant, fixed by the condition $\int W dv=1$. Without specific assumptions as to the equation of state, we may expand $p-p_0$ and write

$$-\int_v^{v_0}(p-p_0)\,dv=\frac{(v-v_0)^2}{2!}\frac{\partial p_0}{\partial v_0}+\frac{(v-v_0)^3}{3!}\frac{\partial^2 p_0}{\partial v_0^{\,2}}+\frac{(v-v_0)^4}{4!}\frac{\partial^3 p_0}{\partial v_0^{\,3}}+O\{(v-v_0)^5\}.$$

We write also $$\gamma=(v-v_0)/v_0,$$

so that γ is the accidental condensation (strictly expansion). Then omitting fifth order terms

$$W(\gamma)\,d\gamma=\mu\,d\gamma\exp\left[\frac{1}{kT}\left\{\frac{v_0^{\,2}\gamma^2}{2!}\frac{\partial p_0}{\partial v_0}+\frac{v_0^{\,3}\gamma^3}{3!}\frac{\partial^2 p_0}{\partial v_0^{\,2}}+\frac{v_0^{\,4}\gamma^4}{4!}\frac{\partial^3 p_0}{\partial v_0^{\,3}}\right\}\right]. \qquad \text{....(2163)}$$

The further development now depends on whether the normal equilibrium state does or does not refer to the critical point. Away from the critical point $\partial p_0/\partial v_0 \neq 0$, and it is sufficiently accurate to take

$$W(\gamma)\,d\gamma=\mu\,d\gamma\exp\left[\frac{v_0^{\,2}\gamma^2}{2kT}\frac{\partial p_0}{\partial v_0}\right]. \qquad \text{......(2164)}$$

If the fluid is a perfect gas

$$\frac{\partial p_0}{\partial v_0}=-\frac{n_0 kT}{v_0^{\,2}},$$

$$W(\gamma)\,d\gamma=\mu\,d\gamma\,e^{-\frac{1}{2}n_0\gamma^2}. \qquad \text{......(2165)}$$

* von Smoluchowski, *Ann. d. Physik*, vol. 25, p. 205 (1908); Einstein, *Ann. d. Physik*, vol. 38, p. 1275 (1910).

In this simple case we find a result which we can also obtain at once from (2153). Putting there $P=n$, the number of molecules in a given small volume, we have $\overline{P}=\mu=n_0$,

$$f(n)\,dn=(2\pi n_0)^{-\frac{1}{2}}\,e^{-\frac{1}{2}(n-n_0)^2/n_0}\,dn. \qquad \text{......(2166)}$$

But $\gamma=(v-v_0)/v_0$ for constant n and therefore equally $\gamma=-(n-n_0)/n_0$ for constant v, so that (2165) and (2166) are identical.

The value of $\overline{\gamma^2}$ is easily calculated from (2164). We find

$$\overline{\gamma^2}=-\frac{kT}{v_0^2\,\partial p_0/\partial v_0}\quad\left(=\frac{1}{n_0},\ \textit{perfect gas}\right). \qquad \text{......(2167)}$$

At the critical point $\partial p_0/\partial v_0=\partial^2 p_0/\partial v_0^2=0$, so that

$$W(\gamma)\,d\gamma=\mu\,d\gamma\exp\left[\frac{v_0^4\gamma^4}{24kT}\frac{\partial^3 p_0}{\partial v_0^3}\right].$$

For the sake of a numerical estimate we may use Dieterici's equation of state (823), which yields after reduction

$$W(\gamma)\,d\gamma=\mu\,d\gamma\,e^{-\frac{n_0}{22\cdot 2}\gamma^4};$$

from this we find

$$\overline{\gamma^2}=1{\cdot}6/n_0^{\frac{1}{2}}. \qquad \text{......(2168)}$$

The next step is to make use of a formula which dates back to Lord Rayleigh,* and states that if light of intensity I per unit area and wave length λ_0 in a medium of refractive index μ_0 is incident on a small volume v_0 in which the refractive index is μ, the dimensions of v_0 being small compared with λ_0, then the intensity of the light per unit solid angle scattered at right angles is

$$\frac{2\pi^2 v_0^2}{\lambda_0^4}\left(\frac{\mu-\mu_0}{\mu_0}\right)^2.$$

To make use of this formula for the light scattered by spontaneous fluctuations we must combine with it a relation between μ and ρ the density of the fluid in v_0. For gases, and probably with sufficient accuracy for liquids also, we may use Lorentz's law of refraction

$$\frac{1}{\rho}\frac{\mu^2-1}{\mu^2+2}=\textit{const.},$$

from which it follows that for small values of $(\mu-\mu_0)/\mu_0$

$$\frac{\mu-\mu_0}{\mu_0}=\frac{(\mu_0^2-1)(\mu_0^2+2)}{6\mu_0^2}\gamma.$$

The intensity of the light scattered by a fluctuation to an accidental condensation γ in a volume v_0 is therefore

$$\frac{\pi^2 v_0^2\gamma^2}{18\mu_0^4\lambda_0^4}(\mu_0^2-1)^2(\mu_0^2+2)^2. \qquad \text{......(2169)}$$

* Rayleigh, *Phil. Mag.* vol. 12, p. 81 (1881); *Scientific Papers*, vol. 1, No. 74.

In unit volume of the liquid the number of such scattering volumes is $1/v_0$ and the average value of γ^2 is $\overline{\gamma^2}$ given by (2167). It follows that the average intensity i of the light scattered by unit volume of fluid per unit solid angle at right angles to the incident beam of intensity I per unit area is given by

$$\frac{i}{I}=\frac{\pi^2}{18\mu_0^4\lambda_0^4}(\mu_0^2-1)^2(\mu_0^2+2)^2\left\{-\frac{kT}{v_0\partial p_0/\partial v_0}\right\}. \qquad \text{......(2170)}$$

§20·51. *Comparison with observation.* We begin with some qualitative remarks. The great difference in intensity between critical opalescence and ordinary fluid scattering follows from the different orders of $\overline{\gamma^2}v_0$ in the two cases (2167) and (2168). Roughly we may say that this factor is greater at the critical point than in the perfect gas state (or any not nearly critical state) by a factor of the order of $\sqrt{n_0}$. For a volume v_0 small compared with the λ_0 of visible light, say a cube of edge 10^{-5} cm., $n_0=10^4$ for a gas at normal pressure and temperature and 10^6–10^7 for a liquid. Again formula (2170) shows that there is a strong selection in favour of the scattering of light of the shorter wave lengths. The scattered light from incident white light should look blue. This is of course in accord with all the facts (blue of sky, *etc.*). The fact that at the critical point the scattered light becomes white shows that the theory is there breaking down because the scattering elements, that is volumes in which the fluctuations are sensible, are no longer small compared with the wave length. We see in fact from (2167) and (2168) that for a cube of edge 10^{-4} cm. comparable with the wave length the root mean square fluctuation $\sqrt{(\overline{\gamma^2})}$ in a perfect gas at normal pressure and temperature is only 2×10^{-4}, whereas at the critical point it is 10^{-2}. A 1 per cent. density change leads to a sensible change of refractive index.

Equation (2170) for a perfect gas in which μ_0 is very nearly unity leads to

$$\frac{i}{I}=\frac{\pi^2}{2\mu_0^4\lambda_0^4}(\mu_0^2-1)^2\frac{kT}{p}. \qquad \text{......(2171)}$$

This formula, generalized for intensities at any angle to the original beam and integrated through the atmosphere, can be applied to calculate the nature of the scattered sunlight incident on the eye at an angle α with the vertical and β with the sun. It is completely successful. It can of course be used inversely, treating k as unknown, to determine k from measurements of i, and has been so used with success.*

A complete quantitative test of (2170) is afforded by the experiments of Keesom.† He studied ethylene ($T_c=11\cdot18$°C.). He verified first that for $p=p_c$ and $T=13\cdot6$° C. the ratio of the intensities of scattered light of two wave lengths was still proportional to $(\lambda_1/\lambda_2)^4$, finding $i_2/i_1=2\cdot00$ for $(\lambda_1/\lambda_2)^4=2\cdot13$, so that the theory should be applicable. For the same wave lengths

* Perrin, *loc. cit.*

† Keesom, *Ann. d. Physik*, vol. 35, p. 591 (1911).

i_2/i_1 falls to 1·18 at 11·43° C., showing that by that stage the scattering elements are, as statistical theory requires, no longer small.

Again near the critical point, since $(\partial p/\partial v)_c = (\partial^2 p/\partial v^2)_c = 0$,

$$-v_0 \frac{\partial p}{\partial v_0} = -v_0 \left(\frac{\partial^2 p}{\partial v \partial T}\right)_c (T - T_c). \qquad \text{......(2172)}$$

Hence near, but not too near, the critical point the intensity of the scattered light should vary like $1/(T-T_c)$. This prediction is well verified.

TABLE 70.

Scattering of light by ethylene near its critical point ($T_c = 11{\cdot}18°$ C.).

Temp. ° C. ($p = p_c$)	i	$i(T-T_c)$
13·53	0·190	0·44
12·54	0·337	0·46
11·86	0·671	0·46
11·61	1*	0·43
11·42	2·61	0·63
11·24	6·11	0·37

* Assumed.

Finally, having thus verified all the details of the theory, Keesom obtained $i/I = 0{\cdot}0007–0{\cdot}0008$ for the D lines of sodium at 11·93° C. The known compressibility of ethylene gives $-v_0 \partial p_0/\partial v_0$. All the data are therefore ready for the use of (2170) for an absolute determination of k, which comes out within 15 per cent. (the accuracy of the measurement of i/I) of its accepted value.

§ **20·6.** *Brownian movement.* The effects of fluctuations are made directly evident to our senses by Brownian movement as well as by opalescence. As is now well known the phenomena of Brownian movement are merely the phenomena of molecular agitation, exhibited on a reduced scale by a particle which is very large on the molecular scale—so large that the light it diffracts at least can be seen in an ultramicroscope—and yet so small that its velocity of thermal agitation in the equilibrium state is sufficient to give it visible displacements in reasonable periods of time. This velocity of agitation may be regarded, as we shall see, as maintained by fluctuations in the collisions with the surrounding molecules. This identification of Brownian movement was finally established by the experimental work of Perrin,† verifying the theories of Einstein.‡ Subsequent investigators have added more accurate measurements in even closer accord with the theory. By the study of suitable particles suspended in a fluid, liquid or gas,

† Perrin, *La théorie du rayonnement et les quanta* (1912), "Les preuves de la réalité moléculaire".
‡ Einstein, *Ann. d. Physik*, vol. 17, p. 549 (1905), vol. 19, p. 371 (1906).

(1) we can see the molecular motions going on before our eyes, (2) we can check the assumptions of statistical mechanics in a rather detailed way by proving that the characteristics of the Brownian movement agree with the demands of the theory, and (3) we obtain a direct, though not very accurate, method of measuring molecular magnitudes.

§ **20·61.** *The particle "atmosphere".* In a liquid under gravity containing particles, but so few that their mutual interactions may be neglected, we should find the atmospheric density law

$$\nu = \nu_0 e^{-M'gz/kT} \qquad \text{......(2173)}$$

obeyed by the particles. The potential energy of any other field of force can of course replace the gravitational energy $M'gz$. M' is the apparent mass of a single grain, that is its mass less the mass of the fluid it displaces. Equation (2173) is almost intuitive, but its formal proof as a theorem of statistical mechanics offers no difficulty. One has merely to contemplate an assembly consisting of the grains and the intergranular liquid and assume that forces between the grains can be neglected and that the energy of the intergranular liquid is independent of its shape, i.e. of the position of the grains. The total energy of the assembly then depends on the position of a grain only through the term $M'gz$. The details are left to the reader.

Returning to (2173) we observe that, if Δ is the density of the grains and δ that of the liquid, we find for spherical grains

$$kT \log(\nu_0/\nu) = \tfrac{4}{3}\pi a^3(\Delta - \delta)\, gz. \qquad \text{......(2174)}$$

In this equation ν_0 and ν can be determined by actually counting the grains visible in the field of a microscope. The other quantities are all easily measurable except a the radius of the grain, but even this can be fixed in various indirect and independent ways, so that (2174) becomes an equation for k, after verification that $T \log(\nu_0/\nu)/z$ is constant. The value of k obtained by Perrin in this manner was $1{\cdot}22 \times 10^{-16}$ in sufficiently good agreement with the correct value.

§ **20·62.** *The diffusion movements of a single grain.* If one attempts to follow as closely as possible the actual movements of a single grain and works out therefrom an observed "mean velocity of agitation", the value so found is always of the order 10^{-5} times the value given by equilibrium theory for the average velocity of a particle of the known mass of the grain. Such calculations however are necessarily grossly wrong. We can never follow the details of the movement of the grain, which has a kink at every molecular collision—about 10^{21} times a second in an ordinary liquid, which is so frequently that it is really wrong to think of separate collisions. What we observe as displacements are of the nature of residual fluctuations about a mean value zero, and have little direct connection with the actual detailed path of the grain. To our senses (pushed to their farthest limit in the form of a cinematograph taking 10^5 pictures a second) the details of the path are impossibly fine. The path may fairly be compared in a crude way to the graph of a continuous function with no differential coefficient. Such a curve has not got a "length", and no idea of length can be obtained from any inscribed polygon.

A more subtle analysis is necessary. Confining attention to displacements in one direction, let the concentration of the grains at any place and time be $\nu(x,t)$. Of those in any interval x, $x+dx$ let the fraction $f_\tau(x'-x)\,dx'$ be found after a time τ in the interval x', $x'+dx'$. This fraction has been given its proper functional form, for clearly it can only depend on $x'-x$ and not on x, x' separately. By integrating the contributions found in dx' at time $t+\tau$ derived from all other elements at time t, we find

$$\begin{aligned}\nu(x',t+\tau) &= \int_{-\infty}^{+\infty} \nu(x,t) f_\tau(x'-x)\,dx,\\ &= \int_{-\infty}^{+\infty} \nu(x'-X,t) f_\tau(X)\,dX. \qquad \text{......(2175)}\end{aligned}$$

This equation must hold for all x', t and τ.* It is satisfied in the equilibrium state in which ν is constant in space and time, since by its definition

$$\int_{-\infty}^{+\infty} f_\tau(X)\,dX = 1.$$

Let us first suppose that τ is small and expand both functions ν. Then

$$\begin{aligned}&\nu(x',t) + \tau\frac{\partial\nu}{\partial t} + O(\tau^2)\\ &\quad = \int_{-\infty}^{+\infty}\left\{\nu(x',t) - X\frac{\partial\nu}{\partial x'} + \tfrac{1}{2}X^2\frac{\partial^2\nu}{\partial x'^2} - \tfrac{1}{6}X^3\frac{\partial^3\nu}{\partial x'^3} + O(X^4)\right\} f_\tau(X)\,dx.\end{aligned}$$

......(2176)

The odd powers of X may be assumed to vanish on integration by symmetry, and we shall assume that the terms $O(\tau^2)$, $O(\overline{X^4})$ are negligible for sufficiently small τ, postponing further discussion of this assumption. We then find

$$\tau\frac{\partial\nu}{\partial t} = \tfrac{1}{2}\xi^2\frac{\partial^2\nu}{\partial x'^2}, \qquad \text{......(2177)}$$

where

$$\xi^2 = \int_{-\infty}^{+\infty} X^2 f_\tau(X)\,dX. \qquad \text{......(2178)}$$

This is Einstein's diffusion equation for the displacements of single grains.

It is easily verified that ξ^2/τ must be a constant independent of τ and characteristic of the grain, a condition necessary to make (2177) significant.

* Compare the similar treatment of velocities in § 19·5.

For if τ is not too small ($\tau > 10^{-5}$ sec. will suffice*) the velocity of the grain at the end of the interval τ will be completely independent of the velocity at the beginning. Displacements in consecutive τ-intervals will therefore be independent. This being so if $\tau' = p\tau$, and $x_1, \ldots, x_p$ are p consecutive x-displacements, then

$$x'^2 = \Sigma x_r^2 + 2\Sigma x_i x_j.$$

If then we average x'^2 over a large number n of grains or displacements, the product term will vanish,† and we shall find

$$n\xi'^2 = np\xi^2$$

or

$$\xi'^2/\tau' = \xi^2/\tau. \qquad \ldots\ldots(2179)$$

Unlike the true velocity of agitation the diffusion constant

$$D = \xi^2/2\tau$$

of a grain is directly measurable. It has been shown by Einstein that such measurements may be made to lead at once to a determination of k. We apply the foregoing arguments to an atmosphere of grains in equilibrium in a field of force. The rate of diffusion under the concentration gradient must then just balance the directed effect of the field of force. If a force F acts on the grains, assumed spheres of radius a, they acquire, by Stokes' law, a steady velocity v, given by

$$v = F/6\pi\mu a,$$

where μ is the viscosity of the fluid. The number crossing unit interface in unit time is therefore $v\nu$ or

$$\nu F/6\pi\mu a.$$

The number crossing by diffusion in the opposite direction is $D\partial\nu/\partial x$. If further the grains obey the atmospheric distribution law, we have

$$\nu = \nu_0 e^{-\chi/kT}$$

or

$$\frac{1}{\nu}\frac{\partial\nu}{\partial x} = \frac{F}{kT}.$$

Combining these results

$$D = \frac{\xi^2}{2\tau} = \frac{kT}{6\pi\mu a}. \qquad \ldots\ldots(2180)$$

This equation has been verified to lead to consistent satisfactory values of k, by observation of ξ^2/τ over wide ranges of T, μ and a.

* This estimate is made by calculating how long the viscosity of the liquid will take to reduce the velocity of a sphere the size of a grain to an insignificant fraction of its initial value See Perrin, *loc. cit.* The generalization necessary when this condition is omitted is given in § 20·9.

† Strictly, from the independence and the definition of f_τ

$$\overline{x_i x_j} = \int_{-\infty}^{+\infty}\int_{-\infty}^{+\infty} x_i x_j f_\tau(x_i) f_\tau(x_j)\, dx_i\, dx_j,$$

which vanishes from symmetry.

§ **20·63.** *Generalizations. Further deductions from the equation* (2175). If we may continue to make the assumption that the terms $O(\overline{X^4})$, *etc.* are all $O(\tau^2)$, equation (2175) will yield much more information than we have yet extracted, and in fact also fixes the form of $f_\tau(X)$. On differentiating (2175) with respect to τ, we find

$$\frac{\partial \nu(x',t+\tau)}{\partial t}=\int_{-\infty}^{+\infty}\nu(x'-X,t)\,\frac{\partial f_\tau(X)}{\partial \tau}\,dX.$$

But as a result of the assumptions just specified we have proved, equation (2177), that

$$\frac{\partial \nu(x',t+\tau)}{\partial t}=D\,\frac{\partial^2 \nu(x',t+\tau)}{\partial x'^2}.$$

It follows that

$$\begin{aligned}\frac{\partial \nu(x',t+\tau)}{\partial t}&=D\,\frac{\partial^2}{\partial x'^2}\int_{-\infty}^{+\infty}\nu(x'-X,t)\,f_\tau(X)\,dX,\\ &=D\int_{-\infty}^{+\infty}\frac{\partial^2}{\partial X^2}\,\nu(x'-X,t)\,f_\tau(X)\,dX,\\ &=\int_{-\infty}^{+\infty}\nu(x'-X,t)\,D\,\frac{\partial^2}{\partial X^2}\,f_\tau(X)\,dX,\end{aligned}$$

assuming the legitimacy of the various inversions of order of limit operations. We therefore find

$$\int_{-\infty}^{+\infty}\nu(x'-X,t)\left[D\,\frac{\partial^2}{\partial X^2}-\frac{\partial}{\partial \tau}\right]f_\tau(X)\,dX=0.$$

This equation is to hold for all x', t and τ, so that it must hold for *any* initial law of density distribution $\nu(x'-X,0)$. It therefore implies that*

$$\left[D\,\frac{\partial^2}{\partial X^2}-\frac{\partial}{\partial \tau}\right]f_\tau(X)=0. \qquad \text{......(2181)}$$

The displacement distribution function is therefore that solution of (2181) which places all the grains near $X=0$ at $\tau=0$ (the point-source solution). Hence

$$f_\tau(X)=\frac{1}{2\sqrt{(\pi D\tau)}}\,e^{-X^2/4D\tau}, \qquad \text{......(2182)}$$

or

$$f_\tau(X)=\frac{1}{\sqrt{(2\pi)}}\,\frac{1}{\xi}\,e^{-\frac{1}{2}X^2/\xi^2}. \qquad \text{......(2183)}$$

The displacement distribution law is therefore the error law, which has been exhaustively tested by observation. Perrin gives the following set of counts of the displacements of a grain of radius $2{\cdot}1\times10^{-5}$ cm. at 30 sec. intervals.

* We may assert that $\int_{-\infty}^{+\infty}\nu(x'-x)\,\phi(x)\,dx=0$ for any given ν and all x'. Let $\nu(x)=1$ from $-\infty$ to 0 and $\nu(x)=0$ from 0 to ∞. Then $\int_{x'}^{\infty}\phi(x)\,dx=0$ for all x' and therefore $\phi(x)=0$.

Out of a number N of such observations the number of observed values of x-displacements between x_1 and x_2 should be

$$\frac{N}{\sqrt{(2\pi)}}\int_{x_1}^{x_2}\frac{1}{\xi}e^{-\frac{1}{2}x^2/\xi^2}\,dX.$$

TABLE 71.

Observations and calculations of the distribution of the displacements of a Brownian grain.

Range of $x\times 10^4$ cm.	1st set		2nd set		Total	
	Obs.	Calc.	Obs.	Calc.	Obs.	Calc.
0 – 3·4	82	91	86	84	168	175
3·4– 6·8	66	70	65	63	131	133
6·8–10·2	46	39	31	36	77	75
10·2–17·0	27	23	23	21	50	44

The agreement is satisfactory.

We observe at this point that the derived form of $f_\tau(X)$ is *consistent* with our preliminary neglect of $O(\overline{X^4})$ and $O(\tau^2)$ in (2176). The assumption however that the terms $O(\overline{X^4})$ are all $O(\tau^2)$ is *essential* to these results, which are not necessarily true without it. The terms in $\overline{X^4}$... *might* contain terms of order τ, in which case (2177) would contain $\partial^4\nu/\partial x^4$, ... If the usual diffusion equation (2177) is exact, then the distribution function is Gaussian, and conversely, as we have shown above. But this is not true on the scale of the eddy diffusion in the atmosphere. It is therefore hardly possible to dispense with appeal to experiment; Perrin's experiments show that for Brownian movement the diffusion equation and the distribution function have their simplest form.*

§ 22·64. *Einstein's extension of* (2180). The argument establishing (2180) can be extended to the displacements with respect to any coordinate α (e.g. an angular one) in which the normal equilibrium distribution is uniform. If A^2 is the mean square displacement in this coordinate in time τ due to the molecular agitation, then by the old argument A^2/τ is constant. The distribution in α satisfies the same diffusion equation

$$\frac{\partial\nu}{\partial t}=D\frac{\partial^2\nu}{\partial\alpha^2}\quad\left(D=\frac{A^2}{2\tau}\right);$$

and the number of particles passing by diffusion across a given value α of the coordinate in unit time is again

$$D\,\partial\nu/\partial\alpha$$

in the direction of ν decreasing.

* Critical comment by J. A. Gaunt.

Now suppose that an external field of potential energy $\phi(\alpha)$ acts on the grains. In the equilibrium state we should have a distribution law

$$\nu = \nu_0 e^{-\phi(\alpha)/kT}.$$

Suppose further that for an individual grain under a force Φ we have a steady velocity controlled by viscous resistances so that

$$\kappa\, d\alpha/dt = \Phi.$$

Then in the equilibrium state when $\Phi = -\phi'(\alpha)$

$$D\frac{\partial\nu}{\partial\alpha} = -\frac{\nu\phi'(\alpha)}{\kappa}.$$

But
$$\frac{1}{\nu}\frac{\partial\nu}{\partial\alpha} = -\frac{\phi'(\alpha)}{kT}.$$

Therefore
$$D = kT/\kappa = \tfrac{1}{2}A^2/\tau. \qquad \text{......(2184)}$$

This is the required generalization of (2180). For a sphere rotating in a viscous liquid about a fixed axis

$$\kappa = 8\pi a^3\mu.$$

Therefore for rotational displacements

$$\frac{A^2}{\tau} = \frac{kT}{4\pi a^3\mu}, \qquad \text{......(2185)}$$

an equation confirmed experimentally by Perrin.

§ **20·7.** *Effective pressure (reaction) fluctuations.* The formulae of § 20·3 give the fluctuations (of the second order) in the instantaneous value of any external reaction, in particular of a pressure. They depend essentially on the law of force between for example the gas molecule and the wall. In many applications however it is not this instantaneous value of the pressure which is important, but rather the integral of the pressure (the momentum transferred) taken over a time interval which is long compared with the time of collision between a gas molecule and the wall, and short compared with the time constants of any response that the wall can make. In particular it is this sort of quantity and its fluctuation which is required in the study of the natural limitations of ultrasensitive measuring apparatus.* A typical problem in which the fluctuations of momentum transfer in "physically short" time intervals are required—to the exclusion of fluctuations in the instantaneous pressure—is the study of the rotational Brownian movement of a small galvanometer mirror suspended on a fine wire, which has been completed in detail by Uhlenbeck and Goudsmit.† We shall therefore

* For a general survey see Barnes and Silverman, *Rev. Mod. Physics*, vol. 6, p. 162 (1934), where numerous references will be found.

† Uhlenbeck and Goudsmit, *Phys. Rev.* vol. 34, p. 145 (1929).

obtain in this section the underlying formula for the fluctuation of the momentum transfer in given small elements of time to a given element of surface. The work is based of course on the formulae of § 20·21 for the fluctuations of the numbers of systems in given states or volume elements. These formulae will here be needed only in their simplest classical forms in which $\overline{a_r}$ always satisfies $\overline{a_r} \ll M$, when we have

$$\overline{(a_r - \overline{a_r})^2} = \overline{a_r}, \quad \overline{(a_r - \overline{a_r})(a_s - \overline{a_s})} = 0. \qquad \text{......(2186)}$$

We may start with the usual formula derived from Maxwell's law (174) for the average number of molecules, in a perfect gas of N molecules of mass m in a volume V obeying classical statistics, which strike an element of wall of area Δo in a time interval Δt with a momentum normal to the wall lying between g and $g + dg$. This number $\overline{a_r}$ is the number of molecules lying at the beginning of the interval in the correct volume element of extension $\Delta o \Delta t g/m$, and is given by the usual formula

$$\overline{a_r} = \frac{N}{V} \frac{g/m}{(2\pi mkT)^{\frac{1}{2}}} e^{-g^2/2mkT} dg \, \Delta o \, \Delta t. \qquad \text{......(2187)}$$

When these molecules have been brought to rest at the wall they have communicated to it a momentum $g\overline{a_r}$. They are then thrown off again by the wall with a further transfer of momentum, but this can only be calculated when the exact nature of the reflection process has been specified.* If we assume for the present that the wall acts like a specular reflector, each molecule is reflected with momentum $-g$ and the transfer of momentum by these g-molecules is $2g\overline{a_r}$. This is of course the average transfer by such molecules. The actual transfer in Δt is $2ga_r$ and the fluctuation $4g^2\overline{(a_r - \overline{a_r})^2}$ which is $4g^2\overline{a_r}$. We can now sum these effects over all values of g to obtain the total transfer of momentum G, its average value $\overline{G}$ and its fluctuation $\overline{(G - \overline{G})^2}$. So long as equations (2186) are valid

$$\overline{(G - \overline{G})^2} = \Sigma_r \, 4g^2\overline{a_r}. \qquad \text{......(2188)}$$

We obtain at once by simple integration

$$\overline{G} = p\Delta o \, \Delta t = \frac{N}{V} kT \Delta o \, \Delta t \qquad \text{......(2189)}$$

which is always valid for a perfect gas, and

$$\overline{(G - \overline{G})^2} = 2m\bar{c}p\Delta o \, \Delta t, \qquad \text{......(2190)}$$

which is valid so long as equations (2186) hold. The corresponding fluctuation in the total force F acting on Δo is

$$\overline{(F - \overline{F})^2} = 2m\bar{c}p\Delta o/\Delta t. \qquad \text{......(2191)}$$

* This point was ignored by Uhlenbeck and Goudsmit.

This condition of validity must be carefully attended to. Equations (2186) always hold for averages over a long time for any specified types of molecule whose average number is small compared with the total number of molecules. They are however here required to hold for a physically short time interval Δt. They can then break down if applied to too large a volume element, for the deficiency of one type of molecule may be correlated with the excess of another. An assembly for which they certainly hold is a highly rarefied gas in which the free paths of the molecules are long compared with the linear scale of any volume element concerned. Different sets of g-molecules and sets of molecules striking different adjacent elements of area are then strictly uncorrelated. It is to such gases that we shall apply the resulting formulae.

If the bombarded element is a small mirror suspended in the gas, both sides are accessible to the gas and the fluctuations are obviously additive. We then find for the momentum transfer $\overline{G}=0$ and

$$\overline{(G-\overline{G})^2}=\overline{G^2}=4m\bar{c}p\Delta o\,\Delta t, \qquad \text{......(2192)}$$

and for the total force

$$\overline{(F-\overline{F})^2}=\overline{F^2}=4m\bar{c}p\Delta o/\Delta t. \qquad \text{......(2193)}$$

A more useful result is the torque and its fluctuations about the axis of suspension of a plane mirror. For the impulse about the axis generated by a momentum transfer G in time Δt to an element of area Δo distant x from the axis of suspension we have the average value

$$px\Delta o\,\Delta t$$

and for the fluctuation in this impulse

$$2m\bar{c}px^2\Delta o\,\Delta t.$$

On summing this over both sides of the whole mirror we find

$$\overline{(P-\overline{P})^2}=\overline{P^2}=4m\bar{c}pO\kappa^2\Delta t, \qquad \text{......(2194)}$$

where O is the area of the mirror and κ its radius of gyration about the axis of suspension. The fluctuation in the average couple Γ exerted by the gas is therefore

$$\overline{(\Gamma-\overline{\Gamma})^2}=\overline{\Gamma^2}=\frac{4m\bar{c}p}{\Delta t}O\kappa^2. \qquad \text{......(2195)}$$

If we now attempt to free these results from the assumption of perfect specular reflection we meet with considerable difficulty in formulating any simple but more general alternative assumption, which can be consistently employed. The following assumption is free from contradiction and reasonably simple, but not very natural physically. We assume that the incident g-molecules can be divided into two fractions f and $(1-f)$ of which the fraction $(1-f)$ are immediately reflected specularly and the fraction f condense on the surface and are re-emitted without correlation to their incident value of g. It is necessary also for the sake of complete consistency to assume that this fraction f is re-emitted on the average at the normal rate in equilibrium with the gas pressure (with normal fluctuations about this rate) but that this rate is unaffected by the fluctuations in the rate of incidence. There must be sufficient storage of condensed molecules on the surface to prevent in this fraction any correlation between the numbers arriving and the numbers re-emitted. Under these assumptions we can discuss the momentum transfer by the fraction f by breaking up both the incident and re-emitted molecules into groups of g-molecules with uncorrelated fluctuations. For this fraction we shall therefore have

$$\bar{G} = \Sigma_r' \, |g| \, \bar{a}_r, \quad \overline{(G-\bar{G})^2} = \Sigma_r' \, g^2 \bar{a}_r,$$

where Σ_r' is a summation over both positive and negative values of g. We obtain at once

$$\bar{G} = p\Delta o \, \Delta t, \quad \overline{(G-\bar{G})^2} = m\bar{c}p\Delta o \, \Delta t,$$

the fluctuation having half its former value. The final correction is therefore to replace (2192) by

$$\overline{(G-\bar{G})^2} = \overline{G^2} = 2m\bar{c}p\Delta o \, \Delta t(2-f)$$

and to make similar changes in (2193), (2194) and (2195).

§ **20·71.** *Campbell's theorem.* Before proceeding to apply these theorems to the study of a suspended mirror system it will be convenient to introduce a theorem first enunciated by Campbell,* which may often be usefully employed in this field. The theorem may be enunciated as follows:

Campbell's Theorem. If in any measuring instrument the quantity θ to be observed is due to the linear superpositions of the effects of a number of independent events occurring at random times, so that

$$\theta(t) = \Sigma_\alpha \Sigma_r f_\alpha(t - t_r^\alpha),$$

where $f_\alpha(t-t_0)$ is the effect of an event of type α occurring at time $t=t_0$, and if the events of type α occur at random instants at an average rate N_α per unit time, then

$$\bar{\theta} = \Sigma_\alpha N_\alpha \int_{-\infty}^{+\infty} f_\alpha(t) \, dt, \qquad \text{......(2196)}$$

$$\overline{|\theta - \bar{\theta}|^2} = \Sigma_\alpha N_\alpha \int_{-\infty}^{+\infty} |f_\alpha(t)|^2 \, dt. \qquad \text{......(2197)}$$

A somewhat superficial analysis of this theorem may be given as follows, in the first instance for events of a single type. A single event P at $t=t_0$ causes the quantity $\theta = f(t-t_0)$ to be observed. A set of such events, a_r in

* Campbell, *Proc. Camb. Phil. Soc.* vol. 15, pp. 117, 310, 513 (1909).

number, in an interval Δt_r at $t=t_r$ short compared with the time constants of the system, will produce the effect

$$\theta=\Sigma_r a_r f(t-t_r).$$

It follows at once that

$$\begin{aligned}\bar{\theta}&=\Sigma_r \bar{a}_r f(t-t_r),\\&=N\Sigma_r \Delta t_r\, f(t-t_r),\\&=N\int_{-\infty}^{+\infty} f(t)\,dt.\end{aligned}$$

If the events are molecular in nature so that $\overline{(a_r-\bar{a}_r)^2}=\bar{a}_r$, $\overline{(a_r-\bar{a}_r)(a_s-\bar{a}_s)}=0$, then also

$$\begin{aligned}\overline{|\theta-\bar{\theta}|^2}&=\overline{|\Sigma_r (a_r-\bar{a}_r) f(t-t_r)|^2},\\&=\Sigma_r \overline{(a_r-\bar{a}_r)^2}\,|f(t-t_r)|^2,\\&=\Sigma_r \bar{a}_r |f(t-t_r)|^2,\\&=N\int_{-\infty}^{+\infty} |f(t)|^2\,dt.\end{aligned}$$

The extension to different sets of events as in the enunciation is immediate.*

A particularly important case occurs when

$$f_\alpha=\alpha f,\quad \Sigma_\alpha \alpha N_\alpha=0,\quad \Sigma_\alpha \alpha^2 N_\alpha=N\overline{\alpha^2}.$$

Then

$$\bar{\theta}=0,\quad \overline{\theta^2}=N\overline{\alpha^2}\int_{-\infty}^{+\infty}|f(t)|^2\,dt. \qquad \text{......(2198)}$$

§ **20·72.** ***Torsional (Brownian) oscillations of a suspended mirror in a rarefied gas.*** The suspended mirror is for our purposes a system of a single degree of freedom defined by its angular displacement ϕ. If it is subject to a linear damping force its equation of motion may be put in the form

$$I\ddot{\phi}+r\dot{\phi}+D\phi=\Gamma(t), \qquad \text{......(2199)}$$

with an obvious notation. When left to itself in the gas it follows from the equipartition theorem that at all gas pressures

$$\tfrac{1}{2}I\overline{\dot{\phi}^2}=\tfrac{1}{2}D\overline{\phi^2}=\tfrac{1}{2}kT.$$

It is however of considerable interest to analyse this relationship in greater detail for a rarefied gas for which all the calculations can be completed. With this object in view we shall require an explicit expression for r which we now proceed to obtain.

Under the condition postulated of a highly rarefied gas the number of impacts in time Δt on any element of wall Δo moving through the gas with velocity u normal to its surface can be calculated from the equilibrium distribution laws of the gas since these are then not upset by the wall move-

* A much more profound study of this interesting theorem is desirable, but cannot be included here. See Rowland, *Proc. Camb. Phil. Soc.* vol. 32, p. 580 (1936).

ment. If the wall is moving to the right with velocity u, the number of impacts from the left with momentum normal to the wall between g and $g+dg$ will be, by Maxwell's law,

$$\frac{N}{V}\frac{g/m-u}{(2\pi mkT)^{\frac{1}{2}}}e^{-g^2/2mkT}dg\,\Delta o\,\Delta t,$$

impacts being only possible when $g/m>u$. For perfectly specular reflection these each transfer the linear momentum $2(g-mu)$. The total average transfer of momentum to Δo is therefore

$$\bar{G}=\frac{N}{V}\frac{2/m}{(2\pi mkT)^{\frac{1}{2}}}\int_{mu}^{\infty}(g-mu)^2e^{-g^2/2mkT}dg\,\Delta o\,\Delta t.$$

Correct to terms linear in u this easily reduces to

$$\left(p-\frac{N}{V}m\bar{c}u\right)\Delta o\,\Delta t,$$

or taking account of both sides of the element of surface

$$\bar{G}=-\frac{2m\bar{c}p}{kT}u\,\Delta o\,\Delta t. \qquad \text{......(2200)}$$

For a rotating mirror the impulse generated in Δt about the axis of suspension by the momentum transferred to Δo is $x\bar{G}$ and $u=x\dot{\phi}$. The average damping couple acting on the mirror is therefore

$$\bar{\Gamma}=-\frac{2m\bar{c}p}{kT}\dot{\phi}\int x^2\,do,$$

$$=-\frac{2m\bar{c}p}{kT}O\kappa^2\dot{\phi}. \qquad \text{......(2201)}$$

The coefficient r in (2199) is therefore $2m\bar{c}pO\kappa^2/kT$.

If we relax the condition of specular reflection by the same generalizations as in § 20·7, the fraction $(1-f)$ of the incident molecules continues to give the same contribution to $\bar{\Gamma}$ per molecule as before. The remaining fractions f on either side are first condensed on the mirror during which process they contribute half as much momentum transfer as before. They are then re-emitted but at a rate in equilibrium with the average gas pressure, and therefore symmetrically from both faces of the mirror giving no contribution to the momentum transfer. The average damping forces are therefore now smaller by the factor $1-f+\frac{1}{2}f$, and

$$\bar{\Gamma}=-\frac{(2-f)\,m\bar{c}p}{kT}O\kappa^2\dot{\phi}. \qquad \text{......(2202)}$$

§ **20·73.** *Torsional oscillations. Application of Campbell's theorem.* It will now be of interest to verify the equipartition theorem by applying Campbell's theorem. We may think of the motion of the mirror as made up by the superposition of a number of free damped excursions each generated impulsively at random times. The $f(t)$ of Campbell's theorem is therefore that solution of (2199) with $\Gamma(t)=0$ which satisfies $\phi(0)=0$, $\dot{\phi}(0)=\delta$. This solution is

$$\phi=f(t)=\frac{\delta}{\mu}e^{-\lambda t}\sin\mu t \quad (t>0), \qquad \text{......(2203)}$$

where $\lambda=\tfrac{1}{2}r/I, \quad \mu^2=D/I-\tfrac{1}{4}r^2/I^2.$

It follows at once that

$$\int_{-\infty}^{+\infty} f^2(t)\,dt=\frac{\delta^2}{4\lambda(\lambda^2+\mu^2)}=\frac{\delta^2 I^2}{2rD}.$$

We can now apply Campbell's theorem and find that

$$\overline{\{\phi(t)\}^2}=\frac{NI^2\overline{\delta^2}}{2rD}. \qquad \text{......(2204)}$$

Now in (2204) we may take $N=1/\Delta t$ and $I^2\overline{\delta^2}$ the mean square angular momentum generated by the molecular impulses in time Δt. Thus by § 20·7

$$I^2\overline{\delta^2}=\overline{(P-\overline{P})^2}=4m\bar{c}pO\kappa^2\Delta t. \qquad \text{......(2205)}$$

Combining (2204), (2205) and (2201) for r, we find

$$\tfrac{1}{2}D\overline{\{\phi(t)\}^2}=\frac{1}{4\Delta t}\frac{\overline{(P-\overline{P})^2}}{r}=\tfrac{1}{2}kT, \qquad \text{......(2206)}$$

in agreement with the equipartition theorem. The form of (2206) shows that this agreement persists when the factors $2-f$ are inserted.

The equality of $\tfrac{1}{2}D\overline{\phi^2}$ and $\tfrac{1}{2}I\overline{\dot{\phi}^2}$ can be verified at once by multiplying equation (2199) by ϕ and averaging each term over a long time τ.

§ **20·74.** *Further study of torsional oscillations by Fourier analysis.* In the investigation already referred to Uhlenbeck and Goudsmit have shown that there is much of interest to be found in a still deeper study using Fourier analysis. Consider a long time interval $0, \tau$ and develop $\Gamma(t)$ in this interval in the Fourier series

$$\Gamma(t)=\sum_{k=0}^{\infty} A_k\cos\omega_k t+B_k\sin\omega_k t, \qquad \text{......(2207)}$$

where $\omega_k=2\pi k/\tau,$

$$A_k=\frac{2}{\tau}\int_0^{\tau}\Gamma(t)\cos\omega_k t\,dt, \quad B_k=\frac{2}{\tau}\int_0^{\tau}\Gamma(t)\sin\omega_k t\,dt.$$

A particular integral of (2199) valid in this time interval is then given by

$$\phi(t) = \sum_{k=0}^{\infty} \phi_k(t),$$

where

$$I[(D/I - \omega_k^2)^2 + r^2\omega_k^2/I^2]\,\phi_k(t)$$
$$= [A_k(D/I - \omega_k^2) - B_k r\omega_k/I]\cos\omega_k t + [A_k r\omega_k/I + B_k(D/I - \omega_k^2)]\sin kt. \quad \text{......(2208)}$$

It is sufficient to consider any particular integral. The complementary functions required to give any other initial values to $\phi(0)$ and $\dot{\phi}(0)$ is damped out and will not affect the results. It follows at once by averaging over 0, τ that

$$\overline{\{\phi(t)\}^2} = \sum_{k=0}^{\infty} \overline{\{\phi_k(t)\}^2},$$

where

$$\overline{\{\phi_k(t)\}^2} = \frac{1}{2I^2}\frac{A_k^2 + B_k^2}{(D/I - \omega_k^2)^2 + r^2\omega_k^2/I^2}. \quad \text{......(2209)}$$

Now $A_k^2 + B_k^2$ can be expressed as the double integral

$$A_k^2 + B_k^2 = \frac{4}{\tau^2}\int_0^\tau\int_0^\tau \Gamma(t)\,\Gamma(t')\cos\omega_k(t - t')\,dt\,dt'. \quad \text{......(2210)}$$

Since $\Gamma(t)$ is the fluctuating couple, there is no correlation between $\Gamma(t)$ and $\Gamma(t')$ except when $t \sim t'$ and the only sensible terms in the double integral come from a region which we may denote by $-\frac{1}{2}\delta + t' \leqslant t \leqslant t + \frac{1}{2}\delta$, where $\delta \ll 1/\omega_k$. It follows that

$$A_k^2 + B_k^2 = \frac{4\delta}{\tau^2}\int_0^\tau \Gamma^2(t)\,dt.$$

We may now put $\delta = \Delta t$ and use for $\Gamma^2(t)$ the value $\overline{\Gamma^2}$ already calculated so that

$$A_k^2 + B_k^2 = \frac{4}{\tau}\overline{\Gamma^2}\Delta t = \frac{16m\bar{c}pO\kappa^2}{\tau}. \quad \text{......(2211)}$$

Using the formula (2201) for r, we can now put $\overline{\{\phi_k(t)\}^2}$ in the form

$$\frac{(\pi m)^{\frac{1}{2}}(8kT)^{\frac{3}{2}}pO\kappa^2}{\tau}\cdot\frac{1}{\pi kT(D - I\omega_k^2)^2 + 32p^2m(O\kappa^2)^2\omega_k^2}. \quad \text{......(2212)}$$

The expression (2212) shows clearly that although $\overline{\{\phi(t)\}^2}$ is independent of p the nature of its time variation is strongly dependent on p. For smaller values of p the Fourier components for which $\omega_k^2 \sim D/I$ become more and more strongly marked in the motion. Such a pressure dependence has been observed and is well shown in Fig. 79 from the work of Kappler.*

* Kappler, *Ann. d. Physik*, vol. 11, p. 233 (1931).

In conclusion we may note that such observations may be used to make a direct measurement of k. Kappler for example determined k in this way with an error of less than 1 per cent.

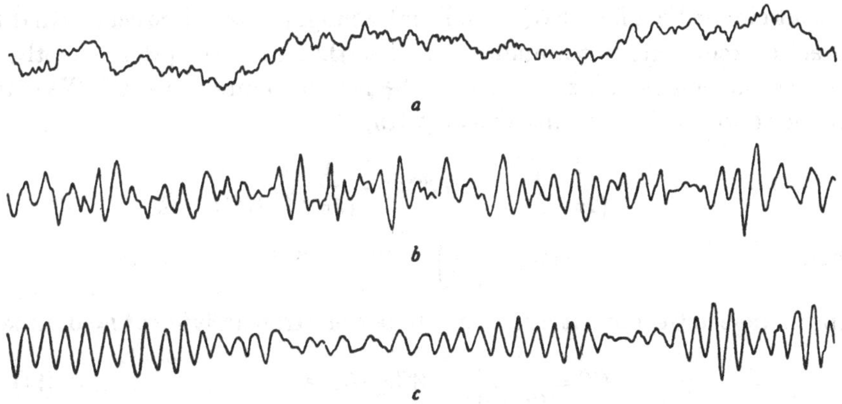

Fig. 79. Registration of an oscillating mirror system under identical conditions except for the pressure, which has the values (*a*) 1 Atmosphere; (*b*) 10^{-3} mm. Hg; (*c*) 10^{-4} mm. Hg.

§ **20·8.** *The shot effect.* The fluctuations in the current from a hot cathode due to the fact that the current is caused by discrete electrons emitted at random intervals of time and is not a continuous flow of electricity is known as the shot effect. It is a well-known source of background noise in telephone circuits of high amplification, and has been studied by a number of authors.* It is obviously an effect of the same nature as those under discussion here, and may be conveniently treated by means of Campbell's theorem.

When an electron reaches the anode in the valve whose effects are to be amplified it will naturally alter the voltage of the anode and thereby affect the anode circuit. It is then necessary as was first pointed out by Fry† to make explicit assumptions about the nature of its effect on the anode circuit which are somewhat difficult to justify. It is assumed that if C is the capacity of the anode system and R the resistance of the circuit and the circuit has negligible self induction, the arrival of an electron of charge $-\epsilon$ produces instantaneously a potential $-\epsilon/C$ which dies away according to the law

$$V(t) = -\frac{\epsilon}{C} e^{-t/CR} \quad (t > 0), \qquad \text{......(2213)}$$

well known to be correct for charges in bulk. We shall accept this assumption here; its justification will appear during the discussion.

* For instance first by Schottky, *Ann. d. Physik*, vol. 57, p. 541 (1918); also by Ornstein and Burger, *Ann. d. Physik*, vol. 70, p. 622 (1923); Fürth, *Physikal. Zeit.* vol. 23, p. 354 (1922); Fry, *J. Franklin Instit.* vol. 199, p. 203 (1925).

† Fry, *loc. cit.*

The effects to be observed are observed after amplification in a system of circuits which need not be specified more closely than by requiring that an oscillating anode voltage of frequency ξ shall be linearly amplified in the output circuit by a factor $G(\xi)$, and that voltages of zero frequency (that is direct currents) are not amplified at all so that $G(0)=0$. [More exactly it will be convenient to assume that $G(\xi)/\xi$ remains finite as $\xi\to 0$.] We start therefore with a Fourier analysis of (2213). If

$$W(\xi)=\frac{1}{(2\pi)^{\frac{1}{2}}}\int_{-\infty}^{+\infty} V(t)\,e^{i\xi t}dt=\frac{1}{(2\pi)^{\frac{1}{2}}}\frac{\epsilon}{C}\,\frac{-1}{1/CR-i\xi},$$

then
$$V(t)=\frac{1}{(2\pi)^{\frac{1}{2}}}\int_{-\infty}^{+\infty} W(\xi)\,e^{-i\xi t}d\xi.$$

The voltage in the output circuit due to one electron arriving at $t=0$ is then

$$f(t)=\frac{1}{(2\pi)^{\frac{1}{2}}}\int_{-\infty}^{+\infty} W(\xi)\,G(\xi)\,e^{-i\xi t}d\xi. \qquad \ldots\ldots(2214)$$

It will be observed that $f(t)$ and $W(\xi)\,G(\xi)$ are Fourier transforms of each other so that

$$\int_{-\infty}^{+\infty}|f(t)|^2dt=\int_{-\infty}^{+\infty}|W(\xi)\,G(\xi)|^2d\xi. \qquad \ldots\ldots(2215)$$

We now apply Campbell's theorem. We suppose that on the average N electrons per second reach the anode in the primary valve, reaching it at times distributed at random. Then the average voltage in the output circuit $\bar{\theta}$ is given by

$$\bar{\theta}=\frac{N}{(2\pi)^{\frac{1}{2}}}\int_{-\infty}^{+\infty}dt\int_{-\infty}^{+\infty}W(\xi)\,G(\xi)\,e^{-i\xi t}d\xi,$$

$$=-\frac{N\epsilon}{(2\pi)^{\frac{1}{2}}C}\int_{-\infty}^{+\infty}dt\int_{-\infty}^{+\infty}\frac{G(\xi)}{1/CR-i\xi}e^{-i\xi t}d\xi.$$

It may be assumed that it is legitimate to express this in the form

$$\bar{\theta}=-\frac{2N\epsilon}{(2\pi)^{\frac{1}{2}}C}\operatorname*{Lt}_{\tau\to\infty}\int_{-\infty}^{+\infty}\frac{G(\xi)}{1/CR-i\xi}\,\frac{\sin\xi\tau}{\xi}d\xi,$$

so that if $G(\xi)/\xi$ is finite $\bar{\theta}=0$. The average squared voltage using (2215) is then

$$\overline{\theta^2}=\frac{N\epsilon^2}{2\pi C^2}\int_{-\infty}^{+\infty}\frac{G^2(\xi)}{1/C^2R^2+\xi^2}d\xi. \qquad \ldots\ldots(2216)$$

In practice the $G(\xi)$ used is only sensible for audio-frequencies, for which $1/C^2R^2\gg\xi^2$. Further, this cut off of the high frequency components justifies neglecting as above the self-induction in the anode circuit and the transit time of the electrons between cathode and anode. Equation (2216) may therefore be simplified to

$$\overline{\theta^2}=\frac{1}{\pi}J\epsilon R^2\int_0^{\infty}G^2(\xi)\,d\xi, \qquad \ldots\ldots(2217)$$

where J is the anode current in the primary circuit, and $\overline{\theta^2}$ the average squared (fluctuating) voltage in the output circuit. This equation can be used to determine ϵ since all the other quantities are directly measurable. It has most recently been used in this way by Moullin* whose values of ϵ so determined agree with the accepted value within about 5 per cent. Much more accurate determinations of ϵ have been claimed by Hull and Williams,† but though there is no doubt that their measurements as analysed determine ϵ with high accuracy it does not appear that the theory of the effect is sufficiently rigorous to support such accurate determinations.‡

§ **20·81**. *Similar phenomena.* Numerous other examples of fluctuations in mechanical and electrical systems have been recognized and studied in recent years. A full account of many of these will be found in the paper by Barnes and Silverman already referred to. A particularly interesting example, akin rather to the shot effect, is the fluctuating voltage across any conductor of resistance R, due to the thermal fluctuations of its conduction electrons, known as the Johnson § effect. A discussion of the Johnson effect, sufficient to be intelligible without auxiliary references, would require too long a discussion of oscillating electric circuits to be included here. The examples discussed in the preceding sections should suffice to show how the statistical calculation of fluctuations may be brought into the analysis of all effects of this type.

§ **20·9**. *General distribution functions in problems of Brownian motion.* In §§ 19·5 *sqq.* we have already given a complete analysis by a special method of the way in which a group of atoms moving originally with velocity u_0 gradually loses its original mean velocity and acquires a Maxwellian distribution depending only on T. The mechanism discussed in those sections is Einstein's mechanism of emission and absorption of radiation of a definite frequency ν, but the discussion of the growth of randomness is quite general in form. The particular mechanism only enters in the assignment of particular values to the coefficients. In § 20·63 we have discussed the similar problem of the distribution in space of a group of free particles initially in a given volume element, but there the discussion was limited to long times.

The restriction to long times can be removed. A general method has been given by Ornstein and van Wijk ‖ by which distribution functions such as $f(u,u_0,t)\,du$ or $f(x,x_0,u_0,t)\,dx$ can be obtained, where $f(u,u_0,t)\,du$ denotes

* Moullin, *Proc. Roy. Soc.* A, vol. 147, p. 100 (1934).

† Hull and Williams, *Phys. Rev.* vol. 25, p. 147 (1925).

‡ The assumption (2213) affects the numerical values and its accuracy is particularly dubious.

§ Johnson, *Phys. Rev.* vol. 32, p. 97 (1928); Barnes and Silverman, *loc. cit.*

‖ Ornstein and van Wijk, *Physica*, vol. 1, p. 235 (1934); see also, Uhlenbeck and Ornstein, *Phys. Rev.* vol. 36, p. 823 (1930).

the average fraction of a set of free molecules, all having the velocity $u=u_0$ at $t=0$, which have a velocity in the range u, $u+du$ at time t, and $f(x,x_0,u_0,t)\,dx$ denotes the average fraction of a set of free molecules, all having the velocity $u=u_0$ and the position $x=x_0$ at $t=0$, which have a position in the range x, $x+dx$ at time t. It is possible also to use the same methods when permanent elastic or gravitational forces are acting on the particles or when the particles in addition are subject to a permanent periodic disturbing force. The problem of Brownian movement and of the oscillating mirror can thus be reduced to a common form. It will be sufficient to illustrate the method by the detailed discussion of $f(u,u_0,t)$ and $f(x,x_0,u_0,t)$.

For a free particle subject to a viscous retardation and rapid fluctuating accelerations of average value zero we can cast the equation of motion in the form

$$du/dt = -\beta u + A(t) \quad (\overline{A(t)}=0). \qquad \text{......(2218)}$$

This equation may be integrated in the form

$$u - u_0 e^{-\beta t} = e^{-\beta t}\int_0^t e^{\beta\xi} A(\xi)\,d\xi. \qquad \text{......(2219)}$$

Now in any interval $\Delta\xi$ $A(\xi)\,\Delta\xi$ will be determined by an expression of the form $X(a_r-\overline{a_r})$. It follows at once that $A(\xi)\,\Delta\xi$ will have the moments characteristic of a_r and given by (2109), so that with sufficient accuracy

$$\overline{\{A(\xi)\,\Delta\xi\}^{2n+1}}=0,$$

$$\overline{\{A(\xi)\,\Delta\xi\}^{2n}}=(2n-1)\ldots 3\,.\,1\{\overline{A(\xi)\,\Delta\xi}\}^2.$$

These are of course the moments of a Gaussian distribution and are by Pólya's theorem (§ 20·4) equivalent to a Gaussian distribution. Moreover if we may further assume that we can take $\Delta\xi$ small compared with the time scale defined by $1/\beta$, but yet still so large that the fluctuations of all orders in consecutive $\Delta\xi$'s are independent of each other, then the moments of all orders for all $\Delta\xi$'s will be additive and the moments of the right-hand side of (2219) will be Gaussian and therefore also those of $u-u_0e^{-\beta t}$. It follows at once from Pólya's theorem that

$$f(u,u_0,t)\,du = \sqrt{\left(\frac{\mu}{\pi}\right)}\,e^{-\mu(u-u_0e^{-\beta t})^2}\,du,$$

and it remains only to determine μ from the value of the second order fluctuation. This gives

$$\int_{-\infty}^{+\infty} r^2\sqrt{\left(\frac{\mu}{\pi}\right)}\,e^{-\mu r^2}\,dr = \frac{1}{2\mu} = e^{-2\beta t}\overline{\int_0^t\int_0^t e^{\beta(\xi_1+\xi_2)}A(\xi_1)\,A(\xi_2)\,d\xi_1 d\xi_2}.$$

We have already seen that the whole contribution comes from $\xi_1 \sim \xi_2$ so that

$$\frac{1}{2\mu} = Ce^{-2\beta t}\int_0^t e^{-2\beta\xi}\,d\xi = \frac{C}{2\beta}(1-e^{-2\beta t}),$$

where C is some constant. Finally if we let $t \to \infty$ we must have $\overline{u^2} = kT/m$ so that

$$\frac{kT}{m} = \int_{-\infty}^{+\infty} u^2 f(u, u_0, \infty)\, du = \frac{1}{2\mu_\infty} = \frac{C}{2\beta},$$

$$\mu = \frac{1}{\dfrac{2kT}{m}(1 - e^{-2\beta t})}. \qquad \text{......(2220)}$$

The complete velocity distribution function in this case is therefore

$$\frac{1}{\left\{\dfrac{2\pi kT}{m}(1 - e^{-2\beta t})\right\}^{\frac{1}{2}}} \exp\left[-\frac{(u - u_0 e^{-\beta t})^2}{(2kT/m)(1 - e^{-2\beta t})} \right], \qquad \text{......(2221)}$$

agreeing with the result of § 19·51. This can at once be adapted to the equation of motion

$$\frac{du}{dt} = -\beta u + g + A(t) \quad (g \textit{ const.})$$

by using $u' = u - g/\beta$.

The distribution law $f(x, x_0, u_0, t)$ is only slightly more complicated. The equations of motion are now two,

$$\frac{du}{dt} = -\beta u + A(t), \quad \frac{dx}{dt} = u,$$

and the proper solution

$$x - x_0 - \frac{u_0}{\beta}(1 - e^{-\beta t}) = \int_0^t e^{-\beta\eta}\, d\eta \int_0^\eta e^{\beta\xi} A(\xi)\, d\xi. \qquad \text{...(2222)}$$

It follows by the same arguments that

$$r = x - x_0 - \frac{u_0}{\beta}(1 - e^{-\beta t})$$

has Gaussian moments and therefore a Gaussian distribution. Therefore

$$f(x, x_0, u_0, t) = \sqrt{\left(\frac{\mu}{\pi}\right)}\, e^{-\mu[x - x_0 - (u_0/\beta)(1 - e^{-\beta t})]^2}, \qquad \text{......(2223)}$$

and as before

$$\frac{1}{2\mu} = \overline{\int_0^t \int_0^t d\eta\, d\eta'\, e^{-\beta(\eta+\eta')} \int_0^\eta \int_0^{\eta'} e^{\beta(\xi+\xi')} A(\xi) A(\xi')\, d\xi\, d\xi'}.$$

Since again the whole contribution comes from $\xi \sim \xi'$

$$\frac{1}{2\mu} = C \int_0^t \int_0^t d\eta\, d\eta'\, e^{-\beta(\eta+\eta')} \int_0^{\eta^*} e^{2\beta\xi}\, d\xi,$$

where C is a constant and η^* is the lesser of η and η'. The integrations are easily completed giving

$$\frac{1}{2\mu} = C' \left[t - \frac{3 - 4e^{-\beta t} + e^{-2\beta t}}{2\beta} \right]. \qquad \text{......(2224)}$$

It remains only to determine C', which can be derived by Einstein's arguments as given in §§ 20·62, 20·63 which here lead to $\overline{(x-x_0)^2}=2kTt/m\beta$. This gives C' and the complete law is

$$f(x,x_0,u_0,t)$$
$$=\frac{1}{\left[\frac{4\pi kT}{m\beta}\left\{t-\frac{3-4e^{-\beta t}+e^{-2\beta t}}{2\beta}\right\}\right]^{\frac{1}{2}}}\exp\left[-\frac{\{x-x_0-(u_0/\beta)(1-e^{-\beta t})\}^2}{\frac{4kT}{m\beta}\left\{t-\frac{3-4e^{-\beta t}+e^{-2\beta t}}{2\beta}\right\}}\right]. \quad \ldots\ldots(2225)$$

Similar methods can be applied to systems for which the equation of motion is

$$\frac{du}{dt}=-\beta u-\omega^2x+P(t)+A(t).$$

§ **20·91.** *Concluding remarks.* In the foregoing discussion, in particular of the shot effect, the fundamental nature of the assumptions underlying (2213) and (2214) has as yet scarcely been sufficiently emphasized. Similar assumptions are really made in the problem of the galvanometer mirror in setting up (2199), and they appear to be unavoidable, or at least never yet avoided, in such problems. In adopting these and similar equations we really assume that we can divide the whole apparatus into two parts, in one of which the discontinuity of the events happening is essential, while in the other it is trivial and may be neglected. This part of the apparatus, therefore, is assumed to obey the ordinary laws of large scale continuous physics. In view of the discontinuities in all matter and all events this division is a drastic assumption, which is made relying only on physical intuition, apart from *a posteriori* justification by success. The correct procedure would be of course to treat the apparatus as a whole, making no use of the laws of continuous physics, except in so far as their use can be explicitly justified. A particular virtue in the use of Campbell's theorem in such problems is that its use emphasizes the dichotomy and does not allow one to overlook the fundamental assumption used in any such treatment.

CHAPTER XXI

RECENT APPLICATIONS TO COOPERATIVE AND OTHER PHENOMENA

§21·1. *Cooperative phenomena. Generalities.* The most striking recent advances in the application of statistical mechanics have been made in the study of "*cooperative phenomena*". These are phenomena which cannot be interpreted in terms of the properties of an assembly of distinct systems only slightly linked to each other, but can only be interpreted in terms of the states of the assembly as a whole, because the states of any distinct system are fundamentally influenced by which states of the other systems are occupied. The classical example of a cooperative phenomenon is ferromagnetism, already discussed in detail in Chapter XII. We have there seen that the states of a would-be ferromagnetic metal can be correctly described in terms of the total magnetization—the resultant spin of all the contributing electrons—but cannot be analysed into states built up out of the states of individual electrons because the energies of the spin states of any one electron depend fundamentally on the directions of the spins of its neighbours. The recently recognized cooperative phenomena which we shall discuss here are (i) the order-disorder transition in metallic alloys, (i) the liberation-rotation transition in certain solids, and (iii) the existence of critical condensation temperatures for the deposition of a metallic vapour on a glass or other metal. We shall discuss the first of these in greatest detail since for it the theory has been most completely worked out. The third is perhaps the simplest of all and the best adapted to form in future an introduction to the subject. The chapter closes with short discussions of the peculiarities of crystals of the paramagnetic salts, and other miscellaneous topics.

§21·2. *Order-disorder phenomena in metallic alloys.** It has been known for a long time that many metallic alloys, which form for a small range of atomic ratios a definite solid phase with a characteristic lattice structure, exhibit anomalous specific heats and associated properties at a certain critical temperature. It has been shown more recently by X-ray analysis that the underlying structural change is a change, as the temperature rises, from ordered to disordered arrangements of the two sorts of atoms in the alloy. The lattice structure of the phase undergoes no significant change, but the degree of order in the allocation of the different atoms to the lattice points is affected.

* Bragg and Williams, *Proc. Roy. Soc.* A, vol. 145, p. 699 (1934); vol. 151, p. 540 (1935); Bethe, *Proc. Roy. Soc.* A, vol. 150, p. 552 (1935); Williams, *Proc. Roy. Soc.* A, vol. 152, p. 231 (1935). Ideas substantially adequate for a correct statistical treatment seem to have first been put forward by Gorsky, *Zeit. f. Physik*, vol. 50, p. 64 (1928).

The types of change that occur are shown below in Fig. 80. The alloy need not contain equal numbers of the two types of atoms nor need all the atoms in the lattice take part in the disordering. It is also not essential that the atoms should be present in the alloy in the exact proportions most typical of

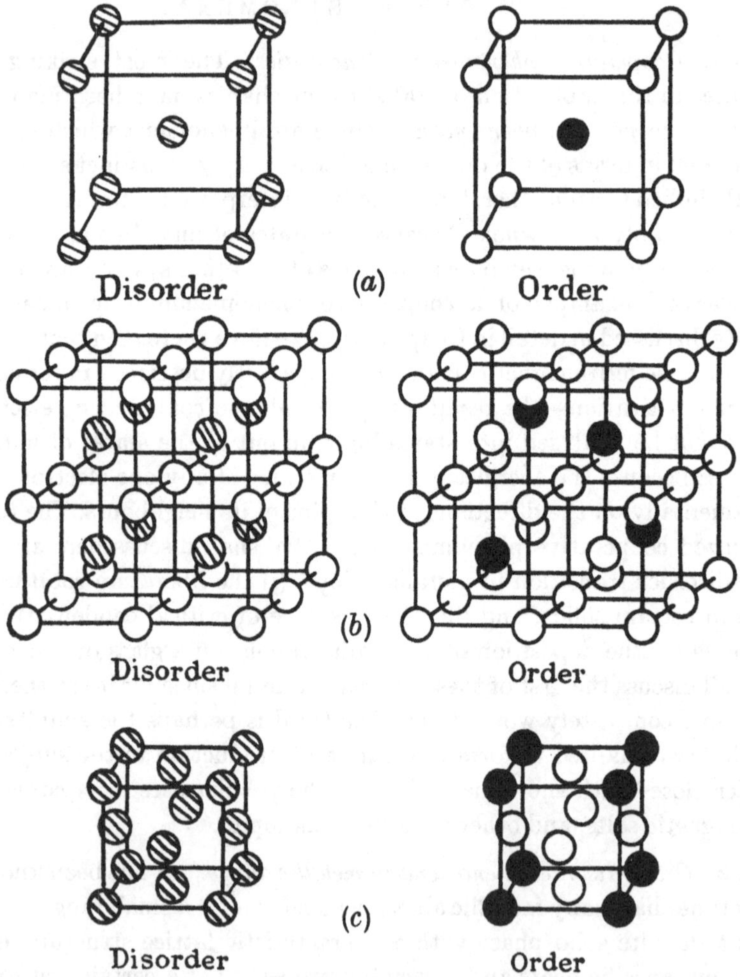

Fig. 80. Showing the arrangements of atoms in the states of complete order and complete disorder for the alloys (*a*) β-brass, CuZn, ○ Cu, ● Zn, ◉ $\frac{1}{2}$Cu, $\frac{1}{2}$Zn; (*b*) Fe_3Al, ○ Fe, ● Al, ◉ $\frac{1}{2}$Fe, $\frac{1}{2}$Al; (*c*) Cu_3Au, ● Au, ○ Cu, ◉ $\frac{1}{4}$Au, $\frac{3}{4}$Cu.

the phase, but we shall not discuss the effect of small changes in the atomic ratios. In β-brass for example, whose typical atomic formula is CuZn, the lattice is body-centred cubic, all the atoms take part in the ordering and the completely ordered and completely disordered states are shown in Fig. 80 (*a*). In the completely ordered state all the Cu-atoms occupy one simple cubic lattice, say the cube corners, and the Zn-atoms the other lattice, the cube centres. In the completely disordered state every lattice point of either kind is occupied indifferently by either atom. In the alloy Fe_3Al on the other hand, Fig. 80 (*b*), only half the atoms take part in the disordering. The lattice is again body-centred cubic, and two-thirds of the Fe-atoms permanently occupy one simple cubic lattice at all temperatures. The remaining atoms, equal numbers of Fe and Al, occupy the remaining lattice points at the cube centres—in the ordered state forming a rock salt lattice, and in the disordered occupying all this set of lattice points indifferently. It will be seen that in this example a larger unit cell is required to describe the ordered state (a cube of twice the edge) than to describe the disordered state. The ordered state is therefore sometimes referred to as a superlattice and the new X-ray reflections which reveal its existence as superlattice lines. In yet another example Cu_3Au, Fig. 80 (*c*), all the atoms take part in the ordering. The lattice is face-centred cubic; in the ordered state the Au-atoms occupy the cube corners and the Cu-atoms the face centres, while in the disordered state every lattice point is occupied indifferently by any atom.

Our problem therefore is to determine statistically the degree of order in the lattice, or degree of perfection of the superlattice, as a function of the temperature. Fundamentally this should be simple. It is clear that we are concerned only with geometrical configurations of the atoms, and that, to a first approximation at least, we may treat the configurations and configurational energy as independent of the lattice vibrations and other contributions such as electronic energy to the total energy of the crystal. The complete partition function for the mixed crystal can to this approximation be compounded of two factors, one, the normal one, independent of the atomic arrangements, and the other depending solely on these arrangements. In more refined treatments, into which we shall not enter here, it may prove necessary to admit that this independence is not complete. Our problem therefore reduces to the construction and evaluation of the partition function for the configurational states of the alloy.

§ 21·21. *The construction of the configurational partition function for the approximation of Bragg and Williams.* Let us suppose that in a given specimen of an alloy there are N atoms in all which take part in the ordering, rN of one kind and $(1-r)N$ of the other. There are then

$$\frac{N!}{(rN)!\{(1-r)N\}!}=q$$

distinct configurational states of varying energies, and the configurational partition function which we have to construct is

$$\Gamma(T)=\sum_{\tau=1}^{q} e^{-W_\tau/kT}.$$

To make progress it is necessary to classify the states in terms of some suitable parameter specifying the degree of order, which at the same time determines the energy of such states with reasonable precision. If s is such a parameter, then

$$\Gamma(T) = \Sigma_s g(s)\, e^{-W(s)/kT}. \qquad \text{......(2226)}$$

Provided a suitable s can be defined and $W(s)$ and $g(s)$ determined with sufficient accuracy as functions of s, $\Gamma(T)$ can be evaluated by searching for the maximum terms in (2226) and the configurational energy content and all other equilibrium properties immediately deduced.

The parameter s proposed by Bragg and Williams is the degree of order or degree of perfection of the superlattice defined as follows. In the state of perfect order there are rN lattice points occupied by A-atoms and $(1-r)N$ occupied by B-atoms. These may be referred to as α-lattice points and β-lattice points respectively. In any state of partial ordering there are, let us say, a fraction p of the α-lattice points occupied by A-atoms and the rest, a fraction $1-p$ by B-atoms. The fraction p goes from 1 to r as the ordering falls from perfect order to perfectly random arrangements (complete disorder). We therefore define s by

$$s = (p-r)/(1-r), \qquad \text{......(2227)}$$

so that s ranges from 1 to 0 over the range from perfect order to complete disorder, and correctly specifies the degree of perfection of the superlattice. The fraction p can of course always range from 1 to 0, and when $p<r$ the corresponding values of s are negative. When $r=\frac{1}{2}$ both signs of s may be included, the negative signs corresponding equally to ordered states with A- and B-atoms interchanged. When $r \neq \frac{1}{2}$ we take $r<\frac{1}{2}$ and the states of negative s as here defined do not correspond properly to a new set of ordered states. No error is made by arbitrarily excluding the states $s<0$ in the simple cases here discussed, which merely means that we confine attention to configurations leading to order with atoms A on the preselected α-lattice points.

One can now see approximately how $W(s)$ must depend on s. Starting from the state of perfect order each new configuration can be formed by the process of interchanging a pair of atoms A and B, and the disorder increases proportionally to the number of such exchanges. Each such exchange will at first require the expenditure of a constant amount of energy. If the state of perfect order is taken as the zero of energy, then the slightly disordered states $s=1-\Delta s$ will have an energy which can be written $W_0 \Delta s \simeq \frac{1}{2}W_0(1-s^2)$. When $s \simeq 0$ the form $W_0 \Delta s$ can no longer be correct. The interchange of a pair of atoms when disorder is practically complete will on the average involve no expenditure of energy and it is natural to assume that in the final stages the energy required for the replacements causing a change Δs varies like $s\Delta s$. If therefore we assume that

$$W(s) = \tfrac{1}{2}W_0(1-s^2), \qquad \text{......(2228)}$$

the s-dependence will be correct for both $s \simeq 0$ and $s \simeq 1$, and it is reasonable to suppose that we shall not be far wrong for all values of s. This is the essence of Bragg and Williams' approximation. They assume that, for given s, W will be a function of s only, effectively independent of local pairings and fluctuations, and given nearly enough by (2228).

To complete the formulation of $\Gamma(T)$ we need only determine $g(s)$, which is the number of configurations in which prN A-atoms are distributed over rN α-lattice points and $(1-p)rN$ A-atoms over the $(1-r)N$ β-lattice points, the B-atoms filling the vacancies remaining. It follows at once that*

$$g(s) = g\left(\frac{p-r}{1-r}\right) = \frac{(rN)!}{(prN)!\,(\{1-p\}rN)!}\,\frac{(\{1-r\}N)!}{(\{1-p\}rN)!\,(\{1-2r+pr\}N)!}, \qquad \text{......(2229)}$$

$$\Gamma(T) = \sum_{prN=r^2N}^{prN=rN} \frac{(rN)!}{(prN)!\,(\{1-p\}\,rN)!}\,\frac{(\{1-r\}N)!}{(\{1-p\}rN)!\,(\{1-2r+pr\}N)!}\,e^{-\frac{1}{2}W_0(1-s^2)/kT} \qquad (2230)$$

In the complete $\Gamma(T)$ which we need not consider the true lower limit of the summation is of course $prN = 0$.

It is evident without calculation that we must assume that $W_0 > 0$. This makes the state of perfect order ($s = 1$) the state of lowest energy and therefore the stable equilibrium state at very low temperatures. At higher temperatures the vastly greater number of disordered states will, as we shall see, allow them to take charge of the equilibrium. For the opposite assumption $W_0 < 0$ the state of perfect order will never be attained and our analysis of the states in terms of s is insufficient for the discussion. It is easy to see that this case would really correspond to a complete separation of the different atoms at low temperatures into as far as possible distinct phases, a problem in solubility with which we shall not concern ourselves.

§**21·22.** *The evaluation of Bragg and Williams' partition function.* The resemblance of (2230) to the Weiss-Heisenberg partition function (1383) for ferromagnetics is at once obvious. Similar methods of evaluation may be used and similar results will be obtained. We write the logarithm of a term $\gamma(p)$ in the series (2230), using Stirling's theorem, in the form

$$\log\gamma(p) = N[r\log r - rp\log rp - r(1-p)\log r(1-p) + (1-r)\log(1-r)$$
$$-r(1-p)\log r(1-p) - (1-2r+pr)\log(1-2r+pr)] - \tfrac{1}{2}W_0(1-s^2)/kT.$$

* Williams, *loc. cit.*

Since $s=(p-r)/(1-r)$, stationary values are determined by the equation

$$-\frac{sW_0}{kT}=Nr(1-r)\log\frac{(1-p)^2r}{p(1-2r+pr)}. \tag{2231}$$

If we write
$$x=\frac{sW_0}{r(1-r)NkT}, \tag{2232}$$

this equation can be reduced to

$$r(1-r)(e^x-1)(1-s)^2=s,$$

whose relevant root is

$$s=1-\frac{\{4r(1-r)(e^x-1)+1\}^{\frac{1}{2}}-1}{2r(1-r)(e^x-1)}. \tag{2233}$$

The roots of equations (2233) and (2232) in s determine the position of the maximum term $s=s_0$ or $p=p_0$ in (2230). It is then as usual sufficiently accurate to take

$$\log\Gamma(T)=\log(\text{Max. term}), \quad\text{......(2234)}$$

and
$$\overline{E_c}=kT^2\frac{\partial\log\Gamma(T)}{\partial T}=\tfrac{1}{2}W_0(1-s_0{}^2), \quad\text{......(2235)}$$

where $\overline{E_c}$ is the configurational energy. The contribution to the free energy of Helmholtz is

$$-kT\log(\text{Max. term}). \quad\text{......(2236)}$$

For $r=\frac{1}{2}$ equations (2232) and (2233) simplify to

$$s=\tanh\tfrac{1}{4}x \quad (x=4sW_0/NkT). \quad\text{......(2237)}$$

These are identical in form with the equations of the Weiss-Heisenberg theory and the same results can be derived. There is a critical temperature T_c given by

$$T_c=W_0/Nk, \quad\text{......(2238)}$$

such that, when $T>T_c$, $s_0\equiv 0$, while when $T<T_c$ s_0 is given by the non-zero root of (2237) shown in Fig. 81. We thus see that s_0, and therefore $\overline{E_c}$, is continuous, but $ds_0{}^2/dT$ and therefore the contribution $C_V{}^c$ to the specific heat is not. For values of T near T_c

$$s_0{}^2=3\frac{T^2}{T_c^2}\left(1-\frac{T}{T_c}\right), \quad\text{......(2239)}$$

$$\frac{ds_0{}^2}{dT}\to-\frac{3}{T_c},\quad C_V{}^c\to\frac{3}{2}\frac{W_0}{T_c}=\tfrac{3}{2}Nk. \quad\text{......(2240)}$$

A comparison of observed and theoretical values of $C_V{}^c$ for β-brass is made in § 21·25.

When $r\neq\frac{1}{2}$ the behaviour of the roots of (2233) is more subtle and s itself is discontinuous at $T=T_c$. The most illuminating method of procedure is to write equation (2231) in the form

$$\frac{sW_0}{r(1-r)NkT}=\log\left\{1+\frac{s}{r(1-r)(1-s)^2}\right\}, \quad\text{......(2241)}$$

and study the intersections of these two functions of s. We shall find that, as T decreases from large values, the first non-zero root of (2241) does not enter at $s=0$ and move continuously away from the existing zero root as it does for (2237). For that equation it was easy to show in § 12·9 that the maximum term in the partition function always corresponds to the non-zero root of s as soon as such a root exists. Here the maximum value of $\gamma(s)$ lies at $s=0$ so long as that is the only root and the new stationary values when they enter must be smaller than $\gamma(0)$ and can therefore not at once correspond to the

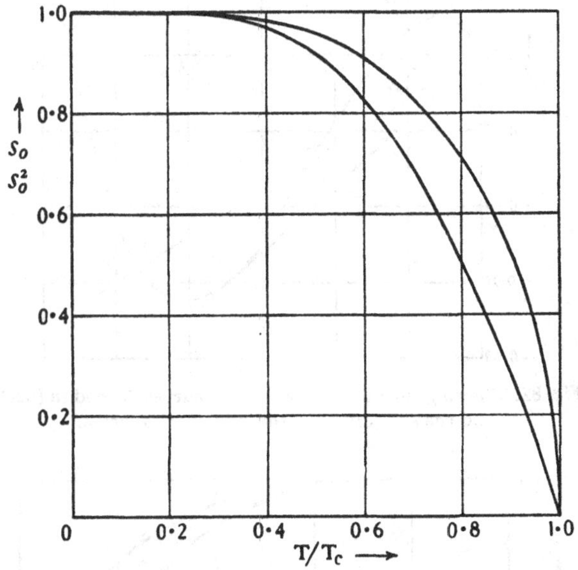

Fig. 81. Showing s_0 and s_0^2 as functions of T/T_c for $r=\frac{1}{2}$ according to equation (2237).

maximum terms. The maximum of $\log\gamma(s)$ corresponds of course owing to the nature of partition functions to minimum free energy. To find when a non-zero value of s first gives the maximum value of $\log\gamma(s)$ one must determine at what point in the three-root range of equation (2241) the two extreme roots (one of them $s=0$) give equal values of $\gamma(s)$ or $\log\gamma(s)$. This can be done conveniently by applying *the rule of equal areas*. Since the difference of the two functions in (2241) is proportional to $\partial\log\gamma(s)/\partial s$ the change from the maximum term at $s=0$ to a maximum at $s=s_0$ occurs when

$$\int_0^{s_0}\frac{\partial\log\gamma(s)}{\partial s}\,ds=0,$$

that is when the curves in the (y,s) plane

$$y=\frac{sW_0}{r(1-r)NkT},\quad y=\log\left\{1+\frac{s}{r(1-r)(1-s)^2}\right\}$$

enclose equal areas between $s=0$, $s=s_0$ and the s-axis. If equation (2241) is transformed, for example by taking exponentials, this simple rule no longer can be applied.

The case $r=\frac{1}{4}$ is illustrated in Figs. 82, 83. Intersections of the curves for

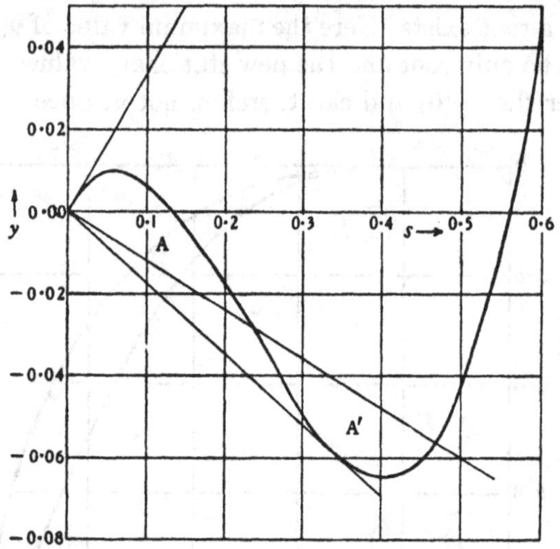

Fig. 82. Showing the intersections of the curves defined in (2242) and the application of the rule of equal areas.

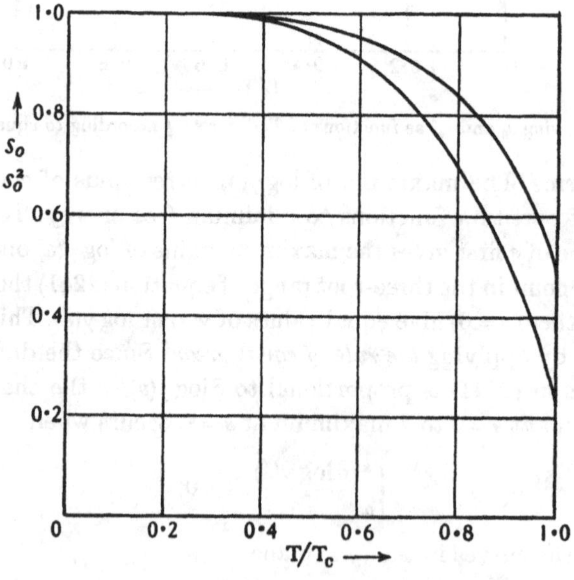

Fig. 83. Showing s_0 and $s_0{}^2$ as functions of T/T_c for $r=\frac{1}{4}$ according to equation (2233) or (2241).

large values of s cause no difficulty and Fig. 82 is confined to the range $s < 0{\cdot}6$. It is convenient to compare the curves

$$y = s\left\{\frac{16W_0}{3NkT} - 5\right\} = \alpha s, \quad y = \log\left\{1 + \frac{16s}{3(1-s)^2}\right\} - 5s. \quad \ldots(2242)$$

The second of these is shown on a large scale in Fig. 82. The adjustment of subtracting a multiple of s from both functions does not invalidate the rule of equal areas. The functions in (2242) have three intersections in the range $-0{\cdot}175 < \alpha < +0{\cdot}33$. The areas A and A' are equal so that the critical value of α is $-0{\cdot}12$. The critical value T_c of T is therefore given by $W_0/NkT_c = 0{\cdot}915$ and the critical value of s to which it jumps discontinuously from $s = 0$ is $0{\cdot}467$. Equation (2235) shows that this corresponds to a latent heat at $T = T_c$ equal to $\frac{1}{2}W_0 s_0^2$ or $0{\cdot}100\,NkT_c$. Corresponding values of T/T_c and s for values of $T < T_c$ and $s > s_0$ are easily derived by following the greatest root of the two equations (2242). The resulting values of s_0 and s_0^2 are plotted in Fig. 83.

Apart altogether from any success which this theory may have in explaining the order-disorder phenomena in alloys, it is of considerable importance in statistical mechanics as the first genuinely molecular model which has been shown to give an energy which is a discontinuous function of the temperature for a single homogeneous phase, thus leading to a latent heat of transition. Such latent heats are of course common-place enough in thermodynamics where they occur only in systems assumed *a priori* to be in two distinct phases; the latent heat is then of course the energy change when the relative extension of the two phases alters. Here the existence of a latent heat is derived from the properties of the single homogeneous phase. One must also ask here what happens for the molecular model when the energy is fixed at some intermediate value between the extremes of the discontinuity? The answer is that the assembly then breaks up into two distinct phases, one ordered and the other disordered.

§ **21·23.** *Refinements of the order-disorder theory due to Bethe.** In the foregoing approximations we have assumed that the configurational energy of the crystal is a function only of the superlattice order s as there defined—all the numerous arrangements leading to the same s are given the same energy. This assumption is however more than doubtful. It is more natural to expect that the atoms in the lattice act on each other with short range forces or even that only the interactions between nearest neighbours are of any real importance. The energy of the crystal will then be determined by the number of pairs of unlike neighbours in the lattice, that is by *the degree of local order* rather than directly by *the degree of long range order s*. It by

* Bethe, *loc. cit.*

no means follows that because we must assume only short range forces or interactions between nearest neighbours we shall therefore not have long range order set up. We have already seen in the theory of ferromagnetism that interaction between nearest neighbours can orientate, i.e. order, the spins of the whole crystal. A theory of order-disorder ought therefore to be possible, based on interactions between nearest neighbours only, and might be expected to give a better physical picture of an alloy than the model of Bragg and Williams. Such a theory has been successfully developed by Bethe for the case $r = \frac{1}{2}$ and for the simplest types of lattice arrangement in which the nearest neighbours of any one atom are never nearest neighbours of each other. Simple cubic and body-centred cubic lattices are of this type so that the theory applies directly to β-brass. Extensions of the theory are now in progress, but we shall only discuss the simplest case dealt with in Bethe's original paper.

§ **21·231.** *Long range order and short range forces in one, two or three dimensions.* In the discussion of ferromagnetism in § 12·9, for low temperatures in the manner of Bloch, we found that ferromagnetism can only exist in a crystal lattice in three dimensions. The short range exchange forces from which it is derived cannot orientate the whole body of spins in a linear chain of atoms or even in a plane array. This importance of the number of dimensions in such problems is brought out even more clearly in the order-disorder example, and it is well to discuss it carefully for its own sake as it promises to be of still greater importance in any advance of statistical mechanics in the direction of a proper theory of liquids.

We shall now show that short-range forces, which we shall take in the extreme form of interactions between nearest neighbours only, cannot order a linear lattice, but can order a lattice in two and *a fortiori* in more dimensions. Consider first a linear lattice of composition AB. Perfect order is represented by the state

$$\ldots A\,B\,A\,B\,A\,B\,A\,B\,A\,B\,A\,B \ldots.$$

At any non-zero temperature there is a finite chance that a state of next greater energy occurs in which one pair of neighbours are alike. This is the state

$$\ldots A\,B\,A\,B\,A_1A_2B\,A\,B\,A\,B\,A \ldots.$$

Since A_1 is irrelevant to the right of A_2 the normal state will be as shown perfectly ordered on A_2; similarly the normal state to the left of A_1 will ignore A_2 and be perfectly ordered on A_1. The chain falls then into two parts each perfectly ordered in itself, but the long range order of the whole is completely destroyed by one break.

In two or more dimensions there must still be a finite chance at any non-zero temperature that the state of next higher energy will occur, in which one atom in the array is wrong. But such a wrong atom no longer can have this catastrophic effect on the long range order. Arrays with perfect order and with one wrong atom are shown below:

$$\begin{array}{llllll} A\,B\,A\,B\,A\,B & \quad A\,B\,A\,B\ \ldots \\ B\,A\,B\,A\,B\,A & \quad B\,A\,B_1X\ \ldots \\ A\,B\,A\,B\,A\,B & \quad A\,B\,A_1A_2\ldots \\ B\,A\,B\,A\,B\,A & \quad B\,A\,B\ \ldots \\ A\,B\,A\,B\,A\,B & \quad A\,B\,A\,B\ \ldots \\ B\,A\,B\,A\,B\,A & \quad B\,A\,B\,A\,B\,A. \end{array}$$

We now consider what ought to happen at X when the wrong atom A_2 is in place. A_2 requires X to be B, but B_1 requires X to be still A, and there are other neighbours as well. Thus the nature of X depends on all the neighbours and no longer merely on A_2, and one mistake like A_2 does not necessarily destroy the long range order. The possible alternatives are

(i) The atoms surrounding A_1A_2 fit A_1 and ignore A_2, so that A_2 is merely one wrong atom and long range order is unaffected.

(ii) The atoms surrounding A_1A_2 fit A_2 and ignore A_1, so that A_1 is merely one wrong atom and long range order is again unaffected.

(iii) There is a real break in the order as for a linear lattice and the lattice falls into two parts perfectly ordered in themselves but perfectly out of phase with each other across an interface, thus

$$\begin{array}{l} A\,B\,A\,B\,|\,B\,A \\ B\,A\,|\,A\,B\,A\,B \\ A\,B\,A\,|\,A\,B\,A \\ B\,A\,B\,A\,B\,|\,B \\ A\,B\,A\,B\,A\,|\,A \\ B\,A\,B\,A\,B\,A. \end{array}$$

Arrangements (i) and (ii) have each four pairs of wrong neighbours, or eight if we include somewhere the displaced B, and require obviously far less energy than (iii), which has a number of wrong pairs of neighbours of the order $N^{\frac{1}{2}}$. At low temperatures therefore there will exist long range order with (until $T=0$) a few wrong atoms, and a negligible chance of long range disorder. This conclusion holds *a fortiori* in three dimensions where an interface requires a number of wrong pairs of neighbours of the order $N^{\frac{2}{3}}$.

At high temperatures when energy differences no longer matter the large number of ways of realizing an interface will of course take charge, and we must ultimately pass to a state of long range disorder—in fact a state of complete disorder, short and long. We shall be able to show that long range order disappears suddenly and completely at a critical temperature, while local order changes much more gradually and is only perfect when $T\to 0$

and zero when $T \to \infty$. For local order there is no overwhelming combinatory factor coming in to wipe out the effect of any energy differences that may survive. We shall therefore distinguish both types of order in the crystal, introduce precise functions for them, and try to evaluate the configurational partition functions on this basis.

§ 21·232. *Local order, or order of neighbours in a crystal lattice.* Consider any simple lattice of equal numbers of two types of atoms, in which in the state of perfect order all the nearest neighbours of any atom are atoms of the other sort. In any configuration let the fraction of unlike (correct) pairs of neighbours be $\frac{1}{2}(1+\sigma)$ and the fraction of like (wrong) pairs of neighbours $\frac{1}{2}(1-\sigma)$. The difference of these fractions is σ and is called *the local order of the configuration.* The parameter σ ranges from 1, perfect local order, to 0, complete local disorder. In these extreme cases there will of course also be perfect long range order and complete long range disorder respectively.

In such a configuration any atom A will have on the average a fraction $\frac{1}{2}(1+\sigma)$ B neighbours and $\frac{1}{2}(1-\sigma)$ A neighbours. The same is true for B, with A and B exchanged throughout. If there are N atoms in all in the lattice and each has z neighbours, the A-atoms will have a total of $\frac{1}{2}Nz$ neighbours of which $\frac{1}{4}Nz(1+\sigma)$ are B's and $\frac{1}{4}Nz(1-\sigma)$ are A's, with similar numbers for B-atoms. We easily deduce that there are in all $\frac{1}{2}Nz$ pairs of neighbours in the lattice of which $\frac{1}{4}Nz(1+\sigma)$ are AB's, $\frac{1}{8}Nz(1-\sigma)$ AA's and $\frac{1}{8}Nz(1-\sigma)$ BB's. We can now give an expression for the energy of the configuration. Let pairs of neighbours AB, AA and BB provide interaction energies V_{ab}, V_{aa} and V_{bb} respectively. We find therefore that E_c, so far as it depends on σ, is given by

$$E_c = \tfrac{1}{4}Nz(1+\sigma)V_{ab} + \tfrac{1}{8}Nz(1-\sigma)(V_{aa}+V_{bb}),$$

$$= const. + \tfrac{1}{4}NzV(1-\sigma), \qquad \text{......(2243)}$$

where $$V = \tfrac{1}{2}(V_{aa}+V_{bb}) - V_{ab}. \qquad \text{......(2244)}$$

We may assume that $V>0$ or else the atoms A and B will tend to segregate at low temperatures.

The energy V introduced here can be related to the energy W_0 of equation (2228). Since when $\sigma=1$, $s=1$ and when $\sigma=0$, $s=0$, the energy difference between perfect local order and complete local disorder is by (2243) $\frac{1}{4}NzV$. Or referring to (2228) we see therefore that

$$\tfrac{1}{2}W_0 = \tfrac{1}{4}NzV. \qquad \text{......(2245)}$$

This relation may also be verified by a study of the effect of replacements when the order is nearly perfect. When s changes from 1 to $1-\Delta s$, the energy of the lattice in Bragg and Williams' model increases by $W_0\Delta s$. Since for $r=\frac{1}{2}$, $s=2p-1$, this corresponds to a change of p from 1 to $1-\frac{1}{2}\Delta s$. Now p

is the fraction of A-atoms on α-lattice points, and therefore $\frac{1}{2}(1-p)N$ is the number of B-atoms on α-lattice points. Each such replacement of an A by a B creates a B-atom with z wrong neighbours (since $\sigma \sim 1$) and requires an interaction energy zV_{bb}. There are an equal number of wrong A-atoms each requiring an interaction energy zV_{aa}. The total interaction energy required is therefore $\frac{1}{2}Nz(1-p)(V_{aa}+V_{bb})$ in place of the interaction energy $Nz(1-p)V_{ab}$ necessary when all these pairs are correct. The total gain of energy is therefore $Nz(1-p)V$, or $\frac{1}{2}NzV\Delta s$. On comparing this with $W_0\Delta s$ we verify (2245).

§ **21·233.** *Long range, or superlattice order redefined.* It is convenient slightly to modify our former definition of long range order in terms of s for the sake of achieving greater symmetry. We divide the lattice points as before into α-lattice points and β-lattice points, and let p be the fraction of α-lattice points occupied by A's, so that $1-p$ is the fraction occupied by B's. We now define s' by the difference

$$s' = p-(1-p) = 2p-1. \qquad \text{......(2246)}$$

This retains the symmetry because both the state $p=1$ and the state $p=0$ are really states of perfect order. In Bragg and Williams' definition $s=(p-r)/(1-x)$, but since for $r=\frac{1}{2}$, $s=s'$ we shall not further distinguish them, and have merely to remember that the range of values of s is now -1, 1, a range which, as we saw, we could have included before in this special case.

§ **21·234.** *The form of the partition function in terms of long and short range order.* We have now to form the partition function. If $g(s,\sigma)$ is the number of configurations with given s and σ, and $\gamma_s(T)$ is the partition function for states of given long range order, then

$$\gamma_s(T) = \Sigma_\sigma g(s,\sigma)\, e^{-\frac{1}{2}NzV(1-\sigma)/kT}, \qquad \text{......(2247)}$$

and $\gamma_s = \gamma_{-s}$ by symmetry. The complete partition function would then be

$$\Gamma(T) = \sum_{s=-1}^{+1} \gamma_s(T). \qquad \text{......(2248)}$$

This could be summed at once if we knew $g(s,\sigma)$ or in fact $\Sigma_s g(s,\sigma)$. This last expression is the number of arrangements of atoms A and B on the lattice points with a given number of pairs of nearest neighbours of type AB (actually $\frac{1}{4}Nz[1+\sigma]$). This number is known for a linear array* but not for any array in two or more dimensions. We are therefore compelled to adopt indirect methods. The type of method adopted is worthy of detailed study, since it is most probable that such indirect methods will become of increasing importance in statistical mechanics.

We may start by observing that at very low temperatures the terms in

* Ising, *Zeit. f. Physik*, vol. 31, p. 253 (1925).

$\Gamma(T)$ have a maximum for $\sigma=1$ and a minimum by symmetry when $\sigma=0$. This implies that then the maximum terms are $\gamma_{\pm 1}(T)$ and the minimum term $\gamma_0(T)$, since $\sigma=1$, $s=\pm 1$ and $\sigma=0$, $s=0$ correspond uniquely. As T increases the long range order s must decrease, so that the maximum terms must shift to lower and lower values of s until by symmetry they coalesce at $s=0$. Provided this coalescence occurs before $T\to\infty$ there must be a critical temperature, above which the long range order is permanently zero. These arguments do not of course constitute a rigorous proof of the existence of T_c, or that the maximum terms move *continuously* in towards $s=0$, but they suggest that we may start by trying to find conditions for the onset of a critical behaviour in this region of s.

If by any method we can determine σ_0 and the corresponding value or values of s_0, which give (perhaps only approximately) stationary values to the terms of $\Gamma(T)$, we can derive all the equilibrium properties that we require without constructing and evaluating $\Gamma(T)$ directly. For given σ_0 we know that the configurational energy $\overline{E_c}$ is $\frac{1}{4}NzV(1-\sigma_0)$. We know moreover that for the corresponding s_0 $\log\gamma_s(T)$ must be stationary for variations of s. But by the thermodynamic properties of a partition function this implies that

$$\frac{\partial S_c}{\partial s_0}=\frac{1}{T}\frac{\partial \overline{E_c}}{\partial s_0}.$$

Since we know S_c for perfect order $s_0=1$, we know both S_c and $\overline{E_c}$ as functions of s_0, and we can therefore follow the changes in the free energy as s_0 varies, thus picking out that value of s_0 (when there are more than one) which gives the free energy its absolute minimum, and fixing the critical temperature. The reliability of such indirect methods which do not actually construct and evaluate $\Gamma(T)$ cannot be directly checked. One can only demand that the condition they set up for determining σ_0 and s_0 shall be a plausible approximation to the direct method.

§21·24. *Indirect evaluation of* σ_0 *and* s_0. The indirect method proposed by Bethe is to construct partition functions for the configurations in limited regions of the crystal in such a way that the σ_0 and s_0 they determine should be adequate approximations to the true σ_0 and s_0 for the complete configurations.

Let us start by considering one lattice point and its z nearest neighbours. For a square lattice $z=4$ and the configurations of perfect local order are:

$$\begin{array}{ccc} & B & \\ B & A & B \\ & B & \end{array} \qquad \begin{array}{ccc} & A & \\ A & B & A \\ & A & \end{array}$$

We know that on the assumption of short range forces the configurational energy depends only on $V=\frac{1}{2}(V_{aa}+V_{bb})-V_{ab}$. For simplicity we shall assume here that $V_{ab}=0$, $V_{aa}=V_{bb}=V$. As a first approximation we shall construct the partition function for configurations on one lattice point and its neighbours ignoring all effects of more distant atoms. The atoms in the first shell are not nearest neighbours of each other, and have therefore no interaction energy. Let

$$x=e^{-V/kT}$$

for shortness. Then the configurational partition function for all the 2^z configurations with a given central atom A, say, is

$$f_a=(1+x)^z. \qquad \text{......(2249)}$$

Similarly

$$f_b=(x+1)^z.$$

Moreover the number of wrong (A) neighbours of a central A or wrong (B) neighbours of a central B is obviously

$$x\frac{\partial}{\partial x}\log f_a=\frac{zx}{1+x}.$$

But on the average this number is known to be $\frac{1}{2}z(1-\sigma)$. It follows that to this approximation

$$\sigma_0=\frac{1-x}{1+x}. \qquad \text{......(2250)}$$

This gives the degree of local order as a function of T to this approximation. It is a smoothly varying function of T and shows no critical behaviour. This however is to be expected owing to the crudity of our approximation, which ignores the ordering effect in the first shell of the second shell and still more distant atoms.

The obvious next improvement is to take into account the second shell explicitly and ignore the third and more distant shells. This approximation has been worked out in detail by Bethe for a plane square lattice. It yields a value of σ_0 of the same general character as (2250) but with a sharper variation. It is clear that repeated steps of this type will eventually solve the desired problem accurately but already the third approximation is very laborious and one must turn to the indirect attack.

We take account of the ordering effects of distant shells on the last shell explicitly included by inserting a factor ϵ in each term in the partition function for every wrong atom in the last shell. This ϵ will take care of the ordering effect of the neglected distant atoms to a very good approximation. Taking merely explicit account of a central atom and its first shell of z neighbours (the first approximation above) we shall now have, for a lattice point whose central atom should be A, partition functions f_a and f_b corresponding to central atoms A and B given by

$$f_a=(1+\epsilon x)^z, \quad f_b=(x+\epsilon)^z. \qquad \text{......(2251)}$$

Of these configurations the average fraction with central atoms B is

$$f_b/(f_a+f_b),$$

and the average number of A (wrong) atoms in the first shell is

$$\epsilon \frac{\partial}{\partial \epsilon} \log(f_a + f_b).$$

We can now determine ϵ by a reflexive argument. For there is nothing to distinguish the central atom from the atoms of the first shell in the lattice as a whole and therefore the fraction of wrong atoms at each point must be the same, or

$$\frac{f_b}{f_a + f_b} = \frac{1}{z} \epsilon \frac{\partial}{\partial \epsilon} \log(f_a + f_b), \qquad \text{......(2252)}$$

leading to

$$(\epsilon + x)^z = \epsilon x(1 + \epsilon x)^{z-1} + \epsilon(\epsilon + x)^{z-1},$$

or

$$\frac{\epsilon + x}{1 + \epsilon x} = \epsilon^{1/(z-1)}. \qquad \text{......(2253)}$$

A similar calculation can be carried through at this stage of the second approximation.

The parameter ϵ_0 determined as the proper root of (2253) will obviously serve in place of s_0. For we have defined s as the fraction of α-lattice points occupied by A-atoms less those occupied by B-atoms. Therefore

$$s = (f_a - f_b)/(f_a + f_b), \qquad \text{......(2254)}$$

from which it follows by using (2253) that

$$s_0 = \frac{\epsilon_0^{-\frac{1}{2}z/(z-1)} - \epsilon_0^{\frac{1}{2}z/(z-1)}}{\epsilon_0^{-\frac{1}{2}z/(z-1)} + \epsilon_0^{\frac{1}{2}z/(z-1)}} = \tanh\left[\frac{\frac{1}{2}z}{z-1} \log \frac{1}{\epsilon_0}\right]. \qquad \text{......(2255)}$$

Again σ_0 can be expressed simply in terms of ϵ_0. For the average number of wrong neighbours of the central atom is obviously $x\,\partial/\partial x\,[\log(f_a + f_b)]$, and also by definition $\frac{1}{2}z(1 - \sigma)$. Therefore after a simple reduction

$$1 - \sigma_0 = \frac{4x\epsilon_0}{1 + x\epsilon_0} \frac{1}{1 + \epsilon_0^{z/(z-1)}}. \qquad \text{......(2256)}$$

Since for any configuration $\overline{E_c} = \frac{1}{2}NzV(1 - \sigma)$, equation (2256) is equivalent to an equation for the average configurational energy.

The next step is to determine the nature of the roots of equation (2253). By inspection we see that

(i) To any value of x there correspond pairs of values of ϵ, namely ϵ' and $1/\epsilon'$ ($s = \pm s_0$), denoting merely that the A- and B-atoms have been interchanged throughout.

(ii) For all x, $\epsilon = 1$ is a solution for which there is no long distance order. The state for which $s = 0$ may therefore be assumed always to correspond to a stationary value of the terms of the complete partition function, but whether maximum or minimum remains to be settled.

(iii) If $\epsilon \neq 1$, we can assume $\epsilon < 1$ and write

$$x = \sinh\left[\frac{z-2}{2(z-1)} \log \frac{1}{\epsilon}\right] \Big/ \sinh\left[\frac{z}{2(z-1)} \log \frac{1}{\epsilon}\right].$$

It is easily deduced from this equation that x is monotonic increasing as ϵ increases and therefore never greater than its limit as $\epsilon \to 1$; that is that

$$x < x_0 = (z-2)/z \quad (\epsilon \neq 1). \qquad \text{......(2257)}$$

If therefore $x > x_0$ so that $e^{-V/kT} > (z-2)/z$, or

$$T > T_c = V/k \log\{z/(z-2)\}, \qquad \text{......(2258)}$$

then $\epsilon = \epsilon_0 = 1$ is the only root, $s_0 = 0$, and there is no long range order.

(iv) When $\epsilon_0 = 1$ equation (2256) shows that σ_0 reduces to its value for the first approximation given in (2250), namely

$$\sigma_0 = (1-x)/(1+x) \quad (x > x_0),$$

while as $\epsilon_0 \to 1$

$$\sigma_0 \to (1-x_0)/(1+x_0) = 1/(z-1). \qquad \text{......(2259)}$$

(v) The roots of equation (2253) not equal to unity enter continuously with the root unity at the critical temperature, the non-zero value of s_0 starting continuously from the zero value. We may assume therefore without further discussion that the non-unit root of (2253), when it exists, corresponds always to the true maximum term in the partition function.*

§ **21·241.** *The configurational specific heat.* The calculation of $C_V{}^c$ presents no particular difficulties though it is somewhat lengthy. Since σ_0 does not vanish for finite T $C_V{}^c$ is never zero, but there is a strong discontinuity at $T = T_c$. We find

$$C_V{}^c \sim \tfrac{1}{4}Nk(z-2)\left(\frac{z}{z-1}\right)^2\left(\log\frac{z}{z-2}\right)^2(\tfrac{3}{2}z-1) \quad (T \to T_c - 0),$$

$$C_V{}^c \sim \tfrac{1}{4}Nk(z-2)\left(\frac{z}{z-1}\right)^2\left(\log\frac{z}{z-2}\right)^2\tfrac{1}{2} \quad (T \to T_c + 0).$$

Thus the ratio of the limiting values is $3z-2$, which is already 16 for a simple cubic lattice.

The limiting values of $C_V{}^c$ according to these equations can easily be calculated for various values of z; results are given in Table 72.

TABLE 72.

Limiting values of configurational specific heat near the critical temperature according to Bethe's theory. Values given are for one mole of atoms partaking in the ordering.

No. of neighbours	Lattice structure	$C_V{}^c(T_c-0)/R$	$C_V{}^c(T_c+0)/R$
2	Linear chain	0	0
4	Plane quadratic	2·14	0·214
6	Simple cubic	1·90	0·119
8	Body-centred cubic	1·78	0·081
12	—	1·68	0·049
∞	—	1·50	0

* This point has been examined more closely by Williams, *loc. cit.* by calculating the entropy and free energy.

§21·242. ***Comparison of the two theories.*** It is interesting to observe that the results of Bethe's theory (first approximation) reduce to the results of

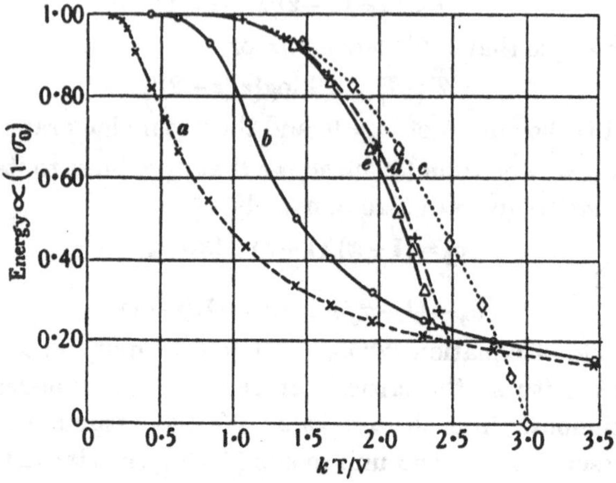

Fig. 84. Local order σ_0 for $r=\frac{1}{2}$ and simple cubic lattices $z=6$ according to (*a*) and (*d*), Bethe's first approximation without and with long range order; (*b*) and (*e*) the same, second approximation; (*c*) s_0^2 according to Bragg and Williams.

Bragg and Williams when $z\to\infty$. This is what one should expect, for when any atom has infinitely many "neighbours" the conceptions of local and long

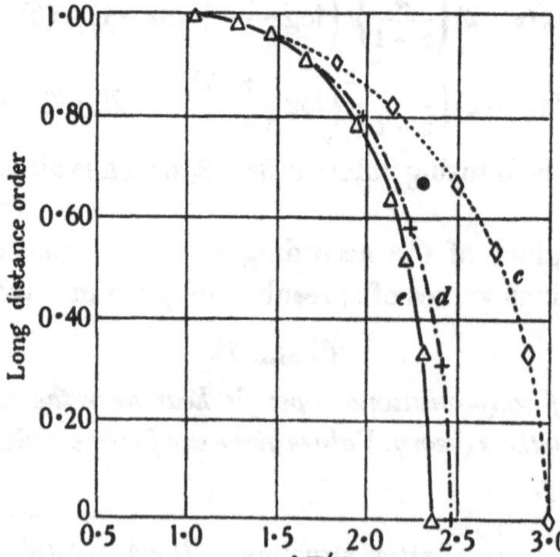

Fig. 85. Long range order s_0 for $r=\frac{1}{2}$ and simple cubic lattices $z=6$ according to (*d*) and (*e*) Bethe's first and second approximations and (*c*) Bragg and Williams.

range order must merge into each other. A proof of this equivalence will be found in Bethe's paper; it can easily be reconstructed by (i) using the substitution

$$\epsilon=e^{-2\delta/(z-1)}$$

and observing that, when $z\to\infty$, $z\delta$ remains finite in the important ranges;

(ii) recalling the relation between W_0 and V, namely $W_0 = \frac{1}{2}NzV$; zV must therefore be assumed to remain fixed.

The results of the two theories for $r = \frac{1}{2}$ and simple cubic lattices $z = 6$ are shown in the three accompanying diagrams, which include also the results of Bethe's second approximation in which the ϵ-factor is brought in in the second instead of in the first shell of atoms.

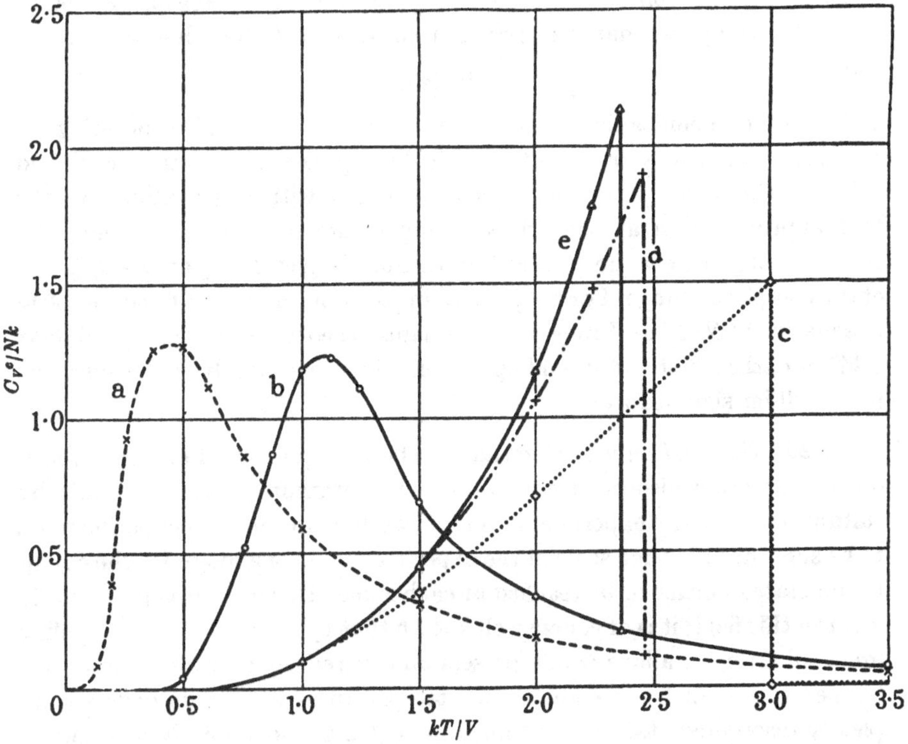

Fig. 86. The configurational specific heat for $r = \frac{1}{2}$ and simple cubic lattices $z = 6$ according to (*a*) and (*d*) Bethe's first approximation without and with long range order; (*b*) and (*e*) the same, second approximation; (*c*) Bragg and Williams.

Fig. 84 shows the local order or configurational energy content. Curve (*a*) shows σ_0 according to equation (2250) ignoring any ϵ-factor and curve (*d*) shows σ_0 according to (2256) including the ϵ-factor. These curves give Bethe's first approximation. Curves (*b*) and (*e*) give similarly Bethe's second approximation. Curve (*c*) gives s_0^2 according to Bragg and Williams as shown in Fig. 81. It represents local order in this theory for comparative purposes in the sense that the energy content is proportional to $1 - s_0^2$.

Fig. 85 shows the long range order s_0 according to the same theories. It will be observed that the onset is most sudden for what is presumably the best of the theories, Bethe's second approximation.

Finally Fig. 86 shows the configurational specific heat. The same tendency will be observed for the presumably best theory to give the sharpest and most concentrated variation of the specific heat near the critical temperature.

An important point of comparison of these theories, which is also an important point for comparison with observation, is the relation between the complete change of configurational energy and the critical temperature, since this relation contains no adjustable parameter. On Bragg and Williams' theory the complete change is $\frac{1}{2}NkT_c$; according to Bethe's first approximation

$$\tfrac{1}{2}NkT_c[\tfrac{1}{2}z\log\{z/(z-2)\}].$$

This however includes the change in the energy due to local disordering remaining to be achieved above the critical temperature. Strictly one should subtract this term particularly for comparison with observation. To the first approximation at the critical temperature $\sigma_0 = 1/(z-1)$. Hence the energy change below the critical temperature is just $1-\sigma_0$ or $(z-2)/(z-1)$ of the complete change. For $r=\frac{1}{2}$ and a simple cubic lattice $z=6$ the complete change is $0{\cdot}5NkT_c$ in Bragg and Williams' theory, $0{\cdot}486\ NkT_c$ and $0{\cdot}48\ NkT_c$ in Bethe's first and second approximations respectively. These numbers hardly differ significantly.

§ **21·25.** *Comparison with observation.* The configurational energy content during the order-disorder transition can be determined experimentally by suitable calorimetric experiments. The great difficulty in such experiments is to be sure that the time scale of the experiment is long enough for the equilibrium state of order to be reached at each stage. This danger is particularly great in this field; it is of course well known that by sufficiently rapid cooling (quenching) most alloys which possess an ordered equilibrium form at low temperatures can be brought to low temperatures in a more or less completely disordered state and retained in that state indefinitely in a sort of metastable equilibrium. It is moreover probably correct to hold that many other alloys, not known to occur in an ordered state at low temperatures, are really *frozen* in a metastable disordered state, because the temperature at which changes of position in the lattice cease to be possible at reasonable speed is higher than the critical temperature for the order-disorder transition. These questions of the rate of attainment of order have been discussed by Bragg and Williams in the papers quoted; we cannot discuss them further here as they lie outside the scope of this monograph.

Careful experiments on the specific heat or energy content of β-brass ($r=\frac{1}{2}$) and Cu_3Au ($r=\frac{1}{4}$) have recently been carried out by Sykes* and discussed by Bragg and Williams. β-brass is particularly suitable experimental material

* Sykes, *Proc. Roy. Soc.* A, vol. 148, p. 422 (1935) and unpublished experiments described by Bragg and Williams, *loc. cit.*

as its relaxation time for atomic rearrangements in the lattice is unusually short. Fig. 87 compares the observed and calculated values of $\{\overline{E_c}(T_c)-\overline{E_c}(T)\}/RT_c$ as functions of T/T_c for (i) β-brass (CuZn) and (ii) Cu_3Au.

The agreement on the whole is very fair. The theoretically more abrupt changes near the critical temperature for $r=\frac{1}{4}$ are well seen in the observed values. There is however a distinct tendency for the energy changes to be more closely concentrated towards $T/T_c=1$ than is allowed for by the theory.

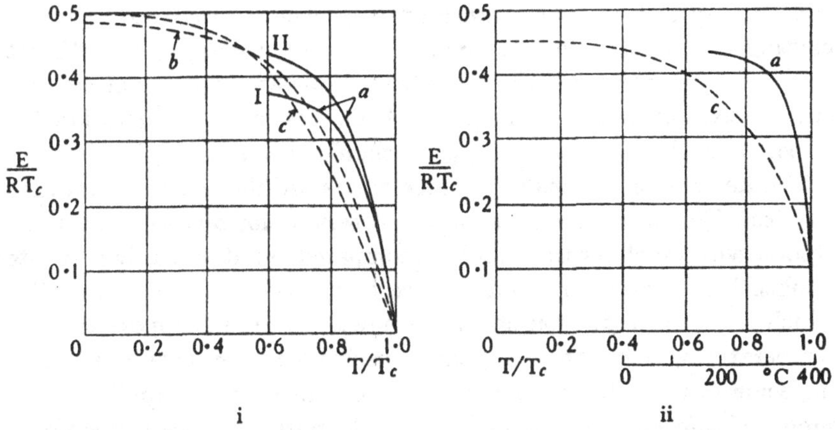

Fig. 87. Observed and calculated values of $\{\overline{E_c}(T_c)-\overline{E_c}(T)\}/RT_c$ for (i) β-brass, CuZn ($r=\frac{1}{2}$), (ii) Cu_3Au ($r=\frac{1}{4}$). The values marked (c) are calculated according to the theory of Bragg and Williams; those marked (b) according to the theory of Bethe (second approximation). The values marked (a) are from Sykes' observations; those marked I for rising, II for falling temperatures.

For β-brass Bethe's theory gives a more abrupt change than Bragg and Williams' but the observations are more abrupt still, and it is unlikely that the observations here are greatly in error, as in β-brass the atoms still move freely in the lattice near the critical temperature. In Cu_3Au the observations, which were made with a rising temperature, must give somewhat too great a value of $\overline{E_c}(T_c)-\overline{E_c}(T)$, owing to the lag in attaining the equilibrium order. It is also likely that Bragg and Williams' approximation is not very close to the true partition function in this unsymmetrical case, and that the true $\overline{E_c}$ will change more abruptly* near $T=T_c$. Making all allowances however, it is not fair to claim more than that existing theory is a decent first approximation; it seems likely that some omitted effect, such as interaction between the ordering and the lattice vibrations, is of importance in sharpening up the variation of $\overline{E_c}$.

* This appears to be borne out by the calculations of Peierls, *Proc. Roy. Soc.* A, vol. 154, p. 207 (1936).

§21·3. *Rotations of molecules in solids.* Other solids besides metallic alloys—notably halogen hydrides and ammonium salts—present the phenomenon of a critical temperature and a discontinuous specific heat, which can now be associated with some confidence with the rather sudden onset of free rotations of the ions or molecules forming the crystal lattice. The general interpretation of these phenomena was first given by Pauling.* In conformity with his views the anomalous specific heat of these substances must be associated with a transition from libration motions to almost free rotations for the ions or molecules in question. The transition may even be described as rotational melting.† Qualitatively this explanation seems completely satisfactory. A tentative quantitative theory has been proposed, which is not yet particularly successful, but is perhaps worth discussion as a suggestion for the direction in which a better description should be sought.

The description we shall give here treats all librations and rotations as classical. This limitation is perhaps not serious for any of the actual substances with which the theory is here compared, but it probably prevents any application of the theory to the interesting case of solid methane (CH_4) for which a quantum version should be used. We must suppose that at low temperatures, before the anomalous specific heat develops, the axis of the molecule can librate about a direction or directions of equilibrium. Such preferred directions for the molecular axis must be imposed therefore by a directional field of force. We shall only attempt to discuss the simplest possible model for the molecule and the field in which its angular motion takes place. We shall therefore assume that the energy associated with this motion is that of a symmetrical rigid rotator without axial spin in a field of force $-W\cos\theta$, where θ is the displacement of the axis of figure from a single preferred direction of equilibrium. The total energy in Hamiltonian form is then

$$H=\frac{1}{2A}\left(p_\theta^2+\frac{p_\phi^2}{\sin^2\theta}\right)-W\cos\theta. \qquad ......(2260)$$

At sufficiently small amplitudes, and therefore at low temperatures, librations of the molecular axes will not be distinguishable from any other small oscillations and will form part of the body of normal modes of the lattice. As the higher states come into play the oscillators will cease to be simple harmonic, and become gradually (for constant values of W) more and more like free rotations with a smaller contribution to the specific heat. These effects may be followed at once by calculating the partition function

$$f(T)=\frac{1}{h^2}\iiiint e^{-H/kT}\,dp_\theta dp_\phi d\theta d\phi,$$

* Pauling, *Phys. Rev.* vol. 36, p. 430 (1930).

† Frenkel, Todes and Ismailow, *Acta Physico-Chem. U.S.S.R.* vol. 1, p. 97 (1934).

where H is given by (2260). This leads at once to

$$f(T)=\frac{4\pi^2AkT}{h^2}\int_0^\pi e^{W\cos\theta/kT}\sin\theta\,d\theta,$$

$$=\frac{8\pi^2AkT}{h^2}\frac{kT}{W}\sinh\frac{W}{kT}. \qquad \text{......(2261)}$$

If $\overline{E_{\text{rot}}}$ includes the energy of orientation in the field $-W\cos\theta$ and there are N such rotating systems in the crystal, *whose rotations do not affect each other*, then

$$\overline{E_{\text{rot}}}=NkT\left[2-\frac{W}{kT}\coth\frac{W}{kT}\right], \qquad \text{......(2262)}$$

$$C_{\text{rot}}=Nk\left[2-\frac{W^2/k^2T^2}{\sinh^2 W/kT}\right]. \qquad \text{......(2263)}$$

If the curve for the rotational energy or the specific heat is plotted, it will be found of course that it is perfectly smooth and gives no sign of any violent variation when kT and W are of the same order of magnitude. We have, however, following Pauling, no reason to expect that the rotations of one molecule are independent of those of the other molecules or in short that W is constant, but rather every reason to expect the contrary. The directional terms in the field to which any one molecule is subject, specified by $-W\cos\theta$, are of course mainly due to the lack of spherical symmetry in the combined fields of the surrounding molecules. This lack of symmetry will be greatly weakened and might in fact be almost destroyed by a sufficient degree of rotation among the surrounding molecules. Thus W itself cannot be a constant but must depend on the degree of rotation already present among the molecules. An approximate partition function for the whole body of libration-rotations for the N systems can therefore in theory be constructed in the form

$$\log L(T)=N\log\frac{8\pi^2AkT}{h^2}+\int_0^1 g(s)\log\left\{\frac{kT}{sW_0}\sinh\frac{sW_0}{kT}\right\}ds, \qquad \text{......(2264)}$$

where $g(s)\,ds$ is the number of systems which are effectively subject to the field $-sW_0\cos\theta$, and $\int_0^1 g(s)\,ds=N$. If $g(s)$ could be formulated, $\log L(T)$ could at once be evaluated by searching for the maximum of the integrand in (2264).

No satisfactory method of evaluating $g(s)$ has yet been proposed and roundabout methods must again be adopted.

§21·31. *A formula for determining the equilibrium value of s.* It has been proposed to determine s by an approximate equation based on the following arguments.* We specify a degree of non-rotation among the molecules by

* R. H. Fowler, *Proc. Roy. Soc.* A, vol. 149, p. 1 (1935).

regarding as not rotating all those molecules for which

$$\frac{1}{2A}\left(p_\theta^2+\frac{p_\phi^2}{\sin^2\theta}\right)<\beta W(1+\cos\theta). \qquad \text{......(2265)}$$

The inequality (2265) asserts that such molecules have at any θ less than β times the kinetic energy required to allow the axis of the molecule to reach the pole $\theta=\pi$. We can easily calculate the fraction of all molecules "not rotating" according to this definition. It is $f_0(T)/f(T)$, where $f(T)$ is given by (2261) and

$$f_0(T)=\frac{2\pi}{h^2}\int_0^\pi e^{W\cos\theta/kT}\,d\theta\iint\limits_{p_\theta^2+p_\phi^2/\sin^2\theta<2A\beta W(1+\cos\theta)} e^{-(p_\theta^2+p_\phi^2/\sin^2\theta)/2AkT}\,dp_\theta dp_\phi.$$

By obvious substitutions this can be reduced to

$$f_0(T)=\frac{4\pi^2AkT}{h^2}\int_0^\pi e^{W\cos\theta/kT}\sin\theta\,d\theta\int_0^{\beta W(1+\cos\theta)/kT} e^{-x}\,dx,$$

from which it follows that

$$f_0(T)=\frac{8\pi^2AkT}{h^2}\left[\frac{kT}{W}\sinh\frac{W}{kT}-e^{-\beta W/kT}\frac{kT}{W(1-\beta)}\sinh\frac{W(1-\beta)}{kT}\right].$$

We therefore find that

$$\frac{f_0(T)}{f(T)}=1-\frac{e^{-\beta W/kT}\sinh W(1-\beta)/kT}{(1-\beta)\sinh W/kT}. \qquad \text{......(2266)}$$

It is now suggested that $f_0(T)/f(T)$ may be used as a measure of the strength of the orientating field. The simplest assumption is to take $s=f_0(T)/f(T)$, $W=W_0s$, so that (2266) becomes an implicit equation for s of the form

$$s=1-\frac{e^{-\beta sW_0/kT}\sinh sW_0(1-\beta)/kT}{(1-\beta)\sinh sW_0/kT}. \qquad \text{......(2267)}$$

The special case $\beta=\frac{1}{2}$ is particularly simple, since then

$$s=\tanh\tfrac{1}{2}sW_0/kT. \qquad \text{......(2268)}$$

For all values of β equation (2267) gives a critical temperature above which $s\equiv 0$. Assuming that (2267) gives an adequate representation of the behaviour of the maximum term in (2264), it follows that

$$\overline{E_{\text{rot}}}=NkT\left[2-\frac{s_0W_0}{kT}\coth\frac{s_0W_0}{kT}\right], \qquad \text{......(2269)}$$

$$C_{\text{rot}}=Nk\left[2-\frac{x^2}{\sinh^2 x}\right]-\frac{NW_0^2}{2kT}\frac{ds_0^2}{dT}\left[\frac{\cosh x\sinh x-x}{x\sinh^2 x}\right], \qquad \text{......(2270)}$$

in which s_0 is the correct root of (2267) and $x=s_0W_0/kT$. If $\beta>\frac{1}{2}$ both s_0 and ds_0^2/dT are continuous through the critical temperature. If $\beta=\frac{1}{2}$, s_0 is continuous but ds_0^2/dT is discontinuous, so that C_{rot} is discontinuous. In this familiar case the discontinuity is easily calculated, since

$$s_0^2\sim\frac{3T^2}{T_c^2}\left(1-\frac{T}{T_c}\right), \quad T_c=\tfrac{1}{2}W_0/k.$$

It follows that

$$C_{\text{rot}} = Nk \quad (T > T_c), \qquad C_{\text{rot}} \to 5Nk \quad (T \to T_c - 0),$$

and the discontinuity is $4Nk$. If $\beta < \frac{1}{2}$, we pass over to the case in which the new roots of (2267) enter discontinuously at values of s_0 greater than zero, and cannot immediately represent the equilibrium state. The exact critical value of s_0 must then be determined by the method indicated at the close of § 21·23. There is now a finite latent heat at the transition point.

§ 21·32. *A generalized model.* A slight generalization of the model which is not physically impossible leads to a libration-rotation transition with even more violent properties. If we may assume that the field of potential energy in which the molecule rotates is not simply $-W\cos\theta$ but $-\alpha W - W\cos\theta$, there is an extra term $-N\alpha s_0 W_0$ in $\overline{E_{\text{rot}}}$, and $N\alpha W_0(-ds_0/dT)$ in C_{rot}. It is easy to show that when $\beta \geqslant \frac{1}{2}$, $T_c = \beta W_0/k$, and that when $\beta > \frac{1}{2}$

$$s_0 \sim \frac{3T}{T_c}\left(1 - \frac{T}{T_c}\right)\frac{\beta}{2\beta - 1}, \qquad -\frac{ds_0}{dT} \sim \frac{3\beta}{2\beta - 1}\frac{1}{T_c}.$$

It follows that, if $\beta > \frac{1}{2}$, C_{rot} has a finite discontinuity which tends to infinity as $\beta \to \frac{1}{2}$. We shall not however give any further details for any of these models, since we shall see on comparing the theory with observation that it is still too far from giving an adequate explanation of the facts.

§ 21·33. *Comparison of theory and observations.* Typical observed values of specific heats and energy contents are shown in Figs. 88, 89 for the halogen hydrides and for some ammonium salts. We see in Fig. 88 that the curves for the halogen hydrides are of the general form that one would expect to be accounted for by such a theory as we have proposed, but that there are for HI two similar breaks, for HBr one single and one double one and for HCl such violent variations that the specific heat was not recorded, but only a latent heat. It appears therefore that the rotational degrees of freedom do not set in together, but that perhaps rotations about the different crystal axes set in separately. This of course makes strict comparisons impossible. It seems reasonable however to assume that these breaks refer to one rotational freedom, and therefore to abstract from the observed curve the abnormal part of the specific heat, call this C_{obs} and compare $C_{\text{obs}}/\frac{1}{2}R$ as a function of T/T_c with the theoretical values of $C_{\text{rot}}(T/T_c)/R$. Points derived in this way from the two breaks for HI are shown in Fig. 90 and compared with possible theoretical curves. One sees that no very good agreement is obtainable. The distinctive feature of the disagreement is that the abnormal specific heat due to the term ds_0/dT is much more sharply confined to the neighbourhood of the critical temperature than the theory suggests. Since the breaks in the specific heat curves for HBr and HCl are still more sharply confined to the

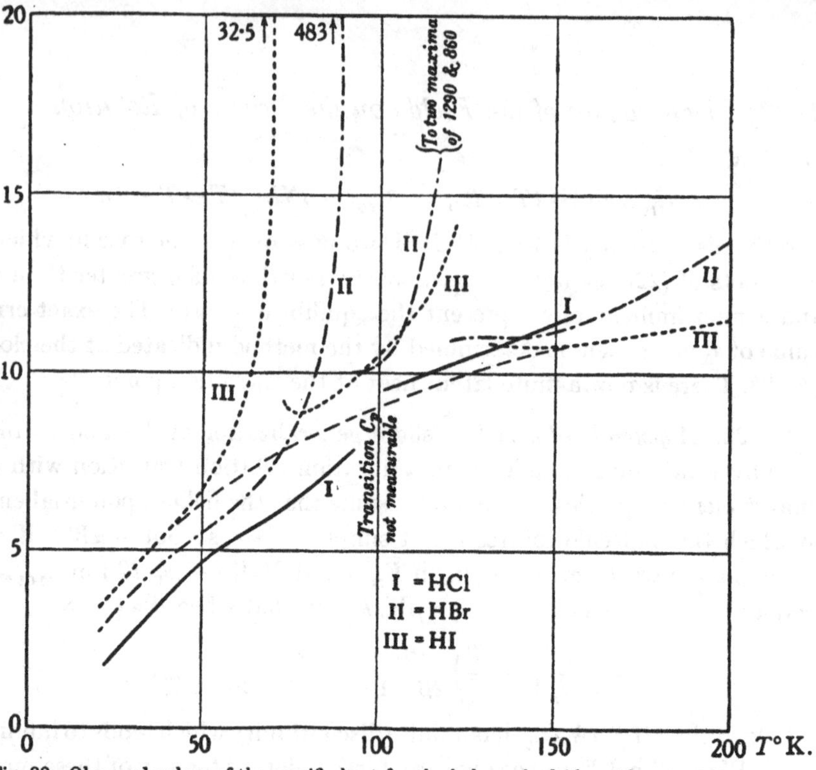

Fig. 88. Observed values of the specific heat for the halogen hydrides at low temperatures.
[Giauque and Wiebe, *J. Amer. Chem. Soc.* vol. 50, pp. 101, 2193 (1928); vol. 51, p. 1441 (1929).]

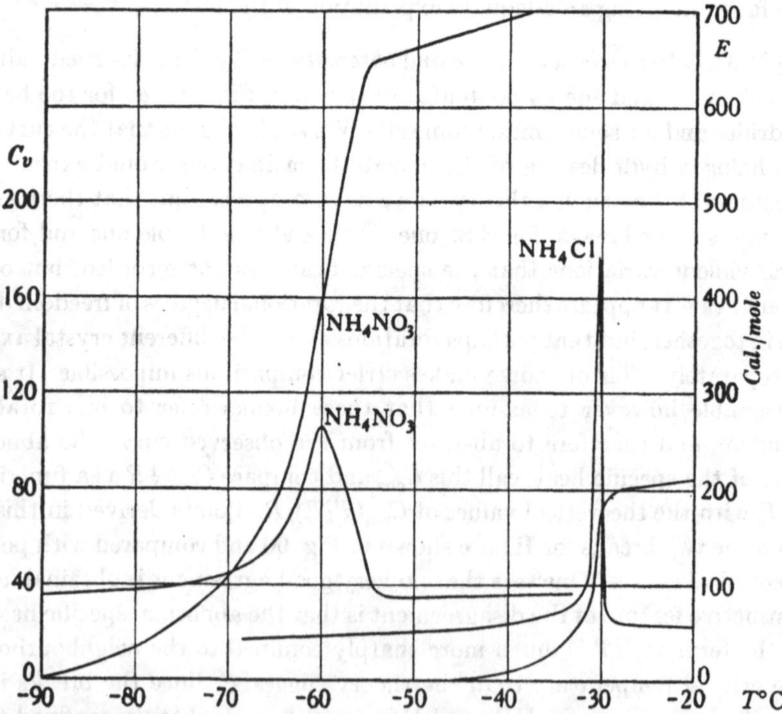

Fig. 89. Observed values of the specific heat and energy content for the ammonium salts, NH_4NO_3 and NH_4Cl.

[For NH_4Cl, Simon, Simson and Ruhemann, *Zeit. f. physikal. Chem.* vol. 129, p. 339 (1927); for NH_4NO_3, Crenshaw and Ritter, *Zeit. f. physikal. Chem.* B vol. 16, p. 143 (1932).]

neighbourhood of the critical temperature, the present comparison will fail still more badly for these substances.

For the ammonium salts and even for HBr and HCl specific heat comparisons are really out of the question, and one must compare energy contents

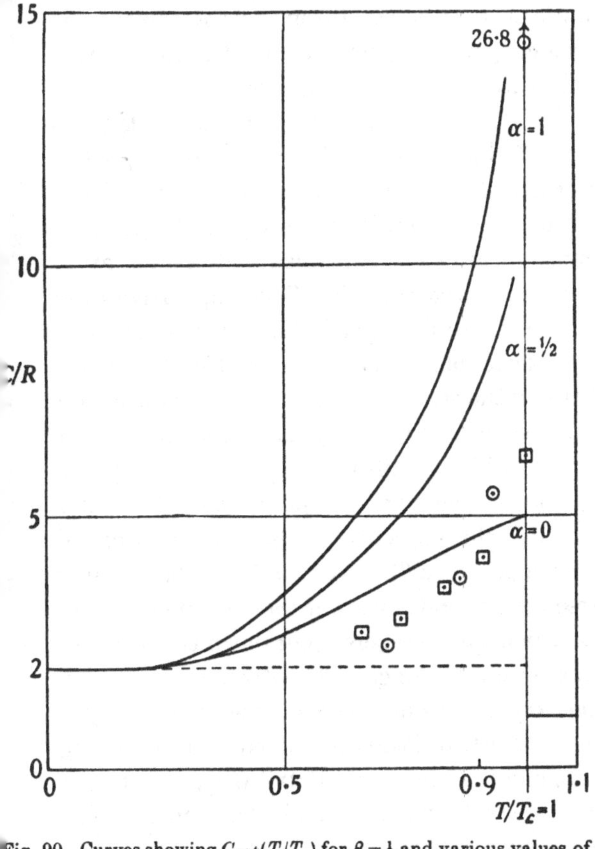

Fig. 90. Curves showing $C_{\text{rot}}(T/T_c)$ for $\beta = \frac{1}{2}$ and various values of the additional binding α. The plotted points show corresponding observed values for HI.

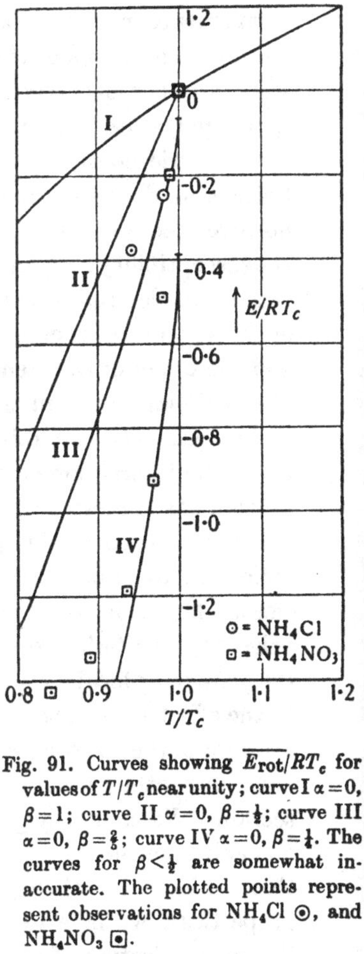

Fig. 91. Curves showing $\overline{E_{\text{rot}}}/RT_c$ for values of T/T_c near unity; curve I $\alpha = 0$, $\beta = 1$; curve II $\alpha = 0$, $\beta = \frac{1}{2}$; curve III $\alpha = 0$, $\beta = \frac{3}{8}$; curve IV $\alpha = 0$, $\beta = \frac{1}{4}$. The curves for $\beta < \frac{1}{2}$ are somewhat inaccurate. The plotted points represent observations for NH_4Cl ⊙, and NH_4NO_3 ⊡.

instead. The theoretical curves shown in Fig. 91 show $\overline{E_{\text{rot}}}/RT_c$ plotted against T/T_c for various values of β. The curves for $\beta < \frac{1}{2}$ are not accurate as the precise critical value of s_0 has not been properly determined. The theory however is hardly sufficiently successful to make this worth while. The observed values for NH_4NO_3 and NH_4Cl have been reduced to these variables and adjusted arbitrarily to agree with the theoretical curves when $T/T_c \to 1 + 0$. It is evident that, while there is general agreement in form, the abnormal variation of $\overline{E_{\text{rot}}}$ is again much more closely restricted to the neighbourhood of the critical temperature than the theory can be made to indicate. Nevertheless the theory proposed goes some way towards interpreting the facts and is perhaps not entirely valueless.

§ **21·4.** *Dielectric constants of solids and liquids containing dipoles.* It is well known that Rochelle salt $NaKC_4H_4O_6.4H_2O$, for a limited range of temperature, may for practical purposes be said to have an infinite dielectric constant analogous to the infinite susceptibility of iron in its ferromagnetic state, another clear example of a cooperative phenomenon. The cooperative (polarized) state in Rochelle salt is limited by an upper critical temperature T_u (or Curie point) such that for $T > T_u$ the susceptibility though large is finite and decreases rapidly as T increases. Unlike the ferromagnetics however there is also a lower critical temperature T_l such that when $T < T_l$ the susceptibility is again finite and decreases (at least initially) as T decreases. It is probable* that these phenomena are to be explained by the orientation of the water dipoles present as water of crystallization. The cooperative state and the upper critical temperature T_u can be explained by an exact analogy of the Weiss-Langevin theory of ferromagnetism expounded in §§ 12·9 *sqq.*, and no difficulties are raised here by the strength of the necessary molecular field. The interaction energy of electrical dipoles is so large that it supplies precisely the necessary energy term which it fails to do for magnetic dipoles. The explanation of this part of the phenomenon requires the polar water molecules to orientate freely under the influence of the effective applied electric field. The lower critical temperature T_l must then be explained, it is believed, by a failure of the free orientations at lower temperatures, due to the growth of local constraints which can so severely cut down the response to the effective field that the material is no longer self-polarizing.† We shall attempt in the following sections to present a quantitative theory of this type based on the theory of libration-rotation transitions developed in the preceding sections. We cannot hope to explain in this way in detail the properties of any actual substance such as Rochelle salt, which are really far more complex than we have described.‡ We can hope only to succeed in showing that such properties as we have described may be expected to find a natural explanation by the application of normal statistical methods to models of the proposed type.

Again the dielectric constant of water or ice is finite at all temperatures, and falls to low values even for low frequencies as the temperature is decreased below 150° K. This can only be understood, assuming that the H_2O molecule in ice or water carries approximately the same dipole as in steam, if its orientations are not free but severely restricted by the fields of its immediate

* See e.g. Kobeko and Kurtschatow, *Zeit. f. Physik*, vol. 66, p. 192 (1930).

† B. and I. Kurtschatow, *Physikal. Zeit. d. Sowjetunion*, vol. 3, p. 320 (1933).

‡ Mueller, *Phys. Rev.* vol. 47, p. 175 (1935).

neighbours even at the highest temperatures for which the dielectric constant of water has been investigated. The water dipoles are so numerous and so strong that water would be in a cooperative state of self-polarization at all such temperatures if the dipole carried were even approximately free. Somewhat similar phenomena occur for other polar liquids such as some of the alcohols and nitrobenzene, which possibly are explicable in the same way. There is probably general agreement about these qualitative explanations, but we are still far from a satisfactory quantitative one. In the following sections a discussion will be given on the basis of libration-rotation transitions, which though still far from adequate, indicates that a satisfactory explanation can perhaps be found in this way.

§ **21·41.** *The theory of the dielectric constant of material composed of or containing librating-rotating dipoles.* We discuss an assembly of N systems in volume V ($n=N/V$), carrying dipoles of moment μ, each subject to a restraining field $-W\cos\theta$. We must not assume that the equilibrium directions of orientation are all the same, or the assembly at low temperatures would be permanently polarized in this direction in the absence of any applied field and would scarcely polarize in any other. For a starting point we must therefore divide the systems at least into two equal groups whose natural equilibrium orientations are in opposite directions. We shall therefore start by assuming that the assembly consists of $\frac{1}{2}N$ systems each in a field $-W\cos\theta$, and $\frac{1}{2}N$ systems each in a field $W\cos\theta$. An effective field F' acts on the systems in the direction $\theta=0$, so that their total potential energies are $-(W+\mu F')\cos\theta$ and $(W-\mu F')\cos\theta$ respectively. This simple arrangement will be generalized later. The partition functions f_1 and f_2 are then

$$\frac{8\pi^2AkT}{h^2}\frac{kT}{W+\mu F'}\sinh\frac{W+\mu F'}{kT},\quad \frac{8\pi^2AkT}{h^2}\frac{kT}{W-\mu F'}\sinh\frac{W-\mu F'}{kT}, \quad\ldots(2271)$$

respectively. By equation (1294) the polarization (induced dipole strength per unit volume) is given by

$$P=\tfrac{1}{2}nkT\frac{\partial}{\partial F'}(\log f_1+\log f_2),$$

from which it follows that

$$P=nkT\mu\left[\frac{\mu F'}{W^2-\mu^2F'^2}-\frac{1}{kT}\frac{\sinh 2\mu F'/kT}{\cosh 2W/kT-\cosh 2\mu F'/kT}\right]. \quad\ldots\ldots(2272)$$

Assuming that $\mu F'\ll W$ this reduces easily to

$$\frac{P}{F'}=\frac{n\mu^2}{kT}\left[\frac{k^2T^2}{W^2}-\frac{1}{\sinh^2 W/kT}\right], \quad\ldots\ldots(2273)$$

which has the expected limiting form $n\mu^2/3kT$ when $W \to 0$. For further study it is convenient to write (2273) in the form

$$\frac{P}{F'} = \frac{n\mu^2}{3kT} g(x) \quad \left(x = \frac{W}{kT}\right), \qquad \text{......(2274)}$$

so that $g(x)$ is a reducing factor which the binding imposes on the effective value of $n\mu^2$. As $x \to 0$, $g(x) \to 1$; as $x \to \infty$, $g(x) \sim 3/x^2 \to 0$.

To deduce the dielectric constant η we have to use the standard relationships of Chapter XII, namely, in the present notation,

$$\eta = 1 + 4\pi P/F, \quad F' = F + \tfrac{4}{3}\pi P.$$

The second of these holds only (with the coefficient $\frac{4}{3}\pi$) for those media for which Lorentz's Lemma applies. This lemma does not apply to Rochelle salt and we shall use this equation therefore in the general form

$$F' = F + \gamma P, \qquad \text{......(2275)}$$

where γ will depend on the direction of P in the crystal. We shall only consider principal directions for which F, F' and P are all parallel. It follows that

$$\eta = 1 + \frac{4\pi n\mu^2 g(x)/3kT}{1 - \gamma n\mu^2 g(x)/3kT}, \qquad \text{......(2276)}$$

provided that F' and therefore P can be regarded as small. We can therefore write

$$\eta = 1 + \frac{4\pi}{\gamma}\frac{T_d}{T - T_d} \quad \left(T_d = \frac{\gamma n\mu^2}{3k} g(x)\right), \qquad \text{......(2277)}$$

provided that $T > T_d$. If $T < T_d$ then the approximation leading from (2272) to (2273) is illegitimate. We have then to solve (2272) and (2275) for P as a function of F. It is easy to see that in this case the equilibrium state, or greatest value of the partition function $\frac{1}{2}N\{\log f_1 + \log f_2\}$, corresponds to a root P of (2272) which does not vanish with F and that we are then concerned with a cooperative state of self-polarization in zero field, which may be regarded as a state with an infinite dielectric constant.

The binding W in $g(x)$ must not however be thought of as a constant. To a first approximation it will be governed by the rotations of neighbouring molecules and have the effective value sW_0, where s is given by (2267) or some more accurate equation of this type. The complete behaviour of the dielectric constant as a function of T is therefore governed by the set of equations

$$\eta = 1 + \frac{4\pi}{\gamma}\frac{T_d}{T - T_d}, \quad T_d = \frac{\gamma n\mu^2}{3k} g\left(\frac{sW_0}{kT}\right) = T_d^0 g, \quad s = s\left(\frac{T}{T_c}\right), \qquad \text{......(2278)}$$

with

$$g(x) = 3\left[\frac{1}{x^2} - \frac{1}{\sinh^2 x}\right]. \qquad \text{......(2279)}$$

It can be seen at once that when $T \to 0$, $s \to 1$, $x \to \infty$, $T_d = O(T^2)$. Hence at low temperatures we certainly have $T > T_d$ and $\eta \to 1$. For large values of T,

$s \to 0$, $g(x) \to 1$, and T_d assumes a constant value, so that η behaves according to the ordinary dipole theory of Chapter XII and ultimately $\eta \to 1$. At intermediate temperatures η will increase to a finite maximum and then decrease again provided that always $T > T_d$. But if ever $T < T_d$ then we must have a range of temperatures within which η is infinite, bounded by upper and lower critical temperatures T_u and T_l. This is exactly what is observed for Rochelle salt along the a-axis, but the assembly discussed here is still rather too specialized for actual applications.

§ **21·42.** *Random orientations of the natural directions of equilibrium.* We have so far only considered a polarizing field P' parallel to the natural directions of equilibrium of the dipoles. Let us now consider the same assembly when F' is perpendicular to $\pm W$. In either case the two potential energy terms combine vectorially so that all the dipoles are in fields of potential energy $-(W^2+\mu^2F'^2)^{\frac{1}{2}}\cos\theta'$, θ' being measured from new positions of equilibrium. Therefore

$$f(T) = \frac{8\pi^2AkT}{h^2}\frac{kT}{(W^2+\mu^2F'^2)^{\frac{1}{2}}}\sinh\frac{(W^2+\mu^2F'^2)^{\frac{1}{2}}}{kT}. \qquad \text{......(2280)}$$

It follows that

$$P = n\mu^2kTF'\left[\frac{\coth(W^2+\mu^2F'^2)^{\frac{1}{2}}/kT}{kT(W^2+\mu^2F'^2)^{\frac{1}{2}}} - \frac{1}{W^2+\mu^2F'^2}\right]. \qquad \text{......(2281)}$$

Assuming that $\mu F' \ll W$ this reduces to

$$\frac{P}{F'} = \frac{n\mu^2}{kT}\left[\frac{kT}{W}\coth\frac{W}{kT} - \frac{k^2T^2}{W^2}\right]. \qquad \text{......(2282)}$$

It is now possible to combine (2273) and (2282) to give the relation for the more general case in which the equilibrium directions of the dipoles are distributed at random in space. It is easily verified that, for any set of dipoles whose equilibrium directions are half and half in opposite directions, the effects of imposed fields parallel and perpendicular to the equilibrium directions are independent of each other to the order of accuracy of (2273) and (2282). Moreover instead of averaging the response for all equilibrium directions we can equally well average for all directions of F'. It is then clear that this average will be obtained by taking for P/F' two-thirds of (2282) plus one-third of (2273), since there are two independent components of F' normal to $\theta = 0$ and only one parallel. Thus in general

$$\frac{P}{F'} = \frac{n\mu^2}{3kT}\left[\frac{2kT}{W}\coth\frac{W}{kT} - \frac{k^2T^2}{W^2} - \frac{1}{\sinh^2 W/kT}\right]. \qquad \text{......(2283)}$$

It is probably therefore correct to discuss the dielectric constant of liquids or of solids of high symmetry using equations (2278) but with

$$g(x) = 2\frac{\coth x}{x} - \frac{1}{x^2} - \frac{1}{\sinh^2 x}, \qquad \text{......(2284)}$$

in place of (2279). For Rochelle salt itself we shall see that (2279) is still probably the more appropriate. The functions $g(x)$ are shown in Fig. 92.

A greater variety of behaviour is now possible as $T \to 0$. Equations (2278) and (2284) do not require that $\eta \to 1$ as $T \to 0$, and do not even require η to remain finite. When $1/x$ is small

$$g(x) \sim \frac{2}{x} - \frac{1}{x^2} = \frac{2kT}{sW_0}\left(1 - \frac{kT}{2sW_0}\right).$$

We may write $W_0/k = T_c/\beta'$, where $\beta' = \beta$ when $\beta \gg \frac{1}{2}$, but β' is somewhat

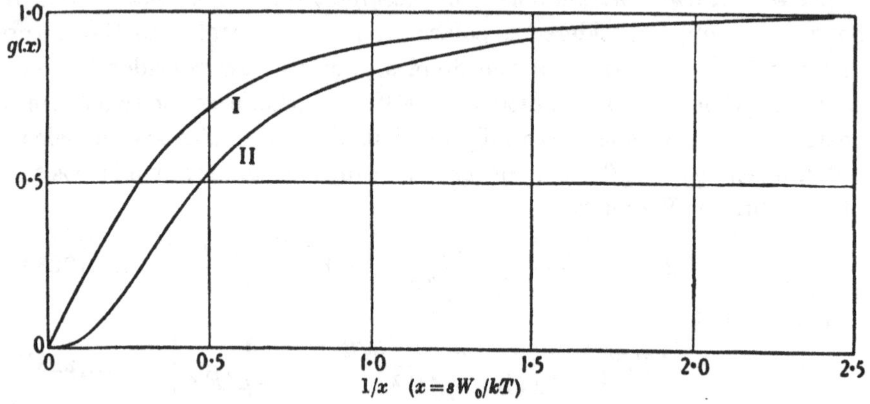

Fig. 92. Curves showing $g(x)$ as a function of x for

$$\text{I.}\quad g(x) = 2\frac{\coth x}{x} - \frac{1}{x^2} - \frac{1}{\sinh^2 x}.$$

$$\text{II.}\quad g(x) = 3\left(\frac{1}{x^2} - \frac{1}{\sinh^2 x}\right).$$

greater than β when $\beta < \frac{1}{2}$. The parameter β was defined in § 21·31. We may also in this region put $s \simeq 1$. Then

$$g(x) \sim \frac{2\beta' T}{T_c}\left(1 - \frac{\beta' T}{2T_c}\right), \qquad \text{......(2285)}$$

$$\eta = 1 + \frac{4\pi}{\gamma}\,\frac{1 - \beta' T/2T_c}{T_c/(2\beta' T_d{}^0) - 1 + \beta' T/2T_c}. \qquad \text{......(2286)}$$

There are two critical temperatures in this formula, T_c the critical temperature for the vanishing of s and the onset of free rotations, and $T_d{}^0$ $(= \gamma m\mu^2/3k)$ the critical temperature for the cooperative state of self-polarization among these freely rotating dipoles. We see from (2286) that η steadily increases as T diminishes and remains finite if and only if

$$T_c > 2\beta' T_d{}^0.$$

A variety of possible curves for η is shown in Figs. 93.

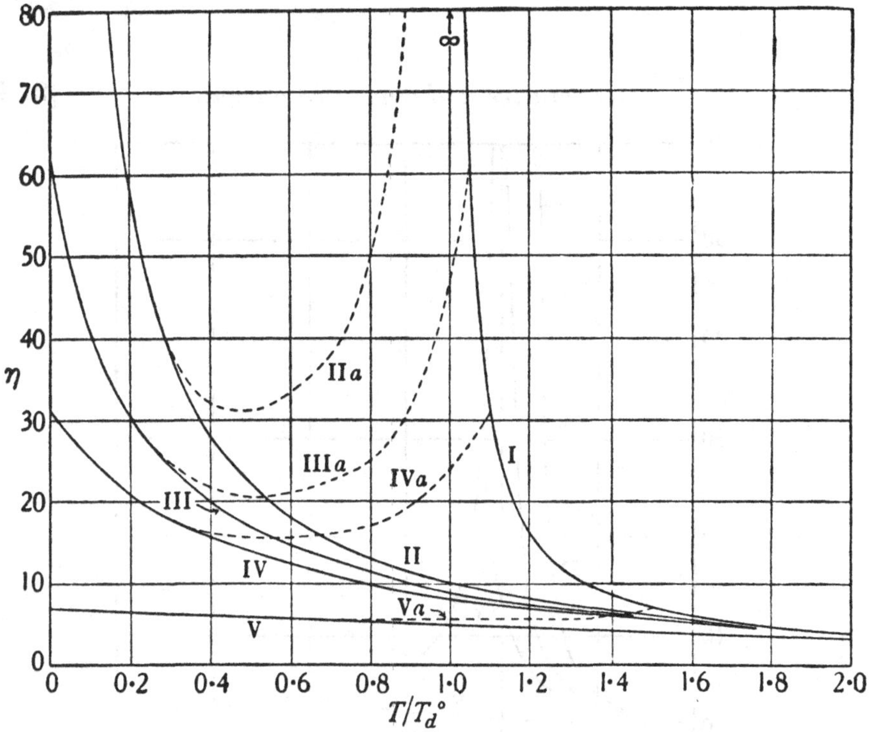

Fig. 93*a*. Curves showing η as a function of $T/T_d{}^0$, using the $g(x)$ shown in I, Fig. 92. Curves are drawn for $\beta = \beta' = 1$, and I, $T_c = T_d{}^0$; II, $T_c = 2T_d{}^0$; III, $T_c = 2{\cdot}1T_d{}^0$; IV, $T_c = 2{\cdot}2T_d{}^0$; V, $T_c = 3T_d{}^0$; also for $\beta = \beta' = \frac{1}{2}$, and I + II*a*, $T_c = T_d{}^0$; I + III*a*, $T_c = 1{\cdot}05T_d{}^0$; I + IV*a*, $T_c = 1{\cdot}1T_d{}^0$; I + V*a*, $T_c = 1{\cdot}5T_d{}^0$. The second set of curves are shown broken only where they are distinct from the corresponding parts of those of the first set.

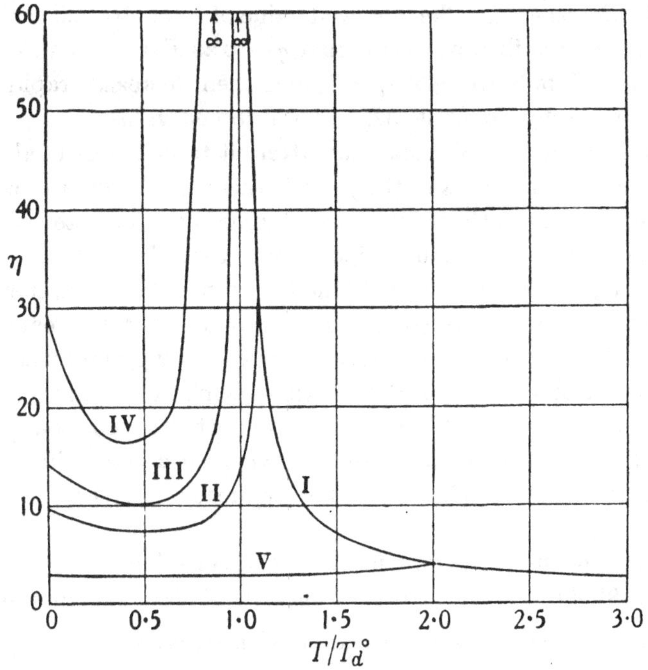

Fig. 93*b*. Curves showing η as a function of $T/T_d{}^0$ using the $g(x)$ shown in I, Fig. 92 for $\beta = \frac{2}{3}$, $\beta' \simeq 0{\cdot}44$, and I + III, $T_c = T_d{}^0$; I + II, $T_c = 1{\cdot}1T_d{}^0$; I + IV, $T_c = 0{\cdot}9T_d{}^0$; I + V, $T_c = 2T_d{}^0$.

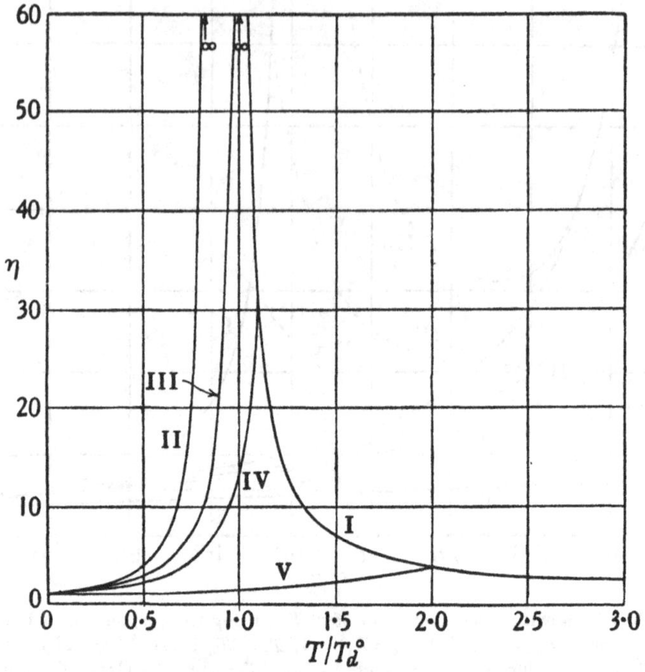

Fig. 93c. Curves showing η as a function of $T/T_d{}^0$, using the $g(x)$ shown in II, Fig. 92, for $\beta=\beta'=\frac{1}{2}$, and I+III, $T_c=T_d{}^0$; I+II, $T_c=0{\cdot}9T_d{}^0$; I+IV, $T_c=1{\cdot}1T_d{}^0$; I+V, $T_c=2T_d{}^0$.

§ **21·43.** *Rochelle salt.* The observed value of η for Rochelle salt along the a-axis may be described by saying that $\eta\to\infty$ as T decreases to T_u, remains infinite until T falls to $0{\cdot}865T_u=T_l$, and then decreases rapidly reaching ordinary values by about $0{\cdot}83T_u$. The observed values at right angles to the a-axis are normal. We shall only attempt to give a general description of this behaviour in terms of the present theory.* In order to compare the observations with the theory we must assume that the water of crystallization provides the orientating dipoles and that $T_u=T_d{}^0$. If their dipole moment is equal to the moment of the H_2O-dipole in steam, then $T_d{}^0$ would have the value 550° K. for $\gamma=\frac{4}{3}\pi$, and a proportionately greater value if γ is greater. We shall see that γ must be appreciably greater than $\frac{4}{3}\pi$ so that $n\mu^2$ for the water of crystallization must be some $2\frac{1}{2}$ to 3 times smaller than the value for steam would lead one to expect. The possibility of such a change in this direction can hardly be excluded *a priori*, but the need to postulate it unfortunately makes the application of the theory still less precise.

The source of the anisotropy must next be considered. The two curves for $g(x)$ in Fig. 92 show that it is always easier to polarize a medium by deflecting

* Mueller, *loc. cit.*; he also gives a much more detailed analysis of the observations.

the dipoles by a force at right angles to their natural directions of equilibrium than by a force parallel to these directions. This result must be generally valid. Thus any anisotropy in η combined with special natural directions of equilibrium must make it easier to polarize a medium in a plane, whereas it is actually easier to polarize it normal to a plane. The anisotropy can in short only be referred to γ, and can be shown by γ, if the dipoles are suitably arranged in the crystal. The actual arrangement of the water molecules is not yet known. It is easy to show that if the dipole carriers are arranged in strings parallel to the a-axis, being separated by a shorter spacing along these strings than the separations of the strings themselves, then $\gamma(\parallel)>\gamma(\perp)$. The natural directions of equilibrium of the dipoles may then be assumed to lie along the strings. We must then use the $g(x)$ of equation (2279) for polarization along the a-axis and a $g(x)$ derived from (2282) for polarization at right angles. This $g(x)$ does not differ greatly from that of (2284). The η-curves of Fig. 93c therefore apply along the a-axis and those of Figs. 93a,b at right angles. If for example $\beta=\beta'=\frac{1}{2}$ and $T_c\simeq 0\cdot 9T_d^0(\parallel)$ while $T_c=1\cdot 5T_d^0(\perp)$, the observed characteristics of η for Rochelle salt are remarkably well reproduced by the theory. This requires a ratio $\gamma(\parallel)/\gamma(\perp)=1\cdot 666$. Such a value does not require anything extreme in the ratio of the spacings of the dipoles along and across the a-axis, though the ratio is difficult to compute exactly. It remains to be seen whether such spacings are confirmed by X-ray analysis.

§21·44. *Polar liquids.* Curves showing η as a function of T for a number of substances are shown in Fig. 94. At first sight these curves look not unlike the curves of Fig. 93c. These calculated curves however are based on the $g(x)$ of equation (2279) and cannot apply to liquids, which must be reasonably isotropic. Similar peaks near $T=T_d^0$ are shown by the curves of Figs. 93 a,b but these are followed at low temperatures by another rise of η to high values, and of this there is experimentally no sign. On the evidence it is clear therefore that the complete curve of η for a substance such as ice-water cannot be accounted for as a whole by any simple form of the present theory. There is another difficulty in such a comparison, namely that the calculated values of T_d^0 using (2278) with $\gamma=\frac{4}{3}\pi$ are in all cases higher and in some cases much higher than the temperatures of the observed η-peaks as is shown in Table 73. Before therefore we attempt to interpret η for these polar liquids in terms of the present theory, we must see whether some further important factor may not have been overlooked. One sees at once from Fig. 94 that these temperatures of maximum η though not identical with melting points are certainly closely associated. It seems therefore that the fall of η for low temperatures should be associated rather with an increased binding due to

TABLE 73.

Temperatures of maximum dielectric constant η and values of T_d^0 for various polar liquids. The values of T_d^0 are calculated from (2278) with $\gamma = \frac{4}{3}\pi$ and the value of μ found for the same molecule in vapour or dilute solution.

Substance	T_d^0	Temperature of maximum η, ° K.
H_2O	1200	273
$C_6H_5NO_2$	920	263
CH_4O	410	163
CH_2O_2	375	286
$C_5H_{12}O$	185	147

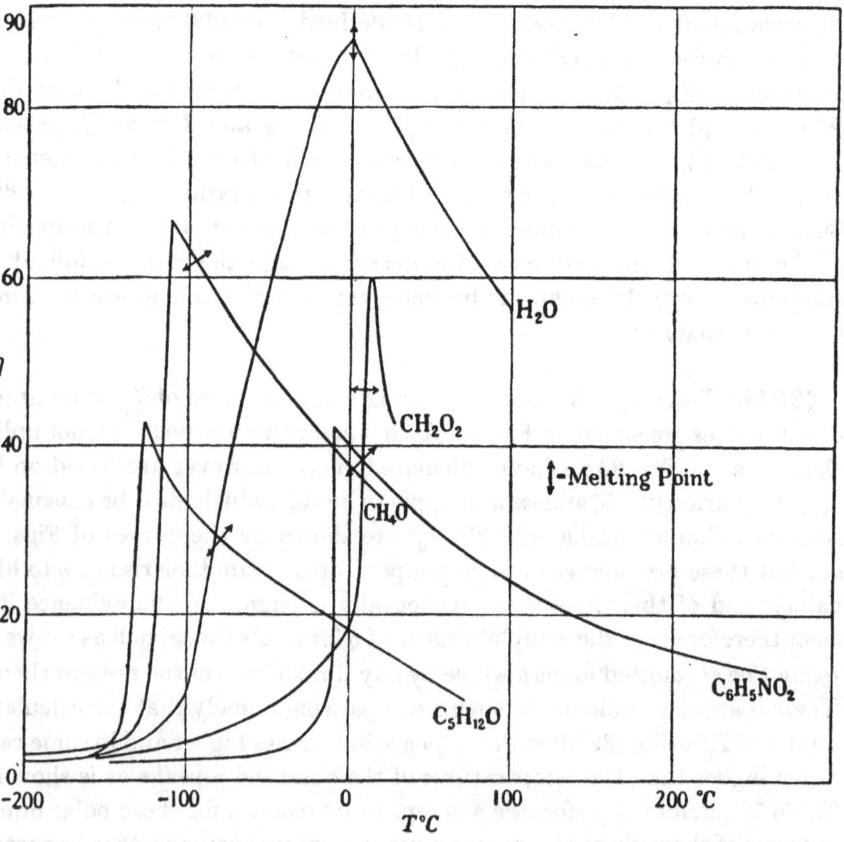

Fig. 94. Observed values of η as a function of T for the substances specified. The cross arrows mark the melting point temperature on each curve for that substance.

ordinary freezing and not merely with "rotational freezing" when the binding is controlled by the rotations themselves. It would then be incorrect to expect the present theory to account for the whole η-curve, but it might remain legitimate to apply it in detail to the form of the η-curve on the high temperature side of the peak.

No satisfactory analysis can be made by assuming that T_d^0 is modified so as nearly to correspond to the temperature of the observed peak, because, if such an identification is made, the greater part of the observed curve must be correlated with part of Curve I, Fig. 93*a*, namely

$$\eta = 1 + \frac{3T_d^0}{T - T_d^0},$$

and no such correlation fits the facts for constant T_d^0. Fairly satisfactory analyses can however be made by retaining T_d^0 more or less unmodified and correlating the observations with parts of the low temperature branches of the curves of Fig. 93*a*. The discussion is as yet too tentative to record in detail here.* As an example we may mention the analysis for water. The range of values of T is $0{\cdot}23 \leqslant T/T_d^0 \leqslant 0{\cdot}31$, so that (2286) can be used with $\gamma = \frac{4}{3}\pi$. This formula then contains the single parameter T_c/β', so that β and β' are scarcely relevant. If $T_c/\beta' = 2336$, the observed values of η are reproduced fairly well, the calculated values ranging from $\eta(273) = 91$, $\eta(373) = 53$. This slightly too great variation of η, if significant, could be removed by somewhat reducing T_d^0 and using a suitable value of β. For other substances similar slight modifications of T_d^0 appear to be essential to a satisfactory analysis.

It will be appreciated that the whole theory is extremely tentative and little better than a suggestion of the type of theory required to explain the effects of the libration-rotation freedoms of molecules in liquids and solids.

§21·5. *Surface phenomena. Properties of adsorbed films.* We have as yet made no reference in this monograph to the equilibrium states of surfaces or surface phases or to the calculation of surface tension, except for a passing reference for crystals in Chapter x. It does not seem possible to hope for much progress in the statistical theory of surface tension for liquids, which is an important field, until the theory of liquids themselves is considerably further developed. We shall make no attempt to discuss this problem here. But there is an allied problem, the equilibrium theory of an adsorbed film on liquids or solids with which some progress can be made by regarding the liquid or solid as providing a given field of force in which the two-dimensional adsorbed phase is established. The equilibrium properties of such adsorbed phases can be discussed by statistical mechanics, and we proceed to do so now. In fact it is not always realized that Langmuir's adsorption isotherm

* See R. H. Fowler, *loc. cit.* p. 811.

for example is strictly a theorem on the equilibrium state, for it is almost always derived by kinetic arguments.*

In general the adsorbed phases to be discussed consist of a layer at most one molecule thick of adsorbed molecules attached to the surface of a solid, or to the surface of a liquid in which the adsorbed molecules are practically insoluble. Such a layer may be conveniently referred to as a *monolayer* for shortness. On a liquid the molecules of such a monolayer must be supposed to be freely mobile except in so far as they get in each other's way or form a condensed or two-dimensional crystalline phase. On a solid they may or may not be freely mobile depending on the temperature. An adequate model for the liquid surface or the solid surface at high temperatures is to assume that some point in the adsorbed molecule (for example its centre of mass) is more or less rigidly bound to the geometrical surface of the liquid but with complete absence of any binding forces in the two directions in the surface. An adequate model for the solid surface at low temperatures when the molecules of the monolayer are immobile or nearly immobile on the surface is to assume that there are definite points of attachment on the surface capable of accommodating just one adsorbed molecule, and that, when so attached, the adsorbed molecule has the usual series of stationary states possessed by any quantum system in a prescribed field of force. It should not be impossible to find a satisfactory mathematical technique for tracing the gradual change over from the immobile to the mobile monolayer, but little work as yet has been done on this problem and we shall not attempt to discuss it here.

§21·51. *Equations of state of mobile monolayers. The perfect gas film.* It is at once evident that the mobile monolayer can be discussed as a two-dimensional gas by methods identical with those used for ordinary gases. When the cross-sections of the adsorbed molecules and the forces between them can be neglected, the molecules are each confined to an area A, the area of accessible surface, and to the stated approximation they do not interfere with each other. They each possess therefore the classical partition function

$$Q(T)=\frac{1}{h^2}\iiiint e^{-\left\{\frac{1}{2m}(p_x^2+p_y^2)+W\right\}/kT}\,dx\,dy\,dp_x\,dp_y,$$

in which W is a boundary field confining the particles to the area A. It follows at once that

$$Q(T)=2\pi mkTA/h^2. \qquad \text{......(2287)}$$

If F is the surface pressure (negative surface tension) in dynes/cm. for

* See for example Langmuir, *J. Amer. Chem. Soc.* vol. 54, p. 2798 (1932). This paper contains much valuable material on all types of adsorbed layers.

an assembly of M adsorbed molecules, then in the usual way it follows that

$$F = MkT\frac{\partial}{\partial A}\log Q(T) = MkT/A. \qquad \text{......(2288)}$$

There are also the obvious formulae for the kinetic energy content and velocity distribution given by Maxwell's law for two dimensions.

§ **21·52.** *The imperfect gas film.* The equation of state of an imperfect gas film correct to terms in $1/A^2$ can be obtained simply by the method of the virial, when we may assume that the interaction energy of two adsorbed particles is sufficiently nearly fixed by the distance apart of their points of attachment to the surface, or at least that we may use a mean value for their energy of interaction which is a function only of this distance.*

The method of the virial given in § 9·7 may be adapted to the present case as follows; we shall not include any frictional resistance term here. The two surviving equations of motion for any particle are

$$m\ddot{x} = X, \quad m\ddot{y} = Y,$$

which, if $r^2 = x^2 + y^2$, $v^2 = \dot{x}^2 + \dot{y}^2$, may be combined to give

$$\frac{1}{4}\frac{d^2}{dt^2}(mr^2) = \tfrac{1}{2}mv^2 + \tfrac{1}{2}(Xx + Yy). \qquad \text{......(2289)}$$

On summing (2289) for all the particles on the surface and averaging over a long time τ, we find

$$\tfrac{1}{2}\overline{\Sigma\, mv^2} = -\tfrac{1}{2}\overline{\Sigma\,(Xx + Yy)}. \qquad \text{......(2290)}$$

This equation applies to the whole surface phase or to any portion of it contained within any geometrical boundary line on the surface. The stress across any such boundary is by definition F per unit length of boundary, and the stress components $-lF\,ds$, $-mF\,ds$ are part of the forces acting on the particles near the line element ds due to the retaining wall or the particles outside the boundary. They contribute to $-\tfrac{1}{2}\overline{\Sigma\,(Xx + Yy)}$ the term

$$\tfrac{1}{2}F\int(lx + my)\,ds.$$

The remaining contribution comes from the interaction of every pair of molecules within the boundary. If $E(r)$ is the potential energy of a pair of adsorbed molecules at a distance apart r, then the interaction of this pair contribute as in § 9·7

$$\tfrac{1}{2}r\,\partial E/\partial r.$$

We find therefore that

$$\tfrac{1}{2}\overline{\Sigma\, mv^2} = \tfrac{1}{2}F\int(lx + my)\,ds + \tfrac{1}{2}\overline{\Sigma\Sigma\, r\,\partial E/\partial r},$$

the double summation being a summation over all pairs. This reduces easily to

$$FA = MkT - \tfrac{1}{2}\overline{\Sigma\Sigma\, r\,\partial E/\partial r}. \qquad \text{......(2291)}$$

* Mitchell, *Trans. Faraday Soc.* vol. 31, p. 980 (1935).

Now the average number of pairs at a distance apart between r and $r+dr$, by an argument analogous to that giving (875), is

$$\tfrac{1}{2}M^2\frac{2\pi r\,dr}{A}\,e^{-E(r)/kT}.$$

Therefore
$$FA = MkT - \frac{M^2\pi}{2A}\int_0^\infty \frac{\partial E(r)}{\partial r}\,e^{-E(r)/kT}\,r^2\,dr, \qquad \text{......(2292)}$$

$$= MkT - M^2\frac{\pi kT}{A}\int_0^\infty (e^{-E(r)/kT}-1)\,r\,dr. \qquad \text{......(2293)}$$

This can be checked by a direct evaluation of the partition function for the complete two-dimensional configuration in the manner of § 8·31.

It is unfortunate that the range of validity of these formulae is too small to reach a region in which they may be adequately compared with experiments such as Langmuir's or Adam's on the expanded phase of a monolayer on a liquid. The problem of extending (2293) to higher powers of $1/A$ is a problem of difficulty comparable to the similar problem for an ordinary gas.

It is also theoretically possible to discuss the equilibrium of a two-phase system consisting of the mobile monolayer and an ordinary gas of the same molecules; we can also discuss a system of two surface phases, one crystalline and one a gas as above. But such discussions are of little or no practical value at the moment. We pass on therefore to the immobile monolayer on a solid for which the theory can be successfully compared with experiment.

§ 21·53. *Langmuir's adsorption isotherm. Molecular adsorption.* The feature of primary interest for the immobile or almost immobile monolayer on a solid is the fraction of surface covered in the monolayer in equilibrium with gas at a given density. We have therefore to study this two-phase equilibrium. It is commonly studied by kinetic methods. The present discussion will serve to show that the results have the usual independence of the mechanism by which equilibrium is attained.

Let us suppose that the states accessible to the adsorbable systems are (i) the ordinary states for such free systems in the gas phase, that is a set of states of weights ϖ_r and energies ϵ_r, and (ii) states of attachment to any one of the surface atoms (or other suitable location) of the adsorbing solid. We shall suppose for the present that the molecule is attached as a whole to a single point of attachment, and that in this adsorbed state it possesses a series of possible quantum states of weights ρ_r and energies η_r with the partition function $v_s(T) = \Sigma_r \rho_r e^{-\eta_r/kT}$ for the whole set belonging to any particular location. We suppose further that there are N_s such locations and that each one can only accommodate one adsorbed molecule. The adsorbed states belonging to any one location are assumed to be independent of whether surrounding locations are holding adsorbed molecules or not. It will not matter what statistics the gas molecules are assumed to satisfy. We shall use the unspecified generating function $g(q)$ for them.

Suppose that the assembly contains M adsorbable systems in all, with a total energy E. Then the total number of complexions can be written down at once as the coefficient of $x^M z^E$ in the product of $\Pi_r [g(xz^{\epsilon_r})]^{\varpi_r}$ for the gas states, and

$$[1+x\Sigma_r \rho_r z^{\eta_r}]^{N_s}$$

for the adsorbed states. The factor for the gas states is that already given in § 2·4. The factor for the adsorbed states contains N_s factors so arranged that the occupation of any quantum state on one particular point of attachment puts out of action all states on that point of attachment for further adsorbed systems. This is what we require. We find therefore that

$$C=\frac{1}{(2\pi i)^2}\iint \frac{dx}{x^{M+1}}\frac{dz}{z^{E+1}}\prod_{r=0}^{\infty}[g(xz^{\epsilon_r})]^{\varpi_r}[1+x\Sigma_r \rho_r z^{\eta_r}]^{N_s}. \quad\text{......(2294)}$$

Having thus composed C the rest of the derivation is simple. We have at once by the usual arguments

$$\overline{M_g}=\lambda\frac{\partial}{\partial\lambda}\sum_{r=0}^{\infty}\varpi_r \log g(\lambda e^{-\epsilon_r/kT}), \quad\text{......(2295)}$$

$$\overline{M_s}=\lambda\frac{\partial}{\partial\lambda}N_s\log[1+\lambda v_s(T)]=\frac{N_s\lambda v_s(T)}{1+\lambda v_s(T)}, \quad\text{......(2296)}$$

where $\overline{M_g}$ and $\overline{M_s}$ are the equilibrium numbers of systems in the gaseous and adsorbed phases respectively. Since the gas will be effectively classical we may simplify (2295) to

$$\overline{M_g}=\lambda\sum_{r=0}^{\infty}\varpi_r e^{-\epsilon_r/kT}, \quad\text{......(2297)}$$

$$=\lambda\frac{(2\pi mkT)^{\frac{3}{2}}V}{h^3}b_g(T)\,e^{-\chi/kT}. \quad\text{......(2298)}$$

In (2298) χ is the energy step from the lowest adsorbed state, assumed to be of energy zero, to the lowest free state in the gas—the heat of adsorption per molecule at the absolute zero—and $b_g(T)$ is the partition function for the rotations and vibrations of the free molecule. Its term of lowest energy is a constant since the energy of this state has already been taken account of in χ. The remaining factor is the usual one for free translations in a volume V.

It is usual to write $\overline{M_s}/N_s=\theta$, so that θ is the fraction of a complete monolayer held by the surface. We have also $\overline{M_g}/V=p/kT$. Combining these relations with (2296) and (2298) we find

$$p=\frac{\theta}{1-\theta}\frac{(2\pi m)^{\frac{3}{2}}(kT)^{\frac{5}{2}}}{h^3}\frac{b_g(T)}{v_s(T)}e^{-\chi/kT} \quad\text{......(2299)}$$

for the relationship between p and θ. This is often written in the form

$$\theta = \frac{Ap}{1+Ap} \quad (A = A(T)), \qquad \text{......(2300)}$$

and known as Langmuir's adsorption isotherm. Current proofs of this important formula are usually based on explicit assumptions as to the *mechanism* of deposition and re-evaporation. We see here that no such assumptions are necessary. This isotherm (2300) must hold whatever the kinetics of the processes, provided only that the molecules are adsorbed as wholes, independently of each other, on a fixed number of definite locations on the solid surface. The use of (2300) is familiar from Langmuir's writings and we shall not examine it here.

§**21·54.** *Langmuir's adsorption isotherm. Atomic adsorption.* A somewhat different isotherm is obtained if for example the adsorbable systems are atoms X, while the gas phase consists overwhelmingly of molecules X_2. We can apply all the formulae of the preceding section if we assume they refer to the atom X and add to the gas phase an extra set of states accessible to the atoms in the form of molecules X_2. These states require an extra factor

$$\prod_{r=0}^{\infty} [g'(x^2 z^{\zeta_r})]^{\sigma_r}$$

in the integrand of C, if the molecules X_2 have a set of states of energies ζ_r and weights σ_r. Besides the equations (2295)–(2298) for the equilibrium state there is now the extra equation

$$\overline{(M_2)_g} = \lambda^2 \frac{\partial}{\partial \lambda^2} \sum_{r=0}^{\infty} \sigma_r \log g'(\lambda^2 e^{-\zeta_r/kT}), \qquad \text{......(2301)}$$

$$= \lambda^2 \frac{(2\pi m' kT)^{\frac{3}{2}} V}{h^3} b_g'(T)\, e^{-\chi'/kT}. \qquad \text{......(2302)}$$

In (2302) m', χ' and $b_g'(T)$ refer to the free molecule X_2 in the gas phase in which the equilibrium number present is $\overline{(M_2)_g}$, and χ' is the energy step from the lowest adsorbed state for two atoms X to the lowest free state for the pair as a molecule X_2 in the gas. Since the number of free atoms in the gas may be assumed to be negligible we have now $\overline{(M_2)_g}/V = p/kT$. Combining this with (2296) and (2302) we find

$$p = \left(\frac{\theta}{1-\theta}\right)^2 \frac{(2\pi m')^{\frac{3}{2}} (kT)^{\frac{5}{2}}}{h^3} \frac{b_g'(T)}{[v_s(T)]^2} e^{-\chi'/kT}, \qquad \text{......(2303)}$$

or

$$\theta = \frac{(A'p)^{\frac{1}{2}}}{1+(A'p)^{\frac{1}{2}}} \quad (A' = A'(T)). \qquad \text{......(2304)}$$

Whatever the mechanism of deposition and re-evaporation Langmuir's isotherm must have the form (2304) whenever a molecule X_2 is to be adsorbed as two independent atoms X requiring two points of attachment.

These formulae are in common use, but it should be remembered that they both have the weakness that they depend essentially on the assumption that the fields holding the adsorbed molecules, and therefore $v_s(T)$ and χ, do not depend on the temperature (except in the usual explicit manner) or on the presence of other adsorbed molecules, and that all locations are equally efficient adsorbers. Any or all of these assumptions may be in error.

§ **21·55.** *Adsorption of competing molecules.* It is simple to discuss by the same methods the equilibrium of an adsorbed layer formed out of two or more competing adsorbable molecules in the gas phase.* The fundamental assumption here is that any location on the surface can be occupied by any one of the available molecules from the gas phase, but not by more than one of any kind. For definiteness we will first consider molecular adsorption as in § 21·53, for two competing gases. The generalization to more gases is obvious. We have to use selector variables x_1, x_2 for the two types of systems and find

$$C=\frac{1}{(2\pi i)^3}\iiint\frac{dx_1}{x_1^{M_1+1}}\frac{dx_2}{x_2^{M_2+1}}\frac{dz}{z^{E+1}}\prod_{r=0}^{\infty}[g_1(x_1z^{\epsilon_r^1})]^{\varpi_r^1}\prod_{r=0}^{\infty}[g_2(x_2z^{\epsilon_r^2})]^{\varpi_r^2}$$
$$\times[1+x_1(\Sigma_r\rho_r^1z^{\eta_r^1})+x_2(\Sigma_r\rho_r^2z^{\eta_r^2})]^{N_s}. \quad \ldots\ldots(2305)$$

The factor $[\ ]^{N_s}$ allows for no molecule or one molecule of either kind but no more to be attached to each location on the surface. From this equation it follows at once by the usual arguments that, if p_1, p_2 are the partial pressures of the two constituents in the gas phase and θ_1, θ_2 the fractions of the surface covered by the two constituents in the adsorbed phase, then

$$\theta_1=\frac{\overline{M_s^1}}{N_s}=\frac{\lambda_1v_s^1(T)}{1+\lambda_1v_s^1(T)+\lambda_2v_s^2(T)},\quad \theta_2=\frac{\overline{M_s^2}}{N_s}=\frac{\lambda_2v_s^2(T)}{1+\lambda_1v_s^1(T)+\lambda_2v_s^2(T)}, \quad \ldots\ldots(2306)$$

$$p_1=\frac{\overline{M_g^1}kT}{V}=\lambda_1\frac{(2\pi m_1)^{\frac{3}{2}}(kT)^{\frac{5}{2}}}{h^3}b_g^1(T)\,e^{-\chi_1/kT}, \quad \ldots\ldots(2307)$$

$$p_2=\frac{\overline{M_g^2}kT}{V}=\lambda_2\frac{(2\pi m_2)^{\frac{3}{2}}(kT)^{\frac{5}{2}}}{h^3}b_g^2(T)\,e^{-\chi_2/kT}. \quad \ldots\ldots(2308)$$

The superior affixes in these formulae of course distinguish the different molecules and do not denote powers. From these equations it follows that

$$p_1=\frac{\theta_1}{1-\theta_1-\theta_2}\frac{(2\pi m_1)^{\frac{3}{2}}(kT)^{\frac{5}{2}}}{h^3}\frac{b_g^1(T)}{v_s^1(T)}e^{-\chi_1/kT}, \quad \ldots\ldots(2309)$$

$$p_2=\frac{\theta_2}{1-\theta_1-\theta_2}\frac{(2\pi m_2)^{\frac{3}{2}}(kT)^{\frac{5}{2}}}{h^3}\frac{b_g^2(T)}{v_s^2(T)}e^{-\chi_2/kT}, \quad \ldots\ldots(2310)$$

or

$$\theta_1=\frac{A_1p_1}{1+A_1p_1+A_2p_2},\quad \theta_2=\frac{A_2p_2}{1+A_1p_1+A_2p_2}$$
$$(A_1=A_1(T),\ A_2=A_2(T)). \quad \ldots\ldots(2311)$$

* Equations equivalent to those of this section or their generalizations were first derived by kinetic arguments by D. C. Henry, *Phil. Mag.* vol. 44, p. 689 (1922).

These formulae can be extended as in § 21·54 to atomic adsorption of molecular gases. A case of particular interest is the competing adsorption of light and heavy hydrogen. Equations (2311) then continue to hold if p_1 and p_2 denote the partial pressures of H-atoms and D-atoms in the gas phase. These partial pressures must then be converted into molecular partial pressures by the usual equations of dissociative equilibrium (Chapter v). In such a case of course the molecular gases H_2, D_2 and HD must all be present in concentrations in full equilibrium with each other. If the gas phase is not in full H_2-D_2-HD equilibrium it cannot be in true equilibrium with an atomic monolayer.

Still more complicated cases can be handled in which the second system can also be attached to an already adsorbed first system and so on. This particular case can be dealt with by using in C the factor

$$[1+x_1(\Sigma_r \rho_r{}^1 z^{\eta_r{}^1})\{1+x_2\,\Sigma_s\,\sigma_s z^{\xi_s}\}+x_2(\Sigma_r \rho_r{}^2 z^{\eta_r{}^2})]^{N_s}. \quad \text{......(2312)}$$

The assumptions underlying such formulae are however probably not a sufficiently good representation of the physical conditions to warrant further discussion at present.

§ **21·56.** *Generalizations of the adsorption problem. Critical adsorption.* At the close of § 21·54 we noted certain important restrictions on which the adsorption isotherm of those sections are based. One of these assumptions, that all adsorption locations are equally efficient, can of course be quite simply removed, by postulating a distribution law for adsorbing locations over a range of values of χ. It is now convenient to take as the energy zero for a molecule its energy in its lowest free state in the gas; its energy in its lowest adsorbed state is then $-\chi$, a variable depending on the particular location of adsorption. It is clear that the result of this generalization is merely to replace the factor $[\ ]^{N_s}$ in (2294) by

$$\Pi_\chi\,[1+xz^{-\chi}(\Sigma_r\,\rho_r z^{\eta_r})]^{q(\chi)d\chi}; \quad \text{......(2313)}$$

we have here assumed that the energies η_r of the excited adsorbed states are substantially unaffected by χ, and that $q(\chi)\,d\chi$ is the number of locations in which the maximum adsorptive binding energy lies between χ and $\chi+d\chi$. We find at once by the use of (2313) in (2296) that

$$\overline{M}_s=\lambda\frac{\partial}{\partial\lambda}\int_0^\infty \log[1+\lambda e^{\chi/kT}\,v_s(T)]\,q(\chi)\,d\chi, \quad \text{......(2314)}$$

$$=\int_0^\infty \frac{\lambda e^{\chi/kT}\,v_s(T)}{1+\lambda e^{\chi/kT}\,v_s(T)}\,q(\chi)\,d\chi;$$

$$N_s=\int_0^\infty q(\chi)\,d\chi. \quad \text{......(2315)}$$

Remembering (2298), equation (2314) can be reduced to

$$\overline{M}_s = \int_0^\infty \frac{pA(T)\, e^{\chi/kT}}{1+pA(T)\, e^{\chi/kT}} q(\chi)\, d\chi, \quad N_s = \int_0^\infty q(\chi)\, d\chi. \quad \text{......(2316)}$$

Equations (2316) give θ ($=\overline{M}_s/N_s$) as a function of p and replace (2330); $A(T)$ is independent of p and χ. No systematic use of such equations appears to have been made at present, but their further study might be profitable.

A quite distinct and much more important generalization of (2300) can be obtained by abandoning the assumption that the energy of binding to any location is independent of the presence of other adsorbed atoms. Adsorption immediately becomes a typical cooperative phenomenon. Instead of the regular (θ, p) relation at constant T or (θ, T) relation at constant p given by (2300) we find critical conditions, and are able to give a satisfactory account of the existence of the well-known critical conditions for the deposition of metallic vapours on a dielectric surface such as glass or mica, or on the surface of another metal. Alternatively, when the interaction of adsorbed molecules loosens instead of strengthens the binding, we obtain formulae, showing no critical conditions, which are applicable to the formation of caesium layers on tungsten and similar phenomena.* We shall not discuss their further applications here.

§21·57. *Critical adsorption phenomena.* The critical conditions just referred to are usually studied kinetically. In the experiments a stream of metal vapour falls on a plate at a given temperature, the metal vapour making a given number of impacts per cm.² per second. It is found that if $T > T_c$ the metal is not deposited, but that if $T < T_c$ the metal is deposited in bulk. What is observed has therefore nothing directly to do with the formation of an equilibrium monolayer, but it is clear that there is an underlying equilibrium problem. Metal vapour at pressure p is in equilibrium with an adsorbed phase. If $T > T_c$, the fraction θ of the surface covered should be small, effectively zero, but if $T < T_c$ the fraction covered should rise effectively to unity. In such a case further layers of adsorbed atoms will actually be deposited on the first layer once it forms and the stable equilibrium state corresponds to the deposition of the metal in bulk. This further stage however can perhaps be ignored in a first discussion, and attention concentrated on the critical behaviour of the first adsorbed layer, because the formation and reasonable completion of the first layer is obviously essential for the later layers to be formed at all.

The existence of such critical conditions has been discussed previously by

* Langmuir, *loc. cit.*

Langmuir* and Frenkel,† from a more kinetic standpoint than that adopted here. They have shown how such critical conditions can arise if the energy of binding of two atoms of the vapour on adjacent locations on the surface is more than twice the energy of adsorption of a single atom. It is only necessary to incorporate the idea of a dependence of the heat of adsorption on the number of adsorbed neighbours in order to be able to construct a partition function for the adsorbed layer showing typical critical phenomena for the equilibrium state of the monolayer for given vapour pressure.

In § 21·53 we assumed effectively that the energy of adsorption of M_s atoms in their lowest states on similar locations was $M_s \chi$, independent of their arrangement on the surface. This assumption we must now abandon. Instead it will be reasonable to suppose that the energy of adsorption of any atom in its lowest state will be determined by the number of neighbouring locations occupied by other adsorbed atoms. This energy may not vary exactly linearly with the number of adsorbed neighbours but we cannot commit any great error in making this assumption. On making it we can express the energy of adsorption of M_s atoms in the form

$$M_s \chi_0 + X\chi_1,$$

where X is the number of pairs of nearest neighbours. The partition function of an adsorbed layer of M_s atoms all in their lowest adsorbed states can therefore be put in the form

$$Q(M_s, N_s, T) = \Sigma_X \, g(M_s, N_s, X)\, e^{(M_s\chi_0 + X\chi_1)/kT}; \qquad \ldots\ldots(2317)$$

heats of adsorption are of course negative energies. The coefficient $g(M_s, N_s, X)$ is the number of arrangements of M_s atoms on a lattice of N_s points with X pairs of nearest neighbours occupied. If we complete $Q(M_s, N_s, T)$ by allowing for a set of excited states for each adsorbed atom which is unaffected by the presence or absence of neighbours, we have

$$Q(M_s, N_s, T) = \Sigma_X \, g(M_s, N_s, X)\, e^{(M_s\chi_0 + X\chi_1)/kT}\, [v_s(T)]^{M_s}. \quad \ldots\ldots(2318)$$

The coefficient $g(M_s, N_s, T)$ cannot be evaluated exactly except for a linear array. To proceed further from (2318) we can make an approximation similar to that made by Bragg and Williams, or to that made by Bethe, in the order-disorder problem. We shall develop here the consequences of the Bragg and Williams type of approximation.‡ We assume that a fair approximation to the energy $M_s \chi_0 + X\chi_1$ for the values of X that matter is to take it equal to

$$M_s(\chi_0 + \tfrac{1}{2}\theta\chi_1) \quad (\theta = M_s/N_s).$$

* Langmuir, *Proc. Nat. Acad. Sci.* vol. 3. p. 141 (1916).

† Frenkel, *Zeit. f. Physik*, vol. 26, p. 117 (1924).

‡ For Bethe's methods applied to this problem see Peierls, *Proc. Camb. Phil. Soc.* vol. 32, p. 477 (1936).

This χ_1 is not the same as the former χ_1 but no confusion will arise. We can then evaluate $Q(M_s,N_s,T)$, for

$$\Sigma_X g(M_s,N_s,X) = N_s!/\{M_s!(N_s-M_s)!\},$$

and therefore, with sufficient accuracy,

$$\log Q = N_s \log N_s - M_s \log M_s - (N_s - M_s)\log(N_s - M_s) \\ + M_s \log v_s(T) + M_s(\chi_0 + \tfrac{1}{2}\theta\chi_1)/kT. \quad \text{......(2319)}$$

§ **21·58.** *Determination of the equilibrium state.* The present assembly consists of the two phases, adsorbed and gaseous. According to the general rules of Chapter VI we can at once write down the free energy of Helmholtz F_a, for the assembly,* namely

$$F_a = F_s + F_g = -kT\log Q - kTM_g\left\{\log\left[\frac{(2\pi mkT)^{\frac{3}{2}}V}{h^3 M_g} b_g(T)\right] + 1\right\}. \quad \text{......(2320)}$$

The formula for F_g follows at once from Theorem 6·31. Equilibrium is attained when F_a has its least possible value, which is therefore given by one of the roots of the equation

$$\frac{\partial F_a}{\partial M_s} = -kT\left[\left\{\log\left(\frac{1-\theta}{\theta} v_s(T)\right) + \frac{\chi_0 + \theta\chi_1}{kT}\right\} - \log\left\{\frac{(2\pi m)^{\frac{3}{2}}(kT)^{\frac{5}{2}}}{ph^3} b_g(T)\right\}\right] = 0. \quad \text{......(2321)}$$

In forming this equation, we remember that $\theta = M_s/N_s$ and that $M_g + M_s = const.$ Provided that the root for θ is correctly chosen, it follows at once from (2321) that

$$p = \frac{\theta}{1-\theta}\frac{(2\pi m)^{\frac{3}{2}}(kT)^{\frac{5}{2}}}{h^3}\frac{b_g(T)}{v_s(T)} e^{-(\chi_0+\theta\chi_1)/kT}, \quad \text{......(2322)}$$

an isotherm identical with (2299) except that χ_0 is replaced by $\chi_0 + \theta\chi_1$, the "effective" value for the next increase in θ. Equation (2322) is valid for χ_1 of either sign.

Equation (2321) can be put in the form

$$\log Ap = \log\frac{\theta}{1-\theta} - \frac{\chi_0 + \theta\chi_1}{kT} = P(\theta,T), \quad \text{......(2323)}$$

where $A\ [=A(T)]$ is a slowly varying function of the temperature. If $\chi_1 > 0$, this equation exhibits the critical phenomena we wish to establish. The behaviour of the function $P(\theta,T)$ for $\chi_0 = 0$ is shown in Fig. 95. We see at once that, if $T > \frac{1}{4}\Theta$, the function $P(\theta,T)$ is a monotonic function of θ and critical conditions cannot occur for any value of p. If T has any smaller

* The argument is given here in terms of the free energy. It could of course be given equally well directly in terms of the complete partition for the assembly as in § 21·22.

value then, when p is sufficiently small, there is still only one root of equation (2323) and one possible (rather small) value of θ for the given p. But as p increases we reach a value $p=p(T)$ above which for a limited range of p there are three intersections. It is obvious that the central intersection represents an unstable state and need not be further considered. For higher pressures there is again only one root, for which $\theta \sim 1$.

To determine the value of p at which the change over occurs from the smallest root giving the least value of F_a to the root nearly unity giving the least value of F_a we can again apply the rule of equal areas. For the condition is that if θ_0, θ_1 are the critical extreme roots of (2323) then

$$\int_{\theta_0}^{\theta_1} \frac{\partial F_a}{\partial \theta} d\theta = 0,$$

or

$$\int_{\theta_0}^{\theta_1} P(\theta, T)\, d\theta = [\log Ap]\,(\theta_1 - \theta_0). \qquad \text{......(2324)}$$

This implies that the critical adsorption isotherm for a given pressure is that isotherm in Fig. 95 which cuts the horizontal line $\log Ap$ in three points in such a way that the areas enclosed above and below the line are equal. If Fig. 95 itself is used, the omitted term χ_0/kT can be incorporated in $\log A$.

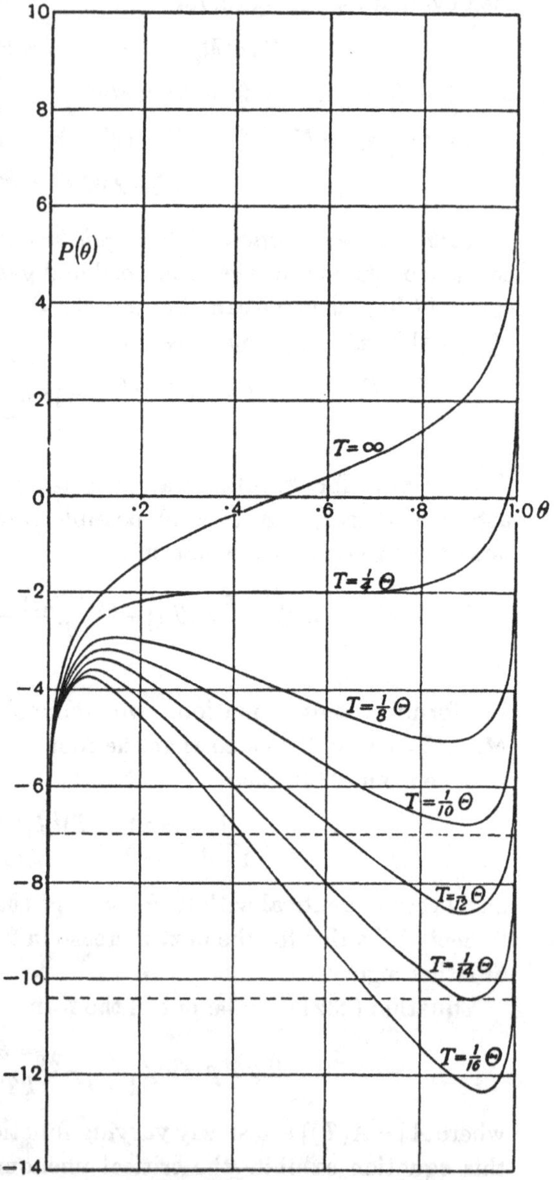

Fig. 95. The function $P(\theta, T)$ for $\chi_0 = 0$, $\chi_1/k = \Theta$.

§21·59. *Numerical values.* Let us now insert rough numerical values which might illustrate the condensation of Cd vapour, so far as orders of magnitude are concerned. We combine $e^{\chi_0/kT}$ with A in (2323), and the explicit expression for $Ae^{\chi_0/kT}$ is then

$$\frac{h^3}{(2\pi m)^{\frac{3}{2}}(kT)^{\frac{5}{2}}}\frac{v_s(T)}{b_g(T)}e^{\chi_0/kT}.$$

To illustrate the case of Cd-atoms we can take

$$b_g(T)\equiv 1,\quad m\simeq 100\times 1{\cdot}6\times 10^{-24}\,\text{gm.},\quad \text{and}\quad v_s(T)\simeq (kT/h\nu_0)^3\simeq (T/50)^3,$$

taking the excited adsorbed states to be those of an isotropic three-dimensional oscillator. Then if p is in atmospheres (10^6 dyne/cm.^{2}) and $T\simeq 100^\circ$ K., we have

$$Ape^{\chi_0/kT}\simeq 10^{-5}pe^{\chi_0/kT},$$

$$\log Ap+\chi_0/kT\simeq -11{\cdot}5+\log p+\chi_0/kT. \quad \text{......(2325)}$$

The exact value of T must be retained in the last term. The value chosen for $v_s(T)$ is not very important. We can then see at once from Fig. 95 that the horizontal for 1 atmosphere cuts off equal areas from the $P(\theta,T)$ curve for $T=\frac{1}{8}\Theta$, if $\chi_0/kT=7{\cdot}5$. The energy required to remove one atom from the monolayer is then

$$k(0{\cdot}94+\theta)\,\Theta,$$

and the critical temperature at one atmosphere pressure is $\frac{1}{8}\Theta$. The discontinuity is a jump in θ from the value 0·03 to the value 0·98.

More extreme conditions are provided by the curve for $T=\frac{1}{14}\Theta$. The horizontal for one atmosphere cuts off equal areas from this curve if $\chi_0/kT=4{\cdot}5$. The energy required to remove one atom from the monolayer must then be

$$k(0{\cdot}32+\theta)\,\Theta,$$

and the critical temperature at one atmosphere pressure $\frac{1}{14}\Theta$. The discontinuity is a change from $\theta=0{\cdot}001$ to $\theta=0{\cdot}999$. The discontinuities as here presented are primarily discontinuities on the *isobar* for one atmosphere as T passes through a critical temperature. They can equally well be presented as discontinuities on the adsorption *isotherm* as $\log p$ (p in atmospheres) passes through the value zero.

The curves of Fig. 95 show that there is a true critical temperature $T=\frac{1}{4}\Theta$ above which critical conditions never arise for any pressure. But the critical temperatures below $\frac{1}{4}\Theta$ which we have discussed above are in no way absolute but are of course of the form $T_c=T_c(p)$. By examination of Fig. 95 we see that an increase of p by a factor 2·7 raises the horizontal line exactly 1 unit. The corresponding decrease in $1/T_c(p)$ is almost exactly $2/\Theta$.

In attempting to compare the theory with observation some care is necessary. The theory applies to equilibrium states and the observations to the impact of a vapour stream with velocities corresponding to an oven temperature T_0 on the condensing surface at temperature T. If we compare the actual vapour stream with gas in equilibrium at the temperature of the condensing surface and at such a pressure that the number of atoms striking unit area in unit time is the same, then the theory would apply strictly if the reflection coefficient of the impinging atoms is not greatly altered by the change in their velocities of impact. Such changes do of course lead to errors in any comparison of theory and experiment but it does not seem likely that they will be large.

Accurate experiments which can be compared with the theory on this basis have been made by Cockcroft.* His results for the condensation of Cd on Cu are shown in Fig. 96 as a relation between $\log_{10} n$, where n is the number of impacts per cm.² per second made by the vapour stream on the plate where it is to condense, and $1/T_c(n)$, where $T_c(n)$ is the critical temperature for condensation at that impact rate. We see at once that the linear relation predicted by the theory is well satisfied. Coming to details the observed linear relation is

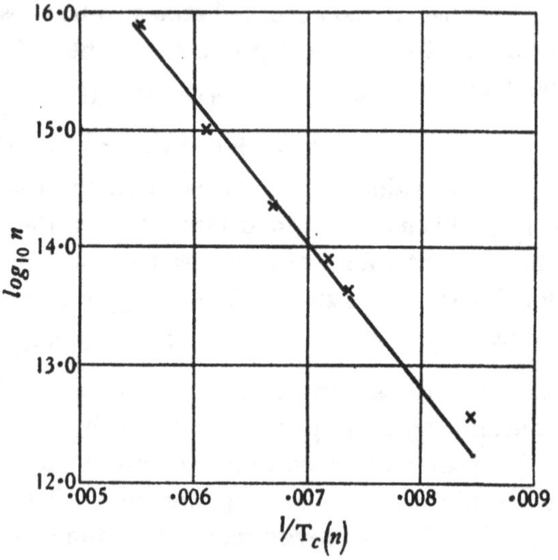

Fig. 96. Showing the linear relation observed between $1/T_c(n)$ and $\log_{10} n$ for critical condensation of Cd atoms on a copper surface.

$$\log_e n = const. - 2840/T_c(n). \qquad \text{......(2326)}$$

If we transform this to equivalent pressures by the relation

$$n = \tfrac{1}{4}\nu\bar{c} = p/(2\pi mkT)^{\frac{1}{2}}, \qquad \text{......(2327)}$$

we have $\log_e p = const. - 2840/T_c(p)$. The equivalent pressure for the observed impact rate of $2{\cdot}4 \times 10^{14}$ at 150° K. is 10^{-9} atmosphere. We obtain the right theoretical slope for the $\log p$, $1/T_c(p)$ relation from Fig. 95 if $\Theta = 2100$. We obtain the correct value of the equivalent pressure for $T_c(p) = 150 = \frac{1}{14}\Theta$ if $\chi_0/150k = 25{\cdot}2$. The observations are therefore well reproduced by the theory if the binding energy for a Cd atom on a copper surface is

$$k(1{\cdot}80 + \theta)\,\Theta, \quad \Theta = 2100.$$

§ **21·6.** *The attainment of very low temperatures by adiabatic demagnetization.* We have already mentioned that crystals of the paramagnetic salts, of which the classical example is gadolinium sulphate $Gd_2(SO_4)_3 \cdot 8H_2O$,

* Cockcroft, *Proc. Roy. Soc.* A, vol. 119, p. 293 (1928).

exhibit special properties of peculiar interest at extremely low temperatures, properties which enable these substances to be used in the effective attainment of temperatures of the order of 0·1° K. and probably considerably lower. The lowest temperature yet claimed is of the order 0·04° K. but the theory used in the analysis of the observations is as yet too inaccurate for the establishment of a reliable temperature scale. The establishment of such a scale however will no doubt shortly be achieved. The special properties of these salts arise of course from the orientational states of the paramagnetic ions Gd^{+++} (*etc.*) which in general when unperturbed have a normal state which is an S state of high multiplicity (Gd^{+++}, ^8S). Such ions contribute an extra orientational factor to the partition function of the crystal, since their orientations are to a high degree of accuracy independent of the lattice vibrations, and therefore make an additive contribution to the entropy and free energy and so to the specific heat and other properties. Their contribution to the entropy as a function of temperature and magnetic field is of particular importance, since a knowledge of this function enables the adiabatic cooling by demagnetization to be calculated. This effect was first explicitly formulated by Debye* and Giauque† and has since been successfully realized, notably by Giauque,‡ de Haas,§ and Simon.‖ It has not however as yet been rigorously discussed using partition functions, and therefore entropies, of a form which agrees in detail with other knowledge of the properties of these paramagnetic ions. The discussion given here is still incomplete, but it is based on work by Van Vleck¶ in which all details known about the partition functions are taken into account. We shall discuss only gadolinium sulphate, but the methods to be used are the same for all the paramagnetic salts.

§ **21·61.** *The states of* Gd^{+++} *ion in crystalline and magnetic fields* The ^8S normal states which the Gd^{+++} ions would possess in the free state are affected in the crystalline salt in the following ways. The field of the neighbours in the lattice, if it is composed effectively of fourth order terms with cubic symmetry, splits the ^8S state into the three states of the partition function

$$2+4e^{-5\delta/kT}+2e^{-8\delta/kT}. \qquad \text{......(2328)}$$

The splitting intervals 5δ and 3δ might be inverted and the partition function become

$$2+4e^{-3\delta/kT}+2e^{-8\delta/kT}. \qquad \text{......(2329)}$$

* Debye, *Ann. d. Physik*, vol. 81, p. 1154 (1926).
† Giauque, *J. Amer. Chem. Soc.* vol. 49, p. 1864 (1927).
‡ Giauque and MacDougall, *Phys. Rev.* vol. 43, p. 768; vol. 44, p. 235 (1933).
§ de Haas, Wiersma and Kramers, *Physica*, vol. 13, p. 175; vol. 1, p. 1 (1933).
‖ Kürti and Simon, *Proc. Roy. Soc.* A, vol. 149, p. 152 (1935).
¶ Van Vleck, Hebb and Purcell, in course of publication.

There is as yet insufficient evidence to prefer one or other form *a priori* and we shall use (2328). Crystalline fields of lower symmetry can split the middle state into $e^{-5\delta/kT}(2e^{-\delta'/kT}+2e^{-\delta''/kT})$, but no further splitting is possible due to this cause.

Magnetic fields can cause a further splitting of the ^8S terms and in fact will completely resolve the multiplicity without help from the crystalline splitting. When there is an effective magnetic field H' acting on each ion and $\delta/kT \ll 1$, the orientational partition function is as usual

$$e^{\frac{7}{2}g\mu_B H'/kT}+e^{\frac{5}{2}g\mu_B H'/kT}+\ldots+e^{-\frac{7}{2}g\mu_B H'/kT}. \qquad \ldots\ldots(2330)$$

In (2330) μ_B is Bohr's magneton and g Landé's splitting factor (here $g=2$). The corresponding value of the small field susceptibility is

$$\chi=\frac{ng^2\mu_B{}^2}{3kT}\frac{7}{2}\cdot\frac{9}{2}, \qquad \ldots\ldots(2331)$$

where n is the number of ions per unit volume.

If this were the whole of the perturbation of the states of the ions, that is if we could identify H' with H, the applied external field, and neglect entirely all other possible magnetic energy terms (spin-spin interactions), matters would be comparatively simple. We shall give such an elementary discussion in the next section where we shall see that for practical purposes a knowledge of these limiting forms of the partition function is then almost sufficient, without a complete knowledge of the dependence of the partition function on both H and T for the region $\delta \simeq kT$. The formulae above can then be supplemented by a more accurate formula than (2331) for the small field susceptibility, valid for all values of δ/kT on these assumptions. This formula is*

$$\chi=\frac{ng^2\mu_B{}^2}{kT}\,\frac{\left(\frac{9}{4}+\frac{6}{5}\frac{kT}{\delta}\right)+\left(\frac{130}{36}+\frac{188}{135}\frac{kT}{\delta}\right)e^{-5\delta/kT}+\left(\frac{49}{36}-\frac{70}{27}\frac{kT}{\delta}\right)e^{-8\delta/kT}}{1+2e^{-5\delta/kT}+e^{-8\delta/kT}}. \qquad \ldots\ldots(2332)$$

Unfortunately the magnetic energies omitted above are not trivial. At the actual concentration of Gd^{+++} ions in $Gd_2(SO_4)_3.8H_2O$ the spin-spin interactions are so large that they should contribute about a third of the whole anomalous specific heat. Any analysis which is made neglecting these terms must therefore be regarded as merely provisional. As yet there are hardly enough observations for well-determined temperatures below 1° K. to enable a proper analysis to be satisfactorily completed. It is also necessary to remember that, before any analysis of the observations can be made the correct demagnetizing factor must be used to calculate H inside the specimen. From this H H' must then be derived by using the

* Van Vleck, in course of publication.

$\frac{4}{3}\pi$ factor. A ferromagnetic state with a very low critical temperature may be reached, but whether or not this occurs appears to depend on the shape of the specimen. From a classical point of view of the interaction of magnetic dipoles the eventual attainment of the ferromagnetic state for bodies more prolate than a sphere appears to be certain. However the true spin interaction states are spread out into something like the Gaussian spread assumed by Heisenberg in his theory of ferromagnetism, and this spread may be sufficient to prevent a ferromagnetic state from ever being attained for $T>0$. At this stage of the development of the theory it is therefore not possible to do more than mention these difficulties. No attempt will be made here at any analysis of experiments beyond a preliminary one ignoring spin-spin interactions. We may however quote for future reference Van Vleck's first approximation to the extra factor in the partition function, which arises from spin-spin interactions. It is in fact to this approximation an extra factor independent of the factor due to the crystalline splitting, and in the form here given should be valid when $H=0$ and the temperature is high compared with the expected Curie point of the spin-spin interactions. For N atoms in zero field we have by (2328) the partition function

$$[2+4e^{-5\delta/kT}+2e^{-8\delta/kT}]^N.$$

The first approximation for spin-spin interactions in the gadolinium sulphate crystal replaces this by

$$[2+4e^{-5\delta/kT}+2e^{-8\delta/kT}]^N\,(1+9\kappa Nn^2\Omega/k^2T^2), \qquad \ldots\ldots(2333)$$

where κ is a numerical coefficient of the order of 1·3–1·5, n (as before) the number of Gd^{+++} ions per unit volume $(=N/V)$, and*

$$\Omega=\frac{g^4\mu_B{}^4}{[1+2e^{-5\delta/kT}+e^{-8\delta/kT}]^2}\left[\left(\frac{81}{16}+\frac{36}{5}\frac{kT}{\delta}\right)+\left(\frac{137}{4}+\frac{623}{30}\frac{kT}{\delta}\right)e^{-5\delta/kT}\right.$$
$$+\left(\frac{49}{8}+\frac{224}{15}\frac{kT}{\delta}\right)e^{-8\delta/kT}+\left(\frac{4225}{324}-\frac{4064}{405}\frac{kT}{\delta}\right)e^{-10\delta/kT}$$
$$\left.+\left(\frac{12985}{324}-\frac{623}{30}\frac{kT}{\delta}\right)e^{-13\delta/kT}+\left(\frac{2401}{1296}-\frac{980}{81}\frac{kT}{\delta}\right)e^{-16\delta/kT}\right]. \qquad \ldots\ldots(2334)$$

§21·62. *Preliminary study of the specific heat and the cooling by adiabatic demagnetization for* $Gd_2(SO_4)_3\,.\,8H_2O$, *neglecting spin-spin interactions.* Using (2328) for the partition function, the preliminary value of the specific heat is given by

$$C_\delta=Nk\sigma^2\frac{\partial^2}{\partial\sigma^2}\log[2+4e^{-5\sigma}+2e^{-8\sigma}] \quad \left(\sigma=\frac{\delta}{kT}\right), \qquad \ldots\ldots(2335)$$

$$=Nk\frac{\sigma^2[50e^{-5\sigma}+64e^{-8\sigma}+18e^{-13\sigma}]}{[1+2e^{-5\sigma}+e^{-8\sigma}]^2}. \qquad \ldots\ldots(2336)$$

* The limit of (2334) as $\delta\to 0$ (free atoms) has also been derived by Waller (unpublished).

This value fits Clark and Keesom's* observations fairly well with $\delta/k = 0{\cdot}170$, as shown in Fig. 97, but no serious test can be made until reliable observations are extended to lower temperatures.

Equation (594) determines the entropy contribution. Using the value $\delta/k = 0{\cdot}170$ just determined, the continuous curve in Fig. 98 shows the entropy for one gram-ion of gadolinium sulphate contributed by the crystalline splitting with the partition function (2328). This is therefore the entropy in zero magnetic field omitting spin-spin interactions. If spin-spin

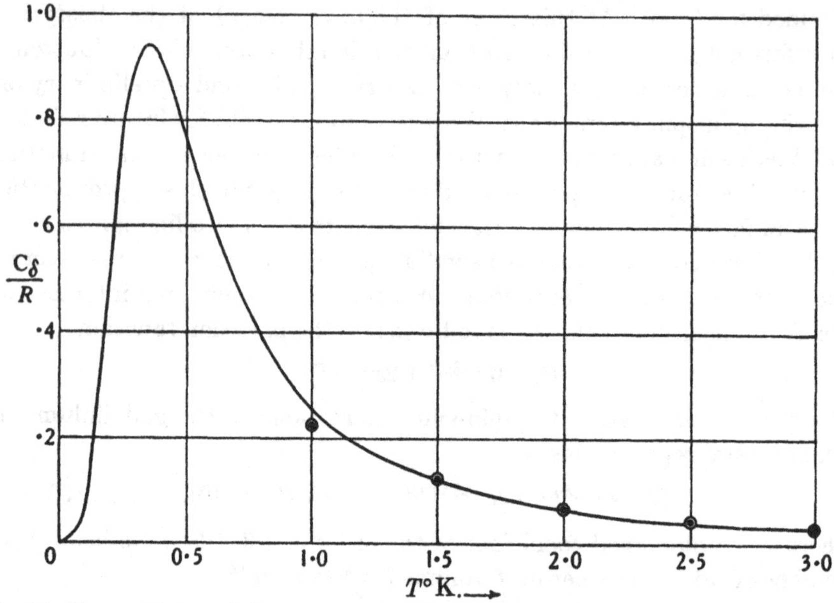

Fig. 97. The specific heat C_δ/R according to (2336) for one half mole (=one gram-ion) of gadolinium sulphate with $\delta/k = 0{\cdot}170$; the observed values of Clark and Keesom are shown thus ⊙, after the normal lattice specific heat has been subtracted.

interactions are included, the entropy will be modified to a form something like that shown by the broken curve. The sharp break at the bottom will occur if the crystal becomes ferromagnetic. If it does not, the entropy curve will be higher and be rounded off like the curve for the crystalline splitting only.

Perhaps above 2° K. and certainly above 3° K. δ/kT is sufficiently small for the crystalline splitting to be ignored in calculating the entropy in the effective magnetic field H'. On any vertical on the right-hand side of the diagram we can therefore calculate the entropy as a function of H' by using the partition function (2330). The entropy according to (594) takes the form

$$\frac{S}{R} = \log\frac{1-e^{-8\gamma}}{1-e^{-\gamma}} + \gamma\left[\frac{1}{e^{\gamma}-1} - \frac{8}{e^{8\gamma}-1}\right] \qquad \left(\gamma = \frac{g\mu_B H}{kT}\right). \quad \text{......(2337)}$$

* Clark and Keesom, *Physica*, vol. 3, p. 1075 (1935).

On inserting numerical values for

$$g\ (=2),\quad \mu_B\ (=9{\cdot}22\times10^{-21}),\quad \text{and}\quad k\ (=1{\cdot}372\times10^{-16}),$$

we find
$$H' = 7{\cdot}43\times10^{3}\,T\gamma. \qquad\qquad \text{......(2338)}$$

The calculation of the H' scales is then easily completed. The scale for $T = 3°$ K. is shown in Fig. 98.

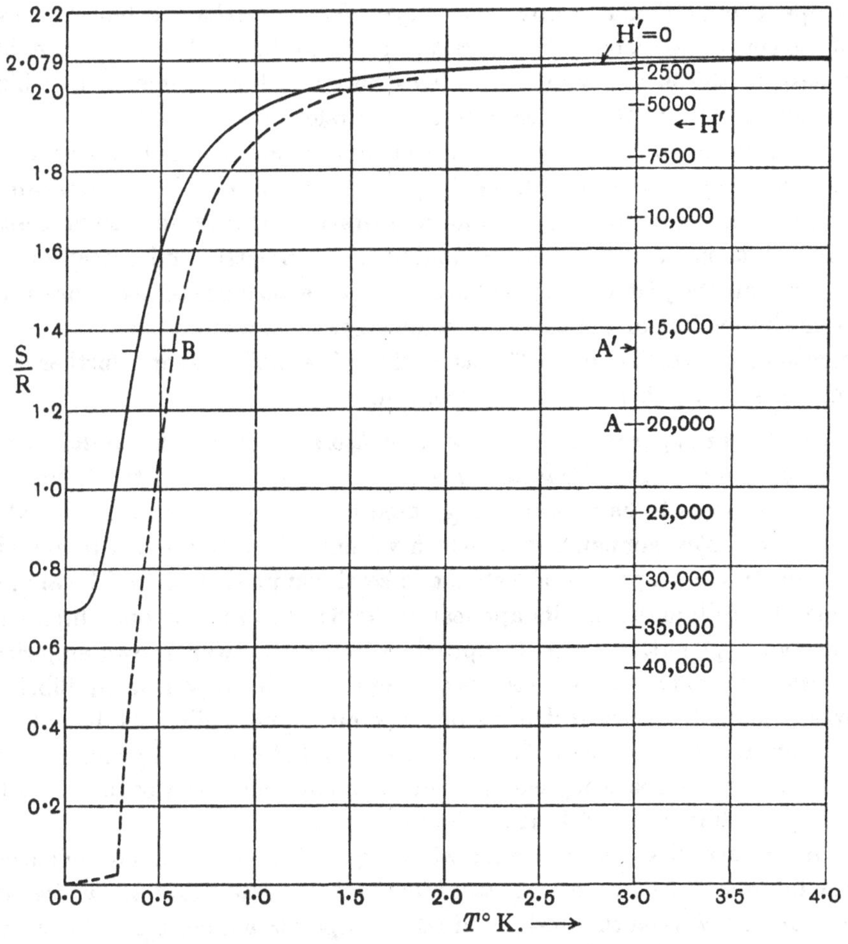

Fig. 98. The entropy S/R for one gram-ion of gadolinium sulphate as a function of T for zero field H' and as a function of H' for sufficiently large T, in particular for $T = 3°$ K.

The method of using such a diagram to study cooling by adiabatic demagnetization is now very simple. Equilibrium is set up for an initial state of say $T = 3°$ K., $H' = 20{,}000$ gauss. The state of the specimen then corresponds to the point A in the entropy diagram, or if allowance is to be made for the residual entropy of the lattice vibrations of the gadolinium sulphate and any other systems in thermal contact with it, to the somewhat higher point A'. The systems are then heat insulated and after insulation the field H' is removed. The entropy remains constant and the state of the system therefore changes to the point B of equal entropy on the curve $H'=0$. If the form of this entropy curve has been accurately calculated, the resulting temperature can be read off the diagram and checked by comparing the observed susceptibility in the final state with the calculated value for this temperature. Calculations by Teller suggest, however, that the temperatures of the spin state distributions so reached will not be reached (in reasonable times) by the lattice vibrations. This may affect experiments in which the exact state of the lattice vibrations is relevant.

Accurate temperatures cannot of course be determined in this way without an accurate theory of the entropy curve in zero field.* It is possible to determine correctly the absolute temperature attained in such processes without a molecular model by combined calorimetric and susceptibility measurements, just as it is possible to set up the absolute temperature scale from the reading of the constant volume gas thermometer combined with measurements of the Joule-Thomson effect. We shall not enter further here into this purely thermodynamic problem.

§21·7. *An appendix to Chapter XIII. A theory of "the rectilinear diameter" for liquid and vapour densities.* An important advance in the theory of liquids appears to have been inaugurated by Eyring† in remarking that a theory of holes, similar to that which we have already applied to explain the electronic properties of metals and semi-conductors, should be applicable also to liquids. By its application he has shown that one obtains at once an explanation (in first approximation) of the well-known empirical relationship between the densities of a liquid and its vapour in equilibrium with it, and the critical density of the same liquid, called the law of the rectilinear diameter or the law of Cailletet and Mathias.‡ Eyring has also shown that the idea is of great value in advancing the theory of the viscosity of liquids, but these applications lie outside our field.

We observe first that if χ is the work required to extract a molecule from the interior of the liquid, *and leave behind a hole in the liquid into which that molecule fitted*, then the work required to evaporate a molecule and leave no extra hole behind—that is the ordinary heat of evaporation—is just $\frac{1}{2}\chi$. For we may regard the work required in the first process as that required to break a number of bonds between the molecule removed and its neighbours, no compensating formation of new bonds being allowed. The work required

* de Haas and Gorter, *Physica*, vol. 2, pp. 335, 438 (1935), have avoided some of the uncertainty by using ions with doublet states, such as Ti^{+++}.

† Eyring, *J. Chem. Physics*, vol. 4, p. 283 (1936), giving also a valuable study of viscosity.

‡ For the most recent experimental study of this law see E. Mathias, Onnes and Crommelin, *Proc. Sect. Sci. Amsterdam*, vol. 15, p. 960 (1913).

is therefore given by $\chi = \Sigma_i n_i \epsilon_i$ summed over all types of bonds or pairs of neighbours in number n_i that contribute effectively an energy ϵ_i to the binding of the particular molecule. If however the whole liquid of N molecules is evaporated then the work required is $\frac{1}{2}N\Sigma_i n_i \epsilon_i$, since each bond is then counted twice over. But by definition this work is N times the normal heat of evaporation per molecule, which is therefore $\frac{1}{2}\chi$. The same result can be reached by considering the energy restored when the hole is allowed to heal itself by the formation of new bonds. The result is exact so long as the binding can be correctly analysed into interactions between pairs of molecules only. It implies that the energy required to form a molecule in the vapour in its lowest state ($\frac{1}{2}\chi$) is equal to the energy required to form a hole in the liquid in its lowest state ($\chi - \frac{1}{2}\chi$).

Now consider the partition function for the motion of the hole in the liquid. So long as the temperature is not too low and the molecules move fairly freely the hole will "move" freely by neighbouring molecules occupying it, so that the hole takes their places in succession. When there are only a few holes, therefore, they do not interfere with one another and a first approximation to their states of motion will be to take them as the same as those of a free molecule of the same mass. This should be a good approximation for monatomic molecules—probably less good for more complex molecules and "holes", but even for these the approximation of taking the partition function for the hole equal to the partition function for the free molecule should be reasonably reliable. Moreover any associated molecules in the vapour and associated or multiple holes in the liquid will possess similar partition functions and therefore be present to similar concentrations in the vapour and the liquid respectively. Since therefore both the lowest energy required to form a molecule in the vapour and a hole in the liquid are equal and also the partition functions for the states of higher energy and the degrees of association are the same, it follows that the equilibrium density of the vapour must be equal (to this approximation) to the equilibrium density of the holes in the liquid. Expressed in symbols this implies that

$$\rho_v + \rho_l = const. = 2\rho_c, \qquad \text{......(2339)}$$

where ρ_v and ρ_l are the mass densities of the vapour and the liquid and ρ_c is their common density at the critical point.

At lower temperatures the partition function for the holes will inevitably be less than that given by the above first approximation, and for more complex molecules than monatomic ones may be expected to be somewhat less at all temperatures. The binding of the liquid molecules to each other makes them not entirely free to move in to occupy the hole and the hole if "diatomic" will not be entirely free to "rotate". The next approximation which we shall not attempt to develop quantitatively will give a value of "hole-density" in the liquid less than that above and less by an amount which increases as the temperature falls. The liquid density ρ_l will be that much the greater, and a more accurate version of (2339) will be

$$\rho_v + \rho_l = 2\rho_c + \rho' \quad (\rho' > 0), \qquad \text{......(2340)}$$

where $\rho' \ll \rho_c$; ρ' increases as T diminishes, and is greater the more complex the molecule. The empirical law of the rectilinear diameter is that

$$\rho_v + \rho_l = 2\rho_c + \alpha + \beta(1 - T/T_c), \qquad \text{......(2341)}$$

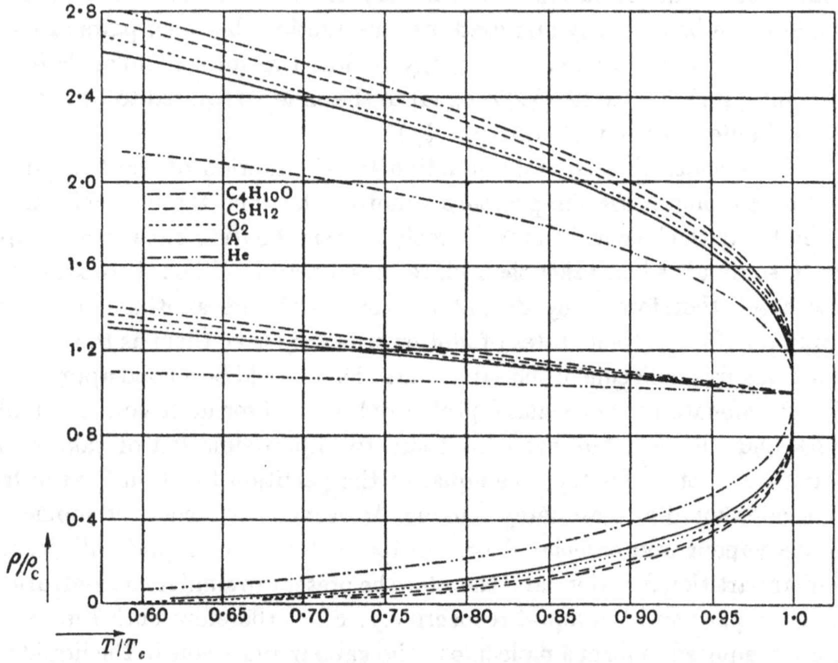

Fig. 99. Showing the empirical law of the rectilinear diameter for a number of substances. The ratios of the liquid and vapour densities and of their mean to the critical density are plotted as functions of T/T_c.

where T_c is the critical temperature and the correcting terms are both positive and small compared with $2\rho_c$ at least until T/T_c has fallen to 0·6. Equation (2340) is thus a long first step towards the successful explanation of (2341).

Fig. 99* shows the experimental results for a number of substances, which are just of the form which the theory requires. It will be observed that (2339) is most nearly obeyed by helium and then by argon, and that the β term is greater the more complex the molecule. Further study of liquids from this interesting starting point appears most likely to be profitable.

§21·8. ***An appendix to Chapter XVIII. An improved formulation of gaseous reaction rates, particularly bimolecular.*** It has recently been shown by Eyring

* E. Mathias, Onnes and Crommelin, *loc. cit.*

and his collaborators in a series of papers* that a formulation of reaction rates for bimolecular reactions can now be made, which depends on equilibrium theory to just the same extent as the formulation in terms of collisions given in Chapter XVIII, but is much more intimate and illuminating than the older formulation and has already almost superseded it. The importance of the new method appears to be great, and we shall devote the closing sections of this monograph to a short account of its foundations.

The ideas underlying the new formulation are as follows. The forces between the atoms of the molecules in reaction are due to the motion and distribution of their electrons and the charges on their own nuclei. These forces can be calculated by quantum mechanics and even in the most violent collisions, at least for many reactions, the relative motions of the nuclei are slow and the forces are those corresponding to the same static configuration of the nuclei. This means of course that we may then assume that the action is reversible† and that no electronic transitions occur with any appreciable probability. In the configuration space of the reacting complex we can therefore set up a potential energy function which defines the configurational energy of the complex at every stage of any collision and from which the forces acting may be derived by differentiation. In this potential energy field the relative motion of the various nuclei must be treated by quantum mechanics when necessary; there may be vibrations of too high a frequency for a classical approximation, but in general all tunnelling effects may be neglected and the treatment of the nuclear states of motion may at least be taken to be quasi-classical. It is however a definite assumption that the potential energy function is unique and that no electronic transitions occur. If any such transitions occur reaction rates of a different order of magnitude may be found. We shall not consider such reactions here. Since at least three atoms, say q, are concerned in any reaction, configuration space is of $3q$ dimensions. But since the position of the centre of gravity of the complex is irrelevant, and since the potential energy function is also independent of the rotation of the complex as a whole, the configuration space required for the representation of the potential energy function and the essential details

* See for example Eyring, *J. Chem. Physics*, vol. 3, p. 107 (1935); Hirschfelder, Eyring and Topley, *J. Chem. Physics*, vol. 4, p. 170 (1936), where many further references will be found. London, *Probleme der modernen Physik* (Sommerfeld Festschrift), p. 104, Leipzig (1928), first suggested that many reactions must proceed reversibly in the manner here investigated. Eyring and Polanyi, *Zeit. f. physikal. Chem.* B, vol. 12, p. 279 (1931), constructed the potential energy surfaces in configuration space for such reacting systems. Pelzer and Wigner, *Zeit. f. physikal. Chem.* B, vol. 15, p. 445 (1932), using the potential energy surface of Eyring and Polanyi, calculated the rate of $H + H_2 \rightarrow H_2 + H$. Wigner, *Zeit. f. physikal. Chem.* B, vol. 19, p. 203 (1932), formulated the quantum corrections required by the earlier classical calculations.

† That is "adiabatic" in Ehrenfest's sense.

of the relative motion reduces to $3q-6$ dimensions, or $3q-5$ if the complex is linear.

Let us now consider the nature of the potential energy function more closely. For certain configurations the potential energy will be low. These regions correspond to the separation of the reactants or resultants to great distances. These regions will be separated by a potential energy mountain range and the lowest pass in this range, owing to the factor $e^{-E_0/kT}$ affecting the probability of occurrence of any configuration, will be the route by which the reaction occurs. When the combined pair of reactants is in a configuration near the pass and in a state of motion in which its representative point in configuration space can cross the pass, the combined pair may be called *an activated complex*. The calculation of reaction rates by application of equilibrium theory proceeds by calculating the equilibrium number of activated complexes present in the gas at any temperature and the rate at which any activated complex gives rise to an effective reaction—the method will be explained more precisely in a moment. The whole method is therefore still based on the assumption that the reaction rate is not fast enough to upset the equilibrium calculation of the number of activated complexes. This necessary assumption naturally survives from the cruder collision theory.

The lowest pass separating the two regions in which the representative point must be, to represent the separated reactants or the separated resultants respectively, may be a simple pass, but more usually is not. It then leads first from the region of separated reactants to a limited high level basin in configuration space, points in which correspond to the formation of *the associated complex*. There must then be a second lowest pass in the rim of the high level basin leading from the basin to the region of separated resultants. When there is no high level basin, the rate of reaction is merely the rate at which activated complexes reach the pass in such a direction that (regarded as classical particles) they can pass over and through it. When there is a high level basin, then this rate is merely the rate of entry to the basin, and the rate of reaction is this rate of entry multiplied by the probability that the associated complex breaks down by exit over the second pass and not over the pass of entry.

It is now possible to give the equilibrium calculation of the number of activated complexes, and hence the rate at which the representative points of the complexes cross the pass of entry, for a volume V of the reacting gas in which there are N_1 and N_2 molecules of the two reactants respectively. Every pair of molecules (1,2) is a complex for which the complete partition function to a sufficient approximation is $f_1(T)\, f_2(T)$, the separate factors being the usual partition functions for the two reactants. There are $N_1 N_2$

such complexes in all. We now require the fraction of such complexes which are activated, that is to say the fraction in the neck of the pass whose position and momentum in a suitable coordinate corresponding to motion across the pass lie in the range dp^*dq^*, all other coordinates and momenta having any values whatever consistent with these. This fraction will be calculable if we can construct the partial partition function for such activated configurations. Let us now consider the nature of the potential energy function for the complex in this region. By definition of the pass the potential energy in this region will be a minimum for variations of any other configurational coordinate, except the special one corresponding to passage across the range for which it is a maximum. We shall assume this maximum to be so flat that motion across the pass is practically a free translation. We can therefore set up a partition function for the activated complex by integrating or summing over all types of motion of the complex in the other variables consistent with the specified values of p^* and q^*, thus constructing a quasi-partition function $f^*(T)$ in $3q-1$ configurational variables and their corresponding momenta and multiplying it by the partition function corresponding to the possession of the specified values of p^* and q^* themselves. This latter factor is of course

$$e^{-E^*/kT}\,dp^*\,dq^*/h,$$

where E^* is the energy in this coordinate which must satisfy $E^* \geqslant E_0$, E_0 being the height of the pass. The equilibrium number of these activated complexes is therefore

$$N_1N_2\frac{f^*(T)}{f_1(T)f_2(T)}\,e^{-E^*/kT}\,\frac{dp^*\,dq^*}{h}. \qquad \ldots\ldots(2342)$$

The energy zero for the calculation of $f^*(T)$ is now reckoned as at the top of the pass.

Representative points corresponding to these activated complexes cross the pass at a number per second obtained by replacing dq^* in (2342) by p^*/m^*; for if m^* is the effective mass of the complex for relative motion in this coordinate, p^*/m^* is then the velocity of approach of the representative point to the pass. The total number of crossings of the pass in unit time in volume V is therefore

$$N_1N_2\frac{f^*(T)}{f_1(T)f_2(T)}\int_{E_0}^{\infty} e^{-E^*/kT}\,\frac{p^*\,dp^*}{m^*h}.$$

If, as will usually be accurate enough, we put $E^* = E_0 + p^{*2}/2m^*$, then this rate of crossing reduces to

$$N_1N_2\frac{f^*(T)}{f_1(T)f_2(T)}\,\frac{kT}{h}\,e^{-E_0/kT}. \qquad \ldots\ldots(2343)$$

If the fraction κ of these crossings results in reaction, or in such a reaction as the system of measurement will detect, for example in the ortho → para hydrogen reaction, then the reaction rate or *number of reactions occurring in unit volume in unit time at molecular concentrations* n_1 *and* n_2 *is*

$$\kappa n_1 n_2 \left[\frac{Vf^*(T)}{f_1(T)f_2(T)}\right]\frac{kT}{h}e^{-E_0/kT}. \qquad \ldots\ldots(2344)$$

The partition function ratio in [ ] does not depend on V. In many simple reactions κ can be adequately estimated and the success of the method then depends merely on ability to calculate $f^*(T)$.

Without going into details for any particular reaction the general nature of the factors in $f^*(T)$ can be made evident at once. From the potential energy function in $3q-5$ or $3q-6$ variables, assumed known, the configuration of the atoms when the representative point is in the neck of the pass can be derived. This configuration being fixed, the complex as a whole is free to move in the volume V. Thus $f^*(T)$ contains the factor

$$\frac{(2\pi[m_1+m_2]\,kT)^{\frac{3}{2}}\,V}{h^3}.$$

The complex is also free to rotate as a whole in this configuration; if the configuration is linear, $f^*(T)$ contains the factor

$$8\pi^2A^*kT/\sigma^*h^2;$$

if the configuration is non-linear, then by (195) $f^*(T)$ contains the factor

$$\frac{8\pi^2(8\pi^3A^*B^*C^*k^3T^3)^{\frac{1}{2}}}{\sigma^*h^3}.$$

The coordinates p^*, q^* have already been allowed for. The configuration is stable in the remaining $3q-6$ or $3q-7$ freedoms, and they may be taken each to supply a normal mode of definite frequency with a corresponding Planck factor in $f^*(T)$ which can be deduced from the potential energy surface. In this way $f^*(T)$ may be completely constructed and the reaction rate (2344) calculated *a priori*, so far as equilibrium considerations allow.

§ 21·81. *The ortho→para transformation in the reaction* $H_2+H \rightarrow H+H_2$ *and the rates of the reaction* $H_2+D \rightarrow HD+H$ *and other reactions involving heavy hydrogen.* Hirschfelder, Eyring and Topley have shown that the various hydrogen reactions conform closely to the theory just developed and we shall describe their results shortly as the simplest illustration of the theory which has already been widely applied. All the reaction rates observed can be successfully interpreted in terms of a single semi-empirical potential energy surface. For the reactants H_2+H (or the corresponding sets in which any number of H's are replaced by D's) the activated complex is linear. The potential energy can therefore be represented in a configuration space of $3\times3-5=4$ variables, but the representation is simplified by the symmetry of the surface. If the distance apart of the H_2 nuclei is fixed at any value, then the equipotential surfaces are surfaces of revolution about the line of the H_2 nuclei. Fig. 100 shows a section of them when the H_2 nuclei

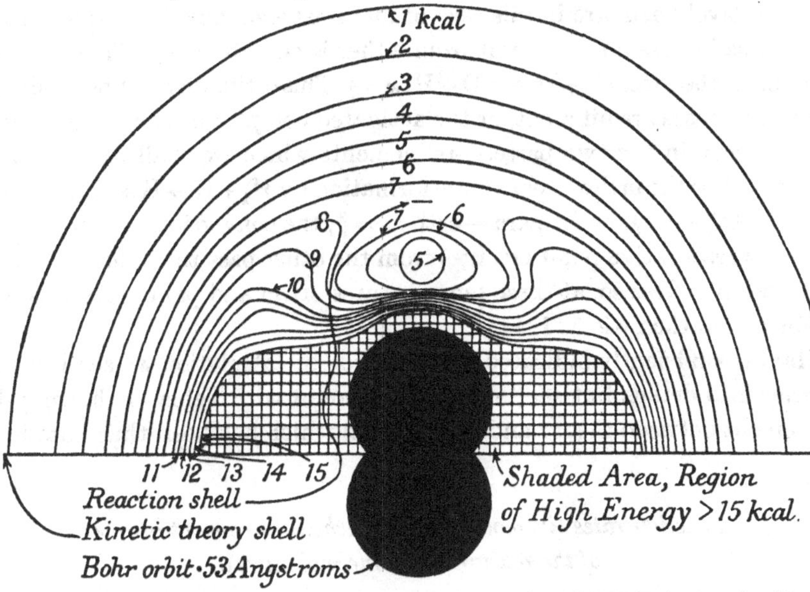

Fig. 100. Showing a set of equipotential surfaces for the three atoms in the complex $H_2 + H$ when the nuclear separation of the H_2 atoms is fixed at its normal value.

are fixed at their ordinary equilibrium separation. These curves combined with the curves of Fig. 101 showing the equipotential surfaces for the straight configuration and variable nuclear distances are sufficient to enable all the features essential to the calculation of reaction rates to be deduced.

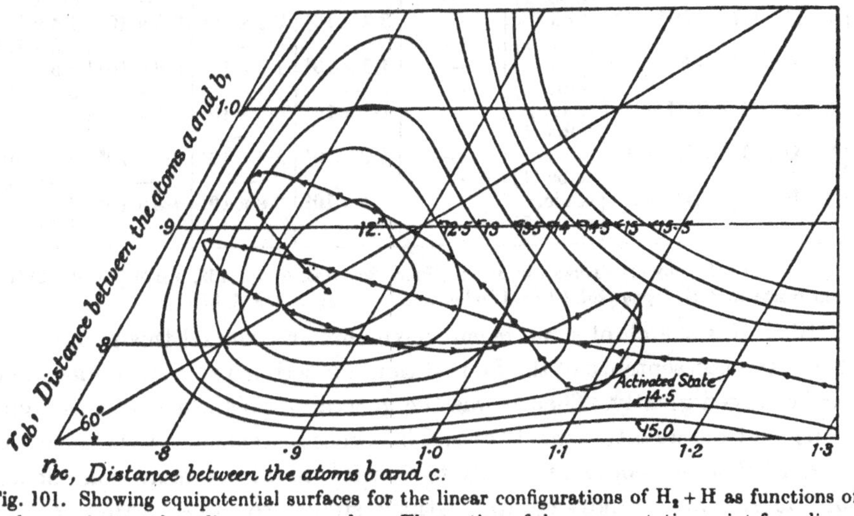

Fig. 101. Showing equipotential surfaces for the linear configurations of $H_2 + H$ as functions of the two internuclear distances r_{ab} and r_{bc}. The motion of the representative point for a linear vibrational disturbance of the complex is shown. [Energies in *k* cal., distances in 10^{-8} cm.]

The trajectory shown in Fig. 101 is of some importance. Its considerable complication, combined with the fact that the passes of entry and exit to the high level basin are in this case of identical level and form, shows that we may safely assume that exit from either is equally likely. Thus for any reaction of the form $H_2+D \rightarrow HD+H$, $\kappa=\frac{1}{2}$. This estimate must be modified when the passes are different, or the activated complex has a different zero-point energy in the two passes—refinements which we shall not pause to describe. For the ortho → para transformation by $H_2+H \rightarrow H+H_2$ however we must take $\kappa=\frac{1}{8}$, or for para → ortho $\kappa=\frac{3}{8}$; the conversion can only occur if the representative point emerges from the other pass than that of entry, and even if it does so, at these temperatures ortho and para molecules will be formed in the ratio 3 : 1.

Having sufficiently indicated the methods of calculation we shall not go further into details but shall close this account with Table 74 showing the agreement between theory and observation which has been thus obtained.

TABLE 74.

Reaction rates (in moles/litre/sec. when the concentrations of the reactants are one mole/litre).

Reaction	T° K.	283°	300°	600°	900°	1000°
1. $H+H_2 \rightarrow H_2+H$*	(Calc.)	—	$7·3\times10^4$	$7·3\times10^7$	$9·2\times10^8$	$1·5\times10^9$
	(Obs. 1)	—	—	—	$1·5\times10^9$	$2·2\times10^9$
	(Obs. 2)	$8·5\times10^4$	—	—	—	—
2. $D+D_2 \rightarrow D_2+D$*	(Calc.)	—	$3·0\times10^4$	$3·5\times10^7$	$4·4\times10^8$	$7·6\times10^8$
	(Obs. 1)	—	—	—	—	$1·2\times10^9$
3. $H+DH \rightarrow HD+H$	(Calc.)	—	$2·2\times10^4$	$2·6\times10^7$	$3·2\times10^8$	$5·2\times10^8$
	(Obs. 1)	—	—	—	—	$6·8\times10^8$
4. $D+HD \rightarrow DH+D$	(Calc.)	—	$2·4\times10^4$	$2·4\times10^7$	$2·6\times10^8$	$4·4\times10^8$
	(Obs. 1)	—	—	—	—	$1·0\times10^9$
5. $H+HD \rightarrow H_2+D$	(Calc.)	—	$1·1\times10^4$	$1·8\times10^7$	$2·6\times10^8$	$4·5\times10^8$
	(Obs. 1)	—	—	—	—	$9·5\times10^8$
6. $D+H_2 \rightarrow DH+H$	(Calc.)	—	$7·1\times10^4$	$6·2\times10^7$	$7·1\times10^8$	$1·2\times10^9$
	(Obs. 1)	—	—	—	—	$2·5\times10^9$
7. $D+DH \rightarrow D_2+H$	(Calc.)	—	$3·0\times10^4$	$2·5\times10^7$	$3·0\times10^8$	$5·0\times10^8$
	(Obs. 1)	—	—	—	—	$7·9\times10^8$
8. $H+D_2 \rightarrow HD+D$	(Calc.)	—	$1·5\times10^4$	$2·8\times10^7$	$4·3\times10^8$	$7·4\times10^8$
	(Obs. 1)	—	—	—	—	$1·2\times10^9$

Observed values by (1) Farkas and Farkas, *Proc. Roy. Soc.* A, vol. 152, p. 124 (1935). (2) Geib and Harteck, *Zeit. f. physikal. Chem.* (Bodenstein volume), p. 849 (1931).

With this sketch of some recent developments we end this monograph. Since the last sentence of the first edition was written statistical mechanics has changed greatly. Then it was the general theory which was emerging clarified from quantum mechanics. This process seems now to be complete, and further developments likely to consist mainly of more complicated and more widely ramifying applications, similar to those treated here.

* Complete reaction rates derived from the measured rates of the para → ortho change for H and the ortho → para for D.

INDEX OF AUTHORS QUOTED

Adams, 549, 560
Akulov, 516
Anderson, 364
Angerer and Müller, 334
Appell, 314
Appleyard, see Skinner and A.
Arrhenius, 293

Backer and Goudsmit, 601
Barker, 455; see Plyler and B.
Barnes and Silverman, 775, 785
Barnett, 484
Bartholomé, see Clusius and B.
Becker, 516, 689, 724
Benedicks, 495
Bernal and Fowler, 452, 540
Bernal and Tamm, 522
Bethe, 6, 397, 789, 797; see Sommerfeld and B.
Bidwell, 416
Birkhoff, 712
Birtwistle, 280
Bitter, 513
Bjerrum, 536, 541, 546, 552
Blackett, Henry and Rideal, 93
Blackman, 118, 120, 137
Bloch, 338, 496, 502
Bloch and Gentile, 503
Blüh and Stark, 335
Bodenstein and Lindner, 707
Bohr, 19, 196, 404, 472, 690
Boltzmann, 15, 189, 284, 666
Bonhöffer and Harteck, 86
Born, 19, 74, 118, 119, 127, 141, 148, 187, 188, 313–320, 331–334
Born and Brody, 314
Born and Fock, 74
Born and Göppert-Mayer, 118, 119
Born and Mayer, 295, 319, 328, 329
Bose, 43, 467
Bothe, 672
Bragg and Chapman, 332
Bragg and Williams, 6, 789, 808
Bridgman, 418
Brillouin, 338, 382, 468
Brody, see Born and B.
Brönsted, 559, 560
Bronstein, 387
Brown, 223
de Bruyne, 356
Bryan, 194
Buckingham, 292
Buisson, see Fabry and B.
Burger, see Ornstein and B.
Burgers, 19
Burkill, 198
Burns, see Curtis and B.

Campbell, 778
Cardwell, 359
Cario, 684, see Franck and C.
Cario and Franck, 684
Chandrasekhar, 584, 636, 652; see Milne and C.
Chapman (A. T.), see Johnston (H. L.) and C.
Chapman (S)., 309, 669; see Bragg and C.; see Topping and C.
Chapman (S.), and Milne, 627
Chapman (S.), Topping and Morrell, 332
Christiansen and Kramers, 718
Clark, see Keesom and C.
Clark and Keesom, 842
Clausius, 286
Clayton and Giauque, 234
Clusius, 218, 320
Clusius and Bartholomé, 87
Cockcroft, 838
Colby, 89, 495
Coleman and Egerton, 218
Compton (A. H.), 730
Compton (K. T.), 451
Compton (K. T.) and Langmuir, 346
Condon and Morse, 9
Constable, 377
Cook, see Hassé and C.; see Lennard-Jones and C.
Courant, 54, 114
Courant and Hilbert, 550
Cox, 217
Crenshaw and Ritter, 814
Crommelin, see Mathias, O. and C.
Curtis and Burns, 632
Cuthbertson (C. and M.), 305
Czerlinsky, see Gans and C.
Czerny, 223

Daily, see Mott-Smith and D.
Dalton, 64
Darwin, 1, 437, 444, 447, 473, 677
Darwin and Fowler, 743
Davidson, 632
Davison and Germer, 354
Debye, 112, 118, 299, 437, 448, 450, 453, 460, 469, 730, 839
Debye and Hückel, 269, 541
Dennison, 83, 96
Dent, see Lennard-Jones and D.
Dickinson, see Tolman, Y. and D.
Dieke, 84
Dirac, 9, 19, 25, 42, 696, 697, 721, 727
Donat, 684
Drude, 338
Du Bridge, 353, 361
Du Bridge and Roehr, 361
Dushman, 345, 352

Eddington, 6, 576, 585, 591, 641, 646, 652, 722
Egerton, 218, see Coleman and E.
Ehrenfest, 19, 196, 200
Ehrenfest and Trkal, 89, 151, 192, 205
Einstein, 43, 118, 207, 720, 728, 753, 766, 769
Elliott, 223
Ellis and Kneser, 106
Emde, see Jahnke and E.
Emden, 265, 585
Emersleben, 317
Enskog, 309, 669
Epstein, 496, 499
Eucken, 77, 127, 218, 223, 225, 230, 279
Eucken and Hiller, 86
Eucken and d'Or, 103
Eucken and Parts, 100, 101
Ewald, 315
Ewald and Hermann, 329
Eyring, 844, 847; see Hirschfelder, E. and T.; see Millikan and E.
Eyring and Polanyi, 847

Fabry and Buisson, 632
Fajans, 333, 334
Falkenhagen, 269
Farkas, 87, 168, 852
Fermi, 42, 259, 574
Fock, see Born and F.
Fokker, 734
Försterling, 129
Forsyth, 14
Fowler, see Bernal and F.; see Darwin and F.
Fowler, Gossling and Sterne, 357
Fowler and Guggenheim, 639
Fowler and Lock, 712
Fowler and Milne, 595, 598, 606
Fowler and Nordheim, 356
Fowler and Sterne, 169
Franck, 334, 685, see Cario and F.
Franck and Cario, 685
Frank and Sommerfeld, 418, 422
Frenkel, 9, 471, 834
Frenkel, Todes and Ismailow, 810
Fröhlich, 359
Fry, 783
Fues, 92
Fürth, 783
Fuoss, 555

Gans and Czerlinsky, 508, 510, 511
Gatty, 546
Gaunt, 576, 614, 774
Geib and Harteck, 852
Gentile, see Bloch and G.
Germer, see Davison and G.
Giauque, 234, 235, 469, 839; see Clayton and G.; see Johnston (H. L.) and G.
Giauque and Johnston, 224, 234
Giauque and MacDougall, 470, 839
Giauque and Wiebe, 234
Gibbs, 7, 65, 186, 187, 195, 743, 750, 756
Glaser, 462
Glasoe, 364
Goens, see Gruneisen and G.
Göppert-Mayer, see Born and G.-M.
Gorsky, 789
Gorter, see de Haas and G.
Gossling, see Fowler, G. and S.
Goudsmit, see Backer and G.; see Uhlenbeck and G.
Griffiths, see Sheratt and G.
Gronwall, 550
Gronwall, La Mer and Sandved, 550, 557
Gross and Halpern, 546
Gruneisen and Goens, 149
Guggenheim, 481, 522, 524, 527, 533–537, 560; see Fowler and G.

de Haas and Gorter, 844
de Haas, Wiersma and Kramers, 470, 839
Haber, 333
Hall and Harkins, 549, 560
Halpern, see Gross and H.
Hammer, 462
Harkins, see Hall and H.
Harteck, 218; see Bonhöffer and H.; see Geib and H.
Hartree, 437, 571, 638
Hassé and Cook, 310
Hebb, see Van Vleck, H. and P.
Heckmann, 148
Heisenberg, 25, 26, 84, 484, 487, 488, 516, 724
Heitler, 654
Heitler and London, 292, 486, 496
Hellman, 305
von Helmholtz, 147
Helmholtz (L.), see Mayer and H.
Helmholtz (L.) and Mayer, 334
Henglein, 325
Henning, 279
Henry (D. C.), 831
Henry (P. S. H.), 93, 94, 97, 99; see Blackett, H. and R.
Hermann, see Ewald and H.
Herzberg, 106
Herzfeld, 1, 294
Herzfeld and Rice, 93
Heuse, 100
Heydweiller, 322
Hilbert, see Courant and H.
Hildebrand, 324, 533, 535
Hiller, see Eucken and H.
Hinshelwood, 700, 701, 704, 706
Hirschfelder, Eyring and Topley, 847
Hobson, 737
Holborn and Otto, 282, 297, 303, 308
Honda, 493, 513
Honda and Kaya, 509
Hori, 84
Hovorka and Rodebush, 549, 560
Huber, 460
Hückel, see Debye and H.
Huggins and Mayer, 329
Hull and Williams, 785

Hund, 26, 84, 94, 317, 465

Imes, 89
Ingham, see Lennard-Jones and I.
Ising, 801
Ismailow, see Frenkel, T. and I.
Ittmann, see Kramers and I.

Jahnke and Emde, 315
Jeans, 11, 57, 77, 112, 241, 275, 278, 284, 294, 298, 584, 585, 620, 704
Jeffreys, 53, 68
Joffé, 419, 436
Johnson (J. B.), 785
Johnston (H. L.), see Giauque and J.
Johnston (H. L.) and Chapman, 104
Johnston (H. L.) and Giauque, 225, 234
Johnston (H. L.) and Walker, 104, 106
Jones (H.), 382; see Mott and J.
Jones (H.) and Zener, 382
Jones (J. E.) (≡ Lennard-Jones), 669
Jordan, 672
Jüttner, 650

Kamerlingh-Onnes, 297, 303; see Woltjer and O.; see Mathias, O. and C.
Kamerlingh-Onnes and Keesom, 275, 277, 303
Kamerlingh-Onnes and Weber, 311
Kappler, 782
Kassel, 82, 98, 99
Kaya, 508, 511, 521; see Honda and K.
Keesom, 277, 294, 299, 768; see Clark and K.; see Kamerlingh-Onnes and K.
Keesom and Clark, 391
Keesom and Koks, 344
Kemble, 83
Kemble and Van Vleck, 91
Keyes, see Kirkwood and K.
Kikuchi, see Nordheim and K.
Kingdon, see Langmuir and K.
Kirkwood, 305, 321; see Slater and K.
Kirkwood and Keyes, 292
Klein and Rosseland, 4, 677
Kneser, see Ellis and K.
Knudsen, 698
Kobeko and Kurtschatow, 816
Koks, see Keesom and K.
Kossel, 293, 312
Kramers, 42, 68, 274, 724; see Christiansen and K.; see de Haas, W. and K.; see Ornstein and K.
Kramers and Ittmann, 42
Kronig, 437, 455, 457
Kronig and Penney, 382
Kürti and Simon, 470, 839
Kurtschatow, 816; see Kobeko and K.

La Mer, see Gronwall, La M. and S.
Ladenburg and Thiele, 218
Landau, 473
Lange and Robinson, 547
Lange and Simon, 218
Langevin, 449
Langmuir, 698, 826, 833, 834; see Compton (K. T.) and L.
Langmuir and Kingdon, 354, 364, 372, 375, 377
von Laue, 366, 370, 375
Lauritsen, see Millikan and L.
Lehrer, 462
Lennard-Jones, 5, 292, 295, 299, 309, 620
Lennard-Jones and Cook, 283, 308
Lennard-Jones and Dent, 332, 335
Lennard-Jones and Ingham, 318
Lennard-Jones and Taylor, 331, 332
Lewis (G. N.), 207, 234, 312, 719
Lewis (G. N.) and Randall, 228, 229
Lewis (G. N.) and Smith, 719
Lewis (G. N.) and Von Elbe, 105
Lewis (W. C. McC.), 77, 280
Lindemann, 711
Lindner, see Bodenstein and L.
Livens, 437
Lock, see Fowler and L.
London, 296, 847; see Heitler and L.
Lorentz, 404, 437, 440, 740, 765
Love, 127

McCrea, 65, 91, 97, 611, 636, 637
MacDougall, see Giauque and M.
MacGillavry, 234
Madelung, 314
Mahajani, 503
Manneback, 455
Massey and Mohr, 310, 311
Mathias, Onnes and Crommelin, 844, 846
Maxwell, 8, 308
Mayer, 329; see Born and M.; see Helmholtz (L.) and M.; see Huggins and M.
Mayer and Helmholtz, 329, 334
Mecke, 99, 223
Mendelssohn, see Simon, M. and R.
Mensing and Pauli, 455
Menzel, 597, 636
Millikan, 698
Millikan and Eyring, 357
Millikan and Lauritsen, 357
Milne, 6, 186, 265, 286, 585, 595, 596, 611, 620, 625, 630, 631, 635, 652, 724, 725, 733, 734; see Chapman (S.) and M.; see Fowler and M.
Milne and Chandrasekhar, 611
Milner, 274
Minnaert, see Pannekock and M.
Mitchell (A.), 827
Mitchell (K.), 359
Mitchell (S. A.), 630
Mohr, see Massey and M.
Molk, see Tannery and M.
Morrell, see Chapman (S.), T. and M.
Morris, 359
Morse, see Condon and M.
Mott, 67, 393, 396, 524
Mott and Jones, 397
Mott-Smith and Daily, 459

Moullin, 785
Mueller, 816, 822
Mulholland, 80
Müller, 550; see Angerer and M.

Nernst, 4, 228
von Neumann, 7
Noddack, 742
Nordheim, 338, 349, 350, 356, 496, 672; see Fowler and N.
Nordheim and Kikuchi, 672, 677
Nottingham, 347, 352

Onsager, 269
Oppenheimer, 727
d'Or, see Eucken and d'O.
Ornstein, see Uhlenbeck and O.
Ornstein and Burger, 783
Ornstein and Kramers, 672
Ornstein and van Wijk, 785
Otto, see Holborn and O.

Pannekock and Minnaert, 636
Partington and Shilling, 77, 91, 92, 94, 275, 278
Parts, see Eucken and P.
Pauli, 28, 456, 471, 727, 753; see Mensing and P.
Pauling, 222, 322, 810
Payne, 594, 597
Peierls, 424, 809, 834
Pelzer and Wigner, 847
Penney, 99; see Kronig and P.
Perrin, 765, 768, 769, 772
Planck, 1, 77, 192, 200, 201, 204, 257, 280, 284, 572, 734
Plaskett, 595, 632
Plyler and Barker, 99
Poincaré, 200
Polanyi, see Eyring and P.
Pólya, 762
Powell, 503, 510, 516
Purcell, see Van Vleck, H. and P.

Randall, 560; see Lewis (G. N.) and R.
Rankine, 294
Rasetti, 223
Rayleigh, 289, 767
Reimann, 346, 352, 364
Richardson, 345, 346, 357, 366, 370, 417
Rice, see Herzfeld and R.
Rideal, see Blackett, H. and R.
Riemann, 314
Ritter, see Crenshaw and R.
Robinson, see Lange and R.
Rodebush, see Hovorka and R.
Roehr, see Du Bridge and R.
Rosseland, 585, 638; see Klein and R.
Rowland, 779
Ruhemann, see Simon, M. and R.
Rupp, 354
Russell, 594, 615

Saha, 594, 595
Salow and Steiner, 106
Sandved, see Gronwall, La M. and S.
Scatchard, 535, 546, 560
Schlesinger, 712
Schottky, 176, 345, 355, 366, 417, 436, 783
Schottky and Waibel, 401
Schrödinger, 17, 125, 724
Schubin, see Tamm and S.
Schultze, see Villars and S.
Schuster, 611
Seitz, 333; see Wigner and S.
Semenoff, 718
Sheratt and Griffiths, 94
Shilling, see Partington and S.
Sidgwick, 312
Silverman, see Barnes and S.
Simon, see Kürti and S.; see Lange and S.
Simon, Mendelssohn and Ruhemann, 222
Simon and Ruhemann, 814
Simon and Swain, 131
Skinner and Appleyard, 679
Slater, 322, 333, 397
Slater and Kirkwood, 292, 305
Smekal, 1
Smith, see Lewis (G. N.) and S.
von Smoluchowski, 766
Snow, 99, 223
Sommerfeld, 72, 338, 404, 416, 467; see Frank and S.
Sommerfeld and Bethe, 338, 397, 404, 406, 409, 416, 496, 526
Spangenberg, 322
Sponer, 95, 97, 99
Stark, see Blüh and S.
Steiner, see Salow and S.
Sterne, 26, 225, 655; see Fowler and S.; see Fowler, G. and S.
Stewart, 612
Stoner, 467, 652
Strömgren, 644
Sucksmith, 484
Sugiura, 724, 727
Sutherland, 77, 99, 103
Swain, see Simon and S.
Sykes, 808

Tait, 621
Tamm, see Bernal and T.
Tamm and Schubin, 359
Tannery and Molk, 54, 81
Taylor, 635; see Lennard-Jones and T.
Teal, see Urey and T.
Teller, 844
Teller and Topley, 101
Thiele, see Ladenburg and T.
Thomson, 690
Todes, see Frenkel, T. and I.
Tolman, Yost and Dickinson, 717
Topley, see Hirschfelder, E. and T.; see Teller and T.
Topping, see Chapman (S.), T. and M.
Topping and Chapman, 332

Trkal, see Ehrenfest and T.
Turner, 220

Uehling, 411
Uehling and Uhlenbeck, 672
Uhlenbeck, 107; see Uehling and U.
Uhlenbeck and Goudsmit, 775, 781
Uhlenbeck and Ornstein, 785
Unsöld, 612, 633
Urey, 259, 260, 574
Urey and Teal, 168
Ursell, 26, 241, 245, 294

Van Velzer, 353
Van Vleck, 19, 447, 449, 455, 458–467, 488, 721, 738; see Kemble and V. V.
Van Vleck, Hebb and Purcell, 839, 840
Verschoyle, 282, 307
Villars and Schultze, 94
Viney, 82
Von Elbe, see Lewis (G. N.) and V. E.

Waibel, see Schottky and W.
Walker, see Johnston (H. L.) and W.
Waller, 841
Wasastjerna, 320–322
Watson, 315, 650; see Whittaker and W.
Weatherby and Wolf, 459
Weber, see Kamerlingh-Onnes and W.
Webster, 509, 511, 518
Weiss, 479, 483, 493, 501
Wentzel, 359
Weyl, 54, 114
White, 101
Whittaker and Watson, 708, 709
Wiebe, see Giauque and W.
Wiersma, see de Haas, W. and K.
Wigner, 26, 847; see Pelzer and W.
Wigner and Seitz, 333
van Wijk, see Ornstein and v. W.
Williams (E. J.), 789, 793, 805; see Bragg and W.
Williams (N. H.), see Hull and W.
Wilson (A. H.), 398
Wilson (E. B.), 94
Wilson (H. A.), 345
Winch, 359
Wolf, see Weatherby and W.
Woltjer and Onnes, 469

Yost, see Tolman, Y. and D.

Zahn, 450
Zeidler, 218
Zener, 93, 381, 389; see Jones (H.) and Z.
Zwicky, 299

INDEX OF SUBJECTS

Initial capitals mark words used in ordering the entries, or headings of alternative entries in the index.

Absorption Spectra, Stellar, 593–619
Abundance lines for Nuclei of even mass number, 657
Accessible states, Defined, 9
  Enumerated, 23, 29
Activated complexes, number of, for Bimolecular Reactions, 849–850
Adiabatic processes, 74
Adsorbed films on Surfaces, 825–838
Adsorption isotherms, Atomic, from diatomic gases, 830–831
  of Competing molecules, 831–832
  Critical, 833–838
  Generalizations, 832–833
  Langmuir's, 828–831
  Molecular, 828–830
Antisymmetrical states, 27, 42
Arrhenius' equation for Reaction rates, 702
Assemblies, of Classical systems, 56–66
  Defined, 8
  of Dissociating gases, 157–168
  of Electrons and positive ions, 574–583
  of Electrons with Relativistic energies, 648–650
  Evaporating, 168–172
  General, of Crystals and vapours, 173–174
  of Harmonic oscillators, 30–35
  with Nuclear transformations, 655–657
  with Positive electrons, 653–654
Atmospheres, of Brownian particles, 770
  Free paths in upper, 625–626
  Ionized, equilibrium of, 585–593
  Planetary, Escape of molecules from, 620–630
Atomic Ions, Overlap and van der Waals energies of, in Crystals (*q.v.*), 320–328
  Paramagnetism of (*q.v.*), 464–469, 840–841
  Partition functions for, 561–562
  Structure of, described, 562–571
  Weights of states of, 569

Berthelot (D.), Equation of state of, 276
Bimolecular Reactions, Activated complex for, 848
  in Gases, 703–706
  Quantum theory of, 847–852
Binary mixtures, of Imperfect gases, 281–283
  Virial coefficient (second) for, 306–308
Boltzmann's Constant, 189
  Distribution law, 63
  Equation, 669
  H-theorem, 666–669
  H-theorem, general form of, 670–671
  H-theorem, quantum form of, 675–677
  Hypothesis, 200–201, 205–206
Boundary density in Imperfect gases, 291
Brownian movements, 769–775
  Distribution laws, general, for, 785–788
  Random displacements of any type, 774–775
  Rotations of mirror in a gas, 779–783

Campbell's theorem on Fluctuations, 778–779
Characteristic Function, for Assemblies of Electrons and positive ions, 574–583
  for Imperfect gases, 257–259, 283–284
  Method of excluded volumes, 574–579
  Method of Planck, 579–583
  Planck's, defined, 192
Chemical constant, defined, 209
Chemiluminescence, 741–742
Chromosphere, Calcium, 630–637
  Fully supported by radiation, 632–635
Classical Rotations, 62–63
Classical Statistics, 43
  Approximated to, 55
Classical Systems, Assemblies of, 56–66
Collisions, Classical, number of given type in gases, 663–665
  Classical, preserving equilibrium, 659
  First and second kind of, 677–682
  of Gas molecules with Surfaces, 697–699
  General 2- and 3-body Dissociating, 691–695
  Inelastic, applications of, 684–685
  Inelastic and superelastic, for Electrons and atoms, 677–682
  Inelastic and superelastic, for heavy particles, 682–684
  Ionization by electron, 685–690
  Quantum reformulation of gas, 672–674
  Target areas for Ionization, 690–691
  Transferable Energy in, 707–710
Complexions, Defined, 9
  Enumerated, 23
  Enumerated for Assemblies of complex systems, 153–157
  Total number of, for Classical systems, 47
  Total number of, evaluated, 37
Compressibilities, of Crystalline salts calculated, 328–332
  of Isotropic solids, 140
Compton effect, 730–732
Condensation of cadmium on copper, 837–838
Configurational Partition function, Bethe's approximation, 797–806
  of Bragg and Williams, 791–797
  Specific heat, 805–809
Contact potential, of Metals, 362–364
  of Semi-conductors, 401–403
Contact transformations, 15
Contacts, Metal, 429–432
  Metal Semi-conductor, current-voltage relation for, 432–433

Contacts, Metal Semi-conductor, of high transparency, 433–434
Metal Semi-conductor, of low transparency, 435
Strongly rectifying, 429–436
Continuity of path, hypothesis of, 8
Continuum, Normal modes for, 112–115
Cooling, by adiabatic Demagnetization, 470–471, 841–844
Cooperative phenómena, 789–838
Correspondence principle, 19
Critical point, for Imperfect gases, 278–280
Crystalline salts, properties calculated, 328–332
Crystals, Cubic, Potential energy constants for, 319
Easy directions of Magnetization for, 502
Entropy of, 191–192
External reactions of, 138
Ferromagnetic, Magnetization curves for, 511–516
General, Equations of state for, 141–150
Free energy of, 147
Homogeneous displacements in, 143
Mixed, 176–182
Order of neighbours in, 800–801
Overlap and van der Waals energies for Atomic ions in (*q.v.*), 320–328
Partition functions for, 118–150
Potential energy per cell for (*q.v.*), 312–319
Specific heat of (*q.v.*)
Strongly anisotropic, properties of, 149–150
Surface forces outside, 335–337
Vapour pressure of, 172–173
Zero-point Energy of, 123
Curie point, Defined, 478
Phenomena of, 491–496

Dalton's Distribution law, 64
Degeneracy, Relativistic, 651–652
Degenerate Assemblies of Electrons, 71–73
Degenerate matter, Equation of state of, 650–652
Degenerate systems, 45
Distribution laws for, 45–46
Demagnetization, Cooling by adiabatic, 470–471, 838–844
Density, great, of Stellar material, 647–648
Detailed balancing, 659–660
in General Collisions, 696–697
Requirements of, in gas Reactions, 716–719
Diamagnetism, Absence of, for classical free Electrons, 472
of Electron gas (quantum theory), 473–475
Dielectric constant, Classical theory of, for Gases, 447–451
General theory of, for gas of complex molecules, 458–459
Isotropic property of, 456–457, 460–461
for Librating-rotating dipoles, 817–822
for Rigid rotators, 455–456
of Solids and liquids with polar molecules, 816–825
Dielectric constant, for Symmetrical top-like molecules, 457–458
Thermodynamic theory of, 453–454
Dieterici's Equation of state, 276–277
Diffusion, of Brownian particles, 770–771
Dipole moments of molecules, 451
Dipole and quadripole Energy of Magnetization, 503–504
Dissociation, by Collisions of heavy particles, 691–695
Fluctuations in, 758–760
Dissociative equilibrium, in External fields, 185–186
in Imperfect gases, 255–260
in Magnetic fields, 475–477
in Perfect gases, 157–168
Distribution laws, Boltzmann's, 63
Boltzmann's, for free Electrons, 64–65
Dalton's, 64
for Degenerate systems, 45–46
for Electrons, free, in Metals, 340–343
for Electrons in overlapping bands, 394–396
for Electrons, free, in Semi-conductors, 397–401
in External fields, 67–69
Fluctuations in, 751–755
for Harmonic oscillators, 33–35
Maxwell's (*q.v.*), 3, 58
Maxwell's, with mass motion, 58–60
Molecular, for Imperfect gases, 253–255
for Quantum statistics, 44–45
Dulong and Petit's law for Solids, 124

Einstein-Bose Statistics, 43
Einstein's law of Photochemical equivalence, 742
Electrical conduction in Metals, Change of, on Melting, 526–527
Formal theory of, 404–418
in Semi-conductors, 418–421
Electrolytes, strong, see Strong electrolytes
Electromagnetic theory of Susceptibilities (*q.v.*), 437–447
Electrons, Assemblies of, with Relativistic energies, 648–650
Atmospheres of, 364–378
Degenerate Assemblies of, 71–73
Distribution laws for, in overlapping bands, 394–396
Distribution laws for, in Semi-conductors, 397–401
Emission of, by cold metals, 356–357
Emission of, cooling effect of, 357–358
Emission of, the Schottky effect, 355–356
Free metallic, defined, 378–381
Free metallic, Distribution laws for, 340–343
Para- and Dia-magnetism of, 471–475
Partition function for, near nucleus, 572–574
with Positive ions, Characteristic function for Assemblies of, 574–583
Specific heat of, in Metals, 343–344
Specific heat of, in Nickel, 391–397

Electrons, States of, in Periodic fields, 379–388
  States of, in Periodic fields; simple bands, 383–388
  Theory of Metals, 338–436
  Thermionic emission of, by Metals, 347–358
  Thermionic emission of, by tungsten, 351
  Vapour pressure of, for Metals, 344–346
Elements of even mass number, Abundance lines for, 657
Emission and absorption of Radiation by fixed atoms, 720–724
Emission of Positive ions by Metals, 370–373
Energy, of Crystalline salts, calculated, 328–332
  of Crystalline salts, from Born's cycle, 333–335
  Fluctuations of, 745–750
  Fluctuations of, in Dissociating assemblies, 760–761
  Interatomic (*q.v.*)
  of Magnetization, Dipole and quadripole, 503–504
  Transferable, in Collisions, 707–710
  Zero-point, of Crystals, 123
Ensemble, Gibbs', 7, 10
Entropy, Absolute, 204, 231
  Change of, in Reactions at zero temperature, 233
  Comparisons of, observed and calculated, 234–235
  Contributions, 191–192
  Defined, 188–189
  Increasing property of, 193–194
  Limit of, for zero temperature, 231–233
  of Magnetization, 469–470
  of Radiation, 116
  Statistical definition of, 203–205
  and Thermodynamic probability, 200–206
Equation of Mass action, 160
Equation of Mean values, 669–670
Equation of state, of Degenerate matter, 650–652
  for General Crystals, 141–150
  for Imperfect gases, representation of isothermals, 297–298
  for Isotropic Solids, 139
  for Monolayers, 826–828
  Reduced, 278–280
  with Relativistic Degeneracy, 650–652
  for Simple gases, Berthelot's, 276
  for Simple gases, Dieterici's, 276–277
  for Simple gases, van der Waals', 274–275
Equilibrium, Independence of mechanisms of, 3
  State, nature of, 658
Equipartition theorem, 60–62
Escape of molecules from Atmospheres, 620–630
Ettinghausen effect, defined, 422
Evaporation, Assemblies with, 168–172
Exclusion principle, 28
External fields, Distribution laws in, 67–71
External reactions, 73–76
  of Crystals, 138
  Fluctuations in, 755–757

Fermi-Dirac Statistics, 42
Ferromagnetics, Block structure of, 501–503
  Specific heat of, 492–494
  Susceptibility of, above Curie point, 480
Ferromagnetism, 477–521
  Bloch's approximation for, 496–501
  Heisenberg's theory of, 484–494
Fluctuations, 109, 741–789
  Defined, 10
  in Dissociation, 758–760
  in Distribution laws, 751–755
  in Energy, 745–750
  in Energy, in Dissociating assemblies, 760–761
  in External reactions, 755–757
  Formal consequences of, 762–764
  in Momentum transfer, 775–783
  Simple special cases, 764–765
Forces of long range, in Imperfect gases, 260–266
Forces of short range, in Imperfect gases, 240–245
Forces, Surface, outside a Crystal, 335–337
Free energy, of Crystal, 147
  of Intermolecular forces, 267–269
Free path, mean, Defined, 622
  of Electrons in Metals, use of, 406–408
  in Uniform or non-uniform gas, 621–623
  in Upper Atmosphere, 625–626

γ-Phases, Hume-Rothery's rule for, 382
Gases, Dielectric constants, classical, for, 447–451
  Dipole moments of molecules in, 451
  Entropy of, 191
  Imperfect (*q.v.*), 236–291
  Paramagnetism of, 463–464
  Perfect (*q.v.*), 77–94
  Reactions, chemical, in (*q.v.*), 700–719
  Scattering of light by, 765–769
Generating function, defined, 43
Gibbs' Ensemble, 7, 10
  Phase integral, 65

H-theorem, Boltzmann's (*q.v.*), 666–669
Hall effect, Defined, 422–423
  in Semi-conductors, composite, 428–429
  Statistical theory of, 423–424
Harmonic oscillators, Assemblies of, 30–35
  Partition functions for, 40
  Schrödinger's equation for, 18, 20
Henry's law, for ideal Solutions, 527
Holomagnetization, Defined, 501–503
  Partition function for growth of, 504–507
Homogeneous displacement in general Crystals, 143
Homonuclear molecules, defined, 84
Hume-Rothery's rule for γ-phases, 382

Hydration of Ions, 539–540
Hysteresis loop, for Ferromagnetics, 477

Image fields, Effects of, 375–378
 of Metals for Electrons, 350
Imperfect gases, Binary mixtures of, 281–283
 Boundary density in, 291
 Characteristic function for, 283–286
 Critical point for, 278–280
 Dissociative equilibrium in, 255–260
 Distribution law, molecular, for, 253–255
 Equations of state, simple, for, 275–280
 Long range forces in, 260–266
 Partition function for Potential energy in, 237–240
  First approximation to, 240–245
  General theory of, 245–253
 Potential energy constants for, 306
 Reduced Equation of state for, 278–280
 Short range forces in, 240–245
 Stress per unit area in, 288–291
 Thomson-Joule effect in, 280–281
 Virial, use of, for, 286–288
 van der Waals' Equation of state for, 244
Insulators, Semi-conductors and Metals, defined, 388–390
Intensive parameters, defined, 52
Interatomic energies in Crystals, classified, 292–293
 Overlap energy, 295
 van der Waals' energy, 295–297
Internal Stresses, 182–186
Ionization, by electron Collisions, 685–690
 Equilibrium, Langmuir's test of, 372–373
Ions, Association of, in Strong electrolytes, 552–557
 Gd^{+++}, states of, in crystalline and magnetic fields, 839–841
 Hydration of, 539–540
Isotopes, mixtures of, in Dissociative equilibrium, 164–168
Isotropic Solids, Equation of state for, 139

Johnson effect, 785

Langevin's formula for Susceptibility, 461
Langmuir's Adsorption isotherm (*q.v.*), 828–831
Last multiplier, 14
Lattice constants, of Crystalline salts, calculated, 328–332
Limiting principle, 19
Liouville's theorem, 11–12
Liquids, Densities of, Rectilinear diameter for, 844–846
 Dielectric constant of polar, 816–825
 Pressure effects on, 529–531
 Pure, approximate Partition functions for, 522–525
 Scattering of light by, 765–769
 Structure of strongly polar, 451–453
Local order, 800–801
Localized systems, Partition functions for, 39
Long range order in a lattice, 792–801
Lorentz's Formula for Susceptibility, 441
 Lemma, 440–441

Magnetic deviation effect, 507–511
Magnetic fields, Dissociative equilibrium in, 475–477
Magnetic fields, Transverse effects of, in Metals, 421–425
Magnetic phenomena, Thermodynamics of, 481–483
Magnetization, of Crystals, (I-H) curves for, 511–516
 Dipole and quadripole Energy of, 503–504
 Easy directions of, in Crystals, 502
 Entropy of, 469–471
 of Ferromagnetics, ideal (I-H) curves for, 478
Magnetostriction, 516–521
Mass action, Equation of, 160
 with Solid phases, 174–175
Maxwell's Distribution law, 3, 58
 General form of, 671–672
 with Mass motion, 58–60
 Preserved by Emission and absorption of Radiation, 733–739
Melting, Change of Electrical conductivity of Metals on, 526–527
 Elementary theory of, 525–526
Metals, Contact potentials of, 362–364
 Contacts of, 429–432
 Electrical and Thermal conductivities of, 404–418
 Electron Emission of, 347–358
 Electron Specific heat of, 343–344, 391–397
 Electron theory of, 338–436
 Electron theory of, elementary, 338–364
 Electron Vapour pressure of, 344–346
 Emission of Positive ions by, 370–373
 Free Electrons in, defined, 378–381
 "Impurity", 390–391
 Insulators and Semi-conductors, defined, 388–390
 Photoelectric effect for, 358–362
 Potential energy step at surface of, 354–355
 Thermoelectric effects in, 411–418
 Transverse effects in, 421–425
 Wiedermann-Franz law for, 410–411
Molecules, classified by last Collision, 624
Momentum of Radiation, 732
Momentum transfer, Fluctuations in, 775–783
Monolayers, Equation of state for, 826–828
 Virial, use of, for, 827

Nernst effect, defined, 422
Nernst's heat theorem, Enunciated, 228–229
 Statistical basis of, 230
Neumann and Regnault's law for Solids, 124
Normal modes, for Continuum, 112–115
 Einstein's approximation for, 118
 for Linear lattice, 132–133
 for Square and cubic lattices, simple, 133–138
Normal properties, defined, 11

Nuclear transformations, Assemblies with, 655–657
Nuclei, Symmetry rules for, 156
  Abundance lines for, of even mass number, 657

Opalescence, critical, 765
Order-disorder phenomena, 789–809
Order, of Neighbours in a lattice, 800–801
  Superlattice, 792–801
Ordering effect of Short range forces, 798–800
Orthohydrogen, 85
Orthohydrogen-Parahydrogen transformation rate, 850–852
Overlap Interatomic energy, in Crystals, 295, 320–328

Parahydrogen, 85
Paramagnetism, of Alkali metals, 475
  of Atomic ions, 464–469
  of Electrons, free degenerate, 471–472
  of Gadolinium sulphate, 839–841
  of Gases, 463–464
  of Nickel above Curie point, 480
  Saturation of, 467–469
Parameters, Intensive, defined, 52
Partial potentials, Thermodynamic, 194
Partition functions, for Adsorbed Monolayers, 834–835
  for Atomic ions, 561–562
  Bloch's for Ferromagnetics, 496–501
  Configurational, of Bragg and Williams, 791–797
  Configurational, Bethe's approximation to, 797–806
  for Crystals, Debye's approximation to, 119–128
  for Crystals, Einstein's approximation to, 118
  for Electrons near a nucleus, 572–574
  for Harmonic oscillators, 40
  Heisenberg's, for Ferromagnetics, 485–491
  for Holomagnetization, process of, 504–507
  for Localized systems, 39
  Long and short range Order in terms of, 801–805
  for Mixed Crystals, 176–182
  for Potential energy of a gas, 237–240
  for Radiation, 115–116
  for Rotations, rigid, 41, 79–82
  for Rotations, rigid, asymptotic formulae for, 81–82
  for Vibrations of diatomic gases, 89
Pauli's Exclusion principle, 28, 293
Peltier effect, Defined, 412
  Statistical theory of, 413–418
Perfect gases, Equation of state of, 77
  Specific heats of (*q.v.*), diatomic, 79–94
  Specific heats of (*q.v.*), monatomic, 78
  Specific heats of (*q.v.*), polyatomic, 94–106
Periodic field, Electron states in (*q.v.*), 379–388
Permeability, 438; see Susceptibility
Persistence of velocities, effect of, 704
Phase integrals of Gibbs, 65
Photochemical effects, 741–742
Photoelectric effect, for Fixed atoms, 724–726
  for Metals, 358–362
  for Solids, 740–741
Photon theory of Radiation, 117–118
Planck's, Characteristic function, 192
  Law for temperature Radiation, 112, 116
  Oscillator, see Harmonic oscillator
Poisson-Boltzmann equation, Established, 263
  Solved for Strong electrolytes, 549–552
Polar Liquids, Dielectric properties of, 823–825
Polarizing force in matter, 439
Pólya's theorem on Fluctuations, 762
Positive electrons, Assemblies with, 653–654
Positive ions, Emission of, by Metals, 370–373
Potential energy, of Crystalline salts, 312–319
  Partition function for, of gases, 237–240
Potential energy constants, for Crystals, cubic, 319
  for Imperfect gases, 306
  for Imperfect gases, from viscosity, 308–312
Pressure, 75
  of Radiation, 116
Pyroelectric effect, 148–149

Radiation, Entropy of, 116, 191
  Momentum of emitted, 732
  Partition function for temperature, 115–116
  as Photons, 117–118
  Planck's law for temperature, 112, 116
  Pressure of, 116
  Stefan-Boltzmann law for, 116
Radiative processes, Emission and absorption, by Fixed atoms, 720–724
    by Free atoms, 733–739
    by Solids, 739–740
  Free-free transitions, 726–727
  General nature of, 720, 727–730
  Photoelectric effect for fixed atoms, 724–726
Raoult's law for ideal Solutions, 527
Reaction isobar, Isotopic effects on, 226–227
  Statistical theory of, 225–228
  Thermodynamic theory of, 225
Reaction isochore, 160
Reaction rates for hydrogen and deuterium, 850–852
Reactions, chemical, Arrhenius' equation for, 702
  Bimolecular, Eyring's theory of, 847–852
  Bimolecular, simple theory of, 703–706
  General nature of, in gases, 700–702
  Order of, 701
  Reverse reactions, effect of, 706–707
  Unimolecular, 710–716
Reactions, Photochemical, 741–742
Rectification, general discussion of, 435–436
Rectilinear diameter for Liquid densities, 844–846
Relativistic energies, Assemblies of Electrons with, 648–650
Richardson-Einstein-de Haas effect, 484

Righi-Leduc effect, defined, 423
Rochelle salt, properties of, 823
Rosseland's theorem on Stellar interiors (*q.v.*), 638–639
Rotational freezing, 824
Rotational Specific heat, 81–82
Rotations, rigid, Classical, 62–63
 of Molecules in Solids, 810–815
 Partition functions for, 41, 79–82
 Weights for, 21
Rule of equal areas, 795

Scattering of light by Liquids and gases, 765–769
Schottky effect, in Electron emission, 355–356
Schrödinger's equation, for Harmonic oscillators, 18, 20
 for Structureless particles, 53
Seebeck effect, Defined, 412
 Statistical theory of, 413–418
Sellmeyer's formula, 441
Semi-conductors, Composite, general theory of, 425–429
 Contact potentials for, 401–403
 Electrical conductivity of, 418–421
 Impurity, 390–391
 Metals and Insulators, defined, 388–390
 Thermoelectric effects in, 418–421
 Work functions for, 401–403
Short range forces, Ordering effect of, 798–800
Shot effect, 783–785
Solids, Dielectric constant of polar, 816–825
 Emission and absorption of Radiation by, 739–740
 Photoelectric effect for, 740–741
 Rotations of molecules in, 810–815
Solutions, ideal, 527–533
 Extreme dilution of, 532–533
 Freezing point lowering in, 529
 Henry's law for, 527
 Osmotic pressure of, 531
 Raoult's law for, 527
 Zero heat of dilution for, 531–532
Solutions, perfect, 532
Solutions, *s*-regular, 533–536
Solutions, of Strong electrolytes (*q.v.*), 536–560
Space charge effects, 364–378
 with Positive and negative ions, 373–378
Specific heat, of Acetylene, 99
 of Air, high temperatures, 94
 of Ammonia, gaseous, 95–96
 of Carbon dioxide, 97–98
 Configurational, 805–809
 of Crystals, Blackman's theory of, 130–138
  from Elastic constants, 127–128
  Försterling's formulae for, 128–129
  Law of corresponding states for, 125–126
  T^3-law for, 125
 Electronic, of Metals, 343–344
  of Nickel (metallic), 391–397
 of Ethylene, 99–101
 Excitational, of Gases, 101–106
Specific heat (*cont.*)
 Excitational, of Nitric oxide, 102–103
  of Oxygen, 104–106
 of Ferromagnetics near the Curie point, 492–494
 of Gadolinium sulphate, 841–842
 at High temperatures, 92–106
 of Hydrogen at high temperatures, 91–92
 of Lithium (metallic), 131
 of Methane, 96–97
 of Nitrous oxide, 98–99
 of Oxygen at high temperatures, 94
 of Perfect gases, 77–107
 Rotational, of hydrogen, 82–89
 of Rotations, rigid, 81–82
 of Vibrations, for diatomic gases, 90
 of Water vapour, 97
Spectra, Stellar Absorption, 593–619
States, non-combining groups of, 26
Statistics, Classical, 43
 Classical, approximated to, 55
 Einstein-Bose, 43
 Fermi-Dirac, 42
Steepest descents, method of, 36
Stefan-Boltzmann law, 116, 200
Stefan's constant, 116
Stellar Absorption lines, Decay of, part maxima, 618–619
 Formation of, 610–617
 Marginal appearance of, 617–618
 Maxima of, 595–597
 Statistical theory of, 597–610
Stellar interiors, Rosseland's theorem on, 638–639
Stellar material, Great densities of, 647–648
 Mean molecular weight of, 643
  Strömgren's calculation of, 644, 646
 Nature of, 639–646
Stellar temperature scale, Russell's, 615
Stresses, in Imperfect gases, 288–291
 Internal, 182–184
Strong electrolytes, Bjerrum's theory of, 552–557
 Debye and Hückel's theory of, 269–274, 541–549
 Dissociated in solutions, 536–539
 Heat of dilution of, 546–548
 Osmotic coefficient of, 541, 548–549
 Poisson-Boltzmann equation solved for, 549–552
 Specific ionic interactions in, 557–560
Structureless particles, Schrödinger's equation for, 53
Superlattice Order, 792–801
Surfaces, Adsorbed layers on, 825–838
 Gas Collisions with, 697–699
Susceptibilities, Darwin's theorem on, 447
 Electromagnetic theory of, 437–447
 of Ferromagnetics above Curie point, 480
 Langevin's formula for, 461–463
 Lorentz's formula for, 441
 Polarizing force for, 439
 Sellmeyer's formula for, 441

Symmetrical states, 27, 43
Symmetry number, 89
Symmetry rules, derived, for Complex systems, 153
  for Nuclei, 156
Systems, Defined, 8
  Degenerate, 45
  Independence of, 16

Temperature, Absolute, defined, 188–189
  Empirical, defined, 187
  Low, reached by Demagnetization, 838–844
  Radiation, Planck's law for, 112, 116
Thermal conduction in Metals, formal theory of, 404–418
Thermal expansion, coefficient of, for isotropic Solids, 140
Thermionics, see Electrons, Emission of
Thermodynamic, Partial potentials, 194
  Probability and Entropy, 200–206
Thermodynamics, of Magnetism, 481–483
  and Statistical mechanics, 187–207
  Third law of, 228
Thermoelectric effects in Metals, 411–418
  in Semi-conductors, 418–421
Third law of Thermodynamics, 228; see Nernst's heat theorem
Thomson-Joule effect, 280–281
Thomson's Specific heat of Electricity, Defined, 412
  Statistical theory of, 413–418
Torsional oscillations of suspended mirror, 779–783
  Fourier analysis of, 781–783
Transferable Energy in gas Collisions, 707–710
Transmission coefficients for Electrons, 349–351
Transverse effects of Magnetic fields in Metals, 421–425

Unimolecular gas Reactions, 710–716
Unit mechanisms, 660–663

Vapour pressure, Electronic, of Metals, 344–346
  Constants, Defined, 209
    Diatomic, 220–224
    Monatomic, 218–220
    Monatomic, from dissociation equilibria, 217–218
    Polyatomic, 225
  of Crystals, simple, 172–173
  Equation, Statistical, 210–215
    Thermodynamic, 208
Vibrational Specific heats (*q.v.*)
Virial of Clausius, 286–288
Virial coefficient (second), for Binary mixtures, 306–308
  for Imperfect gases, 302–305
  for Monolayers, 827
  Theoretical, 298–302
van der Waals' Equation of state, 244, 274–275
van der Waals' Interatomic Energy, 295–297

Weights, Analysis of, 16–17
  of Atomic states, 569
  of Classical systems, 18
  Defined, 10
  Derived from Specific heats, 197–200
  Invariance of, 195–197
  of Isotropic oscillators, 19
  of Rotations, rigid, 21
White dwarf stars, 648
Wiedermann-Franz's law, for Metals, 410–411
Work function, Thermionic, for Metals, 348
  for Semi-conductors, 401–403

Zero-point Energy, 123

Printed in the United States
By Bookmasters